THE SCIENCE OF LIFE

SOME OF THE BOOKS BY
H. G. WELLS

THE OUTLINE OF HISTORY
MR. BRITLING SEES IT THROUGH
THE WORLD OF WILLIAM CLISSOLD
MEANWHILE
TONO-BUNGAY
THE TIME MACHINE
MEN LIKE GODS
THE AUTOCRACY OF MR. PARHAM
THE SHORT STORIES OF H. G. WELLS
THE OPEN CONSPIRACY

BOOKS BY
JULIAN S. HUXLEY

ESSAYS OF A BIOLOGIST
ESSAYS IN POPULAR SCIENCE
THE INDIVIDUAL IN THE ANIMAL KINGDOM
RELIGION WITHOUT REVELATION
THE STREAM OF LIFE

THE TEEMING LIFE of an Australian coral reef. In the foreground is the giant yellow and blue anemone Discosoma, which shelters the little fish Amphipryon in its interior. To the left a large edible sea-cucumber or Trepang crawls over the coral, and to the right are two kinds of sea-urchin. Above, is a hurrying crowd of gaily-colored coral-fishes and at the upper left is a specimen of the great sea-fan Gorgonia, the polyp sky-scraper.

THE
SCIENCE OF LIFE

H. G. WELLS
Author of "THE OUTLINE OF HISTORY,"
ETC., ETC.

JULIAN S. HUXLEY
Author of "THE STREAM OF LIFE,"
ETC., ETC.

G. P. WELLS
UNIVERSITY COLLEGE
LONDON

THE LITERARY GUILD
New York

PRINTED AT THE *Country Life Press.* GARDEN CITY, N. Y., U. S. A.

COPYRIGHT, 1929, 1930, 1931, 1934
BY HERBERT GEORGE WELLS
JULIAN S. HUXLEY
GEORGE PHILIP WELLS
ALL RIGHTS RESERVED

CONTENTS

INTRODUCTION

THE RANGE, NATURE, AND STUDY OF LIVING THINGS

	PAGE
1. The Origin and Aim of This Book	1
2. What Do We Mean by Life?	4
3. The Limitation of Life in Space	6
4. Is There Extra-terrestrial Life?	11
5. The Subjective Side of Life	13
6. A Preliminary View of Living Forms	14
7. The Progress of Biological Knowledge	17

BOOK ONE—THE LIVING BODY

CHAPTER

I THE BODY IS A MACHINE
1. The Fundamental Routine of Two Living Creatures 24
2. Why We Call These Bodies Machines . . 29

II THE COMPLEX BODY–MACHINE AND HOW IT WORKS
1. What Every Schoolboy Knows about the Body 32
2. About Cells; the Lesser Lives within Our Life 38
3. Blood 47
4. The Course of the Blood 52
5. Breathing 60
6. Kidneys and Other Exhaust Organs . . 66
7. How Our Food Becomes Blood . . . 71
8. The Continual Struggle against Infection and Chill 90

III THE HARMONY AND DIRECTION OF THE BODY–MACHINE
1. A Study of Adjustment 97
2. Chemical Messengers 103
3. Man and Mouse as Individuals . . . 104
4. The Controlling System 106
5. Sensation and the Senses 110
6. The Nervous Mechanism and the Brain . . 133

CONTENTS

IV THE WEARING OUT OF THE MACHINE AND ITS REPRODUCTION
    1. Age and Decay 140
    2. Reproduction and Fertilization . . . 144
    3. The Growth and Development of the Embryo 150
    4. Rhythm and Birth 156
    5. Childhood, Adolescence, and Maturity . . 159

BOOK TWO—THE CHIEF PATTERNS OF LIFE

I THE FIRST GREAT PHYLUM: VERTEBRATES
    1. Classification 168
    2. What is Meant by a "Phylum" . . . 170
    3. The Classes of the Vertebrate Phylum . . 174
    3a. Mammals 175
    3b. Birds 180
    3c. The More Ancient Class of Reptiles . . 182
    3d. The Linking Amphibians 184
    3e. Fishes 186
    3f. Cyclostomes, a Class of Degraded Antiques . 187
    3g. Semi-Vertebrates 188

II THE SECOND GREAT PHYLUM: THE ARTHROPODS
    1. A Contrast of Arthropod and Vertebrate . 194
    2. The Arthropod Plan of Structure . . 199
    3a. Shrimps, Crabs, Water-fleas, and Barnacles . 201
    3b. Spiders, Mites, and Scorpions . . . 204
    3c. Insects 205
    3d. Centipedes and Millipedes 208
    3e. Peripatus 209

III FURTHER PATTERNS OF INDIVIDUALIZED ANIMAL LIFE
    1. Other Animal Phyla 210
    2. The Molluscs 211
    3. Echinoderms: Nature's Pentagonal Experiment 217
    4. Segmented Worms 221
    5. Roundworms 223
    6. Flatworms 227
    7. Et Cetera 232

IV LESS INDIVIDUALIZED ANIMALS
    1. A Preliminary Note on Individuality . . 235
    2. Obelia 239
    3. Polyps, Jelly-fish, Sea-Anemones, Corals . 244
    4. Sponges 249

CONTENTS

CHAPTER		PAGE
V	VEGETABLE LIFE	
	1. Stems, Leaves, and Roots	253
	2. Individuality in Plants	259
	3. Flowers and Seeds	261
	4. The Flowering Plants	267
	5. Ferns and Mosses	268
VI	THE LOWLY AND MINUTE	
	1. Amœba	275
	2. The Minutest Animals	280
	3. Plant-animals and Seaweeds	284
	4. Moulds, Toadstools, and Yeasts	288
	5. Lichens	292
	6. Slime-moulds	295
	7. Bacteria	299
	8. The Smallest Living Things	306
VII	IS OUR KNOWLEDGE OF THE FORMS OF LIFE COMPLETE?	
	1. Sea-serpents and Living Dinosaurs	311

BOOK THREE—THE INCONTROVERTIBLE FACT OF EVOLUTION

I	THE FACT TO BE PROVED	
	1. Evolution and Creation	314
	2. The Nature of the Proof	316
II	THE EVIDENCE OF THE ROCKS	
	1. The Nature and Scale of the Record of the Rocks	318
	2. Defects and Happy Finds in the Record	326
	3. A Sample Section in the History of Life: the Evolution of Horses	332
	4. The Continuity of Evolution as Shown by Sea-urchins	346
	5. "Missing Links"	348
III	THE EVIDENCE FROM PLANT AND ANIMAL STRUCTURE	
	1. Structural Plans, Visible and Invisible	356
	2. Vestiges: the Evidence of the Useless	362
	3. The Evidence of the Embryo	366
IV	THE EVIDENCE FROM THE VARIATION AND DISTRIBUTION OF LIVING THINGS	
	1. The Variability of Living Things	374
	2. What Is a Species?	378
	3. The Distribution of Living Things	388
	4. The Evidence Summarized	402

CONTENTS

CHAPTER		PAGE
V	THE EVOLUTION OF MAN	
	1. Man: a Vertebrate; a Mammal; a Primate	405
	2. Fossil Men	411
	3. Man's Body: a Museum of Evolution	415
	4. Man's Place in Time	419

BOOK FOUR—THE HOW AND THE WHY OF DEVELOPMENT AND EVOLUTION

I	THE ESSENCE OF THE CONTROVERSIES ABOUT EVOLUTION	
	1. The Chief Theories of Evolution	425
	2. Method of Treatment	433
II	HOW INDIVIDUALS ORIGINATE	
	1. Can Life Arise Spontaneously?	436
	2. Sexless Reproduction the Original Method	438
	3. Sex Is a Complication of Reproduction	443
	4. The Gametes, or Marrying Cells	445
	5. Some Evasions and Replacements of Sexuality	448
	6. Artificial Propagation	449
	7. A Note on the Regeneration of Lost Parts	452
	8. Grafts and the Chimæra	454
	9. What Is Meant by the Germ-plasm	457
III	THE MECHANISM OF INHERITANCE	
	1. How Our Cells Multiply	459
	2. How the Gametes are Formed	462
	3. Fertilization	463
	4. What the Chromosomes are For	465
	5. Chromosomes in Plants	466
IV	THE A B C OF GENETICS	
	1. The Different Kinds of Variation	468
	2. The Discoveries of the Abbé Mendel	473
	3. Mendel's Two Primary Laws of Heredity	476
	4. Numerical Proof of Mendel's Laws	484
	5. Genes and Their Effects upon Human Characteristics	485
	6. Colour Genes in America as an Example of Multiple Genes	489
	7. Mapping the Chromosomes	491
	8. Plant Breeding—Some Practical Applications	496
	9. Is Inbreeding Evil?	498
	10. Reversion to Ancestral Type	503
	11. How Much of Inheritance is Mendelian?	505
	12. Some Common Superstitions about Heredity	508

CONTENTS

CHAPTER		PAGE
V	THE GROWTH OF THE INDIVIDUAL	
	1. Normal and Monstrous Development	514
	2. The New Individual Settles Down to Development	518
	3. The Moulding of Organs by Use	523
	4. The Transformation of Tadpoles	526
	5. The Rôle of the Genes	531
	6. The Limitation and Control of Growth	535
	7. Age, Retrogression, and Rejuvenation	541
	8. Is Human Rejuvenation Desirable?	550
VI	WHAT DETERMINES SEX	
	1. The Normal Causation of Sex	552
	2. The Proportion of the Sexes	555
	3. Can We Control the Sex of Our Children?	558
	4. Sex-Linked Inheritance and Sex-detectors	560
	5. Gynanders	563
	6. Sex Hormones	565
	7. Reversal of Sex	568
	8. Intersexes	569
VII	VARIATION OF SPECIES	
	1. The Open Questions of Variation	576
	2. Mutation: the Experimenting of the Germ-plasm	577
	3. Are There Forward Bounds in Mutation?	580
	4. Are Acquired Characters Inherited?	583
	5. Artificially induced Mutation	593
	6. Moths and Smoke	595
	7. The Significance of Sex	596
	8. Variation in Plants	598
VIII	SELECTION IN EVOLUTION	
	1. The Vindication of Darwinism	600
	2. Natural Selection as a Conservative Force	606
	3. Natural Selection Under Changing Conditions	609
	4. Selection of Characters Useless to the Species	614
	5. Isolation as a Species-maker	617
	6. Crossing May Produce New Species	621
	7. Failures to Vary and Extinction	623
IX	IS THERE A MYSTICAL EVOLUTIONARY URGE?	
	1. Straight-line Evolution	629
	2. Do Races Become Senile?	632
	3. The *Élan Vital* and the Life Force	637
	4. Is There Purpose in Evolution?	639

CONTENTS

BOOK FIVE—THE HISTORY AND ADVENTURES OF LIFE

CHAPTER	PAGE
I THE PROLOGUE	
1. The Scale of the Universe	644
2. The Setting of the Stage	648
3. The Origin of Life	649
4. Changes in the Terrestrial Scenery	653
II LIFE BEFORE FOSSILS	
1. The Earliest Life	660
2. Union Is Strength: the Buildings of the Plant-body	661
3. Marthas and Marys Among Cells: the Origin of the Higher Animals	664
4. The First Brains	666
5. Blood as a Step in Evolution: the Cœlomates Appear	667
III THE ERA OF CRAWLING AND SWIMMING: EVOLUTION IN THE WATER	
1. The Opening Eras: the Archeozoic and the Proterozoic	675
2. The Age of Ancient Life: the Paleozoic	678
3. The Cambrian Period: Age of Trilobites	680
4. The Age of the Sea-scorpions	684
5. Echinoderms: a Phylum Which Has Never Produced Land Forms	686
6. The Ancient History of Molluscs	688
7. The Earliest Fish	694
IV LIFE CONQUERS THE DRY LAND	
1. The Desert Land of the Early World	701
2. The Curious History of the Land-plant and Its Seed	702
3. The Plants of the Ancient World	716
4. Coal	719
5. The Vertebrata Crawl Out of Water	722
6. A Coal-measure Forest	729
7. An Age of Death	733
V THE FULL CONQUEST OF THE LAND	
1. The Reptilian Adventures	738
2. A Digression upon Adaptive Radiation	743
3. The "Back-to-the Water" Movement	745
4. The Conquest of the Air	749

CONTENTS

CHAPTER		PAGE
	5. Dinosaurs	756
	6. Insects Multiply as Plants Develop	761
	7. The Battle of the Giants and the Dwarfs	769
VI	THE MODERN ERA	
	1. Modern Life	774
	2. The Troubled Earth of Cenozoic Times	777
	3. The Origin and Prosperity of the Mammals	779
	4. The Tangled Skein of Evolution	786
	5. Is There Progress in Evolution?	789
VII	MAN DAWNS UPON THE WORLD	
	1. Why We Say "Man Dawns"	796
	2. The Remote Ancestors of Man	797
	3. A Short History of Pleistocene and Recent Climate	806
	4. Traces of Man Before and In the Pleistocene Period	817
	5. The Advent of Modern Man	820

BOOK SIX—THE SPECTACLE OF LIFE

I	HABITATS	
	1. Ways and Worlds of Life	823
	2. Habitats and Their Inhabitants	827
	3. Ways of Getting a Living	832
	4. The Adjustment of Inhabitant to Habitat	835
II	LIFE IN THE SEA	
	1. Life in the Sea	839
	2. The Surface Life of the Sea	845
	3. The Deep Sea	853
	4. On the Floor of the Abyss	863
	5. Sea-Shore Life	866
	6. Coral Reefs and Islands	871
	7. Holes and Corners of Sea-life	874
III	LIFE IN FRESH WATER AND ON LAND	
	1. Fresh-water Life	878
	2. The Life of Flowing Waters	883
	3. The Life of Standing Waters	889
	4. Land Habitats	893
	5. The Desert	895

CONTENTS

		PAGE
	6. The Tropical Forest	901
	7. Regions of Rock, Snow, and Ice	907
	8. Island-dwellers	913
	9. Cave-dwellers	915
	10. Out-of-the-way Modes of Life	918

IV SOME SPECIAL ASPECTS OF LIFE
1. Partnership and Parasitism 922
2. The Scale of Living Things 936
3. Colour and Pattern in Life 944

V THE SCIENCE OF ECOLOGY
1. Ecology Is Biological Economics . . . 961
2. The Chemical Wheel of Life 962
3. The Parallelism and Variety of Life-communities 967
4. The Growth and Development of Life-communities 973
5. The Grading of Life-communities . . . 982
6. Food-chains and Parasite-chains . . . 989
7. Storms of Breeding and Death . . . 996

VI LIFE UNDER CONTROL
1. The Balance of Nature 1012
2. Pests and Their Biological Control . . 1016
3. The Beginnings of Applied Biology . . 1022
4. The Ecological Outlook. 1027

BOOK SEVEN—HEALTH AND DISEASE

I INFECTIOUS AND CONTAGIOUS DISEASE
1. *Is* Man Particularly Unhealthy? . . . 1033
2. Microbes 1035
3. Insects as Microbe-carriers 1040
4. Immunity 1047
5. Avoiding and Killing Microbes . . . 1050

II THE NOURISHMENT OF THE BODY
1. Mr. Everyman at Table 1054
2. The Six Vitamins 1059
3. Some Possible Poisons 1064
4. Drugs, Their Uses and Dangers . . . 1068

III FRESH AIR AND SUNLIGHT
1. Town Air and Country Air 1076
2. The Air of a Stuffy Room 1080
3. Sunlight as a Tonic 1085

CONTENTS

CHAPTER

IV THE PRESENT HEALTH OF *Homo Sapiens*
    1. The Control of Epidemic Diseases . . . 1089
    2. The Heart and Lungs 1090
    3. Cancer 1093
    4. Tuberculosis 1096
    5. Mr. Everyman and the Fight against Disease . 1099

BOOK EIGHT—BEHAVIOUR, FEELING, AND THOUGHT

I RUDIMENTS OF BEHAVIOUR
    1. The Three Elements of Behaviour . . . 1102
    2. Receptivity 1104
    3. Response 1114
    4. Correlation: the Origins of the Nervous System 1118
    5. Vegetable Behaviour 1128
    6. Instinctive and Intelligent Behaviour . . 1131
    7. The Behaviour of the Slipper Animalcule . 1136
    8. The Different Worlds in Which Animals Live 1143

II HOW INSECTS AND OTHER INVERTEBRATES BEHAVE
    1. The Arthropod Mind as the Culmination of Instinct 1147
    2. An Anatomy of Instinct 1151
    3. Solitary Wasps 1158
    4. Insect Societies 1162
    5. Ways of Life Among Ants 1169
    6. The Parasites of Ant-colonies . . . 1174
    7. Termites 1178
    8. Bees 1182

III THE EVOLUTION OF BEHAVIOUR IN VERTEBRATES
    1. The Vertebrate Nervous System . . . 1200
    2. The Mind of a Fish 1210
    3. The Amphibian Mind 1214
    4. The Brain in Reptile, Bird, and Mammal . 1218
    5. Courtship in Animals 1226
    6. The Evolution of Mammalian Intelligence . 1239
    7. Education in Animals 1247
    8. Play 1252
    9. The Behaviour of Monkeys and Apes . . 1259

IV CONSCIOUSNESS
    1. Objective and Subjective 1270
    2. Is Consciousness Passive or Active? . . 1272
    3. What is the Range of Consciousness? . . 1272
    4. Body-Mind 1275

CONTENTS

CHAPTER		PAGE
V	THE CULMINATING BRAIN	
	1. The Expansion of the Cortex	1278
VI	THE CORTEX AT WORK	
	1. Pavlov	1288
	2. What Is a Conditioned Reflex?	1289
	3. The Dog as a Simpler Man	1293
	4. Inhibition and Control	1297
	5. The World of a Dog	1305
	6. Boredom, Alertness, and Sleep	1306
	7. The Dog Hypnotized	1311
	8. Temperament in Dogs	1314
VII	HUMAN BEHAVIOUR AND THE HUMAN MIND	
	1. Human Behaviourism	1318
	2. The Mind as the Thing That Knows	1323
	3. Emotion and Urge	1330
	4. Hypnosis	1331
	5. The Unconscious	1334
	6. The Splitting of the Self-multiple Personalities	1339
	7. Hysteria	1344
	8. Exaltation	1349
	9. Automatism and Mediumship	1351
	10. Neurasthenia	1353
	11. Repression and the Complex	1356
	12. Psycho-analysis	1362
	13. Minds Out of Gear and In Gear	1369
	14. Differences between Minds	1381
VIII	MODERN IDEAS OF CONDUCT	
	1. The Conduct of Life	1390
	2. What Is the Self We Conduct?	1391
	3. The Primary Biological Duties	1395
	4. Of Self-knowledge and Moods	1397
	5. Candour	1400
	6. Restraint and Poise	1403
	7. Evasion, Indolence, and Fear	1408
IX	BORDERLAND SCIENCE AND THE QUESTION OF PERSONAL SURVIVAL	
	1. The Theory of Body-Soul-Spirit	1411
	2. Dream Anticipation and Telepathy	1413
	3. Clairvoyance, Table-tapping, and Telekinesis	1416
	4. Materialization and Ectoplasm	1422
	5. Mythology of the Future Life	1432
	6. The Survival of the Personality after Death	1433

BOOK NINE—BIOLOGY OF THE HUMAN RACE

CHAPTER   PAGE

I   PECULIARITIES OF THE SPECIES *Homo Sapiens*
1. Fire, Tools, Speech, and Economics . . 1436
2. Origins of *Homo Sapiens*. . . . 1439
3. Primary Varieties of Human Life . . . 1447
4. The Development of Human Interaction . . 1450

II   THE PRESENT PHASE OF HUMAN ASSOCIATION
1. The Religious Tradition 1454
2. The Passing of Traditionalism 1460
3. The Supersession of War 1464
4. The Change of the Nature of Education . . 1465
5. The Breeding of Mankind 1467
6. The Superfluous Energy of Man . . . 1472
7. The Possibility of One Collective Human Mind and Will 1473
8. Life Under Control 1476

INDEX 1481

THE SCIENCE OF LIFE

INTRODUCTION

THE RANGE, NATURE, AND STUDY OF LIVING THINGS

§ 1. *The Origin and Aim of This Book.* § 2. *What Do We Mean by Life?* § 3. *The Limitation of Life in Space.* § 4. *Is There Extraterrestrial Life?* § 5. *The Subjective Side of Life.* § 6. *A Preliminary View of Living Forms.* § 7. *The Progress of Biological Knowledge.*

§ 1. The Origin and Aim of This Book.

A FEW years ago one of the writers of this book made a summary of historical knowledge called *The Outline of History*. He dealt with all history as one process. He displayed it—or, rather, it displayed itself as he gathered his material and joined part to part—as the appearance of life in space and time, and as an achievement of self-knowledge and a release of will; a story unfolding and developing by a kind of inner necessity, until at last man was revealed, becoming creative, becoming conscious of the possibility of controlling his destiny, and groping through kingdoms and empires and wars and revolutionary conflicts towards unity and power. This epic story, presented plainly though it was, without any attempt at literary flourishes or poetic passages, seized upon the imaginations of a great number of people. It placed them in relation to the whole scheme of things. It joined up such historical facts as they knew into a whole. It explained their patriotic feelings, cleared up their conceptions of international relationships, and rationalized their political and social activities. It was something for which they had been ripe, and were waiting, and the book had a success altogether beyond the deserts of its learning or its literary quality.

Men and women turned to it from the previous outline of history upon which, consciously or unconsciously, they had hitherto based their general conception of world events—the historical portion of the Bible, supplemented by some scraps of classical literature, and their own national record. For modern needs a history which opened with barbaric myths, which concerned itself mainly with the tribal affairs of the Jews, which disregarded the past of nearly all the world outside Syria and Egypt, and ended two thousand years ago, had become insufficient. Little by little a vaster Genesis had been written by science, and unobtrusively, detail by detail, the scholarly excavator had opened up mightier Chronicles and pieced together a more splendid roll of Kings. A broader world had grown in human experience and now altogether dwarfs the fortunes of Palestine and the jurisdiction of the Cæsars. For all its faults, its compiled quality, its pedestrian if unpretentious prose and much incidental sketchiness, *The Outline of History* brought something of the greatness of these new vistas before the minds of many for the first time. They lived, they realized, upon a wider stage, and the things of the present in which their activities mingled had a greater significance than they had supposed.

But it is not only in the field of human history that our science has been enormously expanded. Our realization of the nature of life, our knowledge of its processes, has been changed, deepened, and intensified. A great and growing volume of fact about life as it goes on about us and within us becomes available for practical application. It reflects upon the conduct of our lives throughout; it throws new light upon our moral judgments; it suggests fresh methods of human co-operation, imposes novel conceptions of service, and opens new possibilities and new freedoms to us.

Much of this new material is still imperfectly accessible to ordinary busy people. It is embodied in scientific publications, in a multitude of books; it is expressed in technical terms that have still to be translated into ordinary language; it is mixed up with masses of controversial matter, and with unsound and pretentious publications. Complicated with the clear statements of indubitable science there is a more questionable literature of less strictly observed psychology, and a great undigested accumulation of reports and statements turning upon current necromancy, thought-transference, speculations about immortality, will-power, and the like. Mingled with orthodox physiological teaching are the doctrines of dietetic dogmatists, and the prohibitions and injunctions of religious and other regimens. Obscuring the facts of heredity there are heavy accumulations of prejudice and superstition. In the care of his health and the conduct of his life, the ordinary man, therefore, draws far less confidently upon the resources of science than he might do. He is unavoidably ignorant of much that is established and reasonably suspicious of much that he hears. He seems to need the same clearing up and simplifying of the science of life that *The Outline of History* and its associates and successors have given to the story of the past. And the present work is an attempt to meet that need, to describe life, of which the reader is a part, to tell what is surely known about it, and discuss what is suggested about it, and to draw just as much practical wisdom as possible from the account.

Three writers have joined forces in this compilation, this *précis* of biological knowledge. They are all very much of the same mind about the story they have to tell, they have all read and worked upon each other's contributions, and they are jointly and severally responsible for the entire arrangement and text. The senior member of the firm, so to speak, is Mr. Wells, who wrote *The Outline of History*, but equally responsible with him are Professor Julian Huxley, the grandson of Mr. Wells' own inspiring teacher, Professor T. H. Huxley, the great associate of Darwin, and Mr. George Philip Wells, the senior partner's son. The latter is a scholar of Trinity College, Cambridge, who refreshes the detailed research he has been doing at Plymouth and London with this excursion into general statement. The senior partner is the least well equipped scientifically. His share has been mainly literary and editorial, and he is responsible for the initiation and organization of the whole scheme. Together this trinity, three in one, constitutes the author of this work.

The reader may like to know something of their procedure in putting it together. After various preliminary discussions they decided that as the first concrete stage they should sketch out something between a syllabus and a table of contents for the projected book. With that as a guide they would be able to preserve a consecutive flow of interest and maintain a just proportion between part and part. They would map out the work before them, and classify and apportion the very great mass of literature and specialist publications they would have to digest. They would also run less risk of omissions, and a squeezing out of important questions, if, with such a sketch to submit, they consulted authorities less concentrated upon the animal biology in which all three of them were mainly educated. Each made his own sketch, and then came exchanges and discussions, and a wide search among their friends and interested people for comment and advice. A large part of a year was spent in drawing this outline of an outline, and in the end a system of chapters and sections was achieved, a framework sufficiently stable to carry the whole work. Here and there subsequent alterations, extensions, and additions have been made, arising out of the fact that the writers learnt things they had not known and came to realize values they had apprehended imperfectly, but the broad lines, as they were first drawn, stood the test of all their later reading.

The triplex author claims to be wedded to no creed, associated with no propaganda; he is telling what he believes to be the truth about life, so far as it is known now. He is doing exactly what the author of *The Outline of History* attempted for history. But no one can get outside himself, and this book, like its predecessor, will surely be saturated with the personality of its writers. The reader has to allow for that, just as a juryman has to allow for the possible bias in the evidence of an expert witness or in the charge of a judge. This book is written with a strenuous effort to be clear, complete, and correct; each member of the trinity has been closely watched by his two associates with these qualities in view. But they cannot escape or even pretend to want to escape from their common preconceptions. The reader of this book will not have made the best use of it, unless, instead of accepting its judgments, he uses them to form his own.

§ 2. What Do We Mean by Life?

WE WILL begin our summary by asking what is meant by life? What are its distinctive characteristics?

Our answer at this early stage must be provisional. All this work is no more than a descriptive contribution to the complete answer. But when an ordinary superficial man speaks of life he has a certain group of distinctive points in his mind. Firstly, a living thing moves about. It may not be in movement continuously—many living seeds lie immobile for long periods, but sooner or later they will thrust and stir. Life may move as swiftly as a flying bird, or as slowly as an expanding turnip, but it moves. It moves in response to an inner impulse. It may be stimulated to move, but the driving-force is within. It does not move simply like dust before the wind or sand stirred up by the waves

And not only does it move of itself, but it feeds. It takes up matter from without into itself, it changes that matter chemically, and from these changes it gathers the energy for movement. Crystals, stalactites, and other non-living things grow, but only by additions, by the laying-on or fitting-in of congenial particles, without any change of chemical nature or release of energy. This process of taking in, assimilating and using matter, is called *metabolism*. Metabolism and spontaneous movement are the primary characteristics of living things.

In addition, life seems always to be produced by pre-existing life.

It presents itself as a multitude of individuals which have been produced by division or the detachment of parts from other individuals, and most of which will in their time give rise to another generation. This existence in the form of distinct individuals which directly or indirectly reproduce their kind by a sort of inherent necessity is a third distinction between living and non-living things. Waves in water or wind-ripples in sand may be said to reproduce themselves, but not by a detachment and growth of their own substance; drops of oil or water grow and break up under suitable conditions, but not through any innate disposition to do so. Living things display an impulse to reproduce themselves even in adverse circumstances.

So far as our knowledge goes, life arises always and only from preceding life. In the past it was believed that there could be a "spontaneous generation" of living things. Aristotle taught that plant-lice arise from the dew which falls upon plants, and that fleas spring from putrid matter.

Alexander Ross, a charmingly assured writer of the seventeenth century, in sound controversial style, reproved Sir Thomas Browne for doubting this. "So may he," said Ross, "doubt whether in cheese and timber worms are generated; or if beetles and wasps in cows' dung; or if butterflies, locusts, grasshoppers, shell-fish, snails, eels, and such like be procreated of putrified matter, which is apt to receive the form of that creature to which it is by formative power disposed. To question this is to question reason, sense, and experience. If he doubts of this, let him go to Egypt, and there he will find the fields swarming with mice, begot of the mud of Nylus, to the great calamity of the inhabitants."

Very gradually this error was dispelled. The mice story evaporated—the necessary connection of flies and maggots was demonstrated. It was only in the middle nineteenth century that the concluding dispute took place over Bastian's assertion that abundant bacterial life appeared in sterilized infusions of hay and other material. Pasteur and others demonstrated the insufficiency of his sterilization. Germs and protected spores had escaped his boiling. It is accepted now by all biologists of repute that life arises from life and in no other way—*omne vivum ex vivo*. Life as we know it flows in a strictly defined stream from its remote and unknown origins, it dissolves and assimilates food, but it receives no living tributaries.

Yet the distinction of what is living and what is not living is by no means easy. Seeds, small worms and microscopic animalcula can be dried up and left totally inert for long periods of time, so that it is impossible to distinguish them from dead organisms; then at the touch of moisture they will resume the recognizable process of life. Do they *live* meanwhile?

Science has passed through a phase in which there was a sharp distinction between "organic" matter, which was either in or had been produced by a living thing, and inorganic matter. While living things possessed individuality, inorganic matter was supposed to consist of repetitions of identically similar atoms and molecules. Certain modern writers, however, writing from the standpoint of mathematical physics, display a disposition to recognize individuality even in atoms, to speak of each atom as an "organism," simpler in nature, but otherwise not dissimilar to the intricate complexes which make up definitely living tissue. This is certainly pulling the word "organism" down to the level of a much simpler structure than has been customary in its use. These writers believe that a process of complication is all that separates the non-living from the living, and in spite of the want of any direct evidence on the biologist's side, they assert the credibility of spontaneous generation from their point of view. The complicated carbon compounds that are found in living things were at one time thought only to arise under the influence of a "vital force," and their distinctiveness was one of the most vehemently asserted arguments of the vitalists—until, in 1828, Wöhler synthesized urea in the laboratory. Since that time our knowledge of the chemistry and physics of the systems that are found in living tissues has developed rapidly. It has been found possible to imitate a number of the properties of living things by means of artificial models. It is at least possible that the distinctive properties of living things depend simply upon the complexity of their molecular organization—in which case, spontaneous generation, under suitable circumstances, is at least credible. But one fact remains, that all the life we know is one continuing sort of life, that all the life which exists at this moment derives, so far as human knowledge goes, in unbroken succession from life in past time, and that the unindividualized non-living world is separated from it by a definite gap. We will return to the questions of its origin and of its separation from non-living matter at a later stage, after a fuller account of its chemical and physical processes has been given.

§ 3. The Limitation of Life in Space.

So FAR as our certain knowledge goes, life is wholly confined to the surface of the planet Earth and to a few miles above and below its surface. Before we reach the summits of the highest earthly mountains life has practically ceased, and the bottom of the sea is the downward limit of the vital process. Within the earth is lifeless matter, and never in all human experience and observation has there come the slightest intimation of any life beyond our atmosphere. So far as we know, the immensities of space, the other planets, the suns and stars and nebulæ know nothing of this thing we are studying, this something of which we are a part, that feels and moves of itself and reproduces its kind. It is a unique being, it appears, new and strange; it is an aggressive continuing being that does as yet but begin in the limitless universe of matter.

There is reason to believe—we shall give some of the considerations later—that life appeared upon this planet thousands of millions of years ago, and that it was confined to warmish saline water. Since then it has extended its range very greatly, in height, in depth, and into cold, dry, and desolate regions. And it still extends its range. It still presses against its limitations.

Within the last few years men have thrust themselves in aeroplanes far above the level of Mount Everest, and that great mountain itself, more than six miles above the sea, was perhaps conquered by G. L. Mallory and A. C. Irvine in 1924. They started from a camp pitched at a height of 26,800 feet on June 6th and never returned. From this camp they were watched high up on the mountain slopes still struggling onward, and then they were hidden by driving mists and never seen again. Dr. Somerville and Lieut.-Col. Norton, with infinite pain and suffering, had previously got to 28,200 feet. Somerville suffered horribly from the parching of his throat in the intensely cold and dry air, and Norton, after his return to camp, was stricken with snow-blindness. It is probable, but not certain, that Coxwell and Glaisher exceeded these limits in a balloon in 1862, and attained something over 30,000 feet. They were conscious above 29,000 feet. After Glaisher became insensible, Coxwell, by a last effort, pulled the valve cord with his teeth, his hands being too frost-bitten to use. Until recently, the height record for any living thing was the balloon record of Berson and a companion, made in July, 1901. They certainly reached 34,500 feet, but then they became unconscious in spite of their inhalation of oxygen. In the last few years a series of ascents, made in balloons with closed gondolas, have raised this record very considerably. Professor Picard exceeded 53,000 feet in August, 1932. In September of the following year, the Russian aviators Prokofieff, Birnbaum and Godunoff reached 62,400 feet, which is very nearly twelve miles.

They had a spherical gondola of duralumin, large enough for them to walk a few paces, warmed inside and with a supply of compressed oxygen, and they handled the ballast and external instruments by means of electrically operated controls. The record height for a heavier-than-air machine is 44,000 feet, set up by Capt. C. F. Unwins in April, 1933. No other living creature, so far as our knowledge goes, has ever transcended these limits. The condor of the Andes is credited with 23,000 feet by Humboldt. At the higher levels these great birds use the passes, flying low and close over the heads of travellers.

Sixty-two thousand feet, a trifle under twelve miles, not a fifth of what any automobile can do in an hour on the level, but so far it is practically the limit of the upward range of life. Even to win that much demands elaborate preparation and frightful exertion.

Three main wants restrict the living body to the lower levels—deficiency of oxygen to breathe, deficiency of pressure upon the exterior of the body, and intense cold. The Everest expeditions of 1922 and 1924 met the former need by taking up large supplies of compressed oxygen, and the same thing is done in the attempts of aviators to break the height record. Within limits the body of the mountain climber can be gradually educated to the diminished pressure, a fact upon which the Everest expedition of 1933 based their programme, but the aviator runs great danger of suffering from an extreme form of mountain sickness.

Airmen who have attempted height-records describe themselves as feeling swelled, puffed out, and deformed, with a painful buzzing in the ears. The heart beats too vigorously in the attempt to send sufficient oxygen round the body. There may be bleeding in ears, nose, and lungs, and even from the eyes and gums. The mountaineer at high altitude has to exert himself, whereas his flying compeer sits still; he moves with a general feebleness, and is in a continual state of fatigue.

Major Hingston, the medical officer of the Mount Everest expedition of 1924, says that "the very slightest exertion, such as the tying of a bootlace, the opening of a ration-box, or the getting into a sleeping-bag, was associated with marked respiratory distress." At 27,000 feet Dr. Somerville had to take eight or ten deep respirations for every step forward. Lieut.-Col. Norton at his highest ascended 80 feet in an hour of extreme exertion.

Tissandier, who made a balloon ascent to 27,950 feet from Paris in 1875, fainted at 26,500 feet, and when he recovered consciousness the balloon was descending and both his companions were dead. A curious apathy comes on the explorer at such elevations. He no longer sees, hears, or acts so rapidly, and there is great loss of muscular and mental power. Glaisher stared at his instruments, but could not read them. "There is no suffering," wrote Tissandier. "Only an inward joy."

The cold at these heights sinks to levels as low as thirty degrees below zero centigrade and it is extremely difficult to retain the heat of the body. The aviators can wrap themselves amply in wool, paper, and other non-conductors, but the mountaineer must not be too heavily encumbered. Two members of the 1924 Everest expedition died of cold, and seven of the preceding one were killed by an avalanche. The supreme mountain heights are subjected to icy winds and violent snow-storms and to snow and rock avalanches, inconveniences the aviator escapes. But the aviator has to anticipate great stresses upon his machine due to the cooling and consequent contraction of the metal. The task of the engine to sustain the weight of the machine increases with the attenuation of the air. Long before the aviator has struggled to the five-mile limit all other living things have left him. Insects have long since fallen insensible from his planes. No bird can emulate him, though the condor might have kept with him as high as five miles.

Downwards life is also restricted, but downwards it is increasing pressure and, in the solid earth, increasing temperature, against which life has to fight. The forms of life familiar to us do not extend down into the sea for very many fathoms; they are replaced by other species more adapted to cold and darkness and high pressure. The barriers set to air-breathing life are soon reached. A diver in a diving-dress, with skilful management and under good conditions, can go to about 300 feet below sea-level, can stay there for twenty minutes, and return to the surface by stages in about an hour and a half, provided he is constitutionally adapted to the work. A naked diver may get to something like thirty feet and stay at most about a couple of minutes. A submarine is under similar restrictions in its range of submergence.

The limiting factor to men going upward and downward under water is the increasing solubility of the atmospheric gases in the blood with pressure. The respiratory and circulatory system of a creature adapted to ordinary surface conditions works with increasing difficulty under high pressure, and a rapid return to normality produces an effervescence of the gases absorbed. "Caisson disease" is a disease due to this release of bubbles of gas in the blood, and its characteristic victims are divers and workers in the submerged caissons used in bridge and breakwater construction. Its symptoms are various, depending upon the part of the body in which the bubbles happen to be released, and include fainting, vomiting, deafness, inability to breathe, paralysis, pain of the joints and muscles, and sometimes sudden death. At the St. Louis Bridge, under a pressure of four and a half atmospheres, one hundred workmen suffered seriously and fourteen died out of a total of six hundred men employed.

Because of this relationship of living things to limiting pressures, there is a very considerable restriction in the movements of all life in the sea. We are too apt to think of whales, and so forth, plunging into the abyss and rushing up to the surface without inconvenience. Such familiar creatures, however, soon descend to definite limits; they do not go down to the lowest depths. Whales live deep—astonishingly deep, when the pressures upon them are considered. They have peculiar reticulations of their blood-vessels which may serve to ease off the bubbling of gas as they rush up to the surface again after a prolonged submergence. The Greenland whale, says R. W. Gray, can get down to 800 fathoms, the best part of a mile, to judge by the length of line a harpooned individual will reel out. But this depth is not equal to the height of such a winter resort as Arosa.

Anything normally enclosing air or other gas, such as the strands of a rope or a piece of woolly fabric, is flattened out at great depths. If a ship's hawser, says Professor J. Arthur Thomson—a hempen hawser, of course, is meant—is sunk to the depth of a couple of thousand fathoms, it is squeezed to less than the diameter of one's wrist, and a piece of wood that has gone down so far will no longer float. Whatever lives at great depths—if it contains air or gas—must have its internal pressures as great as those about it.

Oceanographic exploration shows us that there is a definite zoning of life in the abyss, the creatures of the deeper levels being adapted to the peculiar conditions of their particular level. They can no more come up than the surface life can go down—until it dies and drifts slowly to be eaten by the inhabitants of the lower darkness. Sometimes, however, some of the deep sea fishes *fall upward*. Many of them, like their surface kindred, have swimming bladders to accommodate themselves to varying levels. These bladders contain gas under enormous pressure. If one of the abyssal fishes rises too high in pursuit of its prey, the gas in its bladder may expand out of control to the muscles. The fish can no longer go down, but continues to rise helplessly, and as it rises to levels of less and less pressure, the expansive force of the gas within is less and less restrained, until at last the creature rushes up to the surface and arrives there distended or burst and dead.

The ocean at its deepest goes down perhaps for seven miles. Life, therefore, so far as we know, is confined to a layer of air and a layer of water, having a total thickness of less than fourteen miles, on this comparatively small planet Earth, and no one single form of life is able to span even these petty limits. Man's vertical range seems to be as great as any creature's and it is no more than eight miles. In all the wilderness of space outside this double film there are only the very faintest suggestions of the possible occurrence of life or of any processes parallel with and comparable to life.

§ 4. Is There Extra-terrestrial Life?

SPECULATION about life in extra-terrestrial space has occupied itself mainly in a search for conditions similar to those recognized as the limiting conditions of life upon earth. Something like a wetter, hotter parallel to earthly conditions may be found upon the surface of the planet Venus, but the atmosphere of that planet is so continuously clouded, that it is impossible to obtain any confirmatory glimpses of vegetation or other living stuff. Slight changes of tint have been observed upon the surface of the moon and might just possibly be due to a transitory growth of plant-like organisms during the long sunlit day of that satellite. Mars displays rectilinear double markings, the "canals," so unlike any other natural fissures in their straightness, that they have been ascribed to the engineering of intelligent beings, and there are also changes of colour upon the Martian surface that may be caused by the growth and ripening of irrigated crops. Mars has snow-caps which expand and contract as the Martian winter comes on and gives place to the Martian summer, and the canals may conceivably distribute the thawing water.

But if indeed there are living things upon these other members of our solar system, they must be profoundly different in many ways from terrestrial life. It is doubtful if they could be transferred to terrestrial conditions and live, or if much or any terrestrial life could be transplanted to and acclimatized on the moon or Mars. The mass of Mars and still more so the mass of the moon is much less than that of the earth, the gravitational energy at the surface of either planet would be less, the atmospheric pressure would be much less, on the almost entirely airless moon it would be practically nil, and so the weight of a living body and of its parts and fluids would be much less there and the pressure on its surfaces very much less. The lungs of a terrestrial animal would be unable to get air enough to breathe, the gases dissolved in the fluids of the body would effervesce and expand and blow out all its closed cavities, our joints would lock, the heart would drive the blood through the membranes that are sufficient to contain it on earth, and we should bleed in our lungs, in our throats, eyes, ears, and be choked with blood. On the other hand, a man reconstructed to walk on Mars would be crushed to death by his own weight on the earth. Life on Mars would have to be so different in its character from the life we know, that one would almost need another word for it. If we called terrestrial life Alpha life, we might call the life-parallel on Mars Beta life, an analogous thing and not the same thing. It may not be individualized; it may not consist of reproductive individuals. It may simply be mobile and metabolic. It is stretching a point to bring these two processes under one identical expression.

Moreover, if once we begin to speculate about the possibilities of something not exactly life as we know it, but analogous, in its complexity or activity, to life, we are at liberty to entertain, so far as the sciences of chemistry and physics are concerned, a limitless variety of imaginary parallels. Life as we know it seems to be attached to and dependent upon, in ways we will presently investigate more fully, complex chemical compounds of carbon, nitrogen, hydrogen, oxygen, and other elements; but we are really not limited to such compounds nor to the limiting temperatures and pressures within which they can play their parts, once we allow our minds to move beyond the life we actually know. We can conceive vaguely of silicon playing the part of carbon, sulphur taking on the rôle of oxygen, and so forth, in compounds which, at a different *tempo* under pressures and temperatures beyond our earthly ken, may sustain processes of movement and metabolism with the accompaniment of some sort of consciousness, and even of individuation and reproduction. We can play with such ideas and evoke if we like a Gamma life, a Delta life, and so on through the whole Greek alphabet. We can guess indeed at subconscious and superconscious aspects to every material phenomenon. But all such exercises strain the meaning of the word life towards the breaking-point, and we glance at them only to explain that here we restrict our use of the word life to its common everyday significance of the individualized, reproductive, spontaneously, stirring and metabolic beings about us.

In a recent survey of the whole universe known to physics, Sir J. H. Jeans, the brilliant secretary of the Royal Society, summarized the relation of life as we ordinarily conceive it to that universe in its entirety. "The physical conditions," he said, "under which life is possible form only a tiny fraction of the range of physical conditions which prevail in the universe as a whole. The very concept of life implies duration in time; there can be no life where the atoms change their make-up millions of times a second and no pairs of atoms can ever become joined together. It also implies a certain mobility in space, and these two implications restrict life to the small range of physical conditions in which the liquid state is possible. Our survey of the universe has shown how small this range is in comparison with the range of the whole universe. Primeval matter must go on transforming itself into radiation for millions of millions of years to produce an infinitesimal amount of the inert ash on which life can exist. Even then this residue of ash must not be too hot or too cold, or life will be impossible. It is difficult to imagine life of any high order except on planets warmed by a sun, and even after a star has lived its life of millions of years, the chance, so far as we can calculate it, is still about a hundred thousand to one against its being a sun surrounded by planets. In every respect—space, time, physical conditions—life is limited to an almost inconceivably small corner of the universe."

It is limited as yet, but it is still premature of us to define its final limitations. It seems that life must once have begun, but no properly informed man can say with absolute conviction that it will ever end.

§ 5. The Subjective Side of Life.

ONE characteristic of life we have not yet noted. We know it directly only in our own individual cases; we infer it in the life about us; we suspect its extension to all life and perhaps even beyond the limits by which we have here defined life. This is feeling. We react to external stimulus, and we not only react but feel. We are conscious and we have every reason to suppose that a great proportion at least of the living creatures about us are conscious too. When we see an oyster close its shell at the jabbing of a stick or a lobster search with its feet and pincers for a fragment of food, it seems natural to assume that these creatures also are feeling, after the fashion in which we feel when we recoil from disagreeable or respond actively to agreeable sensations. We never experience any but our own feelings; all our belief in the feelings of other beings is based on the analogy of their movements to our own. We seem to detect shrinking and appetite even in the microscopic amoeba, and Sir Jagadis Chandra Bose, by a system of ingenious measurements, shows how even plants appear to be elated or depressed by the application of favourable or unfavourable substances, how they will seem to shudder at this or writhe at that in a fashion suggestive of feeling. But even a wire spring can be made to shudder at a touch. For all we know there may be gladness in dancing waves and a real freshness in the sunrise itself.

So far as our knowledge goes all feeling, all consciousness, requires concurrent material changes; we do not know of feeling without material change in ourselves or in those others who tell us of their sensations. We have already hinted that the converse may also be the case, and there may be no material change without feeling. But on one point let us be clear. When we attribute feeling to an oyster or an amoeba we do so because it behaves to a certain extent like ourselves, and we imagine such feeling as resembling our own to the same degree as the behaviour resembles our own. When we attribute feeling to a material change, on the other hand, we are led to do so not by sympathy but by logic. One cannot imagine any sort of intelligible emotion in a piece of zinc that is dissolving in hydrochloric acid; one does not suspect litmus of shame because it blushes at the touch of vitriol. The idea of feeling in inorganic material processes is an almost mathematical abstraction; it involves considerable extension of the usual meaning of the word "feeling." This point will be clearer when we have dealt with the texture and structure of living substance, and particularly of brains, sense-organs, and nervous tissue. We will study them and their behaviour as far as we can before we bring feeling into our discussion. But we note this internal aspect of life here merely to explain that for a time it will be convenient to disregard it and view living things from an entirely external standpoint. When we have built up a coherent picture of the forms and behaviour of life as they appear outside ourselves we shall be in a securer position for the correlation of that visible world with the far more mysterious world, outside the frame of time and space, which sees and knows.

§ 6. A Preliminary View of Living Forms.

LET us now consider how life in general as an external phenomenon presents itself to an ordinary intelligent human being who has had no special instruction in biology. In response to the question, what were the chief sorts of life, such a person would probably answer, plants and animals, and explain that plants were green-coloured in whole or in part, were rooted to one place and fed chiefly upon non-living matter, derived from the atmosphere and the soil, while animals moved about freely and needed to feed upon the substance of other living things. There would probably be a difficulty about sponges and one or two other such things—sea-anemones, for example, and sea-mats—whether they were plants or animals, and an exception to the general greenness of plants would have to be made in the case of fungi. Further questioning would elicit further difficulties in maintaining the distinction of plants from animals, which had at first seemed so plain and clear. It had seemed all right on land amid familiar surroundings, but as odd corners of the world, beaches and rockpools, dark shadows and festering corners of decay in forests, were explored, the first easy assumptions would have to be modified.

Moreover, if the question of what sorts there are of animals were next taken up, there would perhaps come the traditional answer of "Birds, Beasts, and Fishes." Are men animals? A question much debated. And are these three classes inclusive? Some further sorts of animals would be named. Insects would be noted, worms, "creeping things" such as spiders, lice, and so forth. The "so forth" would be found to be a rather expanding and varied division. "Fishes" also would have to be stretched beyond its original intention when one thought of lobsters, starfish, and oysters. And what was a frog or a crocodile? A "beast," perhaps, but lacking the familiar protection of hair or feathers, and with no warmth of blood. To the ordinary man living a commonplace life in a town or in the countryside, the birds, beasts, and finny fishes would still seem to be so largely the bulk of life, that it would be a strange and difficult thing to hear that life had been active and varied for millions of years before ever a bird or a beast or a leafy tree or a flower existed. They are the living setting of our world now, they are an essential familiar part of our existence as those others are not. Our race has built its mind, its fears, panics, fables, and beliefs on fur and feather and fin, and it is hard to think of a whole world of animated existence without them. From them we go to the stranger creatures and to them we return, as one returns to one's home from unfamiliar places.

Aristotle we may take as standing in this matter on the dividing line between the unsystematic intelligent person and the systematic inquirer. He was one of the earliest men to become systematic. He left no complete classification of animals, but he seems to have divided them into those with red blood and those without. The former he subdivided into (1) Mammals (beasts), who bring forth their young alive, into which class he put men; (2) Birds; (3) Four-footed or creeping things (reptiles and amphibia) which lay eggs; and (4) Fishes. The animals without red blood he referred to (1) a soft-bodied group, in which he put the octopus and squid; (2) a "soft-shelled" division of lobsters, crabs, shrimps, and such like, with jointed bodies and limbs;

(3) insects, with which he included spiders and scorpions; (4) creatures with hard, protective outer shells like the snail, murex, oyster, and sea urchins; and (5) zoophytes or plant-animals like sponges and sea-anemones. Later we shall give a contemporary classification which has replaced this preliminary survey of living forms.

What we find as we study the history of animal classification down to the present day is a very slight variation of the original classification of the more familiar animals and a steady increase of classes and distinctions among odd and less familiar forms. The naturalist kept opening up new departments with a sort of reluctance, as odd, intractable, unclassifiable riddles were forced upon his attention. Always until quite recent times in every classification there would be a kind of ragbag class, "Vermes," "Radiata," or "Zoophyta," into which all the residue of "puzzlers" would be thrust. First came a fine array of kindred species, varying agreeably round some well-known central form, hundreds of bird species and hundreds of fish species, clearly named and brightly depicted, and then at the end a deprecatory jumble and muddle of nonconformists. The systematic botanist kept a similar dump. He made a lovely classification of flowering plants, he never tired of arranging them, and then hurled all the awkward cases together into a division of "Cryptogams" which had nothing in common except that they had no flowers to be dried and put neatly between sheets of blotting-paper. It is among these queer, out-of-the-way, low-caste flowers and plants, as we shall see, that some of the most interesting and illuminating work of modern times has been done.

From the green scum on a dank garden path to Solomon in all his glory, from the tree to the tiger, from the swarming millions of germs in a poisoned finger to the tame elephant in the Zoological Gardens, from intestinal worm to rosebud, and from lichen to whale, life plays in endless variations that drama of movement, metabolism, and reproduction which marks it off from the mineral kingdom and from all the interplay of inanimate Nature. And, perhaps, in endless variations it plays also upon the themes of conscious and sub-conscious life, it dreams and slumbers in the plant or in the motionless fish, or drinks deep of contentment or flashes into frenzies of desire and delight and terror in hunter and hunted, in basking snake or playing cub or singing bird. And the writing and reading of this book and the thought-process behind these things are life also. All these aspects must receive our attention in the general review of current biological knowledge we are now undertaking.

§ 7. The Progress of Biological Knowledge.

BIOLOGY, the science of life, was practical before it became systematic. Man was a biologist perforce long before the dawn of history, classifying plants into edible and inedible, and accumulating a lore of the animals he hunted, and perhaps of the animals that hunted him. In the Old Stone Age, perhaps twenty thousand years ago, he was already making very competent drawings of the beasts that concerned him.

The Neolithic Age must have been a time of active biological enterprise. In a few thousand years man carried out a very large amount of experimental work, he domesticated and tamed nearly every animal that has ever been tamed or domesticated, and he began deliberate breeding. Mythology hints at a disposition to attempt the most extraordinary hybridizations. Every conceivable combination of domesticated species was probably attempted by Neolithic man.

The vigorous days of the Greek republics and the Hellenic empire of Alexander saw the dawn of modern science. It was a period of immense curiosity, which declined only as the dark shadow of Roman Imperialism fell upon the Mediterranean world. There was an organized search for facts by Aristotle's agents under the patronage of Alexander. The first museum was, as the name implies, dedicated religiously to the Muses, and the earlier collections of curiosities were in temples. Hanno, the Carthaginian, after his memorable coasting voyage about Africa, hung the skins of his gorillas in a temple with an inscribed record.

Hellenic science during its phase of energy achieved some remarkable things. We hear, for the first time, of dissection and of vivisection at Alexandria. Men were beginning to pry into the mechanism of these sympathetic creatures so like and yet so unlike themselves. The question whether vivisection was actually performed upon human subjects in Alexandria is still an open one. A little later than the Hellenic spurt came a phase of botanical experimentation, and particularly of experiments in acclimatization and plant-collection under Asoka, in India. The four centuries before Christ were years of exceptional stir and expansion in the intellectual world. But as the vast organization of Rome wasted Sicily, North Africa, Greece, Egypt, Asia Minor, Spain, as it staggered through its brief uneasy centuries of feasts, circuses, and general grossness, to collapse at length into barbaric confusion, the impetus of that intellectual dawn died away, and the first harvest of biological knowledge shrank to neglected manuscripts in the libraries of the acquisitive illiterate, and was well-nigh forgotten.

The biology of that first phase of scientific activity was not perhaps so remarkable on the whole as its astronomical and mathematical achievements. The dissection undertaken was not sufficiently systematic to leave the world a thorough collection of drawings and traced connections, and no special results seem to have accrued from the vivisection done. It seems to have been a very occasional effort of curiosity. It is remarkable that no resort was had to any magnifying contrivance. The use of a water-flask for magnifying objects may have been known, and probably was known, to the jewellers of ancient Egypt, but it never seems to have got through to the philosophers. Guild barriers and class restrictions may have prevented the necessary intercourse between learning and technical skill.

The onset of Christianity and Christian monasticism in the declining Romanized world helped to check the development of biological curiosity. Neither secular art nor secular learning had a place in the Christian scheme of salvation. And concurrently with the spread of Christianity came a deepened sense of sin, an enhanced fear of lurking snares and temptations, which cramped enquiry. This or that knowledge might perhaps prove to be forbidden and unholy knowledge. The Greek mind had attained a fairly free, candid, and experimental attitude towards thought in Athens and Alexandria between the fifth and third centuries B. C., but the mind of the decaying Roman Empire, and still more so that of the confused centuries that followed, was obsessed by a sincere terror of the Tree of Knowledge. The dissection of animals could no longer be thought of as a frank examination of something curiously and fascinatingly hidden; it was a process touched with a quality of monstrosity, a vile prying into entrails for base and evil ends, best done in the dark of the moon. These feelings were not in the least humanitarian; they did not deter the medieval world from the most liberal use of rack and thumbscrew and every form of torture, nor from the infliction of death in exquisitely painful forms, but they made men recoil from the cool and deliberate investigations of the anatomist's knife. Why should you cut up a man, unless you hated him and wanted to triumph over his poor remains? That one could understand. To dissect things simply in order to see and know was to defy the powers that had hidden his own interior from man. It was presumption. It was not good, honest cruelty; it was pride and impiety to do such things.

STUDY OF LIVING THINGS

Let us not suggest that the restriction upon intelligent curiosity, the submergence of the first beginnings of scientific biology in Europe and western Asia by the Roman civilization, was the coming of a new and unprecedented darkness upon the human mind. Still less would we attach any specific blame to the spread of Christian doctrine as such. Quite outside the Christian movement, the Roman world was pervaded by a general drift towards magic and superstitious squalour. The Pagans just as much as the Christians felt the fear and fascination of things forbidden. It was a relapse to a congenial obscurity from which man had for a brief period emerged. The great age of Greece was an exceptional release of intellectual courage in the ancient world. Man has always been, and still is, disposed, perhaps instinctively, to suppression and panic at plain statement; it is the hardest task of the educationist to train him to look facts in the face. He is curious by nature—yes, but he is meanly and furtively curious. He does not like to be caught looking or suspected of thinking. About the body a cleanly frankness is the last achievement of our civilization. There are millions of people in the world who have never contemplated themselves unclothed in a mirror, and would not willingly do so. Many religious orders deny their votaries an entire bath. There is a real natural and instinctive resistance to biological knowledge in the mass of mankind, and the nearer that knowledge approaches to our own minds and bodies, the intenser the resistance becomes.

The resistances to knowledge are not merely passive. It is not only that most human beings are indisposed to know and learn; they are afraid of and hostile to all that they do not know and they seek to prevent it in others. The human mind is much more tortuous and indirect than it will consciously admit; it often fails to understand its own motives. Since the great revival of scientific work in the sixteenth century there has been a steady undercurrent of deprecation and antagonism which rises very easily to active obstruction and suppression. The self-love of the ignorant has demanded that the man of science, in play and story, should be caricatured, ridiculed, and misrepresented. From Laputa to the Pickwick Club, British literature, for example, spits and jeers at its greater sister, and to devote a life to science and the service of truth is still to renounce most of the common glories and satisfactions of life for a hard and exalted mistress. But while the pomps and glories of every other sort of human activity fade and pass, the growth of science is a continuing and immortal thing.

The revival of biological enquiry and free biological teaching became remarkable in the sixteenth century. An increasing number of active minds directed themselves to observing, collecting, and classifying. Art was becoming observant and more sedulously representative. Dissection at first was quite as much artistic as medical and surgical in its aims. There is a strong decorative purpose apparent in many early science collections. They consisted of "curiosities," not specimens. The student, turning over the notebooks of Leonardo da Vinci, finds there the artist and scientific man inextricably mingled. From the studio and the private investigations of dissatisfied medical men, far more than from the bookish study, did the modern science of life arise. Modern science owes more to art and natural curiosity and less to literature and philosophy than is commonly understood.

The geographical exploration of the fifteenth and sixteenth centuries was an enormous stimulant of biological work. The invigorating influence of practical benefit was supplied by the acclimatization of new food plants, the application of new medicinal substances, the production and importation of such commodities as cane-sugar, tobacco, tea, coffee, and a great variety of economically valuable novelties.

The development of the microscope in the seventeenth century meant a mighty enhancement of scientific observation. It revealed not only a whole new world of unsuspected life-forms, but a thousand structural secrets too fine for the scalpel. In many of its branches the advance of biology has been determined by the progress of microscopy. A third great mine of contributory knowledge was opened up by the study of fossils, of which Nicholas Steno (1638–1686) was the pioneer.

FIG. 1. One of the first biological 'microscopes—an instrument made and used by Anthony van Leeuwenhoek (1632–1723).

The magnification (up to 160 diameters) is accomplished by a single lens.

Biological science remained largely observational for three centuries, passing slowly from the development of botanical and zoological gardens and the industrious collection of shells, bones, fossils, flowers, fruits dried and preserved, specimens of all sorts, to dissection and the microscopic examination of tissues and living structure, and so on to the subtle and ingenious analysis of all this ordered accumulation of facts and relationship.

Experiment has gone on from the seventeenth century, and since the middle of the nineteenth it has become the leading strand of biological work. Before that there were trials of this or that substance or device upon a living subject, vivisection to test deductions from observations, and a certain amount of exploratory experimenting, but there was no wide progressive continuum of experimental work. In the middle of the nineteenth century the Abbé Mendel made a series of breeding experiments, the full importance of which was not recognized at the time. The work was revived about the beginning of the present century, and became the starting-point for a great activity of experimental breeding of poultry, rabbits, mice, guinea-pigs, flies, and plants of many sorts, types which reproduce with sufficient frequency to make possible the observation of a series of successive generations. Work upon the living subject became more and more frequent and relatively more

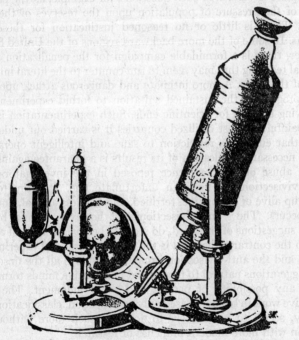

FIG. 2. The microscope with which Robert Hooke (1635–1703) discovered vegetable cells.

The microscope itself is on the right; on the left are a lamp and a glass globe filled with water, for focusing a spot of brilliant light on to the object under examination. (This and Fig. 1 by courtesy of the Royal Microscopical Society of London.)

important. Increasing delicacy of manipulation, particularly under the microscope, has, for example, rendered directive interference with the development of the eggs of many creatures a practicable thing, and much work has and is being done upon detached fragments of still living tissues, carefully kept alive in nutritive solutions, and upon the grafting of tissue from one organism upon another. As the task of observing, classifying, comparing, and recording the forms of life approaches completion, experiment will become more and more the normal method of progress.

Antagonism to biological knowledge is by no means dead. There is a constant struggle to keep physiological or pathological information from people who might put it into beneficial practice, and to prevent the complete discussion of such questions, for example, as the possible control of the pressure of population upon the reserves of the community. There is little or no reasoned justification for these suppressions. In some of the more backward regions of the United States, moreover, there is a formidable campaign for the penalization of any biological teaching that may seem to run counter to the literal interpretation of the Bible. A more intimate and dangerous attack upon biological progress is the sustained agitation to forbid experimentation upon living animals for scientific ends. Such experimentation is very rarely painful; in most civilized countries it is carried out under conditions that ensure its restriction to sane and intelligent operations, and the necessary publication of its results is a guarantee against any possible abuse of the confidence reposed in the investigators. The word "vivisection" is itself an unfortunate one. It intimates the cutting up alive of a sensitive, terrified, and helpless animal, and that never occurs. The "anti-vivisectionists," however, wrung by the horrible suggestions of the word, do seem to believe, in spite of all evidence to the contrary, that this is the normal method of experimental biology, and the anti-vivisection campaign displays all the unscrupulous exaggerations natural to tender and imaginative minds tormented beyond any possibility of patient and sober judgment. The older descriptive work in biology, based only on collection, classification, and anatomy, seems to be unlikely to achieve much more without coöperation with more modern branches of research.

Meanwhile a new province of observational and experimental work, ecology—the observation of animals in their proper surroundings and of their normal interaction and ways of life, and what one may call field physiology—is being very extensively developed. Men of intelligence are taking cameras and building watching-shelters in forest and jungle and prairie, where formerly they took gun and trap and killing-bottle. Zoological gardens are being reconstructed and enlarged, so that, while formerly the animals were exhibited as specimens, they are now watched going about their normal affairs.

In place of the encyclopædic zoological gardens of yore, placed often in the proximity of great cities, with their smoky air and abundant contaminations, governments are now setting up reserves and small stations, here, there, and everywhere, in which particular species and groups of species may be observed less for structure than behaviour. This sort of observation passes insensibly into experiment—experiments in behaviour, experiments upon habit and intelligence, experiments in physiology. Parallel to these modern zoological gardens, the modern botanical garden expands from the old obsession with specimens. There are now even bacteriological zoos, so to speak, where collections of living cultures of this, that, and the other infection are available for the experimentalist. Human psychology has also become observational and analytical in the past quarter of a century, and has been brought into closer touch with nervous physiology on the one hand and with the behaviour of animals on the other. To summarize the accumulating harvest of the army of workers now busy in all these fields is the task we have set ourselves.

We shall begin by giving an account of the structure and general life-processes of the most familiar of living things—man. After that we shall take an ampler view of the forms of life, to revise the popular classification we have cited in the preceding section. We shall then be in a position to study the reactions of life upon life and its general evolution. We shall consider life in health and disease. Then we can come to the finer questions of feeling and thought, to man as a biological type, and to the huge expansion of his range during the last million years, which has made him now the master and main representative of living matter. And so to life as we know it in the city and the street, the home and the laboratory, and its outlook towards time to come.

BOOK ONE

THE LIVING BODY

I

THE BODY IS A MACHINE

§ 1. *The Fundamental Routine of Two Living Creatures.* § 2. *Why We Call These Bodies Machines.*

§ 1. The Fundamental Routine of Two Living Creatures.

WE SHALL begin this work as we shall end it, with man. There our interest in life begins and there it culminates. First we shall review what is apparent and what is known about his life, in broad general terms. All this work, *The Science of Life*, is sustained by the question of what life is and what its possibilities are. And to begin with, we have to consider its material facts, to ask how this body, which is either itself life or the expression of life, moves and acts upon surrounding things. To do so, we may have to go over many facts that will be more or less familiar to most educated people. Nevertheless, they may find it advantageous to run over what we have to say to refresh and define their knowledge. We shall find first that the human body is a machine, amenable to the same chemical and physical laws as most of the machines man makes himself for his own ends. But next we shall realize that it is far stranger and more complicated than any machines that have ever been contrived by human agency. And, thirdly, we shall find it is a machine which does its own repairs, sees to its own fuelling and lubrication—though many automobiles attend to the latter need nowadays—and has the capacity of making new machines to replace itself.

For the most part we shall be writing of Man in the opening book. But there are close similarities to him in most of the familiar animals about us, and occasionally it will be convenient to cite the ape, the dog, the cat, and the mouse, to fill in gaps of our description. As the reader probably knows, these animals are classified together, with man and all other creatures that have hair and warm blood and suckle their young, as *mammals*.

Let us first take a Man, Mr. Everyman, and consider certain familiar but sometimes disregarded aspects of his daily life. We will say nothing here of his loves and hates, his dreams, and his political opinions. That must come later. We will consider him first as a moving body and note little but the facts that illuminate the mechanical processes of his being.

He is, we will assume, a denizen in a modern city, who works as people say for his living. We will commence our observations while he is still asleep. On the stroke of seven he is roused by his persistent alarm-clock. He stirs and grunts and, since it will continue its clamour until he presses the release, and since he has deliberately placed it out of reach at the far side of the room, he is forced to get out of bed. Mechanically he turns to his ablutions. He washes his face and neck and he brushes his teeth; probably, unless it is Sunday or some other red-letter day, he does not worry about the rest of his anatomy, for that will be effectively concealed by his clothes. Perhaps as he dresses he is reminded of that slow, relentless change that no man can stay—he may drop his comb and note that he cannot stoop as swiftly and supplely as he used to do; he may look carefully in the glass and pluck a white sheep out of his flock of hair; if he is very young he may pride himself on the firm, manly lines that are appearing on his brow and round the corners of his mouth.

By the time he is dressed and brushed he is well aware of a clamour from his own inside—from his digestive organs; and they have to be silenced and kept busy before he can attend to anything else. It is a very persistent clamour. It is even more persistent than his alarm-clock. He might stop his alarm-clock and go to bed again without immediate inconvenience, but hunger goes on. It is his machinery demanding fuel. And so he turns to breakfast. For a time his attention is divided between swallowing to satisfy his interior need and reading the newspaper to employ a mind which might otherwise be bored while this primary business preoccupies him. Then he sets out to his work.

It may be that he spends his morning at a desk, or behind a counter, or in some physically arduous employment. He may be one of those rare, fortunate individuals who have an occupation that they enjoy. We will not discuss here how he expends his energy. For most men that takes the form which is called earning a living. He is making his food supply secure. Sooner or later he finds his energy slackening, his attention wandering. His digestive organs are calling for attention again.

So out he goes to lunch and then back again to his desk, or wherever it is that he works. Many people are doing the most various things about him in the City, but the lunch hour calls to them all. At lunch again he may seek the distraction of a book or paper, or talk with companions. There, perhaps, is something not quite like a machine. But whatever else he does, he eats. Before he goes to bed he will have to answer that imperious demand from within again. He may answer it once or twice. Three or four times a day he loads his stomach to keep his wheels going round—and several times, we must note, he has to deposit the residue of their working in appropriate places. There we have the fundamental and unavoidable element of his routine. And of everybody's routine. Saint and sinner, plutocrat and proletarian, judge and criminal must all observe a similar round. All men must eat. All the rest of their activities are secondary to this daily necessity. These activities may vary endlessly with the accidents of private fortune and the chances of the day. On any particular day our Mr. Everyman may work or not; he may exhaust himself in his daily employment or be urged by a superfluity of energy to play, to seek amusement, make love, or quarrel. He may find such a sedative drug as tobacco useful in slowing down the urgency of his bodily engines. He may be pursued by danger or moved to deeds of violence by a fire or a dispute. Still he will eat or want to eat. Under abnormal conditions people have staved off eating for as long as forty days. It has left them exhausted and emaciated and manifestly on their way to extinction. After only a day or so their general activities have been restricted and their energies have been concentrated upon resisting a supreme urgency.

And at length after Mr. Everyman's day of eating, etc.—the etc. may vary but the eating is constant—comes the night and the desire for sleep. There again is another great universal, and our Mr. Everyman, however his afternoon and evening have been spent, realizes at last a growing need for repose. He makes his way back to the bedroom in which we found him, winds up his alarm-clock, puts his head on his pillow and drops back into the comparative inactivity of slumber. The daily round of life has completed its circle. Man must eat and also man must sleep. There may be interruptions and delays, but that is the formula, and to disobey it for any length of time is death.

Now here perhaps in the sleep-necessity the critical reader may ask whether there is not a difference from any mechanism. Machines, he will say, do not sleep. That is almost true of such a simple and unvariable mechanism as a clock, which, regularly wound, may never cease ticking for years, but it is not true of such a versatile piece of engineering as an automobile which is subjected to constantly changing stresses. That must go out of activity periodically, for adjustment, lubrication, and minor repairs. The more complex the machine the greater the difficulty of keeping it in perfect adjustment. And Mr. Everyman, if indeed he is an apparatus, is a far more versatile apparatus than any automobile, and he and his kind and a vast majority of the higher creatures have made the habit of rest and reparation during the most convenient part of the twenty-four hours into a second nature. We shall have more to say of that necessity for readjustment when we consider the inevitability of old age and death. But even when Mr. Everyman sleeps, he is not all inactive. The mechanism is still moving. His heart, his lungs are not sleeping. They tick on just as the clock does, but at a slower, steadier pace than during the day.

Let us now compare the routine of Mr. Everyman, the most interesting of living creatures to us, with another daily round, at a rather lower level than his own. Mr. Everyman is sleeping, silent or gently snoring. If he has not eaten or if he had eaten badly his sleep would be slight and distressful, for digestion is more important than slumber, but we will assume that all is well within him. The house grows still. Presently the silence is broken by little sounds that echo disproportionately loud, a squeak and a little stir behind the skirting, a patter over the floor. It is Mr. Everymouse, another familiar embodiment of life, abroad.

For a mouse the day is a time of peril, a time when the world belongs to active, clumping, dangerous giants. Somewhere in a safe corner— behind a book-case, under the floor-boards, in a hollow wall—he has built a nest, soft and warm, with such materials as he has been able to gather; here he lurks during the day. At nightfall he grunts in his turn, and stretches and sallies forth to satisfy his tiny clamouring stomach. The life of a mouse in its physical essence is entirely parallel with that of a man. The mouse-life is largely spent like the man-life in eating and in toiling—in gnawing holes and runs, in collecting nest-materials, in foraging for crumbs and suchlike remnants. There are intervals for ablution—and although a mouse does not use soap and water he goes over himself with a thoroughness that would put many of our own species to shame. There are social occasions—friendly meetings, fights, and the periodic disturbances of love. There are young mice and old mice, male mice and female mice, helpless pink baby mice to parallel Mr. Everyman's social circle. The chief difference is that the mouse-world is not so well organized as ours is, nor is Mr. Everymouse's day so definitely planned; he has to hunt about for his food instead of going to special eating-houses, and he has to be more on the alert for danger. He probably goes hungry more often than does Mr. Everyman. Presently, a little after dawn the house world from which a sensible mouse retires, becomes a world of loud noises and perils, and so he gives place to the man again.

These two living creatures are as mechanism very much alike indeed. Later on we shall have to consider many forms of life so different from man in form and nature that relatively a kind of cousinship will appear between the two. As the reader probably knows they are classified together, with dogs and horses and all creatures that have hair and

Fig. 3. A highly successful mammal, to judge by its abundance—a mouse at its toilet.

milk, as "mammals." Their bodies are such beautiful and wonderful things that to many people it is almost repugnant to approach these palpitating sensitive structures in a coldly calculating mood. And yet if we are to understand them aright that is the mood in which we must study them. In that mood alone can we seek the explanation of this perpetual and universal desire that makes the stuffing of food into a hole in our faces the first fact in life, and reconcile the higher aspects of existence with even more sordid realities. These things and many others become clear and only become clear if we analyze the body from a mechanistic point of view, if we insist upon knowing why food is thus primarily necessary for life, and if we ask why does life, our life and the life of all the creatures we know and observe, hinge first of all on that? If we refuse to slur over these questions we shall find only one possible answer.

THE BODY IS A MACHINE

§ 2. Why We Call These Bodies Machines.

A MAN is continually giving out energy. He is never completely at rest as long as he lives. Even when he sleeps his limbs twitch, his ribs heave, and his heart continues with regular rhythmical beat. And besides these actual movements, there is another way in which he loses energy: his body is warm. A hot piece of metal or stone will, in a short time, cool down to the temperature of the surrounding air, but a living body is always warm to the touch. A man, like any warm object, loses heat continually to the cooler air which surrounds him, but he has internal sources of energy which compensate for his loss, so that except in extreme tropical heat his temperature is always higher than that of his surroundings.

The need of a man for food and drink is apparent in a great part of his waking activities. On that we have already insisted. Less conspicuous, but in reality even more urgent, is the need for air. Normally, it is unnecessary to seek that. Always, almost unconsciously, a mammal is inhaling and exhaling air. It hardly realizes the necessity of that until, under some exceptional conditions, the supply is cut off and suffocation begins. But you can put a man out, just as you put a fire out, by stopping the air supply. If we take the air a man breathes out and analyze it, we find that he has used some of the oxygen gas that normally air contains, and we find substances, water vapour and carbon dioxide, which are precisely what we would expect to be exhaled if the oxygen were used for disrupting the food in order to yield energy. A petrol motor, or a fire, does exactly the same; in both cases, besides fuel, there must be a supply of air, and in both cases oxygen is used up and carbon dioxide and water vapour are given off.

Finally, we find various matters passing out of a mammalian body, matters other than the mere undigested residue of his food. If our analogy with a petrol motor is sound, we may compare these matters to the more complex ingredients of exhaust gases—both to the more complex products that result from the burning of the fuel, and to the fine particles of steel that wear away during the working of the engine itself. And, indeed, this is what our excreted substances represent.

Now, is this analysis sound? Is a living man fundamentally a machine? That is a question capable of experimental decision. We can measure the amount of food that a man or an animal consumes over a given period of time, and we can measure the energy yielded during the same period. If we burn an equal weight of similar food in a suitable apparatus and find out how much energy its combustion yields, and if this value is equal to the energy yielded by the experimental subject, then evidently the living organism, so far as its energy-output is concerned, is really and precisely a combustion engine.

Such experiments have been performed. A man has been shut up for a time in a small compartment, so constructed that the energy he gave out, as heat or otherwise, could be accurately measured. He was fed on a weighed amount of food of known composition and the energy actually yielded was compared with that which would have been obtained had the food been simply burnt. Such experiments have been made repeatedly both with man and animals. The following figures were obtained by Atwater, who worked with a human subject. A man was shut in a calorimeter (heat-energy measurer) for nineteen periods of twenty-four hours each; on the average, he gave out 2,682 large calories in each period, and ate an amount of food which would yield if burnt 2,688 calories. (A large calorie—there are several sorts of calorie—is the amount of heat required to raise one litre of water from $0°$ to $1°$ centigrade.) The agreement is good. The two sums differ by less than one fourth of one per cent, which is well within the margin of error of the methods employed. Numerous other experiments have yielded equally convincing results. We are, therefore, justified in concluding that man and the animals generally are fundamentally mechanisms, driven by the energy liberated in the oxidation of food. A mouse or a man works in much the same way as a petrol motor; its fuel is its food, and precisely after the fashion of a petrol motor, if it lacks either fuel or air, it will cease to move, and slowly become cold. Moreover, the living engine needs an exhaust; as exhaled carbon dioxide or in other ways it must get rid of the products of this combustion or they will clog its system and impede its working. Indeed, the whole of what we have called the fundamental round of Mr. Everyman is a mechanical process.

Naturally, we do not suggest that this demonstration of our fundamental mechanical nature explains everything; as we have already pointed out, the living body has certain obvious properties that distinguish it from any man-made machine. It grows and reproduces its kind, and it is conscious. In later parts of this work we shall discuss the intensely interesting question how far it has been found possible to bring these higher and more intricate phenomena underlying development and mind into an agreement with known physical and chemical laws similar to that which holds good for the primary round of life. But here we will content ourselves by pointing out that these phenomena do not occur unless that fundamental round is sustained. A brain must have food and oxygen or it cannot think, and an embryo, if it is to grow, must have them too. There is no observable life-process which is independent of a supply of physical energy. When that ceases, growth, movement, and all signs of consciousness cease also—so that in studying the physical mechanism of the body, even if we are not studying development and mind themselves, we are studying conditions upon which they rest and from which they are apparently inseparable.

We have spoken in this discussion of Mr. Everyman and of the animals that are most nearly related to him. But our conclusions are not by any means limited to mammals. It can be shown that they have a very much wider application than that. *There is hardly a living creature that does not depend in the last analysis upon oxidation for the energy by which it lives.* All animals and all green plants are energized in this way. Plants differ from animals in being able to make their own fuel instead of eating it; they can do what human invention has so far failed to do upon a paying scale—they utilize the radiation of the sun directly for making substances on which their engines can run. There are indeed one or two microscopic creatures that have other energy-sources—yeast, for example, can use a chemical reaction that does not involve oxygen. But these are exceptions; we shall tell of them later. For the present it is enough to remember that all animals (including men) are combustion engines of an intricate and curious kind, which live by oxidizing their food. We will now go on to the lay-out of these engines and some interesting details of their structure and working.

II

THE COMPLEX BODY–MACHINE AND HOW IT WORKS

§ 1. *What Every Schoolboy Knows About the Body.* § 2. *About Cells; the Lesser Lives within Our Life.* § 3. *Blood.* § 4. *The Course of the Blood.* § 5. *Breathing.* § 6. *Kidneys and Other Exhaust Organs.* § 7. *How Our Food Becomes Blood.* § 8. *The Continual Struggle against Infection and Chill.*

§ 1. What Every Schoolboy Knows About the Body.

THE writer of a general review of the sciences that concern life such as this we are engaged upon is confronted by certain peculiar difficulties. He is writing for readers nearly all of whom have some knowledge of the subject in hand. He is threatened, therefore, on one hand by the probability that he is boring his readers with what they know already, and on the other by the probability that he is assuming they know too much. The writer of the text-book or the university lecturer is free from these perplexities. His readers and hearers are under compulsion to go on attending. He can say *everything*. He can inform his victims, graciously and elaborately, where their backbones are to be found and impart the stirring discovery that "at the top of the spinal column we find the head, of which the larger part is the cranium or brain case, below which are two passages separated by a horizontal partition, the upper one being the nasal passage and the lower the mouth." And he can go on in that strain for as long as he likes and then jump into technicalities that will make our heads spin.

We cannot do that sort of thing with that much magnificence here. Yet our better-informed readers must allow us to run through, as compactly as possible, a certain amount of knowledge they no doubt acquired at school. If these better-informed readers will run beside us, they may find a fact or two of interest for themselves on the way and arrive at our more detailed expositions with their own knowledge sharpened and refreshed. Anyhow, it has to be done. We are going

COMPLEX BODY-MACHINE AND HOW IT WORKS

to recite here what, in Lord Macaulay's phrase, "every schoolboy" knows about the body.

We will assume that the reader is already aware that every movement described in the First Chapter is caused by muscles and that he knows these muscles are the meat or flesh that clothes our skeleton. We will assume, too, that he knows the general structure of his skeleton. He knows where to find his backbone, the tie-beam of all his bony framework. He knows it is composed of separate bones (the vertebræ) jointed together to make it flexible, while at the same time the nature of the joints and the strong ligaments binding it together limit this flexibility and make it rigid enough for its supporting and strengthening function. It encloses and protects a vitally important nervous organ, the spinal cord. The column is hollow, a flexible tube; each vertebra has the shape of an irregular knobbly ring, and when they are all jointed together the holes in the rings form a smooth, continuous canal in which this spinal cord is securely housed. Above the canal opens into the brain-case and through this opening the spinal cord is continuous with the brain.

All this is matter of common knowledge, and so, too, is the fact that the joint-surfaces, where two bones have to slide over each other, are covered with thin layers of gristle, as smooth as glass. Moreover, the joints are enclosed in close-fitting bags which contain a watery fluid to lubricate the sliding bones. At the joints the bones are lashed together by strong white fibrous bands, or ligaments. These jointed bones are worked by the muscles, which are the "flesh" or "meat" of the animal. Muscles and bones form nearly all the substance of the limbs of mice and men. Muscles are generally band- or spindle-shaped, and they are joined usually at the ends to two separate bones, which they pull together by altering their shape. They have the power of becoming thicker and shorter. No doubt the reader has amused himself by watching his "biceps" muscle thicken up as he bends and tightens his arm. Except for the inside of skull, chest, and belly (*abdomen* is merely the polite word for the latter) we are all bone and muscle. The human arm and hand are worked by fifty-eight different muscles working on thirty-two separate bones (three long ones and the rest in the wrist and hand)—counting the big muscles that go to the shoulder-blade, ribs, and collar-bone. The face is provided just under the skin with the thin slips of muscle that move the features and produce various expressions. Anatomists distinguish thirty-one of these muscles. At the sides of the head there are the eight muscles that move the jaw. Below are the complex muscles of the throat, tongue, and voice-box, and occasionally we meet gifted individuals who have extra head muscles—thin films over the domed top of the skull by means of which they can waggle their scalp, delicate strips that allow them to twitch their ears. The neck is mostly muscle; it is purely and simply an anatomical device which allows the head to be turned, raised, or lowered without moving the trunk. (A frog has no neck; if a frog wants to snap

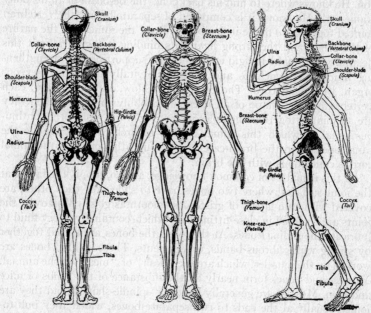

FIG. 4. The human skeleton from the back, from the front, and from the side.

at an insect he has to turn his whole body until his mouth is pointing in the right direction. A dog or a man, on the other hand, can turn his eyes and mouth in a number of directions without moving his body.) In the shoulder and chest there are great muscles that bind the arms to the trunk and move the shoulder joint in various ways; in the hips and buttocks there are others that bind the legs to the trunk, moving the hip-joint and supporting the whole weight of the body. Running over the ribs there is the clothing of muscles that produces the rhythmic swell and ebb of our chests; and wrapping round the abdomen, just below the skin, is another (three-ply) muscular sheet. These are some of the more important muscles. But in the small space at our disposal we can give no adequate idea of the full complexity of the system. The medical student who is studying "myology," as it is called, has to familiarize himself with some two hundred names, and since most of these are paired muscles (such as those of the arms and legs), and some are serial (such as those of the ribs), the actual number of muscles in the human body is very much greater. But let us leave to the medical student the labours peculiar to his profession.

These hundreds of muscles that move the human machine about are controlled by the nerves, tough white cords of tissue, some thick and conspicuous, some so thin as to be barely visible, running and branching among the quivering muscles. As everyone knows, they function as telephone wires; they bring stimuli to the muscles whenever the latter are to move. Their arrangement is quite as complex and intricate as that of the muscular system, for each muscle must have its own nerve-supply. They radiate out from the brain and its continuation the spinal cord. Brain and spinal cord constitute a government, so to speak, which rules the muscles; it is constantly flashing instructions along the nerves and commanding a movement here, a relaxation there. Without this central control—if, for example, its nerve is cut or diseased—a muscle is paralyzed, flabby, and helpless.

The brain and spinal cord do not exercise a wanton or capricious dictatorship; besides sending commands they receive a copious stream of information from all the parts of the body, and laboriously respond to that stream and move the mechanism in appropriate ways. The central government is the brain. Its informants are the sense-organs—eyes, ears, nose, tongue, the organs of touch, warmth and pain, the organs in our muscles and joints that feel at any moment what our limbs are doing, the others in our chests and abdomens. These and many more are continually sending in their messages to the brain and the spinal cord, and the incoming stream, like the outgoing stream of instructions, is also borne by nerves. As our imaginary professor told us at the outset, the brain-case is the upper part of the head and the nose and mouth the lower. In a man, having a powerful brain and feeble teeth, the former part is by far the larger; the brain-case overshadows the face. In a mouse, with a comparatively ill-developed brain but with large and wonderfully specialized teeth, the brain-case is the smaller of the two.

The blood-vessels that supply our muscles are of two kinds—the thick elastic arteries, along which the warm, pulsing stream of blood is pumped, and the thinner veins, along which it flows sluggishly back. Between the two there is a third system of very fine vessels, for the arteries lead into a network of invisibly small tubes, the capillaries, from which the veins lead away. The blood-stream, coursing and filtering through the muscles, keeps them active and healthy: it brings them fuel to burn and oxygen to burn it with, and it brings matter with which they can repair themselves from the slight wearing away that their working entails; moreover, it washes away the exhaust gases, so to speak—the by-products of their activity. The blood-stream, be it

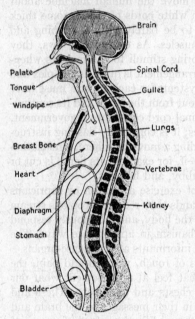

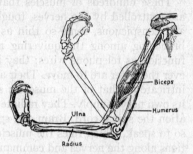

FIG. 6. The movements of the limbs are produced by the pull of shortening muscles on the jointed bones.

The figure shows how the biceps, attached to the shoulder and forearm, produces a bending at the elbow joint.

FIG. 5. A diagram of the human body seen from the side, to show as simply as possible how the more important organs are disposed.

noted, is not a peculiarity of the muscles; it perfuses every corner, every shred of tissue in the body, bringing food and oxygen and rinsing waste matters away.

The circulation of the blood centres in the heart and that is in the chest, where also the lungs are accommodated. Need we do more than name the ribs and the midriff enclosing this cavity? The professor insists on calling the midriff "the diaphragm, a domed muscular partition between the *thorax* (his name for chest) and the abdomen." But he points out that the chest is really a conical rib-cage, narrow above and broad below, and that is worth remarking. What we thrust out as our chests when we want to swagger like an athlete is mainly the big muscles on the *outside* of the thorax which go from the ribs to the arms.

The abdomen is singularly unprotected in front, with nothing but a soft muscular wall. Most of us have wondered at times why Nature never foresaw the kick and the stab at our lower vital centres. The liver and stomach indeed cower up under protection of the lower ribs, being able to do so because of the domed frame of the midriff but the poor bowels are dreadfully exposed in a beast that walks erect. The rest of the contents of the abdomen need but the slash of a sharp knife to be spilled abroad, as happens in *hara-kiri*, that form of suicide so popular in Japan. The abdomen is largely filled with the massive apparatus that digests, absorbs, and handles food—the stomach, the coiling bowels, and so on. Many of these parts are in incessant movement, churning the food inside them, mixing it with the digestive juices, and driving it along their canal by means of forceful rhythmical contractions. At the back of the abdomen, in the small of the back, lie the kidneys, and below is their associate, the bladder; they are the apparatus which weeds out from the blood the poisonous by-products of the activities of our bodies and ultimately expels them. Moreover, the abdomen contains most of the organs of reproduction.

Fig. 7. A dissection of the forearm, in which some of the muscles and related parts are displayed.

A limb consists largely of muscles; branching among them are arteries (seen as greyish tubes), veins, (left out of the drawing for clearness), and nerves (seen as white cords.)

Thus far the body of a man is very like that of any other mammal—a mouse, for example. Both possess limbs and chests and abdomens, hearts and spinal cords and bowels. But when we come to the lower end of the body we find a serious divergence between human and murine anatomy. A man terminates in a smooth, pink expanse of skin, but a mouse continues into a long and ingeniously movable tail. In a fish the tail is even more important; it is the principal organ of locomotion, and includes all the bones, muscles, blood-vessels and nerves that are necessary for movement in a limb. The tail of a mouse is less important. The mouse propels himself chiefly by means of his limbs, and although the tail is used for balancing the body and for getting a firm grip in climbing, it is by no means his chief organ of propulsion. It includes the same organs as the tail of a fish, but very much more poorly developed. In ourselves the tail is represented by a useless little curling end of the vertebral column that does not even project from the surface of the body. At least this is true of our adult selves; as embryos we had perfectly good tails, as we shall tell later.

So, with apologies, we tell over again what everybody knows, and review the general layout of a man. And having made quite sure where our heads and tails are, we will now go on a little more precisely to other matters, also very widely known but not so extensively and not nearly so exactly as what has gone before.

§ 2. About Cells; the Lesser Lives within Our Life.

WE HAVE shown in Chapter One that the human body is fundamentally a machine. We have now to explain that it is not a simple machine. It is a great mechanical system made up of an almost infinite multitude of smaller machines.

We remarked in the introduction that one of the most important and revolutionary extensions of the science of life occurred when the microscope came to bear upon living structures. Before that time the wisest men knew no more about the composition of flesh and blood than the most ignorant butcher who ever cut up a carcass. Many philosophers knew less. Now nearly everyone knows that the blood is full of billions of microscopic bodies called blood corpuscles and that most of our bodies are built up of units called "cells," as a wall is built of bricks or a cotton strand of fibres.

The word "cell" is a most unfortunate word in this connection. That is why the triplex writer has put fastidious inverted commas about it in the last two sentences. He dislikes handling and using it. He has to do so, but thus he shows his reluctance. The old original meaning, the proper meaning of "cell," was a compartment, an enclosure. We still talk of the hermit's cell or a cell in a prison. And many people at the outset of their biological reading are misled, therefore, into imagining that our living tissues have a sort of honeycomb structure. Nothing could be farther from the reality. The proper word should be "corpuscle" (little body) and not cell at all. Only comparatively few of our

COMPLEX BODY–MACHINE AND HOW IT WORKS 39

own cells are imprisoned in a case. Yet the misnomer arose very easily and naturally. Hooke, in 1667, peered down a microscope at a thin slice of dead cork, and found it to consist of an enormous number of

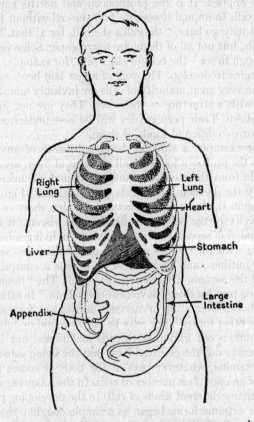

FIG. 8. A diagram showing where some of our internal organs are placed with reference to the ribs, the breastbone, and other easily recognized surface landmarks.

tiny empty boxes fitted close together, and cells seemed to be the only proper word for them. His observation was extended to a considerable number of other vegetable substances, and in all of them he found cells likewise. Very gradually the interest of investigators shifted from the boxes to the contents, but not abruptly enough to rechristen these objects. In living cork, or in the other tissues of a plant, the cell walls are inhabited by pieces of a slimy, gelatinous substance called *protoplasm*, and it is to these contents and not to their prisons, that the word cell is now applied. It is the protoplasm and not its garment which makes the cell. In animal tissues we find the cell without the box. The protoplasm may go bare; "the cell's the cell, for a' that."

Most cells, but not all of them, are microscopic. Some vegetable cells are big enough to see—the cells of elder pith, for example, and any egg before it begins to develop. The yolk of a new laid hen's egg is a single cell. But the very great majority of cells are invisibly minute. And they are things with a structure of their own. They are not just lumps of this protoplasm. Their peculiarities will be best understood if we describe one or two different kinds of them.

Take, for example, a shred of body substance of any kind—any "tissue" as the professor loves to call it—a bit of liver, say, or of brain, or of muscle from the stomach—and put it under a microscope, and immediately the division into cells is quite manifest. Figure 9 shows, greatly magnified, the minute structure of a thin sheet of tissue that lines the cavity of the belly and sheathes the viscera; it consists, we see, of a number of separate minute flat plates of living substance, lying side by side with their edges touching each other, and separated by thin, wavy outlines, and each has in its middle a conspicuous round body called the *nucleus*. That is one sort of cell. The "tissue" of which it is the living part is called "pavement epithelium." In other "tissues" —muscle, bone, liver stuff or nervous substance, for example—the cells assume other forms. They may be flat or round or cubical or star-shaped or drawn out into thin threads and fibres; but it is always possible to make out the essential fact, that the living substance of the tissue we examine, whatever part of the body it comes from, is an aggregate of an enormous number of cells. In the following chapters we shall meet many different kinds of cell. In the developing young every one of these extreme forms began as a simple one, like that to be immediately described. We may compare the body to a community, and the cells to the individuals of which this vast organized population is composed.

It is very important to realize that this is not a merely allegorical comparison. It is a statement of proven fact, for—we resort here to the stress of italics—*single cells can be isolated from the rest of the body, and kept alive.*

This is a branch of biological investigation only recently developed. Its fascination is evident. The procedure is to take a minute fragment of living tissue from an animal, and to put it in a drop of blood-serum (the liquid left over after blood has clotted) from the same species. In order to make microscopical examination of the fragment possible, the drop is suspended from a thin square of glass, a "coverslip," over a little depression excavated in a thicker glass slip, the microscope slide. (It is necessary to make the most elaborate precautions, we may note, against bacterial infection when making such a "culture," and there are various other matters that must be attended to if the cells are to be kept alive and active—for example, if the tissue is taken from a warm-blooded animal it must be kept in an incubator at a carefully-regulated

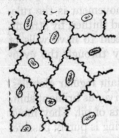 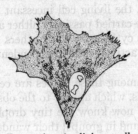 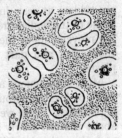

FIG. 9. A piece of the membrane that lines the abdomen highly magnified and revealed as a smooth pavement of thin flat cells.

FIG. 9-A. A living cell taken from an embryo chick, seen from above as it creeps over a sheet of glass in a tissue-culture. The large egg-shaped body slightly to the right side of the cell is the nucleus. (*Courtesy of Dr. H. B. Fell.*)

FIG. 10. A slip of gristle, highly magnified, showing the cells embedded in a hard substance that they themselves have made.

temperature—so that the work demands considerable experience and skill, and elaborate experimental equipment.) After a few hours, if the culture is successful, some of the cells will wander out of the fragment of tissue into the surrounding liquid. They move about in that liquid by themselves, and while they are creeping over the cover-glass it is possible to examine them with the microscope, and to take note of the most intimate details of their structure. The clearest observations are made by letting an intense beam of light fall obliquely on the cell; the surfaces of the various structures inside it reflect the light and are seen as brilliantly luminous objects on a dead-black background.

A cell is a tiny, flat, irregularly-shaped lump of protoplasm. It is not bounded by any visible wall; there is just a simple interface (the meaning of this useful new word is too plain to need definition) separating the cell-substance from the surrounding fluid. The outline is constantly changing. Sometimes it changes so rapidly that one can watch bulges being thrust out of the cell, and withdrawn again; sometimes so slowly that one must make accurate drawings—or, better, photographs—every few minutes, and compare them with each other before any change can be detected. By means of such movements the cell creeps slowly about.

The ground-substance, which is called the *cytoplasm*, is a fluid as clear as glass. Somewhere about the middle of the cell is the large, rounded nucleus; floating in the transparent cytoplasm round the nucleus are numbers of smaller bodies of various kinds. These bodies are seldom at rest; in the living cell incessant movement is the rule. Some of the bodies are carried passively hither and thither, by streaming movements of the cytoplasm, while others seem to have wills of their own, and dart actively and independently through the cytoplasm.

Most conspicuous among those bodies are certain regular globules, more or less numerous, which appear to the observer as brilliant shining spheres. These we now know are tiny droplets of oil. They move about the cell slowly and in groups; their wandering is purely passive, and they simply drift along, carried by currents in the cytoplasm. Scattered among these globules, less brilliantly luminous, and much smaller, is a host of fine granules. These granules show irregular movements which, like those of the oil-droplets, are probably passive and due to currents in the transparent cell-fluid.

Very different from these granules and droplets are certain snake-like threads which writhe slowly through the cell. These threads are exceedingly fine and their length is variable; sometimes a thread can be seen to break into two, or two threads may join together end-to-end to form one. Their movements are apparently spontaneous, and independent of cytoplasmic currents. The threads are called *mitochondria*, and are composed of albumen and a fat-like compound, *lecithin*. Though evidently of importance to the cell their use has not yet been discovered. Lying like a cap over one end of the nucleus, a zone called the *centrosphere* may be distinguished. The principal function of this zone, which is dimly seen as a tangle of filaments and granules, will be described later on in connection with cell-division. For the present its interest lies in the fact that it seems to be a region where mitochondria are made; they can be seen wriggling out from it into the cytoplasm.

Lastly, if the tissue under examination was taken from the eye of the embryo chick, there may be present among these other bodies within

COMPLEX BODY–MACHINE AND HOW IT WORKS

the cell a number of tiny rods of pigment. Sometimes these pigment-rods lie passively still, showing only a barely perceptible trembling movement. This slight quivering is an example of what is known as *Brownian movement;* it is due to the actual impact of molecules upon the sides of the rods. But from time to time these rods exhibit movements which it is hard to attribute to cytoplasmic currents; they dart to and fro "more or less rapidly and irresponsibly, like guinea-pigs in a run." We take our description from a paper by the late Dr. Strangeways and Dr. Canti.

Among this host of brilliantly illuminated bodies the outline of the nucleus is seen. The nucleus is a relatively large oval or spherical mass which drifts slowly to and fro in the cell. The boundary between nucleus and cytoplasm, like that between cytoplasm and culture-medium, has no visible wall. Inside the nucleus there are usually two mistily opaque bodies of irregular contour, the *nucleoli;* these bodies are continually changing their size, shape, and position. Except for the slowly writhing nucleoli nothing can be seen inside the nucleus but a clear fluid.

This, then, is a simple kind of cell, released from the body and leading a life of its own. The size of this object is such that about 2,500 laid side by side would measure an inch. And it is itself separately and independently alive. Such is the stuff that man and all his life is made of. In our bodies there are millions of such individual cells, inherent and necessary parts of us. They are not dead like the bricks in a wall; they are alive like the soldiers in an army. And they can be persuaded by the arts of Dr. Strangeways to desert! Then they will move by themselves, take nourishment, absorb oxygen, exude waste matters. They can be starved or suffocated. Not only will they move about as free individuals, but they will reproduce themselves. After a few hours some of the cells in these cultures will be seen to pull themselves into two parts, each of which lives and grows as an independent unit. They are no longer parts of an organized animal, yet so far from pining away in exile they are absorbing nourishing substances from the liquid about them, growing and ultimately performing an act of reproduction. In two or three days the multiplying and growing cells will have poisoned the serum with their own excretions, and the "culture" will die; but this death can be prevented by putting a little of the failing culture into a drop of fresh serum, when the reproduction of the cells will resume and go on until that drop also is fouled. This administration of these exiled body cells is called sub-culturing. By sub-culturing every few days it is possible to keep strains of detached cells alive for months, even years; indeed such strains may be kept going for periods of time far exceeding the individual life-span of the species from which they are taken. Cell-strains taken from a chick embryo have lived over fourteen years, while the ordinary life of a fowl is not a third of that span. No reason is apparent why we should not sub-culture for ever. If Dr. Strangeways had lived in the time of Julius Cæsar and set a series of sub-cultures growing from a scrap of him, fragments of that eminent personage might, for all we know to the contrary, be living now.

This release of living cells to a free life is not restricted to the primitive connective-tissue cell. It has been possible to get similar strains of cells from muscle, or nerve, or kidney. In such cases the cells lose much of their specialization before they begin to wander about and reproduce: they cease to be typical muscle-cells, nerve-cells, or kidney-cells, and become more like the primitive connective-tissue cell that we have described above. They cease to be parts, they become wholes.

We may say then that every living tissue consists of cells, and that every living cell is a potential individual (or sub-individual), containing in its own body all that is necessary for life. But, be it noted, the cell is a unit; it is not possible to isolate *parts* of cells and keep them alive. If a cell be cut into halves—one with the nucleus and centrosphere and the other without—the first, which contains all the elements of the original cell, will live on and reproduce itself; the defective half, on the other hand, may creep about for a time, but it can neither grow nor divide, and presently and surely it will perish. (The experiments on which this statement is based were made, not on cells from higher animals, but on Amoebae and other Protozoa, which are independent creatures resembling the released tissue-cells of a higher animal very closely.) When cells multiply they divide into two by an elaborate process which ensures that each of the two daughter-cells shall have had the cytoplasm with its various inclusions, half the centrosphere and half the nucleus. Thus each of the two resultant cells has all the essential cell-elements; if this were not the case, they could not live any more than a man who lacked lungs or heart could live.

Naturally, when they are parts of a living body the cells are disciplined, they do not wander about where they like, growing actively and reproducing themselves, as the cells in a culture do. An organ such as the brain or liver is like the City during working hours, a tissue-culture is like Regent's Park on a Bank Holiday, a spectacle of rather futile freedom.

The ordinary activities of organs are summations of the activities of their constituent cells. The pull of a muscle, for example, is in reality the synchronized effort of thousands of muscle-cells, and bile is the pooled outpourings of countless liver-cells. Evidently this co-operation involves control. No muscle would be of any use if its cells were allowed to contract independently whenever they liked; they wait, tense and expectant, until the muscle is required to move, when they all join together in a hearty but disciplined pull. Even growth is a summation of cell-activities, for the increase in size of any part is due to the orderly growth and multiplication of its constituent cells "according to plan."

We shall deal in later chapters with the processes of reproduction and development. We shall see then that the fertilized egg is a single cell with cytoplasm, nucleus, centrosphere—all the essential elements, that this cell divides into two, that each daughter-cell divides in its turn, and that in this way by a process of continued subdivision, all the millions of cells in the body are produced. Naturally, this growth process must be regulated like any other cell-activity. The thumb must not be allowed to grow to the size of a leg. Occasionally a group of cells shakes off the normal controlling influence and grows disproportionately, producing a tumour or a cancer, but normally all the cells are kept well in hand.

We have reached an idea whose importance in biological theory can hardly be overestimated, and we may perhaps be pardoned if we emphasize it. Briefly, it is this. The reader has a feeling of single individuality; he or she feels and acts as one; the various parts of his or her body work smoothly and harmoniously together. But he or she is also a community, a vast assemblage of invisibly small cells. These cells are living together and they are controlled and specialized in divers ways for the common good; nevertheless they are themselves individuals, for under suitable conditions they can be detached from the body and kept alive indefinitely.

At this point the reader may protest. "I am myself an indivisible individual," he or she may say, "for if one of my arms or legs be amputated it cannot survive. How then can my body be a community?" It is of course true that an isolated limb will shortly die; but if while a busy factory was working it was suddenly surrounded by a high brick wall, so that its personnel could not get any food, it would soon be starved out and become still. An isolated limb dies for precisely the same reason. Normally its cells are nourished and supplied with oxygen by the blood; when it is amputated its blood-supply is cut off and so the cells are killed (although as a matter of fact they suffocate before they have time to starve). Nevertheless, if proper precautions were taken, the limb could be kept alive by perfusing it along the arteries with blood. This sort of thing is constantly done in the laboratory with detached nerves or muscles or hearts or stomachs or kidneys, because by using such methods it is possible to study living tissues without causing pain to living organisms.

On one occasion (a long time ago) a physiologist obtained the heart of a criminal aged thirty-five. He removed the heart eleven hours after the execution, and ran a suitable artificial fluid into it through the veins, to take the place of blood; the heart revived and began to beat again, and was then studied and experimented upon for a full three hours. The heart of a cold-blooded animal, such as a tortoise, will live for weeks under such conditions. Similarly the mammalian kidney can be removed altogether and supplied with artificial blood through a glass tube tied into its artery; it will continue for hours to remove any undesirable matter from the fluid that runs through it. It has even been shown that the isolated brain of a goldfish will go on for a considerable time sending out perfectly regular breathing impulses down the stumps of the nerves, that used to go to the gill-muscles.

The indivisibility of the human body is then a spurious indivisibility, resulting from the fact that its cells are specialized, that the cells of a leg, for example, have come to rely on the cells of the heart for a supply of nourishing blood. On the other hand the indivisibility of a cell is absolute, for the cell represents the simplest possible way in which animal tissue can be organized; without its essential parts the exchanges of energy, the complex interplay of chemical reactions that underlie life, cannot continue. It is in fact a primary indivisible unit of life.

The number of cells in the reader's body is staggering. In the blood of an average man there are over fifteen million million cells in the blood alone; his brain system contains nearly two thousand million; and the total number in the human body is over 1,000,000,000,000,000 —a thousand billions (and English billions, not American ones). They serve the body community in various ways and have various appropriately specialized forms. Some are of service because they can actively change their shape—such as muscle-cells; others, the nerve-cells, are drawn out into enormously long, thin threads, and are like living telephone wires; others, more cubical, serve by exuding special chemical substances—such as the cells of the salivary or thyroid glands. We need not catalogue all the possible varieties, but can content ourselves with stating that there are well over fifty distinct kinds of cells to be found in every man's body. But, since we are dealing with the ultimate nature of our tissues, it is interesting to note in passing that there are very important parts of our bodies that are not in any sense alive. Our cells are all alive but we are not all cells.

Figure 10 shows the structure of gristle, or as the biologist terms it, hyaline cartilage. The cells do not lie in contact with others but are separated by a stiff transparent substance. If a fragment of living cartilage be cultured, those cells that can escape will migrate out into the serum, but the matrix between the cells will not move or grow. It is not alive, but an exudation made by the living cartilage cell in much the same way as spittle is made by the cells of the salivary glands. Nevertheless it is to this non-living substance that cartilage owes its stiffness and smoothness, the qualities that make it useful to the organism. Similarly with the other kinds of framework tissue. Bone consists of small cells imbedded in a matrix which has been made rigid by the deposition in it of lime salts, and it is upon this matrix, exuded (or, in biological language, secreted) by the bone-cells, that the strength of a bone depends. The binding connective tissue that permeates organs and holds them together consists of cells surrounded by a web of non-living fibres of great strength and elasticity that they have laid down. Tendons and ligaments are much the same sort of thing, but more so—with a greater proportion of fibres, and smaller and fewer cells. The connective-tissue cells are carpenters, and build the non-living scaffolding that supports the body; but the parallel would be closer if the scaffolding made by human carpenters consisted of their consolidated sweat. Finally, blood consists of cells floating in a non-living watery fluid, and the watery part sustains most of the transport duties of blood.

§ 3. Blood.

"THE life which is the blood thereof," says the Bible, and we may best go on to our examination of the life in a man's body by considering the nature and functions of the blood. Where the blood is not circulating the processes of life, in any mammal, cease almost immediately. Stop the blood-supply to the head and in a moment the brain is vacuous and the body it is directing stumbles and faints. Lungs, liver, kidneys and stomach, all our internal organs, are mere factories and depots to purify and feed this essential vitalizing stream. An almost infinitesimal change in its composition makes us exalted or depressed or fills us with unwonted fears or desires. Manifestly our next step on our way to the understanding of this vast cell-community, an animal body, must be a study of the blood.

It should be realized at once that blood is not a simple lifeless fluid like water or milk; it is, in biological language, an active living "tissue." It is a teeming population of living cells. The cells of blood float in a watery medium, and blood as a whole is therefore a fluid, but it is nevertheless quite as much a tissue as cartilage or bone—it is simply a liquid tissue.

The cells of blood are of two kinds, red and white, and it is to the presence of vast quantities of the former kind that the red colour of blood is due. A human red blood cell is a disc about seven-thousandths of a millimetre across, about one quarter as thick at the rim, and rather thinner in the middle, or, putting it into English measures, 3500 red blood corpuscles side by side would measure about an inch. It has no nucleus. Inside the cell there is an apparently homogeneous fluid in which no structural elements can be discerned, containing as one of its chemical constituents a red pigment known as *haemoglobin*. The whole is bounded by a membrane, so delicate that it cannot be seen even with the highest powers of the microscope, and which is believed to consist of a double layer of fatty molecules arranged side by side in a regular way. The white blood cells are of several different kinds, but they are all alike in being a little larger than the red cells, in having conspicuous nuclei, and, like isolated cells of a tissue-culture, capable of a slow change of shape. The cells are present in enormous numbers; a cubic millimetre of blood contains about ten thousand white and six million red—that is to say, there are more red cells in one drop of blood than there are people in the United Kingdom. The reader contains just about a gallon of blood, so imagine the vastness of this continually circulating population. The red cells are concerned with the transport of oxygen, and the white cells include among their many duties that of combating bacterial parasites.

Together, the red and white cells constitute one third of the volume of blood. The remaining two-thirds is water, carrying a great and varied assortment of chemical compounds. This water with its dissolved substances is called the *plasma;* it forms a clear, nearly colourless fluid in which the blood-cells float. Plasma is the non-living part of blood, but it is nevertheless by no means the least important, for it is as solutions in the plasma that the great majority of transported substances are carried by the blood.

It will be interesting to survey this curious miscellany of plasma-ingredients very briefly, because by so doing one gains an insight into the number and variety of the functions performed by this essential fluid.

In the first place blood fetches and carries for all other tissues. It brings them supplies of oxygen and of various kinds of food they

COMPLEX BODY-MACHINE AND HOW IT WORKS

require, and it takes away and eliminates various unwanted products of their activity. The red cells are solely responsible for the transport of oxygen and they play the chief part in carrying carbon dioxide; the remaining duties are undertaken by the plasma. Dissolved in the plasma are all the many and complex food supplies that cells need, and the waste products are simply shed by the cells into it. These two duties alone necessitate the carrying of a great number of different compounds—some in considerable amount, others only in minute traces.

But besides the bringing and taking away of substances which

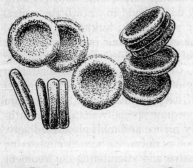
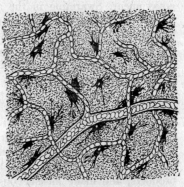

FIG. 12. A fine artery branching out into capillaries, as seen in the frog's foot.

FIG. 11. A few cells from human blood.

Above, some red cells are seen in various positions. Below, three kinds of white cells.

participate directly in the chemical processes of the cell-machines, the plasma carries other matters, not actually consumed by cells, but necessary constituents nevertheless of the fluid which bathes them. A cell needs not only fuel and oxygen to work, but its delicate mechanism is acutely susceptible to the presence or absence of certain mineral salts in the surrounding fluid. Its activities respond to the slightest changes in their proportions. A minute trace of calcium must, for example, be present if our muscle-cells are to obey our will. Withdraw that and they will begin a rhythmic twitching. There is a whole range of such accessory substances in the blood needed for the proper working of the body.

Here, then, we have the primary functions of the blood, fetching and carrying for the tissue cells, for multifarious citizens of our body, and ensuring them the comfort necessary to their activity. These functions alone mean a very great complexity. But superimposed upon them are others, still more intricate and extraordinary. Day by day, and as a general rule quite outside the realm of our consciousness, there is a continual fight against certain evil forces that seek perpetually to disturb our serenity. Generally this conflict troubles us as little as the criminal activities of Pimlico and Shoreditch disturb the mental peace of the Prime Minister in Downing Street. But sometimes the criminal onslaught gets sufficiently out of hand to disturb the central government. Swarming in the air we breathe, lurking on the solid objects we touch, and even on the surface of our own skin, are the spores of minute bacteria, many of them ruthless parasites, waiting for a chance to work us mischief. Constantly, through a scratch in our skin, through the tender lining of our lungs, through the injured tissues of a sore throat, swarms of bacteria invade the blood-stream itself, and finding themselves in a medium specially adapted for bathing and nourishing living cells, they rejoice and multiply exceedingly. The very efficiency with which blood performs its normal ministering function makes it the most attractive and stimulating environment for these parasites. Many of them are extremely troublesome and ungracious guests. They produce their own often very disagreeable by-products, and make mischievous changes in the composition of the blood. They invade and attack the tissues. Unless they are defeated and repelled they produce the colds, influenzas, fevers, epidemic diseases, typhoid, small-pox, and so forth that disturb and may overthrow the central government altogether. The resistance, once skin or membrane is pierced, goes on almost wholly in the blood.

The chief antagonists of the bacteria themselves are the white blood corpuscles. These actually fight the bacteria. That war goes on interminably. It is only in the more serious engagements that painful congestions and inflammations make Mr. Everyman aware of the mischief afoot. In addition, in a manner too complex to explain at this stage, the chemical poisons which the invaders pour into the plasma are neutralized by specially manufactured antidotes. These chemical invaders, these poisons and antidotes, formed only in minute traces but nevertheless physiologically potent, constitute another class of ingredients in that complex fluid, the plasma, and provide another type of event in the swarming highways of the living cell-community which we call a man.

COMPLEX BODY-MACHINE AND HOW IT WORKS 51

The composition of blood is further complicated by a second kind of protective arrangement—not in this case against living invaders, but against a more direct result of mechanical injury to the animal body. This is the provision for forming clots. When the surface of the body is cut or scratched and the injury involves blood-vessels, the escaping blood, as we know, clots into a solid mass and so seals the wound. Were it not for this fact, there would be nothing to prevent blood flowing away indefinitely once our bodies were punctured. Some unfortunate people have blood that does not clot. In the hereditary condition known as *Haemophilia*, in which the blood is incapable of clotting, patients lose large quantities of blood from even trivial cuts, and have been known to bleed to death in the dentist's chair. Normal clotting, which fills the wound and so checks the escape of blood, involves as an essential feature the conversion of a certain substance dissolved in the plasma, *fibrinogen*, into *fibrin*, which is insoluble and which precipitates therefore to form the fibrous clot. This conversion is the result of an extraordinarily complex chain of preliminary reactions, in which a number of different substances participate, some of which exist ready-made in the plasma, while others are contributed by damaged cells at the point of injury. The plasma carries this fibrinogen and various accessory substances, so that this protective process of clotting takes place when occasion demands.

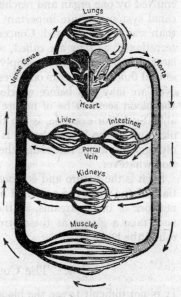

FIG. 13. The course followed by the blood in its unending flow round the body.

Vessels carrying fresh blood are shaded; those carrying used blood are black.

And still there is more to tell of this marvellous fluid, which is, as the Bible has it, the "life" of a higher animal. It feeds, it comforts, it protects. It is also a means of communication between part and part. It carries messages that secure the harmonious co-operation of one organ with another.

Twenty or thirty years ago it was thought that the co-operation of part with part was ensured through the nervous system alone, either consciously through the brain or unconsciously by the subordinate systems of nervous communication through the spinal cord and inferior centres. But nowadays we are beginning to realize that a very large part of the harmonizing task is done through substances emitted by one organ and reaching another by way of the blood. The canal system is more important relatively to the telegraphic system than was once suspected. Concerning these substances, the "internal secretions," as they are called, we shall give some interesting details later. Here we note them simply to round off our account of this essential fluid which bears our lives along. One more fact about it, however, we may add before we close the section. Some constituents of the blood seem to be of no use to us; they happen to be unavoidable. Such, for example, is the dissolved nitrogen which the plasma picks up in the lungs when it is taking up oxygen for the cells. The nitrogen has a free ride round the system, but apparently has no function whatever.

Such is the nature and composition of the blood. About one-twentieth of the weight of a normal man is blood, and in the meshes of a network of blood-streams, all the life in our bodies goes on. We will next take a glance at these streams and note how they are driven round the body.

§ 4. The Course of the Blood.

It is not difficult to see the blood actually at work among the tissue-cells. The delicate web of a frog's foot, for example, can be examined under a microscope, without being severed from the rest of the frog or indeed causing any pain, and we can observe the vital business in progress. The field of vision is occupied by masses of greyish tissue, in which the boundaries of individual cells can be made out only with difficulty. In the frog's foot pigment-cells can be seen dotted over the field—black star-shaped masses with irregular processes radiating out from a common centre. Riddling the sheet of tissue under examination, and apparently dividing it into lobes, is a network of canals, the blood-vessels. In these vessels the course of the blood may be watched by following the oval red corpuscles as they float along.

The blood can be seen hurrying apparently at headlong speed*

---

*Velocities always look deceptively high through the microscope, because the instrument magnifies distance without magnifying time.

COMPLEX BODY-MACHINE AND HOW IT WORKS

along certain larger vessels—highways—which are minute arteries. The arteries branch and divide and their branches divide again, so that the blood passes into smaller and smaller canals. At each division the cross area of the two branches taken together somewhat exceeds that of the original trunk, so that the blood travels more and more slowly as it finds its way into narrower and narrower passages. The final, minutest blood-vessels are the *capillaries;* their diameter is only very slightly greater than that of the red blood-cells themselves, and in these ultimate canals the cells no longer bustle—they crawl. It is through the invisibly thin walls of the capillaries that there takes place that interchange of substances between blood and flesh for which purpose the circulation exists. The whole appearance suggests women coming to a market—the swift stream pours into the busy market-place, it breaks and divides into slower and slower subsidiary streams, until, finally, the housewives are hovering and bargaining in the narrow spaces among the crowded stalls.

But the blood does not stay long in the capillaries. As it loiters, its dissolved food and oxygen are diffusing into the cells, and the unwanted residue of their activity is passing back in exchange. Diffusion, over distances measured in thousandths of a millimetre, is a process of lightning speed, and the blood must not hang about after its function is performed. Hence, after a brief passage through one of the myriad capillaries, the blood reaches a point where the vessels, instead of subdividing, come together again. The capillaries join, and the resulting trunks join each other, and so veins are formed. But now the previous conditions are reversed, for each trunk has an area slightly smaller than that of the two branches which unite to make it, so that as the blood finds its way into larger and larger vessels it travels at a greater and greater speed. Thus, in one of the little veins it hurries out of the picture with the same effect of headlong haste that characterized its arrival. We cannot watch the blood as it travels, ever more swiftly, towards the heart; the frog is too opaque for our microscope to penetrate. The rest of the story has been pieced together by dissection and by experiment.

When it leaves the field of view, the blood continues in the same way. The veins from the web are joined by veins from the muscles of the toes, and later by veins from the muscles and skin of the calves, from the bone and cartilage, and other tissues in the leg. The stream becomes greater and swifter. Still later, in the region of the hips, two great vessels from the legs join each other to form a venous trunk, the *inferior vena cava.* This, as it runs up the back of the abdomen, receives branches from the muscles of the back, from the kidneys, from the intestines, from the liver. In the chest it combines with two other main trunks, the *superior venæ cavæ*, which carry blood down from the head, neck, and arms. Finally, the stream of used blood collected from every part of the body by the union of these three vessels pours into the heart.

When the blood from the tissues reaches the heart through the venæ cavæ it is depleted blood. While passing through the capillaries it has parted with the substances required by the flesh and with its oxygen, and it has encumbered itself with the waste products there produced. It is unfit to be driven round again until it has undergone a process of restoration and purification. Now the various needs of living cells vary in their urgency. The food-supply is necessary, but it need not be constant. Cells are able to accumulate little stores of sugar, fat, or a very important nitrogenous material, protein, in their own bodies, to be used when required, and it is sufficient for the blood to bring them food once in a while, after a meal, for them to replenish their stores. Moreover, the removal of waste products need not be very rapid; they are formed only in small quantities and slowly. But with the oxygen supply things are different.

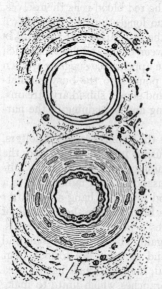

FIG. 14. An artery (below) and the corresponding vein, (above) cut across to show the strengths of their walls.

The need for oxygen is imperative—a man cannot live without oxygen for more than three minutes—and the cells have no way of laying up stores of oxygen. It must be brought to them constantly and abundantly. A fresh charge of oxygen is therefore a primary requirement of this returning blood before it can go back to the tissues.

The circulatory system in a mammal works first and foremost to meet this need for oxygen. The removal of other waste substances and the taking up of more foodstuff is a less urgent matter. Venous blood, returning from the tissues, is driven at once to the lungs, where the oxygen-content is restored. In the lungs the blood travels through another network of capillaries across the wall of which it comes into contact, not with active and hungry flesh, but with oxygen-containing air. Instead of being impoverished it is refreshed. From these lung capillaries the oxygenated blood is collected as before into veins and returns again to the heart. It is now fit to be sent once more to feed, stimulate, and oxygenate the active tissues, and it is pumped by the heart, along a branching system of arteries, into the vast network of capillaries that permeates every organ in the body; there it repeats the exchanges, the marketing of oxygen and nutritive matter for waste products, with which our description began.

The circulation of a man or a mouse is, therefore, a completely double one—it is not so much a circle as a figure of eight. There is the journey through the lungs, during which the blood is charged with oxygen and parts with surplus carbon dioxide, and there is the longer journey, from the heart to a capillary in some other part of the body and back, during which the oxygen is surrendered. The former round is called the *pulmonary* (lung) and the latter the *systematic* (body) circulation.

FIG. 15. The ventricles of the heart cut across to show the relative strengths of their walls.

Manifestly this could be managed by two distinct pumps—a body-pump to disperse the blood everywhere, and a smaller, less powerful pump to send the blood on its special trip to the lungs. But the mammal's heart combines both these pumps into one organ. It is really two pumps side by side. Each has a thin walled upper part, the *auricle*, to receive the inflow of blood, and a powerful muscular part, the *ventricle*, which, as there are valves to prevent any back flow, impels the blood forward into the great arteries of the body or into the lungs, as the case may be. In Fig. 15 we have a cross section of the ventricles, and the reader can see how much feebler and flabbier the right (pulmonary) ventricle is than the left, which has to force the blood to scalp and toes and everywhere.

It is quite easy to hear anyone's heart at work and distinguish the phases of its action. By means of a stethoscope, which is simply a listening tube, one end of which is put to the ear and the other pressed against the left side of the chest, it is possible to hear the sounds made by the beating heart. There is a silence, then a long sound, then a short, sharp sound, then silence again—*Lubb* . . . *dup, Lubb* . . . *dup,*

Lubb . . . dup. The filling of the heart from the veins and the contraction of the auricles to fill the ventricles are noiseless; the sudden, violent rush of blood when, in turn, the ventricles contract gives the long sound, *Lubb.* . . . The short sound depends on a property of the arteries. The arteries have elastic walls. When blood is forced into them under considerable pressure by the ventricles they give, and when the ventricular contraction ceases they recoil. If it were not for the valves at their openings, the result of this recoil would be to drive blood back into the heart, but actually it closes the valves. The *aorta*, the main arterial trunk which supplies all the body except the lungs, is abruptly cut off from the ventricle and the slam of its valves causes that second heart sound . . . *dup*. This sound marks the end of the cycle; it is followed by silence while the auricles fill again. In one minute this pump drives about a gallon of used blood to the lungs and sends out an equal amount of pumped blood to the body.

The heart lies in a bag—the *pericardium*, which has smooth walls and contains a fluid. The purpose of this structure is to lubricate the movements of the heart; lying in this bag of fluid, the body-pump is able to contract and expand (about seventy-two times per minute) without friction against the adjacent organs.

The arteries which take the blood from the heart have to stand a fairly heavy strain and are strong and elastic to resist it. Every time the heart contracts there comes a tidal rush and the pressure rises to a climax and falls. Accordingly the walls of an artery are strong and elastic; they include a thick layer of muscle-cells and elastic fibres. Blood travels in the main arteries at a relatively high speed; in the great aorta, which leads directly out of the heart, it travels at about one-and-a-half to two feet per second, and in the limbs its pace is still considerable. At the wrist a main artery to the hand comes conveniently near the surface and the tidal pulsation of blood can be felt, and here it is that doctors inform themselves of the vigour and excitement of the heart.

As the arteries branch and divide, the force of the pulse becomes less, their walls become thinner, and the blood travels less swiftly and more evenly. Finally, in the capillaries, its velocity has fallen to about an inch a minute—about one-thousandth of its speed in the aorta. The pulse by that time has disappeared and the pressure is much lower. There is therefore no need for a capillary to have strong, resistant walls. The structure of a capillary is determined by its function, which is to bring the blood into the closest possible relation with the surrounding flesh. For this reason the walls are almost in-

COMPLEX BODY–MACHINE AND HOW IT WORKS

calculably thin. Moreover, the fact that the blood, instead of flowing along in one broad current, has been divided into a multitude of tiny parallel trickles, ensures that the surface of contact between blood and flesh is as great as possible. The number of the capillaries is enormous; in a cross-section of muscle, each square millimetre (that is, about double the cross area of an ordinary pin) contains well over a thousand, and all the capillary vessels in a man's muscles put end to end to form a continuous tube would girdle the earth two and a half times! When filtering through this network of fine tubes, the amount of exposed surface of the blood is enormous; a single cubic centimetre of blood will have about ten thousand square centimetres of surface

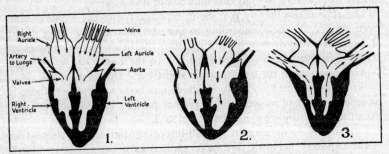

FIG. 16. The Three Phases of the Heart Pump.

(*1*) *The auricles filling from the veins;* (*2*) *the auricles driving blood into the ventricles;* (*3*) *the ventricles pumping blood into the arteries.*

at which chemical interchanges can take place. So we come to what we have already seen in the frog's foot.

Deoxygenated blood flows from the capillaries into the veins slowly and at low pressure. The driving force of the heart is mainly spent. The flow in the veins is helped by various accessory factors. The most important of these factors is a series of valves, baggy membranous projections of the veins' walls into their cavities, which only allow blood to pass them in one direction. These are present even in the smallest end-branches of the veins. The veins are squeezed by the muscular movements of the body, which thus assist in driving blood along them. During inspiration, for example, the veins in the abdomen are pressed upon as the diaphragm descends, and during voluntary movement of the limbs blood is driven out of the capillaries and veins of the contracting muscles. It is also possible that the muscle-fibres in the walls of the veins themselves are capable of slow rhythmic contraction. Because of the valves, the various factors tending to compression of the veins can drive blood in one direction only—towards the heart.

The internal pressure in the veins being much lower than that in the arteries, and there being no violent pulse to resist, the walls of the veins, although they include muscular and elastic elements, are not nearly as thick nor as strong as those of the arteries. The blood-flow in the veins, deriving its principal motive force, not from a special pump but from the incidental action of unrelated organs, is very much more sluggish than that in the arteries. The cross-section of any vein is about double that of the corresponding artery, and inside it the speed of the blood is only about half as great.

The time taken by any particle of blood to round the whole double cycle varies, of course, according to the particular artery along which it happens to be driven; a journey from the heart to the muscles of the ribs is very much quicker than one to the toes. On the average, however, in a man, the time taken is slightly less than a minute.

So it is that the blood travels about our bodies. However, there is another system of vessels which play a part in the circulation, but about which nothing has hitherto been said.

Let us consider once again that all-important part of the circulation—the capillary network. We have spoken as if the capillary blood-vessels were in immediate contact with the surfaces of the cells, but this is not strictly true. The capillary vessels are, roughly speaking, cylinders; the cells, according to the nature of the particular tissue, may be cylindrical, cubical, spindle-shaped, or may assume various other forms. In between the cells and the capillaries there are minute irregular spaces, and these spaces are filled by a clear plasma-like fluid—the *lymph*. In the interchange of substances between blood and tissue, lymph plays the part of a go-between; substances diffuse from the blood and flesh into the lymph, and from the lymph into the blood and flesh. Lymph has been compared to a middle-man, taking substances from the blood and handing them to the cells, and vice versa; but, unlike a human middle-man, it extorts no profit from this turnover, and as a matter of fact it plays no active part therein. Lymph, in the intercellular spaces, is simply a fluid through which substances diffuse. It comes from the blood and it is destined to return to the blood. The capillary walls, as we have seen, are very thin, and the blood in the capillaries is under pressure. The pressure is not great, but so delicate are the walls that it is high enough to drive fluid through them, and there is always a very slow, very steady seeping

COMPLEX BODY-MACHINE AND HOW IT WORKS 59

of blood-plasma from the capillaries into the surrounding lymph-spaces. This is how lymph originates and how the supply of lymph is sustained.

For the return of the lymph to the blood there is a system of vessels, very thin-walled and delicate, which actually open into the lymph-spaces, so that the fluid can escape along them. These vessels are called *lymphatics*. They unite with each other, like veins, to form larger vessels, and, like veins, they guide their contained fluid, which is simply filtered blood-plasma, slowly upward towards the heart. They are indeed shadowy plasma veins.

The flow of lymph in the lymphatics is slower even than that of

FIG. 17. A vein cut open, showing two pocket-shaped valves.

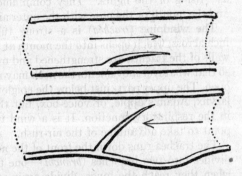

FIG. 18. How the valves prevent blood from flowing in the wrong direction.

blood in the veins, and, as in the veins, it is assisted by valves, by bodily movements, and probably by rhythmical contraction of the vessels themselves. The course of the lymphatic vessels is not so direct as that of the veins. They unite and separate again to form networks, and their course is interrupted by passages through curious spongy labyrinthine filters, the "lymph-glands," but ultimately the lymph is led to a great main lymphatic in the chest, the *thoracic duct*, and this trunk runs up the back of the chest and opens into one of the great veins near the heart. Thus the lymph is returned again to the blood-stream from which it originated.

Besides merely collecting the blood which oozes out of the capillaries and returning it to the circulation, we may note here that the lymphatic system has some interesting special functions connected with the absorption of food, to which we shall return later.

§ 5. Breathing.

WE HAVE already made it plain how and why man must breathe and little need be said here about the machinery of breathing. It would take about three minutes to suffocate Mr. Everyman. We all know that the lungs are two spongy organs, lying to the sides of and behind the heart and that together with that organ they fill the chest. To the cat lover their appearance will be familiar, and they will be better known as "lights." Lights is the old word for lungs; in the bills of mortality of the eighteenth century, pulmonary consumption figures as "rising of the lights." They communicate through the windpipe and the mouth or the nose with the outer air.

The windpipe (*trachea*) is a strong tube, about two-thirds of an inch across, which opens into the mouth at the back of the throat. The walls of the trachea are strengthened and made rigid by rings of gristle, so that whatever movements the neck may make this tube stays widely open. The upper part, just below the epiglottis, is modified to form the larynx, Adam's apple, or voice-box, but this structure plays no part in the respiratory function. It is a wind instrument inserted at this point to take advantage of the air-rush.

The trachea runs down the front of the neck into the chest, where it divides into two branches (*bronchi*)—one to each lung. The bronchi, when they reach the lungs, divide again, and their branches divide; ultimately, by continual division, they lead into minute air-channels, less than a hundredth of an inch across, which permeate the substance of the lungs. Out of these fine channels (the *bronchioles*) there open little clusters of hemispherical bays, the *alveoli*, and it is from these alveoli that air is taken up by the blood. In the narrow spaces between the densely-packed alveoli run the capillaries, the arterioles (little arteries), and the venules (little veins); it is of the alveoli, the blood-vessels, a little packing tissue, and the orderly labyrinth of ducts which bring air, that the substance of the lung is composed.

The thorax (the cavity of the chest) is a completely enclosed space, with the backbone behind it, the breast-bone in front and the sloping ribs on either side. All these parts are movable upon one another, and by their movements the air is brought to and taken from the alveoli. The movements are of two kinds—movements of the ribs and movements of the diaphragm. The ribs, as has already been said, run round the sides of the thorax from the vertebral column behind to the breast-bone in front; in between the ribs, connecting each to its

COMPLEX BODY-MACHINE AND HOW IT WORKS 61

neighbours, there is a sheet of vertically disposed muscle-fibres, the *intercostal muscles*. The action of the ribs depends on two purely structural features. The first is that as the ribs run forwards they also run downwards; the attachment of any rib to the breast-bone is lower than its attachment to the backbone. The second is that the ribs are curved, lying as they do in the curved walls of the chest, and that when the lungs are empty the convexity of this curve points slightly downwards. During inspiration the actual movement of the chest wall is

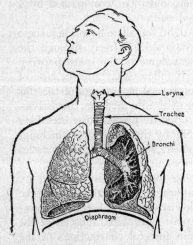

FIG. 19. The Organs of Respiration.

The left lung is dissected to show the bronchi.

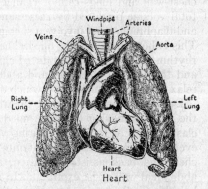

FIG. 20. The heart and lungs, as they lie in the chest.

produced by contraction of these intercostal muscles, which tends to pull the ribs closer together. The uppermost rib, solidly attached to the structures in the neck and shoulder, acts as a fixed point, the other ribs being pulled up towards it. As a result of the anatomical features just noted, this pull has two effects—firstly, because of the downward slope of the ribs, it pushes forwards the lower part of the breast-bone and so increases the front-to-back depth of the chest; secondly, because of the curvature of the ribs, it pulls their middle parts upwards and outwards, so increasing the breadth of the chest from right to left. These two factors combine in increasing the volume of the thorax, and are assisted in so doing by a simultaneous movement of the diaphragm.

The diaphragm is concave downwards, attached round its edges to the ribs, backbone, and breast-bone, but free in the middle. Its central part is of tendon, but its marginal part consists of radiating muscle-fibres. When these muscles contract the domed centre is pulled downwards; the diaphragm flattens, pressing on the liver, which lies immediately beneath it, and on the other abdominal organs, and at the same time it helps to increase the volume of the thoracic cavity. Let us note in passing that contraction of the diaphragm, since it presses on the organs in the abdomen, must be accompanied by a slight bulging forward of the abdominal wall.

Now the thoracic cavity is itself air-tight, but there lie inside it these two elastic bags, the lungs, which communicate freely through the trachea with the outer air. Since the volume of the space between chest-wall and lungs cannot change, it follows that any increase of total volume produced by these simultaneous movements of the ribs and diaphragm must involve a sucking of air into the lungs themselves. This is how the lungs are filled. The opposite phase of the breathing rhythm—expiration—is produced chiefly by the elasticity of the lungs and the walls of the chest and abdomen; when the various muscles relax there is a natural recoil which drives air out of the lungs.

During ordinary, unforced breathing inspiration is produced by the muscles moving the ribs and diaphragm, and expiration by the elastic recoil of the parts. During violent or forced breathing, however, other factors are brought into play—additional muscles of the chest and back assist inspiration, and forceful contractions of the abdominal wall assist expiration. In men, the principal factor in quiet breathing is the movements of the diaphragm. In women, on the other hand, the principal part is played by the upper ribs; this is because at certain times the female abdomen has a precious and delicate charge, which might be injured by rhythmical pressure from the diaphragm.

In this way the air in the alveoli of the lungs is constantly renewed. The alveoli are thin-walled structures; each is bounded by a single layer of flattened cells, extremely thin. Immediately outside this layer and closely pressed against it is the tangled mass of capillary blood-vessels, also very thin-walled, so that in the lungs blood comes into as close a relation with air as it does with hungry tissue-cells in other organs. As in the latter case, the interchange of dissolved substances takes place by a simple physical process of diffusion. The inner surfaces of the alveoli are damp; oxygen from the air inside them dissolves in this layer of moisture, and from this it can readily diffuse through the thin alveolar and capillary walls into the blood.

Ordinary room air is for various reasons unfit for so great an intimacy as this: it is only an improved and chastened air that is admitted to the alveoli. A very considerable process of purification and preparation has had to occur en route from the nostrils. Air carries, for example, floating bacterial spores—minute organisms in a state of suspended animation, waiting for a chance to get into our warm and nourishing blood. The outer surfaces of our bodies are fortified against invaders by a strong, resistant layer of skin, and it is only when this skin is torn that parasites can enter from that quarter. In the lungs things are different, for the delicate alveolar lining could put up no effective resistance against them; they have to be prevented from getting there. It is therefore necessary to filter the inspired air to remove this danger. Moreover, air contains dust, and some kinds of dust (such as flinty particles) can cause serious trouble if they reach the delicate lung membrane. The chief filtering organ, both for bacteria and dust, is the nose.

The cavity of the nose is divided into a maze of passages by shelves of bone, projecting from its walls and twisted into complicated and bizarre shapes. The thin layer of living tissue covering these partitions and the lining of the trachea and bronchi throughout are provided with thousands of glands, each producing a continual trickle of sticky mucus, and also with other cells whose duties are to distribute the mucus evenly over the surface and to keep it moving slowly towards the throat. The latter cells illustrate a kind of motion, very different from muscular movement, of which we have said nothing hitherto. Each cell has, rising from its free surface, a number of fine, whiplike projections—*cilia*, by whose continual lashing it keeps in motion the fluid layer covering it. Air as it enters the nose leaves its coarser particles of dust entangled by the hairs in our nostrils. As it passes among the labyrinthine curlings of the nasal bones, most of the finer particles stick to the layer of mucus as flies do to a fly-paper; finally, in the trachea and bronchi, any particles that have got through the nose are caught in the same way. It will be seen that the nose is the most important part of our filtering system—hence the importance of breathing through the nose.

In fresh country air the amount of dust is small, and practically none of it gets through this protective labyrinth into the lungs. In towns, on the other hand, the air carries soot and various other consequences of human concentration, and the alveoli of town-dwellers are usually definitely grubby. In some of the less popular occupations—coal-mining, for example—the air inhaled is so loaded with particles that the normal protections are inadequate, and the alveoli and finer bronchioles become choked. The lungs of an agricultural labourer are pink: those of a coal-miner are black.

Besides the filtering of dust and bacteria, there are other duties that the nose performs. In the first place, air is cold—imagine an incessant cold draught blowing into a man's chest and against the walls of his heart! But as it filters through the warm channels of the nose the chill of cold air is taken off, and it reaches the alveoli at a more comfortable temperature. More important than heat is the question of water-loss. The alveolar walls are moist and their moisture plays an essential part in the exchange of gases between air and blood. A continuous draught of ordinary room air would rapidly dry up the delicate alveoli by evaporating their water. But the lining of the nasal cavities is also moist, and air is saturated with water-vapour as it passes through the nose; thus in the alveoli it takes up very little.

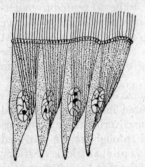

Fig. 21. A few cells with cilia, from the windpipe (greatly magnified).

It is curious to note in passing a system of air spaces running into the bones of the face and communicating with the nasal cavity by means of small openings. Apparently the purpose of these spaces is to lighten the architecture of the skull. Sometimes they may be invaded by bacteria through the openings into the nose, or, in the case of the large sinus in the upper jaw, from the root of a bad tooth, and inflammatory conditions may set in. Owing to their situation the treatment of such inflammations involves very difficult and painful operations.

We have said that the oxygen from the air reaches the blood by diffusion. It is dissolved in the plasma. But the amount the plasma could carry in a state of solution would not be enough for the energy-production of a mammal's body. And here it is that the importance of the red blood corpuscle comes in. They are specialized oxygen-carriers. The *haemoglobin*, the red substance colouring the red blood-cells, combines chemically with oxygen and so increases enormously the amount of oxygen that any given volume of blood can carry. The oxygen diffuses through into the plasma and the red corpuscles pick it up. The haemoglobin-oxygen compound is an unstable one; it can exist only where there is abundant oxygen. If there is much oxygen about it will be formed, and if there is little it will dissociate. Hence haemoglobin combines very readily with oxygen in the lungs and gives it up very easily in the other tissues where there is a want of it.

As has already been said, it is to haemoglobin that blood owes its colour. The colour of oxygen-free haemoglobin differs from that of

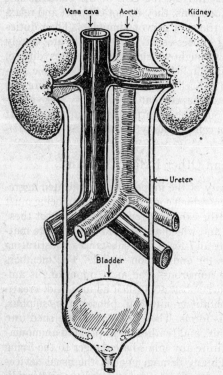

FIG. 22. The organs of excretion.

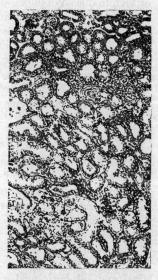

FIG. 23. A kidney consists of a multitude of closely woven tubes. Cut across and examined through the microscope, its porous nature is revealed.

The area shown is about $\frac{1}{15}$th of an inch high.

the haemoglobin-oxygen compound; the former is purple, the latter scarlet. This is why arterial blood is red while venous blood is purplish-blue. If you cut an artery the blood jets quickly because of the pulse of the heart and is scarlet from the outset. If you cut a vein the blood comes in a steady slow stream: it is darker but brightens to red in the oxygen of the air. Anaemic people are people who suffer from a shortage of red blood corpuscles; they are pallid and lack energy because of the insufficient supply of oxygen to the cell engines. "Red-blooded" is not a bad phrase for an energetic person whose tissues are all vividly alive.

In passing, let us note the structure of the cells involved in this respiratory exchange. As we have seen, the processes are very simple. The cells of the alveolar walls and the red cells of the blood do not perform complex and varied functions; they do not contract and relax, or secrete digestive juices, or transmit nervous impulses. Their duties involve no energy-consuming processes at all; the cells are merely water-containing structures through which dissolved gases diffuse. We find corresponding to this simplicity of function an extreme simplicity of structure. A human red-cell is little more than a minute bag containing haemoglobin, and so shaped that it has a very large available surface. A lining-cell of our lungs is little more than a thin, flat plate. They are the extreme of cell simplicity in our bodies.

§ 6. Kidneys and Other Exhaust Organs.

THE next round in our itinerary of the immensely complicated aggregation of living cell-mechanisms which constitutes a human being must be the organs that get rid of the exhaust products of billions of these internal combustion engines, and which also get rid of the waste matters resulting from their wear and tear. The substances constituting the former class are two—water and carbon dioxide. The members of the second class are more numerous. The most important is ammonia, but there are also other substances, such as uric acid, creatinine, and sulphuric acid, containing nitrogen, phosphorus, sulphur, and other elements in various forms. These substances are shed into the blood in small amounts only, and (with the exception of ammonia) their presence does not constitute any immediate danger to the living body. Nevertheless, if through any derangement of the usual devices for their removal they accumulate in the system, quietly and insidiously they work evil, stiffening the connective tissues, hardening the arteries, and producing subtle disorders of nutrition. If, for example, there is difficulty in getting rid of uric acid, it collects in the joints, giving rise in course of time to the stabbing pains of gout. Mr. Everyman as he gets on in life is apt to consume a considerable amount of trustworthy and untrustworthy solvents for his uric acid. If the removal of other waste products is inefficient, there may be such symptoms as headaches, nausea, and vomiting. In general, the results of defective elimination are chronic and cumulative; they represent, not an immediate disaster, but a slow fouling of the living machine.

Excretion is the word used to express the removal of these substances. Carbon dioxide can exist in gaseous form and is therefore excreted by the lungs. It diffuses into the alveoli as oxygen diffuses into the blood. The other substances, however, have to be expelled in watery solutions. The principal agent in this process are the kidneys and liver, although other parts of the body take minor shares. The wise physician watches for clogging of the kidneys and liver as the wise chauffeur watches for carbonization in his cylinders. The kidneys are chemical separators; blood reaches them with various undesirable matters dissolved in it and it flows away cleansed, while the impurities leave along a different channel, which guides them safely out of the body. The liver is a chemical accomplice of the kidneys; it performs certain operations on the waste products in the blood, and converts them into other substances with which the kidney can more readily deal.

The duties of the liver are many; it is an organ to which we shall constantly refer in this and subsequent sections. From our present point of view, its most important function is concerned with the ammonia produced by active tissues. Ammonia is a definitely harmful substance, and has a convulsant action on nerve-centres; there is, therefore, danger in the ammonia which living cells are continually shedding into the blood. It is interesting to note that every living cell in us is doing its best to poison us as a whole and has to be specially restrained. It so happens that as blood flows through the liver the ammonia is converted into *urea*, a comparatively harmless substance, and in this way its sting is removed. This process becomes particularly important just after a meal, for, as we shall see, the digestion of the nitrogenous substances known as proteins involves the entry of large quantities of ammonia into the blood. Since the whole of the blood from the intestines flows directly along a special system of veins (portal veins) to the liver, the ammonia is all dealt with before it can reach and injure any other part of the body. If it were not for this protection the most serious disturbances would occur, for during the digestion of a heavy meal of meat the amount of ammonia produced would be quite enough to derange the nervous system and throw us into convulsions.

Besides ammonia, there are other dangerous products which are made harmless by the liver. For example, bacteria in the intestines produce two foul-smelling poisons called indol and skatol, which find their way in small quantities into the portal blood; but here again the liver intervenes and converts them, before they can reach the general circulation, into harmless substances that the kidneys can remove. The liver may be imagined as a chemical censor of the blood that leaves the intestines, detecting such poisons as result from the digestive process. It does not actually expel them; it modifies them and then returns the disarmed products into the blood stream for the kidneys to expel at their convenience. It has been called the Ellis Island of the body. It arrests and it marks the undesirable immigrant for potential deportation, it manacles him, but it does not actually remove him.

Our kidneys are much the same shape as those of a sheep that figure on the breakfast-table; they are about four and a half inches long and lie just in front of the vertebral column in the small of the back. Their blood-supply is copious; they receive blood by a short branch from the aorta, the great main artery of the body, and return it to the main venous trunk, the posterior vena cava. Since, as we have already pointed out, the pressure in the aorta is higher than that in any other blood-vessel in the body, and that in the vena cava lower, the head of pressure that drives blood through the kidneys is enormous, and the current through them is full and swift. Nevertheless, what they are doing is a work by no means so urgent as the oxygenation of the blood by the lungs, and it suffices that at each heart-beat only a part of the arterial blood goes through them. But it is a large part because of the high pressure and swift return. Like the liver, the kidneys also censor the blood that passes them, but in addition they remove undesirable molecules together from the blood-stream, instead of merely neutralizing them. The liver rebukes and neutralizes, but the kidneys expel. In addition to the blood-vessels, there is another tube leading out of each kidney—the ureter—down which there runs a continual trickle of water, carrying dissolved in it those substances which have been separated from the blood. This trickle of fluid—the urine—accumulates in the bladder, whence from time to time it is expelled.

It should not be thought that the only function of the kidneys is to weed out poisonous substances from the blood. Their importance is very much more general than that. Any substance, even the most salutary, can be harmful if it is present in too great an amount, and the kidneys exercise a general standardizing influence on all the ingredients (except gases) of the blood-plasma which passes through them.

COMPLEX BODY–MACHINE AND HOW IT WORKS 69

They regulate the proportions in which the various blood-constituents are present. As an example of this influence, we may consider the excretion of salt—for there is always a little salt in urine. We saw in a previous section that a certain amount of salt is a necessary ingredient of the fluid that bathes living tissue; moreover, this amount must be accurately maintained, for either excess or lack injures the cells. The kidneys assist in regulating the salt-concentration of blood by varying the proportion of salt to water in the urine it produces. Suppose, for example, that after violent sweating (which involves loss of salt) we drink a lot of water, the blood becomes watery and contains too little salt. To compensate for this deviation, the kidneys produce a large amount of urine which is practically pure water. On the other hand, during dry weather, as a result of water-evaporation from the lungs and skin, there may be too little water (and therefore too much salt) in the blood; under these circumstances, the kidneys produce a urine containing an unusually high proportion of salt. In this way the kidneys control many of the constituents of circulating blood; if for any reason the amount of potassium, or sugar, or sulphate in the blood is abnormally high, the offending substance appears at once in the urine. The kidney is a blood-regulator, and removal of actual poisons is only one aspect of its regulating function.

Evidently the amount and composition of urine produced by the kidneys will vary from hour to hour according to the ever-fluctuating condition of the body, depending upon the amount of water drunk, the amount of exercise taken, the size and nature of the last meal, the stage which its digestion has reached, and so on. On the average a healthy person voids about a litre and a half of urine per day, containing about thirty grams of urea (converted ammonia), fifteen of salt, and ten of other soluble substances. The remainder—more than ninety-six per cent of the total weight—is simply water. It is curious to note this continual loss of water, because water is one of the most necessary constituents in our blood. The reasons for its expulsion are two. Firstly, as we have seen, water-excretion may under certain circumstances make up for lack of salt in the blood; and secondly, the kidneys can only deal with substances in dilute solution. Urea, salt, or creatinine do not form gases and cannot be got rid of in the lungs; they must be expelled dissolved in water.

A kidney consists of a multitude of coiling tubes. It is built up of tubes of two kinds. Firstly, the arteries, the veins, and the capillaries which connect them together; and secondly, the *kidney-tubules*, end-branches of the ureters, in the walls of which the separating processes take place. There are, roughly, a million of these tubules in each kidney of a full-grown man.* The tubules end blindly at one end, at the other they lead into the ureter. The details of their structure and working are complicated and it would occupy too much space to go into them at all fully; the essential fact is that each kidney-tubule is surrounded by a dense network of blood-capillaries, and that a fluid oozes from the blood vessels through the walls of the tubule into its canal and so to the ureter. But the ooze is not a simple filtering like the lymph-ooze that we have already described. It has filtration as its basis, but the cells of the tubule wall interfere actively and modify it in various ways, hurrying up the departure of unwanted substances and delaying or preventing altogether the escape of those that are desirable. Now this needs energy. A busily excreting kidney is working against diffusion just as a busily contracting muscle is working against gravity or inertia or friction, and in both cases the energy required is produced in exactly the same way—i.e., by burning chemical fuels. As a matter of fact, one gram of kidney tissue consumes, on the average, more oxygen per minute than the same weight of the conspicuously labouring heart, and, like the heart, its oxygen consumption rises three or four times when it is given heavy work to do.

This co-operation of liver and kidneys is the most important excretory mechanism with which we are provided. But among the other organs there are one or two which assist in the process to a small extent, although their primary duties are different. The sweat-glands, for example, are primarily organs which regulate the temperature of the body. But since sweat contains a little urea, the sweat-glands assist the kidneys to a certain extent in eliminating this substance. They also eliminate water. Again, the first duty of the salivary glands is to produce a digestive secretion, but they have a subsidiary excretory function. During the chemical operations of our cells minute traces of that deadly poison prussic acid are produced; this is converted by the liver into a relatively innocuous substance, potassium sulphocyanide, which is excreted partly by the kidneys and partly by the salivary glands. In the former case it leaves the body at once, while in the latter it has first to run the whole length of the digestive tube. Similarly bile, the digestive juice made by the liver, has an incidental excretory function, for it carries the products of the continual breakdown of red blood cells out of the body via the bowels.

---

*The number of stars that a good eye can distinguish on a winter night is between two and three thousand.

COMPLEX BODY-MACHINE AND HOW IT WORKS

But the most important accessory excretory organ is the large intestine. Among the end-products of vital activity is phosphate of lime, a very insoluble compound, which would form clogging masses of crystals in the delicate tubules of the kidney if it were excreted by this organ. This substance is excreted into the large intestine from the blood-vessels in the intestinal wall and ejected from the body with the faeces. Other mineral substances are also removed in this region.

§ 7. How Our Food Becomes Blood.

WE HAVE now made a very extensive examination of Mr. Everyman's internal workings. Since we chose him (with occasional allusions to his attendant mouse) for our first examination of a living thing, and noted his extraordinary preoccupation with feeding and his close parallelism to a machine, we have illuminated his structure and the nature of his mechanism very considerably. We now know him and the mouse his parasite to be extraordinarily complex organizations of billions of inferior beings, his cells; we know that within his skin all his body is soaked in blood and lymph; we have magnified his structure, made him, so to speak, transparent, dragged his lungs and heart and kidneys into the light, and exhibited them in illustrations, and shown how part works with part while he (almost regardless of this immense multitude of detailed activities within) goes through the routines of his daily life. But there still remains a gap before our opening review of the routine working this human body-machine is complete. We know that he needs food to supply the blood with the fuel to keep his billions of constituent cell-machines going, but we have still to trace how the food he packs so sedulously into himself becomes the warm, nutritious, comforting blood which bathes all his internal being. To that we will next address ourselves.

It is possible now to draw up a rough list of the different kinds of food that he requires. We know that he needs fuel—matter which is capable of yielding, when oxidized, the supply of energy without which life cannot continue. The substances of this class are not built into his living structure, but are taken up and stored in his cells, to be burnt whenever required; they correspond to the petrol that runs through a motor, rather than to the machine itself. Besides fuel, he needs materials for growth and for making up for the slight wearing away of the living engine that any activity involves. Then there are substances, such as the water or salt of blood, that are neither built up into protoplasm nor used as fuel, but are nevertheless necessary constituents of the bathing fluid of cells. Lastly, we add to this list certain "accessory food factors," the so-called *vitamins*, whose rôle is still obscure, and which we shall discuss more fully in a later book.

Life, however, is not so simple that he can sit down to his meals and take so much fuel, so much tissue material, so much water, and so much of the minor helps and stimulants. Certain types of restaurant indeed mask the dishes with rather doubtful indications that they are flesh-formers or energy-generators. But life is not so straightforward. Mr. Everyman has to take his food as he finds it, and it has not undergone nearly as much sorting out as that. The simplest item in his consumption is the water he drinks, either pure or with various colouring, flavouring, or stimulating additions. Of that, therefore, we will speak first. We have already stressed the profoundly important part that water plays in the living machine. About 59 per cent by weight of the human body is water, the proportion varying from tissue to tissue. Thus bones, which include a large amount of lime between the cells, have only 22 per cent; but the liver contains 69 per cent, muscle 75 per cent, and the kidney, which is comparable to a sponge of tubules whose pores are full of blood and urine, contains as much as 82 per cent of water. The inside of cells is fluid and consists largely of a solution of various substances in water. Moreover, water

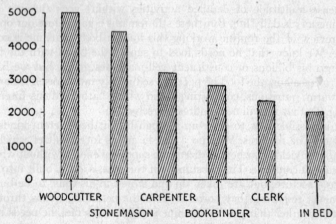

Fig. 24. Energy requirements from food.

The diagram shows approximately (measured in calories) the energy needed in twenty-four hours by men in various occupations.

COMPLEX BODY–MACHINE AND HOW IT WORKS 73

acts in certain characteristic ways on molecules which are dissolved in it, so that the properties of aqueous solutions—upon which vital phenomena depend—are unparalleled by other mechanical systems. Life without water is about as conceivable as music in a vacuum. As Sir Arthur Shipley says in his attractive little book *Life*, "Even the Archbishop of Canterbury comprises 59 per cent of water." Water is continually being lost by the body. It evaporates from skin and lungs, and it is used also to flush away the waste products of cell activity. To make up for this loss, a resting man needs somewhere about three pints of water a day; a man taking considerable physical exercise much more.

In addition, Mr. Everyman must swallow a variety of other substances, chiefly the more or less altered living or dead tissues of plants and other animals, and trust his intricate internal arrangements to assimilate them to his needs. This assimilation involves very complicated changes, for the stuff of tissues of dead or dying animals and plants is quite unsuitable for immediate entry into his blood. To explain how it is made suitable involves a certain use of chemical terms, and it will make things clearer if we remind the reader of the exact significance of a few expressions that are unavoidable in the discussion, that are always being heard when food is under consideration and about which most of us are apt to be a little vague and wanting in precision.

Nowadays we read so much of the "break-up of the atom," and so forth, that some people seem to be in doubt whether they may still think and speak about atoms. Thirty or forty years ago the atom was supposed to be a very simple little particle, the smallest quantity, the unit amount of an element that could go into combination. Today we understand the atom has an extremely complicated structure; but the fact remains that an atom is the smallest quantity of an element that can go into chemical combination. And a "molecule" still means the smallest quantity of a chemical substance that can exist separately. It may consist of from one to a vast number of atoms of the same or different elements; but though there are large molecules as well as small ones, the largest molecule is still ultra-microscopic and minute beyond our everyday imagination. A water molecule contains three atoms; a molecule of common salt two; but in many of the substances we shall next deal with the atoms may run to thousands or tens of thousands.

In these discussions on digestion we are always hearing the terms carbohydrates, fats, and proteins. Carbohydrates and fats are made up of carbon, hydrogen, and oxygen. They differ from each other in the proportions of these three elements that they contain. In carbohydrates there are two atoms of hydrogen and one of oxygen to every carbon atom (hence their name, for water is H_2O), while in fats the proportion of oxygen is much lower. They differ also in the pattern on which the atoms are fitted together. The best-known carbohydrates are sugar and starch. The term fats as used in physiological discussion includes the vegetable oils (such as olive oil) in addition to the more obvious fatty substances. Both carbohydrates and fats are readily combustible—hence their importance to the living body. They are the chief fuels that it consumes. They can be substituted for one another. The Eskimos get their energy chiefly from oils and fats, while the inhabitants of a tropical climate substitute carbohydrates to a large extent.

The third class of food-stuffs, the proteins, are a vast variety of much more complicated substances, containing not only carbon, hydrogen and oxygen, but nitrogen and often other elements, such as sulphur and phosphorus and iron. Their relatively gigantic molecules contain hundreds and even thousands of atoms. Gelatin and egg-white are proteins unmixed with any other food substances; of the commoner foods, meat, cheese, and vegetable seeds (such as peas, beans, and lentils) are richest in protein. Proteins, like fats or carbohydrates, can be burnt as fuel, and a large proportion of the proteins we eat is used up in this way. But they have another and more essential rôle, one which they alone can perform. For the actual fabric of the living cell is built largely of proteins. Not only can they be burnt by our engines; they are the steel of which these engines are made. Herein lies the importance of proteins in our diet. Any of the three kinds of foodstuffs will serve as an energy-source, but protein is the only one that can be used for body building—for growth and for keeping the tissue-machines in repair.

There is just one other chemical term we shall have to use here—amino-acid, because these amino-acids (there are a number of them) lie at the base of protein structure. An amino-acid molecule is simple compared with a protein, complex compared with water. It contains from ten to twenty atoms of carbon, hydrogen, oxygen, nitrogen and sulphur, which may be put together in various ways. The number of different arrangements—in other words, of different amino-acids—that have been found in living tissue is small; there are about twenty-five. Now these amino-acid molecules can combine together, and it is possible for great numbers of them to unite to form a single giant molecule. It is these giant molecules which are the proteins.

COMPLEX BODY-MACHINE AND HOW IT WORKS 75

There are only twenty different amino-acids, but countless numbers of proteins. That is because there is an infinity of ways in which the amino-acids can be built together. Imagine a box of building blocks, containing twenty kinds of bricks, and about ten or fifteen of each kind. How many different houses could you build, using from fifty to a hundred bricks at a time? An incalculable number. The bricks correspond to amino-acid molecules, the houses to protein molecules. Like the houses, proteins differ from each other in two ways—in the proportions in which the different amino-acids are present, and in the geometrical plan on which these acids are arranged. As a further complication, a few proteins include factors that are not really amino-acids at all; thus haemoglobin contains iron and the nuclear proteins contain phosphorus.

And now we are able to describe the essential process of protein digestion. The body-cells require proteins, but they require *particular* ones. The proteins in our food are different from those in our flesh. It is not possible, for example, that the proteins of a cabbage-leaf could play any part in the activity of a muscle-cell; its amino-acids are fitted together in the wrong way. There is indeed no way of eating the correct kind of protein except cannibalism, and even then there exists no way of getting it direct to the blood without unmaking and making it again. It is necessary thus to convert a great variety of different food-proteins into flesh-proteins. The method adopted is as follows: We eat proteins of all kinds; we break them up in our bellies into their constituent amino-acids; we absorb the amino-acids into our blood and thus distribute them to our tissues; finally, the cells take up these amino-acids and fit them together to make up the particular proteins that they require.

The work that is done upon the carbohydrates and fats is quite parallel to what is done upon proteins. The most familiar simple carbohydrate is *glucose*, the chief sugar of many fruits, and especially of grapes, and the sweetening constituent of common jams and sweets. This is the form of carbohydrate which circulates in the blood, and all other carbohydrates—the starch in potatoes, for example, or table-sugar—which are more complex in their composition must be broken down to glucose before they enter our real interiors. Similarly the fats are broken down to their simpler factors glycerine and fatty acids.

Assisted by bacteria in his intestine, the mouse can also turn cellulose, the main substance of ordinary vegetable tissue, into glucose, but in that he has the advantage of man.

In digestion all the three important classes of organic foods—proteins, carbohydrates, and fats—are thus treated according to the same general plan; the molecules are broken up into simpler constituent ones, absorbed as such into the blood, and then rebuilt by the tissue-cells into other arrangements which suit their individual needs. Water and the mineral salts that we require are not dissected in this way; their molecules are already as simple as need be, and no adjustment is necessary.

In addition to taking the food apart into units for protoplasmic re-

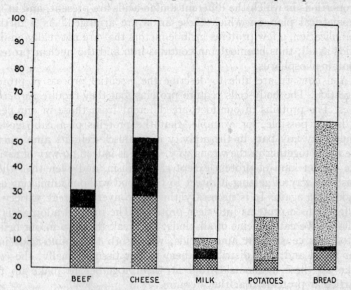

FIG. 25 The percentage composition of some common foodstuffs.

Protein, cross-hatched; fat, black; carbohydrate, stippled; water and mineral salts, white.

construction, digestion also serves to get the food from the open passage of the gut into the real interior of the body. Insoluble food like starch must be turned into soluble compounds to do this; and when molecules are large like those of the proteins they must be broken up into smaller ones, which can pass more easily through the walls of the intestine.

All this disintegration involves operations carried out with great delicacy and precision. It is not a senseless shattering, like the blowing up of a house by a bomb; it is a methodical taking-apart of the molecule, a division of it into certain units that we need, and units which have to be handled with care lest they themselves be broken.

The reader will be familiar with the way in which mass-produced motor-cars are assembled. There is a long moving platform on to which the bare chassis is put. As it is carried steadily along it passes a series of skilled mechanics, each of whom slips one particular part into its place. Each man has his own special operation that he performs on the cars that pass him—a part to insert, a particular bolt to tighten. The process is so organized that finally, at the other end of the platform, the car stands completed. Now imagine the motion of the platform reversed; imagine the finished car standing at one end and moving past a file of highly-skilled dismemberers; each man lifts out one particular part or loosens a particular nut and so helps the disintegration one stage further; finally, when the car reaches the other end, it has been carefully and completely taken to bits. This reversed process would be exactly parallel to digestion. Food is passed from chamber to chamber along the digestive tube, and in each region it undergoes special and appropriate stages of break-down.

The important point about the image is that of the workmen, standing beside the moving belt and each a specialist in some particular operation, for in our digestive tubes the food is attacked and modified by a chain of chemical workmen—substances known to the physiological chemist as *enzymes*—each of which makes its own adjustment, and leaves its own mark on the food.

Consider, for example, the digestion of starch. Starch molecules, as we have seen, are complexes which have to be broken up into the simpler unit called glucose. As a mouthful of bread is chewed it is mixed with saliva, and the saliva contains the first chemical operator, an enzyme known as *ptyalin*. The ptyalin begins to act on the bread at once, and, being swallowed at the same time, continues its work in the stomach. Ptyalin does not completely digest the starch molecule; it breaks it into parts which are simpler than the original molecule, but nevertheless more complicated than the final product. On leaving the stomach, the bread is handed over to two more enzymes, *amylopsin* and *maltase*. The former works on the pieces that the ptyalin left, and breaks them into simpler structures, *disaccharides*, each of which is the equivalent of two glucose molecules stuck together. The latter completes the process by splitting these dual structures into glucose. Thus the starch molecule is not suddenly blasted; it is pulled to pieces with precision and method in a series of graded stages. So with fat, and so with protein.

The enzymes are pre-eminently specialists; maltase, which specializes in the handling of disaccharides, can do nothing to intact starch molecules, and ptyalin, which performs the earlier stages, cannot carry the dissection beyond a certain point. The two have to co-operate. Moreover, like human experts, they are particular about their working conditions. Thus maltase refuses to work in an acid medium, and we shall

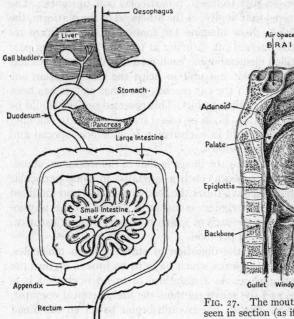

FIG. 26. The organs of digestion.

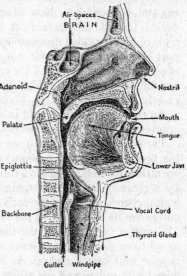

FIG. 27. The mouth, nose, and throat seen in section (as if the head had been sawn into equal halves) to display the parts concerned in swallowing.

later note what precautions are taken to neutralize in the intestine the acidity of the gastric juice. These substances are peculiar in that they act in very small concentration; they are consequently hard to analyze, and, beyond the fact that their molecules rival those of the proteins in complexity, little is known of their constitution.

The story of Mr. Everyman's breakfast may now be summarized very briefly. It is handed from enzyme to enzyme, and stage by stage it is broken up. The digestive apparatus consists of a series of tubes and chambers through which the food passes, and a series of glands

COMPLEX BODY-MACHINE AND HOW IT WORKS 79

which make the disintegrating substances. The stages of the process are as follows:

The first workshop, in which the dismantling is begun, is the mouth. In the mouth food is chewed—that is to say, it is broken up into smaller pieces and thoroughly mixed with saliva. The teeth reduce food to a sort of shredded pulp, which is readily penetrated by the chemical substances responsible for digestion. The purpose of mastication is two-fold. First and most important, saliva is a lubricant, and its sliminess facilitates swallowing and the passage of food to the stomach. After it is swallowed, a well-chewed and therefore slippery mouthful takes about six seconds to glide from the mouth to the stomach; a hard dry object (such as a cachet of bismuth carbonate) takes about fifteen minutes. The second function of saliva is to begin the digestion of starch, which is done, as we have seen, by the ptyalin which it contains. Ideally, when mastication is finished and the bolus is swallowed, the food should be reduced to a finely grained pulp and thoroughly mixed with saliva, but Mr. Everyman is generally too impatient for this to happen.

Saliva is made by the *salivary glands*, of which there are three on each side of the face. Each gland consists of a branching hollow tree of tubes of which the twigs end blindly while the main trunk, or salivary duct, carries the saliva away from the gland to the mouth. The end tubes are lined by thick cubical cells, and it is these cells that actually make the saliva. The saliva contains a little salt, a little protein, some ptyalin, and a little mucin (the substance which makes it slimy). In manufacturing this juice the cells make special substances (ptyalin and mucin); moreover, in separating from the blood a fluid in which the proportion of dissolved substances to water is very much lower than in blood, they do work and use up energy just as the tubule cells of the kidney do. The salivary glands of a man may turn out as much as ten times their own weight of saliva a day.

When the food has been masticated it is swallowed by the gullet (*oesophagus*), a muscular tube about an inch across and ten inches long, which leads down the neck and through the chest into the stomach.

At the back of the throat there are two other openings besides the oesophagus—the opening of the nose-passage, floored by the soft palate, and the opening of the wind-pipe. Swallowing is a complex act involving, among other things, the closure of these openings so that food cannot "go the wrong way." During mastication the food is kept at the front and sides of the mouth; when about to be swallowed, it is collected in a lump on the back of the tongue. There is then a rapid succession of events: first a pause in the masticatory movements, then a slight upward movement of the diaphragm, then a violent thrust upward and backward of the tongue, which shoots the bolus back into the oesophagus. As this last movement occurs the soft palate moves back to protect the nose-cavity, the cartilages of the larynx displace themselves so that the opening of the windpipe becomes a tightly closed slit, and a curious little flap of cartilage, the *epiglottis*, which normally sticks out into the throat, ducks smartly out of the way of the oncoming bolus. In this manner the food is flung into the oesophagus and at the same time it passes out of our ken. It leaves the mouth, which is under the direct control of our conscious minds, for those dark and mysterious regions of whose proceedings we are normally unaware.

The oesophagus (and this is also true of the rest of the digestive tube) is able to perform active movements automatically without any direction from the brain. Its muscles are always slightly contracted, and their contraction is maintained by the agency of a special local nervous system in its tissues. The presence of a bolus in any part of the oesophagus, large enough to stretch its wall, sets these muscles going. The part of the oesophagus immediately in advance of the bolus relaxes, becomes loose and flabby, while the part immediately behind the bolus contracts vigorously. As a result, the bolus is squeezed forward by the tightening muscle into the looser part. But as it moves forward it sets up the same changes; the part immediately behind it contracts and urges it on, while the part in front slackens to receive it. In this way the bolus is driven on by a ring of contraction. This process is called *peristalsis*. In the oesophagus, peristalsis is so powerful that a man can swallow perfectly well when he is standing on his head. He can and most animals do drink "upwards," and in cases of operation in this region the danger of an instrument being swallowed if once it gets into the gullet has to be provided for. Whatever the patient may wish in the matter, his oesophagus will follow the laws of its own being so soon as it gets a grip.

The stomach has a capacity of about three-and-a-half pints. It has two openings, that of the oesophagus, through which the mixture of food and saliva enters the stomach, and that of the duodenum (the first part of the intestine), through which the products of gastric digestion go out. The wall of the stomach is complex in its minute structure, the essential parts to notice being the inner layer, containing thousands of glands which make the gastric juice, and an outer layer of plain muscle-fibres, by means of which the stomach executes slow churning movements during digestion.

COMPLEX BODY-MACHINE AND HOW IT WORKS 81

Food remains in the stomach for at least an hour and undergoes slow permeation by the gastric juice. At the beginning it is hardly altered from its original state; it has been coarsely shredded and mixed with a slimy lubricant and the break-down of some of the starch has begun, but it retains its cellular structure almost intact, and the proteins and fats have not yet been attacked. During its sojourn in the stomach it is reduced to a homogeneous texture, the cells being destroyed and all trace of structural organization eliminated; at the same time, the

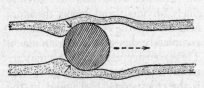

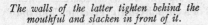

FIG. 28. How food is squeezed along the digestive tube.

The walls of the latter tighten behind the mouthful and slacken in front of it.

FIG. 29. A few of the muscle cells responsible for the movements of the digestive tube, greatly magnified.

digestion of starch continues and that of proteins and fats is begun. The final result is a soft, slightly acid pulp known as *chyme*.

The *gastric juice* is a liquid made by a multitude of very tiny and simple glands situated in the inner layer of the stomach wall. It has several important characters: it is acid, containing a little hydrochloric acid, and it includes enzymes—*pepsin,* which attacks proteins, and *lipase,* which attacks fats. The glands begin to secrete this juice about five minutes after a meal is swallowed, and it soaks slowly into the new mass of food. As it does so it produces certain profoundly important changes. The pepsin begins the dissection of protein molecules. At the same time some of the simpler carbohydrates, such as cane-sugar, are attacked by the warm hydrochloric acid and broken down to glucose. A certain digestion of fats also occurs, partly under the influence of the hydrochloric acid and partly under that of the lipase. It should be noted, however, that the extent of fat digestion in the stomach is very slight, for fats have a tendency to form comparatively large spherical droplets, into the interior of which neither hydrochloric acid nor lipase can penetrate, and the extent to which these substances can attack fats is limited by this factor. To the end the chyme contains little droplets of undigested fat.

Let us insert here a word or so about this characteristic of oils and fats and certain consequences that it entails. Anyone who uses brilliantine or who makes his own salad-dressing knows how hard it is to mix oil with liquids like water or vinegar. Oil and fat are *exclusive:* they keep themselves to themselves. The fats which are stored in most fatty substances (such as meat-fats) are there in the form of droplets, and consequently when the cells containing them are broken down by pepsin they escape in droplet form. This handicaps the enzymes, which can only work at the surfaces of the drops and cannot get at the interior. But in some foods—egg-yolk, for example—the fats are already finely divided, and being therefore accessible to the enzymes are almost completely digested in the stomach. This exclusiveness of fats, we may remark, is probably the reason why fat is the chief stored fuel found in the body. In many of our tissues, and especially in the liver, there is storage of a carbohydrate, glycogen, but only in limited amounts. Glycogen, being soluble in water, cannot be accumulated in a cell without affecting its chemical and physical processes, but a fat remains aloof and does not interfere with the processes about it until it is called into use. A being with fixed meal-times like Mr. Everyman might very well dispense with such storage. But his body will insist upon hoarding droplets of fat in all sorts of odd corners—under his skin, in between his muscles, in his bone-marrow, in his liver, in the membranes of his abdomen—if he gives it a chance. He must diet himself or take exercise to burn up his fuel surplus if he does not want to be encumbered with this provision against a highly improbable involuntary fast.

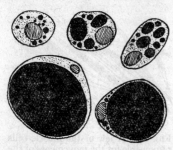

FIG. 30. A few of the cells which accumulate hoards of fat for use in time of need (greatly magnified).

The drops of fat are shown black.

But to return to our story of digestion, the most important aspect of the stomach-phase is the breaking down of proteins by pepsin and the consequent loss of structure which prepares the food for more effective action in the intestine.*

The permeation of the food-mass by gastric juice is a very slow process, so slow that the portion lying in the middle of the stomach

---

*The clotting of milk which occurs in the stomach is due to a change in its protein constituents. This change is often taken as evidence for the existence of a separate enzyme, *rennin:* but it is probably due to pepsin.

may not be reached for over half an hour after swallowing. This fact is an important one from the point of view of starch digestion. Ptyalin cannot work in an acid medium and is therefore inactivated by gastric juice; but during the thirty minutes or so before the middle of the food mass is got at by the juice, digestion of starch by ptyalin continues. At the same time, there is another activity in the central zone. We invariably swallow with our food a certain number of bacteria; these organisms, finding themselves in a warm place with plenty of food material, grow and multiply exuberantly. There is time for at least one complete generation of bacteria before the acid reaches and exterminates the little colony.

As we have already pointed out, the process of gastric digestion is assisted by slow churning movements of the stomach wall. These movements are like the peristaltic waves that drive food along the oesophagus—they have the form of rings of contracted muscle, embracing the stomach and moving downward and to the right in the direction of the duodenal opening. The waves move steadily and slowly. They originate near the middle of the stomach, at intervals of from fifteen to twenty seconds, and they creep along the stomach like ripples moving incredibly slowly over the surface of a pond. The effect of the waves is to mix up the stomach contents and to drive them towards the tube along which they are to depart.

The way out of the stomach, the opening of the duodenum, is called the *pylorus* (the gateway). It is guarded by a powerful ring of muscle which normally keeps the exit tightly closed. The pyloric muscle is a sentinel, and will not allow food to enter the intestine before it has been adequately treated in the stomach. During the early stages of gastric digestion, before the food mass is ready to pass on to the next stage, the pylorus remains shut; the sluggish contractions of the stomach cannot push matter through it, but merely produce eddies and currents in the food mass. As digestion proceeds the stomach-waves become more and more forcible, and the pylorus begins to open; with every wave a little jet of chyme escapes into the intestine, and so, when ready, it is passed on.

It is clear that the time spent by a meal in the stomach will depend upon the ease with which it is digested. Thus veal, particularly minced veal, is easily broken up and is not detained for long; but pork, being more resistant to the gastric juices, remains in the stomach for about four hours. If a draught of water be taken on an empty stomach, to quench thirst, the pylorus relaxes at once, and it reaches the duodenum in one or two minutes. For obvious reasons, the thoroughness of mastication will affect the digestion-time; a properly chewed and shredded meal is much easier to permeate than a meal that is gulped down in lumps. That is why Mr. Everyman does well to read or talk during his meals and take and deal with his mouthfuls slowly. The cult of Fletcherists is a health cult which makes a great point of never swallowing solid lumps of food. Fletcherists just go on chewing until the food is practically liquefied. They claim that they are rewarded by great digestive tranquillity.

The small intestine is a tube about an inch across and some twenty feet long, which forms an intricate coiling mass in the abdomen. The whole tube is very uniform in its minute structure, but physiologically it is possible to distinguish between the first and second halves. In the former the digestive process is completed, while the latter is chiefly concerned with the absorption of food into the blood. We shall deal at present with the first half only.

We have seen how the intestine leads out of the stomach. Some four inches from the pyloric opening there opens into it a narrow tube. This tube has two branches coming from important digestive glands, the *liver* and the *pancreas*, and it serves to carry the juices made by these glands into the intestine. Each has essentially the same structure as a salivary gland, but they are larger and more complicated.

The *liver* is by far the larger of the two organs. Everyone knows its shape and position. It secretes a greenish fluid called bile, which trickles slowly down a tube, the bile duct, into the small intestine. If a meal is being digested, the bile runs straight down to the intestine; if not, it is stored until required in a hollow sac, the *gall-bladder*, which lies at the end of a side branch of the bile duct. The secretion of bile is more rapid during digestion.

The *pancreas* (the sweetbread properly so called of domestic animals, though other glandular matter is frequently dished up under this name) is a smaller gland, pink, about eight inches long, having an irregularly elongated shape, and lying in a horizontal plane behind the stomach. It secretes a colourless *pancreatic juice*, which flows directly along the pancreatic duct to the intestine.

Moreover there is yet another supply of ferments. The intestine, like the stomach, has in its wall an enormous number of minute, simple glands, whose tubules open into the intestinal canal. These glands make the *intestinal juice*, or *succus entericus*, a fluid which plays a part of the utmost importance in the digestive process. These various juices act not only on the food, but also on one another. The presence of bile makes some of the enzymes in the pancreatic juice more active, and the succus entericus converts inactive "enzyme-precursors" in that juice into active enzymes.

The chyme is mixed with the secretions of all these glands as it enters the intestine, and so becomes exposed to the action of a much more vigorous team of enzymes than heretofore. They attack it simultaneously, continuing and finally completing its digestion. It is quite possible for a man to adapt himself to live without a stomach, but his small intestine is essential. Proteins—or rather, the shattered results of the gastric digestion of these substances—are attacked by two enzymes, *trypsin* and *erepsin*, contributed by the pancreas and intestinal wall. Trypsin specializes more in the earlier and middle stages of protein dissection, while erepsin is concerned with the splitting of the penultimate products. Between the two the protein molecules are completely taken apart into their constituent amino-acids. Carbohydrates also have been reduced to assimilable glucose—all, that is, that Mr. Everyman will get out of his meal, for he may well have eaten more than his interior was prepared to cope with. Fats are fallen upon by certain substances present in bile, the so-called bile-salts, which overcome their stubborn tendency to form large droplets and disperse them into a state of minute division, called an emulsion. (In physical language, the bile-salts lower the fat's surface-tension.) When this has been done, the fats are readily assailed by *lipase*, an enzyme made by the pancreas, which splits them up into their last stage of fatty acids and glycerine. Thus for all the main classes of organic foodstuffs is the process of digestion completed.

But besides the various enzymes there is another important substance in this region, contributed by the pancreas. This is carbonate of soda. The enzymes just mentioned will not work in an acid medium, therefore the acidity of the chyme has to be neutralized before their operations can be performed. This is done by the alkaline carbonate of soda.

As in the stomach, digestion in the intestine is assisted by muscular movements. The writhings of the intestinal tube are of several kinds. There is, for example, a slow, regular swaying of whole loops of the intestine from side to side, and there is a sudden chopping up of a straight length of intestine into segments by the appearance of a series of local contractions. These movements can be studied by giving an experimental meal containing bismuth, and examining the subject with X-rays; the bismuth casts a dense black shadow, and so the movements of the food can be watched. The meal is seen to remain for a time in each loop of the intestine, undergoing various rockings and churnings which thoroughly mix the chyme with the digestive juices. After a little time a strong peristaltic wave runs along the loop, and sweeps the mixture into the next, where it suffers another period of mixing. Its progress is therefore irregularly rhythmical; it pauses for a time, to be thoroughly churned, then sweeps on for a foot or so, then pauses again, and so on. In this manner, in some three hours, the meal is worried along the twenty odd feet of small intestine, and its entry into the blood-stream made possible.

Now the inside surface of the intestinal tube is given a velvety appearance by the presence of myriads of finger-like projections, the *villi*, each about a twenty-fifth of an inch long. But it is restless velvet; the villi wriggle about when a meal is being digested, lengthening, shortening, and swaying from side to side. It is through the villi that nutriment enters the blood. Each of these absorbent fingers is in contact outside with the digested food, while inside it passes a copious stream of blood. The cells of the villi are active, like the cells of the kidney or of the salivary glands. They lay hold of nutritive molecules in the gut and force them into the blood-stream. In this way proteins (broken down to amino-acids), carbohydrates (in the form of glucose or other equally simple sugar), and such substances as salt, which, being already simple enough, have undergone no digestion, are passed directly into the blood-stream. Thence they are carried to that very complicated organ, the liver.

This dark-red large portentous organ plays a very central part in the internal activities of Mr. Everyman. We are all familiar with that hackneyed answer to the trite question: Is life worth living?—"That depends upon the liver." And truly the working of this versatile but sensitive viscus is of primary importance to the chemistry of the body, and to the colour of the mood of which that body is either the medium or the material substance.

Let us rehearse the tale of the liver's activities. We have already noted that it is a digestive gland, and described the digestive use of the bile in emulsifying fats and stimulating the pancreatic juice. The bile also contains excretory material, due to the breaking up of old blood corpuscles, about which we will not trouble ourselves here. Since the secretion of bile necessitates obvious anatomical structures— the presence of a bile-duct, tubules in the liver, and so forth—this biliary secretion was the first of the liver-functions to be recognized; later, and only as a result of painstaking experimental research, was the chemical side of the liver's activity brought to light. Now we realize that its chemical activity is even more important than its

COMPLEX BODY-MACHINE AND HOW IT WORKS 87

digestive. We have noted its action upon ammonia in our account of the kidneys. It can also deal with undesirable matters present in the blood from the gut—such as the substances produced by bacteria in the intestine during protein digestion. Its censorship is not exclusively chemical. If solid particles (such as bacteria themselves) make their way into the blood, certain of the liver-cells can actually lay hold of them and consume them. Next the liver performs a great work of adjustment. If, for example, the sugar absorbed by the villi is in excess of

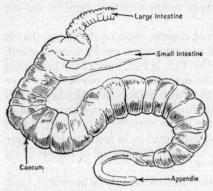
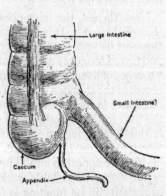

FIG. 31. The organs of Fig. 32 as they appear in the rabbit.

The caecum is large in vegetable-feeding animals.

FIG. 32. An organ that we should be better without—the human vermiform appendix.

that required by the body, it can store it in the form of glycogen, while if, as is often the case, there is insufficient fuel, but a surplus of protein, it can also break down the latter and convert the products into carbohydrate.

Because of its exposed position—exposed to anything that may get into the blood from the intestines, and also communicating with the intestine by means of an open duct—the liver is very liable to be deranged. That is a necessary consequence of its censoring duties. A number of parasitic organisms that cause various troubles, such as dysentery, in the bowels may penetrate to the liver, and produce abscesses. If Mr. Everyman is in the habit of overworking his censor—if, for example, he regularly consumes unnecessarily large amounts of alcohol—its sensitive specialized cells may be damaged and waste away, leaving too high a proportion of connective tissue, a condition known as cirrhosis. And the regular working of the liver may be upset in other ways. For example, inflammation of the bile-duct itself, or of the part of intestine into which it opens, is very easily set up, a chill will do it. Then comes jaundice or a jaundiced condition. The bile cannot pass through the swollen, closed duct, so that the fats in the intestine are no longer dispersed, and therefore they are no longer digested. Moreover, they form films over the other food-substances, and interfere with their digestion also. And the pent-up bile-pigments force their way into the blood, and are manifested by yellow eyeballs and yellow skin. Our complexions go yellow with this pigment, and our faeces white for lack of it.

Nevertheless, the liver does not deserve most of the abuse that is showered upon it. It is imagined as a capricious, temperamental gland, the most delicate part of our bodies. But, in fact, most of the headaches and furred tongues and losses of appetite that we call "bilious attacks" or "touches of liver" have nothing to do with that organ at all; they are disorders of the stomach due to errors in eating and drinking. Even jaundice originates in most cases in trouble in the intestine or bile-duct, not in the liver itself; the yellowness we just noted is, in fact, due to the liver faithfully carrying on its duty, although the avenue of escape of its secretion is closed. Wherefore, if the reader is in the habit of maligning his liver we appeal to him to revise his estimate. It is an ingenious, busy organ, doing responsible work in a very exposed situation, and any irregularity in its function is far more likely to be due to the way he treats it, or has treated it in the past, than to any inherent frailty in its own constitution.

In this manner, then, most of our food-stuffs are absorbed and scrutinized. But there is a curious exception, for digested fats are not sent to the liver; they travel to the heart by another way. We have already referred to the oozing out of lymph through the capillary walls that goes on in all our tissues and its return by the lymphatic system to the heart. Now in the villi this ooze is particularly copious, and it is this stream that carries the fats away when they have been absorbed. This is probably because of the stubborn exclusiveness of fats that we have already noted. As soon as they have been absorbed by the villi, the glycerin and fatty acids form fats again, and these run together into droplets; these droplets, if they were sent by the blood, might clog the liver capillaries, and obstruct the portal circulation. After a meal rich in fats, the lymph contains enormous numbers of these droplets. They give it a milky appearance, which can be traced up to the thoracic duct, and even into the great veins.

When the enzymes have finished their work, and when all that we require has been absorbed, there is still a certain residue of our meal left over. There are substances that are useless for food, and substances that might be useful if we had a chemical apparatus for dealing with them—such as cellulose, the carbohydrate wall of plant-cells, which forms a large part of the food of a cow. Moreover, since digestion is not a perfectly efficient process, there is a small proportion of good food that has successfully run the gauntlet. Before it is dropped overboard, this residue travels about a yard and a half, up the right side of the abdomen, across the top, and down the left, along a broad tube, the *large intestine*. On its way it undergoes further alteration, not always to the benefit or comfort of Mr. Everyman.

The opening of the small intestine into the large is guarded, like its opening into the stomach, by a muscular sentinel, a ring muscle, in this case the *ileo-caecal valve*. From this point the large intestine, or *colon*, pursues its devious course to the anus. It is curious to note by the ileo-caecal valve a small pouch leading out of the colon, known as the *caecum*, and leading out of this a little worm-like tube, three or four inches long, and ending blindly, the *appendix*. In man these structures have no functions that other organs cannot carry out—indeed, it is probably better to be without an appendix, for it may be the seat of acute and even fatal inflammation.

It is improbable that the human large intestine plays any important part in digestion. People can live to be healthy and active after its removal. When the remnants of our meal pass the ileo-caecal valve they have lost at least ninety-five per cent of their fats and carbohydrates, and they contain less water than the amount shed into them by the various glandular secretions they have received. The greater part of this residual water is absorbed in the large intestine. As far as other substances are concerned, although the possibility of further absorption during the slow progress of food along the large intestine (a progress that takes about twelve hours) cannot at present be absolutely denied, it is at least certain that whatever occurs is insignificant. Moreover, the large intestine is a hotbed of bacterial putrefaction. Any bacteria that can successfully resist our digestive processes flourish in this warm, relatively restful situation, and consume the undigested parts of our meal. It has been calculated that as much as fifty per cent of the faeces (exclusive of water) consists of living and dead bacteria. It is unlikely that much of the other half is food residue, since such residue has been used to build up bacterial bodies or as fuel for bacterial activity. Most of it is dead cells that have dropped off the intestinal wall, just as dead cells are constantly peeling off our skins. This dropping off of dead cells has, of course, nothing to do with nutrition, and it is remarkable that professional fasters have produced faeces after taking no food for thirty days, and that faeces can be generated in the colon even when its connection with the small intestine is artificially closed.

It is interesting to point out in conclusion here that Mr. Everyman is fitted up with a food canal which is certainly too long for his present needs, and that even in the case of the healthiest human being a fermenting multitude of these alien bacteria form a frequently mischievous foreign quarter in the intricate community of the body. They increase and multiply in complete disregard of their host and may cause him the distresses and inconveniences of colitis and various other annoyances. In the mouse the caecum and the large intestine have definite and essential uses. To a herbivorous mammal they are indispensable.

FIG. 33. A small part of the inner surface of the intestine, magnified.

The finger-like villi absorb food; among them are the mouths of glands which make a digestive juice. The oval in the middle is a guard-house against bacteria.

We have already pointed out that Mr. Everyman gets his food more regularly and surely than any other mammal. It is also better chosen, and it is prepared in various ways that anticipate digestion. A hungry mouse in hard times will do tremendous feats of digestion, from boots to carpets and grease-paint, that are quite beyond the present powers of Mr. Everyman. For him, under existing conditions, the small intestine seems to be nearly all that he requires. Intestinally he is over-equipped. An eminent London surgeon has declared that Mr. Everyman might with advantage be deprived of stomach, appendix, and large intestine, and benefit (ultimately) by the change. Perhaps as the enlargement of his belt begins to trouble Mr. Everyman he may be persuaded presently to try this heroic way back to the slenderness of youth.

§ 8. The Continual Struggle against Infection and Chill.

BEFORE we conclude this résumé of the Body-machine or Cell-community which constitutes the material substance of Mr. Everyman, we will say a little more about certain of its defensive activities. Apart

COMPLEX BODY–MACHINE AND HOW IT WORKS

from the graver dangers of fire and flood, war and traffic, it is continually being assailed by more insidious and deadly invaders.

In our account of the blood we described the police activity of the white corpuscles (*leucocytes* or *phagocytes*—"white cells" and "consuming cells" are alternative names for them) which fight and devour those bacteria that are swallowed in the food or otherwise reach the blood-stream. At times this police work attains dimensions that make it comparable to warfare. The main frontier of the cell-empire that we know as Mr. Everyman is the skin, and we may devote a little attention to the organization of this first line of defence.

Fig. 34 presents a section through a small part of the skin and the underlying connective tissue. At the upper surface separating the tissues beneath from the outside air there is a layer of special cells, the *epidermis*. The deeper cells of this layer are constantly growing and dividing; because of this continual multiplication they spread outwards towards the surface. But as they spread they become cut off from their blood-supply, for the capillaries do not extend into the epidermis. Only the deepest epidermal cells are properly nourished; the others are all in various stages of starvation. The most superficial cells, being remotest from the blood-stream, actually die, the outermost layer being composed of dead cells.

Now these dead cells are dry and horny; they form a tough rampart between the outer world and the living tissue. These cell-corpses are ordinarily brushed or washed away as they die; but when they are allowed to accumulate, as under a bandage which is not undone for many days, they form dense white masses of flaky scurf or scarf-skin, showing to what an extent our bodies die daily. At the same time, by the active division of the deeper layers, new cells are produced to starve and die so that the rampart is renewed from within. The epidermal cells are the martyr volunteers of the cell community. It is their sole business to multiply and perish so that their corpses may protect the state.

In Fig. 35 the hard impervious outermost layer has a winding appearance which suggests the Great Wall of China. The parallel is a good one; the skin guards the body from the Mongol hordes of bacteria. But if the Great Wall had been built of the mummified bodies of Chinamen the parallel would be closer.

Sometimes, through a cut or a scratch, a breach is made in the wall, and the bacteria lurking in the air or on our skins make an entry, whereupon there follows a battle royal in our tissues. The first response to any injury to the epidermis is a dilatation of the capillary vessels in that region. Chemical products of the injured stuff stimulate them to dilate. By thus increasing the local blood-flow reinforcements of white corpuscles are at once rushed to the affected spot.

If the process be watched in a transparent tissue, such as the web of a frog's foot, the movements of these skirmishers can be traced. At the region of injury these defenders can be seen to leave the capillaries.

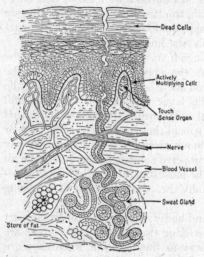

Fig. 34. A vertical section of skin, as seen through the microscope.

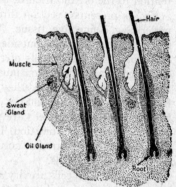

Fig. 35. A thin section of the human scalp, magnified, to show how the "roots" of our hairs are constructed.

They elbow their way, so to speak, between the living cells of the capillary wall and reach the intercellular spaces. Here they fall upon the bacteria and, incidentally, upon any injured or dying tissue-cells, and consume them; they take them into their own bodies and digest them. If the injury be a slight one, and the invading bacteria few and relatively harmless, the leucocytes have no difficulty in cleaning up the situation. A few hours after their entry into the battle-area they have withdrawn, presumably along the lymphatic vessels, and nothing is left but healthy tissue-cells, actively dividing in order to reconstruct the tissue that has been broken down. But if the invasion is more serious the struggle lasts longer. More and more leucocytes are brought to the battle-area, which becomes to the mind of Mr. Everyman "sore and inflamed." The bacteria, keeping and extending their footing, multiply in the tissues and destroy tissue-cells; the leucocytes frantically consume both bacteria and debris, and in excess of zeal, perish themselves of indigestion. At the end of a day or two, if we examine the battlefield with a microscope, we find a mass of dead and dying bacteria and cells, a mass that is disintegrating into the yellowish fluid called "matter," or *pus*, and round this mass we find a cordon of living leucocytes, sheathing it and hemming it in so that there can be no further spread of the trouble. Finally, since the blood-supply of the skin lying over the abscess is now quite cut off, the skin breaks down and the abscess "points," discharging its contents into the outer world. Mr. Everyman washes it, ties it up, and does his best to forget it.

But not always do the leucocytes prevail. And then the bacteria may succeed in invading the blood-stream, and once they have got access to blood-vessels they are swept round the body, and all the tender tissues of Mr. Everyman are at their mercy. This is when the warfare becomes really serious for him. Leucocytes join issue with the bacteria in the blood, fresh armies of leucocytes are formed in the bone-marrow, the lymph glands, and other special centres of proliferation, and at the same time there is a chemical battle between toxins (poisons) and antitoxins. The whole body becomes the seat of war, and the mental serenity of Mr. Everyman has to be subordinated to the urgent struggle. He has a fever, he says, goes to bed in order that his bodily energies may all be mobilized for this war, and he calls in the aid of a doctor, an allied cell-community, so to speak, for advice, comfort, and medical munitions for the struggle.

The power to swallow up and destroy hostile cells is not confined to these phagocytes, which wander about in blood and tissue. In the spleen there are localized cells which take in any undesirable bodies from the blood that circulates past them. During undulant fever, for example, the spleen-cells are found to be full of the organisms responsible for that complaint. The two parts of our interior surfaces which are most exposed to infection are also protected in this way, for cells in the liver exercise a similar filtering power on blood from the intestine, and so do cells in the lungs. Moreover, the walls of our intestines are dotted with special oval areas where leucocytes can be formed when necessary. These guard-houses are called Peyer's patches. Presumably this is because of the bacteria that swarm in our digestive tubes, warmed by our own heat, and taking our food not only out of our mouths but out of our very bowels. To an organism able to resist our digestive juices—and there are many such—the tender intestinal wall affords an easier way of entry into our blood than the horny skin, and it is therefore necessary to garrison strongly that weak point in our defences. The tonsils of the throat, and the appendix at the entrance to the large intestine also, are composed entirely of leucocyte-forming tissue of this kind. They are particularly strong salients on the line of defensive fortification. So at nearly every vulnerable point Mr. Everyman is garrisoned against his smallest (and greatest) enemies.

It is not only against positive infection that his multitudinous commonweal must defend itself. Mr. Everyman is everlastingly resisting certain physical processes, as persistent and insidious as his bacterial foes. First, like the mouse and all other mammals and birds, he has to keep himself warm inside. He belongs to a large and dominant sort of machines called warm-blooded creatures, of a distinctive delicacy and versatility. It is only within a very limited range of temperature that his protoplasm can work at its best. That is probably true of all living things, but while the warm-blooded creatures keep going as a rule in almost all the weather changes to which they are subjected, the cold-blooded creatures lie up and become inactive when it is too cold or too hot. Their bodies acquiesce in the temperature of their surroundings and all that it implies in sluggishness or liveliness, while the warm-blooded creatures are generally warmer than their surroundings and will not consent so readily to follow the intimations of the thermometer. The temperature of a healthy man's blood is about 37° Centigrade (98.4° Fahrenheit) and its fluctuations are fractional. (A moderate and agreeable temperature for a room is about 16° C., or 60° F.) His sensitive body shows signs of distress with a fall or rise of two or three degrees (Fahrenheit) in his internal temperature. A rise of this amount is enough in itself to produce fever symptoms, headache, malaise, and loss of nervous and muscular power. But the essence of the danger of a chilled condition lies in the diminished alertness of the leucocytes. Infections that are quite easily disposed of under normal conditions may now make headway. Colds, influenzas, bacteria always besieging the body, but normally held at bay, get successfully busy as the immediate result of a chill.

The body is continually losing heat to the cooler air which is usually around it, and therefore it must have internal sources of heat to keep its temperature up. This heat-production is the result of the continual chemical activity of our muscles and glands, for example of the processes needed to keep our muscles taut and ready for action. The temperature of the body depends upon a balance between this internal heat-production and the loss of heat from the skin. Both these factors may vary and need to be regulated. The rate of heat-production depends on the activity of our tissues, and of our muscles in particular;

COMPLEX BODY-MACHINE AND HOW IT WORKS

it is very much greater during severe exercise than at rest. The rate of heat-loss depends on the state of the air; on its temperature, motion, and humidity. Therefore, since our own temperatures must be kept strictly constant to secure bodily efficiency, there must be some method of regulating both heat-production and heat-loss, so that the variable factors may be compensated for.

To a certain extent this is done by controlling our own heat-producing processes. Shivering is a way of producing heat by otherwise pointless muscular activity. But such control, since it involves interference with our own life-processes, is inconvenient. The chief part in temperature regulation is played by the skin, which in the case of man is supplemented by clothing.

In most warm-blooded animals the skin is clothed either with hair or feathers. A hair is a rod of dead cells, proliferated from a living root, that sticks out from the skin into the outer air. The part that lies inside the skin is housed in a tube, the hair-follicle. Round the mouth of the follicle there is a sensitive nervous belt which we will consider later, and opening into it there are one or two *sebaceous glands* which secrete a natural brilliantine to keep the hair and skin supple. The outside parts of the hairs form a forest in which imprisoned air stagnates, so that heat loss by convection is reduced. At the same time, since air is a very poor conductor of heat, heat loss by conduction is also diminished. Man, however, does not produce hair or feathers for himself; he steals them from other creatures or replaces them by woven vegetable fibres.

To a certain extent the insulating power of this air-layer in fur, feathers, or clothing can be varied. In very cold weather a mammal's fur stands on end and a bird's feathers are ruffled up in order to increase the thickness of the air jacket. Under the same circumstances a man gets "goose-flesh" and his hair, such as it is, bristles, which is a feeble attempt to do the same thing. He relies, however, more and more on his hat and overcoat to replace Nature's failing gifts.

The chief mechanisms of heat-regulation are two—variation in size of the skin-capillaries, and perspiration. The skin is an adjustable radiator. By dilating the blood-vessels in the skin the amount of blood that is exposed to the cooling action of the air can be increased: therefore we redden when we are too hot. We are increasing the amount of blood in our air-coated radiator. Conversely when we are too cold the vessels of our skins contract; we blench and the skin pales, and may even become bluish from lack of bright red oxygenated blood. Perspiration is due to the action of the *sweat-glands*, which pour out on to the surface of the skin a fluid that consists very largely of water. That evaporates and cools us. It is our most effective method of temperature regulation. By sweating and so cooling his body a man can stand exposure to a temperature at which water would boil. As early as 1775, a Mr. Blagden reported to the Royal Society experiments on this subject. He describes how he and others who stayed for some time in a dry-heated room at the temperature of boiling water found that "the air heated to these degrees felt unpleasantly hot, but was very bearable." They were particularly struck by their power of keeping their own temperature constant in spite of the heat around. "Whenever we breathed on the thermometer, the quicksilver sank several degrees. Every expiration gave a very pleasant impression of coolness to our nostrils."

It is indeed perfectly possible for a living man to remain in a hot chamber long enough to see the dead flesh of a mutton chop cooked by the same heat that his temperature-regulating machinery enables him to withstand.

But to endure such heat the air around a man must be dry. If the air is already charged with water-vapour, or if he is immersed in water, his sweat cannot evaporate and is therefore useless; under the circumstances such a temperature would be rapidly fatal. That is why damp heat is so much more oppressive than dry heat.

And so we realize why it is that Mr. Everyman, as he grows older and wiser, and as his heat-regulating devices lose a little of their elasticity, becomes more and more solicitous about his overcoat, his umbrella, his neck wrap, the soundness of his shoes, and the texture of his underclothing. Next perhaps to his urgent need of air and food is protection from these bacteria which pursue him night and day, and from the chill which enfeebles his resistance to them. These invisible enemies may drive him an exile to warmer and drier climates. They may become the dominant interest in his life. So long as he lives he is never safe from them.

III

THE HARMONY AND DIRECTION OF THE BODY-MACHINE

§ 1. *A Study of Adjustment.* § 2. *Chemical Messengers.* § 3. *Man and Mouse as Individuals.* § 4. *The Controlling System.* § 5. *Sensation and the Senses.* § 6. *The Nervous Mechanism and the Brain.*

§ 1. A Study of Adjustment.

WE ARE "fearfully and wonderfully made," says Holy Writ, and all that has gone before must seem to the reader but an elaboration and filling in of that statement. Probably the thing that will seem most marvellous to him will be the co-ordination of this intricate diversity of cells and organs, all so capable of a measure of independence and all so disciplined towards the common end of being Mr. Everyman and obeying his will. We have already noted that the co-ordination, amazing as it is, is not perfect. We shall be better able to grasp the difficulties in the way of perfection and the extraordinary nature of the adjustments that do occur, if we make a brief study of a particular set of these. We will choose the heart and lungs under varying conditions of exertion and repose for the study, and we think the facts we shall state are sufficient to astonish anyone not already hardened to them by familiarity.

Now the first adjustment that we have to consider is the adequate supply of oxygen to the working cells of the body. In a resting man the heart, the stomach, the kidneys, and the respiratory muscles are active, and their oxygen demands must be satisfied: but the great mass of his muscles—the voluntary muscles—are quiescent, and their oxygen-requirement, although it is not zero, is very small. During violent exercise, on the other hand, the oxygen-need of the voluntary muscles is enormous. A resting man consumes roughly 250 cubic centimetres of oxygen per minute (just under half a pint), but during exercise he may use ten or fifteen times this amount. These are the opposite extremes of his requirements. In a sudden crisis he may have to flash from one to the other in a minute or so. Even during ordinary life, the oxygen requirement varies widely, walking up a staircase requires very much more energy than walking at the same speed on the level. To meet this constant variation in the amount of muscular exertion, there have to be corresponding changes in the breathing rhythm and in the blood-flow. When activity increases, the blood-flow must be accelerated, so that oxygen can be brought to the tissues more rapidly, and the rate at which air is breathed must also be augmented so that more oxygen can get into the blood. Let us take first the regulation of the breathing.

The contraction of the muscles concerned in breathing—the intercostal muscles between the ribs, and the muscular part of the diaphragm—is due to nervous impulses coming from the brain. The impulses emanate from a centre at the lower end of the brain-stem and travel to the muscles concerned along certain nerves—the *intercostal nerves* to the intercostal-muscles, and the *phrenic nerves* to the diaphragm. It is a continual series of rhythmic nervous discharges from this centre that keeps the respiratory system in motion.

Now the activity of the respiratory centre can be modified either by nervous impulses from other sources, or by changes in chemical composition of the blood which bathes it. Of these factors the latter is probably the more important in regulating breathing to correspond with exercise taken. The respiratory centre is very sensitive to the amount of carbon dioxide in the blood, responding to any increase by a more vigorous and rapid stream of impulses. If the carbon dioxide in a man's blood is increased by as little as 3 per cent of its normal value, the rate of breathing is doubled. Similarly if the carbon dioxide in blood is unusually low the rate of breathing falls off. *If it is decreased by only 3 per cent his breathing stops altogether.* On the other hand, the centre is not very sensitive to the amount of oxygen in the blood. Considerable oxygen-lack does produce compensatory effects, but these effects are feeble compared with those of the smallest change in the carbon dioxide. Normally, adjustment is brought about in the following way. If after a period of rest there is a sudden spell of exertion, the muscles will consume more oxygen and produce more carbon dioxide; therefore the blood becomes poor in the former gas and rich in the latter. But it is the excess of carbon dioxide and not the shortage of oxygen which stimulates the respiratory centre. It is the unneeded and not the needed gas that makes the breathing increase and restores the proportion of these gases in the blood to the normal.

This arrangement will work all right if the oxygen lack is accompanied by simultaneous carbon dioxide excess, and this always occurs during exertion. But sometimes the oxygen shortage may be due to causes acting not from within but from without; in that case there is not necessarily any excess of carbon dioxide and the adjusting mechanism may fail. Suppose, for example, that a man goes into a room where the air is deficient in oxygen—containing, say, half the normal amount—but where the carbon dioxide concentration is normal. There is, as has already been pointed out, a slight sensitiveness to oxygen-lack, and he therefore pants a little. But by panting he blows out most of the carbon dioxide in his blood, and since this is the more potent controlling factor, his breathing slows down again. He suffers no further discomfort, but a certain lassitude comes upon him, and presently he falls unconscious from oxygen-lack. Or suppose that a man climbs a high mountain or soars in an aeroplane to a region of low atmospheric pressure. Both oxygen and carbon dioxide will be deficient; because of the lack of the latter he does not respond by increased breathing to the shortage of the former, and oxygen-want may take him by surprise. The paradox presents itself therefore that although carbon dioxide is a poison there must be a certain amount of it present if the respiratory system is to work. This does not injure an animal or a savage living a normal life; any oxygen shortage in the blood of either will be due to bodily effort and therefore accompanied by carbon dioxide production. But it is beginning to endanger man. Exploring and venturing into places with strange atmospheres, into deep mines or above the snow line, is a thing man does against nature and tradition, and it is profoundly interesting to note that he has no certain adjustments to such adventures.

The oxygen demands of an active body are met not merely by increased breathing, they are also supplied by more rapid circulation. The most important adjustment of the circulation is made by the capillaries. The capillaries are not rigid tubes; they are contractile, and can narrow or widen their bore from its normal condition. During exercise the capillaries open more widely, permitting a greater volume of blood to flow through the tissue. The importance of this fact is enormous. It has been calculated that as a result of the simultaneous relaxation of the millions of capillaries when the body is violently exerted, the volume of blood in the voluntary muscles is increased about three hundred times. Moreover, since a large cylinder has a greater surface than a small one, the surface of the capillaries at which exchange of dissolved gases can take place is also greatly increased. The blood-flow through the muscles of a resting man is gentle and restrained; as his exertions increase it becomes a swollen torrent.

Contractility is not confined to the capillaries; the smaller arteries and veins are also capable of marked changes in diameter. The factors which can bring about these changes in bore are very various. Like the breathing centre, the capillaries can be controlled by nervous or chemical agencies. There are nerves which make the blood-vessels contract and others which make them relax, and there are a variety of chemical substances which, if they are present in blood, can influence the diameter of the vessels. With some of these mechanisms we shall deal in a moment; for the present we may note that either an excess of carbon dioxide or a shortage of oxygen can cause the capillaries to widen, and so increase the blood-supply of any part.

This action differs from the responses of the respiratory centre or heart in that its effects are purely local. If carbon dioxide accelerates breathing through the respiratory centre, the result is to put up the oxygen supply in the whole circulation. But if exercise is confined to one small region—to a hand, for example—and the rest of the body is relatively quiet, the carbon dioxide produced by that single group of muscles may not be enough to affect the respiratory centre. It can, however, exert a local action by causing the local blood-vessels to dilate, and in this way the blood-supply to the active part is increased. During exertion which involves the whole body the capillary dilatation is general, and in this case other nervous and chemical agents come into play.

Variation in size of the capillaries may be seen without any elaborate apparatus. The redness of skin depends on the amount of blood it contains, and the flush produced by a hot fire, or the blushing that accompanies various emotional states, are due to this factor.

The essential response of the circulation to muscular exercise is this increase in blood-flow through the tissues. But there are other changes too. It is clear that a mere dilatation of the capillaries, unsupported by correlated changes in other parts of the circulation, would be of little avail. The speed of the circulation as a whole depends on the output of the heart, the pump which keeps it all in motion, and any widespread dilatation of the capillaries would result in a slowing rather than a quickening of the circulation, unless the output of the heart were augmented at the same time. The two factors work together. The disastrous effects of capillary dilatation unsupported by the heart are seen in the condition called surgical shock, which follows any extensive tearing of the flesh. It appears that torn flesh

THE HARMONY OF THE BODY-MACHINE 101

pours a substance known as *histamine* into the blood, and that histamine causes the capillaries to relax completely without having any effect on the heart output. We have already had a glimpse of histamine at work in the early stages of a small wound (Chapter Two § 8). The result, however, of an extensive production of histamine, since the total volume of the capillaries is enormously increased (about seven hundred times more than normal) and the heart is only pumping just enough blood for a normal resting circulation, is that blood stagnates in the bloated capillaries, and the whole blood-stream becomes sluggish. Clearly this is advantageous in one respect, because it will diminish bleeding from the wound, but this advantage is gained by sacrificing the efficiency of the transport system. The oxygen supply of the body becomes inadequate, the delicate brain-cells are soon affected, and the subject suffers from collapse.

When, during exertion, the capillaries enlarge, a similar condition would be produced, were it not for corresponding changes in other parts. Of these the most important is an increase in the output of the heart. Unlike the rhythm of the muscles concerned in respiration, the beating of the heart is spontaneous and automatic, and will continue for days after the organ is isolated from the body. It can, however, be controlled by nervous and chemical influences; there are, for example, nerves which make it beat more slowly, and others which make it beat faster. During exercise the latter come into play; they increase the number of beats per minute and the volume of blood pumped at each beat, so that when exercise is taken and the capillaries dilate, the output of the heart also increases, and the blood-supply to the muscles is duly adjusted.

Besides changes in capillary volume and heart output, there are yet other changes in the circulation during exertion. Among them it is interesting to note a compensatory mechanism, only very recently brought to light, which involves the most mysterious organ in the human abdomen—the spleen.

The spleen is a purple organ, roughly oblong and about five inches long, lying behind the stomach and receiving a copious blood-supply direct from the dorsal aorta. Its minute structure is curious, for in its substance the blood is not confined within definite vessels as it is in any other organ; the arterioles and vessels have open ends, and instead of passing through a capillary network the blood flows freely in the spongy tissue, and directly bathes the spleen cells. The spleen has long been a puzzle to physiologists, partly because it is an organ with many functions. Probably, for example, it is concerned with the breaking down of old and worn blood-cells, and with manufacturing new ones to replace them. But at present we are concerned with another aspect of its activity, which depends on the presence of a well-defined system of muscle-fibres, forming a capsule round the organ and sending interlacing strands through its substance.

During rest, the heart is pumping just enough blood to keep it circulating at an effective speed through the partly-closed capillaries. If, as a result of exertion, the capillaries suddenly open up, the blood-stream will evidently be slowed unless the heart labours strenuously to keep up its speed. But if at the same time some other part of the circulation, which normally holds a considerable volume of blood, reduces its own capacity and ejects its contents into the general circulation, then the expansion of the capillaries will be compensated for. Now there exist in the body certain blood-reservoirs which have precisely this function, and of these the spleen is chief. During rest the muscle-fibres of the spleen are for the most part relaxed; they are rhythmically contractile, beating slowly and indolently, like a very lazy heart, and so keeping a sluggish flow of blood through the organ. In its distended tissue-spaces it holds about a litre of blood—about one-fifth of the total volume in our bodies. During exercise, on the other hand, the muscle-fibres are permanently contracted, and its capacity decreases to only about one-third of the resting value. The reserve blood in the organ is therefore driven out into the rest of the circulation, and in this way the blood-supply to the muscles is increased.

The spleen, then, is essentially a blood-reservoir, and during moderately severe exercise the contribution which it makes to the muscular blood-supply is sufficient. During very severe exercise, when every resource of the organism is called upon, the blood in the spleen may not be enough, and in this case the process is carried a step farther, blood is taken from other organs, such as the digestive organs, and sent to the toiling muscles, the activities of the former being temporarily suspended. In a crisis, when a man or an animal has to struggle for his life the whole circulation is concentrated into a blood-flow through his muscles. The portal veins, for example, are the capacious vessels which carry blood from the intestine to the liver, where absorbed food has to undergo various chemical adjustments before it is fit for general distribution. In an emergency, when the maximum possible muscular work is required, these veins contract; their normal function is suspended, that circuit is all but closed, and the blood-flow thus economized goes to swell the muscle supply. The kidneys, again, hold a considerable volume of blood, and during severe exercise they also contract and, like the portal veins, restrict the demand. In the greatest muscular exertion of which a healthy man is capable, this latter process may be carried so far that the kidney-cells are actually injured from blood-loss—the kidney-cells have been sacrificed so that the whole organism can survive the crisis.

This is simply another reflection of the fact which has already been stressed—that food-supply and the excretion of waste-products are not processes of immediate urgency, for they can be temporarily suspended without serious loss to the organism. The respiratory function, on the other hand, is of immediate and critical importance; when necessity demands, other functions are neglected and all the resources of the circulation are concentrated on this one vital need.

It is clear, then, that the thoracic and abdominal organs are controlled and regulated to suit the ever varying needs of the organism; moreover, we have seen that this control can be exerted in either of two ways—by means of nervous influences, or by means of chemical substances in the blood. As everyone has heard, a nerve is a sort of protoplasmic telephone wire, so that the nerves to the various viscera constitute a telephone system by means of which the brain can communicate with them and modify their activities.

We may, however, look a little farther into chemical regulation before we leave it altogether and turn our attention to the nervous system.

§ 2. Chemical Messengers.

CHEMICAL regulation, as we have seen it operating through the carbon dioxide in the blood, plays a very large part in securing the harmonious co-operation of our viscera. Another example is afforded by the pancreas. This organ does not continually pour its digestive juice into the duodenum; it only produces it when there is food to digest in that part of the alimentary canal. The mechanism by which the pancreas is suddenly activated at that particular moment is as follows: There are cells in the duodenum whose duty it is to secrete a particular substance, *secretin*—not into the digestive tube, but into the blood. This they do whenever food passes from the stomach into the duodenum. The secretin is distributed by the blood all over the body, and in most places it has no effect at all; but when it reaches the pancreas it stimulates that organ and makes it produce its digestive juice.

The process of secreting particular substances into the blood is called *internal secretion;* and the various organs which do this are called ductless glands, or *endocrines,* from the Greek for internal secretion. Many of the ductless (endocrine) glands produce their secretions continuously—we shall learn more about them when we are considering development and growth—but a few, like the ductless gland cells in the duodenum, do so only at particular moments, and are used for regulating other organs. Regulation by means of nerves is comparable to telephonic communications, because particular parts of the body are connected by definite nerve-fibres, and thus the nervous messages are sent only to the particular organs they concern. Regulation by means of an internal secretion is more like stamping "Buy British Goods" on all the letters that go through the British post. The appeal is broadcast to everybody, but only upon those already disposed to respond to it has it the intended effect.

As a further and very remarkable example of chemical regulation we may take the *adrenal bodies*—two small yellowish glands, weighing three or four grams each, and lying just above and in front of the kidneys. They are made of two kinds of tissue, having quite distinct functions, the one forming a thick capsule round the other. It is with the central part that we are now concerned. This tissue is used in mobilizing the resources of the body to meet a sudden emergency. In a crisis when everything depends on muscular exertion—on escape from an enemy, for example, or on fighting and winning—a nervous message is sent to these two glands, which promptly pour their secretions into the blood-stream. This substance, *adrenin,* is swept round the body, and as it goes it affects different organs in different ways. It speeds up the heart and dilates the capillaries, it stimulates the sweat-glands, so that the body may be cooled, it slows the movements of the digestive organs and contracts their blood-vessels, it makes the liver shed its stored glycogen so that the muscles may have a copious supply of fuel, it stands the hair on end, dilates the pupil and bulges the eye, so that the individual may be terrifying to look upon—in fact, it is a chemical Whip, a broadcast S O S, a tocsin which makes every organ play its part in the general mobilization.

§ 3. Man and Mouse as Individuals.

But, as everyone knows, the main unifying organization in the complex activities of Mr. Everyman is his nervous system. So far we have been so busy inside of him, dealing with the details and problems of his continually more astounding mechanism, that we have given little

THE HARMONY OF THE BODY-MACHINE   105

heed to his behaviour as a whole. We have preferred to study how he does it, before we even thought of what he does. Before we have done with *The Science of Life* we hope to illuminate the activities of this gentleman up to their very highest level, to ask how far his comings and goings, his loves and hostilities, his laughter and tears, are spontaneous and how far they are as determinate as the swallowing action of his oesophagus when it is set going by a bolus of food. But here we

FIG. 36. Three vitally important ductless glands,—the right adrenal (above), from above the kidney, the thyroid (centre), from the front of the throat, and the pituitary (below), from the base of the brain.

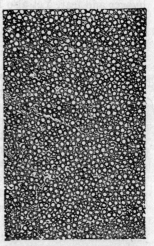

FIG. 37. A nerve cut across and examined under the microscope is seen to consist of an enormous number of fibres lying side by side.

The area shown was, in the actual nerve, about one-fortieth of an inch high.

simply want to consider behaviour so obvious and primitive—running away from a mad dog, or looking for drink when thirsty—that it seems almost below Mr. Everyman's dignity to discuss him. Such behaviour as we are going to study here may be studied equally well in the case of his minor cousin, the mouse.

Most of us have seen that little creature, hunger-driven, foraging in our houses. Its sensitive nose quivers perpetually, and samples and appreciates the surrounding air. Meanwhile, its eyes and ears are alert for the least intimation of cat, dog, or other enemy. It moves about in swift, eager runs or sits up to listen. The warning squeak of a kindred mouse will instantly send it flying to cover. It is the smallest of common mammals, a poor fighter, and its strategy is hiding. Whenever it can, it travels in the dark behind the wainscot or under the floor, feeling its way through the obscurity with its extraordinarily sensitive whiskers. By making it captive we can spy upon its shy existence. It balances between fear and hunger and hunger and fear. Yet also there is something else in its life that can at times prevail over either of these forces. It responds to something, we may call it an obligation to the species; storms of desire impel it to do strange and dangerous things so that the race may continue.

Here, then, we have three imperatives that govern the life of the mouse and to which its conduct must conform. It must fill its belly, it must save its skin, and it must breed; and the penalty for failure is death, either of the sinner or of his stock.

So with any wild mammal that we choose to name, and so, perhaps to a lesser degree, with ourselves. All normally are under these three imperatives. This is why the brotherhood of viscera—heart, stomach, kidneys, lungs—needs other helpers; muscles to move them about, eyes, ears, and so forth to observe their surroundings, a nervous system to govern the community and hold it together. A living body is a co-operative alliance between a system of chest and belly viscera on the one hand and a system of behaviour organs on the other. The former provide the latter with a stream of clean, nourishing blood—that is their essential duty in the partnership; while the latter protect and guard the former and keep the stomach filled. We have studied the former group; we will turn now to the organization of watchful sense-organs and telephone nerves, by which the community is guarded and held together.

§ 4. The Controlling System.

EVERYONE has a general idea what the brain and spinal cord are like and how they are situated in skull and backbone. From them come nerves, spreading and branching out from the brain and passing through bony apertures in the cranium, not only to the face and head generally, but sending branches down even to the neck and heart; and also from the spinal cord, in pairs, one pair between every vertebra and its neighbour, to the limbs and body. We have already compared the nervous system to a telephone system—from which it differs in the fact that its fibres are one-way fibres that either carry outgoing impulses (motor fibres) or bear incoming ones (sensory fibres). There are in addition to this main system (Central Nervous System) subordinate centres and systems in more or less complete communication with it. They perform tasks of co-ordination of a lesser scope, with whose details it is unnecessary to burthen the main system. The solar plexus, a network of cells and fibres concerned with abdominal activities, is chief among these inferior centres.

Now the structure of the brain and most of its relationship to the spinal cord we shall find it better to defer to a later part of the work. Upon it centre some of the most fascinating, illuminating, and controversial questions concerning life in general, and human life in particular that it is possible to raise. In man and mouse the Central Nervous System has a close general resemblance, but the brain of the man is far larger, more powerful, and intricate in its working. There centres the behaviour of the creature as a whole. Whatever is least mechanical, and anything there may be that is not mechanical in life, have their seat there. There it is we must face the problems of consciousness and the freedom of initiative. To understand it as far as current knowledge goes will be to crown and complete the task we are setting ourselves in this compilation, and at present we are only opening up the introductory matter of our subject. Here we will consider the nervous system and its working as the machinery of communication that holds the body together for unified action, as the apparatus through which the whole body, with the billions of cell-denizens and the endless variety of substances and secretions we have reviewed, is made one collective thing with a sense of itself and a purpose of its own.

We have already made use of a well-known comparison—that of the nervous system with a telephone system—and it is important at the outset to see how far the analogy is true and how far it is misleading. What in the nervous system corresponds to the telephone wire? What exactly are the channels along which part communicates with part? To answer that question we must look into certain details of microscopic structure.

The cells of the nervous system differ from other cells in an important respect—their surfaces project into long, slender fibres of living substance. Fig. 38 represents a nerve-cell of the spinal cord. It has an irregular, star-shaped body, which is unusually large for a cell, and radiating away from this centre are a number of delicate protoplasmic arms (nerve-fibres) that spread and branch in the spinal cord. One of these arms, in this particular cell, is longer than the rest; it leaves the spinal cord and runs along a nerve to some organ in a distant part of the body. We shall follow it in a moment. The soft, pinkish tissue of the central nervous system consists entirely of such nerve-cells and

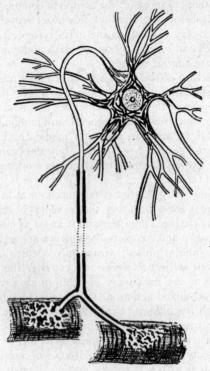

FIG. 38. A single nerve-cell (much magnified).

Above is the cell-body, placed in the brain or spinal cord. Below, the nerve-fibre leads away and supplies two muscle-cells. If the muscle supplied were in the foot the nerve-fibre, drawn to scale, would be a quarter of a mile long.

their fibres, except for a little connective tissue to support them and bind them together, and these hundreds of millions of cells are units— the clerks, so to speak, of the organization. The fibres correspond to the telephone wires in our analogy. Some of them run from nerve-cell to nerve-cell growing out from one and ending in a tiny cluster of finger-

THE HARMONY OF THE BODY-MACHINE   109

like branches which twine around and clasp the fibres of another. Others leave the central nervous system altogether and run to other tissues—to muscles, glands, sense-organs—and along these commands are sent out which control the working tissues, and reports from the sense-organs are carried in.

The telephone wires of the body, then, are living threads of astounding delicacy, finer than the finest gossamer. They are about one-tenth the thickness of a human hair and may be several feet long. The messages which flash along these protoplasmic wires are called "nervous impulses"; they are physical changes which travel at about four hundred feet per second in man. Our bodies are permeated by a network of these fibres, centring in the brain and spinal cord.

Such of the fibres as go out from the brain and spinal cord to the rest of the body are collected together into bundles, the nerves; a nerve is simply an enormous number of nerve-fibres lying side by side like the separate wires in a telephone cable, and bound together by a little connective tissue. As the nerve runs away from its root it branches, and the fibres begin to part company and disperse to their various destinations. Some of them are connected up to sense-organs, and carry to the central organization the information these organs obtain; some go to the voluntary muscles and bear the impulses which make them contract; some run to glands and make them secrete; some plunge into the chest and belly and supply the more independent organs there, bringing impulses which slow them or hurry them up.

It is difficult to realize the full complexity of the nervous system. By dissecting as completely as possible the finer nerves in a limb, for example, one can get a fair idea of it. We will try in another way—by considering the control of the muscles as one only of the many sides of its activity.

The muscles that cause the wilful movements of man or mouse work under the direct impulsion of the central nervous system. To every muscle-fibre in a well-controlled muscle goes a nerve-fibre from a nerve-cell in that system and spreads out at last in a palm-like cluster of branches against its sides. Every movement made by that muscle-fibre is due to an impulse coming from the associated cell in the brain or spinal cord. Every movement made by a muscle depends upon a nerve made up of innumerable fibres. But there is hardly a movement of the body that is not the outcome of the co-ordinated action of a number of muscles, and hardly a moment of our waking life when we are not in movement. As the reader peruses this section he is performing continual slight movements of the muscles of his eyeballs and head and neck, so that his line of sight may run smoothly over the print. Now and again his arm moves up and his fingers turn the page. Perhaps at the same time he smokes and knocks the ash from a cigarette; his lips may flicker in a smile of approval, or curl in contempt, or stretch spasmodically in a yawn. Here are muscles working by the score. And all the time the ninety-odd muscles that move his ribs continue their rhythmical contraction and relaxation. Every one of these movements involves the co-operating action of some thousands of muscle-fibres, and every one is precisely controlled by nervous impulses sent by his brain or spinal cord.

But the activities of the central nervous system extend further than this; they include a supervision of resting muscles. When a muscle is passive it is not allowed to hang limply from its attachments, but is kept braced and ready for immediate action when called upon. In the language of the physiologist it is in a state of slight partial contraction or *tone*. This tone is due to impulses from the central nervous system, and if its nerve is cut the muscle loses its tautness and becomes slack and flabby. Further, besides keeping the muscles keyed up, the central nervous system seems to exert an influence on their general health, for cutting the nerve causes a slow shrinking of the muscle, which in the course of months or years may even disappear and give place to a mass of connective tissue. Perhaps this effect is due simply to disuse, for the size of a muscle depends on the frequency and vigour with which it is used; the point is not clearly established. But, however this may be, it is evident that a muscle depends completely on the central nervous system both for the impulses that control it and for its general welfare, and that the latter organ is to be regarded not as a wilful and despotic government but as a vigilant, untiring leader.

§ 5. Sensation and the Senses.

IF WE could see the whole living web of the nervous system laid out before us, and if a nervous impulse was a visible thing, we should get a picture of continual thrilling and rippling activity. We should see a ceaseless succession of impulses flashing to the muscles, keeping them taut and ready for work, and spurring single muscles or groups of muscles to vigorous, disciplined activity. But also we should see an equally incessant series of impulses travelling in the opposite direction. Before we come to the central working of this marvellously complicated and accurate government we must consider those other impulses which bring information unceasingly from all the quarters of its realm.

THE HARMONY OF THE BODY-MACHINE   111

For the web—unlike the telephone web of a city—is a double one. Any nerve contains fibres of two kinds, *motor fibres* that carry impulses outwards and *sensory fibres* that bring impulses inwards. These two kinds of fibre are bound up together to form nerves, their structure is similar and the physical nature of their impulses is similar, but they

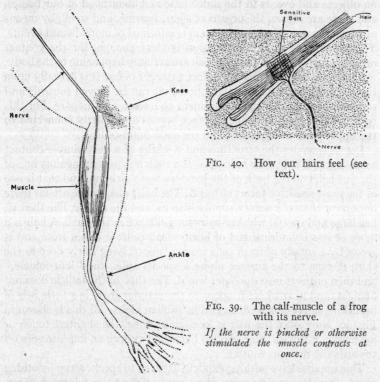

FIG. 40. How our hairs feel (see text).

FIG. 39. The calf-muscle of a frog with its nerve.

If the nerve is pinched or otherwise stimulated the muscle contracts at once.

differ in the direction in which the impulses travel, and in the origins, destinations, and functional meanings of these impulses.

The *sense-organs*, the intelligence agents, in which the sensory nerve-fibres end, are of many and various kinds. There are sense-organs scattered about in the muscles, in the joints, and indeed in every tissue of our bodies, reporting the state of affairs there. So the central government is informed of the positions and movements of our limbs, and sometimes, when there is discontent in the thorax or abdomen, it feels hunger or indigestion. These sense-organs, which watch and report on the other tissues, may be called *internal* sense-organs. Contrasting with the internal sense-organs there is a second sort of sense-organ which takes note of circumstances external to the body state. This second group, the *external* sense-organs, may be further divided into two: there are some, such as the organs of feeling and taste, that report on objects and events in the immediate neighbourhood of our bodies, and there are others, the organs of sight, hearing, and smell, by means of which the central nervous system is informed of more distant events. The importance of this distinction is clear enough, for the contact organs can only detect things which are actually happening to the body, while the distance organs can detect a danger before it is actually upon the organism, or a remote attraction that can be moved towards and secured. They enable the government to foresee and prepare. For this reason, as we shall find, the distance organs are directly connected to the upper more independent and personal brain-centres.

Let us consider the structure and working of a very simple contact sense-organ, a hair. Our hairs feel. If a match is pressed against one of the fine hairs on the back of the hand we feel a tickling, and that is one of the most sensitive forms of touch. The hairs of a mouse are far more important than our scanty supply is to us, and the mouse, like the cat, has large and special whisker-hairs for guidance in the dark. A hair is a more or less cylindrical rod of horny dead cells; it arises from and it grows by a supply of fresh cells from a "root" lying fairly deep in the skin; it runs to the surface along a special tube, the "hair-follicle," and then projects into the outer world. The tube of the follicle does not fit tight, and near where it opens to the exterior there is a little belt of branching nerve-fibres which end in swollen blobs. All this is shown in Fig. 40. Now since the hair is loose in its tube, the slightest touch or push upon it tickles these swollen blobs and at once an impulse goes off towards the nervous centres.

This impulse says nothing explicit. It is, so to speak, a rap—nothing more. Any two sensory nerve-fibres are as like as two telegraph wires, and the impulses they transmit are, as far as we can see, identical. The central government has to interpret these messages, and to discriminate between the various kinds of feeling. How can you do so if the messages are all alike? If we could examine all the sensory nerve-fibres in the body we should find that they vary in one respect only—they have different endings. It is in this difference that the answer to our question lies, for an impulse is interpreted by the central nervous system according to the particular nerve-fibres along which it arrives, and the centre or group of cells that receive the impulse. When they get inside the central nervous system different fibres run to different destinations. If, for example, an impulse comes along a fibre that supplies a hair-follicle, it is led to a centre which interprets it as a sensation of touch, and if it comes along a fibre from the sensitive layer of the eye, it is led to another centre which interprets it as a sensation of light. Because of these facts it is quite easy to deceive the central nervous system, and to send it lying messages. For example, we can press on the eyeball and so stimulate the sensory cells there. The reader can do it forthwith. The brain sees spots, clouds, or flashes of light according to the sharpness of the impact or pressure. Similarly, disease of the nerve of hearing produces sound-sensations, often interpreted as "Voices," and stimulation of the sense-organs of the tongue, by a weak electric current, for example, gives us a taste in the mouth.

A point of general interest is that if the pressure upon that very simple sense-organ the hair is sustained, the stream of messages soon dies away in the nerve. Nervous impulses generally result from a *change* of conditions; permanent conditions do not go on producing them. The desirability of that is evident. Our mouse when it creeps out has to be on the alert for approaching enemies. His eyes are sentinels, on the lookout for danger, and evidently they must not be officious with their information; they must not shout continually to the brain, "The bed is dim and grey and Mr. Everyman is motionless; Mr. Everyman is motionless; Mr. Everyman is motionless." They must be discreet and report only essentials, and so long as Mr. Everyman is motionless he does not matter a rap. The impulses travelling along the optic nerve are more like this, a dying report: "everything quiet—quiet—quiet—quiet," lifting suddenly to an arresting announcement—**"hallo! something moving there on the right!"** A little movement in the room has more interest for the eye of a mouse than a whole roomful of still furniture. Its eyes become intent. The same is true of its ears or its nose; there is little interest in all the customary flavours of Mr. Everyman and his belongings or in any of his familiar breathing exercises, but a sudden ejaculation or a sudden whiff of cat may be of vital importance. It is the changes that matter.

With all the sense-organs that produce precise sensations, a constant unvarying stimulus has little effect, but a change sets the impulses storming. Suppose, for example, that on a cold day we come into a warm room. The sudden rise in temperature affects a set of organs in our skin that are sensitive to warmth; there is a sudden discharge of impulses, and for a few minutes we feel warm all over. But soon the effect wears off. The temperature stays constant and the warmth-organs calm down; the room seems to get cooler. On stepping out into the cold again there is a similar cycle of events; the sudden change of temperature excites another set of organs, sensitive to cold, but they soon calm down at the constant low temperature and we have the illusion of "warming up." On this principle there rests a well known and very instructive physiological paradox. Take three bowls of water, one hot, one cold, and one lukewarm. Immerse the right hand in the hot water and the left hand in the cold and wait for a minute or so. There will be a cycle of sensation like that just described; at first the right hand will feel hot and the left will feel cold, but the sense-organs will soon calm down. Now quickly immerse both hands in the lukewarm bowl. The right hand is suddenly cooled, so its cold-sensitive organs are stimulated; the left is suddenly warmed, so its warm-sensitive organs are stimulated. Therefore one and the same bowl of water feels hot to the left and cold to the right. Our skins are meant to feel not temperature but changes in temperature.

True, our feelings of warmth and cold are not confined to changing temperatures, and we can feel extreme heat or cold even if the temperature is constant. But such feelings are not due to the action of the temperature sense-organs. On a hot day one has the moist sensation of perspiration and sensations due to the expansion of the blood-vessels, and the consequent more violent pulsing of blood in the skin, and on a cold day one has sensations caused by the response of shivering. The continual warm tingle of exposure to strong sunlight is a feeble stimulation of the pain-organs by ultra-violet light. None of these things concerns the temperature sense-organs, which are unaffected by constant temperatures.

Similarly with our other senses. In an aeroplane flight we speedily become accustomed to the steady roar of the engine and ordinarily we are unaware of the constant contact of our clothes, although a slight change of pitch in the former case or a slight displacement of the latter is instantly felt. The principle holds even for our eyes; the images of still objects make little impression, but moving objects catch the attention and are more distinctly seen. For this reason when we think we are looking fixedly at an object to see it distinctly our eyes are not really still at all; they quiver rapidly and almost imperceptibly from side to side so that the image may move about on the retina and always be falling on a new part of its sensitive surface.

But we have dwelt long enough on the general aspects of sensation. We must pass to a more detailed treatment, and examine the various senses one by one.

Pain, Temperature and Touch. The skin is able to feel four different kinds of sensation—warmth, cold, pain, and "touch" or deformation of the surface. It can be proved by a simple experiment that these four sensations correspond to four different kinds of sense-organs. If

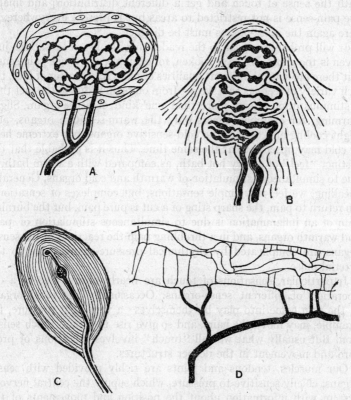

FIG. 41. Four types of "touch" organ (magnified).

A is responsive to changes in temperature, B and C to pressure, and D to pain. Note the relative simplicity of the pain organ.

we dab about on the skin with the point of a warm pin, we find that the warmth can only be felt at scattered points; there are warm-sensitive areas with intervening areas where warmth is not perceived. Similarly with a cold pin; the sense of cold is restricted to definite areas. Now the warm-sensitive areas and the cold-sensitive areas do not correspond with each other. Some points are sensitive to warmth and not to cold, some to cold and not to warmth, some to both, some to neither. Therefore there must be two distinct kinds of sense-organ, one for the warmth sensation and one for the cold sensation. One can proceed similarly with the sense of touch and get a different distribution; and finally the pain-sense is not restricted to areas but is present everywhere, so here again the sense-organs must be distinct.

It will probably appear to the reader that the list of sensations just given is too short. We have spoken, for example, of a single pain-sense, but there are several different qualities of pain. As a matter of fact, the rich variety of feelings that we get from our skin is due to the fact that a stimulus often affects more than one kind of sense-organ. Slight warming, for example, affects only the warm-sensitive organs, and slight cooling affects only the cold-sensitive organs, but extreme heat or cold may upset both at the same time. Thus it is probable that the distinct "feel" of a very hot bath, as compared with a warm bath, is due to simultaneous stimulation of warmth and cold organs. Generally speaking, we feel not simple sensations, but complexes of sensations. To return to pain, the sharp sting of a cut is pure pain, but the burning pain of an inflammation is due to simultaneous stimulation of pain and warmth organs, and in a throbbing pain the reactions of the sense-organs are complicated by rhythmical pressure-changes due to the arterial pulse.

In particular, sensations of touch are nearly always due to a co-operation of different sense-organs. Occasionally the touch-organs of the skin come into play by themselves—a very light pressure, for example, may move the hairs, and so give rise to a pure touch sensation. But usually what we call "touch" involves sensations of pressure and movement in the deeper structures.

Our muscles, tendons and joints are richly provided with sense-organs, chiefly sensitive to pressure, which supply the central nervous system with information about the position and movements of the limbs. These sense-organs also play an important part in feeling the shapes and textures of objects. Suppose, for example, that the finger is drawn along the table, the eyes being closed. The impression of a hard, flat surface depends chiefly on sense-impulses from the deeper structures; we know that it is hard, because we can press on it vigorously without deforming it, and we know that it is flat because in order to draw the finger along it we have to move our muscles in a certain way. The sense of pressing on the table is due to organs lying in the deeper connective tissue in the finger, and that of movement to sense-

THE HARMONY OF THE BODY-MACHINE 117

organs in the muscles concerned—which are placed, not in the finger at all, but in the arm and shoulder. In this way, even the simplest impressions of touch are made up in reality of sensations from a number of different kinds of sense-organ, which are fitted together by the central nervous system, and interpreted as a single feeling.

In structure the organs responsible for sensation in the skin and muscles are comparatively simple. We have already noted that the touch-organs in the hair-follicles consist merely of the branching endings of nerve-fibres. The other organs are generally more complicated than this—they involve not only branching nerve-fibres, but special cells—but they do not compare in intricacy of design with the sense-organs that we shall study in subsequent sections. A great number of different kinds of sensory structures have been found in the skin and deeper structures, but unhappily very little is known at present about the way they work. For this reason we need not describe them here.

We may, however, take note of a kind of sense-organ that is even more elementary in design than the touch-organs in the hair-follicles. Everywhere in the skin there is a network made by the branching ends of nerve-fibres, running among the cells without any evident order, and without any specialized accessory structures. When these fibres are stimulated they give rise to sensations of pain. The pain sense is of more importance to the organism than any other of the skin sensations, for while a message about the temperature of the air or about the hardness and shape of a neighbouring object may or may not be interesting to the central nervous system, a stab of pain always means immediate danger. The pain organs are very much less sensitive than their more elaborate neighbours, but, on the other hand, they are less specialized and can be excited by any stimulus, whatever its nature, if it is of sufficient intensity to injure living tissue. Slight warmth, for example, excites only the special warmth cells, but a temperature high enough to burn or scald is brilliantly painful. Pain is an alarm signal, and its biological importance is reflected in the anatomical fact that whereas warmth and cold and touch are senses confined to particular patches of skin, the pain network is unrestricted and pervades the whole surface of the body.

Internal Sensation. While we are on the subject of these scattered and rather mysterious sense-organs, we may note the feelings that arise from time to time in our thoracic and abdominal viscera. Normally, we are unaware of these independently working parts. There is a close correspondence between the parts of the body over which the sway of the central nervous system extends, and the parts from which it receives precise information. The mouth, for example, is controlled by the central government—chewing and swallowing are the results of impulses from the brain—but the oesophagus is an independent structure, driving mouthfuls down to the stomach by means of its automatic movements. Corresponding with this, we find that the mouth is very sensitive to the temperature, texture, and chemical composition of our food, but that as soon as a mouthful is swallowed it ceases to stimulate sense-organs; it is handed over to the autonomous factories and passes at the same time out of our control and out of our ken. Generally speaking, whenever an abdominal organ sends a vivid sensory message to the brain, it means that something is seriously wrong and that the government is expected to do something about it.

The whole subject of sensation in the viscera is still obscure. We know very little of the elements into which visceral sensations can be analyzed, or of the sense-organs that are responsible for them. It is clear that the arrangements are very different from those in the skin, and the differences seem to depend on the relative independence of the viscera to which we have just alluded. Surgeons find, for example, that the intestine can be cut or sewn without giving rise to any painful sensation whatever, but a strong contraction of its muscular wall or excessive distension of its cavity—the sort of thing that happens when its working is disturbed—may give rise to intense griping pains.

There is, however, a possibility that the sense-impulses received from the viscera include messages other than these cries of distress. We saw that motor impulses to the voluntary muscles are not confined to the moments when contraction is desired; they include continual encouragements, an unceasing influence that keeps the muscle fit and ready. In the same sort of way it may be that the government receives continual intimations from the chest and belly, that all goes well with the kidneys, or that there is a slight unrest in the stomach that may presage more serious trouble, and so on. In addition to our accurately-analyzable sensations there are vague feelings that are at present imperfectly understood—feelings of vigour, appetite, discomfort, restlessness, weariness, oppression, and so on, that lead up to, and are evidently related to emotional states. It may be that such feelings are to a large extent due to impulses from our viscera. It is a familiar fact that the condition of the viscera has a profound effect on the colour, so to speak, of the mind. When the stomach is empty, for example, the organism becomes restless before it feels definite hunger, and when the stomach is filled it becomes indolent and sleepy. On the other hand it may be that these feelings are not depend-

THE HARMONY OF THE BODY-MACHINE

ent upon messages received along sensory nerves, but upon changes in the chemical state of the nerve-cells, due to alterations of the way in which circulating blood is distributed between belly and brain or of the composition of the blood itself. We have at present no precise information about these vague sensations, and therefore we cannot profitably discuss them further; we must simply note their existence and leave it at that.

Smell and Taste. Smell and taste are the chemical senses—that is to say, the stimulating agents in each case are not physical influences, such as temperature, pressure, or light, but chemical substances which act because of their molecular structure. In man, the chemical senses,

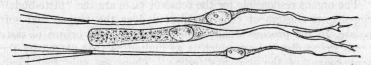

FIG. 42. Three cells from the sensitive membrane of the nose (highly magnified).

Two smell-cells and a supporting cell are seen. Surface is to left.

and particularly smell, are poorly developed, being overshadowed by sight and by hearing, but in the lower mammals the sense of smell is very much more acute. In a dog, for example, the nose is probably quite as important a sense-organ as the eye or the ear. A man recognizes his fellows by their appearance, and by the noises they make, and the store-rooms of his memory are largely filled with labelled records of sights and sounds; the memory of a dog, on the other hand, is more probably a sort of card-catalogue of smells.

The mechanisms of smell and taste are fundamentally alike. In both cases the essential elements are cells, sensitive to particular substances, which dispatch impulses to the brain when those substances are present. In both cases, the stimulating substance has to be dissolved in water before it is effective. Thus, although even a weak tincture of quinine has a powerful taste, dry quinine powder, which dissolves slowly and sparingly in saliva, has hardly any taste at all; and similarly odorous substances have to dissolve in the moisture covering the nasal mucous membrane before they evoke any sensation of smell.

The nostrils lead into two triangular cavities which are broken up and made labyrinthine by the scroll-like turbinal bones. The special cells that perceive odorous substances are found at the upper end of these cavities, and on the thin partition that divides the right nose cavity from the left. Owing to the complicated structure of the nasal cavities, inspired air does not generally blow over the olfactory membranes; if it did so, it would tend to clog them with its suspended dust, and to dry their moisture. The lower part of the nose is swept by this rhythmic gale, while in the upper regions the air is comparatively still. Normally, stimulating molecules can only reach these peaceful places by diffusion. But, when we are especially interested, a brisk sniff, accompanied by dilation of the nostrils, can upset the normal direction of the air-currents and waft a scent-charged breeze straight to the sensitive cells.

The organs responsible for the sense of taste are the "taste-buds," egg-shaped clusters of cells distributed over the tongue and soft palate. In the process usually called "tasting," the organs of taste *sensu stricto* play a part which is as limited as that played by the sense-organs of the skin in "feeling." There are only four kinds of pure taste sensations: Sweet, sour, bitter, and salt. Any particular taste-bud is specialized to respond to one only of these four tastes. The many and varied sensations arising from substances in the mouth are complexes, made up partly of the four elementary taste-sensations, and partly of other kinds of stimulus. The difference between mutton and ham, for example, is smelt, not tasted; it is due to a diffusion of odorous molecules into the hinder parts of the nose, and most of the subtler discriminations usually attributed to the palate are in reality performed by the olfactory membranes. A serious cold, by clogging the nasal passages with mucus, can completely destroy this part of our sense of taste, and a man who shuts his eyes and holds his nose will be hard put to it to distinguish between a bit of apple and a bit of onion in his mouth. Further, substances may stimulate sense-cells which are neither gustatory nor olfactory; thus, mustard stimulates the warmth-ends, peppermint stimulates the cold-organs, and some substances produce a delicious tingling that is in fact a very feeble pain.

Hearing. The sense-organs which have been described thus far perceive stimuli that are known to affect living substance directly. A fresh muscle isolated from the body can be made to twitch by a pinch or wetting with certain chemicals, and in a touch-organ or a taste-cell this natural irritability to mechanical or chemical stimuli is exaggerated and taken advantage of. But the sense-organ to which we are now coming is more complicated than this. Sound consists of rhythmical pulsations of the air, and there is no evidence whatever that such pulsations can act directly on living protoplasm. Indeed, we shall find on examining the ear that the stimulus which excites the auditory cells is not a sound-stimulus but a touch-stimulus, for the ear is an elaborate machine so constructed that whenever sound-pulsations fall upon it, sensitive cells are touched. Herein lies the wonder and ingenuity of the ear; it is a definite extension of the faculties of living matter. A lowly organized creature, such as an earthworm or a polyp, can taste or feel, but it lacks the structures that convert sound-waves into stimuli capable of exciting protoplasm directly, and is therefore ignorant of the rich and varied world of sensations that sounds can evoke. An examination of the anatomy of the ear will make the point plain.

The organ of hearing consists of three parts—the outer visible ear, the middle ear with its ear-drum, and the inner ear. The inner ear is the true sense-organ, for it is here that the vibrations cause the excitation of sensory cells and the initiation of nervous impulses. The outer and middle ears are structures which collect sound vibrations and transmit them to the inner ear, making it possible for that elaborate and fragile device to lie safely embedded in the bone of the skull. Most mammals are provided with a natural ear-trumpet in the external ear, a hollow cone which can be turned about to face the direction from which sounds are coming, and which therefore makes feeble sounds more distinctly audible. From this ear-trumpet a short tube leads to the ear-drum, the boundary separating the outer and middle ears. In man this amplifying apparatus is poorly developed; the external ear is a mere flap of no acoustic importance, and the power of moving it, except in a few gifted individuals, is absent. For this reason, although a man can distinguish sounds varying over a wide range of pitch, he is less able to discern feeble sounds than a dog or a horse, and he cannot get such an accurate idea of the direction from which sounds are coming. The tube leading from this reduced ear-trumpet to the ear-drum is provided with hairs and wax-secreting cells that have a filtering and protecting function similar to that of the hairs and mucus-secreting cells of the nose.

The middle ear is a narrow chamber filled with air and communicating with the mouth by means of a duct, the *Eustachian tube*. The walls of this tube are usually pressed flat together, but when the pressure in the middle ear becomes too different from that outside, they are forced open and a little air passes one way or the other. This causes the click in the ear which we experience after ascending or descending rapidly in a lift or mountain railway. The walls of the middle ear are strong and bony, except in three places. The first and largest of these is the ear-drum, and in the second and third, the *fenestra ovalis* and *fenestra rotunda* (oval window and round window) there are even more delicate membranes which separate middle and inner ear. A chain of three minute bones, prettily jointed together, runs from the ear-drum to the fenestra ovalis; because of their curious shapes these bones are called the hammer, the anvil, and the stirrup.

Sound waves, travelling down the outer ear, strike the ear-drum and make it vibrate; by means of the chain of bones, the vibrations of the drum are transmitted to the fenestra ovalis. This elaborate device is partly protective, for the hammer is so jointed to the anvil that if a violent jar occurs, such as a box on the ear, the two are disengaged and the shock is not passed on to the delicate inner structures. But there is an even more important reason. The inner ear is filled with a watery fluid, and the properties of sound waves travelling in water are very different from those of sound waves travelling in air. The bones act as levers and reduce the amplitude of the vibrations, but at the same time, by concentrating the energy of the vibrating ear-drum on to a window only one-twentieth of its size, they make the sound pulses more vigorous. They are a means of getting over the difficulty of transmitting sound-waves from air to water.

The inner ear is an elaborate labyrinth of passages, embedded in bone and filled with a lymph-like fluid; as we have seen, this fluid is made to vibrate when sounds fall on the outer ear.

A part only of the inner ear is concerned with hearing. We shall see later what the rest is for. The part that concerns us at present is a narrow, tapering tube, about an inch long, which is coiled into a dwindling spiral like the shell of a snail and is therefore called the *cochlea*.

This cochlea is divided into three compartments by a partition which runs along its whole length. On this partition the sense-cells are situated. The sense-cells are represented in Fig. 46; they bear stiff, hair-like projections on their upper faces, and they are surrounded by the endings of nerve-fibres; moreover they are placed on an elastic membrane and overhung by a rigid shelf. The working of the ear depends entirely on these anatomical relationships. When the fluid surrounding the apparatus vibrates, the elastic membrane bounces up and down; this brings the sense-cells into collision with the rigid shelf, and it is their impact with the shelf that stimulates the cells and causes them to dispatch impulses along the nerve-fibres to the brain.

According to the theory most extensively held (for there are differences of opinion here) the method of discriminating between sounds

THE HARMONY OF THE BODY-MACHINE 123

of different pitch is as follows. The diameter of the cochlea decreases steadily from end to end of the spiral; it is a spiral staircase that dwindles to a point at the top. Since all the structures inside it decrease in proportion, the width of the elastic membrane on which the sense-cells are placed also decreases. That is to say, since the period of vibration of an elastic body varies with its size, the different parts of

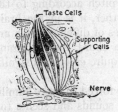

FIG. 43. A taste-bud from the tongue (magnified).

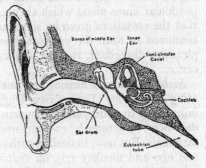

FIG. 44. A section through the ear to illustrate the mechanism of hearing.

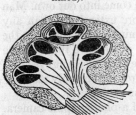

FIG. 45. The spiral cochlea cut across. It is embedded in bone. (Compare Fig. 52.)

this membrane are tuned to different notes. The apparatus is like a piano or a harp, where the wires give higher notes as they decrease in length. Now it is well known that if one sounds a tuning-fork near a piano, with the damping pedal down, the particular note of the tuning-fork is echoed by the appropriate wire of the piano—the pulsations emitted by the tuning-fork make a wire vibrate if it is tuned to the same pitch. And in a similar way rhythmical vibrations of the fluid in the inner ear only shake the particular part of the elastic membrane that is tuned to their own pitch. Notes of different pitch affect different parts of the membrane, and so cause impulses to be sent to the brain along different nerve-fibres. Clearly, there is no distinction at all in the manner in which C and C sharp stimulate cells—both notes cause sensitive cells to be bumped against an overhanging shelf, and the distinction lies in the fact that each note shakes a different group of cells.

Thus we see that the auditory cells are not sensitive to sounds as such, but they are sensitive to being touched; the ear is an ingenious device for touching different cells when notes of different pitch are sounded. The nerve-fibres run from the ear to the brain, and here the impulses are interpreted as notes of varying pitch according to the particular fibres along which they arrive. Further, it is in the brain that the sensations given by simultaneous notes of different pitch are combined to form the complex sensation of a chord, and becoming pleasant, discordant, rousing, and so forth, acquire emotional significance.

Sight. In studying hearing and smell we were forced to make humiliating confessions of inferiority. Dogs perceive sounds and scents too faint for us to notice, and the human nose and outer ear are miserably developed structures when compared with their homologues in other mammals. But now, in dealing with sight, we come into our own. Man and ape and monkey rely on vision far more extensively than any other mammals, and it is probable from its minute structure that the human eye can see more distinctly than any other, save that of some birds. Here, then, is our consolation. We have such excellent eyes that we have no need of noses, and we may interpret the enviably sensitive canine muzzle as a sign of the fogginess of the canine eye.

The structure of an eye is very like that of a photographic camera. We may distinguish two essential parts: a sensitive screen at the back, the *retina*, and an optical system that projects an image of the outside world on to that screen. The retina is the sense-organ proper, for it is here that the rays of light forming the image act on sensitive cells and initiate nervous impulses. The eyeball is blackened within, like a

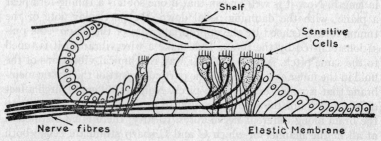

Fig. 46. A few sensitive cells of the ear, to show how they are situated.

THE HARMONY OF THE BODY-MACHINE

camera, in order to prevent reflection and scattering of light. The structures in the front part of the eye are very strikingly like the structures in the front part of a camera. The wall of the eyeball in this region is transparent and bulges forwards as the *cornea;* a short distance behind the cornea there lies a doubly convex mass of transparent tissue, the *lens.* The iris of the eye works like the camera's diaphragm. When illumination is poor—in a dark room, for example, or at dusk—the iris opens widely, so that as much light as possible enters the eye; when the light is very bright it closes down to a pin-point, for if the intensity of light inside the eyeball exceeds a certain value the retinal cells are injured and become temporarily or permanently incapacitated. There is another advantage to the use of an iris diaphragm, whether in the eye or in a camera, depending on the fact that the middle of a lens always focuses light more accurately than the edges. If the light is bright enough it is always a good thing to stop down the iris to a small

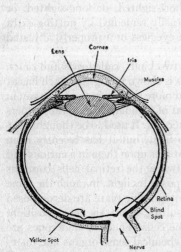

FIG. 47. The eyeball cut across to show its essential parts.

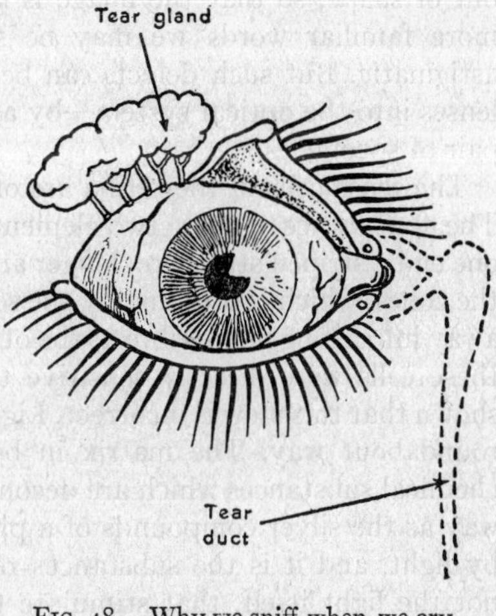

FIG. 48. Why we sniff when we cry.
The gland that makes tear-fluid to keep the eye clean and the duct which drains it away into the nose.

size and cut out the edges of the lens, so that the image formed has the least possible distortion of colour and line. The lens divides the cavity of the eyeball into two parts. Of these, the part in front of the lens is filled with a watery fluid, the *aqueous humour,* and that behind it is filled with a transparent jelly, the *vitreous humour.* The aqueous humour is constantly being secreted by glands in the ciliary region and oozing out through fine pores at the edges of the cornea; by means of this slow, steady stream, food and oxygen are brought to the cells of the lens and adjacent structures, and their waste products are washed away.

The eye, then, is provided with a combination of two smoothly-curved bodies, the cornea and the lens, by means of which light can be focused on to the retina. By far the larger part in image formation is played by the cornea, which is responsible for about two-thirds of the bending of rays of light. The importance of the lens lies in the fact that it is elastic. By means of muscle-fibres which pull on it the convexity of the lens can be altered; the optical system is therefore adjustable, and in this way we can focus our eyes on objects of various distances. Sometimes this apparatus does not work as well as it should. The muscles may not pull properly, or the eyeball itself may be slightly out of shape, so that the image is not crisply focused. In other and more familiar words we may be short-sighted, or long-sighted, or astigmatic. But such defects can be easily remedied by putting extra lenses into the optical system—by an eyeglass or a properly adjusted pair of spectacles.

The sense-cells of the retina are of two kinds, called *rods* and *cones*. The appearance of these two elements is shown in Fig. 49. Each has at one end a striped structure, longer and more slender in the rods than in the cones, which is presumably the seat of stimulation, and each tapers away into a long nerve-fibre at the other end. It used to be thought that these cells were directly sensitive to light, but it has recently been shown that this view is incorrect. Light acts upon them in a curious and roundabout way. The matrix in between the retinal cells contains chemical substances which are decomposed by light, in much the same way as the silver compounds of a photographic plate are decomposed by light, and it is the substances resulting from this decomposition, not the light itself, that stimulate the cells. The rods and cones are not sensitive to light at all; they are chemical sense-organs like taste-cells or smell-cells, and just as sound is converted into touch-stimuli in the ear, so light is converted into chemical stimuli in the eye. Sight is, in fact, an extreme utilization of chemical response. It is the brain which interprets the impulses of the retinal fibres as form and colour.

We have seen that the great variety of touch-sensations are in reality only blends of one or two elementary kinds of feeling. A similar analysis can be applied to colour-sensations. It is known that we can give rise to any desired colour-sensation by projecting into the eyeball the correct mixture of red, green, and violet light. Probably there are three different light-sensitive substances in the retina, decomposed by red, green, and violet light respectively, and their decomposition-products give rise to the three elementary colour-sensations. The process of blending these three sensations into a compound colour such as yellow or brown or white, occurs in the brain and is analogous to the mental blending of different notes into a chord.

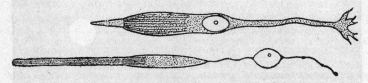

FIG. 49. A rod (below) and a cone (above) from the retina.
The sensitive end is to the left (highly magnified).

The retinal cells communicate by means of nerve-fibres with the brain, and before we consider the mental processes accompanying vision we may take note of a curious anatomical fact about the way in which these nerve-fibres are arranged.

By analogy with other sense-organs one would expect the striped sensitive ends of the rods and cones to point towards the light and the nerve-fibres to lead away in the opposite direction. But, as a matter of fact, the exact opposite is found. The sensory ends of the retinal cells are turned towards the wall of the eyeball and the nerve-fibres form a layer intervening between them and the light to which they are sensitive. One would expect this nervous layer, interposed between the retinal cells and the light, to fog the images that are cast on the cells; and so it does, over the greater part of the retina. But in the eyes of men, apes, and monkeys this arrangement is modified in one region, lying nearly in the centre of the back wall of the eyeball, and called the *yellow spot* because it turns yellow after death. This yellow spot is the spot of distinct vision, and it makes the sight of the animals that possess it a clearer and better thing than that of any other mammals. In this place the nervous layer is reduced to about one-sixth of its normal thickness, so that the fogging of the image is imperceptible. The sensitive cells are very closely packed together, so that the image can be more accurately defined, and there are no blood-vessels in that particular part of the retina. From it the nervous fibres run together to a special region of the brain, and, small as it is relatively to the rest of the retina, the fibres from its sensory elements make up a third of the whole thick stalk of the optic nerve. We can see a thing distinctly only when it is focused upon this spot, and so it is natural to conclude that such animals as the cow, horse, dog, or mouse do not see things with anything like the same definition as ourselves.

FIG. 50. The blind spot.

Close the left eye and look fixedly at the cross with the right, holding the page ten inches or a foot from the face. By moving it nearer or farther a distance can be found at which the circle disappears, its image falling on the blind spot.

A short distance from the yellow spot is the point where the optic nerve leaves the eyeball. The fibres of the retina all converge to this point and are there gathered together and depart. This is the *blind spot;* at this point there are no sense-cells at all, so no appreciation of the image is possible. Under normal conditions the blind spot of one eye is covered by a sensitive part of the other, so its existence is not suspected, but by means of the well-known experiment reproduced as Fig. 50 the reader will be able to convince himself that he possesses a blind spot. King Charles II was so entertained by this experiment that he used to practise taking off the heads of his courtiers by this harmless method.

There is an important contrast between the mental processes accompanying hearing and those accompanying vision. We have seen that the particular part of the cochlea stimulated by a note depends on its pitch. On the other hand the particular part of the retina stimulated by an object depends on its position. The ear is a sense-organ for perceiving the qualities of sounds; our ideas of the direction from which sounds are coming are generally vague, based, for example, on the relative intensities with which they are heard in the two ears, and which ear hears them first. But the eye is an organ for determining the spatial properties of objects—their positions, shapes, and movements. The brain is constantly elaborating the information received from the

THE HARMONY OF THE BODY-MACHINE 129

eye by means of unconscious associations with past experience. In our judgments of solidity and the three-dimensional shape of objects, for example, we rely to a large extent on their shading and on associations of shading with experiences derived from the touch sense. But we cannot go further into this constant unconscious judging and weighing, nor into the many ingenious optical illusions that illustrate some of the errors that both our optical apparatus and our visual brain-centres can make. One or two such illusions are reproduced herewith.

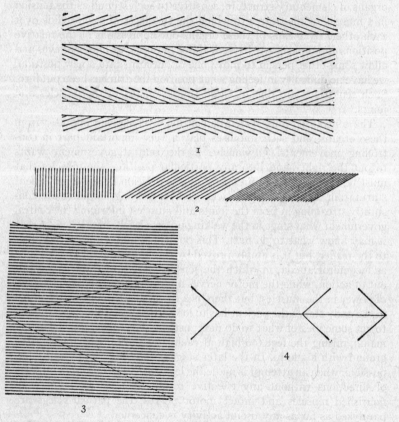

FIG. 51. Illustrating some errors of judgment in our visual brain-centres.

In (1) the four horizontal lines are straight and parallel; in (2), the shaded areas are equal in height; in (3), the sloping lines are straight, not stepped; in (4), the two parts of the horizontal lines are equal halves.

The characteristic thing about the eye is the precision of the spatial information that it can obtain. By means of its nose a mouse can tell that there is a cat about somewhere; by means of its ears, it can hear him moving and locate him roughly; but by means of its eyes it can get precise information of the enemy's whereabouts—and that, after all, is the most important thing to know.

Sensations of Position and Movement. We have already remarked that the muscles, tendons, and joints are provided with minute sense-organs of elementary structure, sensitive to such stimuli as the tension in a muscle, or the degree to which the two faces of a joint slide over each other. By means of these organs our impressions of the relative positions of our parts are derived. If, for example, we close our eyes and allow some other person to move one of our arms into a new position, we have no difficulty in feeling what position the arm has been put into. Such sensations are due to these simple organs in the muscles and joints.

The central nervous system is continually receiving impulses from these organs, and such impulses play a very important part in controlling movements. Obviously, if the central government wants to put the right hand into any particular position, the muscles that must be called into action depend on the position that the movement starts from. During complex acts, such as walking, impulses are constantly streaming in from the joints and muscles, informing the central government what stage in the walking process has been reached, so that it may know what to do next. This point may not seem very evident to the reader, but it is amply proved by the symptoms of diseases such as locomotor ataxia, in which the sensory nerves of the legs are put out of action, while the motor nerves are unimpaired. There is no loss of power in the muscles, but there is a surprising loss of control. Even in the early stages the patient has to look at his legs while he is walking to get some idea of what to do next; usually, he exaggerates his movements, raising the feet too high at each step and putting them to the ground with a stamp. In the later stages, walking becomes quite impossible; when an attempt is made the legs are thrown about in all sorts of directions without any effective result. Thus, in spite of well-nourished muscles and intact motor nerves, the patient's legs are paralyzed as far as any useful activity is concerned.

In addition to these important sense-organs there is yet another group, a group of very elaborate and specialized sense-organs, situated in the head, from which we get impressions of the position and movements of the body as a whole.

THE HARMONY OF THE BODY-MACHINE

The labyrinth of the inner ear, as we have seen, is a hollow bag of complicated architecture, embedded in bone. Moreover, only a part of it is concerned with the perception of sounds. Fig 52 shows the appearance of this labyrinth. On the left is the spirally coiled cochlea, where sound-vibrations are converted into touch-stimuli; in the middle there is a bag, the vestibule; and on the right lie the structures with which we must now concern ourselves.

It will be seen that there are three arching tubes, the *semi-circular canals*, each leading out of the vestibule, sweeping round in a regular half-circle, and opening again into the vestibule. It is important to observe that these three canals lie in three planes at right angles to each other. The posterior canal lies in the plane of the page, while the other two stand out perpendicularly from it; at the same time the planes of the superior and lateral canals are perpendicular to each other.

Each of the three canals has at one end a slight swelling, the *ampulla*. It is here that the actual sense-cells are located. In each ampulla there is a cluster of cells with long hair-like processes projecting into the cavity of the canal, and continuous below with nerve-fibres.

Now, suppose that the head is suddenly moved horizontally from right to left—i.e., in the plane of the lateral canal (a sudden backward glance over the shoulder would involve such a movement). The labyrinth, as we have seen, is full of a lymph-like fluid, and when the head is suddenly moved the inertia of this fluid will make it lag a little behind the movements of the labyrinth itself.

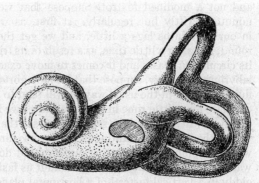

FIG. 52. The labyrinth of the inner ear.

The cochlea on the left; the semicircular canals on the right.

There will be a slight circulation of fluid in the lateral canal, and therefore, a bending of the delicate hairs on the sense-cells in the ampulla. This bending stimulates the cells and causes them to dispatch nervous impulses, which are interpreted by the brain as a sensation of horizontal movement. Now, since the three canals lie in planes at right angles to each other, corresponding to the three dimensions of space, any movement of the head will bring about a circulation of fluid in one or more of the canals, and the direction of the movement will determine exactly the relative amount of disturbance in each of the three. That is to say, any movement, if it is vigorous enough, will cause a stimulation of the sense-cells in the ampullae, and by comparing the strength of disturbance in each of the three the brain is able to judge the direction of the movement. Without this information we should be helpless when we tried to walk or run.

It will be noticed that the working of the apparatus depends on the inertia of the fluid in the canals. It therefore detects, not constant uniform motion, but accelerations and retardations. If we are seated in a railway carriage with the blinds pulled down, as long as the train travels at constant speed we cannot feel whether we are facing the engine or not; but when the train slows down or speeds up the change in velocity is clearly felt. The semicircular canals are like most of our sense-organs; they feel, not motion, but changes in motion. It is possible to demonstrate this fact by a simple experiment, an experiment that was familiar enough in the days when a waltz was a waltz and not a modified foxtrot. Suppose that we spin round for some minutes, swiftly but regularly. At first, as we accelerate, the fluid in our labyrinths lags a little, and we get the sensation of twisting round; but after a little time, as a result of its friction against the walls, its circulation ceases and it comes to move exactly as the surrounding labyrinths do. Now, suppose that we stop abruptly. The inertia of the fluid will work in the usual way; it will tend to go on moving, and will circulate for a little time in the canals after we have come to rest. Evidently, it will now bend the sensitive hairs in the opposite direction, and we get the illusion of rotating in the reverse direction to that in which we were spinning. By putting our forehead down on to the top of a walking-stick, and then moving round it as fast as we can, we can get giddy in a vertical instead of a horizontal plane.

On the other hand, there are sense-organs in the ear that apparently do not conform to the general law of contrast. On the wall of the vestibule there are two patches of sensitive cells, one patch lying in a vertical plane and the other in a horizontal plane. The structure of these cells is essentially like that of the sense-cells in the ampullae, but in among their hair-like processes there are tiny nodules of carbonate of lime, called the *otoliths*. The otoliths are constantly pressing on the surrounding hairs, and they constantly stimulate the sense-cells with an intensity depending on the direction of their pressure. Since this direction depends on the position of the head—i.e., on its inclination to the vertical—the stimulation of the cells will also depend on the position of the head. In this way, by streams of impulses from the sense-patches in the vestibule, the brain is constantly informed about the inclination of the head.

At first sight the inner ear seems a very heterogeneous structure, since it detects factors so diverse as sound and position. But there is a principle common to all these organs—cochlea, semicircular canals, and patches in the vestibule—and one that distinguishes them from all our other sense-organs. We saw that the sense-cells in the cochlea are not sensitive to sounds as such, but that they are touch-cells. Similarly, the sense-cells in the semicircular canals and vestibule are touch-cells, and the parts are so constructed that they are stimulated whenever we move, or whenever our position is abnormal. The inner ear, then, is a complex of several ingenious anatomical devices for stimulating touch-cells whenever certain conditions arise—conditions which themselves do not excite living protoplasm directly. The basal sense in all these cases is touch.

§ 6. The Nervous Mechanism and the Brain.

Now we are in a position to state certain general ideas about the working of the behaviour-system in the animal bodies we have made our opening subjects of study. Later on we shall take up these general ideas again and scrutinize them much more closely. We have considered the motor mechanism and the organs of sensation. Their action is correlated. How closely and exactly it is correlated, we shall defer until that later discussion. We feel something and we act in response; let that suffice for the present. We may hardly observe that we feel something, and yet we may act in response.

It is customary to distinguish and contrast two aspects of behaviour in reply to a sensory impression. As an example of the first kind we may consider what happens when the hand touches a hot object—the red end of a cigarette, for example. We withdraw the hand, and do so very quickly indeed, for the movement begins before our conscious minds are even aware of the pain. Such an action is called a reflex action, or more compactly a *reflex*. It is entirely automatic and independent of the mind; we withdraw our hand, not because we think we are going to be burnt, but simply because we are made that way. It is, so to speak, part of our anatomy.

There is another type of response which does not appear to be so automatic as a reflex. It is more complicated. A certain element of memory and consciousness is involved. The reader hears a gong, for example, and makes appropriate preparations for lunch, or he hears the National Anthem and stiffens up as he has been trained to do. The behaviour of the higher animals is twofold in character. Its acts may be of a simple and primary form, part of an inherited and automatic system of reflexes as much a part of one's constitution as the disposition of one's arteries or bowels, or they may involve a larger or smaller element of acquired response. There is the simple reflex and there is the reflex complicated by something less obviously mechanical.

This duality is reflected in our anatomy, for these two systems of response have different seats. In the accompanying figure it can be seen that the greater part of the cranium is filled with an oval, wrinkled mass, the *cerebrum*. A vertical cleft divides this organ into two halves, the *cerebral hemispheres*, but the cleft lies in the median plane and is not seen in the figure. The hemispheres are the seat of the mind; it is here that those processes of memory and association take place which underlie responses of the second kind. The rest of the central nervous system, the spinal cord and certain lower parts of the brain that are overlapped and hidden by the cerebral hemispheres, are concerned with automatic actions only. Only the activities that come up to the cerebrum, and not all of these, are conscious.

By removing the cerebral hemispheres it is possible to get a mindless mammal, and to see just how far one's many actions are due to the intervention of the higher centres. Such animals have the spinal cord and the higher reflex centres intact, but they lack any power of learning, understanding and memory, and presumably they lack any sort of continuing consciousness. A "decerebrate" frog, for example, can breathe and swallow, it will sit in the usual position, and when pinched or otherwise stimulated it will jump away, guiding itself by sight so that it avoids obstacles; it turns over at once if it is put on its back. If it is put into water it swims about until it finds an island of some sort, such as a floating piece of wood, when it climbs up and then sits perfectly still. A decerebrate dog can do much the same; it can breathe or swallow or walk about—indeed it spends nearly the whole of its time in prowling restlessly and aimlessly around—and at night it sleeps soundly. But these decerebrate animals differ from normal animals in the complete lack of any sign of psychical phenomena. Mentally they are dead. All intimations of feeling cease. They seem

THE HARMONY OF THE BODY–MACHINE

unable to remember or to anticipate. If food is put into their mouths they will chew it, swallow it, and digest it, but they never show any signs of appetite or hunger; even when they are starving they never look for food, and if food is put into their cages they will not touch it of their own accord. They do not recognize it as food. And similarly with painful stimuli; a decerebrate frog will jump away if it is pinched,

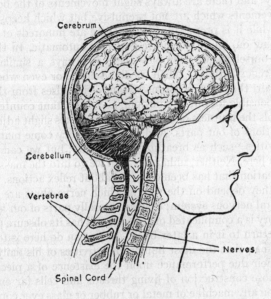

Fig. 53. The brain and upper end of the spinal cord exposed from the right.

and a decerebrate dog will snarl and snap, but neither shows any sign of fear at a threat. They make no response at all unless they are directly stimulated.

Perhaps the most noteworthy point about these animals is not the things they cannot do but the things they can do. Consider simply the fact that without cerebral hemispheres a mammal can walk about and avoid obstacles. Walking is a very complicated activity indeed. It involves the contraction of a number of muscles each at the right moment—moreover, when any muscle contracts those which oppose its action relax their tone; it involves a stream of sensory impulses from the limbs that report progress so that the central nervous system may know what to do next. But it also involves other things, for the organism must keep its balance. There are impulses streaming in from the labyrinth of the ear, from the pressure organs in the soles of the feet, from the eyes, which tell the central nervous system about the position of the body; and there are always slight movements of the body to be made, movements which are not propulsive but which keep the body balanced while it is propelled. Clearly there are hundreds of different reflexes going on, all of which are perfectly automatic. In the intact mammal—ourselves for example—there is always a similar undercurrent of reflex activity. When we stand erect, or even when we sit up in a chair, there is a constant stream of impulses from the sense-organs responsible for our equilibrium, and a constant counter-stream that controls the tautness of our muscles and makes slight adjustments of the positions of our parts. And in this category come many of our other activities—such as breathing movements; but we cannot catalogue them here. Neither, unhappily, can we go into the mass of accurate information that has been collected about reflex actions. Roughly speaking, they depend on the way in which nerve-fibres are disposed in the central nervous system; they are literally parts of our anatomy. But the story is a complicated one, and it still has its obscure passages. We shall return to it in a later book. All we can do here is to remind Mr. Everyman that the most humdrum activities of his daily life depend for their due performance upon the existence of a piece of machinery whose construction of living threads and cells far outdoes in complication any machine of metal or rubber or glass ever constructed by men.

The cerebrum, the organ of the mind, is superposed upon this system of automatic, machine-like responses, and it is capable of interfering with and modifying the activity of the reflex centres. Returning to our comparison of the body to a state and the central nervous system to a government, the spinal cord and the brain-stem may be compared to clerical and administrative departments, dealing with ordinary matters of routine even up to a very high degree of complexity, while the cerebral hemispheres are the executive considering the state of affairs as a whole and making special decisions only when the routine responses are insufficient.

If we cut a slice off one of the hemispheres—the reader may do this for himself whenever sheep or calves' brains come to table—we observe that there is a division into two kinds of tissue; running over the surface of the hemisphere there is a layer of grey matter; and this is where the nerve-cells are seated and where the mental processes actually take place. And inside this there is a branching core of white matter, consisting of the fibres that put the cells of this grey matter into communication with the lower routine parts of the brain and with other parts of the body, and also connect different areas of grey matter one with another. Impulses arriving from the sense-organs come first to the lower receiving centres whence fibres lead away in two directions; some go to the reflex centres and others to the cerebral hemispheres. The impulse may be wholly reflected in a reflex or pass on in part and more or less modified to the hemispheres. Responsively a system of fibres takes impressions down from the hemispheres to the motor nerve-cells in the brain-stem or spinal cord. Thus the executive can consider special circumstances, command special movements, and even inhibit and overrule the normal activity of the automatic administrative centres—as for example the mind of Mucius Scaevola inhibited his reflexes when to impress Porsenna with the fortitude of a Roman he held his hand in the burning coals and kept it there in spite of his pain.

We shall see in a later part that the extent to which this control by the personal executive can be exerted varies in different animals. A fish or a frog is to a large extent a reflex machine, but as we examine in turn the lower and higher mammals leading up to man, we find the lower centres relying more and more completely on the guiding control of the hemispheres. At the same time the hemispheres themselves are better and better developed; in a frog they are relatively small, but in one of the higher mammals, even at the mouse level, they are large and their surface is thrown into wrinkles so that as much grey matter as possible can be fitted into the cranial cavity. There can be little doubt that the distinctively large cerebrum of man is associated with the distinctive complexity and definiteness of his thought-processes.

We may note here, before we break off this brief first account of the working of the brain, a curious part of it lying at the back, behind and below the cerebral hemispheres and having a more closely wrinkled surface than they. This organ, the *cerebellum*, is a sort of private secretary to the cerebral hemispheres. It is discriminating but not conscious. Its activities may best be displayed by a single instance. A very intricate complex of reflex actions keeps our bodies balanced. Suppose that the cerebrum intervenes and sends a command to the appropriate motor cells in the spinal cord so that the right arm is raised. The centre of gravity of the body will be shifted, and in the absence of any compensating adjustments we should sway over to the right side. But the cerebellum sees to this. It watches the acts of the superior authority, and takes unobtrusive but appropriate steps so that the general equil-

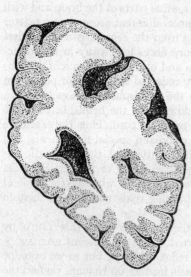

Fig. 54. A slice of the cerebrum, showing the outer layer of grey matter and the core of white matter.

Fig. 55. A bit of the grey matter of Fig. 54 highly magnified.

It consists of nerve-cells with complicated tangled processes. The numbers indicate layers with different characteristics. (Courtesy of Dr. C. J. Herrick and W. B. Saunders Co.)

ibrium of the body may not be upset—in the example chosen, it would call forth a slight compensating movement of the left arm or a slight adjustment of the position at the hip-joints. It is like the official in a cinema studio who specializes in "continuity," it notes the actions of the directors, supplies their omissions, and, without bothering the mind in the least, sees them through.

Occasionally Mr. Everyman may afford an unpremeditated demonstration of the function of the cerebellum. It is sensitive to an excess of alcohol in the system and speedily becomes disorganized under its influence. The cerebrum itself may be even elated by this drug, and may issue many bright and pleasing orders in consequence, but the cerebellum fails in its customary secretarial care. Consequently Mr. Everyman's movements become unbalanced. He walks, but he no longer walks with grace and precision; he may become excited by ideas and garrulous, but his larynx, lips, and mouth fail in the fine adjustments necessary for perfect articulation. He seeks to give playful slaps to his friends and misses them perplexingly. Similar results may be due to organic injury of the cerebellum, and it is sometimes difficult to distinguish the permanent from the transitory derangement.

This is as far as it is convenient to carry our account of the brain at the present time. Throughout the Books that follow this one we shall be gathering the material for a later more exhaustive attack on the mysteries of that coating of grey matter upon the outside of the cerebrum, which is the seat of Mr. Everyman's dreams and desires, the vehicle of his imagination, the medium of his joys and sorrows, the most important stuff in the universe for every human being.

IV

THE WEARING OUT OF THE MACHINE AND ITS REPRODUCTION

§ 1. *Age and Decay.* § 2. *Reproduction and Fertilization.* § 3. *The Growth and Development of the Embryo.* § 4. *Rhythm and Birth.* § 5. *Childhood, Adolescence, and Maturity.*

§ 1. Age and Decay.

IN THE course of time Mr. Everyman discovers that he is growing old. He has avoided complex chills, his leucocytes have defeated a score of dangerous infections, he has had wounds and they have healed, but nevertheless it is borne in upon him that he cannot go on for ever. He realizes a slow rusting up of this strange machine with which he is identified, and neither he nor any help he can call in seem able to arrest that inevitable march towards cessation.

The amount of specialized tissue that a body can lay down is limited. We possess, normally, the apparatus for making two sets of teeth, but no more. When any tooth of the second set is worn out, whether by decay or by inadequate blood-supply, or by sheer grinding away of its hard surface, it leaves a gap that can never be filled again. So with our softer tissues. Our nerves and muscles and glands, just like other delicate machines, are damaged a little by their own working, and we cannot keep them in repair for ever. For this reason after we have lived for some forty years our bodies begin to wear away; gradually, but with gathering speed, we lose weight.

This wearing-away chiefly involves the more specialized tissues. Skeletal or connective tissue, which consists of anatomically undifferentiated cells loafing about among a scaffolding that they have built up, is relatively unaffected. It is the brain and muscles, the kidneys and liver, that shrink. Moreover, different tissues dwindle away at different rates. The brain-tissue goes more rapidly than the cerebral arteries, so that the latter, having their original length but a shortened course, be-

THE WEARING OUT OF THE MACHINE 141

come twisted into curious loops and spirals. Similarly the skin does not shrink at all fast, so that, as the underlying muscles wear away and the fat stores are used up, it becomes thrown into wrinkles. As our specialized cells degenerate our faculties gradually become blunted. We cannot see or hear as acutely as before, our muscles are less powerful, our digestion is less efficient, our brains less active and less accurate.

The chemical basis of this wearing out is at present not understood. Old age seems to be associated in some way with defective calcium metabolism; the brittleness of senescent bones is due to the resorption of lime salts into the blood. Moreover, there seems to be an accumulation of poisonous substances in the blood. It has been shown that if a culture of healthy tissue from a growing chick (Chapter 2, § 2) be mounted in blood-serum from a very old bird the cells cease forthwith to divide and grow.

The slow exhaustion of old age may also be complicated by other changes; in the great majority of old people (but not in all) there is a gradual stiffening (*sclerosis*) of the arteries, and there may be a similar stiffening in other tissues.

Sooner or later one or other of the essential organs fails and the body dies. In most cases of natural death in people over fifty the failure occurs in the circulatory system; the heart stops, or one of the stiffened arteries bursts, and all the other tissues perish because of the cessation of their blood-supply. In other cases the failure occurs elsewhere—in the brain-centre that calls forth the respiratory movements, for example.

It is important to realize that our cells do not die because mortality is inherent in their internal structure. They die because they are parts of a very complicated system based on co-operation, and sooner or later one of the tissues lets the other down. In death from heart-failure, for example, it is wrong to suppose that the heart's muscle-fibres die and therefore stop beating; their stoppage precedes and causes their death. The cells first weary of their unending labour and stop beating, and only subsequently die, like all the other cells of the body, because as a result of their own stoppage their blood-supply is cut off. As a matter of fact, living matter is potentially immortal. If one keeps a culture from the tissue of a young animal and takes sub-cultures regularly, the race of cells can apparently go on growing and dividing indefinitely. Death is a consequence of incomplete organization; the tissues die because they are parts of an imperfectly balanced body.

How far it may be possible for a body to sustain its balance and continue indefinitely, or at least for a much longer period than the normal life of its species, is an interesting matter for speculation. Such a prolongation is not Nature's ordinary way. But there are long-lived species which do remain recuperative for relatively long periods. Parrots will last out eighty years and eagles a hundred, and such a fish as the carp, especially if it is protected from external danger in a garden pond, seems able to go on almost without a limit. There is evidence for hundred-and-fifty-year-old carp and for pike which have lasted for two centuries. Some trees have an enormous span of life. The baobab of Cape Verde is supposed to endure five thousand years, and the *Sequoia gigantea* of California almost as long.

The prolongation of the restless, various human life is a much more complex issue than that of keeping alive a captive bird or fish. Many old people are still hungry for experience and the fact that the intelligence outlasts many other bodily powers gives the death of ripe-minded and balanced men and women a quality of tragic waste. From what we have said and from what we will explain a little later, it will be plain that the process of old age is at least partly chemical and accompanied by defects of internal secretion. But it is not by any means a simple process; this man gives first at one point and that man at another, and the replacement or reinforcement of the secretion that would help in the first instance might simply intensify the want of balance in the second. Ilya Metchnikov was disposed to ascribe many of the phenomena of senility to decay in the large intestine, and sought to correct this by a liberal use of *koumiss*—soured milk, such as the long-lived Tartars consume—which abounds in bacteria whose products check intestinal putrefaction. For a time koumiss was a fashionable drink among the elderly, but after the death of Metchnikov, at the age of seventy-one, faith in his remedy declined.

In the young and developing adult certain secretions from the interstitial tissues of the reproductive glands play an invigorating part in the growth and strengthening of the body, and various experimenters have tried to restore the dwindling supply of these secretions as age comes on. Steinach seeks this end by local surgery, which for a time at least reanimates the declining tissue. Voronoff grafts new tissues taken from young apes, in the effective region. There are other fairly obvious applications, by injection and so forth, of the same idea. Resort is also made to the substance and secretions of the thyroid, the pituitary body and other ductless glands which influence the growth processes. Steinach, Voronoff, and various other workers indubitably produced some remarkable rejuvenescences. But also they have had distinct failures and disappointments. Much more experiment is needed in

THE WEARING OUT OF THE MACHINE

this direction before we can confidently and safely hold back the advance of age.

In age, we are dealing not with an orderly decline, but with an irregular development and progressive exaggeration of what at first were minor maladjustments. The body is like a machine in which parts wear loose and little rifts widen. It was never perfectly adjusted and

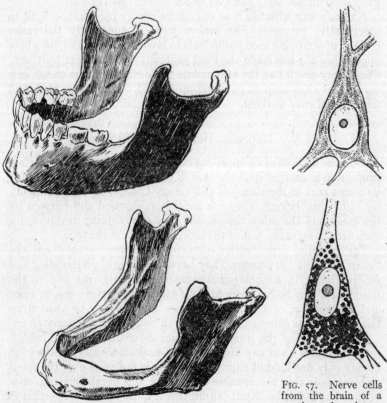

FIG. 56. The lower jaw at the prime of life (above), and at a great age (below).

FIG. 57. Nerve cells from the brain of a youth of sixteen (above) and of an old man (below).

The old cell is filled with droplets of fat (shown black).

the defects appear in use. Each one of us ages after his nature and in his own fashion. A very capable, observant, and devoted medical attendant might by continual alertness check this strain, repair or replace that fagging part, and eke out a life far beyond the normal span. In some cases that might be worth while. And a general prolongation of vigour may be possible with more knowledge and care and improved surroundings. A time may come when men as a race will live more sanely and fully and suffer less. But it seems inconceivable that any individual human body will ever evade its final goal of death.

Nature's way with life is to economize energy by setting a limit to recuperative processes. She prefers what is apparently the easier method of scrapping used individuals in favour of fresh ones, for whose appearance she has made the most generous and elaborate provision. She is very much like the automobile trade which prefers to sell new cars rather than keep on patching up the old, as they wear more and more out of easy working.

§ 2. Reproduction and Fertilization.

AMONG the millions of co-operating cells that build up a body a certain proportion is charged with the task of initiating new lives. They play no direct part in the chemical and energy changes of the community to which they belong. They are nourished, cleansed and warmed by the labours of the other tissues, but they give nothing in return, for they are the servants, not of the individual, but of the race.

The essential reproductive elements are minute; they are the "marrying cells" or *gametes*. They have the power of proliferating and differentiating in suitable circumstances and so giving rise to the tissues of a new body. There are two kinds of gametes—*ova*, or eggs, and *spermatozoa*, or sperms, and a new life can only be created by the union of one gamete of each kind, an ovum with a spermatozoon.

In this fact lies the reason for sexual differentiation. The human species is made up of two kinds of individuals, male and female, each having only one kind of gamete. The female possesses ova, the small globular cells that are capable of growing and developing into new bodies; but before an ovum can do this it must be fertilized—that is to say, it must unite with a spermatozoon. The male possesses spermatozoa, and his part in the reproductive process is the providing of the spermatozoa that fertilize the ova.

The fertilized ovum of all the mammalia grows and develops inside the female and therefore her body contains, in addition to an ovum-

THE WEARING OUT OF THE MACHINE

factory, a special chamber where the embryo is tended and nourished. There is no such special chamber in the bird, which lays its fertilized eggs. Nor is it usual in reptiles, fishes, and the majority of lower animals.

The lowliest and simplest animals do not show this division into two sexes. When we are studying them in a later chapter we shall find quite different methods of reproduction. But in the mammals the creation of a new life necessarily involves the encounter of male and female bodies, and the consequent union of male and female gametes.

The male gametes, or *spermatozoa*, are curiously constructed. They are very unlike most of the cells we have hitherto considered. Each spermatozoon consists of a rounded head—this is the part that fertilizes the ovum—and a long, lashing tail, by means of which it can propel itself in quest of the waiting female gamete. The whole thing is about one five-hundredth part of an inch long.

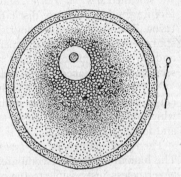

FIG. 58. An ovum and a sperm drawn to the same scale.

The male organs are of three kinds. There is a pair of roundish bodies, the *testes*, where these spermatozoa are manufactured, a series of passages along which the spermatozoa pass to be injected into the female system, and in which they wait until they are wanted, and a series of accessory glands whose secretions mix with the sperm as it is discharged.

A testis consists, in the main, of a close mass of coiling tubes, lined by ceaselessly proliferating cells. Some of these cells are converted into spermatozoa, the nucleus becoming the sperm-head, part of the cytoplasm becoming the sperm-tail, and the rest of the cell dropping off and disintegrating. Others simply multiply, giving rise to new sperm-forming cells. In between the coiling tubes there are cells of another kind, which have nothing to do with sperm-formation, but have quite another function in the body.

From the testis the spermatozoa enter a fine, tortuous tube, the *epididymis*, which, although between twenty and thirty feet long in the human case, is curled up into a mass smaller than the testis itself. In this canal the spermatozoa are stored until required, when they are forced out of it by muscular movements of its wall. From the epididymis the spermatozoa pass along a tube to the *urethra*, a passage that guides them out of the body and which, by a curious economy, they share with the urine. The last part of the urethra projects and the walls about it are spongy in structure. When spermatozoa are discharged these walls are stiffened by distension with blood in order to facilitate insertion into the recipient female passage.

A number of accessory glands open into the urethra—the *prostate*, the so-called *seminal vesicles*, and *Cowper's glands*. These glands manufacture fluids, which are discharged at the same time as the spermatozoa. The fluids serve to nourish the spermatozoa, and they seem to spur them to activity, for the spermatozoa lie passive in the epididymis, and do not begin to lash and swim until they are mixed with the secretions of the accessory glands.

Nature is curiously lavish with spermatozoa. At each human sexual act as many as two hundred million spermatozoa may be put forth, of which only one (in rare cases two or three) may achieve its object and fertilize an ovum.

The female gamete or egg (*ovum*) is very much larger than the male. The human egg cannot compare in size with that of a bird, but it is nevertheless just visible to the naked eye. It is a spherical cell, about a one hundred and twenty-fifth part of an inch in diameter, and surrounded by a transparent pellicle; besides the indispensable nucleus, its protoplasm includes a number of granules of yolk, the fuel that is needed to provide the energy for the earliest stages of development. The female organs (Fig. 61) include the *ovaries*, where the ova are made, female equivalents to the testes, and a series of passages and compartments: the *uterus*, or womb, where the growing embryo is housed and nourished, the two upper *oviducts*, leading ova from the ovaries to the womb, and the passage which receives the male gametes. As in the male, these compartments are provided with various accessory glands.

The ovary is not tubular in structure as a testis is; the main part of its substance consists of a network of connective tissue bearing in its meshes the *interstitial cells*, to whose function we shall return. Dotted about in this mass are the ova in various stages of development. The ova lie in spherical vesicles called, after their seventeenth century discoverer, the *Graafian follicles*. The follicles of the youngest and smallest ova consist of a simple layer of cubical cells surrounding the ovum, and nursing and nourishing it, but as the ovum grows the follicle becomes more and more complicated. The largest ova, which are approaching maturity, are surrounded by two layers of follicular tissue each several cells thick, and separated from each other by a space containing a nutrient fluid.

Thus the female gametes are treated with very much more respect than the male. The spermatozoa proliferate almost untended in the testis, but the developing ova are elaborately cared for by the cells of their follicles. Further, ova are not produced by the million; once every

Fig. 59. Two sperms very highly magnified seen from different angles to show the shape of the head.

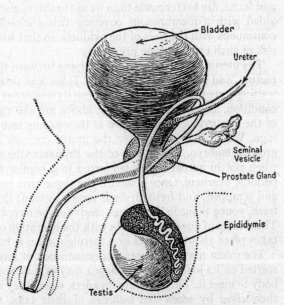

Fig. 60. Diagram of the Male Organs of Reproduction in a higher mammal as seen from the left side.

month an ovum ripens in one or other of the two ovaries, and is discharged at the surface of the ovary by the bursting of its follicle.

The oviducts are narrow tubes opening at one end into the uterus, and having at the other a broad, funnel-like expansion, which wraps round and partly encloses the ovary. Into this funnel the ova drop as they are discharged, and they are passed down the oviduct, partly by the action of cilia (Chapter 3, § 5) and partly by muscular movements of the duct itself. It is curious to note an imperfection in the design of the reproductive system of most mammals: the funnel-shaped end of the oviduct does not completely enclose the ovary, and there is, therefore, a slight risk of the discharged ovum failing to find its way into the oviduct and escaping into the general abdominal cavity. Occasionally (but fortunately very rarely) the ovum does this, and still more occasionally it is fertilized in the abdominal cavity by spermatozoa that have penetrated beyond the ends of the oviducts; the resulting embryo implants itself somewhere among the viscera, and much danger and distress arise. Some mammals, such as the dog and ferret, are better made than we are in this respect, for they are provided with a membranous covering that encloses the ovary and is continuous with the ends of the oviducts, so that with them there is no risk of such extra-uterine pregnancy.

The uterus is a pear-shaped tube, lying between the bladder and the rectum, and having very thick walls. It is about three inches long in its waiting state in the human subject but its size varies with its functional condition. It has three openings; above are the very small openings of the two oviducts, and below is the opening into the outward passage. We shall see later how the uterus encloses and nourishes the growing embryo. The uterus of the Primates (the highest mammals) when not in actual use, is subjected to a regular and very stringent periodic overhaul. Once a month the tissue that lines it is broken up and scrapped, and fresh tissue is grown instead; thus there is always fresh, young tissue in the uterus, ready for the reception of the embryo. This overhaul does not coincide with the liberation of the ovum, which takes place about ten days or a fortnight after it has begun.

The ovum may encounter a spermatozoon at any time after it has started on its journey. Spermatozoa may or may not have entered the body to meet it. These minute invaders, where they are present, propel themselves by screw-like lashings of their tails, and receive some assistance in the earliest stages of their journey from muscular contractions of the uterus itself. For an object as minute as a spermatozoon this journey of several inches is a tremendous undertaking. Only a small proportion of the two hundred million launched by the male reach the upper end of the oviduct. The majority perish on the way. But when the survivors reach that destination, their troubles are over; they rest in the oviduct and retain their capacity for fertilization for at least a week. It is probable that the ovum can live for three or four days in the oviduct after its escape from the ovary. If during this time it meets an active spermatozoon, it is fertilized and can fulfil its destiny. If not, it dies and disintegrates in the uterus and is presently washed away. We will assume that the former alternative occurs, and we will

THE WEARING OUT OF THE MACHINE 149

trace the gradual development of the minute ovum into a young mammal.

The actual fertilization of the ovum is accomplished by one only of the myriads of searching spermatozoa. The successful competitor

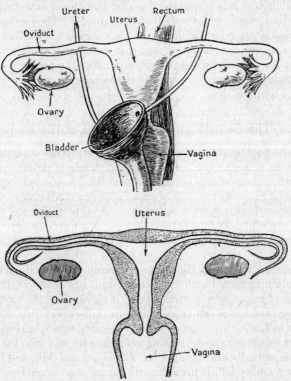

FIG. 61. Above, Front View of the Female Reproductive Organs (the top of the bladder is removed to display the parts more clearly). Below, the same in section, to show their cavities.

presses its head against the side of the relatively colossal ovum, and as a result of that simple contact the latter is activated; it ceases to be sterile, and begins forthwith the series of changes that constitute development.

After the spermatozoon has activated the ovum it does something further. Its head burrows actually into the substance of the ovum, leaving the tail outside to wriggle for a while and then to perish. When it gets inside, the head swells up and becomes a typical nucleus, which approaches and fuses with the nucleus of the ovum. The resulting cell, having the cytoplasm of the ovum and a compound nucleus derived partly from the original ovum and partly from the sperm begins to grow and divide. It is important to realize that this fusion of nuclei has nothing to do with the activation of the ovum; it is the mechanism by which characters inherited from the male parent are transmitted to the offspring. Activation, the change that enables the ovum to develop, occurs at the moment when the sperm touches the ovum. It is a sort of microscopic laying on of hands that awakens the slumbering possibilities of reproduction.

§ 3. The Growth and Development of the Embryo.

DURING its passage down the oviduct into the uterus—a passage that takes about a week—the fertilized ovum develops rapidly. It is, to begin with, a giant as cells go, and the first thing it does is to split itself up into a large number of cells of reasonable size. This process is called *segmentation*. It occurs by a series of divisions of cells into two; the original ovum divides into halves, then each half divides again, giving four quarters, then each quarter divides again into two eighths and so on.

By the time it reaches the uterus the ovum has turned itself into a mass containing some hundreds of cells. But it has done something else as well, for besides merely dividing and dividing, the cells have begun to arrange themselves in a definite pattern. They build themselves into a hollow ball, with a thickened knob projecting into its cavity at one point. At this stage the developing germ eats its way into the wall of the uterus, actually destroying some of the maternal cells, and becomes completely embedded. It lies in a little burrow, surrounded on all sides by uterine tissue and bathed by the blood of the mother. As a result of its implantation the germ has become a parasite upon its mother and is in a position to absorb nourishment. In the oviduct it had no external food-supply and was living on its own store of yolk, so that it was unable to increase in weight; but now it is bathed by blood charged with food substances, so it can absorb what it wants and build up new embryonic tissue. Hitherto it has been simply developing; now it develops and grows at the same time.

For a few days the tiny vesicle differentiates and elaborates itself, but without showing any trace of the organs of the future baby. The first well-defined structures to appear are not parts of the baby at all;

they are the apparatus of membranes and ducts that is to wrap round the embryo and protect and feed it. Towards the end of the second week of human development the germ is little more than a millimetre in diameter, and consists of three hollow vesicles, two lying inside the third; moreover, the cells forming the walls of these spheres have become arranged into sheets or layers. During the third week the whole apparatus grows rapidly, and at this time the first visible organs

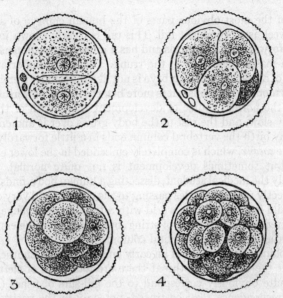

FIG. 62. Segmentation—the first steps in the development of a human egg.

of the embryonic body appear. On the wall that separates the two inner vesicles there is a thickened disc, about a millimetre and a half across; traversing this disc there is a faint furrow, the *primitive streak;* and just in front of this furrow two thickened ridges are beginning to form, ridges that will close together and give rise to the brain.

The tiny disc grows and differentiates very rapidly. By the end of the fifth week of development all the important organ-systems are already laid down. The embryo is now about one-fifth of an inch in length and possesses a beating heart, a definite nervous system, little sprouting buds that will become limbs, relatively large eyes, a rudimentary kidney, a growing digestive tube, and even a well-defined mass of cells that will eventually give rise to gametes, and replace the embryo by new individuals when it has developed and lived and is in its turn worn out. The essential systems are already present. But in its anatomy, in the way these parts are fitted together, the embryo presents at this stage some curious features that are destined to disappear as growth proceeds. We will take careful note of these points, for they shed revealing light on our relationships to other living creatures.

One of the most obvious parts of the human embryo of about the fourth week is a prominent tail. This tail is clearly shown in Fig. 63. It curls forwards and upwards and has a muscular development which rivals in massiveness that of the trunk. That is an important point. The tail of an early human embryo is not like the slender termination of an adult monkey, or mouse; it is more like the broad, powerful tail of a fish. In the normal course of development the growth of this tail is relatively slow, and the rest of the body grows over it and encloses it, so that by birth the vertebral column ends in a little forwardly curving hook, the *coccyx*, which is completely embedded in the lower end of the trunk. But sometimes development is not quite normal. Not uncommonly the medical student, dissecting a human body, finds vestigial and perfectly useless muscles passing to the coccyx. And very occasionally a child is born into the world with a tail—an appendage which is little better than a soft piece of string, but which nevertheless contains the tapering end of the vertebral column.

On looking at the throat of an early embryo we find another curious point. At the sides of the throat there are four pairs of clefts—clefts which quite obviously correspond to the gill-slits of fishes (Fig. 63). This correspondence may be traced, not only in the position of the clefts but also in their skeletal supports—the gill-arches—and in the blood-vessels of that region. As development proceeds this early arrangement is modified beyond recognition. The gill-arches are converted into the ear-ossicles, into the hyoid bone (a bone lying in the throat, to which the muscles of the tongue are inserted), and into the cartilages of the larynx; the arteries are reduced in number, only a few of them persisting as part of the great vessels in the thorax; the clefts are filled up and disappear, except for the first, which gives rise to the Eustachian tube and middle ear. Nevertheless the fact is unmistakable; instead of developing a human throat simply and directly, each of us during our embryonic life lays down a gill-apparatus like that of a fish, and then modifies it and twists it about until it becomes a human throat. It should be remembered that as far as the embryo

THE WEARING OUT OF THE MACHINE

is concerned the gill-apparatus is never used for breathing, and indeed never develops the actual respiratory filaments that are found in the gills of fish.

Further, on dissecting such an embryo, we should find other conditions that resemble those seen in the lower vertebrates. In the arrangement of its principal veins the embryo is distinctly fish-like. It has a simple heart with only one auricle and one ventricle like the

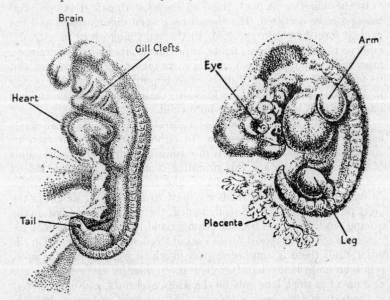

FIG. 63. Human embryo, about three weeks old.

FIG. 64. Human embryo, about four weeks old.

heart of a fish, and in the fact that the urinary, genital, and rectal passages all open into a common pouch, the *cloaca*, which in its turn opens to the exterior, it differs from adult mammals. These points, again, are altered and corrected as development proceeds. Finally, the early embryo has a skeletal structure running down the back—the *notochord*—which is destined to disappear and be replaced by vertebrae. We shall deal with the interpretation of these facts elsewhere; for the present it is enough to note this fish-like stage in our development.

During the second month of development the embryo grows and differentiates further, and straightens out this fishiness through which it has ascended. By the end of this period it is already recognizably human. An eight weeks' embryo is about an inch long; it has already got fingers and toes, and its tail is reduced to a hardly noticeable pimple. There is still a long way to go—the cerebral hemispheres, for example, are very poorly developed, and the face is grotesque and almost horrible—but the general form is laid down. The embryo is beginning to look like a baby mammal.

In the first two months, then, all the most drastic developmental changes have occurred. The tissues have been differentiated and the general form has been assumed. In the remaining seven months the embryo undergoes various minor adjustments; its limbs, for example, grow relatively rapidly and acquire more reasonable proportions, and its face becomes human. But the most important process that occurs in this long period is simple growth—the expansion of an embryo barely an inch long into a new-born child.

It is curious to compare the late human embryo with that of an ape—a chimpanzee, say, or a gorilla. In either case the seven months' embryo has the soles of its feet turned towards each other like the palms of its hands. In the ape this primitive condition is retained; indeed the foot develops very like a hand, and is used for grasping things. In ourselves the foot becomes modified to support our weight in the erect position; the legs are straightened; the sole is planted flat on the ground, and the bones become stronger and less mobile. The foot becomes an efficient support at the cost of freedom of movement. On the other hand there is another respect in which we, and not the apes, retain an embryonic character. The seven months' embryo of an ape or a man has thick hair only on the scalp, eyebrows, and lips; the rest of the body is nude, except for scattered fine hairs or *lanugo*. In both, the embryonic lanugo is succeeded by a general outgrowth of fully developed hair, but in man this outgrowth is much slighter than in an ape. Man's naked skin is less developed than the skin of an ape; it is a case of arrested development.

Before we continue the story, and trace the developmental changes that occur after birth, we must take note of the way in which the embryo is housed and nourished.

We have already noted that the developing germ embeds itself in the wall of the uterus as soon as it reaches that chamber. At that stage it consists of a hollow vesicle, the *chorion*, which contains a little knob where the embryo and certain enveloping membranes will develop. As the embryo grows this chorionic vesicle grows too, so that at a later stage (Fig. 65) it bulges out into the cavity of the uterus, being coated with a thin layer of uterine tissue. The embryo grows and develops, remaining attached to the wall of the chorion by a stalk, the *umbilical cord*, which sprouts out of its belly. Furthermore, the embryo develops a circulatory system very early, and in addition to the heart and vessels in its own body it has arteries and veins running along the

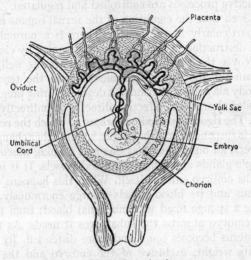

FIG. 65. The embryo in the uterus.

The blood-vessels of the embryo are shown black; those of the mother by close shading.

umbilical cord to the chorion, where there is an extensive network of capillary blood-vessels. Now at the boundary between the chorion, which is composed of embryonic tissue, and the wall of the uterus, which is maternal tissue, the embryonic and maternal bloods come into very close contact, without actually mixing. They are separated only by very thin membranes of tissue. In this region, the *placenta*, there is a continual interchange between the embryonic and maternal bloods; the embryo receives food and oxygen from the maternal stream, and gives up its waste products to it.

It will be seen that the embryo is nourished by a circulatory system of its own, and that the placenta acts as lung, intestine, and kidney combined. This is why the heart is the first of our organs to develop and begin working. From first to last, from the first month to the eightieth or ninetieth year, the incessant heart keeps the body going.

§ 4. Rhythm and Birth.

AT THIS point it will be best to digress for a moment from our account of an individual life-history and to note something of the way in which the reproductive processes are controlled and regulated.

We have seen that the condition of the sexual organs is constantly changing, particularly in the female. There is a normal rhythm of growth and destruction in the uterus, accompanied by changes in the ovary; moreover the rhythm affects other parts as well, for at the periodic clean-out the pulse-rate quickens and the temperature of the whole body may rise a fraction of a degree. During pregnancy there are profounder alterations, affecting directly or indirectly every part of the body. The rigorous periodic overhaul to which the resting uterus is subjected is suspended and that organ suffers other changes. Its wall has two layers, an outer muscular one and a lining mucous one containing simple glands and numerous blood-vessels. It is in this inner layer that the embryo embeds itself. When this happens the mucous layer thickens and its blood-vessels enlarge enormously, so that it becomes like a sponge filled with maternal blood; from this copious supply the embryo absorbs the substances it needs. As the embryo grows the uterus becomes more and more distended. By the end of pregnancy its weight, exclusive of the embryo and the embryonic membranes, is about three hundred times as great as normal. This swelling up causes certain alterations in the disposition of the abdominal organs and may result in disturbances of digestion or the circulation. Besides the changes in the uterus, and the disturbances indirectly produced, pregnancy involves special modification in other organs; there is, for example, the preparation for suckling.

None of these adjustments are due to the action of the nervous system. The nervous system is primarily concerned with rapid responses to one's surroundings, and its part in reproduction is practically confined to the very elementary series of reflexes that bring the sexes together, and to certain other reflexes that occur during childbirth. The slower processes to which we have just referred are produced in another way; they are regulated by internal secretion.

We have already shown how the activity of our organs can be modified as a result of alteration in chemical composition of the blood that bathes them, and we saw that there are special bodies, such as the adrenal gland, which shed substances into the blood-stream for the express purpose of affecting and controlling other tissues. In the regulation of the sexual functions similar processes play the leading part. Our knowledge of this regulation is developing rapidly at the time of writing, and like all rapidly developing subjects it is still a little chaotic. There are brilliant lights and alluring obscurities. It is clear that the monthly rhythm is due to internal secretions produced in the ovary, which control the condition of the uterus, the breast, and so on; moreover, it is at least possible that the rhythm originates primarily not in the ovary at all, but in an anatomically unrelated structure lying at the base of the brain, the *pituitary gland*. This gland apparently produces an internal secretion that varies in amount with a monthly rhythm, and this secretion seems to affect the ovaries, modifying their activity, which, in turn, affects the uterus and breast.

It may be that the mechanism of the sexual rhythm is even more complicated than this. The thyroid gland, for example, which is another organ of internal secretion, often swells slightly during these phases and this may indicate active participation in the cycle. But these are undecided questions. We have said enough to show that the normal sexual rhythm is a very complex thing, depending on minute but regular changes in the composition of the blood. Let us turn to the suspension of this rhythm that follows fertilization, for in the phenomena of pregnancy the part played by internal secretion can be traced more clearly. Every time an ovum is shed from the ovary a special and temporary ductless gland is developed in that organ to see the ovum through.

We noted that in the ovary the ovum is tended and nourished by an embracing capsule of cells, the follicle. When the ovum escapes the follicle cells are relieved of this duty, but they do not lose their devotion to their charge; they multiply and alter somewhat in shape, forming a mass that fills the follicle and is called from its colour the "yellow body" (*corpus luteum*); and prepare to serve the ovum in another way. If the ovum is not fertilized the mass soon breaks down and disappears, but if fertilization occurs it remains and acts as an organ of internal secretion. It manufactures and sheds into the blood substances that are carried all round the body by the circulation and work various changes. The substances stimulate the inner layer of the uterus and prepare it for the reception and nourishment of the embryo; they stimulate the secretion of milk; they soften the junction between the two halves of the hip girdle so that birth may be easy.

The corpus luteum is not the only organ that manifests special activity during pregnancy. It is probable that internal secretions are produced from other parts of the ovary—perhaps even from the placenta—which assist that of the corpus luteum in making the necessary modifications in the female organism. But the secretion of corpus luteum is the most important.

One of the most interesting points about the internal secretions produced during pregnancy is that, like the hormones controlling sex-characters, they can directly influence the working of the brain. Shortly before birth occurs, a bitch prepares a special bed for her whelps and a doe rabbit plucks fur from her own breast to provide a soft, warm nest for her young. It has been proved that these unusual acts are directly dependent upon the presence in the blood of special internal secretions produced by the ovaries. This is one of the clearest examples of the determining influence that the chemical composition of the blood may have on mental processes. We spoke of the mind as the government of the body, but it should be remembered that the government is by no means despotic. Just as human governments may be swayed and controlled by small, active sections of the population, so here the brain is controlled by one of the abdominal organs. Further, it should be remembered that this influence is by no means restricted to pregnant individuals. The non-pregnant female experiences cyclical changes in her mental life, changes in her emotions, her desires, her intellectual liveliness, that synchronize with the rhythm in her reproductive organs. Moreover, it would seem that in animals with a periodic rut the males also have a rhythm that is caused by internal secretions. Even in some men there is said to be a monthly tide of sexual feeling. It is clear enough that the colour and efficiency—the tone, so to speak—of our minds may depend to a very considerable degree on the chemical substances poured into the blood by our organs of internal secretion, including our reproductive organs, and thus indirectly on the condition of those organs themselves.

The corpus luteum acts as a chemical fairy godmother for the unborn child, taking all sorts of steps which ensure its being properly cared for both before and after birth. In particular we may note that the uterus is not a naturally hospitable organ; its monthly habit is to clear itself out; it receives and nourishes the embryo because the corpus luteum instructs it to, by means of its chemical messengers. As pregnancy advances the uterus begins to move and contract restlessly. Its muscular layer is affected by internal secretions from two sources. The substances made by the corpus luteum dispose it to keep quiet and nurse the embryo; but there is an antagonistic substance, made by the monthly periodic activity of the pituitary gland, which stimu-

THE WEARING OUT OF THE MACHINE 159

lates the muscles of the uterus and urges it to contract. For nine months the substances of the former kind prevail, but at the end of this time the corpus luteum begins to degenerate and ultimately it ceases to turn its products into the blood. The uterus, relieved of this controlling influence, revolts at once, and contracts more and more forcibly, until its now unwelcome guest, the baby, is ejected. First the baby departs, and then the placenta and the investing membranes—the *after-birth*—are thrown out. In this process the uterus is not unassisted, for the diaphragm and the muscles of the abdominal wall participate heartily. It is as if the whole abdomen had been chafing at its burden, and as soon as the fairy godmother dies, rises to expel her protégé.

In this fashion the baby finds itself flung into a new and larger world. As soon as this happens the muscles that move the tiny ribs begin to work, air rushes for the first time into the lungs, and the vocal cords are stirred, giving out the awkward, tentative cry with which the newcomer heralds his own appearance.

§ 5. Childhood, Adolescence, and Maturity.

AT BIRTH the whole economy of the child is suddenly changed. It has now to breathe air and collect its own oxygen, and it has to digest its own food. The lungs and bowels begin to work. As a matter of fact the strain on the digestive organs is not very severe, for the baby receives in its mother's milk a diet that requires hardly any chemical adjustment. Moreover, since it is no longer housed in a warm incubator, it has to regulate its own body-temperature.

Nevertheless, as far as development is concerned, birth is an incidental event that hardly affects a slow, continuous process. During the first few weeks of uterine life all the organs of the embryo are laid down in a more or less rudimentary way, and thereafter development consists in the elaboration and perfection of those rudiments. This is a process that goes on for some twenty-five years. Birth is an event for which the organs of the mother, not the baby, are responsible, and it does not correspond to any definite developmental change in the organism of the latter. As soon as the baby has adjusted itself to its new surroundings the processes of growth and differentiation go on in very much the same way as they did before.

One of the most striking changes that occurs during the first years of childhood is the mushroom-like expansion of the brain. By the end of the second year the child's brain has attained more than half the adult size; by the end of the fourth year over eighty per cent of its final bulk is already present. The adult cranium is a strong bony box that protects the brain, but at the same time prevents any further expansion of its precious contents; the brain has therefore to attain its full size early in life while the skull-bones are still only partly formed and before they are solidly joined together. At the same time the human child needs a large brain, for it has to learn an enormous amount during the first few years of life.

As the brain expands there is a similar growth of the legs, accompanied by changes occurring elsewhere in the body, and associated with the assumption of the upright posture. The most important of these changes is the development of the nerve-tracts leading from the brain to the limb-muscles, and enabling the latter to be accurately controlled by the former. At the same time the legs grow longer and stronger, and the lumbar region of the vertebral column—the loins—elongates and consolidates to support the weight of the upper parts. "Learning to walk" is not a process of real learning at all; it follows gradually but automatically with the perfecting of these structures as a reflex follows automatically upon the construction of the sense-organ, nerve-connections, and muscles laid down before birth.

During the first two years or so there is another important process— the eruption or appearance of the milk teeth, the first of the two sets with which we are provided. The second set begins to replace the first at about the fifth year, and it is not complete until the wisdom teeth have erupted in about the twenty-second year.

These are only the most obvious changes that occur during childhood. Throughout the body, in every shred of tissue, development is going on. Moreover, this development continues until the parts actually wear out, for there is no wholly stationary period in the human life-cycle; the body is always either developing or decaying.

It is not possible to name a definite prime, or climax, as it were, to human life, because our different tissues perfect themselves and wear out at different rates. The elastic and muscular coats of our arteries are at their best at about twenty-five; this is when the professional athlete, who has to make sudden physical spurts, reaches his highest point. The brain has a longer life; it attains its most accurate control of the muscles between thirty and forty, and as an organ of thought it is probably at its best between forty and fifty. In this there is a striking disharmony, for the circulatory system may stiffen and fail before the mental powers begin to fall off. The professional athlete has to retire and be forgotten long before his death, but the professional thinker generally has the privilege of dying in harness.

Perhaps the earliest sign of decay in the human body is the appearance of lines on the face—proof that the skin is already losing its elasticity.

Decay, then, is soon upon us; and this brings us again to the need for the revolution that happens in the early teens, the acquisition of the ability to reproduce, and the beginning of the insistent clamour that our reproductive organs set up so that we may be replaced.

In describing the ovary and testis we noted that in both glands, besides the tissue directly concerned with the manufacture of gametes, there is tissue of another kind; there are the *interstitial cells*. These interstitial cells, like the corpus luteum, are organs of internal secretion. They begin to work very early, long before any germ-cells are developed, and their secretions, carried round the body by the blood, determine most of the differences that distinguish man from woman.

The reproductive organs of the male and the female are built upon a common type. To take an obvious example, the male has rudimentary nipples although he will never suckle his young. It is possible to go through the whole system, part by part, and show how male and female correspond. Now the very early embryo has a generalized reproductive system, one that is neither discernibly male nor female but containing rudiments capable of developing into either the male or female form. The first sign of definite sexuality manifests itself in the testis or ovary, as the case may be, and it is the appearance of working interstitial tissue that sheds either male or female internal secretions into the blood. These substances affect the body and, by enlarging some of the parts and inhibiting the growth of others, make them develop into the characteristic type of one sex or the other. The female pelvis is broader than the male, so that it may support the pregnant uterus and allow the head of the baby to pass out. As early as the fifth month of embryonic development it is possible to make out a distinction in shape between the male and female pelvis, and this also is due to internal secretions from the interstitial cells.

After these early changes there is a long period of comparative rest throughout childhood. Sexual differentiation, although it is continuing, does not proceed as violently as it did before. But at puberty—at about fifteen years in the male and fourteen years in the female in temperate climates, and about two years earlier nearer the equator—the gametes begin to ripen, and with this acquisition of the reproductive faculty there is a violent burst of activity on the part of the interstitial cells. At this time the final distinctions between the sexes begin to be laid down. In the female, the pelvis broadens further, a layer of fat is deposited all over the body that gives it a rounded, graceful contour, and the distinctive feminine figure is accentuated. In the male, there is a growth of hair on the face and the larynx enlarges, causing the "breaking" and deepening of the voice—processes which are not completed until about the twenty-fifth year. In both sexes there are correlated psychical changes: the sexual instincts attain full power and the mind deepens and changes its emotional tone.

Somewhere in the early twenties these adjustments are completed and the body is fully mature. For a number of years reproductive capacity is at a maximum, and then it begins to decline. At about forty-five years, in the female, the monthly crisis ceases suddenly; in the male the ability to produce spermatozoa diminishes slowly. By this time the cycle is completed; the individual is wearing out, and new bodies have been made to replace it.

The difference between the internal secretions from the testes or ovaries determine the other differences of the sexes. In this respect sexual development is not unique, for the growth and differentiation of the majority of our parts are regulated, in the later stages at least, by internal secretions from such ductless glands as the thyroid and pituitary.

The study of internal secretion is one of the youngest and, at the present time, one of the most fruitful branches of physiological research. We may well close our account of the body by noting some of the most important results of this work, taking first the two glands that play the greatest part in regulating growth.

The *thyroid gland* lies in the neck. It is a roughly U-shaped body, brownish-red in colour, with two thickened lobes lying one on each side of the larynx and a bridge connecting the lower ends of the two lobes and crossing in front of the windpipe. It contains a number of minute, completely closed vesicles that are full of a yellow fluid. The essential facts about the thyroid have been known for a longer time than about any other ductless gland, owing to the fortunate accident that its secretions pass unchanged through the digestive system into the blood, so that patients can be treated by simply feeding them with dried gland substance. And our knowledge of the thyroid has recently taken an important step forward, because it is one of the two ductless glands whose active principles have been isolated and analyzed, and even successfully synthesized in the laboratory from simpler materials. The other is the central part of the adrenal gland, of which we spoke in § 2 of Chapter Three. The products in question are called *thyroxin* and *adrenin* respectively. The majority of the internal secretions have not yet been isolated, their presence being inferred from the effects they produce.

The secretions of the thyroid affect profoundly a great number of different physiological processes. One milligram of thyroxin injected into a normal man will cause his chemical activity, as measured by oxygen consumption and carbon dioxide production, to go up by two per cent, although the amount injected is only about one seventy-millionth part of the man's weight. Heavy doses of thyroid speed up the vital engines to such an extent that unless enormous quantities of food are provided there is a breakdown of the substance of the tissues themselves in an effort to get more fuel to burn, and the animal or man wastes away. The substance can also affect temperament; heavy doses of thyroxin make a perfectly normal man nervous and irritable.

It is clear that a gland like the thyroid may have very marked effects upon all the bodily processes, including growth, if its secretion is for long periods of time a little above or a little below normal. A disease known as *myxoedema* is due to deficiency of the secretion in the adult. It is characterized by a sluggishness of the various chemical and physical reactions that go on in the body; the pulse is slowed, the temperature is low, the appetite is blunted, speech becomes a drawl, the mind becomes viscous. At the same time there is a thickening of the connective tissues under the skin, producing a puffiness and yellowness of the face and hands and falling of the hair and a considerable accumulation of fat.

In children with defective thyroids similar phenomena appear, but growth, particularly of the brain and skeleton, also ceases. A child so affected may live for many years, but even when thirty years old it retains a distorted childish appearance, and has scarcely the intelligence of a child of four or five. The condition is known as *cretinism*. In most cases cretinism and myxoedema can be cured by administering a regular dose of thyroid extract with the food. It is probable that slight thyroid deficiency is very common, for in many cases administering thyroid to people of middle age has a beneficial effect, reducing corpulence and promoting the growth of hair.

Effects produced by an over-active thyroid are seen in the disease known as *exophthalmic goitre*. There is quickening of the pulse, and even palpitations, there is also wasting, and a curious bulging of the eyes. We do not therefore recommend our readers to try the effect of thyroid extract without proper medical supervision.

The thyroid is not essential to life. Life and its oxidations continue even when the whole thyroid is removed, but they continue very sluggishly; the rate of oxidation falls to about sixty per cent of its normal value, and the temperature tends to be subnormal. The thyroid hormone acts like a draught to the fire of life. The draught is necessary to get a bright fire; but without it, the fire will still smoulder on. From another point of view, the thyroid is Nature's device for drugging the higher animals up to a pitch of activity not found in lower forms. The proper growth of our body, the full activity of our mind, is only possible when the thyroid is pouring its little daily dose into our system.

The *pituitary gland* is a small, reddish-grey, egg-shaped body, about half an inch across, and projecting from the middle of the lower surface

Fig. 66. Three drawings of the same woman showing how over-activity of the front part of the pituitary gland (see Fig. 36) may gradually affect the physique.

Left to right, at twenty-five, twenty-six, and forty-two years of age. (Courtesy of Sir A. C. Geddes and "Journal of Anatomy and Physiology.")

of the brain, with the substance of which it is continuous, into a little cup in the base of the skull. It is made up of two distinct halves, each of which produces its own secretion and exerts its own influence on the working of the body.

The hinder half or post-pituitary is mainly concerned with the body's chemistry and chemical supply. Its secretion excites and augments the movements of the bowels, keeps the blood-vessels toned up, and stimulates the kidneys to do their work. In addition if it is deficient, the utilization of sugar by the body is disturbed. A high proportion of the carbohydrate consumed is turned into unnecessary fat instead of being used as fuel; the result is great obesity and an exaggerated appetite, especially for sweet things.

The front half or pre-pituitary, on the other hand, is mainly concerned with growth and development. Over-activity of the pre-pituitary (such as occurs when a tumour grows in the gland), if it occurs before the bones have "set" and finished their growth in length, results in an overgrowth of the whole body, especially in height, which, however, is far more marked in the long bones of the limbs than in the trunk. Most of the giants of fairs and circuses owe their peculiarity to this cause, while the fat lady in the next booth gains her living through a deficiency of the other part of the same minute gland. If over-activity of the gland does not begin until full stature has been reached, the bones cannot grow in length, and the excess growth takes place mainly in the hands, feet and part of the skull, giving rise to a disease called *acromegaly*. Not only the bones, but the tissues over them enlarge, so that the unfortunate patient gradually acquires a coarse, heavy expression, as well as having to discard his or her gloves and boots from time to time for larger sizes. The bones of the fingers usually develop abnormal bony outgrowths. The pre-pituitary affects not only growth, but also stimulates the development of the reproductive organs.

When the pre-pituitary is deficient, children do not grow up properly; they remain undersized, though of perfect proportions, and they fail to mature sexually, giving one of the most typical pictures of what is called infantilism.

The pituitary also helps to regulate temperature. Human beings with deficiency of the pituitary as a whole are generally sub-normal in temperature, and in addition often somnolent to excess. Add to this the tendency to lay on fat and the enormous appetite, and we realize what an excellent picture of an under-pituitaried individual Dickens gave us in the fat boy in Pickwick.

These effects of pituitary and thyroid have been turned to account by Nature. To prepare hedgehogs and dormice and other hibernating mammals for their winter sleep, thyroid and pituitary become damped down. Fat is put on, the creatures grow sluggish, cold, and drowsy, and finally succumb to sleep. Doubtless other glands are at work, too, but the share of the thyroid is prettily demonstrated by the fact that a single thyroid injection will completely waken a hedgehog from his winter sleep.

Half a century ago, nothing was known of the functions of these ductless glands. They were merely puzzles for the anatomist. To-day we know that, including the thyroid and the pituitary body, there are four, perhaps five different glands in our bodies entirely concerned with this business of internal secretion, and two of these five, the adrenal and the pituitary, have a double function and are formed of two distinct parts, each producing one or more quite different hormones.

Besides these definite anatomical structures we have named there are several organs which have other functions as well, but which also contain masses of internally secreting cells. Altogether we know now of more than a dozen separate hormones each with its own special task within the body.

Of these hormones, two—secretin and adrenin—co-operate with the nervous system in bringing about rapid variations in the state of our organs in adjustment to the fluctuating conditions to which we are exposed. We have already described this action in the second chapter of this book.

Other internal secretions have exclusively chemical functions. The pancreas, for example, has islets of endocrine cells scattered among its digestive tubules, and their secretion, *insulin*, has a very important function in controlling the handling of sugar by our tissues. The *parathyroid glands* are small bodies lying at the sides of the thyroid, and until a few decades ago were regarded as mere appendages of that organ. Now we know that they produce a secretion which regulates the amount of lime in the blood. When this is insufficient, defects appear in bones and teeth; when it is gravely deficient, the patient has an attack of muscle-spasms or tetany. In the last few years a way has been found to prepare extracts of the parathyroids, whose injection will relieve the symptoms of this disorder.

The endocrine functions of various other organs of the body have still to be explored. The *thymus*, for example, is a large, pink gland lying over the heart in very young children but dwindling and disappearing before the age of discretion is reached. Presumably it has an influence on early development.

It should not be thought that the different ductless glands exert their functions in isolation and independently of each other. Often a number of glands affect the same process in various ways—in the case of sugar metabolism, for example, the thyroid stimulates oxidation of sugar, the post-pituitary favours the conversion of sugar to fat, the adrenal medulla brings about liberation of sugar from the liver, and the pancreas controls the taking-up of sugar by the tissues from the blood—and thus the various secretions may interfere, antagonizing or reinforcing each other's activities. Again, the various ductless glands may influence each other directly. There is an intimate interrelation between the adrenals and the thyroid, for feeding dogs on adrenal cortex causes their thyroids to store more thyroxin. Similarly the endocrine parts of the ovary and testis affect other ductless glands, for castration and pregnancy are accompanied by various changes in

THE WEARING OUT OF THE MACHINE

various glands; and X-raying the testes of some mammals, which causes overgrowth of the internally secreting tissue, produces an apparently degenerative decay in the pituitary. Like other tissues, the ductless glands are sensitive to changes in the composition of the blood, and thus, in a variety of ways, they can interact.

Manifestly our present knowledge of internal secretion is still very incomplete, and a good deal of premature speculation—indeed, of nonsense—has been written about it. Nevertheless it is clear that the actions of these glands are of the profoundest physiological importance and the unravelling and mapping of this tangled skein of chemical actions—a process comparable to the mapping of fibres in the nervous system—is one of the most fascinating problems that the physiologist of the next few decades will have to tackle.

So we close this succinct review of current knowledge in this field. Beneath the outer appearances of life this interplay of secretions and counter-secretions goes on, like the authors, producers, prompters and so forth of a dramatic piece. They do not challenge our attention. The play's the thing. The play is a Cat or a Fish or a Mouse or an Elephant according to circumstances. The play they put on the stage when the field of action is *Homo sapiens*, has been called by an eminent observer of the species the Seven Ages of Man. Pituitary, pineal gland, thyroid, the interstitial cells of testis and ovary, all contribute to the drama, whose first act is " the infant mewling and puking in its nurse's arms " and whose last, the "lean and slipper'd pantaloon," belatedly in search of Voronoff, Steinach, Norman Haire or whatever other hope may offer.

BOOK TWO

THE CHIEF PATTERNS OF LIFE

I

THE FIRST GREAT PHYLUM: VERTEBRATES

§ 1. *Classification.* § 2. *What Is Meant by a "Phylum."* § 3. *The Classes of the Vertebrate Phylum.* § 3a. *Mammals.* § 3b. *Birds.* § 3c. *The More Ancient Class of Reptiles.* § 3d. *The Linking Amphibians.* § 3e. *Fishes.* § 3f. *Cyclostomes, a Class of Degraded Antiques.* § 3g. *Semi-Vertebrates.*

§ 1. Classification.

WE HAVE now given an outline of existing knowledge about the structure and working of that most familiar form of living thing, a human being. From the material point of view it is a mechanism, but it has this remarkable difference from all manufactured machines:—it can make other mechanisms in its own likeness. We have shown how wonderfully it is built up of billions of subordinate individuals, its cells, held together in co-operation by the most complex and marvellous means, and made to behave as one whole through the agency of the nervous system. We have considered the remarkable subtlety and also the extraordinary limitations and vulnerability of this living body. In the world there are now perhaps a thousand million of these complex mechanisms going about their business and constituting the species Man, *Homo sapiens*.

In relations of conflict or tolerance with this species there are a very great number of other species, more or less numerous and more or less similar, beasts, birds, and fishes. We have already given some transitory glances at one of these species, *Mus musculus*, the house mouse, of which the terrestrial living population must far outnumber humanity. For our present purpose we may define a species as a kind of living thing which reproduces itself but does not interbreed normally with other kinds of living things, even though they may resemble it very closely. The sort keeps itself to itself.

Since the great days of Greek intellectual activity, men have set themselves to grasp and get in order in their minds the realities of the

THE FIRST GREAT PHYLUM: VERTEBRATES

spectacle of life, and one part of that attempt has been the endeavour to put all its varieties into an orderly classification. Of that effort we have already had a word to say in the introduction. From the outset it was observed that many species fell into groups. There were, for example, the house mouse, the harvest mouse, the wood mouse and the black and the brown rats, all kinds keeping themselves apart as a rule, and following distinctive ways of life. Dog, wolf, jackal and fox fall

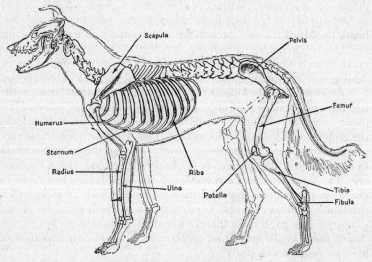

FIG. 67. The Skeleton of a Dog.

If this is compared with the human skeleton in Fig. 4 it will be seen that the two are built on the same fundamental pattern.

into a similar group, and cat, lynx, panther, leopard and tiger into another. Such obviously similar groups of species were early assembled together as genera (singular, *genus*) under a common name. Linnæus in the eighteenth century introduced the modern practice of defining species by using a special terminology which was more exact and explicit than the often very loosely used common name. The name of the genus was first given and then the precise specific name. And instead of using homely words for genus and species it was more convenient to use a rather debased mixture (grammatically) of Latin and Greek, because this prevents confusion arising out of the loose use of popular words. Thus the house mouse is *Mus musculus*, the harvest mouse is *Mus minutus*, the lion is *Felis leo*, the leopard is *Felis pardus*.

And to be uniform, man is called *Homo sapiens*, although no other species of Homo now exists. But once upon a time there were other species of *Homo*. Occasionally the bones and other traces of a long-banished race come to light, heavily built, with differences in the jawbone, neck, legs, and skull so great as to constitute a different species, *Homo neanderthalensis*.

There is no exact definition of what a genus is, and there are wide differences of opinion and practice about genera. Some types of naturalist seem to like to file their facts in big groups; others distinguish and break up. Much toil and interchange of opinion are necessary to secure anything approaching a uniform practice. Even species sometimes break bounds, and there may be differences of opinion, whether we are dealing with distinct species or what are called varieties of one species.

With as much exactitude as possible naturalists marshal genera into families and families into orders, orders into classes and classes into phyla (singular, phylum). Finally there are the two great kingdoms, the Animal and Vegetable Kingdoms. This makes a good framework in which to sort out the immense variety of living things we have now to consider. But the scheme, be it clearly understood, is not an exact one. The facts are often awkward and the terminology has to be as elastic as possible, so that intermediate grades, such as super-order, sub-family, sub-species, and so on, may be interpolated in the series.

We will now go through this classification, and consider the extraordinary range in the types and patterns of living mechanisms that we have set ourselves to study.

§ 2. What Is Meant by a "Phylum."

THE greatest divisions (after the two kingdoms have been separated) used by the modern naturalist are called *phyla*. A phylum assembles all the creatures which possess one common ground-plan. Dog and man, for example, have a common ground-plan; we need only repeat the words skull, backbone, ribs, liver, spleen, forelimb, hindlimb, to illustrate this point. Turn to a fly or starfish and none of these fundamentals in the lay-out of man and dog are to be found. Fly and starfish, like man and dog, are built up of innumerable cells and are bathed internally in a sort of blood and held together for united ends by a nervous system, but all the arrangement of the parts is different. They belong to different phyla.

As the reader knows, man and dog are classified with a multitude of other creatures as *Vertebrata*. They are the beasts, birds, fishes, and

THE FIRST GREAT PHYLUM: VERTEBRATES 171

reptiles; they are the most obviously important living things to us. Let us consider what are the characteristics that bring all these Vertebrata into one phylum. That we may do most illuminatingly by comparing the anatomy of a fish with that of a mammal. We will take the dogfish and ourselves as convenient types.

The dogfish is a miniature shark who prowls round our coasts in search of small crabs and worms—indeed, anything that he can devour—and who appears under a variety of names on the counters of our fried-fish shops. His general appearance is shown in the accompanying

Fig. 68. The Rough Dogfish, another representative of the vertebrate pattern.

(*From the Plymouth Aquarium Guide Book, by courtesy of the Marine Biological Association of the United Kingdom.*)

drawing. Like an adult mammal, he has a head with eyes and mouth and nostrils that lead into an olfactory cavity; like an embryo mammal, he has a series of gill-slits on the sides of his throat. The front gill-slit, or *spiracle*, we may note in passing, is different in structure and position from the rest and different in a very amusing and interesting way. It is small and round and placed just behind the eye, and although it is not used for hearing, it runs very close to the inner ear of the dogfish and then opens into his mouth. In its position, and in the particular nerves and arteries that lie against it, it corresponds to the channel made by our outer and middle ears and Eustachian tubes (which run from our ears to the backs of our noses), and this idea of the correspondence of the two structures is confirmed by the fact that in a mammalian embryo the front gill-slit actually becomes separated from the rest and develops directly into that channel. Biologists call this parallelism *homology;* the two structures are "homologous." And in creatures of the same phylum, but of different orders or classes, we shall be constantly detecting homologies.

The fins of a dogfish are of two kinds—the unpaired fins, including the fins of his back and tail and the hindmost of the fins on his lower surface, and the paired fins, of which there are two pairs. The unpaired fins do not correspond in structure or position to anything that is found in a mammal, but the paired fins are homologous with our limbs; they resemble our limbs in their position and in the fact that they are supported by skeletal girdles inside the body.

Many of the differences in external appearance between a dogfish and a mammal may be directly correlated with the fact that the former lives in the sea while the latter lives on land. The body of the dogfish is beautifully stream-lined, so that he may glide swiftly through the water. His tail is a powerful organ of propulsion. His paired fins do not project far from the trunk, for they would offer too much resistance to the water if they did so; they are used as balancing vanes, partly like the ailerons of an aeroplane, to keep the fish the right way up, and partly like elevating-planes to steer it upwards or downwards through the water. A mammal, on the other hand, living in a medium with less resistance than water, has no need of accurate stream-lining, and can afford the luxury of a neck; at the same time he cannot propel himself by wagging his tail, and so his limbs must form long supporting and propulsive levers. The difference in respiratory mechanism has a similar rationale. A dogfish breathes by gulping a current of water into his mouth and out through the gill-slits. As the water passes over his gills it yields oxygen to the blood in them and washes away the carbon dioxide. A mammal could not do this; if he passed a continual current of air over his respiratory membranes they would dry up, and moisture plays an essential part in the exchange of gases. He breathes by drawing air into those hollow bags, his lungs, and blowing it out again. In the dogfish there are no lungs; we do not even find a homologue here, though many other fishes have it in the form of the swimming bladder. We may note that a dogfish could not inhale and exhale water as the mammal does air, because the inertia of water is too great. Running along the side of the dogfish's body is the *lateral line*, seen as a white streak in the drawing. This is a sense-organ, probably recording changes in pressure; owing to its structure it would not work in air, and we do not find it in mammals.

So far as outward and visible structure is concerned, a dogfish is like a mammal, but twisted about and modified because of the necessities of an aquatic existence. Or we may put it round the other way and say the mammal is after the dogfish pattern but adapted to the air. If we turn to internal anatomy we find that the correspondence of lay-out

THE FIRST GREAT PHYLUM: VERTEBRATES

still holds. The dogfish, like the mammal, has a skull, and a backbone made of jointed vertebræ; the former houses the brain and the latter has a canal running through it which contains and protects the spinal cord. He has a heart in a position corresponding to our hearts, but as he has no lungs he has no double circulation; the dogfish heart, like the heart of a human embryo, is a single pump and drives all the blood forward via the gills to the rest of the body. He has a stomach lying rather to the left of his abdominal cavity and kidneys at the back; he has a liver and portal veins, a spleen, thyroid, pineal and pituitary glands, cerebrum and cerebellum, testes or ovaries. The homology of his organs can be traced even in minute details, in the courses and branching of his nerves and arteries. Upon point after point the differences between a dogfish and ourselves are modifications of a plan common to both. And in many cases where the homologies of the two adults are not apparent a study of the human embryo shows how the two correspond.

The position might be summarized by saying that if we had a model dogfish made of plasticine we could twist and stretch and bend it bit by bit, increasing the quantity at this point perhaps and diminishing it at that, *but adding nothing essentially new*, until we had a very passable model mammal—a modifying process that is crudely reproduced in the stages of our own embryonic development.

If we examined any other vertebrate—a frog, say, or a tortoise, or a sparrow—we should find a similar agreement in plan. They all possess practically the same set of organs, put together in the same way. On the other hand this does not apply to a great multitude of other living animals. A lobster, for example, has organs that do not correspond at all to ours, and the whole architecture of his body is distinct; the same is true of a starfish, or a tapeworm, or an oyster. A plasticine model of any of these creatures would have to be broken down to nearly nothing before you could begin to worry it into the form of a vertebrate. You might perhaps retain mouth and anus, but that would be almost all.

Such then is the import of "phylum." If we have a thorough knowledge of the anatomy of one particular kind of animal we are able to find our way about any other kind of animal belonging to that same phylum with only moderate difficulty. For example, knowing the plan on which a mammal works, we know roughly what organs to expect in a bird or a fish and what part of the body they are in. But if our knowledge is confined to that one kind of animal we are hopelessly at sea if we try to dissect a member of a different phylum—if we tried, for example, to find organs corresponding to the organs of mammals in the inside of a lobster.

In reply to the question, What are the characteristic features of Vertebrates, as distinct from other animals? a trained zoologist could reel off a string of anatomical points, particularly if he was a teacher or a student about to undergo examination. We shall discuss some of these points later; at present it will be more profitable for us to examine one by one the various classes of Vertebrates and so get an idea of what they are like than to plunge into anatomical details. Nevertheless, there is a simple definition; the Vertebrates are the athletes of the Animal Kingdom. They have better-developed muscles, and better devices for supplying those muscles with oxygen than any other creatures. The mammals on land, the birds in the air, the fish in the water—these include in their ranks the largest, strongest, and swiftest of creatures, and those with the greatest powers of endurance. That is why their one phylum thrusts itself upon our attention in front of all the other phyla of the Animal Kingdom. Theirs is the leading pattern; they include all the high-grade makes in the world of life.

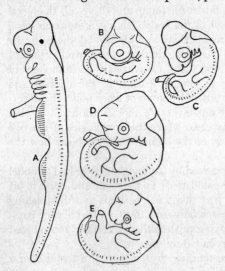

FIG. 69. Early embryos of a dogfish (A), a lizard (B), a chick (C), a rabbit (D), and a man (E) resemble each other very closely; (magnified.)

As development proceeds these creatures arise as divergent modifications of the primitive plan. (From "Vertebrate Zoology," by G. R. de Beer, Sidgwick & Jackson.)

§ 3. The Classes of the Vertebrate Phylum.

AT THE head of any survey of Vertebrate forms must come the mammals and birds. They have dominated this earth for many millions, perhaps for scores of millions of years, and are manifestly the best adapted creatures to the present climatic variety and uncertainties of

THE FIRST GREAT PHYLUM: VERTEBRATES

our planet. But they have not always ruled the terrestrial scene. In the past, stupendous ages passed in which neither bird flew nor mammal ran, and beings less adapted to an active life in a fluctuating temperature prevailed. There was an age when the reptiles ruled, before Nature it would seem had invented the hair or feathers, the double circulation and the blood at a constant temperature, which account for the active presence of the bird and mammal all over the globe and all round the year.

§ 3a. Mammals.

THE class Mammalia includes most of the animals with which we are familiar—ourselves and dogs, cats and horses, pigs and mice. Mammals are warm-blooded animals, a distinction which they share with the birds, and to prevent loss of heat they are more or less completely clothed with hair. This hair, we may note, is one of the distinguishing features of mammals. Moreover they are peculiar in the treatment of their young. For the most part they bring forth their young alive instead of laying eggs after the fashion of most creatures, and they are unique in that their females have special milk-producing glands with which to suckle their offspring. These are the most characteristic points, but there are others. The muscular diaphragm which separates chest from belly is a characteristic feature of the mammals; in other vertebrates the heart and lungs and bowels all writhe and jostle in one large body-cavity. Mammals have better-developed brains, and consequently better minds, than any other creatures. And while the lower jaw of all other vertebrates consists of several bones joined together, the mammals have a lower jaw of one single bone. This fact, which may seem unimportant to the reader, is of very great value in tracing the evolution of the group, because in many extinct species, and par-

FIG. 70. An Ungulate.

ticularly early mammalian species, only the lower jaws have been found, owing to the ease with which lower jaws drop off floating carcasses, to be embedded and preserved in river mud.

Now there are three grades of Mammals, differing in the manner in which they treat their young. The first grade, the *Monotremes*, is a very small one; it includes two curious creatures, the spiny anteater and the duck-bill or water-mole, which show a number of striking resemblances to the reptiles. They stand markedly apart from the rest of the class. They differ from the other members of the class in that they lay eggs instead of bearing young; nevertheless, when the young ones hatch, they are suckled by the mother. In their egg-laying habits they are reptile-like, and they show other reptilian features in their skeletons, in the structure of the brain, and in not being able to keep their temperature as constant as that of other mammals. The duck-bill is found only in Australia and Tasmania; it lives in streams, feeding on insects, snails, small bivalve molluscs and the like, and it makes a burrow in the bank where it lays its eggs. The spiny anteaters, of which there are two genera and several species, are found in Australia, Tasmania and New Guinea; the females carry their eggs in a large pocket in the skin of the belly until they hatch. These creatures are fascinating from the evolutionary point of view; they are survivors of a time when some of the reptiles were turning themselves into mammals by acquiring warm blood and hair.

The remaining mammals fall into two parallel sections, of which one is very much larger than the other. The first and smaller section is the *Marsupials*, or pouched mammals. These animals are intermediate between the egg-laying monotremes and the higher mammals that are familiar to everybody, for they bring forth their young alive, but they bring them forth in a very helpless and undeveloped condition, and then carry them in pouches on their bellies, where they remain glued to the mother's teats until they are able to fend for themselves. They are viviparous, but only just so. An idea of the helplessness of a young marsupial at birth may be gained from the fact that a newly-born kangaroo—of one of the biggest kinds of kangaroo—is about the size of one's little finger. Living marsupials are found only in Australia, New Guinea and the adjacent islands—except for the opossums, which occur in America. The best-known marsupials are the kangaroos and opossums, but there are many other forms, showing a great range of specialized structure and habit. There are wolf-like, cat-like, mole-like, and anteater-like marsupials; and some of them even have parachutes of skin connecting the fore and hind limbs together so that they can

THE FIRST GREAT PHYLUM: VERTEBRATES

plane through the air like a flying squirrel. In fact, the marsupial orders and sub-orders parallel most of the orders of the next section in their diversity of structure and habit.

The third section, the *placental* mammals, includes all the better known mammals alive to-day. They are distinguished from the marsupials by a number of points—their brains, for example, are better developed—but the essential difference lies in the way their young are produced. The placentals are thorough-going mammals; as their name

FIG. 71. An Unguiculate.

implies they possess true placentas such as we have already described in the last chapter of Book I, and for this reason they are able to house and nourish the embryo in the uterus until it is pretty nearly able to run about by itself. The marsupials on the other hand develop no true placenta and so bring forth their young in a state of inconvenient immaturity; hence the need for the pouch. We need not examine the great variety of placentals in any detail (there are fifteen orders living to-day and many extinct ones); we will merely take note of some of the most interesting and important forms and refer the reader to the Zoological Gardens and the Museum of Natural History where their characteristics can be far more profitably studied.

The placental mammals are conveniently divided into four great cohorts based upon their way of life, and particularly upon the uses to which they put their hands, feet and teeth. We may distinguish as a first cohort the great assembly known as *Ungulates* or hoofed mammals, a group of vegetarians having flattish-topped but knobbly or ridged molar teeth to grind up plant tissues; for the most part large beasts using their four limbs exclusively for supporting their bodies and for locomotion. We may note as examples the three largest and most important orders—the *Perissodactyla* (the horses and their close relatives the asses and zebras, the tapirs and the rhinoceroses) which put most of their weight on the middle toe of each foot, the *Artiodactyla* (the pigs and hippopotami and the great cud-chewing assembly of camels, llamas, oxen, sheep, goats, antelopes, deer, and giraffes) which divide their weight equally between the third and fourth toes, and the *Proboscidea* or elephants.

The second cohort is the *Unguiculates* or clawed mammals. To this cohort belong, amongst other orders, the *Rodentia* or gnawing mammals (rabbits, rats, mice, squirrels, porcupines, beavers, jerboas) the *Insectivora* or insect-eating mammals (hedgehogs, shrew-mice, moles), the *Chiroptera* or bats, the *Carnivora* or beasts of prey (cats, dogs, bears, civets, weasels, badgers, hyænas, otters, seals, walruses) and the toothless *Edentata* (sloths, true anteaters, armadillos). They are on the whole smaller than the ungulates, and they are an altogether more varied cohort. Their feet are not so exclusively used for carrying their weight as are those of the hoofed mammals and they may be turned to a variety of other purposes. Many of the clawed mammals can grasp and hold things with their paws (which no ungulate can do) and many of them can climb. In some the claws are strong and the hand is specialized for digging; the mole and the anteater illustrate this. In seals and walruses, which spend a large part of their lives in water, the hands and feet are paddles. And many of the unguiculates can fly. The flying-squirrels and the so-called flying-lemur have parachute-like folds of skin between their arms and legs by means of which they can skim down from the trees on which they live; and the bats, having flying membranes between the enormously elongated fingers of their hands, can flutter and twist and turn as nimbly as any bird. The clawed mammals are also more catholic in

FIG. 72. A Primate.

THE FIRST GREAT PHYLUM: VERTEBRATES 179

their diet than the hoofed mammals; as a general rule (and except for the rodents, which will eat anything) they do not consume leaves and stems as the flat-toothed ungulates do, but feed on other animals or on fruits and the juicier parts of plants.

The third cohort is the *Primates*, and includes the lemurs, the monkeys and apes, and ourselves. It is difficult to separate this cohort clearly from the unguiculates, for the lemurs are in many respects like some of the insectivores and extinct carnivores. The primates may be called the inquisitive mammals; they are characterized first and foremost by their intelligence, and include the animals with the largest brains, the most useful, mobile hands, and the clearest-seeing eyes. The latter point may be associated in some way with another primate character—the presence of a more or less complete bony partition separating the eyeball from the muscles moving the lower jaw.

Moreover (except for one or two lemurs and the marmosets, which are clawed) they are distinguished from all other mammals by having flat nails on their fingers and toes, instead of claws or hoofs. Their hands contrast very strikingly with the feet of ungulates. In the latter group the foot is purely and solely a prop, a single shaft, and there is consequently a tendency to have few toes and to have them all bound together into one column. It is very strong and firm, but its possible uses are severely limited. In the primates the hand is used for grasping and examining things, and here, therefore, there is a full complement of five fingers which can move more or less independently of each other. It is a hand designed for variety and adaptability of movement rather than strength. For the most part the primates live among trees and climb.

The fourth and last cohort of placentals is the *Cetaceans*, the whales, dolphins, and porpoises. They are wholly aquatic mammals. Many other kinds of mammal have taken more or less to water but, except for those strange ungulate-like creatures, the dugong and manatee, none of them have utterly surrendered their land-living habits; even seals come ashore to breed. But the whales never set foot on land. Indeed, they have no feet; their front limbs are small, like the front pair of fins in a fish, and are used as balancing "flippers"; their hind limbs are represented by mere splints of bone, embedded in the body-wall, which are not used at all; their tails have broad, horizontal flukes which are used, like the tail of a fish, to drive the body along. Moreover, the group includes the largest known animals, for the blue whale weighs over a hundred tons. The smallest mammal of all, we may note, is a common insectivore, the pigmy shrew, which is less than a millionth of that weight. Even the smallest Cetaceans, when fully adult, are three feet long. As regards diet there are two kinds of whales. Some, the toothed whales, eat large prey (such as giant cuttle-fish). The common porpoise, which chases and devours fish, is of this kind. The others, the whalebone whales, have a sieve of whalebone round their mouth in the place of teeth and they feed on the tiny creatures that swarm in myriads at the surface of the sea; they take a large mouthful of sea and squirt it out again through their internal moustache of whalebone, and then lick off the masses of tiny creatures that have stuck on the

FIG. 73. A Cetacean.

sieve. Most whales are exclusively marine, but there are one or two species which venture into rivers—the common porpoise, for example, has ascended the Seine to Paris. Moreover, as the only finned mammals, they occupy a unique position in human economics; to the Catholic Church they have always been fish and therefore eatable in Lent. Hence the name of the porpoise which is a corruption of Porkpisce, or pork fish.

All the mammalian cohorts include a great number of orders, families and genera now extinct, and in many cases these extinct forms are interesting because they form links and bridges between groups of which the surviving members are in sharp contrast. But the consideration of these extinct and linking forms we will reserve for a later Book.

§ 3b. Birds.

BIRDS appeared in the world in the same middle period of time as the mammals, and became abundant as mammals became abundant. They are a parallel class of heat-retaining, high-efficiency animals.

Birds obviously are heavier-than-air flying machines. There are indeed one or two birds that cannot fly—the ostrich, for example, or the kiwi—and the penguin, instead of flying in air, flies in water, but the great majority of them are able to take to the air, and the fact is reflected in every bone, every muscle, every internal organ. Moreover, they are the only class of animals—except the insects—the majority of whose members are winged. They include the largest and heaviest flying animals, and their champions can fly higher, more swiftly, and for a longer time than any other unaided living creature.

Birds are characterized by the possession of feathers, just as are mammals by hair; feathers long and used for propulsion and steering on the wings and tail, short and forming an efficient garment on the rest of the body in order to reduce the amount of heat lost to the air that streams past during flight. Their arms are wings, and their breastbone bears an enormous keel in front to which the muscles which flap the wings downwards are attached. Being wings, the arms cannot be used to support the body, and when the bird is not flying its whole weight is poised on the legs; the hip-girdle is therefore very solidly joined to the backbone, the feet are large, and the legs are struts, firm but light. Moreover, except for a few extinct forms, birds have no teeth, and their jaws are ensheathed by horny beaks made of the same stuff as the dead outer layer of our skins, and variously moulded according to the feeding habits of the bird. In their internal anatomy there are several devices to assist the bird in performing vigorous and sustained exercise which necessitates a copious oxygen supply to the muscles. Their hearts are larger in proportion to the size of the body than those of mammals, and their lungs communicate with a unique system of air-spaces which make the oxygen supply of the body more efficient—some running among the viscera and some into the bones, which, being hollow, combine the maximum of rigidity with the minimum of weight.

The mechanical problems of propelling a heavy body through air are so considerable, and involve specialization of structure in so many parts of the body, that bird organization is to a large extent stereotyped. There is little room for variation. To be sure there are wading, swimming, perching, and running birds, each with a different kind of leg and foot; there are soaring, hovering, and fluttering birds, and clean, swift fliers, each with its own shape of wing and tail; there are vegetable-feeders that swallow stones and use them in the gizzard (a special part of the stomach) to grind up their food and there are birds that eat meat and have no need of such a device; there are various shapes of beak and neck, and a tremendous range of coloration. But compared, for example, with the mammals the variation of structure

FIG. 74. The Pelican preens himself.

found among birds is slight. Indeed, the only birds which show really profound departures are those exceptional ones which do not fly through air—the penguins, which fly in a medium having very different physical properties, and the completely flightless kiwis, ostriches, rheas, emus, and cassowaries, in which the wings are small and often useless, and in which, since there is no need of powerful muscles to move them, the keel on the breast-bone is small or absent.

§ 3c. The More Ancient Class of Reptiles.

The three greatest classes of vertebrates to-day are the mammals, birds, and fishes. It is as if these three classes had divided the world between them, the mammals taking the land, the birds the air, and the fishes the water. The fishes rarely leave their proper element, but each of the other two groups has sent a few representatives—missionaries and prospectors as it were—into the allotted territory of the other; the class of mammals, for example, has sent the bats into the air and

THE FIRST GREAT PHYLUM: VERTEBRATES

the whales into the sea. Nevertheless, in the main, each of the three classes rules in its own element. The remaining classes are now less dominant, and of minor importance; but among them the reptiles, at least, have a great history. At one time they ruled both land and sea, and birds and mammals were not.

In a number of points the reptiles are like mammals or birds—their embryos, for example, are surrounded in the egg by an elaborate system of membranes like the membranes of a chick in the egg, or of a

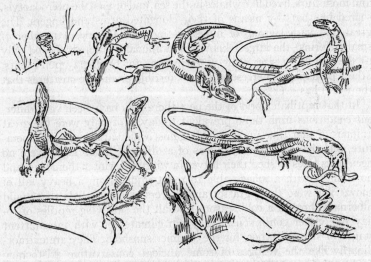

FIG. 75. The Green Lizard (Lacerta viridis).

developing mammal in the uterus, and they breathe throughout their life-histories by means of lungs—but they are cold-blooded, that is to say the temperature of their blood has to be that of their surroundings, and they have scales instead of the hair of mammals or the feathers of birds.

There are four orders of reptiles living. The first includes the crocodiles, alligators and gavials—predatory animals living in tropical rivers, some snapping the heads of mammals that come to drink and holding them under water until they drown, others taking fish and the like. The second consists of the tortoises, turtles and terrapins, land-living and aquatic, their bodies encased in a hard, protective shell into which in some, but not all types the head and limbs can be withdrawn. The third is by far the largest and most varied order, and includes the snakes and lizards—the rattlesnakes, cobras, kraits, pythons, geckos, skinks, chameleons, monitors, iguanas, slow-worms, and a host of others. Finally, the fourth order contains one genus only —the lizard-like Sphenodon, which is sufficiently distinct in its internal anatomy to be put in an order by itself.

With the extinct reptiles—twenty orders of them!—we shall deal more fully later on. Most of the extinct reptiles, like the survivors of the group, were terrestrial, but some, the ichthyosaurs, plesiosaurs, and mosasaurs, lived like whales in the sea, and others, the pterodactyls, soared like bats by means of their elongated arms and fingers. The great and varied group of dinosaurs included (besides a number of smaller forms) the largest known land animals—animals second only to the whales in size; some of them peaceful browsers upon plants, others perhaps the fiercest and most terrible predatory creatures that the world has ever seen.

In that reptilian world of the past there may have been very different conditions from those prevalent to-day. Not only were the great animals of that time unprotected against wide variations of temperature, but the vegetation also was of a different kind. There were no deciduous trees to shed their leaves in winter, if winter there was, and no conifers with a foliage able to resist injury from a heavy fall of snow. Nor were there grasses and flowers ripening seeds that could hibernate through a cold season. And all the surviving reptiles of to-day live either continuously in warm climates or with intermittent stages of winter torpor, for the summer sunshine. They are extraordinarily like the vestiges of some ancient conservative aristocracy living on in a world of harder conditions and more energetic newcomers.

§ 3d. The Linking Amphibians.

FROGS and salamanders are often classified by the uninitiated as reptiles, together with lizards and other cold-blooded land vertebrates. But they are not so. They belong to a very different class, the *Amphibia*—a class which is interesting because it sits on the fence that divides the lunged and limbed from the gilled and finned, the air-breathing mammals, birds, and reptiles from the water-breathing fishes. It is a class of vast antiquity, but never has it displayed the variety and abundance of the mammals and birds of recent times, nor of the reptiles in their prime. Amphibia may be recognized by their smooth, moist skins, which lack hair, feathers, or scales. This naked skin is one of the reasons why (with one or two rare and specialized exceptions)

THE FIRST GREAT PHYLUM: VERTEBRATES 185

an amphibian cannot venture far from moist places; possessing neither the dry scales of a reptile nor the air-jacket of a bird or mammal he has no protection against water-loss. Sometimes on very hot days one finds mummified frogs lying on roads—frogs which have ventured into the arid desert of the road and have dried up before they could find their way off it again.

Two of the three living orders of amphibians are well known. The first, the frogs and toads, are tailless, and have long powerful hind

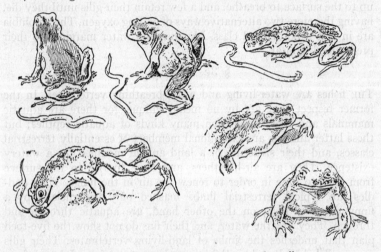

FIG. 76. The Common Toad grabs, overcomes, and swallows a worm.

limbs by means of which they leap or swim. The second, the newts and salamanders, swim by fish-like sideways wriggling of the body, which tapers behind into a long tail; they generally have four limbs, which they use for crawling on land, but in some forms the limbs are very small, or even absent altogether. The third order includes a few rare burrowing forms, having worm-like bodies and lacking both tail and limbs.

Nearly all amphibians spend at least a part of their lives in water. They have a fish infancy and an air-breathing adult life. The reader is probably familiar with the life-history of the common frog. Frog-spawn consists of a mass of clear jelly in which the hundreds of tiny black eggs are embedded; the embryos are not provided with the apparatus of ensheathing membranes found in the three classes with which we have already dealt. The young frogs or tadpoles have no limbs; they have large tails with which they propel themselves, and they breathe as a fish does by means of gills. In other characters also they are fish-like. They have, for example, lateral line sense-organs. But as they grow older they develop lungs and also limbs with hands and feet at the ends; they absorb their gills and tails; they convert themselves from fishes to land vertebrates.

In the great majority of amphibians there is a similar life-history. But some of the salamanders appear reluctant to leave the water; many of them stay in streams or ponds for their whole lives, swimming up to the surface to breathe, and a few retain their gills until they die, having therefore two alternative ways of getting oxygen. The amphibia are in fact a borderland class. Marshes and water margins are their predestined habitat.

§ 3e. Fishes.

THE fishes are water-living and water-breathing vertebrates. In the former respect they are by no means unique, for there are aquatic mammals and there have been many kinds of aquatic reptiles; but these latter animals are exceptional members of essentially terrestrial classes, and their structure of a land animal modified for a watery existence. They are air-breathers, and have to rise to the surface from time to time in order to renew the air in the lungs. Their paddles are typical terrestrial limbs masked only superficially by a finny disguise. Fish, on the other hand, are aquatic through and through. They breathe water, and their fins do not show the five-toed plan that underlies the limbs of land-living vertebrates. Their gills are paralleled by larval amphibians and even by a few adult members of that group; but most amphibians have lungs when they grow up, and their jointed limbs always end in unmistakably terrestrial hands and feet. The most distinctive feature of the fish is the fin—a broad, flat expanse ending in a plate of characteristic "fin-rays." They may be defined as finny vertebrates.

There are two sub-classes of fish—the gristly and the bony fish. In the former the skeleton is composed in the main of gristle; this is strengthened by the deposition of lime-salts in various places—always in the teeth and scales, and often in the vertebræ and jaws. In the second sub-class there is bone in the skeleton, and especially in the skull; it is as if the skull of a gristly fish had been plastered over with a layer of flat bones. The easiest way of distinguishing the two sub-classes is by looking at the gill-slits. In the gristly fish (with one or two rare exceptions) the five or more gill-slits can be seen as separate openings, while in the bony fish they are covered by a flap of bone, the gill-cover, and they open together by a single cleft at its back margin.

The gristly fish include sharks and dogfish, the flat, bottom-living skates and rays, and the weird and rarely-seen "Chimeras." They include the largest of fish, the basking sharks, up to forty feet long. The eggs of dogfish, rays and the like are well-known seaside objects, often thrown up on the beach by storms and called "sailors' purses" or "mermaids' purses." They have tough horny shells and contain a large store of yolk, and the embryo may develop inside them for over a year before hatching. Some species, like the "spur dog," are viviparous. But they are not now the prevalent kind of fish; it is to the second class that the great majority of fish belong.

The bony fish vary enormously in form and habit. Some, beautifully streamlined, swim swiftly and with endurance; a good example is the mackerel, whose clear blue colour, light below and darkly rippling above, makes him almost invisible as he slides through the sea. Others, curiously flattened, lie on the bottom, snapping with their wry mouths at unsuspicious passers-by, and only occasionally swimming to a new pitch; the plaice and turbot, for example, have an astonishing power of changing colour and assuming the pattern of the bottom on which they lie. Some, like the carp, have small, fastidious mouths and swallow worms and such-like delicatessen; others, like the pike, have enormous jaws and cruel teeth, and destroy fishes nearly as large as themselves. There are fish with mouths that shoot out like the latticework "ticklers" of our fairs and snatch unwary creatures; there are fish with lures to attract their prey within reach of the hungry jaws. The flying-fish have large, broad fins so that they can skim for a considerable distance through the air. There are cylindrical fish, ribbonlike fish, globular fish; in the abyss there are fish more grotesque, more horrible than any mythological monster. There are fresh-water fish and salt-water fish, and there are some, like the salmon, that travel regularly to and fro between sea and river. There are also fish that have structures very like lungs and that can, if they choose, breathe air. They are amphibious fish and they are found in tropical rivers subject to summer drought. Then they cease to use their gills, encase themselves in mud and lie breathing gently with their lungs until the waters return.

§ 3f. Cyclostomes, a Class of Degraded Antiques.

AND now we come to another group less numerous now than it was in the remote past. These are the Cyclostomes. Of this class only a few genera now survive. Once they seem to have been the only vertebrata in existence. The living forms are the lampreys and the hag-fish or slime-eels. They live some in fresh water and some in the sea. Their bodies are long and eel-like. They breathe by means of gills as a fish does. But in a great number of respects they are extraordinary. They have no trace of paired limbs, finny or otherwise; the skeletal supports of their gills are unlike those of any of the true fishes; they have only one middle nostril and only one nasal cavity; instead of three semi-circular canals in their ears which all other fishes possess they have one or two; in some species the hollow pituitary pouch of the brain opens into the roof of the mouth. But their most striking peculiarities are in the structure of the mouth itself. They have no jaws and no proper teeth; instead of being a slit that can be open or shut, their mouth is a funnel that can be used as a sucker, and their tongues, armed with horny substitutes for teeth, can be moved like a piston in the mouth-funnel. They nourish themselves by attaching themselves to fishes by means of this sucker and rasping off the flesh with their tongues. They may even bore through to the abdominal cavity of their victim, punching a neat round hole in his body-wall as they do so.

The earliest known fossil vertebrates are certain heavily-armoured fish-like forms more like these Cyclostomes than any other creature. It has been shown that these, too, were jawless, and that they were built in the same sort of way as a lamprey or a hag-fish. Their method of feeding, however, was probably different. Apparently the lampreys and hag-fish are, like the reptiles, the rare survivors of a once dominant aristocracy, but one enormously more ancient than that of the reptile world.

§ 3g. Semi-Vertebrates.

The animals that we have already considered share a common plan of organization. Their most evident distinctive feature is the possession of a backbone, a column of ring-like blocks of bone or gristle jointed together, running down the back and housing the spinal cord. Because of this vertebral column the animals are called vertebrates. The phylum includes ourselves, our domestic animals, and all the largest and most conspicuous wild creatures; it is the most immediately striking and for practical purposes the most important phylum. Since they are less obtrusively important, all the animals that are not vertebrates are often lumped together as the invertebrates, although this negative group, including as it does a number of different phyla, is in reality a much more varied one than the single phylum of vertebrates. It was

THE FIRST GREAT PHYLUM: VERTEBRATES

inevitable that the earliest naturalists should pay more attention to vertebrated animals than to others, and the thorough exploration of the great group of outsiders and the recognition of their variety and systematic importance only followed long after the general principles of vertebrate classification had been recognized. How the invertebrata have been at last sorted out into a system of phyla we shall tell in the following chapters, but here we must note certain odd lowly forms which naturalists, after the most careful consideration, have decided to throw back into the vertebrate basket. They are not full vertebrates certainly but they are put with them as associates, distant cousins, semi-vertebrata.

The lancelet—*Amphioxus* (sharp at both ends) is its systematic name—is an inconspicuous little marine animal, whitish, translucent, shaped like a flat toothpick, and less than two inches long (Fig. 77). It lives an obscure life, burrowing in sand in shallow water round most coasts, and was considered by its discoverer to be a slug. But subsequent investigation has removed it from the low phylum to which it was assigned and placed it with the vertebrates. If it is not legitimately a vertebrate, it is at least a poor relation.

This little two-inch slip of life has a smooth, limbless body. Its mouth is surrounded by tentacles and leads into the pharynx, a perforated, nearly cylindrical structure not unlike a gas-mantle. The fabric of this gas-mantle is covered with cells bearing tiny vibrating hairs or cilia of the same nature as those in our air-passages, and these microscopic hairs are continually working water through the perforations in an outward direction. The space round the gas-mantle opens into the sea by a special pore; thus there is a continual slow current of water in by the mouth, through the slits of the gas-mantle, out by the special pore. This apparatus is not unlike the gill-apparatus of a fish, for here also there are clefts in the side of the throat through which water passes out, and it was for a long time regarded as a similar respiratory device. But it is now known that the chief duty of the apparatus is to collect food. As it passes through this animated sieve the water is filtered; minute organic particles are collected on strands of sticky mucus and passed down to the gullet, which opens at a place corresponding to the blind end of the gas-mantle. On the nutritious matter which is collected in this way the animal subsists.

Amphioxus resembles the vertebrated animals in several important respects. In the first place, it has a tail. In invertebrate animals the anus generally opens at the extreme hinder end of the body, and there is never the muscular backward prolongation of the body-wall, containing the end of the spinal cord, that is found in all embryo vertebrates, and in the great majority of adult members of that group. Secondly, it has a spinal cord down its back, which, like that of a vertebrate, is a hollow tube traversed by a continuous narrow canal. Thirdly, although its digestive tube runs straight from mouth to anus, there opens out of it a blind pouch which seems to be the homologue of our liver. The blood that has supplied the intestine is carried by a

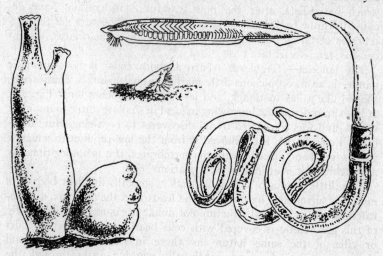

FIG. 77. Semi-vertebrates: on the left, two Sea-Squirts; above, two Lancelets, of which one is buried in sand with only its mouth above the surface; on the right, a Balanoglossus.

portal vein to this pouch and there traverses a second system of capillaries, a portal circulation, such as all vertebrata have; and although the liver of an adult vertebrate is not a simple pouch it nevertheless develops in the embryo from a pouch that grows out from the intestine like the liver-pouch of Amphioxus. And another most important point of resemblance is a structure called the *notochord*. Amphioxus has no vertebræ, but it has a stiff, elastic rod, lying along the whole length of the body, between the spinal cord and the alimentary canal. This notochord, like our backbones, is the mechanical axis of the body on which the muscles play. It is made of a highly characteristic kind of tissue—its cells have large spaces inside them, containing water under pressure, and it is surrounded by a thin, elastic membrane secreted by the cells, so that its rigidity and elasticity are like the rigidity and

THE FIRST GREAT PHYLUM: VERTEBRATES

elasticity of a rubber tube filled with water under pressure. Now, in the embryos of all vertebrated animals a notochord precisely similar to the notochord of Amphioxus appears. In the lampreys and their allies, and in a few fish, this persists and the backbone forms round it and encloses it, but in all other vertebrata it is supplanted during later development by a true backbone that owes its rigidity to lime-salts and its elasticity to binding ligaments.

The notochord is really the distinctive character of the phylum that we are now examining, and which is often called, for this reason, the *Chordata*. In the humblest members of the phylum, such as Amphioxus, the notochord is retained throughout life. In the higher forms, in all the true vertebrata, it is present in the embryo, but becomes replaced more or less completely by a better axial skeleton—stronger and as flexible, and at the same time affording a protection for the spinal cord—the column of jointed vertebræ.

* * * * * *

Another still more remarkable affiliation to the vertebrate phylum in recent years has been that of the sea-squirts and their allies. Sea-squirts are common enough on rocky shores, where they occur, often in large masses, growing fixed to rocks and the like. An individual sea-squirt is a roughly cylindrical, gelatinous, translucent object, often several inches in length; at the unattached end there are two holes from which, when it is touched, it ejects with considerable force two jets of water. In examining the creature more closely we find that its gelatinous exterior is not part of the animal itself but a garment that it secretes around its body. When the sea-squirt is unmolested, a current of water passes slowly in at one of the openings and out at the other; on the way through the body it traverses a gas-mantle filter or pharynx, like that of Amphioxus, and there yields the suspended particles upon which the animal subsists. The whole adult life of the sea-squirt is spent in the same place, and its only outwardly visible sign of animation is the slight stir in the surrounding water produced by its slow, indolent feeding-current—except on rare occasions when, alarmed by the approach of a hungry fish, it contracts for a while into an unattractive gelatinous blob, squirting water out of itself in two violent jets.

There is nothing very like a chordate in the anatomy or mode of life of an adult sea-squirt, nothing to suggest a nearer cousinship to *Homo sapiens* than that which a lobster or spider might claim—except that its food filter is highly suggestive of that of Amphioxus. But when its life-cycle is examined the case is altered. It is upon the anatomy of its early life that its inclusion among the chordata is based. The egg of a sea-squirt develops into a lively little creature like a minute tadpole, which swims along by means of a powerful muscular tail. This larva is quite obviously a chordate. In its tail it has a notochord like the notochord of Amphioxus, and the mechanics of swimming are very similar in both forms; it has a tubular nerve-cord running along its back. But in a little while it settles down on a rock, sitting on its head and fastening itself by means of sticky horns that it possesses there. It now absorbs its tail and notochord, and its nervous system (since it no longer has anything very much to think about) dwindles to a single rounded ganglion with a few radiating branches; at the same time it loses its eyes and other special sense-organs.

The class to which the sea-squirt belongs, the *Tunicata*, includes a number of other curious forms. Many of them are able to reproduce themselves by sprouting out buds; in this way they give rise to extensive colonies of individuals embedded in a common gelatinous mass. Some are not sessile but, beautifully transparent, float about in the sea; of these the most striking is *Pyrosoma* whose thimble-shaped colonies —up to twelve feet in length—are among the most brilliantly phosphorescent objects known. A few never discard the tail and notochord of their youth but spend their lives as active, tadpole-like creatures, inhabiting relatively large gelatinous houses that they secrete and propel through the open waters of the sea.

Balanoglossus is another extraordinary associate of our phylum. It is a white or orange worm-like creature—the length varying in different species from an inch to three yards—that spends its life burrowing in mud at the bottom of shallow seas. It is far less like a vertebrate than either Amphioxus or a young sea-squirt, yet it is more like a vertebrate than it is like the members of any other phylum. It has, for example, a great number of gill-slits leading out of the sides of its throat, and the possession of a perforated throat is apparently a chordate peculiarity; as the animal nourishes itself by swallowing mud or sand and digesting out any organic matter as it passes through the intestine it may be presumed that the slits are respiratory. Again, the front part of the body is very mobile and is used for burrowing; it is supported by a thick-walled tube of narrow bore that runs forwards from the roof of the mouth. The cells of this tube have a structure suggestive of the cells of a notochord, and it lies, like a notochord,

THE FIRST GREAT PHYLUM: VERTEBRATES 193

just below the principal nerve-cord; moreover, in embryo vertebrates the notochord develops as a tube that is pinched off from the roof of the digestive tube. It is at least possible, therefore, that this structure is homologous with the notochord. On the other hand Balanoglossus has a point of resemblance with a very different phylum, the echinoderms (starfishes, sea-urchins, and their allies); in many species the eggs hatch into minute larvæ that are like the larvæ of sea-urchins, except that they grow into Balanoglossus. Thus to link vertebrate and echinoderm is a remarkable achievement.

Classified in the same group as Balanoglossus are two genera of similar creatures that live like polyps in colonies, surrounded by gelatinous tubes, and capturing suspended particles by means of their branching tentacles. The best known is called Rhabdopleura. It marks the extreme limit of the chordate phylum. The chordate ground plan, reduced to the mere shadows of backbone, liver, and gill-slit, is traceable no further among living things.

II

THE SECOND GREAT PHYLUM: THE ARTHROPODS

§ 1. *A Contrast of Arthropod and Vertebrate.* § 2. *The Arthropod Plan of Structure.* § 3a. *Shrimps, Crabs, Water-fleas, and Barnacles.* § 3b. *Spiders, Mites, and Scorpions.* § 3c. *Insects.* § 3d. *Centipedes and Millipedes.* § 3e. *Peripatus.*

§ 1. A Contrast of Arthropod and Vertebrate.

AND now we come to the second great phylum of living things. This includes all the insects which creep and buzz about us, bite and pester us, sting and poison us, the spiders that shroud our rooms and cellars, the lice that infest us, and earwigs that creep and run from every stone we turn. It is a strange, competitive world side by side with the chordate phylum, unsympathetic, alien, and for the most part hostile and malignant to it.

We can best consider its distinctive differences from other phyla by taking first a concrete example, and for that the size and accessibility of the lobster (*Homarus vulgaris*) make it a convenient type.

The general appearance of a living lobster is sufficiently well known. His prevailing colour is dark blue; the familiar red colour of boiled lobster only appears in the boiling. On contrasting him with a vertebrate, two outstanding differences immediately present themselves; first, his whole body is covered with sheet armour; second, in the place of our legs and arms he is provided with a great number of hard, jointed limbs.

The front half of the lobster is protected above and at the sides by a single sheet of stiff armour. This part is called the *cephalothorax* —meaning head and chest in one. The hinder half consists of a chain of seven movable segments; each segment is strongly protected, but in between the segments the layer of armour is very thin—so thin that it is flexible. This part is called the abdomen—absurdly, as we shall explain. Its segments are jointed together in such a way that the abdomen can either be held straight out as in the figure, or curved forward under the cephalothorax; by suddenly flapping his abdomen forward in this way the lobster can jerk himself rapidly backwards through the water, and it is by means of a series of such flaps that he retires when seriously alarmed.

On the lower side of the abdomen there are six pairs of limbs or appendages, one pair to each segment except the last. Below the cephalothorax, there are thirteen more pairs of appendages. The

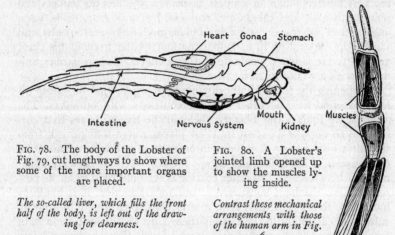

FIG. 78. The body of the Lobster of Fig. 79, cut lengthways to show where some of the more important organs are placed.

The so-called liver, which fills the front half of the body, is left out of the drawing for clearness.

FIG. 80. A Lobster's jointed limb opened up to show the muscles lying inside.

Contrast these mechanical arrangements with those of the human arm in Fig. 6.

structure of these appendages varies with their situation and the duties that they are intended to perform. Most conspicuous are the two pairs of whip-like feelers in front, the enormous claws, the four pairs of walking legs, and the last pair of all which, together with the last abdominal segment, form a flat plate used in the abdomen-flapping mode of retreat. But besides these obvious appendages there are others. There is a battery of six pairs round the mouth, having various shapes and used for smelling, tearing up and sorting food; there is a series below the abdomen of which in the male the first two pairs are specially modified for transferring spermatozoa to the female. Moreover, some of the appendages of the cephalothorax bear gills on their bases, and are constructed in various ways so that a current of water may be passed over the gills.

It is interesting to compare this series of specialized appendages with our own two pairs of limbs. Our legs are designed for one function only—they support and propel us—but our arms are essentially unspecialized structures. From an evolutionary point of view they are primitive. Our hands have not committed themselves, so to speak, to the exclusive performance of any one act; they are plastic, in the sense that they can do a variety of different things. A lobster does not possess anything that could be called a hand. Every one of his nineteen pairs of limbs is designed for the efficient and exclusive performance of some particular duty. They are not so much limbs as tools. A lobster is like an armless carpenter who has his mallets and planes, his awls and chisels and reamers, his saws, bits, drills, axes, augers, and sandpaper—even his knives and forks and spoons and toothpicks—sprouting in a double row out of the front of his chest and belly. He is like a penknife of the pocket workshop variety animated with a will of its own.

On opening the lobster and examining the organs that lie within his hard exterior we find other striking contrasts with ourselves. The armour, we find, is the lobster's skeleton; he has no bones that correspond to ours, and except in a few places where rods project from the surface into the soft interior in order to give firm attachment to the muscles of his limbs, and except for a marvellous chewing apparatus in his stomach, this outer coating is the only rigid tissue he possesses. Like the hard substance of our own skeletons it is made *by* but not *of* living cells; it is dead matter consisting largely of a horny substance called chitin, and strengthened by the deposition in it of lime salts, and it is secreted by a layer of cells that lies immediately below it.

Since the lobster lives inside his own skeleton, the mechanical arrangements of his limbs and other movable parts have to be different from ours. In ourselves, the jointed bones are clothed by a layer of muscles which pulls them in various directions; in the lobster, instead of bones there are jointed tubes, and the muscles that move them lie inside them.

The lobster's main interior cavity in the cephalothorax is not divided into thorax and abdomen cavities like ours; it is one large cavity. Moreover, this cavity is filled with blood. It is not a special cavity as our body-cavities are; it is as if the veins of a vertebrate had swollen enormously and united together, squeezing the true body-cavity (and, incidentally, the whole lymphatic system) out of existence. Further, the colour of the blood is different. Instead of our red oxygen-carrying pigment hæmoglobin, the lobster has a pale-blue pigment called *hæmocyanin*, with copper in it instead of iron, and instead of being contained in special circulating cells this pigment is diffused, dissolved in the blood-plasma.

THE ARTHROPODS

Within this vast blood-space his various organs are disposed, and, as a glance at figures 5 and 78 will reveal, the manner of their arrangement is quite different from that of a vertebrate. The lobster's central nervous system runs along his belly; his heart is in the middle of his back; his kidneys and bladders lie in his head and open at the bases of the antennæ a little way in front of his mouth; his or her testes

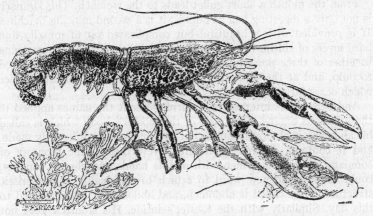

FIG. 79. The Common Lobster (Homarus vulgaris).

(*From the Plymouth Aquarium Guide, by courtesy of the Marine Biological Association of the United Kingdom.*)

or ovaries (as the case may be) are placed between the heart and the intestine.

As we look more closely into the working of the lobster's organs we are impressed ever more clearly by his pervading unlikeness to ourselves. In illustration of this point let us follow the course of his food. Our specimen, we may note, eats anything that he can pick up, with a strong preference for animal food—such as smaller crustaceans, molluscs, and worms, or any kind of carrion.

The lobster has no mouth comparable to ours, with lips, tongue, teeth, salivary glands and so on; the very thorough chewing that his food undergoes is done partly by the jointed limbs, and partly in the so-called stomach. The food is generally seized by the large claws and passed by them to the first or second pair of walking legs; these in their turn hand it to the last pair of mouth-appendages, the third maxillipedes, which tear it to shreds with their rows of sharp teeth. Thence it is handed in turn to the second maxillipedes, the first maxillipedes, the second maxillæ and first maxillæ, all appendages which do but little chewing of the food, and finally it comes to a pair of powerful cutting and grinding jaws, the mandibles. So the food undergoes a thorough mastication, and only when it has been finely minced by this living canteen of appendages is it allowed to enter the mouth.

From the mouth a short gullet leads to the stomach. This stomach is not only a digesting bag as ours is; it is a second mincing machine. It is provided with a beautiful but complicated set of muscles and hard levers of chitin, the *gastric mill*, which bring about the gnashing together of three powerful teeth. By these teeth the food is reduced to pulp, and at the same time it is acted upon by a digestive juice which oozes forward from the so-called liver.

And here let us interpolate a warning about the names applied to these various organs. That part of the lobster's body which is called his abdomen is obviously unlike our abdomens; it is differently made, and it contains different structures. Nevertheless, the pioneers of comparative anatomy, men with nimble hands and clumsy tongues, could think of no new word to attach to this highly characteristic structure. They called it abdomen, and abdomen it has remained to this day. Similarly with the lobster's inside. His stomach does not correspond to our stomachs, and his liver—so called because it is a large gland communicating by means of a narrow tube with the beginning of his intestine—is utterly distinct, both in its minute structure and in its working, from our own. When we compared a fish to a mammal we found that both organisms had on the whole the same set of organs working on the whole in the same way. We can speak, without any fear of misleading ourselves, of the liver or stomach or spleen of a fish, because these organs in a fish are essentially the same as our own. Right down to Amphioxus there is a liver. But a lobster is put together in a different way. He has no spleen. His "stomach" is differently made from ours and does different things. He has no organ that corresponds either in structure or working to our livers, and we have no organ corresponding to his. It would be clearest and best to invent a new name for every organ in the lobster's body, but unhappily this has not been done; we shall have here to follow the general usage and apply the old terms, remembering, however, as we do so that every one of them is a misleading misfit, borrowed from an alien organism.

The lobster's "liver" is an enormous, tubular gland, manufacturing every one of the digestive ferments that have yet been discovered in the animal: in this respect it corresponds to our salivary, gastric, intestinal, and pancreatic glands combined. It also stores glycogen and fat, resembling our liver in this respect at least. But besides these substances it differs from all our organs in storing lime; the lobster, being enclosed in a rigid armouring, has to moult from time to time so that he may grow, and calcium salts required for forming a new shell are stored in readiness in the liver. The gland lacks the most characteristic feature of a vertebrate liver, the portal vein, and is in no sense a scrutineer of blood from the bowels. Lastly, it resembles our intestine, for food actually travels into its ducts and tubules to be absorbed.

From the gastric mill the food is passed into a smaller chamber, provided with an elaborate apparatus of sieves and filters, where it is sorted into two fractions. The coarser particles pass into a long, cylindrical intestine, whence much of their nutritious matter is absorbed and whence their residue is ultimately voided. This fraction is of minor importance. The digestive juice on the other hand, carrying most of the food materials either in solution or in suspension as fine particles, is sucked back into the tubules of the liver, and here the greater part of the absorption of food takes place.

No need to stress the profound dissimilarity between these digestive arrangements and our own. A comparison of any other system of organs would give the same results. We should find just the same distinctiveness throughout. In the disposition and dynamics of the circulation, in the kidneys, in the astonishing compound eyes, even in the way the blood clots we should find this profound and pervading difference from ourselves. A lobster is a living creature; his individual cells work in much the same way as ours do and their chemical and physical needs are the same as ours. But all the arrangements which are made to satisfy those needs are different. He is a cell-community as we are, but he is a community organized *ab initio* on a different plan.

§ 2. The Arthropod Plan of Structure.

A NUMBER of different kinds of animal are classified with the lobster in a separate phylum, the *Arthropoda*, just as the vertebrates are classified together in one phylum, because in the general plan of their organization they resemble each other and are profoundly different from other living creatures. We have introduced ourselves to the group by examining in some detail one of its members, and before passing to a systematic survey of the others it will be best to make clear how far the points we observed are general features of the arthropods and how far they are peculiar to the lobster or to that smaller group, the class Crustacea, to which he belongs.

The abdomen of a lobster, we noted, is composed of a series of hard segments, flexibly jointed together; each segment bears a pair of appendages. The cephalothorax, on the other hand, has a single shield of armour and bears thirteen pairs of limbs. Now, when we examine other arthropods, we find that the segmental plan of structure, as illustrated by the lobster's abdomen, is a very important one. It underlies the organization of the whole group. In many arthropods, such as the centipedes or some of the smaller and less specialized crustaceans, the greater part of the body consists of a chain of hard segments jointed together and each bearing a pair of appendages, and indeed all the members of the group are best regarded as modifications of chains of this type. In a lobster the first thirteen segments are solidly welded together, forming the cephalothorax, only the appendages remaining distinct from each other; in other arthropods the underlying plan has been modified in other ways. Imagine a chain of segments— an enormously extended lobster abdomen; each segment with an armour skeleton outside it and bearing a pair of jointed appendages. That would be a diagrammatic, generalized arthropod. By playing on such a scheme, by varying the number of segments, by welding some together and keeping others distinct, by modifying and adapting different appendages for the performance of different functions, all the various existing arthropods could be evoked.

The idea that the arthropods are variations upon a pervading segmental chain can be applied also to their internal architecture. All arthropods live inside their skeletons; all have a general body-cavity filled with pale blue blood; all have a nervous system running along the lower sides, and if, as is usual, they possess a heart it is in the back. But the various structures may shift to and fro along the length of the body, from segment to segment. The heart, for example, may be a relatively small organ placed in the middle of the back as in the lobster, or it may be a long tube running down most of the length of the body, with a pair of valved openings in each segment. The position of the excretory, respiratory, and reproductive organs varies widely. And the structure of the digestive organs varies—there may, for example, be no gastric mill.

And now for a rapid review of the five great classes of arthropods, to note the main changes that have been rung on this underlying theme.

§ 3a. Shrimps, Crabs, Water-fleas, and Barnacles.

THE first class of arthropods, the *Crustacea*, includes a great and varied assemblage of forms. Nearly all of these animals are aquatic, breathing by means of gills, and most of them are marine—indeed, they are the only great group of marine arthropods. The more familiar forms—familiar because they are the largest and also because they are edible

FIG. 81. The growing Hermit Crab changes into a roomier shell (1—4). Below on the right is another crab with sea-anemones on his shell—a commonly found partnership between two very different kinds of creature.

—are the lobsters, cray-fish, prawns, shrimps, and crabs. These animals are all very similar to each other. A crab is essentially a lobster who has given up the abdomen-flapping method of swimming, and whose abdomen, small and feeble, is permanently bent forward under his broadened cephalothorax. The hermit-crabs also belong to this group; they are lobster-like creatures whose abdomen is not hard, but soft and delicate, and who have therefore to wear the deserted shell of a dead sea-snail on their hinder parts, not for modesty, but as a protection against the many prowling and hungry creatures of the sea.

Other crustaceans, not quite so well known, are the wood-lice and the group including the so-called fresh-water shrimps, which differ in a number of important points from the true shrimps.

Besides these easily visible forms there are several groups of minute crustaceans, seen only as whitish specks to the naked eye, but nevertheless of great economic importance. Cyclops, a one-eyed, literally heartless crustacean, hardly a millimetre in length, swarms in countless numbers in lakes or pools or slow streams; this and the slightly larger Daphnia are among the commonest inhabitants of fresh water. The name "water-fleas," often applied to these animals, is misleading because they feed on even more minute organisms and not as fleas do on the blood of larger creatures. Daphnia is particularly and deservedly popular among amateur microscopists because her body (nearly all Daphnias are female) is so transparent that all her living organs, her beating heart and her churning bowel, can be watched with ease under a low power. In the surface layers of the open sea there are prodigious numbers of minute crustaceans belonging to the same group as Cyclops, the *Copepods*, exceeding in multitude any other kind of marine animal except the unicellular ones, and providing one of the most important sources of food of fishes. The late Sir Arthur Shipley, in his delightful *Hunting Under the Microscope*, says: "If on a sea voyage you tie a piece of bolting cloth loosely to the bath-tap which introduces the salt water into the bath, in a very short time you will find a deposit or paste at the bottom, which in the main consists of copepods." The whole of the sea is like that. There are more copepods in Plymouth Sound than there are people in the world—indeed, there are hundreds of times more.

We may note that crustaceans, like the great majority of arthropods, enclosed as they are by a hard, unyielding coat of armour, have to moult that coating off from time to time so that they may grow. The life of a lobster, for example, is a rhythmic alternation of two states. For a time he lives in his shell, not growing, but storing in his liver the materials necessary for a short but energetic burst of growth; then he retires to a safe crevice in the rock and casts off his armour—not only the outer armour, but even the lining and teeth of his stomach and the lining of his intestine—and, temporarily defenceless and helpless because of the softness of his appendages, he uses the stored material to grow and then to build a new and more roomy skeleton. This done, he emerges from his retreat, bigger and stronger for his reconstruction, into a warring world.

During the earlier stages of their life-histories the crustacea generally undergo one or more profound changes of bodily shape. The young are very unlike the old. We need not catalogue the different larval forms that are found in the various groups; some of them are shown in the illustrations herewith.

THE ARTHROPODS

The most striking changes of shape during the life-history are seen in the group to which the barnacles belong. There is no very evident resemblance between the lobster, with which we began our account of the crustacea, and an adult barnacle, permanently fixed as it is to a rock or to some other hard object, shut up in a little box of skeletal plates, and feeding on minute organisms in the sea-water which it entangles by means of lashing movements of a cluster of long, feathery, whip-like arms. Nevertheless the barnacle begins its life as a free-

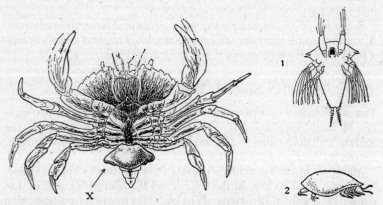

FIG. 82. The parasite Sacculina growing (at X) under the tail of a crab and sending hungry roots into all its organs.

The Crustacean affinities of this creature are evident in its two young stages, seen, highly magnified, at 1 & 2.

swimming larva which is obviously a crustacean, and later settles down and changes its shape to suit its new sedentary life. It does, in fact, very much what the sea-squirt does. In a parasitic animal, Sacculina, which is classified with the barnacles, the change is even more profound. The newly-hatched Sacculina is a free-living creature with outside skeleton, limbs, eyes, heart and so forth; but, a sign of its destiny, it has no digestive tube. It moults and acquires a new form, in which it resembles one of the water-fleas, Cypris, much as an embryo man resembles a fish, and it then swims around looking for a crab. Having found a suitable victim, it settles down somewhere on his body, loses its structural specialization to become a mere mass of hungry cells, and burrows through his armour into his blood-stream. By the host's blood the parasite is nourished, and it rolls hither and thither for a time, moved by the currents of blood in the body-cavity. Finally it settles down in the back part of the cephalothorax, just under the abdomen, and becomes a mass of parasitic tissue (Fig. 82); it sends greedy roots into every corner of the crab's body (excepting only his heart and his nervous system, for no provident parasite kills its host) and sucks nourishment from his tissues as a plant sucks moisture from the soil. Thus all its chemical needs are satisfied, and the only care left is reproduction; it becomes a mere mass of ova and spermatozoa (for it is hermaphrodite) with a halo of absorbing roots, and ends its career by fertilizing its own eggs.

Sacculina is by no means the only parasite among the crustacea. Altogether some 700 parasitic species are known—and these are only a small proportion of an enormous and varied class.

§ 3b. Spiders, Mites, and Scorpions.

THE second class of arthropods, the *Arachnida*, includes a great number of land-living animals. The better known kinds—such as the spiders and ticks—may be distinguished from other land-living arthropods by their possession of four pairs of walking legs.

Many of the land arachnids possess a curious method of breathing, found only in this class, in the insects and the centipedes, and in that curious little animal Peripatus. The body is permeated by a branching network of air-tubes or *tracheæ*, stiffened by rings and spirals of chitin and opening to the exterior at a number of places on the body-wall. Air is pumped in and out of the tracheæ by means of rhythmical movements of the body—movements that can be seen very clearly in the abdomen of a wasp—and it is carried by the fine end-branches of the tubes directly to the tissues themselves. There is no need with a system of this kind for the blood to transport oxygen or carbon dioxide; the tracheæ are a device for putting all the tissues of the body in immediate contact with fresh air. But some arachnids have, instead of tracheæ, a lung-like structure in which the blood is oxygenated, and a few have both mechanisms together.

The carnivorous spiders are the best-known members of the class. They include, besides the cruel, skilful architects with whose snares we are familiar, a number of prowling and hunting forms which capture their prey by stalking or chasing it, and which have no need of webs. The phalangids or harvest-men, little round bodies moving on eight long, stilt-like legs, are put in a separate order from the spiders because their body is not divided in two by a waist, like that of a spider. To the arachnids belong also the scorpions and one or two small and comparatively unimportant orders. But the order which is most important to man is the great army of mites and ticks, some biting us, sucking our blood, burrowing into our flesh and infecting us with various diseases, some attacking our cattle, some producing galls in our crops and forests, some living innocently, as far as we are concerned, in ponds and ditches, blasted trees, old cheese and all sorts of other places.

The class also includes that curious marine creature the king-crab, covered by an enormous circular carapace, and suggestive of a small boy crawling about under a tin bath with a broomstick trailing behind. And there is an extinct group of gigantic marine arachnids, the sea-scorpions, some of which were as much as eight feet long. In the remote geological past, before even the appearance of the earliest vertebrate, the sea-scorpions were the bullies of their world, the strongest and most impressive living things.

§ 3c. Insects.

THE *Insecta* are an enormous group; in the rich variety of their forms they surpass every other class in the Animal Kingdom—indeed, there are more species of insects alone than of all other animals taken together. Butterflies, moths, dragon-flies, beetles, gnats, mosquitoes, house-flies, blow-flies, bot-flies, daddy-long-legs, bees, wasps, ants, fleas, bugs, lice, termites, grasshoppers, earwigs, cockroaches—these are the most familiar forms, but there are many others. For the most part they are mere nuisances. Some, it must be admitted, are æsthetically satisfying, and a few, such as bees and cochineal bugs and those insects which eat weeds or other insects or those which fertilize flowers, have been found actually useful to man; but these are the few bright exceptions in a vast family of bloodsuckers, stingers, spreaders of disease, and devourers of crops. They are a distressful aspect of creation. They are a hard problem for the theologian.

Some of these animals live in water—the larvæ of gnats and dragon-flies, for example, or the air-breathing water-beetles—but the greater number are land-living forms, and like the spiders they breathe by means of a system of air-tubes or tracheæ. They may be distinguished from other land arthropods by their possession of three pairs of walking legs; in addition, the great majority of insects have a highly characteristic feature in their wings. Alone among invertebrates, the insects have conquered the air, unless we consider that the gossamer spider who floats along parachute or balloon fashion on his slender threads is also a conqueror.

The wings do not correspond to any of the appendages of a crustacean; they are different structures, sprouting, not from the belly, but from the back. Typically there are two pairs of wings—as for example in the dragon-flies, where the wings are transparent and delicately veined, in the butterflies and moths, where they are coloured and dusted over with prettily constructed scales, and in the bees, where

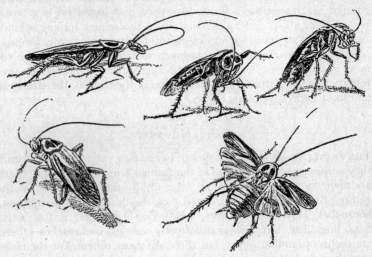

FIG. 83. The Cockroach cleans himself and flies away.

the front and back pairs are fastened together by hooks so that they may beat together more accurately—but this arrangement may be departed from. In the beetles only the hinder pair is used for flying; the front pair of wings forms a hard, opaque casing that covers and protects the delicate second pair when the latter are not in use. In the mosquitoes and commoner flies, on the other hand, the front pair are the ones which are used as wings or should be used; the hind pair are developed as tiny knob-like structures, the so-called "balancing organs." Finally, many insects, such as the fleas and the spring-tails, have no wings at all.

A second respect in which the insects excel over other invertebrates is the surprisingly efficient degree of communal organization to which many of their species attain. There is only one other animal—man—

THE ARTHROPODS

which surpasses them in this respect. With the division of labour and consequent specialization of individuals that is found in bees, ants, and termites, and with the fascinating problem of the insect mind we shall deal in later chapters; as we shall be returning to these and other questions about them, our present treatment of the class may be brief.

There is one consequence of the possession of wings that is worth noting. The arthropods, as we have seen, have to moult in order to grow, and the insects are no exception to this rule. But it is apparently impossible to moult the skeleton of a wing and then to grow it again. An insect's wing is a delicate, tenuous structure; it consists of little

FIG. 84. Peripatus—a link between Arthropod and Worm.

more than a double film of chitin with an invisibly thin layer of living tissue in between. If the stiff outer film were cast off the inner layer would be quite unable to retain any sort of shape. For this reason, the winged stage of an insect's life is invariably its last. For most of its life it simply creeps and eats and grows; once it has developed its wings it cannot moult and its growth therefore ceases. During its long childhood the insect is called a larva; in its often brief final stages, when it flies and reproduces its kind, it is called an adult or imago. The difference of duration between the two periods may be very striking; in some of the may-flies the larva crawls and feeds for two years or more as a preparation for one day of winged love-making.

In some insects, such as the earwig or the dragon-fly, the transition from larva to adult is comparatively gradual. The larva in such forms is often distinguished as a "nymph." The dragon-fly larva lives at the bottom of a pond, snatching and devouring tadpoles and other creatures by means of the hinged "mask" that projects from its face. From time to time it moults, and after the third or fourth moult there appear on its back two small backwardly directed lobes, which are the rudiments of the wings. During the later moults these rudiments increase in size, and at the same time a system of tracheæ develops in the body. Finally, often as long as a year after hatching, the fully-grown nymph crawls up a weed or some other object that projects from the surface of the pond into the air and the final moult occurs. When it has found a suitable place, it pauses for hours, sometimes even for a whole day; then abruptly the nymph-skin splits down the back and the tender body of the imago comes out. There is then a period of almost volcanic expansion before the outer skeleton dries and stiffens. The wings, at first small and moist and crumpled, expand and straighten very quickly; the body alters its shape. In a few hours the body is well-formed and rigid, and after a few tentative movements, the adult dragon-fly sails off to hunt and devour other insects and to breed until the chill of winter puts an end to its life.

In other insects such as butterflies, beetles, bees, and house-flies, the metamorphosis from larva to imago is more abrupt. The butterfly larva, or caterpillar, is a voracious soft-bodied grub with hardly a trace in its body of any of the structures that it will possess when it is adult. The fully-grown caterpillar, when it is ready to moult for the last time, generally encloses itself in a bed of silk and then undergoes a series of profound changes. First it moults and assumes a form resembling more nearly that of the adult. In this stage, as a *pupa*, it remains for weeks or months, and during this time its organs are reconstructed very drastically indeed, undergoing almost complete liquefaction in the process. Finally, as with the dragon-fly, a limp and crumpled butterfly emerges, to expand and dry its wings and then take to the air for its culminating experience of life.

§ 3d. Centipedes and Millipedes.

THE class *Myriapoda* includes a number of land-living arthropods, breathing by means of tracheæ, and having long serpentine bodies with from nine to over a hundred and fifty pairs of legs. In the arrangement of the appendages on the head, in their embryonic development and in other respects they show many resemblances to the insects. The best-known members of the group are the flat-bodied centipedes and the rounded millipedes; they may be distinguished by the fact that in the latter the body-segments are welded together in pairs, so that there appear to be four legs to each segment, while in the former there are only two. The centipedes are fierce, active creatures, stalking small invertebrates and killing them by means of their strong, poisonous jaws. The millipedes are inoffensive plant-feeders, with a preference for decaying plants; their bite is not poisonous, but some of them, if seriously disturbed, can exude a foul-smelling fluid in self-defence.

§ 3e. Peripatus.

FINALLY we come to a class, the *Onychophora*, containing a few species only, all belonging to the single genus Peripatus. The animals are humble creatures, found hiding under stones or damp bark in South America, Africa, Malaya, and Australasia. To the layman they are harmless unimportant grubs, but to the zoologist their anatomy is full of interest.

Peripatus differs in a number of important respects from other arthropods. Its chitinous integument is comparatively thin and beset with little spiny warts; being flexible it shows no jointing on the limb or division into body-segments. The numerous appendages show less specialization of structure than do those of other arthropods. In its general appearance the animal looks half-way between a centipede and a worm.

In its internal anatomy the animal shows many arthropod features. It has a body-cavity that contains blood, simple unbranched tracheæ, and so on. But in other respects it resembles the worms. It has a series of kidney-tubes, one pair corresponding to each pair of legs like the kidney-tubes of an earth-worm, and its central nervous system, instead of lying along the belly, forms two distinct cords running along the two sides of the body and connected together by regular communications like the rungs of a ladder—a condition that is paralleled by the flat-worms and by some molluscs.

And in its development, it, like sea-squirts or Sacculina, drops clues as to its affinities. Early in embryonic life, it has an ordinary blood-system, with veins and arteries, and an ordinary body-cavity with no blood in it, just like a worm. Gradually some of the veins enlarge, compress the original body-spaces into insignificance and become the blood-filled body-cavity of the adult.

Peripatus, then, is an arthropod which shows many resemblances to animals of other phyla, particularly the segmented worms; it seems to bridge much of the gulf between the arthropods and this latter group.

III

FURTHER PATTERNS OF INDIVIDUALIZED ANIMAL LIFE

§ 1. *Other Animal Phyla.* § 2. *The Molluscs.* § 3. *Echinoderms: Nature's Pentagonal Experiment.* § 4. *Segmented Worms.* § 5. *Roundworms.* § 6. *Flatworms.* § 7. *Et Cetera.*

§ 1. Other Animal Phyla.

WE HAVE spoken already in our introduction of the dumps used by the early biological classifiers. At first the only real phylum of animals to be picked out was the vertebrates. The remainder was all a miscellaneous dump labelled (wrongly) "Bloodless Animals." The next great phylum to be separated away from this confused heap was the arthropods we have just dealt with. Linnæus (1707-1778), the father of precise biological classification, made the distinction, using the word "Insect" to cover all the arthropods—but he did not regard the arthropod division as equivalent to the vertebrates; he grouped animals into the six classes—mammals, birds, amphibians (including also the reptiles), fishes, insects, and worms. This last class, the "worms," is his dump; here all other creatures were left together because as yet there was insufficient anatomical knowledge for them to be sorted out.

Gradually we have come to realize that quite a number of other phyla—of other "general ideas" and general plans of animals, that is to say—exist in the Animal Kingdom. We have, we hope, made clear that there is no sort of precedence or subordination between these phyla. The lobster is not lower or higher than the frog or fish. He is altogether different. He belongs to a different series of experiments that Nature seems to have made. So, too, we now recognize another very important and various phylum, another plan of living mechanism, in what are called molluscs (such as oysters, mussels, snails, squids, octopuses), and yet another in the echinoderms (the starfish, sea-urchins, sea-cucumbers, and sea-lilies). There are three perfectly distinct phyla of worms, the segmented worms, the roundworms, and the flatworms, each a large and important group. And there are the phylum of jelly-fish and polyps, the phylum of sponges, and the phylum of animals too small for the naked eye to see. But besides these there are other phyla, which give a less ample spectacle of range and variations. One might speak of them as Nature's less successful or less versatile trials. She is like an enterprising manufacturer who has put a number of devices on the market and followed up those that caught on. The chordata, when it got to the vertebrate pattern, was a huge success, and the call for arthropods was steady and abundant. Both these phyla conquered sea, land, and air. The molluscs and flatworms and roundworms went well, the segmented worms and echinoderms went fairly well. But certain other phyla, we shall note, never ran away with matters in that fashion. The zoologist to-day has still to wind up his tale with a collection of odd and apparently unrelated things, minor phyla, and at last something it will be most truthful to label "et cetera." There are phyla Nature has not gone on with; in the past there may have been endless phyla she has set aside and ceased altogether to produce.

To each of these less exciting patterns of animal construction we will devote only a section or a part of a section here. That must not hide from the reader that we are dealing with phyla, with series of fundamentally different kinds of animal body, each, from the point of view of physiological mechanism, as important as the chordata and the arthropods.

§ 2. The Molluscs.

MOLLUSCS are creatures nearly all of which possess shells—double (bivalve) shells like those of oysters, clams and cockles, single, hollow shells like those of snails, shells embedded in the body-wall like the "bone" of a cuttlefish. Except for these shells they are soft-bodied, without any skeleton. They are not segmented as arthropods are, but their blood is coloured pale blue with hæmocyanin, like the blood of a crustacean. In the great majority of cases (the shell of the bivalves being a notable exception) their muscles do not form accurately defined bands playing on jointed skeletal levers, like the voluntary muscles of a vertebrate or an arthropod; their locomotor apparatus consists generally of an apparently disorderly tangled mass of muscle-fibres, the so-called "foot," an arrangement which leads to plasticity rather than precision.

There are five classes of molluscs, of which only three, the bivalves, the snails and slugs, and the octopuses and cuttle-fish are of sufficient importance to mention here.

The bivalves, or *Lamellibranchia*, are all aquatic and most of them marine. They lie between two hard shells which are hinged together along one side; by means of powerful muscles the shells can be pulled together to enclose the animal completely when any danger threatens. The poor thing's only retort to hostility in most cases is just to "slam the door." If man must go to the ant, that arthropod, for lessons against sluggishness, it is to the lamellibranch he must be sent to calm his violence and anger. Some of these animals, like the fresh-water mussel, creep sluggishly about by means of that lobe of muscular tissue, the "foot," which they protrude from the shell.

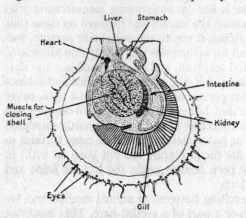

FIG. 85. A Scallop after removal of one valve of the shell, showing where some of the more important organs are placed.

Some, like the heavy-shelled oyster, or the common blue mussel of our shores, that attaches itself firmly to rocks by means of a tuft of adhesive threads, do not move about at all. Some, like the cockle, which can jump by means of a spring-like action of its "foot," or the scallop that swims by rhythmically opening and closing its shell-flaps, are surprisingly active creatures. The scallop is a marine butterfly; to guide it in the long flights which it takes it is provided with a row of glistening eyes just inside the rim of each shell. To ourselves the bivalves are important for three chief reasons; firstly because most of them are good to eat, secondly because the oyster makes pearls, and thirdly because some members of the class bore holes in wood or rock, and can do extensive damage in this way to piles, breakwaters, wooden ships and the like.

The bivalves live by passing a continual current of water through the shell, a current from which they extract oxygen and the minute particles of organic matter upon which they feed and into which they shed their waste-products. On opening a mussel, the most obvious structures that present themselves are the so-called "gills," the engines which keep this slow current of water in motion. The "gills" are not primarily breathing organs at all; they are elaborate filters like the pharynx of Amphioxus or a sea-squirt, through which water

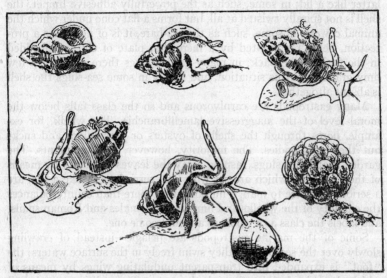

Fig. 86. The Common Whelk (Buccinum undatum) coming out of its shell and attacking a clam with its long proboscis.

The speckled, uplifted tube is the "siphon" through which the whelk breathes water. Below on the right is a cluster of tough, papery whelk's eggs.

is worked by means of cilia, and which collect suspended particles of food and pass them down to the mouth.

Some of the most important points in the anatomy of these creatures may be seen in the figure opposite. Perhaps their most bizarre feature, in vivid contrast to the scheme of the two phyla previously studied, is the fact that the intestine runs actually through the ventricle of the heart.

The second class of molluscs, the *Gastropoda*, is also the largest; it includes the terrestrial snails and slugs and the whelks and winkles of the sea. There are about 6,000 terrestrial and 10,000 aquatic gastropod species living to-day, most of the latter being marine. With them the "foot" is a flat muscular sole to the body lump—hence the name of the class, from the Greek for stomach and foot. They go on their stomachs; and most of them bear on their back a spirally coiled shell into which they can retire when they choose. In some forms the tail of the "foot" carries a hard chitinous plate, so arranged that when the animal retires into its shell the plate closes the opening of the latter like a lid; in some, such as the powerfully adhesive limpet, the shell is not spirally twisted at all, but forms a flat cone under which the animal cowers; in some, such as the sea-hare, it is of no use as a protection, being represented by a mere thin plate of chitin embedded in the flesh of the back; in many garden slugs there are only a few limy granules in this situation, and finally in some sea-slugs the shell is absent altogether.

Many gastropods are carnivorous and so the class falls below the moral level of the inaggressive lamellibranchia. The whelk, for example, bores through the shells of oysters or periwinkles and sucks out their soft bodies. The majority, however, feed on plants. The garden snails and slugs, rasping away the leaves of plants by means of their tongues, which are set with fine, serried teeth like a file, are a serious nuisance to man, a nuisance that more than counterbalances the edibility of the whelks and limpets and winkles and Roman snails, and labels the class as a whole as a troublesome one.

Some of the marine gastropods are pelagic—instead of crawling slowly over the sea-bottom they swim freely in the surface waters; the "foot" is expanded into transparent undulating wings by means of which the animal glides through the water. The ordinary pond-snail can often be seen in a sort of slow, writhing swimming just below the surface of the water. These pelagic gastropods develop this idea.

The third class of mollusca, the *Cephalopoda*, is more exciting, including as it does the cuttle-fish and octopuses. There are comparatively few species, all marine; nevertheless, its members are the most highly organized molluscs, and perhaps the most highly organized invertebrates. Further, they include by far the biggest invertebrate animals, for giant cuttle-fish have been recorded over fifty feet long. Their name means "head-footed"; for the "foot," a tongue-like body in the lamellibranch and flat underneath in the gastropod, here grows forward to surround the mouth and is prolonged into flexible arms.

The majority of living cephalopods capture their prey—which consists chiefly of crustaceans and fish—by stalking and pouncing upon it. The circle of writhing arms surrounding the mouth is provided with powerful suckers, by means of which they catch and hold their victims. Once captured, the prey is bitten by a pair of jaws, sharp and curved like an eagle's beak, and a digestive juice is pumped out into its body from the captor's mouth. The juice paralyzes the victim very quickly, and then it dissolves and partly digests the flesh, the resulting soup being sucked back into the captor's mouth held close to the captured prey.

FIG. 87. The common Cuttle-fish (Sepia).

Above on the right in backward flight; below, capturing a crab.

The octopuses live mainly on the bottom, lurking in caves and crannies, often building a shelter of stones, pouncing out upon unsuspecting passers-by and falling on top of them like a parachute of serpents. In the argonauts, which are closely related to the octopuses, the female carries a delicately built spiral shell, partly in order to protect her own body but chiefly as a sort of perambulator for carrying her eggs. Both octopuses and argonauts have eight long tentacles. The cuttle-fish and squids, however, have ten; they constitute the largest and most varied group of living cephalopods, and swim through the sea like fish, propelling themselves either backwards or forwards by means of fin-like structures that run along the sides of their bodies. Of the ten tentacles two are very long, and are generally kept curled up out of the way, but they can be suddenly shot out to snatch at victims; the remaining eight are shorter, and are used to hold the prey more firmly once it has been caught. Embedded in the flesh of the back of the squids is a shell homologous with the shells of snails; in some species the shell is a thin, horny "sea-pen," and in others it is the white, massive "cuttle-bone" familiar to canary owners. Another kind of cephalopod, the pearly nautilus, is found swimming near the bottom round the shores and coral reefs of the South Pacific—it lives in a beautifully coiled spiral shell; a shell very different in structure from the perambulator-shell of the female argonaut, and its numerous tentacles are not provided with suckers. The fossils known as ammonites are the spirally coiled shells of a vast parallel group of cephalopods that has long been extinct.

The cephalopods breathe by rhythmically inhaling and exhaling water into and out of a large "mantle-cavity" where it passes over their plume-like gills. By suddenly ejecting water from this cavity they can squirt themselves rapidly backwards; this is their method of flight from danger. The cuttle-fish have an ingenious additional method of evading their enemies, depending upon their astonishing powers of rapidly changing colour—powers that put a chameleon to shame—and upon the inky fluid, sepia, which they can emit when they wish. A closely pressed cuttle-fish tries at first to mislead its pursuers by changing as it flees from white to dark brown, then to a mottled colour and so on. If these devices should fail it has another resource. First it becomes very dark indeed, and for a few seconds remains so. Then abruptly it turns off to one side, escaping in a direction at right angles to its original line of flight; as it does so it suddenly turns white and at the same time it belches out a puff of sepia, which travels away in the opposite direction. The pursuer, which has been chasing a dark object, follows the black puff; by the time it has realized its mistake the cuttle-fish has made good its escape.

These cephalopods are undoubtedly the most intelligent-looking invertebrates. They have well-developed and grimly expressive eyes, and the subtle coiling movements of their arms and the rapid and very striking colour-changes that they undergo when stalking their prey or when pursued are suggestive of mental accompaniments. Observers who have watched the very elaborate courtship of these animals, or who have noted the voluptuous writhings of the tips of the arms when a hungry octopus devours its prey, find it easy to believe that these creatures can experience a rich variety of intense emotions—though it must be admitted that we have, as yet, no clear evidence of memory or forethought on the part of a cephalopod.

§ 3. Echinoderms: Nature's Pentagonal Experiment.

MOST of the animals that we have considered so far bear a certain very crude resemblance to ourselves in the way their bodies are laid out. They have front ends where the chief sense-organs and the mouth are placed, they have hinder ends, they have backs, and bellies on which they crawl. A bivalve or a flatworm offers difficulties; but for the most part, by stretching one's imagination just about as far as it will go, one can get a crude idea of what the animals feel like. If, for example, one were covered with close-fitting combinations made

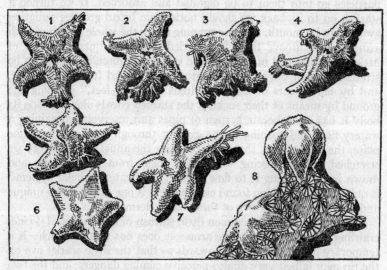

FIG. 88. Asterina, a little starfish, about life-size.

In 1 the animal has been turned upside down; in 2 to 6 it rights itself again, and in 7 it crawls away. In 8 two individuals are browsing over a gelatinous colony of sea-squirts.

of mediæval armour, and if the number of one's limbs were legion, one would feel not unlike a lobster, and if one had no arms or legs and lived in a pitch-dark tank of warm thick soup one would feel not unlike the roundworms to be presently described. But with the phylum *Echinodermata* things are different; with a few exceptions they do not know front and back, right and left. They are not bilaterally symmetrical, as we are. They do not "look before and after." One cannot imagine oneself a starfish or a sea-urchin however hard one exerts one's mind.

A starfish consists of a number of arms—usually five—radiating out from a central disc—but the so-called "arm" of a starfish is no more like our own arms than a lobster's "liver" is like our livers—indeed, hollow as it is and containing viscera, it is more like a belly than an arm. Its mouth is exactly in the middle of that disc, on the lower side on which it crawls; its anus is in the middle of the upper surface. Its mouth opens into a loose, baggy stomach; that leads into a short intestine; from the intestine ten long branches run off, two into each arm—branches not unlike the liver of a lobster, for the finer food-particles go into them to be digested and absorbed. If we turned a starfish on to its back we should notice five broad grooves radiating away from its mouth, and one running along the whole length of each arm; in each groove there is a serried double or quadruple row of little tubes that end in suckers. The tubes are muscular. They can be stretched out or drawn into the grooves or waved from side to side, and by movements of these hundreds of "tube-feet," gripping the ground by means of their suckers, the starfish crawls along. Inside its body it has an elaborate system of pipes and reservoirs containing a watery fluid and opening to the exterior through a porous, sieve-like plate, the *madreporite*, that is placed near the anus; the tube-feet are stretched out by forcing water into them from this system, and drawn in by allowing it to flow back. This "water-vascular system" is quite unlike anything found elsewhere; it is one of the many unique and characteristic features of these echinoderms.

Fig. 88 shows a very common little starfish of our shores, *Asterina*, crawling along. One of its five arms—it does not matter which—is a temporary head; it is curved upwards so that the little scarlet eye on the tip may (albeit very dimly) perceive coming dangers, and the two or three end pairs of tube-feet are extended in the water like the antennæ of an insect. The other arms are subordinate; they are simply walking, and not sitting up and taking notice. If we tap the front end of the leading arm, or if for any other reason it occurs to the animal that the direction in which it is proceeding is not desirable, that arm ceases to act as a head and becomes passive; one of the other arms holds up its eye-spot and waves its tentacles like antennæ, and the animal follows the new head in a new direction. It will be noted that at any one time the many hundreds of tube-feet are all pulling together, some working in a direction parallel with the groove in which they sit, some, in other arms, working across their grooves at various appropriate angles. How much more applicable to the starfish than to the centipede is the celebrated epigram:

> "The centipede was happy—quite!
> Until the toad in fun
> Said, 'Pray, which leg moves after which?'
> This raised her doubts to such a pitch,
> She fell exhausted in the ditch,
> Not knowing how to run!"

The starfish feeds on animal prey—colonial sea-squirts or polyzoa or small worms in the case of Asterina, oysters, mussels, scallops and so on in the case of some larger forms, other starfishes in the case of the seven-armed *Luidia*. They have no masticatory apparatus. A few starfishes take their victims whole into their bodies and there digest them; others hold them by means of their tube-feet and turn their own stomachs inside-out through their mouths on to the prey, digesting it outside the body. Sometimes Asterina may be seen to browse slowly over colonies of sea-squirts by means of its everted stomach.

The starfish has no internal skeleton, but its integument is made hard and stiff by a network of bony plates and spines. Scattered over the surface of its body are little pincer-like structures, the *pedicellariæ* —tiny groups of two or three toothed jaws mounted on a muscular stalk. By means of these gnashing pincers the starfish cleans its body and defends itself against its enemies; in one or two instances they have been seen to be used when attacking prey, and sometimes they secrete poison.

The echinoderms are an exclusively marine group. The sausage-shaped trepangs or bêches-de-mer are an important luxury food among Eastern peoples, and there exists a large industry for collecting them from the reefs of California and the Eastern Archipelago. And the spawn of certain sea-urchins is eaten in Mediterranean countries. The chief trouble caused by echinoderms is that some of the larger and more voracious starfishes wreak havoc on oyster beds.

There are five living classes. The starfishes we have already mentioned. The second comprises the brittle-stars—surprisingly active animals with round button-like discs and very thin, serpentine arms, by lashing movements of which they move along. The third class, the sea-urchins, cake-urchins, and sand-dollars, have no arms. The last two types nourish themselves like earthworms by shovelling mud into their mouths and digesting out any nutritious matter that it may contain, while the prickly urchins are mainly browsers upon seaweed. A sea-urchin is a hollow, bony ball bristling with spines, tube-feet, and pedicellariæ. Below, on the side on which it crawls, is its mouth, provided with five sharp teeth; the teeth are moved by an elaborate engine of bones and muscles called "Aristotle's lantern." Above,

FIG. 89. A group of Echinoderms.

On the left are three Sea-gherkins (Cucumaria) catching and swallowing minute organisms. In the centre are two Feather-stars (Antedon), one swimming and one resting on a stone. Below on the right is a Brittle-star (Ophiothrix), above that a Starfish (Asterias), and above that again a Sea-urchin (Echinus). The characteristic tube-feet can be clearly seen in several of the animals.

at the antipodes, so to speak, are the anus, the madreporite, and the openings of the testes or ovaries.

The fourth class, the holothurians, includes the cotton-spinners (so-called because they exude sticky white threads from the anus when disturbed, in order to entangle and confuse their enemies), sea-puddings, sea-cucumbers, sea-gherkins, and bêches-de-mer, besides a number of curious burrowing, pelagic, and deep-sea forms. They are like sea-urchins that have been softened and pulled out into long sausages, and in the process they have acquired bilateral symmetry, for they creep on one particular side of the body with the mouth in front and the anus behind. Many of them feed like sea-urchins, shovelling mud into themselves by means of a ring of spade-like tentacles round the mouth. Others consume minute organisms that float in the sea water; they wave their sticky tree-like tentacles, entangling the organisms, and then suck them one by one like a small boy sucking jammy fingers.

The fifth class, the feather-stars and sea-lilies, are somewhat different from the rest. They have long, branched arms radiating out from a central disc. The feather-stars swim by sinuous movements of their arms, occasionally anchoring themselves for a shorter or longer period by means of a cluster of claw-like roots. The sea-lilies are permanently fixed, living at the end of long, jointed stalks.

In addition to these five classes there are curious fossil groups that have become extinct.

We have already stressed the fact that in the majority of cases the symmetry of an echinoderm is radial—instead of having two symmetrical sides to its body as we have, it consists of a number of symmetrical structures radiating out from a central disc. It is curious to note that this is only true of the adult. A baby echinoderm is typically a minute, delicate, transparent creature swimming through the water by means of one or more fringes of cilia that run round its body. Moreover, it is bilaterally symmetrical: it has front and back ends, right and left sides. But presently a set of rudimentary organs appears on its left side, rudiments that grow at its expense like a tumour and finally absorb the whole of the larva, becoming the radially symmetrical adult. It is as if a lump appeared on the left flank of a child of six or seven and grew at his expense, developing new organs of its own, finally absorbing all his tissues, and giving rise in that way to the adult man or woman—except that in the case of echinoderms the organization and symmetry are very different in child and adult.

§ 4. Segmented Worms.

The next invertebrate phylum, the *Annelida*, includes such animals as the earthworms, the leeches and the lugworms—a series of animals which are distinctly like the arthropods in their organization. They have, however, no rigid armouring, and their blood does not fill their body-cavities but is confined within definite vessels. It may be red as ours is, or it may be green, but it is never coloured pale blue by hæmocyanin. In these respects they differ from arthropods. But they have a central nervous system running along their bellies, and a long, rhythmically contracting artery that plays the part of a heart running along their backs; in these matters, and in the way the nervous system forms a loop round the mouth, they are like arthropods. Moreover, they are segmented. We noted that the body of any arthropod may be regarded as a modified chain of segments, each of which bears one pair of appendages. This type of structure is even more obvious in annelids. The body shows a series of ring-like grooves that divides it into segments; on opening it up we should find that the body-cavity is divided into a series of compartments by transverse partitions corresponding to the rings. Each of these segments contains a pair of tubes that act like kidneys, each contains a ganglionic swelling of the nervous system, and in many forms each contains a pair of testes or ovaries. Moreover, in all annelids except the earthworms and leeches, each segment bears a pair of appendages, not jointed but bearing tufts of bristles, and used in most species as primitive limbs, which strengthens their resemblance to the segments of arthropods. If the reader will glance back at our description of Peripatus he will probably be disposed to agree with the many authorities who regard the arthropods as a sort of higher, more complex, development of the annelid plan.

There are three main groups of annelids. The first includes the burrowing earthworms and one or two similar but aquatic forms; one species, from Tasmania, attains a length of six feet. Charles Darwin's book on *Vegetable Moulds and Earthworms* has shown what a vitally important part is played by earthworms in loosening and turning over the soil. The animals consume earth in great quantities and digest nutritious matter out of it as it passes down their intestines; they defecate it on the surface of the ground as "worm-castings." In rich garden soil, where worms are especially numerous, there may be over 50,000 individuals to an acre; in these circumstances more than ten tons of earth will pass through their bodies and be brought to the surface in one year, and in ten years this will form a continuous layer of finely divided surface soil at least two inches deep. In Darwin's own words, "The plough is one of the most ancient and most valuable of man's inventions; but long before he existed the land was, in fact, regularly ploughed and still continues to be thus ploughed by earthworms. It may be doubted whether there are many other animals which have played so important a part in the history of the world as have these lowly organized creatures."

Earthworms, we may note, are "hermaphrodite," male and female in one, and their embraces are reciprocal. This is not by any means the only departure from the relentless division of individuals into two sexes that plays so important a part in our own social life. Snails, for example, are both male and female, but in this case the two faculties are not exercised simultaneously; the creature takes turns, being now one sex and now the other. The edible oyster alternates slowly but regularly between male and female, and there are many cases in the invertebrates of animals which normally start their lives as males, become mature, and then at a certain age turn over and become female for the rest of their lives. The vertebrate way is not the only way. Indeed, as we proceed with our survey we shall find increasingly striking departures from our own way of life, even in those matters which we regard as the very warp and woof of it.

The second group of annelids is of less economic importance, but it includes a greater variety of forms, all of which are marine. It includes among its hundreds of species the graceful, sinuous paddle-worms of our shores, the iridescent sea-mouse, the lobworm, which is often used as bait, and a number of animals which live in tubes attached to rocks, sea-weeds and the like, and feed on suspended matter, much as bivalves or sea-squirts do, by means of filtering crowns of delicate tentacles.

The third group comprises the leeches, freshwater, marine and terrestrial. They have suckers fore and aft, and some, like the medicinal leech, get their nourishment by attaching themselves to larger creatures and sucking blood. Others are active beasts of prey, devouring pond-snails, blood-worms and the like. In a number of respects the leeches resemble the earthworms, so that it is customary to divide the annelids into two great groups, one comprising the earthworms and the leeches and the other, probably the more primitive of the two, including the more varied marine group.

§ 5. Roundworms.

THE phylum *Nemathelminthes* or roundworms includes a number of worms of plain, unassuming appearance but nevertheless of great importance to ourselves because a number of them are parasitic. The roundworms often found in the intestines of pigs, horses, and men are well-known members of the group. They have thin cylindrical bodies, up to a foot long, pale buff in colour, and tapering away at both ends. Their anatomy shows well the simplicity of structure that is often found in parasitic organisms. Living surrounded by digested food, all they have to do is to gulp it into their own bowels and absorb it; they have no need of glands to manufacture digestive juices, and their alimentary apparatus consists of two parts only—a muscular pump, the pharynx, which sucks the juices in, and a simple intestine where they are absorbed. Moreover, as they inhabit a warm, sheltered place and have neither prey to seek nor enemies to fear, their behaviour is almost non-existent. They have no eyes or other specialized sense-organs—indeed, eyes would be useless in the darkness of our bellies—and their nervous and muscular systems are very elementary in structure. They can perform simple writhing movements, but it may be

FIG. 90. A group of Marine Worms.

Below, left to right, are two iridescent Sea-mice (Aphrodite), a burrowing Lug-worm (Arenicola) with its tufted gills, and a Scale-worm (Lepidonotus). Above are a bronze-coloured Rag-worm (Nereis), and two different kinds of tube-building worms (Sabella and Amphitrite), spreading their tentacles for oxygen and food.

presumed that they do not do so very often; beyond reproducing themselves and occasionally dodging an exceptionally violent peristaltic wave it is hard to see what movements they require. Correlated with this low level of muscular activity we find that they possess no special circulatory or respiratory organs; their muscles do so little work that all the oxygen they need can diffuse in from outside.

It is at first sight difficult to explain how an animal which spends its life in an intestine can get any oxygen at all, and indeed it used to be thought that the parasitic roundworms depended upon a different kind of chemical mechanism for their vital energy. But the story has now been made clear. When the host-animal is not digesting a meal there is very little oxygen, if any, in his bowels; under these conditions the roundworms become anæsthetized by the chemical products of their own activity and sink into a drugged immobility. But when the host consumes a meal the blood-supply to his gut is increased enormously and oxygen diffuses—accidentally so far as he is concerned—into the intestinal canal. When this happens the roundworms revive and begin to gulp; when his digestion is over they sink again into their torpid sleep. There is a rhythm in the life of an intestinal roundworm as marked as our own rhythm of day and night; a period when there is nothing doing, and then a period of activity, when the world in which it lives begins to churn and writhe, and when oxygen and food are suddenly abundant. Instead of dawning and waning light there is dawning and waning oxygen. It is an easy life; there is nothing to fear (except the death of the host) and nothing to do but sleep and swallow; the weather is always perfect. Fortunately, having no brain worth mentioning, the worm cannot possibly get bored.

But there is another side to the life of a parasite. Before it can enjoy this Elysium it has to get there, and it is no easy thing for a worm as simply made as the roundworm to invade a mammalian intestine. Consequently we find that eggs are produced in enormous numbers —a single Ascaris, it has been reckoned, can produce 200,000 eggs a day—and only very few among these thousands reach their earthly paradise. In some species, such as the comparatively harmless *Ascaris lumbricoides* of the human bowel, the eggs pass out with the fæces, and a few lucky ones are swallowed by a new host—for instance, on salad leaves—and thus win through. It is simply a matter of hit and miss, and the prolific mother makes so many shots that they cannot all miss. In other species, however, there are more elaborate histories. In the dreaded *Trichinella spiralis*, for example —a worm only a few millimetres long—the adult lives and pairs in the intestine of a man, a pig, a rat, or another suitable mammal. The fertilized female burrows into one of the villi and there brings forth living young—sometimes many thousands—which escape into the lymphatic vessels of the host. Thus, reaching the circulation, they are carried all over the body, until they settle down in the voluntary muscles. They burrow into the muscles, shut themselves up in little hard cases and wait. Nothing more happens to them until the host dies; but if after his death his flesh is eaten by another mammal—a man by a rat, or a pig by a man—they revive and flourish in the bowels of the latter. An infected animal may have tens of millions of these parasites patiently waiting in his muscles for his death, and hurrying it up by producing the inflammations or other symptoms of "trichiniasis." It has been calculated that one ounce of infected pork may contain 85,000 of them—if the pork was not properly cooked its consumption would involve the liberation of all these parasites

FIG. 91. Microscopic view of a piece of infected pork, showing white, parasitic worms (Trichinella) curled up among the muscle-fibres.

FIG. 92. Planaria, a kind of flatworm, showing the digestive tubes in black

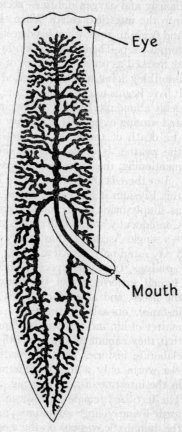

in the bowel of the eater, and if half of these were females, producing on the average five hundred embryos each, the result would be a migration of twenty million young worms into his muscles.

Again in the guinea-worm, *Dracunculus medinensis*, often called *Filaria*—a worm from one to six feet long—the adults live in the connective tissue of man, just under the skin. The young escape into the outer world through abscesses caused by the presence of the parasite, and some of them have the good fortune to get into a stream or a pond.

Here they make their way into the bodies of water-fleas, *Cyclops*, where they wait until the water is drunk by a man careless enough not to filter it first. Thus they reach the human bowel, and thence, presumably by the circulation, they travel to the connective tissue of the skin. In a case of this kind, where the adults live inside a creature of one species, but the young, in order to get there regularly, enter the bodies of some quite different animal, the latter is called an intermediate host.

We cannot enter here into an account of the many peculiar features of the organization of the roundworms which has led to their being put apart in a phylum of their own. In a number of respects they are interesting—their spermatozoa, for example, instead of swimming by means of long, lashing tails, creep slowly like white blood corpuscles; and their epidermis, instead of being divided up into cells, consists of an uninterrupted sheet of protoplasm freckled with nuclei. We may, however, note that these creatures are not all parasitic. A huge number of them live harmlessly in salt or fresh water or in soil or the tissues of rotting plants. Some species are free when they are young, but become parasitic when they mature—such as the dangerous hookworms, *Ancylostoma* and *Necator* which infect man, and *Tylenchus scandens*, which invades the tissues of cereals. In some, on the other hand, the young are parasitic while the adults are free-living; an example of this is afforded by *Mermis* whose young grow in the tissues of insects. Finally, many, as we have seen, are parasitic throughout, although the larvæ and adults may infect different hosts or different parts of the same host; and many of these are very troublesome indeed to man.

§ 6. Flatworms.

THE *Platyhelminthes* or flatworms are for the most part parasitic. Their organization is curious, and we may introduce ourselves to it by examining a particular member of the group.

Planaria lactea is a white, flat, worm-like creature, about three-quarters of an inch long, that leads the life of a scavenger at the bottom of fresh-water ponds, nourishing itself for the most part on dead organic matter. Usually it travels along with a smooth gliding motion; from time to time it varies this somewhat monotonous method of progression by "looping" like a looper caterpillar.

The head of Planaria is interesting. Our own heads are chiefly remarkable for two reasons; first, because they contain the chief sense-organs and the brain, and second, because they contain the mouth and the apparatus for taking in food. Now the head of Planaria is only half a head; it carries the brain and eyes, but it has no mouth. The mouth, instead of being at the front end of the body, is placed at the end of a long tube that sticks out of the middle of the belly; moreover it combines the functions of mouth and anus.

Another curious feature of this animal is that it possesses no circulatory system at all, although, unlike an intestinal roundworm, it is an actively moving creature. The body-cavity is filled with spongy connective tissue, through which the digestive and excretory systems ramify like the branches of trees. Since there is no blood, the branching intestine has itself to carry food into every corner of the body and the branching kidney-system has to run into every corner and drain the waste-products away. Because of this fact the anatomy of Planaria looks very complicated; if, in the accompanying figure of the digestive apparatus, the reader imagines an equally arborescent system of fine excretory canals, intricate reproductive organs (both male and female, in the same body), and a web-like nervous system he will get some idea of the labyrinthine nature of this tiny inside. But the complexity is more apparent than real. The organs are elaborately branched, but there are not nearly so many kinds of organs, or so many kinds of tissues, as in a vertebrate.

Moreover, Planaria has no special respiratory organs; its body is so thin that enough oxygen to satisfy its needs can diffuse in through the tissues.

There are three classes of flatworms. The first class is that to which Planaria belongs; they live for the most part independently in damp places on the land, in fresh water, and in the sea, but one or two of them are parasites.

The second class is the flukes—a group consisting entirely of parasitic forms. These animals show the most ingenious devices for invading and destroying their unwilling hosts; we may illustrate the point by following the life-history of the liver-fluke, *Fasciola hepatica*, found in sheep and causing the wasting disease known as sheep-rot, particularly in beasts that feed in wet, soggy pastures—a life-history that is surprising because it involves a regularly alternating rhythm of several quite different generations.

The flukes that inhabit and devour the liver of an infected sheep are hermaphrodite—any one individual having both male and female organs—but nevertheless they fertilize each other. Like most parasites, they are incredibly prolific. Their microscopic eggs are carried by the bile down the bile-duct of the sheep to the intestine; thence, among the fæces, they reach the exterior. During the next spell of warm, moist weather the hard egg-shell breaks and there emerges a creature very unlike its parents. This creature, the *Miracidium*, has a tiny conical body covered with cilia, two eyes of very elementary structure, and no mouth or stomach. Begotten by parents over an inch long, it is itself less than one two-hundredth of that size. Nevertheless it is active from the moment of hatching. It swims about in the puddles

Fig. 93. The Life-cycle of the Liver-fluke.

or ditches or larger pools in the grazing meadows, or even through the thin surface film of water left on the grass after rain or dew, with every appearance of eager purpose. Now, there are certain kinds of snail which live in ponds or among the damp grass in swampy places, and it is for one of these that the little creature is searching. Its eagerness is excusable, for unless it succeeds in its quest within eight hours, it perishes. But if it does succeed, it bores its way into the body of the snail and alters in shape, becoming a long, hollow bag. At this stage it rests, nourished by the snail's blood and growing slowly, for about a fortnight; then it reproduces itself—it gives rise to a number of embryos and dies. That is the first generation. The offspring of this little adventurer differ from both their immediate parent and from the original flukes. They are elongated and wormlike, about a twenty-fifth of an inch long when fully grown, and they wander about the body of the snail and consume it, showing a particular preference for the liver. These ingenious parasites, the *rediæ*, live and die and reproduce themselves inside the snail, weakening it but prudently not killing it, and in this way there follow several generations more, all very much alike. But at the chill of the following autumn there is another change. New forms appear in the little community—forms called *cercariæ*, somewhat smaller than the rediæ, with big round heads and long thin tails. The cercariæ are not satisfied by the leavings of their parents and—platyhelminth pilgrim fathers, so to speak—they bore their way out of the snail's body in quest of a larger organism to colonize. They swim up wet blades of grass or the leaves of other herbage, and there they fortify themselves; they exude a slime which thickens into a hard white capsule and, thus protected, they wait. If nothing more happens to them they eventually die, but if the grass on which they are waiting chances to be swallowed by a browsing sheep (or other grazing animal, for flukes of the same species have been found in oxen and deer and even rabbits) they make good use of their opportunity. Their capsules are obligingly digested away by his gastric juice, and when he has freed them in this way they creep up his bile-duct to his liver, where they thrive and grow at his expense. In about six weeks' time they are ready to beget eggs to start on the adventurous cycle again. There is, then, a perfectly regular rhythm of generations that differ from each other—first the fluke, a creature over an inch long, living in the sheep's liver, then the barely visible creatures living in the snail, then the fluke again, and so on. It is as if the offspring of men were mice and the offspring of mice were men.

The flukes include a number of different forms with more or less picturesque and adventurous life-cycles. The one most troublesome to ourselves is Bilharzia of tropical countries, which spends its fluke stage in the human intestine or urinary bladder, and its microscopic stage, like that of Fasciola, in a fresh-water snail. It invades us through the mouth or through the skin, and reaches the bladder along the veins.

The third class of flatworms is the tapeworms—also parasitic, and also showing an alternation of generations in two different hosts. The animals have long bodies—sometimes up to eighty feet of them—and they have suckers and hooks on their "heads" by means of which they cling to the wall of the host's intestine. They have no digestive system of their own, but absorb the food at the surfaces of their bodies when it has been digested for them by these hosts. The body consists of a chain of segments—two or three to many hundreds—each with a complete reproductive system in it. As the hinder ones ripen they drop off, and swollen pockets of eggs are loosed into a wider world with the fæces of the host; but they are continually replaced by new

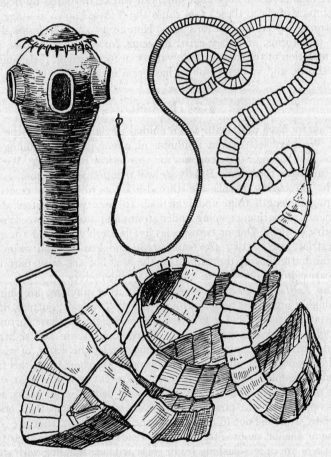

FIG. 94. Tænia, a common tapeworm.

Inset is magnified view of the head, showing the armoury of adhesive hooks and suckers with which it is provided.

segments which originate in a zone of growth close to the head. Of those which infect man some spend their adult stages in our own bowels and their intermediate stages in the muscles of pigs or cattle, entering us with our meat; another, even more troublesome, spends its adult stage in the dog's bowel and its intermediate stage in our own bodies, where it forms large "hydatid cysts" in the lungs, liver, and other organs in the hope that our flesh will be devoured by dogs. In the brains of sheep these cysts cause the disease known as staggers.

Strange indeed are Nature's ways. Here are two special classes, flukes and tapeworms, and they exist, it seems, for no other purpose than the affliction of the innumerable variety of forms which the dominant vertebrate, and molluscan phyla have produced.

§ 7. Et Cetera.

So FAR we have dealt with seven animal phyla. Now comes the rag-bag. We have left over a residuum of minor phyla containing few forms and having little economic or theoretical importance. We will not catalogue them here. Briefly we will mention four groups.

The *Nemertine* worms are thread-like or ribbon-like creatures, ranging in length from under an inch to twenty-five metres. Most of them are marine, creeping under stones and among seaweed; some are able to swim. One or two species live in fresh water and a few are terrestrial; one or two are transparent and leaf-like and swim like marine butterflies at the surface of the sea. For the most part they are carnivorous. They are in a phylum by themselves.

The *Rotifers*, or wheel-animalcules, another phylum, are minute but highly organized animals, found chiefly in ponds, gutters, damp moss and the like, and a great source of delight to those fortunate enough to possess a microscope. "If, retaining sense and sight, we could shrink into living atoms and plunge under the water, of what a world of wonders should we then form part! We should find this fairy kingdom peopled with the strangest creatures: creatures that swim with their hair, and have ruby eyes blazing deep in their necks, with telescopic limbs that now are withdrawn wholly within their bodies and now stretched out to many times their own length. Here are some riding at anchor, moored by delicate threads spun out from their toes, and there are others flashing by in glass armour, bristling with sharp spikes or ornamented with bosses and flowing curves; while, fastened to a green stem, is an animal convolvulus that by some invisible power draws a never-ceasing stream of victims into its gaping cup, and tears them to death with hooked jaws deep down within its body." (Hudson and Gosse in their classical monograph of "The Rotifera," 1886.) The rotifers are neither useful nor harmful to man; their interest lies almost entirely in this microscopic æsthetic appeal.

The *Polyzoa*, including the so-called sea-mats, are aquatic and most of them are marine. At first sight they look like the hydroid polyps to be presently described, for they are small animals living in colonies, and each individual has a circlet of tentacles surrounding its mouth; nevertheless they are very much more elaborately organized than hydroids, although most people take them for plants. Their colonies are supported by horny, slimy, or gelatinous skeletons, generally intricate and often beautiful when seen through the microscope. There are an enormous number of different species. Polyzoa are very common on the seashore, forming encrustations on seaweeds, rocks, crabs and the like; the freshwater forms, by clogging the pipes and filters of water-works, may become a serious nuisance.

FIG. 95. Superficially like a clam, but very different in its inner organization—Terebratula, a kind of lamp-shell (Brachiopod) from the deep water of the Faroe Channel.

The *Brachiopods* or lamp-shells are marine animals having a shell that looks at first sight like the shell of a bivalve mollusc. A lamp-shell leads the life of an oyster or a mussel, lying at the bottom of the sea, collecting suspended particles of food by means of a ciliated sieve that works like the gills of a clam or the pharynx of a sea-squirt, snapping together the two shields if any danger threatens. Nevertheless their organization is very different from that of bivalves. The two halves of the shell, for example, correspond to the back and front of the creature and not to its right and left as they do in bivalves. And the food-filter is made in a peculiar and characteristic way. At one time they were classified with the molluscs, but now they are either put in a phylum by themselves, or with the polyzoa in the phylum *Molluscoidea*. At the present day there are about a hundred species of brachiopods living in the sea at various depths, but they are only the survivors of a group that flourished and attained a rich variety of forms in the Palæozoic era, and has declined steadily in numbers ever since. Slowly, for some reason, as the world aged the oyster and the scallop and the mussel took their place. But one genus, *Lingula*, living in the mud of warm seas, has continued the even tenor of its existence from a nearly incredible antiquity. It is the senior genus of the world of life.

IV

LESS INDIVIDUALIZED ANIMALS

§ 1. *A Preliminary Note on Individuality.* § 2. *Obelia.* § 3. *Polyps, Jelly-fish, Sea-Anemones, Corals.* § 4. *Sponges.*

§ 1. A Preliminary Note on Individuality.

WE ARE accustomed to think of ourselves as individuals, for our bodies act and feel as single undivided things; we are in this sense the units of which our species consists. Nevertheless we are also cell-communities, and these cells, in such circumstances as we have described in Book One, can behave with remarkable individuality and independence. We may indeed distinguish two entirely different grades of individuality—the conscious, actual individuality of our bodies and the potential, suppressed individuality of our constituent cells. Now, as we explore the realms of life farther and farther from our own phylum, this grading is less distinct. All the organisms that are examined hitherto are individualized communities of cells just as we are, but in some of them the organization of the community is a little less definite, less inflexible than it is in ourselves. The cells are not so rigidly subordinated, not so strictly specialized.

Perhaps the most independent cells that we possess, excepting, of course, our spermatozoa, are our white blood corpuscles, ranging more or less freely in our blood and lymph and playing the part of scavengers, consuming and destroying bacteria or other undesirable objects. They are least subject to the disciplining influences that hold most of our cells in place and make them do special and limited activities at appropriate moments. Now, in many invertebrates freely moving cells of this kind play other parts in the economy of the organism, not merely absorbing undesirable matters. In the oyster, for example, the digestion and absorption of food is accomplished in part by creeping cells which elbow their way through the wall of the digestive tube into its canal; they fall upon the food, taking it into their own tiny bodies, and then they creep back through the wall of the intestine into the tissues of the oyster again. They even assist in transporting food round the body, for they creep everywhere and give up the particles that they have taken in wherever they are required. Further than this their liberties extend, for they can leave the body of the oyster altogether and prowl about in the mantle cavity (the space between the creature's body and his shell), where, presumably, they act as outposts against invading bacteria. In the earthworm there is a similar state of affairs; creeping cells leave his body and wander about over his skin, scavenging and keeping it clean. Truly an astonishing degree of freedom for cells that are really parts of a body!

The more flexible organization of the lower animals is clearly seen in the reproductive processes. We ourselves have only one method of reproduction; we have definite cells set aside for the purpose, and the only possible way in which a mammal can originate is by the union of an ovum with a spermatozoon. Moreover, our developmental processes are nicely regulated; a particular group of cells in the embryo is delegated to form an arm, and that is the only way in which an arm can be formed; if we lose an arm the part can never be replaced. In the lower vertebrates, on the other hand, the cells are not so rigidly specialized; a newt, for example, can grow a new limb if one is accidentally lost. Lizards can grow new tails, and they make good use of their ability; if a lizard is pursued and finds itself hard pressed it deliberately breaks off its own tail, which, writhing aimlessly like the body of a newly decapitated man, distracts the pursuer while the rest of the lizard makes its escape. In those very lowly organized members of our own phylum, the sea-squirts, the tissues have a tremendous power of reconstitution. There are kinds of sea-squirt in which the animal can be cut completely in halves, and in which each half then grows the missing organs so that two individuals result. This power which ascidian tissue has of reconstituting itself is used for reproducing the species; besides the normal sexual method the animals can reproduce *asexually* by sprouting out buds. There are sea-squirts, such as Doliolum, that bristle with buds as a pageboy bristles with buttons. Often, the individual sea-squirts formed in the latter way do not part company, but remain close together, so that by several asexual generations extensive groups or colonies of animals are built up.

As we pass from the sea-squirts to ourselves, we see an increase in the specialization of the tissues and a decrease in their power of developing independently and giving rise to new parts. In ourselves, the healing of a wound, the formation of new skin to seal a cut, of new fibres when a muscle is torn, of new bone when an arm or a leg is broken—these are the only representatives, the physiological homologues, so to speak, of the great power of reconstitution possessed by the sea-squirt.

Among invertebrates, the lobster is at about the level of a newt in this matter of reproductive power; it can part company with a limb in order to save itself, and it can slowly grow a new one. An octopus, too, can grow a new arm. And in the starfishes the power of reconstitution is amazingly well developed. If a starfish be cut in halves, each half will sprout new arms and become a whole starfish—if a little of the disc is attached a single isolated arm can grow the rest again. There are species of starfish which normally use this power, like the sea-squirts, for reproduction. After living harmoniously together for some time the arms seem to fall out and to group themselves into two parties; one party walks as hard as it can in one direction and the other party pulls equally hard in the other, so that the central disc with mouth, stomach, anus, and the rest is split right across. Finally, each of the two halves grows the missing parts and becomes a complete starfish. Fishermen, infuriated by the quantities of starfish that appear in their nets and by the havoc wrought by these animals on oyster-beds, often chop them in halves and fling them back into the sea; but since each half then proceeds to develop into a whole starfish, the fishermen are increasing rather than reducing the numbers of the species.

In the segmented worms we find various methods of sexless reproduction. In the little marine worm *Autolytus cornutus*, new individuals are grown at the hinder end of the body, so that there may be a chain of from two to at least forty individuals, each with a head of its own, but stuck together head to tail by a mortar of living tissue until such time as one of the hinder ones becomes rebellious, refuses to follow its leader, and tears itself away. Moreover, in this species we have an example of the phenomenon known as "alternation of generations." The fertilized eggs develop into sexless individuals which swim about sprouting new individuals until they die—but the individuals thus sprouted are male and female (and differ in a number of external characters from the sexless generations); they do not reproduce themselves asexually but produce fertilized eggs—and so the cycle begins again. In another kind of worm, *Syllis ramosa*, which is found inside the cavities of sponges from the deep sea, new individuals are sprouted off as side-branches from the flanks of the old, so that a web of worms is produced from which the sexual individuals snap themselves off and escape.

Finally, similar but even more striking phenomena are seen in flatworms. In most species of Planaria (but not in all) the creature creeps about as we have described, but while it does so there is disaffection, so to speak, in its end; as the animal becomes fully grown the latter half reorganizes itself into a new individual, and when it is ready it suddenly rebels and firmly grips the ground like a resolute donkey, refusing to be pulled along by the front any longer. There is a frantic

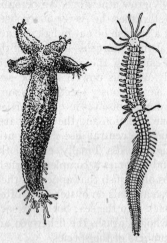

Fig. 97. The hinder half of this marine worm has developed a head of its own and will soon break away.

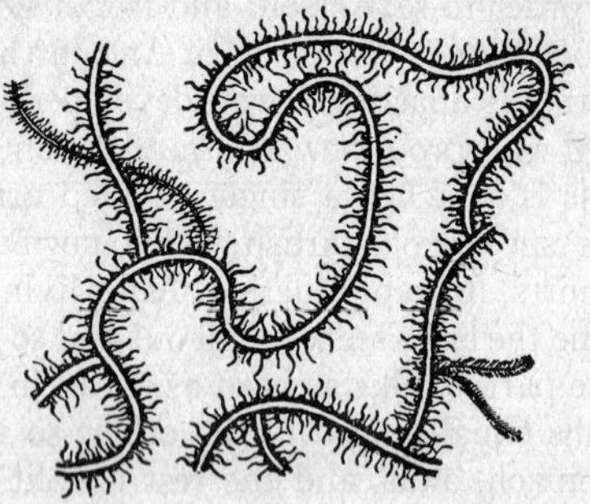

Fig. 96. A "Comet Starfish"—a single starfish arm, accidentally cut off, is growing the missing parts.

Fig. 98. The branching growth of the worm Syllis ramosa.

tug-of-war and a final split; the front end crawls away with an appearance of extreme indifference to grow a new and ultimately equally troublesome tail. Meanwhile, the hind end starts on an independent life, to be rebelled against in its turn by its own posterior. Moreover, in this flatworm phylum we find alternation of generations carried to its highest and most elaborate pitch; we have already seen how in the liver-fluke there is a regular succession of a whole series of generations, externally very unlike each other—the sexually reproducing flukes in the liver of the sheep, and the minute sexless forms in the water-snail.

We are accustomed to think of ourselves as individuals, and of our species as consisting entirely of individuals that are all alike, except for the relatively slight distinction between male and female. But manifestly in these invertebrate animals there are different kinds of individuals, and a different sort of species. We find one half of the body rebelling against the other half, and we find them splitting apart and reconstituting themselves; we find that the liver-fluke, a creature two-thirds of an inch in length, spending its life in a sheep's liver, and the redia larva, about one sixteenth as long, living in a pond-snail and with every organ different from that of a fluke, are both members of one and the same species, begetting each other with a regular, inevitable rhythm. The rigidity and uniformity of organization with which we are familiar in ourselves is not universal. Our ideas of individuality have to be revised and extended. We shall find them still more profoundly shaken by the animals that we are next to examine.

§ 2. Obelia.

It is common to find a light brown fur-like growth just below low-water mark on seaweeds, the wooden piles of piers, and the like. The growth consists of branching filaments of about the thickness of fine sewing-cotton, the branches terminating in tiny knob-like enlargements. The whole thing is suggestive at first sight of a miniature forest of seaweed, but the filaments are animals, not plants; they belong to the genus called *Obelia*.

On examining the filaments with a microscope, we find that they consist of hollow strands of living substance ensheathed by a transparent, horny tube. At most of the knobs the tube opens out as a delicate goblet of simple design, and in each goblet there lives a polyp. The polyps—the individuals of the little colony—are very simple in their structure. They consist of plain cylindrical bags of living tissue; at one end they are joined to the central filament of the branch on which they rest, and at the other they have tiny funnel-shaped mouths surrounded by single rings of writhing tentacles. Scattered all over their surfaces, and particularly numerous on the tentacles, are minute stinging-capsules. Each of these is a tiny oval vesicle with a spiral filament coiled up inside and a sensitive "trigger-hair" projecting from the surface of the animal into the outer world. When the trigger-hair is stimulated, the spiral filament, barbed and poisonous, is shot out with surprising vigour. The flower-like polyp is a living snare; it waits, with its tentacles stretched out, and if a suitable victim—such as one of the microscopic crustaceans—brushes one of the tentacles as it passes by it is suddenly riddled with the stinging, numbing filaments, then grasped in the other tentacles and crammed through the mouth into the hungry bag. The cells lining the bag secrete a digestive juice which disintegrates the prey, and they can also take particles of food into their own bodies very much as our white blood-corpuscles consume bacteria. Finally, after a little time, the indigestible remains of the victim are discharged through the same door that served as an entry, and the net of stinging fingers is spread again. That, in brief, is the life of the polyp; occasionally, if prey is not too abundant, the spread tentacles stir restlessly, and occasionally, when the creature is alarmed, it ducks back, contracted, for a few seconds into the safety of its horny cup, but it never wanders or explores, nor does it reproduce its kind; it leads a life of patient waiting and grabbing and digesting.

Compared with any of the animals with which we have dealt hitherto the anatomy of the polyp is incredibly simple. It consists of two layers of tissue, one inside the other and separated by a thin gelatinous sheet of non-living substance, like the firm matrix of our gristle. It has some seven kinds of cell, as opposed to the seventy or so that are found in our bodies. It has no heart, no eyes, no ears; as far as we know, it has only two senses, taste and touch; it has no brain, its nervous system being scattered through its body and organized about as elaborately as the local nervous system that produces writhing movements in our bowels. Except for the mouth, tentacles, and gut of which it consists, it has no specialized organs of any kind.

At its base the polyp tails away into a narrow tube of living tissue, running down the branch that the polyp crowns, and this leads into a main tube that forms an axis for the colony. Every polyp on the little frond is directly continuous with this central tube of tissue, and, through it, with every other. They are like flowers growing from a common stem, one flesh. Occasionally a side-branch grows out from this main tube, making a clear, horny sheath as it goes, and gives rise to a new polyp. Thus, by a process of budding and branching, the colony grows.

But occasionally, here and there along the central stem, there arise polyps of a new kind. Buds sprout out, at first very much like the usual kind of bud, but later developing into plain, club-shaped cylinders called *blastostyles*. The blastostyles do not develop mouths or tentacles; they are not feeders but breeders. They become covered with tiny wart-like buds which grow and develop into little free-swimming creatures. There is a division of labour between the different parts of the colony; the cavities of the stomachs of the feeding-polyps are in direct communication with the bore of the tubular stem, and some of the products of their digestive processes diffuse along this channel to the other branches. The feeding-polyps nourish the whole colony. The blastostyles, like our own ovaries or testes, do not contribute to the economy of the colony to which they belong, but, nourished by the diffusing juices from the feeding-polyps, they concern themselves with the founding of fresh colonies. The feeding-polyps can move independently, and if they are cut off from the stem the material of which they consist lives on and reorganizes itself into a new colony; nevertheless, biologically speaking, they are parts, not wholes.

The buds that are produced in great numbers by the blastostyles develop into miniature jelly-fish, as clear as glass, and little more than a tenth of an inch across. Their shape is that of an umbrella with a very short, thick handle; they have a hemispherical bell fringed with tentacles, and, hanging from the middle of it, a short stalk which bears the mouth at its end. As they become properly developed they detach themselves from the blastostyles and swim actively and independently through the sea. Like the feeding-polyps, they are predatory; they have stinging-cells and they paralyze and engulf a variety of microscopic creatures. But they are altogether more complicated in structure than the polyps; they have a definite stomach leading into a system of canals which, like the branched stomach of a flatworm, carries food to all parts of the body; they have special sense-organs; they are male or female. Dangling down under the bell there are tiny bags filled with eggs or spermatozoa. The little jelly-fishes, or *medusæ*, swim in shoals at the surface of the sea and, when they are ripe, they shed their spermatozoa or eggs, as the case may be, into the sea-water. The two mix together, and in this primitive manner the eggs are fertilized. They develop into simple elongated larvæ, called *planulæ*, mouthless bags which swim by means of cilia with which they are covered. These larvæ are the founders of new colonies. They settle on pieces of wood, seaweed, and so on, and develop into feeding-polyps; the polyps bud out stems which give rise to other polyps; and thus a new colony is formed. Evidently there is an alternation of phases in the life-cycle of Obelia: first, the fixed, budding colony of polyps, and then the free-swimming, sexual medusa, charged with the duty of founding new colonies in distant places.

We have used the phrase "alternation of phases" rather than "alternation of generations" because it is open to question whether the fixed stage should be called a generation at all. It consists, we have seen, of a central tube with an indefinite number of side-branches; of the side-branches the majority end in those living snares, the tentacled polyps, whose duty it is to nourish the whole colony, while a few end in the blastostyles—polyps which specialize in reproduction. Now, the feeding-polyps can perform independent movements of a considerable degree of complexity—they look like individuals—and the blastostyles have essentially the structure of polyps, so that we may, if we choose, regard the feeding-polyps and blastostyles as

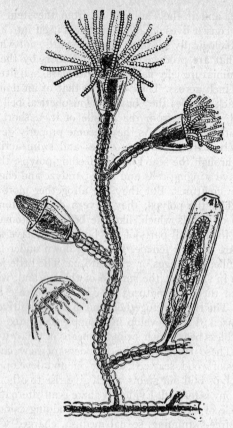

FIG. 99. Part of a colony of Obelia, seen through the microscope.

Above are three polyps in various stages of expansion. Below are a Blastostyle on the right and a free-swimming Medusa on the left.

individuals, specialized in different ways in order to build up a colony, much as our cells are individuals specialized in different ways in order to build up a body. We may regard the polyp as the homologue of a man. Then we should distinguish three grades of individuality—the cells, the polyps, and a higher order of individual, the colony. But the organization of Obelia is very loose, very plastic. There is no rigid specialization of tissues like that found in ourselves. If a small fragment of the colony be cut off from the rest—a single feeding-polyp, for example—it will sprout out a new central stem from which both feeding- and breeding-polyps will arise. What is more, and most vividly interesting, the tissues of the creature can be forced through fine muslin so that its cells are completely separated from each other; nevertheless, they will come together again into masses, and organize themselves into new polyps, new colonies. Imagine the tissues of a man doing this!

In ourselves, the species consists entirely of discrete and highly complicated cell-communities that we call individuals. Moreover, in our case, this kind of organization is necessary, for human tissue cannot live unless it is part of a human body.* But in Obelia the reality is not a number of separate individuals but the plastic living tissue, growing and budding and branching, throwing out now a feeding-polyp, now a blastostyle, now a medusa; not definitely committed to any one of these kinds of body, but sprouting and reorganizing itself as circumstances determine. It is a living continuum; a mob of cells which may be turned to this or that. So long as any part of the mob lives it may continue the race. Even in ourselves there is an underlying reality of this kind, for we, the individuals, die and yet through us the race continues; small portions of tissue, single cells detached from our bodies may hand on the life of the species by growing and building themselves into new individuals. But in our case, the capacity for reproduction is retained only by a few specialized cells. Through them human protoplasm lives on and may live for ever; human individuals

FIG. 100. The tissues of Pennaria (a creature very like Obelia) have been pressed through fine silk and the cells separated from each other; nevertheless, they have crept together again to rebuild polyps (magnified).

The smaller cell-mass (inset), three days old, is beginning to organize itself; the larger one, six days after the operation, has already formed two perfect polyps. (From Wilson, "Journal of Experimental Zoology," 1911.)

---

*Unless it is elaborately nursed in a tissue-culture—made to believe, so to say, that it *is* part of a body.

are the temporary shapes that it assumes—excrescences, so to speak, that the race throws out as the Obelia stuff throws out polyps. We are so accustomed to thinking in terms of individual experience, to regarding a species as merely a convenient group-name for a number of separate persons, that the idea of the living, material race as a more enduring reality than the individual is at first sight altogether strange; nevertheless, it is clear when we come to examine these lowly-organized creatures that the elaborate, self-centred individual is not by any means an indispensable part of the scheme. Life can go on without his or her intervention.

§ 3. Polyps, Jelly-fish, Sea-Anemones, Corals.

Now Obelia belongs to a phylum, the *Cœlenterata*, which includes a large number of different forms, all aquatic and nearly all marine. Like Obelia, their bodies are always simply organized, without elaborate organ-systems and with little differentiation of tissue and their symmetry is nearly always radial like the symmetry of a starfish, and not bilateral like the symmetry of a vertebrate. For the latter reason the Cœlenterates used to be classified with the Echinoderms (Fig. 89) in the phylum "Radiata"—but it is now recognized that they are utterly distinct from each other; the Echinoderm has an altogether more highly organized and specialized interior than the Cœlenterate. The bodies of Cœlenterates are soft, although they often make rigid houses to live in, or frameworks to live round, and the group is characterized by their peculiar stinging-capsules. Many of them are sessile and flower-like, so that they have been called the Zoophyta, or plant-like animals.

There are three classes of Cœlenterates, of which the *Hydrozoa* includes Obelia. The members of this class are all variations, so to speak, upon the Obelia theme; the kind of life-history that we have described, with its alternation between the free-swimming sexual medusa and the colony of budding polyps, underlies the group, although it may be more or less profoundly modified. Best known are a great number of forms, growing like Obelia on hard objects round our coasts and often mistaken for small seaweeds; they differ from each other in the structure of the polyps and medusæ and of the horny tubes and cups in which the colony lives. The structure of our common representatives of the group can be seen clearly only with a magnifying glass, but in some forms the polyps may be brilliantly coloured flower-like objects two or three inches across, and the medusæ, as transparent as glass, may be fifteen inches across the bell. In some species the life-history is simpler. There may be no colony stage at all; there are forms in which sexual medusæ develop directly from the fertilized egg—medusæ otherwise very similar to those of Obelia. On the other hand there are forms without medusæ, and we may pause for a moment to examine one of these.

Hydra is a minute creature found commonly enough in fresh-water ponds; it may be a third of an inch long. It is like a single, free-living polyp of Obelia; it has a hollow, cylindrical body, with a mouth surrounded by tentacles. It usually remains fixed, catching such prey as Cyclops with its tentacles, which, when the animal is hungry, trail in the water like a driftnet. Sometimes it moves from place to place, gliding along on its base or looping like a looper caterpillar. Like Obelia, it is a loosely organized cell-community; if it is cut into halves, or even into small bits, each fragment can reconstitute itself and form a whole Hydra. Its methods of reproduction are two. It may produce buds—a lump of tissue appears somewhere on the side of the creature, grows, develops a mouth and tentacles, and breaks away. Sometimes, when food is very plentiful, the bud may start to grow new buds from its own flanks even before it breaks away from its parent; this is a hint, so to speak, of the colony-formation seen in Obelia. Normally, however, Hydra is not colonial. Sometimes little swellings may appear on the sides of Hydra—not in definite places, but anywhere—which do not form buds. These swellings produce eggs if they are near the base of the body, spermatozoa if they are near the top. The eggs are relatively large, passive cells like our own eggs; the spermatozoa are tiny, active tadpole-like creatures as our own are. The sperms swim up to and fertilize the eggs, which then develop into little embryo Hydras without any intermediate stage.

Hydra, then, contrasts with Obelia in two respects; first, there is no medusa and the polyp is itself sexual; second, there is no colony-formation and no division of labour between polyps like that between feeding-polyps and blastostyles in Obelia.

On the other hand, there is another group of Hydrozoans, the Siphonophores, in which colony-formation is carried to an extreme degree. They are found floating in the open sea—chiefly in warm waters. The "Portuguese man-of-war," *Physalia*, is a good example; it has an elongated bladder filled with air, from one to twelve inches long, by means of which it floats, and having above a raised crest or sail; the whole is peacock-blue or orange. From the lower side of this float there hang bunches of miscellaneous-looking objects—blue, violet, and carmine, with a silvery sheen. Each object is a unit, either a polyp or a fixed medusa, and they specialize in the performance of various duties; some, not unlike the Obelia polyps, are feeders; others, having no mouths and a single long tentacle very richly provided with stinging capsules, are soldiers; the medusæ, clustering like dark blue

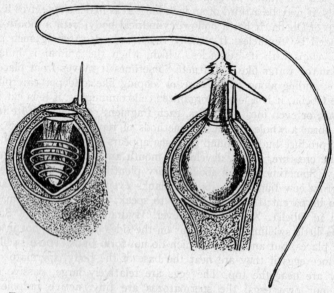

FIG. 101. Two of the characteristic stinging-capsules with which a Cœlenterate benumbs its prey, very highly magnified.

The one on the left is undischarged; the barbed thread is still coiled up within it. The one on the right has shot out its thread.

grapes, bear sexual cells; from the whole there depend particularly formidable tentacles, up to sixty feet in length and with a sting that can raise weals on a human being, like the sting of a giant nettle, and give him fever for several days. Other Siphonophores are differently planned and coloured. We may note that although a Siphonophore is conveniently described as a colony of specialized individuals, those individuals are not nearly so sharply defined as in Obelia—they are continuous with each other, and it is often very hard to say where one begins and the other ends. The problem of assigning to any part of the colony an individuality comparable to our own individuality is much harder even than in Obelia.

The second class of Cœlenterates, the *Scyphozoa*, includes all the larger and more familiar jelly-fish. The medusa of a Hydrozoan is like a miniature jelly-fish in its general appearance, but the Scyphozoan medusa is more complex in its organization. The Scyphozoa are all marine, and most of them float at the surface of the ocean; they are generally transparent, with patches of purple, blue, or orange, especially on their reproductive organs; the deep-water forms are often red or brown. Many are phosphorescent. The greater part of their bodies consists of a structureless jelly that is in fact mostly sea-water; only about one per cent of the weight of a jelly-fish consists of organic matter. One or two species are curious, not swimming but living on the sea-bottom; *Haliclystus*, shown in figure 102, affords a good example —it has the structure of a jelly-fish, but leads the life of a polyp, attaching itself to weeds and the like.

Many jelly-fish develop directly from the egg, but in some species there is an intermediate stage reminiscent of the alternation found in Obelia. In *Aurelia*, for example, the commonest jelly-fish of our shores, the eggs hatch into tiny planula larvæ like the larvæ of Obelia, which settle and give rise to a single feeding-polyp. The polyp later divides itself into a series of discs, like a pile of crumpets sprouting tentacles, which detach themselves one by one and grow into jelly-fish.

The third class, the *Actinozoa*, includes the sea-anemones and corals. These animals exist only in the polyp phase, no medusæ being known, and their polyps are more highly and elaborately organized than those of any other group. Some, like the sea-anemones of our shores—the beadlet, the plumose anemone, the sea-dahlia and the rest—remain solitary through life, but in most cases, notably among corals, there is extensive budding, which results in the formation of more or less enormous colonies. The sea-anemones, like Hydra, have a mouth surrounded by stinging tentacles; in many species peculiar threads covered with particularly active stinging cells can be shot out through the mouth, or through special port-holes in the side of the body. The larger kinds—such as the dahlia anemone—can even capture prawns and small crabs.

The corals make a hard skeleton of lime, round which, and sometimes inside which, they live; they are nearly always colonial, and may build fan-like structures, rods like the pink "dead man's fingers" of our coasts, or massive reefs. The Great Barrier Reef of Australia, over a thousand miles long and fifty miles across, is the work of myriads of tiny coral-polyps. The Actinozoa are also remarkable for the brilliance and intensity of their coloration. Anybody who has prowled a rocky shore at low tide will know how varied the colours of sea-anemones can be. The tropical corals are even more striking; there are pink,

FIG. 102. A group of Cœlenterates.

Below, from left to right, are four Plumose Anemones (Metridium), two of the green polyp-like jelly-fish Haliclystus, two Sea-dahlias (Tealia) capturing prawns, and two kinds of coralline growth (Sertularia and Gorgonia), each the tubular dwelling of a colony of tiny Polyps. Above, on the left, is a clear, milky-blue Jelly-fish (Aurelia), and, on the right, two glassy Sea-gooseberries (Pleurobrachia), of which one has just caught a young pipe-fish.

yellow, green, and purple corals; corals with green polyps and crimson skeletons; bright blue corals; corals with scarlet, finger-like skeletons fringed with white polyps.

A fourth class, the *Ctenophores*, is sometimes classified with the cœlenterates and sometimes put in a phylum by itself. The sea-gooseberry, *Pleurobrachia*, is a delicate transparent globe of jelly-like consistency that floats in the open sea. Running down the sides of its body there are eight comb-like structures, each tooth of the combs being a gigantic cilium; the combs are in incessant motion, giving the animal a beautiful iridescent appearance, and thus it swims through the water. Trailing downwards (but capable of being drawn in) are two long tentacles set, not with stinging-capsules, but with special adhesive organs. By means of these tentacles the creature catches its prey—small fishes, crustaceans and the like—and pulls them into its mouth, which is on the lower surface of the globe.

Most Ctenophores live, like the sea-gooseberry, at the surface of the sea; most are transparent, tinged with pink or yellow, and some are brilliantly phosphorescent. The wonderful "Venus' Girdle" of the Mediterranean is a rippling, band-shaped animal, delicate violet in colour, with a brilliant greenish fluorescence, and may attain a length of five feet. One or two exceptional members of the group do not float but creep over the sea-bottom.

Ctenophores develop directly from the egg without any intermediate polyp-stage.

§ 4. Sponges.

It has probably never occurred to many of us that we get into our baths in company with a skeleton. The sponge that the reader employs—unless, of course, he or she prefers the flaccid rubber substitute that is attempting to oust the natural article—may be variously formed, for there are several species on the market, but it is most likely to be a more or less rounded object, a flattened globe. Perhaps it has been trimmed a little by the dealers; most of the sand that had to be washed out of it after it was bought had been artificially introduced, for sponges are sold by weight. There is nothing very suggestive of a living thing about the domestic sponge; and yet when it is examined with care it reveals some degree of organization. The caverns that riddle it are of various kinds; there are conspicuous round holes, often large enough to admit the finger, and leading into neatly trimmed, more or less straight cylindrical shafts that sink deep into its substance, and there are smaller, less regular holes, leading into a labyrinth of passages some of which open into the main shafts. Moreover, the substance of the sponge is itself a network of branching fibres, a network so fine that its meshes are only just visible to the naked eye. These fibres are the picked and whitened bones of a living thing.

At one time the sponge was covered with flesh, and it lived at the bottom of a warm, shallow sea—round the shores of the Mediterranean, the West Indies, Australia. It was a yellowish, brownish, or dark, purplish object, smelling faintly of garlic, elastic and slightly slimy to the touch; living bath-sponges have been compared, in colour and consistency, to fresh liver. Most of its holes were closed with a thin skin of tissue. Some of the larger ones—called *vents* or *oscula*—were open, although most of these could be closed at will by means of films of tissue round them, and scattered freely over its surface there were microscopically fine pores.

The creature did not move from place to place, it remained rooted firmly to a rock, and, except for occasional slow changes in the diameter of its oscula and pores, it made no visible movements. Indeed, sponges are such passive creatures that for a very long time their animal nature was hardly even suspected. They grow, and therefore there must be life in them of some sort; but many biologists considered them to be plants, and a few even believed them to be the dead by-products of living things, secreted by the worms that are invariably found crawling about in their cavities, much as the gelatinous investment of a sea-squirt is secreted, or the tubes and cups of a coral. But just over a hundred years ago Robert Grant made an observation of fundamental importance. He noticed that there was always a slow, steady current of water through a sponge—passing in through the microscopic pores, out through the vents. That current is the essential vital activity of the sponge: its means of livelihood and almost its only means of self-expression.

Sponge-cake, spongy iron, spongy platinum—the name has come to mean anything riddled with holes, for a sponge consists in the main of a tangle of aqueducts guiding its precious stream. Its cavities are of three kinds: there are the "flagellated chambers" where the current is both caused and exploited, there are inward passages guiding the current from the surface pores to the chambers, and there are outward passages guiding it away to the oscula. A flagellated chamber is a microscopic cavity lined with living cells possessed of long whips like the tails of spermatozoa; by lashing their whips they keep the current moving through the chamber, and indeed through the whole sponge. Round the base of its whip each cell has a transparent collar, a collar which has the power of grasping suspended organic particles as they drift by and passing them down to the cell below. They are then taken into the cell itself, and there digested. The current, then, is like the current that flows through the pharynx of a sea-squirt, or the gills of a clam; it brings food and oxygen, and washes waste matters away, but sponges are unique in having an enormous number of separate openings through which they take in their food; they have nothing that can be called a mouth.

There are many different kinds of sponges, all rooted to, or in, the bottom of the sea, river or pond that they inhabit, all making this slow, creeping current, and giving no other sign of life. There are common sponges that form green and yellow encrustations on rocks, crabs and so on, between tide-marks, others that occur in fresh water —sometimes in such enormous masses that they may stop up water-pipes; some, inhabiting the deep sea, weave glassy baskets of beautiful and symmetrical shapes instead of the horny web we wash with.

The sponge is a very loosely organized community of cells. It has a certain definiteness of architecture that varies from species to species, but its shape is by no means stereotyped; individuals vary so much that the classification of the group is a matter of extreme difficulty. Besides the myriads of milling collar-cells that keep the current moving there are one or two other kinds of tissue— the muscle-like cells that control the diameters of the canals, wandering cells that drift about the gelatinous body of the sponge as they choose, recalling our leucocytes, the flat cells that pave its outer surface, and other cells that make the skeleton. This skeleton may be of various kinds; there are sponges whose skeletons are made of tangled horny fibres, sponges whose skeletons are made of spicules of carbonate of lime, or of silica. These spicules are generally microscopic, and present a great variety of beautiful and symmetrical forms in different species. But the tissues of a sponge do not compare in complexity or diversity of organization with our own, or even with those of Obelia. It is even more undisciplined, even less individualized than the latter creature.

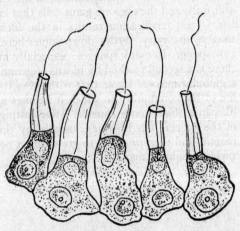

FIG. 103. Five of the myriads of collared cells which keep the water in a living sponge in motion (highly magnified.)

Like a Cœlenterate, a sponge has great powers of reorganization. A small piece cut away from a sponge will live and grow—and indeed, there is no very evident reason why it should not, for the parts of a sponge are not differentiated into organs, and a fragment a tenth of an inch across would contain all the different kinds of tissue that form a sponge. When it is isolated it will simply continue to grow. Moreover, as with Obelia, a sponge can be forced through gauze so that all its cells are separated from each other; nevertheless, they will clump together again and rebuild a shoal of little sponges.

There are two methods of reproduction in this phylum. There is the sexual method: The spermatozoa are tiny cells with short, blunt, stiff tails, and the eggs are giant cells that creep about the sponge like leucocytes; fertilization results in the formation of a ciliated larva that swims away, settles down somewhere else and develops into a new sponge. In many sponges—especially in the fresh-water species—there is a second method by means of gemmules. A gemmule is simply a globular mass of tissue that withdraws itself, so to speak, from the life of the community to which it belongs and encloses itself, hermit-like, in a thick horny capsule. After resting for a time in the tissues of the sponge it is liberated, usually by the death and decay of the parent, and then proceeds to grow into a new sponge, just as any detached fragment may do.

V

VEGETABLE LIFE

§ 1. *Stems, Leaves, and Roots.* § 2. *Individuality in Plants.* § 3. *Flowers and Seeds.* § 4. *The Flowering Plants.* § 5. *Ferns and Mosses.*

§ 1. Stems, Leaves, and Roots.

AND now we come to a vast group of living things laid out for the most part upon a common plan—the plant world; its importance in our daily consciousness is second only to that of our own phylum of vertebrated animals. Its structural conception is still an aggregation of cells; it displays the phenomena of sexual reproduction, but in nearly everything else it is so different from the lay-out of any phylum hitherto considered that it is difficult even to put types side by side for comparison.

A plant—our crude description will apply to any plant that bears flowers—consists of three essential parts; the stem, more or less rigid and holding the plant in shape, and at the same time guiding sap from place to place, a system of green leaves spread like a net to catch as much sunlight as possible, and a system of roots, branching eagerly in the soil, anchoring the plant, and sucking up moisture and other substances upon which it thrives. We may leave out for the present the reproductive parts, the flowers and seeds, and discuss very briefly the physiological processes underlying the ordinary day-to-day life of an individual plant, just as in the first Book we began by discussing the day-to-day processes, the digestion and breathing and excretion, that underlie animal life. The life of a plant is so different from that of an animal, different even in its chemical basis, that we shall have to start again nearly from the beginning before we can understand the vegetable world: but the effort is well worth while, for we shall find in a simple physical fact a clue that explains and clarifies all the mysteries of vegetable life.

An animal, we noted, consumes food and uses it in various ways; with parts of it he repairs and adds to his own living fabric, while part of it is burnt in order to give the energy upon which his life depends. The food that he consumes consists of complicated substances—substances with large and intricately constructed molecules—which can be broken up and decomposed, and therefore used as fuel. Now, a plant never takes in such substances. The things upon which it subsists—water, carbon dioxide, and various mineral salts—have extremely simple and elementary molecules, and can under no circumstances be burnt. They contain no bottled energy, as the food of an animal does. Whence then does a plant get its energy, the energy for its growth and breeding, and for the hardly perceptible movements that it performs?

If we place a plant in a compartment so contrived that we can, from time to time, analyze the gas around it, and if we keep the plant in darkness—if, for example, we experiment at night—we discover a profoundly important fact; a plant consumes oxygen and gives out carbon dioxide. It breathes just as an animal does. Its leaves are covered with minute, barely visible pores, thousands of tiny nostrils, as it were, and through these pores the gases diffuse in and out. The plant, then, is breathing, and therefore it must be burning something as fuel—but what is it burning? There is nothing in the substances absorbed by its roots that could be burnt.

Supposing now that we repeat our experiment by daylight, or even in the presence of an electric light; we get a very different result. We find that things are just the other way round, for now the plant is absorbing carbon dioxide and giving off oxygen. And in this fact lies the whole secret of the apparently foodless plant existence, for the plant makes its food in its own tissues—makes it out of water, carbon dioxide from the air, and a few simple salts from the soil giving off oxygen in the process, a thing beyond the power of any animal.

The leaves of a plant are green. This is because the tissues of the plant contain a characteristic green pigment called *chlorophyll*. Chlorophyll is a unique substance as far as the chemistry of animals and plants are concerned, for when it is present in a living cell that cell can absorb and make use of the radiant energy contained in light itself. That, in a word, is the secret of the plant. It can drink the pure energy of sunlight, and it can use that energy to build out of the elementary substances that it absorbs the higher complex molecules of which its tissues consist.

The plant breathes, as we have already noted, and it breathes continually, both by night and by day. (By day-time, however, its breathing is not noticeable because it is masked and overweighted by this process that works in the opposite direction, involving the absorption of carbon dioxide and the storage of energy inside the molecules of chemical compounds instead of its liberation by their disruption.) Nevertheless, it breathes, and underlying the whole physiology of its tissues is a scheme like the chemical scheme of an animal, an oxidation

Fig. 104. Part of the surface of a leaf, magnified, showing the outlines of the cells and the pores or stomata, the microscopic nostrils through which the plant breathes.

A single oak-leaf has several million of these pores.

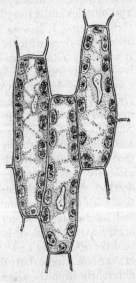

Fig. 105. Three typical plant cells from a moss leaf.

Note the tough boxes in which they are imprisoned and the oval green plastids ranged along their walls.

of complex substances in order to yield energy. The difference between animal and plant is that the latter has something in addition. Over and above this scheme it has another scheme founded upon sunlight and chlorophyll, by means of which it can synthesize its food out of the most elementary ingredients instead of having to hunt about and look for it. Take the most famous chef in the world; give him water, a handful of nitrate of lime, a teaspoonful each of nitrate of potash, phosphate of potash and sulphate of magnesia, and a trace of iron (such as the rust from an old nail); what sort of dinner could he turn out for you from that? Give them to a plant, let it absorb them into the kitchens of its own cells, and if there be light it will turn them into delicious tissues, into asparagus or lettuce or beetroot, into spinach or turnip or artichokes or potato or wheat or peas, according to its particular bent.

Naturally, the organization of a plant is very different from that of an animal—a mouse, let us say. A mouse has nerves and sense-organs and muscles and a busy little brain, but what should a plant want with these things? It has no food to seek, and that is the chief thing that an animal does with its brain. Indeed, if a plant could walk about, it would have to tear up its roots at every step! It sits still and spreads its leaves towards the light, and its roots into the soil. It lacks the complicated digestive apparatus of a mouse, for the mouse consumes energy-yielding fuels that have to be adjusted before they can be fitted into his tiny body. But the plant has no need to consume fuels; it contents itself with the elementary molecules that it finds in the air and soil. It does not rush about from place to place and has no muscles that toil strenuously, therefore it has no need of a heart-pump and violently pulsing blood; the slow, steady ooze of sap, upwards and downwards along the stem, suffices to keep its parts in communication with each other, and is quick enough for its slow, dignified chemical interchanges. Nevertheless, it should not be thought that the organization of a plant is simple. It has a very great variety of specialized tissues, struts and girders, pipes, living laboratories, storehouses of various kinds, but they are planned and specialized altogether differently from the tissues of a mouse.

The plant is built up of tiny organized protoplasmic specks just as we are—indeed, if we look into these intimate details of their organization the gulf between animal and vegetable, between man and cabbage, let us say, seems to narrow. In many plant-cells active movements can be seen in progress; the impassive stillness that distinguishes cabbage so markedly from man breaks down. These movements can naturally be best seen in transparent plant tissue—in the leaves of the Canadian waterweed, in the stinging hairs of a nettle, or in the thin fronds of mosses—and in many of these the protoplasm can be seen to stream regularly round the little box in which it is enclosed, almost as if it were trying to find a way out. It is obviously and actively *living*, just as our protoplasm is. But there are general differences between animal and vegetable cells. In the first place, the cells of the green parts of a plant contain, in addition to the nucleus, mitochondria, and so forth, that we noticed in the animal cell, little round spherules within which the vital chlorophyll is confined. After what has been said there is no need to stress the importance of these *plastids*, as they are called. The second point is of minor importance; plant cells almost invariably secrete little boxes round themselves, boxes of a complex substance related chemically to the sugars and called cellulose. Because of these myriads of microscopic but indiges-

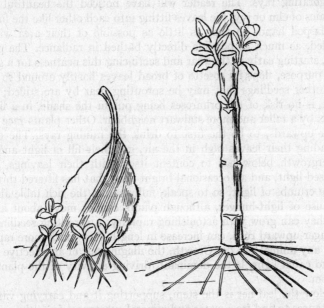

FIG. 106. Regeneration in Bryophyllum calycinum, a plant from Bermuda.

A leaf and a piece of stem, cut off and dipped in water, are sprouting out new roots and shoots. (After Loeb.)

tible boxes in which the protoplasm of a plant is enclosed, mammals like cows or horses, which lead an exclusively vegetarian life, have to carry longer and more complicated digestive tubes than we do. Moreover, their own digestive juices are incapable of attacking the cellulose and thus these herbivores are dependent, every time they digest a meal, on microbes that live in their bowels—but that is another story, that we shall recount later.

The urgent need of an animal is food; that of a green plant is sunlight—and herein lies the clue to above-ground vegetable architecture. Air is plentiful enough, and, once its roots have struck in a suitable soil, a supply of moisture and the necessary mineral salts is guaranteed; but the third vital necessity, light, is a thing that has to be fought for. All over the world, wherever roots can get a grip and an adequate supply of nourishing moisture, plants are wrestling silently with their neighbours for light. Leaves are a net, spread to catch the invigorating rays. The reader will have noticed the beautiful leaf-mosaics of elm or ivy, the leaves fitting into each other like the fingers of clasped hands so that as little as possible of their area will be shaded, as much as possible directly bathed in radiance. The primrose, starting early in the year and sacrificing this neatness for a grimmer purpose, flops its rosette of broad leaves hastily around so that any other seedlings that may be sprouting near by are stifled; thus there is no risk of the primroses being put in the shade, in a literal sense, by a taller and more stalwart neighbour. Other plants race each other upwards, to be the first to drink the raining rays. The trees, spreading their leaves high in the air, get their fill of light and the undergrowth below has to content itself with their leavings, with diffused light, and an occasional bright fleck that has filtered through —the crumbs of light, so to speak, falling from the rich table above. Because of light-hunger, although plants cannot move about as we do, they can grow with astonishing rapidity; a sprouting seedling, in its eager upward rush, can increase in length very much more rapidly than any animal can do. Growth, the flinging out of an effective leaf-net to catch light, is the dominant, driving activity of the plant organism.

Below the leaf-net is the stem, supporting it and carrying various substances to and fro—water and mineral substances from the mining roots to the light-driven factories in the leaves, finished chemical products from the leaves down to the toiling tissues of the roots. And the roots, sinking hungrily into the soil, sucking up raw materials for the green factories to work upon, branch and branch again into a net of another kind, a net spread not for light, but for moisture.

These are the broad essentials of plant anatomy. There are, of course, many other features, many variations on the scheme. There are, for example, parasitic plants, that find it easier to abandon the strenuous light-race and to suck the nourishing juices of their neighbours. The dodder is an example of such a plant and the mistletoe is an interesting half-way house, battening on the labours of the roots of other plants, for mistletoe perches at the top of a high tree, and its roots, penetrating into the tissue of the latter, suck out the upward stream of moisture and salts for the mistletoe leaves to work

VEGETABLE LIFE

upon—an ingeniously effortless way of coming out top in the light-race.

And the organs of plants may be turned to other purposes. The leaves may be hard and sharp to repel vegetarian animals, as in gorse, or the twigs may serve the same purpose, as in hawthorn; there may be stores of food, like the swollen roots of carrots, beetroots and turnips or the bulbs, made of fleshy subterranean leaves, of onions and tulips. But these are matters that we shall leave aside here; the essential thing to remember is that the primary life of a plant is not a food-hunt but a light-hunt, that instead of moving about it grows, that instead of our elaborate apparatus for muscles, nerves and sense-organs it has an elaborate architecture of stems, leaves and roots, and that all these things are simply results of the fact that plants are green.

§ 2. Individuality in Plants.

WHAT does it feel like to be a plant?

In the first place, it is very doubtful whether a plant feels anything at all. In our own bodies sensation depends on the presence of specialized sense-organs and nerves, and there is no trace of either of these structures in a plant. There are philosophers who argue, without too closely defining their terms, that some sort of consciousness must be associated with all matter; that is as may be. In watching the behaviour of plants one gets little evidence of any sort of conscious sensation. It is, of course, true that many plants grow towards the light, or away from it, or towards moisture, or in a direction opposed to gravitational pull, and in that one discerns a responsiveness to external agencies. But Hammond, an American inventor, once constructed a little machine on wheels that would follow a light as a dog follows its master, and could be led along any desired path in a dark room by simply holding a flashlight in front of it, so that these tropisms, as they are called, do not necessarily imply a considering and deciding mind. The movements that plants perform—such as the turning of flowers and leaves towards the sun, the opening and closing of daisies which depends upon the light, the opening and closing of crocuses or tulips which depends upon temperature, or even the sudden swoon of the sensitive *Mimosa pudica* when it is touched—are brought about by processes very different from those occurring in our own muscle-fibres. True, there are simple plants, as we shall shortly learn, which skip along by means of lashing tails like animalcules, and Sir J. C. Bose, by using very sensitive apparatus, has detected quivers and shudderings in injured plants; but we may safely conclude that even if a plant is conscious its mind is something very different, very much more elementary than the highly organized individual consciousness that we ourselves possess. Calves' foot jelly can be made to quiver, an automobile started on high gear shudders; shuddering by no means implies pain.

Looked upon as a cell-community the organization of a flowering plant is somewhere about the level of Obelia in the animal kingdom. Obelia, we noted, is not a discrete, highly specialized individual, but a mass of living tissue, growing and branching and throwing up polyps of various kinds without any very rigid order as occasion demands. Vegetable tissue behaves in much the same way. The structure of a plant-body is by no means as stereotyped as the structure of one of the higher animals; a primrose does not have a fixed and invariable number of leaves, and the branching of elm-trees is very much more variable than the branching of our own arteries and veins. There is an evident reason for this, for the plant can adapt itself to the particular conditions of illumination under which it finds itself, and grow in such a way as to secure as much light as possible.

Many plants grow and spread by means of creeping underground stems; couch-grass, for example, has a tough, rapidly growing underground stem that sends up a bunch of leaves and flowers from time to time, much as the hollow main stalk of Obelia throws up polyps. The runners of a strawberry are similar, and the reader, if he or she has ever attempted to weed a garden, will be able to think of many troublesome examples of the same thing. Moreover, it is a well-known fact that many plants can be reproduced by means of cuttings; in Begonias, for example, a single leaf or even a part of a leaf, detached from the plant, will sprout out other leaves and rootlets, and reorganize itself into a whole plant, just as an isolated Obelia polyp will sprout out a new axis-tube and found a new colony. Mosses are celebrated for their great power of regeneration from cut portions of all the organs, and they can reproduce themselves by means of gemmules like sponges, by separating off little bits of themselves which grow into new plants. The bulbils of some of the flowering plants are devices for doing the same thing.

To sum up, then, it is difficult and probably wrong to think of conscious individualized plants. One should think of plant tissue as organized in quite a different sort of way from our own, as an unconscious race, rather than a number of rigidly differentiated persons.

VEGETABLE LIFE

§ 3. Flowers and Seeds.

LIKE animal tissue, vegetable tissue has to reproduce itself, to spread and colonize new localities, and although, as we have seen, it has all sorts of sexless ways of spreading and multiplying, the principal method employed, at least in the higher plants, is sexual; as with ourselves, it involves the union of gametes of two kinds.

Now, there are great variations between the different kinds of plants in the method of reproduction and in the life cycle. We shall start by describing the reproduction of the flowering plants, and afterwards, in reviewing the vegetable kingdom, we shall note what arrangements occur.

The chief differences between the sexual process in animals and that seen in the higher plants are due simply to the fact that animals move about while plants are rooted. Among higher animals, male and female see or smell or hear each other, are smitten with desire, move towards each other, make love, and so their gametes are brought together. But plants have no eyes to see with, no brain to lust with, no muscles to move with, and they have to rely on outside agencies to bring about the union of their gametes. Plants woo each other "by correspondence"; they are provided with elaborate devices for utilizing the transport possibilities of winds, streams or insects, and inducing them to act as go-betweens in this necessary transfer.

Let us consider for a moment a buttercup flower, if possible, in the presence of the real thing. There are five brilliantly yellow petals, the most obvious parts of the flower; outside them are five little green sepals, the leaves that enclosed and protected the other parts of the flower when it was a tender, forming bud. If one of the petals be pulled off, a tiny yellow scale will be seen at its attached end; under the scale is a moist spot, moist with a droplet of the sugary fluid called nectar. This nectar is a bribe for the black insects that are generally found crawling over buttercups, and the shiny golden petals are nothing more or less than advertisements, shouting to the insect world that there is nectar to be had. Inside the petals is a cluster of delicate structures, a ring of yellow stamens surrounding a bunch of green ovaries. The stamens are little stalked knobs, and when they are shaken each of them emits a few grains of the yellow dust called pollen; this dust is the male element of the plant. A pollen-grain is something much more than a spermatozoon, but that we will discuss later. The ovaries are green capsules, each containing a little round white body which is the female element. From the free end of the ovary a short neck projects, ending in a sticky yellow spot called the stigma. Just as in most cases the eggs of an animal cannot develop until they have been fertilized, so this waiting globule can do nothing unless a pollen-grain should fall on the end of the stigma. This is called "pollination."

When a grain of pollen gets on to a stigma it behaves like a minute

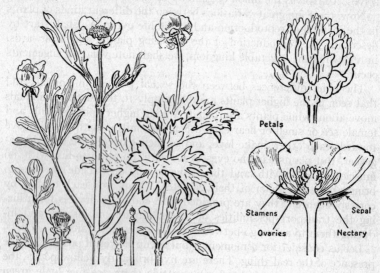

FIG. 107. The Buttercup flower.

Above, right, a cluster of fruits; below, right, a single flower cut across to show its construction.

plant in favourable circumstances; it sprouts out a little stalk or filament, the pollen-tube, which burrows down towards the egg-cell like a rootlet growing down through the soil; the tip of the stalk ultimately touches the egg-cell and fertilizes it so that it may develop into an embryo.

The manner of operation of the whole affair is this. Insects are attracted by the petals and come to seek the nectar; the stamens shed their pollen outwards into their circular groove at the bases of the petals, so that the insects, scrabbling for the nectar, get dusted with the tiny yellow grains. The insects, not satiated, fly to another buttercup to continue the meal; looking for more nectar, they rub up against

VEGETABLE LIFE

one or more of the stigmas and leave a few of the precious yellow grains upon them—and thus the egg-cells are fertilized.

Most of the gaily coloured flowers with which the reader is familiar work in the same way as the buttercup; their sepals, petals, nectaries, stamens, ovaries, and stigmas can be readily identified. The flower may be twisted up into a more or less elaborate shape, as in snapdragon or monkshood or columbine. And the non-botanical reader may have trouble in finding the parts of daisies and dandelions, which are not single flowers at all but bunches of a great number of curiously made little flowers crowded together. But with a hand-lens even this mystery may be unravelled without serious difficulty. There are many ingenious modifications of the scheme; flowers may specialize, for example, in the attraction and utilization of particular kinds of insects. In general, the colours and perfumes of flowers play the part of advertisements, the honey that of a fee, and the insects are busy go-betweens between stamens and stigmas, collecting a heavy bribe at both ends of their fertilizing journey.

But insects are by no means indispensable to the vegetable kingdom. It is just a way in which immobile plants have utilized the abundance and restless prevalence of arthropods. Most of the flowers that attract casual notice are pollinated by insects for the very good reason that the petals and perfume that attract us are actually intended to draw the six-legged carriers, but there are other ways of securing pollination, and many plants could get on perfectly well if there were no insects at all in the world. Grasses, for example, are pollinated by wind; their graceful spikes are in reality clusters of small flowers, but they have no conspicuous petals or perfume, because the light, dry pollen is simply blown from plant to plant. They have no use for insect visitors. There are many other examples of wind-pollinated flowers, generally small and greenish or brownish in colour because they need not attract attention. The dangling catkins of hazel are clusters of tiny male flowers, that is to say, flowers containing stamens but no ovaries, and pollen is blown from these to the female flowers, containing ovaries but no stamens, which form bud-like clusters only distinguishable from ordinary hazel-buds by the tiny brush of soft crimson stigmas with which they are tipped. Moreover, water is used as a medium for this transfer by many of the plants that live therein. In the Canadian waterweed, to which we have already made reference, the male flowers break bodily away from the plants and float about on the surface of the pond until they come up against female flowers, when pollination can occur.

And here we may mention a curious and important point. The buttercup, we have seen, is provided with an elaborate apparatus of petals and nectaries so that insects may carry pollen from one flower to another. Why should this be? Why should not the pollen from the stamens drop directly on to the stigmas, and so bring about fertilization without the bother of hiring an insect intermediary?

Throughout the plant and animal kingdoms we find a general preference for cross-fertilization; an avoidance of the union of eggs with male elements from the same individual. In most animals and a few plants the individuals are sexed, either male or female, so that here there is no possibility of self-fertilization. But in those animals such as snails and earthworms and liver-flukes, and in the great majority of plants, which are hermaphrodite (male and female in one), there are usually precautions to restrain self-fertilization. The male half of one individual seldom fertilizes its own female half, but is mated with the female half of another.

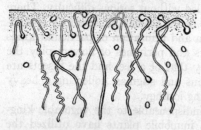

FIG. 108. Pollen grains sprouting in a drop of water.

They grow away from the edge (above), i. e., away from oxygen. This "growth instinct" would guide them in towards the egg-cell if they were on a stigma (highly magnified).

As a matter of fact there is no immediate physiological reason why self-fertilization should not occur. Indeed, in a number of plants it does. Primroses, and other flowers which open early in the spring before many insects are abroad, fall back on self-pollination if they have not been cross-pollinated. In the violet and the wood-sorrel, which, being modest flowers, are apt to be overlooked, special flowers are developed towards the end of the season that do not open at all, but are regularly and automatically self-pollinated. And many of the humbler flowering weeds, such as chickweed, shepherd's purse, and groundsel, nearly always pollinate themselves. Apparently self-pollination is perfectly possible and workable, but cross-pollination is in some way more desirable; the former is a second-best thing, to fall back upon if the latter fails.

In contrast to the modest weeds that put up with this second-best is the upright meadow buttercup, the ovules of which cannot be fertilized by pollen from the same flower. They just do not respond. This is not in any sense a sign that self-fertilization is generally an impossible method; it is a device for securing that only the preferable method, cross-pollination, shall occur. An enormous amount of unfounded belief has accumulated round the process of self-fertilization and the closely similar process of inbreeding. They have been accused of producing a loss of vigour, degeneracy, and a number of vaguely defined phenomena of that kind. It is the exact study of heredity on the basis of Mendel's laws—the most important advance that biological science has made in the twentieth century—that has made the story clear, and put it into precise terms. We cannot go into the details of the problem here; they are best deferred to a later chapter when the essential principles of heredity will be set forth. But we may say that there is nothing inherently devitalizing in inbreeding or even in self-fertilization itself. The chief reason for its unpopularity, so to speak, lies in the fact that self-fertilized populations are more homogeneous, less variable than those in which cross-fertilization is the rule. As we shall learn when we treat of Evolution, variability of the species is a profoundly important thing, for it gives a basis for natural selection to work upon; in the long run variable, cross-fertilized species will tend to supplant homogeneous self-fertilized species merely because they are variable. There will come to be more of the former. It is not that inbreeding is inherently bad and outbreeding inherently good; it is merely that the latter method affords more opportunity for Evolution, for the adaptation of the race to the ceaselessly changing conditions of the organic world, that determines its predominance.

But we have dwelt long enough on flowers; let us turn to fruits and seeds, to the embryos that result from successful pollination.

We have described the process of pollination in the buttercup; let us proceed with our description, and trace the subsequent events. The little bunch of green ovaries, each with a waiting egg-cell inside it, has been dusted with pollen; the sepals, petals, and stamens, their purpose discharged, have withered and dropped off. The ovaries swell up and ultimately become dry and hard; inside them the egg-cells are developing into tiny embryos, and being fitted out with little stores of food, like the yolk of a bird's egg. The embryo, together with its food store and the tough membrane enwrapping both, is called a seed, and the enlarged ovary that contains it is called a fruit. So in the buttercup each fruit contains one seed.

At this stage the plant is faced with another problem, a problem analogous to that of pollen-transfer already discussed. Supposing that the fruits were simply to drop off and that the seeds were to germinate in the ground beside the parent plant, what would happen?

There would result a bitter struggle between parent and offspring, their leaves wrestling and fencing for light, their roots for moisture. It is better for the seeds to be sown in distant places, so that they do not interfere with their parent or with each other, and so that the species may extend its range. They have, to use the technical term, to be dispersed.

The fruits of plants are often variously and ingeniously contrived so that their contained seeds may be dispersed. The reader will be familiar with a number of examples. The plumed fruits of dandelions and thistles and the winged fruits of sycamore and ash sail in the wind for considerable distances. The dry poppy-head rocks in the wind, and as it sways it scatters little puffs of seed, like a sower scattering handfuls of grain. It is indeed a better and more even distributor of fine grains than an ordinary saltshaker or pepper-pot; and a household shaker deliberately designed from the model provided by the poppy has recently been patented. The fruits of violet and wood sorrel and the pods of gorse and the sweetpea burst with surprising violence and eject their tiny offsprings into the world. The seeds and fruits of water-lilies have spongy coverings filled with air, by means of which they float for a time on the surface of the pond, later to become water-logged and sink. The fruits of goose-grass and agrimony and burdock bristle with tiny hooks that catch in the fur, feathers or plus-fours of passing creatures, and so are carried for great distances before they are dropped, brushed off or noticed, pulled off and tossed aside. And there are many other ingenious devices to reward the observant naturalist.

Finally, there are the fleshy and succulent fruits—blackberries, yew-berries, elderberries, currants, apples, cherries, plums, oranges, bananas, grapes, peaches, figs. Here, again, we find plants overcoming their own immobility by bribing animals to act as their postmen. The flesh of an apple, a strawberry, or an orange is not a store of food for the growing embryo, as the flesh of a bean is; it is a free gift, deliberately set out for birds or mammals to consume. In return, the consumers unconsciously distribute the plant. They often carry the fruit a little way before attacking it; if the hard part within is large, like the stone of a cherry, it is dropped uneaten to the ground, and has thus achieved distribution. If it is small and inconspicuous, as in a blackberry, it is swallowed with the rest of the fruit, and being indigestible emerges unchanged from the other end of the alimentary canal; in the latter case it is not only distributed but manured. The conspicuous appearance of most fleshy fruits is again an advertisement, drawing attention to the gift. The advertisements of fruit are mostly directed at vertebrates, those of flowers mostly at insects.

We may close by referring to the seeds of gorse, which employ humbler carriers; they have a small, fleshy appendage that is appetizing to ants, and in addition to being exploded out of their pods they are therefore dragged about by the ants, often for some distance.

§ 4. The Flowering Plants.

It is by no means true that all plants reproduce themselves by means of flowers and distribute themselves by means of the resulting seeds. These matters are characteristic peculiarities of one division of the vegetable kingdom. It is true that division includes by far the greater number of the plants that now cover the earth with meadows and jungles and forests, and nearly all that are grown for beauty, or for food, or for timber; nevertheless, there are other plants with other methods, and with these we shall have to acquaint ourselves.

But before we do so we may note briefly that the division with which we have hitherto concerned ourselves, which may be called either the flowering plants (phanerogams) or seed-bearing plants (spermatophytes), is divided into two classes.

The first class includes all the coniferous trees and shrubs—fir, pine, larch, cedar, cypress, juniper, yew, and the rest—and also the peculiar tropical cycads and one or two other curiosities of which we shall learn more hereafter. The conifers bear their pollen or egg-cells on little catkins, and the female catkin grows into a woody cone, each scale of which bears two seeds, but we may note two well-known exceptions to this rule; in yew and juniper, which are classified as conifers for other and weighty reasons, the fruits are fleshy and distributed by birds like other berries. The definitive characteristic of this class of seed-plants is that the seeds are not enclosed in special ovaries but sit exposed at the bases of the scales; for this reason they are called *gymnosperms*, which means "naked-seeded."

In the second class the seeds are housed in ovaries, and the members of the group are therefore called *angiosperms*, which means "with seeds in containers." The class includes nearly all of the herbs and trees with which the reader is familiar. They belong to the contemporary world. There are no phanerogams in the most ancient rocks; gymnosperms appear in association with reptiles; and the angiosperms, flower and grass and leaf-shedding tree, only became dominant as the mammal and bird became prevalent in the world.

§ 5. Ferns and Mosses.

WE TURN now to plants that do not possess flowers, and do not distribute themselves by means of those elaborate and sexually-produced structures we call seeds.

The reader will be familiar with the appearance of one or two kinds of fern, but he or she has probably only met the comparatively unassuming members of the group that inhabit temperate countries. For the ferns attain their fullest luxuriance in tropical climates. In Central America the common bracken fern towers to double the height of a man, and there are species peculiar to the tropics that grow like trees, with erect cylindrical stems sixty feet or more in height, bearing crowns of spreading fronds. The group is large and varied; it includes over three thousand species, ranging in size from these tree-ferns to the minute filmy ferns, hardly larger than mosses.

Instead of bearing flowers and seeds, ferns reproduce themselves by means of spores. A spore is a tiny particle of living substance, enclosed in most instances in a more or less resistant shell, and capable of growing directly into a plant without any sort of fertilizing process. It is a sexless production. Fern-spores are produced by proliferation of cells in the pale brown areas that form such conspicuous patternings of circles or stripes on the lower sides of the fronds of most ferns. The spores are very small and light; indeed, they are single cells, and when they are ripe they escape and are blown about, often for considerable distances, by the wind.

If a spore drops on suitable soil it germinates and begins to grow, but it does not develop into a new fern. It grows into a structure called a *prothallus*, which, instead of bearing conspicuous erect fronds, is, in most cases, a flat leaf-like plate of green tissue, less than an inch across, and sending a few tiny roots into the soil. This simple object is not a young fern—indeed, it is not a young thing at all, for it is sexually mature; it reproduces and dies without further development. It is a generation in itself. Scattered over its under surface there are tiny reproductive organs of two kinds, male and female. The female organs are shaped like long-necked flasks, and waiting in the belly of each flask is a single egg-cell. The male organs are simply globular masses of tissue that is proliferating and giving rise to the male elements. Now the male elements are not passive, like pollen-cells, but they swim actively like spermatozoa. They have little bodies, shaped like a corkscrew and with a tuft of lashing tails at one end; at the other they usually carry hollow, bladder-like structures which may come off and be left behind without in any way disconcerting the little swimmers. These spermatozoa make their way towards the female organs through the moisture that usually collects under the prothallus, and they pass down the necks of the flasks to fertilize the waiting egg-cells. All this is more like animal reproduction than anything we

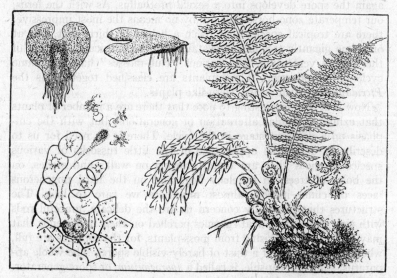

FIG. 109. The Life-cycle of the Fern.

On the right is a Fern-plant (Aspidium). Above, on the left, are two Prothalli—the sexual generation which the sexless fern begets. Below them is a highly magnified view of the tube, in which the egg-cell awaits fertilization. The egg is at the bottom of the tube; passing up the neck is a drop of an attractive fluid round which three sperms are hovering.

have hitherto related of plants. As the result of fertilization the egg-cell begins to grow, and it develops into a fern-plant like the one we started from. Evidently, we have here in the fern an alternation of generations, like the alternation that we noticed in the liver-fluke. The fern-plant produces spores; the spore grows into a prothallus; the prothallus produces spermatozoa and ova; the fertilized ovum grows into a fern-plant, and so on. We may note that there is no very definite purpose in this arrangement, as far as can be ascertained; it is not in any sense better than the arrangement seen in the higher plants. It is an arrangement that becomes intelligible, as we shall learn, from the evolutionary point of view; the ferns, like the amphibians, are transitional forms, caught in the act of passing from one style of existence to another. Like the amphibia, the ferns cannot get far away from wetness. They have not fully attained the dry land.

There are one or two other kinds of plant that show a similar life-history. The horse-tails bear their spores on erect cones, and here again the spore develops into a sexual prothallus. As with the ferns, our temperate zone horse-tails are by no means the most impressive; there are tropical species that reach a height of thirty feet, and at one time gigantic tree-like horse-tails were a conspicuous feature of the world's vegetation. The curious "club-mosses" have the same cycle of reproduction. These plants are classified together as the *Pteridophyta*, which means "fern-like plants."

Now it is very interesting to note that there are a number of plants that exhibit a similar alternation of generations, but with the emphasis on a different stage in the cycle. There is no need for us to describe in detail the appearance of the little tussocks of various species of moss that are found growing on walks and stones, on the boles of trees, in squelchy bogs and on the hard ungenerous faces of cliffs—indeed, almost anywhere we care to look. The structures that especially concern us are the delicate, graceful rods with tiny globular or oval capsules perched on their upper ends that may often be found rising from moss-plants, for the capsules are full, when they are ripe, of a dust of barely visible spores. The whole apparatus, stalk and capsule, is called a *sporogonium*, or spore-generator.

The spores are shed from the sporogonium and are blown about, much as they are in ferns. When they germinate they give rise at first to a *protonema*—a branching feltwork of extremely delicate green threads having no very definite arrangement. This elementary, groping web is moss-tissue feeling about, so to speak, for the best and most comfortable situation in which to settle down. Here and there the protonema strikes lucky; it throws special branches up into the air to become stems and leaves of a primitive kind and sprouts root-like hairs down into the soil. In this way the protonema gives rise to several moss-plants, after which it generally dies, and its offshoots, the moss-plants, lose their organic connection with each other; they become discrete and separate things although they originated as parts of the same living body. So the moss as we know it is formed with simple roots and stems and leaves, and proceeds to live on water and salts and air and light in the manner that we have already described.

After a time this moss-plant develops sexual organs, very like the

sexual organs of a fern prothallus and quite unlike the flowers of a
higher plant. They appear in tiny clusters at the ends of the main
stem or of its branches. As with the prothallus there are male organs
producing active spermatozoa and female organs shaped like tiny
flasks, with necks down which the spermatozoa swim to the waiting
egg-cells. Then the fertilized egg-cell, like the egg-cell of a fern and
unlike the egg-cell of a flowering plant, does not become part of a

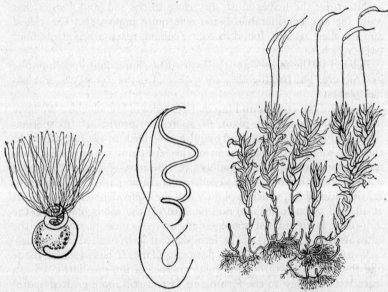

FIG. 110. More animal than vegetable in their restless activity—two sperms from a Fern (left) and a Moss (right).

FIG. 111. A Moss-plant, showing the slender spore-bearing generation rising parasitically from the leafy sexual stems.

hard seed but begins to grow and develop at once. It grows into the
stalk and capsule that we have already noted—the sporogonium. It
gives rise to this mass of tissue that sprouts up parasitically from the
moss-plant, sucking nourishment from it, but which is nevertheless
not a branch but a child of that plant.

Here, again, then, we have an alternation of generations—the sporogonium gives rise to spores, each spore gives rise through a protonema to one or more moss-plants, the moss-plant gives rise to spermatozoa and egg-cells, and the fertilized egg-cell gives rise to a sporogonium again. The chief difference between a moss and a fern is that in the latter the spore-bearing phase of the cycle is larger and more elaborately organized than the gamete-bearing phase, while in the former the reverse is the case. But there is another important difference, for mosses lack the "vascular tissue" of the stem—the installation of pipes along which fluids are driven from roots to leaves and from leaves to roots. Their transport arrangements are very primitive and inefficient, and for this reason they never develop any impressive architecture; the leaves must stay close to the soil, and a stem that soars for any considerable distance is quite impossible. The tallest and stateliest mosses, found in New Zealand, reach a height of about twenty inches.

Related to the mosses are the liverworts, similar but even humbler in structure. The two together are spoken of as the *Bryophyta*, or moss-like plants.

Now here is a curious and fascinating series of homologues. We have in the fern-cycle (a) fern-plant, (b) spores, (c) prothallus, (d) spermatozoa and ovum, (e) fertilized ovum and so back to (a) fern-plant. In the moss we have the equivalent of the fern-plant in (a) the subordinated sporogonium which produces (b) the moss-spores and so leads with the insertion of a little vegetative spreading and breaking up to (c) the moss-plant, which is plainly not the equivalent of a fern-plant at all, and so on to (d), (e) and back to (a) the sporogonium. So the fern-plant is an exalted sporogonium and the moss-plant is not the equivalent (homologue) of the fern-plant; it is an exalted prothallus.

Now let us turn back to the flowering plant. It has vascular tissue like the ferns, and through a great number of intermediates, we can trace its homology to the fern-plant. It is a still more exalted sporogonium. Where is its spore?

Well, it happens that among some of the so-called club-mosses, which are related to ferns, there are two sorts of spores and two sorts of prothalli. There are female prothalli producing egg-cells only, waiting in their tiny flasks, and male prothalli producing the active spermatozoa. Moreover, the spores that grow into female prothalli are bigger than those that give rise to the male prothalli—indeed, the latter are just about the size of a pollen-grain. And the lives of the prothalli are short, they are even smaller and more insignificant than the prothalli of ferns. The plant-tissue is, so to speak, going all out for the sexless generation; it bustles perfunctorily through that ancient and venerable formality the prothallus and gets it over as soon as possible, like a worldly chaplain saying a rudimentary grace before he gets down to the business of eating.

And now let us turn to the more primitive seed-plants—to the conifers and cycads and the like. Here there is an even more swiftly gabbled prothallus. A pollen-grain, we noted, does not fertilize an ovum simply and straightforwardly as a spermatozoon does; it settles near the ovum and germinates and sprouts into a little mass of tissue, growing like a thirsty root towards the ovum, and only a bit of it, becoming detached from the rest, actually accomplishes fertilization. This looks

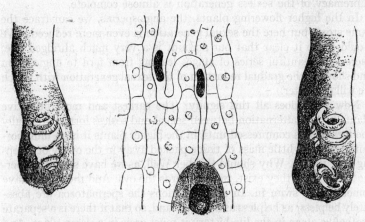

FIG. 112. Fertilization in the higher plants.

In the center a pollen-tube is growing sinuously through the ovary tissue and nearing the egg below. The tube is a vestigial generation; it contains two fertilizing nuclei (shown black). The grey mass below is a vestigial female generation; it contains an egg-cell (shown very dark). This is the typical condition. But in some lilies the fertilizing nuclei in the pollen-tube are spiral like the nuclei of fern or moss sperms. This is drawn (on a larger scale) on the right; compare Fig. 110. And in the primitive Cycads the pollen-tube produces active sperms (left). These are evolutionary reminiscences, like the transient gill-clefts of a human embryo.

suspiciously like the growth of a male prothallus. And the comparison is clinched by a curious fact: whereas in most seed-plants the fertilizing elements are very different in appearance from the spermatozoa of ferns there are one or two exceptions—the ginkgo or maidenhair tree and the cycads—in which the end of the growing pollen-tube gives rise to spermatozoa for all the world like the spermatozoa of ferns or mosses, which swim up to the ova and fertilize them. The pollen-grain of all these plants then is not a gamete but a spore, and it gives rise to a rudimentary male prothallus. And on the female side there is a similar story. In the young female catkin of a coniferous plant certain special cells, which we may call "spore-cells," are set aside. Each of these spore-cells divides and proliferates and produces by its own unaided efforts a special mass of tissues, and in this mass a little cluster of egg-cells is developed. The spore-cells are indeed spores, but they are not allowed to escape from the parent plant; the masses of tissue that they give rise to are female prothalli, reduced and insignificant, which stay imbedded in the substance of their parent. The supremacy of the sexless generation is almost complete.

In the higher flowering plants, the angiosperms, we can trace the same story, but here the sexual generation is even more reduced. (We would make it clear that our account is a very much abridged one; there is a beautiful series of stages leading from fern to angiosperm and showing the gradual reduction of the sexual generation with which we will deal later.)

Now what does all this signify? The earliest and most primitive plants had an alternation between sexual and sexless forms; gradually the former was compressed until in the higher plants it is an unimportant gesture, a little mass of tissue packed away in the ovary or creeping over its top. Why should this be? Well, as we have seen, the spermatozoa, when they exist, *swim* towards the ova. And they must have something to swim in. In dry conditions the spermatozoa are absolutely helpless, as helpless as a fish on land, so that if there is a separate prothallus stage in the life-history it must grow in a damp place. And since the sexless stage sprouts directly out of the prothallus such a plant can never get very far away from moisture. So that the cutting down of this stage, first the tucking away of the whole process into the moist cracks of a cone on the parent plant and then the elimination of a free spermatozoon altogether, is a process of emancipation from water. It is quite parallel to the way in which, in all vertebrata above the amphibia, the larval gill-breathing phase of development is reduced and hurried through in the egg-cell or uterus as the case may be. In both instances, in higher plant just as in higher vertebrate, a progressive emancipation from water and an adaptation to dry land is traceable. And with the plant just as with the animal, it was the water-inhabiting forms which preceded and produced the dry land ones.

VI

THE LOWLY AND MINUTE

§ 1. *Amœba.* § 2. *The Minutest Animals.* § 3. *Plant-animals and Seaweeds.* § 4. *Moulds, Toadstools, and Yeasts.* § 5. *Lichens.* § 6. *Slime-moulds.* § 7. *Bacteria.* § 8. *The Smallest Living Things.*

§ 1. Amœba.

AS YET we have found little employment for the microscope in our survey of the forms of life. Here and there we have met creatures too small to be easily examined, tiny worms and crustaceans and mites and rotifers, but for the most part they were miniatures of the larger animals that we can see around us; we have used the microscope only for prying into the finer details of the anatomy of larger living things. But we must plunge now into a world of creatures which are nearly all minute, attaining only in exceptional instances such dimensions as to be seen with the naked eye.

It is a strange and exciting world. Its existence was hardly suspected until the seventeenth century, when Anthony van Leeuvenhoek, a Dutch draper, peered through lenses of his own making at drops of water and bits of cheese and the scrapings of his own teeth—indeed, at anything he could lay his hands on—and beheld for the first time their curious inhabitants.

An enormous and vitally important branch of biological science has sprung from these observations. Nowadays everybody has heard of "microbes," tiny insidious creatures, populating the earth with unseen multitudes. Because of their minuteness the smallest drop of water is a habitable world for them, and a few particles of decaying organic matter are enough to support a thriving microbe population. They swarm in pools and puddles, in the moisture of the soil, in the moisture that clings to mosses and other plants. Their spores float everywhere in the air. If a few blades of straw or a little pepper or some other organic substance be added to water, and if it be left in a glass, it will be found in a few days to be teeming with these tiny creatures. Moreover, their species may attain to inconceivably vast numbers. The number of men in the world to-day is about seventeen hundred millions, and the numbers of individuals in most species of the larger mammals are probably lower than this. But microbes attain numbers almost beyond estimation, numbers that make such figures as these negligibly small. The green colour of stagnant ponds is due to microscopic organisms, individually too small to be visible; the colour of the Red Sea is due to untold myriads of microscopic plants, "making the green one red." There are microbes which invade our bodies and rack us with a variety of diseases, there are others which are not enemies but indispensable friends. And aside from its practical importance there is beauty and wonder in the micro-world. Few poems have been written about the green soupiness of stagnant ponds or about the green on the boles of trees or about the powdery granules of flinty earths; nevertheless, as we explore we shall find creatures beautiful and fascinating and of the profoundest interest to anyone of a philosophical turn of mind.

Abundant in the greenish, apparently lifeless scum that accumulates at the bottom of most ponds, water-butts and the like is the curious, elementary little creature known as Amœba. It is a speck of a transparent jelly-like substance, about one-hundredth of an inch long, showing very little organization in its frankly visible inside. The particular individual in Fig. 113 is seen from above, crawling—or rather flowing—over a flat surface; from the bird's-eye point of view of the artist it is proceeding towards the left. It is a creature vastly more simple than any that we have hitherto considered. It has no tissues, no definite organs; no heart, no brain, no stomach, no kidneys, no bones. Except for two special structures that will be considered in a moment, it is nothing but a spot of jelly without enduring form—but a jelly that lives—breathing by absorbing traces of oxygen and giving out traces of carbon dioxide, and moving about and feeding itself in a peculiar and characteristic way.

In this jelly two layers may be discerned, a comparatively opaque and very granular inner layer and a perfectly transparent outer layer, more solid in consistency than the inner. This layering is not permanent; the substances of the two layers mingle and are turned into each other. The creature moves by the contraction of the outer layer, which slowly squeezes in at the sides and back and forces the inner layer forward, so that as we watch the granules may be seen to stream along in the direction of the arrows in the figure. When it gets to the front end, the inner layer becomes converted into the outer layer, and when it gets to the back the outer layer is absorbed into the inner layer, so that as the animal proceeds it is constantly being urged forwards by the contracting outer layer. It crawls, so to speak, by being everlastingly squeezed out of its skin. Its method of feeding is as elementary as its locomotion; it consumes microscopic plants, other creatures like itself, and so on, and, lacking a definite mouth, it takes them in indiscriminately at any point on its surface; it simply flows over them and round them and, thus embracing them in three dimensions, gets them into its own body. And when it has exhausted their nutritive possibilities it flows away from them, abandoning them from any point at the surface of its body. The individual in our drawing is prowling about among a number of crescent-shaped microscopic plants; it has recently consumed three of them, and is dragging them along in its hinder part and slowly digesting them.

There are numerous species of amœbæ, living for the most part in fresh- or sea-water or in damp soil. Some of them are parasitic, living in our bowels (which seem to have an irresistible attraction for microscopic organisms, for they are populated by a great variety of different kinds), and although most of the parasitic amœbæ are harmless, there is one species that causes a serious form of dysentery. The amœba represented in the drawing belongs to a kind found in ponds; it is called a "limax amœba" because it crawls smoothly and steadily forwards in one direction like a slug (limax is Latin for slug). The "proteus amœba" has an even less definite shape; its body-jelly can flow out into lobes at any point over the surface, and by throwing out a lobe and then flowing forwards into it, so to speak, the creature moves from place to place.

In this liquidly gelatinous body two definite structures are suspended. The first is an obvious, transparent sphere, situated near the hinder end of the body—a hollow sphere, filled with a watery fluid. If it is carefully watched, this sphere can be seen to grow slowly larger and larger until ultimately it bursts like a bubble, squirting its contents through the clear skin into the outer world. Then, in a little while, a tiny drop reappears, grows again, bursts again—and so on, with a rhythm of several minutes from burst to burst. This structure is called the contractile vacuole. There is an apparent parallel between its behaviour and the slow accumulation and roughly periodic expulsion of fluid that occurs in the human bladder, and for a long time the contractile vacuole was regarded as a special excretory organ for getting rid of the waste-products of the creature's activity. But, as a matter of fact, it is an organ for getting rid of water. The inside of the amœba, like our own insides, contains salt and other dissolved substances, and these substances, by a simple physical process, are constantly sucking water in through the surface of the amœba from the outside. In technical language the dissolved salts set up an osmotic pressure inward. The water has to be everlastingly baled out again by the contractile vacuole, because otherwise the animal would swell up and burst. Most of the chemical interchanges between amœba

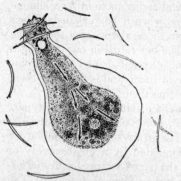

Fig. 113. A Limax Amœba—one of the smallest and simplest animals, built like a single cell.

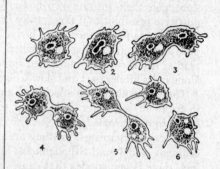

Fig. 114. Amœba multiplies by simply tearing itself into two Amœbæ (highly magnified).

and its environment—the absorption of oxygen, the emission of carbon dioxide and other matters—occur indiscriminately over its whole surface and need no special organ.

The second structure is one we have already remarked in the tissue-cells of the higher types—the nucleus.

It completes a very striking resemblance that may already have become apparent to the reader between the anatomy and general behaviour of amœba and that of a white blood corpuscle, or of an isolated cell creeping in a tissue-culture. Both are of a size and both are masses of living protoplasm without enduring form. It is possible, because of this great similarity of organization, to look upon an amœba as a single cell which is living by itself instead of being part of a body-community, and for this reason amœba is generally called a single-celled or "unicellular" organism.

This view is not unanimously held among biologists. Some point out that amœba is more like a whole human body than one of our constituent cells, for it can do a number of things that our own cells cannot do. It can, for example, live in a pond, finding and securing its own nourishment. No single human cell could do that, nor could a cell from any of the higher animals or plants. Tissue-cultures have to be carefully nursed and grown in elaborately standardized and sterilized fluids containing appropriate foods. Moreover, the life-history of amœba is more complicated than that of a cell in a tissue-culture, for, besides simply tearing itself into two halves, it has other methods of reproduction, which we shall describe later. Amœba, in short, is a whole and independent organism, while the tissue-cell is only a subordinated organism.

Yet it is an attractive comparison. Our white blood cells creep about our bodies and consume things very much as amœbæ creep about and feed. And if our own cells lack the self-completeness of amœba, that is closely approached by the cells of many lower invertebrate forms. We have already noted what happens when Obelia or a sponge is passed through gauze. We can see the cells that originally belonged to a many-celled creature wandering about by themselves, lost and disorganized like a routed army. They creep about in the manner of amœbæ, and they can be seen to clump together gregariously in enormous flocks; they arrange and specialize and organize themselves so that the clumps, at first formless, gradually shape into individuals of a higher order, into tiny polyps or sponges. Moreover, we can trace a series of animals leading up by gradual stages from these feebly organized cell-communities to our own strictly disciplined bodies. These are facts, and basing ourselves upon them we make the statement that amœba is an unicellular organism—unicellular because it is like a single cell, organism because it is a complete living thing in itself—a statement which implies that a man is a great number of amœba-like creatures, their independence sacrificed to the common good.

We may most clearly understand this statement by comparing amœba to a naked and solitary savage on a desert island, and a human tissue-cell to a trim, bowler-hatted clerk, trotting with umbrella and attaché-case to his appointed desk in the city. Both are built in much the same way, and their corresponding parts work in much the same way. They are homologues. But one is without the versatility of the other. The clerk is a specialist in his own particular job, but in specializing himself he has lost touch with many of the basal activities of men. Remove him from the civilization to which he belongs, isolate him on a desert island, and although he can write and calculate more efficiently than the savage, you will have to take the most elaborate care of him, providing him with shelter and clothing and cooked food, or he will surely perish. The savage, on the other hand, can build a hut and scrape together a meal, and his tough skin can survive exposure to the elements. He is a jack-of-all-trades, albeit doing nothing very efficiently, and it is precisely because he is unspecialized that he is independent. One can trace just the same distinction between a free, unspecialized amœba and an efficient but limited muscle-fibre or nerve-cell or gland-cell; the latter three, like the clerk, are accustomed to a division of labour, and have come to depend on other units for the satisfaction of many of their needs. Extend the parallel one degree further; contrast the day-to-day routine of that one-man community the savage, with the enormously complicated and varied activities of a whole civilized nation, and you get a valid parallel to the difference between amœba and your own highly organized body.

§ 2. The Minutest Animals.

AMŒBA belongs to an exceedingly large and varied phylum of minute animals, the *Protozoa*. It is the simplest, the least organized member of the group. Most of the protozoa have definite and enduring shapes, and special organs such as lashes or vibrating hairs with which to propel themselves; usually they take in food at some special mouth point on their surface. But amœba presents in its nakedness the essential features of protozoan organization; for nearly all members of the group are invisibly small so far as the naked eye is concerned, and their structure is comparable to the structure of a single cell of the higher animals.

The protozoa, having for the most part unprotected surfaces over which their chemical interchanges take place, are confined to wet or damp situations; they cannot exist in the absence of moisture (except as encysted or resting forms). But nearly everywhere where moisture collects they swarm, and they may produce the most far-reaching effects.

The reason why such minute organisms can be so very important, both in the general economy of Nature and to ourselves, is at first sight a paradoxical one. It is because their lives are so short. To show how this may be, let us return to amœba. We have seen how it is built, how it eats and grows; let us see how it increases and multiplies.

The simplest method—for there are several—is, by our standards at least, a very curious one. It is like the multiplication of cells in one of the higher animals. When the creature has lived for a time and grown to a sufficient size it simply pulls itself into two halves, each part getting half the nucleus, and the resulting fragments creep away and live free and independent lives of their own. They wander about and nourish themselves and grow, and in course of time, when they are big enough, they divide in their turn. And so on. This simple splitting, this tearing of oneself into two halves is the chief method in which the protozoa multiply; it is, so to speak, their routine method of reproduction. They have other ways. They may vary the monotony of continual splitting by a primitive kind of sexual union, or sometimes instead of dividing into halves they may chop themselves up into a greater number of bits, into a sort of living mince each particle of which can grow into a new individual. But that need not concern us here; for the present we will confine ourselves to the commonest method of splitting into halves.

The time that intervenes between successive divisions varies in the different kinds of protozoa. In the larger species an individual may live for days or months before he feels the call to tear his body into living bits. But in the smallest species the time is shorter; it may be as brief as an hour. Let us consider one of these short-lived kinds; let us start with a single individual and imagine that it has no enemies and all the food it wants. It will flourish and grow, and at the end of an hour there will be two. At the end of two hours there will be four. At the end of three, eight. And so on, the number increasing ever more rapidly. At the end of thirty-six hours there will be two raised to the thirty-sixth power, which is sixty-eight thousand five hundred million —over thirty times the number of people in the world. This at the end of a day and a half! Imagine that the creature was one of the smallest free-living protozoa, one of the monads that can only be clearly seen with the highest magnification of the microscope, this thirty-six hours' proliferation would be enough to form quite a respectable heap on a shilling.

Let the experiment continue. In another thirty-six hours the number will be squared—there would be enough to make heaps on seventy thousand million shillings. In a week there would be a number so inconceivably vast that it would be waste of space to write it down, and the monads would together weigh many times more than the whole earth.

This is, of course, an impossible experiment. It is true that most protozoa are short-lived, and that they tend to multiply at this rate, but under natural conditions their numbers are kept down by factors outside themselves, by limitation of the food-supply, by hungry enemies. Nevertheless, the calculation shows the almost explosive expansion of a protozoan population under favourable conditions. And in this fact lies the power of these microscopic creatures. The countless swarms of micro-organisms in the sea are eaten by minute crustaceans, and these in their turn are eaten by fish, and the fish by other fish or by ourselves; constantly and by the million they are being devoured, but the survivors can multiply so rapidly that the supply is inexhaustible. And the entry of a few parasitic protozoa into the blood of a man may fill him in a very few days with pullulating millions of invisible enemies —enough to interfere seriously with his vital processes, perhaps enough to destroy him.

There is another important aspect of the curious methods of multiplication that are found among these tiny creatures, another reason why they can increase so rapidly. We have noted that the individuals are very small and that they do not survive for long. But let us consider again the way in which they cease to be. A protozoon ends its individual existence by dividing its own body into living bits—into two halves, or into a greater number of fragments. Every one of these bits grows into a complete new individual. Try to imagine that in terms of human experience. Imagine that when a man or a woman reaches a ripe old age he or she simply divides into two halves, each of which is a growing child! If all goes well with a protozoon its tissue never dies; every bit of its body is handed on for the use of future generations. Our own method is very different from that. Little bits of our bodies do indeed live on and grow to be our heirs, but most of our substance dies when we as individuals cease to be. With every generation there is an apparent wastage, a scrapping of great quantities of material. In the protozoa, on the other hand, there are no corpses; every bit of protozoan tissue continues to multiply and grow. It is potentially immortal; although the individuals exist only for an hour or so their tissue lives for ever.

The protozoa, then, put their whole weight into the business of reproduction. They do not relegate that all-important task to a small part of their bodies. The method that we have just discussed is a sexless one, but we may pause for a moment to consider the sexual methods that they sometimes employ, for in this they are equally wholehearted. The details of sexual reproduction vary in the different kinds of protozoa. But the essence of the process is nearly always this: a complete mixing together of the substance of two originally separate individuals. They meet and come together in pairs, and they blend as completely as drops of water run together. They achieve an intimacy of union that we can never aspire to, becoming in a perfectly literal sense one flesh. And the resulting blend lives on as a single thing, after the manner of protozoa.

THE LOWLY AND MINUTE

These protozoa are divided into four classes. There are first the *Rhizopoda*, which include amœba; their name means "root-footed," and is an allusion to their power of everlastingly changing their shape by protruding and withdrawing blunt lobes or fine branching threads of living substance. The other kinds of protozoa are more definitely and permanently shaped. Two orders of this class are important—the *Foraminifera* which build shells of carbonate of lime and the *Radiolaria* which build flinty shells of silica. When these creatures multiply in the protozoan way by division of their bodies into a family of children their offspring abandon the parental shell and make new ones of their own. All the Radiolarians and many of the Foraminiferans live floating in the waters of the sea. There is consequently a perpetual soft rain of abandoned shells dropping on the ocean floor. The creatures are incredibly numerous and they can multiply rapidly, so it will be manifest that infinitesimal though they are their accumulations may be immense. In some places the sea-floor is covered deeply with a greyish mud called "globigerina ooze," which consists largely of billions and billions of the abandoned shells of a foraminiferan called Globigerina. In the deep sea, where the physical conditions are such that abandoned limy shells are re-dissolved in the water, only the siliceous shells of the radiolaria remain and form a deposit called "radiolarian ooze." In the course of past ages such oozes have hardened into earth and rocks, and as, during the evolution of our globe, the sea has advanced and receded over most of its surface, they have been left as geological strata on dry land. Huge thicknesses of white chalk and limestone consist very largely of the microscopic skeletons of foraminifera. The rock of which the Pyramids of Egypt are built is composed largely of the shells of Eocene foraminifera—and the earth known as "tripoli stone," used commercially as an abrasive or an absorbent, is composed mainly of radiolarian skeletons, the variety found in the Barbados containing at least four hundred different species.

In a second class of the protozoa, the *Flagellata*, the individuals are characterized by the possession of one to four long lashing tails or flagella (singular *flagellum*), and flog themselves along through water much as spermatozoa do. The group is a large one and profoundly important, both from the practical and the theoretical point of view. They are of practical importance because they swarm in incredible multitudes in the sea, forming the chief food of a host of crustaceans, young fish and other larger creatures and thus indirectly nourishing man himself. The diffuse phosphorescence of the sea is due largely to swarms of flagellates. Moreover, there are kinds of flagellate which live parasitically in our bodies or the bodies of our cattle and cause grave tropical diseases such as the sleeping-sicknesses of Africa, and the Indian scourge kala-azar. Theoretically they are interesting because some of them are coloured green with chlorophyll and lie on the border-line between animal and plant, and because others combine together to form little colonies, a link between the independence of amœba and the communal life of our own body-cells.

A third class are the *Sporozoa*. They are all parasites. The most troublesome kinds are those of the various mosquito-borne fevers and agues, ranging from the comparatively mild intermittent fevers, which have been abolished by the drainage of our own fen and marsh districts, to the deadly malarias of tropical countries.

The fourth class, the *Infusoria*, includes some of the largest and most elaborately organized protozoa. They have definite mouths; their bodies are covered with hard pellicles; instead of one nucleus they have two, of different kinds, one for the chemical business of every day, the other reserved for sexual union and other rare reproductive occasions. Instead of one or two propulsive whips, like the whips of flagellates, they possess whole batteries of vibrating hairs or cilia by means of which they swim. The cilia may be of various shapes and sizes, and are variously arranged on the body, and the modes of progression of their owners are correspondingly diverse. Some infusorians glide more or less smoothly through the water; some crawl and then dart; some spring by means of catapult-like cilia at their hinder ends. Among the commoner fresh-water forms well known to microscopists are Paramecium, the slipper animalcule—quite large for a protozoon, nearly a hundredth of an inch in length and visible as a tiny speck to the naked eye—Vorticella, the bell-animalcule, a graceful living goblet fastened to some solid mooring by a long stalk, which can shorten to a tight spiral when the animal is alarmed and so pull it back out of danger—and a host of others. Attractive though the infusoria are to the possessor of a microscope, they are of little direct importance to man. They do him no good, and, except for one species that lives in the intestines and causes dysentery, they do him no harm.

§ 3. Plant-animals and Seaweeds.

MOST of the protozoa are quite obviously animals; they wander about and feed in the animal manner. But there are curious exceptions. There is, for example, a tiny creature known as *Euglena*, one of the flagellates possessed of a lashing whip by means of which it swims about. Now Euglena has got two perfectly distinct ways of nourishing itself and growing. It has a mouth and can take things into its body and there digest them after the manner of animals. But it is also green in colour; it possesses chlorophyll and can build up its tissues out of the simplest ingredients after the manner of plants. It can do whichever it likes. Here then is a problem: is Euglena an animal or a plant?

We noted in the last chapter that chlorophyll is the characteristic feature of vegetable organization. It enables plants to dispense with the complex food that animals require, and a number of the most striking features in plant anatomy depend simply upon that fact. It accounts, for example, for their lack of muscles and nerves. But it is not an absolute criterion; there are organisms, such as the fungi, which are more like typical plants than animals in their organization, but which do not possess chlorophyll. Indeed, it will become more and more apparent as we proceed with this chapter, that there is no absolute criterion. Typical plants such as buttercups and oaks and cabbages differ strikingly enough from typical animals such as mice and men and lobsters; they are green and feed on substances of simple constitution; they do not move about and apparently they do not feel; their cells are enclosed in tough boxes of cellulose. And as long as we confine ourselves to the more obvious organisms the line is easy enough to draw. But when we peer through our microscopes at the curious world of micro-organisms we find the distinction breaking down. We find, for example, that there are colourless flagellates which whip about and feed like animals; we find very similar flagellates, whipping about in much the same way but green and building up their substance in the manner of plants. We find some of the filamentous algæ, delicate green threads of protoplasm, motionless and nourishing themselves like plants; we find the moulds, which have very similar threads but colourless, motionless also but nourishing themselves in darkness on substances of complex constitution. And we find other filamentous algæ, with chlorophyll, which are not motionless but sway and glide about like microscopic serpents by some means that is at present entirely mysterious. What then are we to do? If we regard chlorophyll as the definitive feature of plants, then mushrooms are animals. If we regard immobility as the criterion, the green flagellates are animals—and what of the streaming protoplasm of many plant-cells and the active animal-like spermatozoa of ferns and mosses? We are forced, in fact, to abandon the old categorical distinction between animal and vegetable, to recognize that there are transitional forms lying between the two. There is a curious assembly of such forms—the green flagellates, the "slime-fungi," and one or two others—which are described in both zoological and botanical text-books—a sort of no-man's-land (or rather, both-men's-land) between the animal and vegetable kingdoms.

And now we must point out the indefiniteness of a second categorical distinction. We described the structure of amœba and pointed out that it is in all essentials like a single cell from one of the higher organisms; upon that fact we based a distinction between single-celled and many-celled organisms. Now, the line between the two, like the

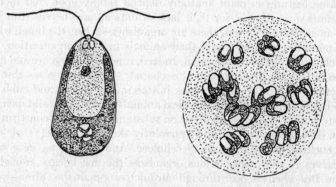

FIG. 115. Left, a single individual of the active green flagellate Chlamydomonas, half animal and half plant. Right, less highly magnified, several individuals temporarily associated into a colony and living in a gelatinous mass.

line between animals and plants, is not so much a fine black line as a broad grey smudge, not so much a boundary as a neutral zone. Here again there are transitional forms.

We noted that the commonest method of reproduction in protozoa is simple fission. The creature pulls itself into two halves, two cells, which creep away in different directions and live independently of each other. Now, suppose that instead of parting they remained stuck together, suppose that they divided and divided and built up a colony of many adhering cells instead of a population of free individuals, the result would be a step in the direction of the many-celled body of the higher animals. And, indeed, such a process as this actually occurs during the life-history of many flagellates. The single-celled forms may remain together more or less permanently as colonies, and thus they lead up to the many-celled plants, and by a series of stages to the familiar seaweeds. Because of this relationship the whole series, from Euglena to the gigantic seaweeds of Antarctic seas, is grouped together and spoken of as the *Algæ*. Let us examine the various kinds and note this grading in more detail.

The simplest plants include the green flagellates that we have already noted, together with several other single-celled kinds. There is, for example, the green coating on the shady sides of trees, rocks, fences, and the like, which consists of millions of tiny, globular, thick-walled algæ. And there is an enormous group of forms, the diatoms, which are characterized by the siliceous shells in which they live, shells consisting of two valves which fit together like a pill-box and its lid. There are very many species of diatoms, showing a wide variety of forms; their shells are patterned with tiny pits and grooves so fine as to be standard objects for testing the high-power lenses of microscopes. They live wherever there is moisture and light, but the richest variety of forms is found in the surface layers of the sea; their skeletons may sink to the bottom, as with marine protozoa, and form oozes and earths. In some places whence the sea has receded layers of fossil diatom shells over a hundred feet in thickness have been found. Such siliceous earths may be used commercially; the kieselguhr, which is mixed with nitroglycerine to make dynamite, is an example, and diatom skeletons are responsible for the abrasive properties of certain silver polishes and tooth-pastes. Also conspicuous among the microscopic vegetation of ponds and the like are the desmids, algæ lacking siliceous shells but showing a variety of beautiful and symmetrical forms.

Now, many of these single-celled plants have a habit of associating themselves into chains or masses, sometimes temporarily and sometimes for the whole of their life-history. In this way they lead up to the so-called filamentous algæ, which consist of chains of cells stuck together end to end. Spirogyra is an example well known to microscopists; it forms slimy, felt-like masses on the surface of ponds, and its cells are marked with characteristic green spirals. And from these simple filamentous forms there is an equally gradual transition to the largest and most elaborate algæ, the seaweeds of our shores.

A few of the seaweeds are green in colour—the sea-lettuce, ulva, is an example—but in most of them other pigments are present in addition to chlorophyll, so that the colour of the latter is masked and obscured. Their classification is based to a large extent on these extra pigments; thus we speak of the blue-green algæ, the red algæ, and the brown algæ, as distinct from the green algæ already mentioned.

The blue-green algæ are simple in structure, some single-celled, some filamentous, some aggregated into gelatinous globules, and do not reach the size or elaboration found in the red and brown algæ. They are widely distributed in fresh- and salt-water. They multiply and swarm like other microscopic organisms, and in fresh-water reservoirs their rapid development, followed by death and decay, may cause serious trouble by giving rise to unpleasant tastes and odours. The plants which give the Red Sea its name belong to this group. The brown algæ, represented by the very common bladder-wrack, with its gas-bladder floats, that is exposed in masses on our shores at low tide, and by the ribbon-like fronds of Laminaria that grow by the extreme low-tide mark, include the largest members of the group. The Laminarias that grow on the Pacific Coast of America may be over a hundred feet in length; those from the Antarctic may be twice as large as this. And there are other kinds in the Antarctic which grow like drooping submarine trees, several yards in height and having a main trunk as thick as a man's thigh. The brown algæ are among the few members of the group that are of use to man; some species are eaten in China and Japan, and before the development of modern industrial chemistry they were used as sources of iodine and potash—indeed, iodine is still extracted from them in Ireland—and they are often used as fertilizers. The red algæ (whose colour may be red, violet, or purple), like the brown, are nearly all marine; they are a varied group, but do not attain the great size or the economic usefulness of the brown.

§ 4. Moulds, Toadstools, and Yeasts.

IT IS not true that all plants possess chlorophyll. Most plants possess it; but there are exceptions, organisms with an anatomical structure like that of the humbler green plants but not themselves green, living on the organic substances that their green neighbours have built up.

Everywhere on the face of the world, except in lifeless regions and in a very few places where civilized man has built, there are the corpses of living things. In a wood, for example, there are masses of dead leaves, which, with other organic matter, make a rich rotting layer on the ground. And wherever the corpses are found we find also a host of creatures that feed on their luscious decaying tissues, animals and plants that specialize, so to speak, in well-hung food. There are beetles and millipedes and worms and grubs of many sorts, and there are bacteria, all fattening on the dead material; but the guests that concern us now at this banquet are the delicate white or bluish felts that are known as moulds.

An attractive provender for moulds is stale, moist bread. If a bit of bread be left about somewhere exposed freely to the air, there will soon settle on it a number of the invisibly small, dust-like spores of mould-fungi. Finding themselves well situated, the spores will hatch, and each will give rise to a very thin, white thread. The threads grow and branch and tangle together into a soft feltwork, on the surface of the bread and penetrating into it. They are not subdivided into cells; their protoplasm forms a continuous cylinder, with numerous nuclei dotted about. Seen through a microscope, this protoplasm is revealed to be in continual restless movement, streaming about inside the walls of the delicate threads. This tangle of white threads is a simple kind of fungus. It has no chlorophyll. It needs organic food, which it gets from the bread in a curious way—by exuding a digestive juice and sucking in a nutritious, digested soup in return.

In fact, it digests like an animal, and the whole world is its stomach, into which it pours enzymes and from which it absorbs the products of their action.

From time to time a branch of one of the threads rises up into the air, and it grows at its free extremity a little globule—black, in the commonest bread-mould. This globule is a spore-case, and in it hundreds of tiny spores are developed, to ripen and be freed and to float in the air in the hope of finding a new piece of dead organic matter to live on. And occasionally there is a primitive sexual process. Two threads, growing from different spores, meet at the tips and flow together to form one mass, a mass which surrounds itself with a thick, hard case and rests for awhile, ultimately to give rise to a number of spores. It is curious to note that in these lowly, creeping threads there is often a division into sexes. There are two kinds of threads, and they will only unite with members of the opposite kind. There is no outward sign of any difference—except that one kind often grows more actively and vigorously than the other—neither do they play distinguishable parts during the actual fusion; nevertheless, the fact that there are two strains can be made out clearly.

There is no exact distinction between fungi and algæ, except that the former do not possess chlorophyll. As far as form goes, some fungi and some algæ are exactly alike. The fungi may be considered as algæ which have given up chlorophyll, and the independence which it confers, and taken to digesting and consuming the organic substance of other creatures instead of making their own. Bread-moulds feed on dead matter, but all moulds are not so inoffensive; some prefer their food fresh. They turn their juices against living things and become parasitic digesters of the tissues of living plants or animals.

We may note that both saprophytism (living on the decay of living things) and parasitism (living on live things) occur among plants other than fungi. The disposition to throw up honest productive work and live on the activity of other beings is as strong in the vegetable as in the animal and human worlds. Among flowering plants, for example, the bird's-nest orchid is a saprophyte and the dodder and broom-rape are parasites. Such plants, having little or no chlorophyll, are unpleasantly

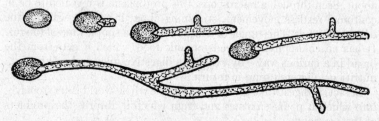

FIG. 116. The germination of a mould—five stages in the sprouting of the first hypha-thread from a spore. Very highly magnified.

pale. But these are shameful exceptions in a group of green, independent plants. The fungi, on the other hand, specialize in dependence, every one of their multitude of species (there are tens of thousands) is either a saprophyte or a parasite.

In nearly all fungi the essential living thing is a web of delicate threads, burrowing inconspicuously into the material upon which it thrives. The most conspicuous parts are the spore-bearing structures. In the common toadstools and puffballs and mushrooms, for example, the umbrella or globe or hat with which we are familiar is not the whole plant any more than an apple or a blackberry is a whole plant; it is merely the fructification, the spore-producing organ, and the essential living thing from which it sprouts is a white, silken web that creeps and digests unseen in the rich soil below. The "mushroom spawn" from which edible mushrooms are grown consists of lumps of richly manured soil with this living web in it. And even a mushroom with its definite shape is not so firmly built as the body of a higher plant. There is no brick-laying, so to speak, of its component cells: instead of the cells being mortared together into a solid mass, it consists of an interlacement of separate hypha-strands, like a hut of interwoven branches instead of a brick house.

On the whole the fungi like the arthropods are malignant towards mankind. Many of the parasitic forms cause havoc among cultivated plants. The rust, smut, and bunt of wheat, the diseases of potatoes, sugar-canes, and larches, the destructive mildews of roses and hops and grapes—these and many other plant plagues are due to fungi parasitic in the tissue of the sufferers. A few attack our own bodies, producing ringworm and other itching affections of the skin and hair— even in civilized countries mouldy children are quite common. The saprophytic forms may be troublesome, too; there is, for example, the dry rot that undermines our buildings, as there are the moulds that spoil our food. In favour of the fungi it may be said that they are among Nature's undertakers, helping in the decay and disintegration of dead things, and that one or two of them have edible, even delicious fructifications. But this scarcely balances the evil they do.

There is, however, a group of organisms, the yeasts, which play an important part in human economy and are highly valuable to us. They differ in structure from other fungi and stand in an order apart. Their chemical reactions are distinctive and extraordinary. They are single-celled moulds; they thrive on various organic matters and they are cultivated because of their residues, their chemical excretions, of which the most important ingredients are carbon dioxide and alcohol.

Yeast, as usually purchased, is a moist powder. Every granule of that powder is a microscopic globule of life. An ounce of brewer's yeast contains about five thousand million of them. It has been known since the dawn of history that if dough is left about for a time before it is baked, the resulting bread is quite different from the unleavened bread produced by baking the dough as soon as it is mixed. And it was known, too, that a little unbaked dough which had already fermented, leaven that is, if it was mixed with newly made dough, leavened the whole lump. But it was reserved for Pasteur to make these facts intelligible. The leavening of bread is due solely to yeast; it nourishes itself on the bread and the changes produced are manifestations of its life—the most important being that the carbon dioxide it releases gives the bread its light, frothy, aerated consistency.

Again, if fruit juices are left undisturbed for a time they undergo changes of flavour and become alcoholic. This also is due to the action of yeasts (of other organisms, too, but chiefly yeasts), which live on the sugar in the juice and produce alcohol and carbon dioxide. If the liquor is bottled while it is still raw the carbon dioxide, unable to escape, dissolves under pressure and a sparkling drink is produced. If not, the carbon dioxide escapes and a still wine or beer or cider results. The various subtle flavours of wines depend partly on the kinds of grapes employed, but much more on the fact that each vineyard has its own peculiar mixture of fermenting organisms, its own stud, so to speak, whose combined excreta have a distinctive bouquet.

From the physiological point of view yeasts are very remarkable because they can live in the absence of oxygen. Nearly all other organisms have to breathe; they can no more live without air than a fire can burn without air. But yeasts, and various other humbly organized plants, have different devices. True, if a yeast cell is supplied with oxygen it will breathe, taking in the oxygen and giving out carbon dioxide just as we do. But if it is denied an oxygen supply it falls back on another device; it turns sugar into alcohol and carbon dioxide, a process which yields energy just as combustion does. It is because of this peculiar method of avoiding suffocation that yeasts are of importance to man. Buried in the dough of his bread they lighten it and make it easily digestible; drowned in flat fruit juice, they change it to sparkling wine.

§ 5. Lichens.

AND next we must notice a widespread series of curious forms that are really double plants. There are fungi that capture and domesticate other plants, as we domesticate other animals.

It was thought until about the middle of the last century that lichens, the familiar encrusting growths that are found on the trunks of trees, on walls, on rocks, and similar places, formed a natural group of plants, equivalent, for example, to the algæ, or the fungi or the mosses and liverworts. But, as a matter of fact, they are not single plants at all. If we examine a shred of lichen through the microscope, we find that it is made of two very different kinds of tissue intimately mixed together. There are little green cells, and surrounding and entangling them there is a closely woven mass of colourless threads. The colourless threads are mould-like fungi; the green cells are single-celled algæ that have been captured by the fungi and are held prisoners in the tangled web. A lichen is no more a single organism than a dairy-farm is a single organism.

Each of the participants in this curious union derives some benefit from the other. The fungus, not being green itself, is able to exploit the synthetic powers of the algæ and to use the sugars and other matters that they build up, with the help of light, from air and moisture. In some species of lichen the fungi penetrate into the algal cells and actually consume them, oozing digestive juices on to them and absorbing the resulting broth. On the other hand, the alga is sheltered within the mass of mould-threads and has more chance of getting the moisture and mineral salts that it needs than it would if it lived exposed on a rock or on the bark of a tree. It is noticeable that the algæ that have been enslaved by lichen-fungi are larger than free-living individuals of the same species, much as our domesticated cattle and sheep and pigs are fatter than their wild relatives. As a result of this mutual assistance the lichens can grow in places which are too barren to support any other form of vegetable life.

The fungi are the dominant, active members of the association; it is they that capture the algæ that are enslaved. The specific characters of a lichen—its form and colour and so forth—depend on the particular kind of fungus that it includes. Moreover, the lichen-making fungi are completely dependent upon their captives, and they cannot live by themselves (unless they are carefully nursed in the laboratory in a nutrient solution). The algæ, on the other hand, are involuntary partners; they belong to species which normally live independently by themselves. Often several different kinds of lichen-making fungus make use of the same species of alga; there is one kind which uses two different species.

Perhaps the relations between fungus and alga will be most clearly understood if we consider the methods of reproduction of lichens. Of these there are three. The first is simple fragmentation; if any part of a lichen be broken off and if it blows or tumbles or rolls into a suitable situation, the fungal threads will sprout and grow and the algæ will multiply, so that the fragment gives rise to a new lichen plant. The second method, slightly more complicated, is by means of special and characteristic buds called *soredia*. A soredium is a tiny spherical particle, consisting of a number of fungus-threads surrounding and holding one or more of the algæ; it is like a party of settlers moving to a new region, the fungi being the active members of the expedition and the algæ corresponding to such livestock as they take with them to ensure a food-supply when they arrive and settle. Soredia are liberated in thousands from the lichen as a powdery mass, and are distributed by wind. In the third method the fungal part of the lichen forms spores after the usual manner of fungi, without the participation of its algal associates. Such spores, when they settle and germinate, are faced with the immediate necessity of finding and entangling appropriate algæ; unless they alight in a place where such algæ may be had in plenty they perish of starvation. Reproduction by soredia is evidently the safer method, since the fungal emigrants take a stock of algæ with them; on the other hand, soredia are large and heavy compared with fungus spores—although to our eyes they are minute powdery granules—so that the spores can be carried for much greater distances by the wind, and spread the species through a wider range.

We have noted that the association of these two very different kinds of plants enables its participants to live under conditions which would be impossible to either by itself. This fact is clearly shown by the geographical distribution of lichens, which are found almost everywhere on the face of the earth. The largest and most elaborate lichenous plants are found in the tropics; from thence they range to the otherwise bare rocks of Arctic and Antarctic coasts, and to mountain heights uninhabitable by other plants. They are even found on rocks far above the snowline. In Arctic regions lichens are more plentiful than any other form of plant, and such forms as the "reindeer moss," "Iceland moss," and "tripe de roche," are important articles of food, both for reindeer and for hard-put men. In the barren steppes stretching from Algeria through Palestine to Tartary the manna-lichen, growing on the ground in little, pale, greyish lumps, is eaten by men and beasts—indeed, by saving the Israelites from starvation during their wanderings this species may claim to have had more influence on the present political and economic situation of the world than any other vegetable organism. Before the extensive use of aniline dyes the lichens were used as sources of pigment; litmus, the acid-alkali indicator of chemists, is an example.

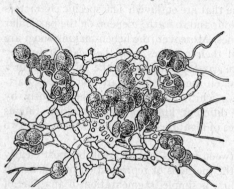

Fig. 117. The birth of a Lichen—fungus-threads, which have just grown out of spores (Fig. 116), are surrounding and capturing globular algæ (highly magnified).

The lichens are the classical example of the biological phenomenon known as *symbiosis*. This term (which is Greek for "living together") is applied to cases when two organisms of different kinds live in intimate union, and to the benefit of both. In the lichen, as we have seen, both the alga and the fungus derive some benefit from their union, although the fungus is the active member in bringing it about

THE LOWLY AND MINUTE

and certainly profits most. It is interesting to find that lichens can be divided into two groups, each derived from a different fungus type. The two groups must have taken to domesticating green algæ quite independently of each other. There are other cases where single-celled algæ are captured and domesticated in this manner by plants, and even by animals. A characteristic feature of the radiolaria, for example, is the presence in their bodies of tiny globular "yellow-cells." These cells are algæ, entangled and held in the protoplasm of the radiolarian; they can survive the death of their host and live and multiply independently. Here, again, the radiolarian benefits by the synthetic powers of the algæ, and the algæ benefit by being protected. There are similar organisms in other protozoa, and even in the tissues of certain many-celled animals. Some species of the common freshwater polyp Hydra (Chapter 4, § 3) are coloured a brilliant green by plant-cells in their tissues, and nearly all corals have yellowish algæ in the inner layers of their bodies and apparently cannot live without them. And there are flatworms and sponges which do the same thing.

This relationship provides an amusing armchair speculation. Should the algæ in a radiolarian, say, be called part of the organism or not? *Pro:* They are apparently indispensable to the organism, an integral part of his body. *Con.:* They can survive his death and lead happy, active lives of their own. The triplex author leaves this nice question unanswered; he states the facts and leaves the interpretation to his reader.

§ 6. Slime-moulds.

WE COME now to an exceedingly curious group of creatures lying on the broad frontier between the animal and vegetable kingdoms—a group of creatures of no importance whatever to the practical man but of the utmost philosophical and scientific interest, common enough, sometimes even conspicuous, and yet hardly ever noticed except by naturalists.

For most of its life a slime-mould (the technical name of the group is either *Myxomycetes*, meaning slime-fungi, or *Mycetozoa*, meaning fungus-animals) is a naked, slimy mass of protoplasm, an almost formless mass without cells or special tissues. Its appearance varies from species to species. It may be translucent, or white, or pink, or green, or yellow, or purple; it is usually large enough to be easily seen, and in extreme cases such as the bright yellow "flowers of tan" it may be as much as a foot square and two inches thick—one can fill a bucket with this formless creature. It is found in such situations as decaying logs, old tree-stumps, heaps of rotting dead leaves, and the like. Usually it is spread out as a thin creeping film of living substance without enduring shape. It is nearly as structureless as a dropped egg. To the naked eye it presents a veined pattern, having a network of thickened strands connected together by thin films in the meshes. Seen through the microscope, it is a simple sheet of protoplasm without nerves or muscles or digestive organs or any such specialized parts—not even divided up into cells. It is dotted with thousands of nuclei and contractile vacuoles; the protoplasm consists, like the protoplasm of an amœba, of a clear outer layer and granular inner mass. The creature is, in fact, like a gigantic amœba.

The substance of this plasmodium, as the large, creeping stage in the life-history of a slime-mould is called, is in continual sluggish motion. It ebbs slowly hither and thither, often with a regular rhythmic motion, flowing in one direction for a minute or two and then reversing. Generally the streams in one direction are longer and stronger than those in others, and so the mass of thin fluid jelly surges along. Its motion is so slow as to be barely perceptible; nevertheless, it creeps with an appearance of appetite and purpose. Normally (unless it is about to form spores) it avoids light and makes its way into damp, dark corners. At its advancing fringe it puts out and withdraws lobes, like the lobes of an amœba, and it can flow around and thus consume the matters upon which it lives. It will turn aside from its course to flow over an attractive lump of food. In most cases only dead and decaying things are taken into this shapeless interior to be digested, but there are species which feed on living vegetable prey—on fungi, for example.

A botanist once made a slime-fungus flow through a dense pad of wet cotton wool. As it progressed its substance broke up into a multitude of parallel streamlets to pass through the interstices of the wool, and as it emerged on the other side they joined up again. Previously it had contained great numbers of the dark brown spores of the fungus on which it was feeding, but now it was colourless; they had all been filtered out and left behind as the protoplasm oozed through the wool. Many slime fungi spend most of their lives thus permeating through rotten wood.

Sometimes, under adverse conditions (such as drought), the creature passes into a resting condition. Part of its substance hardens to a horny consistency in order to protect the rest and forms a number of hollow capsules, each capsule housing a mass of fluid protoplasm with ten or twenty nuclei in it. This mass of capsules can live, motionless, without giving a sign of life, for as long as three years. When conditions become more favourable—when the resting mass is moistened, for example—the hard cyst-walls are absorbed and the contents flow together into a single sheet again, and continue their slow, creeping life as if their long inanition had been only the sleep of a night.

So far our slime-mould has behaved more like an animal than a

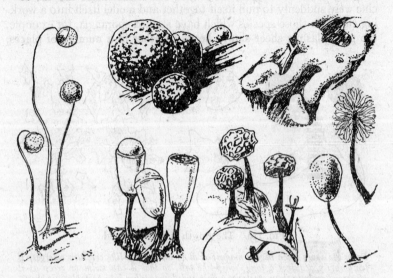

FIG. 118. The graceful spore-cases of six species of slime-mould—magnified to many times life size.

They are common on rotting wood, dead leaves, and so on.

plant. It creeps about like an extremely sluggish animal; not being green, it feeds like an animal. But its reproductive processes are definitely plant-like. Because of this alternation between an animal mode of life and a vegetable mode of reproduction the group is one of those frontier forms that have found a place both in zoological and botanical text-books. The slime-mould reproduces itself by means of spores, contained in fructifications that remind one of those of the true fungi. These sporangia, as they are called, are of various shapes, depending on the species to which they belong. Fig. 118 gives an idea of the range of form that is found. They may be spherical or oval or cylindrical; they may rise on short stalks as knobs or cups or mushroom-like umbrellas. Usually they are extremely small and have to be examined with a low-power lens. Besides containing spores, they may have tangled fibres in them, which uncoil like springs when they are moistened, thus jerking out the spores, which are then distributed by wind.

The process by which the inchoate, creeping film of protoplasm converts itself into a number of highly specialized and elaborated sporangia is extraordinarily interesting. It is almost as if a lump of plasticine were suddenly to pull itself together and model itself into a working clock. In those species which have stalked sporangia, for example, the protoplasmic sheet slowly heaps together in a number of places,

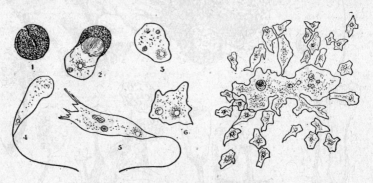

FIG. 119. The growth of a Slime-mould.

From the spore (seen highly magnified at 1) an Amœba-like creature emerges; it temporarily acquires a long, whip-like tail, so that it can swim or creep as it pleases. Presently these creatures collect and flow together into protoplasmic sheets. This union is drawn to a smaller scale at right.

acquiring a warty surface; the warts grow slowly upwards and a solid stalk is formed at their core by a hardening of their own substance; the viscid, living fluid climbs up this stalk as it makes it and, at the top, it hardens and elaborates itself into the complicated structure that we have briefly described. In this manner the whole plasmodium turns itself into a tiny forest of sporangia—the process taking somewhere about a day.

The ripe spores, a few thousandths of a millimetre across, are blown about by the wind, and if they settle in a damp place with a supply of decaying organic matter for food they hatch and develop. The creature that escapes from the hard spore-shell is of microscopic dimensions; it is like a combination of an amœba and a flagellate, having a plastic body and long whip-like tail. It can either creep like an amœba or dance like a flagellate. Moreover, it reproduces by division as the protozoa do, and so may give rise to a great number of similar "swarm-cells." After a time these swarm-cells show a sociable tendency; they collect together in groups, and they fuse with each other in pairs, completely and unreservedly, like the spermatozoon with the ovum. The result of this simple sexual process is a microscopic plasmodium, a miniature of the protoplasmic sheet with which we began our description.

This tiny film creeps about, feeding on bacteria and the like and slowly growing, but it does not develop directly into the fully-grown plasmodium—at least, not by its own unaided efforts. It takes more than one to make an adult. Whenever the tiny creature meets another such sheet belonging to the same species, it mixes eagerly with the new acquaintance, the two forming a single living mass. And so on. So that the final large plasmodium is in reality a union of hundreds of dancing swarm-spores that have completely merged their individuality into one shapeless gelatinous sheet. Imagine that whenever two people meet each other in the street they run together into one blob, as drops of water run together, so that ultimately the whole population of a town is rolled up into a gigantic mass of living substance that creeps about like a single creature; that is the sort of thing that happens as a matter of course in the life-history of a slime-fungus.

§ 7. Bacteria.

THE smallest and most humbly organized creatures that we have hitherto considered are the single-celled animals and plants—amœba, the flagellates, the lower algæ, and the yeasts. But there are smaller and simpler organisms even than these. For the bodies of most bacteria are incredibly small. The common bacillus that is found in infusions of dead organic matter the world over is a short rod about one-thousandth of a millimetre (one twenty-five thousandth of an inch) across and five or eight times as long. That may be called an average bacterium. Some kinds of bacteria are longer than this and some are very much smaller, but most are of about this size. Imagine such a bacillus enlarged to the size of a cigar, half an inch across by four inches long; a man magnified to the same extent would be fifteen miles in height.

There is little room for specialized structures in such a tiny interior, and indeed its organization is far simpler than that of any of the cells that we have hitherto examined. It has none of the various writhing bodies that we noted in a tissue-culture cell. There is not even a nucleus. The chemical substances that characterize the nucleus in animals and plants are diffused through the whole bacterial body, mixed in with the rest of its substance, instead of being confined within a special vesicle. A bacterium is little more than a microscopic blob of protoplasm having a definite shape. Moreover, the life-history of our bacillus is as simple as its anatomy; the things it can do are very limited. It can grow and proliferate (by simply breaking into two halves) with incredible rapidity. The average length of life of a bacterium that finds itself in favourable surroundings, the time between successive divisions, is twenty minutes, so that in rapidity of multiplication they leave even the protozoa behind. And in many cases it can weather difficult periods, when conditions are unfavourable for its growth, by turning into a resistant, tough-walled spore. There is an important difference, we may note, between the spores of bacteria and those of other creatures. The spores of amœba, for example, are reproductive devices; each amœba produces a great number of spores at a time. But the spores of bacteria are in no sense reproductive. Each bacterium converts itself into a single spore. They are devices for resisting unfavourable conditions.

In bacteria, then, the life-history (or, better, race-history, because it is difficult to interpret the things that bacteria do in terms of our own personal experience) consists in an alternation of two phases—the active phase proliferating with astonishing rapidity, and the passive, resting spore with equally astonishing powers of resisting adverse conditions. Active bacteria are killed at once by the heat of boiling water, but their spores may survive boiling for several hours. Indeed, in order completely to sterilize surgical instruments or fluids for use in bacteriological work it is necessary to expose them to moist steam at about double atmospheric pressure and a temperature of 120° Centigrade in an instrument known as an autoclave—and even under these scalding conditions, some spores survive for a quarter of an hour. At the other extreme, bacterial spores have survived prolonged exposure to the temperature of liquid air ($-190°$ C.) for six months. The creatures can live for very long periods in this spore stage, perfectly passive, waiting as it were in a trance for an opportunity to emerge and grow and divide. They have been kept for three years in this condition and have been fully capable of revival at the end of that time. Taking the normal life-span of a bacterium as twenty minutes, these spores slept during the time of over a hundred thousand generations of active bacteria. It is as if a man had passed into a trance ages ago, before ever the Pyramids were built—even before anything at all was built, for he might have been one of the earliest flint-chipping men—and woke blinking and rubbing his eyes at the present day to go about his accustomed occupations.

In this quiescent, resistant state bacterial spores wait on the ground, in the sea, in the air for an opportunity to hatch and grow. They have been found frozen in icebergs. They have even been found in hailstones, which shows that they float very high in the air. If they are lucky enough to settle in a suitable spot they grow and proliferate, in some species with almost miraculous speed. A flask of sterilized broth, inoculated with bacteria by dipping an infected needle into it, may become visibly turbid in twenty-four hours' time. Every cubic inch of the broth will now contain some eighty thousand million organisms.

A striking peculiarity of bacteria is that they have no trace of a sexual stage in their life-history. Even in the lowest protozoa and algæ there are primitive kinds of sexual union, blendings of cell with cell. But nothing of the sort has ever been seen in bacteria. They proliferate by dividing, and then can do so vigorously and indefinitely without ever betraying that need for the refreshment that sexual union affords.

Considering the great number of different kinds of bacteria that are known, the forms that they may assume are very limited. Their names are based in part on the shapes of their bodies. A rod-shaped bacterium, for example, is called a bacillus (or stick), a spherical bacterium is a coccus (or berry), and a curved or wriggling bacterium is a spirillum. Some bacteria have long, lashing whips—either one, or a tuft at one end of the body, or a number bristling all over the surface of the creature. And they may stick together in bunches or clusters or chains whose shape is characteristic of the species. But, nevertheless, their range of form is very limited. Often quite distinct species of bacteria look so much alike as to be distinguishable only by the effects that they produce. There are, for example, three kinds of bacteria almost indistinguishable in appearance, with globular bodies that form little chains; the first causes the souring of milk, the second flourishes in our blood and causes a rapidly fatal infection, while the third is a perfectly harmless creature nearly always present in the healthy human mouth.

It is because of their physiological diversity that the bacteria are remarkable. There are bacteria which show the profoundest departures from the normal chemical method by which animals and plants get their energy. Their engines run on unusual fuel. Some get energy by oxidizing simple inorganic compounds—for instance, by turning the inflammable, foul-smelling gas sulphuretted hydrogen into sulphuric acid. Others have to use sodium thiosulphate to make sulphuric acid. Others, again, can energize themselves by burning ammonia. And by means of these devices the creatures can build up the complicated organic substances of which their bodies, like any other living bodies, consist, from such simple ingredients as water and carbon dioxide. As a further curiosity, there are bacteria which can live, like yeasts, without oxygen, and there are even bacteria which shun oxygen and are poisoned by it. But most bacteria are more conventional than this; they burn fuel and need oxygen and nourish themselves like animals or fungi on the living or dead bodies of other creatures.

We noted in an earlier section how difficult it is to classify micro-organisms into categories. The separation of bacteria from animals and plants is another example of this difficulty; here, again, there are border forms. There are, for example, the so-called filamentous bacteria which are very like the humbler blue-green algæ, except that they are not blue-green. And there are the spirochætes, a group which includes the parasites responsible for syphilis, that are in many respects intermediate between bacteria and the flagellate protozoa. It is still uncertain whether they are most nearly related to the former or to the latter. Moreover, in some ways, bacteria are very like yeasts. The bacteria, then, merge into the simpler animals and plants just as the latter two groups merge into each other. Usually they are regarded as being less animal-like than vegetable-like (they are described, for example, in standard text-books of botany but not of zoology), but this again is a nice question; there are bacteria that wriggle very much like animals and nourish themselves on organic food not at all after the typical plant manner. Indeed, the line between animal and plant is such a vague one that the question has little meaning. At this level of being, the distinction of plant and animal, like the distinction of sex, has practically disappeared.

It would be difficult to overstate the practical importance of these creatures—invisibly small, but swarming in incalculable multitudes. There is hardly a human activity that is not affected in some way by bacteria, acting either as friends or as enemies. There is hardly a nook or cranny on the surface of the earth that they do not inhabit. They are found in the air, in the soil, in the waters. The bacteria in the air are chiefly carried on particles of dust; the air of an assembly hall in Paris was found to contain nearly ten thousand bacteria and two thousand five hundred mould-spores per cubic yard. The decay of dead things is in large measure due to bacteria. They penetrate into the bodies of living creatures and there behave in various ways, some being harmful to their hosts and some, as we shall learn, doing good. They teem in the soil, and especially in the upper few inches of it; the smell of a newly-tilled field is due to the products of countless soil-bacteria. A lifeless world, we may note, would be completely odourless except perhaps for such cold, unemotional scents as chlorine and ozone and the oxides of nitrogen. And most of the odours in this world are due to microscopic creatures. There are, of course, many flowers and animals with characteristic smells for which they are directly responsible, but the smells of decaying things (not really a sign of death, but a sign of teeming bacterial life); the smell of a sea beach, the smell of a wood in winter, the smell of a stable, the smell of the earth after rain—these are all bacterial odours, produced by creatures too small to be seen. The most subtle and complicated and delicate perfumes that we know are produced in this way. The bouquet of a wine is due to the swarms of micro-organisms that have thrived in it, and it is at least possible that the subtler flavours of choice cigars are due to bacteria that have lived on the heaps of tobacco leaves when they were put out to dry.

The bacteria, indeed, are on the whole a much-abused group. They are usually spoken of simply as creatures which cause disease; their more beneficial activities are less widely noised. It is, of course, true that there are disease-producing bacteria, and very important ones—cholera, typhoid, diphtheria, plague, pneumonia, tuberculosis—these are only a few of the major disorders for which bacteria are responsible. We shall return later to this side of their activities. But for the present we will note the existence of other kinds of bacteria, even friendly bacteria. For it is no exaggeration to say that, without bacteria, life on this world would be quite impossible.

Bacteria, for example, are very important in agriculture. There are a number of kinds of bacteria living in the earth which lay hold of the nitrogen gas in the air and turn it into substances in the soil which can be absorbed and used by plants. The fertilizing action on the soil of crops of leguminous plants, such as beans and clover, is due to the fact that their roots are always swollen and warty from the presence of swarms of symbiotic bacteria—parasites, if you like, but parasites which benefit their host more than they harm it, for they use nitrogen from the air, which no higher plant can do, and build from it organic substances that the hosts absorb. Again, bacteria are both enemies and allies to the dairy farmer. The taste of butter depends to a large extent on the bacteria that inhabited the milk in the time between milking and churning; there are some bacteria that make pleasant flavours and others that make foul. For this reason it is customary in scientific butter-making to sterilize the milk as soon as possible after milking and then to add a "starter"—a culture of the right kind of organism. In this way one is certain of getting a pleasant flavour. Moreover, the "starter" usually includes the bacteria that turn milk sour, because it is found that the acidity of sour milk prevents the growth of undesirable bacteria. Further, the flavours and distinctive properties of cheeses (even their contained air-bubbles) depend on the organisms (chiefly bacteria and yeasts) that have inhabited them during the ripening process. The characteristic flavours of cheese from different districts, like the characteristic flavours of wines, are due to the fact that each district has its own races of the appropriate micro-organisms. Vinegar is made by the action of bacteria which turn alcohol into acetic acid; the slimy mass known as "mère de vinaigre" consists of these organisms embedded in the slime that they produce. And many of the processes in leather manufacture, such as the "sweating" process, which softens the hairs so that they are readily scraped off the skins, are really carefully watched bacterial decay.

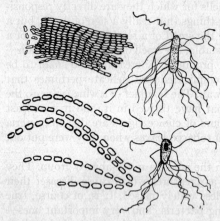

FIG. 120. Three kinds of bacteria of economic importance.

The bacilli forming chains on the left are two of many kinds that turn wine into vinegar. On the right are two specimens of a bacterium with lashing flagella, which causes butter and other fats to turn rancid.

On a higher plane, bacteria have played a part in strengthening religious faith. The prodigy of the bleeding Host, in which bread broke out into scarlet spots, was due to a bacterium which grows on bread and manufactures a red pigment; the same organism may infect human skins, if not too frequently washed, and thus cause miraculous "sweatings of blood." There are other bacterial curiosities, too. The phosphorescence that appears in rotting wood, fish, and the like is due to bacteria; and so is the heat of manure heaps and moist haystacks.

But the outstanding thing about bacteria, perhaps the most important of all, is that they are the chief organisms responsible for putrefaction. Usually one regards putrefaction as a nuisance—if anything in the larder goes bad or if a rat dies under the boards of the floor one resents this apparent tendency of dead organic matter to deliquesce and stink. But decay is not a natural property inherent in dead and abandoned things. It is a sign that they are being inhabited and consumed by busy millions of microscopic creatures. And decomposition is a vitally important process. If everything that died were perfectly preserved and lay for ever as a corpse on the surface of the earth we should be hard put to it to find room for our feet. But there is a better reason than that. A living body, a man for example, is a more or less definite weight of organic matter set apart for special use; when the creature dies its body is finished with, and in the usual course of Nature its substance is melted down, so to speak, and used again, built up again into other bodies. For the amount of available carbon, for example, in the world is limited. Equally important, there is energy even in a dead body; it is combustible, a potential fuel. A preserved dead body is so much matter, so much energy withdrawn from the general interchanges between living things; it leaves life as a whole so much the poorer. But normally, as soon as anything dies it is discovered by the floating spores of putrefactive bacteria, Nature's housebreakers, which proceed to demolish it.

There is a great variety of these bacteria. There are, so to speak, a number of rival gangs of housebreakers working without any sign of discipline or co-operation. Some attack the proteins and fats and break them up into fluid substances; others, rivals of the first, fall upon the same material and turn it into stinking gases. Some fall upon the nitrogen containing molecules and turn them into ammonium carbonate—a substance which is wrestled for by two other kinds, the first turning it into atmospheric nitrogen and the second into nitrites. Then the nitrites are seized upon by yet other bacteria and converted into nitrates. The confusion is increased by other creatures which butt in—moulds and grubs of various kinds. But the net result of this swarming, struggling activity is good; the useless body is converted into molecules of simple constitution, gases and simple compounds in the soil, which can be used to nourish green plants. The chemical capital, which would otherwise be locked up and unproductive, has been brought back again to play its part in the commerce of living things.

For the material of living things seldom rests; it is kept in continual circulation from body to body. Consider, for example, the muscles of a man. Their material may be used up—burnt or worn away—during his life, in which case it will be excreted in his breath or in his urine. Or it may be present at his death, in which case it will be fallen upon by bacteria and demolished, and the chemical bricks of which it is built make their way as before either into the air or into the soil in the form of comparatively simple molecules. In either case, sooner or later, it will be built again into a living thing. Carbon dioxide in the air is absorbed by plants to be built up into their substance; nitrogen is absorbed by special bacteria in the soil and turned into nitrogen compounds. And the substances in the soil are sucked up by the roots of plants. The plants are eaten by animals, the animals die or are eaten by other animals—even by men—and thus the material is handed on from one form of life to the other. Life is a continual commerce. There is a rhythm, a cycle, from inorganic substances in the air and soil and soil to plant-tissue, thence to animal-tissue, from either of the last two stages via excretion or death and decay back to the air and the soil.

And this everlasting rotation of life is kept in motion by the radiant energy of the sun. For the green plants build up complicated substances out of simple ones by means of solar energy, the vegetarian animals use and burn these compounds and the carnivorous animals, consuming the bodies of the vegetarians, use them, too; true indeed it is that "all flesh is grass." And, finally, the last drop of energy is wrung out of these substances by the putrefactive bacteria. So we can compare the life on our planet to a wheel; matter going round and round, solar energy streaming through it and keeping it in motion.

§ 8. The Smallest Living Things.

THE invention and gradual improvement of the compound microscope has led us stage by stage into a world of tiny wonders, a world that becomes more and more mysterious as one proceeds to smaller creatures. In the first excitement of this discovery it was thought that our power of magnification was, in theory at least, unlimited. It was thought that one could go on combining lenses for ever, that if only one could make lenses without aberrations of line and colour there would be no secret of Nature, however minute, that one could not magnify and make visible. But, unhappily, that is very far from true. There is a limit to the smallness of the things which can be seen, a limit imposed not by any fault of the lens-grinders but by a property of light itself. We know that we shall never be able to see distinctly anything less than one four-thousandth of a millimetre across (one hundred-thousandth of an inch) however powerful our lenses may be, because of the unalterable wave-length of visible light. Smaller objects can indeed be made perceptible by means of optical tricks, but they cannot be examined; they appear as spots or blurs without definable shape. Now, that limit has already been reached. Microscopes have been made which clearly reveal objects of that size, and it has been proved that there are living things even smaller than this.

It may sound paradoxical to say that we are sure of the existence of living things which cannot possibly be seen, but, as a matter of fact, our belief rests on solid ground. But before we go into this matter we may pause for a while and define a new word. For it is inconvenient to measure these minute distances in such comparatively vast units as millimetres or inches; it is like describing the dimensions of a postage stamp in fractions of a mile. The microscopist, in studying these matters, uses a measure of his own invention called a micron (usually written μ); a micron is one-thousandth part of a millimetre, or about one twenty-five thousandth of an inch. To the unaided eye it is an invisibly small distance. The smallest object that we can observe with any microscope is about a quarter of a micron across—and only the very best instruments can take us as far as this. Most bacteria come pretty near this limit; the diameter of a coccus is usually somewhere about a micron. And now for the evidence on which our belief in these unmagnifiable creatures rests.

The argument is very simple. Suppose we pass a handful of coarse sand through a sieve whose pores are a twentieth of an inch in diameter; we know that none of the grains that are bigger than this can get through. It is possible to get filters whose pores are about a quarter of a micron in diameter—they are made of unglazed porcelain or of similar material—and we know that if we pass a fluid through such a filter anything bigger than a quarter of a micron across will be held back. This is the limit of microscopic vision. Therefore, we may infer that nothing visible through the microscope will be able to pass such a filter—an inference that can be tested by direct experiment. Now there are several disorders that can be transmitted to healthy subjects by injecting them with fluids that have passed through such a filter. Foot-and-mouth disease was the first to be discovered. The characteristic feature of this cattle scourge is the appearance of sore blisters, especially on the hoofs and lips. If a little of the fluid from these blisters be passed through the finest porcelain filter, so that all traces of visible bacteria are eliminated from it, it will, nevertheless, infect a healthy beast if it is injected into its blood; moreover, a third beast can be infected from the blisters of the second, and so on. Evidently there must be a living thing in this fluid, a thing that can multiply and proliferate. If it was simply a poisonous substance its effects would become weaker and weaker as it was transferred through successive beasts, but this does not occur. Like a bacterial infection, the invisible creature proliferates inside its host.

A living thing of this kind, too small to be seen with the highest powers of the microscope and passing even a porcelain filter, is called an "ultramicroscopic" or "filter-passing" organism; since all the known kinds cause diseases they are also spoken of as viruses.* There are other diseases with similar causes. Smallpox, typhus fever, trench fever, yellow fever, measles, mumps, the distemper of dogs, and a number of vegetable afflictions causing the leaves of potatoes, beans, turnips, and the like to come out in white spots—these are examples. Perhaps the common cold is another. There is a possibility—a hotly contested possibility—that an epidemic destruction of bacteria themselves that sometimes breaks out in cultures is due to a living thing called bacteriophage (or devourer of bacteria), which, if it exists at all, is about a fiftieth of a micron across.

This, we may note, is getting within measurable distance of the sizes of the ultimate units of which living things are made up. A single molecule of protein is somewhere about a thousandth of a micron across, so the body of a single bacteriophage organism cannot contain much more than a thousand protein molecules. But the living nature of bacteriophage is still a matter of controversy. We may note that all the ultra-microscopic organisms whose existence has been proved are causers of diseases, and this is but natural; the pain they cause has spurred us to their discovery. It is perfectly possible that there are other kinds living independently in the soil, for example, just as visible bacteria teem everywhere (although the tentative search that has already been made has as yet revealed none). It may be that there is a world of these incredibly minute forms of life, as little suspected as was the world of microbes three hundred years ago.

A tantalizing situation! Broad hints of the existence of these creatures, even proof in some cases, and yet no instrument to see them with! But happily a new method is being worked out, a method by which we can pry into the anatomy of some at least of the larger of these ultra-microscopic creatures. It has been shown that although we cannot see them with any distinctness we can take their photographs.

In 1925, Mr. J. E. Barnard devised a method, demanding

---

*Strictly speaking, the terms "ultramicroscopic" and "filter-passing" are not synonymous. There are slender, serpentine creatures large enough to be seen through the microscope, yet able to twist and wriggle through a porcelain filter.

the most elaborate apparatus and an infinity of patience and experimental skill, by which we can extend our knowledge a little farther. It rests on the fact that ultra-violet light has a shorter wave-length than visible light, therefore if objects are illuminated under the microscope with ultra-violet light their images will show finer detail than is possible with visible light. Now ultra-violet light has no action on the retina, so these images cannot be seen; nevertheless, it blackens a negative, so they can be photographed. The first organisms to be examined in this way were those that cause a disease of the lungs in cattle. The result was surprising. The organisms were found to be not miniature bacteria but creatures as different from bacteria as bacteria are from protozoa or moulds. Bit by bit, by photographing cultures in different stages of growth, the story of their life-cycle was pieced together.

In the first stage in the cycle, the stage that escapes direct observation and passes through filters, the creature is a very minute particle, about one-fifth of a micron across. Apparently these particles can proliferate directly. But they also have a method of multiplication utterly unlike that seen in bacteria. They swell up into hollow spheres, six or seven times as large—big enough to be seen with a microscope; then little wart-like thickenings appear dotted over the walls of the spheres and detach themselves. These are the particles with which we started. Sometimes the particles remain connected for a time to the parent sphere by means of an incredibly fine thread before they break away. Sometimes a number of particles and spheres may be found thus connected together. And that is pretty nearly all we know. It is a fascinating story because the creatures are so unlike anything that has yet been seen, and fascinating also because it is incomplete, the first step in what must become a profoundly important branch of investigation.

A second step also has been taken: Barnard has photographed similar but even smaller organisms—about a tenth of a micron across—from certain types of cancerous growths of fowls and of men. But whether these organisms are the cause of cancer is a matter that we shall discuss in a later Book. It will be noticed that the creatures which have been photographed in this way are only just beyond the margin of visibility; how far it will be possible to develop the method and reveal yet smaller organisms it is difficult to say. A corner of the veil that hides these tiny creatures has been raised for an instant, an instant long enough for the click of a camera; the rest, at present, is darkness.

And so, on a note of uncertainty, we close our account of the known forms of life. There are living things, as we have found, that are out of reach of microscopes, at least when we use ordinary light; they are almost entirely mysterious as far as our present-day knowledge is concerned. How far we shall be able to investigate them depends on the methods that we are able to devise. There is no need for us to despair. Bragg, by passing X-rays through crystals, has been able to find out how their very atoms are arranged—it is as if he could see them. It may be that the biologists also will arrive at roundabout but nevertheless reliable ways of exploring the anatomy and physiology of these specks of life and so "by indirections find directions out."

VII

IS OUR KNOWLEDGE OF THE FORMS OF LIFE COMPLETE?

§ 1. Sea-serpents and Living Dinosaurs.

BEFORE we leave this survey of the forms and phyla of life, we may perhaps give a little space to the question how far this survey may be considered comprehensive. Does much remain unknown? In the fossil record certainly we may almost count upon remarkable types in store, but here we are concerned with things still living. That there is still much to be revealed by ultra-microscopic methods is manifest from the concluding section of the previous chapter. Apart from that, is it possible that a number of large, exciting creatures still live undiscovered in remote and shadowy and inaccessible corners? The exploration of the land surface of our globe is proceeding more and more rapidly, but there are still places, barren heights and luxuriant, pestilential jungles, where the describing and classifying white man has hardly penetrated. And in the sea there are vast regions that have been barely sampled by the drift-net, the trawl, and the dredge.

It is from the sea particularly that rumours of strange beasts come, glimpsed momentarily by sailors, and then plunging down into the unsearchable deep. The favourite marvel of the nineteenth century was the "sea-serpent" of enormous proportions. This reappeared annually and was annually discussed and disposed of. Now it was a school of dolphins swimming end to end and now a sulphur-bottom whale seen by unaccustomed observers; now a great basking shark, now an actual sea-snake (for such there are as big as ten feet long) with its proportions all exaggerated, and now a brilliant effort of the unaided, creative imagination. Did such an undescribed monster exist it is incredible that no bone, no scrap cast up upon a beach, no floating corpse of it has ever been recorded. A single tooth, a single vertebra, said Professor Owen long ago, would suffice to establish its identity scientifically. A certain number of species of giant squid and octopus may still be eluding the naturalist in the middle deeps in which these cephalopods abound. Several species and one genus are known only from fragments ejected by sperm whales in their death flurries. But there is nothing very sensational for the general reader in the discovery merely of new genera.

There is, however, another possibility of strange things in the sea. The discovery of living creatures in the extreme depths of the ocean is very recent. It was held as recently as 1840 that there was no life, animal and vegetable, deeper than three hundred fathoms. But since then there have been a number of successful experiments in deep-sea collecting. The three-and-a-half years' expedition of the British ship "Challenger" (1872–6) is the classic piece of ocean exploration; it laid the foundations of the science of oceanography which many other expeditions have since expanded and elaborated. In the first excitement of the discovery of living things even in the deepest abyss it was thought that we should find new phyla altogether, new plans of animal organization, and that in these deep, secluded places some of the ancestral forms of life might have been preserved from great antiquity. The intensive researches that have since taken place have made such hopes untenable. The creatures that come up from abysmal depths are weird indeed, but they are not fundamentally different from those that live in more accessible places. They are grotesque distortions of the fishes, crabs, sea-cucumbers, and so on that we already know. They are not in any sense primitive; they are, on the contrary, highly specialized. Indeed, it is becoming more and more evident that the abyss is a region as different from the original home of living things as dry land. There is no light in the abyss and therefore there are no plants; the creatures that inhabit it live on each other or on the corpses that float down from higher layers, from which it follows that they cannot have been the first forms of life. They are a comparatively recent evolution, members of the phyla that are already established above, adapting themselves to that new and at first quite inhospitable environment.

On land, it is equally improbable that we shall discover anything strikingly new. There are jungles that have hitherto proved too lush and too pestilential to be systematically explored, and here it may be that unknown creatures lurk. There was a time when tropical forest and jungle occupied much more of the surface of the world than they do now, and were inhabited, as the fossils tell us, by curious mammals that are for the most part long extinct. Here and there a relic of that forest age still survives. The tapir, for example, is an anachronism. At one time it had a wide and extensive distribution, but now it lingers on insecurely and prowls by night in one or two marshy patches of jungle, in South and Central America, in Malaya, and Java and Sumatra. And the okapi, lurking in the deepest recesses of the forests of the Belgian Congo, evaded scientific description until 1899. It may be that other forest-dwelling mammals lurk on in the same way. But these creatures we may note are not very striking departures from those that are more widely distributed; the tapir is a distant poor relation of the rhinoceros and the okapi is a closer poor relation of the giraffe. Even here, unless some novel ape or monkey amazes the world, there is little chance of a really striking departure from the forms of life we know. In passing, we may set aside as quite untenable the very wild rumours of gigantic survivors from the reptile age that invade the Press from time to time. The reptile age was very much more ancient than the forest mammal age to which we have just referred, its climatic condition may have been very different from any now on earth, and it is so improbable as to be inconceivable that any of the dinosaurs, for example, should have lingered on unseen to the present day.

To sum up, then, it is probable that the range and variety of life as it exists to-day is already pretty well charted. There are certainly many details to fill in, many unrecorded members of existing phyla, but it is in the highest degree improbable that we shall find any fundamental departure from the plans of organization that we already know.

Perhaps an unromantic conclusion. One likes to dream of strange, evasive creatures that are still at liberty, that have eluded the cold, systematic docketing of science. But there is wonder enough in what we already know. In our brief review we have noted something of the variety of life, of the astonishing diversity of its forms. Consider, for example, the life of an amœba, a life where an individual tears himself without residue or wastage into two and in which two individuals blend permanently and without reserve to form one. Consider the reconstitution of a sieved and scattered polyp, or the alternation of forms in the life-story of a liver-fluke. The world as we know it is surely strange enough to satisfy the most curious; the fabled sea-serpent and the fabled African brontosaur are banal imaginings compared with such mysteries as confront us already.

BOOK THREE

THE INCONTROVERTIBLE FACT OF EVOLUTION

I

THE FACT TO BE PROVED

§ 1. *Evolution and Creation.*   § 2. *The Nature of the Proof.*

§ 1. Evolution and Creation.

HITHERTO we have described the forms of life, avoiding as far as possible the introduction of any controversial matter. We have now to take up aspects of our subject that have been the centres of great controversies. We have first to tell of the gradual recognition of a great reality—the Evolution of Life. And having done that as clearly and plainly as we can we shall have to discuss in the Book that follows certain theories about which opinion still varies widely and at times violently.

Now, until a century or so ago it was commonly believed that the world as we know it to-day had begun suddenly. It had been created, with man and all the species of beings as we know them to-day. Great numbers of people, including most educated people, held to the view with great tenacity. They had adjusted their moral and religious ideas to that view, and they did not realize that these ideas were not inseparably dependent upon it. All of us are prone to resist changes in our fundamental ideas. We feel instinctively that it may mean a disturbance of our way of living and the abandonment and change of objectives; it is a threat to our peace of mind and our satisfaction with our lives. The idea of the earth's going round the sun was considered to be just as impious in its time of novelty as was the idea of Evolution by the Fundamentalist of the backward States to-day.

Then steadily and more and more abundantly came evidence to show that the existing forms of life were not all the forms of life, and that there had been a great variety of animals and plants which had passed away, a greater variety and multitude indeed than that which still exists. The science of geology became a new region of intellectual activity, and in the study of the earth's crust the traces of a past infinitely longer than men had hitherto suspected were unfolded.

THE FACT TO BE PROVED

Varied and wonderful as was the present spectacle of life, the series of faunas and floras that had preceded it and passed away was found to be more wonderful. Life had a past, a stupendous past. So far from it being a thing of yesterday, the creation of a few thousand years ago, it had a history of enormous variety and infinite fascination. We can still imagine something of the excitement of our grandfathers when the fantastic and marvellous dinosaurs, the vegetation of the coal measures, the flying dragons of the Mesozoic Period were revealed to them. Continually now that once incredible catalogue is expanded. Every year the palæontologist, the seeker and student of fossils, adds fresh details to this history of living forms.

Faced with these marvels the Creationist at first denied and then, no longer able to deny, declared that these extinct forms of life had been created in the past, tried out for some unknown end, to be extinguished in favour of fresh creations. They were but the prelude of these later creations. They had no clear rational relationship to living things and living things had no clear rational relationship to them. But a bolder school of interpretation appeared. These ancient forms were not so strange and incredible as they seemed. Life had produced them on its way to its present state. Generation by generation it had changed from the wonder it was to the wonder it is. There had been no Creation since the beginning of life. Life had unfolded—or, to latinize unfold, it had been "evolved"—from some remote and very simple beginning.

What weighed with the Evolutionist in his denial of successive creations was this that the abounding and continually accumulating record of past forms of life is not a disorderly multitude, not a confusion of inexplicable "wonders" but that it falls into shape, it has a plan, and every fresh discovery drops into place in that plan. *All* of these forms fall into the scheme of a common tree of descent. That is the plan of it. If there was no other evidence to sustain it, we should still have to believe that Evolution has occurred on the strength of the plan of the fossil record alone.

Let us be very clear here. We are telling the reader in this chapter that this later view is the sound one, that Evolution has occurred. But we are making no suggestion as yet as to how it has been brought about. We are simply declaring that life has come to its present variety through the modification year by year, and age by age, of simpler and less various ancestral species. In making this declaration we are denying a belief, formerly very prevalent, the belief that animal species, as they are now, came into being suddenly, through some abrupt act of Creation. That belief has now become impossible in the face of an assemblage of countless known and established facts. On the other hand all these contributing facts build themselves up into the comprehensive vision of Evolution as the fact of facts, the quintessence of the whole display.

But we are not attempting any *explanation* of this fact of Evolution here. We are not attempting any account here of why species have changed. We will write later of the various theories by which an explanation of this central fact is attempted. We are not discussing here the Theory of Natural Selection, or the Theory of Creative Evolution, or any theory at all of how Evolution has been carried on. First the facts and then these more stormy issues may be faced. Here we traverse ground upon which scientific men of every creed and school are now agreed.

We make this distinction between fact and theory here and, so to speak, underline it, because we know there is still a considerable confusion in the public mind between the fact of Evolution and the conflicting theories about how it works. Dishonest Creationists, narrow fanatics, and muddle-headed people attempt to confuse the very wide diversity of opinion among scientific men upon the questions of how and why with their assertion of established fact. Through this confusion it is suggested that the hated fact is still unproven. It is, on the contrary, proven up to the hilt, and here we shall unfold as much of the evidence as is necessary for conviction.

§ 2. The Nature of the Proof.

BEFORE we go on to the evidence, however, let us consider what our evidence must show if Evolution is to be accepted as the general process of life.

First, then, all things living, or once living, must fall into a branching plan. Everything in the past must be reasonably shown to be either ancestral to a living thing or else without descendants; there must be no renewal of the process, nothing in the past must be plainly derived from some later form. Every mammal, for example, is held to be descended from a reptilian ancestor. Suppose in the early Coal Measures, before ever a reptile existed, we found the skull of a horse or a lion. Then the whole vision of Evolution would vanish. A single human tooth *in situ* in a coal seam would demolish the entire fabric of modern biology. But never do we find any such anachronisms. The order of descent is always observed.

Next there must be an orderly sequence in fossil forms, so far as they are found. We must see very distinctly that form passes into form. In the days of Darwin such sequences were hard to find. In those days there were probably not a hundredth part of the present multitude of fossils that are now collected and arranged. The Creationists pointed triumphantly to a gapped and fragmentary story, sustained by hypothesis, broken up by "missing links." Darwin was challenged to show anywhere in the fossil record the steps by which one species has passed into another. It was then quite a difficult challenge. To-day we have an answer, a score of answers, to that challenge, beyond Darwin's utmost hopes.

Then if animals have been specially created just as they are to fit special conditions, it is reasonable to suppose they are perfectly and completely adjusted to those conditions. There is no reason why any animal should fail to have any structure that might be helpful in its way of life, or possess any structure it has no need for. If a cat lives on birds and a tiger on ground game, is there any reason why a cat should not have wings because a tiger has not? But if the diverse species have been evolved step by step, a certain disharmony is to be expected between inherited structure and reactions, and the full possibilities of the life a creature leads. The second section of our evidence then will be an examination of plant and animal structure to see how far animal and vegetable organs are special to their needs, and how far they have the air of being primarily an inheritance merely fitted to those needs and limited in that fitting by conditions of descent.

And then the way animals and plants are scattered over the world will not be haphazard if Evolution is really the truth of life. If we found a region where an animal might live abundantly and that animal is not there, but somewhere else in the world, then if we are to believe in Creation we have to find Creation very remiss upon the distributive side; but if we believe in Evolution, then it is quite reasonable to suppose that an animal evolved in one part of our planet may never get to another for all the fitness of conditions there. All that also we will illustrate and weigh.

Finally we will take up a question that was once a burning question and is still regarded as smouldering. Does man come into this process of Evolution or is he in some strange way outside general biology, following laws of his own? We will show that there is no exception in his case. We hope to show the reader convincingly that Evolution is the form of all life in time, man and his acts included. Evolution is, in fact, *the* life-process.

II

THE EVIDENCE OF THE ROCKS

§ 1. *The Nature and Scale of the Record of the Rocks.* § 2. *Defects and Happy Finds in the Record.* § 3. *A Sample Section in the History of Life: the Evolution of Horses.* § 4. *The Continuity of Evolution as Shown by Sea-urchins.* § 5. *"Missing Links."*

§ 1. The Nature and Scale of the Record of the Rocks.

IN ORDER to make the nature of the record of the rocks perfectly clear it is necessary to remind the reader of certain elementary geological facts. They will probably be familiar to him, but we do not want to have any "missing links" in our chain of argument. We warn him of this beforehand, so that if the note of the professional lecturer creeps into our discourse he will forgive it—for the sake of its explicitness.

It is only after decades of patient work, we must remember, that the fact of Evolution obtruded itself as a necessity in the face of the palæontologist. Right up to the end of the eighteenth century the comparatively few fossils then known were almost universally regarded as mere curiosities; many dismissed them as sports of Nature, freaks of the earth, and not really the remains of flesh-and-blood, while at most they got credit for being witnesses to the universal biblical Deluge. Nothing more could be expected until geology had made her profound advance of introducing time, and time on a vast scale, into our ideas about the earth's crust.

Let us recall how that extension of time dawned upon the human intelligence.

Everybody knows that flowing waters bring down sediment and deposit it in layers on the floor of seas and lakes, or on flood-plains. Sometimes the deposit reaches the surface, as in deltas. Consider the great tongue of the Mississippi delta, built out into the sea for sixty miles and more, the Mississippi brings down every year over four hundred million tons of sediment. At other times the deposit spreads over the bottom, as when a pond is gradually filled up, or an alluvial meadow built, layer upon layer, from the silt laid down by successive floods. Layers of material may be laid down in other ways; the great spit of Dungeness, on England's south coast, grows out to sea at the rate of over five feet a year, from shingle brought along the coast by the waves and currents. In moor country, deep layers of peat are formed by the successive death of the bottom parts of the bog plants. Currents and waves deposit stretches and banks of sand in quiet bays. After a volcanic eruption vast quantities of dust and pumice and rock fragments fall in the neighbouring sea and sink to the bottom. And from the surface layers of the ocean a constant rain of billions of skeletons of animals and plants, many of them microscopic, but as we have already shown, incredibly abundant, is always falling softly towards the depths.

In these and other ways new materials are to-day being accumulated in the form of sheets or layers of varying extent in innumerable regions of the globe, and they must obviously have been accumulating in the same sort of way through all the ages since liquid water has existed on our planet. These accumulations of slowly deposited layers are what we call sedimentary or *stratified rocks** (as opposed to the *igneous rocks*, forced in among the other layers of the crust in a molten state from below, or belched out over the surface by volcanoes), and the sheets themselves are technically called by the Latin word for layers—*strata*.

In these stratified rocks fossils are often found entombed—the remains and traces of dead animals and plants which were often strikingly different from any creatures alive to-day. We need give but a couple of examples. If you happen to spend your holiday on the Dorset coast near Lyme Regis and search the crumbling cliffs near by (or, indeed, if you explore any clay quarry across from Dorset to Peterborough), you will be pretty sure to find some bi-concave bone discs, the vertebræ of some large animal. Persistent and lucky hunters have found whole skeletons containing such vertebræ. They are of giant reptiles, christened Ichthyosaurs, or "fish-lizards," wholly unlike anything existing to-day, obviously aquatic, for their limbs are converted into paddles.

In the same clay you will be likely to find other fossils, the spiral ammonites, often beautifully patterned. These, too, though their shells are to be found in millions embedded in various rocks, have

---

*By the geologist, all constituents of the earth's crust, except the actual soil, are called *rocks*, whether they are hard granite or limestone, friable chalk or sandstone, or soft clay or loess.

never been discovered alive. But we know that these shells were inhabited by creatures not unlike the pearly nautilus of to-day, and more distantly resembling the cuttlefish and octopus.

It would not take much reflection, one might think, to realize that in any such sedimentary deposit, whether thick or thin, the lower layers must have been laid down before those above them. But such were our ancestors' prejudices and preconceptions, based for the most part upon the belief in the sudden creation of the world at a not very remote period, that it was not until the turn of the eighteenth century that this fundamental but elementary idea was properly put forward, to become from thenceforward the basis of geology. William Smith, an English surveyor, as his work took him from one part of the country to another, noted that a number of characteristic rocks, such as chalk, oolite limestone, red sandstone, or gault clay, occurred as layers which covered large areas of country. Moreover, wherever these layers occurred, they were always in the same order. The gault clay, for instance, was always close below the chalk, the green sand always immediately below the gault, the oolite limestone many layers below the chalk, the red sandstone several layers below the oolite, and so forth. And, a third point, each layer of rock was characterized not merely by the material of which it is made, but also by the fossils which it contains. This last was of vital importance; often two layers of clay or of sandstone may be nearly indistinguishable in their consistency and materials, but easily distinguished by their contained fossils. For instance, the London clay, over which London is built, lies above the chalk layer. It contains fossil fruits of palms and conifers, some nautilus shells, numerous characteristic sea-snails, and a few mammals. No ammonites or ichthyosaurs have ever been discovered in it. The gault clay, on the other hand, from below the chalk, has no plant fruits, but does contain ammonites, often uncoiled in a peculiar way instead of regularly spiral; while the Oxford clay, a thick layer close above the oolite limestone, has huge numbers of ammonites, almost all built as regular spirals.

Such facts as these obviously mean that we ought to be able to arrange all the sedimentary rocks of the world in a series, according to their age; and, this once accomplished, all the fossils in the earth's crust will fall into their time-sequence, too. The task has been accomplished for the great majority of layers. As a result, we can say that one kind of fossil belonged to an animal which lived and died before another kind of animal found fossilized in another layer; and the bewildering variety of life becomes more orderly through receiving an arrangement in time. To take merely the same examples we first mentioned, ammonites are found to be absent from all layers below the coal measures and from all above the chalk, but present in all congenial layers between these limits; while ichthyosaurs, though they also have the chalk as their upper limit, only extend downwards through about two-thirds of the layers in which ammonites are found.

This fundamental principle, that the different layers of the earth's crust can be arranged in a time-sequence, is the basis of that department of science known as palæontology. These sheets of inert matter are the pages of the book of our planet's history. They lie scattered over the globe, often torn, defaced, or crumpled. But patience and reason combined have been able to reconstruct whole chapters and sections of that great book. In it we can read not only the physical changes that the world has experienced—when the Rockies were built, or the Scottish Highlands worn down to mere stumps of their former grandeur, the date of great Ice Ages, æons before the last Ice Age, or of the appalling flow of lava which overwhelmed a quarter of a million square miles in North-Western India—but also the history of Life, printed on the pages of the book in the form of fossils, hieroglyphs which to persevering study reveal readily enough the secret of their picture-writing.

The principle is both fundamental and simple; but there are sometimes difficulties in applying it. Part of the record may have been destroyed or defaced till the life-story it contains becomes illegible; or a whole set of pages may have been crumpled or turned upside down into reverse order (as in the upthrust of some mountain ranges), so that their proper rearrangement is a matter of the greatest difficulty; or an isolated page or chapter from some out-of-the-way corner of the globe may be hard to place.

Happily such difficulties only concern parts of the record; whole chapters of it have the pages all tidily in order, the fossil-writing abundant and easy to interpret. In most cases these confusions have been analyzed and overcome, and the result is that, with rare exceptions, the fossil-bearing rocks of the world can now be assigned to their proper position in the time-scale—their right place among the pages of the Book of Earth.

As a result, the earth's history has been divided up, as this book is divided into books, chapters, sections, and so forth, into subdivisions of various grades. Two sets of terms are used, according as we are thinking in terms of geological time, or in terms of layers of rock.

The main divisions of geological time are usually called Eras; within each Era a number of Periods (or Epochs) are distinguished, and they are further divided into sub-periods.

When, on the other hand, we are speaking of rock-layers, we refer to a System as our main unit, roughly corresponding to the rocks laid down during one Period of time. The Systems are divided into Formations, and so on down to the narrow Zones, which may be likened to the paragraphs of a book. The Cretaceous System is thus an actual set of rock layers laid down during the time of the Cretaceous Period.

For our purpose, the Eras and Periods are all we need trouble about; when we need to go into further detail, we can specify sub-periods by simply using the words "upper," "middle," or "lower." For instance, the Carboniferous Period, during which the world's coal was deposited, is the fifth of the six Periods of the third main Era. The Upper Carboniferous, then, is its latest sub-period, for obviously the uppermost layer must have been the last deposited.

The names of the Periods are at present, unfortunately, unfamiliar to the majority of people. They should be as well known as the names of the continents and main countries of the world, or as the great dates of human history. We give a set of diagrams here, geological time-maps (Fig. 122), and we suggest that the reader make himself familiar with the divisions of these diagrams if he does not know them already. Very roughly the divisions of these diagrams are spaced out in the proportion of time assigned to each Period. We shall note later in this Book how the lengths of these Periods have been determined.

Now, the fact of primary importance in the history of life displayed by these geological Periods is the orderly succession of living forms. They *progress*. They progress from simple beginnings to more complex and versatile types. At the bottom (earliest) of our rock series come rocks with barely a trace of life and then in succession life unfolds. Comes first the ARCHEOZOIC ERA, with the dawn of life. Then the PROTEROZOIC ERA, with creatures as highly organized as worms. Then the vast PALEOZOIC ERA. There are no vertebrata at all and no evidence of land life in its opening period, the CAMBRIAN. Then in a second period, the ORDOVICIAN, is the dawn of vertebrate life. Then comes the SILURIAN, in which fishes and some land plants and invertebrata appear. Then DEVONIAN and CARBONIFEROUS, with· an ever-increasing amount of land forms, and the whole of the era closes with the PERMIAN, in which reptiles first appear.

Above these comes the MESOZOIC SYSTEM of rocks, that gigantic volume which tells of the Era of mighty reptiles and coniferous and cycad-like plants. Its formations (which like the Periods of the Paleozoic derive their names either from the districts in which the rocks are well-developed or from well-defined physical characters) are the TRIASSIC, the JURASSIC, the CRETACEOUS.

Finally comes the CENOZOIC ERA, the age of modern life, of mammals, birds, grasses, flowering plants, and trees. And here we warn

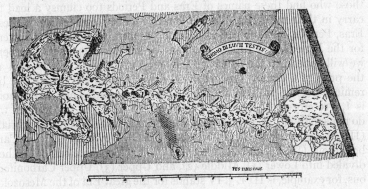

FIG. 121. When the remains here shown were unearthed in 1726, they were described by Professor Scheuchzer of Zurich, who published this figure of them, as "the damaged skeleton of a poor sinner drowned in the Deluge."

As a title to the picture is written "Homo Diluvii Testis"—man, witness of the Deluge. A century later Cuvier showed the remains to be those of a giant Salamander, now called Andrias scheuchzeri; it came from strata of Miocene Age (V C) in Baden.

the reader of one of those exasperating indistinctnesses of nomenclature in which the scientific mind at its worst seems to delight. The Cenozoic Age is subdivided into the EOCENE (dawn of recent life) and so onwards with progressive modernization of animals through the OLIGOCENE (slightly recent), MIOCENE (less recent than the next division), the PLIOCENE (more recent), the PLEISTOCENE (most nearly recent)—the Period of the last great Ice Age—and the "RECENT" Period, since the retreat of the ice, in which we live. The use of local and mineral names for the formations is suddenly abandoned for these kindred confusable names. For some reason quite a number of slightly inattentive students get "mixed" with Miocene and Mesozoic, just as it is the commonest misapprehension in the world to substitute Paleolithic (a stage in human development) for Paleozoic. The scientific systematist has never grasped what any novelist can tell him, that names must be distinctive if they are to be remembered. How would he like to struggle through a story in which Tompkins, Tomlins, Tomkinson, Robert Thompson, Robins and Robinson were the names of the principal characters? Unhappily the present writers have no power to rechristen the geological formations, as they would gladly do, for the ease and comfort of their readers.

We will, however, attempt something that may be of service to those who find these names of Eras and Periods too clumsy a load to carry in their memories. We will attach numbers and letters to these Eras, Periods, and sub-periods, Roman numerals for the Eras, letters for the Periods, and Arabic numbers for sub-periods. These numbers we will append to the names as they crop up in what follows, and at the price of a certain typographical disfigurement the reader will be reminded of the position of each Age as it is named. Thus, Archeozoic is I, Proterozoic II, and the Paleozoic Era III. The two former we do not subdivide for our purposes. But III falls into divisions, Cambrian (III A), Ordovician (III B,) Silurian (III C), Devonian (III D), Carboniferous (III E), and Permian (III F). Each of these can be further divided into Lower (1), Middle (2), and Upper (3). Upper Carboniferous, for example, is III E 3. IV stands for the great Era of the Mesozoic, with its divisions, Triassic (IV A), Jurassic (IV B), and Cretaceous (IV C). Finally, most modern of all, our present Era, the Cenozoic, is distinguished by V. That again subdivides into Eocene (V A), Oligocene (V B), Miocene (V C), Pliocene (V D), Pleistocene (V E), and the current Period, the Recent (V F). With the help of the printer's reader these numbers shall as a rule appear after each repetition of these geological names.

We may add one further word of elucidation. In the earlier days of geology only three great Eras were distinguished, instead of the five we recognize now. These were called Primary, Secondary, and Tertiary. Primary was our Archeozoic (I), Proterozoic (II), and Paleozoic (III) together, Secondary was what we now call Mesozoic (IV), and Tertiary, the Cenozoic (V). Later a fourth term Quaternary was added to distinguish the most modern deposits—those now called Pleistocene (V E) and Recent (V F). The word Primary is now rarely used, but Secondary and Quaternary turn up at times, and Tertiary (because it is an easier word, perhaps) has more than held its ground against Cenozoic. The reader is likely to find us falling rather frequently into the use of that more familiar word.

In column 3 of our diagram the maximum thickness of the layers is given, Period by Period. Of course, the total thickness, or anything

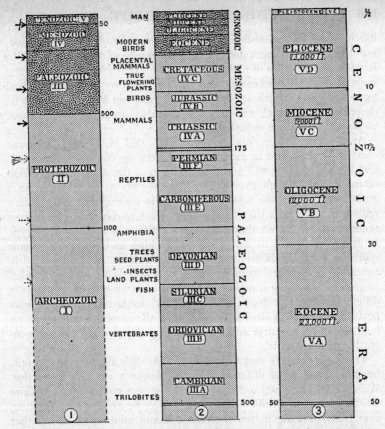

FIG. 122. The time-scale of the record of the rocks.

The whole of Geological Time is represented in (1), except for the earlier Archeozoic (dotted), whose history is not yet known. The darker part of (1), from the first well-preserved fossils to the present day, is shown on a larger scale in (2). The relative lengths of the periods can only be approximate. The period of active Mammalian evolution, darkened in (2), is shown still larger in (3). Here the total thickness, in feet, of the strata laid down during the various periods, is added, to indicate the speed at which deposition occurs. The recent period since the last Ice Age is too short to be visible in a diagram of this scale; it is hidden by the printer's ink of the upper line. It can be seen in a less compressed diagram of the Pleistocene and recent periods, which appears in Book 5. The figures to the right hand of the columns show the time-scale in millions of years, measured backwards from the present day. They are based on analyses of radioactive minerals. (Some authorities would make the Proterozoic of shorter duration.) The arrows in (1) indicate the great "revolutions" or times of violent mountain-building. The dates of the earliest of these disturbances (dotted) are not certainly known; is generally assumed that they took place at about the same rhythm as the later ones. The first appearance as fossils of a few dominant groups of animals and plants is indicated to the left of column (2); but it should be remembered that in several cases the first stages in the evolution of a group have not yet been discovered.

like it, is not to be found piled up at any one spot of the earth's crust, but the total thickness gives a rough measure of the time taken to lay down these miles of rock, film upon film, year after year. The time must evidently have been prodigious; and when we look at the actual figures in years (which are determined by another method, later to be described, and which can now be taken as accurate within about 10 per cent) they are indeed staggering.

Archbishop Ussher, less than three hundred years ago, dated the creation of the world in 4004 B. C. (and gave the day and hour, too!), and his calculations still adorn the margin of the Authorized Version of the Bible. Though they are wrong, they are at least wrong on a grand scale, since the age of the earth is somewhere about half a million times greater than he supposed. Even to-day the average man tends to think the six-thousand-year antiquity of Babylon or Egypt enormous. But just as astronomy is teaching us to think of cosmic space on a wholly different scale from geography, to be measured in terms of "light-years" running into ten thousands of millions of miles, so geology is making it necessary to think of earth-history on a wholly different time-scale from human history, in terms of million-year periods, to which a decade bears almost the same proportion as an hour does to a century, and a century as a day does to a whole generation of human life.

To think in such magnitudes is not so difficult as many people imagine. The use of different scales is simply a matter of practice. We very soon get used to maps, though they are constructed on scales down to a hundred-millionth of natural size; we are used to switching over from thinking in terms of seconds and minutes to some other problem involving years and centuries; and to grasp geological time all that is needed is to stick tight to some magnitude which shall be the unit on the new and magnified scale—a million years is probably the most convenient—to grasp its meaning once and for all by an effort of imagination, and then to think of all passage of geological time in terms of this unit.

§ 2. Defects and Happy Finds in the Record.

THE principles on which the geologist relies in his attempt to decipher the past history of the earth and the life upon it are clear and simple —so simple and so clear that at first sight it would seem that the task of the palæontologist, apart from the physical labour of finding and digging out fossils and the mental labour of studying them, should be easy. But Nature rarely reveals her secrets cheaply, and we have not yet told of the complications of the task.

In the first place, even when a large thickness of rock shows every evidence of having been laid down steadily and continuously, year after year, it may well change its character. For instance, clay, being composed of fine particles, will only be deposited farther out to sea or in quieter water than the more coarse-grained sandstone; but a layer deposited off a coast which happened to be slowly rising (as for instance Spitsbergen is rising to-day) may easily begin as mud and end as sand, the one deposit gradually hardening into clay, the other into sandstone. And if it change its character, the character of the animals and plants which live on and in it will change, too. Mud-dwellers will give place to sand-dwellers. If the change is too rapid for adaptation the fossils will not show a gradual evolution, but there will be an invasion of new creatures from other parts of the sea-bed as the conditions alter. The old forms are extinguished and drift off elsewhere, and the palæontologist is left with his story broken.

Still more frequently there is a break in deposition. The layer perhaps comes within the range of scouring currents, which prevent deposition; or it is shoved out of water by some upward movement of the crust of earth and has its newly deposited sheets removed. Later it sinks into favourable conditions and deposition begins again. But now, whether the new material be the same or quite different from that laid down before, there is a gap, during which life has bequeathed no record of itself. In general no widespread deposition will occur offshore except when the coast is sinking, and very few animals will be preserved except when deposition is rapid.

Difficulties in some ways more serious confront us in studying the sequence of the rocks in many mountainous regions. Anyone who uses his eyes and opportunities on a railway journey through hilly or mountainous country will see that it is rare for the layers of rock exposed in the cuttings to be horizontal; they are usually tilted, and sometimes tilted at sharp angles. This tilting is due to movements of the earth's crust, such as its shrinkage and consequent bending and folding during periods of cooling. Over all of midland and southeastern England the tilting is usually slight but definite. This was very favourable to William Smith, the British pioneer of geology, for, although it left no doubt as to which layer was above which, it caused new layer after new layer to come to the surface, open to investigation, as he passed across country upon his work of canal-making. His genius worked under the luckiest conditions.

But in mountain regions the disarrangement may be much more serious. In the Alps it is common enough to see layers of rock standing on edge, or even turned quite upside down; and in some places, as in parts of the Scottish Highlands, the pressure has been so great that the rocks have been what we may call accordion-pleated—thrown into a whole series of deep folds, so that when seen in section on an exposed face they look like a fan. Not only that, but what the geologists call faulting may take place. Under the stresses and strains of mountain-building or of earthquake, great layers of rock may crack across, and one side slip down or be forced up and made to ride over the other side. At the time of the San Francisco earthquake a fault hundreds of miles long was produced, in which the whole country to one side of the crack fell suddenly from five to ten feet; but in mountain regions areas of rock may be faulted down hundreds of feet.

During these magnificent crumplings whole pages of the record have been made altogether illegible. Even when life has succeeded in writing its story on the rocks, the writing has too often been obliterated again. The rock-layers may be subjected to colossal pressures under great depths of newer deposits or scorched by contact with huge intruding lakes of molten material from below, so that any contained fossils are squeezed, distorted, or baked out of recognition or even out of existence. The rocks themselves change their very character. Such transformed rocks are called *metamorphic;* marble, for instance, is thus metamorphosed out of limestone, quartzite or gneiss out of sandstone, and so forth.

This difficulty becomes more serious as we go farther back in the record. For rain and wind and frost never stop their scouring and splitting and wearing, and the younger sediments must all be derived from the débris of the old. Whole mountain ranges, with their contained fossils, have been destroyed, worn down to level plateaus to furnish material for new layers, often hundreds or thousands of feet thick, which in their turn will be upheaved, and in their turn eroded away. Entire chapters of the Book of Earth have thus been pulped to furnish material for new pages. Luckily, however, the making of the book went on simultaneously in many regions, each one often of huge extent, and it is rare that all the records of a whole Age have perished.

Then we must remember that fossilization is the fate of very few animals and plants. Only one in a million makes its mark in the Book of Life. The great majority of dead things simply decay and disappear and their material is returned to the general circulation of Nature to be built up into the bodies of new organisms. But once in a while a

THE EVIDENCE OF THE ROCKS

corpse is preserved more permanently; it falls into mud or silt or some place where the bacteria of decay cannot get at it. Insects have been caught and sealed and preserved for countless years in the fossilized resin we call amber. The bodies of mammoths can be dug out of frozen mud-cliffs in Siberia with the skin and flesh still preserved. And even when the flesh decays away the bones may escape the dissolving action of rain or other waters and ultimately find their way into the palæontologist's cabinet.

But it should be remembered that these direct preservations of the material of extinct creatures are the rarest accidents. Most fossils (for any dug-up trace of an organic being is called a fossil) are not actual surviving bits of corpses at all, but bits of corpses which have been changed into rock by a slow translation and replacement. They are copies at second hand of the original writing. As the bones or other enduring fragments lie buried they slowly dissolve away and are replaced more or less completely by mineral substances. In a word they are "petrified." In this process, of course, they undergo varying degrees of distortion, although in one or two exceptional cases the translation is astonishingly accurate, each little difference of texture in the original being faithfully reflected in the mineralized fossil. The record may be even more indirect than that; it may be a dried footprint or the hollow impress of a bit of skin or a shell. From these scattered and accidental remains the palæontologist patiently reconstructs his picture of the world as it used to be.

Occasionally a lucky chance makes us realize how scanty our information really is.

In California, for instance, a pool of water with sticky margins impregnated with tar proved a death-trap to thousands of creatures as they came down to drink, and to hordes of carnivorous mammals, like wolves and sabre-toothed tigers, which endeavoured to catch the drinkers when they stuck fast. Now the palæontologist finds a hoard of treasure in that one pool. In France, in the drought of 1911, it was noted that all the fish in a pool burrowed into the mud when the water dried up, and were eventually baked hard in their hundreds. Unless the geologists of future ages happen to hit on such a patch of trapped life, they are not likely to find more than isolated bones of these kinds of fish.

Almost the only complete skeletons of the extraordinary giant reptile, Iguanodon, are those of a whole troop, twenty-nine of them, old and young, which were found in a Belgian cave, obviously entombed by some accident.

Fossil ants, we note, are often found in amber. Amber is the fossilized gum which once exuded as resin from long decayed pine-trees. Only an insignificant fraction of the ant population of the world gets trapped in resin; and only an insignificant fraction of the resin is hardened and preserved as amber; and yet almost all our knowledge of the ants of the past is derived from specimens in amber.

The amount of detail preserved to us is also very much a matter of luck. As a rule, nothing survives but the skeleton, and even that may be distorted and sometimes partly rotted. Now and again, however, happy accidents have caused traces of the softer parts to survive to our day. For example, one or two specimens of the strange, duck-billed dinosaur, trachodon, in the Upper Cretaceous (IV C 3), died and fell on a patch of soft mud. They decayed and the place where they had been became covered with sand. But the mud beneath held the impression of their skins with surprising fidelity—so that now we have the form of their skins preserved, with a mould in hardened mud and a cast in hardened sand. Some of the dolphin-like ichthyosaurs left records of their flesh and fins; some even of their fæces, marked with a spiral twist by the folds of their intestine, and containing undigested remains of the fossil squid-like creatures called belemnites.

The very first known land beasts left footprints in the mud across which they lumbered; the bird-reptile Archæopteryx stamped its feathers with astonishing detail in the fine-grained lithographic limestone of Solenhofen.

But there are even older happy chances than these. Some Devonian plants (III D) are so minutely petrified that we can study the precise shape of the microscopic hairs protruding from their leaves. And recently Walcott has discovered a wonderful array of invertebrates imprinted on Middle Cambrian shales (III A 2), nearly as clear to view as when they were swimming about, ages before the first fish, almost five hundred million years ago. The soft mud, now pressed hard, reveals to us the outlines of jelly-fish, annelid worms, with their appendages and bristles, small crustacea, with all their soft leaf-like appendages, even the outlines of their stomachs, and arrow-worms, just like those that swim in our modern seas, with their transparent fins. Walcott has also found swarms of bacteria from still earlier rocks. It is doubtful whether this is to be considered the most extraordinary case of fossilization; it is certainly rivalled by some Paleozoic (III) fishes (to be seen in New York) which have been so delicately petrified that thin slices of their muscles, ground down to transparency, and looked at under a high power of the microscope, show the cross-sections of the muscle-fibres as clearly as a fresh-made preparation from a modern dog-fish. How the microscopic structure of living tissue came to be thus translated into stone we do not fully understand; but at least the rarity of such lucky finds brings home to us the multiplicity of what is lost for ever.

The various difficulties thus put in the way of palæontologists are of two main kinds. There are those which lead to gaps and imperfections in the fossil record, whether through the fewness of animals and plants which became fossilized, the washing away of large sections of the crust when brought above water, for wind, rain, and frost to destroy, or the destruction of fossils by the heat and pressure of metamorphosis. And there are those which make it difficult to arrange what fossils we have in the right order, whether the difficulty springs from the turning of layers topsy-turvy and from faulting, or from the fact that two contemporaneous layers might show quite different fossils, either because they were laid down in quite different situations, or because, though comparable in the environment they provide, they were situated far apart on the earth's surface.

The imperfection of the record is unfortunate, but nothing more. It makes the labours of fossil-hunters greater, and should warn us against clamouring for the immediate discovery of this or that "missing link." But patience and the exploration of more and more of the earth's surface are bringing their own reward. Every year the record becomes less scrappy, and in many groups of animals, what fifty years ago seemed impossible to hope for has been achieved, and an unbroken series discovered, leading through ages of time from simple to highly developed types. The difficulty is not one of principle.

But the other difficulty seems at first sight more serious, and the opponents of Evolution, anxious to find any stick to beat a dog with, have tried to make out that it is a defect of principle. The Evolutionist, they say, pretends that he dates his fossils by the order of the rock-layers in which they are found; but in reality he very often dates his rock-layers by the fossils found in them—when, for instance, they are distorted in mountains, or when only a single fossil-bearing layer can be found, or when new quarters of the globe are explored for fossils; and this, they proclaim, is arguing in a vicious circle.

As a matter of fact, however, the palæontologist does not fake his results with the naïveté which these assertions ascribe to him. He first of all examines the earth's crust in some region, like the south-eastern half of England, or the Bad Lands of Wyoming, where the strata lie plainly one above the other and are not crumpled or distorted. He collects series of fossils from as many as possible of such undisturbed layers, examines them, determines to what group they belong, and whether there exist any others of exactly the same species from other layers. By so doing he can provide for each main layer and for each kind of deposit—sandstone, shale, clay or limestone—within each main layer, a list of fossil species which are not found in any other layers, and of others which are especially abundant in the layer, or very scarce or absent therein. It is only after he has thus dated his fossils by reference to an undisturbed succession of deposits that he uses them to help him in dating layers whose age cannot be determined by these straightforward methods.

The only really serious difficulty in dating rocks is the problem of relating the rock-systems of distant regions of the globe. The succession of rocks in China may be known, and also the succession in Western Europe; but the evidence may be lacking which enables us to say exactly how the one should be fitted against the other. This may lead to uncertainty in points of detail; but even in such cases it is always possible to date within geologically moderate limits, and every scrap of new knowledge helps to narrow those limits down. Here again the difficulty is not one of principle.

The essential fact is this: fossils are not used to date rocks of doubtful age until they themselves have been dated by their position in rocks whose order is not doubtful but obvious. And the proof of the correctness of the method is that its results are coherent and intelligible. We do not find ammonites appearing haphazard in the earth's crust, but in a definite set of its layers; mammals do not appear and disappear sporadically through geological time, but come on the scene when they are expected, and from then onwards show a steady development; four-toed and one-toed horses are never found in the same stratum, and so forth. If each fossil is a word in the Book of Life's history, then a century ago these words made only a few scattered but promising sentences; to-day, thanks to unceasing exploration, they have fallen into place, and have told man a new and clearer story of life's past and life's destiny.

§ 3. A Sample Section in the History of Life: the Evolution of Horses.

AND now let us take one of the better preserved sections from this vast, confused autobiography which life has written in the rocks. It is a section of which we shall give a considerable amount of detail and

THE EVIDENCE OF THE ROCKS

rather a bothersome multiplicity of generic names. The reader is under no obligation to remember these names, but they have to be "produced in court" for the purposes of our proof.

It is loudly argued by many Creationists and semi-Creationists

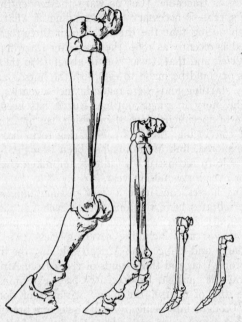

FIG. 123. Stages in the evolution of the horse's hoof; a series of left fore-feet.

On the right, Eohippus with three of its four toes visible—the fifth, fourth, and third. Next to it, Mesohippus with fifth digit reduced to a splint-bone, and considerable lengthening of the toes. Then Merychippus with fifth digit small and lifted off the ground, and enlarged central digit. On the left the modern horse, Equus, with big central hoof, fifth digit quite disappeared, and second and fourth digits reduced to splint-bones.

that there is no fully-worked-out pedigree of any existing forms of life, and that there is nothing to dispose of the view that at irregular intervals creative forces intervene in the evolving process and make life take a convulsive stride forward. This, however, is not the case. So far from there being no well-worked-out pedigree, in which the successive forms in some group of animals are seen visibly modified and differentiated, there are now several such family trees in existence. We are giving here the past record of the existing horses. They have been evolved from a small, four-toed Eocene mammal and every step in the process is traceable. (Let us warn the reader that the time-diagram—Fig 122—is necessary to the reading of what follows.)

It is worth noting that the earliest known three-toed fossil horse was described as recently as 1860, the year after Darwin published the *Origin of Species*, and that it was not till about 1870 that any serious attempt was or could be made to establish the horse's ancestry from fossils. Many startling finds were made in the 'seventies and 'eighties presenting the story in rough outline. But it has been the patient accumulation of specimens since then which has filled in the details and made it convincing. As the fossil-bearing rocks have been more intensively explored, link after link has been brought to light, until now we are able to reconstruct an almost unbroken chain of change extending for over forty million years.

The existing horses constitute a very distinct family of animals. No other vertebrates have but one toe to each foot; and no other animals have quite similar teeth. The family now comprises one genus only, called *Equus* and including seven species—two Asiatic wild asses (the onager and kiang), one African wild ass, the little Przevalsky's horse from Asia, and three kinds of zebra from Africa. A fourth kind of zebra, the quagga, has recently been exterminated. In addition, there are, of course, the many varieties of domestic horse and donkey, brought into being by man's selective breeding. Now all these wild horses and asses and zebras live in much the same way. They run swiftly over hard, level plains (wild asses in the Mongolian desert have been timed doing their forty miles per hour), and they feed on a food that is difficult to chew—on the tough leaves and stems of grasses, which are often hardened by a certain amount of flinty matter. And corresponding with this hardness of ground and of food, we find special provisions in their feet and teeth.

The single toe of a horse's foot corresponds to the third or middle toe of the more ordinary five-toed foot. Only the last joint of the toe touches the ground; the hoof in which this last joint is encased is the exact equivalent of an overgrown toe-nail. The horse's wrist and ankle are far above the ground, forming the joints commonly called "knee" and "hock"; the true knee is what is styled the stifle. The region corresponding to our palm or sole contains the single elongated cannon-bone. But the other fingers or toes are not completely absent,

THE EVIDENCE OF THE ROCKS

for attached to the hinder angles of this cannon-bone are two little splint-bones; and that these are the remains of the second and fourth fingers and toes is amply proved not merely by their position, and by the fact that one or both of them occasionally develop the missing joints and a miniature hoof, but also, as we shall soon see, by their development in the embryo.

The horse's limb, then, is a specially modified limb; it is a limb in which one toe is enormous and strong, and in which the others have dwindled more or less completely away. It is a limb devoted to a special function. On open, grassy plains the best means of escape from enemies lies in speed; and it is for speed on comparatively level and hard ground that the legs of the horse are suited. Everything else has been sacrificed to that. The elongation of the actual foot-region gives a better leverage; the concentration of all the limb-muscles in the upper part of the legs allows for rapid swing; a limb which consists of a single pillar, joined so as to move only fore-and-aft, transmitting all the weight downwards to a single expanded hoof, is stronger than one which can be moved in all directions, or than one in which the lower arm or leg contains two bones (as in ourselves), and there is less "give" than in a limb ending in several toes. Accordingly the horse, though it can only execute a very few kinds of movements with its limbs, though it sinks in soft ground owing to lack of spreading toes, though it is not well adapted to broken country, triumphs in speed on dry and rolling plains (Fig. 123).

Incidentally, length of leg makes length of neck a necessity; without that the horse could not reach down to its food.

The grinding teeth of a horse are beautifully and elaborately adapted for dealing with the tough grasses that he consumes. They are peculiar in three ways. First, they are all alike, instead of the true molars being more complex than the premolars as is usual. The premolars, we may remark in passing, are those grinding teeth which have predecessors in the milk dentition; the molars have not. Secondly, these quite similar molars and premolars of the horse all have an extremely complicated surface-pattern. Before they cut the jaw-bone they are covered with hard, glossy enamel, rising up in ridges round a couple of deep cavities. These cavities later get filled up with cement, a substance not quite so hard as enamel, which is secreted by special glands as the tooth breaks through the gum. Under the enamel is the softer dentine. Use soon grinds off the top of the teeth, with the result that their tops become nearly flat; but as the three materials wear down at different rates, sharp edges of enamel stand up a little beyond the cement, and a little higher above the dentine. The whole forms a re markably effective miniature millstone, with the advantage over our millstones that it keeps its grinding-ridges sharp as it is worn down (Fig. 124).

The third point about the teeth is that they are of remarkable depth, and that during the first eight years of life they have no closed roots,

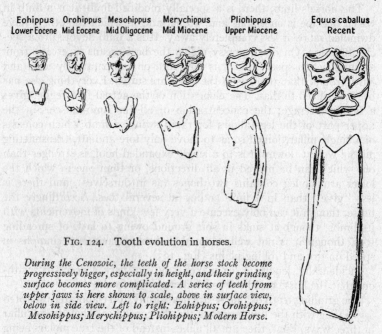

FIG. 124. Tooth evolution in horses.

During the Cenozoic, the teeth of the horse stock become progressively bigger, especially in height, and their grinding surface becomes more complicated. A series of teeth from upper jaws is here shown to scale, above in surface view, below in side view. Left to right: Eohippus; Orohippus; Mesohippus; Merychippus; Pliohippus; Modern Horse.

but go on growing from below, like a rabbit's front teeth, as they are worn away above. After this roots are formed, new growth ceases, and the teeth are simply pushed up to compensate for the wear at the surface till they are all worn away, and the animal dies because it cannot chew its food. Our own teeth, and most mammalian teeth, are finished and complete as soon as they have erupted, but in a horse completion is delayed for eight years to give a longer working life.

The horse's teeth, then, are admirably adapted for chewing up grasses; and the size and peculiar shape of his head is due to the need for finding room for these powerful and deep-rooted living millstones, and for the muscles to work them.

Now, when we look back to the fossils of the Eocene (V A), the earliest period of the Cenozoic, we find nothing resembling a horse. We find no mammal so specialized as a horse. In the earliest Eocene, almost all the mammals were small, they all had either four or five toes to each foot, and their teeth were short, low-crowned, and provided with more or less rounded or conical cusps, never with grinding ridges. Obvious carnivores with teeth for slicing and cutting, like those of lion or wolf, did not exist, nor obvious herbivores, with grinding and chewing teeth like those of cow or elephant. In the later stages of the Eocene Period, definite carnivores and herbivores can be recognized, together with other well-marked types; but these early forms are all extremely different from any living animals. When Owen in 1856 described Hyrocotherium—for that is the first name in our history—he never even guessed at the relationship between it and the horse; now not only do we know that the one is ancestral to the other, but we can fill up all the gaps between the two.

If we go back stage by stage through the rocks of the whole Cenozoic Period (V), we find that the horse has recorded its pedigree in fossils. There are four main stages. In the last, the fossil horses resemble the living forms in all essentials of teeth and feet, differing only in details of proportion. They are all grass-plain animals.

In the stage before this there are no one-hoofed horses. Instead we find smaller creatures, of obviously horse-like type, but with three hoofs on each foot; the two outer hoofs, however, are small and must have been useless in running, since they did not reach the ground, but hung in the air like the dew-claws of deer and other animals. The teeth had a less elaborate grinding system, and were much shorter. Fossil bones of these—shall we call them the fathers of the horses?—are not uncommon in China, and (with others) are dug up to be sold to apothecaries as "dragons' teeth," that essential ingredient in the Chinese pharmacopœia. Professor Watson tells us that even the Chinese labourers employed to dig them up recognize the skulls as like those of their donkeys.

In the next, and still older, assemblage of forms, the grandparents, so to speak, the ancestral horses were no larger than a large dog or a small Shetland pony. They also were three-toed; but all three hoofs touched the ground, and in addition they possessed on the fore-foot the trace of a fourth toe, in the form of a little splint against the cannon-bone; the tooth-pattern again was less elaborate, the whole tooth shorter, and there was no trace of the cement which in all later forms filled up the valleys between the ridges of enamel and dentine and so ensured a flat grinding surface throughout. Yet one can easily recognize the skeletons even of this stage as those of horses—three-toed and rather lumpish horses, but horses.

Finally, in the earliest stage, the far ancestors, in which we can still definitely detect the tendencies which culminated in the modern horse, none of the animals were bigger than a medium-sized terrier; there were four little hoofs on the fore-foot and three on the hind;

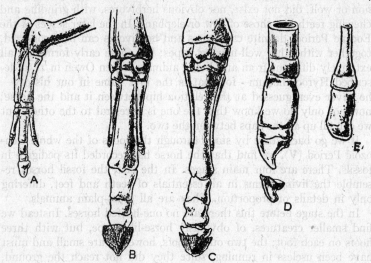

FIG. 125. The modern horse recapitulates its own evolution.

(A) The skeleton of the limb of an embryo horse six weeks old, showing three toes. (B) The same, eight weeks old, showing the side toes much reduced relative to the middle toe, on which the hoof is forming. (C) The same in a five-months embryo, showing side toes reduced to splints. (D) The end of a side toe of (C), much enlarged, showing its rudimentary hoof. (E) The end of a side toe at birth; the various bones are still separate, not joined as in adult horses. (Modified from drawings by Professor Cossar Ewart.)

sometimes the hind-foot also showed two splint-bones representing the missing first and fifth digits; the teeth were very short, with only indications of the system of grinding ridges, and the premolars were not so large nor of so complicated a pattern as the molars.

Besides the teeth and toes, other characters, too, show steady parallel changes as we go back through time. The earlier forms had shorter necks and faces, less tightly fitted wrists and ankles, two separate bones instead of one in the lower arm and leg. In a word, as we go back we find horses less and less adapted efficiently for swift running and for grinding hard vegetable food and more and more like the other generalized mammals of the early Eocene (V A1).

The existence of these three-toed horses in the past acquires a double significance when we remember that every individual one-toed horse of to-day actually passes through a three-toed stage in its embryonic development. In a six-weeks embryo the limbs are short; two separate bones are present to make the skeleton of the forearm; the wrist and foot are short, and the middle toe is flanked by two other smaller but perfectly formed toes, each complete with the same number of joints as the centre toe. At eight weeks the two side toes are on the way to become mere splints, but they still show the full number of joints. They eventually develop a cross between a nail and a hoof at their tip, and this later grows into a regular hoof. Not until some weeks after birth do the separate joint-surfaces disappear, and the three end joints coalesce with each other and with the elongated first joint to form the little buttons that tip the splint-bones. (Fig. 125.)

Thus the three-toed state occurs twice, in two different ways, in the history of horses. It occurred as a stage in the past history of the race; and it occurs as a stage in the development of every modern one-toed individual: the current stage recapitulates the three-toed past in its own embryo person.

All this in broad outline was known more than half a century ago, when isolated representatives of all our four main stages had already been described, and the general trend of horse evolution correctly deduced from them. But to-day we have more than the broad outline; we are able to fill in the details. It is amazing how fine the detail has now become.

The amount of material at our disposal is enormous. In this single horse-stock, beginning with the little four-toed creatures from the Eocene (V A) and ending with the forms alive to-day, a total of over two hundred and sixty species have now been distinguished and named. It is, of course, true that in a continuously changing life-flow like this, species, genera, and families become merely arbitrary, since no sharp lines can be drawn. But even a continuous melody is divided up into bars: and the naming of two hundred and sixty species of fossil horses means roughly that we may chop up the continuously written record into two hundred and sixty pieces, making the difference between each piece and the next about equal to that found between forms recognized as species in an average abundant and variable family living to-day. Actually, however, in many cases the species and genera of fossil horses which were readily separated in early days have simply blended into each other as new material came to light. Sharp breaks in fossil history appear always to be the result of poverty of finds. Furthermore, not only are there these hundreds of types, but of many single types hundreds or even thousands of specimens have been unearthed. Occasionally the specimens are complete, but more often they are fragmentary, the skull (and especially the lower jaw) being perhaps most commonly found. All together, the specimens of evolving horses now in the museums of the world run into several tens of thousands. Even of the Lower Pliocene (V D 1) three-toed forms alone, at least ten thousand have been collected; and though they come from many localities in Europe, North Africa, Asia Minor, India, Persia, Mongolia, China, and North America, all of them agree with one another in their testimony with regard to the evolution of teeth and feet.

There are two complications of the horse's history. First, though the main trend is always on towards the modern horse, many of the fossil types represent side twigs which have died out sooner or later, leaving only the central branch to grow on to culmination. Second, the horses, being a very mobile species which can travel great distances rapidly, have a wide arena for their development. The earliest forms so far found are from the Western United States; but from North America they soon invaded the Old World across a land connection where is now Behring Strait; and the living tide flowed back and forth between the Old World and the New, or was dammed back, according as this bridge emerged or was sunk under the waters; it invaded Africa, and flowed down into South America, when the Central American connection came into existence, in the late Miocene Period (V C 3). Thus, as climate changed and barriers were bridged, the various types of horse would move from place to place of this wide scene, so that sometimes quite new types suddenly appear in the local record—invaders which were evolved in some other locality where perhaps fossils have not yet been discovered. But in spite of these obscuring factors the story has been clearly worked out; the fossils are so numerous that it can be construed without any doubt at all.

It will repay our trouble if we try to penetrate a little deeper into this representative chapter of life's history. First we may ask what could have been the reasons for this steady evolution in one direction. The answer appears to lie largely in a climatic change. The evidence of fossil plants makes it clear that during the Cenozoic, from Eocene (V A) to Pliocene (V D), the climate of the north temperate and subtropical regions became progressively drier, and that over much of

North America and Eurasia forest gradually gave place to glades, and these to open plains, on which the newly evolved grasses flourished and spread. It was but natural that a few of the animals from the already crowded forests ventured out to try their luck in the new world that thus presented itself. It was a new, exacting life; the new plants of the plains were wiry and hard to chew, and in a plain, since effective concealment is difficult, one has to run fast. Some of the adventurers chewed well enough and fled fast enough to survive; the pursuing carnivores had yet to achieve the speed of the wolf. The earliest four-toed horses were forest-dwellers; they must have lived as their closest surviving relations the tapirs live to-day, moving slowly and warily over soft ground and eating comparatively soft vegetation; but the three-toed forms had for the most part taken to the grazing life of the plains. The progressive spread of the plains during the rest of the Cenozoic up to the Glacial Epoch (V E) put a premium upon continued improvement in the same direction. Thus the early horses increased and flourished. Their more conservative forest-dwelling cousins dwindled in numbers as their range was restricted, and in most cases became altogether extinct, while the primitive wolf and the primitive horse were teaching each other speed.

The various stages of this main line of evolution have been classified into ten genera, of which we append the following thumb-nail sketches. Number one is Eohippus from the American Lower Eocene (V A 1), the earliest known four-toed horse; of this genus thirteen species are now known, ranging in size from a cat to a medium-sized terrier. It represents the more primitive type of the earliest of our four main stages, still with rudiments of the first and fifth toes on the hind-foot. There are also three or four European species of the same age and general character, but with even more primitive teeth, which have been given the name of Hyracotherium. Orohippus comes second, from the Middle Eocene (V A 2), with ten known species. This lacks the rudimentary hind-toes, and its premolar teeth gradually become more molar-like as we pass up in the rock layers from species to species. In both these types the lower arm and leg contained two separate bones, and the fore-limb could still be twisted and turned about in a way impossible for the modern horse. Number three, Epihippus, with only two species hitherto described, is from the Upper Eocene (V A 3). It is still four-toed, but the fourth toe is smaller, the central toes larger, and two of the three premolar teeth are just like the true grinders. There is no real break between it and Orohippus.

That closes our Main Stage I. Mesohippus is the first type of Main

Stage II, the horses with three usable toes. Between it and Epihippus there is a slight gap, during which the final stages in the reduction of the fourth front toe to the merest splint must have been run through, and during which there was also a considerable increase in size. Thousands of specimens of the genus have been found, the earliest from the Lower Oligocene (V B 1), but running on through Middle into Upper; these can be separated into no less than eighteen species. All three of the actively-grinding premolars are now like the molars, and in the fore-arm and lower leg one of the two bones is enlarged to take nearly all the weight, while the other is on the way to disappearance.

The fifth genus is called Miohippus. It is first found in the Upper Oligocene (V B 3), and is scarcely to be separated from Mesohippus, save by its rather larger size (up to that of a sheep) and the better mechanical construction of the skeleton of its wrist and ankle. It also is abundant and variable, with seventeen known species. Parahippus, first found in the next higher rock formation, the Lower Miocene, is a real connecting link between Main Stage II and Main Stage III. The earlier of the eighteen species so far discovered are on the whole very like Miohippus, while some of the later forms have the outer two hoofs and digits so much reduced as to be useless. The teeth, too, are interesting; the cement filling appears for the first time, but only in some of the later species is it anything but a very thin coating.

Parahippus runs on right through the Miocene, but already by the Middle Miocene (V C 2) a more progressive type had appeared, which is called Merychippus. This type, abundant and wide-ranging, with some twenty-five species, definitely initiates our Main Stage III, for its side toes never touch the ground. But it makes a greater advance in the depth of the teeth. In the earliest species of Merychippus the grinders are only as high as wide; in some of the latest, the height is two-and-a-half times the width. The cement is always abundant, though it was not deposited until just before the tooth emerged from its bone socket, instead of some time before, as in the living horse. The ulna, the dwindling second bone in the lower fore-limb, is a separate splint in colts of Merychippus, but in the adults is fused with the main bone or radius, as it is from birth onwards in modern horses: here we get a second glimpse of recapitulation, that widespread phenomenon of which the three-toed embryos of modern horses have already given us an example. Much increase in size also took place within the group before it faded out in the early Pliocene (V D 1), the later forms being often as big as small ponies.

Our eighth genus, Pliohippus, grades back insensibly into some of the Merychippus forms, first emerging distinctly towards the very end of the Miocene (V C 2) and continuing through the Lower Pliocene (V D 1). As regards feet, Pliohippus bridges the gap between Main Stages III and IV, some of its seventeen species possessing tiny but perfect side toes complete with miniature but useless hoofs, while in others they were reduced to single splints as in modern horses. The chief point of progress again concerns the teeth. In earlier genera, the molars led the way; then their more elaborate pattern was gradually adopted by the premolars too. However, this applied to the permanent premolars; their milk-teeth predecessors in the colt were always of simpler pattern. But in Pliohippus the elaborate pattern has been thrown back into the milk-tooth stage, and from now onwards all the grinders at every stage of life share equally in every advance made in the grinding machinery.

The ninth genus is Plesippus, which comes definitely into the last of our four main stages. So far, only one species has been discovered, from the Upper Pliocene (V D 3), but of this one species the anatomy is known in full detail. These creatures were very much like a smallish horse, but the hoofs, as in Pliohippus, were still much smaller than in a modern horse, and the teeth show but little advance on those of Pliohippus. The splint-bones representing the outer toes are very interesting, for they are longer than those of living horses and more expanded at the tip. The fifth digit of the fore-foot, which, as we saw, became useless in Mesohippus, hung on for a long time as a remnant, and was apparently loth to disappear, for it is still represented in Plesippus by a tiny nodule of bone. Indeed, in living horses, though usually wholly absent, it is still to be found in a few individuals as a still smaller nodule. The shape of the skull is like that of our modern horses.

From this type to Equus, the true horse, is but a small step, which was taken at the turn from Pliocene to Pleistocene (V E). Even among the various species of Equus, however, evolution can be seen at work, for many of the earlier species are both smaller and more primitive in tooth-pattern than any existing modern horses. On the other hand, one or two recently extinct forms considerably exceeded any living wild species in size, and were probably even larger than our domestic draught-horses; their size is perhaps to be correlated with their living in an interglacial period when the world was warmer and richer in vegetation than it is nowadays. As many as forty-five fossil species have been unearthed and described, so that less than a sixth of the total known variety of the genus Equus is in existence to-day.

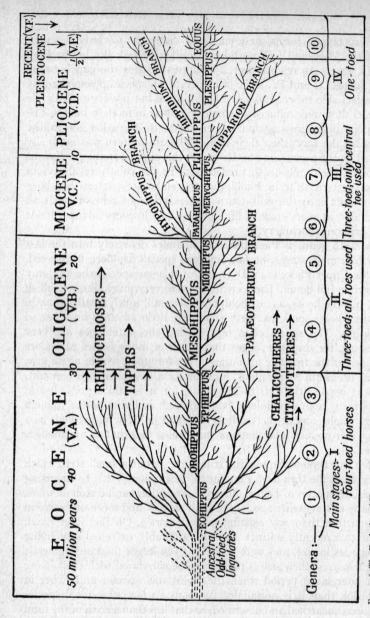

FIG. 126. The Evolution of Horses. The different geological periods are all represented on the same time-scale, except for the Pleistocene and Recent Periods, whose scales have been magnified five and twenty-five times respectively. In the early Eocene the various stocks of odd-toed Ungulates are scarcely to be distinguished, but soon the separate stocks (Rhinoceroses, Tapirs, Horses, etc.) diverge and become markedly distinct. The horse stock alone is here followed throughout. The names of its ten successive genera mentioned in the text are given, with (below) the numbers corresponding to them. Below this again the duration of the four main stages is indicated. The initial stock is all the time throwing off side branches which become extinct; most are short and unimportant, but some produced distinct types of horses, such as Hypohippus and Hipparion.

Thus has the change been brought about, from the tiny, furtive, forest-haunting, browsing Eohippus to the swift, strong, grazing Equus of the open country. But the study would not be complete without a mention of certain side branches which grew out, turning aside from the main trend of horse evolution, and sooner or later came to nothing. We have this evolution of their extinction, step by step.

The first of these branches started in the early Eocene (V A 1). Some of the Old World, four-toed forms wandered off on their own, but became extinguished in the Lower Oligocene (V B 1) after advancing a certain way parallel with the main horse-stock. Palæotherium, three-toed but somewhat heavy and tapir-like, was the best known, but others were lighter and better adapted to swift running. Over twenty species are known to have been evolved in these sidelines; what brought their career to a close is uncertain.

The next divergent branch began at the close of the Miocene (V C). Its peculiarity consisted in its retaining three good toes on each foot, very much splayed out, and teeth less well adapted for grinding grass, but suitable for browsing leaves. While some species of Miohippus form a connecting bridge to Parahippus and the rest of the main horse-stock, others connect equally insensibly with the base of this side-branch. The tendency, first revealed in them, runs its course through a couple of genera, culminating in the early Pliocene (V D 1) in Hypohippus. In this branch, with at least ten known species, we can trace a progressive adaptation to forest life, the animals apparently eating more juicy food and supporting themselves on softer ground by the aid of their spreading toes. Here we have, so to speak, a line of quitters, of animals which shirked the more arduous life of the open plains that was the goal of their race and turned back into the woods. For a time they throve, but only for a time. It was not in this direction that the horse-stock was destined to win through.

In the late Pliocene (V D 3) the horse-stock invaded South America, and here was evolved the Hippidium branch, possessing extremely short legs and a strangely-constructed nose-region. These creatures were a group of three genera, of which only four species have as yet been unearthed, only extinguished within the last million years. Their short legs probably indicate that they were adapted to mountain life. The origin of this aberrant line is traceable through Protohippus, a group of thirteen somewhat earlier species.

The fourth and largest side-line is the Hipparion branch. It came off from Merychippus in the late Miocene and ran on to the close of the Pliocene, when it was extinguished, but not before it had given rise to some thirty-five species. It is interesting as retaining the two outer toes, though not touching the ground, long after the main-line had reduced them to splints; while, on the other hand, some of its species improved their teeth beyond anything known even in present-day horses, the grinding pattern being more complex in some, the height of the teeth greater in others. It is likely that this increased specialization of teeth was an adaptation to a desert life, where the hard, dry vegetation needs more grinding. It may be that these horses fell out of the battle for life because of this too exclusive dependence on the special virtues of their teeth and were caught by a change of circumstances that made speed of greater importance.

That is a condensed summary of the story of the horse and its ancestors and vanished cousins as we know it to-day. It is a tale of adventure and arduous conquest, of steady and successful adaptation of a race to new surroundings. But it is more interesting as a part of a vaster drama. It displays one streak of the process of Evolution very completely and convincingly. Step by step, variety by variety, the progressive changes can be traced. One can hardly say where one species ends and another begins. Doubtless our knowledge of fossil horses will be further filled in and rounded off in the future, as new specimens turn up; but new discoveries can do no more now than fill in a little gap here, correct a minor error there. The essential facts are already before us in their fullness. In one long gallery one might assemble all these stages. We have here in a crushing multitude of steadily progressing specimens just that complete, continuous exhibition of Evolution in action the Creationist has demanded. He is answered.

§ 4. The Continuity of Evolution as Shown by Sea-urchins.

ONE great merit of the horse's evolutionary record is that the animal is familiar and that we can readily understand the biological meaning of the main trends in its long, ancestral development. Its only defect as a demonstration is that the record is nowhere continuous *in one single locality*.

If we could find a considerable thickness of rock, all deposited under approximately the same conditions, we should expect to find an absolutely unbroken sequence, a still more unbreakable evolutionary chain, in the fossils which it contains. Such large thicknesses of one kind of deposit are naturally, though unfortunately, rare; for usually, as deposits pile up they are brought nearer to the surface of the sea by the mainland rising, or by their own gradual accumulation, or they are submerged deeper by the land's sinking. In either case the character of the sediment, and therefore of the animals which can live at the bottom, will change, since, for instance, sand particles will sink before fine mud, and future sandstones therefore be laid down nearer inshore and in less quiet waters than future clays.

But the chalk of the Mesozoic Epoch (IV) happily serves our purpose. Up to a thousand feet of it were continuously laid down in large but shallow seas, originally as a limy mud, largely formed of the tiny skeletons of single-celled animals rained softly down from the waters above. This went on during much of the Upper Cretaceous Period (IV C $_3$), for at least ten million years; and through much of this vast lapse of time the conditions of life on the bottom of the chalk-depositing sea continued so similar that many of the same kinds of animals are found in every layer.

Among the most abundant and the best studied of the chalk fossils are the sea-urchins known as Micraster. These are found, and found abundantly, throughout most of the lower half of the chalk. In Southern England, for instance, they persist through 450 to 500 feet of chalk, the total thickness of the deposit varying from nearly 1,300 to nearly 1,500 feet. Translated into time, this means 35 per cent to 40 per cent of the total chalk period, certainly over four million years.

Throughout this long period the fossil Micrasters are so abundant that hundreds of thousands can be collected and a gradual evolution can be traced as we pass upwards. The changes are apparently trivial. There is a slow alteration of shape from rather flattened to rather arched, and from rather elongated to about as broad as long. The mouth creeps steadily forward, its distance from the front border of the lower surface decreasing from about a third of the body-length in the early types to a sixth in the latest, on a total length of fifty to seventy millimetres. A low ridge, totally absent at first, appears and grows slowly higher along the hinder part of the upper surface. The grooves from which the tube-feet emerge grow longer, and their surface, smooth at first, becomes sculptured. The mouth gets more and more overhung by a protruding lip of the hard skeleton; the little knobs on the skeleton become in some regions gradually more prominent. There are other changes, but these are the most obvious.

Opinions differ as to the number of separate species into which Micraster should be divided during this period. Since the series is continuous this question can only receive an arbitrary answer; but it is worth recording that the most conservative estimate is half a dozen.

The changes are infinitesimal, both in extent and biological meaning, compared with those in the horses. The explanation doubtless lies in the fact that at the beginning of the chalk period the sea-urchins were already an old and well-differentiated stock. Even the earliest Micrasters were highly specialized for life in sand or mud, and there was neither need nor room for any radical improvements. The mammals, on the other hand, were one and all primitive and small in early Cenozoic times; there was obvious necessity and opportunity for their improvement, and the horses shared in the great evolutionary movement which completely remodelled the whole mammalian stock between Eocene (V A) and Pliocene (V D).

Steadfast as it is, Micraster answers our present purpose, for though the changes involved are small, they are absolutely continuous, the urchins found at one level grading quite imperceptibly into those of the rest; a single specimen, indeed, may show characters of one "species" in some of its tube-feet grooves, characters of another in the rest. There is no question but that the Cretaceous sea persisted without any notable change in its conditions throughout the whole time, and that our urchins lived, died, and reproduced upon its bed in a continuous succession, so that the fossils from the lower layers were actually the parents of those embedded higher up. Slowly the race modified itself; by almost imperceptible degrees Micraster changed its shape. The evolutionary movement which Micraster demonstrates to us is only small and sluggish, but the demonstration which it gives is complete, with not a chink or loop-hole in it.

§ 5. "Missing Links."

IT WOULD be possible to give a number of other evolutionary series, almost or quite as perfect as those we have described in detail. The Titanotheres, strange, horned, extinct mammals, rival the horses in fullness of record. So do the camels. The tapir and rhinoceros branches are not far behind. An extraordinary parallel to the later development of the horse-stock is afforded by the evolution of the Litopterna, a group of South American animals, all now extinct. Although many characters prove that they were not horses, and not even closely related to horses, they responded to the world-wide change of vegetation in the same way as the horses did. Evolving from a different

FIG. 127. A Reconstruction of Archæopteryx in the Jurassic woods.

It must have used its wing-claws to scramble among the branches.

five-toed ancestor, they gradually increased in size, lengthened their legs, reduced their outer digits, and from three-toed became one-toed. They grew regular hoofs and their teeth became progressively longer and longer and of more complicated pattern. It is interesting that while their side toes were finally reduced farther than in any horse, to mere stubs of bone, their teeth remained less efficient, and never formed cement. Grassy plains developed in South as in North America, and since for long before the Pliocene (V D) the southern continent was wholly cut off from North America, no true horses could then invade it; but the Litopterna developed along a parallel line to fill the same niche in Nature.

The elephants, too, provide us with a fine evolutionary series, which is peculiar in that its trend was first towards the development of a four-tusked creature with long lower jaw. Later, however, with increasing bulk, the further elongation needed, if the animals were to continue rooting in the ground with their lower jaw, became mechanically impossible; and at the same time the head, with its great tusks, was too heavy to be borne on any but a stout and short neck. Accordingly, evolution changed its direction and pushed forwards towards a short, tuskless lower jaw, while the development of a trunk kept the animal in touch with the ground. And other series almost equally perfect can be found among various groups of carnivores. In reptiles, the fossil crocodiles gradually shift the internal openings of their air-passages farther and farther back along the palate, making it easier for the animal to breathe while holding prey under water in wide-open jaws. Moreover, series of smaller scope but greater continuity, like that of Micraster, are now available for certain star-fishes, lamp-shells, ammonites, pond-snails and other invertebrate forms.

Sometimes the fossil record is not so complete, but yet discoveries, though isolated, may be of startling interest as supplying the "missing links," as our grandfathers called them, between hitherto isolated groups. Typical of such linking types is the primeval bird, Archæopteryx. Two almost perfect skeletons of this creature are known, both from the Jurassic Period (IV B) in the middle of the Age of Reptiles. The rock in which their form is preserved, at Solenhofen, in Bavaria, is so fine-grained that it is used for lithographic stone and has retained the smallest details, down to those of the delicate feathers.

Birds, as was suspected even before the discovery of this missing link, are descended from reptiles; they are reptiles which have been specialized for an aerial life.

THE EVIDENCE OF THE ROCKS

In Archæopteryx we see the specialization in progress, incomplete. In a modern flying bird, for example, the reptilian fore-legs are turned sideways to serve as wings, and this has necessitated profound changes in their structure. Several originally distinct bones are welded together for rigidity, and the claws are lost (except in the young South American Hoatzin, which uses its wings to clamber about in the bushes). But in Archæopteryx, although the fore-limb is very certainly a wing, the welding of bones had not yet been brought about and there were still three well-developed and movable clawed fingers protruding from the wing, used presumably in climbing. Again, in a modern bird, the tail-skeleton is short, a stumpy little support for the fan of tail-feathers, but Archæopteryx had a long lizard tail with twenty vertebræ or more, and a row of large feathers on each side along its whole length. And instead of the horny, toothless beak of a modern bird, this winged, feathered lizard had ordinary reptilian jaws with a fine array of socketed teeth. Had it not been for the happy accident by which the feathers were preserved, it is doubtful whether the skeleton alone would have warranted us in definitely calling the creature a bird. As it is, Archæopteryx is in its general construction a perfect link between the two great groups of birds and reptiles, though more than half-way to modern birds.

FIG. 128. A bird with teeth like a reptile. The extinct diving bird, Hesperornis, as it probably appeared in life.

Its remains are found in rocks of the Cretaceous Age (IV C). It was over five feet in length.

Its wings were too small to sustain its whole weight, and so the hind part of the body had to be supported (as in almost all aeroplanes to-day) by a large tail-plane. Archæopteryx must have lacked the power of rapid, controlled flight that came only with stronger arm-skeleton and larger wings, which allowed the tail to be reduced and so permitted the bird to turn, check, and drop suddenly, instead of planing along and coming to grief below a certain speed as an ordinary aeroplane will do.

But there are other linking fossil forms in the bird stock. In the Cretaceous Period (IV C) birds have been discovered which were essentially similar to living birds in wings and tail, but still had teeth on the jaws (Fig. 128). Before the Middle Eocene (V A 2) these toothed birds had disappeared, and from thenceforward all bird fossils are of modern type. A somewhat similar example comes from reptiles. All modern tortoises and turtles are toothless and beaked like modern birds; but far back in the Permian Period (III F) there lived a tortoise, Eunotosaurus, with well-formed teeth like the majority of reptiles.

The mammals, too, are now linked by fossils with their reptilian ancestors. We knew already that mammals must have sprung from reptiles. Apart from all other lines of evidence, the discovery of those "living fossils," the Platypus and the Echidna, clinched the matter. Had we nothing but the skeletons of these animals, it would be very doubtful whether we should call them reptiles in the last stage of becoming mammals, or mammals which had just ceased being reptiles. But the fact that they nourish their young by a milky secretion, possess a coat of hair, and have a more or less constant temperature, stamp them as true mammals, even though their egg-laying habits and many other peculiarities show by how little they have crossed the boundary.

This is convincing proof; but all the same it is only indirect. However, in the Permian (III F) and Triassic (IV A), as the Age of Reptiles was dawning, a group of creatures existed, whose name of Theromorphs implies a likeness to mammals. Their teeth were already beginning to differentiate into incisors, dog-teeth, premolars and molars, and the whole form of the skull was approaching the mammal. They did not crouch on their belly, but ran with body lifted off the ground. But they still retained the little hole in the top of the head, the pineal foramen, below which, in many modern reptiles, the third or pineal eye is still to be found; this has been lost in all known mammals, with the transformation of the third eye into the pineal gland.

And they link reptiles with mammals in yet another way. One of the most important differences between existing mammals and all other vertebrates is the fact that in mammals the lower jaw consists of but one bone on each side, instead of several distinct bones stuck together. At the same time, there are three little bones in the mammal's middle-ear, transmitting the vibrations of its ear-drum to its organ of hearing, while in all other land vertebrates there is but one. It had been suspected for a long time that these two extra bones, the so-called "hammer" and "anvil," correspond to the two bones which in a reptile make the hinge-joint of upper with lower jaw; for in a mammal embryo before they are built into the machinery of the ear they are actually nipped off from this very region of the developing

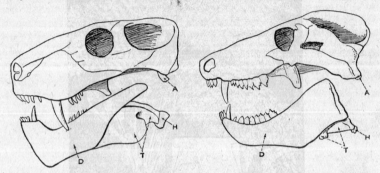

FIG. 129. Steps in the evolution of ear-bones from jaw-bones in the mammal-like Theromorph reptiles.

Left, skull of Scymnognathus from the Upper Permian (III F 3). Right, Cynognathus from the Lower Triassic (IV A 1). (D): the bone which forms the jaw in mammals. (A) and (H): bones which in most reptiles form the joint between upper and lower jaws, and in mammals become the little Anvil- and Hammer-bones in the ear. (T): the bone on which, in mammals, the eardrum is stretched.

jaws. And the tympanic bone, on which the ear-drum is stretched, can similarly be homologized with another of the multiple bones of the reptile's lower jaw. Now, in the Theromorphs most of the steps in this process can actually be traced. In the earliest fossils, the arrangement is like that of other reptiles. But as we go forward in time, the two hinging bones of the jaws gradually release themselves from this duty, grow smaller and are to be found in the region of the ear; from this condition, only a small step would be needed to convert them into ear-ossicles. Meanwhile the future tympanic bone also was becoming less concerned with biting and more with hearing (Fig. 129). Archæopteryx had just raised itself to bird status; the Theromorphs were still reptiles but were on the verge of climbing out into a higher stratum of biological society.

If we want an example of such a transition among the invertebrates, we can go to the brittle-stars. These Echinoderms are very like starfishes, but their central body or disc is much more sharply marked off from their wriggling arms. In all living brittle-stars the grooves along

Fig. 130. A vegetable missing link. Lyginopteris, one of the Pterido sperms or Seed-ferns, from the Carboniferous Period (III E).

A plant is shown bearing seeds on one of its fronds. Three fronds are expanded, the others still curled up. Below, on the left, a bit of a frond bearing seeds is shown enlarged; on the right, a young, unexpanded frond. The reconstruction was made by Miss F. Robertson. (From "Extinct Plants and Problems of Evolution," by Dr. D. H. Scott. Courtesy of Macmillan & Co., Ltd.)

the lower surface of the arms, so prominent in starfishes with the tube-feet arranged along them, have disappeared below the surface, roofed over to form tunnels; further, the main skeleton of each arm

THE EVIDENCE OF THE ROCKS

is a chain of little ossicles, beautifully jointed together. But in the Silurian (III C) and Devonian (III D) Periods there existed obvious brittle-stars which possessed open arm-grooves and arm-skeletons of much less elaborate construction: in both these ways they link the existing brittle-stars with the true starfishes. They were starfishes becoming brittle.

There are plant missing links as well as animal ones. The Pteridosperms, or "seed-ferns" which flourished during the latter half of the Paleozoic Era (III), are linking types of rather a different description. They appear not to provide a *direct* connection between seed-bearing plants and the true ferns, seedless and spore-producing; but rather to be a side-branch (like the Hipparion branch of the horse stock), which, while progressive in respect of seed-evolution, remained primitive in its general fern-like form and growth. But if the true link between ferns and seed plants must be sought in some type ancestral to both modern ferns and seed-ferns, none the less the seed-ferns decrease the gap (Fig. 130).

Once more, examples could be multiplied; but these suffice, since all we are concerned with here is to show that missing links turn up in the most diverse groups of animals and plants, and from all periods of the earth's long history. Steadily the gaps are filled and the ramifications of the tree of life mapped out with ever-increasing confidence and precision.

III

THE EVIDENCE FROM PLANT AND ANIMAL STRUCTURE

§ 1. *Structural Plans, Visible and Invisible.* § 2. *Vestiges: the Evidence of the Useless.* § 3. *The Evidence of the Embryo.*

§ 1. Structural Plans, Visible and Invisible.

IN THE previous chapter we have reviewed the geological facts which constitute the direct evidence for Evolution. They have shown us Evolution as actually taking place among living things: they have demonstrated that it has needed enormous lapses of time for its operations; that it is a gradual, steady process; and that it operates to produce, not merely progressive change, but also divergence and variety.

But although this transformation made visible in the past history of life is by itself sufficient to establish the fact of Evolution and to intimate something of its *modus operandi*, yet the indirect evidence must not be passed over. For one thing, it is impressive to see how each line of evidence confirms the same story; and for another, the indirect evidence throws light upon many of the facts and methods of Evolution which the direct evidence does not touch.

The first line of indirect evidence is comparative anatomy. This is often very similar to the direct evidence derived from fossil forms; in a number of cases the linking forms between groups are not wholly extinct, but a few of them linger on to the present day. We have already seen how the mammals and reptiles are linked by the duck-billed Platypus and its allies; worms and arthropods by the grub-like Peripatus; vertebrates and sea-urchins by Balanoglossus. So also the dog-faced but tree-living lemurs link monkeys with insectivorous mammals, the tailless apes are half-way in structure between monkeys and man, the lungfish help bridge the gap between fish and terrestrial vertebrates.

But the chief evidence from comparative anatomy comes from the broad study of structural plan. In Book II we saw that each of the great phyla of animals and plants is characterized by a common plan that underlies the construction of its members. That is fact. But why should it be fact? What is the sense of flying bat, swimming whale, burrowing mole, and jumping jerboa, all being built on one plan, while a wholly different plan runs through flying butterfly, swimming waterboatman, burrowing mole-cricket and jumping grasshopper?

Long before the time of Darwin, naturalists had recognized these underlying similarities of plan; Cuvier endeavoured to explain them by asserting that each main plan, or archetype as he called it, corresponded to an idea in the mind of God, who had rung changes on it in the process of creation. It is difficult to understand why only a small, definite number of archetypal ideas should have been thus divinely conceived. Why should God be limited in his ideas? And, further, as we shall see, the facts of embryology make that conception unacceptable. In any case, the idea of Evolution provides a more natural and much simpler presentation of the reality. If bats, whales, moles, jerboas, and the rest of the mammals were all descended from some common stock, then it would be *expected* that they should all show the same general plan, that they would start with that and vary from that to meet the demands of their distinctive ways of life.

What could be more different at first sight than whale's flipper, human arm, horse's foreleg, and bat's wing? Yet the skeleton of each is very plainly built on the same plan—a plan originally comprising one long bone in the upper arm, two in the lower, ten little knobbly ones in the wrist, and five jointed fingers to end up with.

This original model is distorted, cut about, modified. Sometimes one or two parts are enormously enlarged, like the two bones of the horse's fore-arm; sometimes parts are shortened and broadened, like the humerus, radius, and ulna of whales; sometimes they shrink almost to nothing, like the second and fourth toes of the horses, or wholly disappear, like their first and fifth toes; but the general plan remains as the common point of departure (Figs. 131, 134).

Even in limbs all serving the same function, the plan may be treated very differently and yet survive; in the wing of a bird, the bones of wrist and palm are fused into one solid mass, and only three tiny fingers are retained; the bat enlarges all the fingers except the first; while pterodactyls, the extinct flying reptiles, enlarged only the "little" finger; or again, the flippers of some among the ichthyosaurs, though at first sight very like those of a whale, achieved their paddle-like shape by a new kind of variation on the original plan; they broadened themselves by adding to the original number of fingers until these numbered seven or even eight. If all higher vertebrates, from Amphibia up, are descended from one common stock with a five-fingered hand, all these curious details are illuminatingly sane; if otherwise, they are incomprehensible (Fig. 132).

In precisely the same way, the jaws and mouth-parts of all insects conform to a single plan—the piercing, blood-sucking tube of the mosquito, the butterfly's coiled miniature trunk for sipping nectar, the house-fly's licking proboscis with its expanded lobes, the stag-beetle's formidable weapons of attack, the ant's chewing apparatus—all can be reduced to a simple plan such as is most clearly seen in a grasshopper or cockroach, with an upper lip, one pair of strong mandibles, and two pairs of weaker maxillæ, the second pair united to make a single lower lip (Fig. 133). Once more, if all insects are blood-relations, with bodies basically similar but specialized in divers ways to suit their diverse habits, the common plan of their jaw parts is easy to understand; if they are all separate creations, then only the supposition of a monstrous pedantry in creation seems to afford a glimmer of elucidation.

Examples could be multiplied almost *ad infinitum*—the appendages of lobsters, crabs, and other crustaceans; the teeth of mammals; the skull or the brain through the whole vertebrate series—all tell the same story.

A handful of different flowers gathered in a country walk would suffice for demonstration, if looked at searchingly with the eye that can pierce below the surface; for, as T. H. Huxley wrote, "Flowers are the primers of the morphologist; those who run may read in them uniformity of type amidst endless diversity of plan with complex multiplicity of detail. As a musician might say, every natural group of flowering plants is a sort of visible fugue, wandering about a central theme which is never forsaken, however it may, momentarily, cease to be apparent." Only descent seems able to explain that unity.

Of recent years fresh support for the evolutionary idea has been forthcoming from a new quarter in the proof of the chemical resemblances and differences between animals and plants. This is really evidence of the same kind which we have just been examining, save that the witness speaks a different language. The evidence still concerns resemblances in construction, but they are the invisible resemblances of chemical structure instead of the visible ones of anatomy.

Let us be a little more explicit about this new line of evidence. Many of the triumphs of modern medicine, as the reader probably knows, such as the antitoxin treatment of diphtheria, or preventive inoculation against typhoid and paratyphoid fevers, are based upon the fundamental principle that when any foreign substance belonging to the chemical group of proteins gets into the circulation, the body manufactures an "antibody"—a substance which in some way neutralizes

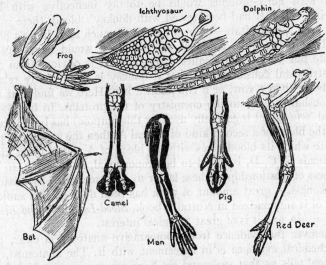

FIG. 131. The structural plan of the vertebrate's fore-limb is exemplified by that of man.

The general plan remains the same throughout the vertebrates, from amphibians up, though the details may be altered. In the Frog, the two bones of the lower arm are fused. In the Ichthyosaur, they are extremely shortened, and extra rows of finger-bones are added. In the Dolphin, two of the fingers are elongated; in the Bat, four. In the Pig, the thumb has vanished. In the Red Deer the second and fifth fingers are on the way to disappearance, and in the Camel only the third and fourth are left.

the foreign protein or puts it out of action; and the antibody acts with full force only against the particular protein introduced, with less force against proteins of similar chemical structure, and not at all against the rest.

Now, one of the chief ways in which the foreign protein may be put out of action is by precipitating it in solid form. If, for example, a rabbit be poisoned by a few injections of horse's blood it produces an antibody which gives its own blood the power of precipitating the blood of horses. If measured quantities of the bloods of treated rabbit and of a horse be mixed in a test-tube, a cloud appears and settles to the bottom—the horse-proteins have been eliminated.

If, however, the same amount of the treated rabbit's blood had been mixed with blood from a hen, there would have been no precipitate, not even a trace of cloudiness. The rabbit antibody which was efficacious with horse-proteins would be totally ineffective with hen-proteins. But if it had been mixed with donkey's blood, there would have been a precipitate—only not quite so much as if horse's blood had been used; while cow's or sheep's blood would have given a definite but very much smaller precipitate. It is obvious from general structural considerations that a donkey is more nearly related to a horse than a cow, or a cow than a hen. Here we find that this relationship extends to the chemistry of the proteins. In brief, after animal X (a rabbit is generally used for this purpose) has been treated with the blood of a second kind of animal A, then the amount of precipitate which its blood gives with the blood of A and of other kinds of animals B, C, D, is found to be proportional, so to speak, to the closeness of relationship of these latter to A, as measured by anatomical likeness. A great amount of work has been done on this subject, much of it summarized in Nuttall's book, *Blood-Immunity and Blood-Relationship*, and it is of great technical interest.

Wherever the evidence from comparative anatomy is clear, this new chemical evidence is in agreement with it. The anatomist, for instance, tells us that seals and sea-lions are carnivores which have taken to life in the sea: and their blood-proteins are chemically more like that of dogs, cats, and bears than of any other creatures. The anatomist puts man in the same group with the apes and monkeys, and tells us that he is more like apes than tailed monkeys, more like tailed monkeys than lemurs, and more like any of these than he is to all the rest of the mammals. His very blood-proteins reinforce this conclusion. The blood of a rabbit previously inoculated with human blood gives a heavy precipitate when tested with chimpanzee's blood, less with a baboon, or an organ-grinder's monkey, still less with a South American spider-monkey, but next to nothing with any animal from other mammal groups.

Occasionally anatomy gives a dubious verdict; and then the blood-test may throw fresh light. The whales, for instance, have points of resemblance both to carnivores and to ungulates, and anatomists have hesitated between the two alternatives. Blood-tests seem to show that they are more akin to the ungulates—a valuable classificatory indication.

PLANT AND ANIMAL STRUCTURE

FIG. 132. The fore-limbs of three flying vertebrates, showing different modifications of the same structural plan.

Above, a Pterodactyl. The wing-membrane is stretched on the enormous "little" finger, three other fingers are left as claws. In the centre, a Bat. Only the thumb is left as a claw; the other four fingers are used to spread the wing on. Below, a Bird. The thumb is a mere vestige, and besides this only the second and third digits remain.

We have spoken so far only of animals; but antibodies are produced equally well against plant-proteins, and the method can be, and has been, successfully used with plants. The rabbit can be injected with an extract of some plant tissue, and its blood later mixed with similar extracts from other plants. The results, from plants equally with animals, can be summed up in a sentence: likeness of chemical plan goes hand-in-hand with likeness in anatomical plan. Likeness is intrinsic and touches every aspect of the living thing. Living things resemble or differ from each other in thread and texture as in plan and form. This falls in with the idea of Evolution, but it is reasonless on any other assumption.

§ 2. Vestiges: the Evidence of the Useless.

THERE are certain facts of anatomy which have proved not merely difficult, but impossible to explain on any other assumption than that of Evolution. These are what are called vestiges—organs which are useless to their possessor, but resemble and correspond to useful organs in other creatures. Such organs are often loosely called rudimentary organs; however, since they seem definitely not to be the beginnings of something better, but rather to represent the ruins of past usefulness, it is better to style them vestigial.

Perhaps the most striking vestigial organs are the legs of whales. Whales have, in their flippers, well-developed fore-limbs; but externally they show no trace of hind-limbs. However, if they are dissected, one or two little bones are to be found embedded in the flesh in the region of the hind-limb. In some whales, a pair of long rods is all that remains, representing the vestige of the limb-skeleton, while the limb is altogether gone; in others, the vestige of the hip-girdle has a vestige of a thigh-bone attached. These bones are wholly useless—there is no trace of limbs for them to support, and they have not been turned to other uses. If we believe in the special creation of each kind of whale, or even of whales as a group, we must confess that these limb-vestiges spell nonsense. But if whales have evolved from land mammals, their presence is not only natural but full of significance. Leviathan we realize is not a perfect, immaculate whale, made as a whale and as nothing else, but the descendant of a land animal doing its best to swim.

Very similar vestiges of limbs are found in some snakes. No snake has any trace of a fore-limb, and most lack hind-limbs, too. But in the boas, pythons, and one or two others, vestiges of hip-girdle and hind-limbs are to be found. Sometimes these seem to be wholly useless, while in other cases, although they have no use as limbs, they protrude as two claws, which doubtless serve some new if minor function. If, as their general anatomy indicates, snakes have evolved from lizards, these vestiges make sense; without the background of Evolution, they are inexplicable.

PLANT AND ANIMAL STRUCTURE 363

The Duckbill Platypus has no teeth; like a bird, it uses a horny bill instead. But it is a mammal, and on evolutionary principles must have descended from toothed forbears. We look for evidence on this point—and there the evidence is, carried about by the baby Platypus in the shape of teeth which, though they are complete with dentine and enamel, never cut the gum. They have no function whatever, save that

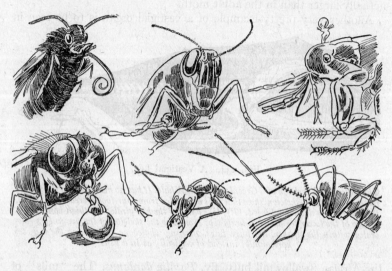

FIG. 133. Six insect faces.

Above: A Moth (Poplar Hawk Moth); a Locust; a Flea. Below: A House-fly, sucking a drop of liquid; an Ant; a female Malarial Mosquito, with the parts of its proboscis separated. The mouth-parts, whether built for sucking, licking, chewing, biting, or piercing, are all built on the same general plan—unpaired upper lip and three pairs of jaw-appendages. The last two pairs usually bear little feelers or palps, as shown in the locust; one pair of these is seen in moth and fly.

of reassuring the Evolutionist: and the same is true of the whalebone whales, for here also the embryo develops teeth and then changes its mind and absorbs them again, all before it is born.

The vestiges of toes preserved to us in the horse's splint-bones have no sense save an evolutionary one; and here the fossil record clinches the matter by showing that this sense is the true sense.

In the common Vapourer moth, which of recent years has become such a pest to trees in London parks, the females are wingless. Where they burst out of their cocoons, there they stay, and are there sought out and fertilized by the winged males. Most moths, of course, are winged in both sexes; thus on the theory of Evolution we should expect that the wingless female Vapourer had arisen from winged ancestors. That she has done so is shown by the fact that she still bears the vestiges of wings—mere buds, wholly useless for any purpose; and, interestingly enough, in the chrysalis stage the wing-rudiments are actually larger than in the adult moth.

Another very pretty example of a vestigial organ is to be seen in

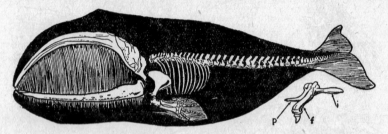

FIG. 134. A Vestigial Leg.

The outline and skeleton of a Greenland Right Whale. It has no hind-fin, but the useless remains of a limb-skeleton show that it is descended from four-footed creatures. Below, the vestiges are drawn enlarged. (P), (i), parts of the hip-girdle; (f) thigh-bone with a vestige of the lower leg attached to its tip. Note also the skeleton of the five fingers concealed within the flipper; the huge jaws carrying whalebone plates; and the tail, set horizontally instead of vertically as in a fish.

the African Swallowtail butterfly, *Papilio dardanus*. The "tails" of Swallowtails are prolongations of the hind wings; and in order to make room for these during the resting stage between caterpillar and butterfly, their chrysalis-cases possess special little pockets into which the tail-rudiments project. Now, *Papilio dardanus* differs from most Swallowtails in being tailless in the female sex; for the females, in shape, colour and pattern, mimic other butterflies which happen to enjoy immunity from attack by most enemies. But the chrysalis-cases of the females possess tail-pockets just like those of the males, although they are obviously useless and not to be explained, unless we suppose that *dardanus* is descended from an ordinary Swallowtail with tails in both sexes.

Sometimes one only of a pair of organs becomes vestigial. Female birds have only one ovary and oviduct—the left—doubtless to provide against the accidents that might occur if two large and brittle eggs were to knock about simultaneously in their insides; but the right ovary and oviduct are always present as miniature and useless vestiges.

PLANT AND ANIMAL STRUCTURE   365

The reptiles, from which birds undoubtedly have evolved, possess a pair of functional ovaries and oviducts: so that the presence of the vestigial right set of organs is perfectly intelligible to the Evolutionist.

Vestiges may, of course, also occur in plants. The well-known Butcher's Broom (*Ruscus*) gives us an example. In this plant, what appear to be the leaves are really flattened-out stems, as is shown by the fact that on them are born the flowers, and by other anatomical details. This curious arrangement is an adaptation to a dry soil; the leaf-like stems are flattened vertically, instead of horizontally like ordinary leaves, and are accordingly not so much heated as leaves would be, and so lose less water from their pores. However, leaves are not absent in the Butcher's Broom; they are still to be found, but only in the form of vestiges, mere scale-like organs below the leaf-like stems and the flower-stalks. These contain very little chlorophyll, and in any case soon wither and fall off, so that they are quite useless for the leaf's prime function of food-manufacture.

Very similar vestiges of leaves, though often still more reduced in size, are to be found in many of the cactuses and prickly pears, which, too, have taken on the function of food-manufacture by their thickened stems, and for the same reason of economizing water.

In flowers, too, vestigial organs may be found. The flower of the

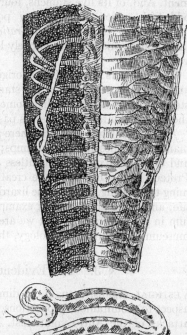

FIG. 135. Another example of a Vestigial Hind-limb in vertebrates.

Below, on Anaconda. Above, a view of the region of its vent from beneath. On the right the surface is shown, with a single protruding claw, sole external remnant of the hind-limb. On the left the skeleton is drawn, with the vestigial skeleton of the hip-girdle and hind-limb lying outside the ribs, and not connected with the backbone.

common figwort, *Scrophularia*, has changed from its original five-rayed symmetry, so common among flowers, to a bilateral arrangement. And, of its five stamens, four are grouped in two pairs, and the odd fifth never develops any pollen; it is purely vestigial, and quite useless. In the related *Gratiola*, only two of the stamens produce pollen, one has been entirely lost, and two have been reduced to vestiges.

These flowers have lost their original, perfect, five-rayed symmetry; and in the process some of the stamens have become useless. But instead of being discarded they sometimes linger on as vestiges to help the botanist unravel the plant's past evolution.

And so we might proceed. There are the sightless vestiges of eyes in many cave-fishes and cave-shrimps; the feeble vestiges of wings in kiwi and dodo, and in many flightless insects; the poor useless limbs of Proteus, and other newt-like creatures, which have preferred swimming to crawling, and of some lizards which have taken to a burrowing life; and there are a host of examples which man carries about with him in his own person. But we are reserving for a later section that museum of evolutionary biology, the human body.

§ 3. The Evidence of the Embryo.

VESTIGIAL organs, actually so diminutive, swell in their theoretical aspect to mountainous proportions, forming impossible barriers to the attacks of the Anti-Evolutionist. But obstacles almost or quite as formidable await him in the facts of embryology.

About a hundred years ago, von Baer, the great embryologist, omitted to label some specimens of embryos which he had put away in spirit. When he came to examine them later he found—but we will quote his own words:—"I am quite unable to say to what class they belong. They may be lizards, or small birds, or very young mammals, so complete is the similarity in the mode of formation . . . of all these animals." Thinking over this he came to formulate a general law—that animals resembled each other more and more the farther back we pursued them in development. This law in general holds good, and this resemblance of embryos or larvæ is a very striking fact, very difficult to explain save on evolutionary lines. A child of two can tell a pig from a man, a hen from a monkey, an elephant from a snake. But these animals are only easy to tell apart in the later stages of their development. When they were early embryos, they were all so alike that not merely the average man but the average biologist would not be able

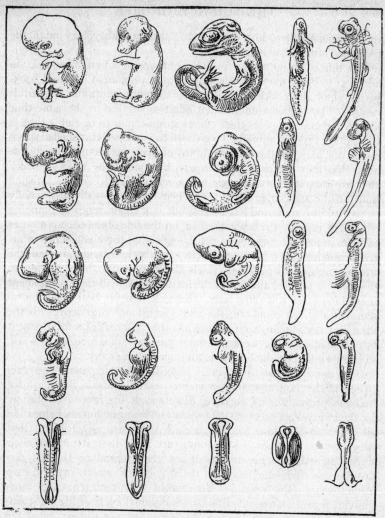

FIG. 136. A collection of vertebrate embryos.

Each upright column represents the development of a single type—left to right, Man, Rabbit, Lizard, Newt, and Dogfish—the earliest embryos being below and the latest above. Note that the early stages are very like each other, and that the animals diverge as they develop. In the first stage the nerve-folds are closing in to form the brain. Then the gill-clefts appear. In the land animals these close later; in the newt and dogfish feathery gills appear. The human tail and its gradual shortening are clearly seen.

to distinguish them, and even a specialist in embryology might be pardoned a mistake (Fig. 136).

But this is by no means all. The embryos of different animals, in addition to being more like each other as development is traced backwards, show also a widening contrast with their parents and their adult destiny. They become unlike their adult selves, but at the same time and in the same respects their construction comes to resemble that of quite other types of animals. To go back to von Baer's unlabelled specimens, not only are the early embryos of man, cat, hen, and snake so alike that they are hard to tell apart, but one of the ways in which they are alike is in having their heart, main arteries, and neck-region built on the same plan as in fish. Their heart is not divided, wholly or partly, into right and left halves, but is a single series of pumping chambers, just like the heart of a fish; on the side of the neck is a series of clefts in just the position of a fish's gill-slits; there is a series of arteries running down between the clefts just as in fish; and, indeed, the whole arrangement of blood-vessels and nerves and their relation to the clefts is piscine, and not in the least indicative of the arrangement they themselves will show later. These clefts never bear gills; the resemblance is not complete; the reader must not run away with the idea that the human embryo ever has the "gill-slits of a fish." But it has a transitory rude passage through that type of structure. It is not, so to speak, a reproduction; it is an imperfect memory.

Once more this means nothing—indeed, makes nonsense—if we are to believe that land animals were created as land animals. But it at once becomes pregnant with meaning if we accept the fact of Evolution, for then we can understand that snakes and hens and human beings and all other air-breathing vertebrates are fundamentally fishlike, that they start on the fishward road and turn away from it towards their higher structural achievement. When they reproduce the old disposition asserts itself; they start towards the old water-way and turn aside towards the uplands. Because of that recurrent urge each individual animal repeats within its individual cycle of life these uneffaced tendencies from the remote part of its race. In Amphibia the recapitulation is much more thoroughgoing; they have not only the clefts but the gills, and most of them actually do breathe by means of gills in their early tadpole stages, and physiologically are indeed fish.

Nearly half a century later, Haeckel, looking at the facts of embryology in the light of evolutionary ideas, broadened and reformulated and perhaps rather exaggerated von Baer's law. Haeckel's revision was this, that every animal in the course of its individual development tends to recapitulate the development of the race; and from this time onwards the facts on which the law is based have been called the facts of recapitulation. But it is a general and not a complete recapitulation. Evolution can affect every part of a life-cycle, and if a stage wastes much time or energy, Nature, who is no historian, will abbreviate it or cut it out quite ruthlessly.

Exactly how far Haeckel's law takes us—what are its limitations; whether recapitulation ever shows us an animal's adult ancestors; whether present development ever does more than recapitulate ancestral development; what is the cause of recapitulation; and why some of the characters and structures are regularly recapitulated in development, others only occasionally or not at all—all this we cannot here discuss. What we are here concerned with are the positive, visible facts. Tens of thousands of animals do recapitulate the past during their development—do, without any apparent advantage in so doing, show organs and constructions which occur elsewhere in the adults of less specialized creatures: and in none of these tens of thousands of cases is this departure intelligible save on the view that in so doing they are repeating phases that were once final forms in the earlier evolution of the race.

There is probably no single case of development among many-celled animals which does not show some recapitulatory feature. Even the origin of all sexually produced individuals in a single cell, the fertilized egg, in a certain sense recapitulates the origin of many-celled from single-celled forms. But there are plenty of more definite examples. Every human being and every other vertebrate at a certain stage of his or her development has an unjointed notochord in place of the future jointed backbone, even though in all save the lampreys and certain fish this notochord-rod vanishes entirely. And the most primitive chordate, Amphioxus, has a notochord all its life long and never develops a backbone.

We have already spoken of the extraordinary transformation that overtakes the young sea-squirt. Here the sedentary adult animal, bearing only faintest indications of its real relationship, passes its early life in a tadpole-like form which shows all the salient points of the chordate plan—the notochord, the tubular nervous system along the back, and the gill-slits. If the gelatinous sea squirt is not a degenerate chordate, why should its larval form be so entirely chordate? What creative idea is served by this hesitation in development?

A much less all-pervading example, but none the less a very pretty one, we have already cited in the early horse-stock. The three-toed

Merychippus had an elaborate grinding pattern on its permanent premolar teeth; but the pattern of the milk-teeth was simpler, like the pattern of the permanent teeth in geologically earlier and more primitive horse-ancestors. Here we can actually put side by side the adult

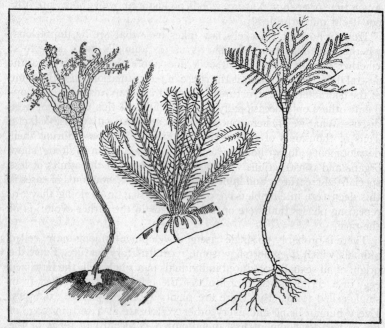

FIG. 137. Ancestral reminiscence in the Feather-star's life-history.

The adult Feather-star (Antedon) can swim from place to place by waving its arms, or anchor itself temporarily by a set of jointed claws, as the specimen in the centre is doing. On the right is an adult Sea-lily (Rhizocrinus) from a depth of about a hundred fathoms in the North Sea. This animal is permanently fixed by a stalk which ends in a root-like tuft. On the left, a young Feather-star in the stalked stage. After growing for some time in this form, the animal breaks off and swims away, leaving its stalk behind. Thus the Feather-star passes through a fixed stage which is like the final form of its more primitive relatives, the sea-lilies.

ancestral form and its young descendant recapitulating it before passing on to the more highly evolved structure.

A beautiful example of recapitulation comes in the life-history of the common feather-star. This is an echinoderm belonging to the crinoid class. The great majority of the class spend their adult lives rooted to the bottom; they are the stalked sea-lilies which wave their graceful, branching arms far down in the abyss, and in the past they were so abundant as to have built up whole layers of rock with their skeletons. From such forms as these the freely swimming feather-stars have evolved. But the feather-star, too, begins its adult existence with a fixed stage. For some weeks it lives and grows rooted to the bottom like a sea-lily. Only later does it abandon this sea-lily guise, break off from its stalk (which is left to die) and become a free-creeping feather-star (Fig. 137).

An interesting case among plants of the preservation of phases of the ancestral life-cycle which are now unnecessary is that of the maiden-hair tree or ginkgo. This lovely tree, with its leaves like little fans, is extinct in the wild state, but has been preserved to us by being cultivated as a sacred tree in the gardens of Chinese temples. It is a naked-seeded plant, related to the pines and firs. As in all other seed-plants, its fertilization is effected by means of pollen. The pollen-grain, as we have already explained in Book II, if it alight on the pistil of a female flower, sends down a long tube towards the egg-cell embedded in the ovary. From its original single nucleus, three nuclei are produced by division; two of these nuclei act as male gametes, either of them capable of fertilizing the egg. In almost all seed-plants these male gametes are merely more or less ordinary nuclei which pass down the tube to the ovum; but in the ginkgo and the cycads the nuclei, associating themselves with some of the surrounding protoplasm, become transformed into two actively swimming, ciliated sperms, like the sperms of fern, moss, or seaweed, which swim on within the tube to fertilize the egg (Fig. 112). They would get there just as surely, we judge from the higher plants, if they had no cilia. These tiny sperms are as revealing as the gill-slits in our own embryonic neck. They show that once upon a time, a very remote time, the ancestors of seed-plants lived in the water, where free-swimming sperms provide the natural method of achieving fertilization. And here, long after the pollen-tube had been evolved to provide a dry method of fertilization for dry-land plants, the motile character of these transitory sperms survive to recall the watery past. Even in some much more modern plants, like the sunflower, though cilia are no longer formed, the fertilizing male nuclei still assume corkscrew shapes in recapitulation of the coiled, swimming sperms of many lower plants.

Two final examples we may note, one from fish, the other from Crustacea. Everybody knows what soles and plaice and other flat-fish look like, and a great many people are aware of their peculiarity in having both their eyes on the one side of their twisted head and in lying flat, not on their belly, but on one side, the opposite side to the eyes. But not so many know that when they hatch out of the egg, flatfish are symmetrical like other fish, swim about in the ordinary upright position, and have their two eyes on different sides of their head like any ordinary vertebrate. It is only after several weeks that a

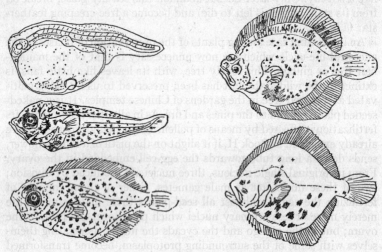

FIG. 138. A Turbot recapitulates its symmetrical past.

First, a newly hatched larva, symmetrical, with organs incompletely developed, and a large yolk-sac. Next, about one-eighth of an inch long, symmetrical, independent, swimming upright. Next, one-fifth of an inch long; the right eye is beginning to shift upwards. Above, right, four-fifths of an inch long; the right eye has grown nearly round to the left side, the animal swims almost flat instead of upright. Below, right, fully transformed; the fish spends most of its time lying on the bottom.

symmetrical growth distorts the head and eyes and the animal gradually settles down to its sideways existence. This is natural enough in the case of a fish which was once free-swimming like other fishes but which, generation by generation and age by age, has taken to life upon the bottom, but it is fantastic if we suppose that the plaice and soles were specially created as they are now. Why should they not spawn themselves in miniature?

The most startling example has been reserved to the end. Portunion is the name given to a repulsive parasite found in the gill-cavities of crustaceans, where it devotes all its energies to sucking its host's blood and maturing its own reproductive cells. In the female sex especially this creature looks much less like a whole animal than like a detached piece of somebody else's internal anatomy. There is nothing to show what sort of a creature it really is, from what kind of animal it has degenerated. But in its development it lets the cat out of the bag—or perhaps we should say it lets the sea-woodlouse out of the bag; for when young it is obviously a crustacean, the details of whose anatomy place it at once in the order which contains the familiar slaters and woodlice. In Book II, Chapter 2, we traced a similar story in the parasite Sacculina, which is shown by its larva to be a curiously disguised and degenerate barnacle.

The evidence of the fossils in the rocks is direct evidence for Evolution. But the evidence of embryology, though indirect, is more immediate. You can watch the individual animal indulge in these amazing reminiscences, and pass almost before your eyes from ancestral primitiveness to adult modernity. All the facts have a simple and straightforward meaning if Evolution be a fact, while a denial of Evolution leaves them unexplained and apparently inexplicable. It is plainly essential to a parasite like Sacculina that it should have a free-swimming larva, but it is hard to see why that larva should be built so exactly upon the crustacean plan if Sacculina has no crustacean ancestry. And a colt, safe in its mother's womb until it is born a well-nigh complete miniature of its parents, has no sort of advantages in preceding its backbone by a notochord or producing and reabsorbing gill-arches. Nor is it some mysterious, widespread harmony which requires this rhythm of repetition, because the amount of recapitulation varies with different animals and the story is often blurred and abbreviated. Recapitulation occurs like something done under a powerful and unavoidable inertia of tradition, like something deep in the nature of living creatures. They recapitulate because they reproduce and because they have been evolved through an infinite series of reproductions from simpler things.

IV

THE EVIDENCE FROM THE VARIATION AND DISTRIBUTION OF LIVING THINGS

§ 1. *The Variability of Living Things.* § 2. *What Is a Species?* § 3. *The Distribution of Living Things.* § 4. *The Evidence Summarized.*

§ 1. The Variability of Living Things.

THE rocks tell us that the forms of living things have changed slowly but steadily in the past; careful comparison of the structure and mode of development of creatures living to-day is in accordance with this fact. We will now open up a new series of facts that harmonize and complement those that have gone before. If Evolution is the form of life's process, it must still rule life. Life must still be evolving to-day. Is that so? We do not see striking metamorphoses happening; Evolution is an extremely slow process, its changes in the case of the slower-breeding organisms take hundreds of thousands of years to accomplish, and it would be strange indeed if any profound alteration in the form of a living thing had occurred in the couple of hundred years during which animals and plants have been carefully and systematically observed. To expect rapid changes of this kind is like expecting visible movement in the hour-hand of a clock. Nevertheless, we can detect slight changes in progress, sufficient to convince us that Evolution still continues.

One of the clearest and most striking proofs of the plasticity of living things is the extraordinary variability they display under domestication. Consider, for example, the dog. Here we have an enormous assemblage of forms, the extremes differing from each other far more strikingly than many natural species, ranging in size from the St. Bernard and the Great Dane to the toy Lapdogs, in proportion of parts from the slender-limbed Greyhound to the low-hung Dachshund, from the long-nosed Collie to the snuffling Pekinese, and showing an enormous variety of colours and coat-patterns. Yet they seem to be all

VARIATION AND DISTRIBUTION

of one kind; they recognize each other as like creatures, and if the physical disparity between them is not too great they breed freely together. In a word they are all dogs. They show to what an extent a living form may vary.

There are plenty of similar instances among domesticated animals. Compare, for example, a cart-horse, a race-horse, and a Shetland pony; a carrier-pigeon, a tumbler, and a pouter; or the multitudinous fancy breeds of rabbits and guinea-pigs and mice. And consider also the enormous richness of varieties that is found in cultivated plants—in roses or primulas or cereals.

It is unfortunate from the point of view of evolutionary science that the mode of origin of most of those special domesticated breeds has not been recorded. They are in many cases very ancient. There were domesticated dogs in the Bronze Age cave and lake-dwellings of Central Europe; in Egypt there were several distinct breeds, including a greyhound, as early as 3000 B.C.; a dog very like the St. Bernard appears on Assyrian bas-reliefs. Nevertheless, it is clear that man has been the primary cause of this extraordinary divergence. He has kept and bred from those animals which best suited his fancy, and he has drowned or starved or given away the others. So he has gradually moulded the breeds.

FIG. 139. Two fully adult dogs, a Saint Bernard and a Toy Black-and-Tan Terrier, affording an extreme instance of variation within an interbreeding group of animals.

There are plenty of cases in which the gradual changes have been recorded. The greyhound of to-day is more slender-legged than the greyhound that appears in Egyptian paintings; a specialist in hunting hares, he is smaller and lighter even than his Elizabethan ancestor, which was sent after deer and all sorts of game. And the bull-dog has been modelled about almost like a lump of plasticine. He was bred first for the special purpose of bull-baiting—hence his short, stocky build, which enabled him to dodge the swing of the bull's horns, his underhung jaw and strong gape, and his method of attack, which is to come boldly from the front, to seize the muzzle and hold on instead of dancing in from the rear and slashing after the manner of deerhounds and other elegant dogs. When bull-baiting was made illegal in England the breed was kept on as a curiosity and was changed from a fighter into a caricature; its face became so short that it could hardly breathe through the nose, its legs were absurdly bowed and twisted, while its general physique and stamina deteriorated. To a large extent this change for the worse was due to selective breeding, although that process was assisted by such tricks as keeping the puppies in cages so low that they could barely stand up, to exaggerate the curve of the legs, and keeping them in harnesses to hollow their backs. Recently there has been a reaction in the other direction, and the bulldog is becoming less grotesque. Thus, now in one direction and now in another, the form of the breed has been moulded.

Presumably all of the domesticated animals and plants have been derived by the selective breeding of wild species. The various sorts of domesticated pigeons have all been evolved from the wild rock pigeon, *Columba livia;* and an astonishing diversity of domesticated forms has been produced. But it is not true that in all cases the races of a domesticated animal spring from a single wild ancestor. In dogs it is probable that more than one ancestral species have been involved. There are several kinds of wolves and jackals, which will breed freely with our dogs, and many of these can be tamed. It is rather the rule than the exception that the tame dogs of any region carry an obvious dash of blood from their local wild kindred. The highly civilized countries are exceptional in this respect, for we select and mould our dogs to an extraordinary degree and we destroy our strays. But from Central Europe, across Asia, and down through Africa the common dogs of the people show evident resemblances to the local kinds of wolves and jackals; and it is the same in America, for the Eskimo dog is like the North American wolf and the Hare-Indian dog is like the coyote. Apparently the habit of domesticating dogs is ancient and widespread and it has entangled a number of different species into one interbreeding complex.

But whether a domesticated race involves one species or many, the essential fact is this—that divergence of form has been brought about by the selection of variations, and that such variations do occur in the wild forms from which our tamed or cultivated races have been derived. Indeed, it is not possible to study intensively any single group of

VARIATION AND DISTRIBUTION

animals without finding abundant variations of type. Living tissue is always groping about in search of improvements. Naturally enough we do not find such *outré* creatures in the wild as we do in our kennels or

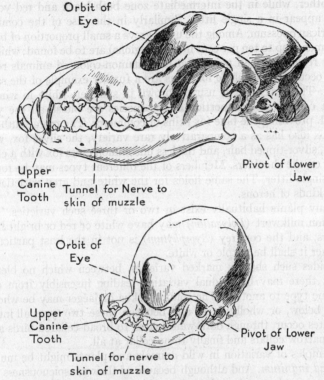

Fig. 140. The skull of a King Charles' Spaniel (below), contrasted with that of a primitive, wolf-like pariah dog (above).

Both are drawn from the left side and to the same scale. One or two points are labelled correspondingly in the two skulls to bring out more clearly how the nose of the fancy breed is shortened.

farmyards or gardens: we do not find pug-dogs wheezing through the woods after their prey or ravenous cart-horses scouring the plains for grass to build up muscles quite unnecessarily large for their own needs. These are creatures fitted to man's special requirements, for his amusement or for his service, and they would be unable to survive or breed in a wild state. But we find plenty of variations among wild species all the same.

For example, our common European squirrel exists in two varieties, red and black. In some regions only one kind is found, in others only the other, while in the intermediate zone both black and red young may appear in a single litter. Similarly in the case of the common American opossum. Among the usual greys a small proportion of black specimens (up to ten per cent in some regions) are to be found; while in other regions a small percentage of cinnamon-coloured animals regularly occur. The valuable fox skins are a further example of the same thing. The Arctic fox is usually brown in summer, white in winter; but a considerable proportion of the species are bluish all the year round, providing the blue fox skins of commerce. The more southerly red fox also has as a comparatively rare variety, the silver fox, with black, silver-tipped hair, and the less uncommon cross fox with a cross mark on the shoulders. Members of the different types may be found in a single litter. The same holds for the white and grey varieties of some kinds of herons.

Many plants habitually exist in two or three such varieties. The common milkwort (*Polygalum*) may have white or red or bright blue flowers, and the comfrey (*Symphytum*) is not in the least particular whether it shall be purple or white.

Besides such sharply marked varieties, between which no blends occur, there may be gradual variation leading insensibly from one extreme type to another. The common skua or jaeger may be wholly white below, or wholly dark; and between these two types all intermediates occur, through dark-waistcoated to broad-collared birds and so to narrow collars and finally to no collar at all.

Examples of variation in wild animals and plants might be multiplied *ad infinitum*. And although because of their conspicuousness we have confined ourselves here to colour-varieties, equally marked variations are found in other points, in shape and size, in the internal organs, even in such invisible physiological characters as disease-resistance or longevity. All kinds of living things vary; the differences are merely a matter of more or less, some varying more strikingly than others.

§ 2. What Is a Species?

IN THE foregoing chapters we have made frequent use of the accepted classification of living things; it is a chart that shows us our way about the Animal and Vegetable Kingdoms and the indeterminate zone between them. But we have said nothing about the way in which that chart has been drawn. The classification of living things is a laborious task; it involves the collection of great numbers of specimens from all parts of the world, and their careful, indexed accumulation in central museums where they can be examined and compared; it involves the full-time labour of collectors and of museum experts in the various groups. It will be worth our while to glance for a moment at the work of these systematists, as they are called, because the way in which living things resemble and differ from each other is in itself a strong piece of evidence in favour of Evolution.

Naturally we cannot examine the drawing-up of the whole map; we shall have to confine our attention to a corner of it. We shall speak only of animals in what follows, although the classification of plants is based on similar principles and entails similar difficulties, and we shall begin with one particular group of animals, the wild dogs.

Running wild and hunting over almost the whole land surface of the globe—excepting only New Zealand, Madagascar, the ice-caps at the Poles, and those intensively civilized regions that man has made wholly his own—there is a tribe of closely related animals, the wolves, jackals, and foxes—in biological language, the family *Canidæ*. The appearance of a wild dog, such as a wolf, is well known enough. He has a bushy tail, erect ears, a pointed muzzle, a shrewd expression, and sharp, cruel teeth. He is lightly built; the dogs are a family of swift runners with exceptional powers of endurance. In general they are carnivorous, i.e., they kill and devour living prey, and for the most part they run their prey down and do not pounce upon it after the manner of cats. But dogs are by no means particular in their diet; when fresh meat is scarce they are willing to take invertebrate animals, vegetable food, or carrion.

Now, there are a number of kinds of wild dog; they vary in size, colour, and proportions of the parts, in habits (some hunting by night and others by day, some hunting singly and others in packs), and so on. And they have to be classified. Their popular names—wolf, jackal, fox, and the like—are unsatisfactory, because they rest on the superficial appearance of the creatures and do not give any idea of their true relationships. It is the business of the systematist to classify these different kinds of dogs properly; he has to examine them and note how they resemble and differ from each other and give them unambiguous names which will define as clearly as possible their true relationships.

How then does he proceed? We can most clearly understand the process by first giving a somewhat idealized account, and then noting some of the difficulties that stand in the way.

The unit of biological classification is the *species*. If we find a number of animals, resembling each other and differing distinctly from all other animals, breeding freely and fruitfully together and recognizing each other as kin, then those animals constitute a species. In Britain, for example, the common fox is one species and the wolf, which was exterminated a few centuries ago, is another; in North America the coyote, the timber wolf, the common fox are examples of separate species. In this manner all the known Canidæ are grouped into species.

Now, just as individuals may be grouped into species by considering their resemblances and differences, so may the species be grouped into assemblies of a higher order. There is, for example, a large group of species—the wolves and jackals—that are obviously very like each other; there is another group—the foxes—which also resemble each other, but which differ in a number of respects from the members of the first. Each of these species-groups is called a *genus* (plural *genera*). The wolf and jackal genus is called *Canis* and the fox genus is called *Vulpes*. Sometimes a single species is so strikingly different from all other kinds of dog that it is put in a genus by itself; thus the long-legged, foxy-red "maned wolf" of Brazil and Paraguay constitutes the genus *Chrysocyon*. Usually, however, a genus contains a number of species.

In speaking of animals, the zoologist finds it convenient to give both the generic and specific names—very much as in speaking of human individuals, we often give both Christian and family names. In order to avoid confusion, certain conventions are adopted; the name of the genus is put before the name of the species; the name of the genus is written with a capital letter, that of the species with a small letter. Thus the common European wolf is called *Canis lupus*, the Indian jackal is *Canis aureus*, the common fox *Vulpes vulpes*, the long-eared fennec fox *Vulpes zerda*, the maned wolf *Chrysocyon jubatus*, and so on.

Carrying the classifying process a step farther, we group the genera of dogs together to form a family, the *Canidæ*. This is brigaded with a number of other families, with cats and bears and hyænas and weasels and seals and a host of others, to form an order, the *Carnivora;* this, with a number of other orders, forms a sub-class, the placental mammals, and so on, as we have already seen.

So much for the ideal. Now for the difficulties.

VARIATION AND DISTRIBUTION

We noted in the last section that the individuals composing a species are never exactly like each other. Even in the clearest, most sharply defined species the individuals show slight variations in their colour, size, instincts, and so on. In the common European wolf, *Canis lupus*, for example, animals from different localities vary in length of fur, the coat being thicker in northern wolves, and there is a tendency for the latter to grow to a larger size than the southern individuals. This particular case does not present a very puzzling problem to the systematist. There is no question of dividing the European wolf into two or more species, for the extreme types are connected together by a grading series of intermediate forms, all breeding freely together. They are simply variations within the limits of a single species.

But when we take into consideration all the wolves in the world, the problem of variation becomes more serious. The true wolf ranges over the whole of the northern hemisphere from the Arctic Circle to south temperate latitudes, and apart from such "sports" as complete blackness, which may turn up anywhere, the characters of wolves vary in accordance with the climate of the particular region they inhabit. Northern wolves, for example, are on the average larger than the southern and have a thicker and whiter coat, and in temperate zones the wolves inhabiting comparatively dry country are on the whole paler in tint than those from districts where the rainfall is heavy. And there are variations in build and proportion. Now, sometimes these differences are considerable enough to make it doubtful whether the forms should be grouped in the same species or not. There are, for example, the American timber-wolf, the pale wolves of North-Western India, the small, short-legged Japanese wolf, the little red wolf of Texas —these and many others, all fairly distinct from the common wolf of Europe, but just similar enough to leave us in doubt whether they should be separated or left together. Confronted with this problem, the experts differ among themselves. Some prefer to distinguish these various kinds as separate species; some include them all within the species *Canis lupus* and distinguish them as local races or sub-species. Moreover, this variation is not a canine peculiarity; the foxes show it, too, and indeed nearly every wide-ranging kind of animal or plant has its local varieties and therefore its problems for the systematist.

This sort of difficulty is not confined to the first step in classification. It attends also the formation of groups of higher grades. It is often very hard to tell whether species are sufficiently like each other to be put in the same genus, or whether genera are sufficiently like each other to be put in the same family.

It will make the point clearer if we take another example, not this time from the dogs but from their cousins the cats. As is well known, the lion, tiger, leopard, lynx, and the rest are all plainly related to the smaller cats, and until recently they have always been called different species of one genus; the lion, for example, was *Felis leo*, the tiger *Felis tigris*, the leopard *Felis pardus*, the tame cat *Felis catus*,

FIG. 141. Two freely interbreeding species of crows.

The Hooded Crow, Corvus cornix, is above on the left; the Carrion Crow, Corvus corone, is above on the right. The other birds are hybrids, showing the hooded pattern more or less distinctly.

and so on. But, as a matter of fact, the group is divisible into two subdivisions, on the one hand the lion, tiger, leopard, and jaguar, and on the other the puma, lynx, and all of the smaller cats; their most striking distinction is that, while only the former group can roar, only the latter can purr (limitations which have an anatomical basis). Here arises the first difference of opinion: Some authorities call the two groups sub-genera, while others call them genera with the names *Panthera* and *Felis* respectively. Now each of the commonly accepted species shows considerable variation, and here the second difference arises. The tiger, for instance, is big and long-haired at the northern end of its range, smaller and short-haired in India, very small indeed in Sumatra, unusually closely striped in Turkestan. Some authorities regard the various forms as local varieties or sub-species of the species tiger and denote them by adding a third name, the sub-species name, to the already elaborate titles—thus the Manchurian tiger is *Panthera tigris longipilis*, the Sumatran tiger is *Panthera tigris sondaica*, and so on. Others, however, consider that they are distinct enough to be called so many species, in which case the old tiger species is made a genus, the different kinds becoming *Tigris longipilis*, *Tigris sondaica*, and so on.

These differences of opinion, be it noted, are not signs of incompetence on the part of the systematists; they result from the nature of the facts, *which do not admit of categorical classification.* It is written that Noah collected "every beast after its kind, and all the cattle after their kind, and every creeping thing that creepeth upon the earth after its kind, and every fowl after its kind, every bird of every sort," and even nowadays most people share the delusion implicit in this passage—that if we could assemble together every individual beast and creeping thing and fowl that lives in the world to-day, and if we could examine and compare them all, it would be possible to group them with neatness and precision into definite kinds, or species. But that is a myth. What has been said of dogs and cats is true of other animals—and of plants and microscopic creatures as well. There exists every conceivable grade of difference between organisms. There are animals which are very much alike, so much alike that they are obviously members of the same species. There are animals which are so different that they are obviously not members of the same species. And there are animals just different enough to make one wonder whether they should be called different species or not—and this is where the nice problems arise. Authorities vary widely in the freedom with which they erect new species; some group animals together whenever there is any doubt at all, while others put two specimens in separate species almost whenever they can see a noticeable difference between them. In an extreme instance, the British brambles and roses have been classified as sixty-two species by one authority and as two by another of equal eminence.

For there is no absolute criterion of species, no feature that will stamp living things definitely as the same or different. It used to be thought that there was a distinction between creatures which would breed together and produce fertile offspring and creatures which would not; the former were members of the same species and the latter were not. But in fact there is every conceivable stage, in some organism or other, between mutual fertility and complete mutual sterility. At one extreme the males and females breed freely together, and their offspring are fertile; in this case they are generally called the same species— mankind will serve as an example. At the other extreme the male and female do not recognize each other as similar creatures; a lion and a cow will illustrate this. But in between those two extremes there are various links; there are cases where a male and a female are friendly together but show no desire to breed, cases where they come together

Fig. 142. Some Typical African Mammals.

Top left: Giraffes. Top right: Baboon. Centre (from left to right): Four kinds of Antelope —White-tailed Gnu, Eland, Oryx, Greater Kudu. Below (left to right): Aard-Vark; Cape Hyrax (Coney); Burchell's Zebra.

but are completely sterile, cases where they produce a few weakly or abnormal offspring, cases where they produce healthy but sterile offspring, cases where their offspring are healthy and fertile but their grandchildren are weak and unhealthy. Where in this series are we to draw the line? Which of the graded stages are sufficiently fertile to be called members of the same species?

Indeed, there are plenty of examples of animals and plants which are considered to be different species interbreeding freely in Nature. The case of the carrion crow and hooded crow is classical. These birds inhabit the greater part of Northern Europe and Asia, but the two species divide this vast area between them. The carrion crow (or common crow) is found in England and South-Western Europe, and also in the eastern half of Siberia. The hooded crow is found in North-Eastern Europe and from thence across to the middle of Siberia and down as far as Egypt. There are therefore two lines along which the species meet: the western boundary runs through Scotland, Denmark and the Elbe Valley to Northern Italy, while the eastern boundary runs down through Siberia. In these boundary zones the two species frequently breed together and produce every possible intermediate stage between the coal-black carrion crow and the hooded crow with its grey body and black head, wings, and tail. A case with specimens of the two species and their intermediates may be seen in the Natural History Museum, London. A similar example is afforded by the flicker, a common North American woodpecker. There are two perfectly distinct kinds of flicker, the common flicker on the eastern side and the western flicker on the west; they differ conspicuously in a number of points in their coloration. But along a broad zone running from British Columbia to Galveston the two kinds mix and interbreed freely, and in this boundary region every conceivable kind of intermediate is found. One could multiply similar examples indefinitely; one more must suffice. In East Africa there are two species of antelopes of the hartebeest kind which, until recently, inhabited different districts and also were always perfectly distinct in such points as the shape of their horns; but a few years ago they spread towards each other and began to interbreed—presumably some sort of barrier that had been keeping them apart broke down—and now there are all sorts of intermediates between the two. So that the old fertility-criterion of species can no longer be said to work.

All of this was very perplexing to biologists in the days when they believed living things to have been created in a fixed number of immutable kinds in the garden of Eden. The Creator, they thought, had made so many different species, and it was the business of the systematist to recognize and identify those species and to base his classification upon them. The great Linnæus, for instance, laid down as biological dogma that "the number of species is as many as the different forms created in the beginning." Even in the nineteenth century Cuvier, as Professor J. W. Gregory remarks, "believed that species are as distinct as the different makes of boots sent out from a factory."

But it came to be realized that in many cases it was extraordinarily difficult to recognize absolute distinctions of this sort; to tell, for example, whether there were both Sumatran and Manchurian tigers in Eden or whether there was one tiger which begat them both. The early systematists blamed their own incompetence. They believed in species, even where they could not recognize them, and they spent an enormous amount of time and energy in the quest of universal diagnostics of species (such as the test of mutual fertility that we have just discussed) and to devising definitions of species which would apply to every case. But the nineteenth century saw the passing of the idea of separately

FIG. 143. Some Primitive Mammals from the Australian Area.

Left, from above downwards: Vulpine Phalanger; Rabbit-eared Bandicoot; Wombat (see also Fig. 147). Center: Kangaroos; (Below): Spiny Ant-eater (Echidna). Right, from above downwards: Tasmanian Devil; Koala, or Native Bear; Duckbill (Platypus).

created "kinds." When the fact of Evolution was clearly stated and gained general acceptance, the confusion and difficulty of systematic work suddenly became luminously intelligible. It is precisely what one would expect. It is in itself evidence that Evolution is taking place. Our ideas of species have undergone a change and it is important to realize just how profound that change has been.

We have already spoken of life, seen as a whole in time, as a tree; the vertical height of the tree represents the time-dimension, and its branches, forking and multiplying from a common trunk, represent the various lines along which living things have evolved, diverging

VARIATION AND DISTRIBUTION

and spreading away from some common ancestral form. The horizontal distance between any two twigs represents the difference between two races, the degree to which they have diverged from their common stem. Clearly the spectacle which life presents at any particular time—the present, for example—will be represented by a horizontal slice through the tree at an appropriate level. Now, what would such a slice reveal? It would pass through a great number of separate twigs; they would appear on the section as circles, and in our analogy they represent those species that are sharply marked off and distinct—such for example as the so-called Maned Wolf of South America. Some would be close together—if they had recently branched from a common stem—and others would be far apart; the former are closely related, the latter distantly related species. But here and there our section would pass through an actual fork—it would appear as an ellipse, or a figure of eight, or two circles just touching each other—and these correspond to the doubtful cases, the border-line cases, like the case of the tigers, that may with equal justice be regarded as several species or as one. In a word, the twigs in our section would show every sort of relation to each other, every conceivable grade from twigs that are just branching to twigs that lie far apart. And that is how living things present themselves to-day.

Note the importance of this idea from the point of view of the systematist. In ancient, pre-Evolution days it was believed that living things had been created in definite "kinds" according to divine, but nevertheless presumably intelligible plan. It was our business in classifying living things to detect that plan; one started by assuming that species existed and then tried to find out what they were. Nowadays, on the other hand, we know that living things are a slice through a tree, showing every imaginable degree of cousinship, and not falling tidily and infallibly into "kinds." We know that their forms are not constant but changing. And in classifying this assembly, in trying to reduce it to some sort of order and describe it in unambiguous terms, we may choose what conventions we please. Phylum, class, order, family, genus, species, variety—these are words used by general consent to denote the *degree of relatedness* between forms; they are not clearly and categorically distinct from each other, but merge together, arbitrary divisions of what is really a continuous series. It is like saying of our tree-section: "If two twigs are over an inch apart we will put them in different families; if they are over two inches apart in different orders; if they are over three inches apart in different classes." And a species, in particular, is no longer a unit created by God, nor is it a natural unit at all like an atom, or a quantum; it is an arbitrarily defined grouping set up by Man for his own convenience.

From time to time many definitions of species have been put forward, tested and rejected. One only is unassailable. It was proposed by Dr. Tate Regan at a recent meeting of the British Association, and it runs: "*A species is a group of animals that has been defined as a species by a competent systematist.*" Taken in conjunction with what we have said that definition is perfectly sound. It brings out two essential points—first, that a species is an arbitrary convenience; second, that an enormous amount of toil in huge numbers of specimens is necessary before a judgment of any value can be reached on a question of classification.

§ 3. The Distribution of Living Things.

EVERYBODY knows that different animals come from different countries—the platypus from Australia and Tasmania, the zebra from Africa, the marmoset from South America, the musk ox from Greenland and Arctic Canada, and so on; and the same is, of course, true for plants. But it is not always grasped that different regions differ in respect of whole groups of their animal and plant inhabitants. Contrast the three southern continents: Africa south of the Sahara (that sea of sand which is as much a barrier to life as any sea of water), South America, and Australia. All comprise both temperate and tropical regions; all have their mountains, forests, and open plains. But their animal populations are extremely unlike. If, for the sake of brevity, we restrict ourselves to the mammal population, we find that Africa is characterized by an abundance of antelopes of many kinds, by rhinoceroses, giraffes, elephants, wart-hogs, zebras, lions, leopards, baboons, and buffaloes, and in the rain-forests, by gorillas and chimpanzees, okapi and many kinds of monkeys. Farther south the coneys or hyraxes and the extraordinary aard-varks are very characteristic. The whole giraffe family, with both giraffe and okapi, is found in no other region, nor is the aard-vark family.

Just as characteristic as the presences are the absences. There are no deer, no beavers, no field-mice or voles, no shrews, no bears, and scarcely any goats or sheep (we are, of course, speaking only of animals found wild).

Contrast this assemblage of mammals with that found in South America. Here live llamas and their relatives, edentates like the sloths, the true ant-eaters and the armadillos, primitive monkeys with prehensile tails, vampire bats, peccaries, tapirs, guinea-pigs, vizcachas and agoutis, opossums. None of these occur in Africa; and most of them are either wholly restricted to South America, or at the most penetrate a little way into Central or North America. The whole order of the true edentates is confined to this region (Fig. 145).

Finally, Australia (with which for our present purpose we must include the neighbouring islands of Tasmania and New Guinea) is more peculiar still. Before the advent of white men, it contained none of the higher placental mammals whatever, with the exception of bats, whose wings, of course, give them facilities for spreading denied to mere land forms, with a few ubiquitous mice and the dingo dog, both probably introduced by the early human immigrants of the country. But by way of compensation Australia possesses a unique menagerie (now, alas, rapidly dwindling, with many species in danger of extermination if protective measures are not introduced) of the two lower sub-classes of mammals, the pouched marsupials and the egg-laying monotremes.

In point of fact, no egg-laying mammals occur outside this area, and no marsupials, except the American group of opossums, and one curious little creature called Cœnolestes, with teeth in some ways recalling those of kangaroos, from South America. All the rest are Australian—kangaroos and wallabys, cuscuses and phalangers, wombats and bandicoots, marsupial wolf and Tasmanian devil, pouched ant-eaters and pouched moles and pouched mice—some forty genera, with hundreds of species. Add to this the Platypus and its egg-laying confrères, the spiny ant-eaters, and you have indeed a strange zoo.

Now, it might naturally be supposed, and in the past often was assumed, that each species and each group lived in the region best suited to it. But such is demonstrably and obviously not the case. New Zealand has no native mammals save a bat or two and possibly one species of rat—and yet introduced mammals thrive and multiply. Rabbits, for instance, have run wild over large areas, and red deer introduced from Scotland have not only thriven, but have grown much larger than they ever do in their native land. Then the house-sparrow has spread and the starling is spreading over the whole of North America, in spite of the competition of the hundreds of kinds of native birds. The few horses introduced by the Spanish conquerors of South America multiplied and ran wild in huge herds over the pampas. Far from the native Australian birds and animals being especially well adapted to Australian conditions, they are no match for the species that have been introduced from other regions. The mere mention of rabbits will make an Australian farmer cross. And when we come to plants we find that one of the gravest problems of agriculture in various countries, notably New Zealand and Australia, is to prevent introduced species like the prickly pear and blackberry from overrunning the country and ousting not only the native plants, but man and his agricultural efforts as well.

Why then are whole groups of related animals tied down to limited

FIG. 144. Some extinct South American Mammals of the later Cenozoic Period (V), as they probably looked when alive.

Top left: Macrauchenia, one of the Litopterns. Unlike most of its relatives, which took to a horse-like life, this animal was bulky and tall, and probably browsed on the branches of trees. Top center: Megatherium, the Giant Ground Sloth. This creature closely resembled the existing sloths in its anatomy, but lived on the ground and must have weighed over a ton. Top right: Toxodon, a large herbivore of a unique type. Bottom right: Pyrotherium, representative of another exclusively South American group of herbivores. Bottom left: A form of Glyptodont, very similar in construction to a large armadillo, but with carapace all in one piece, and a knob of heavy spikes on the end of its tail, which it doubtless used as a club. Lower right-hand corner: Outline of a collie dog, to give the scale.

regions of the world? What meaning is there in the restriction of the giraffe family to Africa, the whole of the edentate order of sloths and armadillos and ant-eaters to Southern America, all the monotreme sub-class and almost all the marsupials to Australia?

The answer is to be found in the past, in the history of evolving life in relation to the history of the seas and continents. Through fossils we

VARIATION AND DISTRIBUTION 391

are able to discover not only the past development of existing groups, but also their past distribution in each epoch. Geology, on the other hand, can tell us a great deal about the extent of sea and land in past periods. It can do this by studying where in each epoch marine deposits were laid down, where there were deserts, or evaporating inland seas

FIG. 145. Some Characteristic Animals from South America.

Top row: Tamandua (Tree Ant-eater, an animal probably subsisting in the main upon tree-living Termites); Humboldt's Woolly Monkey (with prehensile Tail, like most New World monkeys); Great Ant-eater (with huge claws for tearing down the cement-like walls of the nests of ground-living Termites). Centre: Huanaco (related to the Llama); Three-toed Sloth (an animal which spends most of its life hanging upside down from the higher branches of trees; it resembles its surroundings by being coloured green through the growth of a species of single-celled green plant in special crevices in its hairs); Marmoset (one of the smallest and most primitive of New World monkeys). Bottom row: Nine-banded Armadillo (a type which has spread from South America to the southern United States); Opossum (a species in which the young are not carried about in the pouch, but anchored by their tails to the tail of their mother); Vizcacha (a large rodent).

which produced beds of salt, where the invading ocean had carved beaches, where ice-sheets had passed or mountain-ranges had been elevated. Through such evidence, geology is able to say definitely that the present distribution of land and water is in no way permanent. In the past, the main land-masses of the world have been connected and disjoined in many other ways; and geology can often tell us just when and where the connections and partings were made, and what was the distribution of seas and continents during a particular geological period.

Animals such as land-mammals can and do migrate slowly until they are spread over the whole of a land-mass. But there are barriers which they cannot cross. The sea is the most formidable of such barriers, ice-sheets are another, and broad deserts may be nearly as bad. Thus, the distribution of any group of land-animals will depend upon three factors—first, upon the region where the group happened to originate; second, upon the connections which this region then and later happened to have with other land-masses; and third, upon the fate of the group in the different regions to which it obtained access.

If mammals first evolved after New Zealand had been cut off by a barrier of ocean from all the continents, we should not expect to find any land-mammals in New Zealand. If lung-fish were once widely distributed all over the world, but later were all but extinguished in the struggle for existence, we should expect to find the few existing lung-fish scattered in isolated regions which happened to favour their survival; it is along such lines that our reasoning must run.

The key to the present distribution lies in past distribution. When palæontology and geology are able to provide us with evidence, the distribution of animals and plants ceases to be a puzzle and becomes a simple matter of history. In the same way the distribution of human races, often so puzzling at first sight, clears up directly we know the history of their movements. The Mongol Turks in Asia Minor, or the fair-haired Lombards in Northern Italy, are at first sight anomalies; but with a knowledge of their migrations, the problem disappears. The only difference with animals and plants is that the periods of time involved are so huge that transformation as well as mere migration of stocks comes into play.

With these ideas in mind, we can turn back to the three southern continents and their three wholly different sets of animal inhabitants. First comes Australia and its marsupial menagerie. During most of the late Mesozoic Era (IV), Australia was connected with the rest of the world. Whether there existed a land-bridge to Asia, or, as many are inclined to believe, to Antarctica and thence again up to Cape Horn, or whether, as Wegener thinks, it once formed a part of a great southern continent which was later broken up, we must leave to the geologists to settle; in any case it is immaterial to us. We know that true placentals did not appear on earth before the later Cretaceous (IV C), that the earlier, Mesozoic (IV), mammals were akin to monotremes and to marsupials, though mostly even more primitive in type, and occurred all over Europe and North America. All of them, however, were very small, and they showed little variety in their ways of life. Some time in the Cretaceous (IV C) these primitive marsupial and monotreme mammals penetrated into Australia; the land-connection between it and the rest of the world was broken, before any placentals could enter. Australia thus became the marsupial's Ark—with the important difference that they did not stay merely a year and ten days in it like the animals in Noah's Ark, but over fifty million years. Indeed it was to them like a combination of Ark and Promised Land. For during that long time they flourished and were able to give rise to new and varied forms of life not found in any other region of the world. These possibilities of the pouched mammal were never realized elsewhere, since in all other regions the marsupials were kept from rising or exterminated by the competition of the placental hordes, biologically more efficient in the protection of their unborn young and in the construction of their brains.

Now that modern man has introduced placentals into Australia, the marsupials are no match for them, and are dying out. The development of the varied marsupial life of Australia was due to biological protection. Competing imports were barred. Home industries flourished, but their products never came up to those resulting from a more rigorous competition.

South America, on the other hand, was open to the North in early Eocene times (V A 1), but only for a short time. This allowed representatives of some, but not all, of the early placental mammals to enter, and the connection was then broken, not to be re-established till the close of the Miocene (V C 3), perhaps thirty million years later. Thus the inhabitants of South America, like those of Australia, were for long preserved from the full stress of competition; the difference lying in the fact that it was not marsupials, but early and primitive placentals, which there found an Ark.

Similar results followed in both regions—the development of creatures elsewhere unknown, the flowering denied to primitive types in more strenuous regions.

The chief factor contributing to this local evolution of South American types was the total absence of carnivores and of true ungulates. The only flesh-eating mammals were primitive marsupial types, all now extinct. The absence of more specialized beasts of prey permitted a great number of sluggish, large, and inoffensive creatures to come into existence, such as armadillos and their allies, sloths, big rodents like the tree-porcupines and vizcachas, opossums, and ant-eaters; while the absence of those best-developed of herbivorous running machines, the true ungulates, permitted other stocks more or less adequately to fill the gap thus left in life's economy.

We have already mentioned the remarkable if slightly inferior imitation of horse-evolution achieved by what we may call the pseudo-horses, the Litopterna. Among other remarkable ungulates not found elsewhere were the ponderous Toxodonts, with their peculiar incisor teeth, many of them outdoing the rhinoceros in bulk; the equally ponderous Typotheres, with almost rodent-like chisel-edged front teeth; other great herbivores with toes retractile like a cat's; the Astrapotheres, with two pairs of enormous tusks like exaggerated boar's tusks, and very stumpy limbs and neck. Finally, we have the extraordinary Pyrotheres, which not only grew as large as elephants, but paralleled certain features of the elephant stock in their huge teeth and their projecting tusks; they were unlike all other mammals in having the fore-arm and lower leg disproportionately short, scarcely more than half the length of the upper part of the limb, which must have given them a very grotesque appearance.

Fig. 146. How the Marsupials colonized Australia.

The probable distribution of land and sea about halfway through the Cretaceous Period (IV C). In the shallow seas extending over what are now Western America and much of Europe, chalk was being deposited. Not only was the main block of North America connected with Europe, but the Cretaceous marsupials were able to pass across from the Eurasiatic land-mass to Australia. This land-bridge was soon afterwards broken, and Australia isolated. (Modified from Schuchert.)

These were all flourishing, with many other groups such as the sloths and armadillos, when in the late Miocene (V C 3) the corridor was again opened to the north. It let in dogs and foxes, cats great and small, sabre-tooths, tigers, and bears, together with horses and deer, peccaries and llamas, mice and squirrels.

From this time on there was competition between the two sets of animals, the old and the new. The old types were not beaten at once,

VARIATION AND DISTRIBUTION

for many of them did not attain their maximum size or abundance until later. But the final result was decisive, and the great majority of them perished wholly from the face of the earth.

It is interesting to remember that it was perhaps the South American fossils which first turned Darwin's thoughts to the idea of Evolution. When he was going round the world on the "Beagle," and occupying himself with everything from coral islands to the problems of structural geology and from the habits of savages to the structure of extinct animals, he excavated a number of the abundant fossils found in the wide-spreading Pleistocene beds of the South American pampas.

Among the skeletons there preserved are those of such remarkable creatures as the gigantic Megatherium—bulky as an elephant—whose construction made it able to pull down the branches of trees to browse on, and the eight-foot Glyptodon, a veritable animal tank, protected by a dome-shaped cuirass of heavy bone. As soon as their structure is examined it becomes obvious that the Megatherium was a sloth adapted to ground life, and that the Glyptodon was simply a giant armadillo which could not roll up.

What struck Darwin's imagination was the fact that, while these and other fossils belonged to the characteristically South American group of edentates, they were different from any living edentate. If the edentates had from of old been confined to South America and there had been able to evolve into many different types, some of which were extinguished to leave their bones as fossils, the facts would be understandable. Otherwise the existence of fossils similar in all general points, but dissimilar in detail to the living animals, and the fact that all are found in one region of the world and one only, become very difficult to understand.

About the same time, though quite unknown to Darwin, a precisely similar story was being unfolded in Australia. Clift and Jameson had been collecting and studying the fossils of Australian caves and breccia beds dating from the Pleistocene (V E); and these creatures, often now wholly extinct, all revealed in their structure that they belonged to the marsupials, and only to those groups of marsupials still living in Australia. We may mention here the Thylacoleo, as big as a leopard, which is a phalanger adapted to a flesh diet; and the gigantic Diprotodon, almost as large as the Megatherium, which was closely akin to the kangaroos, though far too big to hop. Clearly the facts are parallel to those which impressed Darwin on the pampas. The marsupial stock which we find in Australia to-day (V F) must have been there for long periods, and it has evolved and changed its composition since Pleistocene (V E) times.

Africa is characterized by no such primitive types of animals as either South America or Australia. It seems to have been cut off from the main centres of mammalian evolution by the Sahara for a long time. It received its first land-mammals in the Oligocene (V B), after marsupials and the first clumsy Eocene (V A) placentals had disappeared from Asia and Europe. After this first irruption the way was

FIG. 147. Diprotodon, the extinct giant wombat, from the Pleistocene (V E) of Australia, as it probably appeared in life.

Like so many large animals in different parts of the world, this creature died out during the last Ice Age. Inset, a small spaniel, to give the scale.

again closed, to open again (probably on the eastern side of the Sahara, across to Asia by Arabia and Syria) only in the Pliocene (V D). The second and larger irruption which then followed gave Africa the bulk of its existing animal types; since then there has been evolution in many details, but no great changes. The door was then again closed, or at most left ajar, so that Africa thus became an Ark for a large sample of the Pliocene Old World mammals. Some, like deer and bears, had failed to find the door before it shut again, and there are none of them in Africa; but the rest throve and multiplied in the broad, equatorial stretches, while the drought and cold of the Ice Age dealt hardly with their congeners who had stayed in the north.

Just as stretches of sea act as barriers to purely terrestrial animals, so stretches of land bar the migrations of the inhabitants of the water. The upheaval which in Miocene times (V C) created the Isthmus of Panama and the land-bridge between the two Americas, put an impassable barrier between the marine animals of Atlantic and Pacific, at least between those which could not face the cold of southern seas. As a result, no fewer than six hundred cases are known of pairs of fish-species from the coasts of Panama, one member of each pair from its Pacific, the other from its Atlantic shores, the two closely related, but different in some trivial but constant character. Original identity explains the similarities; independent evolution for some twenty million years has produced the differences.

Another example, very similar in its watery way to the case of the Australian land-masses, is provided by Lake Baikal, the great, isolated sheet of fresh water that lies in Southern Siberia. Since it was formed, possibly in the Mesozoic Era (IV), certainly before the earlier part of the Cenozoic (V), it has been without any close connexion with any other large body of water, fresh or salt. The huge lake, over four hundred miles long and in some places nearly five thousand feet deep, holds out the most varied opportunities to water-living animals; but most of the kinds of animals which elsewhere take advantage of similar opportunities were absent in it from the start. As a result, other types, which elsewhere remain monotonous and feebly developed, have here blossomed out in extraordinary variety and fill the most important places in the economy of the lake.

This is so with certain kinds of fish, but pre-eminently so with a particular crustacean family, the Gammarids. To this belongs the familiar sand-hopper which swarms under moist seaweed on our sandy beaches, together with a goodly number of other types, freshwater as well as marine. Members of this group were in Baikal from its beginnings. Finding there what was denied them in all other parts of the world, a large and friendly home where there was no competition from more developed crustacean types, such as crayfish and crabs, and shrimps and prawns, they have done their best to take the place of higher Crustacea. In this single sheet of water they have evolved into over three hundred different species—more than as many as are to be found in all the rest of the world—and many of them belonging to purely Baikalian genera. They have launched out into the most varied occupations. There are deep-water Gammarids, blind but with long feelers to compensate for the loss of sight, living at three hundred fathoms; there are Gammarids which swim all their life in the open water deep below the surface, and are transparent as any jelly-fish;

there are large shore-living Gammarids, four inches long—sand-hoppers doing their best to be lobsters; and so on. Here again, freedom from competition has allowed the surprising evolution of a group which elsewhere has had to keep its potentialities locked up, unrealized.

Isolation of a piece of land or a body of water from the rest of the world always permits its animal and plant inhabitants to evolve along their own peculiar lines. This is not only true for large groups, like the sand-hoppers in Baikal or the marsupials in Australia, but also for genera and species and varieties. We shall have more to say on this subject when we come to discuss the machinery of Evolution. Here we will content ourselves with but two examples, one of which, however, is of considerable historical interest. Those who are interested in the subject can pursue it in Wallace's famous book *Island Life*.

If a find of fossil animals on a large continent first put Evolution as a seed of thought into the fertile soil of Darwin's mind, the germination of that seed was brought about by a problem of present-day distribution on an isolated archipelago. The "Beagle" visited the Galapagos Archipelago, a collection of some fifteen volcanic islands, separated from each other by distances ranging from a mile or so, up to nearly one hundred miles, lying on the equator in the Pacific. The nearest mainland is the west coast of South America, six hundred miles away. The account Darwin gave of them has been supplemented, though not supplanted, by Dr. Beebe's beautifully illustrated book, *Galapagos*. These islands, as Darwin pointed out, resemble the Cape Verde Archipelago, off the African coast, in soil, climate, height and size of the islands. In a world deliberately planned and created they would be populated by the same kinds of creature; what suits one suits the other. But they are not. Their animal and plant inhabitants are totally different. The inhabitants of the Cape Verde Islands are related to those of Africa, those of the Galapagos to those of America. On the Galapagos "there are twenty-six land-birds; of these twenty-one, or perhaps twenty-three, are ranked as district species, and would commonly be assumed [when Darwin published this passage, in 1859!] to have been here created; yet the close affinity of most of these species to American species is manifested in every character, in their habits, gestures, and tones of voice"—and so with the other animals and plants of both archipelagos—they are closely related to, though often slightly different from, those of the nearest mainland.

Thus, here we have the same fact, of difference in species but resemblance in general type, which emerged from the fossils of the other side of the South American continent, only now the differences and resem-

VARIATION AND DISTRIBUTION

blances concern two sets of creatures separated in space instead of in time.

Such facts at once receive an explanation in terms of Evolution. Chance immigration of storm-pressed birds, wind-blown seeds, tortoises or their eggs drifted in logs or brushwood, would people the archipelago from the continent; this, followed by new evolution in the new and isolated home, would account both for the resemblances and the differences between the inhabitants of the archipelago and those of the

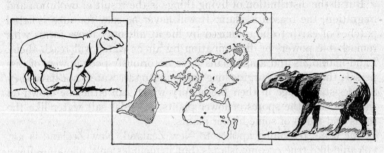

FIG. 148. The regions inhabited by Tapirs are shown stippled. Four species are found in South America and a fifth, with characteristic colouration (below) in Malaya. What is the reason for this wide separation? (The 100-fathom line is indicated as well as the coast-line.)

neighbouring mainland. But in terms of the Creationist view, there is no explanation.

As our second example we choose St. Helena. St. Helena is perhaps the most isolated spot on the globe, the most insular of all islands. Well over half of its two hundred species of insects are to be found in no other region, and three-quarters of its thirty snails, and four-fifths of its flowering plants: and it boasts no mammals, no land-birds, no reptiles, no amphibia, no fresh-water fish and no fresh-water plants. In other words, animals and plants have either wholly failed to reach this water-girt speck of land; or, if they have succeeded, have usually evolved and changed into something new.

But it is not only the living things which are present in a given region which have testimony to offer us. The absences may be as significant as the presences, just as silence may sometimes be as eloquent as speech. On this score the Galapagos and St. Helena and other oceanic islands—islands, that is to say, of volcanic origin, which are separated by many miles of deep water from the continents, and apparently have never had any connection with the mainland—have much to tell us.

If different animals were created and placed in the countries best suited to them, why is it that oceanic islands never possess more than a sprinkling of land-animals and birds and flowering plants, and never any amphibians or land-mammals (except sometimes the too-readily transportable mice)? It is emphatically not because such animals are unsuited to life on islands, for rabbits and goats, cats and frogs thrive well enough when introduced, and often so much too well that they become a pest.

But if the distribution of living things is the result of evolution and migration, the reason is plain. It will never be easy for such isolated patches of earth to be colonized by life at all; only those forms with remarkable powers of dissemination by air or water will reach them. Land-mammals and amphibians have notoriously poor powers of dispersal; they cannot survive long exposure to salt water, nor have they any resistant stage in their life-history which can either be blown on the wind like the spores of lower plants, or resist salt water like the seeds and fruits of some higher plants.

The same reasoning applies to New Zealand. New Zealand is not volcanic like true oceanic islands, but it includes only one amphibian and, apart from bats, which can fly, only one land-mammal—a rat, which may very likely have been introduced by the Maoris. It is separated from the rest of the continents by such distances, and has been separated for so long a time, that its animal population, at the time of its discovery, was in most ways like that of a real mid-ocean volcano. Yet all the time it was most admirably suited to support those very forms of life which it lacked. The zeal of its acclimatization societies has stocked it with all kinds of European birds and animals and plants, many of which have found the country so much to their liking that they have become most abundant nuisances.

Finally, we have a third set of facts, the facts of discontinuous distribution, when almost identical animals are found only in two or three widely separate regions of the world's surface. Why in the name of all that is reasonable are tapirs found only in South America and Malaya? Why is one branch of the camel family, the camels themselves, found only in Asia and North Africa, while the other, the llamas and their kin, grows only in South America? Why are the lung-fish found only in Australia, tropical South America, and tropical Africa? Here geology solves our riddle. The discontinuity did not always exist: the type was once widespread, but to-day has been exterminated save in a few patches. Lung-fish were among the most abundant of fishes in Devonian (III D) and Carboniferous (III E) times, and were then spread over the whole world; the competition of later-evolved and more efficient fish extinguished almost all of them, and only three representatives have managed to survive, in three patches of tropical fresh water.

The camels can be traced back to the Eocene (V A). Their remains are found at first only in North America. Thence, during the Miocene and Pliocene (V C and D), they spread across two newly-emerged land-bridges both southwards into South America and westwards to Asia.

 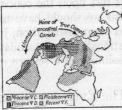 

FIG. 149. The present distribution of the Camel family explained by their past.

Remains of ancestral camels of Miocene Age (V C) and earlier are found in North America. In the Pliocene (V D) they are found also in the Old World. By the early Pleistocene (V E) the family is known from most of the world. Widespread extinction in the late Pleistocene left the Llama branch in South America, the true Camels in Central Asia and North Africa. (Outline of land-masses shown at the 100-fathom line.)

The two emigrant stocks went their own evolutionary ways to what we see to-day, while in their original home the family was abruptly extinguished in Pleistocene times (V E). Finally, the tapirs too, like the lung-fish, were once widely spread. They are known as fossils from various Cenozoic (V) beds in both North America, Europe, and Asia. Doubtless they early penetrated into India and Malaya, and invaded South America by the Miocene (V C) land-bridge from the north. They have been reduced to their present distribution simply by extinction in the regions between.

Discontinuous distribution; the predominance of different groups of animals in different continents; the existence of unique species on remote islands; the total absence of many creatures from countries where they are able to thrive and multiply when introduced—these and many other facts of distribution are concordant.

If the assignment of different kinds and groups of animals to different regions of the earth's surface was made by a Creator, one is forced to admit not merely that these facts are unintelligible and even meaningless, but that the assignment was often definitely unfortunate. But if we are Evolutionists, the chaos becomes order, the mob of facts becomes a marshalled army, held together in complete consistency by that dominating idea.

The perplexities and apparent paradoxes of geographical distribution attenuate and vanish in the light of two chief principles of interpretation: the principle that connects isolation with divergence of type, and the principle of looking in the past for the explanation of the present. Both principles rest upon the concept of Evolution as their basis. Without that basis we should have to relinquish all hopes of rationalizing animal and plant distribution, just as we should have to give up all hopes of rationalizing recapitulation, or vestigial organs, and to abandon the possibility of a science of comparative anatomy.

§ 4. The Evidence Summarized.

WE HAVE now passed in brief review a small fraction of each of the main kinds of evidence for Evolution. We have found that the rocks of earth's crust make a book where our planetary history can be read, and that they contain fossil remains which, when deciphered with the same care which an antiquary bestows upon his inscriptions, yield knowledge as definite as any which he obtains, though it refers to epochs a hundred thousand times as remote. The fossils, dated by the rocks, reveal the actual story of life's past changes. The story does not go back to life's first beginnings, nor does it cover all kinds of living things; but for the latter half of life's existence, and for the most elaborate and interesting of life's children, it is reasonably complete. And the story the fossils reveal is one of steady evolution, of progressive change, of multiplication and divergence of forms of life, of extinction of one type and its replacement by another. The fossils are the remains of creatures other than those of to-day, which once were alive, living a different kind of life in different surroundings. They testify to the past of animal and plant life in the same direct way as do the mummies of the Pharaohs, or the baked-clay bills and receipts of prehistoric Babylon, or the slaughtered retinue of the King of Ur, to the past existence of human beings who lived very different lives from ours; they testify to the evolution of life as directly as do the discoveries of archæology to the evolution of human culture.

Then there are the indirect evidences. There are the similarities of general plan which are not to be accounted for save by the descent from a common ancestor of all the animals showing the plan. The similarity

VARIATION AND DISTRIBUTION

of plan implies common descent; the differences in detail imply descent with modification. There are the useless vestiges which yet correspond rigorously with organs that are indispensable in other animals, inexplicable if their possessor be not descended from some ancestor in which the now useless organ had its use. There are the extraordinary phases of the individual's development in which it passes from one strange likeness to another—likenesses to other creatures, often remote and primitive, to which the adult animal no longer betrays any resemblance, whether in plan or in mode of life. These resemblances are meaningless, and, indeed, deceptive, if they are not recapitulations of ancestral phases in which the race once continued for long periods of past time, though our modern creature hurries through them on its way to its own new and different adult life.

All these facts are inexplicable on any theory of special creation. Save Evolution, no rational explanation of them has ever been put forward; and on the evolutionary view not only are they explicable, but full of meaning.

We reviewed the evidence from life's variability. If we laid more emphasis upon it than has often been done in the past, this is because the evidence from variability, like that from fossils, has of recent years increased enormously, both in account and still more in cogency. Now that a number of groups like birds and mammals and butterflies have had their minutest varieties classified and the details

FIG. 150. A man at about two months of true age.

Front view of a human embryo, four-fifths of an inch long, about seven months before birth.

of their distribution tabulated, the old idea that species are the most real and definite units of life, or even that they are real and definite units at all, sharply marked off from other kinds of units, has gone by the board. There do exist some sharply circumscribed species-units; but other such units intergrade or interbreed with one another. There is no crucial test by which we can distinguish between a local race or a sub-species and a species, or between a species and a genus. There is often disagreement among systematists themselves as to whether a particular kind of animal or plant shall be classified as a full species or a mere variety. There exist interbreeding groups so variable that we would regard the extremes of variation as different species did we not know of the existence of all the intermediates. All this lack of sharp lines and clear limits is to be expected if life's method is Evolution; but on the Creationist assumption it is chaos and confusion.

The considerable degree of variability to be found in all wild forms of life was emphasized, and the conclusions to be drawn from this fact were driven home by an appeal to the astounding changes which man has been able to bring about in his domestic animals and plants. If greyhound, bulldog, toy terrier, and St. Bernard can all be formed out of wild-dog material in a few thousand years, then that living material is of an extraordinary plasticity, and will lend itself willingly enough to change and evolution.

And finally we have recited some of the facts of the distribution of animals and plants, and have shown that they, too, fall into place and become intelligible to the Evolutionist, but remain stubbornly meaningless on any other view.

All these lines of evidence lead to the same conclusion. The way in which each one corroborates all the others is impressive enough, but let it not be forgotten that the actual examples we have chosen are but a fragment of the mass available. If an idea is true, it will apply in every part of its domain. The domain of the idea of organic evolution is the whole domain of life; and the final evidence for Evolution is that throughout the whole domain the idea of Evolution helps our comprehension. It explains old discoveries and leads us on to new; it draws order out of confusion; it gives meaning to what is otherwise meaningless, and brings thousands of isolated facts into a single related whole. There is not one single character or quality of human beings, from the construction of their skeleton to the flush on their cheeks, from their embryonic development to their moral aspirations, which does not become more comprehensible, more interesting in itself, and more significant for the future, when viewed in the light that Evolution sheds upon it. Whether we are dealing with the cone of a pine-tree or the skull of a bird, the fertilization of a flower or the instincts of an ant, Evolution illuminates why they are what they are. And there is no other imaginable illumination.

THE EVOLUTION OF MAN

§ 1. *Man: a Vertebrate; a Mammal; a Primate.* § 2. *Fossil Men.* § 3. *Man's Body: a Museum of Evolution.* § 4. *Man's Place in Time.*

§ 1. Man: a Vertebrate; a Mammal; a Primate.

THERE is no need to stress the physical likeness of the higher apes, chimpanzees, gorillas, or orang-outangs to ourselves. The crowds which gather round their cages in the Zoo are a testimony to this resemblance and the interest which it inspires. Nor is the likeness one of physical structure only: we have but to watch a mother orang with her child, or a young chimpanzee at play, to realize how deep the similarity of behaviour goes. The mother dandles her baby in her arms, kisses it, strokes its head; her gestures and the play of expression on her face have an often pathetic likeness to a human mother's.

But it is perhaps not often realized how extremely close the resemblance is, and how, if a Martian scientist, with no personal prejudices on the subject, had been given the task of classifying the animal inhabitants of our planet, he would at once have put man in the same small group as the tailless apes with the same lack of hesitation with which he would have classified the hive-bee and the solitary bee together; and how quite unintelligible he would have found the long failure of human naturalists to take this step, and the storm of protest which arose when at last a few bold and logical spirits dared to take the objective view of man's place in Nature!

The closeness of our likeness to the apes may best be realized by measuring it against our likeness to other vertebrates. Take a man's body and compare it with a frog's. A man is really very like a frog in the general plan of construction: both have internal skeletons made of bone; a spinal cord running along the back, enclosed in the backbone's tunnel; brain in a brain-box; eyes, ears, nose, mouth, and teeth in the same general relation; two pairs of limbs, with skeleton corresponding almost bone for bone; and a close resemblance in the plan of their internal anatomy. Their chemical arrangements, too, are quite alike. Both have livers which break up amino-acids and store sugars as glycogen; both have a pancreas which secretes trypsin; the adrenalin manufactured by the frog's adrenal glands is not only like but chemically identical with the product of the adrenal glands of man.

In these and scores of other ways a man is not only like a frog, but unlike the vast majority of animal types. The man-frog plan of structure and working is definitely unlike that of a cuttlefish, or an ant, or a crab, or a leech, or a sea-urchin. In zoological terms, men and frogs are vertebrates, and these other animals are not.

On the other hand, if we draw a few other vertebrates into our comparison, we see at once that there are degrees in their likeness. Man is more like a frog than a fish, for frog and man both possess lungs and fingered limbs, and a fish does not. But man is less like a frog than a dog, for man and dog both have hair, and divided hearts, and warm blood, and teeth of several different sorts, and milk, and young that are nourished in the womb; and frogs have none of these things. In terms of classification, men and dogs are mammals, frogs are not.

But again, a man is less like a dog than a chimpanzee. For man and chimpanzee have nails, not claws, and grasping hands with the thumb opposable to the rest of the fingers, and limbs for walking and climbing; they have no tail; their females have monthly periods and a single pair of breasts; their brains are large and deeply furrowed and convoluted. And in all these ways and many others, notably in their teeth, they differ from dogs. In zoological terms, men and apes are primates, dogs are carnivores.

But the resemblance between man and chimpanzee is much closer than these facts alone would indicate. We have already seen how close is their invisible, chemical resemblance. Then the visible resemblance permeates every detail of their anatomy. Their skeletons are not merely alike in general plan, but correspond actually bone by bone; the grinding teeth are extremely similar in pattern; the ape's hands and face and expression are all but human. In a number of points the chimpanzee or the gorilla differ more from the other great apes than they do from man. It used to be asserted that there existed definite structures in man's anatomy, especially in his brain, which were absent in apes. T. H. Huxley, as far back as 1863, demonstrated in his classical essay, "Man's Place in Nature," that this was not true. The differences between human structures and ape structures are only differences of degree. Man's brain and brain-case are proportionately larger; but, as

Elliot Smith has shown us, this increase is due to the enlargement of parts of the brain already present in the ape—the parts concerned primarily with the faculty of association—and no brain-organs are to be found in man which are not also to be found in apes. The dog-teeth of us men are smaller, and so are our big toes, but our thumbs and chins are larger. We have adopted the upright position, and our anatomy shows minor changes as a result. Our legs are straighter than apes'; our spinal cord is bent in a different way; our pelvis, besides still afford-

Fig. 151. Orang-outangs, young and old, drawn from life.

At home in the family circle; mother transporting her baby; young hopeful resting, hanging by one leg. Lower right, an old male of the variety which has a fleshy fold round the face in place of whiskers.

ing attachment to dozens of muscles, has been turned into a flat basin which supports the internal organs of our abdomen; a new development of the gluteal muscles is needed to hold us upright, with the result that our "lower back" protrudes in a way unknown among animal buttocks; our head is poised on a straight pillar of a neck, instead of protruding forward. Our jaws and the bony ridges over our eyes are smaller, our noses and chins larger. Apart from brain-size, the reduced amount and length of hair on the human body and the less hand-like construction of the human foot are perhaps the most obvious differences between man and ape, but even they are only differences in proportions and amounts, no whole structures being gained or lost or transformed into something radically new.

And when we come to behaviour and the mind at the back of behaviour, the chimpanzee is far more like a man than like a frog. To begin with, the actual raw material of an ape's experience, the data provided by its senses, differ from a frog's but resemble our own in many points. The frog lives in a world of black and white; apes, like us, in one enriched with colour. The apes, like us, have a "yellow spot" in their eyes, making them capable of exceptionally accurate vision; the power of discrimination in a frog's eye is much poorer and it pays no attention to any but moving objects. The frog can hear, but its miserable little bag of a hearing organ compares very poorly with an ape's long, coiled cochlea. The range of sound which our ears permit us to hear and the delicacy with which we can discriminate between different tones are almost identical in ourselves and in apes; but the sound-world to which the frog has access is limited and crude in comparison. In one respect, however, the frog has the advantage over us and over the apes. Its whole skin possesses organs of chemical sense, which in us are confined to the moist coverings of nose, eye, mouth, and other mucous membranes. Thus it is able, albeit it would seem in a very crude way, to smell or taste or at least to be stimulated by various chemicals, all over its body. Finally, what emotions a frog possesses appear to be few in number and low in intensity—wholly unlike the extremely human passions of apes.

But if the very bricks out of which it builds its mental life are different from ours, its ability to build with them is no less different. The frog has powers of learning and association, but so feeble that we find it difficult to realize the extent of a frog's inability to profit by experience. Almost the whole of its actions are reflex, predetermined for it from the start by the inherited constitution of its nervous system. The ape, like ourselves, has its due share of reflexes and its complement of inborn instincts with their accompanying emotions; but, like ourselves, it can not only learn, but learn rapidly. Though its actions, like ours, are always built on foundations of reflexes and instincts, yet the great majority of them are what they are because of the animal's individual experience. In a frog's life, learning by experience plays an insignificant part: it plays a preponderant part in a chimpanzee's.

This would be true of a dog as well as an ape but the ape can go farther than this. One chimpanzee studied by Professor Koehler, though wholly untaught, had the idea of fitting one stick into the hollow end of another in order to get at a banana which was out of reach of either stick by itself; and to do that is to anticipate experience by thought. This observation is only one of many, but it must suffice us for the present. What concerns us here is the fact that chimpanzees and other true apes do have this faculty of anticipating experience, of putting two and two together so as to deal with a new kind of situation in an intelligent way, and that this power, in spite of the most careful tests, has never been detected in any lower animal, even in tailed

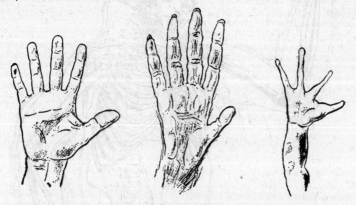

FIG. 152. A Comparison of Hands. Left, Man; centre, Chimpanzee; right, Frog.

The hands of Man and Chimpanzee are much alike, the main differences being differences of proportion and of hairiness. The Frog's hand is built on the same general plan, but differs in many important respects. It has only four instead of five fingers, the true thumb having been reduced to a vestige, so that the apparent thumb is really equivalent to our index finger. This false thumb cannot be bent round and opposed to the fingers for grasping purposes; there are no nails; and the skin is hairless and moist.

monkeys. Many animals have that form of intelligence which we may call intelligent learning, but no others show deliberate invention. It is true that the ape's free ideas are very rudimentary when compared with ours. None the less, our chimpanzee who, instead of doing nothing at all, or of aimlessly fiddling with his two sticks until one chanced to fit into the other, saw beforehand that they would fit and would then serve his purpose—he was, albeit in a humble way, showing the same power which enables an engineer to design a bridge on paper instead of putting something up and trusting to luck that it will stand, or a physicist using his mathematical faculties to calculate how his atoms and electrons should behave if his assumptions are right, so that he can plan the crucial experiment which will tell him if they are right

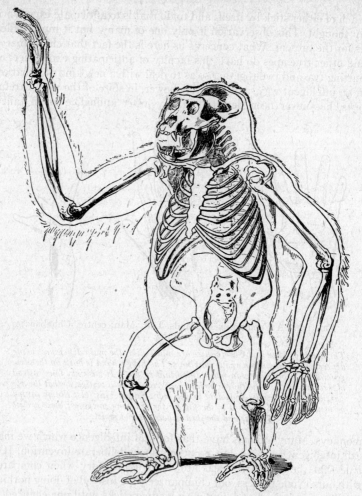

Fig. 153. The skeleton of a full-grown male Gorilla.

Compare it with that of a man (Fig. 4). The two correspond bone for bone.

or no. In mental life and mental powers a chimpanzee is less like a frog than he is like a man.

When we compare our human and our simian, not only when full-grown, but during their respective developments as well, the resemblances increase. Compared with a man, an adult chimpanzee has proportionately longer arms and shorter legs; so has an unborn chimpanzee compared with an unborn human being, but the difference is less; the ape fœtus is more human, the human fœtus more simian in its proportions. During most of the later half of its prenatal life, the human embryo, like the ape's, is covered all over with a coat of short, downy hair; so is an ape embryo of corresponding age. Before birth both ape and man shed this short hair and develop long hair on the head while remaining almost hairless on the body. Man retains this condition throughout life, while the new-born ape soon acquires its thick permanent garment.

The skull-shape of a chimpanzee is much more human before birth; and even the characteristic ape-foot is in the fœtus much less like a hand and much more human than later, while even after birth the human baby's foot, with its inturned sole and eagerly prehensile toes, is charged with hints of a racial past spent in the trees.

In fine, man's structure and development reveal him as zoologically close kin to chimpanzee, gorilla, and orang-outang. Through the invention of language he is made free of a new mental country, to which they have no access; but he does not for that reason cease to be their close cousin, any more than a mentally defective child ceases to be the son of his father, any more than Shakespeare ceased to stand in normal blood-relationship with his cousins, because he entered realms of thought and expression of which they never dreamed.

§ 2. Fossil Men.

THIS testimony from structure and development is confirmed by the authentic voice of the past. Unfortunately, neither apes nor men happen to be preserved as fossils save with extreme rarity, so that we have as yet no such pretty series in our own ancestry as in the ancestry of the horse; but the few remains which have been found tell an unequivocal story.

In the first place, we possess a true missing link between men and apes. He was discovered at Trinil, in Java, in 1892, and christened Pithecanthropus, or the Ape-Man. In the weight of his brain (and brain-development is by far the most important difference between ape and man) he is almost precisely half-way between the largest ape brains known, those of large gorillas, and the smallest, normal human brains, to be found among Australian aborigines. He possessed huge brow-ridges over his eyes, like a gorilla, but walked upright, like a man.

Then there is Piltdown man, unearthed in Sussex, obviously a man and not an ape, but so different from ourselves as to demand being put in a new genus, Eoanthropus or Dawn-Man. His eyeteeth were large and savage, his lower jaw almost wholly ape-like, and his brain both small and primitive. Implement-like objects have been found in association with him.

Quite recently, remains of another still more primitive genus of men have been found in China. A few isolated teeth were first discovered, which, though clearly human, were so distinctive that Professor Davidson Black boldly said they must belong to a different genus of man, which he called Sinanthropus. His boldness was justified. In 1928 parts of two lower jaws were found; and in 1929 Mr. W. C. Pei of the Chinese Geological Survey found a brain-case, the only complete cranium known of a subman outside the genus *Homo*. Altogether, remains from at least ten individuals have been discovered up to the date of this printing. Like Piltdown man and the Java Ape-Man, Sinanthropus dates back several hundred thousand years, to the early part of the Pleistocene Period (V E). His jaws present many resemblances to that of Piltdown man. His skull, however, is closer to that of Pithecanthropus, but definitely more human, with higher vault and better-developed forehead. Crude flint tools and traces of fire have been found in his cave.

Four other extinct types of men which, though admitted to the same genus as ourselves, clearly belong to different species, are Heidelberg man (*Homo heidelbergensis*), without a chin and with a jaw of extraordinary massiveness (Fig. 155), Rhodesian man (*H. rhodesiensis*), with brow-ridges heavier than in any existing human beings, and a brain that to-day would be definitely subnormal, the recently discovered Kanam man (*H. kanamensis*) of which only part of a lower jaw is known; and Neanderthal man (*H. neanderthalensis*). This last is the only extinct species of man of whom as yet we possess abundant material. The reason for this was that he buried the bodies of his dead. Other species of bygone men, it seems, were left to moulder where they died, like animals; but the Neanderthalers laid out some at least of their dead in their caves and put tools and implements beside them and buried them, presumably because they did not believe life was wholly ended, and so put these things for the use of the departed if and when he or she awoke again. It is rash to guess too precisely what ideas led to these interments. They had ideas and doubts about death, no doubt, that resulted in burial. That is as much as we can say of creatures so remote from ourselves. The majority of

THE EVOLUTION OF MAN

Neanderthal skeletons so far discovered owe their preservation to this disposition.

In spite of this human habit of interment, the Neanderthalers were

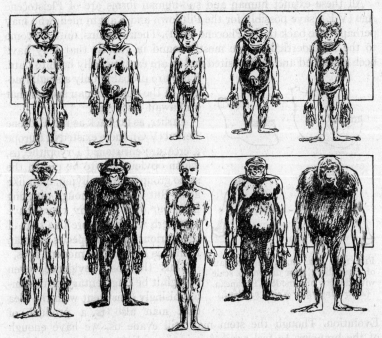

FIG. 154. The five kinds of man-like apes, as adults (below), and as late embryos (above), to show the differences in proportion of the parts.

From left to right: Gibbon, Chimpanzee, Man, Gorilla, Orang-outang. The human embryo is in its fourth month, the others at about the corresponding stage of their development. The figures are not drawn to scale, but the sitting height has been made the same for all. They are all drawn in the same position, to illustrate clearly the proportions of trunk and limbs. Man has relatively the longest legs and the shortest arms. Of the others, the gorilla has the most human proportions. The embryos differ in their proportions in the same kind of way as do the adults, but not so much, so that the five creatures are more alike before birth than when grown up. (Modified from Prof. A. H. Schultz.)

distinctly less human than we; they had heavy brow-ridges, no chin, large teeth—though their canines were less ape-like than ours—head thrust forward on a thick neck, and their short, bent thighs compelled them to walk bandy-legged, with their weight on the outer side of their feet. Many of the implements ascribed to extinct human species are so big as to be unwieldy by the hands of any living race of men.

All these extinct human and sub-human forms are of Pleistocene Age (V E), save possibly for the Piltdown and Kanam men, who may perhaps date back to the Pliocene (V D). Their remains, until we come to the Neanderthalers, are mostly found in gravels that may have been disturbed and re-deposited and so are exceptionally hard to date. They are probably only a small fraction of the quasi-human species that still await discovery.

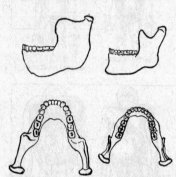

FIG. 155. The huge, chinless jaw of Heidelberg man drawn to scale with the jaw of a modern European. (*After Sir Arthur Keith, modified.*)

Earlier, as far back as the Miocene Period (V C), there existed in Europe a creature christened Dryopithecus, which obviously is to be put in the same group with the living, man-like apes. But its teeth, though differing from our own in many ways, have a pattern which is more like that of the various extinct species of man than it is like that of modern apes.

Thus the fossil evidence, even though it be fragmentary, is uncompromisingly consistent with the idea that man also is a product of Evolution. Though the stem may still evade us, we have enough of the branches to feel assured it is there. Every extinct species of man is in one way or another more ape-like and less human than any living race of men; in Pithecanthropus and Sinanthropus we have true linking forms between men and apes, only slightly, if at all, on the human side, while in Dryopithecus we have a true ape, some of whose characteristics are more human than those of any modern apes—a pier from the apeward end of the bridge from not-man to man.

The evidence therefore is clear not only that man is closely akin to the apes, but that he is actually descended from an animal which, though without doubt different from any living gorilla, chimpanzee, orang-outang, or gibbon, would obviously have to be classified as an ape, which was covered with hair, provided with formidable teeth, had a brain not above half the size of ours, and spent most of its life on the ground or among the branches in the still prevalent but dwindling forests of the Pliocene.

§ 3. Man's Body: a Museum of Evolution.

Man's likeness to the apes shows us clearly enough in what direction to look for his evolutionary origin. But the proof that he has had an evolutionary origin and was not specially created—that rests on a much larger mass of evidence. Our adult human bodies are among the best proofs of Evolution; and the private development of each one of us is an affidavit swearing to the evolutionary history of our race.

Wiedersheim, the celebrated German anatomist, enumerated in the body of man no less than one hundred and eighty organs which are vestigial—wholly or almost useless to us, though useful in other species of animals—each one of them a stumbling-block to the believer in special creation but an ally to the Evolutionist. We may note one or two examples.

The body-hair of men and women is purely vestigial; it no longer serves to prevent us losing heat. And yet each of these tens of thousands of useless hairs possesses a useless muscle by means of which it can be, quite uselessly, raised. For a furry creature to bristle up its hair when the weather grows cold is useful enough—more air is entangled in its coat, and it loses less heat. In the same circumstances we also erect our futile little hairs; but the resultant goose-flesh condition is of no value whatever—we have performed a vestigial action. Even the arrangement of the hairs on our body may recall the past. The hairs on our upper arms point downwards; those on our forearms run upwards and outwards to our elbow; precisely the same arrangement occurs in the orang-outang, and it has plausibly been suggested that in this animal, which often sits with its arms clasped over its head, the arrangement may serve to shed rain off the body, down the spouts of long hair projecting at the elbow. But whatever its significance in the apes, its existence in man is yet another proof of their kinship to him.

A few talented human beings can move their ears. Apart from being a minor social accomplishment this has no value; but for a wild creature, like a rabbit or a zebra, whose safety may depend upon its power of detecting faint sounds and the direction from which they come, the power of moving its ear-trumpets is vitally important. The human ear-mover is indulging in a vestigial action, but the ears of the rest of humanity are one step more vestigial, for the power of moving them has been lost. In spite of this, however, a whole set of muscles to move them is still present, though never called upon for action, since for some reason we are not able to control them. This, by the way, is also true of the great apes, which, like us, have useless vestigial ear-muscles. In tailed monkeys, on the other hand, the muscles are large and can be used, although they are not as strong or supple as those of, say, a dog.

Another interesting vestige in the ear is a little conical projection from the inturned margin of the ear, usually called "Darwin's point," since he showed that it was the remains of the tip of the pointed ear of lower forms, now folded downwards and inwards. It is only found in a certain proportion of human beings (and, curiously enough, is stated to occur more often in men than women); but when present it is, as Darwin wrote, "a surviving symbol of the stirring times and dangerous days of his animal youth."

Again, the little fleshy fold in the inner angle of our eyes (between the openings of the tear-duct) seems to have no function whatsoever; but in most lower vertebrates, including many mammals such as the cats, this same fold is a veritable third eyelid, which can be rapidly swept across the eye from one side to the other. As further proof of man's simian relationship it may be noted that apes and monkeys, too, have their third eyelid reduced to a vestige.

Our wisdom-teeth are on the way to becoming vestigial. In most of us they only appear between the ages of twenty and twenty-five. In quite a number of people, however, they are never cut at all, but remain, useless or even the cause of disagreeable inflammation, within the gums.

Nor is any exception made by man to the animal practice of recapitulating the past of the race during individual development. We have the same family secrets in our embryonic cupboards as the rest of our mammalian relatives. The gill-clefts, the tail, and the furry coat with which our persons were once adorned have already been mentioned (Figs. 63, 64, 136). We may add that the human tail is formed complete with all the muscles for wagging it; later, as the tail fades into insignificance, the muscles degenerate or are turned to other uses.

The early human embryo has nostrils connected with the mouth by a deep groove on either side. Sometimes, through a failure of development, this condition remains throughout life, and we call it hare-lip. It is a reminiscence of the way in which the nostrils were formed in our early fish-like ancestors, and you have only to look at a dog-fish or a skate to see that they still show this construction (Fig. 68). Similar abnormalities of development sometimes allow the prenatal hair to persist, giving us the dog-faced men and hairy women of our fairs and shows; or the embryonic tail forgets to shrink and a baby is born with a little pink tail like a sucking-pig's; or the closure of the gill-clefts is arrested and we have adult human beings with actual slits on the side of their neck, or with white patches of skin, marking the thin places where they closed just before birth.

Then there is the extraordinary capacity of the new-born human babe to support its own weight for several minutes at a time when hanging by its hands alone; indeed, in most cases the child can hang by either hand singly. This capacity persists for a month or so after birth, but then normally fades away, and only after several years is the child again capable of such a feat. All monkey and ape mothers travel through the branches with their babies. But as they need their own arms for travelling, the baby, even from birth, must have the capacity for holding on tight by its tiny hands to its mother's fur. There can be no doubt that the presence of this power in human infants is a survival, now wholly useless, of what was once a matter of life and death. It is interesting to find, as Mr. Kallen has demonstrated with his daughter, that if the baby be repeatedly stimulated by putting graspable objects into its hands, and tugging upwards when the little hands close round the object, the capacity may be reawakened even after it has died out naturally, and may then be made to persist for many months. The inborn, automatic nature of this power is shown by the fact that it was present in a child born without any forebrain (cerebrum), and persisted in full force until the defective infant died at eighteen days of age.

Another recapitulation, which only fades after birth, is the greater prehensile capacity of the human baby's foot, and the fact that its big toe is much more widely separated from the rest than a grown man's. Why should a baby's foot be half-way to an ape's if there is no real relationship between ape and man?

Those who still doubt or reject the truth of Evolution should ponder their own case. For their own private and particular evolution, even though it was all compressed into a nine months' span, was just as spectacular as the slow evolution of life as a whole. Even that valiant apostle of Fundamentalism, Mr. William Jennings Bryan, began his existence as a single cell, passed from this stage of protozoan resemblance through the stage of a cell-colony; hinted at ancestral polyps as he became two-layered; revealed himself akin to Amphioxus in producing a notochord, only to destroy it later in favour of a backbone; indulged in reminiscence of the sea-life led by his fishy forbears by constructing with his amnion a little "private pond" of fluid in which he might embryonically float, and by piercing his neck with gill-clefts, only to do away with them when he subsequently recapitulated his ancestor's greatest feat, the conquest of the land; recalled the furry, four-footed stage of his genealogy by his tail, all ready to be wagged, and his coat of flaxen down; and even, after birth, was unable to help recalling what he later regarded as a blot on his escutcheon —his simian past—by the active, semi-prehensile big toes on his babyish feet and his soon-lost ability (probably never exercised, but undoubtedly present in the first weeks of his free existence) to support his own weight when hanging with his hands.

We are thus no exception to life's rule. The human species has come into being like the rest, not from a sudden act of creation, but slowly, laboriously, by gradual and often devious ways. And having been evolved, man, we are bound to suppose, must still go on along the lines of biological change. So far as our scientific data go, we are bound to believe his present lordship is a precarious one. He may become extinct, like the great beasts of the Age of Reptiles; nothing in his past or in his structure assures us against that; he may linger on in subordination to some new type evolved from another line of life, as the crocodiles and turtles and other reptiles of to-day are subordinated to us mammals; or he may become transformed age by age into something wholly and unrecognizably new, something more powerful or more specialized, as the Eohippus was changed into the horse. In any case let us remember this simple fact—an elementary corollary of Evolution, but never seriously considered before Darwin's time— that there is not the slightest reason for supposing that the powers, intellectual, spiritual, and emotional, which we human beings happen to possess, are the highest of which this planet is capable. Our amphibian ancestor, which certainly had no more brains than a frog, could give rise to descendants with the brains of men. There is no reason whatever for supposing that another such stride, and yet further strides in mental possibility, may not occur. The one sure thing of which the spectacle of Evolution convinces us is that things will not remain as they are. And since there is a vague persuasion very widely diffused at present that in a few million years this planet of ours will "freeze up" and the evolution of life cease, we will conclude this chapter upon the fact of Evolution with a brief note upon Man's place in Time.

§ 4. Man's Place in Time.

EVOLUTION, we now perceive, is a present reality, a going concern. There is no sign in man's incomplete being that it has culminated or is in any way arrested. So that a vividly interesting question opens before us: How long can the evolution of life go on? To that question it is now possible to give a tentative answer.

The dating of fossils by rocks, as we have explained it in an earlier chapter, can be at best only a relative dating; it tells us that one kind of animal or plant lived and died before a second kind and after a third; but it tells us nothing as to any absolute date in years. Yet the reader will have remarked that we have been giving the absolute age of different rocks with very considerable assurance: Early Cenozoic fifty million years ago, and so forth. It is time to explain how this is possible.

It is possible because, as the work of recent years has shown, some of the rocks of the earth's crust contain what we may call, without any fantastic exaggeration, geological clocks. They contain timekeepers that were set chemically ticking when the rock was first formed, and have gone on ticking at the same rate, without once needing to be wound or regulated, ever since. These clocks are what are known as the radio-active elements, radium, uranium, thorium, and others, which exist in certain minerals. The work of Becquerel on uranium, which speedily led to the discovery of radium by Madame Curie in 1898, was the starting-point for a series of researches which have altered all our ideas on the constitution of matter, and, incidentally, given earth-history a measurable chronology. These radio-active elements, as everybody knows nowadays, shoot out particles of matter from their atoms, and in so doing transform themselves into different elements. They do this in such a way as to become effectual chronometers.

The particles given off are sometimes electrically charged atoms of the light gas, helium (which is a stable element and shows no further change), sometimes the far tinier electrons, the spinning bricks of which the atoms, save for their cores, are built. Uranium is the parent of radium. After shooting off three atoms of helium and several electrons, it becomes radium. Radium then continues the process of change; it first discharges a gas, radium emanation, and finally, after five helium atoms have been shot away, becomes plain lead. This, having stable atoms, does not change any more; the clock, so far as the lead atom is concerned, has run down, and we know no means by which it can be wound up again. The lead thus produced, though in its chemical behaviour quite normal, differs from ordinary lead in having an atomic weight of 206 instead of 207.2; we can call it uranium-lead. The whole process may be summed up by an equation: 1 atom of uranium = 1 atom of uranium-lead + 8 atoms of helium + energy.

Thorium, another radio-active element, goes through a similar series of transformations, and after shooting out six atoms of helium, also ends up as lead. This thorium-lead, however, differs from uranium-lead in having an atomic weight of 208.

Now, the bearing of all this on the computing of geological dates is very simple. This disintegration does not go on haphazard and wildly. It is timed. Each of these transformations takes place at its own definite rate, as regularly and inevitably as the swing of a pendulum. The physicist assures us that if we took a definite quantity of radium, say a milligram, then after 1,700 years there would be only half of it left—exactly half—the rest would have turned into helium, lead, and traces of the intermediate steps. Uranium, on the other hand, is a slow disintegrator; you would only lose half your uranium after 4,500 million years. Thorium, too, has its own particular speed of metamorphosis. But since the helium and the lead which are thus generated do not undergo any further changes, we can calculate precisely how long it would take for any given proportion of lead to be accumulated in a mineral containing uranium or thorium. And so, if we can find a rock of, say, Lower Carboniferous Age (III E 1) in which, when it was first formed, radio-active minerals containing uranium or thorium were crystallized out, we can tell its age in years by measuring the proportion of helium and of lead which has been produced in these minerals since they were first locked up in their rocky prison.

Lord Rayleigh measured the amount of helium produced by uranium-bearing minerals, and found that a gram of uranium would take nine million years to produce a cubic centimetre of helium, which agrees with calculations based on the rate of step-by-step transformation. As regards lead, it can be calculated that a ton of uranium gives rise to 1/7,400th of a gram of lead every year. Thus, if on analysis we find 1 per cent of lead in a uranium-containing mineral, this proportion must have taken 1/99th of 7,400 million years to accumulate; while if there had been 10 per cent of lead, the time needed would have been 1/9th of 7,400 million years. The rate for thorium is slower, but the principle is the same.

The one assumption made is that the radium clocks never change their rate of going; and physicists, after trying in vain to alter that rate by every conceivable means, are agreed that they never do. There are various small mathematical corrections to be made, and various precautions to be taken in the practical procedure; but all such difficulties can be readily overcome. How they are overcome is clearly set forth in Professor Holmes' little book, *The Age of the Earth*.

There, too, will be found an account of other methods of estimating geological time—as by the rate of erosion of the land, by the rate of deposition of new sediments, by the rate at which the sea, which is always receiving dissolved salts and always evaporating, grows saltier, by timing geological processes against periods of known length, such as the precession of the equinoxes with its 21,000 years' cycle, and so forth. These all supply checks upon our radio active estimates.

In addition there are one or two methods by which, at certain pages of the record of the past, we can get even an estimate in single years. For example, trees form annual rings, and some of the Big Trees of California take us back several thousand years. When ice-sheets cover part of a country, the flood produced by each summer's melting lays down a layer of clayey deposit over the neighbouring regions; and Baron de Geer has shown, by counting these layers in Southern Sweden, that the retreat of the last ice-sheet in Scandinavia took over 10,000 years. But these more detailed methods commonly apply only to the immediate past, and to periods of time that are but one or two ticks of geology's clock.

There is also a further method applicable to any period where marine shells were fossilized. It is based on the fact that many bivalve shells show annual growth-rings like trees, and that the annual growth-rings are divided up into minor rings, each new addition to the shell being made after a feeding period. These minor rings vary much in breadth, according to outer conditions. Thus each annual growth-ring bears its own individual stamp in the number and size of its separate feeding-rings; and the same stamp will be impressed on all the shells in the same neighbourhood. By comparing a whole series of shells in the successive layers from one locality we can identify the separate years by the pattern they have left on the growth-rings, and so count their succession. This method is being worked out by Prof. W. M. Winton of Texas; it should give valuable results whenever thick layers of one type have been laid down in one place, and contain the shells of bivalves.

Ultimately, by such means as these, we may carry a surprisingly detailed calendar back for hundreds of thousands and even millions of years into the past from the present time. And even for the remoter periods we are already approaching a sufficiency of knowledge whereby the margin of error can be narrowed down to ten million years or so, which on the geologist's scale of time is not a large figure, for it only means being out by about five or ten per cent.

Such are the data upon which we base the figures we have inserted upon our diagram of geological formations. If the reader will turn back to that he will find one or two points of very great interest in relation to our curiosity about the future course and duration of Evolution (Fig 122).

In the first place he will remark that there is a huge disproportion between the lengths of the great eras of geological time. The Age of Mammals seems hardly to merit the name of Era at all: it has only endured for about 50 million years. The Age of Reptiles lasted about 125 million years, while the Paleozoic droned on, without even a bird or mammal or flowering plant, for over 300 million. That puts the date of the Early Cambrian back to 500 million. Below these lowest Cambrian rocks there is still a huge thickness of earlier rock-deposits —at least 180,000 feet of them—in which, however, save in the uppermost 10,000 feet, no fossils have been found. The time taken for the formation of these Pre-Cambrian rocks was longer than for all the later eras together; for one of them is radium-dated back to over 1,200 million years.

The earth came into being when the ancestral sun was disrupted by the too-near approach of some other star. Of that stupendous birth you can read in Jeans' *Astronomy and Cosmogony*. The date of that remote event cannot yet be estimated with such accuracy as can the ages of rocks, though the limits are rapidly being narrowed down. As Holmes shows, it cannot be less than 1,600 nor more than 3,000 million years ago; probably the lower estimate is nearer the truth.

Let us provisionally take 2,000 million years as the age of Earth. Life, as revealed by actual fossils, has been in existence for nearly a third of this time. Without any doubt the real age of life is greater, since the first living things would be soft and squashy and very seldom fossilized, and even if they were preserved would, in most cases, have been later baked and crushed out of recognition.

It is only in the last fifth of the earth's history that we know of vertebrates, and in less than a sixth of it that we know of land-vertebrates. Mammals have been on the scene for only a thirteenth of the time, and modern placental mammals for only a thirtieth. And as for man, he is a mere upstart. The earliest creatures that could be called men, or at least not apes, cannot possibly have been on earth's stage for more than a paltry ten million years, or one two-hundredth of the total, and it may well prove necessary to halve even this estimate; while our own species of man boasts perhaps a million years of history—one-twentieth of one per cent of the earth's full record.

Time is telescoped, and the centuries of history shrivel. Fifteen thousand years ago man was still in the Old Stone Age. Civilization, in the sense of a stable social life based upon agriculture and metal-working, dates back to less than ten thousand years. Ten thousand years—when it took at the very least ten millions to generate man from the first tailless ape, and a hundred million for that ape to be brought into being from the first mammal!

With this revelation of the huge spaces of earth's past the doors of the future, too, seem to open. We know a good deal now about the rate of cooling of the earth and of the sun. That sort of knowledge also grows more and more exact. The discoveries that heat may be generated by radio-activity, by shrinkage, and by the actual transformation of matter, have enormously enlarged the possible future of life upon earth. There is every reason to suppose that conditions on our planet will continue to allow life to flourish in the future for as long as they have allowed it to flourish in the past; indeed, this trifle of a thousand million years or so for the future of terrestrial life is almost certainly an under-estimate, granted of course that no unforeseen catastrophe breaks in upon it.

Man is part of an unbroken stream of life. That same stream in the dawn of life on earth manifested itself in the form of single microscopic cells; hundreds of millions of years later, after transformation through forms we dimly guess at—forms of polyps, of worm-like creatures, of headless things like lancelets, it flowed through thousands of generations in the form of fish; it emerged on land, it learnt to be a reptile, it covered itself with hair and warmed its blood, and fed its young with milk. Still without break of continuity, it transformed itself to become fully mammalian, its young to grow as parasites upon its life. Four-footed, tailed and hairy, it took to the Eocene forests; it grew into lemur, into monkey, into ape; and finally ape turned man-ape, and man-ape grew to man.

If that self-same stream of life that flows through our human generations and that we call man was once fish, and if those fishy ancestors could be transformed into our present selves in three hundred million years, without the aid of conscious purpose in any of the pre-human forbears, who shall prophesy what our race may not achieve and into what it may not transform itself before another such period in the history of life on earth has passed?

To grasp the full implications of this estimate of available time is to realize that we are still only in the dawn of consciousness and thought, and that all human will and wisdom has ever done is no more than an augury of what it may yet achieve. Evolution presents itself as an accelerating process, gathering momentum and hardly yet beyond the beginning of its revelations.

BOOK FOUR

THE HOW AND THE WHY OF DEVELOPMENT AND EVOLUTION

I

THE ESSENCE OF THE CONTROVERSIES ABOUT EVOLUTION

§ 1. *The Chief Theories of Evolution.* § 2. *Method of Treatment.*

§ 1. The Chief Theories of Evolution.

IN BOOK 3 we have been dealing with facts beyond any reasonable controversy. We have put them plainly and we hope convincingly before the reader. If we have failed to convince, the fault lies in our writing and not in the facts. Life, we have shown, has appeared not multitudinously and again and again in the world, but at one particular stage in our planet's history, and from that one beginning it has developed like a branching tree. It has not been multifariously and repeatedly created in its present or kindred forms; it has unfolded from lowly and simple beginnings, through a vast variety of species, to all the animal, vegetable and other organisms the biologist contemplates to-day. Special creation of each animal and vegetable type is a parable or a myth. Evolution is the shape of life and a fact as well-established now as the roundness of the earth or the relative immensity of the sun. All these three facts have been disputed in the past. To-day controversy about any of them is dead.

Equally dead we shall find is the older and really more plausible belief that life has had numerous origins and, even now, can at times be created afresh. That, it seems, is not so. *Life is one thing.* Every living thing is related through a common descent to all the rest of life. There is no reason a priori why this should be so. But all the evidence is that it is so.

In all the three instances we have given—the round earth, the larger sun, and the evolution of life, men have disputed these great generalizations because they had started in life with contrary assumptions and found the shock of the new idea too great. They had intermingled their moral and religious ideas with the notion of a special creation of each kind of animal at a certain date, or with the notion of a flat earth, or with the notion of a small subservient sun going about our planet, and it seemed to them that if these notions were destroyed their very heavens would fall. But new generations have followed them, have accepted the new ideas and found the heavens of religious feeling and moral impulse none the worse for a broadened and enlightened outlook. To-day there is no denial of the fact of organic evolution except on the part of manifestly ignorant, prejudiced and superstitious minds.

But here we enter upon a less certain and established region of biological study. In this Fourth Book we are going to discuss *how* individual development is carried out, and, further, *how* Evolution has occurred. There we find active and intelligent minds still differing very widely. What are the relations of individual development to the development of the species? There is no question any longer that Evolution has occurred, but our question is now, what has been its method? Or its methods?

This is a field where the débris and glow of recent controversies are still evident and where wide and often flaming differences of opinion are still found. And, just as in the opening of Book 3 we made it quite plain what fact we had to prove, so here it will enable the reader to understand the full significance of what follows if we give first the broad questions our chapters are designed to illuminate, and point out what is still arguable and what is the present state of the discussion. What are the Theories of Evolution between which we are asked to decide?

The fact of vital Evolution has gleamed upon intelligent minds at various phases in the world's history, but the modern revival of biological science had been going on for some time before it rose again to recognition. Linnæus (1707–1778) seems to have had no doubts of the fixity of species. It was only towards the close of the eighteenth century and with the increasing study of comparative anatomy and fossils that the fixity of species began to be questioned.

At first the fact of Evolution was seen piecemeal, as a possible change of one species into another within the boundaries of this or that restricted group of allied forms. It was not apprehended as a process comprehending all living things. Perhaps all the carnivores were genetically related, for example, or all the horned cattle. It was then generally called Transformism. The word Evolution came later. And the question whether the process included man was either not raised, or plainly or tacitly answered in the negative.

The first attempt to explain Transformism was to ascribe it to the effort of the living being to adapt itself to the often difficult conditions under which it had to live. The French naturalist Lamarck (1744–1829) pointed out that the individual was responsive to its circumstances, that it used and developed this organ and made little use of and therefore did not greatly develop that, that within limits need and exercise called forth structure; and he supposed that these individual adaptations were in a measure inherited. The three-toed horse—if we may use an example unknown to Lamarck—which under changing conditions was always scampering on firm prairies and scarcely ever going on soft ground, made no use of its once useful side-toes and so they were not stimulated to develop, while the business toe got all the work and all the benefit. The foals, according to the Lamarckian idea, inherited the enhanced main toe and the reduced side ones. This line of argument was made exceedingly plausible by the known fact that we all develop best the organs we use most; the rower his biceps, the singer his chest. The weakness of the Lamarckian case, or at least the unproven assumption of it, was that the individual development was in any degree inherited.

In ordinary biological discussion the individual development is called an "acquired characteristic": the size of the rower's biceps, for instance. Lamarck assumed the inheritance of acquired characteristics and found in that a partial explanation of Transformism. To this day the belief in the inheritance of acquired characters is called Lamarckism. With the inclusion of an involuntary response to the environment (such as the response of growing corals to currents or the darkening of some birds' feathers when they are reared in a warm and humid atmosphere) and the inheritance of this response, it is called Neo-Lamarckism—Lamarckism modernized.

Lamarck's realization of at least a limited evolution of species, Transformism, was based on an infinitely smaller knowledge of fact than we have to-day. He relied chiefly on fossil shells, rudimentary structures, and the manifest anatomical resemblances of animals for his belief that Evolution occurred. It was only later (1828) that Geoffroy St. Hilaire called attention to the embryological evidence for Evolution.

Several distinguished living biologists are Neo-Lamarckians. And the view has appealed to many people because of the moral attractiveness of the idea of effort achieving enduring consequences. Master what you can of mathematics and your child will compute with greater ease; be merciful and your children will find it less difficult to practise mercy. One likes to think in that fashion. And with various additions and improvements Lamarckism is to be found vigorously paralleled in much modern thought outside the world of biological specialists. There has been added to the individual effort the idea of an upward driving force of a general sort. Bergson finds an *élan vital*, George Bernard Shaw a *life-force*, both mystical drives towards adaptation, coming from or acting through the organism. Both owe something, no doubt, to Schopenhauer's idea of a driving Will in things. Whether such an hypothesis is necessary or even harmonious with the facts of the case we shall leave the reader to judge at the end of this Book 4. We give it here as a second theory, which must be treated with respect, the theory of an upward *drive* in life.

Now, while Lamarck was elaborating his transformist ideas, an English clergyman, Dr. Malthus (1766–1834) was developing certain views that did not at first sight seem to have any bearing upon natural history and Transformism at all. His preoccupations seem to have been purely social. He was struck by the rapid increase of the human population about him—and he lived in a prolific age. It was increasing, he thought, much faster than was the food-supply. Consequently there was already a harsh struggle for subsistence going on. Mankind was breeding its way towards starvation; the weakest would go to the wall. Famine and the check of pestilence were the natural counters to this drift towards over-population and an unendurable poverty, and he urged his fellow-creatures to avoid such miseries by restraining their increase through late marriage and through continence. The artificial interference with conception known as birth-control or Neo-Malthusianism, we may note, had no place in his philosophy. That was as far as he got; he betrayed no consciousness of the bearing of his observations upon the ideas of Transformism, of which indeed he may have been quite unaware.

It happened, however, that his writings were read by two scientific travellers and naturalists who were both coming to believe in the fact of Evolution but by no means satisfied with Lamarckism as an explanation of it ("Creative Evolution" with its *élan vital* was still to come). These were Charles Darwin (1809–82) and Alfred Russel Wallace (1823–1913).

It is well to note here that Darwin did not "discover" Evolution, as many people suppose. Evolution is not Darwinism and Darwinism is not Evolution. The idea of Evolution is not only at least as old in modern thought as Lamarck, but adumbrations of it are clearly traceable in such ancient writers as Lucretius and Empedocles. But in the minds of Darwin and Wallace, looking for operating causes for the evolutionary process, the phrase of Malthus, "the struggle for subsistence," found a fruitful soil. Both realized a second great fact— for fact it is—in the general conditions of life, namely Natural Selection. Every living species is continually producing a multitude of individuals, many more than can all survive, varying more or less among themselves, and all competing against each other for food and a place in the sun. On the whole, Nature will let the better fitted ones live more abundantly and she will kill off the less happily constituted. The weaker will go to the wall; they will not breed so much; the stronger and their offspring will prevail. Assuming that weakness and strength and, in general, fitness and unfitness are hereditable qualities —and that is the general persuasion—a species must be always on the grindstone, having its unsuitable strains eliminated and its suitable strains left in possession.

Now, let us be quite clear here; speaking with precision, Natural Selection we say is not a theory but a fact. But does it, in connexion with the small differences that occur between every individual and its peers and the distinctive resemblance of parent and child, suffice to account for the whole spectacle of Evolution? With or without that element of effort and hereditable acquirement which Lamarckism asserts? There we come to speculative matter, to theories. Darwin thought it did. He did not contradict the Lamarckian hypothesis, but he added a new factor in the process, which factor he drew from Malthus. In 1859 he published a book which made an immense stir in the world, and he called it *The Origin of Species by Means of Natural Selection*. We have insisted that Natural Selection is not a theory. But, on the other hand, this appeal to the fact of Natural Selection and the fact of hereditable variations as giving between them a full and sufficient explanation of the fact of Evolution, is a theory; it is the Darwinian Theory. To the majority of even highly educated people at that period, educated for the most part upon lines of a narrow religious orthodoxy, it brought home for the first time the neglected and repudiated fact of Evolution, and made it seem credible. Explanatory theory and fact to be explained appeared together in their minds, and so to this day, in common talk, Evolution, Darwinism, and Natural Selection are hopelessly mixed and muddled. It became the custom to speak of the Darwinian Theory, the Theory of Natural Selection, and the Theory of Evolution indifferently.

Moreover, Darwin and his associates drew attention to the particular aspect of the question of Evolution that had hitherto been in the background. He followed up his *Origin of Species* by a book upon *The Descent of Man*. He insisted that man was an animal and that if the facts of Evolution were true they applied to man. If other living things had not been specially created but evolved, so, too, man must have been evolved. To do this was to challenge and bring into the discussion the whole world of contemporary theology. What had been a field of interesting speculation for naturalists became an arena of intense interest to the ordinary man.

Darwin's publication was followed by furious controversies, in which Thomas Henry Huxley (1825-1895) and Ernst Haeckel (1834-1919) played notable parts in championing the evolutionary cause and defending Darwin and his views from misrepresentation. Huxley liked to call himself "Darwin's bull-dog." But the controversies did much to darken counsel in these matters. The fact of Evolution had to be proved to most people, and many were only too eager to suppose that the defeat or qualification of the theory would abolish the fact. To many of them to the end of their days it remained a theory, and an unsound one at that. All sorts of secondary considerations have played their part in these disputes. There is, for example, a real dislike of the fact of Natural Selection on the part of such a fine and sympathetic nature as Mr. G. B. Shaw's. It seems to him unchivalrous and vile for science to recognize that the weakest do go to the wall. It is hitting the fellow who is down. In the philosophy of a wilful life-force it is natural the wish should be father to the thought. He wishes things were not so, and therefore he declares they are not so, and he does it with great charm, confidence, and conviction. It pleases Mr. Shaw to tell the world at regular intervals that Natural Selection has been "exploded," and it does not hamper the operation of Natural Selection in the very least that he should do this. But Natural Selection has been no more "exploded" by recent research than the rejection of underweight coins at the Mint has been exploded by the doctrine of relativity. Wherever there are favourable or unfavourable heritable variations Natural Selection must be at work.

Nearly three-quarters of a century have passed since the controversial cataclysms of the mid-Victorian period, and Darwinism has been criticized in every conceivable way. It cannot be said that it has been destroyed, but it has undergone restatement in certain respects.

The modification of a species by the natural selection of variations is still an undefeated theory. That idea from Darwin's writings lives and flourishes. The remoulding of Darwinism has concerned the part of it which deals with the mechanism of heredity and the intimate nature of variations. For in Darwin's time hardly anything was definitely known about the inheritance of individual differences. The chromosomes to be presently described had not yet been seen; the essential facts of fertilization were unknown; most important of all, experimental breeding had not drawn a clear distinction between variations which are inherited and those which are not.

Since that time accurate knowledge has accumulated on those questions. The microscopic changes in the germ-cells that accompany fertilization were observed; the chromosomes, the bearers of the physical basis of heredity, were discovered, and their complicated but regular movements were traced as they passed from one generation to the next. Moreover, the study of inheritance was attacked experimentally. Even in Darwin's time the Austrian Abbé Mendel (1822–84) had experimented on the interbreeding of varieties of plants and had discovered the two most fundamental laws that govern hereditary transmission. But the significance of his work was not recognized at the time. His communication to the little Natural History Society at Brunn (now called Brno) dealt chiefly with peas and arithmetic, not the sort of things that cause excitement and clamour, and in the confused tumult of the nineteenth century Evolution controversy they passed unnoticed. Only in the opening years of the twentieth century was his work disinterred and brought to bear on the discussion of Evolution.

This rediscovery was the stimulus for an enormous amount of careful experimental breeding of animals and plants. New conceptions arose of how variations originate and are handed on from generation to generation, and of how they may play their part in the struggle for existence.

Here we are merely revising the broad issues before us. Later on we shall discuss the question whether variations occur through a "life-force" driving the whole species in a definite direction, or through some blind disposition to vary evoked by the action of external conditions upon the reproductive process. That is an interesting issue of profound importance. And we shall have to consider the still-vexed question whether the inheritance of individually acquired characteristics, which was the essence of the original theory of Lamarck, occurs or no. This was flatly denied by Weismann (1834–1914) and disproved in many instances. He and his followers are sometimes spoken of as Neo-Darwinians. They believe variation to be a purely random process, resulting neither from a persistent urge nor from the Lamarckian moulding of the individual body; the direction of Evolution being determined entirely by Natural Selection. Their complete denial of the evolutionary value of individual experience gives a flavour of hard predestination to their views. Unless it vary by the grace of unknown forces, a struggling lineage is doomed; no individual luck or effort can save it. And clearly no education, no social protection can cure the innate defects of any inferior human family. Its rôle is extinction.

A very fine and curious issue to which we shall later direct the reader's attention is that of *Orthogenesis*. It is alleged that in a certain number of cases species, even though fairly well adapted to their conditions, and without experiencing any change of conditions, have by virtue of a sort of inner drive, an innate destiny of the species, gone through considerable evolutionary change. Professor Henry Fairfield Osborn finds this drive convincingly displayed in various fossil series, such as the horses, camels, and titanotheres. Of course, where the changes produced by the orthogenesis have been disadvantageous, Natural Selection has at last arrested the drive and extinguished the line. But the supposition that the new form, new structure or other characteristic is not advantageous or is insufficiently advantageous to have "survival value" is difficult to establish. It may involve an advantage which the observer has not recognized. And it is still more difficult to show that the drive towards variation in a definite direction was innate. There may be a thousand outer influences working upon the minute elements in reproduction of which as yet we know nothing, from obscure chemical factors in the food to electro-magnetic radiations from outer space. There may be a thousand subtle selective influences rejecting this variation and preserving that. Here again our duty will be to sum up the known facts and views and leave the conclusion to the reader.

We have then still active in the biological field (1) the *élan-vitalist*, (2) the Neo-Lamarckian, and (3) the Neo-Darwinian; three distinctive schools as to the origin of variations. They are by no means mutually exclusive. Any follower of either school may attach more or less importance to the variational drive or conversely to the action of Natural Selection, and Orthogenesis may be used to account for more or less or none of the changes that have occurred. In a world of earnest workers these various shades and blending of opinion are hotly debated. Biologists can be as sensitive to heresy as theologians or any other sort of men in deadly earnest, and it is quite easy for dull-minded or dishonest controversialists delving in the literature of the subject to clip out such phrases as that "Darwin has been disposed of," or that "Natural Selection is inadequate," and pretend that a conviction or refutation in some particular is an absolute reversal of view. But, indeed, no such collapse has occurred. Difficulties in "accounting for" variation are minor difficulties in face of the invincible facts of the evolutionary process. Every day the form and details of Evolution, the life process, are seen more clearly, solidly, certainly, and coherently.

We need scarcely point out to the interested reader how temperamental disposition and philosophical and moral preoccupations may dispose men's minds towards one or other of the three types of opinion. This possibility gives this branch of our subject an interest and excitement far beyond the strictly biological field. There is something cold and stern, very attractive to a certain hard, clear type of mind, in the Neo-Darwinian attitude. There is something heroic in the obstinate advance of Orthogenesis. The Neo-Lamarckian view appeals most to those combatant spirits who would figure man in a Promethean and finally hopeful conflict with the universe; again, the mystic and the believer in a continually directive divinity incline very naturally towards the hidden upward urgency of the *élan vital*. For him it becomes the finger of God. It is a return towards the idea of creation, as Bergson's popular and attractive phrase *Creative Evolution* reminds us.

There remains one other temperamental type which has found expression in these discussions, and that is the brilliant sceptic as typified by the late Professor William Bateson. He accepted the fact of Evolution, if only on the palæontological evidence, but, as the outcome of a life spent largely in the study of variation and especially of Mendelism, he developed an increasing inability to satisfy himself how any progressive variation could ever occur. He crowned his scientific career by various lectures and addresses in which he reiterated his imaginative failure. This type of agnosticism was probably the negative aspect of a passionate and unquestioning faith in the implacable unteachableness and integrity of certain Mendelian units of heredity we shall presently describe and discuss. Later work has removed much of the point of his criticisms.

§ 2. Method of Treatment.

THE threefold author of this work must admit that the writing and arrangement of this Book has cost him more trouble and effort than any other portion of the undertaking. A word about his difficulties may be of help to the reader who is steadily following this exposition of the Science of Life and who has now to read what has been so painstakingly written.

The broad facts and the consequences of inheritance are of universal interest; the minute study of the mechanism of individual development and variation is, on the other hand, very obscure, outside everyday experience, and can easily be made very complicated. Yet the broad questions we have posed in the previous section are only to be grasped soundly after the nature and bearing of this minute mechanism has been understood. The problem has been to give a full and exact account of this difficult subject, an account which a lawyer, let us say, or a schoolmaster or anyone of general intelligence but without special biological training, might be expected to follow with interest and understand, and at the same time not to confuse, weary and defeat the natural widespread curiosity about these things.

We decided to eliminate every avoidable technical term, and those we have found unavoidable—*genes, germ-plasm, chromosomes,* for example—we have explained with sedulous care and much mutual criticism. Much of the current literature of heredity resorts to symbolism and formulæ that suggest a mathematical text-book. It sends inquiring spirits weary and empty away. Some passages here were first shaped in that fashion and then, when their meaning had been completely worked out, very carefully rewritten in the English language. We fought each other not indeed for compression, but for simplification and compactness. This Book 4 was at one time, we may mention, twice as long as it is at present. We hope after all this labour to have produced a story as readable as the rest of this work and certainly, we believe, more stimulating to the imagination. For the implications of these minute particulars are immense.

For a time, then, we shall set aside the great interrogations of the previous section as if they were quite open questions and give the essential facts of individual development and reproduction. It may come as a novel idea to some of our readers that sex, identified in their minds with reproduction, probably had nothing to do with it in the first place and that the two were only slowly entangled. We shall then describe the minute mechanism of sexual union and inheritance. The reader will follow the work of the Abbé Mendel from its modest beginnings to its broadening and developing consequences. Then, stripped of all jargon, the central concepts of Genetics will be displayed in their essential beauty and importance. Some further sections of interest about sex will follow, and then we shall take up again those notes of interrogation with which we ended the previous section.

We shall then find that we are able to estimate the factors in the evolutionary process far more surely than we are now able to do. We shall put the Neo-Lamarckian on trial and come to grips with that widespread and popular idea of an upward evolutionary urge, of which, as we have said, Professor Bergson and Mr. G. B. Shaw are perhaps the best known exponents. And then, emerging from such controversial matters, we shall treat in the next Book of the marvellous growth of the Tree of Life from its beginning.

II

HOW INDIVIDUALS ORIGINATE

§ 1. *Can Life Arise Spontaneously?* § 2. *Sexless Reproduction the Original Method.* § 3. *Sex Is a Complication of Reproduction.* § 4. *The Gametes, or Marrying Cells.* § 5. *Some Evasions and Replacements of Sexuality.* § 6. *Artificial Propagation.* § 7. *A Note on the Regeneration of Lost Parts.* § 8. *Grafts and the Chimæra.* § 9. *What Is Meant by the Germ-plasm.*

§ 1. Can Life Arise Spontaneously?

WE HAVE asserted already several times that all life derives from preceding life. But we have given very little to sustain that assertion. Here we will deal rather more fully with this very important issue. Our proposition is that life is a branching tree. We want to dispose of the possibility that it is several branching trees or that new life-trees have arisen or can originate. We have to answer the question whether under existing conditions living things may not still be able to rise from lifeless material?

Right down to the middle of the last century it was thought, even by many biologists, that living creatures could so arise—"spontaneously." It was believed in not as a miracle, a rare and occasional wonder of nature, but as a fact of everyday experience.

The swarm of busy maggots that appears whenever anything is left to decay was supposed to be directly generated from the putrefying material. In antiquity it was believed that frogs and reptiles could be generated out of mud and slime: Virgil in the Georgics gives a recipe for producing bees from the carcass of an ox; and Samson's riddle (Judges XIV) seems based on a similar misconception. He saw some bee-like flies emerge from the decaying body of a lion and put his riddle: "Out of the eater came forth meat and out of the strong came forth sweetness." It was a perfectly natural and reasonable error in the days when the habit of interrogative observation had not yet been developed. But the Bible narrator has gone a little beyond the possibilities of the case by adding a honeycomb to the story.

HOW INDIVIDUALS ORIGINATE

It was not until the seventeenth century that Redi exploded the belief that blowflies and their maggots are produced by decaying meat itself. He marked the blowfly at its work. When he prevented flies from having access to the meat by covering it with gauze no maggots were generated. It was a simple, clear-cut experiment, and yet no one had previously displayed the intelligent scepticism needed to attempt it.

By the seventeenth century the discovery of the microscope had opened up a new world of living creatures, whose life-histories were hard to trace out owing to their complicated life-cycles, to their minuteness, and their liability to be blown from place to place in a condition of nearly-suspended life. These creatures had an air of turning up out of nothingness. Thus they were supposed to be "spontaneously generated."

In the middle of the nineteenth century the genius of Pasteur, with a combination of rigorous experiment and patient perseverance, finally clinched the matter and proved that all visible living things, at any rate in the conditions which now obtain in nature, arise only from others of the same sort. Soups and such-like infusions were up to his time supposed to be natural environments for spontaneous generation. But the most nutrient of soups will stay pellucid, without a trace of decay, if sealed off from the air after being sufficiently heated to destroy all microbes and their spores. Even if air be later admitted, no decay will set in, provided no bacterial spores actually fall into the soup. They may be filtered off by passing the air through cotton-wool; or by letting the air pass very slowly along a series of U-shaped bends in a narrow tube, when they will settle at the bottom of the U's. Or if uninfected air be admitted to sterilized soup, as for instance by break-

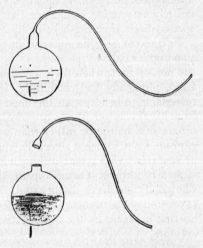

FIG. 156. One of Pasteur's experiments.

When yeast infusion is boiled in a swan-necked flask, no putrefaction follows, even though it is left open to the air, because any living germs in the air settle in the bend of the U or are caught on the moist sides of the neck. If the neck is broken off, putrefaction sets in, since spores and cysts now have ready admission.

ing open the sealed flask at the top of a high mountain, and the flask then resealed, again there will be no decay. Nothing decays unless bacteria or their spores settle upon it. The same boiled infusion which, left exposed to the air, within a week teems with many kinds of bacteria and protozoa, will remain untenanted indefinitely if their germs and spores and cysts are debarred entrance. We can say now with an entirely reasonable confidence that all life which exists to-day has sprung direct from pre-existing life.

But, of course, this apparent impossibility of spontaneous generation applies only to the world as we know it to-day. At some time in the remote past, when the earth was hotter and its air and crust differed, physically and chemically, from their present state, it seems reasonable to believe that life must have originated in a simple form from lifeless matter. It was presumably a fairly gradual change, a slow progressive synthesis, rather than a sudden leaping into being of organisms from formless slime. To that problem we shall return. Our present conclusion is that, although life could be generated under those strange conditions, it never appears spontaneously at the present time. The conditions have disappeared. The animals and plants that we know to-day are branches of a single Tree of Life, growing out of pre-existing branches by the reproductive processes that we have now to examine.

§ 2. Sexless Reproduction the Original Method.

AMONG most of the higher animals reproduction, the origination of new individuals, is inseparable from sex. It takes two to make any addition to the race. But that is by no means true of life generally. Nearly all living things have some kind of sexual process, but most of them have also sexless ways of increasing their kind. In biological language, reproduction is often *asexual*. Our own strict adherence to a single generative technique is in fact to be regarded as the exceptional thing.

The most primitive way of multiplying, the method used by the smallest and simplest creatures, is by splitting the whole body into two halves, each of which grows into a complete new individual. This is known as *binary fission*. (Figs. 114, 157.) In such cases the offspring is not merely a detached part of the parent; the whole substance of the parent becomes offspring. The parent leaves no corpse. It ceases to exist; but for it there is no death—only duplication.

This reproduction by splitting, this multiplication of substance

HOW INDIVIDUALS ORIGINATE

passed on into ever new individuals, is most spectacular in the bacteria. A simple bacterium in a congenial nutrient soup may easily accomplish its whole span of individual existence in half an hour. At zero hour, we have one specimen; in half an hour, two specimens—five hours, a thousand and twenty-four—ten hours, over a million—twenty-four hours (if the food holds out) hundreds of billions.

When the dividing animal has a complex structure, remarkable rearrangements must occur at each act of fission. Thus the complicated protozoan Stylonychia (Fig 157), although it is only a single cell, has a definite shape, with special bristles which act like legs, and a gullet armed with vibratile plates. During fission, many of the old organs disappear, new organs appear in each future individual, and, while division is being effected, develop into their definitive forms and migrate to their definitive stations.

Binary fission, we may note, is by no means unknown in manycelled animals. Not only do many sea-anemones reproduce by splitting themselves slowly through from the mouth downwards, but this longitudinal fission is one of the main methods by which the huge colonies of corals are built up from a single original polyp Transverse division, on the other hand, occurs in various worms, especially Planarians.

From this method, two other main methods of asexual reproduction have been derived. One is multiple fission. In this, the single individual divides into many small ones, either by a series of binary fissions following immediately one upon the other without time for any growth between, or, more frequently in single-celled forms, the central nucleus alone divides repeatedly, and the cell as a whole then splits up simultaneously into as many miniature cells as there are nuclei. The parent individual becomes a mass of smaller cells. This method is frequent with single-celled creatures which grow to a—comparatively speaking—large size, such as Foraminifera, or the common large Amœba, and is found also in the malarial parasite.

The other method consists in making the two products of division unequal. Sometimes, as in various Flatworms and Annelid worms, reproduction is by transverse division, but the hinder individual is at first small and unformed, and only gradually grows and develops. This method has the biological advantage that the original individual retains its individuality, its head and general alertness, while giving rise to a new specimen from the less specialized tissues of its hinder parts (see Fig. 97).

Sometimes we find this process several times repeated, and then whole chains of incompletely developed individuals may arise behind the original head, trailing after it until growth has developed them far enough to be detached.

Another variation of this method is adopted by the polyp form from which, in many species of jelly-fish, the adult medusæ arise. Instead of a string of offspring trailing behind a moving parent, the new individuals appear upon the parent as a fixed stem. The advan-

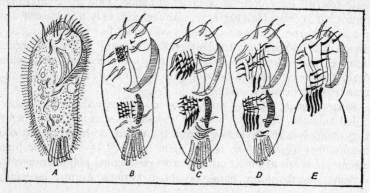

Fig. 157. How a Complicated Cell Divides.

(*A*) *Stylonychia*, one of the most elaborately organized ciliate Protozoa (much magnified). Besides cilia it has various large bristles which serve instead of limbs, and (upper right) a row of triangular vibrating plates to make a food-current. (*B-E*) Various stages in its division (diagrammatic). The bristles of the old animal are dotted, those of the two daughter-animals are black. The current-producing region is transversely shaded. The new bristles appear close together, enlarge, and move to their right places, while the old bristles shrink and disappear. The old current-producing region is kept by the front daughter-cell, while the hind daughter-cell forms a new one.

tage here lies in retaining the fixed attachment during the whole process. The original fixed polyp (which develops from the fertilized egg) becomes divided by a series of transverse grooves into what looks like a pile of miniature saucers, frilled and lappeted at their edges. The top one of these is the most advanced in development, and is the first to free itself and swim away as a baby medusa, the rest one by one following suit, leaving the parent stalk behind to divide and divide again.

When the size of parent and offspring is markedly unequal we speak of *budding*. This is rare in single-celled types, but is the commonest form of sexless reproduction in many-celled plants and animals. Its adoption confers the advantage of leaving the original individual fully formed and active all the time, dispensing with any need for reorganization, by putting all the excess material of growth direct into rapidly developing buds, which can be thrown off when they are fit to look after themselves. Budding is widespread among sea-anemones, jellyfish, sea-squirts and many other invertebrates.

Even when sexless reproduction has been abandoned by the adult, as in all the highest animals, it may persist unsuspected in early stages

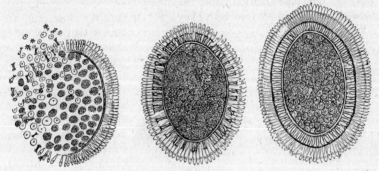

FIG. 158. How a gemmule is formed in the freshwater sponge, Spongilla (highly magnified).

Left: Cells packed with nutriment are gathering at one spot; other cells are beginning to secrete a horny case round them. Centre: The first horny case is finished, and microscopic spicules have been transported to make a further protective layer round it. Right: The horn-secreting cells have retreated and made a second horny case round the layer of spicules. The gemmule is now complete.

of life. Generally, every fertilized egg means a new individual, and if two or more individuals appear at a birth, then two or more eggs were fertilized. The triplets, quartets, quintets and so forth of cats or rabbits arise from groups of separate eggs, three, four, five and so on in number. But there are exceptions. In the Texas armadillo, whose bony carapaces are so often made into baskets to tempt the tourist, four young are always born at a birth, and they are peculiar in sharing a common set of embryonic membranes. These quadruplets are produced by the early division into four parts of a single fertilized egg. Similarly all the cases of human identical twins (twins which are so closely alike as to cause confusion) are due to the splitting of a single embryo at a very early stage. This splitting is, in a perfectly legitimate sense, sexless reproduction. In some parasitic insects this budding of the ovum occurs on a grander scale. Only one egg is laid in the destined victim of the parasite. This divides not into two or four, but into hundreds of cell-masses, from each one of which a whole insect eventually grows.

A very peculiar form of sexless reproduction is found in many sponges. Sponges possess large numbers of wandering cells, not unlike our own white blood corpuscles. A number of these collect together in one spot, much as our white corpuscles collect where there is inflammation. However, their function in this case is not defence against intruders, but reproduction, and they have previously accumulated food-stores within themselves. They pack themselves close, and are then sealed up by other cells in a protective case. In a single sponge there may be a multitude of these agglomerations. When the parent sponge dies down under stress of bad weather or other discouraging conditions, these gemmules, as they are called, fall to the bottom; when conditions are favourable again, they burst their case and develop into new sponges. The interesting fact about these gemmules is that they do not originate at a single center of development, but are built out of scattered cells which come together from different parts of the body (Fig. 158).

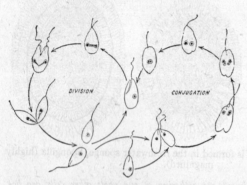

FIG. 159. The life-history of Copromonas, a simple flagellate Protozoan, highly magnified.

Centre: A typical Copromonas with nucleus, flagellum, and gullet. Left: A series of stages in its reproduction; one Copromonas divides into two. Right: Stages in its sexual process; two Copromonas conjugate to form one. (Modified from Dobell.)

Plants, from the lowest to the highest, possess the most varied methods of sexless reproduction. Ferns and fungi, mosses and seaweeds, detach myriads of single cells or spores, each capable of growing directly into a new plant. The thread-algæ, and most fungi through the major part of their life-history, consist of a feltwork of filaments, any bit of which may become detached and continue the race. So with many flowering plants which have creeping underground stems: they send up shoots at intervals, and any one of these, if detached from the rest, can lay claim to be a new-reproduced plant. The strawberry grows new plants at the end of its runners; poplars and many other trees send out suckers from their roots, each of which can grow into a whole new tree. The banyan tree lets fall aerial roots from its branches, each of which on touching the ground takes root and becomes a new trunk, so that a many-trunked tree, a grove in itself, is produced.

§ 3. Sex Is a Complication of Reproduction.

A CONSIDERATION of these strange, asexual methods of reproduction leads to some interesting and important generalizations.

It is evident in the first place that reproduction is essentially nothing more than a special kind of growth. It is growth accompanied by detachment. A bit of the parent-body is split off; a piece of living substance, instead of growing on as a part, grows into a new whole. Reproduction is not, in any strict sense of the word, creation. Nothing is suddenly called into being. It is simply a separation and a remodelling of part of the parent organism.

We come now to a difficult and fascinating problem, the problem of sex. What has sex to do with this reproductive process? Why, if creatures can multiply by mere division, should the complication of sex intrude into the life-cycle?

If we trace very briefly its evolution, and show how it presents itself among the simpler forms of life, we shall discover a fact which to us vertebrates seems startling—that essentially sex is not reproductive. *It is a different thing from reproduction.*

In the bacteria there is no sex. There is simply sexless proliferation. The creatures divide and divide by binary fission, by tearing themselves into halves, and as far as we can see they get on perfectly well without any form of sexual union. Thus the simplest living things today; and thus, presumably, life began.

Among the microscopic single-celled animals and plants we see the beginnings of sexual union. We see it appearing as a new intrusive process, perfectly distinct from reproductive proliferation, interrupting and delaying the latter and in its essence antagonistic to it. In the simplest flagellates, for example, the organisms multiply by binary fission just as bacteria do, but their life-history is complicated by a contrary tendency. Occasionally, if we are watching the creatures through a microscope, two individuals may be seen to come together and to melt completely into one. It is, of course, a much rarer event than normal fission—otherwise the species would not increase—and the individuals taking part in conjugation, as this union is called, often come from different and not very closely related stocks. And it is as obviously unrelated to reproduction as are feeding or excretion (Fig. 159).

In other kinds of protozoa the process is varied in divers ways. In the highly organized Ciliates, for example, two individuals come to lie side by side and then exchange bits of their nuclei. Here the fusion does not involve the whole organism, but, nevertheless, there is a definite mixing of material from different strains. Here also the process is anti-reproductive, because it occupies time which might be spent in the normal rhythm of growth and fission; actually the act of union takes about as long as would three generations of ordinary fission.

In the simpler many-celled animals and plants—the slime-fungi, for example, or the seaweeds—the gradual entanglement of sex with reproduction can be seen. They reproduce by liberating clusters of tiny dancing flagellated cells whose business is to grow into new individuals. But before they do so these cells generally come together in pairs and melt into one. Here then the business of sexual conjugation is relegated to the reproductive cells. Their reproductive value is evidently diminished by this process which halves their number; but it has another compensating purpose, for the members of conjugating pairs are usually from different parents, and so it affords a method for the actual blending of living material from different stocks.

From these lowly plants we can trace a series of stages leading up to the state of affairs that is found in ourselves. In all the higher animals and plants the essence of the process is the same—the reproductive cells have to melt together before they can give rise to new individuals; they are called, therefore, *gametes*, or marrying cells. These gametes are usually of two kinds—active smaller male gametes or sperm-cells and passive larger female gametes or egg-cells. There is every gradation in this inequality. We may have quite equal cells conjugating or we may find very considerable inequality in the size of the conjugating cells. Thus, in the simplest phase of the sexual process there is neither male nor female; there is a sexual process without distinction of sex.

But as a further development of the appearance of the differentiation of the conjugating elements into larger and smaller there presently appears a differentiation of the parent bodies into those producing active and those producing larger conjugating cells. These are the incipient phases of sexual differentiation.

In plants, and in the lowest animals in which sex is thus entangled with reproduction, the sexual method exists, as an alternative to asexual, but in higher forms such as vertebrata reproduction is exclusively sex-ridden. Before we can produce new individuals there has to be this actual mixing of the substance of two parents. Even in our own species this is plainly an anti-reproductive thing, for if we could proliferate asexually it would take only one to do what now needs two, and we could multiply twice as fast. Thus, in the course of Evolution the two originally distinct and antagonistic processes have come together and become inseparably blended.

Evidently there is a riddle here, and one of very profound importance. There must be something very important about this sexual process, this mixing of the stuff of different stocks. What it is we shall presently discover, when the basal principles of heredity have been set forth.

§ 4. The Gametes, or Marrying Cells.

IN SUCH great animals as we are there is a rhythm in the life-history, an alternation between two phases. First there is the human body that we know as a familiar fact of every day. Then, proliferated inside the body, are the gametes, that is to say the sperms and eggs, which are to unite in pairs to form new people. The proliferation of the gametes is our method of multiplying; their subsequent union is the anti-reproductive process which, as we have just seen, is here entangled with propagation.

In many other animals the independence of the gametes is more obvious than it is in ourselves. They are almost like a separate generation alternating with the principal body-phase as the various phases of the liver-fluke alternate. In many marine invertebrates the sperms and eggs are spawned in vast misty clouds into the sea, where the sperms swim about and unite with the eggs in a perfectly independent manner.

The product of the union of two gametes is called a *zygote*. Mr. Everyman, for example, is a zygote, and so are his dog and his canary and the flies on the window of his study and the cabbages in his back garden. All of the familiar living creatures of every day are zygotes. Let us turn to the less familiar gametes with which they alternate.

The egg, the female gamete, is almost invariably a large cell. Even the smallest eggs, such as those of placental mammals, are large as cells go. The human ovum is almost a thousand times the bulk of an average human tissue-cell. Most eggs when ripe are passive and spherical, their only visible activity being to send out a welcoming cone of protoplasm which engulfs the first-come sperm and draws it into the egg, and many lack even this.

The size to which the single egg-cell may grow is enormous. There are small eggs like those of some mammals and flowering plants, under a tenth of a millimetre (1/250th of an inch) across; those of frogs and newts, the size of small shot; others grow as big as peas, like those of trout or some land-snails, as big as damsons, like those of cuttle-fish or pigeons; and so eventually to the huge eggs of dogfish and crocodiles, of swans and ostriches. In all eggs of birds and reptiles the true ovum is the yolky part; this is a cell packed so full of the stored food we call yolk that practically no living protoplasm remains save for a thin film on the surface, and a concentration in the form of a delicate cap at one pole: the white and the shell are not part of the true egg-cell, but merely its wrappings. In Æpyornis, the giant ostrich-like bird, now extinct, from Madagascar this one bloated cell attained a volume of about a gallon!

In shape some eggs depart from the spherical, but usually stop short at oval. Comparatively few, like the eggs of mosquitoes and cockroaches and a number of other insects, are elongated or flattened.

The sperm is much more variable in appearance than the egg. It is nearly always active and seeks restlessly for the waiting ovum. Though the actively swimming sperms of all many-celled animals, from polyps to men, are built on the same plan, with single propelling whip-lash behind, a condensed nucleus forming the bulk of the "head," a middle-piece between the two to carry the centrosome, and a cap covering the front of the head, and probably containing the substance needed to galvanize the egg into development, yet this plan is much varied in its details.

The head may be rounded as in the sperm of bats or starfish, or flattened symmetrically as in man, or flat and set slantwise as in squirrels; it may be a long, thin cylinder as in tortoise, toad and spider; a mere ball like that of amphioxus, or a corkscrew as in finches. The middle-piece may be tiny, or longer than all the rest of the sperm put together. The tail may be a simple whip-lash, long or short, or be equipped with a vibrating fin-membrane of transparent protoplasm. The only exceptions to this general plan are found among flatworms, some of which have sperms with two whip-tails instead of one.

In plants, however, the single-tailed sperm seems never to occur. Those of many lower plants, including mosses, have two flagella, and in some seaweeds these two-lashed sperms are not to be distinguished in appearance from free-living single-celled flagellates. Fern-sperms have usually a corkscrew nucleus, and a great many flagella instead of only two; often they carry with them that part of the general protoplasm which is not needed for production of head and whip-lashes, only to cast it from them before entering the egg. Animal sperms have gone a stage beyond this, for they shed all their unwanted protoplasm before embarking on independent life.

Both among animals and plants, however, are to be found sperms

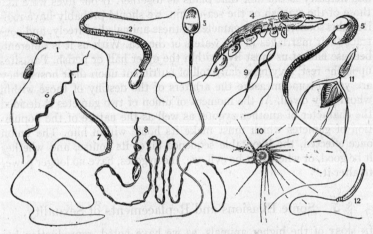

FIG. 160. Some different kinds of male gametes.

1. Sea-urchin. 2. Bat. 3. Man. 4. Fire-bellied Toad. 5. Field-mouse. 6. Spade-footed Toad. 7. Tortoise. 8. Greenfinch. 9. Flatworm (Procerodes). 10. Spider-crab. 11. Squat-lobster. 12. Frog. 13. A Seaweed (the bladder-wrack). All very highly magnified.

which have abandoned their primitive motility. The gamete-nuclei of the flowering plants, which are merely floated along the pollen-tube into the egg, are the most familiar examples; they have abandoned both motility and independence. But all the higher Crustacea have unmotile though independent sperms, strange stiff structures which look like tripods or calthrops or catherine wheels. In some water-fleas the sperms cannot swim but crawl like amœbæ. The roundworms have almost motionless sperms, mere lumps of protoplasm, and mite-sperms and millipede-sperms have also lost the power of swimming.

Thus these gametes present themselves in various forms, but always doing the same things, alternating with zygotes, begotten of zygotes and uniting to form zygotes again. We cannot better summarize the position than by quoting Professor Punnett's little book on Mendelism, in which he says:

"People generally look upon the human species as having two kinds of individuals, males and females, and it is for them that the sociologists and legislators frame their schemes. This, however, is but an imperfect view to take of ourselves. In reality we are of four kinds, male zygotes and female zygotes, large gametes and small gametes, and heredity is the link that binds us together. If our lives were like those of the starfish or the sea-urchin, we should probably have realized this sooner. For the gametes of these animals live freely, and contract their marriages in the waters of the sea. With us it is different, because half of us must live within the other half or perish. Parasites upon the rest, levying a daily toll of nutriment upon their hosts, they are yet in some measure the arbiters of the destiny of those within whom they dwell. At the moment of union of two gametes is decided the character of another zygote, as well as the nature of the population of gametes which must make its home within him. The union once effected, the inevitable sequence takes its course, and whether it be good, or whether it be evil, we, the zygotes, have no longer power to alter it."

§ 5. Some Evasions and Replacements of Sexuality.

IN MOST of the higher animals, as we have noted, reproduction becomes inseparably entangled with sex. But Nature is a tricky worker; as living things have evolved she has vacillated and often enough she has changed her mind and gone back on her previous acts. A certain number of animals are curious in that they have returned to sexless methods of reproduction; evolved from purely sexual stocks, they have found ways of dodging the entanglement.

In many species there exist females which produce eggs as if for sexual reproduction, but nevertheless dispense with males. The eggs develop without fusion with a sperm; the offspring are fatherless. This "virgin reproduction" or *parthenogenesis* is found in a number of organisms, such as greenfly, water-fleas, rotifers, and occasionally flowering plants. From one point of view it is like spore-production, since a single cell is detached and grows into a new organism. But it is different in its origin. Spore-production is among the main original methods of reproduction. Parthenogenesis is secondarily simple, having arisen from true sexual reproduction by dispensing with the need for spermatozoa and with the male that bears them. The spore-producing fern or fungus is asexual, a neuter creature; but the parthenogenetic greenfly or rotifer on the other hand is definitely a female, even though she lacks a mate.

Moreover, the human experimenter has found ways of getting round the need for sexual fusion. One of the most striking discoveries of modern biology was that of Jacques Loeb in 1899, that eggs which normally needed to be fertilized could be made to develop by substituting some man-devised treatment for the natural stimulus of the sperm. In other words, he caused artificial parthenogenesis. Artificial parthenogenesis was first induced by chemical treatment of sea-urchin eggs; later it was found that almost any eggs which are easily accessible to the experimenter by being laid into the water—including those of sea-urchins, starfish, worms, snails, and even frogs—can be made to develop unfertilized; and that the fatherless young thus produced can be as healthy as those arising in the ordinary course of Nature—fatherless starfish and sea-urchins and tadpoles have been reared to the adult stage. The agencies needed to lift the ban from the unfertilized egg and make it develop differ from animal to animal. Heat or shaking will do it in the starfish, while pricking with a needle dipped in blood (a recipe reminiscent of magic!) is needed for frogs.

In mammals the ovum is inaccessible to the experimenter, so that we do not know whether artificial parthenogenesis is possible. There is no reason to suppose that it is not. Surgery and experimental technique are advancing rapidly, so probably we shall find out very soon.

From these strange facts an important point emerges. It is that a spermatozoon does two things to the egg-cell when it fertilizes it. First of all it *activates* the egg-cell and makes development possible; this is the step that Loeb and his followers can imitate. Secondly, it, or part of it, blends physically with the egg, contributing material from another stock and so, as we shall learn, affording a basis for father-to-child heredity. The second is the essential sexual process. The first is simply a trick, devised by Nature for *preventing* reproduction until this all-important fusion shall have taken place. Once more it becomes evident to us that sex is imposed upon reproduction, and is in its essence a different thing.

§ 6. Artificial Propagation.

CLEARED of the complication of sex, reproduction is seen to be simply the detachment of living bits of the bodies of one generation, which grow up into the next.

Now if this goes on under natural conditions, can we imitate it artificially? If the experimenter can make his needles and solutions play the rôle of a sperm—(or, rather, one of the two rôles of a sperm) can he not make organisms reproduce by simply cutting them up? As a matter of fact he can.

Many segmented worms, especially the little freshwater kinds like Lumbriculus, can be chopped transversely into numbers of pieces,

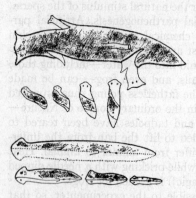

FIG. 161. Flatworms Regenerating

Above: A Flatworm has been cut so as to make it grow two extra heads and two extra tails. Centre: A small oblique piece cut out of the middle of another worm grows head and tail and straightens itself out. Below: A Flatworm has been bisected lengthwise (dotted lines show what has been cut away). Bottom row, left to right: Three stages in the half-worm's regeneration. As the regenerated tissue (white) grows, the old tissue (shaded) shrinks to provide the requisite material.

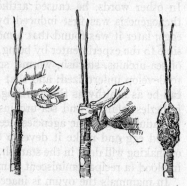

FIG. 162. Two twigs of one variety of tree being grafted on to a stock of another variety.

Left: Insertion into a cut in the stock. Centre: The graft is tied tight. Right: The region of the graft is cemented with grafting-wax.

and each piece will grow a new head and tail and become a new whole worm; and most Planarians among the flatworms have this faculty in even greater degree, for they can be cut in all directions, and yet any piece above a very tiny size will turn into a whole worm again. So with many Cœlenterates; we can multiply them at our pleasure And so with Protozoa; provided they are big enough to chop in pieces, and provided there is a bit of nucleus in every piece, we can force reproduction on them at the edge of the knife.

In plants, not only are such methods possible, but they have been

HOW INDIVIDUALS ORIGINATE

used for practical ends time out of mind. A mere post of green willow wood may take root and sprout and turn into a new willow bush, and a willow twig will even produce leaves and roots when suspended in damp air. Gardeners reproduce a chosen plant by taking slips or cuttings—shoots bearing one or a few buds—and simply planting them in soil. If undue evaporation is prevented above-ground, and the parts below-ground are warmed a little to expedite root-production, many plants can be successfully propagated in this way.

Plants which happen not to grow well from cuttings may often be propagated by what is called layering. A branch is bent down and part of it lightly buried in soil; when it forms roots, it is cut through near its attachment to the parent stem, and grows into—or indeed is already—a new plant.

Or the method of grafting may be employed. Here the detached bit of living tissue (the *scion* as it is called) is not planted in the soil, but is made to unite with the tissues of another plant, the host or *stock*. In grafting, the whole plant is not reproduced, since the graft is served by the roots of the stock and never forms any of its own; but it grows and forms its own leaves and flowers and fruit. Grafting ensures a real though partial reproduction. Sometimes only a single bud is grafted, as with roses or peaches, and is then inserted under the bark at the side; or whole stems are made to unite end-to-end at their cut surfaces, as with apples.

In many animals, artificial propagation could be carried out on a similar scale; but they are not usually of any commercial value. The only exceptions are bath-sponges. These can be cut into little pieces, each of which will grow into a new sponge; and experiments are being conducted to see whether this method can be made commercially useful.

But the most extraordinary method of artificial reproduction is that which we have already mentioned in Book 2—reproduction not by any procedure so mild as cutting or grafting, but by the really drastic method of rubbing the animal through a fine-meshed sieve of finest bolting-silk until it is totally disintegrated into single cells and small cell-groups. The mere cell-sediment which is first produced begins to organize itself into separate balls and lumps. Finally, each lump, provided it is neither too large nor too minute, will organize itself into a miniature animal once more. Before this power the fable of the Hydra pales; an organism which can have every vestige of its organization destroyed and whose scattered débris can reorganize itself into hundreds of new and perfect organisms, is indeed hard to destroy. In the hydroid Clava, which normally has the sexes separate, it has been found possible by this method to make hermaphrodite individuals; such will sometimes grow out of the cell-débris formed when a male and a female individual are sieved and dissociated together.

Artificial reproduction is possible even in land vertebrates, though only in their earliest stages. A fertilized newt's egg, up to the time when it forms definite cell-layers, can be constricted into two parts by a fine hair, and each of the two halves may survive and grow into a whole newt: we can produce artificial twins. If we could but get at the earliest stage of our own development, doubtless we could equally well persuade a human egg to do at our will what it occasionally does at its own—divide into two and give rise to identical twins.

§ 7. A Note on the Regeneration of Lost Parts.

WE TURN now to a series of phenomena which are not in themselves reproductive but which are very evidently related to the reorganization of detached bits of one individual to form others, which is the essential reproductive process. It is an all-too-familiar fact that if we lose a limb or even a tooth or finger, we cannot replace it; and all the creatures with whom we have everyday dealings suffer the same limitations. In fact, Mr. Everyman would be likely to assert that the capacity for regeneration was one of the points which separated animals from plants. This was firmly believed in the eighteenth century; and when the Abbé Trembley discovered the little green polyp Hydra, and was at a loss to know whether it was an animal because it moved, or a plant because of its being fixed and green and flower-shaped, he used regeneration as a test. If he cut it in half and it regenerated, it would be a plant; if it did not, it would be an animal. He cut it in half, and both halves regenerated into perfect little wholes. However, common-sense prevailed over preconceived theory. He had seen the Hydra capture living prey, swallow and digest it. He pronounced it an animal, but an animal which could regenerate.

Starting from this discovery, biologists eagerly tested all kinds of creatures for their regenerative capacities, and found that regeneration is by no means a rare or isolated phenomenon in animals, that worms can grow new heads or bodies, lobsters and crabs replace a missing leg or feeler, and even air-breathing vertebrates such as newts regenerate limbs and tail.

Regeneration in action still remains one of the most striking spectacles which biology can offer. A good-sized lobster loses its claw; and at successive moults we see a little bud of tissue grow on the

HOW INDIVIDUALS ORIGINATE

stump, take shape, become formed into a miniature new claw (still wholly useless for the big creature on which it is growing), and enlarge step by step until the animal is once again completely equipped. A newt has a leg bitten off; it, too, will bud out a whitish lump from the cut surface, the bud will elongate and constrict itself into joints, bones and muscles will form inside it, and toes sprout out at its end, until the newt has achieved what is impossible to us lords of creation. It will restore just what was lost, and no more; if only the hand had been bitten off, the bud will turn into a new hand; if the bite was at the shoulder, into a whole new limb.

We take a planarian worm (Fig. 92), and cut a small oblong out of its body. The fragment cannot eat, for it has no mouth. And yet it throws out an army of new and active cells on both its new frontiers, and these, dividing and growing and differentiating at the expense of the rest, form themselves into a new head and hinder end. At first these new parts are too small for the body; and accordingly remodelling goes on not only in these new-produced regions, but in the original fragment as well. They grow, it shrinks; and both they and it change shape, until in place of the original helpless fragment there is a new and well-proportioned little flatworm (Fig. 161).

Regeneration is so queer and unfamiliar a process to the human investigator that it was thought at one time to be a special protective device evolved by creatures which are particularly exposed to injury. It was supposed, for example, that newts, being peculiarly liable to have their limbs bitten or torn off, had therefore evolved this wonderful aptitude for re-growth in order to save themselves from extinction. But, as a matter of fact, regeneration is a very surprisingly widespread thing. Moreover, it is apparently a primitive thing, a power possessed by all the simplest forms of life, and one which we have lost, not one that the newt and the lobster and the flatworm have gained.

Worms, polyps, sponges, and so forth, are capable, we have seen, of total regeneration. Fragments of the body, torn off, can build up a whole new individual. It is evident enough that this power is related to sexless reproduction; both are aspects of a tendency which primitive living matter has of rebuilding and reorganizing itself. As we pass the Animal Kingdom in review, turning from the simpler to the more complex, we see that the abilities to regenerate and to reproduce asexually fall off as the body gets more elaborately specialized. A newt or a lobster can regenerate limbs but not much more than that. We ourselves can regenerate very little. The healing of a wound, the growth of new skin over a cut, the setting of a broken bone, the repair of a torn muscle or tendon—these are the last vestiges in our own bodies of this strange regenerative capacity. This is all that remains to us of the reproductive power that can make the severed flatworm into two individuals.

§ 8. Grafts and the Chimæra.

MAN, we noticed, can do curious things with the normal rhythm of life. He can imitate fertilization and sexless reproduction. We turn now to an even more striking interference—to the mixing together of organisms of different kinds to form one.

It is well known that branches of plants can be grafted on to other plants, and that they will take and "become one flesh." But grafting is possible with animal tissues, too.

John Hunter in the eighteenth century succeeded in grafting a cock's spur and a piece of chick's leg into another fowl's comb, and had the satisfaction of seeing them continue to live and grow. Almost any organ of an embryo or any young and soft-bodied animal can be grafted anywhere else on the anatomy of the same or another creature of the same kind, and will "take" and grow. A tadpole's growing limb or the regenerating leg-bud of a newt will grow after grafting on to the head or the flank, and will sprout out toes and turn into a well-formed leg quite happily. An eye-rudiment taken from one salamander embryo and stuck on to the flank of another grows into a normal-looking eye (though, of course, the animal can never see with it) and will even change its colour and shape, as do the animal's own eyes, when the host changes from tadpole to land-salamander. And with very young embryos, whole regions, or even whole animals, can be grafted. Double tadpoles stuck back to back or belly to belly have been artificially produced, and will feed and grow and turn into double frogs. Moths too can be grafted while in the chrysalis stage, and each part will go on growing in its appointed way. By this means the most bizarre monsters can be created, such as two tail-ends joined in the middle, or a tandem moth with four pairs of wings.

The intimacy of union effected by grafting is shown very vividly by the mutual reaction of scion and stock in grafted plants. Identical applestems, for example, grafted on to different root-stocks behave very differently. For one thing, the colour and size of the scion's apples is modified by the root-stock. On one stock, the scions show quick growth but fruit sparingly and late; on another they fruit early and well but stay small; and some root-stocks seem to promote both rapid

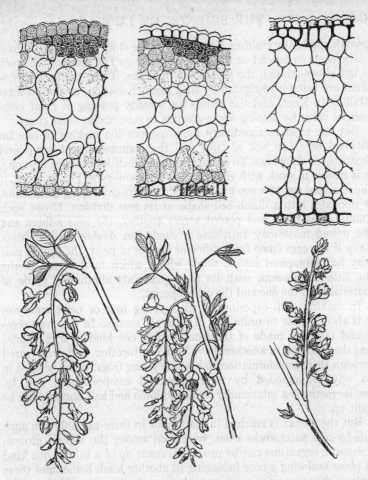

FIG. 163. A Plant Chimæra.

Above, slices, seen under the microscope, through petals of the common Yellow Laburnum (left), the Purple Broom (Cytisus purpureus), right, and the graft-hybrid between them, Cytisus adami (centre). The surface cells of the purple broom are coloured purplish (shaded). Those of the laburnum are yellow (dotted), with occasional patches of brown cells (heavily dotted) which give rise to the dark streaks on the petal. Cytisus adami consists of a skin of purple broom one layer thick over a core of laburnum. A sprig of each kind of plant is drawn below. The laburnum sprays have large yellow flowers and are pendent; those of purple broom stand up and have smaller purple flowers. The graft-hybrid has pendent sprays with medium-sized light purplish flowers.

growth and heavy fruiting. The degree of the stock's influence in these cases seems to depend on the place of grafting. The more of the stock is left above-ground, the greater its influence. These and other mysteries are being investigated by such research stations as that at East Malling in Kent, and the results are already proving of great commercial value by helping apple-growers to standardize their crops.

But the most extraordinary facts concern the making of one individual organism out of two; and the manufacture of compound creatures or chimæras. To make two individualities coalesce into one, it is easiest to work with eggs or very early embryos. For instance, if the fertilized egg of a newt is removed from its enveloping membrane, it elongates into a dumb-bell shape at its first division. If one such dumb-bell is lifted and placed across another, the cells adhere and the united mass may turn into a single but double-sized embryo. Even if the eggs come from different species of newt, the united pair may be reorganized into a single whole, which in this case merits the title of chimæra, with its four quadrants alternately made of material from the one and the other species.

In making graft-experiments with young frog or newt embryos, it is also possible to unite bits of different species. In this way compound tadpoles, made of the front half of one kind of frog-embryo and the hind half of another, have been put together, and even reared through their transformation to become young frogs, which seemed in no way incommoded by their unnatural compositeness, and by similar methods a salamander leg, half albino and half normal, can be built up.

But the climax is reached in plants; for in them not only can such side-by-side patchworks exist, but, even among the highest groups, composite organisms can be produced, made up of a skin of one kind of plant enclosing a core belonging to another kind. Sometimes these strange chimæras arise spontaneously from the junctional tissue where a graft has been made. The ornamental shrub *Cytisus adami*, often cultivated in gardens, apparently arose in this way, and consists of a skin, one cell-layer thick, of the purple broom, *Cytisus purpureus*, over a core of the ordinary yellow laburnum, *Cytisus laburnum*. Occasionally one of the parts gets the upper hand: the core bursts through, makes a bud, and a strong-built, yellow-flowering branch of pure laburnum results; or skin shoulders core out of its proper share in a bud, and there grows a branch of quite different character, bushy and purple-blossomed—pure *purpureus*. Such hand-in-glove unions are possible even between tissues of different genera of plants.

HOW INDIVIDUALS ORIGINATE

Old stocks of hawthorn (*Cratægus*), grafted with medlar (*Mespilus*), have sometimes given rise to branches consisting of a core of hawthorn with a tissue-glove of medlar fitting over it.

Recently a way has been found for manufacturing chimæras at command. The stem of one kind of plant (tomatoes and other species of *Solanum* have chiefly been used, but it has succeeded with other plants such as poplars) is cut across obliquely, and the top of the stem of another species grafted on. After the two have united, the stem is cut straight across the joint; from the cut surface new stems and buds are formed, and a certain proportion of these give skin-and-core chimæras, or graft-hybrids, as they are generally called.

We reproduce some pictures of these strange vegetable monsters herewith. Sometimes the skin is only one cell-layer thick, sometimes two; and the graft-hybrid will differ in appearance accordingly. The growth-rates of the two components may be different; and in such a case the graft-hybrid will either have a stretched, tense look, or its leaves will be all folded and rucked up because their skin is growing too fast for their central parts.

§ 9. What Is Meant by the Germ-plasm.

It is evident from what has gone before that, whether reproduction is sexual or whether it is not, there is material continuity between one generation and the next. The individual is a detached bit of its parent.

Weismann, the great German zoologist, pointed out in the latter half of the last century that that part of a man which grew into his children generally survived the rest of him. And a small part of him, growing into his grandchildren, survived still longer. Indeed, unless a man is childless, he does not wholly die. A bit of him, part of his living substance, is handed on, generation after generation, for ever.

For this immortal bit of the organism, that is to live on after the rest is done with, Weismann coined the phrase *germ-plasm*. He set it in contrast to the *soma*, which is the mortal remainder. More than 99.9 per cent by weight of the reader is soma; a fraction of an ounce of material in his testes or her ovaries is germ-plasm. The soma is the individual, who will live and die; the germ-plasm may go on indefinitely.

The germ-plasm is potentially immortal. Generation after generation it lives on, sprouting out bodies to house it and feed it and keep it warm, driving them with strange appetites and lusts so that it may get release from them and start again. Clearly it is the germ-plasm which evolves, not the ephemeral individual bodies that it throws out. The horse evolution was a slow modification of equine germ-plasm; the Micraster evolution a steady change in sea-urchin germ-plasm; man's rise from the apes is also a stir in the germ-plasm.

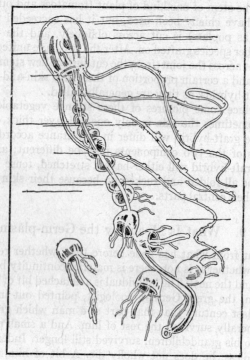

FIG. 164. The Continuity of the Generations.

The little jelly-fish Syncoryne has its mouth on the end of a long proboscis, from which it buds off new jelly-fish which are eventually liberated to grow up and repeat the process.

For in a sense there is only one germ-plasm. Presumably life had a single origin; the living things we know to-day are divergent branches of one stock, twigs of one tree of germ-plasm. In the frame of space-time there is actual material continuity between Mr. Everyman, his wife, his cat, and his aspidistra.

On the workings of this germ-plasm we must for a while concentrate our attention.

III

THE MECHANISM OF INHERITANCE

§ 1. *How Our Cells Multiply.* § 2. *How the Gametes Are Formed.* § 3. *Fertilization.* § 4. *What the Chromosomes Are For.* § 5. *Chromosomes in Plants.*

§ 1. How Our Cells Multiply.

WE ARE now going down into almost ultra-microscopic mysteries. We are dealing with things quite outside and quite unlike the ordinary experiences of Mr. Everyman. Any analogies we can draw from the incidents of daily life are likely to be misleading. And yet what we have to tell here is among the most fundamental stuff of Mr. Everyman's life. All his innate character, all that much of his personal destiny, is determined by these hidden, microscopic, inhuman interactions we shall now detail. What we have to set forth, indeed, takes the place in modern knowledge of the dark spinning of the Fates. It is impossible to grasp modern concepts of the development and destiny of life unless a broad understanding of this, the fundamental texture of the life-process, is attained.

We have already made it clear that all living things take their origin in pieces of living substance detached from the bodies of other living things. This can now be extended by another statement—that every living cell arises from a pre-existing cell.

There are a few organisms, it is true, such as the slime-fungi and some seaweeds, in which cell-boundaries are lost or never developed. They form an exception to the statement, but an exception more logical than real. In the overwhelming majority of organisms, reproduction results from a continued division and growth of cells. Each of us originated, as most of the living things we know originated, from a single living egg-cell. That cell divided itself into two, and then into four, and then into eight, and so on until every one of the millions of cells in our bodies was produced. Cells are never generated independently, never (as was once imagined) arise spontaneously out of some formless matrix, any more than whole animals or plants arise spontaneously from formless slime or decaying corpses. That is why, in order to understand growth and reproduction, we have to think in terms of cells, and to study their methods of multiplication.

Let us turn back to the first chapter of Book 1, when we watched the creeping movements of a single mammalian cell in one of Dr. Strangeway's tissue-cultures. We saw then that the cells multiplied, as Protozoa do, by tearing into halves; let us note as carefully as possible the details of the process.

From time to time one of the creeping cells can be seen to stop and to withdraw its temporary limbs of out-thrust protoplasm until it is a rounded blob. It is about to divide. Soon preparatory changes are evident in its interior. The nucleus has hitherto been a clear globe with one or more slowly stirring nucleoli inside it, but now the nucleoli disappear and at the same time a swarm of faint grey granules looms into view. The granules come together to form a mass of thin threads, some longer, some shorter. These begin to twist and wriggle, more and more actively, until the whole thread-mass looks, in Dr. Strangeway's words, "like eels in a box." Then the writhing slows again and the threads shorten and thicken until they become stubby rods.

These bodies, at first writhing threads and then stubby rods, are called *chromosomes;* they play vitally important parts in our lives.

Meanwhile, the boundary of the nucleus has disappeared, so that its contents mingle with the cytoplasm; and two centres of activity have appeared at either end of the cell, from which dim grey fibres spread out like the lines of force between two unlike magnetic poles. The star-shaped ends of this fibrous figure are called the asters, the central part the spindle, from its shape; and it is across the spindle's equator that the chromosomes now take up their posts. To reach this stage takes about eight minutes from the first appearance of granules in the nucleus.

The changes that we have just described have been preparatory to the actual division, which now begins to take place. First each chromosome can be seen to tear along its length, and the two halves to be pulled apart towards the two ends of the spindle; in another five or ten minutes the halves have been completely separated into two groups at the two ends of the cell. Meanwhile, the cell has elongated, and now begins to show a strange activity. All over its surface little bubble-like outgrowths are thrust forth and after a few seconds drawn in again, so that the observer is reminded of a much sloweddown process of boiling. This goes on more and more actively for about six minutes. Then abruptly a waist appears, dividing this boiling cell into two halves. They move straight away from each other, as if mutually repulsive, until they tear the final connecting strand; then, having won their freedom, they calm down and begin wandering about in the usual way. Meanwhile, round the chromosome-bunch in each cell a clear zone forms, and a membrane round this; the chromosomes grow hazy, they swell and become transparent, and so an ordinary resting nucleus is produced. The actual process of *mitosis*, as this type of cell-division is styled (from the Greek for thread, with reference to the thready chromosomes), takes about half an hour, and the reconstruction of the nucleus one to two hours more.

If the cells are killed and stained, further details are revealed. The chromosomes are seen to differ from each other in shape and form, their longitudinal splitting is more clearly visible, and two little dots are revealed at the centre of each set of radiating fibres. These dots are the centre-bodies or *centrosomes*. Their activity in a few minutes conjures up the spindle and its star-shaped anchoring ends, by means of which the split half-chromosomes are guided to their destinations. What can be the meaning of these strange manœuvres, this "paroxysm of activity," as Strangeways calls it? It cannot be for nothing that the nucleus abandons its ordinary task of regulating the cell's chemistry, dissolves its own boundaries, and generates this cohort of living chromosome-threads; not for nothing that these threads are divided along their narrow length, and that the centrosomes fill the cell with a tackle of microscopic ropes along which the half-chromosomes are transported.

The most striking feature of the process is the careful bisection of the chromosomes. As soon as mitosis had been accurately described, it was pointed out that *if* the chromosomes were the bearers of something essential for life, and *if* this something existed in the form of units strung along the chromosome like beads along a string, then this longitudinal splitting would ensure that each daughter-cell inherited a complete set of these essential somethings.

To these somethings we shall very shortly return. Let us remind ourselves once more that when growth takes place in our bodies, whether it be an increase of the whole body, or the thickening of a muscle as a response to regular exercise, or the growth of new skin on a cut, that growth is due to the active proliferation of thousands of tiny living cells. Normally, at every one of the divisions of these cells, mitosis takes place. At every division there is this sorting out of the chromosomes, this jealous and accurate sharing between cell and cell.

§ 2. How the Gametes Are Formed.

But there is one important exception to the statement made at the end of the last section. Most divisions of cells are indeed as we have described, but there is a peculiar difference in those processes which are concerned with the formation of sperms and egg-cells.

In the normal process of cell-division, we noted, a little bunch of chromosomes appears in the nucleus. These escape into the body of the cell when the bounding membrane of the nucleus breaks down and they arrange themselves at the middle of the spindle. Each of the chromosomes then splits into two, and half of it goes to each end of the spindle. Thus the chromosomes are equally divided between the daughter-cells; each daughter-cell has the same number of chromosomes as had the original cell.

This question of chromosome number is very important indeed. Every cell in the body has a definite and fixed number of chromosomes. Moreover the number is characteristic for the species. Every one of Mr. Everyman's cells has in its nucleus forty-eight chromosomes; Mr. Everymouse's cells have forty; and so on.

Now for the exception. During the formation of sperms and eggs there is an important division when the chromosomes behave differently. Before arranging themselves on the spindle, while they are still writhing about in the nucleus, like eels, they come together in pairs. Then the pairs move on to the spindle. When the sorting-out begins, the chromosomes do not divide into halves, but the pairs simply separate again, one member of the pair going to either end of the spindle. Thus each of the daughter-cells has only half the number of chromosomes possessed by the original cell. For this reason, this particular stage in the preparation of the gametes is called the *reduction division*.

In ordinary tissue-growth the chromosomes are shared out between the daughter-cells by actually halving them and giving a half to each cell. In the reduction division they are not split up but shared out as wholes into two heaps, much as cards are dealt.

The gametes, then, having exactly half as many chromosomes as the cells of the zygote. Mr. Everyman's spermatozoa carry twenty-four chromosomes each; Mr. Everymouse's, twenty. In biological language, the gametes are "haploid," with a single set of chromosomes, while the zygote-cells are "diploid," with two sets; these two words we shall use very frequently. Men and women are diploid; sperms and eggs are haploid.

§ 3. Fertilization.

WE TURN now to the next stage in the reproductive rhythm—the conjugation of sperm with ovum. What exactly happens when the gametes unite to form a zygote? Let us study the process in the sea-urchin, because there it can be watched—even cinematographed—without any difficulty. The stages of the microscopic drama are the same in higher animals and in ourselves.

We take a number of urchins during the breeding season, open their shells one by one, and separate them into males and females. From a female we take out a number of eggs into a flat glass dish, and let them settle to the bottom as a sparse single layer of just-visible grains of life. Then we take a single drop of milky fluid from a male's testis, shake it up in a test-tube of sea-water, and add about half a dozen drops of this to the dish with the eggs. In these few drops there will be ample sperms to ensure that every egg is found and fertilized. We peer into the dish with a microscope.

We see the egg-cells, now appearing as gigantic grey spheres, and the tiny tadpole-shaped sperms fussing about among them. The sperms are attracted by the eggs and run their heads against them. As soon as an egg is touched by a sperm it protrudes a little cone of protoplasm and sucks it in. Immediately afterwards a transparent membrane appears round the egg, a defensive shell which prevents other sperms from approaching. The egg has got all it needs.

Only the head of the sperm-cell enters the egg. The tail is left curling outside, to die very shortly. The tail was simply a device for getting the head to the egg-cell, and now its part is played.

When the head is sucked into the egg it undergoes an important change. It swells up like one of those Japanese toy flowers that expand when they are dropped in a glass of water, and it reveals itself as a nucleus, which has hitherto been very tightly packed and compressed to facilitate its journey. Slowly it creeps towards the other nucleus, which belongs to the female cell. The two come together in the middle of the cell and unite completely to form one. Henceforward the fertilized egg and every cell in the individual into which it develops—urchin or man, for the same is true of ourselves—has a compound nucleus, derived half from the original egg-cell and thus from the mother, and half, via the sperm, from the father. The egg which was haploid has become diploid and continues to divide into diploid cells.

Now we can see the reason for the reduction division, which occurs

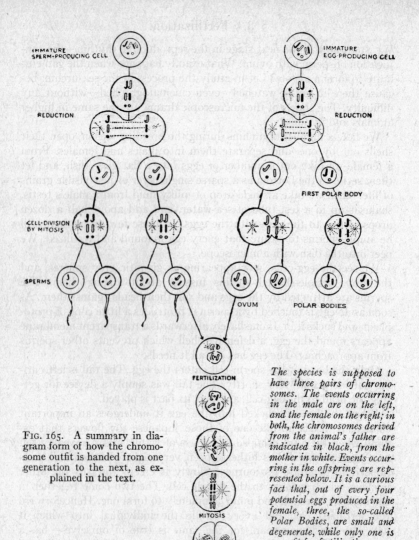

FIG. 165. A summary in diagram form of how the chromosome outfit is handed from one generation to the next, as explained in the text.

The species is supposed to have three pairs of chromosomes. The events occurring in the male are on the left, and the female on the right; in both, the chromosomes derived from the animal's father are indicated in black, from the mother in white. Events occurring in the offspring are represented below. It is a curious fact that, out of every four potential eggs produced in the female, three, the so-called Polar Bodies, are small and degenerate, while only one is fit for fertilization.

when the sperms or egg-cells are being prepared for their adventure. If each of the gametes had the full number of chromosomes, then the zygote (and therefore the next generation) would have double the number, the generation after that would have double again, and so on. Which would be absurd. So each of the gametes is reduced to half its normal outfit, and thus the constancy of chromosome number is preserved.

§ 4. What the Chromosomes Are For.

WE HAVE seen that in every kind of animal and plant, wherever sexual reproduction exists, the nuclei of the cells contain sets of chromosomes. The actual constitution of the set varies from species to species. Some creatures are provided with many chromosomes and some with few, some with small and some with large (there are roundworms, for example, which have only two in each nucleus, while some crayfishes have over two hundred). Moreover, when they escape from the nucleus and collect on the spindle, the chromosomes have characteristic shapes—they may be straight or curved, stout or slender—and in these details also there is variation even from species to species. Each kind of living thing has its characteristic chromosome-outfit, and takes the most minute and elaborate precautions to preserve the constancy of that outfit at every cell-division.

Looking more closely into the matter we find that every cell of the zygote stage (with certain exceptions to be noted in the next section) has a *double* set of these precious chromosomes. Moreover, in species whose chromosomes vary in shape it can be seen that the nuclei have each of the kinds in duplicate. In the Fruit-fly *Drosophila melanogaster*, for example, each nucleus contains a pair of very little blobby chromosomes, two pairs of much longer V-shaped chromosomes, and a pair of stout straight rods. (This, by the way, is only strictly true of the female insects; the males show a slight variation that will be fully discussed in the chapter on Sex.)

When the sperm and egg-cells are formed this double set is separated into two single sets. Thus each egg-cell of the Fruit-fly has one little blobby chromosome, two long V-shaped chromosomes, and one of the stout rods. When the sperm and egg-cells unite in fertilization their nuclei join, and thus a double set is founded again for the resultant individual. It gets a set from each of its parents. Now what is the meaning of these facts?

We know as a matter of everyday observation that a child or a young animal or plant is just as likely to take after its father as its mother. In some ways it will be like one parent, in some like the other And yet the contributions of mother and father to the offspring, the egg-cell and the sperm-cell, are as different as two cells could well be. The one is enormous and inert, the other ceaselessly active and exceptionally small. In one respect only, are the bequests of father and mother alike: each provides exactly half the nucleus, exactly half the chromosome-outfit of the new individual. Here, then, we detect the only possible explanation of the fact that fathers and mothers influence the constitution of their offspring in roughly equal measure. The chromosomes are the material basis of heredity, the vehicle by which the characteristics of one generation are handed on to the next.

This conclusion was arrived at provisionally, based as we have based it on a study of the chromosomes themselves, long before Genetics became an exact science. To-day it has been tested and proved beyond reasonable doubt. It explains at once why every kind of creature should have its own characteristic chromosome-outfit, and why it should be so elaborately and delicately handled. The very word "chromosome" is still comparatively unfamiliar to the general reader—as unfamiliar as was "atom" a century ago or "electron" twenty-five years ago. But the importance of atoms and electrons as the ultimate bricks and units of all matter is so great that they have now forced their way into everyday speech. And chromosomes will do the same, for they are no less important. These little flecks of viscid protoplasm, so small that they can only be seen with the highest powers of the microscope, are the bearers of hereditary destiny. Following the laws that we are about to examine, they form our bodies and our characters. Each of us human beings is what he is, himself and no other, because of the particular outfit of chromosomes with which he was equipped at fertilization.

§ 5. Chromosomes in Plants.

IN THE next chapter we shall study this hereditary function of the chromosomes more closely.

But before we do so we must turn aside for a moment for the sake of completeness and discuss the state of affairs in plants.

We have described the manner in which the chromosomes of all the higher animals behave. There is a zygote stage, the animal that we see and know, which is diploid; this alternates with a brief gamete stage which is haploid. In plants there is a somewhat different rhythm.

In Book II we discussed the phenomenon of Alternation of Generations in the vegetable kingdom. We saw that in certain lowly plants, in mosses and ferns and the like, there is a regular alternation between a spore-bearing and a gamete-bearing generation. We traced the alternation upwards and saw how in the higher plants the latter phase had been very considerably abridged, but how it still nevertheless existed. The two phases also differ in their chromosome constitution. The gamete-bearing phase, whether it is the moss-plant or the little leaf-like fern prothallus, or the pollen-grain of a higher plant, is haploid; the spore-bearing phase, whether it is the spore-capsule of a moss or the main part of a fern or a flowering plant, is diploid. The reduction-division occurs during spore-formation.

In the higher plants this cycle does not differ very markedly from our own, for in both the diploid stage is long and the haploid stage is brief. The strange contrast is in the mosses, where the principal stage in the life-history is haploid.

In the simplest animals and plants there are curious jugglings with the chromosome cycle. For example, in the red seaweeds, in the green thread-like alga Spirogyra, in the exquisite little single-celled Desmids, and in the parasitic Sporozoa, almost the whole of the life-history is spent in the haploid stage, the diploid stage occupying but a few minutes. Reduction occurs at the first division after fertilization. But in all sexually-reproducing organisms the diploid and haploid phases exist and alternate.

IV

THE A B C OF GENETICS

§ 1. *The Different Kinds of Variation.* § 2. *The Discoveries of the Abbé Mendel.* § 3. *Mendel's Two Primary Laws of Heredity.* § 4. *Numerical Proof of Mendel's Laws.* § 5. *Genes and Their Effects upon Human Characteristics.* § 6. *Colour Genes in America as an Example of Multiple Genes.* § 7. *Mapping the Chromosomes.* § 8. *Plant Breeding—Some Practical Applications.* § 9. *Is Inbreeding Evil?* § 10. *Reversion to Ancestral Type.* § 11. *How Much of Inheritance Is Mendelian?* § 12. *Some Common Superstitions about Heredity.*

§ 1. The Different Kinds of Variation.

THE word "variation" covers a diversity of instances and meanings. And it needs watching if we are to escape the logical traps due to its shifting implications. For example, the progeny of a fancy mouse are born of divers colours and patterns; that is variation. Some are fed well and some badly, so that they come to differ in size and vigour. That can also be called variation; but it is a different sort of variation. A pure-breeding strain of sweet-peas suddenly throws a "sport," a new colour or shape, and the new form may be known as a "variety" (that is to say a variation). The word is evidently a general term for many widely differing things; it is not a technical term, and in the past this breadth of meaning has been responsible for much confusion of thought in discussions of heredity and also of evolutionary change.

In order to make the point clear we shall analyze variation as it appears in a particular species. For this purpose the bean-plant is admirably suited. The bean-plant differs from most plants and animals in that it normally and regularly fertilizes itself, so that there is no mixing of two heredities when new generations are begotten. Each plant has only a single parent. As we shall presently see, inbreeding results in a uniform strain, so that in the bean we can be sure that

THE A B C OF GENETICS

any plant has exactly the same hereditary constitution, exactly the same chromosome-outfit as its parent. In biological language a bean-plant, its parent, grandparent and great-grandparent, its children, grandchildren and great-grandchildren, and so on, are spoken of together as a *pure line* because they all have the same inward constitution, handed on unchanged from generation to generation. If they vary, we know that the variation comes not from innate, but from external causes. The sort of variation that comes from the inward constitution we call "inherited," the latter "environmental" variation.

The results of a classical series of experiments on beans were reported by Johannsen early in the present century. He started by trying to modify the hereditary constitution of a pure line by selection. The character he selected for experiment was weight, because it is accurately measurable and because it is of practical importance. His first step was to grow a number of beans from a single plant. Then, after the seeds had been weighed, the heaviest and the lightest were taken as the beginnings of two lines, the lines of plus and minus selection. The plus line arose from the heavy seeds; and in all further generations was again continued only from the heaviest beans, all the rest being rejected. And the minus line received the reverse treatment. But originally they were started from brother-seeds, picked from the same pure stock. The biggest brothers were bred from, and the smallest.

One would expect that this would have a marked effect upon the course of events, and that the average weight of the beans of the plus line would go up rapidly, that of the minus line as rapidly decline. But, as a matter of fact, nothing of the sort occurred. Whichever way the selection was made, the average remained the same, generation after generation. These results can only be interpreted in one way—that the variations in size seen among the beans of a single plant have, in these experiments, nothing to do with heredity. Their origin must be sought in the conditions to which they were exposed during their formation. The position of the bean within the pod; the position of the pod upon the branch; the position of the branch upon the main stem; the amount of light received by the branch—these and other factors will each of them exert an influence upon the size of individual beans. When all the separate conditions happen to be favourable, the bean will be exceptionally large; when all are unfavourable, exceptionally small. But neither the help nor the hindrance has had any effect upon the actual constitution within. When the bean is sown, it is its interior constitution which asserts itself, not the accidental excess or defect of its weight; and the new plant develops as do all other plants of that particular constitution. Its inner destiny was not altered by its own good or bad fortune or by the favourable or unfavourable conditions in which its parents lived.

In addition to the variations in size among the beans on a single plant, or on all the plants of one and the same strain grown in the

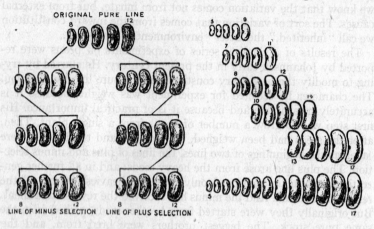

FIG. 166. A diagram of Johannsen's experiment on selection in a pure line of beans.

A pure line under standard conditions produces seeds ranging from eight to twelve units of weight. From this, two lines are started, a line of plus selection, bred only from the heaviest beans, and a line of minus selection, bred only from the lightest beans. In neither case does the selection have any result on the average weight of beans produced.

FIG. 167. Different lines may differ in average seed-weight.

Five such different pure lines are here set forth in diagram. If you mixed seed from the five pure lines and sowed it, your plot would give the range of seed-weight shown in the lower row.

same plot during one season, the average size for different seasons may vary, and also the average size for beans of the same strain grown under different conditions of cultivation. But these differences too, when the same accurate methods are applied to them, reveal themselves as non-inheritable.

Here are environmental variations due to the outside influences acting on the plant; they have nothing to do with heredity. Warmth, light, moisture and so on are moulding the individual plant and the individual bean without having any effect on the germ-plasm within it. The germ-plasm keeps at the point from which the individual started.

This much Johannsen, working from a single bean, found. But when he compared the offspring of different original plants—that is to say

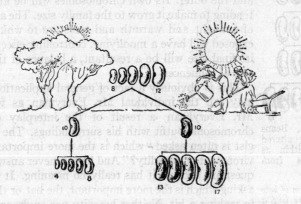

FIG. 168. Outer conditions may modify the pure line's achievement.

Bean-plants from the same pure line grown in unfavourable conditions will produce small seeds, and in favourable conditions will produce big seeds. But these differences are not inherited.

members of different pure lines—he found another face upon things. In the first place, many different pure lines, each with its own characteristic average seed-weight, could be distinguished. Plants of different pure lines were grown side by side under identical conditions and yet would continue to produce beans of different average weight. The differences in such cases obviously depend upon differences between the hereditary constitutions of the strains. Commercial lots of beans will consist of the produce of a number of pure lines mixed together; accordingly, if we start with an ordinary bean-field, and start selecting for seed-weight, using, let us say, the heaviest five per cent of beans produced in each generation as the parents of the next, we shall raise the average seed-weight of our stock rapidly. But we shall do this not by changing the maximum weight of any line, but by eliminating the pure lines of low seed-weight. Of the many original pure lines we shall finally be left with but one, the one of highest average seed-weight.

And, after this, our selection will cease to have any further effect. Here then is a second kind of variation, due to the inherited constitution of the lines and independent of outside influences.

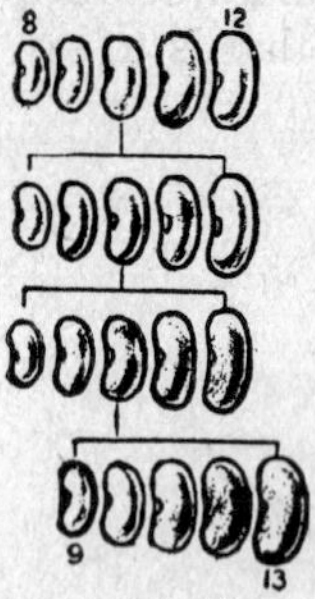

It is clear that the size of any single bean will depend upon the interplay of two tendencies, the inner and the outer. Its own chromosomes will be at work, tending to make it grow to the family size. The amount of sunlight and warmth and moisture to which it is exposed will have a modifying, controlling effect. And its final size will be a resultant, a balance of the two sets of influences.

This is obviously a fact of general application. We may call the individual Mr. Everyman, as well as Mr. Everybean, a result of the interplay of his chromosome-outfit with his surroundings. The biologist is often asked "which is the more important, environment or heredity?" And he can never answer the question, because it has really no meaning. It is like asking which is the more important, the bat or the ball in making a hit. Neither heredity nor environment is the more important—because both are essential. The best constitution in the world, of bean or man, would come to nothing at a temperature 50° C. lower or 50° C. higher than usual. And not even the best imaginable conditions could force a wild bean to yield seeds as large as those of a good cultivated variety, could make an African pigmy grow six foot high, or turn a child with a constitution making for mental defect into a mathematician or a successful lawyer.

FIG. 169. Beans may be altered from within as well as from without.

A diagram of how in a pure line a mutation may occur, changing the average seed-weight, though the outer conditions remain the same; the mutation here occurs in the third generation. Changes like this are inherited.

We turn now to a third kind of variation. We have written as if the germ-plasm were absolutely fixed and unalterable in these pure lines. And indeed heredity is from its very essence a conservative force. It goes back to the old, old story. Beans beget beans, dogs beget dogs, men beget men, generation after generation. But heredity is not absolutely stable. The germ-plasm is ever so slightly restless; it shifts now and then and changes, and the race alters a little. As we shall make plain in a later chapter, these alterations are vividly important, for they provide the raw material on which Evolution works.

Twice, while he was carrying out his elaborate painstaking experiments, did Johannsen detect such a change in the chromosomes: twice did he witness a little jerk of evolutionary progress. On each occasion, the progeny of one bean gave an average bean-weight a little different from that of the pure line to which it belonged. In one case the new beans were lighter, in the other heavier. Though the difference was not great it was constant in the resulting strain, generation after generation. Something had happened in the chromosomes, and the germ-plasm was permanently changed. This kind of variation biologists call *mutation*. Mutation is here a scientific term, and not an ordinary word; it is the special name for a variation of the germ-plasm itself.

In the following chapters we shall analyze further the working of the chromosomes. We shall see how they wrestle with their environment and with each other to build up the organisms. We shall learn more of the appearance of mutations. But before we leave our beans let us pause for a second and point a moral. For they teach us something about method. They show the value of experiments with individuals as opposed to experiments where a number of random individuals are treated together and their produce averaged. If Johannsen had worked with a handful of commercial bean-seed lumped together, it would have appeared as if selection did produce a hereditary effect; and the fact that single mutations occasionally cropped up could never have been established.

It was precisely a realization of the value of this same individual method, which, together with patience, clear thinking, and a certain good fortune in the choice of material, led the Abbé Mendel to the discoveries on which are based all our modern and intimate knowledge of heredity. With those discoveries our next section will deal.

§ 2. The Discoveries of the Abbé Mendel.

THE pure line experiments tell us a good deal about what is *not* inherited, and about the precautions needed in making experiments on inheritance; and they make it clear that what *is* inherited is something very definite, capable of remaining unchanged over a long series of generations. The microscope has, moreover, made it practically certain that the hereditary constitution is contained in the chromosomes. But neither pure line experiments nor microscopical observation tell us anything important about the actual nature and composition of this inheritance. The clues to that problem come from another direction; they were put on record by the Abbé Mendel and in a quite different fashion.

The story of his discovery and its fall into half-a-century's oblivion is a strange one—how, working through a sequence of years in his garden at Brno, in Moravia he hit upon vital facts as regards inheritance in peas, and propounded equally vital theories to account for them; how, in 1865, he published his results in the ordinary way, and was surprised at the scant attention paid them; how in attempting to apply his ideas to inheritance in other plants, such as hawk-weeds, he met with difficulties which he could not overcome—difficulties which we now know were due to unusual reproductive behaviour in these plants and not to any defect in his thinking; how he died, a perplexed and disappointed man; how, in 1900, three botanists, searching through old scientific periodicals for facts bearing on their own researches, independently came across his papers and immediately realized their importance; how the enthusiastic discipleship of Professor Bateson led to the world-wide recognition of his work; how it was taken up and tested by a little army of investigators in every civilized country; and how his principles were speedily extended until we now recognize that almost all inheritance, plant, animal, and human, is Mendelian.

The intensive study of heredity which owes so much stimulation to the rediscovery of Mendel is usually called Genetics and its devotees are spoken of as Geneticists.

The expansion of knowledge in this field has been so rapid, and the opposition to the new ideas from many biologists who have been concerned with other branches of their science has been so great, that the extent of the change wrought by Mendel's work is still perhaps not fully appreciated by the world at large. It constitutes the most important single step forward taken by biology during the last half-century. It has the same fundamental importance for our understanding of heredity and evolution as had the cell-theory for our understanding of the construction and reproduction of organisms, or the atomic theory for our understanding of matter and its chemical transformations. To-day the theory of the atom is undergoing profound revision, but its share in the development of science has been of fundamental importance. To-day we can still criticize our definition of a "cell," and yet all biological work rests on that still-developing conception. Like the cell-theory or the atomic theory, Mendelism has shown us that what we were used to think of as single indivisible wholes are in reality composed of discrete and definite units, and that the properties of the wholes—in this case, the hereditary constitutions of animals and plants—result directly from the particular combination of their component units.

According to the atomic theory, the atom-units of matter persist unchanged through all the strange material transformations which occur in Nature or in the chemist's laboratory; when a candle burns, its matter is not destroyed, but the special arrangements of carbon, hydrogen, and oxygen of which it consists are taken to pieces and, after combining with atoms of oxygen from the air, are liberated in the form of new atom-arrangements which we call water-vapour, carbon dioxide, and so forth. According to the Mendelian theory, the germ-plasm of animals or plants consists similarly of units. These units (apart from the "mutations" of which we have already made mention in our account of Johannsen's work) persist indefinitely without traceable change. When, as we may actually do, we cross a black mouse and an albino, and obtain a progeny consisting of grey creatures exactly like wild mice, the units in the constitution which determine black hair and albinism have neither been lost nor merged together (like drops of black and white paint) to form grey, but remain as distinct as the atoms in a molecule; and each is able to emerge unchanged and to produce its original effect in later generations.

Since the re-discovery of Mendel's work in 1900, the greatest further discovery has been that these units of his have a visible habitation—within the chromosomes; and that there they seem to be arranged in a definite and orderly way. Thirty years ago we knew literally nothing of the solid facts of heredity. To-day there are organisms, like the Fruit-fly Drosophila, which have been so thoroughly and industriously investigated that the Geneticist can take his friends to the microscope, show them this fly's chromosomes—so small as only to show clearly under the highest power—and make such apparently incredible assertions as this: "You see that rod-shaped chromosome? That is the chromosome which determines whether the fly shall be male or female. Close to one end of it, where you see the spindle-fibre attached, is situated the hereditary unit which may turn our fly's body yellow instead of grey. In one of these other V-shaped chromosomes is situated a unit with the power of reducing wings to vestiges, and about one-twentieth of the chromosome's length farther along, a unit which turns the body black. In the other big V live other units with which I do not propose to bore you; and, finally, this tiny dot of a chromosome houses its own few units, of which one prevents the fly from developing eyes at all."

And then he will show them on the wall an actual map of Drosophila's chromosomes, on which are marked the position of all the four hundred and more determining units which have been distinguished in this little animal. The map, of course, is not enormously reduced in scale, like ordinary maps, but enormously magnified. The map which we should take for a walking tour is reduced about 100,000 times; these chromosome maps have to be magnified to about the same extent. Later on we will give this map of the infinitely little.

It is the map of the Geneticists' peculiar Empire. Drosophila is to him like the Eagle to Jupiter, the symbol of his rule. It makes peculiarly convenient material because it breeds every ten days, it produces several hundred offspring in each generation, it is so small that the whole of one of these generations can be successfully reared in a pint milk bottle. Like the fruits of Paradise, it is in season all the year round. Thus a graduate student in the two years allotted for his higher degree can work out a problem covering seventy generations, equal to decades in the selection of rabbits, centuries in the case of cattle, and millennia of human reproduction. Its possibilities have been exploited to the utmost by the magnificent team-work of Professor T. H. Morgan and his school at Columbia University. How far-reaching are the implications of their work, we shall show in subsequent sections.

§ 3. Mendel's Two Primary Laws of Heredity.

But we are running ahead of our exposition.

The method of Mendel in his studies of the workings of heredity was to hybridize varieties and species of plants and observe the consequences. That is still the normal experimental method in Genetics. We hybridize—we make crosses between parents which differ in their hereditary constitution—and study the characters of the following generations. For example, we mate a white rabbit with a black one or a short-horned heifer with a long-horned bull and see what turns up. Or we work with plants. We fertilize one plant with pollen from another, taking care, of course, to keep out insects or wind or any agency which might bring in other pollen and so confuse our experiments.

First, as a very simple and illuminating instance of such results, let us consider the garden flower called *Mirabilis jalapa* or "Four-o'clock." This exists in numerous varieties, two of which are distinguished by the white and red colour of their flowers. Suppose that we cross a red-flowered plant with one of a white-flowered strain. All the offspring will have pink flowers. If now we inbreed this first hybrid generation, fertilizing them with each other's pollen or their own, we get a second generation showing three distinct types. There are red-flowered plants precisely like those of the original red strain, white-flowered plants like those of the pure white strain, and pink-flowered plants like the first hybrid generation resulting from the cross (Fig. 170).

If the experiments are carried a generation further, another striking fact reveals itself. If the reds are self-fertilized or pollinated by other reds, all of the progeny will also be red; thus, these plants, although only in the second generation from a cross, yet breed true for the character that was crossed. This fundamental fact was in total disagreement with the prevalent view of animal and plant breeders in the last century, that crossing so contaminates the "blood" or constitution of both strains used in a cross, that it is impossible to obtain purity of breed again, at least without many generations of rigorous selection.

The whites of this second generation, when tested, prove to be as pure for flower-colour as the reds. But the pink plants all behave like their parents of the first hybrid generation; they do not breed true but throw reds and pinks and whites. And so on for as many generations as we like to raise. Reds and whites will always breed true, while pinks will throw reds and whites as well as pinks. A pink that breeds true is an impossibility.

In order to explain facts like these, Mendel assumed that somewhere in the plant's hereditary constitution there were units busy with the control of this or that character. In our particular example, there must be units whose business is with the control of flower-colour. And these units, he supposed, were self-perpetuating, and did not mix with each other. To-day we call these units by the technical name of *genes*. All the countless breeding experiments that have gone on since Mendel's work was re-discovered have confirmed his original idea, and in addition have established a new fact—that these gene-units are carried in the chromosomes. Like an atom, a gene is something that has never been seen or felt. It is inferred. But its existence is none the less real for that. A century ago, scientific men were forced to believe in the existence of atoms by the chemical behaviour of different substances when they combined and separated. To-day, scientific men have been forced to believe in the existence of genes by the reproductive behaviour of animals and plants when they are forced to breed together. "Gene" is simply a shorthand phrase for "that something in the chromosomes, whatever its nature may be, which is responsible for a particular character and its particular way of transmission from parent to offspring." We have gradually got to know a great deal about the nature of atoms, besides the mere fact that they are the units whose existence we have to assume to explain the way of chemical combination; and we are beginning to learn something about the nature of genes, besides the mere fact that they are the units of whose existence we must assume to explain the way of hereditary transmission. But the main point

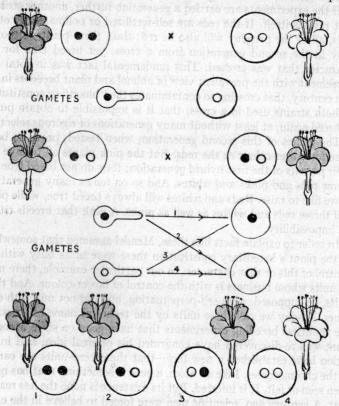

Fig. 170. The Mendelian explanation of the breeding behaviour of Four-o'clocks.

A white flower with two "white" genes in each nucleus (upper right) is pollinated by a red with two "red" genes (upper left). The resulting generation (middle row) have one gene of each kind and are therefore pink. When these fertilize each other, reds, pinks and whites are produced (below) as explained in the text.

about them is that, whatever their nature may be, we have to assume that they exist, or else the facts of inheritance are inexplicable.

Since every cell of the plant (except for the actual reproductive cells) contains a double chromosome outfit, it must contain two of these colour-producing genes. In the original red plant from which we started these genes were red-producing; in the white plant they were different and turned the flowers white. When the cross was made the hybrid plants of the first generation got one of their two chromosome-outfits from each of their parents. Whence it follows that their cells contain one colour-controlling gene of each kind—one red-producer and one white-producer. So that the resulting flowers are a compromise between red and white—that is, pink.

Now when these mixed plants became mature and produced reproductive cells, the result was more complicated. Manifestly when their chromosomes were sorted out in the reduction division there must have been a separation of the red gene from the white; there must have been red-carrying pollen and white-carrying pollen, red-carrying ovules and white-carrying ovules. So that when these pink plants were bred together four kinds of fertilization ensued so far as colour-genes were concerned. These may be catalogued as follows: (1) a red-carrying pollen-grain might fertilize a red-carrying egg—result, a red-flowered plant; (2) a red-carrying pollen-grain might fertilize a white-carrying egg—result, a pink-flowered plant; (3) a white-carrying pollen-grain might fertilize a red-carrying egg—result, a pink-flowered plant; (4) a white-carrying pollen-grain might fertilize a white-carrying egg—result, a white-flowered plant.

As we have seen, all these possibilities have happened. The second hybrid generation was mixed and contained red-, pink-, and white-flowering plants. And so will any generation bred from pinks.

We can go farther than this. By following a similar line of reasoning (the details of which we leave to the reader's ingenuity) we can prophesy that on crossing a pink with a red the first hybrid generation would include both pinks and reds, but no whites. Similarly on crossing a pink with a white we should get pinks and whites but no reds. And this is what actually does happen when the experiment is tried. The gene hypothesis is thus tested and confirmed.

This example well illustrates the fundamentally important conception of the independence and mutual aloofness of the units. When red-producer and white-producer exist together in one plant the visible result is a blend of red and white. But this is only a resultant of two independent and antagonistic influences: the genes themselves neither blend nor contaminate one another. When reduction brings the time for parting, each goes its solitary way, bearing no trace of having been associated for months or years with the other within the microscopic chambers of the cells. Moreover, it evidently makes no difference whether we use a red-flowered plant as our original male parent and a white for our female, or vice versa. In both cases the hybrids will be pink. All that matters is the kind of genes in the resulting mixture.

A widespread and important complication of this scheme is the

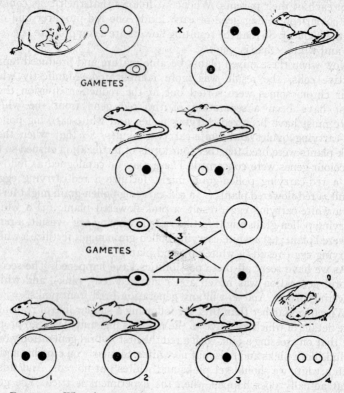

Fig. 171. What happens when a "waltzing" mouse (upper left) is mated with a normal one (upper right).

Two doses of the "waltzing" gene produce a waltzer, but one dose of it together with one of the corresponding gene for normal progression leaves the animal normal, so the hybrids (second row) are normal. But if they breed together their offspring include some waltzers (lower row) as explained in the text.

phenomenon known as dominance. In our example the joint effect of the two different kinds of gene, when both were present together, was intermediate between their effects when present in a pure state; pink is intermediate between white and red. But that is not always the case. One gene may be stronger, more effective than its correlative, and, indeed, such inequality of effect in the genes is the rule rather than the exception.

Suppose we cross a mouse of the Japanese "waltzing" type with one which is normal and comes from a perfectly normal stock. The former has an inherited malformation of the inner ear which upsets its sense of direction and sets it spinning round, sometimes for hours on end, like a puppy trying to catch its tail. The first hybrid generation of such a cross will show not a trace of this waltzing habit. It will consist of perfectly normal mice, progressing with irreproachable directness, their inner ears admirably formed. Nevertheless, genetically they are mixed, just as are the pink Four-o'clocks. Every cell in their bodies contains one "waltzing" and one "normal" gene. But the latter overpowers the former. In the next generation, obtained by breeding the first together, it becomes manifest that we have been dealing with mixed, hybrid animals, for we get three kinds of mice. There are true-breeding waltzers and true-breeding normals, corresponding to the reds and whites of the second generation of the Four-o'clock cross, and, corresponding to the pinks, there are mice like those of the first hybrid generation—normal in behaviour and anatomy, but nevertheless carrying the "waltzer" gene hidden in their chromosomes and bequeathing it to some of their offspring.

This case is obviously parallel to that of the Four-o'clocks except that here the hybrids are not intermediate, but indistinguishable in their characters from one of their parents. In such a case as this the gene which wins, so to speak, in the hybrid is said to be *dominant* to its partner, while the one which is concealed is called *recessive*.

"Dominant" and "recessive" are relative terms. To forget that is to drift towards very nonsensical ideas. An example of incomplete dominance appears when a white leghorn cock is mated to a brown hen of the same breed; the hybrids are white, but with one or two brown feathers here and there on the body. White in this case is dominant to brown, but incompletely so. There are degrees of dominance, ranging from the apparent equality of red and white in the Four-o'clock to the complete submergence of waltzing by normal in mice. We stress this point because in some of the popular writing one meets about Genetics, dominance and recessiveness are treated as things as absolutely opposite as the poles.

And let us emphasize, because it is the very quintessence of Mendelism, that the recessive gene, however obliterated it may be in appearance, is in no way destroyed or altered. It is masked, it is eclipsed, but it is still there. In every cell of the hybrid mice—as distinguished from the pure-bred ones—the recessive gene is present, like a quiet, henpecked husband with a wilful and noisy wife who must be obeyed. In another generation he will find his chance in a few individual cases, get together with another of his own kind, and be as distinctive in his influence as his nature requires.

Herein lies the importance of this discussion. A recessive gene may be carried in a stock, completely hidden by a dominant partner, and may therefore be difficult to detect and to weed out. And it may be a very sinister gene. In our own species, for example, there is a kind of inherited deafness, so complete that it leads to dumbness—deaf-mutism—which is recessive. If a person, after the manner of the hybrid mouse with one imperceptible waltzing gene, carries one gene of this deaf-mute type, he or she will behave quite normally. His "taint" will not matter—to him. He will hear and speak, just as the mouse will run straight. If such a person marries a normal person without this hidden taint some of the children will be pure normals and some will be hybrid normals like the first individual. All of these again will hear and speak. It will not affect them personally at all. But if two such hybrid "carriers" marry—and there is no way known at present of telling the pure normal from the hybrid—there is a fair chance that an abnormality-carrying sperm may fertilize an abnormality-carrying ovum, producing a pure recessive, a deaf-mute. It is (as we shall shortly see) a chance of one in four; a risk, not a certainty.

Similarly with other unpleasant recessive genes—one, for example, which produces a kind of insanity coupled with blindness, and another which produces a hideous abnormality of the embryo, which has one large eye instead of two normal ones. They may be carried for generations undetected, and then, after an unlucky mating, their effects may come suddenly to full realization.

The fact that the genes in these gene-pairs are independent entities, emerging separate and unchanged in the gametes after their co-operation in the zygote, is known as Mendel's First Law. It is a profoundly important law. In popular language, when one is talking of crosses and hybrids one speaks of "mixed blood," and the phrase admirably expresses the belief which lurks at the backs of many minds. If we take two kinds of blood and mix together a pint of each we will obviously get a fifty-fifty mixture. If we mix a pint of this with a pint of one of the original kinds we shall get a twenty-five-seventy-five mixture. If we mix a pint of this with a pint of the same original blood we shall get a twelve-and-a-half-eighty-seven-and-a-half mixture. And so on. This is parallel to our almost instinctive belief about heredity. One expects races to mix as liquids mix. One expects red crossed with white to give pink, red with pink to give a darker three-quarters pink, red with this pink to give an even darker seven-eighths pink, and so on. One expects that no amount of breeding with red will altogether remove the whiteness from the tainted strain. But all this is wrong. Slowly the biologist has come to know better. He thinks not of mixed blood, but of assorted unit-genes—a very different conception, born of the facts that we are here detailing.

Mendel's Second Law, to which we now proceed, is easy enough to understand. It is called the Law of Independent Assortment and concerns the outcome of complicated crosses when the original parents differed in two or more distinct respects.

We may illustrate by one of Mendel's own instances. Mendel worked with peas. He crossed together two strains; one gave taller plants than the other, and at the same time the seeds of the first were yellow, while the seeds of the second were green. He found that he was dealing with two separate and independent gene-pairs. "Tall" was dominant to "Dwarf" and "Yellow seed" to "Green seed." Every plant in the first hybrid generation was tall and had yellow seeds. In the second generation the two pairs sorted themselves out independently of each other. He got pure tall, hybrid tall, and pure short plants; of each kind some were pure yellow, some hybrid yellow, and some pure green as far as seed colour was concerned. Similarly with three or more gene-pairs. The first hybrid generation after a cross between two pure-breeding stocks is always uniform, while in the second we get a mixed batch showing every possible combination of the original genes. In brief, *separate gene-pairs are shuffled and dealt independently of each other*.

It follows from this Second Law of Mendel's that in a cross, so long as the strains are pure and the same genes are involved, it makes no difference how they are combined in the original strains. Instead of crossing a tall yellow and a dwarf green strain, we might have crossed a dwarf yellow and a tall green; but the result, both in the first and in all subsequent generations, would have been precisely the same. Once more, it is the individual genes which count, not the way they happen to be grouped in the parents' bodies. So long as all the gene-ingredients get into the first zygote in the right proportions, it makes no difference from what abode they come, any more than it makes a difference to the taste of the stew whether the pepper-pot happens to have stood next to the rice or the potatoes on the shelf.

These details may seem a little tedious, but they are the very grammar of this subject, the skeleton needed to support its living flesh.

Once they are mastered, our attention can be given to various matters of general interest and far-reaching practical consequences. But they must be properly mastered before they can be used.

§ 4. Numerical Proof of Mendel's Laws.

IN THE last section we outlined the two essential principles of Mendelian Inheritance. But we should be doing scant justice to this great and patient investigator or to his followers if we failed to note the enormous mass of numerical evidence in support of his theory. We have contented ourselves so far by noting the kinds of animal and plant that turn up after a cross and by giving a qualitative explanation. But Mendel's Laws will carry us farther than that. They enable us to estimate with a very reasonable prospect of success the numerical proportions in which the different kinds will appear.

Turn back for a moment to the second cross-bred generation of the Four-o'clock plant (Fig. 170). We catalogued the different kinds of fertilization which occur. Now it follows from the way in which the reproductive elements are made that among the pollen-grains those which carry red will be just as numerous as those which carry white, and the same is true of the egg-cells. Moreover, fertilization is a perfectly random process; there is no reason to suppose that red-carrying pollen-grains have a particular fancy for white-carrying eggs or anything like that. So we can say that the four kinds of fertilization will occur in roughly equal proportions—in the next generation half the plants will be pink, one quarter will be red, and one quarter white.

If the reader has any doubt of this reasoning let him take a pack of ordinary cards, shuffle it and spread it face downwards over a table. Let him draw the cards at random in pairs. He may happen to draw a red and a red, or a red and a black, or a black and a red, or a black and a black. The chances of these various combinations are equal. So, in the long run, he will get about twice as many mixed pairs as double reds or double blacks.

We are dealing, be it remembered, with chance and the result is never perfectly exact. We say "in the long run." But chance is a fickle jade, and often the run is uncommonly long before the mathematical rectification is achieved. The first two or three pairs to be turned up may happen to be pure red or pure black, but as the experiment is continued the proportions will tend to straighten out as we have prophesied. The more pairs one turns up, the more closely will the proportions obtained approximate to those we have deduced.

THE A B C OF GENETICS

So with numerical experiments in heredity. If one wants reliable results one has to work with large numbers of individuals. In the case of the Four-o'clock, if one were to take half-a-dozen seeds from a pink plant and rear them, anything might happen. They might even all be red (although the odds against that are about four thousand to one). But if one rears large numbers of the hybrid generation—fields of it—the proportions straighten out and assume the estimated values within a striking degree of accuracy.

How close actuality may be to expectation when really large numbers are bred is shown by the following example. Mendel and six subsequent investigators have all counted the second generation from crosses between yellow and green colour in peas. Their actual figures totalled together, are: yellow, 152,824; green, 50,676. The theoretical expectations with this number of plants are: yellow, 152,625; green, 50,875. The actual ratio is 3.01 to 1, instead of 3 to 1—a divergence of about one-third of one per cent.

Using similar methods one can anticipate the results of other Mendelian crosses, and then confirm them experimentally. Thus the second hybrid generation of the waltzing mouse cross Fig. 171 will give three times as many normals as waltzers. In the second generation of the peas cross (§ 3) there will be nine tall plants with yellow seeds, three short with yellow seeds, and three tall with green seeds to every one that is short with green seeds. And here again the prophecies can be confirmed with a high degree of accuracy.

Sometimes these calculations are upset. In some cases the mobility of the pollen-tubes is affected by the genes they carry, so that one kind is at a disadvantage, reaching the ovules too late, and the ratio is altered. Or one variety may be more delicate and have a higher mortality than another: we shall see in the following section how sometimes genes may kill their bearers. Sometimes there is a hitch in the delicate chromosome machinery; for example, two chromosomes may stick together during the reduction division and one of them may get pulled into the wrong gamete with strange and unexpected results. But these are exceptions. In the vast majority of cases, whenever sufficiently large numbers of animals or plants are reared, Mendel's Laws have ample practical confirmation, quantitative as well as qualitative.

§ 5. Genes and Their Effects upon Human Characteristics.

THE inner ears of mice or the colours of garden flowers may seem trivial in themselves, but they illustrate clearly a mode of inheritance of the greatest and most general importance. The chromosomes are an aggregation of an enormous number of genes, each playing its own little part in the production of the individual organism, each obeying Mendel's Laws. Of this gene-army our germ-plasm consists.

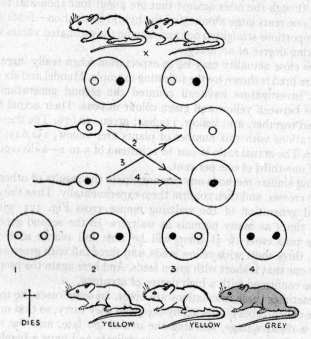

FIG. 172. Why no yellow mice breed true.

The gene for yellow is dominant, and also lethal. When two "yellow" genes are present, the mouse dies while still an embryo. Yellow mice therefore contain one gene for yellow and one for grey, and when two such mice are mated together (above) the offspring (below) always include a considerable proportion of greys.

We cannot catalogue here all the various actions that have been traced to these genes. Let us consider a single species, the Fruit-fly, which, as we have remarked already, is particularly well suited for genetic experiments and has therefore been extensively and intensively studied. In this insect it has been shown that one gene reduces the wings to vestiges, another cuts them off square, others slightly alter their vein-pattern; still others alter the shape and colour of the eye, others the colour of the body; one makes the feet grow double; one remarkable recessive gene not only turns the grey body tan-colour but abolishes the fly's normal instinct to go towards the light; others influence fertility and others again the length of life. There is not a corner of the Fruit-fly's anatomy, not an aspect of his activities that has not been shown to be controlled in some way by these all-important genes.

To take a few examples from other organisms, single genes exist which convert horned into hornless cattle, bearded into beardless wheat, normal snap-dragon flowers into flowers with five-rayed symmetry. All the colour-varieties of domestic animals depend on genes. So do the plumage-varieties of poultry, and here also there are genes influencing the shape of the comb, the size of the body, and even the egg-laying capacity. We shall discover many other interesting effects of genes in later sections.

Among the most curious genes known are the so-called "Lethal Genes," which kill their possessors. For example, there is a gene in the mouse which, when present in one only of the chromosome-sets, produces a yellow animal, perfectly healthy, though with a tendency to fat; but when it is present in double dose the animals die in the uterus at an early stage in embryonic development. Thus a yellow mouse which breeds true cannot exist. There is a similar case in the Fruit-fly, a gene which in single dose produces a scalloping of the margins of the wings and in double dose kills the developing embryos. Again, another gene in the Fruit-fly causes fatal tumours to grow in the young grubs. In many plants there are genes which kill the seedlings by preventing the formation of the vital green pigment, chlorophyll.

So far we have treated of the genes as completely independent entities. But their effects may interfere; they may modify, counteract and reinforce each other. There is, for example, the phenomena of Multiple Genes, when a troop of genes combines for the production of a single effect, each contributing a little push in the right direction. Such gene-troops are responsible for the inheritance of characters such as size in animals, or skin-colour in man, which show whole series of graded degrees of difference. We will illustrate the Geneticist's conception of these Multiple Genes more fully in the following section.

In our own species, plenty of Mendelizing characters have been made out. The difference between ears with free lobes and ears with attached lobes depends, it would seem, on a single gene-difference. So, it is alleged, does the difference between blue and brown eyes, blue being recessive to brown. Two brown-eyed parents can have blue-eyed children, but two blue-eyed parents should not, in the present phase of genetic science at least, have a brown-eyed child. It might cause unpleasant accusations.

Albinism—pink or scarcely coloured eyes and white or almost white hair—which affects about one person in every ten thousand, is much more certainly a simple recessive to normal pigmentation. So we may expect to find albinos cropping up as the result of a marriage between two quite normal people who happen to be carriers of the albino gene. But if two albinos marry, all their children will certainly be albinos. On the other hand, the malformation known as brachydactyly or "short-fingeredness," when the fingers have one joint too few, appears to be a Mendelian dominant. Thus, except in the event of both parents carrying a recessive normal gene, in which case one child in four will have normal hands. If two normal people marry, the children cannot possibly be brachydactylous. If a brachydactyl marries a normal, the children will either be all brachydactylous, or half of them brachydactylous and half normal, according to whether the malformed parent was pure-bred or hybrid for the malformation-producing gene; if two brachydactylous people marry, the children will all be brachydactylous if one or both of the parents were pure-bred for the gene, while if both were hybrid, containing a recessive normal gene, there will be an approximation to the proportion of three quarters brachydactylous to one quarter normal children.

We can also say with a reasonable confidence that special genes control such diverse characters as hair-colour and hair-shape, the tendency to produce twins, and the pitch of the voice. Moreover, it has been shown that such a trivial peculiarity as the presence of a white lock in one region of the head can be due to a dominant gene, and so it seems is the "Habsburg" lip—the fleshy lower lip and protruding lower jaw which Haecker has traced back in portraits of the Habsburg royal line from the present day to the fifteenth century. The kinkiness of negroes' hair seems to be a simple dominant to the straight or wavy hair of other races.

The shape of the nose is controlled by at least four pairs of genes; convex bridge is dominant to snub. The property which causes the blood of some people to clot other bloods, making them useless for blood-transfusion, is also inherited in a Mendelian way. Certain forms of dwarfism are due to a single recessive gene, and so are some congenital forms of short-sightedness and of club-foot. In some cases, at least, left-handedness is inherited as a single recessive character, though the picture is complicated by the tendency to force right-handedness on children by training. The "almond eye" of many Asiatic peoples is also a recessive.

THE A B C OF GENETICS

But besides such human characters which obviously obey Mendel's law of segregation, there are many others which, though less clear-cut, we can affirm to be inherited, without yet being able to assign them to the action of definite genes. This, however, is merely due to the fact that the biologist cannot conduct properly planned breeding experiments with his own species. Wherever experiments can be carried out, differences which are inherited turn out to be due to differences in one or several gene-units; and man in his heredity undoubtedly works like the rest of life.

For determining whether such less sharply-marked human characters are inherited we often have to fall back upon the statistical method known as correlation; we cannot go into the details here but must refer our readers to books like Carr-Saunders' *Eugenics* or Karl Pearson's *Grammar of Science*. But the results of this analysis are very definite. All kinds of differences between human beings may owe their origin to innate differences of constitution: height, proportions, temperament, disease-resistance, intellectual ability. Environment also has its moulding effect, but *within a single social stratum and on the average* the differences between one human being and another are due much more to heredity than environment, much more to genes than to home and school and the accidents of life.

§ 6. Colour Genes in America as an Example of Multiple Genes.

HERE perhaps as an example of the way in which the Geneticist deals with more intricate pedigrees than those we have hitherto considered we may note some interesting studies worked out by C. B. Davenport upon the mixture of the white and black strains in the United States.

He starts his investigations from the mating of "pure" whites with "pure" blacks. His inquiries show that the first hybrid generation, the true mulatto, is very uniform and intermediate in type between the parents. The next generation, however, if mulattos intermarry, is surprisingly varied. Single families of course will give purely chance results, but taken in the large the following consequences have been established. In this second generation there are no such gradings of cream and coffee as the blood-mixture theory would require. On the contrary the offspring are conspicuously diverse and all sorts of gradations between extreme black and extreme white may occur, with half-way mulattos in a majority. If the resulting population continues to interbreed the same confusion persists; there will be everything from coal black to white.

Now at first sight this does not look as clear as the case of the Four-o'clock (Fig. 170); nevertheless the same principles can be traced in it. We remark, for example, that in both cases the first hybrid generation is uniform, while the second is varied and may include the parental types. The difference is that in the second generation of mulattos, instead of the three simple classes, reds, pinks and whites of the Four-o'clock, we get every conceivable shade of blackness. The shades vary widely in the same family; there are never any whole large families simply intermediate between the tints of the parents. How is this to be explained?

Let us suppose that instead of one gene-pair, as in the Four-o'clock

FIG. 173. The Habsburg Lip

In the Habsburg dynasty, a prominent lower lip and a protruding lower jaw recur repeatedly, and seem to be due to a single dominant gene. (1) The Emperor Frederick III (1415–93), from a medal; (2) his son Maximilian I (1459–1519), by Dürer; (3) Maximilian's great-great-great-grandson, Philip IV of Spain (1605–65), as a young man, by Velasquez; (4) Philip IV's son, Charles II (1661–1700), by Juan Careno. The dominant gene still persists to-day, as shown in (5) the former King of Spain, Alfonso XIII, whose mother was a Habsburg, and who is descended on his father's side both from a daughter of Philip IV, and from a daughter of Philip IV's father, Philip III.

the original black and white parents differed in three or four independently assorting skin-colour gene-pairs. Let us suppose also that each gene contributes a little quota of blackness, so that the colour of an individual depends on the number of black-producing and white-producing genes he carries. The first cross-bred generation will be exactly half way between the two parents in tint, because half his colour-genes are black, from the negro parent, and half are white, from the one hundred per cent "Caucasian" parent. In the next generation, by the usual Mendelian reasoning, it can be shown a priori that there will be all sorts of mixtures and combinations; and that is exactly what is found. There will be a few with white genes only—a very small minority. And an approximately equal minority with black genes only. The majority will have black and white mixed in various proportions, the likelihood of any tint turning up being the greater, the nearer it is to half-way mulatto colour. And this is what an extensive statistical examination of coloured population shows to be the case.

We need not go into all the exact technical details of this research. We will only note this: that Mendelian principles can still be detected at work here, and that the black and white genes do not mix together but retain their independence in the mixed stock. What we are dealing with is shuffled genes, not with "mixed blood." It is perfectly possible that a pure white, or a pure negro, could spring from one of the early generations of such a cross. It is, of course, extremely unlikely. In the matter of skin colour alone about one in sixty-four (or if four instead of three separate genes are at work, about one in two hundred and fifty-six) of the second cross-bred generation will be pure white and one pure black. Moreover, there are other genes at work controlling the shape of the features, the voice, the characters of the hair and so on, so that the chance of a full-blooded negro, or of a pure-bred white turning up is exceedingly slight indeed. But it exists for all that. The marriage of two individuals near the same end of the scale increases this likelihood. One of the most active members of the movement in defence of the rights of coloured people might pass anywhere for a Scandinavian blond. He has an unquestionable coloured pedigree on both sides, but in the South, where various racial social distinctions are made between the races, he has great difficulty in refusing treatment as a "White." Hotel clerks and train conductors will not believe it of him.

This sort of inheritance, by teams of co-operating genes, or Multiple Genes, as they are called, is very important in Nature. Wherever a series of graded types appears in a cross we are dealing with Multiple Genes. In our own species a number of characters are inherited in this way. For example, the inheritance of stature and weight, of the tendency to corpulence, even perhaps of temperament, are due to Multiple Genes.

§ 7. Mapping the Chromosomes.

WE HAVE already written of the remarkable confidence with which the Geneticist can display a map of the chromosomes of such a creature as Drosophila and localize and identify the genes thereon. Genetics is a science with all the attractiveness and perhaps some of the defects of heady youth. It has been developed with extreme enthusiasm in the last quarter of a century and it has achieved very brilliant results. But there is a Wonderland touch about some of its claims and assertions that holds back some of the maturer biologists from complete acquiescence in all its enthusiasm.

Meanwhile we record the marvel of this mapping of the chromosomes of Drosophila. Marvel it is. These things we describe here with such detail and exactness are visible only with the highest power of the microscope. We are dealing with forms that would make the finest fairy that was ever imagined seem a colossal monster of clumsiness in comparison. Never has the reader's unassisted eye seen anything so fine and small as the reality we here enlarge for him.

In the various species of the Fruit-fly Drosophila, easy to breed, easy to observe and rich in varieties, Morgan and his indefatigable associates have been able to explore the behaviour of over five hundred genes. In addition there must be an unknown number of true-breeding genes which have as yet given no variations to betray their nature. In the cells of this genus of fly there are only four pairs of chromosomes. It has been shown, by an elaborate process of experiment and reasoning, that these five hundred and more genes are crowded in a chain along the chromosomes. They are not strung together anyhow, it would seem, but arranged along the chromosome in a fixed definite order.

The reduction-division of the germ cells of Drosophila to form the gametes has been watched and recorded in great detail. As we know already, the chromosomes then come together in pairs instead of splitting. Now the pairs that so unite are always the corresponding pairs. The description of the reduction-division in § 2 of the previous chapter may seem marvellous enough, but we did not go then into the full intricacy of this delicate process. In this Fruit-fly the two little blobby chromosomes pair, the two shorter rods pair, and the long V-shaped chromosomes pair, each with its corresponding fellow of the other set. At the reduction-division each daughter-cell gets one member of each pair and thus it is ensured that each of the resulting gametes shall get one complete outfit. The same is true of the chromosomes of many other creatures. Moreover, the chromosomes during the pairing process lie in very accurately dovetailed apposition, even the individual corresponding genes of the members of a pair being opposite to each other. And sometimes they twist a little round each other, like Hermes' snakes, while they are writhing in the nucleus in the early stages of the reduction-division.

Even the minutest details of these microscopic juxtapositions are accurately reflected in the subsequent inheritance. In the first place, if two genes are placed in the same chromosome they will tend to be handed on together from generation to generation. The units

THE A B C OF GENETICS

which are handled and dealt out on the spindle of a dividing cell are chromosomes, not genes. The genes go in bundles. In the case of the tall yellow-seeded and short green-seeded peas (§ 3) the height gene and the seed-colour gene are in different chromosomes. If they were in the same chromosome we should expect the original pairs with which we started to reappear in the second cross-bred generation, and that the tall plants with green seeds and the dwarf plants with yellow seeds would not appear.

However, as a matter of fact we learn from this Drosophila work that this linkage of genes which are carried by the same chromosome is by no means complete. When the pairing chromosomes curl round each other they may exchange bits of themselves. They may come unstuck along a rather different line from that in which they met. Thus the genes are shuffled around and appear in ever-changing combinations even when they are carried by the same chromosomes.

This brings us to the line of reasoning upon which the position of the genes on the chromosomes has been actually mapped out. The probability that two genes on the same chromosome will be separated in the curlings and writhings of the reduction-division depends on the distance

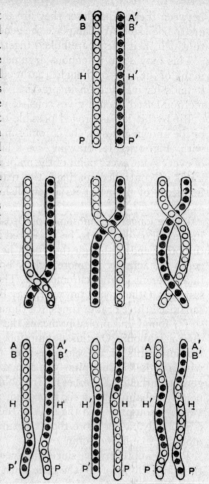

FIG. 174. A diagram of how the two members of a chromosome-pair may exchange bits with one another.

When the two corresponding chromosomes pair up (top figure) they may twist round each other in various ways (middle row). They appear to break at the points of crossing and to join up again in new arrangements (lower row), but always so that each chromosome contains one of each kind of gene

between them. The closer they are to each other, the less likely they are to be parted. We can place the genes therefore by noting the degree to which their effects are linked together in inheritance.

This may seem a tenuous argument, but it is no finer spun than many of the tools of thought that modern physics is forced to use, and it finds its justification in practice. The idea can be tested, and the method based upon it gives consistent results.

Bit by bit it has proved possible on these lines to map the places of the known genes on the chromosomes. The map is reproduced herewith (Fig. 175). Its drawing was a laborious process, for the placing of every gene, every point of the map, necessitated the making of many kinds of controlled matings, the rearing and examination of many hundreds of flies. But its accomplishment is a crowning triumph of modern Genetics.

In the longer chromosomes of Drosophila, a gene nearer one end is just as likely as not to be separated by crossing-over from one near the other end; although they both lie in the same chromosome, such genes are actually no more tightly bound together than if each were in a different kind of chromosome. These chromosomes are loose, long sausages. Other genes, however, may lie so close that they only become separated three or four times in a thousand. This is the tightest linkage so far found and probably means that the two genes are actual next-door neighbours. On this assumption, and with the aid of microscopic measurements of the chromosomes themselves, we can calculate—within rather wide limits—the actual amount of space occupied by a gene, and the total number in Drosophila's germ-plasm. The total number is probably about six thousand, and certainly between two and twenty thousand. And the bulk of the gene, whatever a gene may be, is probably rather more than ten times as large as that of a protein molecule like hæmoglobin.

Put in another way, since the average protein molecule contains a thousand atoms or so, the gene must contain about a hundred thousand. The chemistry of life is evidently complex. The little girl who asked if God made flies as well as people and was told yes, remarked: "H'm—fiddling work, making flies!" The discoveries we have been narrating reveal that the work of making a fly is even more fiddling, or rather more astonishingly delicate and complicated, than we could ever have imagined. If a fly an eighth of an inch long needs several thousand separate and different units to conspire in building up its structure, how many will prove necessary for a man?

We do not know; it may be not a very much greater number, perhaps

THE A B C OF GENETICS 495

the same. For if we are fearfully and wonderfully made, so is an insect. Let us remember the words of Sir Thomas Browne on "Ants, Bees and Spiders": "Ruder heads stand amazed at those prodigious pieces of Nature—Whales, Elephants, Dromedaries and Camels; these, I confess, are the Colossus and Majestick pieces of her hand: but in these narrow Engines there is more curious Mathematicks."

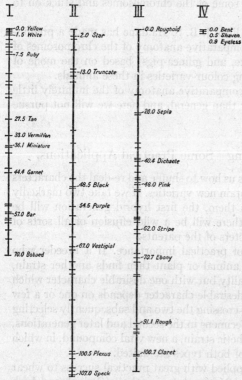

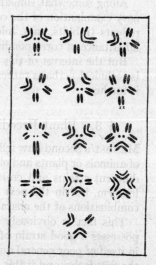

FIG. 176. Diagrams of the chromosomes of various species of Drosophila.

The left upper group is Drosophila melanogaster (Fig. 175). Sometimes, however, the dot-like chromosomes are absent; sometimes the place of one or both of the V-shaped chromosomes is taken by two rods corresponding to the two arms of the V; and sometimes the rods are bent up into small V's. In addition, the proportions of different chromosomes and parts of chromosomes may differ. XX denotes the sex-chromosomes (see Chapter 6). (After Metz and Moses in "Journal of Heredity.")

FIG. 175. A map of the four kinds of chromosomes in Drosophila melanogaster, showing by cross-lines the positions of a few of the genes on the different chromosomes, and the names of some of the more important ones.

I is the rod-shaped chromosome of Fig. 176. *II* and *III* are the big V-shaped ones, and *IV* is the little dot-like chromosome. The numbers refer to the distances of the genes from the end of the chromosome. The names of the genes refer to some characteristic effect which they produce. The map is magnified many thousand times. (After T. H. Morgan.)

Besides the common Fruit-fly *Drosophila melanogaster* other species of the genus have been investigated, and they show a very similar arrangement of the chromosomes, though with certain modifications. In the species *D. simulans*, for example, it is as if a bit had been taken out of one of the long chromosomes of *melanogaster*, turned round, and put back with its ends the wrong way. In other species bits have been taken off the ends of some of the chromosomes and stuck on to others. And so on.

Along somewhat similar lines J. B. S. Haldane has made a preliminary examination of the comparative anatomy of the chromosomes of rodents (rabbits, rats, mice, and guinea-pigs) based on the mode of inheritance of corresponding colour-varieties in these animals.

But the interest of this comparative anatomy of the infinitely little is technical perhaps rather than general, and here we will not pursue it further.

§ 8. Plant Breeding—Some Practical Applications.

MENDEL's Second Law tells us how to shuffle and re-deal the characters of animals or plants and obtain new varieties. If we take two markedly different strains and cross them, the first hybrid generation will be uniform, but in the next there will be a wild effusion of all sorts of combinations of the characters of the parents.

This fact is obviously of practical importance. If a breeder who possesses a good strain of animal or plant then finds another strain, in itself of poor general quality but with one desirable character which the first lacks, and if this desirable character depends on one or a few Mendelian factors, then by crossing the two and subsequently selecting from individually tested specimens in the second and later generations, he can build up a new synthetic strain, a new vital compound, in which all the desirable qualities of both types are united.

This method has been applied with great practical success to wheat by Sir Rowland Biffen. The National Association of British and Irish Millers, dissatisfied with the wheats at their disposal, requested Biffen to try to improve them. They demanded a beardless wheat, immune to the fungus disease known as rust, with a high yield per acre, and with "hard" grain, rich in gluten, and so of good milling and bread-making quality. No known variety combined all these characters, so Biffen set himself to make one. He crossed a rust-resistant grain of low yield and poor grain-quality with a wheat of good quality but great susceptibility to disease. All the first generation were susceptible to rust. This was no discouragement to a Mendelian, so Biffen obtained a large second generation by self-fertilization, and his expectations were justified by finding, among the large number of recombinations which appeared, some plants which combined rust-resistance with the desirable qualities of the susceptible stock. The two main strains which he manufactured by Mendelian principles, "Little Joss" and "Yeoman," though only put on the market in 1912 and 1917 respectively, by 1927 occupied about a third of the wheat-growing area of the world.

The American plant-breeder, Luther Burbank, who died in 1926, employed similar methods. He used to cross two widely separated varieties or species, from the disappointingly uniform first generation raise a huge second, and then select the best types from this. His white blackberry was one success among 65,000 unsuitable combinations; the parents of his edible spineless cactus were isolated from among hundreds of grandchildren of a cross, some of which were more hopeless in appearance, more spiny and more woody than either parent. Among two thousand in the second generation of a cross between Oriental and Opium Poppy, no two were alike. The variety of forms available as the result of this method is very considerable. This, of course, is entirely in accord with Mendelian principles; the variety of new types is due to new recombinations having taken place among the many different genes brought in by the two parents. Curiously enough, Burbank did not seem to realize this. He was a Mendelian without admitting it.

But the most remarkable success so far achieved in applying genetic principles to commercial practice has been the work of that practical people, the Dutch. In 1928, two-thirds of the total sugar crop of Java (a sugar crop second only to that of Cuba) was grown from a pure stock of sugar-cane known as "P.O.J. 2878." The next year over ninety-five per cent of the crop was being grown from it. This cane (we are utilizing the information in Mr. Ormsby-Gore's interesting report on his visit to Malaya, Ceylon, and Java in 1928) has a yield per acre nearly twenty per cent greater than any other known variety: and in addition is more resistant to the numerous ills to which cane is heir in the East Indies.

Only in 1921 was the possibility of creating such an improved strain first seriously envisaged. A high-yielding variety of cultivated sugar-cane was crossed with a wild Javanese cane which, though very resistant to disease, *contains no sugar at all*. Selection among the descendants, together with the judicious admixture of genes from other cultivated varieties, led by 1924 to the production of this new and really marvellous P.O.J. 2878. Its field tests took two more years, and it then began to be grown commercially on a large scale. Incidentally, it is very interesting to find that the improved variety has one hundred and twenty chromosomes instead of the forty of its principal cultivated ancestor, in this respect providing a striking parallel with wheat, whose high-yielding varieties all have forty-two chromosomes instead of the fourteen of the most primitive wild wheats (see page 598).

Java is smaller than England, but, though almost entirely agricultural, has more people in it. The feeding of these multiplying millions is becoming a serious problem. . . . Mendel in his monastery garden; Hertwig and Boveri and the other cytologists straining their eyes down their microscopes to look at chromosomes; Morgan in his "fly-room" at Columbia—who could imagine that their work would so soon, and on such a grand scale, help to solve Java's social problems?

§ 9. Is Inbreeding Evil?

INBREEDING has been looked upon with suspicion by many men of many races. There may be religious taboos upon its practice among human beings, such as are evidenced in the Tables of Consanguinity to be seen on the walls of our churches; or there may be rational (or at least rationalized) objections urged against it, as in the widespread opposition to cousin-marriages on the ground that they will produce defective children, or the reluctance of many animal breeders to risk degeneration of their stock by close inbreeding.

Now it is perfectly true that in general if an animal or plant stock is consistently inbred (for example, by means of brother-to-sister matings) its vigour, its size, its fertility and its economic worth will slowly fall off. But this is not by any means a universal law. There are plenty of cases where inbreeding is not harmful in the least. Many breeds of domesticated animals and plants are very closely inbred, and show no evident degeneration; this is true, for example, of the Saint Bernard dog, of Jersey and Guernsey cattle, and indeed of many breeds of cattle, horses, sheep, and dogs. Pigs, on the other hand, are not usually so closely inbred; and they "degenerate" much more readily on inbreeding. In Nature, there are many species which as a rule and even invariably fertilize themselves. Among plants, such familiar forms as beans, wheat, and tobacco may be cited; among animals certain parasites such as Sacculina and tapeworms—which, in spite of their ethical level, are very efficient parasites! A classical example of actual *improvement* of a stock by inbreeding is the experiment of Miss King, at the Wistar Institute, Philadelphia. Miss King practised brother-and-sister matings on a strain of albino rats for generation after generation, selecting vigorous individuals. After twenty generations of this her rats were all the better for it; the average weight was higher, the fertility (number of young per litter) had increased, and even the individual life-span had lengthened a little. To-day, after more than fifty generations of inbreeding, the strain is still going strong, a standing proof that inbreeding is not harmful *per se*.

In general, nevertheless, it is true that an inbred stock degenerates. And the opposite is also true; a cross between separate strains usually produces strong, vigorous offspring. Everybody has heard of the stimulating effect of "fresh blood" on a race. Pig-breeders who want rapidly maturing animals for bacon are in the habit of crossing two different breeds, for they know that the hybrid animal will grow up more rapidly than either stock. The same method is often practised, for the same reason, with cattle, sheep, and poultry. Even when the result of a cross is sterile it may exceed its parents in vigour. The strength, endurance and resistance to disease of the mule are well known; the sterile hybrid of a radish and a cabbage grows into a rampant bush ten feet in height. But here also there are exceptions, and a cross may not exceed its parents as conspicuously as is expected.

A curious and, at first sight, rather lawless mass of facts! But this mystery of inbreeding and outbreeding is one of those compartments of biology which have been swept and cleaned and put in order by geneticists in the last thirty years. To-day, the rationale of the subject is plain enough.

Consider a family of brother and sister animals or plants, whose stock shows no inbreeding for several generations back. They will have two parents, four grandparents, eight grand-grandparents. If, on the other hand, one of the grandparental couples had been a brother-and-sister mating, the family will only have six great-grandparents, or if their father and mother had been brother and sister, they would only have had four. So that inbreeding diminishes the number of ancestors contributing to a race—the number of different strands of germ-plasm that tangle and blend within it. And indeed it is found in practice that a strain which has been long inbred is more homogeneous, less variable than one which is cross-bred and mixed.

To make the idea clearer, suppose that we start with a mixed stock (let us say, for simplicity, of pink Four-o'clocks) and leave them alone to interbreed without any form of control. Pinks, whites and reds will appear and mate; the genes will mix and separate and mix again. And nothing very much will happen to the proportion of plants with differently-coloured flowers. But suppose that we force the plants to fertilize themselves—the extremest form of inbreeding. The pure reds will produce pure reds and the whites whites; but the pinks will produce reds, whites, and pinks. So that the proportion of pinks will diminish generation after generation. In a word, such an inbred stock tends to sort itself out into a number of races, each pure-breeding and

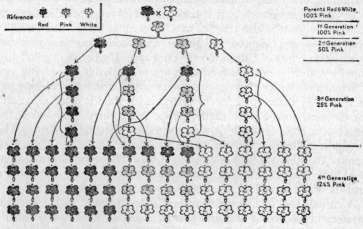

Fig. 177. The Effect of Inbreeding.

A pure red strain of Four-o'clocks is crossed with a pure white strain; result, all pinks. The pinks are inbred; result, 25 per cent. pure reds, 25 per cent. pure whites, 50 per cent. pinks (see Fig. 170). The separate strains are again inbred; each generation of inbreeding halves the percentage of mixed (pink) plants, and doubles that of pure red and white strains.

homogeneous. Now we noted that genes may be dominant or recessive, and that a recessive gene may be carried hidden in a race, mastered by its dominant companion, and may continue so for generations without showing itself. This is true for example of albinism in rabbits, of hereditary deaf-muteness in man. The effect of inbreeding on a stock will be to unearth these hidden recessives. They will be brought together in pairs, freed of their dominant neighbours, and so made to reveal themselves. In rabbits, if there is albinism in the strain it will appear: in the human example deaf-mutes will turn up. And the opposite is also true; cross-breeding will conceal the recessives by increasing the chance of their being submerged by dominants (Figs. 177, 178.)

Clearly the relative merits of inbreeding and outbreeding will depend simply on the recessives that the strain in question is carrying.

If they are harmful recessives, inbreeding will be bad and outbreeding good; if they are desirable recessives, the reverse will be true. In the case of the human stock tainted with a concealed hereditary defect, any form of inbreeding is to be avoided; brother-and-sister marriage would be very dangerous indeed, and even cousin marriages would be running a considerable risk. Consorts should be chosen from quite unrelated families. But in the albino-carrying rabbits the reverse might be the case; if their breeder wanted albinos he should inbreed them, and the closer the better.

If it be true (and, as a matter of fact, it is) that this idea of the concealment or exposure of recessive genes is enough to explain the results of crossing and inbreeding, we must find in it an explanation of the fact that in the great majority of cases close inbreeding is quite obviously bad and crossing is quite obviously invigorating. But to do this we must anticipate the conclusion of a later chapter.

A species is not a static thing but an evolving thing. It is constantly being polished and improved by its surroundings, by a selecting action of the environment which favours some variations and strikes out others. The environment may be said to pick over the germ-plasm of a species and weed out the genes to which it is antagonistic. The process is always going on; there is no perfectly adapted species. Living things are continually adapting themselves more and more closely to the conditions under which they live. Now it is clear that in every generation the dominant genes will be scrutinized and either struck out or allowed to pass. But the recessives may slip by. They may lurk in the race concealed by dominants and only present themselves for selection once every three or four generations. In a word, they will be less intensively picked over than the dominants. So that the dominants, on the whole, will be a better, more tried bunch than the recessives; the recessives, on the whole, will be the skeleton in the nuclear cupboard of the race. This, in a nutshell, is the reason why inbreeding, by bringing a scum of recessives to the surface, will in most cases lower the average standard of the stock.

And now let us note a case in which these principles were used deliberately for the improvement of a stock. East and Jones, working in America with maize, secured intense inbreeding by making the plants fertilize themselves for twelve generations. Out came the recessives, and the plants separated into a number of true-breeding lines. There were strains with round or flat, coloured or colourless cobs, with long or short ears, with more or less seeds on the ear, with large or small grains. And the skeleton in the race-cupboard came out also;

there were stunted strains and infertile strains, yellowish and albino strains, strains particularly susceptible to fungus diseases, strains unable to stand upright because of their malformed roots, strains with distorted ears. Thus the genes were sifted out and isolated. After this had been done the most desirable strains were picked out and recrossed with one another and thus a breed was obtained with all the fine characters of the original stock, but purged of the concealed recessives which made them throw up the sporadic abnormalities that deface almost any field of maize.

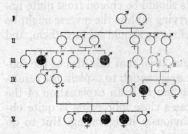

FIG. 178. How cousin marriages may reveal recessive defects.

A pedigree of deaf-mutism. One of the original couple (generation I) must have carried the recessive gene for deaf-mutism, as must individual X, who comes in from another tainted family. A and B were first cousins, C and D were second cousins. In both cases the marriage produced deaf-mute children. (After Albrecht.)

Thus inbreeding may be used to purify a stock of the harmful cargo of recessives that it is smuggling. It is a long and expensive process, but it can be justified by the good quality and uniformity of the final product. It has already been found to pay commercially with maize and sugar-beet, an advantage both to farmer and seedsman. We can be sure that in the near future the method will be applied to many other plants, and in a modified form to animals, with altogether profitable results.

With regard to the question of human marriages, there is little to say. The whole business of human breeding is at present so haphazard that it is difficult to write with any precision of what we should or should not do. Man has evolved recently and rapidly, and is so mixed and cross-bred that we may expect his stock of undesirable concealed recessives to be considerable. From this point of view inbreeding looks dangerous. When there is any suspicion of a harmful recessive in a family, inbreeding should certainly be avoided. But it is not in any sense harmful in itself. For several generations, cousin marriages between the closely related Darwins, Wedgwoods, and Galtons were frequent. Charles Darwin himself married his cousin; and among his children and his grandchildren alike there have been several distinguished bearers of the same penetrating yet patient scientific capacity which characterized him, his cousin Francis Galton, and his grandfather, Erasmus Darwin.

Perhaps in years to come our descendants will look with intelligence over their pedigree, and if there is a probability of recessive genius in a family and no reason to suspect a grave recessive taint they will deliberately encourage inbreeding. A rather grim Utopia might be devised in which for some generations, on the pattern of East and Jones' maize, inbreeding would be made compulsory, with a prompt resort to the lethal chamber for any undesirable results. A grim Utopia, no doubt, but in that manner our race might be purged of its evil recessives for ever.

Even to-day, Mr. Everyman, faced with the issues involved in a cousin marriage, is no longer standing in fear before a superstitious mystery. He can begin to weigh the risk and the advantages in the rational spirit in which he would approach an investment or a business undertaking. He can consider his forefathers and weigh the quality of his prospective relations-in-law.

§ 10. Reversion to Ancestral Type.

ANOTHER matter that we may profitably discuss in the light of recent discovery is the question of Reversion. If a number of fancy breeds of an animal or plant be turned loose, the population tends to revert to the ancestral wild type from which they have been derived. The pigeons of London and Venice are mostly coloured more or less like the blue rock pigeon, although they were originally liberated as fancy breeds. Sweet-peas may revert to the wild purple flower-colour, and so on. This tendency, being in apparent opposition to evolutionary progress, has been made much of by the opponents of evolution. "It has been found that every 'variety' produced artificially, if left to itself for a few generations, reverts to the original type"—we quote from an American Fundamentalist work, published in 1922.

Let us point out at once that this latter statement is a distortion. Reversion only occurs when *different* varieties are crossed. It never occurs in genetically pure stocks.

Let us analyze a simple example. It is sometimes found that when a black rabbit is mated with an albino, both from pure-breeding stocks, the offspring are neither black nor white, nor are they an intermediate grey or piebald; they are a brownish colour like wild rabbits. This occurs sometimes, not always. In such a case we are simply dealing with the shuffling of two gene-pairs. There is one recessive gene which prevents any kind of pigment-formation in the rabbit, even in its eye, and which produces albinos. There is a second and quite unrelated recessive gene which blackens any pigment that may be formed. Now, since both these genes are recessive, the first cross-bred generation is normal, i.e., grey-brown; the black parent has introduced the dominant gene for pigment-formation, while the albino has introduced the dominant gene for brownness.

If these so-called reverted brown rabbits are inbred they will produce a mixed generation including blacks and albinos in the proportion of nine browns and three blacks to every four albinos. The brown colour, being dominant to the other two, will be in a majority in the resulting population.

Evidently there is nothing very mysterious about this "tendency" to reversion. The ancestral type simply turns up as one of the possible arrangements of the genes which the two parents have contributed. Moreover, it has little to do with evolutionary progress, because the crossing of distinct varieties is a rare occurrence in Nature. Reversion is only observable under highly artificial conditions.

The case of the London pigeons is similar. A number of different fancy varieties were liberated and they bred together and one of the shufflings of their genes happened to produce the wild type. Since most of the variations were recessive, the wild type increased rapidly and submerged the others. And so with the sweet-peas, and with many other cases of reversion when domesticated races of animals or plants are allowed to interbreed freely.

One more point about the rabbits. The cross of black with albino does not invariably produce brown. The albino gene prevents the formation of pigment altogether, so there is no telling (except by crossing with other varieties, or from the pedigree) what colour-genes an albino may not smuggle into our experiments. If he is a bleached brown he will produce the original type when crossed with black, but he may be a bleached black or blue or chocolate or lilac, or even a bleached piebald, in which case the result of the cross will be very different. Thus crossing does not necessarily produce reversion. We mention this to remind the reader how complicated the interplay of the genes may be.

The term reversion is sometimes loosely used to cover another quite different class of happenings, where the appearance of an ancestral type is due not to hereditary recombination, but to an arrest of development. This is well illustrated by certain kinds of hare-lip, in which a groove connects either nostril with the mouth. This state of affairs is permanent with dogfish, and is passed through by the embryos of higher animals. Sometimes the grooves fail to close, and then the primitive condition persists into adult life. It is, however, much better to label all such cases as *arrested development*.

§11.   How Much of Inheritance Is Mendelian?

THE Mendelian laws, we have seen, are of service both to the practical man and as cleaners of cobwebs from our general ideas about heredity. But are they the only laws? May there not be other modes of inheritance besides the Mendelian one?

There is a small group of cases in which inheritance is very evidently not Mendelian. It includes a few plants—of which a variety of our friend the Four-o'clock is one—which have variable patches of green and white on their leaves, and it concerns the inheritance of those patternings. The green pigment of plant tissue is contained in special round bodies that float in the cell, outside the nucleus, and are called plastids. These plastids grow as the cell grows and they proliferate in an independent manner of their own, dividing by pulling themselves into halves. Their reproduction has no relation to the rhythm of division and growth in the cell that contains them, and during the tumult of cell-division they are swept, at random, some into one daughter-cell and some into the other. It has even been suggested that these plastids are organisms, independent lives within the cells, just as the algæ in a lichen are captive lives within a larger organism; but nobody has yet isolated a plastid and seen it grow apart from the containing cell. These plastids, then, are the bearers of chlorophyll from a cell to its daughter-cell, and it is because of their presence in the seed that a plant inherits its vital greenness from its parents.

Now in variegated plants like the Four-o'clock some of the plastids are abnormal and contain no chlorophyll, though behaving in all other respects like their green neighbours. And in their generations these plastids, these independently multiplying bodies inside the cell, breed true. If a cell contains only green plastids, all the cells that proliferate from it will also contain green only. If it contains white plastids, all its descendants will be white, and its proliferation will produce a white patch on the plant. If a cell contains some of each they will be divided between the daughter-cells at random; one may chance to get all the white and one all the green, and one will get a few white and a lot of green and so on. There is little orderliness in this method of inheritance. From these facts one can infer the results of sexual and asexual propagation in the plants in question. An egg-cell, or a shoot from a "pure-green" part of the plant will itself be pure green; one from a "pure-white" part will be pure white; one from a mixed part may be anything. And the male parent has no effect whatever on the colour of the offspring because there is no room for plastids in the pollen-grain. Thus there arise various permutations and combinations, but without the disciplined regularity that characterizes Mendelian inheritance.

In these plants, then, there is inheritance in which the nucleus is not immediately concerned. It is inheritance through the rest of the cell. And it also differs from Mendelian inheritance in that the male parent has very little say in the matter because his gamete is practically all nucleus. The question naturally arises, do such phenomena exist elsewhere, outside this little group of patchwork plants? In animals, is the whole of heredity indeed due to the nuclei of the sperm and egg? Does the rest of the egg-protoplasm (cytoplasm) play no part at all?

We cannot answer with an absolute No. Recent research has produced instances that seem to establish cytoplasmic inheritance in certain types, but the weight of the evidence seems to be that these instances are very exceptional and that the bulk of racial qualities, in all many-celled creatures at least, are transmitted by genes according to the Mendelian principle.

We must remember that in the simplest living things, the bacteria, there is no definite nucleus and no sexual process, so there cannot possibly be Mendelian heredity. Here then is a very large group of organisms, in which apparently inheritance must be non-nuclear. It is possible to look upon the chromosome mechanism as a later evolution, a thing added to this non-nuclear inheritance, co-existing in the higher animals and plants with a more ancient mode. There are authorities (of whom the late Jacques Loeb was the most distinguished exponent) who hold this view. They believe in two different methods of inheritance—the "species" method and the "variety" method; the former, they suppose, is something more basal, co-existing with the latter, and presumably carried on by the cytoplasm. According to these dualists, it is usually by this former, the cytoplasmic or "species" method, that the race evolves.

A similar dualism, it may be, presents itself in the development of the embryo from the fertilized egg. It is probable that the nuclei do not begin to play a part in shaping the growing embryo until the first stage in development is over and the rough form of the future person is already planned out. The egg-plasm forms itself into a crude embryo and then the genes take control and put in the details. This is essentially the former view put rather differently; the egg-plasm is the basal formative agent, the designer and builder of the future organism, while the genes are merely painters and decorators, putting finishing touches to the work.

Now let us consider what objections can be raised to this dualism.

Firstly, in the matter of inheritance in the bacteria. It is perfectly true that bacteria have no definite nuclei. But it has been proved that they have nuclear material, the same substance of which the chromosomes of higher organisms are composed, scattered through their tiny bodies.

May it not be that this scattered chromatin is the only hereditary material that they possess? Presumably when the first bacteria evolved upwards into creatures with nuclei, an elaboration that we must stress as the greatest stride ever made in the evolutionary march, they did so by a process of concentration—by collecting all this scattered stuff together into a single central blob. If this process was at all complete, no hereditary material would be left in the cytoplasm outside the nucleus.

For the terms "species" and "variety" inheritance, there is indeed little justification. Either they express a bad idea or they seek to express a new idea badly. They suggest a revival of the old belief in species as something qualitatively distinct from variety, which, as we saw in Book 3, is unsound. Whenever we find fully fertile hybrids between species, the result follows the Mendelian principles. And in all the careful breeding work that has been done on animals and plants, except for the patchwork plants that we have already considered, there is hardly a case which is demonstrably non-Mendelian. Everything points to the predominant importance of Mendelian heredity in the great majority of living forms to-day, whatever vestiges of cytoplasmic reproduction may still linger in obscurity.

Finally, in that matter of the developing embryo. It is apparently true that the embryo begins to develop without the controlling guidance of its own chromosomes. But before it was fertilized it was made in the maternal ovary, and then the maternal chromosomes were fully active. So it is perfectly possible that the properties of the egg-cytoplasm are in the last analysis controlled by the chromosomes of the mother; and not only is this possible, but it has been proved in a number of instances that they are so controlled.

We conclude then by setting aside non-Mendelian heredity, that is to say, heredity through any other agency than the genes, as a process trivial in extent and at present too obscure for practical consideration. There are intimations, as we shall see, of other forces at work, but at present they are very elusive. They demand further study. In the following chapters we shall be guided in our interpretations constantly by Mendel's two laws. Assuming that the genes are the normal means of heredity, we shall study their method of working. Perhaps the most illuminating study we shall make will be that of the determination of sex; there we shall find the most interesting examples of the struggle between dominant and recessive gene-partners, and the way in which they are affected by other and less easily defined influences arising out of the activity of the rest of the genes.

§ 12. Some Common Superstitions About Heredity.

This is an appropriate place to speak of certain beliefs which are widespread enough among the general public. They may be truly called superstitions, since they are without any of the ample basis of facts which underlie breeders' beliefs on inbreeding and outcrossing.

Of these superstitions, the most important is technically called telegony or "infection of the germ." Those who believe in telegony maintain that the first male to which a female is mated has an influence upon her subsequent offspring, even if these are sired by another male. This belief is so firmly rooted, even among practical men, that some sheep-breeders' associations refuse to register any lamb whose mother was ever mated to a ram of another breed than her own; and it often occurs that fine pedigree cows or ewes are slaughtered or sold for a song because they were accidentally covered by a male of another variety. Dog-breeders share these beliefs. On the other hand, curiously enough, the belief does not seem to be strongly or widely held about the matings of the human species. But then human beings, both male and female, with the very rarest exceptions, trouble hardly at all when they marry about the possible quality of their offspring.

To-day it is possible to assert without any question that telegony is a mere fable, which could only have gained ground in the days when men were ignorant of the true mechanism of fertilization and reproduction. The supposed instances of telegony which are constantly being reported even to-day invariably disappear when critical analysis is brought to bear upon them.

Perhaps the most famous example is that of Lord Morton's mare. The mare, a pure Arabian, was mated with a zebra stallion, and produced a hybrid foal. On two later occasions, she was bred to a black Arab stallion, and gave birth to two further foals. These had legs, which were striped even more definitely than those of the hybrid foal or the zebra sire himself, and one had some stripes on parts of the neck also. In addition, they had a stiff mane of very zebra-like appearance. Darwin himself accepted the evidence as sufficient proof of telegony. But when definitely planned and long-continued experiments were made, the proof escaped. Cossar Ewart, for instance, made a number of horse and zebra crosses to test the validity of the belief. When mares previously bred to zebras were afterwards mated with horse stallions, their colts were often without the least trace of zebra characters. In other cases, colts with some degree of striping were produced. But one mare gave birth to a striped colt as a result of her first mating, which was with a horse stallion; while two later matings with other stallions, made after she had been successfully mated once and three times respectively with a zebra, gave unstriped offspring. In other cases, when striped colts were born to a mare and stallion after the mare had been previously mated to a zebra, Ewart took other mares, closely related to the first, bred them to the same Arabian stallion without having mated them previously with a zebra—and they, too, produced striped foals.

In short, the production of striping (and also of erect mane) in foals is not a very uncommon occurrence in horses; it may appear whether previous impregnation by a zebra has taken place or not. The stripes of Lord Morton's foals were a mere coincidence, well illustrating the danger of drawing conclusions from single and therefore possibly exceptional cases, and the need for systematic and repeated experiments.

Most of the examples adduced to prove telegony have the same basis as Lord Morton's. They are coincidences, the resemblance of the later offspring to the first sire being usually due to the emergence of some unsuspected hereditary combinations as the result of the second mating. As a good example, an actual occurrence in poultry deserves citing. A Wyandotte hen was successfully mated to a Leghorn cock. Three months later, on being mated to a cock of her own breed, she produced a number of clutches, in which some of the birds had the single comb of the Leghorn instead of the rose-comb of the Wyandotte. But single comb is recessive to rose; and we also know that there has been a considerable infusion of single-comb Leghorn blood into many modern strains of Wyandotte; so that the occasional mating of two Wyandottes both carrying the gene for single comb would not be surprising, and would account for all the facts.

One or two cases, however, may have a different explanation. In many animals, the sperm of the male continues active for a considerable time in the genital tract of the female. In bats, for instance, impregnation takes place in the autumn, but the sperm lives on in the female until the following spring before fertilization takes place. In domestic fowls sperm can continue to fertilize eggs for three weeks after insemination, and in turkeys one impregnation will suffice for a whole summer. In such circumstances sperm from a previous sire might persist and fertilize an ovum even after a mating with a second male; but the apparent telegony would then be the result of the ordinary processes of heredity, not of any new or special influence. This explanation, however, will not hold for mammals if pregnancy and birth follow the first mating.

In these ways the phenomena attributed by superstitions to telegony arise. But think how rarely they arise. We must not forget the enormous mass of negative evidence. In mule-breeding establishments the mares are usually put to horses after three or four impregnations by asses; and yet no trace of ass-characteristics then appears in their offspring. And in experimental breeding it is the commonest thing to mate an animal of new or unusual type with a number of different mates, yet no cases of telegony have turned up in the laboratory.

The converse belief to telegony—that a male's subsequent performance will be impaired by a mating with an inferior female—is also still held, though by no means so widely or tenaciously. This idea of "infection of the male" is even less credible than telegony, and the evidence in support of it is utterly unconvincing. It is to be hoped that both these beliefs will rapidly die a natural death, as in many ways they hamper the progress of scientific breeding, as well as involving breeders in considerable but unnecessary financial loss.

Another very widespread but baseless belief is that in maternal impressions. This dates back to very early time, as is evidenced by the account in Genesis of how Jacob (somewhat dishonestly, it must be confessed) utilized his knowledge of the process to procure greater riches for himself. Laban having promised him all the cattle and sheep and goats that were "ring-straked, speckled and spotted," Jacob "took him rods of green poplar and of the hazel and chestnut tree; and peeled white strakes in them, and made the white appear which was in the rods. And he set the rods which he had peeled before the flocks in the gutters in the watering-troughs when the flocks came to drink. And the flocks conceived before the rods, and brought forth cattle ring-straked, speckled and spotted. . . . And it came to pass, whensoever the stronger cattle did conceive, that Jacob laid the rods before the eyes of the cattle in the gutter, that they might conceive among the rods. But when the cattle were feeble, he put them not in: so the feebler were Laban's and the stronger Jacob's."

THE A B C OF GENETICS

To-day it is in the field of human reproduction that this superstition is strongest. In spite of popular credulity and scriptural authority, however, there is no warrant for the belief. There exists no possible machinery by which a pregnant mother, shocked by the sight of some deformed cripple or frightened by a negro, could influence her unborn child so that it should grow crippled or acquire a dark skin. Not only is there no nervous connection between the mother and the child in the womb, but no connection by way of the blood. All that passes between them passes by indirect channels, dissolved or secreted through the vessel-walls.

How strong the superstition may be is shown by an incident that actually occurred quite recently. A breeder of Black Aberdeen-Angus cattle was disturbed at the number of red calves born in his herd. He accordingly went to the expense of erecting a huge palisade around his paddock, which he then painted jet-black, so that the cows at and after conception should never have any colour but black before their eyes. But the explanation of the red calves was in reality very different and very simple. In all mammals this type of red colour is recessive to black; and his black bull and some of his black cows happened to be carrying this recessive red, with the obvious result that one-quarter of their offspring were pure red. A knowledge of Mendel's first law and a little patience in testing his breeding stock would have saved him that expensive stockade.

The followers of M. Coué, the apostle of auto-suggestion, have asserted in print that auto-suggestion by the mother can influence the character and talents of an unborn child; that therefore, we may presume, properly conducted meditations on Greek sculpture will make him beautiful, and that musical talent, intellectual gifts, or strength of character can be induced by the same means. This, of course, is mere magic. To repeat: "Every day in every way I am growing better"—or cleverer, or stronger—may have an effect upon one's own mind, but cannot conceivably have any upon the embryo parasitic within one.

There is a tale of a Catholic Couéist, a lady of uncertain age, who during confession said she must avow one sin, that of looking in the glass every morning and repeating: "Every day in every way I grow more and more lovely." To which her confessor drily replied: "Madam, that is not a sin, that is a mistake." But to make such an assertion about the influence of suggestion on the unborn child is not only a mistake: it begins to be a sin, for there is no excuse in the present state of knowledge for making it, and it rouses hopes that cannot be fulfilled.

It is right and proper that an expectant mother should take every possible care of her general health, but hygiene should be rational and as free as possible from superstition. The sooner the belief in maternal impressions and the suggestibility of embryos disappears, the better; it can result in nothing but wasted time, and a crop of false hopes, unnecessary fears, and imaginary troubles.

Breeders cherish many other unwarranted beliefs. There is an amusing tangle of myths about sex-determination which will be touched on later. Another very widespread belief is that there exists a quality called "vigour," a vigour that can be seen in the health and strength and activity of the individual creature, which enables it to stamp its own personal characteristics more indelibly on its offspring. Suppose, for example, we mate a black bull with a red cow; this belief would have it that if the bull is strong and vigorous, the chance of black calves will be greater than if he is weak and ailing. And all sorts of further embroideries of the theory have been made. Here again, however, careful experiment has shown that there is nothing in the idea. So long, it appears, as the organism is not so feeble as to be incapable of reproduction, the genes which it transmits are working genes and their shuffling and assortment goes on in the usual way.

We do not, of course, imply that all sires have equal powers of affecting the offspring. Some have a higher proportion of dominant genes than others, and will therefore be responsible for more of the characters that appear in the offspring. This is the explanation of what breeders, in pre-Mendelian days, called "prepotence." But there is no reason whatever why such prepotence should manifest itself as unusual physical well-being in its possessor.

A somewhat similar idea is that individuals which are ill or underdeveloped or past their prime will give less vigorous offspring than their more healthy or youthful rivals. In females there is an obvious basis for this belief, for the female has the nursing of her unborn young.* But in males it appears simply not to be true. However ill or old a male may be the character of the offspring which he sires will not be affected, provided he is able to sire them at all.

If he can still produce spermatozoa capable of fertilization, they will, it appears, still be normal, equipped with all the proper hereditary outfit in full working order; if they are abnormal, they will be incapable of effecting fertilization. The unwillingness of breeders to use old sires

---

*The offspring of very young females, not yet arrived at full size, may be undersized or weakly at birth. In these, and similar cases, there is no evidence that the weakness or smallness is transmitted to later generations: but the question deserves careful investigation.

has meant that there is little animal evidence bearing on the question; but there is plenty among human beings, in the shape of the children of marriages of old men and young wives. There is no evidence whatever that such children are less vigorous or healthy or intelligent than the children of younger men, any more than that the children of women over forty are in any way below the level of the children of younger mothers. So long as the human frame can go on producing reproductive cells, the reproductive cells continue eternally the same, themselves neither young nor old.

Such views as we have dealt with were excusable in the days when men were still groping in the dark for any clue as to the mechanism of heredity. They are so no longer. To-day the main principles of that mechanism are clear. Practical breeding will never cease to be an art; but it now has the opportunity of becoming a science as well, and is doing so by shedding the superstitions and often meaningless traditions with which it is still encumbered.

V

THE GROWTH OF THE INDIVIDUAL

§ 1. *Normal and Monstrous Development.* § 2. *The New Individual Settles Down to Development.* § 3. *The Moulding of Organs by Use.* § 4. *The Transformation of Tadpoles.* § 5. *The Rôle of the Genes.* § 6. *The Limitation and Control of Growth.* § 7. *Age, Retrogression and Rejuvenation.* § 8. *Is Human Rejuvenation Desirable?*

§ 1. Normal and Monstrous Development.

WE HAVE described the machinery of heredity—the chromosomes and the chains of unit-genes along them. We turn now to an account of individual growth and development, which is their concrete realization. We propose to watch the mechanism at work and to glean what we can of how it works.

The science of Developmental Physiology is still largely in the stage of preliminary enquiry. Its known facts are scattered and somewhat fragmentary. Particularly is this true of invertebrate animals. But there are signs of a coming synthesis. In the last thirty years astonishing progress has been made in our knowledge of heredity, of vital physical chemistry, of internal secretion, of the forces that determine the early shaping of the egg, and the various branches of investigation are reacting upon one another even now. In this chapter and the next we shall note some of the main lines of work in this field and the way they are beginning to interact and to illuminate one another. Much of the work on vertebrates has been done on the very accessible eggs of frogs and newts, and in what follows, where we do not state the contrary, the reader can assume that we have these Amphibia in view.

The fertilized egg divides and subdivides again (Fig. 182). The mass of cells thus accumulated forms and organizes itself; the rudiments of various structures appear; gradually the parts are differentiated, finished and perfected. Finally, from a single protoplasmic globule there has arisen an elaborate, functioning living creature (Fig. 98). What are the laws that control this orderly self-modelling?

THE GROWTH OF THE INDIVIDUAL

The first stage of development is the establishment of a *direction* for the future embryo. When an artist is about to draw a complicated picture he roughs it in, very faintly, in pencil, to decide where the main masses and lines will lie. So with the egg. The first thing it has to do is to make up its mind, so to speak, which parts of it will be right, and left, front and back. All higher animals and plants have a definite main axis about which their bodies are symmetrical. In man and the vertebrates generally, it is the line of the backbone. And human symmetry, like that of most many-celled animals, is bilateral; the body consists of two more or less equivalent sides. In a polyp or a sea-urchin, where symmetry is radial, the main axis runs from the mouth down the stem. In a tree it runs up the trunk.

The first step in the organization of the egg must be then the achievement of this main axis. In some eggs this is decided only after they are laid. In the bladder-seaweed Fucus, for instance, it is the direction of the light falling on the fertilized egg which normally determines what position the two ends of the future plant shall have; and in the laboratory weak electric currents can be made to do the same for it.

Usually, however, it is not the outer world, but the mother that determines her offspring's front and hind end, by impressing an axis on the unfertilized egg during its growth. In the simplest cases the end of the egg-cell which comes to the surface of the ovary becomes the future front-end. When there is much yolk in the egg, more is generally deposited near the future hind-end, and the amount gradually diminishes towards the front. Yolk is food laid aside for the growing embryo, an inert, cumbersome substance, and it delays the inter-play of physical and chemical processes that constitutes the life of the cell. In the front end where head and sense-organs and controls are to be developed, where the finest work has to go on, it would be most in the way. It is thrust to the less important hinder regions. So that there is a gradient, so to speak, of activity from front to back, along the main axis; and even when no graded arrangement of yolk is visible, the gradient in activity seems to be there. The fact that there are fore and hind ends is indeed only an expression of such a gradient.

Next comes two-sidedness. In most higher animals, this is determined at and by fertilization. When the sperm first pricks the egg there is great excitement within it; the protoplasm swirls and eddies slowly round, and it is this streaming which draws the sperm-head in and sweeps the two nuclei together. It is during this excitement that the bilateral symmetry is determined. In the frog's egg, for instance, on the side opposite that on which the sperm happens to bore its way in, a pale patch appears (owing to the sucking-in of pigment from the surface). This "grey crescent" marks the middle of the future back. Before this moment, the egg had fore and hind ends, but was without back or belly, right or left. Where these shall appear is imposed upon it from without by the mere accident of where the sperm makes its attack. In insects, however, the egg is already bilateral when laid. Here conditions within the mother's body decide all the embryo's future direction-lines.

The egg directly after fertilization has usually no more than this preliminary ground-plan marked out. The only internal organization it possesses consists of graded differences in the distribution of yolk and graded differences in vital activity. In the frog's egg, for instance, there is one point of high activity at the head end, another on the future dorsal side. The chief activity at this stage is cell-division (segmentation), and the faster it goes on, the smaller the resulting cells must be. Accordingly, the points of high activity are revealed as regions of small cells.

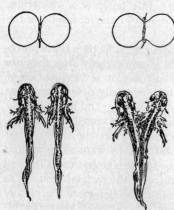

FIG. 179. Artificial control of development.

Left: above, a developing newt's egg has been separated into two halves by a fine hair; the result (below) is a pair of twin newt-tadpoles. Right, the egg has only been partially constricted; the result is a two-headed monstrosity.

The most interesting fact about the egg at this stage is that it behaves as a single whole and is capable of all sorts of adjustments and regulations if it is tampered with —a faculty destined later most conspicuously to disappear. A newt's egg, for instance, can in this early phase be divided into two by tying a fine hair tight around it, and both halves will reorganize themselves to wholes, and each produce a normal animal. If we do not pull the hair quite tight and so leave the two halves connected, we obtain partial twinning—a two-headed monster. Most human double monsters are due to a similar incomplete accidental division of the segmenting egg. And the converse experiment is also practicable; two segmenting eggs can be put together and made to join and form a single animal.

It is the activity-gradients which tie the whole together at this stage. If we alter them, we alter the proportions of the future embryo. A series of experiments by the American workers, Stockard and Child, have demonstrated this. By taking segmenting fish or frogs' eggs and immersing them for a short time in various dilute poisons and narcotics, strange monsters can be produced which lack the front-most parts of the head. When, as often happens, all the region in front of the eyes fails to develop, the two eyes appear united into one at the extreme front end—we have made a cyclops. The explanation is simple. A high degree of activity is necessary for the elaborate head-region to develop. When the activity is reduced by drugging the egg it is insufficient to perform its task. What develop in the place of the head are those structures which demand a slightly lower level of activity for their formation.

But if we can cut out the "high-level" organs by flattening the main activity-gradient, we ought to be able to encourage them by steepening it. Some recent experiments seem to indicate that this is so. Strychnine and caffeine are stimulating drugs; and fish and frog eggs treated with them in early stages grow into embryos with abnormally large heads and abnormally small tail-regions. Another method is to expose segmenting eggs to high temperature at one end, low at the other. When the future head, the region of high activity, is cooled, its activity is depressed and rather small-headed tadpoles are the result; and vice versa. Side-to-side treatment alters the proportions of right and left body-halves. For the present we are confined to animals which lay their eggs free in the water. Could we but get at the early stages of mammalian development, what vistas of strange achievement would open before us!

This minute study of the developing egg throws suggestive light on the monsters which from time to time appear. Everybody has heard of the four-legged ducks, of the two-headed snakes, calves, and tortoises that crop up at times even upon the best regulated countryside. We give some drawings of monsters here. These strange freaks, ranging from nearly separated Siamese twins to double-headed and double-tailed creatures, to double bodies in which one individual sprouts parasitically out of the belly of a second, to heartless and brainless and single-eyed embryos—are in the main the result of accidental interference with the invisible activity-gradients of the recently fertilized egg.

§ 2. The New Individual Settles Down to Development.

HITHERTO the embryo has been quite plastic material; fragments of it could reorganize themselves into wholes. But there now follows a stage in which the various parts of the egg have become specialized.

FIG. 180. Some monstrosities due to abnormal development.

Above, a two-headed calf (cf Fig. 179); a sheep with an extra pair of legs; a turtle-monster with two front halves and no hinder parts. Below, a duckling which trailed an extra pair of legs behind it; a lamb born with another attached to its head; calves with double fore-quarters and with double hind-quarters.

They have been told off to play their several parts; this piece is to grow into a leg, that into a stomach, and so on. An invisible chemical determination has taken place. Each part is set in its own specific direction.

By ingenious experiments upon suitable creatures, biologists can remove small parts of the developing embryo and graft them on to others. Such a piece taken from an embryo newt just before the first rudiments of the nervous system appear, and grafted on to another newt of the same age, will go on developing and will produce what it would have produced if left in its original position, quite irrespective of its new surroundings. The piece that ought to have formed an eye will still do so, even if it finds itself in the flank or even the interior regions of another embryo.

THE GROWTH OF THE INDIVIDUAL

But if the grafting is done only a few hours earlier, we get the opposite result. The group of cells which if left in place would have turned into an eye, would in this earlier phase, when grafted into another embryo's flank, turn into ordinary epidermis. In the first stage the early grafts obey their environment instead of overriding it; they are so much raw material, subservient to their surroundings. In the second stage they have achieved responsibilities of their own.

By a series of beautiful experiments with developing newts' eggs, Spemann has shown us what is responsible for this sudden change from versatility to specialization. It is a region of high activity over the egg's equator (on the embryo's future back). This, for reasons we need not go into, is called the *dorsal lip;* and its remarkable powers become manifested just when the plastic period closes. If at this stage a piece of dorsal lip be transplanted into the flank of another embryo, it not only continues to develop as it would have done if left in place, but it causes the surrounding tissues of its host to grow into an orderly arrangement of nervous system, notochord, muscle-segment, and the rest; in brief, it makes its host produce a second supernumerary embryo. And its action is not specific. The dorsal lip from one kind of newt will set off the tissues of another kind of newt, and can even activate those of a frog. In this universality of action outside the range of its own species it resembles the hormones. On what this strange power of the dorsal lip cells depends, we do not yet know.

There are parallels in other organisms. One of the most interesting concerns fir-trees and other conifers. If the "leader" or single upright shoot at the very top of a growing fir be killed or cut off, one of the circle of horizontal branches just below will grow upwards and take its place. The leader keeps the other branches growing horizontally; it influences the way other parts shall grow—not, however, in this case, the kind of organs they shall produce. In flatworms, however, the parallel with the newt is even closer. If a piece cut from the hind end of a Planarian worm (Figs. 92, 161) regenerates a new head, a gullet and a mouth and protrusible proboscis will grow where, in these curious creatures, they ought to be—i.e., in the middle of the piece; but if for any reason no head is produced, no gullet or proboscis will be formed; the new head evidently organizes the old tissues of the rest of the piece. Like the dorsal lip, "it makes things make themselves."

In vertebrates the dorsal lip region may thus be called the *organizer,* for through its action the embryo first becomes possessed of the rudiments of organs. Before this moment the developing egg behaved as a whole; from now on, until unity is again re-established by outgrowing nerve-fibres and circulating blood, it is a mosaic of parts, each working out its own destiny and fitted together like a chemical jigsaw puzzle.

In this second stage the local independence of each part is amazingly complete. Limb-buds grafted on to back or flank produce limbs; a newt embryo whose tail-bud is cut off remains forever tailless; if the embryo be cut in half when the nervous system first becomes visible, there is no readjustment and no regeneration beyond the mere healing of the wound-surface—the front half develops into head and gill-region, the hind half into trunk and tail. Small bits of a chick embryo can be cut off and grafted on to the embryonic membranes of another egg, which is then sealed up aseptically. The grafts are nourished by the blood-vessels of the membrane, but are not in contact with the host-embryo. And yet in these surroundings they continue their differentiation as if at home: tiny limb-bud grows into well-proportioned limb, and eye-rudiment becomes eye—an eye attached to nothing! The chick's embryonic kidney normally degenerates before hatching, and is replaced by the definitive kidney, which develops farther back in the body. The rudiment of the embryonic kidney grafted on to another egg not only differentiates to form characteristic tubules, but degenerates after the correct lapse of time. It is like a chemical clockwork which, being set, grinds out its result and can do no other.

Even if organ-rudiments (like the early optic cup) are cultivated in tissue-culture, they will still achieve their characteristic differentiation within their glass prison as if in the body. The development of such grafted and tissue-cultured rudiments, however, is often not typical as regards their precise shape, for this depends on purely mechanical, not on chemical, considerations. But the chemical differentiation runs on unaffected, so that, in spite of mechanical distortions, all the characteristic kinds of tissue are produced.

It is often asserted by vitalists that the capacity for regulation—the adaptive adjustment of parts to whole—is a peculiar and universal character of life. Vertebrate development is proof to the contrary. We all of us pass through a stage in which we are little more than a chemical mosaic. During this stage our separate parts are not bound together by any unifying principle, for if one is removed it is not replaced; if one is isolated, it can still pursue its differentiation; if the pieces of the mosaic are rearranged, each still continues its predestined development. The arrangement of the pieces is an outcome of what happened in the stage before: it is without influence on their further differentiation.

THE GROWTH OF THE INDIVIDUAL

But for the full intricacy of the chemical mosaic to come into being, time is needed. There are phases in specialization. Take the vertebrate limb. There comes a moment when a definite region of cells, the fore-limb area, has its fate fixed as future fore-limb. Cut it out and graft it elsewhere and it still will grow into a fore-limb. But all that is fixed is that it must turn into a limb: the different parts of the limb are not yet rigidly predetermined within it, any more than the different systems of the embryo were rigidly predetermined in the egg. As with the egg, if it is cut in half each half can give rise to a whole arm; and two limb-areas properly joined together will coalesce and give but a single limb. From the first moment of its existence its fore-and-aft axis is determined (determined by the existing axis of the body, as the egg's axis was determined by the conditions in its mother's ovary); if you graft it elsewhere its thumb will grow out of what was originally its front side, just as if it had been left in place. And if you wait a little longer its other main axis will have been fixed, also, and the position of its palm and upper surface irrevocably determined, too.

But later, when the flat limb-area has grown out into a little conical bud, the separate main parts of the future limb become also determined. The specialization has gone on another step. Then, though the bud looks uniform throughout, the tip of it, if grafted elsewhere, will give only a hand and wrist, the base only an upper arm: and still later the details within each main part are fixed; within the tip-region, for instance, one group of cells, which could up till then have been turned into wrist or palm or finger, can now become nothing else than a particular predestined finger-joint.

Thus the acquisition of complexity, that greatest marvel of development, is bound up with a narrowing-down of the possibilities open to each region. The same group of cells which, when the limb-bud is well-grown, can do nothing but turn into a finger-tip, a little earlier could have turned into palm or wrist, but not into upper arm. Twenty-four hours earlier, again, they could have been turned into any part of a fore-limb, but not into anything else but fore-limb. But in the earliest stages of all they could have been switched over to becoming any organ whatsoever, from liver to eye, from brain to bladder.

The whole egg, in other words, behaves first as a single field of organization, if we may use the term; and it gradually becomes divided up into smaller and smaller fields, each different from the rest and with a large measure of independence. The splitting up of the singleness of the early embryo into this patchwork is touched off by the action of the dorsal lip "organizer." The influence thus set in motion spreads at a definite rate from region to region. For instance, the influence which determines the formation of the eye-lens from a particular area of outer skin has to travel outwards from the brain through the optic cup or first main rudiment of the eye. If, while this influence is traversing the optic cup, the cup be cut out and grafted elsewhere, it still has the power of transmitting the influence to the skin, and so can call out the production of a lens from skin of back or flank.

That the chemical differences thus established persist through life is shown by some remarkable facts of regeneration. If a newt's tail is amputated the cells restore just what is missing. Some influence, peculiar to the region of the cut, determines what shall be formed there.

If the bud growing out to replace the amputated tail be grafted to some other part of the body, say, on to the fresh-cut stump of a leg, it will go on developing into a tail, provided that it is more than a few days old. If, however, it is grafted on to the leg-stump earlier, while still a mere boss of cells, its destiny is not yet fixed; it will now turn into a leg instead of a tail. The very early bud is like a very young embryo. The active cell-division at the tail's cut surface has produced a mass of undifferentiated cells, capable of turning into arm or leg as readily as tail. Which they shall turn into is at this stage decided by the influences of the area where they happen to grow. Soon it settles down as an embryo does and its parts assume definite responsibilities.

In the developing organism, as plasticity gives place to fixity of direction, the unified whole becomes the mosaic of chemically independent parts. Different creatures differ markedly from each other concerning the time at which this transformation is effected. Just as with the determination of the direction-lines of the body, some eggs have all their symmetry ready-made for them in the ovary, others wait and have it decided by outer agencies after fertilization, so here. In frogs and newts and many other creatures the irrevocable chemical decisions are not taken till late, well after the close of segmentation. In other eggs, however, like those of sea-squirts, the corresponding step has been pushed backwards to just after fertilization; and in still others the egg may already have become a chemical mosaic even before it is shed from the ovary. In such cases, of course, there is little or none of the plasticity found in vertebrates during segmentation. If the two first cells into which such an egg divides be separated from each other, each turns into a half-embryo and does not regulate itself to a whole. But in all the principles remain the same, it is only that the time-relations are shifted.

So the main lines of the new individual are planned and its chief structures built. But soon a new set of forces comes into play. Hitherto development has been preparation, a passive growth, but now the organs begin to work and, working, mould themselves further, as we must now examine.

§ 3. The Moulding of Organs by Use.

NO ORGANISM, and especially no higher animal, could work if it remained only a mosaic of independent parts. And so the next step is to give back to the embryo the unity it has so lately lost, to make its parts interlock and co-operate again. But the machinery by which the new adult unity is achieved is quite different from that underlying the unity of the segmenting egg. Instead of mere gradients of activity tying the whole together, we have now coming into play the ubiquitous streams of blood and lymph and the fine branching of the nerves. Of their unifying functions we have already spoken in Book 1; they provide a common internal environment, a postal delivery system, a telephone system, and a controlling headquarters.

Once this stage begins, new agencies can be set to work to put the finishing touches to the animal's construction. One of these is the battery of hormones which regulate growth and development; of these we have already said something. The other, equally important, is the actual working of the separate organs for the good of the tissue-commonwealth as a whole.

It was the German zoologist, Wilhelm Roux (1850–1924), father of the science of developmental physiology, who half a century ago pointed out the vital importance of what he called *functional differentiation*. Functional differentiation means differentiation brought about through use and disuse, and through the strains and stresses to which growing tissues are subjected. The size of heart, muscles and glands depends upon the calls made upon them. In our bones there is a fine internal architecture of struts and stays beautifully adapted for resisting, with the greatest economy of material, the mechanical strains to which the bone is subjected. In its construction the head of the thigh-bone is like certain cranes; the inner architecture of the heel-bone is not unlike that of a roof, with lines of bone transmitting the pressure from leg to ground, and others at right angles as tie-beams.

All this delicacy of adaptation is called into existence by the actual stresses and strains to which the bone is exposed. If the mechanical conditions alter, the struts rebuild themselves in responsive adaptation. This is seen, for instance, when a broken bone is set crooked, or when two bones which ought to be free to move become united to each other, as in an ankylosed stiff knee. Again there exist people born with deformed feet which can never be used for walking. They cannot begin to walk and so the normal stimulations are withheld. The bony structure of their heel-bone remains poorly developed, and what there is of it is more like an ape's than a man's. The fine structure in our heel-bones is one of the characteristic points in which we differ from the apes; and yet we achieve most of it anew in each generation by walking upright.

Indeed, the whole size and shape of bones is plastic. If a puppy be prevented from using one of its legs, by tying it in a sling, after six months the bones of the unused limb will be only about half the thickness of those which have had to carry weight, as well as being a little shorter. And when an irregular fragment of bone from some other part of the body is grafted to take the place of a damaged or diseased finger-joint, the mechanical forces in the finger will remodel it towards the shape of a finger-joint within a year or two (Fig. 181).

Then there is the moulding of the connective tissues to form tendons. Each tendon in the body runs just in the direction of the greatest tension to which it is exposed, and its size is proportional to the amount of that tension. This beautiful adaptation again is brought about directly by the tension. If the Achilles tendon in the heel of an animal is cut, it is gradually repaired. New fibres grow in between the cut ends. At first, they are quite irregular in arrangement; but gradually they grow in number and all become parallel. The Achilles tendon's attachment is to the calf-muscles; and if these are removed the new-formed fibres never grow parallel, and remain sparse. It is thus the pull of the muscle which normally both guides the fibres into place and stimulates them to grow until the bundle is big enough to resist the highest tension the muscle can exert. In the same way, experimentally setting up a pull in connective tissue (as by exerting gentle tension on a silk thread healed into a wound) will bring about the artificial production of tendons—tendons that do not run from muscle to bone, but simply follow the pull. And recently it has been shown that even in cultures of connective tissue-cells outside the body, if a tension is set up locally, the fibre-forming cells not only grow parallel with its pull, but multiply more vigorously here than elsewhere. Here

THE GROWTH OF THE INDIVIDUAL

then the chain of proof is complete. The original chemical differentiation brings into being that general kind of tissue we call connective. Out of its pervading meshwork, the mechanical forces of the living body themselves elicit all the multifarious details of particular tendons and sinews.

So too the size of blood-vessels and the thickness of their walls depend chiefly upon the volume of blood poured through them, as is strikingly shown by those cases where a big artery is gradually blocked, and the blood, finding its way round through small byways, forces them to enlarge and assume the structure and duties of major vessels. The digestive system alters with the diet. John Hunter, the great anatomist, made a seagull produce a passable imitation of a gizzard by feeding it for many months on grain only. Few organs escape some degree of modelling by their own activity.

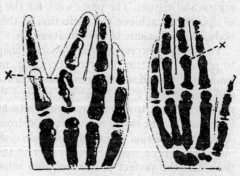

FIG. 181. The Moulding Power of Use.

Two X-ray photographs of a boy's hand in which the basal joint of the middle-finger became diseased. It was removed, and an irregularly-shaped piece of living bone (x) from another situation was grafted in to take its place. Left, immediately after the operation. Right, two years later (the position of the hand has been reversed); the grafted piece has been moulded into a very good imitation of a normal finger-joint. (After Timann, from "Animal Biology," by J. B. S. Haldane and Julian Huxley, Clarendon Press.)

Development, we see, is a step-by-step process. In the bit of living matter that is to develop, each difference between one region and another is the parent of further differences. The first regional differences come from without—the egg has its main axis imposed upon it, and then its two-sidedness. Later the influence of the active "organizer" region spurs it into chemical activity. As a result, the segmented egg, while it is still little more than a mass of cells of different degrees of activity, becomes parcelled out into a series of separate areas, each with its own peculiar chemistry. Each of these areas has its own axis, contains its own local differences in activity, and thus within each area new sub-zones come into being, each characterized by its own chemical peculiarities. The processes they set in train run their appointed course

The chemical mosaic becomes a mosaic of visibly different structures. Ultimately this mosaic becomes a whole again. All its parts have become interconnected by blood and nerves, and are falling into subjection to the government of the brain. Hormones make widely-separated parts grow or shrink in synchrony, the nervous system throws into simultaneous action the most widely diverse kinds of organs and tissues. The parts work for the good of the whole; and in so doing they achieve the perfection of their own structure, and the body its full measure of co-ordination.

To achieve success in this elaborate shaping of a new individual, to prevent monstrosities from appearing, marvels of adjustment and timing are needed, certainly no less accurate than those in the engine and magneto of a car. And the more delicate the adjustment, the more need for it to be removed from outside interference in its plastic early stages. The rarity of human monstrosities is in part due to ages of natural selection by which heritable disharmonies and maladjustments have been weeded out, in part to the rich and uniform conditions and the perfect protection from outside interference provided for the developing embryo in the maternal body.

§ 4. The Transformation of Tadpoles.

EVERY spring tens of thousands of tadpoles become transformed into little frogs and leave the water for the land. This change, so trivial-seeming, so familiar that most of us do not even pause to think about it or its underlying machinery, is in reality pregnant with interest, and from its study far-reaching ideas have been born.

The tadpole's metamorphosis involves a radical change in the whole structure and working of the organism as it passes from one element to another. To mention only the major alterations, gills and tail degenerate, intestine, brain and skin are remodelled, and lungs and limbs, which have been growing slowly, come to full size. How is it that all these changes are so nicely synchronized?

This has only been found out in the last twenty years, beginning with Gudernatsch's discovery in 1911 that tadpoles could be made to turn into frogs at any time by feeding them on thyroid tissue. The frog's metamorphosis is primarily due to the thyroid hormone circulating in its blood and so affecting all the organs of the body at once. By dosing young tadpoles with thyroid, bull-frogs no bigger than flies can be produced; by cutting out a tadpole's thyroid, it can be made to continue growth in tadpole form until it attains sizes

THE GROWTH OF THE INDIVIDUAL

previously unknown in tadpole annals, and never metamorphoses at all.

Most spectacular of all, animals such as the Mexican Axolotl, which normally lives all its life and reproduces itself as a gilled eft in the water, can be made to turn into a land salamander at will by a single

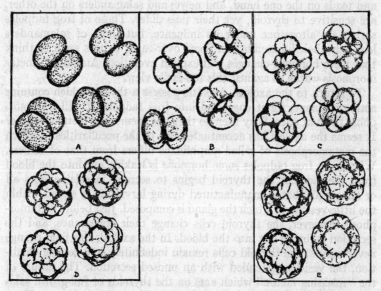

FIG. 182. Development in action; the sea-urchin becomes many-celled. Photographs of the same group of four sea-urchin eggs at different stages of segmentation.

(A) The fertilized egg has divided into two cells; (B) the two cells have become four; (C) one egg (lower right) is in the 8-cell stage; the others are on the way to the 16-cell stage; (D) about 32 cells have been formed; they enclose a space in the centre; (E) between 64 and 128 cells; the central space is larger. (From the film Cosmos; courtesy of Wardour Films, Ltd.)

dose of thyroid. On the arid Mexican plateau, life in water is preferable; and the axolotl has had its metamorphosis suppressed and lived out its life aquatically for thousands of generations. But a fraction of a milligram of thyroxin (Book 1, ch. 4, § 5), even from a sheep or a fish, will bring out the latent salamander in a couple of weeks.

Other normally water-living amphibians, however, such as Necturus or Proteus, which also have lost their land phase, do not react to thyroid, but remain obstinately aquatic. Here we are introduced to an important principle. As it takes two to make a quarrel, so it takes two to give a hormone-reaction—the hormone itself, and the capacity of the tissues to react. The gills and tail fin of the axolotl are sensitive to thyroid; those of Necturus are not. Similarly, although both frogs and toads on the one hand, and newts and salamanders on the other, are sensitive to thyroid, yet their tails differ. Those of frog tadpoles shrivel up altogether under its influence, but those of salamanders lose only their fin-membrane. Some over-eager writers seem to think that mere hormone-changes will explain evolution. Amphibian metamorphosis warns us against this too-facile view.

To return to the axolotl. Axolotls possess a thyroid which contains active hormone, for if engrafted into frog tadpoles it will metamorphose them at once. Why then do they not themselves metamorphose? It seems the axolotl has accentuated one of the peculiarities in which the metamorphosis of tailed amphibians differs from that of tailless. Whereas in frog tadpoles some hormone is leaking out into the blood from the moment the thyroid begins to secrete, in salamanders all or almost all that is manufactured during larval life is stored within the little vesicles of which the gland is composed. Just before metamorphosis, however, the thyroid cells change their appearance, and the secretion is forced out into the blood. In the axolotl, this final change never occurs; the thyroid cells remain indefinitely in a resting condition, the vesicles distended with an unused secretion. The nature of the "releasing factor" which acts on the thyroids of full-grown salamander tadpoles, but is lacking in axolotls, we do not know. If we could find out, it would be of great importance, for in our own thyroids phases of rest and storage alternate with those of activity and discharge.

Subtler differences between tissues are revealed within the simple body of the tadpoles. The destruction of the gills and tail takes place only at the moment of metamorphosis; but the growth of limbs, though also dependent on thyroid (for it is enormously slowed down if the thyroid is cut out), begins from the moment when limb-buds first appear. This is due to the fact that limb-tissue is sensitive to the least traces of thyroid hormone, while gills and tail do not react until a certain concentration is present.

The transformation of amphibians reveals other interesting secrets. Why is it that while ordinary frogs need almost three months to their metamorphosis, some toads take only a month, some bull-frogs over two years? The difference seems due to the relative rate of thyroid growth. If it grows fast relatively to the body, the necessary concentration of hormone in the blood will be reached when the animal is small; if its growth is slow, the tadpole's existence will be prolonged.

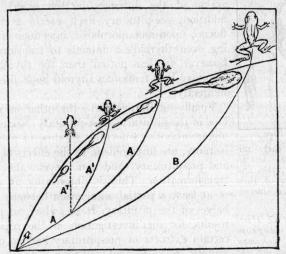

Fig. 183. A diagram to explain why tadpoles turn into frogs at different ages and sizes.

The upper thick line is supposed to represent the growth of the tadpoles; the lower lines, the development of their thyroids. In (B), the thyroid grows slower than in (A), and so the metamorphosis takes place later. If extra iodine is provided, the animals' own thyroids grow more quickly (A^1). If thyroid is administered, the animal metamorphoses immediately (A^T).

We can prove this by actually altering the thyroid's relative growth-rate. The thyroid hormone seems to owe many of its properties to the presence in it of iodine in much greater abundance than in any other part of the body. Iodine is very scarce in nature, and the growth of the thyroid is limited by the amount it can pick up. Give it more than usual, as by immersing tadpoles in weak iodine solution, and it will pick up more, grow faster, and so cause metamorphosis at a lower body-size than usual. And the precocity of metamorphosis is roughly proportional to the strength of the iodine solution. (Fig. 183).

The thyroid is the main gland concerned with metamorphosis; but the pituitary—the ductless gland that lies under the brain and just over the roof of the mouth—is also essential. In the absence of the pre-pituitary or anterior lobe, tadpoles will not metamorphose; and examination shows that their thyroids have failed to grow to a tenth of their normal size. The pre-pituitary is needed to make the thyroid grow—a striking example of the interlocking action of the different ductless glands. But in addition, pre-pituitary itself exerts a direct influence upon metamorphosis, injections of it causing even thyroidless animals to transform; it is, however, far less potent than the thyroid. Thus pituitary here reinforces thyroid both directly and indirectly.

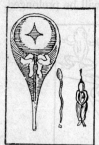

FIG. 184. The zeal of some early microscopists outran their discretion and the power of their primitive instruments.

Left, part of Hartsoeker's diagram (1694) of a human sperm, showing the imaginary homunculus sitting in the head; he also figures a cord extending from it down the tail, which, he supposed, was later converted into the umbilical cord of the embryo. Centre, a human sperm as observed by Dalempatius (1699); Right, the homunculus which he imagined he could see within its protective envelope.

Finally—a strange fact—the other or posterior lobe of the pituitary (post-pituitary) works against metamorphosis. Injections of this, in axolotls for instance, are antagonistic to the effects of thyroid and pre-pituitary, and can prevent the animal's transformation. This is interesting as evidence of at least a partial antagonism between the two halves of the pituitary. It has also had practical results, for the investigation of the reason why certain extracts of pre-pituitary had no effect on axolotls, revealed the fact that if the two parts of the pituitary are not separated as soon as possible after death, post-pituitary hormone leaks into the pre-pituitary, there to work against its hormone. And this has led to improved methods and the preparation for the first time of really pure pre-pituitary extracts for medical use.

The transformation of pollywogs is not so trivial a process as it seemed. Processes are there at work which are of the utmost importance in our own lives; but in the tadpole they are more readily accessible to experiment. The specific sensitiveness of different tissues, their reactions to different intensities of one and the same substance; the synchronizing of many processes by the aid of one hormone; the shifting of the time of some developmental crisis by altering the relative rates of two processes; the interaction of different ductless glands, different parts of that "chemical skeleton" whose general plan runs through all vertebrates just as definitively as does the plan of their true mechanical skeleton:—in all these ways the study of tadpoles has shed new and valuable light on general problems.

§ 5. The Rôle of the Genes.

LET us review our account of individual development thus far:

We have traced the main stages in the evolution of an individual body from an egg. First, the very young embryo is a plastic thing, making the most elementary arrangements, deciding in which direction its head and its tail will be produced and which will be its right and which its left. Then it begins to specialize itself, to crystallize out into tissues and organs. Rudiments appear and shape themselves and differentiate until they become workable parts. And, finally, there is the third stage of action, of pulsing blood, of contracting muscles and hardening bones, of messages flashing along the newly drawn-out nerve-fibres, of ductless glands pouring their controlling secretions into the blood-stream. In this stage the tissues settle down together, so to speak, and knock each other into shape until the body becomes a harmoniously working whole.

Now all through this orderly development the genes in the chromosomes are exercising a regulating, supervising action. As we noted in the first chapter, there is not a corner of the body that cannot be altered and modified in some way by these active little units in the chromosome outfit. The time has come for us to look a little more deeply into the way they produce their effects.

We must remember what a gene is. We grow so accustomed to using convenient but in a sense misleading shorthand phrases like "the gene for blue eyes" or the "albino gene," that we tend to think of the genes as in some way little replicas of the characters with which they are concerned, and of the gene-outfit as being a sort of compressed organism, with a point corresponding to each part of the body. But that idea is wholly false; it is really a survival of the preformationist ideas of the eighteenth century, which so worked on the imagination of one microscopist that he actually drew a human sperm with a *homunculus*, a miniature man, squatting within the head! (Fig. 184.)

The genes are something structural and chemical in the chromosomes, and that is all we know about them materially. The rest of our knowledge is based on their observed effects. The characters they produce are the end-results of physical or chemical influences that they exert. The gene for blue eyes has nothing of blueness about it, nothing suggestive of eyes. It is a chemical unit which modifies the train of events leading to pigment-formation in the eye. Moreover, the genes are scattered about on the chromosomes; there is not a particular region where all the eye-genes are, another where all the foot-genes are, and so on. In the fruit-fly the colour of the eye can be modified by any one of over twenty different genes, which are scattered about at random over the chromosomes; and the same can be said of other thoroughly-investigated cases, like the coat-colour of rodents. The

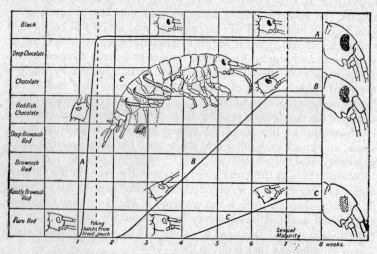

FIG. 185. Genes that act on the rate of development.

In Gammarus chevreuxi (centre, magnified about six times) the normal adult has black eyes (A). Other varieties are known in which the eyes are chocolate (B), or almost pure red (C). The differences between these are due to single genes; and the genes act by altering the rate at which the eye, originally always red, grows darker. This is indicated by diagrams of the heads of the different varieties at different ages. (Left to right, time in weeks: bottom to top, stages in darkening of eye-colour.)

relation between gene and character is by no means a simple matter.

It is an interesting fact that many genes produce their characteristic actions by influencing the *speeds* of processes occurring in the body. For example, the little brackish-water shrimp *Gammarus chevreuxi* is normally black-eyed. But various recessive varieties are known with other eye-colours, and here the effect depends upon the rate at which black pigment is deposited. The normal shrimp has red eyes at first, but black pigment is very rapidly formed in them so that even before hatching the eyes are black. Then there is a variety which produces black much more slowly; for several days after hatching the eyes are red, then they darken slowly to a chocolate colour. In another variety the eyes darken still more slowly and only reach a brownish-red, and, finally, there are individuals whose eyes never darken at all. Now Gammarus is an animal which hatches as a miniature replica of the adult and can be watched as it grows up. In the case of the fruit-fly the development occurs in the pupa stage; when it emerges the fly is a finished product, and does not grow or develop any more. So it may well be that the great variety of eye-colours known in the fruit-fly are really due to the speeds at which the eye-pigments are formed in the pupa (Fig. 185).

Much of the variation in the colour of human hair and eyes seems to be due to similar controllers of speed. Many children change their eye-colour as they grow up, and the earlier the change shows, the more quickly it runs its course and the deeper is the final shade reached. Really dark eyes are, on this view, eyes in which dark pigment had formed very rapidly and had reached its final condition before birth; and in true-blue eyes the rate of pigment production is slowed down by the influence of a special gene to so little that it never appears appreciably during the period when change is possible.

And now we will take up again the question of how much of development is due to the genes? Are these genes directly responsible for all the stages of development, even for the earliest formative steps in the egg-cell? In our opinion it is manifest that they are not. In the finished product, the new individual, there is a complete chromosome-outfit in every cell of the body; yet in one place the outfit produces a blue eye, in another blond curly hair, in a third a snub nose, and so on. The gene controlling eye-pigment has nothing to say on the bridge of the nose, and vice versa. There is obviously a very important controlling factor at work—the place in which the genes find themselves.

A neat demonstration of this was made by Hertwig, a German embryologist. He compressed the fertilized eggs of a frog between glass plates so that at the eight-cell stage the cells were all forced into one layer instead of being arranged in two tiers of four. He found that if the resulting plate of cells was released from pressure, at the next division it formed two tiers of eight cells each. And from such eggs, perfectly normal embryos grew. A glance at the diagram (Fig. 186) shows that some of the nuclei which in the normal course of events would have stayed near the upper pole, and so formed part of the front end or outer skin, in the compressed and released egg went to the other pole and formed part of the hind-end or of the internal organs. The cells' nuclei must therefore be all alike, the difference of their fate and what genes are called into operation is decided by agencies outside themselves—the region in which they come to lie.

An example of how the genes require special local conditions before they will work is afforded by the so-called Himalayan breed of rabbits, whose pattern is recessive to the normal. These are white in colour except that the tips of the nose, ears, tail and feet are black. Now, why should the recessive gene in question blacken parts of the body and whiten others? If we shave part of the skin we find the answer; for the

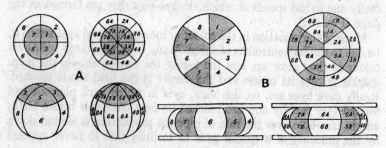

FIG. 186. Disarranging nuclei makes no difference to early development.

A diagram of frog's eggs segmenting normally and under pressure; in each case a top view is drawn above, a side view below. (A), segmenting normally; left, in the eight-cell stage; right, in the sixteen-cell stage. (B), flattened between glass plates, eight-cell and sixteen-cell stages. The nuclei that in the normal egg go into the topmost cells (shaded), in the compressed egg are forced into quite different regions. And yet the compressed eggs produce normal tadpoles.

fur will grow white if the rabbit is kept warm, and black if it is kept in the cool. The action of the gene depends on temperature. At the normal body-temperature it causes the fur to grow white, while at the tips of the extremities, which are slightly but measurably cooler, it causes the fur to grow black.

During early development the embryo becomes parcelled out into different regions with different local conditions. Then the genes get to work, moulding and finishing the ground-plan. But the first rough laying out of the future body is apparently not due to the genes; it is due to the protoplasm of the egg, or, as we have seen, to the more or less accidental impact of external conditions on the egg protoplasm.

A large and tangled subject, the Physiology of Development! Soon, we hope, it will straighten out, and the interplay of gene and protoplasm, or organizing zone and encumbering yolk, of mechanical stress and internal secretion, will be better understood. In one particular field the relation of the various influences that bear on development are becoming clear—in the study of sexual differentiation. There one can see how the different factors work together. To that unusually lucid branch of developmental physiology we shall devote our next chapter. But first we will turn aside to discuss certain very interesting side-issues of this study of individual development.

§ 6. The Limitation and Control of Growth.

For men and most higher animals, growth and development have very definite limits. Why is this?

As we have already seen, the cells of a bird or a mammal in a tissue-culture continue to grow indefinitely, long after they would have stopped growing if left in the body (Book 1, ch. 2, §2). The great riddle of growth is not why we grow, but why we stop growing. Many, perhaps most organisms, never stop growing. Our greater familiarity with higher vertebrates and insects misleads us in this matter. Such diverse creatures as trees, fish, molluscs and crustaceans go on increasing their dimensions as long as they live, though at a steadily diminishing rate. The female American lobster, for instance, may produce eggs when under eight inches in length, and eleven or twelve ounces in weight, and then continue growth until death overtakes her, which may not be until she has reached sixteen inches and over five pounds in weight, while a male may reach twenty-two inches, and nearly twenty-five pounds.

Many fish, such as the ordinary trout, show similar disparities. And then extremes may be even greater; in the common Hermit Crab, the largest breeding females are about four times the length of the smallest, and so about sixty-four times their weight.

In spite of their continued growth, however, there is a limit in size in these animals, albeit it is set not at maturity as it is with us, but at death. The limit is set by the relative rates of growth and of ageing; if growth slows down quickly and ageing is rapid, the animals will be cut off by death before any of them have passed even a moderate size. Even among mammals this undoubtedly primitive method of ending growth is to be found, for many voles and other rodents increase in size every year that they live, even after sexual maturity. This method, once we reflect a little, is what we should expect. What needs explanation is not the continuance of growth throughout life, but its suppression even before the full vigour of life is attained. The higher vertebrates and the insects are the most important groups with limited growth, though the little rotifers are in some ways even more remarkable, arriving at a definite fixed number of cells (often only about a thousand) and then stopping short.

Fixed adult size obviously confers certain advantages, in standardizing the species, making it more delicately adjusted to its mode of life, and stabilizing all relations, social, parental, or sexual, between different individuals of the species. We have only to reflect how awkward family life would be if after we had started to produce children we proceeded, like female hermit crabs, to quadruple our height, or like male lobsters to multiply our weight by over twenty, to realize the biological value of checking growth during the adult phase. In part such considerations doubtless apply to insects, too; but in addition the impossibility of moulting their wings makes it a necessity for them to cut short their growth immediately the winged phase is reached.

In higher vertebrates, the cessation of growth in height is brought about by the solidification of the epiphyses or end-pieces of the long bones and vertebræ, and their fusion with the central part of the bone; the gristly zone of cell-multiplication between them and the main shaft is obliterated, turned into bone-tissue, and after that further growth in length is impossible. We know that the time of this fusion is, in part at least, regulated by the ductless glands; but just how these act, and how the other tissues of the body are stopped from growing too—of this we know next to nothing; and we are equally ignorant of what brings the growth of insects to a close, just as we are ignorant of why many of them, such as the stick-insects, and many crustaceans, double their weight with reasonable exactitude from one moult to another.

We must confess our ignorance of what stops growth. But besides the stoppage of growth there is its slowing; and this is universally found in all many-celled forms, in those with continued growth as much as in those with fixed final size. Our simplest measure of growth is the percentage amount put on each hour, or day, or month; if we plot out curves of percentage increase, we find that almost every organism grows steadily less and less rapidly from the earliest stages onwards. We are apt to think that our own growth after birth, multiplying our original seven pounds or so by about twenty, is considerable. But it is our pre-natal growth which is really both so vast and so rapid. The human fertilized egg weighs two or three hundredths of a milligram. It thus multiplies its weight by about a hundred million times in the nine months before birth; and the egg of a large whale multiplies itself some three million million times before the whale has reached full weight.

THE GROWTH OF THE INDIVIDUAL

The average increase of boy children for each year, expressed as a percentage of their weight at the beginning of the year, is 200 per cent in the first year of life; in the second year only 20 per cent; and by the seventh year it has fallen to 10 per cent. There is a slight increase before and during puberty, but then it falls off again and declines to one or two per cent by twenty years of age. If we had taken months instead of years as our standard, the average for the first year after birth would be under 17 per cent; in the last month before birth it is over 25 per cent, and mounts rapidly as we go back, being over 100 per cent for the sixth month of pre-natal life, and over 200 per cent for the fifth.

Essentially similar curves, though on a different scale, are found for guinea-pigs, silkworms, rabbits, sunflowers, and indeed for every organism which has been accurately measured. The smoothness of the curves may be interrupted by illness or underfeeding, and also by biological crises such as metamorphosis (when weight is almost invariably lost) and sexual maturity.

Thus, broadly speaking, animals and plants are increasing by compound interest all the time; but the rate of interest is falling off progressively throughout life.

Final size depends upon two separate things—the rate of growth, and the length of time which it continues. For instance, a man has a post-natal growth-period of over twenty years, a rabbit of about thirteen months, a guinea-pig of about fourteen months. Man and rabbits put on about the same average amount of weight per day—6 grams as against 6.3 grams—but a guinea-pig only 1.8 grams. Thus rabbits are larger than guinea-pigs because they put on more weight per day, but men are larger than rabbits because they grow for a longer period. If we take the average *percentage* increment as our standard of growth-rate, however, we find that man has the slowest growth of the three, his mean percentage daily increase for his whole growth period being only 0.02 per cent, while that of the rabbit is 0.50 per cent, and of the guinea-pig 0.47 per cent.

It is more likely that some at least of the universal slowing in compound interest rate of growth is due to the difficulties arising out of mere bulk. The animal or plant must draw its food and oxygen through some surface; the living substance that is to breathe or to be fed is a solid mass. And everyone knows that if you double the linear dimensions of any object, a ball or a model yacht, you increase its volume eight times, its surface only four times. Automatically therefore the increase of an organism's bulk slows down its capacity for supplying the cells with materials for growth. Caterpillars, for instance, alter but little in shape and proportions during growth. But they may increase a thousand times in bulk; and this will mean that the absorptive surface of their digestive tube has only been increased a hundred times—the proportion of the absorptive surface to the bulk to be nourished is only one-tenth of what it was when they hatched. Small wonder that their growth slows down as they enlarge.

But this is assuredly not the only force at work in restraint of growth.

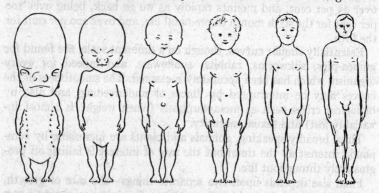

Fig. 187. How human proportions change during growth.

(1) *a two-months fœtus;* (2) *a six-months fœtus;* (3) *a baby at birth;* (4) *a two-year-old child;* (5) *a six-year-old* and (6) *an adult. The arms and legs grow faster than the trunk, the trunk faster than the head.*

It has been found that from old tissues a substance can be extracted which will damp down the growth of tissue-cultures, from embryo tissues a substance which will speed them up; and it has been shown by Warburg that actively growing cells, like those of embryos and still more those of cancers, have different chemical habits from slower growing cells; they need more sugar, and indeed in some ways resemble yeast-cells in their metabolism. Connect these facts with the further fact that certain extracts from yeast-cells will increase growth in many animals, and you will feel that we are on the verge of important discoveries concerning the control of growth. But these discoveries have not yet been made; meanwhile we must be content with these scattered intimations of their approach.

Let us add here a remark or so upon the growth of different parts of the body relative to one another. This differential growth is of the greatest importance, since if one region did not differ from another in growth-capacity and growth-rate we could never achieve characteristic form, but should remain as undifferentiated in shape as we began. Such differential changes may go on all through growth. We are all born with snub noses. A Roman nose is Roman because of the relatively high growth-rate of its bridge, causing the baby's infantile snub to straighten and bend over; a less eagerly growing bridge would have made the nose straight, a yet slower-growing one have left it snub through life.

Typically indeed an organism has no fixed form; it is a bundle of organs and tissues all growing at slightly different rates, and so constantly and regularly changing its proportions. By cutting growth short, a term is set to this flux of growth-change and an adult phase is achieved which is, within minor limits, fixed and stable both as regards its size and its proportions; but this regulation of size and shape, like the regulation of body-temperature or blood-composition, has only been achieved by the higher types of animals.

One of the most obvious examples of the changes in shape brought about by exaggerated local growth of a single organ is seen in Fiddler-crabs. These are so called because full-grown males have one enormous claw which they generally hold in front of them as a violinist holds his fiddle. The other claw of the male is tiny, and is just like both claws of the female. The big male claw may grow to weigh almost as much as all the rest of the body. But this striking masculinity is brought about during the animal's growth. The youngest crabs, both male and female, are all alike, with two equal small claws. And the difference between the sexes is entirely brought about by one simple fact—that while both the female's claws and one of the male's grow at the same rate as the animal in general, the other male claw grows half as fast again. This claw and the rest of the body are like two sums of money put out to compound interest at different rates, the claw's rate being the higher, and so, as the animal increases in bulk, the proportionate size of the claw becomes bigger and bigger. If growth were to go on long enough then the large claw would eventually grow larger than the whole of the rest of the body. But before that, death in some form overtakes this over-balanced animal. Fiddler-crabs, like all Crustacea, go on growing as long as they live. So a male fiddler-crab has no fixed or final shape, since it changes the relative size of claw and body as it grows. All that is constant is the proportion between the claw's rate of growth and that of the rest of the body. A further interesting fact is that the male's big claw is growing fastest of all near its tip, the growth-rate grading down towards the body. As a result the claw changes the proportions of its parts as it grows. This does not happen in the female's claws; not only are they as a whole growing at the same rate as the rest of the body, but all their parts are growing at the same rate. This local exaggeration of growth in a particular region, a "growth-centre," seems to be universal for organs that grow faster than the body as a whole.

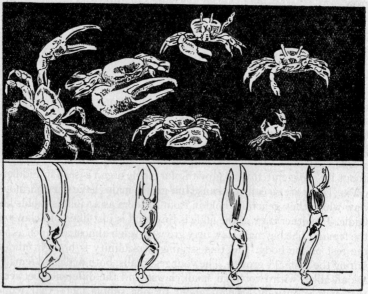

FIG. 188. Exaggerated local growth.

Above, a group of Fiddler-crabs (Uca pugilator). The female (upper right) has two small claws, which she uses in feeding. The rest are males of various sizes; one of their claws stays small while the other grows steadily bigger in proportion to the body. Below, the claws of a female (right), of a quarter-grown, a half-grown and a full-grown male (left), all brought to the same absolute size. The proportions of the joints change; the nipping end part of the male claw grows faster than the rest.

In ourselves and other higher vertebrates, just the same differences in growth-capacity exists between different parts. From very early embryonic stages onwards, for instance, our arms and our legs are growing faster than our trunk. The difference between us and a crab is that in us, at a more or less definite time, growth is mysteriously arrested. Our proportions have been changing as steadily as those of the crab; but now change is stopped. We have attained a more or less fixed final shape. If we could solve the riddle of that arrest and defer the time at which growth ceases, human beings would not simply be larger than they are; their proportions would be different; their shape would be changed.

§ 7. Age, Retrogression, and Rejuvenation.

The experiments of Voronoff on gland-grafting and the exploiting of them by a section of the daily press have made rejuvenation a subject of general interest. But all too rarely has the public discussion of the matter treated it as more than a stunt, a nine days' wonder, a subject of music-hall jokes, or, even if it has got beyond this stage, done more than debate the *bona fides* of the experimenters and the genuineness of the facts. It may serve a useful purpose to discuss the question on the broadest biological scale; against this background the results on man and higher animals will show up in their right proportions.

First of all, we must make quite clear what we mean by age and youth. Are we to define them in terms of time, or as physiological conditions? When we say that one animal is younger than another, do we mean that it has lived fewer years, days and minutes, or that it is not so far along the road to senile breakdown and natural death? Only one answer is possible: age must be measured physiologically, not by time. Even among human beings in whom the life-cycle has a relatively fixed framework, that is obvious enough. Which of us does not know men and women who at sixty are younger in every quality both of body and mind than other people are at fifty? Men and women, we say, and at the bottom of our hearts we firmly believe, are as old as they feel. The clear skin and eye, the wrinkles and stiff joints, these are the things that count; the birth certificate is a mere formality in comparison with these physiological truths.

Then there are the half-abnormal cases of precocious senility; a man of fifty is overtaken by the sudden onset of those symptoms of real senility which in the bulk of humanity are postponed until twenty or thirty years later. And there are definite diseases of the function of ageing. The most spectacular, and luckily one of the rarest, is that called *progeria*, or anticipated senility. An apparently normal baby is born. At two it already has a careworn expression. Like the precocious infant in the Bab Ballads, it is "an old dotard at five," or at least is losing hair and showing other signs of age. By twelve the child is bald and looks like a miniature old man; and it generally dies of old age before "coming of age" in the ordinary acceptance of that term.

When we consider the generality of organisms, our contention becomes even more obvious. Since they have no temperature-regulating machinery, their activities go up and down with the outer temperature, and their rate of growing old is no exception. Most animals and plants only thrive over some part of the temperature range between 5° and 40° C.; between these limits the rate of ageing is increased by temperature about two-and-a-half times for each rise of 10° C.—a rate of increase which holds for most other functions, too, such as breathing, walking, or digestion. Some of the most accurate experiments have been done on the fruit-fly Drosophila. The total span of life at 30° C. is 21 days; at 15° C. it is 124 days. After a female fly had laid her first few eggs, one could, by keeping the young mother at 15° and her offspring at 30°, readily ensure that she should be younger than her own great-great-great-grandchildren.

The same is true of cold-blooded vertebrates. Fig. 189 shows a number of developing frogs. They were fertilized at the same moment; but since that moment they have been kept at different temperatures. By the clock, they are all of precisely the same age—three days. However, their age must clearly not be measured by the clock, but by the changes that have been going on within them, changes which have made those kept in the warm to advance farther along the path that all of them must tread, which leads first up to maturity, then slowly down towards death.

By age, then, we mean physiological age, measurable in terms of the organism's own state, not time-age, measured by any outer time-

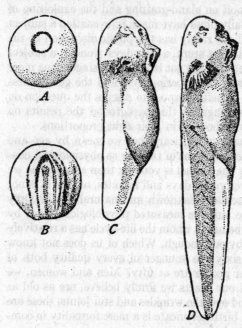

FIG. 189. Four Developing Frogs; are they all the same age or not?

All were from the same batch of eggs, fertilized three days previously. But while (D) was kept at 24 degrees C. and is ready to hatch, (C), kept at 20 degrees, has only begun to grow gills and tail; (B), kept at 15 degrees, is still forming its nerve-tube; and (A), kept at 11 degrees, has only got a little beyond the close of segmentation. (After O. Hertwig from "Experimental Embryology," by Dr. J. W. Jenkinson, Clarendon Press.)

keeper; and by rejuvenation we mean a return to a more youthful physiological state.

In general, it is found that young animals grow faster, move faster, react with more alacrity, and that therefore their oxygen-consumption, which is readily and accurately measurable, is more rapid. It appears probable also that the viscosity of protoplasm increases with age—it becomes thicker, less liquid, as it grows older. Old tissues are thus more solid and more sluggish. Further, while young animals are less resistant than old animals to markedly unfavourable conditions, they are better able to overcome and adapt themselves to slightly unfavourable conditions. Young planarian worms, for instance, will die a good deal sooner than old if put in solutions of poison which will kill even the old worms within twelve hours; but in very weak solutions, though all the old worms die after a week or so, young ones can survive indefinitely. We find a rough parallel to this chemical behaviour in the different nervous resistance of young and old among ourselves. A grown man can stand fatigue and hardship under which youth will succumb; but youth is more likely than middle-age to adapt itself quickly to unpleasant minor changes.

Now that we have some idea of what age means in physiological terms, we can go on to our main problem. And what we find is quite definite; there is no question but that rejuvenation, in one form or another, can and does occur in many lower forms of life.

For example, consider the little sea-squirt *Clavellina*, an elaborately built creature with an intricate food-filtering throat, with gullet, stomach and bowels, heart, reproductive organs and ducts, muscles, nerves. Put one of these animals, when still fairly small, in a small volume of sea-water and leave it to be drugged by its own accumulating waste-products. In a few days it closes its two wide openings, through which it drew water in and passed it out, and begins to shrink and lose its beautiful transparency. Gradually it dwindles from the glassy creature, unceasingly sweeping a stream of water through itself, to a mere lump, irregularly rounded, condensed and opaque, its apertures not merely closed but grown over, and showing no trace of life except a slow and irregular pulsing of the heart, which can barely manage to pump the thick soup of blood-corpuscles through the diminished spaces of the body. Finally, the heart stops and there remains only an inert whitish blob (Fig. 190).

From this extreme stage of reversed development or *de-differentiation*, as the process is called, there is no recovery. But at any time before the heart stops we can revive the animal by putting it into clean water.

Surprisingly swiftly it organizes itself again, and in a few days the shapeless blob is transformed into a working Clavellina, a little smaller than the one with which we started, but otherwise healthy. Moreover, it is a newer, younger animal than the original one. The process can be

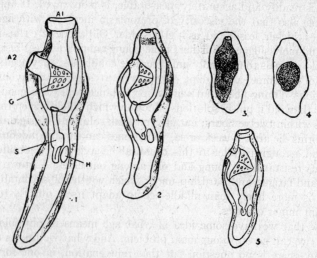

FIG. 190. Living backwards.

(*1*) *A young Clavellina.* (*A1*) *and* (*A2*), *apertures for taking in and passing out water;* (*G*) *gill-slits;* (*H*), *heart;* (*S*), *stomach.* (*2*) *The same beginning to de-differentiate.* (*3*) *Three weeks later; only traces of its organization are visible.* (*4*) *Eleven days later; it is a mere opaque blob.* (*5*) *Replaced in clean water, it has re-differentiated into a normal sea-squirt again.* (*After J. S. Huxley.*)

repeated and the animal made to degenerate and revive again until it reaches a certain minimum size; then it will not work any more.

But here there is a difficulty. Has a single individual been rejuvenated here, or has one piece of matter doffed an old individuality to become clothed later in a new one? Is it the same Clavellina? Whichever way we answer the question, the fact remains that the animal has passed through an obliteration, temporary or definitive, of its essential character.

However, there are a number of examples where this difficulty does not confront us. The best studied are planarians (Fig. 92). It has long been known that these little flatworms can survive prolonged starvation by living on themselves, growing smaller the while. They

THE GROWTH OF THE INDIVIDUAL 545

can draw on their own tissues and convert them little by little into food to satisfy their daily needs. Men or dogs or hens can only do this to a limited extent. They can use up perhaps half their total capital of living substance before they are reduced to skin and bone, as the saying goes, and succumb. The first drafts are made upon the reserve stores of fat and glycogen (the readiness with which fat breaks down makes it possible for hibernating animals to survive the winter) then upon the actual living substance of glands and muscles, and to a less extent of connective tissue. But there are certain essential parts which cannot or must not be liquidated. The heart is little affected. The brain-cells must not be destroyed, even if their destruction were to provide a week's more food for the starving body, since with their destruction all that the animal had learnt by experience would disappear, and higher vertebrates must rely upon what they have learnt in almost every action of life. The skeleton, on the other hand, is not liquidated because there is no fuel value in the lime-salts that stiffen our bones. Thus, as the other tissues shrink, there finally comes a time when the muscles are spread with a lamentable attenuation over the undiminished bony framework; they are unable to move it properly, and the glands are too much wasted away to produce their necessary secretions; the creature dies.

No such limitations beset the planarian. He has no skeleton, and if he can profit by experience at all, the profit is negligible compared with the value of his straightforward reflexes and instincts. Every one of his tissues can be liquidated, and when he is starved he keeps a sense of proportion by destroying all his tissues to the same extent. It is as if a starving ship's crew ate one mate, one deck-hand from each watch, and one engineer, so as still to be a working unit if their circumstances improved. So the whole planarian can continue to shrink without marked change in shape or in behaviour, until it is reduced to less than one-hundredth of its original bulk, to a size even below that at which it hatched from the egg. Such diminished animals still work and can be made to grow again at any moment by supplying them with food.

The reduced worms are different in proportions from adults; their shape is like that of young worms. They look rejuvenated. To see whether they not only looked young but had really become young again, the following experiment was undertaken by Professor Child of Chicago. Two lots of full-grown worms were taken, of a species which normally reproduces asexually by a form of fission. One lot was regularly fed, and grew and reproduced in due course. The others were starved until quite small, and then fed until they reached their original size, then starved again—and so on. By this means they were kept between definite limits of size, while the other lot ran through twelve generations—a period which if translated into human lives would take us back to the Elizabethan age. Meanwhile, those on half-time starvation had never reproduced at all, had remained the same individuals throughout, and had shown no signs of progressive ageing. Whenever they were large, they were as old as ordinary worms of that size, whenever they were made small they were as young as ordinary young worms. If we chose to take the trouble, we could doubtless keep a single flatworm going up and down the hill of life and never transgressing a certain limit of age for periods which would make Methuselah look ephemeral.

The actual experiment has not been tried on other species, but we have every reason to suppose that it would succeed with any organism which while living on itself can keep a just proportion between its various tissues. Such ought to be the case with starved sea-anemones (up to the time when they absorb their tentacles), with certain worms, and even the highest insects in their growing stages, for starved beetle-larvæ can shrink almost as much as Planarians.

But if this intermittent starvation can serve as the fabled *elixir vitæ* to lower animals, it unfortunately will not work with the higher. It is quite possible that occasional holidays from food, such as form the basis of certain cures, would do good to many members of our over-eating modern civilization, and might somewhat prolong their expectation of life; but this is a different matter; it is because their bodies are cluttered up with stored food, and has nothing to do with rejuvenation.

Only two methods have yet been found to work with higher animals—one based on gland-grafting, the other on diet. However, this latter is rather a prevention of growing old than a return to youth. If different proteins are used separately as the main or sole source of nitrogen in the diet, it is found that while some are altogether useless, there are others which contain all the amino-acids (the chemical units of which proteins are built up) needed both for maintenance and growth, and others on which the animal can maintain its weight and health, but not grow. Such a one is zein, one of the proteins of maize. Young rats fed on this zein diet have been kept at half-weight until long after the time at which rat-growth is normally over, and yet begin to grow again perfectly well when put on normal diet. One may presume, though the experiment has not yet been tried, that they could be kept both small and young for a time well exceeding the normal life of their species by prolonging the zein dietary.

It would certainly be interesting to try a similar experiment on human beings, but it would not be easy, and it may be doubted whether any growing boy or girl would be willing to endure the monotony of the diet even to secure an extra twenty years of existence. And even if it were feasible, this would be mere retardation, a holding-up of our developmental processes, not rejuvenation.

Finally, we come to what the daily press usually refer to as "gland" experiments, although most glands have no connexion with rejuvenation.

Steinach was the first to attack the problem. He reasoned as follows. The most obvious sign of old age in mammals is the loss of sexual power and the failure of the sex-instinct. We know that the ductless glands together form an interlocking system. If the reproductive hormone fails, will not the others suffer? And will they not be stimulated to renewed activity if a new supply is provided? Finally, in view of the vital importance of the ductless glands in keeping life's activity up to the mark, might not this renewal of their activity abolish the senile symptoms?

The most obvious methods of renewing the supply of sex-hormone is to graft in bits of reproductive organ from young animals. There is also a second method—to tie a ligature round the *vas deferens* (the narrow tube along which spermatozoa leave the testis). It had previously been found that the back-pressure of the sperms thus prevented from escaping will cause the sperm-producing tissue lining the testis-tubules to degenerate. This meant that there was more space available for the interstitial tissue between the tubules; the interstitial tissue therefore grew, and, since it produces the male's sex-hormone, more sex-hormone was accordingly produced.

Steinach tried both methods, and with both he achieved what he had hoped for. The successes were not invariable: grafts do not always take, and the vas must be tied in just the right place if there is to be effective back-pressure. But they were numerous enough. Old male rats successfully treated in either of these ways threw off their senility. Their appetite improved, their coat took on its old gloss, their weight went up. Instead of sitting apathetically in one spot, like a decayed old villager on the ale-house bench, they began to display that incessant activity and inquisitiveness so characteristic of adult rat life. Potential mates were no longer treated with indifference. In brief, the aged beast reverted to a condition of normal maturity.

This renewal of vigour lasted up to six months or so, and then old age appeared a second time. The twice-senile animal could be again rejuvenated. Indeed, one rat was rejuvenated four times. First one vas was ligatured, then the other, and then it received testis grafts from its younger relatives; and by this means its span of life was lengthened by thirteen months. These prolongations do not sound large to us; but they mean a good deal to rats. The normal span of life among Steinach's untreated rats is about twenty-seven months; thus the maximum prolongation which Steinach's rats achieved was nearly fifty per cent of this, which would be over thirty years in terms of human lifetime.

The experiments with repeated rejuvenation yielded a new fact. Each spell of renewed vigour was shorter than the one before, and was terminated by a more abrupt relapse into senility, until at last the animal collapsed all at once, like the deacon's "wonderful one-hoss shay" in the poem. Thus rejuvenation, of this type at least, is at best a temporary setting-back of the clock. There is some part of the body which is progressively ageing all the time, in spite of the best possible provision of hormones. We do not know for certain what this organ may be, but may reasonably guess that it is the central nervous system. For in higher animals, brain-cells, unlike gland-cells or skin-cells, never divide. We end with the same equipment of brain-cells with which we were born. And it is a fact that brain-cells show marked shrinkage and other degenerative changes when they grow old.

Whatever it be which thus inevitably ages within us, its rate of ageing is very different in different animals. Very aged mice attain four years, while elephants do not begin to be old till they have passed the half-century. But on the other hand mice live faster than elephants. The pulse-rate of a mouse is over ten times that of an elephant, so measured in heart-beats the lives of mouse and elephant come to very much the same total. And mice move and therefore presumably think more rapidly than elephants do. If we could discover the reason for this difference, the reason why a mouse gets through its life more than ten times as quickly as an elephant, we should perhaps have a method of prolonging life, by simply slowing it up, which would be more effective than merely tinkering with the glands as they give out—although it would not necessarily be more satisfactory, since we should merely have the same amount of real living expanded to fill a longer span of years. But that belongs to the speculative future; we must return to the actual present.

THE GROWTH OF THE INDIVIDUAL

Steinach's results have been confirmed and extended by others such as Sand. Dogs, sheep, goats, and cattle have thus been successfully rejuvenated.*

What of the application of the method to man? Here, too, many successes have been claimed, and Voronoff has introduced a new method by using bits of chimpanzee testis for grafts, instead of waiting for the rare accidents when a human being has to have his testis surgically removed. Such grafts from one species to another are always resorbed within a month or two, so that whatever lasting rejuvenation they may cause must be due to some effect exerted on their host's system during the first few weeks.

Much less has been done with women. X-ray treatment of the ovaries appears to have something of the same stimulating effect as tying the vas in men. And now that one at least of the female sex-hormones has been isolated, we may perhaps be able to use it for rejuvenating the whole feminine system. Already a preliminary experiment on female rats has proved successful, and we may expect events to move fast.

In judging the results, we must remember that man is a suggestible animal; we do not know how much of an improvement could be produced solely by belief in the efficacy of the treatment. But when we have discounted this, there remain some cases which seem to have achieved an often striking success. Several patients have come back after a spell of renewed maturity for a second treatment; and one was so overcome by the restoration of vigour after years of decrepitude that he drank himself into a *delirium tremens*. And there are cases in which the effect of suggestion has been as far as possible eliminated, since the patient has not been told what the operation is for.

But before we can come to a sober judgment on the value and the limitations of the treatment, much further work is needed, and the introduction of a more scientific spirit into the whole subject.

Even if the work on rejuvenation has not opened up any rosy vista of perpetual youth, it holds forth promises of a reasonable extension of healthy middle-age. There is in any case no basis for the protests that have been raised by certain well-meaning people against rejuvenation on the ground that grafting chimpanzee glands into human beings is disgusting or immoral. The graft is resorbed within a few weeks; and in any case it cannot possibly alter any of the patient's characteristics,

---

*The claims made by Voronoff that testis-grafts in rams induce not only an increase of vigour in the animals themselves, but a hereditary improvement in the quality of the wool of their descendants, are apparently quite without confirmation.

bodily or mental, in the simian direction. It does no more than provide a convenient source of male sex-hormone. Every time a man or woman takes a thyroid tablet, he is using thyroid hormone prepared for him by a sheep; but he does not become sheeplike in consequence. If the male sex-hormone could be injected or given in tablet form, grafting would be unnecessary; but until it has been chemically isolated grafting is one of the few available methods. Those who would like to prevent testis-grafting should also start a campaign against the use of insulin, thyroid, pituitrin, and adrenin.

§ 8. Is Human Rejuvenation Desirable?

ARISING out of this discussion is the very interesting problem whether there would be any advantage at all, either for the individuals concerned or for the species, in prolonging the lives of men and women of the current type very much beyond eighty years. Health, vigour, and happiness up to seventy-five or eighty seem to be a possible thing now for many people with a good start. At the end of that time it is open to question whether there is anything more for the contemporary old to do that could not be done better by younger people, or whether the pleasure and entertainment they will still draw from life will compensate them for the labour of continuing. Most old people become very philosophical about death; few regard the final cessation of experience with the terror it has for the young. They tire. They have seen the show.

Mechnikov, who paid some special attention to the subject, concluded that in the very old, those who were dying not of disease but of sheer old age, there existed a desire for death as natural and normal as the desire for sleep at the end of a long day.

We have already noted that the fully developed brain and nervous system have a fixed number of nerve-cells. We know no way of increasing them. As in its youth the organism learnt, connexions were established between these cells, paths of association were set up and the main substratum of the mental existence was established. On this fresh associations and persuasions are being perpetually built, but there is no going back to childhood and youth to reconstruct those foundations. The brain has the framework of its time and circumstances set up once and for all. As life goes on it becomes more and more difficult for the brain to accept new fundamental ideas. And not only ideas but feelings and affections become fixed. How much of an old brain must be occupied by tender thoughts and perhaps hostile thoughts, rivalries, considerations, loyalties, about the dead? For the values of ten thousand things it goes back to the dear old home and to the beloved school that time has long since swept away. The brains of many ageing people must be like cemeteries at evening time. They are overcharged with past things, and no sort of hormone will ever alter that.

It is interesting and plausible to argue that the individual life of the larger animals, the soma, is a sacrifice of physical immortality in exchange for power and achievement. The individual has, so to speak, made a bargain. For the individual comes out of the germ-plasm and does and lives and at length dies for the sake of life. It is a bit of the germ-plasm which has arisen and broken away, in order to see and feel life instead of just blindly and mechanically multiplying. Like Faust it has sold its immortality in order to live more abundantly. A bacterium, as we have shown, is *all* germ-plasm, all reproductive material, and there is no soma to die. But its activities are very limited. In a man what remains of the immortal germ-plasm is a mere scrap of material hidden in his body. It has been, so to speak, thrust out of the way, so that he can grow; so that he can use hands, feet, eyes, and brain; so that he can run, swim, climb, fly, and fight his obstinate battle against the universe. He sees, he hears, he thinks, he puts things on record, stores resources, marks dangers and prepares a path for his sons. But he subordinates the germ-plasm only in order that he may serve it, willingly or not. He uses himself up, but gloriously. Until the dawn of greater capacity and the onset of greater tasks, his eighty-year life seems to be enough for what he has to do. Much more, and his brain, clogged with its own private memories and conclusions, would cumber the progressive and impatient germ-plasm. For the individual there is a time for work, there is a time for rest, there is a time to go.

VI

WHAT DETERMINES SEX

§ 1. *The Normal Causation of Sex.* § 2. *The Proportion of the Sexes.* § 3. *Can We Control the Sex of Our Children?* § 4. *Sex-linked Inheritance and Sex-detectors.* § 5. *Gynanders.* § 6. *Sex Hormones.* § 7. *Reversal of Sex.* § 8. *Intersexes.*

§ 1. The Normal Causation of Sex.

WE PROPOSE to devote a whole chapter here to the determination and development of sex, for three reasons. First, because it throws much light on developmental physiology: it illustrates with exceptional clearness the way in which genes, internal secretions and so on work together in individual development. Secondly, because it is one of those fields which have been enormously illuminated by recent work. And, thirdly, because it is good to tell simply, clearly and cleanly in terms of genes and gametes, matters too often treated in an atmosphere of shame and dingy heat.

First, we will consider that central problem, what determines whether a child or a young animal shall be male or female? A rich variety of views has been put forward on this question; the sex of one's offspring is due to the quantity or quality of the mother's food, to maternal impressions, to auto-suggestion; one ovary makes boy-eggs and the other ovary makes girl-eggs—these and many others. Some of these notions still have their advocates. We need not examine these ideas in detail, for the essential mechanism is now known, and it has nothing to do with any of them.

About thirty years ago it was discovered that in certain plant-bugs the number of chromosomes differs in the two sexes. The female has an even number, a pair of each kind; the male has an odd number, one less than the female. In the males, a member of one of the chromosome-pairs is absent. The peculiar chromosome, paired in the female, but single in the male, was christened the X-chromosome. The question at once arose, what happens in the male in the reduction-division, when the sperms are being made and the chromosomes have to be sorted out into two equal sets? The answer may be guessed. The ordinary chromosomes pair up, but the lonely X stays by itself; remaining undivided, it goes to one or other of the two resulting haploid cells. So that half the sperms, in these plant-bugs, have X-chromosomes and half have none. In the female the reduction-division is, of course, perfectly normal, since the number of chromosomes is even; thus every one of her egg-cells contains an X-chromosome.

The eggs, then, are all alike, but the sperms are of two kinds. Now, when a sperm with an X-chromosome fertilizes an egg, the resulting zygote will have two X-chromosomes—that is to say, it will be a female. If the fertilizing sperm lacks the X-chromosome the zygote will have one only, and so it will be a male. Clearly then the sex of the offspring depends on the chances of fertilization, on which kind of sperm meets the egg first. After fertilization all is settled.

There is the secret of sex-determination in the plant-bug. Since the plant-bug is not a unique organism, but simply an example of the general rule, there in its essence we have the secret of sex-determination in ourselves. It is a simple secret. Sex is no more than a Mendelizing character dependent upon a chromosome. And the primary decision, whether the future embryo will be male or female, occurs therefore at the moment of fertilization.

But in minor details the process varies from species to species, from group to group. Sometimes the X-chromosome is not alone in the male; it may have a very much smaller companion. In this case the sperm which gets the full-sized X-chromosome is the female-producer, while that which gets its dwarf companion (or Y-chromosome, as it is called) is a male-producer. Sometimes the X- and Y-chromosomes are very nearly equal in size, but nevertheless distinguishable by a slight discrepancy in length, or by some twist or trick of shape. Sometimes they are exactly alike in visual appearance, but even in these cases their existence and their differences can be demonstrated by breeding experiments along lines to be discussed in § 4 following.

To-day, visible differences between the chromosomes in the two sexes have been detected in insects, centipedes, and spiders; in mammals such as rat, horse, opossum, dog, cat, monkey, and man; in birds, reptiles and amphibians; in sea-urchins and roundworms and molluscs; in some liverworts and in flowering plants with separate sexes such as docks, campions and the Canadian water-weed Elodea. In the vast majority, as in the plant-bug, it is the male who has the unsymmetrical chromosome-outfit, and the sperms therefore which are the decisive begetters either of males or of females. The eggs have no voice in the matter. Very strangely, however, two groups show the opposite condition; in the birds and in the butterflies and moths the male is chromosomally symmetrical and the female unsymmetrical, so that here it is the eggs which are of two kinds. This fact is all the more

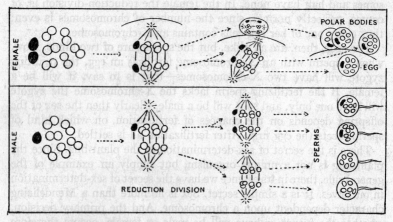

FIG. 191. Diagram of determination of sex in an insect.

Above left is the arrangement in a female germ-cell before reduction. It has 14 chromosomes including two X-chromosomes (black in diagram). This divides into two without any splitting of chromosomes; One nucleus is expelled as a polar body and the other divides again giving off a second polar body (the first polar body also divides). One nucleus is thus left in the ovum (the boundary of which is indicated by a line). All these results of the subdivision of the female germ-cell obviously contain an X-chromosome. Below left is a male germ-cell with only one X-chromosome. It divides into one nucleus containing 7 chromosomes (one X) and another without the X-chromosome. The ultimate result of its division is two sperms with X-chromosomes and two without. If either of the two former fertilize the egg a female will ensue, if either of the two X-less do so, a male.

remarkable since in other vertebrates and in other insects it is the male who is unsymmetrical, as usual.

We may note that the discovery that sex is determined at the moment of fertilization explains an odd little fact about twins. There are, as we have seen, two kinds of twins. Sometimes two children are born together, no more alike than brothers and sisters usually are. This is due to two eggs descending into the oviducts and being fertilized instead of the normal one. But real twins are of the Tweedledum and Tweedledee type—practically indistinguishable from each other. In this case a single early embryo has divided into two halves and each has grown into a complete child. Now it has long been known that whereas the first kind of twin may or may not be of the same sex, identical twins are invariably both boys or both girls. Why?

In the first case there is evidently no particular reason why the two should be of the same sex, any more than there is why two consecutive children should be of the same sex. It is pure chance. But in the second case the twins are really parts of the same zygote, with the same chromosome-constitution—hence their close resemblance to one another, and, since sex is a Mendelian character, hence their identity of sex.

§ 2. The Proportion of the Sexes.

Once the underlying chromosome-machinery of sex-determination had been understood, the remarkable fact that in most animals the two sexes are produced in roughly equal numbers was explained. Hitherto it had been entirely mysterious. But since one sex produces equal numbers of male-determining and female-determining gametes, and the gametes of the other are all alike, it follows that in the long run the offspring must be male and female in about the same proportion.

This equal ratio may be upset, however, by various influences. One sex, for example, may be less resistant than the other. In man, for instance, more males die than females; the difference is greatest in early life, but continues up to old age. The result is that old women outnumber old men; among people over seventy in England there are roughly three women to every two men. That this is a constitutional difference is shown by its being even more pronounced before than after birth. In the United Kingdom, the proportion of male to female deaths before birth is about 150 to 100; for still-births it is 135 to 100, and for the first year of life it is down to about 120 boys to 100 girls. And just the same happens in other mammals, such as cattle.*

There are some further interesting consequences of this lesser resistance of the human male. For one thing, the percentage of female children goes up slightly with the mother's age and the number of her pregnancies. This, it seems, is not due to any change in the eggs she produces, but is a direct consequence of the fact that the internal

---

*The probable *explanation* of this lesser constitutional resistance of the male mammal is also concerned with his sex-chromosomes. As he has only one X, any recessive sex-linked (X-borne—see Fig. 195) genes will show their effects at once in males, while in females they will often be masked by the effect of dominant partners. Thus, all recessive sex-linked genes with weakening effect will be able to exert this effect much more often on males than on females. By selective breeding, we should be able to produce a stock in which no harmful sex-linked recessives were present; and in this the males should be just as resistant as the females.

environment provided by the mother to her unborn children becomes less favourable with age and previous child-bearing; and the males, being less resistant, suffer more. There are more still-births and more

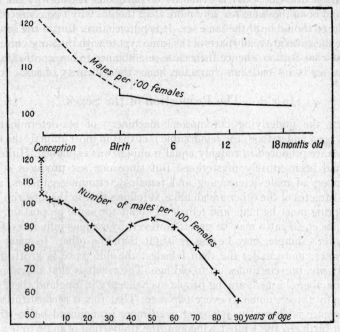

FIG. 192. How the proportion of the sexes changes with age.

Below, the proportion of the sexes at different ages in the population of England and Wales, from the Registrar General's Report, 1913. Apart from an irregularity in late middle-age, the proportion of males drops steadily from conception to death. Above, the same for the first years of life. Before birth, dotted; the exact figures here are uncertain. The upper point for the moment of birth is the figure for all births; the lower, for children born alive. The figures on the left give the number of males to every 100 females.

miscarriages, a high percentage of which are of males, and so the proportion of female children rises.

Indeed, whatever favours pre-natal health and successful childbirth causes a rise in the proportion of boys. In every country legitimate children show a higher male percentage than illegitimate, because they are better looked after before and at birth. Jews have a high male ratio as a race, probably because the Jewish expectant mother pays more attention to hygiene than her Christian sister. In America, white births show a higher male percentage than negro births, largely, it appears, for the same reason.

There has long been a legend that the male birth-ratio rises markedly in war-time. It does rise, as the statistics of the Great War show, but only to the tune of less than one per cent; and this is not due to privation, as used to be supposed, since it did not occur in any neutral countries, even those where the blockade was causing great hardships. It appears, on the contrary, to be due to *more* favourable pre-natal conditions. It only occurred where a great body of the male population was away at the front, save for rare leaves, and this, it seems, meant for a number of women a freedom from the sexual demands of their husbands that was favourable to a slightly higher survival of their unborn children, and so a slightly increased male ratio.

But even when all differences in the resistance of males and females are allowed for, we do not always find exact equality. In most white races the number of boy babies born alive for every hundred girls lies between 101 and 108; and if we take the still-births and the pre-natal deaths into account, we find that the proportion of male to female conceptions must be about 120 to 100. This preponderance of males at conception can only be accounted for by some advantage possessed by Y-bearing sperms which enables them to fertilize more eggs than their X-bearing brethren.

Presumably the male-producers have greater powers of endurance than the female-producers and can better withstand the arduous journey up the uterus and oviducts. We may guess from the statistics that for every six male-producers only five female-producers get to the top of the oviduct; that six out of every eleven among the millions that lose themselves and perish on the way are female-producers.

The machinery of chromosome-distribution inevitably produces a precise equality of male-producer and female-producer sperms. The female-producers are apparently handicapped in some way and a high male conception-ratio is the consequence. But the males are constitutionally less resistant; and so the preponderance of males is gradually reduced and may even turn into a preponderance of females.* Conditions within the uterus and oviduct may change the handicap on the

---

*There should be mentioned the extremely high male sex-ratio, up to 160 males per 100 females, found in some primitive peoples who are decreasing in numbers as a result of contact with civilization, e.g. the Melanesians and some North American Indians. This condition of affairs does not seem to be due to female infanticide, and for the present remains wholly unexplained. Whatever its cause, the paucity of women naturally helps on the disappearance of the tribe.

female-producers, and so alter the sex-ratio at conception. Favourable conditions before and after birth will prevent the percentage of males declining so rapidly. At present in England at least one out of every five fertilized eggs dies before birth; over three per cent more die at birth; yet another fifteen per cent or so in the first year of life—a wastage of nearly forty per cent of all conceptions before one year old. And of these deaths, many more are of males than of females. Thus, increase of pre-natal care and infant hygiene would soon reduce the surplus of women over men.

§ 3. Can We Hope to Control the Sex of Our Children?

Now we may discuss the bearing of these facts on a much-debated question. Is there at present any way of deciding what sex our children are going to be?

Various methods are buzzed from ear to ear, and delicate issues of time and attitude are discussed. None of them seem to have the slightest practical value. The sex of our children is at present a matter of chance; half of the millions of spermatozoa which are injected at any sexual act are male-producers and half are female-producers, and the sex of the child depends on which kind happens to meet the egg first.

It has been suggested that the diet of the pregnant mother can affect the sex of her child, a copious diet for a girl and a lean diet for a boy, and even to-day educated people are trying to realize their wishes in this way. There may be something in this idea, as our later discussion of intersexes will show, but there is probably not enough for any perceptible result, and if any were produced it would be much more likely to be an abnormality than for the diet to carry things off in the face of the gene-determination. There seems to be a confusion here with the effect of external conditions on the sex-ratio that was discussed in the last section. But it should be obvious that there is really no parallel between the action of external conditions on the relative mortality of the sexes and the action of diet on a single embryo.

At present, it must be admitted, we have no practical choice between girls and boys; we have to take what comes, with the odds about fifty-fifty. But this may not always be true. Some day we shall probably work out a real method of control. Perhaps it will be along the lines of the following tentative suggestion.

The head of a spermatozoon, as we have already noted, consists of little else than condensed and closely packed chromosomes. Accord-

WHAT DETERMINES SEX

ingly, when the male has but one X, and the Y is absent or small, the two classes of sperms which he produces may be expected to differ in head-size, the female-determiners or X-bearers having bigger heads than their X-less and male-determining brethren. This expectation can be tested by measurement, and measurement shows it to be correct. There is a considerable range of variation in the measurement of sperms—their heads are no less variable in size than our own. In sperms such as those of fowls, all of which possess an X, this variation varies round a single average size. But in species with marked difference in size between X- and Y-chromosomes there are two high points; in the bug Corizus, for instance, there are a great many sperms with heads close to 27μ long, an equal number with heads about $29\frac{1}{2}\mu$ long, and only a few with intermediate lengths. These two high points represent the average head-lengths for male-determining and female-determining sperm respectively.

This measureable size-difference between the two kinds of sperms has also been proved to exist in several mammals, including man; in man, the average head-length for male-determining sperms is just over 4μ, for female-determiners, just under 5μ. It is of considerable interest because it opens the door to a possible method of controlling sex.

If male-determining and female-determining sperms differ in size, there must be some method of separating them, whether by centrifuging, or if their speed varies, as it probably does, by making them swim long distances in nicely warmed fluids, or in some other way. It will be a ticklish business, but sooner or later the right technique may be found. And artificial insemination has long been shown to be practicable. The Abbé Spallanzani first achieved it with dogs and cats in the eighteenth century, and now it is used on a commercial scale in horse-breeding in Russia and elsewhere. We can accordingly look forward to the installation in animal-breeding establishments of "sperm-separators," whose products, labelled "X" or "Y," will be injected into the females to beget the precise proportion of male and female offspring needed in our domestic animals.

There we have one remote and grotesque possibility by which the sex of individuals may ultimately be controlled from the phase of fertilization onward. Later on we shall produce facts which, while they do nothing to encourage any hope of controlling the sex of the young under present conditions, may point to the possibility of an occasional post-natal reversal of sexual structure.

§ 4. Sex-linked Inheritance and Sex-detectors.

The genes which the X-chromosome carries are not all concerned with the determination of sex. It bears others as well. The genes are crowded on to the chromosomes in serried thousands, and it is hardly to be expected that the X (which is in some species the largest of the outfit) should have only the genes which determine sex.

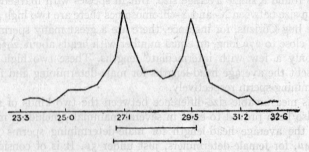

Fig. 193. A diagram showing that two size-classes of sperms exist.

The heads of 500 sperms from the bug Corizus were measured. The graph represents the number of sperms of each head-length. There are two high points, corresponding to the averages for the small-headed male-determiner sperms and the large-headed female-determiners. Below, the comparative lengths for the two averages are given. (After Zeleny and Faust, from the "Journal of Experimental Zoology.")

Now, when a gene is lodged in the sex-chromosome, certain complications of the ordinary Mendelian scheme, as we have stated it, appear. Most genes are handed from parent to offspring quite irrespective of sex, but those which are placed on the sex-chromosomes cannot help but behave differently. As an example of the sort of thing which may occur, we take the plumage-colour of two breeds of fowls.

The Light Sussex and Rhode Island Red are two breeds of poultry differing in colour, the first being white, and the second reddish-yellow. This difference is due to a single gene-pair which happens to be located in the X-chromosome. White is dominant over red. Suppose then that we cross two pure-bred birds, a Light Sussex hen with a Rhode Island cock. Let us remember that in birds it is the female which has only one X; the male has two. The sperms, therefore, will all be alike; they will each contain one X, and in that X there will be the recessive red gene. The eggs on the other hand are of two kinds; some contain an X and in

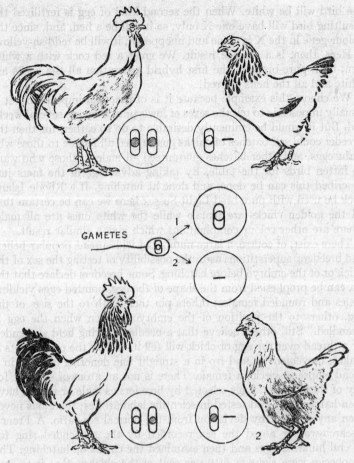

Fig. 194. Sex-linked inheritance.

A Rhode Island Red cock is mated with a Light Sussex hen. All the male offspring resemble their mother in being white, all the females are red like their father. The genes concerned are in the X-chromosomes, and the diagram shows how the criss-cross inheritance is brought about.

it the dominant white gene, while others contain no X, and therefore no colour-gene at all. When one of the former kind is fertilized, the resulting bird will have two X's; it will accordingly be a cock. And it will have one red and one white gene; so, the latter being dominant, the bird will be white. When the second kind of egg is fertilized the resulting bird will have one X only, so it will be a hen, and, since the colour-gene in the X is alone and unopposed, it will be reddish-yellow.

Here, then, is a strange result. We cross a red cock with a white hen, both pure-bred; in the first hybrid generation all the cocks are white and all the hens are red!

We choose this example because it is of practical importance. It is usually impossible to tell the sexes of fowls until they are several weeks old. But it would be eminently desirable to do so earlier, for then the breeder could sell batches of chicks guaranteed all female to those who want eggs, and other batches guaranteed all male to those who want to fatten birds for the table. By taking advantage of the facts just described this can be done, and done at hatching. If a Rhode Island cock be used with pure-bred Light Sussex hens we can be certain that all the golden chicks are female while the white ones are all male. There are other colour combinations which give a similar result.

There exist, of course, a large number of widespread popular beliefs and breeders' superstitions as to the possibility of testing the sex of the chick or of the embryo before hatching. Some breeders declare that the sex can be prophesied from the shape of the egg, pointed eggs yielding males and rounded females; others pin their faith to the size of the egg, others to the position of the embryo as seen when the egg is "candled." Still others believe that a needle or a ring held suspended by a thread over the egg or chick will tell its sex by the movements it makes, movement to and fro in a straight line denoting a male, in a broad ellipse or circle, a female. There is not an atom of support for any of these beliefs. Eggs selected by breeders as male or female have been hatched out and tested in scientific laboratories; there has never been any noteworthy deviation from the normal sex-ratio. A French research-worker noted the sex recorded by the suspended ring for several hundred eggs and then examined the chicks on hatching. The prophecies were right in fifty per cent of the chicks—that is to say, in the proportion expected for blind guessing. "Sex-testers" of various types are still advertised in the poultry papers; since the true facts have been known for years, the sale of such implements is a mere vulgar fraud, and the advertisements should not be accepted.

We need not go into all the complications that sex-linkage, as it is called, introduces into the Mendelian scheme. The reader who likes problems can worry out what would happen if one crossed a Light Sussex cock with a Rhode Island Red hen. Let us simply stress the fact that these curious sex-entangled inheritances are not in reality exceptions to the Mendelian laws; they are special cases of those laws, due to the fact that sex is itself a Mendelizing character.

There are many human characteristics which are inherited in this sex-linked way. Two striking abnormalities may be mentioned. One is night-blindness, the inability to see in the twilight; the other is hæmophilia, a very unpleasant condition, when the blood is incapable of clotting and the smallest cut may cause dangerous bleeding. As their genes are carried on the X-chromosome, these defects zigzag to and fro between the sexes in a peculiar way. Almost without exception they are seen only in males. An affected man never transmits the complaints directly to his children; but all his daughters (and on the average half of his sisters), though themselves healthy and normal, will hand the defect to half of their sons. Hæmophilia occurs in this fashion in some of the European royal families. It was the peculiar affliction of the last Tsarevitch; and the distress of his parents and their resort to the quasi-magical influence of Rasputin made this gene a determining force in the history of Russia. There is a well-known belief that sons tend to resemble their mothers, and daughters their fathers. This finds some justification in the behaviour of the X-chromosomes. For our sons receive all their X-borne genes from their mothers, while our daughters get a set from their fathers as well. As there are twenty-three other pairs of chromosomes, however, the difference between sons and daughters thus brought about will not extend to more than a fraction of their qualities.

§ 5. Gynanders.

COLLECTORS of insects have for many years prized the curious rarities known as *gynanders*. These are mosaic individuals: they are part female, part male. The commonest and most striking types have their male and female areas delimited with considerable exactitude along the middle line of the body. There are also specimens in which a quarter or some smaller fraction of the body is of different sex from the rest; and in yet others little patches or islands of male and female tissues interlock promiscuously as in a jigsaw puzzle. When, as in some social insects, the male is winged but the neuter female wingless, the most bizarre effects may be produced. Modern Genetics enables us to understand their origin. The story of the commonest method might be called "the chromosome that took the wrong turning." Gynanders usually arise from a fertilized egg which, having two X's, should develop into a female; but after one of the early embryonic cell-divisions an X-chromosome gets lost in one of the daughter-cells, and fails to become incorporated in the nucleus. This daughter-nucleus will therefore have but one X-chromosome, while the other will preserve the normal two. In other words, the one nucleus will now be male in constitution, the other female; and they will transmit their sexual constitutions to all their descendants. If the chromosome should stray

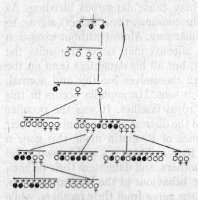
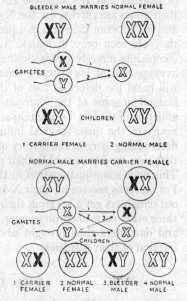

Fig. 195. Above, a pedigree of hæmophilia—"bleeder's disease," ♂ males, ♀ females; affected persons black. All the affected individuals are males. They do not transmit the disease to any of their children; but their daughters and some of their sisters will be carriers, and will transmit the disease to (on the average) half of their sons. This is due to the gene for hæmophilia being recessive, and lodged in the X-chromosome. (After Wickham Legg.) The diagrams (on right) show the method of its transmission (the X which carries the hæmophilia gene is drawn black).

at the first division of the fertilized egg, the male section will comprise half the individual; if at the second, one-quarter, and so on.

It follows that gynanders arising thus must always take their origins in eggs of the two-X sex, and that never more than half of the individual can be of the other or one-X sex.

In silkworms, other types of gynander have been found. One owes its origin to the abnormal production of eggs with two nuclei instead of one, which are then fertilized by two separate sperms. As a result, the two sides of the body can differ in regard to many inherited characters, just as can ordinary brothers and sisters. In bees, strains are known which throw a definite percentage of gynanders.

One recent observation, though as yet quite unexplained, must be mentioned because of its sheer strangeness. It has been found that if the chrysalids of swallow-tail butterflies, just after they have been formed from the caterpillars, are dropped on the floor, or jarred by tapping the box containing them with a hammer, they produce a considerable proportion of gynanders. These are of unusual type, in that they are always patchy and that individuals of either sex may exhibit small regions of the other sex. The cause of the phenomenon is as yet completely obscure.

But in all the other cases the cause is always some little miscarriage of the normal chromosome machinery. A chromosome is delayed in its movements on the spindle; a female germ-cell fails to divide its cytoplasm after dividing its nucleus; an egg divides a trifle too fast for the sperm that fertilizes it—little accidents, but irrevocable. There is none of that mysterious "regulation to the normal" here, which we often hear of as a universal quality of developing life. The little accident leads to a wasted life, a monstrous individual, a biological absurdity. The process, for all its intricate complexity, is mechanical and regardless of any ideal of harmony or perfection in the result.

§ 6. Sex Hormones.

We come now to the various kinds of sexual abnormality found in the higher animals and in mankind. The abnormal may illuminate the normal; by studying how the machinery goes wrong we get a better idea of its normal working.

So far as structure goes, we vertebrates begin our embryonic lives not as males, nor as females (nor even as neutrals), but as both. Our chromosome constitution is determined, it is true; the double X or single X has already decided for female or male. Nevertheless, the sexual organs themselves remain still male-female—hermaphrodite. When the sexual apparatus first develops it consists of a single pair of gonads (which will become either ovaries or testes), but for the rest, in the ducts, for example, between gonads and exterior, it has the rudiments of both sexes. It has a foot in both camps.

But soon it becomes evident to which sex the embryo really belongs. If it is going to be a boy it enlarges the male rudiments while the others remain undeveloped. And the other way round if it is going to be a girl. So it comes about that the adult male has rudimentary bits of femaleness here and there on his body, rudimentary nipples, for example, and deep in his pelvis a very small rudimentary uterus. And the characteristic male parts can be traced in the female. These facts are by no means proof, as has been sometimes suggested, that the mammals are descended from a hermaphrodite ancestor. They are simply the result of the way the body develops.

Now the chief agents in the moulding of this bisexual ground-plan into the form of one sex or the other are the sexual hormones, substances shed into the blood by the ovaries or testes. Everyone knows that a castrated male animal differs from a normal male both in temperament and structure. The ox is dull and quiet, has horns intermediate in shape between those of a bull and a cow, is larger-boned than a bull and puts on fat more readily; the gelding is structurally very little different from the stallion, but their temperaments are poles asunder; the capon preserves the cock's plumage, but his comb and spurs remain rudimentary and he never crows or pursues hens. The same is true for females with their ovaries removed, though the visible changes are generally less. The spayed sow, however, is not only neuter in behaviour, but fattens more quickly than the normal sow. In birds remarkable changes may follow the removal of the ovary, hens and ducks changing into the plumage of cocks and drakes (though without assuming masculine behaviour) at the next moult. The plumage of male birds is thus unlike most male secondary sex-characters; it can develop perfectly well in the absence of the testis, but the ovary produces something which prevents its growth.

Grafting experiments give confirmatory results. A capon's comb, for instance, can at any time be made to grow to full size, and the bird revived into full activity of crowing, fighting, courting and mating, by grafting a piece of cock's testis anywhere into the body-cavity. The grafted testis cannot function in its reproductive capacity, since there is no connection between it and the reproductive ducts. But its glandular function continues, for a grafted piece of tissue always acquires a blood-supply, into which the internal secretion can be shed. In the same way a hen, the removal of whose ovary has brought about the development of male plumage, can be made to redevelop henny feathering by a grafted ovary.

Still more remarkable are the deliberate reversals of sex-characters. Steinach was the pioneer in this field, and his results have stood firm against much hostile criticism. If a male baby guinea-pig, a few days after birth, be castrated, it grows up into an animal not unlike an ordinary male, save that it is completely neuter in instinct and behaviour. But if at the same operation an ovary from a young female is engrafted the creature grows into a passable imitation of a female.

It is smaller than a normal male; it is pursued with sexual intent by males; the milk-glands and nipples enlarge as in a female, and the animal may even produce milk and, if a foster-child is provided, suckle it and care for it like a normal mother.

These "feminized males" are, of course, not complete or normal females: the physical differentiation of the internal organs of reproduction had proceeded so far before birth that it cannot be completely reversed. The feminization is, however, complete for all characteristics developing after the operation. Equally successful results were obtained in the converse experiment of masculinizing young females. And precisely similar successes have been achieved with fowls.

By grafting both ovary and testis into a guinea-pig Steinach even produced an animal which had a double personality—it oscillated rapidly between masculine and feminine instincts. It is one of the most remarkable experiments to see an animal thus made over in its most intimate anatomy, its outward appearance and its mind, by the substitution of one small piece of tissue for another. We mammals are indeed the creatures of our internal chemistry.

Sometimes nature makes these experiments. When twins are conceived by a cow their embryonic membranes meet and join and the blood-vessels in them unite; they acquire a common circulation and the hormones of one can flow through the other. Apparently the male hormones are earlier formed and more potent than the female, for when the twins are of opposite sex the male is born normal, but the female suffers more or less transformation towards maleness. She is known as a free-martin, and is always sterile.

Now, this is a very interesting qualification of our idea of the ruling influence of the gene. In that male guinea-pig of Steinach's every living cell (except those of the graft) had its complete chromosome-outfit unimpaired, and normally it would have produced masculine characteristics wherever they ought to have appeared. Similarly with that free-martin, which ought to have been a proper cow. But sexual hormones, from the graft in the one case and from the bull twin in the other, come in and swamp the influence of the original genes. In these higher vertebrates the machinery controlling the development of sex-characters is a double one, a two-switch mechanism. The chromosomes and their genes set the first switch. If it is set one way the animal travels on to produce a reproductive organ that is male; if the other, to produce one that is female. Here is situated the second switch. The male gonad turns on the male hormone, and it is to its influence, not to the influence of the genes, that the remainder of the male sex-characters are sensitive; and similarly with the female gonad and its hormone. We have learned how to tamper with this second switch, and so to send the animals over on to the opposite sex-track.

In insects, however, we cannot reverse sex-characters in this way. Insects, evolving along another path, have not developed the second switch. They seem to lack the sex-hormone mechanism entirely. Castration or spaying of an insect makes no difference whatever to the other parts of the body. That is why gynanders and sex-mosaics can occur in insects: the sex is determined by the sex-chromosomes in every separate tissue, and whatever chemical substances are needed for the development of sexual characters are brewed locally under their supervision.

§ 7. Reversal of Sex.

IN THE year 1474, a remarkable trial took place in the Swiss town of Bâle. A barndoor cock had laid an egg. With all due formality, it was tried for witchcraft, sentenced, and publicly burnt.

Such sexual aberrations in fowls are not excessively rare. Crowing hens are sufficiently well-known phenomena to figure in a rhymed proverb—

"A whistling maid and a crowing hen
Are neither fit for God nor men."

A bird ceases to lay, her comb enlarges, and she begins to crow. Some go farther and exhibit male behaviour—they fight with cocks and pay court to hens. This is not uncommon.

In very rare cases, however, the transformation proceeds to its logical conclusion, and the bird changes its sex completely. Professor Crew, for instance, reports on a three-year-old Buff Orpington which, after being a good layer and a mother of chickens, ceased to lay, grew spurs and practised crowing. Next season her plumage became male, her crow lusty, her behaviour masculine. Isolated with an unmated pullet, the transformed hen became a father. On "her" death, autopsy revealed that the ovary was diseased, and had become almost completely transformed into a testis. Apparently the disease had altered conditions in the ovary, and stimulated the production of male tissue. But whatever the cause, the fact remains: sex-reversal is possible to a warm-blooded vertebrate, even after it has become adult.

In frogs the normal chromosome-machinery can be over-ridden in various ways. Keeping the early embryos at a high temperature increases the proportion of males; and so does acid water; while a water-shortage (e.g. keeping the eggs only just moist) increases the proportion of females. Most effective of all is making the mother retain her eggs for three or four days after they are fully ripe. Such "over-ripe" eggs, when fertilized, produce ninety to a hundred per cent of males instead of the usual fifty-fifty ratio.

These cases differ from the guinea-pigs of the last section because the reversal is complete. It affects the gonads. In the guinea-pigs the hormones were upset; it was the so-called accessory sexual characters that were altered. But in these fowls and frogs the action is on a different link of the casual chain. It is the development of the gonad itself which is attacked.

However, even if we reverse the gonad of a fowl or a frog and turn it from an ovary to a testis the chromosomes remain unchanged. And the proof of this fact lies in the next generation. Suppose that we turn a female frog by over-ripeness into a male, it will still have the female chromosome-outfit, with two X-chromosomes. So that every one of its sperms will have an X-chromosome in it. When such an animal is mated with a normal female, who therefore is also XX in regard to her chromosomes, the next generation will be exclusively female, instead of male and female in equal numbers. By similar reasoning it can be shown that if a male turned into a female be mated with a normal male the next generation will include two males to every female. And these expectations are confirmed in practice. Thus by breeding from the metamorphosed animals we can demonstrate the immutability of their chromosomes.

In certain invertebrates, we may note, sex-reversal normally and regularly occurs in the life-cycle. As Dr. Orton of Plymouth has shown, the adolescent oyster is male; then it becomes female; and thereafter it swings slowly from sex to sex with a yearly rhythm. But this has little to do with the reversals that we have been considering. There is apparently no chromosome-difference concerned with sex in the oyster any more than there is in those plants which bear male and female organs on one flower. Sex in such cases is a matter of local influences in the process of development.

§ 8. Intersexes.

This brief study of sex-determination has done much to enlarge our minds from the rather crude gene-determinism with which we necessarily began. Normally, as we showed at the outset of this chapter, sex is determined at fertilization and is plainly a Mendelizing character.

But since then we have gradually unfolded a number of qualifying forces that do much to modify the results of that primary decision. Let us take stock of our position.

The egg is an almost unformed sphere of living substance. When it is fertilized it begins to organize and shape itself, and soon it comes under the direction of the genes, the little parliament within the nucleus. The parliament is not altogether harmonious, for every character of the future individual is decided by debate between two rival parties of genes, contributed one by the father and one by the mother. On some points the parties are in agreement. On others they differ, and the result is either a compromise (as in the pink Four-o'clock), or the more or less complete over-ruling of one gene, which, however, is not destroyed, by its dominant rival. And even the majority-vote of the genes is not always the decisive factor.

We can interfere with the developmental processes experimentally, and our results can illuminate those intranuclear operations, and can show in some cases how the chromosomal decisions are put into operation.

Recent genetical work is modifying our ideas about the procedure in this microscopic parliament very considerably. It used to be thought that each gene-pair was independent of the rest of the nucleus; that in the nuclei of the hybrid peas, for example, the rival genes for tall and dwarf and those for yellow and green seeds fought out their little battles independently and without interfering with the others. But the matter is more complicated than that.

We propose to make the point clear by considering in some detail the determination of sex in the fruit-fly. We shall make it clear that other genes may take quite an important share in the sex-determining work. In this most instructive insect, the males, as we have already noted, have one stout, rod-shaped X-chromosome and one hooked Y-chromosome in each nucleus; the females have two X-rods and no Y. These are normally the sex-determining chromosomes. The other chromosomes, which have been called "autosomes" to distinguish them from the sex-determiners, are the same in both sexes. We have told that Drosophila contains four pairs of chromosomes, so that each nucleus contains six of these autosomes, which form two sets, each set including a little round chromosome and two long V-shaped chromosomes. One autosome set, of course, and one of the sex-determiners, comes from each parent. All this we have displayed already in the pictures and map of the genes of Drosophila (Fig. 175).

Now, during the intensive study of variation and heredity in this creature a strain turned up in which the sperms and ova were abnormal, the sorting of the chromosomes at the reduction-division being inaccurately performed. So that among the zygotes, individuals with all sorts of abnormal chromosome-outfits appeared. Some had one X, some had two, some had three and even four. And similarly with the Y's, and with the other chromosomes.

It was found first that the Y contained very few active genes and played the rôle of an inert dummy; it was a mere dancing-partner for the X; at any rate, it had no say whatever in sex-determination. Then it was found that the sex of an individual depended not, as had hitherto been assumed, on the number of X-chromosomes but on the proportion of X's to autosome-sets. For example, a fly with one X and only one autosome-set was female; so was a fly with three of each or with four of each. The normal female fly, of course, has two of each. So that when they are equal a female is produced. On the other hand an abnormal fly with two X's but four autosome-sets was male; so that when the latter outnumber the former by two to one a male is produced.

Various other combinations appeared. In some the X's actually outnumbered the autosome-sets; in these "super-females" the insect was female in type but sterile. In others the X's were overwhelmed by autosomes in a ratio greater than two to one; these "super-males" were male in type but sterile. Most curious of all, there were flies in which the proportion was intermediate between two to one and equality. There were, for example, flies with two X-chromosomes and three autosome-sets. And in their bodily characters and instincts these flies reflected the abnormality of their chromosomes, for they were intermediate in type between male and female—to use the technical term, *intersexes*.

Thus, sex is determined in these fruit-flies not by the X's alone but by a reaction between the X's and the rest of the chromosome-outfit. It is determined by a sort of debate in which all the chromosomes participate. But normally the deciding factor is the number of X's. Both sexes have the same autosome-outfit, and its balance is in each case in favour of maleness; its influence is strong enough to outweigh a single X and produce a male, but is overcome by two X's, which then produce a female. It is, however, only in abnormal flies that the votes, so to speak, of the autosomes can become casting votes. This, by the way, shows how abnormalities can illuminate the normal processes, for without this particular strain of curiously developing flies the true relations between the chromosomes would never have been discovered.

This struggle between male-producers and female-producers in the nucleus is, we infer, a general phenomenon in sexually reproducing creatures. It occurs in the development of each one of us. Normally one set has a sufficient majority over the other, and the individual is definitely male or female. But not always.

Some interesting facts were discovered by Goldschmidt, the German zoologist, when he was crossing different species of gipsy-moth, of the genus *Lymantria*. Here, also, there are rival male-producing and female-producing genes; normally one set outweighs the other completely. But in different species of gipsy-moth these genes seem to work at different rates. In the European species they are fairly slow. But in the Japanese species the whole tempo of the life-cycle is swifter, and the genes are more active. Let us, for the sake of simplicity, begin by considering female moths. All females will have male-determining as well as female-determining genes at work, but normally the female ones are in the majority, and whatever female-determining substances they are generating is in excess over the opposed substance produced by the male genes.

But by crossing different species curious results can be obtained. Female moths can then be produced in which the female genes are slow workers, the male genes quick workers. The moth begins to develop as a female, because of the female genes' initial majority. But, meanwhile, the male genes have been producing their male-determining substance at a great rate; and their production catches up with, and eventually surpasses, that of the female genes, so that the moth finishes its development as a male. The result is an intersex in which those organs or parts of organs which are laid down early are female in type, while those which appear later are male. And the reverse creature, starting off as male and then becoming female, is also obtainable by appropriate crossing. Indeed, by choosing his strains of moth, Goldschmidt has been able to produce any degree of intersexuality he likes. The earlier in development the moment when the original under-dog genes catch up with those of the opposite sex and become top-dog, the more the creature is transformed away from its original sex. And when the catching-up occurs before the rudiments of adult organs are laid down in the chrysalis, there is a complete reversal of sex. The division of the race into two equal battalions, one solidly male, the other solidly female, seems at first sight fundamental and inevitable. But the distinction can be set at nought not only by the slipping of a chromosome-cog, but by a slight defect in timing. Alter one speed, and you throw the whole harmony out of gear.

Various rather similar phenomena have been observed in mammals.

They are not uncommon in domesticated pigs and goats. Animals classed at birth as females, some of which have even won prizes in shows as immature females, develop various masculine characters and instincts later. Internally they contain testes and not ovaries, but they have a uterus besides their male ducts. Again they seem to be animals in which the relative timing of the sex-genes has got out of gear. In goats, in some cases at any rate, they seem to have been caused, like the moth intersexes, by crossing different races.

That the existence of these sex-intermediates is determined by heredity is shown by some strange facts which Dr. J. R. Baker has brought back from the New Hebrides. In the northern island of this Melanesian archipelago, intersexual pigs are extremely abundant, making up perhaps ten or twenty per cent of the whole male porcine population. They are called by a special name (*rawé* or *ndré*) and are employed in religious and social ceremonies. At every step a chief makes in rank, for instance, so many have to be sacrificed. Now, if in Britain the acceptance of a knighthood were necessarily bound up with the ritual slaughter of a number of intersexual pigs, there would be a great demand for sows which produced such valuable monstrosities. This is precisely what has happened in Melanesia. The *rawés* are modified males, which owe their existence to a single recessive sex-linked or X-borne gene. Sows carrying this gene will produce some intersexes in every litter; and such mothers are eagerly sought after and command high prices. As a result of this artificial selection, the proportion of intersexes has grown until it is hundreds of times greater than in the pigs of any other country. This is one of the prettiest examples of the power of selection, given an initial variation, to build up fantastic types. Here artificial selection has turned an abnormality into a normal component of the race.

In our own species quite parallel phenomena may occur. There are about two thousand well-documented cases of children which inclined to female in their external characters but had male ducts and testes (which, however, are retained in the belly), and sometimes a more or less well-developed uterus. These children have often been brought up and even married as girls, but masculine instincts appeared after puberty.

Thus, beneath the calm orderly development of the individual there is war in the nuclei, and in all of us, male or female, the rival forces are present, one outnumbered by the other. In rare cases the battle nearly results in a draw, and that draw marks itself in indeterminate structure. And even when there is a definite win the defeated side may regain a little ground. Various things may go wrong. The chromosomes may not be quite correctly balanced, or they may get to work at slightly different rates, or there may be influences disturbing the ductless glands.

There are feminine men and masculine women, in body and in mind. The interplay of internal secretions in our bodies is extraordinarily complicated; thus, if the outer peel of the adrenal gland is too active in a woman, it may deepen her voice and give her masculine tastes and instincts; too much pituitary may give her abnormally large hands and feet; a deficient pituitary may prevent sexual maturing and keep an individual of either sex permanently infantile; and so on.

The reader, we may remark here, must not confuse these genuine intersex cases, in which structure and function are truly intermediate, with the alleged cases of which he may find accounts in pseudo-psychological works, of perfect men with women's souls and instincts and perfect women with men's. There is a literature and discussion of this subject too abundant to ignore, it is associated with queer little cults and practices upon which we will offer no comment, and so it may be well to point out that the facts we are here recording fly right in the face of any such assumptions that there are ladies sadly misplaced in whiskered bodies or vice versa.

Clearly in all that we have told, disposition follows structure. The hormones keep body and dispositions consistent. *Homo sapiens*, however, is extraordinarily curious, plastic and suggestible in sexual matters, and through bad influence or unfortunate accidents, he may develop the strangest, most pitiful perversions of desire, matters for the parent, guardian, friendly doctor or mental home rather than for publication and sympathy. The importance of these unfortunate eccentrics is enormously over-rated, both by their adversaries and themselves, and more than half the evil of their misfortune is due to such exaggeration. On the one side a little abnormal thought and behaviour is represented as a terrible crime, on the other hand it is elevated to the dignity of a marvellous distinction. Sexual fuss is rather characteristic of the primates, as any cage of baboons will testify.

There is no reason, however, why biology should be invoked to bolster up perverted ideas or why it should tacitly allow its facts to be misapplied.

The true intersex is physically malformed, a thwarted being in that respect but not a perverted one. The organisms with adrenals or pituitaries or some other bit of vital chemistry disorganized are in another case. They are not full intersexes, in so far as their reproductive organs themselves are definitely of one sex. But in their temperament and physical quality they are not as fully sexed as the normal. It is such combinations that give us the masculine women and feminine men of whom we have spoken. Very possibly such organisms are more readily suggestible in the matter of sex abnormality. That we shall only know when the analysis of motives has been pushed farther than its present elementary stage.

Meanwhile let us beware of this grotesque and very unbiological mythology of purely female souls in purely male bodies and all that it implies.

VII

VARIATION OF SPECIES

§ 1. *The Open Questions of Variation.* § 2. *Mutation: the Experimenting of the Germ-plasm.* § 3. *Are There Forward Bounds in Mutation?* § 4. *Are Acquired Characters Inherited?* § 5. *Artificially Induced Mutation.* § 6. *Moths and Smoke.* § 7. *The Significance of Sex.* § 8. *Variation in Plants.*

§ 1. The Open Questions of Variation.

OUR preceding chapters have, we hope, cleared up the general facts of individual development with a sufficiency of detail and explanation for us to resume now those broader questions we had to leave open at the beginning of this Fourth Book. We then explained the main interrogations that are still undergoing discussion to-day about the method of the evolutionary process. We ended our first section with a group of four, not mutually exclusive, alternative possibilities of the way in which species may have varied, concluding with a brief paragraph upon the agnosticism of Professor William Bateson. In what follows we hope to supply the reader with sufficient material to obliterate the scepticism of that eminent authority, with an entirely reasonable understanding of the possible How of the matter.

Our alternatives—or rather our group of possible origins of heritable variation—were firstly, what we may call mechanical or chance changes of the germ-plasm, in which the Neo-Darwinians (Weismannists) find a sufficient cause; secondly, heritable variations due to striving and the acquisitions of the struggling individual (asserted by the Lamarckians and denied by the Weismannists); thirdly, variations sustained in a definite direction, due to an innate drive in the germ-plasm of a group of species (Orthogenesis); and fourthly, variations through some mystically conceived force within or without the substance of the species in a "creative" or "upward" direction (Bergson and Shaw).

VARIATION OF SPECIES

We did not note in that section, but we may note here, how easily this latter belief may relapse from Evolution to the older conception of Transformism by accepting every favourable upward movement as a relapse into creative activity—and by ignoring the frequent degenerative processes that occur.

We shall deal first of all with the fortuitous mechanical changes of the germ-plasm. We shall show that such changes certainly do occur. But we shall not go all the way with the Neo-Darwinians. We shall proceed to discuss in a far more open spirit but without absolute rejection the Neo-Lamarckian claim for the inheritance of acquired characteristics (somatic changes). Then we will take up Orthogenesis (on which we have already thrown the shadow of doubt), and conclude with an examination of the Creative Evolution idea. In that way we hope to make this Book of our work a complete summary, from our point of view be it remembered, of the state of evolutionary discussion at the present time.

§ 2. Mutation: The Experimenting of the Germ-plasm.

HEREDITY is in its essence a conservative tendency, binding each generation to the organic semblance of the last, and in the preceding exposition of its principles we have treated of the genes as immutable units, shuffled and dealt out in various combinations but unaltered by their treatment, sometimes lying concealed by dominant partners for generations but always emerging unchanged, as atoms emerge unchanged from the molecular combinations in which they take part, or as the cards lie on the table unchanged at the end of an evening's bridge. But now we must tackle an important complication of the story. Sometimes the genes themselves may undergo alteration. Let us consider four concrete instances to show how living things may inherit, and at the same time evolve.

In the year 1791 a Massachusetts sheep-farmer of the name of Seth Wright was greatly surprised when one of his ewes bore a lamb with the build of a pekinese, having a long back and short, bowed legs. Nothing like it had ever appeared in his flock before. Now it happened that he was sorely perplexed because his sheep used to roam into other people's fields; the fences of Massachusetts, it seems, were low in those days. So he noted with pleasure that the new arrival was a male. He killed his old ram and reared and bred altogether from the new; and he found the new shape to Mendelize, for thereafter his sheep were always either completely Ancon (as the new shape was called) or completely ordinary. By picking those of the short-legged type he got in a few years a very considerable flock of animals which were unable to jump over fences—and the variety spread throughout Massachusetts.*

In the laboratory of the Carnegie Institute at Cold Spring Harbour, a strain of water-fleas (*Daphnia*) had been bred for 363 generations. The animals of this strain thrive best at 20° C., and die if the temperature goes above 26° or below 11°. After fourteen years, a mutation arose which adapted its possessors to different conditions of temperature. The young of this new type die if kept in the standard temperature conditions. The mutant strain does best at 27°C., dies at 20°, and will tolerate 32°. Apart from this altered reaction to heat and cold, no other difference could be found between the two strains. Here we have an excellent example of the germ-plasm throwing up a new type which might be of the greatest service to the species in helping it to spread to warmer climates.

The Florida Velvet Bean could only be grown in Florida and the Gulf States. Suddenly a variety appeared which would flower and set seed in less specialized weather and which could therefore be profitably planted over the whole cotton belt of the United States.

A variety of Tobacco appeared which would not flower in Northern U. S. A. (where its parent plants had successfully existed) but which could be made to flower and seed by keeping it under artificial illumination which corresponded in length of day and night with a sub-tropical summer. Suddenly at a bound the strain had fitted itself for an entirely new climate.

Here we have four examples of mutations. A stock has bred more or less true for generations; suddenly something happens in its chromosomes, some innovating change, and an individual with an entirely new feature, a "sport," appears.

Let us be clear about this process. Sometimes a recessive will lurk concealed in a strain for generations and turn up abruptly. Thus, and in other ways, the common shuffling of genes may imitate mutations. But these four novelties recalled no known precedent. True innovation is a quite different thing from the release of a recessive; the change produced by an actual alteration in one of the genes, not by merely shuffling the old ones about. It is a genuine innovation, an enduring change in the germ-plasm, and it need not show the slightest disposition towards recessiveness under hybridization.

These little gene-leaps that we call mutations are certainly one very

---

*It was subsequently displaced by the Merino.

VARIATION OF SPECIES

important way in which the race changes. They are the normal ticks of the evolutionary clock. By mutations, now in this part, now in that, races slowly modify themselves.

The classical studies in mutation have been made with various species of that most convenient of insects, the fruit-fly (*Drosophila*). Every now and then, among hundreds of normal fellows, a "sport" appears. The novelty may differ from the rest in eye-colour or in wing-shape, or indeed in the appearance of any part of its body, in length of life, in its reaction to light, and so on. Often the same mutation turns up independently in several different stocks. Altogether in the course of this Drosophila work over five hundred mutations have been carefully recorded. Thus every now and then the germ-plasm experiments and tries something new.

By skilful crossing it has been possible to pile together a great number of these mutants into a single strain. The resulting insects are amazingly unlike the "normal" parent type. They look more like ants than flies, and it is highly unlikely that if an entomologist were confronted with specimens without explanation he would identify them as *Drosophila melanogaster*. He would think them at least a different species. There are many other well-established cases of mutation in domesticated animals. Among plants, maize and the garden flower called *Primula sinensis*, the Chinese Primrose, are classical examples, but there are plenty more.

FIG. 196. A Drosophila which no longer looks like a fly.

A specimen of a strain combining two mutations, vestigial wings and black body. (From "The Physical Basis of Heredity," by Prof. Thomas Hunt Morgan. J. B. Lippincott Co.)

It is of course impossible to *observe* mutations directly in wild nature, for one has no pedigrees and cannot tell whether a variety appears merely by shuffling of pre-existing genes or whether it is really a mutant. But there is every reason to believe that mutation is frequent in wild forms. There are plenty of clearly demonstrated cases where wild animals and plants differ in characters which are inherited according to Mendelian laws, and which we may presume therefore to have arisen by mutation. Thus the grasshoppers *Paratettix* and *Apotettix* of the Middle Western U. S. A. are found in an astonishing range of colours and patterns, all of which Mendelize. In guinea-pigs and rats, the coat-colours of different wild species have been shown to differ by Mendelian genes. In pheasants and in snap-dragons and primulas, where fertile species-crossing is possible, Mendelian species-differences are also known (Book 3, ch. 4).

Note that these mutations that we have been discussing, the Ancon sheep and the Florida Velvet Bean and the Tobacco and the rest, all produce pretty considerable results. They are rather abrupt jerks of progress. But all mutations are not like that. Some are so small as to be barely perceptible, and probably these are the commonest. An abrupt jerk will obviously stand more chance of being noticed by the human observer. Occasionally in the course of very delicate experimentation such inconspicuous mutations have been recorded; Johanssen (Chapter 3) observed two in his beans. Castle (Chapter 8) saw some in rats, and quite a number have been found in the fruit-fly. Such minute changes are probably widespread in animate nature.

Thus, with apparent spontaneity through these greater and lesser mutations, the race makes its experiments. In the remaining sections of this Book we have to see how, out of these mutations, the steady progressive changes of Evolution may be wrought.

§ 3. Are There Forward Bounds in Mutation?

HERE perhaps we may find space for an idea that had some currency in the beginning of the century through the work of de Vries (born 1848). It is that there are spurts, jolts, storms of mutation in the case now of this species and now of that. A stock was supposed to evolve in a series of violent efforts, first a rush of variation, then a rest, then another rush, followed by another rest, and so on. These onward pushes he thought supplied the greater part or all of the variational factor in Evolution. They were considerable enterprises, aggressions, novelties. They might yield new species at a bound. It is as if a species or group of species, going along in a humdrum way, were to be suddenly pricked by an unaccountable whipper-in. Recently this idea has been revised in an exaggerated and purposive form by Mr. Hilaire Belloc in one of his all too rare contributions to biological discussion. Unable to imagine how a fore-limb could be transformed from a walking limb into a wing, convinced of the fact of Evolution and equally convinced by the assurances of an eminent Montpellier biologist that the commonly accepted pedigree of the birds is unsound, Mr. Belloc resorted to the suggestion of a rapid series of large mutations, during which the developing birds must have led a retired and embarrassed existence, providentially escaping extinction until the new gift was attained. Contemporary biology has no need and no evidence for this grotesque hypothesis.

The work of de Vries was based almost entirely on the behaviour of one of the Evening Primroses (*Œnothera lamarckiana*) which he found growing ready to hand in Holland. It seemed to be in the throes of an evolutionary burst, throwing out all sorts of strange varieties among its offspring. His work was already shaped out before the revival of Mendel's ideas or the realization of the full significance of the chromosomes in development, and on the face of it seemed to justify his claim that a species in a stage of violent evolutionary activity might by large and abrupt mutations throw off new forms, some worthy to be classed at once as new species while others constituted marked new varieties.

Unhappily he had hit upon this *Œnothera lamarckiana* for his material, and the species happens to be a very exceptional one. It is loose and inaccurate in its chromosome process to an extraordinary degree. One of de Vries' new types was due to a duplication of all the chromosomes, giving an exceptionally large plant with four instead of two chromosome-sets. The majority, however, owed their origin to the slipping of a cog in the machinery of chromosome-reduction, a fault to which Œnothera, with some other plants, is particularly prone. Sometimes both chromosomes of a pair stick together at reduction instead of separating, so that one gamete-nucleus receives both, the other none. All those with none inevitably die—they have not got a complete chromosome-pack. When those with two members of a pair unite with a normal gamete, the resulting zygote has an extra chromosome, being provided with three instead of the normal two of this particular kind. And this, of course, has an effect upon its appearance. For instance a triplication of one particular chromosome gives rise to plants with very broad leaves.

These types, however, have no evolutionary significance, for they cannot breed true. At reduction, the three chromosomes separate into two and one. If at fertilization, two meet two the resulting zygote, over-balanced by two extra chromosomes of one kind, is unable to live. If one meets one, the normal original type is reconstituted; only when two meets one is the new type repeated, and this in further generations must again always throw a certain number of normals.

Then owing to an extraordinary complication of lethal factors, recombinations occasionally can crop up in a previously unsuspected way and were mistaken by de Vries for new mutations; and finally some new mutations and recombinations appeared as they do ordinarily in other plants and animals.

The net result of twenty-five years of research, summed up in Renner's monograph, is that no special storm of mutation is agitating Œnothera's germ-plasm. So far as we can see, the kind of sports it is throwing now it will continue to throw as abundantly as ever, so long as it exists; and most of them are accidents to the chromosome-machinery, of no significance for Evolution. As regards true evolutionary material, Œnothera is producing no more gene-mutations than do the general run of plants and animals; as regards its other sports, there is no reason to suppose that it is at the moment indulging in a burst of them. De Vries was as unlucky in Œnothera as his material for the study of Evolution as Mendel was lucky in his choice of the pea for his study of heredity.

But even though we dismiss the Evening Primrose as evidence for Evolution by mutational booms, it may still be true that organisms are occasionally subject to fits of change, their germ-plasm to storms of mutation. It is a possibility; there is no reason why this should not be. Few facts, however, are to be found in support of such a view. Every animal and plant which has been carefully and intensively studied—wheat and flies, maize and mice, shrimps and peas—is found to be throwing gene-mutations. Further, the rate at which genes mutate seems roughly the same in all of them. There is some evidence to show that in Switzerland the Bee-orchid is started to produce numbers of varieties in the last fifty years, while the same varieties were scarce or absent earlier. But many more records and experiments on animals and plants, both in a state of nature and under cultivation, will be needed before we can say if such mutational storms are as isolated as they now seem to be or whether they do indeed bear upon the evolutionary process to any notable extent.

One tenaciously-held idea at least seems to be without foundation. It used to be implicitly believed by practical breeders and scientists alike, that the bringing of an animal or plant into new and favourable conditions would cause an epidemic of variations. Rich food, protection, and warmth for animals; cultivation, good soil and plenty of manure for plants—these were asserted to bring on mutation-storms. But modern experience is rendering this conclusion very doubtful. We see variations more readily when they are under our eye and when we are on the look-out for them; and in cultivation we preserve many that would go under in nature. That seems to be all.

This idea of evolutionary spurts was much more reasonable when fossil records were incomplete—for instance, when only half a dozen steps in the Horse progress were known—than it is to-day. Now, with the fossil records of the horses and the sea-urchins, of the pigs, camels and elephants, of the pond-snails (*Paludina* and *Planorbis*), ammonites, and many others at our disposal, we can feel reasonably sure that on the whole Evolution progresses slowly and steadily. Whenever we get a reliable fossil series we find that to be the case. The crises of Evolution when they occur are not crises of variation but of selection and elimination; not strange births, but selective massacres. The germ-plasm, it would seem, has gone on throwing up mutations at about the same rate age after age.

§ 4. Are Acquired Characters Inherited?

Now we come to a knotty and controversial question, and the best way to get clear about it will be to attack it historically.

Lamarck, in his discussion of Transformism in the later eighteenth century, believed that a species modified itself as a result of the cumulative inheritance of the efforts of its constituent individuals. The ancestral giraffe stretched out its neck to reach the leaves of high trees; this elongation was transmitted to some extent to the offspring and thus, by generations of stretching, the species produced an innate elongation. The blacksmith wields his hammer and his muscles swell and harden; as a result of his exertions his children (on a Lamarckian view) are the more muscular. White men venture southward into the tropics; as a protection from the blazing ultra-violet light their skins become tanned, and thus by generations of cumulative tanning the coloured races of to-day are supposed to have arisen.

Lamarck held that, in the race as well as in the individual, exercising a faculty led to its further development. He did not invent this idea. Its truth was generally assumed at this time, as applied to human pedigrees or the breeding of domesticated races of animals, and had been so assumed for centuries. He simply adopted and extended the idea and showed how it might be imagined to play a part in Evolution. Moreover, he believed that neglected or unused parts shrank away and finally disappeared; the classic example of this was the hind legs of the whale, which were reduced to mere vestiges when these aquatic mammals gave up walking about on land and took to swimming in the sea (Fig. 134).

In addition, the modern followers of Lamarck stress the inheritance of the direct efforts of the outer environment. Everyone is familiar with such direct modifications of the individual. Animals on a low diet grow up stunted; trees exposed to a steady coast wind grow all in one direction; plants grown in darkness never form green chlorophyll. More subtle reactions are the individual immunity acquired after an attack of measles or small-pox, or the succulent fleshy leaves produced on geraniums and many garden plants by watering with salt water. The Neo-Lamarckians suppose that such effects, as well as those of use and disuse, are also inherited in their degree, and contribute to evolutionary change.

There are two ideas to disentangle here. One is the strictly biological problem: Do modifications imprinted on an organism during the rough-and-tumble of its lifetime affect the characters of its offspring at all? This is the problem with which as biologists we shall concern ourselves here. The other is the philosophical idea of the race perfecting itself by striving. We need not concern ourselves with this side of the question; plants and bacteria evolve as well as we do, and they cannot be said to strive without distorting the word beyond recognition. And in any case, unless the biological question is answered in the affirmative, the philosophical does not arise.

It is clear enough that the problem is of vital interest, for if acquired characters can be inherited we would reasonably expect education and the effects of occupation to be inherited; we could hope that by nursing our children carefully we were imprinting a permanent improvement on the race and there would be no limit to the improvement which education and social betterment could make on the human species, however carelessly it bred.

Let us remember that there is a strong "common-sense" bias in favour of the inheritance of acquired characters. The tales of the giraffe, the blacksmith and the sunburnt white sound entirely plausible. In the often-quoted example of the pointer dog it is natural to assume that the dog stands rigid at the scent of game because he has been taught for generations to do so by his masters, and the habit has become an instinctive character of the race. Yet at the present day the majority of trained biologists show a definite bias against Lamarckian interpretations. Let us note their reasons for this opposition.

When Lamarck put forward his views, the chromosome machinery of inheritance was still unknown. The profoundly illuminating nature of its discovery is not often realized. Darwin, who believed in the inheritance of acquired characters as a subsidiary cause of evolution, suggested a purely theoretical mechanism to explain it. A number of tiny particles, which he called gemmules, were supposed to be formed all over the body, and to swarm into the testes or ovaries by way of the blood-stream. From every organ, every tissue, they were supposed

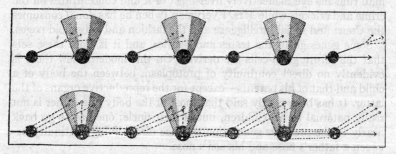

FIG. 197. Above, a diagram of the relations of germ-plasm and body (soma). The germ-plasm is immortal, but in each generation produces mortal bodies, containing many kinds of tissue. Thus the body of the parent is not directly ancestral to the body of the child. Below, a diagram of Darwin's theory of Pangenesis. He supposed that gemmules immigrated from each tissue of the body into the germ-plasm, whence in the next generation they could emigrate to the tissues of the offspring. In this case there would be some kind of continuity between the bodies of parents and children.

to come—from the head and the stomach, the ears and the feet—and they crowded into the germ-cells like the animals into the Ark. Then, in the embryos of the next generation, they were supposed to supervise growth and development, each going to the appropriate part according to its origin, back to the head or stomach or ears or feet, and thus the children were formed after the pattern of their parents. Now it is easy to see how on such a basis the inheritance of an acquired character could be explained. An injury to an arm, for example, by affecting the arm-particles, might at the same time damage the arms of unborn and unbegotten children. This view, with minor variations, was held widely before the chromosomes were discovered.

Contrast with this the modern gene theory. In the view we have just examined the determining particles were dispersed in the embryo, some going to the head, some to the feet, and so on. Like letters in the post, each had its appropriate destination. But on the modern view the determining particles (the genes) are not thus separated; as a result of the delicately adjusted processes of cell-division the whole complex is distributed together, much as a number of copies of a newspaper are printed, all alike, and distributed. Some sets go to the head, some to the feet, and they work according to their situations. It is like copies of one and the same popular newspaper going here into the hands of the sporting man, who reads merely the racing news, here to the speculator, who concentrates on the City column; Mr. Everyman runs his eye ineffectively over most of it and concentrates on the crime and cricket, while Mrs. Everyman (when he has done) consumes the Court and social intelligence and the fashion and household pages.

Some genes go to the testes and ovaries, and it is from these sets that the future germ-cells are derived. On this modern view there is evidently no direct continuity of protoplasm between the body of a child and that of his parents—except for the reproductive organs of the latter. It has been neatly said that most of the body of a father is not really paternal to his children, but is their uncle; one has to go back nearly to the previous generation to trace direct living continuity between a father's nose and his son's nose.

This point of view was stressed by Weismann (1834–1914). We have already noted his distinction between germ-plasm and soma (Chapter 2, § 9). When the chromosome mechanism was discovered he pointed out that it made the inheritance of acquired characters difficult to believe. And he tested the point by direct experiment. If the view, which Darwin propounded, of a contribution from every part of body to reproduction, is true, then local mutilation of the body, by injuring some of the hereditary particles, will produce corresponding injury in the next generation; if the later view is correct, and the germ-plasm is separate from the beginning, mutilations of the soma will have no effect upon it. So he tried the effect of cutting off the tails of mice, with entirely negative results. Generation after generation of mutilation did not curb the tendency of the young mice to grow tails. The biologist might say with the poet: "There's a divinity that shapes our ends, rough-hew them how we will."

There is ample corroborative evidence on this point. The Jews, for example, have practised circumcision for at least four thousand years, and yet they are still born of the same structure as Gentiles. And there are plenty of examples of similar failures to influence the shape of the race. The flat-head Indians and various African tribes compress the heads of their children into the most abnormal shapes. Chinese women of the upper classes were for many centuries compelled to compress their feet to mere trotters. Many races distend their earlobes with pieces of wood until they reach to their shoulders; and the women of some African peoples enlarge their lips in the same way until they protrude like fantastic beaks. But in not one of these cases is there any trace of an inheritance of the effect, whether it lead to a stunting as with the Chinese women's feet, or to over-growth, as with distended lips and ear-lobes.

Most familiar of all such failures is the absence of any hereditary effect of language. Not only do English children have to learn their own language, but they learn it no quicker than they would learn French if brought up from the first in a French-speaking household. Here is a habit practised daily and hourly through all but the first year of life; and nothing of it, neither words nor ideas, neither grammar nor accent, has become engrained in the hereditary constitution.

By these facts the older view of the hereditary mechanism was undermined, and the probability of a complete independence of the germ-plasm was strengthened. If became extraordinarily hard to see *how* an acquired character could be inherited. Weismann proceeded to look into the evidence for the phenomenon. Is there any need for us to admit this theoretically incongruous hypothesis into the harmony of our ideas? In a word, is there any clear case of such inheritance on record?

Many supposed examples, he found, could be dismissed as the result of mere chance, or of the natural but unfortunate tendency of human nature to remember and exaggerate striking coincidences, while allowing the much more numerous cases where nothing striking occurred to fade out of mind. A man has a finger amputated; his son is born with defective fingers. A woman bears a scar on her face as result of a wound; her child has a birth-mark in "exactly" the same place. Such facts may be perfectly true (although we often find we have to discount some of the exactitude of the resemblance on account of this tendency to make the best of a good story), but they do not prove anything. Isolated cases never prove anything, because they may be due to chance and we do not know the odds. Children with defective fingers or marks on their faces are also born from perfectly normal parents. Only a systematic collection of observations, selecting nothing and omitting nothing, or a systematic set of experiments, in which the results of treatment can be compared with those of a control experiment in which no treatment is given—only these will prove anything, whether positively or negatively. They will show us what is due to chance alone, and what to the treatment. To trust to these anecdotes of isolated cases is like backing a horse odds on when his real starting price is completely unknown.

Many other cases Weismann found to be based on misapprehension. The facts were correct, but the interpretation was wrong, or at least not necessarily right. Stock-breeders, for example, tell us that when barbed-wire fencing was first introduced in the western United States there were many cases of cattle injuring themselves on the new-fangled fences, but that after a comparatively few years such accidents became extremely rare. This is often cited as a case of inherited experience; but it could equally well depend not on the young animals inheriting anything, but on their learning directly from the old, and this is on general grounds more likely.

Again, it is perfectly true that many seashore plants produce unusually fleshy leaves. So do many garden plants, such as geraniums, when they are watered with salt water. But in the shore plants the character is inherited; it appears even if they are grown miles away from the coast. May it not be that they owe their fleshiness to the cumulative effect of generations of exposure to the salty spray? Perhaps, but not necessarily. Succulence appears to be a useful quality in such situations, for the presence of salt makes it more difficult for the plant to withdraw water from the soil—the soil is, physiologically speaking, dry; and the fleshy leaves provide a reserve store of water in case the roots fail to keep up the supply. It is just as possible to suppose that mutations making for fleshiness occurred among the salt-plants' genes, were preserved because of their utility, and thus fixed in the hereditary constitution.

And so with the dark-skinned tropical races of man. Pigment is very necessary where the sun's rays are intense to protect the tissues from over-dosage with ultra-violet rays. There are some very fair people who never tan; they burn and peel, and suffer a great deal in low latitudes. In the tropics this protection is so essential that those who happened to have a hereditary constitution making for dark skin survived better than those of lighter colour. It is so essential that any further mutations towards dark skin were seized upon by natural selection. This is in itself just as good an explanation as that the actual tanning effect is cumulatively handed on by heredity. Only further evidence can decide between them: and as a matter of fact the constancy of negroes' skin-colour in such high latitudes as that of New York or Chicago is sufficient proof that full sunburning is not necessary to keep the colour deep.

Similarly with the pointer dogs. The instinct of pausing a moment at the scent of game is found in the wolf and is not uncommonly seen in many breeds of dogs, some individuals showing it more markedly than others. By breeding in each generation from those animals in which the trait was best developed, man could as easily have built up the breed of pointers by selection of suitable variations as he did the breed of bulldogs or pouter pigeons.

An idea first developed in detail by Professor Mark Baldwin must not be overlooked. He pointed out that individual modifications,

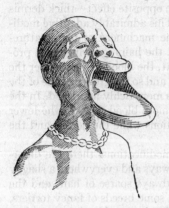 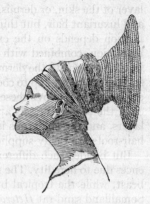

Figs. 198–199. Distortion practised by African tribes: the women of the Sarajingi tribe in the Lake Chad province distend their lips with wooden plugs; the women of the Mangbetu (north-eastern Belgian Congo) compress their heads into an elegant pointed shape. Though these practices have gone on for generations, the children are born normal.

(*From "La Croisière Noire," by G. M. Haardt and L. Audouin-Dubreuil. André Citroën, Paris.*)

though not inherited, might co-operate with inherited mutations in helping the course of evolution. Suppose that the conditions of an organism's life are changed, whether through a change in outer conditions such as climate, or, in animals, to a mutation in instinct, forcing the creature to new modes of living. The new conditions modify the structure of the organism in each generation, so that it manages to survive—a sheep, for example, taken to a colder climate grows longer wool. But if a mutation crops up which alters structure in the same direction, this will be selected—the sheep might mutate so as to produce still longer hair under the same degree of cold, and this would adapt it still better to the new conditions. Modifications will often enable the species to hang on, so to speak, until the right mutation turns up.

Let us consider in more detail what happens in the case of heavily furred animals. There is a direct individual relationship between temperature and thickness of coat. It is no good establishing fox-farms in too mild a climate; the fur is short and sparse, the pelts of little value. Now exposure of most mammals to warm conditions, while it brings about an increase in thickness of the outer skin or epidermis, diminishes the length of the hairs and also the thickness of the lower layer of the skin, or dermis. Cold has the opposite effect—thick dermis and luxuriant hair, but thin epidermis. This admirably adaptive modification depends on the existence of the machinery of temperature-regulation, combined with the fact that the hairs' growing roots project far down into the dermis. In the heat, the blood-vessels nearer the surface are opened up to cool the animal, and so the growing zone of the epidermis, just above these capillaries, is more richly nourished. In the cold, the surface vessels are narrowed; more blood flows in the lower layers, and the epidermis is on short rations while the dermis and the hair-roots are richly supplied.

But besides such differences due to modifications, there are differences due to heredity. The musk-ox is always and everywhere a shaggy beast, while the tropical buffaloes are always sparse of hair, and the Somaliland sand-rat *Heterocephalus*, like some breeds of fancy terriers, is all but hairless. The power of modification enables a mammal to penetrate into colder regions than it could otherwise tolerate, and there the species has to await the necessary mutations before extending its range farther into the cold.

There are many cases where the apparent inheritance of an acquired character is due to a confusion of cause with effect. For example, it is probably true (though statistics do not appear to have been collected) that the sons of blacksmiths are on the average physically stronger than the sons of, say, tailors. But is it not likely that men who inherit strong constitutions will be more inclined to become smiths than sit cross-legged and ply a needle, while tailors will be largely recruited from those whose poor physique rules out heavy labour?

A somewhat similar example is the supposed inheritance of the effects of alcoholism. In spite of numerous experiments, the matter is not yet fully settled. One thing is certain, that some species of animals can be made dead-drunk day after day and generation after generation without showing any deleterious effects whatever on health, length of life, fertility, capacity for learning, or any other measureable characteristic. This is true for mice, rats, and fowls. Indeed, in fowls, the average vigour and weight of the strain was increased by such treatment. Pearl, who made these experiments, suggests that the effect was really

VARIATION OF SPECIES

selective, the weaker eggs being prevented from developing. In guinea-pigs, on the other hand, Stockard found that the result of alcoholic treatment was, in some strains, distinctly bad, fertility being diminished and the percentage of monstrosities increased. Possibly, alcohol-

FIGS. 200-201. One of the hairiest and the least hairy of land mammals—the shaggy musk-ox of the Arctic (Courtesy of the American Museum of Natural History), and the sand-rat Heterocephalus (about five inches long), which burrows in the hot sand of African deserts.

The one has abundant hair even when brought into warm climates, the other is always almost hairless.

damaged germs are killed in fowls, but just pull through in guinea-pigs. For practical purposes what we want to know is whether man in his reactions to alcohol is more like a fowl or a guinea-pig. So far as the statistical evidence of Karl Pearson goes, he is more fowl-like—at least there is no evidence of inherited alcoholic degeneration in drunken stocks. The frequent occurrence of epilepsy, criminal tendencies, feeble-mindedness, and other undesirable traits in the same families where alcoholism is prevalent seems certainly to be not so much an effect of alcoholism as a cause. The unbalanced and the weak-minded take to drink more readily and under less provocation of outer circumstances than do men and women of better stock. The drunkards begin from a bad strain. There are two ends to this stick, and the end which it seemed simpler and more natural to grasp may well be the wrong one.

One more example of the sort of trap that awaits the incautious adventurer in this field. Man is unique among living things in possessing language and a method of transmitting the results of experience from one generation to the next. The handing on of tradition, in the widest sense of the word, from father to son may be looked on as a special method of inheritance, which in many ways stimulates the inheritance of acquired characters, for by its means a son may inherit his father's opinions, or his refinements, or his special techniques. Moreover, the human child is very plastic and subtly absorptive, so that it is often difficult to discriminate between this sort of transmission and true inheritance through the germ-plasm. The sons of a blacksmith make better blacksmiths than the sons of a tailor. Is this because of inherited muscularity or because they were brought up in a smithy?—because of their germ-inheritance or because of their tradition-inheritance? Here then is another source of confusion, one of the many muddy sources from which the idea of inheritance of acquired characters draws refreshment.

We see that indubitable proof of the inheritance of acquired characters is unobtainable from mere observation of the animals and plants and human beings that live their lives and propagate themselves around us. Always there is the possibility of error. The world is so complicated a place that other influences may always have been at work.

It is, of course, possible that experimental evidence of such a mode of inheritance could be obtained. In animals or plants under domestication the circumstances can be controlled and simplified, and the possibility of such rival influence as selection, for example, can be removed. A number of experimental researches have indeed been made, which seem at first sight to give evidence in favour of Lamarckism, but not one of them is in any sense final. Many of them, when repeated by other investigators, have failed to give the original results. Many others have given negative results from the outset. We will not trouble the reader with a catalogue of these experiments. Not one but has some possible alternative explanation, some little slip that invalidates its testimony.

So that the reader, in this disputed matter of the inheritance of acquired characters, is free to choose which side he pleases. He can believe in it if it suits his philosophy. But he has to admit that it is at present quite impossible to see how such a mode of inheritance could work. There is certainly no inescapable argument in its favour. A large majority of contemporary biologists rejects the inheritance of acquired characteristics because it is a totally unnecessary hypothesis. They do

FIG. 202. Variations, profitable and useless, and a failure to vary.

Left to right, a Dipper, a Redshank, and a Phalarope, with larger drawings of their feet. The third of these water birds has evolved webs on its toes, but the Dipper has none and the Redshank hardly any. On the extreme right is a farmyard mutation of no biological utility—a webbed hen's foot.

not find it explains anything that variations due to mutation, whose occurrence is known and proven, cannot explain equally well.

This is perhaps best for humanity. It would doubtless be very desirable if the benefits of healthy upbringing, good education and mellow experience could be entailed upon our descendants. But if they could be entailed, so could the effects of disease and unhealthy homes, the cramping effects of bad education and the fruits of vice and laziness and degenerate living. And it must be confessed that over the greater part of history, the bad has outweighed the good in the conditions under which most human beings have had to live. Disease, drink, excessive toil, slums, semi-starvation alternating with gross excess—the machinery of our germ-plasm is mercifully such that no taint rests upon us through the inheritance of these degradations.

§ 5. Artificially Induced Mutation.

IT SHOULD not be inferred from our refusal to accept the inheritance of acquired characteristics that we think it impossible to modify the germ-plasm from without. It is entirely possible to do that. Mutations can be produced at will. But the thing is done by going straight for the germ-plasm and tackling that, and not by worrying with the individual body which bears it.

This work on the production of mutations is a product of the last few years. Until very recently no way of deliberately altering the germ-plasm was known. It seemed that mutations were like the transformations of radio-active elements—something truly spontaneous, in the sense of being determined from within, not to be influenced in their rate of occurrence by any treatment which could be devised. But during the last few years we have at last received proof that this is not the case, and with this proof a new and vitally important field of research is opened up—the investigation of how to control the variation of animals and plants, not by the tedious and indirect methods of selection, but directly and immediately.

The most spectacular success is that of H. J. Muller, of the University of Texas, who chose the fruit-fly for his experiments, because its genetic constitution is already so well worked out. Briefly, his method was to bombard the fly with X-rays. He employed pretty heavy doses, enough to induce a slight degree of sterility in the flies, and he found that some wave-lengths were better than others. By this means he succeeded in altering the germ-plasm directly. There was no effect on the flies actually exposed to the X-rays, but among their offspring mutations turned up, literally by the hundred. He got flies with white eyes instead of the usual red, flies with unusually small wings, flies with their tiny bristles forked instead of straight, and all sorts of other strange forms, and the alteration in the genes was permanent, for the new varieties bred true after Mendel's laws. Some of them were dominant and some recessive. Some were altogether new; some were identical with mutations that had already turned up spontaneously in untreated fruit-fly cultures. To put it crudely, the germ-plasm had been thoroughly shaken up.

This line of work is being extended to other kinds of living things, and positive results have already been obtained with maize.

An exciting speculation, based on Muller's results, is that some at least of the natural mutation which is always in progress may also be due to X-rays. For X-rays do occur in the radiance that falls on our earth. True, they are very scarce; but then the mutations of wild things are much rarer than the artificially produced mutations of Muller's flies. And the two variables roughly cancel out, so that the suggestion is entirely plausible. A disturbing idea, that life has evolved and is still evolving under the spur of those strange rays, shot casually into our world from unknown corners of the universe!

§ 6. Moths and Smoke.

BESIDES X-rays, it seems that chemical influences can produce mutations. An interesting example of this was found a few years ago by Heslop Harrison, an English biologist. It concerns the black or "melanic" varieties of moths that are occasionally found wild.

About a century ago the first recorded examples of a dark or "melanic" variety of the peppered moth *Amphidasys betularia* were caught in England, and for years afterwards the variety remained rare. Collectors were keen, so that it is unlikely that the variation first appeared much before the date of its discovery or that its rarity was only apparent. It gradually became commoner, until by the beginning of this century it was in some districts more abundant than the normal type. Where it was common was in industrial regions and near big cities; in the countryside it remained rare. Other species of moth in similar localities gradually followed suit, until to-day in regions like the manufacturing areas of Northumberland or Durham scores of species are usually found of dark type instead of with their original white or brown or grey ground-colour. Sometimes the transformation has been amazingly rapid. In North Durham, for instance, Professor Harrison records that the moth *Ypsietes trifasciata* was transformed from the normal light type to a dark one within twelve years; in 1898 no black mutants had been seen, while by 1910 the original light type was hardly to be found. Rather later than in Britain, melanic moths began to appear in other countries, but always in areas where industrialism had reached a certain pitch. They are now known from Germany, France, Belgium, and the United States. Moreover, wherever the breeding behaviour has been tested, the two forms have been found to differ in one Mendelian gene only, dark being almost always dominant to light. No help in escaping the notice of enemies seems to be conferred upon these moths by their dark colour; it is not a protective variation.

Now all green things in industrial districts are coated with a grime that is rich in poisonous metallic salts. It occurred to Harrison, struck by the coincidence between the distribution of black moths and of industrial smoke, that this might be the cause of the change. Accordingly he made the caterpillars of various moths eat tiny quantities of heavy metals, especially lead and manganese, with their food. His suspicion was justified; in the metal-fed cultures a few mutants with black wings appeared. Moreover, the colour, once it had been produced, bred true even without further metal-feeding. The chemical agencies had induced permanent changes in the germ-plasm.

Here, then, we have proof that environmental factors, X-rays or chemical substances, can actually induce mutation in animals. This must be distinguished very clearly from the Lamarckian method of inheritance that we discussed in the last section. The Lamarckian method was directly *adaptive;* the individual responds in a favourable way to some factor in its surroundings and the response is supposed to be inherited. But in these actually proven methods of modification there is no direct adaptive response. The experimental agents simply knock the germ-plasm about, so to speak. They produce random mutations, not adaptive responses.

And, indeed, mutation seems in general to be a perfectly random process. There is no evidence of adaptive striving or set purpose in the way new varieties turn up. The race varies in all directions at random, in a manner as casual and apparently purposeless as the desultory X-ray bombardment that may be the variation's cause. Out of these random, purposeless gropings of the race Evolution builds. But how? To that question we shall very shortly turn.

§ 7. The Significance of Sex.

But here we can at last tackle a problem that presented itself in the first chapter of this Book and has been running like a counter subject of a fugue beneath our argument. What is the advantage of sex? We saw that reproduction is primarily a sexless thing; that living creatures can proliferate and multiply actively without sex. The origin and purpose of conjugation (of which sex is a special case) are still mysterious.

Sex intrudes, an essentially anti-reproductive process, and forces itself on the life-cycle; it entangles itself more and more closely with reproduction until in ourselves the two are inseparable. In the single-celled protozoa two individuals sometimes come together and blend into one. But this conjugation is not essential in these forms. It happens occasionally, but it appears from the results of experiment that strains can be kept going indefinitely, proliferating sexlessly without conjugation, if they are carefully tended. In ourselves this blending of the substance of two individuals is made compulsory; we cannot reproduce without it. This is the rule with the higher animals. In the higher plants also sex is closely entangled with reproduction. But even in these cases there are exceptions—there are parthenogenetic insects, and there are flowers which normally and regularly fertilize themselves. To summarize the situation, sexual union seems to be an eminently desirable thing from the point of view of the species, and yet not absolutely necessary; at a pinch, in exceptional cases, it is dispensed with. Now why should all this be?

We are beginning to see in this chapter how living things evolve. They are not adaptively moulded by their surroundings. They produce mutations at random; their germ-plasm gropes about in the dark and makes experiments, trying now this innovation and now that. And the formative agent, which acts upon these chance mutations and builds out of them the progressive changes of evolution, is, as we shall shortly see, Natural Selection. Natural Selection is like a sifting machine and the mutations are the raw proposals that come to it for consideration, rejection, or justification. Undesirable variations are sifted out and thrown aside; successful ones get through and continue in the germ-plasm of the race.

It is clear enough that a race will stand a better chance of evolving and adapting itself if it supplies plenty of variations to the mill of selection. The more numerous and diverse its varieties the better it will get on. So it pays a species to exert itself and keep the mill well-fed.

Now for the secret. Imagine two allied species of plant, each producing mutations at about the same rate. One normally fertilizes itself, like Johanssen's beans (Chapter 3, § 1) so that it tends to sort out into pure lines. The other is regularly cross-fertilized. It is obvious enough that the mutations will be shuffled about and recombined in the second species. Suppose that two good mutations appear in different individual plants. If they belong to the first species they will always stay apart. If they belong to the second they will stand a chance of being combined together in one of the next generations, and of meeting any other mutations that may be about. If, for instance, in a given space of time each species has thrown up ten mutations, then the self-fertilized species will be in possession just of these ten new types and no more, while its cross-fertilized rival will have the possibility of over a thousand new types built out of the recombinations of the ten.

In a word, sexual union is good in plants and animals alike, because it affords a method of variation. By shuffling about the mutations that appear, by combining them in various ways and presenting the results to Natural Selection, it lends efficiency and speed to the evolutionary process.

Sexual union is essentially the pooling of the mutation-experience of two lines of descent. A useful mutation is a precious event, not too frequent; the sex-machinery keeps up a continual mixing and interweaving of germ-plasm strands so that these treasures can be preserved and combined to the best possible advantage.

Such seems to be the clue to our riddle. Sex is forced upon organisms because they reproduce so fast, because of that Malthusian population-pressure that compels them, as we shall shortly learn, to compete and seize every advantage. Moreover, it is precisely because mutation is a casual random process that the continual juggling machinery of sex is so important. It allows the race to get the most out of its mutations. That is why conjugation interrupts the normal proliferation of protozoa every now and then. That is why in large and elaborately organized individuals like ourselves, which reproduce themselves comparatively infrequently, sexual union is entangled with every reproductive act.

§ 8. Variation in Plants.

So FAR we have dealt only with so-called gene-mutations, i.e., changes involving a single gene. But there are other ways in which the germ-plasm can be modified. Mutation is apparently something of an accident, the result of diet or the impact of an X-ray, and the accident can affect more than a single gene.

Often if the chromosomes of a number of closely related plants are counted they turn out to be simple multiples of each other. Thus, different kinds of wheat may have fourteen, twenty-eight, or forty-two chromosomes in their body-cells—all multiples of seven. Then roses and brambles may have fourteen, twenty-one, twenty-eight, thirty-five, forty-two, or fifty-six. Bananas may have sixteen, twenty-four, thirty-two, or forty-eight. And similarly with a great number of cultivated and wild plants. This phenomenon is called "polyploidy," the presence of multiple chromosome-sets, and is apparently of widespread importance in the vegetable world.

We can presume that these races (which generally show evident differences from each other) arose by sudden doubling, trebling, and so on, of the chromosomes of the member of the series which has least. This can be imagined as happening in various ways. For example, a dividing cell might go wrong and fail to pull itself into halves after the chromosomes had been split. But we have no space to go into the details of polyploidy here.

The fascinating thing about these phenomena is that sometimes they seem to show us species evolving before our eyes. In the garden flower *Primula sinensis* a race suddenly sprang into being (during the last hundred years) with double the normal number of chromosomes. It differs in flower-colour, stem-colour, leaf-shape and growth-form from the parent variety, and because its chromosome-outfit is so unlike that of its parent form they are almost completely sterile together. About one in a thousand attempts to cross the two succeed. Such hybrids as are obtained are intermediate in type between the two parent plants, and are even more sterile. Surely, by all the accepted definitions of species, a new species has sprung into being?

Moreover, in this sense one can actually make species. If a tomato plant is repeatedly cut down always at about the same place, its cells begin to go wrong, and often doubling of the chromosomes appears. If this occurs, then a new kind of stem may grow out and produce leaves and flowers, the whole being quite different in appearance from the original plant from which it is growing. And the new variety is sterile with the old.

We must note that this sort of thing does not happen in animals, except in very rare exceptional cases. Plants have more varied methods of evolutionary progress than animals, corresponding with their more various methods of reproduction. But these strange polyploids share with gene-mutations their strikingly accidental nature. Both turn up, apparently by chance. Both are sporadic happenings, and yet they provide the raw material for evolutionary progress. To the way in which steady, orderly improvement of the race results from these random variations we must now turn our attention.

VIII

SELECTION IN EVOLUTION

§ 1. *The Vindication of Darwinism.* § 2. *Natural Selection as a Conservative Force.* § 3. *Natural Selection Under Changing Conditions.* § 4. *Selection of Characters Useless to the Species.* § 5. *Isolation as a Species-maker.* § 6. *Crossing May Produce New Species.* § 7. *Failures to Vary and Extinction.*

§ 1. The Vindication of Darwinism.

IT MUST be clear already to the reader, after these seven chapters upon individual development and the variation of the germ-plasm, that, as we foreshadowed in our opening section, the broad propositions of Darwinism re-emerge from a scrutiny of the most exacting sort essentially unchanged. Charles Darwin propounded his view—which he applied at first only to animals and plants and extended later to man—that in general the evolution of species was due to the Natural Selection of Variations. What has three-quarters of a century of subsequent criticism done to modify that view? We have now scrutinized all the main lines of research during that period, upon the development of individual characters and the details of heredity, and our answer is, "Practically nothing." Instead of "Variations" simply, we may prefer to write "Mutations" or "Variations of the Germ-plasm." We have already dealt with the questions of convulsive mutation and the improbability (but not the impossibility) of the Lamarckian factor playing any large part in the process. We will deal with the Elanvital group of ideas, which does not so much combat the fact of Natural Selection as ignore or subordinate it, in a subsequent chapter.

Let us be perfectly clear, even at the risk of repeating one or two things already said, what this phrase "the struggle for existence" means. The essential fact on which it lays stress is that the power of living things to multiply is so great that every living species is constantly tending to press upon its means of subsistence. The daily life of man and mouse alike, as we pointed out in our opening Book, is primarily a food-hunt. In that food-hunt, the less capable, the less well-equipped hunters are pushed to the wall. They are pushed out of the game, they fail and die, and their sort dies with them.

Every animal and every plant produces offspring in such numbers that many must die if the numbers of the species are not to increase in every generation. The elephant is the slowest breeder among animals; Darwin estimated that an average pair produces only six young in ninety years of reproductive life. But even so, if all the young survived, a single original pair would in five hundred years become fifteen million. Most animals and plants produce their eggs and seeds by the hundreds or thousands, and yet their average numbers remain steady throughout the generations. This can only happen if, out of all the offspring produced by each couple, on the average only two survive to reproduce their kind again. All the rest must die—it is a question of the simplest arithmetic.

The numbers which must normally die are vividly brought home to us when we see a species temporarily released from pressing on its means of subsistence, by being imported into a new country where it happens to thrive. In a few years it fills the whole area. The Canadian water-weed was accidentally introduced into England in the middle of the last century. Within ten years it was clogging the waterways and holding up canal traffic all over the country. The English sparrow is to-day as common and as much of a pest in America as it is in England, but it filled the United States to saturation point in the geologically negligible time of less than a century.

The struggle may act at any period of life; it affects every department of existence. There may be a struggle within the womb; in most mammals more eggs are fertilized than there can be young born. Under every forest tree there is a struggle among the seedlings—a struggle for air, for light, for nourishment from the soil. There is a struggle between closely related species: the brown rat when it invaded Europe almost exterminated the black rat. There is a neverending struggle between eater and eaten: among the eaten to escape, among the eaters to secure enough prey. There is a struggle among plants to escape being browsed out of existence, a struggle among seeds for wide dissemination. There is a struggle against the forces of nature: of the migrating bird to survive the gale, of the flower not to go under in the rain-squall, of the reindeer not to be frozen in the arctic winter. There is a constant struggle against disease and against parasites. There is a struggle for mates and breeding-places in animals, a struggle for cross-fertilization in flowering plants.

The struggle for existence, as Darwin was careful to point out, is in a sense a metaphorical struggle. It is rarely a conscious effort; there is automatic competition, and some competitors are automatically crowded out. The fast-growing embryo in the womb does not know that it is causing the death of its less-favoured brother-embryo. The black rat was not engaged in conscious warfare with the brown rat, but it simply throve less well and multiplied less rapidly. The bird knows that it is being beaten down by the wind or caught by the hawk; but the true struggle in Darwin's sense lies between it and its stronger or swifter congeners that weather the storm and escape their enemy, and of this the bird knows nothing. But if the struggle is metaphorical in this sense, its results are real enough. The better-equipped survive, the worse-equipped die: that is no metaphor.

Without variation, however, the struggle would bear no fruit for life. If all individuals in a species were exactly alike, then it must be a mere question of luck which failed, and the struggle for existence could not alter the characteristics of the species. Or if they differed but their differences were not inherited, it would not matter from the evolutionary point of view which went under. But since many differences of an advantageous or disadvantageous sort exist and are inherited, the struggle for existence acts on a species like a filter or a sieve. It selects types of success and failure, sets a premium on advantageous variations and continually removes a large majority of the disadvantageous ones, so that the average of the species moves in the advantageous direction.

To picture this selection, this combined effect of struggle and variation, at its work, let us go back in thought to the fir-trees in a forest, each letting fall its thousands of winged seeds. Many fall under their parent; many others just outside its shelter; others are carried farther. Everywhere in the forest and for some distance beyond its boundaries the ground is strewn with seeds every season. The struggle begins at once. There is much purely accidental destruction; as in the parable, some fall on stony ground and cannot germinate, some where they are choked by other plants. But there is a selective struggle, too. An old tree has fallen; there is a vacant spot in the forest. This, too, is covered with seeds. But some germinate faster or better, and so gain a start. Among the horde of seedlings some send leaves up or push roots down more slowly; they will invariably lag and be overshadowed and killed out by their brothers. Some will have less green chlorophyll; some will have leaves less well arranged to catch the light. Some will be particularly efficient at drawing water and salts out of the soil, others at turning the raw materials into new substances for growth. The net result is that some grow faster than others. "From him that hath not shall be taken away," and those that are backward become still more handicapped as their competitors overtop them; the disproportion in growth increases and the innate vigour of the new finally involves the death of the less well-equipped many. There is struggle, there is variation—and so there is selection.

We may perhaps quote Darwin himself to show how he envisaged the struggle and its results in one of the higher animals. "Let us take the case of a wolf, which preys on various animals, securing some by craft, some by strength, and some by fleetness; and let us suppose that the fleetest prey—a deer, for instance—had from any change in the country increased in numbers, or that other prey had decreased in numbers, during that season of the year when the wolf was hardest pressed for food. Under such circumstances the swiftest and slimmest wolves would have the best chance of surviving and so be preserved or selected—provided always that they retained strength to master their prey at this or some other period of the year, when they were compelled to prey on other animals. I can see no more reason to doubt that this would be the result than that man should be able to improve the fleetness of his greyhounds by careful and methodical selection, or by that kind of unconscious selection which follows from each man trying to keep the best dogs without any thought of modifying the breed."

It was this sifting of variations—the automatic preservation of the favourable and elimination of the unfavourable—which Darwin meant by Natural Selection. It would be quite inoperative without inherited variations. But given that, it explains the process of Evolution with a completeness approached by no other explanation.

Opponents of Darwinism make a great difficulty of the correlation of variations, the way in which a change here must correspond to a change there, if a new proportion of feature is to be established. If a stag is to carry twenty pounds weight of antlers on his head, not only must the skull have a different construction from that of a hind, but the tendons which hold up the neck must be stronger, the muscles more powerful, the blood-supply richer, and so on. How is it to be imagined, they argue, that all these variations came concurrently into being? For the antlers would be impossible without the rest.

Our study of development has enabled us to see that the difficulty is only apparent. The tissues of the body are so responsive to the demands made upon them that all the needed adjustments are simultaneously and automatically called out during individual development by the change in weight. The adaptations are made to build themselves anew in each generation; they are not fixed by heredity, and so mutation and selection are never called upon to help produce them. A vast amount of the detailed adjustment of the body is of this sort, dependent not on racial adaptation but on the functional adaptations of the individual. And the existence of all this functional adaptation means that there is so much less for mutation and natural selection to do (Chapter 5, § 3).

Mutation and selection have less to do than was originally imagined, because of the way growth works. We saw how the male fiddler-crab's claw enlarges more rapidly than the rest of his anatomy, and how this is due to the presence of what we may call a growth-centre near the tip of the claw, a region where growth is most intense, and from which it fades gradually down to the normal as we pass down the claw toward the body (Fig. 188). Wherever a part is to grow smaller or larger than usual, it seems that a whole region is affected in this way, with growth grading down in this orderly way from a local growth-centre. Thus to enlarge the proportionate claw-size of the race of fiddler-crabs it is not necessary that mutations should crop up to affect each of its joints separately. One mutation affecting the intensity of growth in its growth-centre is all that is wanted. Again, as D'Arcy Thompson has pointed out in his *Growth and Form*, to make a sun-fish out of a more normal-shaped relative like Diodon, it is not in the least necessary that each part of the body should be separately moulded by natural selection. The development of one very active growth-centre near the hind-end of the body will automatically bring about the bulk of the changes, and selection need only polish, so to speak, and modify detail. Thus a mere change in the amount of growth in one region or in one direction can wholly transform an animal. The power of such differences to alter the look of an organism can be personally studied by looking at one's reflection in a set of distorting mirrors. Extreme distortion turns one into another sort of creature. But mild convexities and concavities only alter the type. It is surprising to see an oval-faced man transformed by a broadening mirror into a passable likeness of Mr. Winston Churchill, while a lengthening one makes him look quite like Lord Birkenhead.

Here we propose to study the process of selection itself in its chief modes and varieties of action. Natural selection, as we have said, is a fact in the process of evolution and not a theory. It is not in itself the *cause* of anything whatever. Professor Fairfield Osborn in a recent enumeration of the factors in the evolutionary process set it aside from the others as "non-energetic," an excellent distinction. The others, the various causes of variation, were "energetic," but natural selection was simply a passive stop or release of what the others had produced. And that is the essential conception of its action. It is a filter; it is a sieve; it is a balance to reject or accept. Either a variation in the germ-plasm qualifies for success or survival, or it disqualifies. The disqualification of a variation may be absolute—as when two recessive lethal genes get together to kill a yellow mouse in the uterus—or relative, when there is merely a disadvantage that enfeebles the new type, puts it at a disadvantage, or prevents it breeding as freely as the species in general. The advantage of a variation, on the other hand, is always relative. It gives the individual an ampler life and fuller opportunities of distributing the new advantageous gene.

We lay this much stress on this fact that selection *is* selection and not production, not because we lack any respect for our readers' intelligence, but because there exists a voluminous foolish literature of controversy in which Darwin is alleged to have taught that Natural Selection, in Heaven knows what inconceivable way, *produced* variations. Thereupon he is trounced, disposed of, burnt in controversial effigy and generally made an end to. Some victim of such mephitic controversy may chance to breathe the purer air of this work, and for him it is that we underline the obvious and explain to him that A means A, and not X, Y, Z. All about us and at every moment Natural Selection is going on. Mr. Everyman slaps at a fly, and it moves or does not move quickly enough to avoid him. He catches a glimpse of Mr. Everymouse, who was not smart enough to keep out of his way and he decides to put down a trap. He strokes his cat, which is neither clever nor patient enough to get Mr. Everymouse, and goes out into the street, so preoccupied with the painful question whether he will keep his cat or send it to the chemist, that he is nearly knocked down by an automobile he has failed to observe. He stumbles back clumsily, bruises his hand against a wall, and is invaded by some myriads of bacteria of a devitalizing and dangerous type. Will his white corpuscles do their duty?

All living individuals are being thus tested in every moment of their lives by little trials that may or may not put them aside among the "rejects" of life. With man, mouse, cat, or bacterium goes so much distinctive germ-plasm, to pull through or to be wiped off forever from the possibility of further development, according to their individual reactions.

§ 2. Natural Selection as a Conservative Force.

WE ARE apt to speak and write of the factors of the evolutionary process as though they were driving us on to incessant fresh developments, new things and strange things, but that is by no means always the case. The action of Natural Selection is probably on the whole conservative, except during periods of marked change in the meteorological or biological environment. It has no bias for wild-eyed novelty. It is just as effective in keeping things in their places.

We have compared it to a filter. But it is, we may say, a directive filter. And the direction in which variation is guided by selection is determined by the environment. While that remains stable, selection will be a stabilizing force, a conservative influence. If a species is well adapted to its rôle in life, Natural Selection will be busy pruning the variants that depart too far in any direction from the temporary ideal. But the environment may change; and it may offer inducements to responsive change. This revolutionizes the selective policy: and Natural Selection in such a changing environment becomes a radical influence in the politics of life. It is now all for new ideas.

A good instance of Natural Selection as a conservative force is supplied by the common sparrow. The English sparrow, as proved by its rapid spread over the world, is a species excellently adapted for ordinary all-round activities. After a severe storm in the United States a number of sparrows were found in distress, beaten down to the ground. They were taken indoors and tended; some revived, some died. Measurements showed that those which died comprised fewer specimens with wings of average length and more with wings unusually long or unusually short. Selection by storms was evidently preserving the central type.

We have further the remarkable fact that no measurable differences, whether in size, proportion, or plumage, can be found between the sparrows of Britain and those of various localities of the United States, where the bird is an alien intruder, and this although many of the environments which it has colonized (such as parts of the American desert) are different from anything to be found in its original home. This is indirect evidence of selection keeping a well-balanced type stable.

To jump from birds to molluscs, snails preserve a record of their growth in the inner whorls of their shell. When measurements of proportions were made on the inner whorls of a number of adult land-snails, and compared with the proportions of young snails of size corresponding to their inner whorls, it was found that the young snails had a greater range of variation. The old snails represented the shots on the central part of the target; the fringe of scattering shots had been eliminated. Natural Selection was refusing the novelties.

And just as Selection the Radical seems capable of guiding variation along paths of far-reaching change, as in the stock of horse or elephant, so it appears that Selection the Tory may keep organisms in a state of evolutionary immobility over awe-inspiring periods of time. The classic example is Lingula, a mud-burrowing lamp-shell. Examples of this identical genus, differing only in trivial details from living species, flourished in the Cambrian (III A) seas, five hundred million years ago. We do not know the length of a generation in Lingula: ten years is certainly well over the mark, but even this would give us fifty million generations of stability—fifty million generations during which all essential changes in the constitutions and habits of the creature have been prevented. It is, of course, possible that the absence of change is here due to the failure of variations to occur rather than to the success of Natural Selection in keeping them pruned down. We can only say that mutations have been found in every organism in which they have been carefully looked for, and that their total absence in this or any other case is unlikely.

The pruning effect of selection is also exerted in another way. Whether change or stability is being encouraged, the organism must be kept up to the mark. Wherever mutations have been studied, many of them are found to be deleterious. Mutations are random changes, and random changes in such complex machinery as that of life will often inevitably be changes for the worse. Natural Selection will always be occupied in ousting these from the germ-plasm of the species. Drosophila keeps on throwing mutations with striking effects; often the same one is repeated again and again, yet they are scarcely ever found in Nature—the reduced vigour which they entail leads to their automatic elimination. They fail at the Natural Selection entrance tests.

Per contra, the absence of selection will allow types to persist and spread which normally are kept down to a minimum. Black rabbits (whose black colour is due to a single recessive mutation) are not uncommon as a natural "sport." In most places but few are found. They arrest the gunner's interest, and are easier marks; as a result their numbers are kept down to a low level. But in some parks in England where no shooting is allowed, black rabbits are unusually numerous. Since the larger birds and beasts of prey have been exterminated in England, the gun is the main agent of selection; and this is here absent. Again, the imported rabbits which were turned out or escaped in New Zealand were of all possible coat-colours; and these survive in the large wild warrens which they have established, giving them a strange appearance. New Zealand possesses no land mammals to prey upon rabbits, and very few enemies of any other type. Rabbits are killed off as vermin by poison and other wholesale methods. It is safe to prophesy that if such a parti-coloured population were turned out in Europe, the percentage of greys in it would rapidly increase; it is the absence of any selective need for a coat of invisibility which allows the other types to persist in New Zealand.

On the Galapagos islands the traveller, so Beebe assures us, can tell without any trouble which birds are resident and which are migrants; the migrants are all as shy as are the run of birds in inhabited countries, while the residents have no fear of man. If the migrants had no fear of man, they would (alas for human nature!) have been exterminated in the inhabited countries where they pass the rest of their life; but on the uninhabited Galapagos there has been no such selection. Thus favourable and unfavourable are often relative terms. What is favourable enough on a desert island would not pass muster elsewhere.

The preservation of "unfavourable" varieties is seen most obviously in domestic animals and plants. It tickles our fancy to conserve hairless lapdogs and pouter pigeons, albino rabbits, and double-tailed goldfish; it pays us to breed impossibly large bulls and pigs that can scarcely waddle; it delights us to have double flowers, even though they be sterile. We alter the incidence of selection, we pet what the natural process destroys, and these types which could not for a generation hold their own in nature, are made to abound and multiply.

As an illuminating example of the way in which mutation and selection co-operate we may consider water-birds' feet (Fig. 202). The Dipper is a water-bird, but with no special adaptations of structure to aquatic life. It is kept a water-bird by its instincts. However, if a mutation for webbing or lobing of the toes were to appear in the Dipper stock, it is difficult to believe that it would not be fixed there by selection. This indeed seems to have happened among the wading birds (*Limicolæ*). Most of these can swim, though without a trace of webbed feet; and this faculty must often serve them well in getting across deep channels or securing a tit-bit otherwise out of reach. The Phalaropes are birds of this sub-order, but in them broad lobes are developed on the toes, making it possible for them to swim much more efficiently, and indeed to pass whole months on the open ocean. In them, the needed mutations have cropped up. Incipient webbing has also repeatedly appeared as a mutation in domestic fowls and pigeons. But as they do not normally swim, it is of no use to them; whereas as waders were already in the habit of swimming, webbing or lobing could be at once seized upon by selection.

The germ-plasm is like a garden, and Natural Selection in many respects like its gardener. Weeds are always cropping up in it, and threatening to swamp the cultivated plants. Selection, as well as sometimes helping in the creation of new types of flowers or fruit, is forever busy with the humbler task of weeding. But there are gardeners and gardeners; some have not the time or energy for weeding of a professional standard; others may even prefer an untidy garden. Thus, dropping our metaphor, sheltered conditions often allow variations to persist which more rigorous selection would eliminate. It is commonly believed that this applies particularly to human heredity. Modern civilization is said to be lightening the severity of the selective process upon our race. It is also commonly believed by the same people—but usually at different hours of the day—that modern civilization is more exacting upon nerves and health than any previous state of society.

§ 3. Natural Selection Under Changing Conditions.

But now passing from the consideration of natural selection as a species-conserving and species-regulating influence, let us look at it in operation as a fosterer of variations and so as adapting species to new conditions. Here is a case we quote from J. B. S. Haldane's admirable *Possible Worlds*.

"The assertion is still sometimes made that no one has ever seen Natural Selection at work. It is therefore perhaps worth giving in some detail a case recently described by Harrison. About 1800 a large wood in the Cleveland district of Yorkshire containing pine and birch was divided into two by a stretch of heath. In 1885 the pines in one division were replaced by birches, while in the other the birches were almost entirely ousted by pines. In consequence the moth *Oporabia autumnata*, which inhabits both woods, has been placed in two different environments. In both woods a light and a dark variety occur, but in the pine wood over ninety-six per cent are dark, in the birch wood only fifteen per cent. This is not due to the direct effect of the environment, for the dark pine-wood race became no lighter after feeding the caterpillars on birch-trees in captivity for three generations, nor can the light form be darkened by placing this variety on pines. The reason for the difference was discovered on collecting the wings of moths found lying about in the pine wood, whose owners had been eaten by owls, bats, and night-jars. Although there were more than twenty-five dark living moths to each light one, a majority of the wings found were light-coloured. The whiter moths, which show up against the dark pines, are being exterminated, and in a few more years Natural Selection will have done its work and the pine wood will be inhabited entirely by dark-coloured insects."

There is a simple and pretty instance of the rôle of Natural Selection in bringing about adaptation. Another, equally simple, is known from agricultural practice. Various strains of cereals imported into Scandinavia, in the course of a few years changed their flowering and seeding period in adaptation to the shorter summers of their new home. This was for long supposed to be due to an inheritance of acquired characters. But careful experiment at the famous research station of Svalot showed that it was a simple effect of selection. As wheat is normally self-fertilizing, a wheat-field consists of a mixture of pure lines. The imported seeds belonged to a number of pure lines, differing in the rapidity of their maturing. The rigorous selection of their new northern environment weeded out all save the most rapidly maturing, and so the average of the stock was changed. There is here no actual record of new mutation in the same direction; but any that occurred would have been seized upon by selection and fixed.

Of course, these are small changes, but then, if we are to produce larger ones we must either alter the time-dimension of our experiments and observations or do something to accelerate the process. We cannot, unless we go back to Methuselah and a long way beyond, produce any current instance to set beside the long-continued, steadfast evolution of the horse. But evidently that evolution went on by steps, each individually as slight as these two. Nature is leisurely and works with a vast profusion of material. We can, however, work faster than she can, by cutting out all the futile trials and vain repetitions in which she indulges. Let us, for instance, put the biological experimentalist, Professor Castle of Harvard, in the place of Nature, and Castle's selection in the place of the loose, wide, hit-and-miss of Natural Selection, and let us see what can happen to certain rats.

Castle worked with piebald "hooded" rats. The standard type of this breed has black head and fore-quarters, with a black stripe along the back; the rest of the coat is white. This pattern is fairly variable, and Castle used it as material to study what effect could be produced by stringent selection for more or for less black. In one lot he bred only from those with most black; in another, only from those with most white. The results were striking: selection achieved types far beyond the range of variation normally found in the piebald strain. Selection for more black drew the "hood" along the back and spread the black line along the spine right down on either side. Meanwhile selection for more white in other rats from the same original stock had reduced the hood to a mere smudge on the nose and ears, and removed every trace of black from the body. (Fig. 203.)

At first sight these results appear to contradict Johanssen's pure line results on beans, and to imply (as Weismann and the neo-Darwinians postulated) that selection could stimulate new variations to come into being; and so, indeed, they were at first interpreted. But while Johanssen's bean-strains, being naturally self-fertilized, were pure lines from the start, the hooded rats, being promiscuous breeders, were not. The germ-plasms of the beans were like so many samples of chemically pure substances; those of the rats like a number of chemical substances being constantly stirred and mixed together.

The results were, as a matter of fact, mainly due to new recombinations of already existing factors. The hooded pattern, whether exaggerated or reduced, always behaves as a simple recessive to normal; but the degree of hoodedness, or in other words, the amount of black, was affected by quite a number of what we may call modifying genes. These by themselves cannot produce a piebald pattern; but once the hooded gene is at work, they can modify the extent of its influence. Each of them singly has only a very small effect; in the medium-hooded rat there are many present, some making for the extension of black and others for its restriction; in the very black hooded rat a large number of extenders of the black patches are present, and so on. In the ordinary mixed stock it is thus very unlikely that all the plus or all the minus modifiers will ever become assembled in one individual as the result of chance mating. Selective breeding, however, will sift the random heap; selection for more black will accumulate more and more of the plus modifiers, the black-extenders, in the germ-plasm; and vice versa. Once the combination of all the plus modifiers has been thus brought into being, a pure line has been produced and selection will have no further effect, unless the germ-plasm itself changes by mutation.

But changes in the germ-plasm can and do occur. In the early stages of such an experiment, while the stock is still very mixed, it will be impossible to distinguish the effects of a mutation from those of a better combination of previously existing modifiers, so that many mutations may pass unnoticed. But once approximately pure lines have been established, the appearance of new gene-changes can be spotted; and in this way at least two mutations which modify the hooded pattern have been identified by Castle. One of these was in the same direction as the selection that was going on; selection in this line was being made for more black, and the new mutation was a black-extender. It was incorporated by selection in the germ-plasm of this line; and thus the range of variation was still further enlarged in the black direction.

Exactly the same sort of thing happens in Nature, only on a grander, looser scale and over vastly longer periods. Castle's work teaches us two vital facts. First, selection in a cross-bred stock can bring about the recombination of existing genes, and so cause the stock to overstep its old limits of variation. Secondly, even though mutations be rare, yet selection in a given direction acts as an automatic trap for all mutations whose effects are in the same direction; thus, if it continues for a long stretch, it may accumulate plenty of these rare visitors, and so in time wholly alter the racial constitution. If mutations go on appearing, the amount of change that can be wrought is unlimited. In a word, it gives an explanation for the steady change of a race in a given direction—and that is precisely the sort of thing that the fossil record shows.

There are many who cannot bring themselves to believe that such trifling alterations, even if accumulated over the generations, can ever give rise to the broad and striking changes of large-scale evolution. They forget the extreme slowness of the change revealed whenever we trace Evolution in action. Only because palæontologists are thinking on a different scale of time from ordinary mortals can they speak of bursts of rapid evolution and the like. During such periods, change may be faster than at other times; but judged by our ordinary standards it is still of an appalling slowness. The Cenozoic Period (V) from its beginning to the Pliocene (V D) was a period of remarkably "rapid" evolution among the mammals. But, all the same, it took about forty million years to make a horse out of an Eohippus; and yet the changes involved, important though they are, are changes of detail, not of essential plan, such as were needed to transform reptile to mammal, or fish to amphibian.

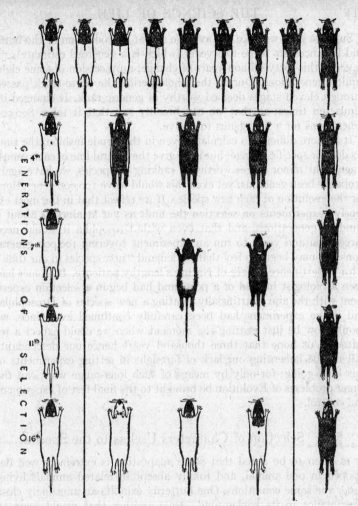

Fig. 203. How selection may gradually change a race.

The effect of selective breeding for sixteen generations, choosing always the whitest rats as parents is shown on the left. The effect of selecting for black is shown on the right. After sixteen generations the averages of the two strains are outside the range of the original strain. In the center, above, the average and extremes of the black strain; the lower surfaces of two are also drawn. The darkest has only a tiny patch of white on the throat. Below, the average and extremes of the white strain. The lightest has only small spots of black on the head. (Diagram constructed from Prof. Castle's tables.)

Suppose that we take the average age at reproduction of the horse and its ancestors to be five years (which is a generous estimate), we have, in this forty million years of change, a procession of some eight million generations. During the whole period the horse-stock passed through eleven stages deemed worthy of generic rank. It changed its genus ten times; so that we can broadly say that it takes 800,000 generations for a new genus to evolve.

It is more difficult to calculate, even in this crude fashion, the time needed for specific change; but if we give the central line of each genus-stage eight minor stages worthy of ranking as species, we have again probably been generous: yet even this would leave 100,000 generations for the evolution of each new species. If we reflect that in the most extensive experiments on selection the limit as yet attained is about a hundred generations, and that even with Drosophila it would need three thousand years to run an experiment covering 100,000 generations, we may begin to feel that to demand "new species in our time" is an impertinence in face of Nature's lengthy patience. If Homer had been a biologist instead of a poet, and had begun a selection experiment with the aim of artificially creating a new species of Drosophila, and if the experiment had been carefully continued ever since, we should now be just nearing the moment when we could expect a result. Let us hope that three thousand years hence our descendants will not be lamenting our lack of foresight in setting experiments of this kind going; for only by means of such long-range work can the larger problems of Evolution be brought to the final test of experiment and control.

§ 4. Selection of Characters Useless to the Species.

IT IS often to be noted that some adaptation is extremely well developed in one animal, and totally absent in related animals living under the same conditions. One butterfly exhibits an amazingly close resemblance to its background, while another that would seem to stand in equal need of such protection is without it altogether. One flower (like the common spotted orchid) possesses elaborate devices for securing cross-fertilization; another (like the bee-orchid) modifies the same mechanism to ensure self-fertilization. Often, too, the adaptation seems more delicate than we should imagine necessary.

From such facts it has not infrequently been argued either that apparent adaptations are really not adaptations at all, since their absence in other species shows them easily dispensed with, and therefore, presumably not at all a matter of life or death; or, on the other hand, that though they may be useful, selection plays no part in their origin, but that some inherent tendency to particular adaptations exist in some species, but is lacking in others.

But such arguments are not cogent. Take a butterfly as example. The struggle for existence is at its hardest during its caterpillar and chrysalis stages; and in any case a limit is soon set to the numbers of the species by its food-plant; however many eggs may be laid, there is subsistence only for a limited number of caterpillars. Thus variations which merely help or hinder the survival of the butterfly stage itself can have no effect on the future of the race, unless they actually prevent all, or almost all, the butterflies from reproducing. Even if two adults survive where only one survived before, and twice as many eggs are successfully laid, no more than the fixed maximum of caterpillars can find food. In the same way, a diminution in adult numbers, if not too extreme, will be compensated for by the survival of a greater proportion of the caterpillars that hatch; again, if the caterpillars are overabundant, they fall a readier prey to ichneumons and other parasites, and vice versa. The caterpillar stage acts as a buffer, taking up all minor fluctuations in the numbers of the adults. And yet it is perfectly possible for variations affecting the adult to be selected. If a variation crops up which gives its possessors greater immunity from attack by making them look more like a dangerous wasp or a nauseous butterfly, a larger proportion of these mimics than of the remainder will survive to reproduce their kind. The caterpillars will be kept down to their maximum numbers, but there will be a greater proportion of those which inherit the genes for the mimicking resemblance. Thus, when the butterfly stage of the next generation is reached, there will be a greater proportion of mimics than in the last. And so the process will go on, generation after generation, until the better mimics have wholly replaced the worse; and it will be repeated every time that there arises a new mutation making for greater resemblance to the "model." Yet, as Dr. Nicholson of Sydney has pointed out, the species as a whole will be neither worse nor better off than before. There has never been any danger of the race going under in the struggle against parasites or elements, or in the competition for food with other species; the struggle has been wholly within the species; it has concerned the survival of one variety in place of another, not that of the species as a whole.

Something very similar is to be seen in our human affairs. Shopkeepers can perfectly well make the same profits by opening for eight hours as for twelve hours—provided that all do the same. But if, when eight hours' opening is the rule, some take it into their heads to stay open for nine, the others must follow suit or lose money to their rivals. There is, of course, an alternative, which is that there should be a mutual agreement to open for a fixed time. The existence of this alternative is a reminder that comparisons of animal and human affairs, while often illuminating, must never be pressed too far. For man, with his affairs based on easily-changing tradition instead of the more slowly plastic germ-plasm, is able to cut the Gordian knot by just such conscious methods as mutual agreement, which are outside the range of animal possibility.

Another human example is seen in the results of unlimited competition in armaments. When one competitor adds a battleship, the others must do so too; the same happens with cruisers and destroyers and submarines; the armaments go piling up, and yet no nation is relatively better protected than at the beginning. Indeed, all are worse off, since all are spending more than they need. Something of this sort too may occur in animals. For example, the brilliant but cumbersome plumes of polygamous male birds like the peacock or argus pheasant must have been developed, like the mimicry we have been discussing, without any benefit accruing to the species as a whole. In some ways they must actually be disadvantageous, since they utilize a disproportionate share of the energy and material resources of their possessors, and make them so much more conspicuous and less agile that they must incur greater mortality. They confer advantage simply on one male as against another male.

If we like to coin a special term for the selection which can thus transform all the individuals of a species (or of one sex within a species) without conferring any advantage on the species as a whole, we can call it *intra-specific selection.* Darwin showed himself acutely aware of one aspect of this process produced by competition, namely, that between rival males or between rival females, and called it Sexual Selection. It has given us the vast size and strength of the bull bison and the elephant seal, the mane of the lion (apparently a defensive protection in conflict), the vast antlers of a stag, and the brilliant plumage of birds to which we have just alluded. Darwin's theory of Sexual Selection has undergone more adverse criticism than any other part of his work, but recent research has largely rehabilitated it.

There is no validity in the argument that adaptations of this sort are not really adaptations at all if other species get on quite happily without them. It is of no biological concern to the species whether they are present or no; and yet whenever the necessary variations turn up, they will be automatically preserved by selection. And so long as such variations arise, they will go on being selected; until the adaptation may become astonishingly perfect. It is characteristic of this intra-specific selection that it will neither extinguish an unadapted type in which no variations making for the adaptation have occurred, nor check the adaptation, once started, from going on to exaggerated perfection; while ordinary natural selection, in which the character selected is of value to the species as a whole, will achieve both.

It seems certain that a great deal of selection is working to produce adaptations of this curious sort. It is equally certain that in the human species similar results, equally obligatory for individuals or nations who wish to keep their heads above water, but equally useless to the community or the world as a whole, will be produced by the struggle of a purely competitive society, or of purely competitive international relations—unless man does what the animals cannot do, and, by deliberate agreement and convention, destroys the need for that vast amount of merely useless but otherwise inevitable competition.

§ 5. Isolation as a Species-maker.

NATURAL SELECTION gives a satisfactory explanation of all evolution that is advantageous. But many of the differences between closely related creatures do not appear to confer any biological advantage at all. How have they come into existence? It may be that further discovery will show that this is due to our ignorance, and that all these characters are really useful in themselves, or necessary accompaniments of some useful property. And if so, our second problem is only part of the first, and the origin of divergence is one and the same question as the origin of adaptation. But meanwhile we must treat the question on its merits.

In the first place, mere geographical separation certainly helps in the development of difference. Many wide-ranging species can be divided into geographical races or sub-species, inhabiting different areas. When there is no geographical break, one sub-species generally fades gradually into another across an intermediate zone. This is beautifully illustrated by the wrens of South America. There is an almost continuous wren-population over the continent; the wrens of different regions differ markedly in size, in colour, in pattern, in proportions, but the various types grade into each other.

Here and there, however, an isolated patch of wrens is separated from their congeners by some barrier, such as a high mountain-range, a stretch of sea, or a desert, and then we find no transitional forms linking them with other races. We can if we like call these groups species and the grading divergences sub-species; but apart from such verbal quibbling, it is clear that isolation has produced sharply marked types.

This effect of isolation in helping to bring new types into being is one of the everyday facts of systematic biology. It is most obvious where the isolation is most complete, for example in the plant and animal inhabitants of islands. Both St. Kilda and the Shetlands, for instance, have a distinct sub-species of wren. The puffins of Spitsbergen are all a little larger and have beaks a good deal bigger than those of the rest of Europe; and we know no adaptive meaning in the divergence.

In other cases, the divergence that is helped and hastened by isolation may be a useful one. We have, for instance, the case of the Florida Deer-mouse, recently investigated by Sumner. This animal is normally dark grey-brown, blending nicely with the colour of the soil. On a little island off the coast, all made of snow-white sand, exists a sub-species whose colour is nearly white—again blending beautifully with the background. On the coast opposite the island a third type is found which is intermediate in colour between the other two, and this, although it inhabits the same white sands as the island race.

Selection of light variations must have been at work both on the island and on the shore; but while the island race is isolated, there will be some intercrossing between individuals from the interior and from the shore, and this will hinder selection at its work.

In our chapter on geographical distribution we spoke of the many animals and plants peculiar to the Galapagos islands, and how they were related to those of South America, the nearest mainland. But that was only part of the story. There is the further remarkable fact that of the archipelago's land animals and resident birds most types are represented by separate species on each of the main islands. This is notably true of the extraordinary giant tortoises; of the mocking-birds; and of the various genera of peculiar Galapagos finches, found nowhere else. It is equally true of the plants. As Darwin wrote: "We have the truly wonderful fact, that in James Island, of the thirty-eight Galapageian plants found in no other part of the world, thirty are exclusively confined to this one island"; and many genera are represented by a single different species on each island.

Chance immigration from the continent, followed by new evolution on the archipelago, would clearly account for the animals and plants of the archipelago as a whole. But what about the differences between the inhabitants of the separate islands—differences which become still more remarkable when we realize that on the American continent single species of similar creatures—mocking-birds, for instance—have ranges extending over hundreds of miles? The environments provided by the various islands are very similar, so that it seems most improbable that the conspicuous differences between the island-species are adaptations to conditions of life. It would rather seem that the prevention of intercrossing between the populations of the various islands has left each stock free to develop along its own line, and that each has happened upon a slightly different one. Mutation is a matter of chance, and here one mutation to which Natural Selection had no objection arose, and there another. Life was easy and adverse forces not very stringent. This extraordinary blossoming of distinct species, each on its little island patch, so impressed Darwin that he wrote, thirty years later: "When I visited the Galapagos archipelago ... I fancied myself brought near to the very act of creation."

Equally striking are the snails of the Philippine Islands. The Philippines contain about 115,000 square miles of land, divided up among thirty large- and medium-sized islands and over three thousand small ones. In spite of the difficulty which land snails find in getting carried overseas, the Philippines possess over a thousand species of these molluscs. The nearest land area in the same latitude is that of French Indo-China and Siam. This includes close on half a million square miles, but only six hundred kinds of land-snail. The Philippines with their isolated islands must have been a veritable factory for snail-evolution. And we could multiply examples from every archipelago.

In fact, the degree of isolation of an island runs closely parallel with the number of types which are peculiar to it. Great Britain's isolation from the Continent is too slight and too recent for many of its birds to have developed into new species—the Red Grouse is its only undoubted case. On the other hand, it does enjoy some measure of isolation, and as a result, many of its birds show slight but constant differences from their Continental relatives; they have evolved into new sub-species. The long-tailed tit is a familiar instance, the bullfinch another.

The Canary Islands, on the other hand, though some of them lie only about fifty miles from the African coast, are separated from it by a channel over 5,000 feet deep, and have never formed part of the continent. And on them half-a-dozen birds have evolved into distinct species.

And for the extreme of isolation, let us take St. Helena, perhaps the most isolated spot on the globe, the most insular of all islands; we have already spoken of it in Book 3. Animals and plants have either wholly failed to reach this water-girt speck of land, or, if they have succeeded, have usually evolved and changed into something new.

Patches of water can be isolated by land as well as patches of land by water; and in the lakes of Ireland and Northern Britain there has been a remarkable divergence of fish belonging to the trout tribe, quite like that of birds on the Galapagos islands. Charr of various kinds, vendace, gwyniad, and pollan—there are a dozen species of the two genera, *Salmo* and *Coregonus*, each restricted to one or a few lakes, while those species of the family which go down to the wide sea instead of living all their life in an isolated lake range over very large areas.

Thus isolation promotes divergence. It seems indeed to promote divergence irrespective of any advantage which the divergence might bring with it. This seems to be due partly to isolation preventing new types being swamped by intercrossing with the old, partly to the release of mutational forces from urgent competitive and destructive conditions. Oddities flourish in out-of-the-way places. The primarily conservative disposition of Natural Selection spares the corners. But the advance of knowledge will probably reveal other agencies at work. It is worth noting that however great the variety which isolation brings into being, it never seems to promote steady advance or long-continued specialization. To achieve that, stringent selection seems to be necessary.

But sharply distinct types may also come into existence by another method of isolation. If a species with many local varieties inhabited a large area, and then some change of conditions killed off the inhabitants of the central belt, the two end-groups would be separate with their differences ready-made. The shallow-water octopus, *O. apollyon*, was first described from the western coast of America. On the opposite side of the Pacific, off China, another form was found, differing markedly from the first in the size and shape of the peculiar arm by means of which the male transfers sperm to the female, and also in proportions and colour. It was naturally assumed to be a distinct species, and called *Octopus hongkongensis*. Later, however, it turned out that these two types formed only the two ends of a continuous chain of forms which extended from China round the northern Pacific, across by the Aleutian islands, and down to California. The two ends of the chain are very distinct, but all kinds of intermediates occur in the northern Pacific zone. If the chain were broken, as it might be for instance by the deepening of the sea between America and Asia, there would exist two distinct types with an unbridged gap between them. And we could equally well have cited such chain-types from among other groups, such as butterflies and birds.

§ 6. Crossing May Produce New Species.

NEW types can come into being through crossing. In the most striking examples, the cross at once gives what must be called a definite new species. For instance, when the two kinds of poppies, *Papaver nudicaule* and *P. striatocarpum*, were crossed, the offspring were quite distinct from either parent, were fully fertile and bred true. It is as if the cross between horse and ass were to give us, not a sterile hybrid, but mules that were fertile and could perpetuate their kind indefinitely.

This was possible owing to their chromosome behaviour. Poppy chromosomes go in sets of seven. One of the parents had two such sets, the other ten; their gametes therefore had one and five sets respectively. Fertilization gave six sets, or forty-two chromosomes. At reduction the six sets paired up, three against three, and all the pollen-grains and eggs had three sets; thus a true-breeding new type was made with six sets or forty-two as its ordinary chromosome-number.

In other cases the cross gives a type in which chromosome-sets do not find themselves able to pair up in this tidy way. It is therefore usually sterile; but in some specimens a doubling of the whole chromosome-number occurs (usually by the division of the chromosomes without division of the cell); chromosome-pairing can then occur, and so the new type become fertile. This is known among Primulas, for instance, and in Evening Primroses; and sometimes the new type is relatively infertile with the old. Here again something like a species is generated all at one bound.

There are still other ways in which new types can be created by crossing. Two distinct species may have the same chromosome-number, and also be fertile when crossed. Mendel's two Laws will then see to it that a great number of new types appear by recombination. Professor Baur of Berlin, for instance, crossed the garden snapdragon (*Antirrhinum majus*) with a wild species, *A. molle*. The first hybrid generation was, as we should expect on Mendelian grounds, uniform. It was also fertile, and in later generations produced a host of new types by segregation. Of some ten thousand plants raised there were not two exactly alike. This is precisely what was to be expected if the differences between the two species depended on a fair number of Mendelian gene-differences, probably between fifteen and twenty. If the two differed in only ten gene-pairs, segregation could produce over 1,000 recombinations; if in fifteen genes, over 30,000; if in twenty, over 1,000,000.

Some of the types of flowers thus produced by recombination were quite different from anything seen in either parent, or indeed within the snapdragon genus, but resembled other genera of the family *Scrophulariaceæ*. We should like to know, of course, whether these types could hold their own in Nature, or might even be better adapted than their parents to special conditions; and investigations on this point are being carried out. Very similar results, as we mentioned in an earlier section, were got by Luther Burbank among his plant hybridizations.

But there are cases where crossing certainly does produce new types capable of holding their own in Nature. The most striking are found among those plant-groups which seem made to puzzle the classifiers. Roses, brambles, willows, hawkweeds—those, as any botanical amateur knows, are a terrible tangle of overlapping types. In every case free crossing between distinct species is the root-cause of the diversity and the tangle.*

Where animal species-crosses are fertile, as in various pheasants and in the natural crosses between two species of the American woodpeckers known as flickers, curious irregularities of pattern often crop up in the recombinations. Each species has worked out its own delicate balance between the different genes responsible for plumage-pattern; when the two are crossed, unbalanced combinations appear which are very easily upset by modifying factors or by environment. We may presume that the same sort of thing holds good with the genes responsible for other and more deep-seated characters. Thus in higher animals, which are far more complexly regulated bits of mechanism than any plant, after a wide cross, the majority of recombinations are likely to be a little shaky in their adjustments, and therefore of no evolutionary account.

Nevertheless, there is one animal species in which this sort of thing has happened on a minor scale—our own. The inhabitants of Northern India are almost all brown-skinned and dark-eyed, although many

---

*Not only segregation, but complications due to parthenogenesis and to difference in number of chromosome-sets may be involved; but they all result from the fact of crossing.

of them have features closely resembling those of fair-haired European people. There is every reason to believe, however, that the invaders who burst into these plains from the north in the middle of the second millennium B.C. were fair-haired and blue-eyed, and that the dark-skinned population they conquered had features not at all of European type. The fair complexion has gone, the features have been preserved. We know that intense sunlight is harmful unless prevented from penetrating by an absorbent layer of pigment. During the hundred generations or so that have elapsed since the conquest and the resultant intermarriage of the two stocks, the fair-skinned must always have been at a disadvantage compared with the dark-skinned. But the European types of features were at a premium, as being the hall-mark of the conquering race. The genes for fair skin have disappeared, but the new combination of aristocratic features with dark skin has survived.

Almost exactly the same thing appears to have happened in Greece. The Dorian irruption, somewhere about the twelfth century B.C., was of fair northern men; and yet there are scarcely any blue-eyed people in modern Greece. Some of their genes, but not others, have been selected out of existence, and so new types have been created.

To sum up, then, new types, often worthy, in plants, at least, to be called species, may spring fully-formed from a cross, like Minerva from the head of Jove, without a long process of selection behind them. But though crossing, like isolation, may sometimes increase the variety of evolving life, like isolation again, it seems incapable of producing evolutionary advance. These agents of Evolution are only subsidiary. Mutation remains the great driving force. Even if crossing between distinct types never took place, even if we could eliminate all the effects of isolation, Evolution, with Natural Selection at the wheel, would still go on so long as animals and plants continued to vary and the conditions of the world about them to change.

§ 7. Failures to Vary and Extinction.

SOME of the more ardent followers of Darwin went a good deal farther than the master. Natural Selection made it unnecessary to do as Paley did, and ascribe every adaptation to the direct agency of the Deity; but they proceeded to make a deity of Natural Selection. Weismann, for example, went so far as to maintain what he called "The Omnipotence of Natural Selection." Such exaggerations gave excellent occasion to the antagonists of Darwinism. But, as a matter of fact, few naturalists failed to realize at any time that the action of Natural Selection is a limited one. It is limited both by the quantity and the quality of the variations supplied to it.

Every case of extinction of a species is a failure to provide any variations which could have been used to meet and overcome the circumstances which were leading the species downhill. Sometimes we can see precisely what was lacking. The Great Bustard, for instance, that magnificent bird which once nested commonly enough in East Anglia, has now disappeared from almost all of North-Western Europe. The cause seems to lie in its excessive instinct of wariness. It leaves its eggs readily at the sight of man, and refuses to return to them so long as he is still in view. The result is that as soon as cultivation invades its haunts, the hen bustard is so often and so long off her eggs that they grow addled. If some mutation had but arisen to reduce this inherited cautiousness, the bird could have survived longer—perhaps long enough to reap the advantages of the modern sentiment in favour of bird-protection. As it was, the species cut its own throat by its shyness, for the right mutation never appeared.

The Great Auk, on the other hand, was doomed by its inveterate tameness. This bird had next to no fear of man and was massacred by sailors for food and oil in such numbers that it became wholly extinct during the nineteenth century. Could but a fraction of the Bustard's shyness of temperament have been granted to a few Great Auks, selection would have preserved them, and in them the species.

In many other cases we cannot put our finger so accurately on the cause of extinction. We know that when the Brown Rat, disseminated by trade, spread over Western Europe, the indigenous Black Rat quickly fell on evil days and was in many places brought to extinction on land (though it survived by taking to a life aboard ship). But we do not know what failed the Black Rat in the struggle, what variation would have saved it. Nor do we know why, of the different kinds of Zebra, Burchell's and Grevy's should have survived the opening-up of Africa, while the Quagga became extinct. Our absence of detailed knowledge does not, however, make the principle less cogent. One species or one group becomes extinct, another in the same conditions survives. The former could have escaped its doom if its germ-plasm could have produced the right variations and produced them with sufficient speed—but it did not.

This leads us at once to ask whether some species or whole groups may not be much less prone to vary than others; unfortunately, we cannot as yet answer this with any certainty. It is often asserted that highly specialized creatures, such as elephant or ant, fully evolved trilobite or dinosaur, lack the power of variation. This, however, seems to rest upon a misapprehension. Dog, horse, and pigeon are all highly specialized animals; and yet sufficient mutations have appeared in them in a short five or six thousand years to allow man to produce such extreme types as bulldog and dachshund, percheron and racehorse, fantail and pouter. Variation thus may be abundant enough in specialized creatures. In Nature, however, the very fact of high specialization handicaps many variations which might have been useful enough to a more all-round and generalized creature. Just because a horse is so well adapted to running, it is further removed from any possibility of becoming adapted to climbing, or flying, or burrowing; and a variation, say, towards more carnivorous teeth, which might have been profitable to an animal not yet wholly committed to one particular line, would, in the horse, merely find itself in opposition to the majority of its other characters. Natural Selection the conservative frowns on such innovation. But alter the tendency of selection, as man the radical does when he breeds his stocks, and the versatility of the creatures is at once revealed.

A highly specialized animal is precisely one which has been already forced by selection about as far as it can profitably go along one particular adaptive line. A horse cannot reduce its toes to less than one; nor improve the mechanical construction of its limbs much beyond their present pitch; nor grow larger without sacrificing speed. Specialized adaptation to one particular mode of life is thus in its very nature a cul-de-sac; variations may occur, but they can only rarely be of service to the specialized animal or plant. The specialized animal is *in its own way* almost as perfectly adapted to its mode of life as is possible; if some other creature should find a better way, it is helpless. Thus Natural Selection has relatively little power to modify highly specialized creatures; but this lies more in the nature of specialization itself than in any failure of the specialized organism to produce mutations as fast as more generalized types.

But some kinds of animals and plants do seem to have a smaller capacity for variation than others. Almost every species of duck, for instance, is relatively very constant; even when its geographical range is a wide one, few or no variations are to be found. Wrens, titmice, and sparrows, on the other hand, are much more variable; each widespread species is divisible into a number of distinct sub-species, which usually grade into each other on the borders of their ranges. The same is to be seen in plants. If examples are needed, the wild pansy will serve, with its dozens of varieties in flower-shape, and flower-pattern. On the other hand, however, that beautiful Alpine plant, the eight-petalled Dryas, is always very much the same, whether on the Alps or the tops of Scottish mountains, on the Great Divide in Colorado or at sea-level in Spitsbergen, nor does the ubiquitous bracken-fern show any noticeable variation from England to its antipodes.

It is certainly true that in the brief space during which Drosophila has been investigated, quite a number of mutations, such as that from red to white eye, have appeared in more than one species, and often repeatedly, a fact which argues the limitation of possibilities; and a similar recurrence of mutations has been found in most other organisms that have been carefully studied. It looks as if their germ-plasm had only a limited repertory of tricks. Limitation is also argued from such facts as the restriction of colour or pattern in various groups of animals and plants. Woodpeckers run to reds, blacks and whites, with occasional outbursts into green, yellow or brown; but blue and its combinations seem to be unknown among them. Gulls, on the other hand, show almost exclusively a combination of white with grey-blue or black; reds, yellows or greens are never found in their plumage. Among plants, the tendency of composites (like dandelion and daisy) to have either yellow or white flowers is familiar.

FIG. 204. A very variable species. The wild pansy of Europe, Viola tricolor, in four of its numerous variations.

The first is deep blue, the second light violet, the third white with a dash of yellow, the fourth yellow with a little white; the shapes, too, are markedly different.

But such facts may be misleading. The prevalent colours may provide some advantage we do not yet understand; or their prevalence may mean only that they constitute the path of least resistance, not that variation cannot produce other colours. The Pierine butterflies or "Whites" give us a useful warning in this respect. As their popular name implies, the prevailing colour is white, usually combined with black; yellow, orange and greenish are also not uncommon; we might well suppose that their variation was restricted to these colour-themes. But in South America a number of Pierines have become mimetic: they mimic butterflies of the Heliconiid family. And among these mimics we find colours and patterns wholly unlike anything

SELECTION IN EVOLUTION

found elsewhere among the Whites. Thus an undoubted tendency to keep to certain colours need not prevent others being produced if need arises.

More to the point are some of the failures of breeders to obtain from Nature something which they particularly desire. Blue roses are the most famous example. For centuries a blue rose has been one of the goals of every ambitious rose-grower; and yet, though bluish-purple has sometimes been obtained, true blue is as far from achievement as ever. It is quite easy for roses to be all shades of red, yellow or white; but, while it is perhaps not impossible, it is clearly very difficult for them to become blue. Black tulips are another case in point. During the tulip mania of the seventeenth century, every effort was made to breed a black-flowered tulip, and huge prices could have been obtained for such bulbs. The flowers were darkened to a deep bronze, but real black eluded the grower.

There are other limitations upon variation to consider. Natural Selection has never been given the chance of refusing or accepting a wheel. Not only has the principle of the wheel never been utilized in the bodies of living things, but it is impossible for it to be utilized. It is impossible, because a living organ rotating like a wheel could not be supplied with blood vessels and nerves. The "Wheel Animalcules" or Rotifers, we may note in passing, simply have circlets of cilia in their structure and not wheels.

Important, too, are some of the broad restrictions to be found in larger groups. No arthropod and no threadworm possesses cilia. No insect has developed any form of skeleton save of pure chitin; no mollusc and no vertebrate has any chitin at all. No vertebrate has any metal but iron in its respiratory pigment; while sea-squirts can use nothing but the rare element vanadium. Chlorophyll, wherever found in plants, varies but little in its chemical composition and mode of working; but it is never found in animals, even though many animals, such as reef-building corals, reveal their need of it by taking green plants into partnership with them. The reason for such restrictions we can only surmise. The insect constitution may be incapable of giving rise to cartilage or bone; or else, insects having happened to make a start with an external skeleton of chitin, the production of substances suitable for an internal skeleton would be valueless without an impossible remodelling of their whole organization. But whatever the precise cause, the fact of restriction remains.

In other cases we can better understand the limitations. No insect weighs more than a few ounces, although if they could have attained the size of horses or even rabbits they would almost certainly have prevented the rise of vertebrates to pre-eminence on land and have become the world's dominant animals. No flying creature, bird, bat, or pterodactyl, has ever reached a hundredth of the bulk of a large elephant, for they all are well below a hundred pounds in weight.

This latter restriction depends on aerodynamic laws; with wings as propulsive machinery and muscles of power, larger creatures could not attain sufficient speed to avoid stalling. And the fortunate restriction of insect size is due to their having embarked on the method of breathing by air-tubes; this method, the most efficient of all ways of breathing for quite small land animals, becomes rapidly more and more inefficient with increase of bulk, since oxygen must diffuse along their tubes to reach the tissues; and if these are too long, diffusion takes so much time that the supply of oxygen is inadequate. Advantageous as size would have been to insects in other ways, Natural Selection could not give it them, because their plan was fixed in a mould which automatically made size above a certain small limit a handicap to efficiency.

Thus, to sum up, the power of living things to change is definitely restricted. Sometimes it is restricted through the limitation of variation: one germ-plasm may be much more stable than another, or one type of constitution may readily produce variations in some directions, but be debarred by its own nature from producing them easily or even at all in others. Sometimes it is restricted by the mere fact of previous evolution; for specialization, without necessarily restricting the supply of variations, automatically makes the great majority of them less advantageous. The specialized animal is committed to a certain line of advance: variations that would take it along other lines can only be useful if it can manage to destroy or modify the plan it has already built up; and even to its advance along its own chosen line a term is eventually set—it reaches the limit of efficiency prescribed by mechanical or chemical laws.

But when all is said, the liberty of change open to evolving life is much more impressive than its restrictions. Here a door is shut, there a limit imposed, but the range of variety and height of attainment is prodigious. When one kind of creature goes under and becomes extinct, it is often, perhaps usually, because another has varied in new and more successful ways. The single type pursuing a particular direction of specialized advance is restricted, but the group of which it forms part is evolving in many and diverse directions. This or that line of advance, this or that change has been barred; but life as a whole has never ceased to experiment and discover.

IX

IS THERE A MYSTICAL EVOLUTIONARY URGE?

§ 1. *Straight-line Evolution.* § 2. *Do Races Become Senile?* § 3. *The Élan Vital and the Life Force.* § 4. *Is There Purpose in Evolution?*

§ 1. Straight-line Evolution.

WE HAVE treated of mutation as a random, sporadic event happening without design or any appearance of persistent aim. This is how it presents itself to the working biologist. In the living creatures, fruit-flies or maize or whatever they may be, in which he studies variation, new mutant forms of many and varied kinds appear; now one feature of the organism is changed and now another; there is no observable tendency for the germ-plasm to improve itself progressively in a particular direction.

Now at first sight this fact of random variation disagrees with the testimony of the fossils. The palæontologist, deciphering his vast histories, finds Evolution pursuing definite and steady trends. The horse-stock, for example, proceeded onwards through millions of years, always steadily horsewards. At the same time such creatures as camels, rhinoceroses, and pigs for certain, and in all probability the rest of the modern mammals, were slowly and minutely perfecting themselves in their own directions. And as we have already assured ourselves, there are plenty of examples of such steady, progressive evolution among other groups of organisms. Out of these facts has grown the idea of an inward directive force somewhere inside the germ-plasm, that makes it vary always in one direction. This is the theory of *Orthogenesis*. On this view the evolution of the horses was due, not to their having taken to the open plains and to any individuals with long legs and complicated grinder teeth having therefore an advantage over the others, but to some innate, independent, orthogenetic tendency of horse-protoplasm to lengthen its legs and complicate its molars.

Now it is obvious that we ought not to assume these mysterious inward directive forces unless we are compelled to do so, unless the facts do not admit of a more intelligible explanation. And in these cases of horse and pig, camel and rhinoceros, they certainly do. In each example the animal is adapting itself very admirably for some particular way of life; and since the changes are manifestly advantageous, Natural Selection acting on random variation will account for them perfectly well. Each successive stage in the horse evolution was an improvement on the last because it adapted the stock still further to a running and grazing existence. Moreover, as we have already seen, once an advance has been made in a particular direction there will naturally be a bias that way, for variations which happen to be on that line will be especially useful, while those that are off it thereby lose their value as weapons in the struggle for existence. A mutation making for millstone ridges on the teeth would have been perfectly useless to the whale-stock, which was evolving a method of straining the nutritious sea through whale-bone, and the millstone would have weighed as heavily round the neck of an evolving tiger. But such mutations were extremely useful to the developing herbivores of the Cenozoic; they had already begun to adapt themselves for grinding grass, and any such improvement of their molars was of great assistance. So that on the whole, other things being equal, animals and plants will naturally tend to evolve in straight lines.

But there are exceptions. There are plenty of cases of animals which have successfully turned aside during their evolutionary histories. Some change in themselves or in the outer world has made a new line more profitable. The evolution of elephants, mentioned in Book 3, ch. 2, § 5, is an excellent example of an evolution which did not keep to one direction. Then many land types have reversed their evolution by going back to water; whales, seals, penguins, ichthyosaurs and plesiosaurs, water-tortoises and turtles, water-insects and some kinds of water-snails are all proof that there is no rigid orthogenetic tendency in living things. The race is always on the lookout, so to speak, for avenues to explore; when a new way of life presents itself it is seized upon. We ourselves are examples of the possible tortuousness of Evolution. One of the most distinctive features of mammal evolution has been the development of hair, and yet we ourselves are hairless. But it can safely be said that we should never have evolved if our ancestors had not warmed themselves with hair. Again, our later ancestors until quite recently were specializing themselves more and more perfectly for an adventurous clambering monkey-life in the trees, but we have broken away and deserted the arboreal habit. Nevertheless it is to this habit in our ancestors that we owe the delicate plasticity of our hands, perhaps even the keenness of our sight and our upright posture.

There are, however, examples which afford stronger evidence of an obstinate orthogenetic tendency than do the horse and other ungulates. There are cases in which a race slowly and obstinately develops a character which, as far as we can see, is of no very vital use.

To illustrate this we may consider the Titanotheres, a group of clumsy, browsing mammals having the build and general appearance of a rhinoceros, that rose very early in the history of mammals, and rapidly attained formidable dimensions and widespread abundance before their extinction in the Oligocene period. At first small, they approached and nearly attained the size of elephants; and all the larger, later types have a pair of horns, diverging like the limbs of a V and placed near the tip of the nose. Within this curious group there are several distinct lines of evolution. Osborn, the well-known American authority, distinguishes at least four. One line retained its front teeth and remained stocky. Another lost its incisors, became more of a grazer, and grew longer-limbed and speedier. In both these two trends the horns remained fairly small. In the other two the horns were better developed; the third, like the second, lost its incisor teeth; while the fourth, of which the gigantic Brontotherium was the final representative, kept its incisors and went in for great bulk and slow, browsing habits.

Now in all these four lines the horns show an independent but parallel development. The earliest representatives, the founders of the four lines, were all hornless. Then, as the lines evolved, each developed horns quite independently of the others, but all four in a parallel way. First a horn-rudiment appeared, a mere thickening of the nasal bones. Then little bosses grew. Then the bosses sprouted into V-shaped horns. Finally in the culminating animals of all four lines the horns had the same general type, a type found in no other mammal, and were seated on the same part of the skull. Believers in orthogenesis say that this independent development of the same type of horn in four parallel lines is impossible to explain on the Selection theory. It demands, they say, the idea that the germ-plasm of the original Titanotheres was predetermined to vary in a definite way, wound up, so to speak, like clockwork, and that in the different stocks it simply unfolded itself in its predestined fashion. The horns *had* to appear and grow at a certain stage in the development of each of the four lines, quite apart from any question of advantage or disadvantage. This is one of the classical examples of orthogenesis.

There is, however, a more intelligible explanation than that. It is well known that in horned or antlered animals (such as sheep, goats, cows, antelopes, deer) the proportionate size of the horns depends on the size of the individual. The bigger an individual of any such species, the greater are his horns in proportion to the rest of him. And it so happened that during their horny unfolding all the four lines of Titanotheres increased steadily in size. Suppose that all the Titanotheres, the earliest as well as the last, had a general tendency to produce horns of their characteristic type. Suppose that the manifestations of this tendency depends on the size of the animal—as it does in deer and sheep and the like. The earliest Titanotheres were too small for their horns to appear at all (just as a young male red deer is too small for his antlers to appear), and as they increased in weight their horns were thereby allowed to develop more and more magnificently.

Thus the steady evolutionary growth of horns in Titanotheres turns out to be an incidental consequence of increase in size. And increase in size in such animals seems definitely to be of advantage. The majority of lines of mammals (and indeed of most vertebrates) show this tendency to grow as they evolve; it gives them greater strength and sometimes greater speed, to help them in their struggle for existence. So the supposedly orthogenetic horn-growth is bound up with biological advantage after all. It is what Darwin called a "correlated variation"; not itself of direct advantage, it is the automatic consequence of a change which has other advantages. Even here there is no need to invoke the inner guiding principles; Natural Selection and random variation will do all that is necessary.

§ 2. Do Races Become Senile?

NATURAL SELECTION will account for the great bulk of straight-line evolution. But when all the cases are eliminated in which an obvious biological advantage plays the rudder that keeps the evolutionary course straight, there are still some examples to explain. The most striking come from among the Ammonites. These animals, so abundant in the seas of the Mesozoic or Secondary Era (IV), grow progressively more richly sculptured, with successive chambers more intricately dovetailed into each other, as the millennia pass. Finally, not long (geologically speaking) before their final and complete disappearance, a number of quite new types appeared. The beautiful spiral became wholly or partially unrolled, and the regular gave place to the fantastic.

IS THERE AN EVOLUTIONARY URGE?

It has been found difficult even to guess at any biological advantage which might explain the existence of many of these later and stranger forms; and Alpheus Hyatt, the American palæontologist, followed by many later authors, has sought to explain them as an expression of what he called "racial senility," due to some kind of inevitable degeneration of the germ-plasm of the group, comparable to the slow individual ageing that Mr. Everyman cannot avoid, and similarly heralding an inevitable and internally determined extinction. The previous trend towards elaboration of sculpture and dovetailing he and his followers would interpret as due to the age-changes of the germ-plasm during its maturity.

There are really two ideas here combined—that of racial ageing and that of orthogenesis. If it could be definitely shown that these latest types of Ammonites derived no biological advantage from their bizarre shapes, we should be forced to accept an orthogenetic explanation of internally determined evolution. It must be admitted, however, that we know next to nothing about the mode of life of Ammonites, or about the changes which may have been taking place in the late Mesozoic (IV) seas. The example reminds us of our ignorance and warns us to keep the idea of orthogenesis in reserve in case it prove necessary after all. But our experience of the Titanotheres and their horns also gives us a warning—that we may discover some advantage lurking in the background of apparently quite useless happenings. It is perhaps wise to suspend judgment. It would be rash to demand a wholly new principle of evolution to account for what is almost an isolated case, before we are certain that other methods of proved efficacy are entirely ruled out.

Racial senescence is another matter. The idea of racial senescence is only a metaphor, taken over from the individual life-cycle; and to this the self-reproducing stream of germ-plasm presents no similarities; the phrase is not really anything more than an analogy, and a loose one at that. But the idea has often been toyed with, usually by palæontologists. According to the upholders of this view, the signs of racial old age include the development of bizarre shape and of great bulk, and especially the production of horns and spines and other excrescences. Indeed there is example after example where such exuberance in a race heralds its extinction.

But when we look into the matter, what do we find? We find that these characteristics are often displayed by stocks which soon afterwards go down the evolutionary hill and become wholly extinct. That is all. It is an interesting fact; but that the specialization or the spininess are expressions of anything comparable to senility—of this no proof is forthcoming.

The explanation of the actual facts seems to be quite different. Specialization, as we have seen, tends to be pushed on by Natural Selection towards a limit. When that limit is reached or neared, various things may happen. Quite often specialized types continue to thrive without essential change. The fully specialized horses appeared in the Pliocene (V D), and have remained with little change ever since. The same is true for many other of the most specialized mammals, such as the whales. Crocodiles, the still more specialized tortoises, and many families of flowering plants, have scarcely changed since the Cretaceous (IV C); lobsters and many fish, including such very specialized types as skates and eagle-rays, date back still earlier; and among the lamp-shells with their elaborate current-producing mechanism, some forms, as we have seen, have persisted from the earliest Paleozoic (III A).

In other cases, a change of outer conditions sets in, and the specialized type, which was well adapted before, can no longer maintain itself, and becomes extinct. Slow and gigantic herbivores that could thrive well enough when plant-life was luxuriant are liable to die out if the climate grows drier and vegetation more sparse. So, too, die creatures who are tied to water for their reproduction. And drought seems thus to have extinguished the giant Amphibians of the Trias (IV A). Forest dwellers must go if their home is all replaced by prairies. This appears to have happened to Hypohippus, the side-branch of the horse-stock which specialized its teeth for browsing on succulent forest-vegetation. As the Miocene (V C) climate grew drier, the forests shrank, and it perished. Cold is an equally potent cause of extinction; the Glacial Period gives us abundant evidence of that. But while some whole groups were then frozen out altogether, like the "tank-armadillos" or glyptodonts, the giant sloths, and the mastodon type of elephants, in other groups which appear to be equally specialized, like the rhinoceroses, camels, or the true elephants, only some species became extinct while others persisted—further proof that specialization in itself need not be the cause of extinction.

In perhaps the majority of cases, however, extinction is not due to so simple a cause. The specialized type is confronted with the competition or the attacks of new types which start from a higher level of organization or possess some special advantage. And it is this competition which, often in conjunction with changed and less favourable conditions, brings about the old types' downfall. Some of the Dinosaurs might well have survived the drought and the cold of the Cretaceous

(IV C) if they had not had to compete with the evolving mammals. Within the mammal stock itself the extinction of so many specialized groups in the first half of the Tertiary (V A and V B), including such formidable creatures as the four-horned Dinoceras and all its kin, was undoubtedly due to the rise of new mammalian types with better physical construction and, in particular, more elaborate brains. And if we need another example, we have only to think of the extinction of modern mammals which has been caused by the coming of man.

An example often invoked to support the idea both of orthogenesis and of racial old age, is that of the wonderful sabre-toothed cats which, first found early in the Oligocene (V B), became extinct only during the Glacial Epoch. Their most striking character—to which they owe their popular name—is the elongation of their upper dog-teeth into a pair of exquisitely constructed and formidable daggers, almost as long as the whole jaw, thin blades of living ivory with fine saw-teeth on their cutting edges. This construction is fully developed only in the latest-evolved species types, such as Smilodon; in earlier forms the "sabres" are smaller, and the whole skull more like an ordinary cat's.

It is often asserted that the gigantic daggers of the latest sabre-tooths not only could not have been useful, but must have interfered with the capture of prey, and so have directly brought about the animals' extinction. Their long-continued enlargement would then be due to orthogenesis; while their final exaggeration would be a symptom of racial senility. Both the advance and the extinction of the stock would be due to inner causes.

But the truth seems to be quite different. The huge teeth of the sabre-tooths are not biting organs in the ordinary sense—almost uniquely among teeth, they are stabbing daggers. While the true cats were becoming adapted for chasing and devouring the fleet grazing ungulates, the sabre-tooths were evolving along other lines. They were the enemies of the huger and slower browsers (the lumbering, tough-skinned giant sloth seems to have been a favourite prey of Smilodon) and their dagger-teeth were evolved to pierce the tough hide of pachyderms. Their whole structure is beautifully adapted for clinging with the feet and then delivering downward stabbing blows, not for gripping with the jaws.

Now one of the most noteworthy events of the Pleistocene (V E) was the change of climate, with the consequent elimination of so many kinds of big, heavy mammals and the restriction in range and numbers of the few survivors. The sabre-tooth of the Pleistocene, far from being a senile type or one that had been forced to overshoot the evolutionary

Fig. 205. Straight-line evolution in the Sabre-toothed Cats.

Right, skulls; left, reconstruction of heads. The Sabretooths gradually increased in size, enlarged their upper canines, made their jaws able to open more widely, and reduced the size of their lower jaw teeth. From below up: Dinictis and Hoplophoneus (Oligocene, V B); Machærodus (Miocene, V C); and Smilodon (Pleistocene, V E). Top, above the line, a true cat (Jaguar), to show the very different construction, with upper and lower canines about equal, and large flesh-cutting molars.

mark, was a beautiful piece of living mechanism evolved to cope with the increasing size and thickening skin of its special prey. And it became extinct not because it ceased to be adapted, but because its prey died out.

Thus, specialization *need* not be the preface to extinction. And when it is, the extinction is not brought about by some inner ageing of the germ-plasm, but by changes in outer conditions or in living rivals and enemies. Very specialized types are more easily extinguished than others: but this is because they have moulded themselves too perfectly to the passing world, not because they are "racially old." In their speciality they have gone too far to turn back before extinction comes.

Racial senility is thus on the same footing with orthogenesis as an explanation of evolutionary happenings. In the great majority of cases there is no need to invoke it, because a simpler and more intelligible explanation will work as well or better. A few examples remain in which we can as yet assign no reasonable outer cause for extinction. The disappearance of the Ammonites is one; the extinction of the great fish-lizards, the Ichthyosaurs and Plesiosaurs and Mosasaurs in the seas, of the late Mesozoic Period (IV), long before the appearance of whales as competitors, is another. As with the few possible cases of orthogenesis, these remain to remind us of our ignorance. With the growth of knowledge, we may find out that there was an outer cause for their extinction after all; or we may discover that occasionally the germ-plasm does fail the race. But even if the latter alternative prove right, it too can have had but a minor importance in Evolution. Extinction is for the most part brought about not by any internal wearing-out or ageing of the germ-plasm, but as a natural consequence of the universal struggle.

§ 3. The *Élan Vital* and the Life Force.

The theories so far discussed are none of them what philosophy calls "vitalistic." To account for Evolution they demand no special and mysterious qualities in living matter. Granted that it can reproduce and vary, a few simple forces come automatically into play, and Evolution by means of selection is the result. There is a machinery underlying Evolution as there is a machinery behind chemical combination or the laws of gaseous pressure. But as we have noted from the outset of this Book, there are other theories in existence whose proffered explanations are not mechanistic at all, but vitalistic. They ascribe Evolution to some directing force which is supposed to be purposive, or at least of the same nature as purpose, and to reside in life, but not in matter which is not alive.

Of such theories there are many. The most celebrated creative force is Bergson's *élan vital* or vital urge, which reappears, expounded in exaggerated form, a pantomime giant, as the Life Force of Mr. Bernard Shaw. The Shavian Life Force need not detain us long. It is Lamarckism in caricature. Life evolves, says Mr. Shaw, by trying, by more or less conscious effort. Evolution takes place because all life is purposeful in its degree. In ourselves the Life Force may be at cross purposes with our conscious but more superficial selves; racial purpose may conflict with individual purpose. But however disguised, however deep below the level of ordinary consciousness, the essence of the Life Force is purpose. Quite apart from the difficulty of ascribing even rudimentary purpose and foreknowledge to a tapeworm or a potato or collective aspiration to the tapeworm race or the potatoes of the world, there remains the impossibility of transmitting the results of this purposeful striving to posterity. If, as we have given ample reason for believing, acquired characters cannot be inherited, Mr. Shaw's Life Force does not exist.

Bergson's *élan vital* is in a rather different case. For him the vital urge is something inherent in the very nature of life, but absent from dead matter. It is in some not very comprehensible way intermediate in its properties between the blind activities of lifeless matter and the conscious and purposeful activities of mind. In operation, it reveals itself as a constant tendency towards adaptation. For M. Bergson it is this tendency towards adaptation which accounts for the existence of adaptations, and not such material considerations as the struggle for existence and the elimination of the less fit by Natural Selection; and equally without aid from struggle or selection, it makes tactfully but firmly for movement onward and on the whole upward in Evolution. The *élan vital* itself causes evolution. It is orthogenesis translated into vitalistic terms.

But when we begin to look into the argument closely, we see that the *élan vital* is a mere metaphor. It is in reality not a new and mysterious creative principle, but the elementary chemical properties of living matter, idealized and personified. Living matter has, as its basic property, the power of metabolism and self-reproduction; and it varies. From these two properties there follow over-multiplication, the struggle for existence, the survival of the fitter, and the constant stress of natural selection. If we like to call the combined effect of the struggle and the variation, *biological pressure*, we can do so. Then why, it may be asked, should we not be poetical and call it *élan vital?* But biological pressure is a resultant of simpler forces, all unconscious, none vitalistic; while the *élan vital* is defined as a property in its own right, not further analyzable, and partaking in some obscure way of such mental properties as purpose. Bergson's *Creative Evolution* gives one of the most vivid pictures ever painted of evolution at work, but as an exposition of evolutionary method it is valueless. It is a brilliant and poetical description, and it has very properly won its creator the very highest literary distinction. But it is not a scientific explanation.

§ 4. Is There Purpose in Evolution?

BERGSON and Shaw, in common with many others, have been so much impressed with the apparent purpose revealed in the slow, stupendous drama of life, that they have not hesitated to make purpose the key of Evolution. They see purpose not only in its results, but would make it the very heart of its method. This question and purpose in Evolution is a crucial one for biology and for the contribution which biology is to make to general thought. Let us briefly summarize the conclusions of this chapter; they will help us to decide if evolutionary purpose is real or only apparent.

Our main conclusion is that the chief agency of evolutionary change is the sifting action of natural selection upon practically random variation of the germ-plasm. Lamarckism will not work, because neither the direct effects of the environment, nor those of conscious or unconscious effort are normally inherited; and orthogenesis is, in most cases at least, a quite unnecessary hypothesis. But the theory of Natural Selection provides an adequate explanation for the great majority of the facts of Evolution; it can explain the detailed adaptations of animals and plants and their long-continued trends of specialization, the rise of new types and the extinction of old, the progress of life, its retrogressions and degenerations, and much at least of its variety. The implications of this are far-reaching. Without constant struggle and competition, Evolution could not have occurred; without the failure and death of innumerable individuals, there could have been no gradual perfection of the type; without the extinction of great groups, there could have been no advance of life as a whole.

And then we see Evolution as a response of life to its environment. It is as much a response as a child's drawing away of a burnt finger from the flame, or the growing of thicker fur by a fox exposed to cold;

but the response is effected by a peculiar and roundabout method. It is effected by a method of trial and error. The environment of a species changes. The species is constantly throwing up new variations. Most of them are no good and are eliminated; a few are improvements and are preserved and incorporated in the evolving flow of life. The type is thus changed, and changed so as to better fit the new conditions of its environment.

Let us not forget that the environment of any living thing includes not only the lifeless environment, but also the living environment of enemies and prey, competitors and parasites. A great deal of evolution is a response to change in that living environment. The horse's speed is a response to the increasing speed of its devourers: its tooth-structure a response to the spread of tough grasses over its feeding-grounds. Environment, in this extended sense, determines evolution—indirectly, by the sifting, trial-and-error method of selection, but none the less surely. An organism apart from its environment is meaningless; if the environment had been different, so would the organism. If this earth were bigger, all the mechanical construction of land animals would be different, for the force of gravity would be greater. If the surface of the earth were perpetually shielded by clouds from the sun's rays, as on some other planets, the human inhabitants of the tropics would not be black. Before the power of colour-vision was evolved there can have been no animals which practised concealment by adopting the colour-pattern of their surroundings, as chameleons or leaf-butterflies do to-day.

The result of Evolution and Natural Selection is a constant increase in fitness. But there are limitations to the perfection of fit attained. Trial-and-error is a rough-and-ready method. What it produces is something that will work, by no means necessarily something that will work perfectly. The creatures that exist are those that happen to have survived: taken together they represent an equilibrium which manages to be more or less stable, rather than life's best possible way of utilizing and sharing out the resources of earth.

But—and now we begin to touch on the question of purpose—as far as we can see, the variations which alone make evolution possible are random variations. That is not to say that they may not be limited in quantity and quality; but that from the point of view of evolution they are at random. In every organism they take place in many directions; the environment, acting through struggle and selection, picks out those that are headed in one particular direction. The identical

IS THERE AN EVOLUTIONARY URGE?

variation that is selected and kept in one environment may be rejected in another. To take but one example. So long as great swamps abounded and the amphibians that lived in them had not evolved to their limit of size and power, variations making for the capacity of living and reproducing on dry land would be of less value than those promoting success in the ample watery environment. But once this environment began to shrink and dry, those same variations would go up in biological value and would be selected, and so impetus be given to the evolution of reptiles.

Variation is at random; selection sifts and guides it, as nearly as possible into the direction prescribed by the particular conditions of environment. Once we realize this, we must give up any idea that evolution is purposeful. It is full of apparent purpose; but this is apparent only, it is not real purpose. It is the result of purposeless and random variation sifted by purposeless and automatic selection.

The term purpose has a very definite meaning. It is a psychological term, describing a certain familiar state of our own consciousness: it implies the prevision of an end, and a determination to reach that end. For evolution to be purposeful, one of two things must be true. Either living things themselves must be purposive in their evolutionary changes—the flower must somehow want to attract bees, the horse intend to lose all its toes but one; or else, although the living animals and plants themselves betray no purpose, purpose must exist in the mind of a divine Being, who is manipulating life and its environment to bring about His purpose, as we manipulate matter and events to bring about our purposes.

The first alternative is that of Bergson and Shaw; and, as we have seen, we must dismiss it. Variation and selection in themselves are blind. Life is by them passively moulded towards adaptation and perfection, and is not the arbiter of its own destiny. The forms of life are as much the automatic product of outer forces as the forms of mountains, lakes, and valleys, although the machinery of their production is different.

The second view is the view of a number of more or less modernist theologians. It is Creationism up-to-date. Instead of creation of the whole scheme of things ready-made, it implies the gradual working-out of a preconceived plan, the gradual realization of a divine purpose. This may, of course, be true. Science cannot say yes or no to such ultimate questions. She can, however, point to facts and ideas which bear on their reasonableness.

She can, for instance, justifiably turn to the theologian and reason as follows. If the natural selection of random variations is the main agency of Evolution, then Evolution receives a scientific explanation in terms of known natural forces. It receives as natural an explanation as does the pressure of air through the impact of its myriad separate molecules, as envisaged by the kinetic theory of gases; or the formation of sulphuric acid from heated iron pyrites, steam and the fumes of nitric acid through the atomic theory and the laws of chemical combination. Not only is the fact that Evolution looks purposeful no proof that there is true purpose behind it; but if there is real purpose behind it, there must be real purpose behind the transformations of dead matter, too.

The biologist can also point out that Evolution, whether looked at as a whole or in detail, is very far from coming up to what we might expect if it were in truth the realization of some exterior cosmic purpose. Firstly, it is extremely slow. Then the method of selection is not only slow but wasteful and, in higher animals at least, involves great suffering. Worst of all, it has achieved much that seems definitely bad. Evolution has deprived barnacles and oysters of movement and brain; it has produced the female mantis, who begins eating her mate during the act of pairing; it has generated the blood-thirsty land-leeches and mosquitoes, and fitted the ichneumon-fly grub to devour its living caterpillar prey slowly, from the inside; it has brought into being not only strong, intelligent and beautiful creatures, but also degenerate parasites and loathsome diseases. In brief, we are confronted with the gravest theological difficulties if we too light-heartedly set out to see purpose in Evolution. The wiser and saner course is to acknowledge our ignorance of ultimate causes and designs. If we do so, then, while frankly confessing that we do not understand what may be behind Evolution, we can yet appreciate at fullest value what it has achieved and seek to turn its further progress into channels that seem good to us.

For when we reach man, Evolution does in part become purposeful. It has at least the possibility of becoming purposeful, because man is the first product of Evolution who has the capacity for long-range purpose, the first to be capable of controlling evolutionary destiny. Human purpose is one of the achievements of Evolution.

Human purpose has arisen as a product of the mechanical workings of variation and selection. But now that consciousness has awakened in life, it has at last become possible to hope for a speedier and less wasteful method of evolution, a method based on foresight and deliberate planning instead of the old, slow method of blind struggle and blind selection. At present that is no more than a hope. But human knowledge and power have grown very marvellously during the last few hundred years. The multitude of our race living to-day still does not know of more than a minute fraction of what is known to man, nor dream yet of the things he may presently do.

BOOK FIVE

THE HISTORY AND ADVENTURES OF LIFE

I

THE PROLOGUE

§ 1. *The Scale of the Universe.* § 2. *The Setting of the Stage.* § 3. *The Origin of Life.* § 4. *Changes in the Terrestrial Scenery.*

§ 1. The Scale of the Universe.

THE time has now come to give an account of the actual history of life upon the earth, the procession of events which, never repeating themselves, have led onwards until this present instant where the past is busy eating into the future. Of this colossal drama no man knows the scheme. Some prophesy catastrophic end; others a slow and gloomy deterioration: still others see no necessary full-stop to the achievements of life, or if they do, comfort themselves with the reflection that life may prove itself abundantly well worth living in the long epochs still to come, itself its own justification. Man at least believes that he has been called to take the leading part in it; and he is not unnaturally interested in finding out all that has gone before in the story.

Goethe's *Faust* and the *Book of Job* both have a Prologue in Heaven, and without attention to the prologue we could not grasp the full meaning of their main story. The same is true of our drama of organic evolution. The history of living matter grows out of the history of planet earth, of which it is but a part; and the history of earth in its turn grows out of the history of the star we call the sun, which, again, far back, merges into the history of a great nebula.

We may be tempted to use theatrical metaphors, speak of the earth as life's stage and the heavens as its background. But this is not really accurate. Living matter is but a special arrangement of ordinary matter, the evolution of life but a local and peculiar eddy, so to speak, in cosmic evolution. Players, stage, and background are all of one substance together.

And before going farther we had better be sure that we have grasped the scale on which time is employed in the making of worlds.

THE PROLOGUE

The matter of this amazing universe of ours is grouped, so the astronomers assure us, into a huge number of island-universes. We live in one, and see others as spiral nebulæ, star-clouds, or star-families. These island-universes drift sparsely in space like jelly-fishes in the sea. Their number runs into millions of millions. Somewhere between five and eight million million years ago (we are following Sir James Jeans) the gassy nebula which was the beginning of our particular island-universe was in process of condensing into those separate and more solid knots of matter we call stars. When it had finished condensing, between twenty and thirty thousand million stars were in existence in our own star-family. Our sun was one of this numerous progeny. It, like all stars, had its own history and life-cycle, whose course is of astronomic rather than biological interest; we must leave our readers to find out about it for themselves in the pages of Eddington or Jeans. But to the sun, after a great part of its history had run, there happened a very rare accident: it was disrupted by the approach of another star. This, the astronomers tell us, took place about 2,000 million years ago, certainly not less than 1,600 or more than 3,000 million years back.

If, to get the whole of our sun's history on to a page, we represent one million million years by one inch, the time during which earth has existed as a separate planet is so small in comparison, that it cannot even be indicated on this scale. (Fig. 206.) Multiply the scale of our time-diagram by a hundred, and the history of earth appears as a space about a quarter of an inch long. Multiply our scale another hundred times, and we get on to our page rather more than half of geologic time since the first stratified rocks were made. About four inches covers all known vertebrate history, and less than a quarter of an inch the whole existence of man.

Again we apply our hundredfold multiplication. Say one inch now represents only a million years. Our page takes in most of the time since the beginning of the Pliocene (V D); the whole Ice Age (V E) occupies about half an inch. Magnify once more, again by a hundred, and an inch represents ten thousand years. Within two inches is included all the time there has been since the final retreat of the ice of the last great Ice Age, and within this space man, as we shall describe, has evolved from his Old Stone Age habits to modern civilization. The time since the discovery of America by Columbus is represented by about one-twentieth of an inch. Magnify by a hundred once more, and our inches become centuries. The time from the discovery of America to the present is represented by nearly five inches; the time since Darwin made it possible for us to see our own place in Nature aright is compressed into less than an inch, while the new discoveries as to the extent and nature of the universe (which we may date as beginning with the discovery of radium) take up only a quarter of an inch.

To arrive at a scale on which we can represent the events of human history, we have had to magnify our original scale, needed to show the life of a star, ten thousand million times. And even to convert a time-scale convenient for geology into one suitable for history, we must use a magnification of a million. We are a long way from the old view (which dominated Western thought so recently that there may still be found persons who cling to it in spite of all evidence) that man's home is the centre of the universe, and that his creation, only a few thousand years back, was simultaneous with the beginning of the whole world.

The great bulk of the physical universe exists in the form of wandering radiation, not of matter; of matter all but a tiny fragment is in the interior of stars, at appalling temperatures in the neighbourhood of forty million degrees centigrade, and quite incapable of supporting life of whatever description, since, as Jeans says, "the very concept of life implies duration in time; there can be no life where the atoms change their make-up millions of times a second, and no two atoms can even become joined together. Primeval matter must go on transforming itself into radiation for millions of millions of years to produce an infinitesimal amount of the inert ash on which life can exist." Man is an inhabitant of a thin rind on a negligible detached blob of matter belonging to one among millions of stars in one among millions of island-universes.

And his insignificance in time is as overwhelming as his insignificance in space. The time of the universe is almost all spent in what to us seem wholly meaningless activities. Stars shrink and dissolve into radiation. The matter of which they are composed is engaged in an atomic and electronic dance, frenzied beyond belief, but persistent through periods which make even the whole past of terrestrial life quite negligible. Man is so far from being central or essential that the tale which the rest of the cosmos has to tell seems meaningless in the light of all his ideas and aspirations. If he is to find justification for these ideas and aspirations, he can no longer seek it in the outer universe, but must look within himself. Human dignity rests upon nothing but itself, and man's activities must have value in themselves and for their own sake if they are to have value at all. That is the outcome of modern astronomy's impact upon the complacency of ordinary thought.

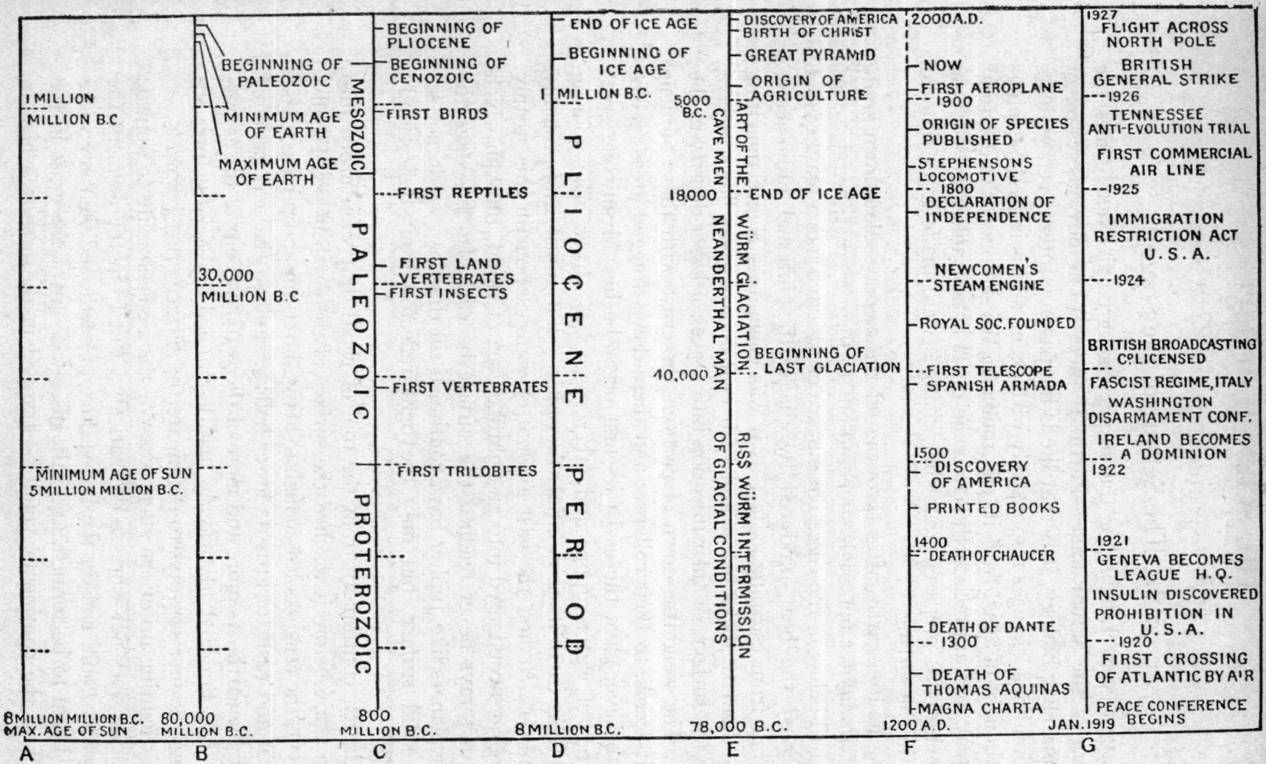

FIG. 206. A Diagram of Stellar, Geological, and Human Scales. Each column is on a scale magnified 100 times more than the one to its left hand. Column (A) gives the lowest and highest possible dates for the origin of our sun from a spiral nebula (according to Jeans). On this scale the age of the earth is so small that it cannot be indicated. (B) gives the last one-hundredth of the time included in (A). The lowest and highest possible age of the earth's origin from the sun are given (according to Holmes). Column (C) gives the last one-hundredth of (B). A few main events in the history of life are indicated on it. Column (D), again magnified 100 times, does not include quite all the Pliocene, which we may presume to have begun about 10 million B.C. The dates of the beginning and ending of the Ice Age are indicated Column (E) gives the last one-hundredth of the time in (D). This includes the last and shortest of the Ice Age's four periods of intense glaciation. The dates of the three historical events are given. Column (F), again magnified 100 times, takes us back to Magna Charta. A further magnification of 100 gives us the period shown in column (G), for which the eight years from the end of the War have been chosen.

§ 2. The Setting of the Stage.

So much, very briefly, for the cosmic background. Next comes the setting of the actual stage—Earth. If there is one obvious lesson of evolutionary biology, it is that life is inseparably interwoven with its surroundings, changes responsively with them, and is indeed meaningless thought of apart from its environment. We shall never understand the play and appreciate the *crescendo* of its action unless we see how the stage was prepared and what changes were made from time to time in the setting.

The molten earth, after throwing off the moon, cooled down gradually, its lighter materials squeezed out to form a crust, heavier ones clenched below in the core. The surface crust acted like a blanket and prevented the heat produced by the earth's internal radio-active minerals from escaping as fast as it was generated. Hence remelting of the heavier rocks below and their eruption as great flows of lava on to the surface, or their injection into spaces of lesser density in the crust. The weighting-down of parts of the crust with these burdens led, it seems, to their collapse; and these down-dragged areas, geologists believe, gave the earth its first ocean-basins. Meanwhile great volumes of water-vapour and other gases must have been pouring out, giving earth its primordial atmosphere—hot, moist, and probably sulphurous. At first the vast amount of water-vapour in this steamy atmosphere condensed into an unbroken sea of cloud, through which the sun's rays never reached the surface beneath. In this stage the larger planets like Jupiter have continued till this day; we do not see their solid surface, but only the light reflected from their cloudy mantle.

Rain fell continually from the roof of cloud, but was turned into steam again before it reached the surface of the earth below. Finally, however, the crust cooled enough to allow the still liquid rain to splash upon it and rush towards its lower hollows; and so the seas gathered about the earth. As more and more of the water-vapour of the atmosphere was liquefied, the cloudy mantle thinned and finally tore, admitting the first sun-rays to illuminate the surface crust below.

This lighting up of the stage was not the signal for life to appear: Evolution does not work so dramatically. But from the moment the first water could run on the hot surface of earth, sedimentary rocks could begin to be formed, and salt, dissolved out of the crust in the run-off of the continents, could be trapped in the oceans, which in the beginning must have been practically fresh. In all probability a very considerable thickness of rock was laid down before the slightest intimation of life appeared; for all the evidence points to primitive life having been adapted from the first to a sea already salt. But the sun-rays, all the same, seem to have had much to do with the origin of life so soon as the conditions became favourable.

§ 3. The Origin of Life.

THE actual origin of life must always remain a secret: even if man succeeds in artificially making life, he can never be sure that Nature did not employ some other means. Some thinkers have supposed that life was carried to this earth in a dormant state within meteorites. But this is to think timorously and to balk the issue; it only removes the problem of life's origin one step farther back. It does not absolve us from asking how and when life originated, but merely introduces an extra difficulty.

It is much more likely that at one moment in earth's cooling-down, the warm seas provided an environment never afterwards to be repeated, an environment differing in temperature, in pressure, in the salts within the waters, in the gases of the atmosphere over the waters, from any earlier or any later environments. The earth at that moment fulfilled all the conditions which the alchemists tried to repeat in their crucibles. It was a cosmic test-tube, whose particular brew led to the appearance of living matter as inevitably as an earlier and different set of conditions led to the formation of rocks and seas and clouds.

Let us remember that there are no elements in living matter which are not found in its lifeless environment; that the energy by which life is operated is not any mysterious "vital force" but is the same energy (we have produced the energy balance-sheet for our readers' inspection) by which the simplest physical and chemical transformations are worked; and that the chemical compounds found in living bodies and as yet unsynthesized seem to differ only in their complexity from those we can already put together in test-tubes and those that exist as not-living matter. The one distinguishing feature of living matter is its capacity for self-reproduction. But the chemist can tell us of numerous chemical reactions which, given proper conditions, are self-continuing in the same way; the only difference is that the chemical transformations of life can reproduce themselves over a wider range of outer conditions than can any of those lifeless reactions.

Let us also remember that the state of not-living matter which we meet with in our earth's crust is altogether exceptional. In the first place, the physicists are finding out that matter and radiation are up to a point interchangeable, both aspects of a single physical reality. The main quantity of this reality does not even exist in the form of matter at all, but of intangible radiation streaming in all directions through space. When we come to matter, the great bulk of it is in the unbelievably hot interiors of stars, and here, as Eddington tells us in his *Stars and Atoms*, exists for the most part in the form not of atoms as we know them, but of what he calls "unclothed atoms," with their rings of electrons stripped off. Even at the surfaces of most stars, matter is so hot that atoms can never stay combined. And, as L. J. Henderson points out in his book, *The Fitness of the Environment*, most of the substances familiar to us on earth can only be formed once liquid water is there to act as chemical go-between in the task of building them up—and liquid water can only exist in a very few and isolated spots in the whole universe.

Matter in the form of naked atom-nuclei and free gadabout electrons—the commonest state; matter in the form of atoms; matter in the form of simple compounds; matter in the form of the special compounds that need water for their formation—a rare state; matter in the form of the self-reproducing and very complex units that we call alive—it seems to be a continuous series, each term in it coming inevitably into existence as the conditions in the cosmic test-tube dictate.

In any case, the great majority of biologists agree in thinking that probably all the life upon the earth had its origin from the matter of the earth at a definite time in the earth's history. Where opinions differ is as to the form in which this primordial living matter appeared. Some scientists, like Sir Ray Lankester, have suggested that the conditions were such as to call forth sheets and blobs of protoplasm-like substance, some of which just failed of self-reproduction and lost life, but served as food for the one or the few which could be called actually alive. Others have supposed that the earliest life was plant-like in its nutrition, either equipped with chlorophyll from the start, or, like many modern bacteria, capable of living on air and salts and water without being green.

However, of late years two discoveries have been made which seem to shed light on the problem, and suggest a third solution. One is the fact that light, even without chlorophyll to act as a transformer, can effect various chemical syntheses. Under the influence of light, small quantities of sugars and other organic substances, some of them nitrogen-containing, are generated from a mixture of such simple substances as water, carbon dioxide and ammonia, as Professor Baly of Liverpool has experimentally shown.

Such substances are presumably being manufactured to-day in sea-water, but in much smaller quantities. For it is the ultra-violet waves of light which are active in this chemical transformation, and most of them are stopped in our present-day atmosphere by the oxygen in it. In those primeval times, the oxygen-content of the atmosphere was certainly lower, perhaps almost absent, and so the light could get to work to some purpose. But to-day any of these substances that may be formed are quickly absorbed by the multitudes of living things that everywhere exist, or got rid of by decay, which is our way of saying that they are broken down by bacteria. But before there were any living things to absorb them or break them down, they must have accumulated until, as J. B. S. Haldane puts it, "the primeval oceans reached the consistency of hot dilute soup."

Any chemical compound which had reached the borderland between dead and living and had acquired the property of chemical self-reproduction would have found in this soupy sea abundant food and stores of potential energy to support it, while it (or rather, no doubt, one survivor among a host of descendant streams, most of them unsuccessful) evolved into something really alive.

And the other discovery is that which we have already mentioned in Book 2, of things which are in truth on the borderland between dead and alive, the bacteriophages or bacterium-consumers. These ultra-microscopic units, it will be recollected, are able to grow and multiply so long as they are given a supply of living bacteria to consume. Dead bacteria are, however, no good. They can be filtered off from the relatively gigantic bacteria and then made to attack other prey of other kinds, if so desired; and the numbers of the bacteriophages increase rapidly (they can be counted by certain ingenious methods) in the process of killing bacteria. D'Herelle who discovered them believes they are alive, because they multiply. Others say they are only an exceptionally active kind of ferment, which happens to be knocking about in the outside world, but is helpless to make more of itself except out of matter which is truly alive.

The truth may lie between these two views. If living matter has originated from dead, then we shall expect that intermediate conditions should exist. In these bacteriophages we have perhaps discovered a "missing link" between the two states of matter, as in Archæopteryx we discovered a missing link between two other states of matter, the reptilian and the avian. If some very recent work is confirmed, it will suggest that these entities may be half-alive parts of cells which have got free of the co-operative restraint of life out into the great world; for it seems that the ultra-microscopic bits into which bacteria are broken up by bacteriophages may occasionally persist and join up again to make full-size bacteria later. And in any case, as Muller suggests, a bacteriophage is in many ways like a gene (a lethal gene, it is true) got loose.

These are but hints; but they are very encouraging hints. They help confirm our opinion, based on a general weighing of alternatives, that, as a matter of history, life on this planet originated from not-life, that it originated at one phase and at one phase only, that it probably originated in the surface waters of the warm early globe, and that sunlight, that "only begetter" of all our terrestrial activities, played a necessary part in its origin. They also hold out a hope that we shall one day be able to make living matter artificially. But that, if ever it arrive, may be a long time coming. To be impatient with the bio-chemists because they are not producing artificial microbes is to reveal no small ignorance of the problems involved. Living matter is matter: but it is quite appallingly complicated matter, many times more complex in its construction than any other substance known anywhere in the universe. It has been evolved through billions of generations under the filtering action of Natural Selection which has rejected every false try and unsuccessful experiment. We rightly praise the skill of the chemists who build up dyes and drugs to order,

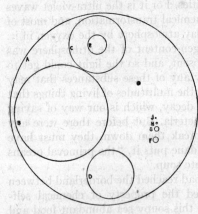

FIG. 207. The sizes of some ultramicroscopic particles; all magnified many diameters.

(A) A particle at the limit of microscopic vision, 0.2µ across. (B) A filter-passing organism, 0.1µ across (Book 2, Ch. 6, § 8). (C and E) A single gene of Drosophila cannot be larger than (C) or smaller than (E), according to our present knowledge; (E) is also, according to Svedberg, about the size of the biggest protein molecule known, that of the copper-containing substance in snail's blood. (D) Approximate size of a bacteriophage (Book 2), a particle on the borderline between living and non-living. (F) A molecule of hæmoglobin. (G) Smallest protein molecule (according to Svedberg). (H) A simple organic molecule like chloroform, with five atoms ($CHCl_3$). (J) A simple inorganic molecule—hydrogen gas.

but to build up living matter, substances as complicated as their highest achievements in synthesis would have to be used as the basic bricks. In any attempt at making living matter, we begin about where the modern organic chemist leaves off, and we begin more than a thousand million years of evolution behind contemporary living cells.

§ 4. Changes in the Terrestrial Scenery.

LIVING matter once formed, it must very speedily have begun to change its environment and render a repetition of its own appearance improbable. Whatever life's original nature, the true green plant seems to have been developed very soon, and once developed, its action upon the atmosphere was very important. Most of the oxygen in the original atmosphere had been in all probability consumed by the unoxidized materials of the great lava-flows from the interior. But now green plants, which liberate more oxygen than they use for respiration, were coming into the play. As they multiplied and as volcanic violence diminished, the air grew richer in free oxygen. Most animals need free oxygen, and roughly speaking the higher the animal in the evolutionary scale, the more oxygen it needs. So that those great early players in life's historical drama, the green plants, prepared the stage for higher and more active organisms.

The chemical balance seems to have been struck pretty early. The early land-plants, at least as far back as the Devonian, had pores on their leaves for the exchange of gases which were very similar in their arrangement and their number to those of modern plants. This would scarcely be so if the proportion of carbon dioxide in the air was very different from what it is to-day.

But if the atmosphere changed in this manner, rapidly at first, to settle later into a relatively stable phase, the seas also were changing, slowly and progressively changing and changing by growing more salty. Salt is removed from the water by being turned into solid layers of the earth's crust when seas are trapped and dry up; but even so the general tendency of things as between land and water is inevitably towards greater saltiness of the water.

A phase of comparative stability in this planet's story was beginning at the very outset of life's career. Life as we know it could not exist at all if the earth's average temperature altered by 50° C. either way, and could not flourish if the alteration up or down were more than 20°. It could not exist if the amount of oxygen in the air was divided by four, or if it were multiplied by four. If life is to go on, there are limits which the carbon dioxide in the air and the salts of the seas must not transgress. Yet these narrow limits of physical and chemical conditions have never been overstepped for perhaps a thousand million years; and, so Sir James Jeans tells us, are likely to continue for anything up to a hundred or even a thousand times as long in the future, so steady is the supply of heat from the sun, so nicely regulated the circulation of water and heat in the currents of atmosphere and ocean over the surface of the globe, and the circulation of matter. solid, liquid, and gaseous, through the bodies of living creatures.

But though the lighting, temperature, humidity, saltness and pressure of life's environment seems to have varied only between comparatively narrow and definite limits since life's story began, that does not preclude immense changes of scene from our consideration, changes from torrid heat to icy desolation, from widely prevalent rains to dry and chilly deserts, from towering mountain-masses to world-wide sunlit shallows.

It has long been known that some geological periods have been characterized by violent outbursts of mountain-building, while in others the lands have been quiet and undisturbed. The earth was comparatively quiet through almost all the Mesozoic Era (IV), but in the much shorter time from the late Cretaceous (IV C_3) to halfway through the Cenozoic Era (V), the Alps, the Rockies and the Coast Ranges, the Pyrenees, the Caucasus, the Apennines and the Himalayas arose in their majesty.

Such times of great mountain-making are sometimes called Revolutions, geological revolutions, and there have been at least six recorded in the history of earth (see Fig. 122). The Cenozoic Revolution we have just spoken of: closing the Paleozoic (III) came another which threw up the Appalachians and the Hercynian chain whose stumps still appear in the Taunus of Germany and the Ardennes of Belgium. Half-way through the Paleozoic, at the end of the Silurian (III C), the world passed through another set of throes, the best known product of which is the Caledonian chain whose worn-down remains extend from Scotland into Scandinavia. Again, just before the dawn of the Cambrian (III A) another spasm raised a huge chain of mountains, now planed down, reaching down from Canada into what is now the Middle West of America, and others in Europe and elsewhere. And geologists tell us of at least two others at equally long intervals in the dimness of Pre-Cambrian times (I and II).

These revolutions are always accompanied by great emergence of the lands and are succeeded by periods when much land is high and

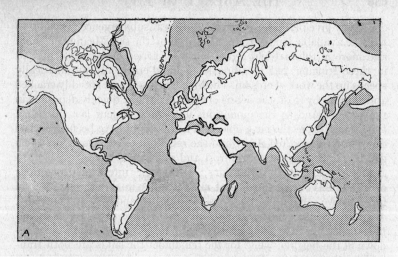

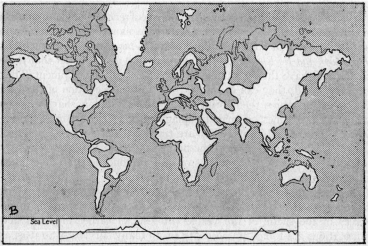

How the outline of the continents would look—Fig. 208 (A) If the lands rose two hundred metres; Fig. 209 (B) If they sank two hundred metres. (Inset) A section, running along the equator, across the Indian Ocean from Africa to the Malay Archipelago, to show the way in which the edges of the continental masses fall rapidly away to the ocean deeps.

dry. They give place slowly to times of great submergence, when up to half the available land-surface is covered by the sea. Meanwhile, more and more of the material of the lands is carried down and deposited in these fringing and invading seas. Land-emergence automatically speeds up the work of erosion. The high lands are more speedily cracked and chipped by frost, the débris of the land is more quickly deposited in the waters. And so the mountains and the grandeur last but for a time; long before the next upheaval is due, they have been smoothed down and transported away, and the relief of the lands is reduced to a gentle undulation, with little frost and sluggish streams.

These great earth-revolutions, with their intermediate periods of tranquillity, seem to come at fairly regular intervals. The interval for those we can date is roughly some hundred and fifty million years, of which twenty or thirty million may be devoted to a series of mountain-births, perhaps as much again, or rather more, to the wearing-down of the heights, while for the remainder the lands are both tranquil and low-lying.

What causes this colossal rhythm of recurrence is not our concern here. The geologists are beginning to offer explanations, and those who are interested can consult Holmes' *Age of the Earth*, or Schuchert's *The Earth and Its Rhythms*. What does concern us is the fact that these revolutions have the profoundest effect upon life's development. They cool the world; they can cause ice ages; and the elevation of the land which accompanies them brings local aridity and a sharply zoned climate. Such times, as may be imagined, are critical times for the world's living inhabitants. They are times both of destruction and of progress. The specialized and the bulky and those that are pleasantly adapted only to the long epochs of smooth conditions are overtaken by disaster and extinguished or brought low. But their very destruction gives opportunity to smaller and less specialized creatures, which have been hardy or quick-witted enough to make a place for themselves in the shade of the vested interests of earlier life; and new adaptations are forced by necessity on to many survivors. So it is, as we shall see illustrated again and again in the course of this Book, that these rhythms are always followed not only by widespread extinction but also by the rapid advance of some new and abler type of animal or plant machine.

But besides this greater rhythm, there is another, of smaller periods, and more irregular, but also of the greatest importance in its effects on life. This is the rhythm of slight sinking and rising of the lands. The main ocean basins and the main continental masses that stick up from their depths seem to have kept their positions without much shifting during earth-history, save perhaps that fragments of the continents have occasionally foundered and become ocean-floor for ever. But much of the continental masses is close to the water-line; and this marginal region is always changing, so that at times a very considerable area of the continents may be flooded with shallow seas, while at others almost the whole of their masses may stick up above water (Figs. 208, 209).

When the land is up, the life of shallow waters is put through a time of reduced territory; when it is down, the same holds for animals and plants that are fully adapted to dry land. When lands are up, there are bridges from one continental mass to another, for land creatures to pass over; when they are down, the bridges disappear and each separate area becomes a centre of local evolution. It is possible, too, as Joly suggested in his *Surface History of the Earth*, that Wegener's theory of Continental Drift may apply during the revolutions, and that at such times the continents may not be fixed in place, but may be shifting gradually, floating so to speak, upon a quasi-fluid mass below, like rafts floating upon extremely viscid tar; and if so, that they may shift slowly from one position to another upon the face of the earth, so that an old connexion may be severed and a new one made. But that is another and more speculative story.

We live now in a time of almost maximum emergence. We can picture to ourselves what submergence must mean to land life, if we try to imagine what would happen to humanity if the land-area of the world was reduced by one quarter. And let us remember that when the next cycle of submergence falls due, the land-area of the world undoubtedly will be reduced by at least a quarter. That will not be for a time to be reckoned in millions of years, and the nibbling away of the land will be very gradual. But the resultant competition will be keen.

In Evolution it is the changes of climate which are important rather than the absolute temperature or rainfall; and these changes, according to such careful investigators as Brooks, are due in the main to the earth-movements that bring about changes in the distribution of land and sea.

When much land is out of water, the general temperature of the world will be lower and the climate of the lands will be of the continental type, cold in winter, hot in summer, and comparatively dry. The greater the elevation of the land and the higher the latitudes in which it happens to be, the greater the effect on climate; Brooks, in his

Evolution of Climate, gives some figures showing how great it can be. If the whole great belt between 50° and 70° North latitude were ocean, then, it can be calculated, the average January temperature in its centre would be 29° F.—only just below freezing; and the average July temperature 41°—only 12° higher than in winter. If that whole belt could be made land, the July average would go up about 30°, the January average fall by about 60°, giving 72° and $-30°$ F., or a range of over 100°. The effect, too, is greater as you go from equator to pole; that is to say, a world with much land submerged will have a comparatively uniform climate, while one with much land out of water will be sharply zoned into tropic, arid, temperate, and polar belts.

The climatic effect of great land-emergence is largely due to interference with the world's heat-circulating system. It alters the circulation of the air—a complicated affair whose explanation we must leave to the meteorologist. And it alters the circulation of the waters. When seas are broad and there exist neither polar continents nor land making a ring round an enclosed polar sea, there is constant circulation, warm water flowing polewards along the surface, there being cooled, sinking to the bottom, and drifting back along the ocean floor. No sea-ice can form near the poles, and warmth-loving creatures can venture right up into the seas of the Arctic Circle.

But when, as in the northern hemisphere to-day, the polar waters are hemmed in by land, the circulation is impeded, and ice can form over the Arctic Sea. A large extent of high land at the pole, as in the Antarctic to-day, will also inevitably get ice-covered; and ice, whether on land or sea, at once lowers temperature. If our imaginary belt from 50° to 70° were entirely land, but also entirely ice-covered, this would have little effect upon its winter temperature, but would bring the summer temperature down by some 50° to 23° F., or 6° lower than its winter temperature would be if it were all water.

Thus the surface of earth is constantly changing in a slow and not irregular rhythm. The waters invade the lands; the lands emerge. The climate passes from mild and oceanic to severe and continental, from wide uniformity to sharp zones and back again. And every change has its repercussion upon life.

But space forbids our elaborating this record of the changing scenes and changing conditions which have tested the versatility of life and obliged it to develop its possibilities. It is still premature to think of a detailed account of the changes of climate which have befallen the earth. For our purpose it will be enough if we draw attention to the occasional times of stress which befell the planet, times of cold or drought which

THE PROLOGUE

spelled either extinction or progressive evolution for life. For the rest, let us remember that these periods were rare, and that for most of geologic time the climate has been warm, with equable conditions extending much farther polewards than to-day; and that the scenery, compared to that which we enjoy, has mostly been what we should consider dull, with low-lying lands, mountains worn down to stumps, sluggish rivers and often great lagoons and marshes.

And now, having discussed the main changes of scene, we are going to trace some of the main ventures of living matter in adaptation to these changes. Life began in the sea and we shall begin with a discussion of its opening phases. Thence we shall proceed to the history of the higher groups of sea creatures. But life touches its highest levels on land; we shall go on to an account of the invasion of the land—a very slow but curiously interesting process accomplished by a number of independent invaders. Next will follow an account of the Middle Ages (IV) of life—its first great blossoming on land and in the air, and then of the rise of Modern Life (V), and how it replaced the old and worked up to its own new strange climax—ourselves. There came a time when one line of evolving creatures developed a new way of life as different from all that had gone before as is life on land from life in the water. This new way of life was based on reason, speech, and tradition, and the creature that took to it is called man. Our last chapter will deal with the dawn of man upon the marvellous ascendent drama.

II

LIFE BEFORE FOSSILS

§ 1. *The Earliest Life.* § 2. *Union Is Strength: the Building of the Plant-body.* § 3. *Marthas and Marys Among Cells: the Origin of the Higher Animals.* § 4. *The First Brains.* § 5. *Blood as a Step in Evolution: the Cœlomates Appear.*

§ 1. The Earliest Life.

THE later and grander chapters of the history of life are based directly upon the testimony of fossil remains, which are like so many stepping-stones leading us back into the past. But in ages before the Cambrian, as we have seen, these fossil facts are very scarce, and in the earliest times they fail us altogether. Yet, though we have no direct evidence bearing on life's early history, that is not to say that we can form no idea of it. We can do so by falling back upon indirect evidence, such as we set forth in Book 3, especially the evidence of comparative anatomy and that of embryology. With these clues to guide us, we can deduce a great deal of what happened in those blank ages when the fossil record is silent, just as a skilful detective can reconstruct the events leading up to a crime from circumstantial evidence. Thus tentatively and rather speculatively let us sketch out a bridge to span the formidable time-gap between the first appearance of life and its varied abundance in the Cambrian (III A)—a gap of at least five hundred million years.

Whatever the form in which life first appeared, we can be reasonably certain that most of the life-stream soon organized itself into small units, reproducing by equal division. For a large part of that monstrous interval of time the structure of the cell and the mechanism of its nucleus was being steadily evolved. Every few hours a new generation of these units was subjected to the approval or condemnation of Natural Selection. These earlier ancestors of existing organisms must have been either cellular in their construction, like the Protozoa, or what we may call sub-cellular, without any definite nucleus, like the bacteria and spirochætes and filter-passing viruses.

We can be fairly sure that there very soon arose those main differences in feeding habits which have ever since marked the lines of cleavage between the major groups of life. One main type exploited the sun's energy by means of chlorophyll, the earliest and simplest green plants. A second group exploited the first, devouring what had already been manufactured instead of setting up manufacture for themselves; and these were the pioneer animals. A third, comprising bacteria and their predecessors, lived in all sorts of strange ways exploiting—or, to be more exact, inventing—decay; or thriving in out-of-the-way chemical environments, such as those rich in iron or sulphur, some living without oxygen, others extracting nitrogen directly from the air. To them must soon have been added the first simple fungi, plants which apparently gave up their original green independence to absorb the rich products of decay or to suck food from other living bodies. We can guess that in those early experimental times all kinds of intermediates existed, combining different methods; and indeed some of these have representatives still living, half plant and half animal, like Chlamydomonas (Fig. 115), or half green plant and half fungus, like some other flagellates.

Doubtless this primitive cellular and precellular life radiated out slowly, age by age, into much of the variety we see to-day, and there appeared amoeba-like forms and radiolaria, flagellates and ciliates, diatoms and blue-green algæ, and many kinds of bacteria. And doubtless a huge variety of unicellular experiments of the most extraordinary sort were evolved and perished again. After the account we have given of mitosis, of chromosomes and genes, it should be clear to the reader that this phase of the evolution of existing cell-structures, minute as they are, such as we find in all the multicellular creatures, was as great a step from the first unorganized specks of living matter as the step from the finished unicellular organism to man.

§ 2. Union Is Strength: the Building of the Plant-body.

For all its variety, life that reaches no higher than the cellular level must remain primitive and feeble, for the reason that its units are, and must remain, so small and so simple in their reactions. The next great step onward was the establishment of larger units. Undoubtedly life made experiments along many different lines towards this end. Sometimes the cells themselves were enlarged, and their nuclei multi-

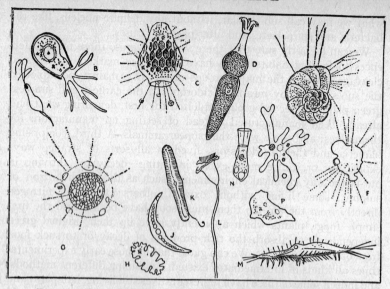

FIG. 210. Some of the forms into which single-celled life has evolved.

(A) A Flagellate, Trichomonas, from the mouth of man; it has three whip-lashes (one two-thousand-five-hundredth of an inch long). (B) Cryptodifflugia, a creature like Amœba, but with a horny shell (one two-hundredth of an inch long). (C) A Radiolarian from the open sea, with a flinty skeleton and a network of fine threads of living substance (shell one two-hundredth of an inch long). (D) A Sporozoan, Corycella, a parasite in the intestine of the whirligig water-beetle; it hangs on with its hooks. (E) A foraminiferan spreading a net of protoplasm. (F, G) Two kinds of Sun Animalcules surrounded by rays of living substance on thin spikes (magnified about 350 times.) (H, J) Two single-celled plant organisms, of the group Desmids. (K) Another single-celled plant, a brownish Diatom, moving by protruding a microscopic "punt-hole" of thick slime that hardens on contact with water. (L) A Ciliate, Vorticella; body about one five-hundredth of an inch long. (M) Another Ciliate, Stylonychia, crawling with its bristles composed of cilia joined together; about one three-hundredth of an inch long. (N) A Flagellate with a collar. (O, P) Two forms of Amœba. For fuller references to all the above consult index.

plied, as may be seen to-day in many Foraminifers and Ciliates; but there is a mechanical limit to such a process, and only a few cells have reached a size to be measured even in millimetres. Or, as in Slime-moulds (Book 2, ch. 6, § 6), they merged together and lost their individuality in a common mass of living protoplasm. But this too leads to a dead-end—organization is impossible in such a mass, and nothing more than a sheet of slime has come out of it.

But union is strength, and cells can unite without sacrificing their individuality. After cell-division, the daughter-cells can remain attached, and so large groups of cells can be built up. The simplest of such groups are the chains made by many kinds of bacteria and blue-green algæ and the flat irregular cell-plates produced by various green flagellates. Besides these, sedentary protozoa often form branched colonies with a cell-unit at the end of each twig, and a few active flagellates also stick together to build free-swimming colonies. And there are flagellates with a rhythm in their life-histories—first free and then fixed in colonies, then free again, and so on. In figures 115 and 211 the reader will find some of these primitive aggregations.

Higher plants in all probability arose from chain-like colonies. The Thread-alga Spirogyra (Fig. 211) is a lowly plant whose cells have thoroughly adapted themselves to a permanent end-to-end existence. In most lower seaweeds and fungi, a filament such as this, built of a single row of cells, is the fundamental unit of life; and it often remains so even when the outer appearance of the plant betrays no trace of it. The elaborate fruiting arrangements that we call mushrooms and toadstools are mere felt-works made only of an interweaving of separate filaments, and the same is true of almost all red and many brown seaweeds. In all higher plants, on the other hand, from mosses upwards, the single cells are cemented together, and there is no trace of separate filaments. A woven wattle construction, so to speak, has been abandoned for solid brick-and-mortar. Only in the higher seaweeds, such as the bladder-wracks and Laminarias, is a beginning made with this new type of construction. In green plants, the passage from one method to the other clearly took place in the seaweed stage, and was a necessary preliminary to the attainment of great size and to success on land.

The early water-plants soon began to show division of labour among their parts. In most seaweeds, the plant-body, no longer confined to the surface between rock and sea, sticks boldly out into the water. But to do this it must develop organs to attach itself. These are not, strictly speaking, roots, for they are not concerned with the absorption of food materials; they serve only the mechanical function of anchorage; and may be called hold-fasts. A stem too may be evolved to connect the hold-fast with the expanded food-manufacturing part of the plant nearer the light. And there are special regions set apart for reproduction.

No longer are all the parts of the plant actively engaged on the task of making food; and so some means of internal transport must be evolved. This is already present in some of the higher seaweeds like Laminaria, in the shape of so-called sieve-tubes—minute cylinder-cells joined end to end, with perforations like a sieve's, putting each one into communication with its two neighbours. Along these living tubes food-stuffs can be passed slowly from one end of the plant to the other.

Of wood, of true roots, of bark, of real leaves, of flowers or seeds, no trace was developed in the sea. Their evolution took place in what we may call historic times. For all that, the plant-body found in some of the higher seaweeds is fairly elaborate, and may be very large (the giant kelps of the Pacific may reach six hundred feet); and to this pitch it is probable plants attained before the full fossil record begins.

§ 3. Marthas and Marys Among Cells: the Origin of the Higher Animals.

BUT chains and filaments, though they serve a plant's turn, are no use to the average animal, with its need for activity and accurate movement; and only in a few parasitic protozoa are such chains of animal cells to be found.

We are very much in the dark as to how the many-celled animals arose from their single-celled ancestors. Presumably the first step was some kind of colony-formation. But we cannot be sure whether these first aggregates were fixed to rocks or weeds, or swam freely in the open water. Probably the former is true. But in either case the salient feature must have been an incipient division of labour among the members of the little cell-community. Even the simplest sponge has at least three main kinds of cells—outer cells forming a protective layer, flagellated cells sweeping in food, and amœba-like cells from among which are recruited the reproductive gamete-producers. If we may judge by protistan colonies, the first step in the division of labour is usually between the soma cells, the Marthas that do the daily work of the individual's body, and the germ-cells, the Marys, tended and looked after by their sisters, which serve the more remote ends of the race.

A beautiful example is seen in the green flagellate Volvox, whose globular colonies contain over a thousand cells, and swim in a co-ordinated way, with one part always foremost. But this creature, though it helps us to visualize the sort of way in which the gulf from one-celled to many-celled must have been bridged, is probably not very like the actual ancestor of either higher animals or higher plants.

What is reasonably certain is that of all the many attempts of animals to solve the riddle of a larger life, only two succeeded—those

that originated the race of sponges, and those that gave rise to Cœlenterates, and so to all other Metazoa. Apparently these evolved
quite separately. Sponge did not give rise to polyp nor polyp to
sponge; the two sprang independently from their single-celled forbears.

Of these two experiments the sponges were the less successful. They

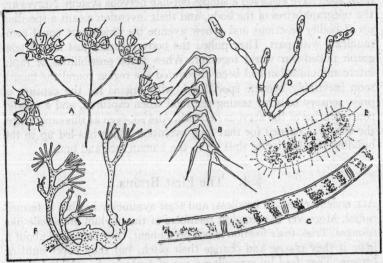

FIG. 211. Co-operations of cells.

(A) A colony of collared flagellates (each one like a single collared cell of a sponge—
Fig. 103), much magnified. (B) The marine flagellate Ceratium multiplies by fission;
sometimes the individuals stay temporarily connected in chains (magnified about 80
times). (C) A portion of a filament of the thread-alga Spirogyra. The cells are joined
end to end. They have a spiral green band running round them. A single thread is just
distinguishable to the naked eye. (D) A half-way stage to the thread-alga. Chlorodendron
is a colony of green cells that live at the end of empty jointed tubes and sometimes break
off and move away. (E) A floating colony. Collozoum consists of hundreds of Radiolarians jointed together into gelatinous sausages up to two inches long. (F) Part of a
colony of Dendrosoma. Each "individual" has a crown of suckers, and many grow
nearly one twenty-fifth of an inch high. They are all connected by living tissue.

are one of life's blind alleys. They have succeeded in colonizing fresh
water as well as all parts of the ocean, in assuming an extraordinary
diversity of shape, in building skeletons of lime, of flinty spicules fine
as glass, of horn so inimitably delicate that we choose it above all other
substances to wash our faces with; but they have remained permanently rooted to the bottom, they have never achieved even the most
rudimentary nervous system or any effective unification.

This doubtless is largely because they have no single mouth, but only a multiplicity of current-orifices. The cœlenterates, on the other hand, the rival experiment in animal-building, were more progressive. They specialized their tissues and anatomical parts more thoroughly. They acquired a true mouth opening into a digestive cavity, and they must soon have achieved a simple net-like nervous system. Nerves are the telegraph wires of the body, and their invention meant a speeding up of bodily reactions and a new avenue by which part could communicate with part. They pulled the body together just as the telegraph pulled our world together. When these possibilities of rapid intercommunication had been discovered the region round the mouth soon inevitably became specialized, with organs for the capture of prey, sensory cells for testing the food when captured, and a greater concentration of the nervous network. And so, even at this early stage, the way was opened for that long evolution which has led up to the building of a head, and thus up to the human face and brain.

§ 4. The First Brains.

ALL cœlenterates are headless; and their symmetry is what is termed radial. Moreover the simplest members of the phylum are sessile like sponges. True, their fixation is not permanent, they can crawl about a little if they choose and change their pitch; but they never hunt or browse. They feed by spreading a net of tentacles to catch whatever chances by. Thus, motionless, the early cœlenterates must have fed. Presumably they took smaller fry than their descendants do to-day— single-celled forms or early experimenting free-swimming colonies— for the crustaceans and so on which nourish a modern polyp had yet to appear.

The next great steps to be taken were first to unstick the animal from the bottom, to emancipate it from its sedentary habits, to send it freely adventuring through the sea; and secondly to confer upon it a bilateral symmetry like our own, with back and belly, right and left— for without such symmetry vigorous locomotion is apparently impossible. All the swiftest machines of travel—ships, aeroplanes, motor-cars—are bilaterally symmetrical, and so are the most active animals.

Jelly-fish and siphonophores, some sea-anemones and the extinct Graptolites are all cœlenterates which emancipated themselves from the bottom; but they show no signs of bilateral symmetry, and are side-branches which have never given rise to anything above the cœlenterate level.

Probably the first two-sided animals were not unlike the flatworms of to-day. They took to crawling on one side only of the body and with one end always in front. This was accompanied by further advances in the specialization of the parts. Thus, as a concentration of sense-organs and nervous tissue in front, there began the evolution of the head and brain.

But this process of concentration and unification we shall consider at length when we come to the behaviour of animals. Here, after reminding ourselves how vitally important the unification was, we will consider only the main steps taken in the general construction of animals and the ways of life they opened up.

§ 5. Blood as a Step in Evolution: the Cœlomates Appear.

The flatworms, for all their new-developed head and nervous system and middle layer of cells, had no blood-system. It is because of this that they have to be flat—so that no part of the interior shall be too far from the life-giving oxygen of the water round them; it is because of this that all their organs must be branched—so that there shall be always a bit of every kind of organ near bits of all the others with which it may have to carry on chemical trading. However, the flatworms are the simplest known creatures to have a regular excretory system. The animal is drained as you would drain a field, by running in drains from the outside into the soggy interior.

The next great step was the development of blood and the tubes for it to flow in—transport system and marketing facilities in one. Equipped with a blood-system, animals can grow bulky in three dimensions, and yet have their organs compact and solid; the blood can bring oxygen to their most interior regions as readily as trains can bring fresh fish from the sea to Paris or Berlin, and the all-penetrating blood-vessels can act as go-betweens even for the remotest organs. So we get the possibility of solid animals like the roundworms.

In roundworms another innovation appears. In lower animal types, the undigested residues go out by the same door through which the food is taken in. But soon the digestive tube acquires a second opening, the anus. The original orifice for the first time becomes only mouth. This has obvious advantages, making it possible for the animal to feed continuously, and to have its orginally simple intestine specialized into a succession of regions, each with its own particular use in digestion or storage.

Next came the formation of the body-cavity or cœlom—the watery space in which the organs of our chests and bellies are suspended. This water-jacket made the intestine and body-wall more independent of each other's movements; we, like other cœlomate animals (and the great majority of many-celled animal groups are cœlomate) can have our intestines writhing away within us, churning the food and passing

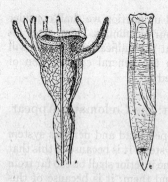

FIG. 212. Improvements in the construction of the nervous system.

Diagrams of the nervous systems of: (A) a Polyp such as Obelia (Fig. 173); it has a nerve-net of interlacing fibres. (Part of the body-wall has been represented as cut away). (B) Of a Planarian; it has a centralized nervous system, consisting of two main strands of nerve-tissue, the nerve cords, which connect with the brain at the front end. From the central nervous system springs a branching and often interlacing system of nerves.

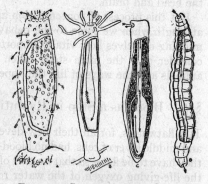

FIG. 213. Improvements in the constructional plan of animals.

(A) A primitive sponge with many small apertures for intake and one large one for outflow. (B) A simple polyp such as Hydra has a digestive cavity with only one aperture and no packing between its wall and the skin. (C) A simple flatworm. The digestive cavity still has only one aperture, but it is separated from the body-wall by a packing of connective tissue with various organs embedded in it. (D) A segmented worm. The digestive tube has two apertures, and is separated from the body-wall by a space, the coelom (which in these animals is subdivided by membranous partitions).

it on its way, while preserving our outer abdominal wall impassive. Moreover, the fluid in the cœlom, with its numerous scavenger-cells like white blood-corpuscles, can act as a sanitary cordon to prevent any bacteria escaping through the gut-wall into the rest of the animal's tissues.

From some worm-creature such as this, with central nervous system, blood-tubes, and tubular gut open both before and behind and running a large part of its course through a cœlom-jacket, all the higher groups of animals must have branched out.

One main line gave the world such diverse creatures as earthworms, molluscs, all the arthropods, and probably the tiny wheel-animals or rotifers as well. The molluscs and the worms, for all the differences between them, show such extraordinary and detailed resemblance in their larvæ and early development that they must have sprung from a common stock. The primitive mollusc's chief biological inventions (we use this convenient phrase as a metaphor, without meaning to imply any fore-thought or design on the evolving animal's part) were the flat muscular foot, whose waves of contraction sent the animal gliding along over the bottom, and the protecting shell over the hump, into which the compact body was up-arched; and they had developed special plumy gills to oxygenate their blood. In a stage scarcely more advanced than this we find their fossils in the Cambrian rocks; and in the Cambrian we will take up their further history.

The bristle-worms, on the other hand, set out along a different road. They became segmented: in other words, they multiplied the middle part of their bodies over and over again, fitting in a whole set of trunk-regions between the little brain-containing tip in front of their mouth and their hinder end.

Many worms propagate by transverse fission, and, as we have seen (Fig. 97), sometimes the separation of the halves is retarded so that temporary chains are produced and often the development of heads in the hinder members of these chains is partly or wholly suppressed until after they become separated. We may hazard the guess that the segmentation of bristle-worms arose by still further retarding the separation of the members of such chains and still further suppressing the formation of heads in them. Meanwhile, these worms developed the chitinous bristles from which they take their name. These, protruding from special lobes on each segment, served as miniature punt-poles by which the wriggling animal could push itself along, or in some cases as oars for swimming.

In the repetition of parts which we call segmentation lay great evolutionary promise. It gave the animal many of the advantages held out by colony-formation without its disadvantages. A colony has as many heads as it has members; in it the work of unification has to begin all over again. But a segmented worm keeps a single head on its shoulders; the multiple trunk is always under unitary control. By this means the segmented creature not only gains size; it is also provided with a multiplicity of similar parts, among which a division of labour, like that between the different cells of a cell-colony or the many-celled members of a Siphonophore, can arise.

From segmented worms, without doubt, the great phylum of the arthropods is descended. To effect the transformation, only two main changes are necessary. You have to turn the delicate cuticle of the worm into a heavier sheet of chitinous armour, capable of acting as a skeleton and providing rigid leverage for the muscles; and you have to convert the bristle-bearing outgrowth on either side of each segment, often an elaborate enough organ even in the worms, into a jointed appendage, a true limb. You have then a creature with endless possibilities of mechanical elaboration. The chitin armour, though conferring great powers of resistance, is yet flexible since it lends itself readily to jointing; even in a tiny arthropod like a sandhopper there are some five hundred separate jointed parts in the skeleton. The large and often indefinite number of segments of the worm, usually all very much alike, give place to a smaller and definite number, each with its own fixed function. Several segments are unified to form a head, some appendages become jaws, others sense-organs, others limbs. The details of the division of labour have been worked out in a different way in each main group of arthropods, just as the actual details of government and administration are worked out differently in different states, but that we need not discuss here.

The transition from worm-type to arthropod-type was of fundamental importance, since it gave to life new possibilities of elaborate behaviour and of speed, and so made better sense-organs a biological necessity. And the tough outside armour made it easy for arthropods, when the time came, to emerge into the air without being promptly desiccated.

With this enlargement of possible activities many more water-arthropods than worms have abandoned the sea-bottom for an active swimming life. But these are for the most part small creatures like copepods and other whale-food; the greatest triumphs among aquatic arthropods were crawlers still. The ideal of construction for combined strength, size, and speedy swimming had not yet been evolved.

The rotifers, or wheel-animalcules, are a strangely twisting branch of the evolutionary tree. They seem to be precocious worm-children, in which the bottom-living adult stage has been cut out, and the reproductive organs made to mature in the tiny larval phase, in order to take fullest advantage of the teeming surface life of open waters. But the experiment had only a lukewarm success. They specialized in various ways. They swarm to-day in ponds, streams, damp moss, gutters and suchlike places, but withal they are tiny and insignificant. They never rose to any sort of dominance in the world of life.

A number of other types of cœlomate creatures went down the evolutionary primrose path towards a more sedentary life, fixing themselves temporarily or permanently and adopting the method of ciliary feeding to take advantage of the rich but tiny débris of the bottom zone. Many of these developed some sort of tentacles, richly beset with cilia to sweep the food towards their mouths. One common result of such a way of living is the bending round of the digestive tube so that the anus opens near the mouth, where it can freely discharge its refuse into the open water; and another is the enclosure of the body in some sort of protective shell. Sometimes, too, the animals become colonial. The most familiar examples of phyla which have pursued this path are the Lampshells and the colonial Bryozoa.

Finally, there remains another main line which early branched, it seems, to give rise to two very different products, the Echinoderms and the Vertebrates. As finished products these are even more different from each other than are snails and lobsters; however, not only do both of them share the unique character of having skeletons of lime manufactured below the outer skin by cells of the middle layer, but the resemblance in various important features of early development between Echinoderms and primitive chordates, such as Amphioxus, is so striking as to make most biologists believe that the two groups must have had a common ancestry—though their divergence must be dated long before the first legible pages of the fossil record.

Both, too, seem to have begun as ciliary feeders. The Echinoderms early became fixed, and swept in their food by food-grooves converging upon the mouth. In such a state we find them in the Cambrian, and the subsequent appearance of the sluggish but freely motile forms we know to-day is a matter of geological history, not of speculation.

The chordates, on the other hand, took a quite different path. They developed a current-sifting machinery which is unique in being internal, a "gas-mantle" distention of the front end of the digestive tube perforated by slits opening to the outside. At first this was a feeding device of the current-sifting type and, as we saw in Book 2, it presents itself in the most primitive chordates to-day, but later it was turned to other uses. And at the same time these pioneers developed an unique and very important set of features. They became extremely muscular, they developed a long elastic skeleton-rod, the notochord, down their backs on which the muscles could pull, and they grew a projection of the body behind the anus—a tail—which helped them in swimming.

It would be hard to exaggerate the evolutionary importance of this early vertebrate invention of swimming by side-to-side undulation of the whole body, like a fish, or a lamprey, or a lancelet. It is by far the swiftest and strongest method of progression that aquatic life has discovered. Its effects are written throughout the vertebrate frame—even in those vertebrates, like ourselves, which left the sea ages ago. The notochord, and therefore its replacement, the backbone, the segmental pattern of muscles and nerves, the lateral-line sense-organ and its

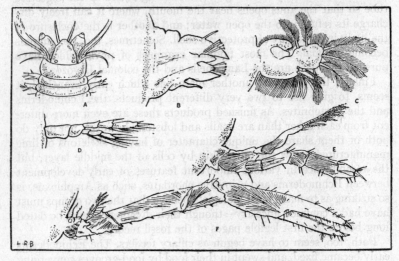

Fig. 214. The evolution of segmental limbs.

(A) *The front of a bristle-worm (the sandworm, Nereis), showing two pairs of its primitive limb-like organs on the body segments behind the head. (B) The limb of another bristle-worm (the lugworm, Arenicola), magnified. (C) A limb of the water-flea, Simocephalus. (D) The swimming foot of the copepod, Calanus. (E) The future big claw of a lobster larva. (F) The same of a nearly adult lobster larva.*

specialization the ear—these are all results of our ancestral undulation.

You have only to look at a cross-section of the best-developed segmented worm and of the lowliest free-swimming chordate like Amphioxus to see how far more muscular the latter is. Moreover, the worm, like nearly all invertebrates, is tailless. The early, experimenting chordates shifted the belly-viscera and anus forward and confined them to the front part of the body so that the hind part could be exclusively used for swimming. This, the tail, speedily became nothing else but muscle, together with its necessary supports and protections and the supply of blood-pipes and nerve-cables needed to nourish it and regulate its actions. One line of worms, the carnivorous arrow-worms, represented to-day by Sagitta and its relatives, developed a tail and musculature rather similar to that of the chordates, but it never produced the mechanical support of a notochord.

The trunk and tail of a water chordate, a fish, for example, is a single flexible unit, and when it swims, the whole body comes into action; the animal does not merely protrude punt-poles or oars, like the rigid-bodied crustaceans. Lobsters and crayfishes swim, but backwards and without precision, by flapping the whole hinder part of the body, and some worms swim with their whole body, but ineffectively; their muscles are far more feebly developed, and their bodies do not run out into a tail of pure muscle, designed to transmit to the water the body's side-to-side undulations with maximum efficiency.

This it was that made the vertebrates, as we have said, the athletes of the animal world; the speed and the power thus conferred upon them called for, promoted and encouraged the evolution of better brains and sense-organs, better co-ordination, better stream-lining of their form, and made it possible for them to reach sizes unattainable by the arthropods, weighted down as these were with armour, and with armour that must be shed and re-manufactured periodically to allow for growth. When it moults, an arthropod is soft and defenceless, and such a necessary periodic retirement must always have been a grave handicap to these possessors of external skeletons.

The larger vertebrates reach sizes unattained even by the giant cuttle-fish. Cuttle-fish, squids and octopuses are in some respects the proudest masterpieces of invertebrate organization, but they do not apply their muscular power so efficiently as the fish. When they swim by vibratile side-fins, their speed is low; when they go fast, by shooting out water through an opening under their heads, they retire in backward spurts, a mode of progression so blind and risky as to be reserved largely for panic emergencies.

It is very probable that the first vertebrates were fresh-water forms. The earliest vertebrate fossils, from the Middle Ordovician (III B 2) right through the Silurian to the Lower Devonian (III D 1) are found in fresh-water or estuarine deposits. One cannot watch river fish hanging almost immobile in a swiftly flowing stream without realizing that this invention of swimming by sinuous lateral movement is uniquely adapted for withstanding a current and preventing an active river creature from being swept out to sea. In this way, perhaps, our ancestors learnt to swim. If this be correct it is but one of several examples of structures or habits which, evolved to overcome some local or temporary difficulty, have later carried their possessors to widespread and permanent success.

But whether the tail was invented in a river or not, its importance to our evolution cannot be denied. Nowadays, we human beings, like the fox in the fable, are inclined to find tails disgraceful and to think it ignominious that we should possess such appendages even in unconscious embryohood; but let us not forget that our stock swam to supremacy before the fingered limb had been invented. In a very real sense it was our tails that made us what we are.

III

THE ERA OF CRAWLING AND SWIMMING: EVOLUTION IN THE WATER

§ 1. *The Opening Eras: the Archeozoic and the Proterozoic.* § 2. *The Age of Ancient Life: the Paleozoic.* § 3. *The Cambrian Period: Age of Trilobites.* § 4. *The Age of the Sea-scorpions.* § 5. *Echinoderms: a Phylum Which Has Never Produced Land Forms.* § 6. *The Ancient History of Molluscs.* § 7. *The Earliest Fish.*

§ 1. The Opening Eras: the Archeozoic and the Proterozoic.

THE record of the rocks opens like many great books, with endpapers, flyleaves, and blank pages. The oldest sedimentary rocks known to us are enormously changed from the state in which they were first deposited, and they give us no clear evidence of organized life, if indeed it existed in that hot and stormy phase of our planet's history. Huge deposits of very early sedimentary rocks, now baked and squashed and crystallized into new mineral forms, exist in Canada and Greenland, with outlying bits elsewhere. At least two great cycles of mountain-building took place while those sediments (there are over 100,000 vertical feet of them in some places) were deposited, and of the vast stretch of time from the earliest known stratified rocks to the present, this period took up nearly a third. We call it now the Archeozoic Era (1), the age of earliest life, since as we have given reason to suppose life may have existed through all or most of its hundreds of millions of years.

For reasons we have already made clear the actual traces of life are of the slightest description in these Archeozoic rocks. There are some doubtful evidences of primitive algæ, some possible wormburrows. There exist great beds of carbon in the form of graphite, and these, as far as our chemical knowledge goes, must be derived from the remains of living things, most probably aquatic plants. That is nearly all the positive fact we have to go upon. We can hazard a reasonable

676 THE SCIENCE OF LIFE

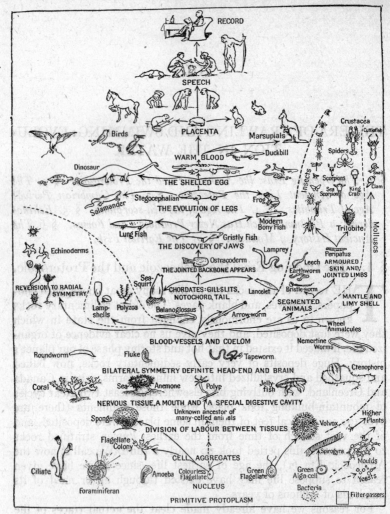

FIG. 215. The Ascent of Life.

guess that bacteria and seaweeds of various types were steadily making the world a place fit for higher animals to live in, and that single-celled animals and, it may be, sponges and Cœlenterates and worms were in existence before the period closed; but that is all. If we walk over the great stretch of Archeozoic rocks of Eastern Canada, we have beneath our feet all the earliest pages of the book of life, but they are almost blank, any history they may once have carried now destroyed for ever.

The record is hardly richer when we come to the next great era, the Proterozoic (II), which includes about a third of all time subsequent to the Archeozoic. On the rocks of this Era, however, the metamorphic forces have been less effective—largely, no doubt, because they have had less time, and because there was less weight of superincumbent material—and many of the sedimentary rocks preserve their original character; right at the base of the Proterozoic, perhaps a thousand million years ago, the sea made ripple-marks which we can look upon to-day. And in general, the character of those early deposits is very like those of more modern times, indicating that the general conditions of the world were becoming stabilized and were no longer markedly different than from those obtaining now.

Late in this Era there was a huge deposition of iron compounds on a scale unknown at any other period, over the bottom of shallow seas; the great bulk of iron mined in the United States to-day was laid down in the Proterozoic (II). Possibly, though we cannot be sure, this iron was deposited with the help of special bacteria and algæ.

Then there occurred a much more remarkable event—an Ice Age. Great stretches of boulder-clay, the product of melting ice-sheets, with ice-scratched stones and boulders embedded in them, have been found in Canada in deposits of early Proterozoic age. Similar deposits from Australia, Norway, China, and elsewhere, seem to belong to a second glacial period right at the close of Proterozoic times, probably separated from the first by a two-hundred-million years' interval. Thus early in the book of earth we receive a comforting assurance that our own particular ice age, in whose aftermath we are still living, does not herald the rapid cooling down of our planet, but is one of a series of recurrent chills to which, at rather long intervals, it is subject.

In the last thirty years a number of definite fossils from these beds, previously thought barren of life, have come to light. Calcareous masses possibly deposited by algæ are not uncommon; among animals Radiolaria, sponge spicules of many kinds and unmistakable worm-tubes have been found; and there are some remains which may prove to be an early sea-scorpion, and have been christened Beltina. But other animals, we believe, with every reason for our assurance, must have existed, and any day a lucky chance may bring their fossil remains to light.

§ 2. The Age of Ancient Life: the Paleozoic.

THE next pages in the book of earth history are those of the Cambrian rocks (III A), first of the volume we call the Paleozoic (III), the Era of Ancient Life, as it was christened in days when no earlier life was known. Abruptly, without warning, they reveal a world of living things whose abundance and diversity is in striking contrast with the poverty of the preceding records.

In the Lower Cambrian (III A 1), sponges and lamp-shells were abundant, and so were the strange and primitive Arthropods known as Trilobites. Jelly-fish, like those of our own day, have left their impress on the shales, though true corals had not developed. But enigmatical creatures with limy skeletons, called *Archeocyathus*, were abundant, sometimes building veritable reefs. They take their name, which means "ancient cup," from their vase-like shape. They have a curious construction of one cup within another, the space between filled by a rafter-work and communicating by pores with the inner cup. This is unlike the construction of any known Cœlenterate or of any known sponge, and some authorities believe them to represent a third type of many-celled animal independently derived from single-celled forms, which failed to hold its own and soon petered out to extinction. Various snails and their relatives, though still rare, are known. The Echinoderms had, it seems, already differentiated into several groups, including the sea-lilies and the sea-cucumbers.

Most Cambrian animals that happen to have been preserved as fossils were bottom-dwellers of the inshore waters, but by a happy chance the Burgess shales of the Middle Cambrian (III A 2) of Canada reveal a rich assemblage of animals which when alive led a floating and swimming life in the open sea. These open-sea animals include jelly-fish, active swimming bristle-worms and arrow-worms, some surface-living Trilobites, and a large number of different types of true crustaceans, mostly not unlike the water-fleas and fairy shrimps of our modern fresh waters. In the mud into which the corpses of these animals fell, there lived burrowing worms of the same type as our mud-burrowers of to-day. Thus not only had animal life already diverged into many groups, but within each group it had split up into many types adapted to distinctive ways of living.

In all subsequent periods this abundance of fossils is maintained. The contrast with the earlier rocks is startling. Passing from the Proterozoic (II) to the Cambrian (III A) is like passing in human history from a time of shadowy legend to one of well-preserved written record. What is the explanation of this sudden abundance of the remains of life? There seem to have been a number of causes. Most important, a long gap of time seems to have elapsed between the laying down of the latest Proterozoic and of the earliest Paleozoic rocks. In no part of the world have any sediments been found which bridge this gap, and that it was long is shown by the fact that wherever Proterozoic rocks are found with rocks of the next period deposited on top of them, they have been first tilted and then worn down to a flat surface again. The material thus eroded away must have been carried off and laid down somewhere in the seas of the ancient world: and yet no trace is known of the layers thus deposited.

Perhaps—the suggestion has been made by competent geologists—perhaps there was a folding of the crust which deepened the ocean basins and so caused the sea-level to fall and the very foundations of the lands to be exposed. If so, the lost sediments are lost because they are all now far out beneath the ocean. Later, there was presumably some readjustment of the ocean floor; the sea filled up with waters liberated from the earth's interior by volcanic action; and the water-line was raised again to within its normal limits. Be that as it may, the gap is there. During the interval of millions of years which it represents, Evolution was busy and many new types came into existence. But this will clearly not account for more than a portion of the contrast.

To account for the new abundance of fossils, it has been plausibly suggested that during this gap in history, animals had succeeded for the first time in constructing strong skeletons with the aid of lime, and so for the first time rendering their remains easy of preservation as fossils. In support of this idea, we find that many creatures with limy skeletons—the molluscs, for example, and the Echinoderms—have them poorly developed in the Lower Cambrian (III A 1), and do not produce them of what we can call normal thickness for another fifty million years or so. Perhaps this was a new acquisition of life; perhaps the lime-supply of the earlier seas was low, or it was carried out of the reach of animals by lime-depositing Algæ. Whatever the truth may be, it certainly looks as if a real advance in the construction of hard skeletons was made by various separate phyla of life shortly before the beginning of the Paleozoic (III)—an advance as important to the modern man of science anxious to unearth the history of life as to the long-dead creatures which achieved it.

§ 3. The Cambrian Period: Age of Trilobites.

BY THE beginning of the Cambrian Period (III A), over half of geologic time had elapsed—over half the time between the deposition of the earliest sedimentary rocks and the present day. And yet what an abundance of living things was still to come upon the record, how many of the types familiar to us were not yet in existence!

There were no land animals. There were no vertebrates at all, not even fish; no insects nor spiders; none of the highest group of molluscs, the Cephalopods, nor of the degenerate lazy-living Bivalves. Although all the main invertebrate phyla were in existence, they were often represented by strange forms, now long extinct. It is highly improbable that any animal then in existence could hear, or could see colours, or had the power of learning by experience in any but the most limited degree. There were no land plants with roots or woody stems, let alone seeds or flowers. Yet, as we have seen, life in the seas was varied and abundant, and each main group had already branched out into many ways of life. There are only four groups of marine animals whose destiny we need to follow in this general review of life's history. The others have achieved all the important part of their evolution before the curtain goes up on the Cambrian (III A), or they are mere supers in the play, or they are soft-bodied and have left no remains. The groups whose history we shall trace in outline are the Echinoderms, the Arthropods, the Molluscs, and the Fish.

Among the Arthropods, the Trilobites were the dominant group of Cambrian animals. These creatures—all of them marine—looked something like large aquatic wood-lice or slaters. But though they have compound eyes and the usual outside skeleton of all Arthropods and a whole series of limb-appendages, segment by segment, the likeness is a superficial one. Wood-lice are Crustaceans, with a regular head and a wonderful division of labour between the different members of their long series of limbs—first a couple turned into sense-organs, then a battery of jaw-appendages, then admirable walking legs, and so forth. But these Trilobites were much more primitive, much nearer the worm-like archetype from which all Arthropods have sprung. They had a single pair of sensory feelers; but otherwise all their limb-appendages were of the same construction—the primitive forked plan still characteristic of crustacean larvæ to-day. None of these appendages had been wholly converted into jaws as has happened to the mouth-appendages of all present-day Arthropods, whether Crustaceans, Spiders, or Insects. But the first step in that direction had been taken; the possibility had been indicated; those near the mouth possessed little toothed lobes which projected inwardly and helped to hold and grind the food. But no Trilobite could chew, it seems, without going through the motions of crawling or swimming, nor crawl or swim without going through the motion of chewing.

The earliest Trilobites seem to have been scavengers, crawling over the muddy bottom and picking up their food under cover of the heavy armour of chitin that covered their bodies above and projected spikily

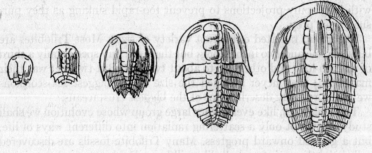

FIG. 216. The life-history of a Trilobite.

The animal added new segments behind its head-shield as it grew up.

on either side over their limbs. Like all modern Arthropods they began their lives with few segments and added new ones at their hind end. So abundant are the fossils of many Trilobites that we can piece together their life-history, all but the earliest stages, as well as if we were able to keep them in an aquarium and watch them grow. The youngest forms are tiny little tadpole things, all head and no body; some of these certainly were active surface-swimming creatures like crustacean larvæ to-day, serving for the dispersal of the species. But they soon descended to the bottom and lived and died after the fashion of their parents.

Scavenging is all very well, but there are other ways of life open to creatures of this construction, and soon there arose a variety of types. There were burrowers, often blind or with reduced eyes like the burrowing mole to-day. Some of these lay in wait in their burrows for passing prey, while others ate their way through the nutritious mud as an earthworm eats its way through the soil. There were others that seem to have taken to a vegetarian mode of life, chewing up the Cambrian seaweeds. And, most adventurous, there were free-swimming Trilobites, some wholly emancipated from the bottom.

For the crawlers, any shadow between them and the light above might mean the passage of an enemy; and they accordingly had their eyes on top of their heads. The swimmers needed better eyes because their greater activity took them more quickly through the changing world, and they needed to see what was happening on every side. Accordingly their eyes were for the most part much larger and lapped over the side margins of the head. And, like so many creatures of the free open water to-day, these free-swimming Trilobites were provided with long spiny projections to prevent too-rapid sinking as they pursued their prey.

The group radiated out into a variety of sizes. Most Trilobites are between one and two inches long; but there are tiny species only a third of an inch long and others up to about two feet. But two feet was their maximum; they never attained the size of the biggest sea-scorpions we shall presently describe, nor of the biggest crustaceans.

The Trilobites, like every other large group whose evolution we shall study, show not only a scattering radiation into different ways of life, but a general onward progress. Many Trilobite fossils are discovered rolled tightly up into a ball, like pill-bugs, head, eyes and face interlocking with tail-flap, and the spiky margins of the upper armour fitting together at the side so that no chink is left for an enemy to reach the limbs and vulnerable belly-surface. They perhaps died of asphyxiation, or of exposure to the air; in any case we find them to-day preserved in the defensive spasm into which they threw themselves as death began to overtake them, half a thousand million years ago. But we never find these rolled-up kinds in the early days of Trilobite history. Rolling up was an accomplishment whose evolutionary acquisition we can trace step by step as we can trace the acquisition of a single hoof by horses.

Most of the later Trilobites also had the hind part of the body turned into a regular tail-flap, with which they probably jerked themselves rapidly backwards out of danger, like a lobster; and some grew fantastic spines all over their bodies.

In early Cambrian times (III A 1) none of these back-flapping or excessively prickly types existed, any more than did the rolling-up types. Very probably the slow improvement in these directions was a progress towards better defence, necessitated by the rise to power and abundance of their enemies and rivals, of whom we shall

EVOLUTION IN THE WATER 683

soon speak, the early molluscs and the sea-scorpions and king-crabs.

This took place during the next period, the Ordovician (III B), and from then on to the long-delayed time of their extinction near the close of the Paleozoic, the Trilobites became gradually less abundant, less characteristic and dominant in the life of the times.

There are three striking facts about the life of the early Paleozoic (III). One is the presence as the most abundant and the most successful creatures of the age, of groups that are now long extinct. Besides the Trilobites, for instance, there were the Graptolites, whose fossil remains are so widespread and so abundant for a spell of geological time from the end of the Cambrian (III A) onwards that they serve as markers by which the geologist can date rock-layers in different regions of the world. There were Cœlenterates, colony-forming polyps, mostly free-floating, with horny skeletons like those of Obelia. The colonies were often fitted out with the most elaborate structure of floating bells, presumably gas-filled, to which were attached straight scaffolding rods bordered by polyps on either side. There were special reproductive polyps which added to the colony by producing new polyp-rows, or founded new ones by budding off single floating polyps, which then grew up into colonies on their own account.

They took the place in ancient seas of the modern Siphonophores, the marvellous glassy polyp-colonies of which we have spoken in Book 2. Perhaps the Graptolites' decline in the Silurian (III C) and extinction in the Devonian (III D) was due to the rise of this new rival type. But this we may never know, for the Siphonophores have no skeleton to be preserved, and so far we have no fossil trace of them.

The second of our striking points about this early epoch is the presence of many lower types of animals which have preserved their general construction almost unchanged through the five hundred million years from then till now.

And the third is the absence of the more developed forms in every higher group—vertebrates, arthropods, molluscs, even echinoderms and corals.

Evolution does not destroy or transform the whole of life. It preserves much of the old, but it extinguishes many of the more elaborate constructions of each age, and in their place is always adding the new and the improved. If we were to judge the Cambrian only by its jellyfish and worms, it was a modern world. But in making our estimate of Evolution we must take account of its most finished products as well; and a world in which Trilobites were the highest achievement of life had still a long road to travel.

§ 4. The Age of the Sea-scorpions.

IN THE Trilobites Cambrian life reached its highest point. By the late Ordovician (III B), however, their age of security and predominance (they had had well over a hundred million years of it) drew to a close. From the Silurian (III C), only thirty-five different genera are known, as against a hundred and twenty during their climax in the period before. They were being pushed aside. Novel, more powerful animals swam through the seas they had ruled. As they declined, their rivals, the sea-scorpions or Eurypterids, rose.

These creatures were aquatic relatives of the true scorpions, the spiders and the mites. They had already come into being in Cambrian times, and rapidly surpassed the Trilobites in size, some of them reaching six and even eight or nine feet long. With their human or even superhuman size and their outstretched limbs they are imposing fossils. Hugh Miller tells us that in his day the Scottish quarryman of the Devonian Old Red Sandstone (III D) baptized them *Seraphim*. Like the Trilobites again they originated as heavy crawling creatures, but many seem to have been capable of prolonged slow swimming, using their expanded hind pair of legs as clumsy oars. Others again were mud-burrowers, and in some the small first pair of limbs, pincer-clawed as in all modern scorpions and spiders, enlarged into passable prototypes of a lobster's powerful claws. We know little about their feeding or their reproduction. But there is one interesting point about their habits. All known Trilobites were confined to salt-water, but while sea-scorpions seem to have originated in the sea, the majority or possibly all of them in the declining half of their evolutionary career took to fresh or brackish water, living in rivers, lakes, estuaries and lagoons. For they, too, in their turn, were giving way to a fresh onset of life.

The Eurypterids continued right through the Carboniferous (III E), but all were finally extinguished before the Permian (III F) brought the close of the Paleozoic. Contemporary with them throughout, however, were some smaller creatures, more trilobite-like in general appearance but of true scorpion-spider construction. Most of these died out at the same time as the sea-scorpions, but one genus, *Limulus*, the King-crab, survives right down to the present day. This veritable living fossil, though all its relatives have been dead for hundreds of millions of years, is still so abundant off some parts of the American coast that King-crab corpses are used to manure the fields.

These occasional successful survivals of antiquated types of life, unburied ghosts of a long past age, are puzzling riddles. It is not always easy to discover what compensating specialization of detail had made up for the animal's general primitiveness and enabled it to survive. But if they propound one evolutionary riddle, they help us solve many others. By investigating their construction and way of working, we can understand their dead relatives and reconstruct the life of the past far more accurately than would otherwise have been possible.

In the Upper Silurian (III C 3) the first true scorpions appear, bearing the closest general resemblance to those alive to-day. They are the first heralds of animal life on land. But they are only its heralds; they show no trace of the openings into the breathing-chambers, prominent in all modern scorpions, even in those of the coal-measure period (III E 2), and they cannot have been fully free of the air and the dry land. It may be hazarded that these Silurian scorpions led a marginal life between water and air, like many modern crabs or the sea-slater Ligia. Inhabitants of the sea-margin, they were probably able to breathe air, but still by means of their old water-breathing organs, which, like all water-breathing organs, needed to be kept always moist. They had to resort to periodic re-immersion in their ancestral element, and could never venture far beyond high-water mark. But probably there was as yet very little to attract them above high-water mark.

In this same period centipedes have also been found. These, too, are to-day exclusively land animals. Thus the later Silurian (III C) witnessed the beginning at least of the emergence of animal life out of the waters, a tentative invasion of the huge uncolonized domain of the dry land. As might be expected, it was among arthropods, with their hard external armour and their gill-covers to keep them and their breathing organs from being dried up by the new and dangerous element of air, that the first pioneers of the land were found.

Up till a few years ago, no completely terrestrial arthropods had been discovered in any rocks earlier than those of Middle Carboniferous Age (III E 2). Recently, however, the huge gap of over a hundred and fifty million years between the Silurian scorpions and the big coal-measure insects has been partly bridged by the discovery of undoubted insects in the same Lower Devonian (III D 1) layers as those which contain the earliest known land plants, which we shall shortly describe. They were all tiny, and all as yet wingless, closely resembling the little springtails of the present day. These insignificant crawling creatures, which have been christened *Rhyniella*, must have been either the ancestors or something very like the ancestors of all the myriad beetles, butterflies, bees, and other winged insects of to-day.

§ 5. Echinoderms: a Phylum Which Has Never Produced Land Forms.

THE Echinoderms also launched out upon a number of evolutionary experiments. If by success we mean survival these were successful enough, but, as we saw when we studied these very curious creatures in Book 2, not a single Echinoderm has escaped from water to land, not one has risen above its primitive headlessness.

All the evidence goes to show that the ancestral form was a small somewhat sluggish creature, not five-rayed in structure but two-sided like ourselves, which browsed slowly over the bottom. Later, the development of a skeleton-armour just below the skin was achieved, perhaps as a protection against the reigning arthropods of the period. About the same time this ancestor took to attaching itself to the bottom, perhaps by sucking on with part of its mouth. Adhesive glands and, finally, a stalk gradually took over the business of attachment, while the mouth was shifted up, away from the region of fixation. This sedentary stage which all the group went through (it has left its traces on even the modern free-living types like starfish and sea-urchins) led to the adoption of a new means of gaining food. Grooves came to radiate out from the mouth, lined with cilia to sift small particles of food from the water and to transport them mouthwards. Originally there seem to have been only three such grooves, but the two hinder ones later forked; and so the five-rayed symmetry, which meets us in almost all the living forms, was established.

The early Cambrian (III A) Echinoderms are all fixed to the bottom; and during this and the Ordovician (III B), a great number of strange and unfamiliar types had been brought to birth, some looking like short-stalked pears, others elongated and with a single food-groove extended within a spout-like tube, others with their food-groove made more efficient by being extended on to arm-like outgrowths, or bordered by a double row of tentacles doubtless covered with actively beating cilia. From some such forms as these the sea-lilies arose, first found in the Ordovician (III B) and rapidly becoming the most successful echinoderm type through Paleozoic time, and the only living order to continue to practise, as its routine method, the group's original mode of feeding (although ciliary feeding is to be found sporadically among star-fishes and sea-cucumbers). Other strange creatures, such as Edrioaster and its relatives, suggest a starfish built into an armoured ball or bun. But, unlike starfish, their mouths faced upwards, and the arm-like structures were not free or movable, but were fixed food-grooves.

The familiar sea-urchins and starfish also arose in the Ordovician,

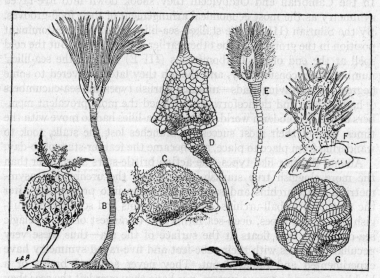

FIG. 217. A group of primitive Echinoderms from the Paleozoic (III).

(A) Echinosphæra, a round animal attached by a short stalk; it had three food-grooves on tentacles round the mouth. (B) Orophocrinus, member of a group which resembled the sea-lilies of to-day, except in having very numerous tentacles. (C) Aristocystis, a creature resembling (A) except that it was bent over on to one side. (D) Dendrocystis, a stalked animal with a single tentacle and food-groove. The anus is at the left lower corner of the body. (E) Lepadocrinus, with three food-grooves along the body, each bordered by tentacles. (F) Protocrinus, with numerous tentacles, each with a food-groove. (G) Edrioaster, a bun-shaped creature with the mouth on the upper side, and five twisted food-grooves from which tube feet were protruded. The anus is near the lower edge.

but long remained comparatively unimportant groups. They differ from all the most primitive Echinoderms in not being fixed, in having their mouth below instead of above, and in having given up ciliary feeding for grosser methods. How they turned themselves upside-down from their original position is one of the unsolved riddles of evolution. All the Echinoderms which crawl mouth downwards to-day do so by means of the highly characteristic "tube-feet" that we noted in Book 2. We may guess that before the creatures turned over these organs were mere tentacles grouped along the food-grooves, for that is how they appear in living sea-lilies and feather-stars.

Though a sea-urchin or a brittle-star represent the climax of Echinoderm activity, yet progress of a sort is registered in their long history. In the Cambrian and Ordovician they shook down into five-rayed symmetry as the most economical arrangement of their food-grooves. By the Silurian (III C), the stalked sea-lilies took over the dominant position in the group from the other earlier stalked types. But the cold spell at the end of the Carboniferous (III E) reduced the sea-lilies' numbers very considerably, and though they later recovered to some degree, the free-living kinds—urchins, starfish types and sea-cucumbers—had become and thenceforward remained the more prevalent members of the Echinoderm world. Even the sea-lilies had to move with the times. One of their most successful branches lost the stalk, took to wandering from place to place, and became the feather-stars of to-day.

Among the star-like types, the active brittle-stars arose later than the more sluggish true starfish, and among the urchins the asymmetrical heart-urchins and sand-dollars came into prominence after the symmetrical ball-urchins. Burrowing forms and scavengers, sluggish and active types, deep-sea forms and (strangest of all) a pelagic sea-cucumber that floats at the surface of the sea—thus these very peculiar creatures with their tube-feet and five-rayed symmetry have experimented and radiated out. They never acquired brains or a highly organized nervous system, nor have they ever left the sea; they lack the vigorous adventurousness of arthropods, molluscs, and vertebrates. Nevertheless, they have played their own distinctive game. Patiently they have evolved and perfected an elaborate internal architecture of water-carrying tubes, tubes of various kinds, quite unlike our own circulatory system, of whose functions we know next to nothing. Within the limits and along the lines prescribed by their own nature, evolution has gone steadily on.

§ 6. The Ancient History of Molluscs.

MOLLUSCS to-day have an extraordinary range of organization—from insignificant worm-like creatures up to gigantic cuttle-fish, from creeping slug to active squid, from sedentary current-feeding clams and oysters to adventurous land-snails that crawl amidst high mountain scenery.

In the Lower Cambrian (III A 1) we find the earliest known mol-

EVOLUTION IN THE WATER

luscs. They all approximate to the snail type, but instead of having most of their internal anatomy neatly done up in a spiral inside a coiled shell, like the great majority of modern snails, most of them had merely a little conical cap-of-liberty shell sitting over the hump into which (unlike the camel's) the vital entrails project.

These early fossils confirm what we deduce from other evidence—that the first molluscs to be evolved were bottom-living creatures that had taken to crawling with the aid of a flattened and muscularized expansion of their bellies—the "foot"—and to protecting their vulnerable upper surface with a limy shell. Another characteristic feature which the earliest molluscs in all probability possessed, for it is found in members of all existing groups save the current-feeding bivalves, is the marvellous "tooth-tongue" or radula. This ribbon of teeth, constantly worn away by use, is as constantly renewed by growth from a formative pocket in the floor of the throat. With the aid of this horny file the first molluscs would have browsed by rasping off bits of solid food. From this ancestral browsing belly-crawling type, the true snail-like forms or gastropods soon evolved; and these have remained, from those Cambrian days till now, the most varied and the most adaptable group of molluscs.

One of their early progressive steps was the evolution of the spiral shell. Originally the gastropod's "foot" was merely the body-wall of its lower surface. To make it more efficient as a locomotor organ it needs to be enlarged sideways into a flattened, projecting sole; it needs enhanced muscularity. As the foot thickened and grew, the viscera, it seems, expanded into a big hump on the back. But the growth of this "visceral hump" and its covering shell brought new problems. If it grows into a flat broad cone, a big surface is always left below open to attack. It is only a few types, like the limpets with their special faculty for holding on tight to the rocks, which have adopted this method. If it grew into a long narrow cone, it would project and flap about awkwardly. But if this long sack were curled spirally, it could be still fairly compact and handy, the whole body could be withdrawn inside a suitable shell, out of harm's way; the entrance could even be sealed with a horny "door" secreted by the part of the foot which is last to get inside, and such is the device found in most modern gastropods.

To make the spiral, one side of the body grows faster than usual, the other less fast. In all snails with spirally coiled visceral humps, the organs of the left side are over-developed, those of the opposite side reduced or absent. The familiar snail is, when we come to look into its history, a remarkable animal. It has succeeded in keeping to symmetrical behaviour—crawling straight, for instance, instead of moving in a spiral path like the asymmetrical Paramecium and many other Protozoa—although in the bulk of its internal anatomy it is in the highest degree one-sided. It is asymmetrical in structure, but symmetrical in behaviour.

We find this change recorded in the rocks. The twisted shell, rare in the early Cambrian (III A 1), grew much commoner in the Upper Cambrian (III A 3), until by the late Ordovician (III B 3) it was as common as to-day.

The gastropods never attained any marked brain-power, or any notable strength or speed. But in quantity and versatility they have always been the leading group of molluscs. They alone of their phylum succeeded in invading the land; and beside giving us the strange and beautiful shell-less sea-slugs, as well as all the shelled snail-like forms, they have produced the Pteropods to colonize the surface waters of the sea, flapping themselves along with their expanded foot turned into a pair of sea-wings.

From these early molluscs, the ancestors of the snails, two other main lines diverged, one by our standards downwards, the other upwards. The downward line led to the bivalves. Imagine a primitive snail with a flat-coned shell that had taken to slowly crawling over and feeding upon mud. It still needs protection from enemies, but it no longer needs to move so rapidly. The foot can dwindle. Enlarge the side margins of the shell and the flaps of living tissue that secrete it until they reach down on either side of the foot and eventually meet below it. Into the space between shell-flaps and foot, make the little gills of the snail grow out into great leaf-like structures richly beset with cilia for producing a current. Reduce the size of the head, which is no longer needed for direction and control, to a minimum. With these changes you have turned a gastropod into an ancestral bivalve. All it has to do further is to develop a line of weakness in its shell along the middle of the back, turning the one shell into two half-shells that can move independently, and to improve the arrangements for admitting and ejecting from the space between shell and foot the current of water produced by the gills, and the bivalve is in its essentials complete. It can now live *in* the mud, as do so many clams and cockles and mussels, instead of *on* it, and so live a safer and more sheltered life. The "gills" not only produce a current, but strain out the microscopic débris therefrom and conduct their sifted treasure along ciliary paths to the mouth and down the gullet. The onward-browsing gastropod has evolved into a sedentary creature with cilia-driven current-feeding.

The earliest bivalves appear only at the end of the Cambrian (III A 3), at least fifty million years later than the gastropods—additional evidence that their sluggish mode of life is not a primitive thing, but a degenerative specialization. Of their subsequent history we need mention but one or two points. During the course of the Paleozoic (III) there was a gradual evolution of an elaborate hinge between the two valves of the shell. The original line of flexibility was converted into a beautiful apparatus of interlocking teeth and sockets, giving great strength as well as perfect freedom of hinging movement. And during the same time there developed asymmetrical bivalves, which lay flat on one side, using one valve of the shell as a foundation-bed, the other as a roof and door combined. These included the ancestors of the oysters and the scallops. The new development made it possible for bivalves to colonize mud and sand surfaces. Previously they had been forced to embed themselves in mud or sand. The new forms lay out and fed more widely as the oysters do. Presently the sea-mussels appeared with a tuft of adhesive threads, the byssus, with which they stuck themselves even to vertical rocks. The scallop, continuing this return from buried obscurity to a more open life, developed a fringe of eyes round its mantle and learnt to swim vigorously. In quite another direction, highly specialized rock-boring forms, like Teredo, the so-called ship-worm, arose.

Finally, we must note the highest of all mollusc radiations: the cephalopods (the octopuses, the cuttle-fish, and their relatives) which include in their ranks the most active molluscs, the largest, the best provided with sense-organs and brains. Their chief advance over gastropods lay in the conversion of the single foot, with its primitive mesh of muscles and its simple, crawling action, into a set of separate and highly-specialized organs, the tentacles or "arms" around the mouth, which are used in the capture of prey and lead to much greater variety of movement and nicer power of discrimination. The foot grew forward, so to speak, round the mouth, and was elaborated to serve it and its sense-organs; hence the name Cephalopod—"head-foot." Another characteristic early development was this, that instead of filling the whole of their shell with their visceral humps, they moved on as they grew larger and left the infantile shell empty, cutting off the abandoned first cavity by a partition. Then again moving on to another as growth proceeded, and so on, they inhabited only the last chamber and carried about with them all the empty rooms they had lived in from their birth onwards. And they seem early to have developed the power of moving much more rapidly than any other molluscs, by squirting out a powerful jet of water from the cavity which encloses their gills through a spout just below the head-region.

Hardly any of the cephalopods that survive to-day are housed in such compartmented shells; they are now altogether shell-less (save for minute vestiges) like octopuses, or have their shell grown over and turned into an internal skeleton, like squids and cuttle-fish. But this is a modern development. From the Cambrian (III A) to the Triassic (IV A) every known cephalopod lived in a shell, and right on to the end of the Cretaceous (IV C) this was true for a large contingent.

The cephalopods, like the gastropods, can be traced with great probability to little cone-capped ancestors. Then, it seems, arose the distinctive habit of partitioning off a living-room from the rest of the cone. This involved the elongation of the shell, which at the same time seems to have been lightened by the secretion of gas into its inner uninhabited air-tight chambers. At first these compound shells were straight, and the original flatly conical caps were elongated in the course of evolution into huge, slender cones up to twelve or fifteen feet long. Even when buoyed up by gas, they protruded awkwardly; and so here, too, coiling came into fashion, and the straight-shelled forms dwindled, to vanish in the Trias (IV A). The coiling, however, instead of being all on one side as in snails, was symmetrical; the shells coiled upwards over the body. All stages in this coiling can be traced, beginning with slightly bent shells, on through curved shells and shells that come round full circle, until in the most specialized forms the earlier parts of the shell are wholly hidden by the embracing wings of the last chamber. One such spiral cephalopod still survives to-day in the Pearly Nautilus of the tropical seas. It is the only living cephalopod with a chambered shell of all the hosts that have been.

In middle Paleozoic times (III) a group branched off from the Nautilus stock to give us the Ammonites. These differ very little in shell-construction from their parent group, and it is one of the puzzles of Evolution why they rose to such abundance in the Mesozoic (IV) while the Nautiloids declined. It is another puzzle to know why all their numerous families save one died out at the end of the Trias (IV A), and why this one surviving stock then blossomed out into a marvellous variety in the Jurassic (IV B). But we still know very little of the physiology and difficulties of the Nautilus, and still less of the forgotten life of the Ammonites. We can only guess rather dully at what might be good for them and what might be bad.

These shelled cephalopods often had a hard head-shield which acted as doorway when they withdrew into their shells. They were always

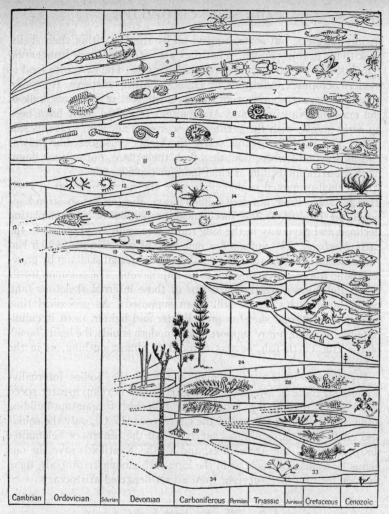

FIG. 218. An outline of fossil history. The main groups of organisms known from fossil remains are shown on the geographical time-scale (cf. Fig. 122), expanding and contracting according to their numbers and importance. Where the evolution of a group is doubtful, the outline is dotted.

1. *Spiders, etc.* 2. *Scorpions.* 3. *Sea-scorpions.* 4. *King-crabs.* 5. *Insects.* 6. *Trilobites.* 7. *Crustacea.* 8. *Ammonites.* 9. *Nautilus and its Allies.* 10. *Belemnites, Cuttle-fish, and Octopuses.* 11. *Lampshells.* 12. *Graptolites.* 13. *Ancestral Echinoderms.* 14. *Sea-lilies* (*Crinoids*). 15. *Primitive Stalked Echinoderms.* 16. *Free-crawling Echinoderms (Starfish, Sea-urchins, Sea-cucumbers, etc.).* 17. *Ancestral Vertebrates.* 18. *Jawless Vertebrates (Ostracoderms).* 19. *Fish (with four main bursts of evolution).* 20. *Amphibians.* 21. *Reptiles.* 22. *Birds.* 23. *Mammals.* 24. *Horse-tails.* 25. *Earliest known land-plants* (*Rhynia, etc.*). 26. *Ferns.* 27. *Seed-ferns (Pteridosperms).* 28. *True Flowering Plants.* 29. *Cordaites (early Conifers).* 30. *Cycad-like Plants (Cycadeoids).* 31. *Cycads.* 32. *Modern Conifers.* 33. *Maidenhair Tree (Ginkgo).* 34. *Club-mosses.*

A number of groups (e.g., *Snails, Clams, Corals, Algæ, Fungi*) have been omitted.

mainly crawlers, but some, buoyed up by the gas within their shells, took to life in the open sea. What they ate we do not know from actual remains, but we can hazard a reasonable guess. The favourite food of modern octopuses and squids are crabs and other crustacea. It is probable that sea-arthropods were the favourite diet of the great bulk of the crawling Nautiloids and Ammonites. Trilobites must have been an early staple, with the smaller sea-scorpions and king-crabs as side dishes, and as these became scarce the higher crustacea, like prawns and lobsters and crabs, doubtless took their place, not only as dominant sea-arthropods, but as chief food for cephalopods.

The shell-protected cephalopods were the most abundant of their kind for some three hundred million years. But, as has so often happened in Evolution, the mechanical protection of armour-plating declined and gave way in the long run before agility and speed. In the early Mesozoic (IV) appeared a new type of cephalopods, which had succeeded in subordinating the cumbersome external skeleton by growing round it and converting it into an internal one. The common fossils called Belemnites are the hard tips of these internal skeletons (and not "thunderbolts" as is still often supposed). As geological time passed this internal skeleton grew lighter and lighter, until it culminated in the thin, horny support of the modern squids, the light "bone" of the true cuttle-fish, or was reduced to almost nothing, as in the octopus.

Freed from their coat of mail and with the bodies internally-scaffolded, these new-model cephalopods could develop greater speed and better stream-lining. When the world-crisis of mountain-building and cooling came at the end of the Cretaceous (IV C), only the squids and cuttle-fish and octopuses survived. All the bearers of belemnites disappeared, all the Ammonites, and all the Nautiloids save the one genus Nautilus, in which, as in the king-crab among arthropods, there lives on for us a lone survivor from a vast departed aristocracy.

§ 7. The Earliest Fish.

So FAR we have done little more than speculate in general terms about the nature of the first vertebrates, pointing out merely that they must have been muscular, tailed, gilled, and that they had given up cilia-driven current-feeding for what to us at least are more ordinary modes of eating. At first blush, such a combination of characters suggests a fish, but if we imagine that all early vertebrates were like the familiar fish of to-day we shall be mistaken. The actual history of the verte-

EVOLUTION IN THE WATER

brates, as given by fossils, began with creatures elaborately but very clumsily built, judged by the standards of the best-constructed of their descendants, and they early branched out into all kinds of experiments, many of them fantastic and over-armoured, many of them only temporary, destined to early extinction. This is so with most groups at the beginning of their main burst of evolution. We have seen it with the Arthropods. The early Trilobites and Sea-scorpions and tank-like King-crabs are elaborate enough, but are primitive and awkward compared with crab or crayfish; and the contrast between the earliest land arthropod and the finished elaboration of spider or insect workmanship is even more striking. It is just the same with human inventions: first the crude experiment, then improvement and refinement. Many of us have watched that happen in the evolution of the automobile during the past thirty years.

The history of vertebrates begins later in the record of the rocks than does that of the other great phyla. Vertebrates are built on a more generous scale than other animals. Not only have they the promise of greater strength, speed and brain, but they run larger in bulk. If we were to make a list of the Hundred Biggest Beasts they would all (save for some of the largest cuttle-fish) be vertebrates. And the smallest vertebrates are tens or hundreds of times bigger than the smallest representatives of the other animal phyla (see also page 936). As one would expect, this elaborate, large-scale construction of theirs took longer to work out than that of mollusc or arthropod.

It is not until the late Silurian (III C 3) that we find well-preserved vertebrate fossils. Among them is one family of true fish—primitive creatures related to the sharks. But the great majority of these Silurian vertebrates are of an extinct type called Ostracoderms. Though well armoured, often with a heavy cuirass of bone, they were of the same general construction as the Cyclostomes, the naked soft-skinned lampreys and hag-fishes of to-day. They were armoured Cyclostomes. Like all Cyclostomes, they had never taken the evolutionary step which gave vertebrates their jaws. Though they had progressed far from the mild and often degrading practice of current-feeding, they had still to achieve the vigorous, efficient *bite* of a higher vertebrate.

It is a curious fact that the jaws of the two most successful animal groups, the Vertebrates and the Arthropods, are in both cases organs converted from their original uses. As we have seen, those of insects and spiders and crustacea were limbs once upon a time. And once upon a time the jaws of vertebrates were merely supports for the gills. The gills of a fish are supported by a series of gristly or bony arches, running round like ribs but jointed, which allow the floor of the throat to rise and fall like a human chest and the gill-slits to open and close rhythmically. With these the early jawless vertebrates were provided. It is the first pair of these throat-supports or gill-arches, lying in front of the first gill-slit, which in the stretch of evolutionary history between Cyclostomes and true fish were converted into supports for the side of

Fig. 219. A group of jawless vertebrates or Ostracoderms from Paleozoic seas.

Upper left, two Pteraspis, with strong bony shield prolonged into a snout. Upper right, three Cephalaspis, with smaller head-shield. The speckled patches on its borders are electric organs. The view from below shows the nine pairs of gill-slits and the mouth. Right centre, Pterolepis, with no shield, but a covering of scales. Its mouth was probably overhung by a flexible lip. The other four animals are Drepanaspis, with big bony plates here and there among the scales.

the mouth. Up to that time the vertebrate mouth must have been for the most part a sucking, mud-grubbing aperture, although certain Ostracoderms made an attempt at mastication with the aid of some of the plates in their outer armour. These became slightly movable, and had tooth-like projections; but the arrangement was a very poor one compared with that of a hinged pair of jaws. The gill-arches were already jointed at the side of the throat (as they are in all existing Cyclostomes and fish) to permit the throat's pulsation; and so it came about that the fish, and all vertebrates that came after them, found themselves provided with biting jaws that moved up and down, instead of across as in arthropods and, indeed, all other creatures with jaws.

Some of the Ostracoderms from the next period, the Devonian (III D), are so minutely petrified that Dr. Stensiö of Stockholm has been able to work out their anatomy almost as well as if he had living specimens to study and dissect. Like the living Cyclostomes, they had not reduced the number of their gill-slits to the five pairs possessed by all modern fish (except two curious sharks, one with six and one with seven); and the first gill-arch was still a gill-arch and not a jaw.

Besides this, they never had the full complement of vertebrate limbs. Modern Cyclostomes have no limbs at all and no traces of limbs; and this limblessness is shared by some of the long-dead Ostracoderms. Others have the tiniest pair of flaps on the side of the body, and there is every gradation from these to well-developed fin-like organs in the position of fore-limbs.

These Ostracoderms were all small; none of them grew more than eight or nine inches long; but they were a varied group. Among them were creatures studded with a shagreen of sharp denticles like a modern dogfish, others with an armour of small mosaic plates, others with wonderful head-shields. It is among these last that the best-preserved specimens have been discovered, and we know the shape of their primitive brains, the position of their nerve-trunks, the course of their main blood-vessels. Some of them had special organs embedded in their head-shields and supplied by very large nerves; these, it seems likely, were electric organs, like those of the Torpedo or Electric Eel. Already in the Silurian (III C 3) several quite differently constructed families of Ostracoderms were in existence. So we must, if we are Evolutionists, believe that they already had a long history behind them.

Before this entry the record is obstinately blank about our ancestors, except for some tantalizing fragments of vertebrate bone from the Middle Ordovician (III B 2), in all probability belonging to Ostracoderms, but Ostracoderms more primitive by fifty million years of evolution than any we know well (what would we not give for a well-preserved fossil of this date!). Before vertebrates developed denticles or bone, they could scarcely ever be fossilized. These heavily armoured bottom-feeding Ostracoderms were probably a side branch of the main vertebrate stem, but traces of that in the Silurian (III C) still elude us.

An important and interesting point in which true fish and all higher vertebrates differ from the Cyclostomes is that they have real teeth. But vertebrate teeth had in the beginning no more to do with the vertebrate mouth than had the vertebrate jaws. The ancestral teeth were developed as armour and not as weapons. If you stroke a shark, a dogfish, or a skate, the skin feels hard and rough like a file, because it contains thousands of spiky scales or "denticles." The more primitive Ostracoderms had similar skins. Each of these denticles is made exactly like a tiny tooth, with enamel, dentine, pulp-cavity, and the rest, anchored in the skin by a little expanded base, so that although our remote ancestors lacked proper teeth in their mouths, they had them in miniature all over their skins. The generality of them in the case of a shark act merely as protection; but those on the skin that grows into the mouth-cavity can help in holding food, and so have been enlarged to form real teeth. In the shark their points are directed stomachward, and greatly assist in retaining recalcitrant prey. Through the accidents of evolution it is only these mouth-denticles—enlarged, elaborated in construction and socketed into jaws—which have survived in land-vertebrates, while the original armour-denticles have wholly vanished. They give the best bite Nature has yet produced. When Mr. Everyman chews his breakfast bacon he does it with teeth which are lineally descended from the shagreen of a cartilaginous fish in the Silurian waters.

But there is something still more curious about these scales and the cartilaginous skeletons of our Devonian ancestor. In the Lower Devonian (III D 1) the very highest and most efficient skeleton inside of any creature was the early vertebrate skeleton of gristle. At first it had no bone in it at all. The only bone in the vertebrate body was on the surface; it was at the base of the armour-denticles. Then by degrees (we find it already in some of the Ostracoderms) bone, which is lighter and harder than gristle, began to *infect* the cartilaginous skeleton. This has never happened in the whole great series of the sharks, rays and other Elasmobranchs—except for their skin they are cartilaginous to this day—but along the ancestral line of the bony fishes and all the land Vertebrata, the process of ossification went on. Bone, harder and stiffer, and when efficiently built, lighter, replaced the more flexible gristle. In this first great age of fishes this process had still to be worked out: but, as we have noted, it had begun. Mr. Everyman in his development repeats this curious advance. As an embryo his skeleton is still essentially cartilaginous and only becomes good hard bone later, some of it not till well after birth. If he is put to walk at too early an age his insufficiently ossified legs give and he becomes bandy, thus demonstrating the biological advantage of the change.

We have then in the Silurian (III C) and early Devonian (III D) rocks two branches but almost certainly not the main stem of verte-

EVOLUTION IN THE WATER

brate evolution preserved. First there came those feeble-mouthed Ostracoderms, betraying their existence by an excess of armour, and then in the Devonian an outbreak of armoured cartilaginous fishes with real jaws and also with efficient limbs. They radiated in great profusion and variety. All the earliest vertebrates we have already noted

FIG. 220. Restorations of some Paleozoic Fish.

In the centre is the huge predaceous Dinichthys. Below it, to the right, is a group of the extraordinary armoured fish, Pterichthys. Just above its tail and by its snout are two specimens of Bothriolepis. The other fish on the left, from above downwards, are Holoptychius; Osteolepis (closely related to the ancestor of land vertebrates); and Climatius, an early shark-like form with finlets between main fore- and hind-fins.

are found in fresh-water or estuarine deposits; Ostracoderms and fishes alike, they are all probably river animals. But in the Middle Devonian unequivocally marine fishes appear in the record. Apparently the discipline of river life, with its currents to contend against, was bearing fruit and the vertebrate stock was already able to make successful sallies into the open sea.

In structure and appearance, too, a great abundance of new types was produced. There were little dogfish-like creatures with huge defensive spines and others with primitive fins. There was one group of true fish so fantastically armoured as to look almost like strange arthropods. They have only front limbs, and these are jointed almost like a crab's; and their eyes are close together, looking upwards on the middle of the head. They probably led a very crab-like life. There were also early lung-fishes in this period, and there were the Arthrodires, large active flesh-eating fish with a formidable cutting beak instead of teeth. Like so many early fish most of them were comparatively unprotected aft, but had their head and the front of their trunk enclosed in a heavy cuirass; these forms appear to have possessed only hind limbs. One of the Arthrodires, Dinichthys, rivalled the biggest modern sharks, attaining a length of twenty feet and more. And there were a number of bony fish, like Osteolepis, more in the main line of onward evolution, but all cumbered by heavy shining armour of large, close-fitting scales, bony within, enamelled without. Probably there were also many small and active vertebrate creatures in this first age of fishes which had not taken to the new fashion of bone and so escaped preservation.

No wonder that with this new array of competitors in all branches of marine existence both Trilobites and Sea-scorpions were on the down-grade from the end of the Silurian (III C 3) onwards.

It is in the late Devonian (III D 3) that the first signs of land Vertebrata appear. As startling to those who explore the recesses of past time as were the tracks of Man Friday to Robinson Crusoe, a few fossilized footprints inform the geologist that land vertebrates trod the Devonian mud. One of the stocks of early fish had thrown up this new kind of creature. But how and why the new construction of lungs and legs was evolved, we must leave for a later section.

IV

LIFE CONQUERS THE DRY LAND

§ 1. *The Desert Land of the Early World.* § 2. *The Curious History of the Land-plant and Its Seed.* § 3. *The Plants of the Ancient World.* § 4. *Coal.* § 5. *The Vertebrata Crawl Out of Water.* § 6. *A Coal-measure Forest.* § 7. *An Age of Death.*

§ 1. The Desert Land of the Early World.

IT IS difficult to imagine this world with all its land surfaces lifeless; yet for more than a half of its history Life played out its drama under water, and the continents were practically barren. They were stark and bare, starker and barer than the utmost desert of to-day. Over the bare cliffs and desolate plains the sole breath of movement came from the wind and rain. A certain margin there may have been of faintly vitalized soil. From comparatively early times, a few simple algæ may have trailed their filaments over the seashore or the moist borders of rock pools, or a few bacteria invaded the crumbling earth surface. But for the rest the earth was innocent of plants, and therefore of animals as well. Even when the earliest true land-plants came into existence, they too (like the earliest land-animals) were restricted to moist places. Not until Middle Devonian (III D 2) times did the earth begin to be amply clothed with the familiar green of vegetation and the stir of life.

The face of the land was like nothing we know to-day. There could have been no real soil, for soil is largely a product of plant action. There was no carpet of plants and felt-work of roots to hold water like a sponge, preventing rapid run-off, and to blanket the ground from excessive gain and loss of heat; and so the work of frost and wind, rain and sun, was much more active. The heights of the land were worn down quicker and sediments more actively deposited, and the scenery was more angular and forbidding. Even long after the first appearance of land-plants vast regions of the land which would now be covered with vegetation remained desert or semi-desert, since all the earliest plants demanded a good deal of moisture. Plants like grasses, which can thrive on dry steppes and prairies, are comparatively modern things. The Paleozoic Era (III) knew nothing of them and the Mesozoic (IV) comparatively little. There was a desert flora and fauna in the Trias; but possibly at least, the regions it inhabited would to-day be steppes or savannahs.

This should be in our minds when we read that such and such a geological deposit was laid down in desert conditions. Geologists, for instance, tell us that much of the land surface in Devonian (III D) and Triassic times (IV A) was more or less desert. But we should remember that much of the desert of those times would to-day be more or less carpeted with plants. There are abundant forms of life now to mitigate what was then a hopeless severity. And the same is true of cold lands and of high mountains. In each case the fact that plants did not grow there meant more violent extremes. The aridity of those remote ages was increased by the failure of plants to conquer it: it was a vicious circle.

Plant-life, like human civilization, has gradually extended its boundaries. Under the pressure of the struggle for existence, its beneficent exploitation of the sun's energy has gradually enriched the world, softened its contours and its climate, and reduced the extent of its waste space. And since the plants opened the door to the animal invasion, their invasion of the land was a necessary pre-requisite to human evolution.

It is a good exercise of the scientific imagination to picture this desolate and desert earth, its continents cyclically rising out of the waters and submerging themselves again, occasionally undergoing a spasm of mountain-building or an ice age, but remaining essentially lifeless for well over five hundred of our million-year periods, in spite of the abundant presence and notable progress of life in the waters. Through all these ages, the lands remained unconquered and must have seemed unconquerable. It is of the invasion of this territory, not only huge in extent but a veritable land of promise from the evolutionary point of view, that this chapter will tell.

§ 2. The Curious History of the Land-plant and Its Seed.

WE ARE so familiar with the plants of the modern world—grass or oak-tree, rose-bush or bindweed—that we forget what evolutionary triumphs they embody, what difficulties surmounted, what adjustments perfected. Before plants could successfully colonize the land, their whole structure and mode of reproduction had to be altered.

LIFE CONQUERS THE DRY LAND

In an earlier chapter we left plants, towards the middle of the Paleozoic Period (III), arrived at the state of higher seaweeds, with holdfast, rudimentary stem, and the beginning of a transport system in the shape of sieve-tubes. What was necessary before such a floating sprawl of greenery could become a land plant? One great change needed to make a fully terrestrial plant was structural: on land its food was no longer all about it, it must seek the two main constituents of its food in two different places. To get carbon and to use it, the plant must reach upwards into the air and light; to get water and mineral salts, including all its nitrogen, it must drive down into the soil where alone these things are to be found. The primitive single factory had to be divided into two: the green leafy shoot concerning itself with the chemistry of air, the colourless absorptive root-system still dealing with the absorption of water and watery solutions.

But this division of functions at once made a better transport system imperative, since the materials secured by the root had to be brought together with the carbon derived from the carbon dioxide of the air before anything that can be called food could be produced. The green chlorophyll-tissue became the region for assembling these materials. It is necessary to have water before even simple sugars can be made, and water is easier of transport than their other constituents. Accordingly, a system of pipes had to be developed to connect root to leaf and arrangements made to set a current going to bring the water and its contained salts up in the pipes from root-hair to chlorophyll-containing leaf-cells. This current is called the transpiration current; so long as the plant is alive and actively at work, it never ceases.

What the forces are which push or pull the column of water upwards through the stem of a land-plant, even to heights several hundred feet above the ground, is not yet certainly and exactly known, but manifestly they exist and evidently pressure exerted by the roots contributes its share to them. This you can vividly see if you cut a stem across in spring when the sap is rising most vigorously. The water oozes out and trickles off the cut surface in large drops. As gardeners say, the stem bleeds. The pipes in which the current mounts are part of the wood—made of cells which have solidified their walls, died for the plant-community's good, and left their cavities unfilled. The empty cells may communicate with each other by diffusion through thin places in their walls, or even coalesce end-to-end to form long microscopic tubes, sometimes strengthened by a close spiral of woody material, as an indiarubber hosepipe is strengthened by a spiral of steel wire.

Once the food has been manufactured up aloft, it has again to be distributed to stem and leaves and other colourless parts of the plant. To do this, a second, downward, system of transport is needed, and this is provided by sieve-tube cells just like those already seen in higher seaweeds, only more numerous. These, unlike the wood pipes, remain alive and their protoplasm actively hands the food along. Such is the system of the land-plant, and in that direction evolution had to go for the land-plant to be attained.

Then comes a second difficulty the seaweed had to surmount before it could pass on to the grade of a land inhabitant. In the sea you never dry up; in land you are always drying up. How not to be dried up was a problem every living creature which pushed out of the waters on to the land had to solve. A land-plant must be protected from desiccation and so its outermost skin has to develop thick cell-walls, or be varnished over, or whole layers have to be turned into dead but protective cork or bark. But it must have some place where carbon dioxide from the air can be brought into contact with the living chlorophyll. This in the normal land-plant is done in the large spaces full of moist air which ramify among the inner tissues of the leaf; they communicate with the outer atmosphere by the myriads of little openings, called "plant nostrils" or stomata (there are a hundred thousand of them on a single big cherry-laurel leaf) which pierce one or both surfaces of leaves. These serve also to let the excess of water brought up by the transpiration current evaporate into the air, and they perform this duty with great delicacy, since the size of each opening is adjustable, closing right down or opening wide as required. In sunlight the stomata open and the upward movement is brisk; the plant draws upon and requires more water at its roots. In the shade or at night the stomata close, the evaporation and the consumption of carbon dioxide diminishes.

Then a third great structural necessity must be met by a land-plant. No longer is its weight supported by the medium in which it lives. It must grow a stiff support for itself. This is provided by pillars or bundles of cells which have died after loading their cell-walls with a heavy deposit of woody substance or *lignin;* and these columns may branch and join up with each other to make an elaborate skeleton, strong yet flexible. Some of these cells serve a double purpose in also providing a channel for the upward course of the water. Finally the plant with its root and its thickened stem must have some means of adding to all these specialized and often dead cells of its trunk; this it does by retaining reserve patches or a whole ring of embryonic undifferentiated cells (sometimes arranged in the form of a complete belt, sometimes consisting of a series of isolated strands) within the stem and root, a living growing tissue (called *cambium* by the botanists) constantly multiplying and adding on new water-tube cells and skeleton fibre-cells and sieve-tube cells as the plant grows.

These main distinctive structures are necessary to any land-plant that is to grow erect: difference of root and stem, woody skeleton, and upward and downward transport systems. And we find that they have all been developed by the Middle Devonian (III D 2) allowing large trees to be evolved even in that remote epoch, only a few million years after the first invasion of the land. Then they appear, essential land-plants; and all the subsequent lapse of time has brought only minor variations or improvements of detail.

It is interesting to imagine the slow stages by which this escape from the waters was attained. Great seaweeds would find a definite profit in every strengthening of the stem that protected them from being torn to pieces by the waves. Every development towards cuticle or bark was a help against drying up at low tide when the waters receded. The inter-tidal region was the school in which the sea-plant learnt the secrets of aerial life. Plants must always have been pressed nearer and nearer the high-tide mark by the supreme advantage of getting the food-building sunlight. And the more the root penetrated down and gripped, the less the risk of the plant being washed away and destroyed. Salt marshes, lakes and lagoons were the training ground of the future conquerors of the dry land.

Still another great adjustment, however, must be made by land-plants if they are to push far from the water, and this concerns their reproduction. In their sea-ancestors, the reproductive cells, both sexual and sexless, had been little, naked, actively swimming specks of life. Discharged at any point into the water they lived and played their part. Such cells, discharged into air, would merely dry up and die. Plants, therefore, to sustain themselves on land, had to have their reproductive cells as well as their bodily structure proof against drying up. It proved a much more complex problem to solve than the problem of support and transportation. The solution was a gradual one. It was worked out in four stages—the spore was the first, provisional response, then came the seed, then the pollen-tube, and finally the flower. Each gave a further extension of range beyond the water, or a further perfection of land-plant life, because each reduced the need for moisture in reproduction or decreased the wastage involved in out-of-water reproduction. It took two hundred and fifty million years before the plants had achieved their final improvement in this matter.

Among the earliest known land-plants were two inhabitants of early Devonian peat-bogs (III D 1) called Rhynia and Hornea, so generalized that some botanists have wished to place them with the ferns, others with the mosses, others with the algæ. But every detail of their structure has been beautifully preserved, and we will let them speak for themselves. The most remarkable thing about them is that, though regular upright land-plants, from four to eight inches high, they had

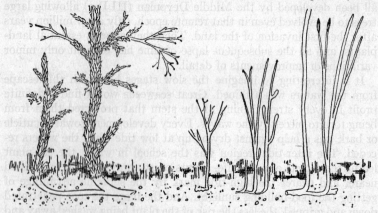

FIG. 221. Reconstructions of the earliest known land-plants.

From left to right, Asteroxylon, Hornea, and two species of Rhynia, all from the Lower Devonian (III D 1). None but Asteroxylon had leaves. The spore-cases are the dark objects at the end of some of the branches.

neither roots nor leaves. There was a creeping underground stem with absorbent hairs on it, but no branching root; and a forking aerial stem. This was doubtless green in life, for it had stomata, or "breathing mouths" of the same type as any modern plant, though very few in number. Internally it had its regular conducting pipes of wood, though here again their number is small. Thus in their construction these plants were obviously, though not very adequately, equipped for land existence. They were land-plants in reproduction, too, for they produced spores with tough thick membranes to resist drought. Their spores were made in mere swellings at the end of some of the branches, which were without any arrangement for bursting open at a particular spot, like a fern's spore-sac.

Another plant called Aster-oxylon ("star-wood") found in the same deposits is related to the other two, but is the first known plant with leaves. The leaves, however, are very simple, and have no "veins" of supply-pipes radiating through their blade. In some ways this seems to mark a half-way stage to club-mosses, for the leaves, like theirs, are small and cover all the stem and branches.

These true land-plants were found associated with many algæ, including a remarkable plant called Nematophycus, found also in the Silurian (III C), with gigantic stems sometimes two or three feet across, and yet all built up at fine interlacing filaments as in modern seaweeds. Such huge algæ are now found only in the sea; but here was one growing in an inland bog. Another remarkable fact is that some true algæ of about the same period had drought-resistant spores and so were adapted to dry-land reproduction, though they had not evolved a woody stem and the dry-land method of growing erect. None of these strange types have survived; they could only thrive when there was little competition from plants better adapted to land.

But it seems clear that in the mountain-building time that brought the Silurian (III C) to a close, when continents were raised and waters shrank, adaptation to land-life must have been at a premium. In those times of change many seaweeds must then have been driven to struggle in various half-way habitats—salt-marshes, peat-moss and fresh-water swamps and bogs; and from one or more of these plant adventurers the ancestors of true land-plants must have sprung. Their actual lineage has not yet been vouchsafed to us. All that we have as yet is a tantalizing collection of scraps and hints. But there is every reason to suppose that the actual transmigration across the water-line took place at the turn of Silurian and Devonian times (III C and D), and speedily gave us plants with root-hairs but without roots, with green stems pierced with air-holes but without leaves, with desiccation-resisting spores but without elaborate spore-sacs—the bare minimum, so to speak, of equipment for land-life. But this once achieved—after hundreds of millions of years of water-life—it took but ten or twenty millions to evolve huge trees and elaborate seeds.

As people say, the plants had "broken the back" of their problem. The rest of the conquest of the land for them was a question of detail and improvement.

And now we come to the most subtle and curious part of this story, the story of the gradual release of the reproductive process in plants from the need of more than a minimum of moisture.

In Book 2 we compared the life-histories of various land-plants, and showed what strange identifications we were forced to make—how the whole moss-plant corresponds with the insignificant prothallus of a fern or the still more insignificant tube of a germinating pollen-grain;

how the green spore-bearing leaf of a fern corresponds with the stamens and ovaries of a buttercup. And we hinted that the transformations that greet us as we pass up the vegetable scale from fern to flower had something to do with plant's progressive adaptation to dry land.

Then we were concerned mainly with forms and the comparison of forms. Here we come to the explanatory story. We could not tell it, however, until we had justified our belief in Evolution; for without Evolution, biology is only disjointed facts; with Evolution, it is seen to be one great dramatic history, formed of thousands of interwoven adventures.

Let us tell the story of the Land-plant as it ought to be told, beginning, as is right and proper, with a humble beginning, tracing our hero's many vicissitudes, and following the tale out to its happy ending. It is a story of emancipation; of life rising above old handicaps. It is not always an easy story to follow as it makes its turns and twists; but it is worth following, since to grasp it gives a new meaning to every tree and flower.

The seaweed from which all land-plants sprung must have had an alternation of generations, a sexless generation reproducing by spores, and a sexual generation reproducing by gametes, such as we have described in the fern; for all land-plants, from mosses up, go through this double cycle. The first step towards successful reproduction on land was to turn the spores from active swimmers in water into tough-walled cells that could lie and drift about unharmed in air. The sexual cells, on the other hand, continued for ages true water-livers. But although the sperm still had to swim to the egg, like Leander to Hero across the Hellespont, all the swimming could be done in microscopic films of water on the plant's surface. For that, however, it was necessary that the sexual generation should still live in moist places, and on this account we may be sure that all the earliest land-plants were inhabitants of the border-zone between land and water. And in all probability both their sexed and sexless generations were, to begin with, mere flat and creeping plates of green tissue.

From this condition, two main lines seem to have branched out. One of these erected the sexual generation into the main plant, and provided it with stem and leaves and root-like organs; the sexless generation being reduced to a mere stalk bearing a capsule full of spores, parasitic on the other. This line produced the liverworts and culminated in the mosses. It got no higher. It is represented in our larger diagram by Stage A. It failed to produce proper conducting pipes in

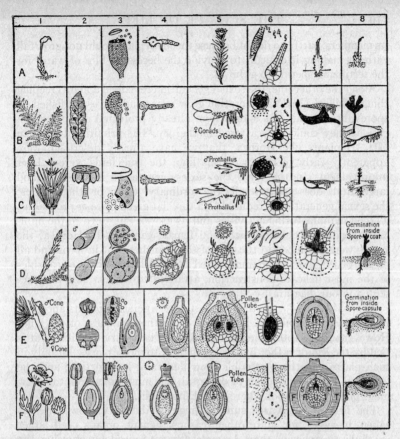

FIG. 222. A Diagram of the Evolution of Seed and Flower.

In each case (1) is spore-producing generations; (2) is spore-leaf; (3) is spore-case; (4) is germinating spore; (5) is gamete-producing generation; (6) is gonads, gametes and fertilization; (7) is development of embryo; (8) is young spore-producing plant. Thus the various corresponding stages are in vertical columns. In all, spore-leaf ▬▬ ; spore-case ////// ; spore-coat (when present) ○ ; gametes and developing embryo ■ . (A) Moss. Spore-producing generation is parasitic on gamete-producing generation. (B) Fern. Spore-producing generation large but (B 8) parasitic on gamete-producing generation for a short time. (C) Horsetail. Like fern, but has spore-leaves unlike foliage leaves, and small male and large female prothalli are often produced. (D) Club-moss (Selaginella); Male and female spores, spore-cases and spore-leaves as well as prothalli. Only four female spores. Spores form prothallus inside spore-coat. (E) Conifer (pine). Male prothallus reduced to pollen-tube. Only one large female spore formed. Female prothallus never escapes from spore-case (lightly indicated in E 4 and 5). A true seed is formed. (F) True flowering plant. Male spore-leaf is a stamen. Female spore-leaf is a carpel and encloses spore-case. Female prothallus never escapes from spore-case or spore-leaf (lightly indicated in F 4 and 5). It forms only eight cells before fertilization. The seed is inside a fruit made from the spore-leaf.

709

its members, partly no doubt because the main plant could not grow tall without making it difficult to provide the necessary film of water for the fertilizing sperms to swim in.

Along the other line, the sexual generation remained a mere creeping thing, while the sexless generation grew stems and leaves, and shed its spores into the air. Its spores were already protected against desiccation; they could be scattered in the dry. The primitive Devonian (III D) plants like Hornea, that we have already described, were apparently early members of this line; the true ferns were a later improvement. In the ferns the sexless generation is elaborate and large and does almost all the ordinary work of the species; the sexual generation is reduced to a tiny flat green plate or prothallus, its food-manufacturing activities and its growth cut down to the minimum needed to develop male and female gametes. In Hornea, the spores were produced in the tips of the branches; but in ferns and all higher plants they are manufactured in special little boxes or capsules of the most delicate construction, attached to leaves. This is represented in our diagram by Stage B.

The existence of a prothallus, obstinately persisting in its ancestral ways of fertilization, is a hindrance to further conquest of the land. However well adapted the main plant may be to dry conditions, it is tied by the leg, or rather by the prothallus, to moist places, for only in such places can the prothallus grow and fulfil its function; and the fertilized egg has no power of dispersal, but must grow up into a new main plant on the spot where it took its origin.

The first step towards surmounting the difficulty, however, was taken, it seems, for quite other reasons. Female organs need larger prothalli to grow on than do male organs: indeed by stunting the growth of a fern prothallus you can make it produce exclusively male gametes. A division of labour came into being. Two kinds of prothalli developed, one small and male, the other large and female. As the male prothallus could produce sperms after very little growth it could take its origin from a small spore; while the more ample growth of the female prothallus was best provided for by making it grow from a large spore. Thus sex was thrown back, so to speak, from the gametes on to the prothalli, and from them on to the spores from which they take origin. The reduction of the male prothallus soon went so far that it lost all its chlorophyll and became little more than a thin layer of cells round a single male organ. And the increase of size in the large spores destined to grow into female prothalli continued until each spore-capsule produced only a few female spores, sometimes only the single group of four produced after a single reduction-division, and each so large that to make a prothallus out of it little growth was needed, the big cell merely segmenting like a large fertilized egg into many little cells. This was the first step in the loss of size and independence by the prothallus. It is Stage C in the general diagram.

This accomplished, the second step, Stage D, could be taken— the first definite step towards making a seed. The female spore-case bursts open as usual, but in such a way that instead of its large spores being ejected on to the ground, they stay where they are and germinate into prothalli while still within its protecting walls. They become in fact parasitic on their parents and refuse to go out into the dry and dangerous world. The little male spores are shed as usual in huge quantities, most of them are wasted and perish, but a few drift and sift into the gaping female spore-sacs. Moisture can drift in here, too; and when it does, the male spores germinate and form their vestigial prothalli and their bunch of active sperms, which swim to fertilize the eggs that the female prothallus has now produced close at hand. This state of affairs we find in some of the higher Cryptogams above the fern level, in the club-mosses for example.

This parasitic female prothallus of the club-moss, once fertilized, is nearly a seed, but not quite. The spore-case soon breaks off, and the four embryos, each sucking nourishment from the remains of the prothallus in which they are still embedded, are tipped out to go on with life on the ground.

This is the phase in which we find a contemporary club-moss like Selaginella. By this stage the sexual generation has suffered a radical telescoping. Compare Selaginella with a fern. In the fern, the job of the spore-capsule is over when it has shed the spores, the job of the spore's own wall is over when it has protected the delicate contents until the moment of germination. But here the embryo may be found absorbing food from the prothallus, which is still inside the female spore's coat, which in its turn still lies within the spore-capsule. One sexless generation has thrown a bridge to the next across the gap of the sexual generation, and the originally separate and consecutive stages now lie one within the other, like Chinese boxes.

But we go on to a further obliteration of the prothallus generation. There can be a division of labour between ordinary leaves and spore-bearing leaves. In most ferns all leaves are alike, and all produce spores. In most higher plants, however, the spores are manufactured on a few leaves only. These special spore-leaves gradually grow less and less like ordinary leaves; they grow smaller, they lose their chlorophyll and give up the business of manufacturing food. They become adapted to one piece of work only—like a workman in a modern factory. Most frequently the spore-leaves are arranged in little groups at the end of special shoots; and these are the first beginnings of cones.

This differentiation of leaves into true leaves on the one hand, and spore-leaves on the other, we may call the third flower-ward modification, although it may just as well precede as follow our second flower-ward modification.

Now comes a fourth modification (Stage E of our diagram). The essential abbreviation here is that the embryos embedded in their parent prothalli are not tipped out of the female spore-capsule soon after fertilization; but the spore-capsule is adapted from the outset, not only to produce large female spores, but to be a protection to the embryo until it has sucked the prothallus dry and is ready for independence. The female spores are never shed: the prothallus undergoes all its development inside the spore-capsule, and never sees the light of day. It has lost all its green pigment and with it all power of manufacturing food for itself. Any growth which it makes is achieved in true parasite style at the expense of the surrounding tissues of the spore-capsule.

The sexual generation is now completely protected within an organ of its sexless parent. The wall of the spore-capsule not only protects the spores, but the prothallus and the embryo as well: it has become the seed-coat, and the whole structure, from its coat outside to the embryo at its core, is a true seed.

But how is fertilization effected? How under these conditions of enclosure are the sperms to swim? The original device for securing fertilization in these seed-plants was what is called a pollen-chamber. This was a cavity in the top of the female spore-capsule to catch the male spores which we shall call pollen-grains from now on for that is what they were. The identification is quite plain. This cavity was often further protected by the growth of a rim round it, which might even shoot up into a spout with a narrow central tube leading from exterior to pollen-chamber. When this was the case, the pollen-grains were caught by a drop of sticky fluid which this tube exuded, and then later sucked down into the chamber. Each male spore, once trapped, grew out into an excessively vestigial prothallus, which we now call a pollen-tube. It wormed its way through the tissues of the spore-capsule's wall, formed sperms, and then burst and broke down into a microscopic drop of fluid, so releasing the sperms to swim to the egg.

This still happens in the most primitive seed-plants now existing, the Cycads.

After fertilization, the embryo grows a bit, and then passes into a resting condition. Meanwhile the spore-capsule becomes tough and woody and turns into a hard seed-coat. The whole structure is now a ripe seed and eventually falls off. What we call the germination of the seed follows in due course: it is the embryo resuming its arrested growth.

In the club-moss Selaginella there are only four female spores to each capsule. In all living seed-plants, the four have been further reduced to one—or rather, though four begin to be formed, all but one degenerate. And though several eggs may be produced by the captive sexual generation, only one grows into an embryo. A somewhat similar reduction is seen, not in the number of male spores to which each male spore-capsule gives rise, but in the number of gametes which each male prothallus forms: in all the highest plants there are but two.

In all living seed-plants, too, another elaboration has occurred, the fifth phase (Stage F) in this strange eventful history. This consists in extending differentiation to the spore-leaves that carry the spore-capsules. Even in Cycads the male spore-leaves differ from the female in their smaller size, in the number of spore-capsules they produce, and in various other ways. Once this step has been taken, their homology is manifest and we can call them by more familiar names: the male spore-leaf is a *stamen*, the female is a *carpel*.

A sixth improvement has also been taken by almost all living seed-plants. It is the suppression of active sperms. Instead of the pollen-tube performing only its original functions of first anchoring the male prothallus and then breaking down to release the sperms, it becomes an organ for penetrating long distances through the tissues of the female spore-capsule and into the female prothallus, and so for transporting its contained gametes all the way to the egg's front-door. Relieved of any need for activity, the male gametes never develop cilia, and indeed become mere nuclei, floated passively along to their destination.

There are yet other steps which had to be taken before the perfection of flowering plants was reached; but these were not achieved until late in the earth's geological history. The six advances we have described have brought us only to the naked-seeded plants, not to the true flower. The flower is in a sense a refinement, not essential to the fulness of land-life. It was not the flower but the seed with its pollen-chamber which liberated land-plants from their subservience to ancestral dampness. All the same, we will pursue our story to its end, for, if not essential, the flower is at least a great improvement in terrestrial reproduction, since in flowering plants the seeds are given extra protection, and waste is cut down all round.

This final step that gave the world its flowers is this. The female spore-leaf or carpel changes its original function entirely, to arch round and enclose the spore-capsule in a little hollow box. Sometimes several carpels conspire to make a single box, sometimes one grows right round. And the spore-leaves, now made protective, do what the spore-capsules did before them in evolutionary history — grow up in a sort of spout. This is the "style" or pillar, with the sticky stigma on its top to catch pollen. This time, however, the outgrowth is solid, not hollow, since the pollen-grains are by now old hands at the job of burrowing their way through solid tissue by means of their pollen-tubes. This extra protective layer made possible the development of fruits; for a fruit is, morphologically speaking, nothing more than one or many seeds enclosed in a container derived from the spore-leaves (and sometimes from other neighbouring parts of the mother-plant).

This change from naked to protected seed was accompanied by two other equally striking developments. The spore-bearing cone was compressed from its original elongated condition, and some of its spore-leaves (or possibly some of the ordinary leaves below it—but that we must leave to the botanists to decide) were converted to still fresh uses: the petals for advertisement, the sepals for protection. The cone has become a true flower. The flower, in other words, is a leaf-bearing shoot like the other shoots of the plant, but first its leaves have become turned to various un-leaflike uses, and secondly it has been so changed in shape that during its development the central part of the flower, which is structurally and was once actually the free tip of the flower-shoot, is arched over by the most basal leaves of all, the sepals, and so the whole is turned into a prettily closed and well-protected bud.

The second change is even more remarkable. Instead of the female spore germinating into a regular prothallus, albeit a small and colourless one, as in the naked-seeded plants like pine or fir, its nucleus divides three times, forming eight nuclei. Two of these move to the middle of the cell and fuse, in just the same way that the nuclei of egg and sperm fuse of fertilization. Now there are seven nuclei; and these all surround themselves with walls and become cells. Of these seven cells one becomes the solitary egg which the vestigial prothallus produces; five come to nothing; and the central cell with the fused nucleus, called the endosperm cell, awaits the event. The event is provided by the pollen-tube burrowing its way in. When it reaches the prothallus, there are strange doings. One of its two gamete nuclei unites with the nucleus of the egg; and the other, instead of resigning itself to degeneration, which is its fate in fir or pine, makes for the nucleus of the endosperm-cell and unites with that. The fertilized egg-cell grows, as per rule, into the embryo; the fertilized endosperm-cell grows into the endosperm or main reserve-material of the seed, destined to be used up later by the embryo. There is thus a double wedding within the walls of the female spore-capsule. Two brothers, we might say, are united with two sisters: but there is a complication, for one of the sisters is herself a double product, two in one, and is later sacrificed to the welfare of the senior couple.

What the meaning of the two-in-one endosperm nucleus may be, we do not understand; but the meaning of the double fertilization is clear. It is to prevent waste. In all our series of plants, from fern up, the prothallus has served not only to produce the gametes, but to help nourish the embryo. Now that it is wholly parasitic on the sexless generation, and produces but one egg, gamete-production need take very little of its energies. Accordingly, in order not to use material in producing a mass of nourishing tissue which would be wasted if the egg were not fertilized, the development of this main part of the prothallus has been made dependent, like that of the egg, on the vital kindling of an act of fertilization.

As further prevention of waste, we find that in flowering plants the development of the fruit too has been made dependent upon fertilization. If the egg is not fertilized, the female spore-leaves or carpels do not enlarge, and harden or become fleshy as the case may be. How precisely this is accomplished, we do not know. It is not due to any extra act of fertilization, but apparently to some chemical influence spreading out from the growing embryo and endosperm. And of course a still further prevention of waste was the substitution of insects for wind as transporters of pollen; but that story we are reserving for a later chapter.

Thus in the higher flowering plants the telescoping of stages has gone far beyond that attained by the gymno-sperms. Spore-leaf, spore-capsule, spore and prothallus—all come to surround, protect and minister to the embryo of the next spore-bearing generation; and the spore-bearing generation can no longer justly be called sexless, since sex has been thrown back from gamete to prothallus, prothallus to spore, spore to spore-capsule, and spore-capsule to spore-leaf; indeed in some plants, like the bisexual hemp or dog's mercury, on to the whole cluster of spore-leaves that we call a flower, and back on to the whole spore-bearing plant. And the originally independent sexual generation has been trapped, imprisoned, reduced, made to serve new ends. Were it not for the complete series of transitional stages, we should never be able to recognize in the pollen-tube or the few cells round the egg of a flowering plant the equivalent of the fern's green prothallus or the moss's leafy shoots.

§ 3. The Plants of the Ancient World.

This story of the perfecting of a land-plant was not so straightforward in reality as our narration might make it appear. We have cut out everything but the bare final essentials. Even the first dispensation of land-plants, the vegetable dynasty that was dominant for some hundred and fifty million years from the Mid-Devonian (III D 2), showed the greatest variation in their progress towards seed and flower.

Of higher woody-stemmed land-plants (which alone can grow tall and alone are likely to be preserved fossil) there were five great groups, all established already in the Devonian. First there were ferns. The vast majority of these have never taken even the first step towards making a seed. Their spores remained all alike, their sexual generation always free.

If the ferns were the least advanced, the most advanced were the Cordaiteans, a group including many fine trees. These, called after the botanist Corda, were an early branch of the true naked-seeded stock, and so related to our pine and other conifers. But they had long strap-shaped leaves in place of needles, and their seeds were provided with a large pollen-chamber, showing that they had not developed proper pollen-tubes.

Then there was the club-moss tribe. These are in some ways the most interesting of the ancient plants, since they were the most successful, the tallest and the most advanced of the earliest large trees. Some of them reached a hundred feet. But, like all their contemporaries, their mechanical construction was not so strong as that of modern trees. With their insufficiency of strong wood, they could never have carried the broad, spreading crown of an oak, or soared up to heights of two and three hundred feet. Among them we find all kinds of attempts, so to speak, at seeds. But the attempts were never wholly satisfactory. Some remained in the primitive state of producing only one kind of spore; but perhaps the majority had taken the first step towards seeds, and many the second step, having had large female spores which often germinated into prothalli within the parent spore-capsule, and there were sometimes fertilized. Finally, in some of the giant club-mosses, what we must call a "false seed" was evolved. Here, as in true seed-plants, the number of female spores in each spore-capsule was cut down to one, and the spore was never shed from the capsule. But instead of the seed-coat being made by a thickening of the spore-capsule's wall, this remained thin, and protection was afforded by a special leaf or bract which grew up round the capsule and later became rough and woody. And further the "seeds" were shed, in many cases at least, before fertilization instead of waiting till an embryo had developed, so that the ancestral need for moist places was not eliminated as it was by true seeds.

Next came the horse-tails or Jointed Plants (Articulates), which, though some grew sixty feet high, never reached the dignity and strength of the club-mosses. All the modern horse-tails are primitive in producing spores all of one kind, without difference of sex, like ferns. But a few of the ancient forms had taken step one and produced small male and large female spores. None, however, seem ever to have gone farther on the road to seededness, though their spore-leaf cones were sometimes of elaborate construction. They are one of the best examples of a relatively primitive form of life, achieving a temporary success by rapid specialization in one or two directions, and then being eclipsed by competitors which had gone in for slower but more solid and more all-round advance.

Finally comes a remarkably interesting fossil type—that of the *Pteridosperms*, which being translated means Seed-ferns. These had foliage so fern-like that until about twenty-five years ago they were universally regarded as ordinary ferns. In 1903 a pretty piece of detective work showed that the fronds of a "fern" called *Lyginopteris*, and certain seeds found separately in the same strata, both possessed little glands, stalked and with a round head, unlike anything known in any other plant. And a year later another fossil "fern" was discovered with seeds actually attached to its leaves. A picture of this plant is given in Fig. 130. Now we know that at least six families of these seed-ferns are to be distinguished and that the group differed from true ferns in the internal construction of its stems as well as in its reproduction. The resemblance between the two groups is a superficial one, like that between a porpoise and a fish.

Some of them were very tall, but the majority grew into smaller trees or bushes. The seeds of some seed-ferns were tiny, in others as big as a hen's egg. All had an elaborately constructed pollen-chamber, in which pollen-grains are sometimes preserved for our microscopes just as they drifted in some two hundred and fifty million years ago, and the walls of the chamber we sometimes find supplied with an elaborate system of little pipes that must have pumped up water to make a pond for the spermatozoa. In one called *Heterangium*, the pollen-chamber had been brought to a higher pitch of specialization than in any other known plant. There was an outer ante-chamber and an inner chamber in the form of a circular groove arched over by a thin, movable flap or lid. At the bottom of the inner or true pollen-chamber were the female sex-organs of the imprisoned prothallus. Apparently this flap was first held open to catch some of the pollen-grains shed from the male spore-capsules, and then grew over to trap them so that they could germinate undisturbed. The male spore-capsules of many seed-ferns have also been found.

The interesting thing about these plants is that they had developed real seeds, but grew them on their ordinary leaves instead of on special spore-leaves: nothing in the least like a cone or a flower was evolved in the group. The only movement in this direction was taken by plants like Telangium, in which each branch divided into three fronds, the two on either side being green and leafy, the one in the centre, though branched and frond-like, devoted entirely to carrying the pendent seeds. The seed-ferns had taken steps one, two, and four, but had omitted step three.

Thus, of our five great groups of ancient plants, two had clung to their old ways and always passed through a free sexual generation on the ground; two had developed seeds; and one, though seedless, had arrived at a half-way house. All these groups can be traced back into the Upper Devonian (III D 3) and most of them, including a primitive Cordaitean and some seed-ferns, as true seed-plants, to the Middle Devonian (III D 2). Though we see many examples of plants which have persisted with partially evolved seeds, the actual evolution of these seeds has not yet been revealed to us in fossils. Presumably it took place with some rapidity (geologically speaking) in Lower Devonian times (III D 1), prompted by the cool and arid conditions which then prevailed.

The truth of the matter may be a little more complicated than the story we are telling here. Many botanists believe that the threshold between sea and land has been crossed not by one stock only of plants, but by several, and that the Jointed Plants, the Club-mosses, and the

Ferns, for instance, are not relatives all descended from a common land-ancestor, but travelling companions who have all happened to take the same road, like land arthropod and land vertebrate. Against this it may be urged that the earliest land-plants known—Rhynia and its relatives—show extremely primitive characters from which the other, later, types could well be derived. The answer lies somewhere in the rocks of the Lower Devonian.

In any case, it is certain that by the Middle Devonian (III D $_2$) the earliest forests were in existence, and that for the first time there was abundance of green life out of water as well as in it. From the point of view of animals, this spread of vegetable life meant so much available food; animals doubtless could have emerged on land long before, but there was no incentive for them to do so, and no possibility of them supporting themselves there, any more than there is the possibility of an agricultural civilization colonizing the Antarctic continent to-day.

§ 4. Coal.

THESE swamp-plants and land-plants which first appeared in the Devonian Period (III D) found, in the subsequent score of millions of years, phases of great encouragement. The world passed through periods of warmth and moisture which released these new possibilities of increase to an enormous expansion. This expansion gave the period next after the Devonian its specific character and its name, the Carboniferous Period (III E). A huge accumulation of vegetable matter occurred and became fossilized as the combustible black rock we call coal. The great part of the world's coal derives from this age. These Upper Carboniferous deposits have played so great a part in the economic development of mankind, especially in Britain and the United States, that the period is known as the Coal-measure Period.

For the best part of twenty million years these Coal-measure forests throve. The lands where they grew were subjected to a general sinking, accompanied by repeated small oscillations of level. The sinking was so prolonged that the total thickness of sediments laid down along the shores piles up to over ten thousand feet. The oscillations of level drowned forest after forest where it stood; sealed its remains away under a lid of sandy or muddy sediment, then raised the whole area above water again for another forest to grow and in its due time to be flooded, killed, and preserved. The results of these regular ups and downs can be read to-day in the way the layers of rock generally lie in a coalfield, the coal seams alternating with bands of other rocks like layers in a Neapolitan ice. Often the layer below each seam is clay, formed of fine consolidated mud in which impressions of tree roots are frequent, and the lid over the seam is generally stratified shale or sandstone.

Often the clay contains the rush-like subterranean stalks of calamites or the forking buttress-stems of giant club-mosses with their roots still attached, showing that the trees grew where we find them to-day, shedding their leaves and branches to contribute to the coal above. But though much of the raw material of coal was doubtless piled up, just as it is in the Great Dismal Swamp of Florida to-day, by the accumulated remains of generations of plants growing in the swamp and dying where they stood, yet it seems clear that this was not the only way. Sometimes true land forests were slowly invaded and drowned; and much material must have been floated down in the sluggish rivers. Some coals were formed in estuaries and sea-swamps, others in freshwater lakes and peat-bogs.

The actual process by which a mass of leaves and stems and spores is turned into a hard black material like household coal or anthracite, eighty or ninety per cent pure carbon, is a matter for the chemist and the geologist. Those who are interested will find an excellent discussion in Arber's little book, *The Natural History of Coal*. Let us remind ourselves that there exists every gradation between the hardest anthracite, through Household and Cannel and Brown coals to Lignite and to Peat, which has scarcely more carbon in it than ordinary wood; and that the raw material for coals of every description is still being accumulated to-day, chiefly in sub-tropical and tropical swamps and peat-mosses. What will happen to this potential coal depends on the way it is accumulating, the sealing-off that it may get, and the weight and the squeezing to which it will be subjected in the next score million years or so. It will never become anthracite, for instance, unless it is subjected to the enormous pressures and temperatures which exist deep within the crust of the earth.

There is plenty of other excellent coal in the world besides the coal made by giant club-mosses and calamites and other Carboniferous plants. In the adjacent arctic lands of Spitsbergen and Bear Island, for example, you find Devonian coal (III D), Carboniferous coal (III E), Jurassic coal (IV B) and Eocene coal (V A), proof that four times in its history this polar region had a warm climate and supported great swampy forests. But the Upper Carboniferous (III E 3) was *the* coal period. Well over half the world's present supplies of good quality coal were laid down in this Coal-measure time. When we remember that it takes about ten years to produce an inch of good peat, and thirty or forty for an inch of coal, we may be thankful for the climate of the Upper Carboniferous and the long-continued oscillating indecision of its sinking shore lines which conspired to give us the great beds of black stone on which nineteenth-century civilization was founded.

The total amount of coal in existence is enormous—some seven million millions of tons, according to the careful estimate made by the Geological Survey of Canada in 1913. This contains over four times as much carbon as the whole of our atmosphere to-day. What effect the withdrawal from the atmosphere of such a vast bulk of carbon as that represented by the Carboniferous Coal-measures might have, we do not yet know. There is a profit and loss account continually going on, carbon dioxide leaking out from the interior of the earth in volcanoes and hot springs, being withdrawn from the air to make wood and coal, from the water to make corals and shells and limestone, and passing from water to air if too much is taken from the atmosphere, back again into water if the atmosphere's share goes up too high. And we do not know enough about this give-and-take to know how big or how lasting an effect such a long-continued drain of carbon may have produced.

It cannot have had any marked permanent effect, for the number of "plant-nostrils" or stomata on modern leaves is not very different from what it was on the pre-Coal-measure plants, showing that the amount of carbon dioxide available in the air is not very different now from then. But it is quite possible that some temporary shortage was brought about; and if so, the coal-plants themselves helped to bring their own luxuriance to an end.

Whether or not we ever succeed in reconstructing the balance-sheet of remote epochs, two items of the economics of life stand out clear enough. Coal to be formed in considerable amounts demands a particular combination of circumstances—the energy of sunlight, the chlorophyll-machinery of luxuriant air-breathing plants, a warm climate, and swampy sinking coasts, and these conditions must be continued for geological periods. The result is bottled energy.

But we are using up that energy many thousand times faster than it was bottled. In 1905 man dug up over 900 million tons of coal; in 1910 he dug up nearly 1,150 million tons; in 1923 over 1,300 million tons. Europe has reasonably available coal to last her, at the present rate of output, about 500 years; the United States, the richest coal country in the world, has only about 2,000 years' supply.

Coal took millions of years to accumulate: we are spending it in a few centuries.

§ 5. The Vertebrates Crawl Out of Water.

THOUGH it is true that animals could never have colonized the land until the green plants had gone ahead to cover the barren rock with nourishing vegetation (just as it is necessary for an advance-party to establish supply-dumps along the route of a polar expedition), nevertheless the evolution of legs and lungs in vertebrates does not seem to have been a simple response to the new possibilities thus opened up, but was forced upon them by an entirely different influence—drought. Once they had acquired legs and lungs and the possibility of land-life, they proceeded to the exploitation of the vegetable food-supply whenever and wherever it became available. But had it not been for drought and oxygen-starvation and the drying-up of their ancestral pools, they might never have left the waters. In much the same way, had it not been for the necessities of war, the aeroplane would never have developed so quickly; but once it was developed, the fields of the air were open to man and advantage was speedily taken of the new opportunities for commerce and peaceful communication.

Lungs and legs are the characteristic mark of the land vertebrate. But the first step towards lungs had been taken long before land vertebrates were possible, when the ancestral bony fishes developed an air-bladder. There can be little doubt but that the air-bladder was an adaptation to life in stagnant waters, enabling its possessors to supplement their water-breathing by gulping air at the surface. Originally perhaps a mere recess in the gullet, richly supplied with capillaries, it must have soon developed into a primitive lung, a sack with abundant blood-vessels into which air could be forced through a primitive wind-pipe. To start with, the bladder-lung probably lay below the stomach and gullet; but in most fish it has been shifted round to the upper side so as not to make the animal turn belly-up when the bladder is full of air or gas.

Nowhere but in fresh-water is there likely to be any long-continued unescapable shortage of oxygen; and we can be reasonably certain that bony fish evolved in fresh-water and not in the sea, so continuing that fresh-water evolution which we found probable for the early vertebrates as a whole.

This evolution of bony skeleton and air-bladder probably befell one of the stocks of primitive gristly shark-like fishes in some corner of the late Silurian world (III C 3); for by early Devonian (III D 1) times several types of true bony fish were in existence. Almost all these early creatures were heavily armoured, their trunk with a chain-mail of beautiful enamelled scales; their head with a plating of bones, also often enamelled, that ran back to join the scale-armour behind; they had rather long bodies, clumsier and less well stream-lined than most modern fish; their fore- and hind-limbs were far apart, and were of primitive types, with a central lobe ending in a fringe of fin-rays. Most of their bone was in their outer armour, but boniness was invading their inner skeleton as well, and a few parts of the gristle of their vertebral column were already replaced by bone.

Those primitive fish, still mainly cartilaginous within but bony outside, diverged into two main lines. One became more fish-like, the other less; one recolonized the seas, the other eventually invaded the lands. In the fishier fish the armour of heavy bony scales was progressively lightened, until it was transformed into a tough but light and wonderfully flexible scaly garment. But the skeleton within was more and more fully ossified until the animal was provided with that elaborate and delicate set of little bony girders which embarrasses us when we eat a herring or a sole; and the armouring of the head was sunk below the surface until it coalesced with the original brain-box below. The group radiated out time and again into every conceivable shape and form; but the bulk of actively swimming fish developed a very perfect stream-lining. They withdrew their limb-bones within the body, leaving the fin more flexibly supported by a set of horny rays. Many moved up their hind pair of fins until they were on the same level with the fore-limbs, or even in front of them: this apparently makes the limbs more efficient as stabilizing rudder-planes; and the air-bladder, transformed to new uses, had a great part to play.

As has so often happened in life's history, an organ evolved by adventure in some new and unusual environment has enabled its possessor to go back to whence it came and beat the stay-at-homes at their own game. Only in the lung-fish and a few survivors of ancient armoured fishes is the bladder still used as a lung. In all the rest it has been turned into a hydrostatic organ—partly a sense-organ, an organic barometer for detecting changes of pressure, and so giving the fish a "depth-sense"; and partly a gas-ballast tank for taking weight off the fish, and so for adjusting to different depths below the surface, just as submarines are adjusted for depth by pumping water or compressed air in or out of compartments in the hull. In the modernized fish the primeval lung has become a gas-filled bladder whose walls secrete gaseous oxygen out of solution in the blood and pass it back again as needed; and in one large group it has even lost its original connexion with the exterior to become a completely closed organ.

The new capacities of adjusting weight thus conferred on aquatic life meant economy of muscular power, and a more delicate adjustment to particular levels in the sea. And these advantages undoubt-

FIG. 223. The Mud-hopper Periophthalmus.

In water, half-way out, scrambling about on mud and mangrove-roots, snapping at a fly, lubricating an eye by withdrawing it into its socket, and looking in two directions at once.

edly had a good deal to say in securing to the bony fish, when they went back to ocean after their long fresh-water stay, their rise to predominance in competition with their gristle-skeletoned competitors. Indeed, when we survey the two great groups of fish, gristly and bony, it is striking to see how many more of the bony and gas-ballasted fishes live a permanently active swimming life; how many of the gristly fishes, without a gas-bladder, are completely adapted to bottom-life, like rays and skates, or spend much of their time resting on the bottom, like most dogfishes.

The full perfectioning of the bony fish for water-life took nearly as long as the full adaptation of land vertebrates to their much more difficult environment. Most of the types that arose in the first outbursts of evolution in the early Devonian (III D 1) died out with the beginning of the Carboniferous (III E 1), which witnessed the beginning of a second radiating-out of new and improved fish-types. If the reader will turn over the pictures of fishes in this book, or in any well-illustrated geological text-book, he will have just the same impression of gradually increasing neatness and smartness that he would get from a museum collection of the automobiles of the last thirty years. Continually he would mark increased efficiency and finer lines. The Permian (III F) saw a second batch of extinctions and a third sprouting of new types; and the process was repeated still once again in the Cretaceous (IV C), when the great bulk of the modernized fish we know to-day—salmon and herrings and conger and the rest—first put in their appearance, only a few million years before the modern mammals and birds.

But we are anticipating. Mammals and birds could not be thought of until some ancestral vertebrate had emerged from water on to land. We must go back to the parting of the ways and discover the origin of legs and how the vertebrate crawled out upon the land.

We take our legs very much for granted. But they have a long history of slow improvement behind them. Before life, now embodied in us, could stand upright on two legs it had to climb with four. Before it could climb it had to run; before it could run, it had to lift its belly out of the dust on four pillars and walk; before it could walk it had to be able to crawl. And to crawl is precisely what the ordinary fish cannot do. Out of water, its weight no longer supported by the element in which it swims, it can only wiggle and jump and lie on its side until it dies. To crawl it needs legs, and it has only fins.

To understand how fins were turned into legs, we must ask the geologists to tell us something of the conditions which prevailed in Devonian (III D) days. For it was these conditions which, for the fishy ancestors of all land vertebrates, made legs the alternative to extinction.

The Devonian Age was a time of great lagoons and lakes. All the vast deposits that we call the Old Red Sandstone (over thirty thousand feet of them in some parts of the world) were laid down in such spots. Conditions in these bodies of water varied cyclically, it would seem. At one period they were brackish embayments connected with the sea by a few irregular channels; then for a few ten-thousand-year periods they would be huge fresh-water lakes; and then a time of elevation would cut down their extent, separate them into smaller bodies of water, and dry many of them up. And so over again, in a cycle of recurrent change.

In the open sea beyond, quite different deposits were forming; and it is one of the most interesting features of the Devonian that for the whole of its perhaps seventy million years, we have two parallel records in the book of the rocks, one entering up marine history on pages of fine-grained blue limestone, the other recording the life of fresh-water on the layers of red sandstone that testify to arid conditions on the high lands beyond the coast.

Every period of severe drying-up of the land waters must have witnessed the extinction of huge numbers of fresh-water fish. Two evolutionary courses were open to them; to go back, or to go on. To go back meant adapting themselves more perfectly to water-life and invading the seas which their remote ancestors had previously abandoned. To go on meant overcoming the difficulties of stagnancy and drought, not merely by taking occasional gulps of air, but by some new construction which would enable them to survive desiccation and, if possible, to leave the water, even if but temporarily, and transport themselves across country.

Most of the fish, as we saw in our last section, went back. Two groups remained and went on. One of these was the lung-fish, which we have mentioned in Book 1. These creatures have turned their air-bladder into a perfectly good lung, and they can live for long periods out of water in a "nest" in the mud. But walking over-land is beyond them. They provide an excellent example of parallel evolution. They were closely allied to the fish which did turn into land vertebrates, and they went part of the same road; but they never got all the way. Such half-way advance is not much good in the long run. The fish that turned back have peopled the seas; the fish that went on have colonized all the lands; but the lung-fish that went half-way and stopped are now extinct save for three scarce inhabitants of tropical swamps. The fish that went on, as we can tell by comparing the pattern of their skulls with that of the earliest Amphibians, were another group of long-bodied, fringe-finned, scaly-armoured, bony-headed creatures, called Osteolepids.

But the actual steps by which fin was made over into leg have not been preserved to us, though we can prophesy with some confidence that some lucky geologist will one day find their traces in the Middle Devonian (III D 2).

However, some modern fish, like the Mud-hoppers, Periophthalmus, get on famously on land with the aid of their fins, which are very muscular and bent under the body like a land-limb. Anybody who has tried to catch one knows that the fins are efficient at their new job (and indeed so well adapted are these fish to land-life, that they die if kept under water for more than a short time). The fins of these amphibious Devonian fish must have been employed in a somewhat similar way, though from various lines of evidence we can conclude that they were not specialized for hopping, but were spread out sideways, flat on the mud, and served to pole the body onwards. A glance at most modern salamanders, with their awkward clockworky movements, reveals essentially the same type of motion. The bends at wrist and elbow, ankle and knee, must speedily have developed as mechanical aids in leverage, and to facilitate keeping the hand or foot firmly and flatly where it was put down in the mud; and of the bones supporting the fin membrane, five were selected for preservation as fingers and toes. It is a far cry from the clumsy bent fin of this creature, just more than fish, but less than salamander, to the leg of a racehorse, the wing of a bird, the shapely human calves that post-War fashion has decreed shall be visible in our streets. But the bent fin was the beginning of them all.

By the Upper Devonian (III D 3) this transformation was complete: we have the land-creature's own "hand and seal" on a page of the record—a veritable foot-print, toes and all, in the mud. By the Lower Carboniferous (III E 1) the legged vertebrates were plentiful. But there is one remarkable fact about them; the majority, as Professor Watson has shown from various peculiarities of their construction, normally did not use their new-won legs, but lived on in their ancestral aquatic home. The reason for this is to be sought far off, in the waters of the sea. In the sea there are very few real vegetarian animals of any size. Why? Because the great bulk of marine plants are floating and microscopic. They cannot be browsed at, but must be eaten whole. Many are swallowed by very small animals, which in their turn fall a prey to larger, others are filtered out of the water by the sedentary army of current-feeders. Thus almost every sea-animal of any size is either a current-feeder, a mud-eater, or a carnivore.

All the early fish were carnivores; and every one of their amphibian descendants has remained true to a meat diet. Now the later Devonian lands were well covered with plants, but very few land-animals had succeeded in exploiting these new resources, which for the most part simply died and rotted unconsumed. They were as useless to the earliest amphibians as gold to a man on a desert island. The only possible animal food was provided by the few early scorpions and myriapods and ancestral insects which must have existed in the early Carboniferous woodlands.

But even this diet cannot have been abundant, and in any case it had to be hunted and captured in new ways. Most animals are as conservative about their feeding as most men; and so it came about that the bulk of the earliest amphibians went on living, eating and breeding in the waters, with legs reserved solely for the occasional necessity of moving from one pool to another. Thus to the early amphibian, meals on land were like what sandwiches on a railway journey are to us—with the important difference that the picnic dinner was growing by the amphibian's wayside. The few who did become adapted to eating their meals out of their old home prospered and multiplied; their legs, evolved for quite other purposes, stood them in good stead; and their seed has inherited the earth. It is not the only time in the history of life that some device, evolved only to meet emergencies, has been turned into an every-day necessity of fuller living.

But even when the adult amphibian emancipated himself fully from the water, in his youth he remained essentially a fish. The young amphibian is a larva, a tadpole, which breathes by gills and swims by a broad tail-fin. We have already spoken of the chemical machinery which effects the transformation from aquatic larva to land adult. Here we will add that a tadpole stage has remained from the remote Carboniferous time to the present a characteristic badge of the great majority of amphibians. It is this, more than anything else, which has kept the group from achieving the abundance or the importance of reptiles, birds or mammals; for it has kept them tied to water. There are a few frogs which have cut the aquatic stage out of their life. The Surinam toad lays her eggs all over her back by means of a long protrusible oviduct, and the skin of the back grows up to surround the eggs, which develop there up to the toad stage. Nototrema grows a huge pouch all over its back and flanks, and broods its babies there. In other species the male turns his vocal sacs into nurseries, or the eggs are laid in a foamy mass. But these are all makeshifts. In none of them has independence of water-breeding been achieved in that simple and radically efficient way that characterizes the reptilian egg.

The first precursors of the land vertebrates, which we find in the Carboniferous, are called Stegocephalians, or animals with roofed-over heads, from the fact that their flat skulls are made of a single roof of bones, continuous save for their nostrils and the orbits of their eyes. In this they differ from their descendants, the modern Amphibia, but resemble the ancestral fish from which they took their origin. Many of them also differed from the slimy-skinned amphibians of to-day in being plated here and there with bony armour on their bodies.

And most of them had a curiously complicated and inefficient construction of their backbones. In all existing land animals, each vertebra of the backbone is an elaborate but single bony unit. But in most of these Stegocephalia, each vertebra was compound, a number of bits wedged together. This condition was inherited from the dim past when the original notochord rod first began to be strengthened by blocks of gristle around it. The multiple construction still survives in water vertebrates like the dogfish, but it has had to be unified into strength in all animals that live on land, and so must carry their own weight. Evidently these Stegocephalians were weak-backed crawlers with no intimation of the bounce of a modern frog. In one respect, however, they achieved something of which no modern amphibian has proved capable; they colonized salt-water. One small group of these from the Trias (IV A), so far discovered only in Spitsbergen, were all marine. They were long-snouted fellows, and probably fed on fish after the manner of the long-snouted fish-eating crocodiles of to-day.

These primitive Amphibia endured from the Upper Devonian (III D 3) to the end of the Trias (IV A 3); and in the meanwhile they and their relations gave rise not only to the ancestors of modern Amphibia, but also to the first reptiles, and so to the ancestors of all higher vertebrates.

§ 6. A Coal-measure Forest.

BY THE time of the Coal-measures, both plants and animals were perfectly at home on land. The conquest of the hinterlands and highlands had still to be made, but life had won the shore and the lowlands and the swamps. But the picture which land-life presented even there was entirely and amazingly different from the picture we are familiar with to-day. If we could be made free of time, and be suddenly transposed some two hundred million years or so into the past, we might be pardoned for imagining we were in another planet, even though we were still on the same spot of our old earth's surface.

Suppose that at the moment when this time-transformation was effected, we had been walking the streets of Newcastle or Pittsburgh. Suddenly the houses and streets and factories are gone. In their place is a moist forest, sloping down through swamps and marshes to the shores of a shallow sea. The forest, it must be admitted, does not rival the tropical forests of to-day, either in the variety or in the size of its trees. But fine trees there are, running up to a hundred feet or more; and there are enough kinds of strange plants to keep curiosity busy for many hours. The giant club-mosses and the seed-bearing Cordaites are the masters of the wood. Among the former, the handsome scale-trees, Lepidodendron and its relatives, raise pillar-like trunks embossed with a wonderfully regular scaly pattern, each lozenge marking where a leaf has fallen; for these plants, like the modern club-

FIG. 224. Some Stegocephalians.

Left, Archegosaurus, about five feet long, whose elongated snout indicates fish-eating habits. Right, top, Mastodonsaurus, a lumbering ten-foot Amphibian. Right, lower corner, Microbrachis, about six inches long. Centre, two specimens of Diplocaulus, with flattened head, huge gape, and body about six feet long; and one of Batrachiderpeton, with skull two-and-a-half feet long.

mosses, grow leaves all up their stems, but resemble many palms in shedding the lower leaves as they get cut off from light by the new growth above. Near the top the Lepidodendrons branch into a fine crown, each of the branches set thick with little leaves and ending in a spore-bearing cone, sometimes a foot or more in length; or, in some cases, the cones are gathered into clusters near the base of the branches like dates on a date-palm. Below, the stem is buttressed in a remarkable way. It diverges downwards into four separate pieces, each of which forks regularly and repeatedly as it runs into the ground. These are really creeping stems, and from their under-side the slender true roots penetrate the soil.

Then there are trees with long leaves, stiff band-shaped twigs up to a yard long and gathered into extraordinary clusters sticking up like fat gigantic paint-brushes, sometimes only one brush to each tree, sometimes two or four or eight. And below each brush of leaves is a thick clustering ring of cones. These are seal-trees, such as Sigallaria, so called from the beautiful pattern of leaf-scars on the stem, each scar often looking like the imprint of a seal. These too are club-mosses grown to the estate of trees, though they never reach the size of the great scale-trees.

Besides these there are the Cordaites, predecessors of our conifers, tall and slender, with long strap-like or even grass-like leaves. Some are over a hundred feet high, with beautiful leafy crown, others more conical, others mere shrubs. Ferns are more abundant and varied than any dweller in our temperate regions can easily imagine, though these grand Carboniferous tree-ferns with pillared stems, sending up their raying crown of fronds to where it can compete for light and air on equal terms with the other forest-trees, would not surprise an inhabitant of the tropics.

But the majority of the ferny plants in this old woodland are seed-ferns. Here is one with huge feathery fronds, their stalks sometimes five or six inches across, with big berry-like seeds borne among the leaflets; there another with tiny winged seeds carried on a frond very like that of an ordinary bracken-fern; there one with huge thick-walled seeds as big as hen's eggs. Many of them are small, but others grow to bushes or even trees, and some have taken to climbing like lianas.

Here are other climbing plants, luxuriantly scrambling over the branches and trunks of their fellows—the Sphenophylls, taking their name from their wedge-shaped leaves. These were climbing horse-tails, a group now long extinct. And there are tree horse-tails, too— perhaps the most fantastic of all the trees in the coal forest. They never rival the largest among the other kinds of trees; but some touch the respectable height of sixty feet. Here is one species of Calamites, the Giant Rush; it creeps about the bog with a subterranean stem from which at intervals shoot up jointed pillars, just like magnified horse-tail stems of to-day, with a frill of leaves round each joint. Each pillar ends in a great brush of upward-pointing cones. And there are many other types, some looking like caricatures of a Lombardy poplar, others like mere magnifications of modern leafy horse-tails, others with jointed branches as well as jointed stems. But all have a curiously inefficient look as trees; and indeed they are a mushroom growth of plant evolution, all the larger types destined soon to disappear.

As the wind blows, clouds of golden spore-dust drift down through the stems. The moist ground is green with mosses and liver-worts, and the innumerable prothalli of all the seedless plants. Prostrate stems lie rotting across each other in the shade and there is a thick carpet of fallen leaves.

On the ground and in the swamps crawl and swim the Carboniferous vertebrates. Most of them are amphibians, but amphibians that will surprise anyone who thinks of amphibians solely in terms of frogs and salamanders. True, there are little salamander-like fellows, in the hollow bogs and through the undergrowth, snapping up insects; but there are also fish-eating creatures with long snouts like alligators; there are legless wrigglers and swimmers; some have fantastic flat heads nearly as big as their bodies, others are plump and hideous, with patches of armour-plating here and there on their heads and backs. Many of these are permanent water-dwellers, having sunk back to their ancestral element after the briefest taste of the land-world. But some of them, as T. H. Huxley put it, "pottered with much belly and little leg, like Falstaff in his old age, among the coal-forests." These are true creatures of the land, even though it be moist land only.

Among the land-folk are also a few little ancestral reptiles, insignificant in themselves, but rich with promise. In and out of the vegetation there run cockroaches in plenty, large and small, with mites and scorpions and centipedes and ancestral spiders and a few land-snails to keep them company; and through the air flaps and flits an abundance of winged things, the primitive net-winged insects and their relatives. Some of these are of alarming size, the largest, not unlike a very awkward dragon-fly, measuring nearly two-and-a-half feet across the wings. Many of these insects live their early lives in the pools and creeks, which teem with invertebrate life as richly as such sluggish waters always do.

It is a strange world, and a varied world. But after the strangeness has worn off, how much we miss! No birds nesting or singing in the trees; no bellowing, roaring, squeaking, or howling of mammals, savage or small; no flowers anywhere, no fruits; no caterpillars to eat the leaves; no bees or butterflies; no four-legged creatures that can do more than crawl. The forest, without colour of flowers or of autumn leaves (for there is no change of seasons here), is a symphony in brown and green, with yellow veils of spore-dust sinking into black pools below.

If there were any sounds beyond the wind in the branches, the occasional fall of a trunk, the humming of the insects' wings, and the splashes of the primeval mud-puppies sliding into the water, they were the raucous mating-calls from those first air-breathing throats.

But somewhere among those swamps there crawled the ancestors of all birds, of all mammals, of us who write, and of you who read this book.

§ 7. An Age of Death.

THE coal-swamps could not last for ever. Another spasm of mountain-building was being prepared in the earth's crust; and the climate began to change. The swamps shrank, the sea retreated, the coal-forests dwindled. The Appalachian mountains were thrown up in a series of violent movements, which, speeded up in the mind's eye, look like a paroxysm, but in reality were spread over a vast space of time—probably ten or twenty million years.

The crust of Europe heaved in sympathy, and the Hercynian chain, now worn down to such stumps as some of the Irish mountains, the Eifel, the Taunus, and the Ardennes, was synchronously brought into being.

Meanwhile in the Southern Hemisphere a great ice age was brewing. What caused it is not our business to explain, but the geologist's and the geographer's. Most probably its onset was due to the remarkable elevation of the southern lands, which first produced a huge antarctic continent, and then cut most of it off from the approach of warm currents from the tropics by joining it with Australia and possibly with South America too, and throwing a vast transverse bar of land across from Australia through Malaysia and India to Africa, and perhaps beyond. This transverse super-continent has been christened Gondwanaland by the geologist. Antarctica, high and with its central-heating system of warm currents cut off, covered itself with a great ice-cap; this infected and cooled all the neighbouring inland sea and filled it full of icebergs; and then the heights of land in the continent of Gondwana also began accumulating their own cap of snow and their own glaciers.

Whatever the cause, it is certain that the whole of what is now South Africa was submerged by the sea of ice, and that the same sheet buried the southern tip of Madagascar; huge areas of Australia and South America were glaciated, and bits of India and possibly Central Africa, right up to the equator. The Northern Hemisphere largely escaped the ice, but was subjected at about the same time to the first onset of a long spell of desiccation, which continued through most of the Permian (III F). This is evidenced by the huge accumulations of salt and gypsum laid down as some of the seas over the continents were trapped by the lifting of the land, turned into basins with no escape, and dried out.

As the southern half of the earth grew colder, a new set of plants was evolved to meet the new conditions—smaller and less luxuriant, but hardier with thicker, tougher leaves and often bushier in growth.

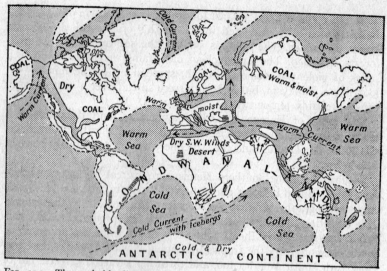

Fig. 225. The probable distribution of land and sea in the Ice Age of the early Permian (III F 1).

A huge continent, Gondwanaland, probably spread across most of the Southern Hemisphere. Arrows denote ice-sheets whose direction of movement can be determined; small parallel markings denote other traces of ice-action. Modified from Schuchert, with the kind assistance of Dr. C. E. P. Brooks.

This is generally called the Glossopteris flora from the prevalence of Glossopteris, a shrubby plant with long tongue-like leaves; this was probably a seed-fern or Pteridosperm, though we cannot be sure until we find its seeds indubitably attached to its remains. And there were Cycads, and a few ferns and seed-ferns, and small horse-tails, and Cordaite trees of different type from those that flourished in Europe and North America. This assembly of plants is found in Africa, in India, in Australia; this practical identity in these distant countries is one of the most important items of evidence which dispose us to believe in that wide connecting Permian (III F) continent of Gondwanaland.

LIFE CONQUERS THE DRY LAND

As the southern ice-cap spread, so this plant-army was forced north, till it reached beyond the equator. When the Permian Ice Age faded, the new plants spread back, following the retreating ice-sheet down on to the Antarctic continent. Among the specimens found in the tent with the dead bodies of Captain Scott and his fellow explorers in 1912 were rock-specimens containing Glossopteris which they had discovered far up on the Beardmore Glacier and judged to be of such interest that they had pulled them through all the weary miles until overtaken by the blizzard that finally overwhelmed them.

This new set of plants seems to have been better fitted than the old to cope with the general aridity which succeeded the cold spell. Accordingly they continued their successful northward extension, replacing the doomed plants of the coal-swamps.

Thus what the cold began, the drought continued. There can be little doubt that it was the stress and pressure of increasing aridity which forced some seed-bearing members of the Glossopteris plant-army on to the further pitch of dry land adaptation, which from the Trias (IV A) onwards, embodied in the Cycads and Conifers, dominated the vegetable world. Unfortunately, dry surroundings are the worst imaginable for preserving plant-remains, so that we have tantalizingly little record of the detailed steps taken by Evolution during this critical time.

But a critical time it was, as evidenced not only in the virtual disappearance of the first dispensation of land-plants and its replacement by the second, but by equally great changes in other branches of the tree of life.

Even in the oceans it was a time of extinction and rapid change. The sea-lilies (Fig. 89) reached their climax in the warm Carboniferous seas (III E), contributing by their skeletons a very considerable bulk of the mountain limestone which then accumulated to the thickness of half a mile or more. Of all their vast variety, only one family withstood the chilling of the sea, and survived into the Mesozoic Era (IV).

The old types of corals and lamp-shells and sea-urchins went through a similar catastrophic time of wholesale extinction; the last of the trilobites disappeared, and only a few sea-scorpions lingered on. Taking advantage of this wholesale killing, other and more resistant forms established themselves. Most notable among these new successes were the Ammonites; and a group of fish whose only living survivor is Amia, the American Bow-fin, which represent a half-way stage between the ancient bony fish, poorly strengthened within and ponderously armoured without, and the modernized fish which for the most part have given up armour-plating in favour of speed, good mechanical construction and flexibility.

But changes of climate and variations of temperature are felt more intensely on land than in the water; and it was on land that life suffered the most revolutionary transformations. We have seen how seed-bearers triumphed over the spore-bearing plants. Meanwhile the insects, as we shall set forth in a later section, were forced into new and more efficient ways of living; and among vertebrates the reptiles rose into the position hitherto occupied by mere amphibians.

The shelled reptilian egg is to land-vertebrates what the seed is to land-plants: it breaks the last link, in the chain which ties half-emancipated life to ancestral water. Just as the earliest seed-plants followed quick on the heels of the first true land-plants, so reptiles of primitive type had arisen before the Coal-measure Period began, quite soon after vertebrates appeared on land at all.

The reptiles had a great advantage in their elimination of the tadpole stage from their life-cycles. But it needed the violence of outer change to give the new types their chance and bring them to the top, much as in England it needed the extraneous accident of the Great War to turn the idea of woman's suffrage, which had long been alive in the political thought of the country, to its successful realization.

The shelled egg has had other important consequences. Whereever it was evolved it made internal fertilization a necessity, in this giving us an interesting parallel with the internal fertilization effected by the deep-burrowing pollen tubes of seed-plants. But besides imposing the need of internal fertilization, it reduced infant mortality. Now internal fertilization makes it necessary for the female to be courted and won; a lower infant mortality means that fewer young need be produced, and so opens possibilities of intensive parental care. Thus the shelled egg played its part in the evolution of courtship and the development of family life.

The Permian (III F) sealed the fate of those primitive amphibians, the Stegocephalia. It is true that they continued to live for some tens of millions of years more; it is true that their largest and most formidable representatives were only evolved late in the Trias (IV A). But from the early Permian they were fighting a losing battle.

Size and armour and teeth have never been enough in Evolution. What counts in the long run is general efficiency in the business of life. In the reptiles, the triumph of vital organization was the shelled egg, with amnion-pond, and tight-packed yolk, extra food wrapped

LIFE CONQUERS THE DRY LAND

round it in the shape of the white, and final material protection in the tough, drought-resisting shell. In the new conditions of wide dry lands, the Stegocephalians, lacking this, lacked all; after their final burst of evolution, they went out like a snuffed candle, and only little salamanders and the precursors of the frogs and toads were left to carry on the amphibian stock. Seed and egg are the badges of Mesozoic (IV) life, as flowers and warm-blood are those of their Cenozoic (V) successors, and brain of the present day (V F). By the Trias (IV A), the Age of Reptiles had definitely begun.

V

THE FULL CONQUEST OF THE LAND

§ 1. *The Reptilian Adventurers.* § 2. *A Digression upon Adaptive Radiation.* § 3. *The "Back-to-the-Water" Movement.* § 4. *The Conquest of the Air.* § 5. *Dinosaurs.* § 6. *Insects Multiply as Plants Develop.* § 7. *The Battle of the Giants and the Dwarfs.*

§ 1. The Reptilian Adventurers.

"REPTILE"—the word itself has a nasty sound. The idea of crawling is in its derivation; and to that has been added a flavour of unpleasantness derived from the only two reptilian groups which the general public habitually thinks of as Reptiles—the snakes and the crocodiles. The snakes supply instinctive horror and the idea of poison; crocodiles are not only dangerous, but enjoy a mythical reputation for hypocrisy. Add further the wholly undeserved notion of cold sliminess, derived from the popular error which persists in classing newts and frogs as reptiles, and the common view of the order is complete.

This picture, however, is an altogether erroneous one. It comes partly from zoological confusion of the lowly Amphibia with their higher cousins; partly from forgetfulness that such pleasant, harmless, and dry-skinned creatures as lizards and tortoises are reptiles; but mainly from the fact that the triumphs of reptilian evolution lie in the past, and its most remarkable products are long extinct.

The clamminess of frogs and newts is due to the fact that they breathe partly (and in one or two cases entirely) through their thin skins, which must, therefore, be continually lubricated and kept moist. The Reptiles owe much of their success to their having once and for all rid themselves of this handicap, and completely adapted themselves to the new medium of air by providing themselves with a tough, dry covering. Most modern reptiles crawl and creep; but it is an irony of fate that the whole group should be christened the Crawlers, for it was they who first raised the vertebrate belly off the ground.

THE FULL CONQUEST OF THE LAND

In the water, animals are almost weightless, because their specific gravity is only slightly greater than that of the medium in which they live. But on land, the problem of supporting weight must at once be faced. The primitive Amphibia, freshly emerged into the strange medium of air, never got farther than resting this embarrassing and new-found weight on the lower surface of their bodies; their limbs protruded sideways and their function was more that of land-oars than of true supporting legs. You have only to watch a salamander to see a creature which has retained for three hundred million years this primitive method of support and progression.

But far back in the Permian (III F) there were reptiles which had got their limbs under their bodies and turned them into pillars; these supports were heavy and clumsy at first, but their efficiency and lightness was speedily improved. That is typical of the real Reptile. Reptiles were the first full conquerors of the land and were, as a group, the greatest evolutionary adventurers which vertebrate life has ever produced, invading and exploiting a new world with something of the same spectacular effect with which the Spaniards explored and conquered the Americas.

Let us chronicle the Reptiles' major adventures, which begin about half-way back between the present time and the Cambrian (III A) period. The true Reptile had arisen from the Amphibian. There was at first a gradual evolution, which we can trace in the skeletons preserved to us, from the lower to the higher type; but before the Permian (III F) it was completed, and the Reptile was ready. His advantages were two: the tough skin that allowed him to disregard the danger of desiccation; and the amnion, his embryo's protective water-cushion within the egg, that enabled him to reproduce his kind without resorting to water. The transition from Amphibian to Reptile was like the transition from a civilization cumbered by a too-elaborate ancestor-worship to one which, throwing tradition overboard, turns its face from the past towards the future.

The earliest reptiles were for the most part sluggish creatures, all, it appears, carnivores or insect-eaters, not differing markedly either in build or habits from many of the contemporary Amphibians. The main stimulus to their advance doubtless came from outside, in the shape of the rise of the land which preceded the mountain-making revolution at the end of the Carboniferous (III E), and the cold, dry times which accompanied and succeeded it. The swamps shrank, the steaming forests largely disappeared. The opportunities and the space available for the half-and-half life of Amphibia dwindled, and a

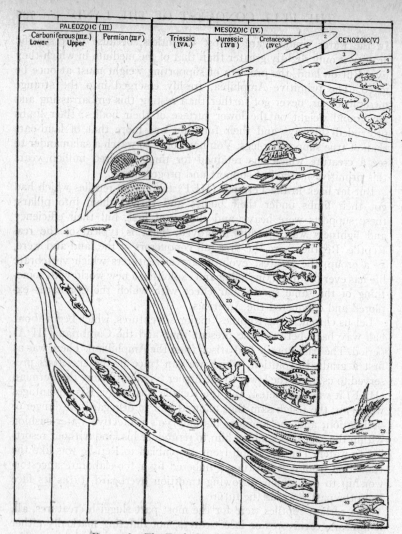

FIG. 226. The Evolution of Land Vertebrates.

1-7, Mammals: 1, *small ancestral mammals* (Trituberculates); 2, *Triconodonts;* 3, *Multituberculates;* 4, *monotremes* (duckbill); 5, *marsupials;* 6, *archaic placentals;* 7, *modern placentals.* 8-36, *reptiles and birds:* 8, *ancestral stock of mammal-like reptiles;* 9, *Pelycosaurs* (Naosaurus); 10, *Theromorphs* (Inostransevia); 11, *Cotylosaurs;* 12, *tortoises and turtles;* 13, *Plesiosaurs;* 14, *false crocodiles;* 15, *long-tailed flying reptiles.* 16, *tailless reptiles.* 17-25, *dinosaurs:* 17, *running dinosaurs* (Ornitholestes); 18, *flesh-eating biped dinosaurs* (Tyrannosaurus); 19, *huge quadruped herbivore dinosaurs,* or *Sauropods* (Diplodocus); 20, *biped herbivore dinosaurs* (Camptosaurus),

740

THE FULL CONQUEST OF THE LAND

premium was set on the development of full adaptation to dry land.

The result in the Permian (III F) was a rapid evolution of the Reptiles into many and varied forms, a number of them going in for vegetable-feeding and bulk. Vegetarians will note with satisfaction that their distinctive practices were of very great advantage to these conquerors of the dry land. Most of these Permian vegetarians continue evolving through the Trias (IV A). They were the Reptile's first big adventure in Evolution.

The most primitive Permian reptiles are the Cotylosaurs, survivors, it would seem, of the group which served as ancestral stock to all the rest, and so to birds and mammals and our human selves. Then there were animals like Elginia which had their heads defensively beset with spiky horns. The still bigger Pareiasaurus—also, it seems, a vegetarian and one of the earliest vertebrated creatures to raise its belly off the ground—reveals the novelty of its achievement in the clumsiness of its limbs, heavy as pillars in an Early Norman cathedral, and there were many active and often fierce animals of the general type of dragon or giant lizard.

But the most remarkable creatures of the age were the Theriodonts or Mammal-toothed forms, so christened because they were the first vertebrates to show in well-developed form the division of the teeth into different groups, molars, canines, incisors, each with its own form and function, so characteristic of most mammals. In the earlier types these differences are usually not great. There are one or more great dog-teeth above and below on each side, and all the other teeth are peg-like. But in some of the later, Triassic (IV A) forms, there are only four dog-teeth in all, and the teeth in front of these are regular incisors, while those behind have developed crowns and cusps like those characterizing the molars of flesh-eating mammals, These, apparently with the long ancestral tail as much reduced as in most mammals of to-day, were active, running creatures, and the most obvious distinctions between them and a modern carnivore, such as wolf or marten, are the minute size of their brains and the still-

---

with two branches; 21, duckbill dinosaurs (Trachodon), and 22, Iguanodon; 23, armoured dinosaurs (Polacanthus), with two branches; 24, Stegosaurs, and 25, horned dinosaurs (Triceratops); 26, ancestral bird stock; 27, Archeopteryx; 28, toothed birds (Hesperonis); 29, modern birds; 30, true crocodiles; 31, Ichthyosaurs; 34, ancestral stock of snakes (32), lizards, (33), and Mosasaurs (35); 36, ancestral reptiles (Seymouria). 37-44, amphibians: 37, ancestral amphibian stock. 38-44, stegocephalians: 38, Loxomma; 39, Mastodonsaurus; 40, Eryops; 41, Branchiosoma; 42, tailed ancestors of modern Amphibia; 43, frogs and toads; 44, newts and salamanders. (A diagram drawn on the same principles as Figure 218.)

reptilian construction of their jaws. Of their manner of reproduction we know nothing. In all probability the mammals took their origin from some of the smaller Theriodonts.

Besides the dog-like forms, the same group also produced extraordinary creatures called Dicynodonts, some of them as small as water-rats, others almost as big as hippopotami, which gradually lost all their teeth save the canines (and in some cases abandoned even these), substituting a horny beak like that of a turtle. Some of them were desert-dwellers, but others were semi-aquatic creatures which must have browsed on water-plants in swamps. Another side-branch, the Pelycosaurs, gave rise to extraordinary creatures, up to eight feet long, with a great frill of bony spikes along the back, each spike sometimes even equipped with side-spikes like yard-arms on a ship's mast. Other water-reptiles of the time had an extraordinary resemblance to crocodiles, but were in reality products of an independent evolution, having a differently constructed jaw, and nostrils on top of the head like a whale's blow-hole.

One often hears the Golden Ages of Ammonites, of Reptiles, of Mammals and so forth discussed as if each great group had had a steady crescendo, a climax, and a wane. But this is over-simplification. As a general rule each of the great groups shows several successive outbursts of evolution. The Reptiles are no exception. This first phase of the reptilian adventure, beginning with the late Carboniferous (III E), came to an end with the end of the Trias (IV A); by then all the specialized creatures we have just described had become extinct. During the Trias, however, a second outburst had begun. This ushered in the grand period of reptilian life, stretching through in the Jurassic and Cretaceous (IV B and C), during which the creatures reached their greatest elaboration and their greatest size. But actually their greatest variety of types was attained in the late Trias, when members of the two waves of evolution overlapped.

The tortoises and probably the true scaly lizards began their careers with this second wave. The main land group, however, was the Dinosaurs, which themselves branched out into almost as many types as their successors the land-mammals—large and small, active and ponderous, vegetarians and beasts of prey. Flying reptiles launched out into the air, gaining a start of perhaps twenty million years over the birds—a handicap which did not prevent the birds from catching up and passing them in the course of sixty or seventy million more. And a varied collection of reptiles took to the water.

THE FULL CONQUEST OF THE LAND   743

Such an outburst of evolution is generally called, in Osborn's phrase, the *adaptive radiation* of a group, because the group as a whole radiates and scatters in many lines, and each line leads towards fuller adaptation to one particular way of life. We have met with such radiations before—in the marsupial of Australia, described in Book 3, in the Trilobites, in the Molluscs, in the Echinoderms. But this is the most striking we have yet encountered, and we must look at it a little more closely.

§ 2. A Digression upon Adaptive Radiation.

EACH stock within a radiating group seems generally to be evolving towards greater specialization, like the horses we discussed in Book 3, or the elephants or sabre-tooths in Book 4; and the evolution of each such line can, as we have seen, be accounted for on the principle of Natural Selection. But Natural Selection will account too for the radiation as a whole, for the simultaneous divergence of all these lines. When a new group, capable of playing a leading rôle, appears on the world's stage, it has before it a territory to exploit. No one stock can exploit more than a small region of this, or fill more than any particular rôle in the general economy of life. And indeed the more efficient one stock becomes at its own particular job, the narrower its rôle, the more room for another line to exploit slightly different collateral modes of life in its own way. Not only that, but the specialized type often creates itself, so to speak, a new rôle to be filled: the development of a stock of gigantic vegetarians is an evolutionary incitement to the development of correspondingly powerful carnivores which may kill and eat its members—creatures so large that they could not live on smaller prey.

The radiation of a group is thus from one aspect the sharing out of the different opportunities of making a livelihood among the different lines of which it is composed, so that the world's resources are ever more abundantly and efficiently utilized. It is on Evolution's stage what the division of labour is in the affairs of man. The number of trades and professions and of specialist jobs and positions within each increases as human civilization progresses; and so, automatically, the number of animal types and the degree of their specialization increases as a group of animals or plants evolves.

Now the different rôles available remain roughly the same through long periods of geological time. We might suppose that the animal types which are the embodiments of ways of exploiting these various means of livelihood would repeat themselves over and over again with each new radiation.

Up to a point this is true. Evolution has produced a long series of examples of what is generally called *convergence*—the moulding of unrelated stocks into wonderfully similar forms by the needs of their way of life. We may point, for example, to the extreme resemblance of some of the South American grazing mammals to the wholly unrelated horses which we noted in Book 3, and we may mention one or two other examples, like the Pouched Mole, which in response to the needs of burrowing life has been moulded out of the marsupial pattern into a startling likeness of the ordinary placental mole; the slow-worm which is a lizard turned into the likeness of a snake; and the amazing resemblance of the social organization of ants to those of termites —both with sterile workers and soldiers and a winged royal castle, in spite of the fact that one springs from a stock allied to wasps, while the other belongs to a wholly different order of insects related to the cockroaches.

But though these resemblances mask they never obliterate the essential community of structure due to kindred.

Any group of land vertebrates, for instance, is likely to radiate out into certain main types—plant-eaters and flesh-eaters and insect-eaters; climbers, walkers, runners and jumpers; those which go back into water, and those which push on into the air. But the exact way in which the various lines achieve their results is often very different; and sometimes whole ways of life are passed over. The Australian marsupials have never produced a true flying stock, nor, surprisingly enough, any which has taken, even incompletely, to the water. The Placental Mammals, on the other hand, though they have produced bats and whales and sea-cows, have never given rise to giant jumpers like the kangaroo. Frogs and toads have never produced any salt-water types, apparently because of their soft and permeable skin; and no insect has even begun to take advantage of abundant plant-food by being bulky like a cow—for the physical and mechanical reasons which we discussed in Book 4. And the set of creatures produced by the first and second great reptile radiations were very different, and both differed from those of the mammals.

Human affairs sometimes shed light on animal evolution. We may remind ourselves that in different countries, even at the same general stage of civilization, the necessary business and administration of the nation may be carried out in very different ways. The exploitation of the ether by wireless is carried on by a single huge state concern in Great Britain, by innumerable public and private organizations in the United States. People get transported rapidly from one part of Chicago to another as they do in London; but in Chicago it is largely done by overhead railways, in London by subterranean tubes. So the business of turning grass into flesh is carried on by life both in Africa and Australia—but is undertaken by zebras and antelopes in the one country, by kangaroos in the other.

Then circumstances may differ from age to age. The vast bulk of some of the Jurassic Dinosaurs would have been impossible in a drier period; while the abundant spreading of cold-blooded reptiles far towards the poles had to come to an end when the climate grew cooler. And finally, both chance and the accidents of history play their part with animals as with men. From the point of view of the Reptiles, it was an accident that the climate grew cold and dry at the end of the Mesozoic (IV). But for that, the Dinosaurs might have survived and evolved for many millions more of years. And if the traditions of the past can help change the destiny of a nation, the accidents of ancestry can alter animal evolution. We have already seen how the fishy conservatism of Amphibia kept them tied to moisture; in the same way the land ancestry of the seals and the turtles ties them to breeding out of water, and the air-breathing past of whales forces them to return to fill their lungs at the surface.

§ 3. The "Back-to-the-Water" Movement.

ONE of the most interesting facts about the great Reptilian radiation is the number of stocks which went back to life in the water. This is a frequent phenomenon in the evolution of land-animals, the insects and the mammals providing many other examples. Land-life demands greater strength and activity to deal with weight—that embarrassing property which appears in full intensity when life leaves water for air; it demands tough impermeable skin to resist drying, greater general resistance and greater powers of self-regulation to cope with the far greater extremes and more rapid changes of temperature and other outer conditions that have to be faced on land; it demands specialized methods of reproduction and efficient protection for the tender young stages. Armed with these new qualities, the land-animal can descend into the waters again to compete successfully with the creatures who have never left their ancestral home, beating them fairly in their own sphere.

It is not for nothing that the roll of such "secondary aquatics" who once were inhabitants of the land includes the whalebone whales, largest animals in the world, and the toothed whales and killers, largest and most ferocious water-carnivores of their times (V) as Ichthyosaurs and Mosasaurs (IV) were of theirs; that water-birds are almost as abundant and quite as successful as land-birds, and that the water-insects dominate all small bodies of fresh-water.

It is of the greatest interest to find in each of the reptile stocks that thus took to the water a progressive adaptation to their new life. The typical Ichthyosaurs of the Jurassic (IV B) were dolphin-like in shape, with the close rows of small, sharp teeth characteristic of all fish-eating types. What once were legs had by then again become the most admirably constructed fins. Inside, the skeleton of arm, wrist and hand had been turned into a veritable mosaic of little bones, in which even the radius and ulna, which make up our lower arm skeleton, can only be distinguished from the rest by their position, and often extra rows of bones are added in the position of new fingers (see Fig. 131). The animals swam by means of a powerful tail-fin, placed vertically like that of a fish, but with the backbone bending down into its lower lobe. In the earliest known Ichthyosaur, however, from the Trias (IV A), the teeth are fewer and less sharp, the tail runs out almost straight, and must have been far less powerful as a propeller, and there are but the normal five fingers in the flippers; the arm-bones are still quite different from the wrist and hand-bones. It is interesting to find that young specimens of later forms recapitulate a stage half-way between the primitive and the well-developed tail-fin.

Ichthyosaurs, like whales, were so perfectly adapted to their way of life that they did not have to come to land to lay eggs, but brought forth their young alive at sea, as we know by the evidence of the skeletons of a litter within the ribs of their mother. Other skeletons, however, reveal to us that Ichthyosaurs, like many fish to-day, would devour young ones of their own species as gladly as any other prey. Their fossilized stomach contents tell us, too, that they fed abundantly on cuttle-fish and their allies.

Another great group of water-reptiles, the Plesiosaurs, did not swim fish-like, with their tails; they rowed themselves with their four great paddle-limbs. Most of them caught prey, after the fashion of a grebe or a cormorant, with sideways darts of their long sea-serpent neck; others were more stocky in build, with stouter and less flexible neck and heavier head. Of the earliest known Plesiosaur, from the Lower Trias (IV A 1), it is hard to say whether it lived a terrestrial or an amphibious life. Then came animals like Nothosaurus, much larger, and clearly aquatic, but with somewhat pulled-out neck and enlarged hands and feet, doubtless fully webbed. These creatures must have come ashore, like seals, to sun themselves on those beaches of a hundred million years ago; and perhaps, also like seals, they still bred on land. But by the Jurassic (IV B) the Plesiosaurs, some of them well over thirty feet long, had a diminutive tail, a huge and flexible neck, and

FIG. 227. Various reptiles which have returned from land to a sea life.

Left, Metriorhynchus, a sea-crocodile of the Upper Jurassic Period (IV B 3) with flippers and tail-fin. Next, a huge mosasaur from the Cretaceous (IV C). Above, on the right, two ichthyosaurs, adult and young. The tail-fin enlarged, especially in its upper lobe, as the animal grew older. Below them is a plesiosaur. Below that again are two turtles and a sea-snake (Platurus) with expanded fin, found living to-day.

for limbs, enormous sweeps quite useless for walking with. They had become, like Ichthyosaurs and whales, creatures of the sea alone.

The remains of Plesiosaurs' meals have also been found, turned into stone. They ate fish, ammonites and cuttle-fish, and small specimens of their own kind; one has been discovered who had succeeded in devouring a Pterodactyl. Unlike the Ichthyosaurs, but like many seals, they gulped their prey down whole and ground it up in their gizzards by the aid of pebbles, which they swallowed and which are often found, nicely polished, between their ribs. Like Ichthyosaurs, their skin was bare of any covering of scales.

The crocodiles are a third group of reptiles which took to life in the water; in the Mesozoic (IV) many of them became marine, and much more highly adapted to a fully aquatic life than any now living. Some of these had tail-fins like the more primitive Ichthyosaurs, and with undulations of these and their elongated body swam like a more powerful newt. They used their hind-limbs as a paired rudder; but their degenerate fore-limbs were often of such diminutive size that the animals could never have come on land.

Nor must we omit the water-tortoises and turtles. The tortoises apparently began their career after the fashion of the burrowing armadillos of to-day, and like them developed armour and had their teeth reduced. In both respects, however, they went farther than most armadillos. The armour became a veritable strong-box, unique in its solidity among land-vertebrates. Even the Tank-armadillos, the Glyptodonts (Fig. 146), had no armour-plated flooring, nor was their roof of bone firmly cemented with their vertebral column, as in the tortoises. And the teeth were altogether lost by tortoises, and sharp, horny beaks substituted. One tortoise from the Trias (IV A) has a few small teeth on its palate—ancestral mementoes. The earliest tortoises were all land creatures; but by the Jurassic (IV B) some of them had taken to the water, to adopt carnivorous habits and became the progenitors of the modern producers of tortoise-shell and turtle soup, increasing the length of their great flippers and developing great speed by reducing their heavy carapace of bone almost to nothing.

There are also a number of true water-snakes. But these are sporadic ventures; no important group of aquatic reptiles has taken origin from among the serpents. Finally come the true lizards. Several of their relatives took to the water; but only one stock need be mentioned here—the Mosasaurs. They appear much later than all these other water-reptiles, not till the early Cretaceous (IV C 1). The earliest types, though obviously water-animals, are still very like the living lizard Varanus, which includes the Komodo "Dragon." The later forms were much larger—some of them forty feet long, with long, undulating body and tail-fin, four smallish paddles, and formidable teeth. Like snakes, they could dislocate the two halves of their lower jaw, and so could swallow prey apparently bigger than their mouths. Whether this or some trick of speed or resistance was the cause, they speedily became the monarchs of the Cretaceous (IV C) seas. The marine crocodiles of the Jurassic (IV B) did not survive their coming; the Plesiosaurs diminished in numbers; the Ichthyosaurs took a still more rapid plunge downhill, and all became extinct well before the end of the Cretaceous.

§ 4. The Conquest of the Air.

LIFE in the open ocean moved in three dimensions. When it emerged on land, it became restricted to two; up-and-down was removed from its repertory. But above the surface on which it crawled or fixed itself, there was another ocean, the ocean of air. In this ocean, too, it was possible to swim, if life could find out the way. The way was not easy, for the fluid of which the ocean of air is composed is eight hundred times more tenuous than salt-water, and eight hundred times lighter than the living substance that demands support from it. None the less, life surmounted these difficulties to the full no fewer than five times, besides making a score of partial explorations of this new territory.

The partial explorations are those of passive floating, as in dandelion or milkweed seeds, in wind-borne pollen, in gossamer spiders; and those of parachuting as in "flying" squirrels, lemurs, phalangers, frogs, and lizards. And even water-animals have exploited the air, shooting out of water for a hundred yards or more on outspread planes, like the flying-fish. The five active conquests, in historical sequence, have been those of insects, flying reptiles, birds, bats, and man— man being the only organism except the gossamer spider to use a vehicle to take him aloft.

Man's aerial achievements are mechanical rather than biological in their interest, and we cannot touch on them further here. The active invasion of the air, with the exception of man, has always been accomplished on true wings, which, though they may be temporarily used as gliding planes, are essentially movable organs, supplying motive power as well as support, unlike the planes of an aeroplane. In the three groups of vertebrate air-invaders it is the fore-limb which has been turned into at least the main part of the wing (see Fig. 132). But in insects the wings have nothing to do with limbs—they are outgrowths from the side of the so-called thorax. No insect, so far as we know, has ever had more than two pairs of wings sprouting from the second and the third of the thorax-segments. But in some of the earliest fossil insects a rudimentary flap, obviously corresponding with the true wings, is found on the first thorax-segment, too.

Many suggestions have been made as to the possible origin of insect flight; but it must be confessed that they all remain very hypothetical. At least, it seems probable that their wings were first mere flap-like outgrowths which helped the creatures to parachute from tree to tree or from rush-stems to the safety of the water; and that gradually the second two pairs became large and movable, while the first degenerated. Their later evolution consisted chiefly in a rearrangement of the chitinous "veins," which act as their tiny but indispensable girders and stays, a rearrangement which gave greater strength; and in a substitution, in the better fliers, of a screw-motion for the heavy up-and-down flapping of the earliest forms. Parallel with the evolution of the wings has gone an evolution of the muscles. The striped muscle-fibres of insects have much bigger light and dark bands in them than our own, and this seems to make them capable of more rapid contraction. Rapid contraction is certainly necessary, since a bee in flight beats its wings over ten thousand times a minute, and a house-fly nearly twice as often.

The insects, though of course they started as mere crawlers, became flying animals very early in their history; the group as a whole is a flying group, and when we find wingless forms, they are almost all, like fleas or many wingless beetles, creatures which have lost their wings and gone back to legs alone. But with the vertebrates it is very different. Compared with insects, they are big, heavy creatures, and it was much more difficult for them to embark upon the thin air-ocean. With the insects, all the attempts at flight, all the half-and-half successes, are things of the remote geological past. But with vertebrates full success in the air has been sporadic, confined to three groups; and among other groups we find to-day many examples of incomplete exploitation of flying.

In practically every case the incomplete vertebrate flier is a parachutist. The commonest method is for the animal to be flattened, with a ridge of skin extending along the sides of trunk and limbs. This may remain very rudimentary, or it may enlarge until fore- and hind-limbs are joined to each other and to neck and tail by a regular membrane; in the flying lemur the parachute-membrane reaches almost to the finger-tips. When the creatures jump into the air with outspread limbs they are living parachute-gliders, steerable by movements of the tail. The chief variant on this is given by the flying-frogs, in which the webbed feet have been utilized as the basis for the parachute, and enlarged until they are big enough for the little beast to ski through the air on them. And there are the flying lizards, Draco, which blow themselves up with air when they "fly," and combine the properties of parachute and balloon.

All these creatures are tree-livers, and their flight is always a parachuting glide which merely extends the distance to be covered by a

THE FULL CONQUEST OF THE LAND 751

leap. It is thus natural to imagine that birds, bats and pterodactyls originated in the same way, as jumping parachutists, and although we have no actual intermediate stage, it is reasonably certain that bats and pterodactyls, at any rate, did start their flying career thus, for the resemblance of their construction to that of a good parachutist

FIG. 228. Two ways in which the flight of birds may have evolved.

Left, the ancestral bird may have parachuted from tree to tree with a fringe of feather-like scales on legs as well as arms. Right, it may have run rapidly over level ground with fore-arms extended and a fringe of feather-like scales on them alone.

like the flying lemur is striking, and their great membranes, stretched from limbs to neck and tail, could hardly have originated in any other way. All that is needed to convert the one type into the other is an elongation of the fore-limb's fingers, and a strengthening of the muscles needed to flap it.

But this we may note about the parachutists, all four legs are always involved in the vane, hind as well as fore. This is true also of bats and of pterodactyls. But it is not true of birds. The bird flying-apparatus involves the arm only; the leg is free to be used as a leg should. Herein lies the success of the birds, as opposed to the bats, for example; a bird can fly and run as well. Moreover, another point of difference, the vane itself in birds is not a membrane of stretched skin, as it is in parachuting beasts or bats or pterodactyls; it is made of those marvellous and unique constructions we call feathers.

We still lack a continuous series of fossils to give us the story of the origin of the bird's wing. It is possible that feathers were not evolved stage by stage with the wing, and as part of the wing, but separately and in the margin of the Jurassic world, as a temperature-protecting covering, an improvement on the simple scales from which they must have been elaborated. Since the larger part of the world in the Jurassic Period (IV B) was under warm and equable conditions, it seems reasonable to suppose that the early evolutions of this warmth-retaining covering occurred in the circumpolar regions and that they crept into vogue only with the increasing climatic extremes of the Cretaceous (IV C). Once they were developed their lightness would open up possibilities of gliding and flight hitherto unknown in the world. Just how these possibilities first came to be exploited is not yet clear. It may be that the birds developed from running reptiles of the open plains, or from clambering reptiles that leapt and then glided from branch to branch. In support of the first suggestion we have the skeletons of numerous running biped Dinosaurs with hind-legs and a bodily poise extremely like that of birds, and with their fore-limb in a reduced state and evidently available for any adaptive utilization. There are authorities who believe that the wing arose from the development of the fore-limb into flappers and sails more and more adequately feathered, taking more and more weight off the ground as the creature ran. If that is true, it explains the admirable development of the bird leg, as contrasted with the leg of a bat or a pterodactyl. In support of the second view, we have the skeleton of the first known bird Archæopteryx, beautifully engraved on the lithographic Solenhofen slates of the Upper Jurassic (IV B 3), with a large wing more primitive than any existing bird's wing, but built on the same plan, and a long tail with well-developed feathers on either side of it quite unlike the rump of any existing bird (see Fig. 127). This creature seems clearly to have been a forest dweller, fluttering and planing from bough to bough; thus, perhaps, the bird stock learnt to use its wings. Meanwhile, the pterodactyls prevailed exceedingly.

Fossil remains of bats are very rare; but in both birds and pterodactyls the earliest types known are, as flying machines, of poor design compared with the later representatives. The long-jointed tail of Archæopteryx only differs from a lizard's in being fringed with large feathers. (It is worth noting, by the way, that this row of big feathers continues right on to the body, up to the hip-region, much as do side rows of large scales on the tails of various reptiles.) This is evidence that it must have kept above a certain minimum speed if it was not to "stall." It must have been plunged rather headlong into the branches where it wanted to land, instead of being able to poise and drop accurately, like a modern bird with its tail reduced to a fan of feathers radiating out from a mere stump of bone. The development of the "bastard-wing," a tuft of feathers attached to the thumb, and movable separately from the wing as a whole, seems also to have been a great aeronautical improvement.

Flying reptiles effected many improvements in their design throughout the Jurassic-Cretaceous Period. The earliest forms known, and those most like ordinary reptiles, had long tails and must have practised a gliding, planing flight, with occasional bursts of wing-beats. Their wing-fingers cut through the air with a stiff and slightly concave edge, their wings were long and pointed and when the animal walked must have been cocked up over the back.

The earliest forms had a membrane stretched between tail and hind-legs, as specially bent bones of the hind-foot testify; but by the later Jurassic (IV B 3) the much elongated tail was probably bare save for a diamond-shaped horizontal plane at its tip.

Flying reptiles of this tailed type died out at the end of the Jurassic, soon after those of more advanced construction had appeared. These had vestigial tails and began their career with the short rounded wings that seem to indicate flapping flight by rapid wing-beats. Their skull was much more bird-like and, like true birds, they gradually rid themselves of the excess weight of teeth, substituting for them a regular horny beak.

This new and more successful type radiated out in several directions. It produced flying-lizards no larger than a lark; and at the other extreme gave the world one of the most extraordinary creatures it has ever seen, the great Pteranodon of the late Cretaceous (IV C 3), with a strange crest of bone protruding back from its skull and doubtless serving as a vertical fin. This creature was in its linear dimensions by far the biggest-living flying-machine ever constructed, measuring up to twenty-five feet in wing-span, twice as much as an albatross. But in weight it was probably no greater than an albatross, reaching something like thirty pounds. As Schuchert says: "Their body was but an appendage to a pair of wings." The remains of these creatures have only been found in formations of pure chalk, which were deposited far from land. They must have lived a life like the albatrosses of to-

THE SCIENCE OF LIFE

day, creatures of gliding flight, searching the open sea for food, perhaps resting on its surface, and only returning to land to breed.

Pterosaurs evolved many characters adapted to air-life which were also later quite independently evolved by birds. The horny beak and the toothlessness we have already mentioned; their cerebellum, brain-organ for balance and quick adjustment, was large as in birds; still

FIG. 229. Various kinds of flying reptiles.

Left, a number of Pteranodons, the biggest known flying vertebrate, which was toothless and probably caught fish far out to sea. On the rock are Rhamphorhynchus and Dimorphodon, two pterosaurs with a long tail ending in a membranous rudder, and, below them, Pterodactylus, a small, almost tailless form.

more remarkable were the hollow bones, filled with air instead of marrow, serving for lightness. But they never developed a warmth-retaining covering; though some have been found with a long crest of hair-like plumes, these must have been for adornment only; the body and leathery wings, even in the most exquisitely preserved specimens, are bare.

This nudity undoubtedly was an important cause of their ultimate downfall. It did not matter so long as the temperature was high and the climate mild. But when the sharp change of climate brought the Cretaceous (IV C) to a close, the birds began to reap the full advantage of their feathers and warm blood. They carried their warmth about within themselves, and so could live at such a rate, with such a formidable energy, that they quite outstripped their leather-winged rivals.

There was also another reason for their success. Birds are the only vertebrate fliers who have kept their legs from being mixed up with their wings (though insect legs have also kept free of such "entangling alliances"); even in the tailless Pterosaurs the wing-membrane had to be spread across from the finger and made fast on the knee. And they have evolved by far the neatest way of folding their wings when not in use, so that, unlike those of Pterosaur or bat, they do not interfere with other activities. The bird, thus unencumbered, can turn its legs to running or swimming, it can pass from air to land or water and be adapted to all, while bat and flying-lizard are mere shufflers on land, and bats at least must roost by suspending themselves head downwards.

It is no wonder, therefore, that after the great geological revolution that closed the Mesozoic Era (IV) the birds became the dominant form of aerial life. They have never succeeded in developing their brains to the same degree as certain mammals; but in some regions of the globe their mobility combined with their warm blood makes them the dominant vertebrates. Thus, in the arctic lands, the land-mammals, tied to the soil, are few in numbers and in kinds; but tens of thousands of sea-birds come each summer to nest in cliffs and exploit the teeming waters for food, making off through the air at the approach of winter.

Flying-birds, however, reach a limit of size much sooner than land-mammals. This is because, as any text-book on aeronautics will explain, the bigger a heavier-than-air machine, whether living or not, the faster it must fly to keep itself from crashing. In order to develop this speed, more power is needed; and it can be calculated that with increasing size the power needed goes up faster than the weight if the proportions of the flying-machine remain the same. But the only way for a bird to secure increased power is by increasing its wing-muscles; and increase in dimensions thus means a disproportionate increase of these, its engines, and the time must come when further enlargement is quite impossible. In flying vertebrates the crucial point is reached at about fifty pounds weight. Were it not so, as J. B. S. Haldane reminds us, "eagles might be as large as tigers and as formidable to man as hostile aeroplanes."

The same author calculates that in order to fly a human being would have to have such enormous wing-muscles that the keel of the breast-bone, to which they must be attached, would have to be over four feet long, his legs would have to be reduced to spindly stilts to economize weight—and even so his biological success would be dubious in the extreme. Thus while the mechanics of animal flight, luckily for man, keeps birds down to a harmless size, a knowledge of their principles explodes a religious fantasy. It demonstrates the impossibility of angels, or at least of angels of the accepted pattern.

§ 5. Dinosaurs.

ON LAND, reptilian life had embarked on equally striking adventures. We have already mentioned the land-tortoises. Everyone has heard the story of Sidney Smith and the kind little girl he found stroking a tortoise. "Waste of time, my dear," said he. "You might just as well stroke the dome of St. Paul's and think you were pleasing the Dean and Chapter." The stimulus was all too far away. Sidney Smith was not quite accurate; as a matter of fact, there is an abundance of pressure-sensitive touch-organs between the thin horny part of a tortoise's carapace and the inner bony dome, and the animal doubtless was aware of the little girl's caresses. But for all that, the thick carapace has meant a loss of contact with the environment, and has kept the creatures rather insensitive; while its great weight has prevented them from achieving speed, and its protection has allowed them to dispense with mental activity. Once they had achieved full tortoise-hood, which they did before the end of the Jurassic (IV B), once they were, so to speak, established and endowed, they could survive or they could die out, but they could not very well evolve; and they have remained, relatively unchanged, ever since.

It was the Dinosaurs, alone among reptiles, who were really successful on the Mesozoic lands. Their dominance lasted a hundred million years; but finally every one of them was wiped off the face of the earth. Some eight or more sub-orders of this wonderful stock are to be distinguished, radiating out until their members differ from one another as radically as a giraffe from a rat or an elephant from a lion.

A striking difference between Dinosaurs and mammals is that a large proportion of the former walked mainly or solely on their hind-feet. It seems that the ancestral Dinosaurs, in striking contrast to the ancestral mammals, must have supported themselves entirely on their hind-legs, and only among the bulkiest types was there a reversion to a four-footed mode of locomotion. All the earliest known types were bipedal, well suited to the open dry plains that covered so much of the early Triassic landscape (IV A 1), either hoppers, like the kangaroo, or two-legged runners with tail outstretched behind for balance, like some of the running lizards of to-day. Some of them seem to have taken to the trees; others, growing more bulky, used their tail as a support as they shuffled over the ground or sat and munched. One of these early lines retained the light build and active habits, and, after giving the world some little hoppers like modern jerboas and wallabys in the Trias (IV A), later on produced some remarkable running creatures, such as Struthiominus, extremely like enlarged and reptilian ostriches in all but their long tail and their apparent lack of feathers.

None of the hoppers ever grew very bulky; two related lines, however, both achieved record size, the one a record for land-carnivores, the other, vegetarian, a record for all animals not wholly aquatic. The carnivorous line became larger and larger to cope with the increasing bulk of its vegetarian relatives and prey, its teeth became more formidable, its fore-arms, at first provided with vicious claws to dig into victims, later degenerating into almost useless vestiges in favour of the huge skull with its ferocious teeth. It culminated in the Tyrannosaurs, whose knee was a tall man's height above the ground, whose head was twenty feet up in the air—formidable, if ponderous, engines of destruction.

Meanwhile the vegetable-feeding line, growing more and more immense, had come down on all fours and given us the only higher vertebrates which begin to be built on the same scale of bulk as whales, the Sauropods or lizard-footed Dinosaurs. We can calculate that several of these gigantic creatures must have reached over thirty tons and a few over forty tons in weight. Gigantosaurus, the largest of them all, is an African form with stocky body and fore-limbs longer than hind. All the rest have the body nearly horizontal on the legs. The American Brontosaurus, the "Thunder Dinosaur," has the body of a much enlarged elephant, but (as in almost all these creatures) a huge neck and tail, "as though," writes Lull, "an elephant were deprived of its normal terminals and provided with those of an enormous snake." Diplodocus was slenderer, with a veritable whip-lash of a tail, forty feet long. This it probably used in true whip style to defend itself. Fractures in the tail-vertebræ of one or two specimens perhaps mark where they struck the bodies of their avid flesh-eating enemies.

Popular fancy persists in imagining prehistoric man engaged in a struggle with creatures of this type. As our readers will have realized, this is an anachronism; popular fancy is out by at least fifty million years.

It is doubtful, on purely mechanical grounds, whether any animal living altogether on land could exist of the size of those huge Sauropods. The supporting power of bone increases as the area of its cross-section. Therefore the cross-section of land-animals' limb-bones must increase proportionately to the weight to be supported. This, as a little calculation will show, means that the bones' own weight must go up

FIG. 230.   Various Dinosaurs from the Cretaceous (IV C).

Above, from left to right: the vegetarian Duckbill Dinosaur, Trachodon, a hopper; its half-aquatic relative Corythosaurus, the Crested Dinosaur; the rapid-running Ostrich Dinosaur, Struthiomimus. Below, Polacanthus, with two rows of defensive bony spikes along its back, and two specimens of Ornitholestis, an active predaceous animal, catching primitive birds.

faster than the weight of the active tissues by which they are moved and nourished. A stock of animals evolving towards greater bulk is thus faced, metaphorically speaking, with a dilemma. Either it must reduce its skeleton below the margin of safety, or it must devote an ever-increasing proportion of its substance to the task of supporting the rest, until the amount of skeleton becomes wholly uneconomic. These handicaps set a limit to the size it can attain.

In the water, however, weight is negligible; size is limited only by considerations of food and reproduction, and life can produce creatures well over a hundred tons in weight, like the great whales. On land, on the other hand, the present maximum attained by elephants is well under ten tons; and the limit, as given by certain gigantic fossil creatures called Baluchitheres, seems to have been under twenty tons.

The Sauropods were in all probability half aquatic, the hippopotamuses of their age, living in rivers and lagoons and marshes, using their long necks to browse on the banks, and perhaps, swan-fashion, on the bottom. This conclusion, based on mere weight, is supported by their having their nostrils turned into a blow-hole high up on the tip of the head. It is a curious fact that they have no grinding teeth; but they had stones in their gizzards, and doubtless, after bolting their food whole, used these instead of molars. These half-aquatic giants had, unlike their carnivorous cousins, already reached their maximum size and specialization in the Jurassic (IV B). In the Cretaceous (IV C) they dwindled in numbers, apparently supplanted by the creatures we are now about to describe.

These fall into three main groups. There are, first, the beaked and toothed forms, like Iguanodon, celebrated because of the discovery of a whole band of twenty-nine specimens, entombed together with many other creatures, such as crocodiles and tortoises, apparently as a result of being overwhelmed by a flood. The remains were found at a depth of over a thousand feet in a Belgian coal-mine; much of the deposit is still unexplored, and would doubtless yield many more Mesozoic reptiles. The Iguanodons combined two distinct species, one a good deal smaller than the other; but no young ones were found, and apparently the two sexes were alike in their skeletons. These animals must have lived much like the giant ground-sloths of a later age, but they were provided with a remarkable battery of grinding teeth, multiple rows of them above and below, and the shape of their jaw makes it almost certain that they had a long prehensile tongue like a giraffe's for pulling off leaves and branches. A closely related line comprises the duckbill Dinosaurs, with their upper-jaw expanded into a gigantic replica of the beak of a shoveller duck or a spoonbill. This, with the horns and crests that some of them carried, together with their often great size, must have made them look very queer creatures indeed. They went in for quantity instead of quality in their grinders. Their teeth were poor in enamel, but as soon as one row wore out, another came on to take its place; the tooth-magazines of Trachodon comprised a total of 2,072 teeth! The imprints and even fossilized remains of the skin have been preserved in a few cases.

Then there was another branch, the Stegosaurs, which went about on all fours and developed the most astounding armour on the back

—great upstanding two-foot plates of bone in one, in another a dense bone-shield over the loins. The tail in general turned into a club or mace, often with spines on it, to lash out at their carnivorous relatives, and the head and brain were almost unbelievably small.

And finally, another quadrupedal branch, the Horned Dinosaurs, anticipated in some ways the heavy horned Titanotheres and Rhinoceroses among mammals. They had, however, one unique feature—a huge neck-shield of bone, sometimes as a solid frill, sometimes in the form of a *cheveux de frise*, growing back from the skull, over the nape; and the skull was generally equipped with several powerful horns. In one species the skull, from tip of nose to end of neck-frill, was eight feet long. These creatures, bracing themselves on their bandy fore-legs, seem to have taken saplings and trees in their powerful horned beak and broken them off by a violent twist of head and neck; the neck-vertebræ are left open above, as Professor Tait has pointed out, to avoid damage to the spinal cord from this twisting. Thus, if we may be Irish, they carried their Achilles' heel in their neck; and in all probability the bony frill was evolved to guard this vulnerable spot.

FIG. 231. Above, restoration of Stegosaurus, a vegetarian Dinosaur which ranged from twenty to nearly thirty feet in length. Below, an outline of the animal with its brain and spinal cord in black. The brain was tiny. In the hip region was an enlargement of the spinal cord far larger than the brain, for reflex movements of tail and hind-legs.

These were the latest Dinosaurs to evolve; and it was one of them that left the eggs with the fossilized remains of embryos inside them which the American Museum Expedition found in Mongolia. Many eggs of extinct reptiles have been discovered, but these were the first found in a nest, still in the position in which they were laid.

The later Dinosaurs, we may note, included a great number of plant-eaters, often highly specialized for a vegetarian diet. In the Trias, and earlier, both among Dinosaurs and other reptiles, the number of plant-feeders was much less; and the most primitive reptiles were all flesh-eaters. We have already seen that all amphibians, living or extinct, are or were flesh-eaters, and have traced this back to its ancestral cause in the economics of marine life, to the fact that the great bulk of plant-food in the sea exists in such small units that only tiny creatures and current-feeders can utilize it. The reptiles were the first to break through the carnivorous vertebrate tradition, and among them the Dinosaurs did so most effectively. This marked a new step in the exploitation of the resources of the land by animal life.

But though the Dinosaurs assuredly made evolutionary history, they remained very backward in one respect—in the development of brains. The average weight of a white man's brain is about three pounds. No Dinosaur, not even the forty-ton Sauropods, had a brain weighing over two pounds. That means a ratio of one pound of brain to about fifty pounds of body in man, of one pound of brain to forty-five thousand pounds of body in a big Sauropod. The Horned Dinosaurs, though weighing only about ten tons, had brains as big as Diplodocus or Brontosaurus, so that they were probably the cleverest of the bigger Dinosaurs. The palm for brainlessness goes to the Stegosaurs. Professor Lull has shown that the weight of their brain was about two and a half ounces. As in a number of other Dinosaurs, the enlargement of the spinal cord in the hip region, which regulates the automatic movements of tail and hind-limbs, is many times bigger than the real brain. Memory and thought must have been very dim in these creatures. It was not until the next era, the Cenozoic (V), that the lagging evolution of the vertebrate mind began to catch up with, and eventually to outstrip, the evolution of the vertebrate body.

§ 6. Insects Multiply as Plants Develop.

It was not only the vertebrates which perfected their land-life during the Mesozoic; their rivals the arthropods did so too. We left the land-arthropods in the Coal-measures (III E) at a crude and lowly phase, a condition about corresponding to that attained by land-vertebrates at the end of the Permian (III F). Of their two highest groups, spiders and insects, the spiders were scarcely differentiated from ticks and mites; the insects were all of the lower orders, without full metamorphosis; caterpillars and grubs there were not, nor any resting stages of chrysalis or pupa.

All these early insects seem still to have possessed the ancestral type of simple biting jaws; the ladle and nectar-sucking tube of the bee, the piercing mouth of the mosquito, the house-fly's proboscis, the moth's coiled trunk—none of these were in existence, though, perhaps, in the ancestors of the plant-bugs the jaws were being modified for piercing and sucking, thus making the rich store of vegetable sap available to their owners. And these earliest known insects were not merely all four-winged, the two-winged forms like flies and beetles not yet having been evolved, but their wings were of primitive construction, permitting gliding and up-and-down flapping, but none of the screw motion which gives the modern fly or dragon-fly its amazing aerial ability, or allows the tiny bee, a hundredth of an ounce in weight, to shoot by at twenty or thirty miles an hour, in spite of the enormous air-resistance experienced by such a tiny body.

The cold, dry spells of geological time which came after the Coal-measure Period must have been hard times for the early insects. In any case, they left a measurable effect on insect size. Warm conditions are so much more favourable to insect-life that among modern insects the average wing-spread of tropical forms is more than double that of the insect population of Central Europe. The average wing-spread of the early Coal-measure insects was enormous—over two inches. In the glacial times of the Permian (III F) this fell to about three-quarters of an inch, to rise again to nearly a full inch in the warm late Jurassic (IV B $_3$).

But cold and the drought, though they stunted the insects, obliged them to progress in other ways. In cool climates insects must grow as fast as possible in the brief summer; in arid ones they must make the best possible use of the diminished and short-lived vegetable supply. This gives an advantage to those forms which during their time of growth doff their adult guise and turn themselves into the larval feeding-machines we call grubs and caterpillars, and this, of course, makes necessary a full metamorphosis through the coffined resting state of the pupa, which then can be used to tide over times of winter and drought.

While no groups of these higher, metamorphosing insects are known from the Coal-measure forests, the Permian (III F) shows us a number of such stocks, including beetles and caddis-flies, and a primitive type, represented to-day only by the scorpion-flies, from which most of the highest modern groups seem to have evolved. This change from the state of insect affairs found in the Carboniferous was undoubtedly due to the climatic upheaval. Suckers of plant-juices, plant-bugs, some of them very similar to modern frog-hoppers, leaf-hoppers, and cicadas were already abundant, together with mantises and modern-type dragon-flies.

In the Trias (IV A), linking forms leading on to the gnats and true two-winged flies and to the Hymenoptera are to be found, and in the Jurassic (IV B) these orders, the highest development of insect-life, are abundant. By the earliest Jurassic termites were present, showing that social life had been developed. It is believed that ants, too, had begun their existence at that time, probably forced by the arid times of the Trias into underground nests and communal association. Certainly by the Jurassic all the main orders of insects had arisen, though the higher orders were as yet represented chiefly by primitive forms.

But now a series of new reactions were coming into play between the evolving insects and the evolving land-plants. Each series was destined to react upon the other with the most intricate consequences.

You have only to shake a male pine-tree on a warm, still day of late spring to stand appalled at the wastefulness of coniferous life. The yellow cloud that floats slowly away, dusting you and the surrounding herbage, is nothing but microscopic pollen-grains, not thousands, nor millions, but hundreds of millions, and all but a few tens or units destined to come to nothing.

This wastefulness was the penalty automatically imposed on plant-life by the air which it invaded. The resistant spores were the best disseminators of the life-stream in this rarefied and desiccating medium. Even later, after seeds had been evolved and the need for an independent sexual stage or prothallus abolished, the male spores, now to be styled pollen, remained as the best bearers of the male life-stream on its search for union with the female, and the wind was still the only means of transport available.

This had been so from the time of the first forests. In the Devonian, in the Carboniferous, in the Permian (III D, E, and F) wherever there were forests and whenever it was breeding-time for trees, "though the green gloom of the long-vaulted colonnades of naked tree-trunks" (as Mary Borden writes in *Jehovah's Day*) "slowly, languidly, as snow-flakes do, a pale golden shower of spores floated ceaselessly down from the leafy roof to the ground." Some seams of coal are almost wholly made of microscopic spores from the Carboniferous trees, so abundant was this golden rain in Coal-measure time. For over two hundred million years this wasteful scattering had continued, most of the innumerable spores and pollen-grains doomed to speedy death and decay. But pollen is rich and nutritious, and it cannot be supposed that the early insects neglected this abundant source of food-supply. Some of them doubtless became occasionally or habitually "pollinivorous." And equally without doubt their visits from plant to plant in search of this food would result very often in transferring pollen to the female organs.

This, however, would only happen if male spores and female spores were borne on the same cone or flowering shoot, and, since the great majority of naked-seeded plants like conifers and maidenhair trees produced their pollen and ovules on different cones, any pollen-eating visitors they may have had remained (from their point of view) robbers.

The case was different with the Cycads of the times. The true, surviving Cycads obstinately stuck to cones not unlike those of Conifers. But an extinct group known as Bennettites after the botanist Bennett, pursued an evolution along quite other lines. As some of the ancient club-mosses of the first land-dispensation evolved false seeds, so these evolved in the direction of what we must call false flowers. In many of them these reproductive shoots, half-cone, half-flower, were two-sexed; and their general resemblance to modern true flowers was often striking. One called *Williamsoniella* had a row of stamens round a central knob which bore the female parts, the stamens were very large, and perhaps were bright-coloured to make up for the absence of petals. *Bennettites* had an even more elaborate flower, with quite enormous stamens, and long plumy-haired bracts in place of a modern flower's calyx. Even forms with separate sexes, like *Williamsonia*, had star flowers as conspicuous as a modern celandine or anemone, though built on a very different plan.

We do not know—and perhaps never shall know—but it is a by no means unreasonable guess that in some of these Jurassic Cycads there had been developed glands secreting sweet juices to supplement the attraction of the pollen for insect-guests. And whether this was so or not, this step in evolution must have been accomplished in the stock of true-flowering plants (which branched off apparently from some Cycadeoid type) before the early Cretaceous (IV C 1), for by then flowers almost identical with some that still exist to-day are to be found.

This could have been prophesied on the basis of knowledge of fifteen or twenty years ago. A very recent discovery illustrates both the tricky nature of the fossil record and the beautiful fulfilment it makes of the biologist's prophecies when it does yield its treasures. The essential step by which the true-flowering plants, the Angio-sperms, differ from all lower forms, is not the showiness of their reproductive shoots, but the enclosure of the whole spore-capsule and the seed into which it develops, inside the leaf on which it grows. This leaf, it will be remembered, is then styled a carpel, and the hollow structure which it and its fellows produces has so long been miscalled *ovary* that we cannot rid ourselves of the term.

Now in 1917, in Middle Jurassic rocks (IV B 2), a plant was found which was in the evolutionary act of turning spore-leaves into carpels. In this plant, christened Caytonia, the spore-leaves bend right round the spore-capsule; but the enclosure is not yet quite complete. Caytonia may not be on the direct line to the modern flowering plant, but it cannot be far removed from that line.

Actual flowers in the true sense had come into being by the Cretaceous (IV C).

This was the plants' and the insects' opportunity. The wasteful method of scattering the solid nitrogenous material of pollen broadcast on the wind could be abandoned by the plants; and the much more direct and economical facilities for cross-fertilization provided by insects could be exploited by the expenditure of a few drops of sugar-water. For the insects, new ways of life were opened up, and a whole new winged population came into being entirely nourished by flowers and their secretions. In the flower-visiting insects, the main changes took place in tongues and instincts. The honey-bees, while keeping their mandibles to help in manipulating their waxen cells, had the rest of their mouth-parts modified into a longish suction-tube with a sort of spoon at the end; the butterflies and moths developed a remarkable tubular trunk for nectar-sipping, while in all save one or two primitive forms the biting mandibles have been reduced to vestiges or to nothing.

Though there are a few butterflies and moths which have taken to fresh sources of liquid food, like the juices of fruits or even putrid meat, the vast proportion of them are still tied to flowers for their food as closely as a whale is tied to the sea, and we can assert without fear of contradiction that without the evolution of flowers the group would never have come into existence. And the same is true of honey-bees, with their whole economic life based, like their anatomy and their behaviour, upon honey and pollen.

But once the flower-insect partnership was set up, the destiny of the flowers was as much changed as that of the insects. In the provision of nectar, the unconscious forces of evolution anticipated human economics, the recompense of goods for services rendered. But the matter did not stop here, and advertisement was brought in to push the plant's wares. Bees, as we now know from accurate experiment, perceive colours, can appreciate differences of shape, and are possessed of a delicate sense of smell. Accordingly in plants the pressure of selection caused the flowers to develop a whole set of new characters to attract the attention of their insect visitors through their senses of sight and smell.

Petals are new organs, neither leaves nor bracts nor stamens, whose prime function is to stimulate the sense of sight by their colour and pattern; other neighbouring parts may sometimes supplement and even take over this duty, as when sepals add to their original function of bud-protection by becoming brightly coloured. Nor must it be forgotten that for most purposes the advertisements of plants, like those of men, must be distinctive. It is not much good to Mr. X, the manufacturer of X's soap, if his advertisements merely stimulate a desire to buy soap in general; and it is not much good to a plant if its floral advertisement is so like that of others that insects visit them all indiscriminately; for then most of the pollen they bring will be that of the wrong species and their visits will be, from the plant's point of view, wasted. Accordingly not only is it generally of service for each flower to have its own trade-mark, so to speak, of form and colour, but special scents are often developed, each one characteristic and unique, and each announcing to the questing insect the presence of its possessor as unmistakably as wireless signals tell the listener-in what station is transmitting.

The size of flowers or flower-clusters, culminating in the yard-broad Rafflesia, the five-foot sprays of Wistaria, or the displays of tropical orchids; the brilliance of the gentian's intense blue or the Indian Paint-brush's scarlet; the fields of poppies and the gold of the golden-rod in waste places; the conspicuousness of the white and light yellows of the flowers that open in the twilight, like the Evening Primrose and Tobacco, to attract night-flying moths; the vivid shape of the Star of Bethlehem or the magnolia flower set among dark green leaves; the scent of heather and wild rose and honeysuckle—these are neither accidents nor properties designed for the delectation of human beings, but are evolutionary products of the partnership between insect and plant, with as material a meaning as that of an insertion in the advertisement column of a paper, as solid a commercial basis as a poster-campaign to launch a new brand of cigarettes.

In a certain real sense, insect-pollinated flowers and flower-pollinating insects together constitute a single biological group. They are as mutually interdependent as the fungus and alga which together make up a lichen biologically if not physically interwoven.

Like other biological groups, this partnership group progresses and evolves and shows adaptive radiation. We can only mention two of

THE FULL CONQUEST OF THE LAND

the chief ways in which their evolution moved. In the first place, to preserve the nectar-bribe from casual visitors not likely to carry out their part of the biological bargain, it was more and more securely hidden away at the bottom of long corolla-tubes, as in tobacco plants and many lilies, in special flower-spurs, like those of columbine or honeysuckle, or behind closed flower-mouths like those of snapdragon which only the weight of some creature as big as a bee could open. And these changes, of course, necessitated corresponding changes in the flowers' regular visitors. To reach the concealed honey, the insect proboscis became longer and longer, till it culminated in the trunk of butterflies and moths, packed in a coil when not in use, but often longer than the animal's body when extended. When a Madagascar orchid was discovered with a honey-spur eleven inches long, Darwin prophesied that an insect with a trunk of this unlikely length would be discovered too in the same region; a few years later the discovery was made—a hawk-moth with an eleven-inch proboscis.

Then secondly it was often to the plant's advantage to become more and more closely partnered with one or two kinds of insects, which in their turn visited few or no other flowers, rather than relying on many and more catholic species, since by this means the wasteful visits, each a dead loss in nectar, when an insect arrives with a load of pollen from another species of flowers, can be cut down. Accordingly, many flowers develop in peculiar ways so that their honey is only available to insects of a particular size or bodily construction. The long-tubed Madagascar flower with its long-trunked moth that we have just mentioned is a good case of such specialized partnership; so is the red clover which dies out without bumble-bees. So is the lovely Yucca or Spanish Bayonet with its associated moth Pronuba. The Yucca transported to Europe from its American home flowers every year; but all in vain—there is no Pronuba to visit the flowers and they never set seed.

Thus, for these two reasons it came about that from the primitive open and symmetrical type of flower, like that of Magnolia or buttercup, all kinds of more elaborate and specialized kinds came into existence during the late Cretaceous (IV C 3) and early Cenozoic (V A). All those with petals joined together to make a single cup, all long-tubed flowers, and those with honey-spurs, all the asymmetrically-built ones like the mints and snapdragons and the pea-flower tribe, often provided with special aerial landing-stages or hips, all the flowers, like the sage, with pretty mechanical tricks for dusting bees with pollen in the right place; all the orchids, whose almost incredible devices for securing cross-pollination so intrigued Darwin that he devoted a fascinating book to them—these and many other elaborate constructions are specializations which grew out of the earlier and simpler insect-visited flower as the Pterodactyls and Dinosaurs and the Ichthyosaurs grew out of the earlier and simpler types of reptiles. And the response in insect evolution was seen in longer and more elaborate tongues, greater dependence on a particular kind of plant-partner, or, as in the honey-bee, by elaborate instincts which ensure that, though many kinds of flowers are visited, usually only one sort is visited by one group of individuals on any one day.

Once the flower as we know it had been brought into existence, other animals besides insects began to enter into partnership with the flower-plants. The golden saxifrage is said to be pollinated by snails; in Java and Trinidad the fertilization of some trees is effected with the help of bats; and, in the tropics particularly, there are a number of tiny birds like humming-birds and sun-birds and honey-suckers which specialize in visiting flowers. It is interesting to find that most flowers which depend on birds for pollination are red, while pure red is a very rare colour among insect-fertilized flowers. We know from experiment that bees are incapable of seeing red, while the eyes of birds are blind to blues, but are stimulated by red; so that the flower-colours are what we should expect. In addition, birds have a notoriously poor sense of smell; and the bird-pollinated flowers are practically all scentless. The insect-visited flower, too, does not always advertise itself by sweet scent or attractive colouring, nor are its wares always nectar or pollen. There are a number of flowers with rank smells or smells as of decay, which are mostly visited and fertilized by carrion-feeding flies; such flowers are often dirty purple or brownish instead of having brilliant or pure coloration, like the buttercup, rose, or gentian. Most extraordinary of all, there exists a species of orchid which gets fertilized by smelling like a female ichneumon-fly. The flowers are visited by the male ichneumon-flies (never by the females), which attempt to mate with them, and in so doing, transport the pollen of one flower to the stigma of another. Paley and his school maintained that every adaptation implied a conscious designer; one wonders what he would have said of this orchid.

The strange and beautiful adaptations thus produced we can only mention; they are described in many books. Here we want to emphasize the fact that the adaptations were mutual, and that the whole evolution of flowering-plants and higher insects is a co-operative evolution. The particular rôle which the flower and its insect-visitors now fill in nature's economy is one into which they have crept jointly. If there had not happened to be small winged animals in the woods and fields of the Jurassic world, we should have had no flowers save green cones or yellow catkins. If there had not happened in that same epoch, to be land-plants with open cup-shaped inflorescences, we should have no honey or humming of bees, no butterflies, no moths, no humming-birds or sun-birds.

§ 7. The Battle of the Giants and the Dwarfs.

IT IS customary to speak of the reptiles as the dominant land-animals of the Secondary Era (IV). An insect, however, might well ascribe this to mere prejudice. So far, man has written all the histories of life and, being a vertebrate, has awarded the palm for the Mesozoic (IV) to his near kin, the reptiles, and for the Cenozoic (V), to his still nearer relatives, the mammals. Our insect philosopher could point to the fact that there are now fifty times as many kinds of insects as of mammals in existence, and that in Cretaceous times (IV C) they undoubtedly rivalled the reptiles both in abundance and in variety.

And yet he would be wrong. It is not only a vertebrate race-prejudice that has made man relegate the insects to second place. There do exist evolutionary differences, and very important ones, between the two groups. No insect has ever grown big; the biggest insects weigh about as much as a smallish mouse. And the evolution of insects, or at least their progressive evolution, came to an end in the early Cenozoic, when that of the mammals was in full swing and that of man was not yet thought of. It has never been, and, so far as we can imagine, it can never be resumed.

These two facts are in reality both connected; and it turns out that the evolutionary struggle for domination of the land is essentially a battle between giants and dwarfs. In Book 6 we shall look into the matter of animal size a little more closely. Here we may anticipate what is perhaps our chief conclusion.

The vertebrates, besides being the most active and vigorous of creatures, are the big animals *par excellence*. The smallest vertebrate weighs several hundred times as much as the smallest crustacean or worm or insect or cœlenterate. All the largest animals in the world have been vertebrates; their average size must be at least about ten times as great as that of any other phylum. In the insects, on the other hand, there is a ban on size. In maximum bulk they are far exceeded by every other main group of many-celled animals—from the humble sea-anemones and tape-worms to snails and lamp-shells, sea-urchins and crabs. The only exceptions are those other land-arthropods, the spiders and the myriapods, both of which have their size cut short at about the same maximum as the insects.

The explanation of this curious dwarfing is to be looked for, as the Danish physiologist Krogh has pointed out in the land-arthropods' breathing-machinery. As we saw in Book 2, insects do not use their blood to carry gases to and from their oxygen-hungry tissues; they take them direct to their cells by means of branching air-tubes. This is an admirable method—so long as the animal is quite small. But the supply of oxygen to the tissues, and the removal of carbon dioxide from them, has to be worked by diffusion along the narrower branches of the air-tubes; and diffusion is a slow creeping compared with the rush of blood along vertebrate blood-vessels. So, as an insect increases its size above a quite small limit, it has difficulty with its aeration; and the self-same air-tube machinery of oxygen supply which alone makes the activity of the bee or tiny ant or fly possible would give out completely if installed in a creature even of the modest weight of a pound. As it is, the bulkiest insects are on the whole sluggish types; as we mentioned in Book 4 they are nearing the frontier of biological possibility open to the air-tube method of making a land-animal. In a not dissimilar way, the existence of very large ships is only possible by the aid of some system of forced ventilation.

The insects were thus condemned by their essential construction, by the primary pattern of their ancestry, to a very small maximum size; while the vertebrates, with their powerful muscles and skeleton, could go on provided the food-supply was rich enough, to the limits imposed by purely mechanical reasons.

The struggle between land-vertebrates and land-insects was thus a struggle between the giants and the dwarfs of life. At first sight, moderately small size does not seem to be a serious handicap. A bee has room for a whole set of the most elaborate tools at the outside of its little body, a well-equipped laboratory of vital chemistry within; it can fly as fast as most birds or bats, and carry out operations more remarkable than those possible to the average land-vertebrate. A small animal can reproduce and multiply with extraordinary rapidity, as the social insects prove; it can make up for size by force of numbers. There is more termite protoplasm in a big white ants' nest than there is lion protoplasm in the body of a lion.

Even so, the dwarf is at a disadvantage in various ways. For the most part dwarfs can only annoy giants. Even when they combine in

FIG 232. Brontosaurus, one of the largest of the Dinosaurs.

It was about sixty-five feet long and weighed nearly forty tons. It almost certainly lived in swamps, browsing on the aquatic vegetation. Its brain weighed about a pound. Inset, a spaniel to the same scale.

bands, like the ferocious driver ants, they cannot kill any large animal unless it is fettered. But the gravest defect of small size is that it is hostile to the development of individual intelligence. In Book 8 we shall have repeated occasion to see how the behaviour of insects, for all its marvellous precision and uncanny adaptiveness, is in the main mechanical, instinctive, unintelligent. Among insects there is little learning and no education. The history of vertebrates, however, from the reptiles on, is largely a history of the improvement of intelligence and the power of learning from experience. The difference seems definitely due to the difference in size. For clockwork action, however elaborate, only a quite limited number of nerve-cells and nerve-paths are needed. But the essence of learning is that a great quantity of permutations and combinations of nerve-paths should be available, so that from the great range of possibilities experience and habit should be able to pick out the best and most suitable; and for this a far larger number of nerve-cells is required. It is no coincidence that there are no animals one can call even moderately intelligent under the size of a mouse, and none of high intelligence below that of a dog.

Thus the small size imposed on insects by their breathing-machinery in its turn restricted the development of their intelligence. And once they had reached a certain mechanical perfection of structure and instinct, they remained stationary. If active insects could have reached the size of rats, the vertebrates would have had a hard time; if they could have reached the size of wolves, their rivals might never have gained a foothold on land at all, or, even if they had, their evolution would have been stifled and they relegated to a subordinate hole-and-corner existence, few and small.

The battle of the giants and the dwarfs was fought out through the Mesozoic (IV). It was decided, like most such evolutionary battles, not by any actual fighting, but by the inherent qualities of the competing stocks, which enabled one or the other to insinuate itself into a more favourable position. It is worth remembering that its closing phases took place in strangely modern surroundings. By the later Cretaceous the majority of the second dispensation of land-plants were gone; the third dispensation that we have still with us had taken their place. But against this modern background of palms and magnolias, maples and walnuts, with bees and butterflies in their branches, Duckbill Dinosaurs wandered, Ostrich Dinosaurs ran, fantastic horned reptiles laid their eggs, the great Tyrannosaurs leapt on their prey. A mediæval play with modern scenery.

THE FULL CONQUEST OF THE LAND

By the end of the Cretaceous (IV C) the issue between arthropod and vertebrate was decided. The dwarfs, the insects, were at the end of their evolutionary resources. The types they had brought out to fight with were the same, in all but details of uniform and equipment, then as now. But the giants were at that very time training up a wholly new kind of troops—the placental mammals—which by their progressive increase of brain were destined to change the whole higher strategy of Evolution. It is true that during the Cretaceous these new troops of the vertebrates were very small compared with the vast bulk of some of their reptilian cousins; yet not only were they destined to increase their size further, but from the standpoint of general biology they were still giants, ten thousand to a hundred thousand times as big as their insect competitors.

We may amuse ourselves by speculating on the beings which the insect stock might have produced but for its limitations of size. Six-legged creatures, clad in flexible mail or chitin, as big as dogs and cleverer than we human beings? They would have been strange, but perhaps no stranger to an impartial eye than men and women. But if there is one evolutionary event for which we may especially render our selfish human thanks, it is the development of a tracheal system of respiration in terrestrial arthropods; but for that, we vertebrate intelligences would in all probability never have existed.

VI

THE MODERN ERA

§ 1. *Modern Life.* § 2. *The Troubled Earth of Cenozoic Times.* § 3. *The Origin and Prosperity of the Mammals.* § 4. *The Tangled Skein of Evolution.* § 5. *Is There Progress in Evolution?*

§ 1. Modern Life.

THE Cenozoic or Tertiary Era (V) has often been called the Age of Mammals. This, though a good enough title in its way, is apt to blind us to the fact that the mammals are but one among its many characteristic life-forms. The Age of Fur, Feather, and Flower would be a better phrase, for it reminds us that the place of the dominant reptiles was taken by birds as well as mammals, and that the earth changed its old plant-dress for one more varied and more lovely. But even this is not broad enough. If we wished to do justice all round, we should have to include snakes and lizards, frogs and toads, higher insects, crabs and teleost fish in our list of new dominant animal groups, and make special mention of herbaceous plants and grasses among the flowering plants. But this list becomes unmanageable, and so we had better call the Age briefly but comprehensively the Age of Modern Life.

First of all we must remember that, in spite of all the destruction of reptilian groups which took place at the end of the Mesozoic, the reptiles as a whole were far from defeated. Most of the readers of this book will be inhabitants of temperate countries. It will come as a surprise to them to learn that there are over five thousand species of reptiles alive to-day—nearly half as many as those of mammals. The scaled reptiles, lizards and snakes, were as relieved by the extinction of the Dinosaurs and Pterodactyls and the rest as were the early mammals. With their formidable competitors removed, they too throve, multiplied, and radiated into a thousand forms. There are running-lizards (some even running upright, as some of the Dinosaurs did, on two legs only), climbing-lizards, geckos with adhesive pads

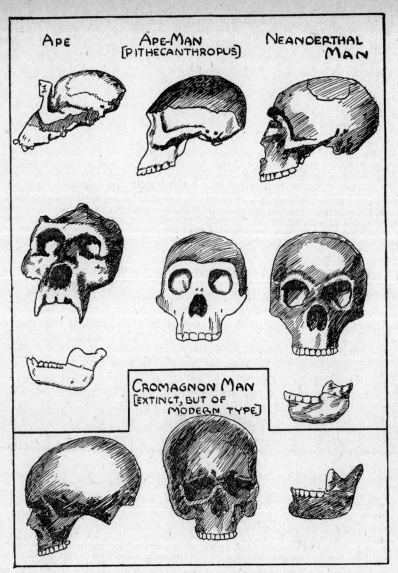

Stages in the evolution of man's skull and jaws.

At the left, Ape; in the centre, Ape-man (Pithecanthropus); at the right, Neanderthal man; at the bottom, Cromagnon man. The front view of the ape's skull is that of an adult gorilla; the side view and lower jaw are of a chimpanzee. (From an exhibit in the American Museum of Natural History.)

on their feet, burrowing-lizards, chameleons with enormous tongues shot out to capture prey, the venomous gila monster, legless-lizards, desert-lizards prickly all over, and lizards that parachute from tree to tree. There exist plenty of lizards heavier than a man, but even the Komodo "Dragon," a fifteen-foot species of Varanus, is a mere dwarf compared to its Austrian relative extinguished in the Pleistocene (V E) which apparently attained fifty feet.

There is almost the same variety among snakes. Some are completely marine, swimming like fish with the aid of a vertically-expanded tail-fin, and even breeding at sea; and other water-snakes live in freshwaters. There are two groups of poisonous snakes; there are the boas and pythons that crush their prey to death, snakes that only eat eggs, snakes that burrow, snakes that climb trees. And again among them we find huge creatures, like the pythons and anacondas which grow to thirty feet and more.

But this great variety and great size can only be attained in the tropics. The lower the temperature, the lower sinks reptilian activity, so that as we pass polewards from the equator, reptiles become less and less serious as competitors of warm-blooded creatures. It is in temperate zones that the Cenozoic (V) is truly the age of mammals and birds; in warmer regions, not only have they all the lizards and snakes to compete with, but tortoises, turtles and terrapins, and the great crocodiles and alligators and gavials to boot.

The reptiles of the present are often stigmatized in biological and semi-biological writings as the degenerate, lingering survivors of a stock that has had its day. But it is worth emphasizing that some of them are very definitely up and coming. The crocodiles, the tortoises and turtles, and the family which includes the strange Tuatera, are three very ancient groups; they are all found in the Trias (IV A) and the second perhaps even in the Permian (III F), and they witnessed the great days of reptilian ascendancy. But the first known lizard comes late in the Jurassic (IV B 3), and the first snake is from the late Cretaceous (IV C 3). The radiation of these creatures has been a Cenozoic radiation, contemporary with that of the mammals.

On a more humble footing are the frogs and toads; but they comprise at least a thousand species, and their main rise also seems to have taken place in early Cenozoic times (V A and B). The earliest fossil frogs so far discovered are from the Eocene (V A). Let us not forget that in this radiation these strange, slimy creatures, tail-less and with enormous hind-legs, have reached the size of a terrier, become wholly aquatic, taken to burrowing, colonized the tree-tops and even the deserts, and that they are still among the most successful exploiters of moist places and small ponds in all save the coldest regions.

Of the rise of the higher insects we have already spoken. Here we may perhaps recall the facts that more species of insects are known than of all the other groups of animals put together. The insects we have always with us. They are an unpleasantly dominant group; and, as we have seen, it is only the providential fact that they have been unable to grow big that has stopped them from rising from being a serious nuisance to complete dominance.

Of the modern plant too there is a word to be said. We have seen how the variety and complication of insect-pollinated flowers went on increasing in early Cenozoic times. Another important change was the rise of herbs. There seems every reason to believe that most of the early flowering plants were either regular trees or woody shrubs. The present abundant supply of herbaceous plants came later. This it is probable had something to do with the progressive sharpening of the climate. For one thing, much land previously covered by forest became vacant. For another, in cold and in dry places, where trees, if they could grow at all, would have to relapse into quiescence for long seasons of the year, the herb has great advantages. At the first breath of warmth and moisture, it can shoot rapidly up from seed or from a sheltered subterranean root-stock, take advantage of the brief flowering time, get rid of its above-ground chlorophyll-machinery and its advertisements to insects, and pass into a snug resting stage once more. Small profits, quick returns; in some regions the slower-growing vegetable organizations we call trees cannot compete with these smaller businesses that go in for quick turnover.

Among the most specialized herbs are the grasses and cereals. Of their rise, in response to the spread of great dry plains, we have spoken already.

The study of fossil plants enables us to trace the gradual change of climate. During some of the Cretaceous (IV C), Greenland grew magnolias and monkey-puzzles, cypress and sassafras, poplars and figs. In the succeeding Eocene (V A), it was still supporting some of these, but maples and pines and other trees indicate a slightly colder climate—a climate also confirmed by the fossils of Spitsbergen in the same period. Europe in the Eocene, however, was still tropical. On the site of London, in the estuaries where the London clay was being deposited, the Nipa palm then flourished, as it flourishes to-day in the estuaries and lagoons of the tropics of Asia.

The Oligocene (B V) was colder, but cinnamon and camphor still grew in France, and palmettos in the Isle of Wight. Miocene (V C) and Pliocene (V D) tell the same tale of increasing coolness. The world-wide semi-tropical flora of the Cretaceous was driven out of our northern hemisphere southwards until it survives to-day only in the warmest parts of earth; and new types, adapted to a harsher contrast of seasons, took their place. These, too, were driven southwards by the increasing cold, and replaced by cool temperate and sub-arctic forms. And so the process went on until it culminated in the Ice Age.

§ 2. The Troubled Earth of Cenozoic Times.

THE mountain uplift that brought the Mesozoic (IV) to an end was the herald of a long series of geological convulsions. This epoch of disturbance (IV to V) left the lands high and the climate sharp, with local glaciers here and there. But the mountains were slowly worn down, the lands sank, and the climate gradually became milder and more oceanic. The tale of life expands again. All the later Eocene (V A 3) and the Oligocene (V B) seem to have been a resting time of warm, moist and equable climate. But in the Miocene (B C) came another upheaval of the lands, another outburst of mountain-building. Many regions grew arid and dry, the climate cooler, and there were phases of extreme volcanic activity. The Pliocene (V D), after opening with a brief subsidence of the lands, resumed what the Miocene had begun, until at its close the ice had already started to invade the lowlands of the north, and one of the two greatest ice ages in the world's history was creeping upon the world.

These earth movements diverted the course of Evolution by their effect on climate; they also affected it by continually opening and closing the gateways of migration between the continents. Near the close of the Cretaceous (IV C) all the great land-masses were, either simultaneously or in quick succession, in communication with one another. Then first Australia was shut off, and soon afterwards South America; while the land-bridge between North America and the old world was not broken till half-way through the Eocene (V A 2).

The gateway between Asia and North America, and that between Africa and Asia continued this slow opening and shutting during the rest of the Cenozoic Era. The gap to South America was bridged again somewhere in the Miocene (V C), but Africa eventually was almost shut off from the north by the development of the great desert barrier of the Sahara; while through the whole of the Pleistocene (V E) the

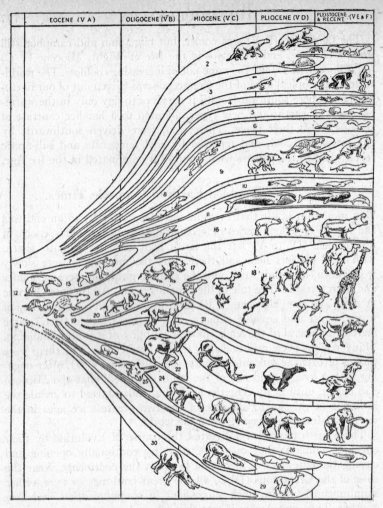

FIG. 233. The Evolution of the Placental Mammals during the Cenozoic Period.

1. Primitive Insectivores. 2. Edentates (ground sloth, armadillo, anteater, Glyptodon). 3. Primates (lemur, baboon, ape, early man). 4. Chiroptera (bat, flying-fox). 5. Insectivores (mole, hedgehog, tenrec). 6. Rodents (beaver, squirrel). 7. Primitive Creodonts. 8. Specialized Creodonts (Hyænodon). 9. Creodont-carnivore stock (sabre-tooth, bear, wolf, lion, weasel). 10. Aquatic carnivores (seal, sea-lion). 11. Cetaceans (Zeuglodont, whalebone-whale, sperm-whale, dolphin). 12. Primitive ungulate stock. 13. Amblypods (Coryphodon, Dinoceras). 14. Condylarths. 15. Primitive even-toed ungulate stock (Phenacodus). 16. Pigs (babirussa, wild boar, hippopotamus). 17. Giant pigs (Entelodon). 18. Ruminants (Protoceras, Syndyoceras, deer, Sivatherium, camel, giraffe, kudu, ox). 19. Primitive odd-toed ungulate stock. 20. Titanotheres. 21. Clawed ungulates or northern ice-sheet was a more effective bar to migration than any sea. On the other hand, there was no ice over Alaska and Kamchatka; so while the bridge from Europe to eastern America was barred, that from America into Asia, though chilly, was left open.

During the great bulk of geological time, the earth has been wrapped in the veils of a warm and mild climate, much more equable from equator to pole than the climate of to-day. The lands have usually been low, the fringes and even the centres of the continents often covered with shallow seas. When high mountains have been formed, they have been worn down to low ranges or plateaus long before another breastwork of mountains was built up in their place. Uniformity of climate and flatness of scenery are the rule; diversity is the exception, widespread earth-activity comes but seldom, and ice ages are among the rarest of incidents.

In all these ways the Cenozoic Era (V) has been exceptional. It has suffered a succession of the crustal throes that gave birth to mountains, with the result that the scenery of the earth to-day is probably far grander than it has been in most periods of its history. The Cenozoic had a cold spell at its beginning, another far more severe at its close; its latter half was a time of high plateaus, widespread activity, sharply-zoned climate. The million years or so of the Ice Age was one of the most exceptionally severe periods in the world's history, and let us not forget that these times in which we live, though not so grim as the Ice Age, are far cooler and stormier, and have far bigger deserts and ice-caps, than the average earth period. Man has lived so far in an exceptionally exacting age.

§ 3. The Origin and Prosperity of the Mammals.

THE early stages of mammalian development, which must, we infer, have gone on from the late Permian (III F 3) or early Triassic (IV A 1) to the Cretaceous (IV C), are not so poorly documented by fossils as the evolution of the birds, but they are by no means fully recorded. Throughout the Mesozoic, traces of mammals are the most precious and welcome of finds.

In both cases we seem to have a scheme of life for which the rich, easy, equable environment of the Mesozoic Era (IV) had little use.

---

chalicotheres (Moropus). 22. *Baluchitheres*. 23. *Existing odd-toed ungulates (rhinoceros, tapir, horse)*. 24. *Proboscideans (Mœritherium, Palæomastodon, Tetrabelodon, Elephas)*. 25. *Arsinoitheres*. 26. *Sirenians (Dugong)*. 27. *Primitive stock of South American ungulates*. 28. *Typotheres and Astrapotheres*. 29. *Litopterna (Macrauchenia)*. 30. *Pyrotheres*.

The early mammals were small, hardy, obscure creatures, whose peculiar qualities only became decisive advantages as conditions grew rigorous towards the close of the Cretaceous Period (IV C). We have already noted the Theromorphs, that series of reptiles in the Trias (IV A) with many anatomical features otherwise characteristic of mammals. To this group the ancestral mammals seem to have belonged. Some jaws and teeth from the early Trias (IV A 1) are on the borderline of mammal and reptile; they might be classified with a note of interrogation on either side of that line. But some from the late Trias (IV A Z) are definitely mammalian.

Small infrequent jaw-bones constitute most of the mammalian traces in the Jurassic. Some of these jaws are more simple than those of any surviving mammal. Some have features that foreshadow the marsupials, but it is too bold for us to assume from that that these creatures had already mitigated or superseded egg-laying by carrying their young in pouches. A third line, though now extinct, survived to witness the end of the great reptiles and well on into the Eocene (V A). These were the Multituberculata, so named because of the rows of ridges of tubercles on their strange grinding teeth. Like all Mesozoic mammals these were generally small beasts, but in the Eocene, before their final extinction, they produced some good-sized genera.

These Mesozoic mammals were the prey of larger reptiles, a series of lurking, hunted races. Particularly difficult was it for them to nest and preserve their eggs. It is easy to see that for innumerable species, the Jurassic Era was one glorious egg-hunt. For the mammals it was a desperate struggle to save their eggs and helpless young. This they were able to do first by the retention of the egg in the body so that they became viviparous, and secondly by continuing a protective association with their young after birth. The equally hard-driven birds developed the second disposition but not the first. As we know, two genera of egg-laying mammals still survive in the world and their rare ancestors struggled through the Mesozoic and Cenozoic, leaving few fossil traces; but the other mammalian lines completed the development of the ovum in the body and either had an early birth of immature young which were then taken into a pouch, or kept them longer and longer in the uterus and developed the placenta so that at last they would be born ready to run and eat and live. The former method was the easiest to evolve, and there are grounds for supposing that the marsupials got the start and for a time were the prevalent pattern. Then the more elaborate but more efficient method of the placental mammals enabled them to recover ground and draw ahead.

So incomplete is the Mesozoic record of mammals that, at the date of writing, the whole of the Lower Cretaceous Period (IV C 1) has to be reported blank. Experts prophesy that if presently we come upon land-deposits of that period with mammalian remains we shall find they are marsupials. When at last we begin to find a supply of fossils again in the late Cretaceous (IV C 2 and 3), marsupials almost identical with modern opossums occur. The placentals had meanwhile become distinctive and varied. While the marsupials were dominant, however, they succeeded in getting into the Australian region, unaccompanied by any placentals, some time in the Cretaceous, before it was severed from the rest of the world by the sea; and there, as already narrated in Book 3, they blossomed out into the wonderful variety of types, from kangaroo to marsupial mole, from flying phalanger to Tasmanian devil, which are found there to-day (Fig. 143). They also early penetrated in force to South America, and there, in its partial isolation, managed to evolve into a number of types. But later almost all of these perished, after South America was thrown open to the tide of animal immigration from the north. In the rest of the world they never accomplished much. As Dr. W. D. Matthew writes: "By the Lower Eocene the marsupials were already a defeated group which took to the trees"; and the opossums are the only survivors which have held their own against full placental competition.

Early placentals seem to have originated in some northern land. The American Museum expedition to Mongolia found there in mid-Cretaceous rocks (IV C 2) the remains of a number of small creatures not unlike a cross between a shrew and a very primitive weasel, with long skulls and tiny brains. They were found in what was then a sand-dune area, together with the nests and eggs of Dinosaurs, probably attracted from near-by woodlands by the eggs of the big reptiles, or the small fry of insects and mites that would haunt these nurseries. They doubtless stole eggs, but left no eggs about to be stolen.*

By the earliest Cenozoic times, Basal Eocene (V A o), the placentals are the dominant group of land-animals in all the main continental masses and have already differentiated into several stocks. The Dinosaurs had all disappeared, at any rate from the great land-areas of the northern hemisphere. The placentals now show every sign of successful

---

*Recent geological discovery has considerably extended the series of Eocene rocks back into the past. To avoid confusion, these oldest rocks of the period are called *Basal Eocene*, followed by Lower, Middle, and Upper in the ordinary way. We will here designate Basal Eocene V A o.

adaptation to the changed world. They radiate. They do as did the reptiles vast ages before them in the Trias—they branch out into a number of lines, many of which, especially among the earlier more experimental ones, presently become extinct. Through the Eocene (V A) and Oligocene (V B) a succession of new stocks are budded out, like whorls of branches successively budded by a growing tree; and the main extinction of the unsuitable branches takes place a little later, from Middle Eocene (V A 2) to Miocene (V C).

There were two main bursts of placental evolution, if we may use the phrase of such a slow process. The first must have begun in the Cretaceous (IV C); it culminated in the Basal Eocene (V A o). Among the primitive ungulate types of this first radiation, some, called Condylarths, are small and extremely like some of the flesh-eaters of the same period, save that their teeth reveal inclinations to a mainly herbivorous diet. These creatures were still clawed. Others, the Amblypods and their relatives, were larger, had developed hoofs, and were the most abundant creatures of the time. In the Unguiculate group the insectivores had already appeared in forms not very unlike those of to-day. The early flesh-eaters, styled Creodonts, had not advanced from the early omnivorous type, snappers-up of unconsidered trifles, from which they and the ungulates as well as modern insect-eaters seem to be descended. Ground-sloths also had appeared.

This group of creatures has been called the "archaic mammals"; and it is they that blossom out to dominate the later Eocene (V A 3). The little Condylarths ran more and more swiftly and browsed more and more efficiently. The Amblypods became huge and clumsy, finally producing Coryphodon, as big as a great ox, and tusked and four-horned creatures like Eobasileus, heavy as a small elephant, with pillar-like legs to support its bulk. The flesh-eaters became more specialized to cope with the greater development of defensive weapons or the greater fleetness or bulk of their prey. Just as wolf type and wolverine type, in the shape of marsupial wolf and Tasmanian devil, were evoked in Australia among the marsupials, so among the Creodonts we find an adaptive radiation which produced creatures more or less resembling hunting dogs, and wolves, and hyenas, and martens, and primitive heavily built cats, and bears, precursors though not actually progenitors of the parallel modern types.

Meanwhile a second radiation was going on in some unknown region. In the Lower Eocene (V A 1) we meet for the first time with the main groups of modern mammals—the gnawing rodents, the modern ungulates, edentates foreshadowing the armadilloes and their relatives, and true primates of the lemur group. These appear suddenly in Europe and North America. They do not seem to have originated there and we do not know where they originated. They must have originated in some other region during the Basal Eocene (V A o) and then flooded out in a wave of migration.

The "archaic fauna," brought into competition with this new series of forms, was for the most part found wanting. Condylarths all died out by the end of the Eocene (V A 3); so did the Amblypods. Their final changes before extinction were towards still greater size and more fantastic development of horns and tusks, but their teeth and their brains were scarcely altered at all. They were cell-communities which increased their population and their armies, but neglected their industrial machinery and their form of government, and their fate is a useful moral lesson for the contemporary statesman.

A similar increase of brain relative to brawn is seen in almost every case when we compare members of the archaic placental fauna with creatures of corresponding size and habits which were evolved in the later radiation. There was little difference in the bodily machinery of the two dispensations; the improvement lay in the improved way in which that machinery could be used.

The successors of the Amblypods and Condylarths, the modern ungulates, fall into two main groups, odd-toed and even-toed. The odd-toed began their main evolution first. In Book 3 we have already traced the evolution of their most successful branch—the horses. It was, as we have seen, a slow and steady perfectioning of their machinery for feeding and for speed. The tapirs and the rhinoceroses are the only other surviving branches of the odd-toed trials, and both are dwindling groups. But once there were hornless running rhinoceroses about as speedy as the half-evolved horses of their day, and another side-branch gave rise to Baluchitherium and its allies. These, the heaviest purely terrestrial animals which ever existed, were also hornless, had huge necks and legs, and apparently browsed on the upper branches of trees, combining the bulkiness of elephants with the shape and habits of a more massive giraffe. The largest of them all comes from Mongolia, with skull over five feet long, fourteen-foot fore-legs, and a twelve-foot neck above; it must have weighed nearer twenty than ten tons. We may surmise that the huge Mongolian Creodont, Andrewsarchus, with skull a yard long, was evolved to prey upon it.

One odd-toed stock that has long died out we have already mentioned—the Titanotheres. This too put its trust in size and formidable horns, and has found that these alone do not pay in the long run. Finally, there are the Chalicotheres, a branch that lasted on to the Ice Age (V E). These are the creatures which disproved the rule laid down by Cuvier, the great French zoologist, that grinding teeth were always associated with hoofs, and flesh-eating teeth with claws. For they had horse-like skulls and teeth but also long claws, three to each foot, and retractile rather like those of cats. Probably they were steppe-dwellers and used their claws to scratch up succulent roots.

FIG. 234. The biggest land-mammal.

A Baluchithere whose skeleton was found by the American Museum Expedition to Mongolia in strata of Oligocene Age (V B). It must have been able to browse on leaves more than twenty-five feet off the ground.

The even-toed ungulates were later in starting their main evolution; but, once they had got under way, they ousted most of the odd-toed forms and rose to be the dominant planteaters of the modern world. A few side-lines early died out, some of them combining characters of pigs and ruminants, while others, like the Entelodonts, were pig-like creatures of giant size.

The two surviving groups are the pig-like forms, with pigs and warthogs, peccaries and hippopotamuses; and the ruminants. From this latter stock the camels and llamas branched off early. The remainder were among the latest of mammals to become specialized, since the cattle, the antelopes and sheep, the giraffe and the modern deer arose only in Miocene (V C) times. All these can be traced back to an ancestral stock of hornless, deer-like animals. The rodents, with their little, simple brains but wonderfully specialized gnawing teeth, were another group which met with constantly increasing success: not a single rodent family has become extinct since their appearance in the Eocene, and they are among the most abundant and widespread of living mammals.

The hosts of plant-eaters were so much raw material for the carnivorous creatures; and these were not behindhand in their evolution. All but one of the several Creodont families died out; but this one, called the Miacids, lived on and gave birth to all other modern flesh-eating mammals. It is no coincidence that this family of the Creodonts was the biggest-brained. What its relatives gained by brute size and strength it achieved by skill and cunning; and it had its reward. The others were too clumsy or too stupid to compete with it. Near the end of the Eocene this single family repeated the adaptive radiation of the original Creodont stock. It gave rise to the bears and the raccoons, the dogs and wolves, the weasels and martens and otters, the civets and mongooses, the seals and walruses, the hyenas, and the cats, great and small. Only one of its descendant families has died out—the sabretooths, whose rise and fall we have discussed in Book 4.

The bats had taken the air as early as the end of Basal Eocene, and at the same time some little creatures called Plesiadapis marked a halfway stage between an insectivore of tree-shrew type and primate-like lemurs. But the history of these must be deferred until we discuss the fascinating riddle that surrounds the ancestry of man.

The known history of whales and dolphins begins in the Middle Eocene, with the creatures known as Zeuglodonts. Like the early Ichthyosaurs, though they had wholly abandoned the land, they were not nearly so fish-like in shape nor so beautifully adapted to aquatic life as their later descendants. They were all actively carnivorous; the whale-bone whales did not develop until the Oligocene (V B). The earliest whales were found in Africa. In the same region originated the sea-cows, the elephants, and the extraordinary horned creature, Arsinoitherium, the only known representative of what must have been a separate African group of herbivores.

Of the history of those long-isolated regions, Australia and South America, we have already spoken in Book 3. Here we must add that recent exploration has yielded abundant remains in the Basal Eocene (V A o) of Mongolia of a little creature closely related to the strange extinct herbivores that once characterized South America, the Typotheres and Toxodonts and their kin (Fig. 146). It would seem that this group, arising in Cretaceous times (IV C), was soon extinguished in the regions of more active competition, but, slipping into South

America with other primitive creatures like the ancestral edentates before the isthmus was submerged, it found its chance, throve, evolved and radiated.

§ 4. The Tangled Skein of Evolution.

THE story of Evolution is so often told as a simple progressive unfolding, as a triumphant march from the early creeping life-stuff through fish and theromorph to man, that the bewildering complexity of the actual facts is in grave danger of being forgotten. For Evolution is not by any means a plain unfolding. While our ancestors were improving their hands and eyes and brains a host of other weapons were being tried and sharpened. Frogs were developing their catapult thighs, birds their wings, rodents their chisel teeth, liver-flukes their tortuous life-stories. Evolution is the sum of a swarm of processes, now independent, now mutually interfering. The plot of the drama is not a single thread but a tangled skein of hundreds of threads of which our own is only one.

Let us look back at the mammals' evolution with this idea in mind. We first disentangle it into a series of lines of specialization—horse, elephant, tiger, whale and the rest—each line steadily becoming better adapted to some particular mode of life. If they do not die out first, each specialized line finally reaches a limit, like that of the Pliocene horse, and changes no further save in minute adaptations of detail.

But these specializations are not independent. Groups evolve by what we have called adaptive radiation, bursting forth into a number of specialized lines, which, evolving together, cover between them all the main ways of life possible to the type. In the long history of a large group like the mammals, however, there are not one but many such bursts of spreading evolution. Every more or less isolated area, like Australia, or South America, has had its separate radiations producing types in its own way. Then there is a succession of radiations in time, each starting from a different evolutionary level. There was a radiation of the first-evolved mammals far back in the Mesozoic, succeeded probably by an early-Cretaceous (IV C 1) radiation of early marsupials. Then came the first radiation of the placentals in the late Cretaceous (IV C 3), giving us the "archaic mammals," and finally the radiation of more modernized mammals in the late Eocene (V A 3). Each of these started from a higher level than the one before: egg-laying—early birth into a pouch—late birth of placenta-nourished young—those were the first three steps. The fourth concerned brain, the second burst of placental radiation starting from animals with greater brain-size than the first. The products of the various radiations commingle and compete. Most of the earlier types die out, but some survive, and even progress. Our main groups of modern beasts are not all products of the latest radiation; the insect-eaters and the carnivores, for instance, had their roots in the "archaic mammals."

Then within the limits of a main radiating group each minor group itself shows radiation—the ungulates, for example, and the carnivores; and within the ungulates the earlier radiation of the odd-toed was followed by the later and more successful radiation of the even-toed.

Apart from radiation, though akin to it, is a process we have not yet discussed, to be found within each single advancing line. Within the horse stock we distinguish between the central evolutionary trunk and the side branches which came to a dead end. But the central stock itself is not really a single smooth trunk; it is, if we pursue our metaphor, a trunk thickly beset with side-twigs. Horse evolution, it will be recalled, was an affair mainly of the improvement of grinding teeth and of running limbs. Whenever numerous species of horses have been unearthed at one period of evolution, there will be found some which have gone farther than the average in respect of their teeth, but less far with the improvement of their legs, and vice versa for others. More haste, less speed. These creatures, a little over-specialized in one feature, a little under-specialized in another, could not in the long run compete with the more all-round improvement of the central types, and so the central trunk grows on, surrounded by the thick fringe of short-lived twigs.

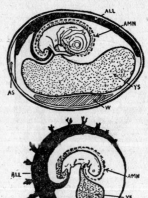

FIG. 235. Evolution of vertebrate reproduction.

The embryonic membranes in a bird (above) and a placental mammal (below). The embryo is enclosed in the protective amnion (AMN). It is attached to the yolk-sac (YS) which in the bird contains yolk, but in the mammal is vestigial. From it springs the allantois (ALL). In birds this is plastered against the inside of the shell and acts as a breathing-organ. In mammals (cf. Fig. 65) it is nutritive and grows tufts of blood-vessels which fit into the wall of the uterus of the mother. (AS) Air-space. (W) White (albumen).

This seems to be a general rule; during the main evolution of any stock, for each type that lives on to be ancestral to the next evolutionary phase, there are a dozen thrown off to live a few tens of thousands of years and die without descendants. A great deal of parallel and independent specialization is going on as regards separate characters; but only those with certain combinations of improvements survive. This we shall find well illustrated in the early evolution of man from ape.

But in the evolution of one group the competition of other groups may also have a great deal to say. All the radiation achieved by mammals in Mesozoic times (IV) was limited and feeble, owing to the presence of the powerful reptiles, like so many vested interests, already occupying all the main positions open to life, crowding out the less specialized newcomers. Here it is that the pressure of environment on life, a pressure quite external and fortuitous, makes itself felt. The great climatic revolution that killed off the dominant reptiles opened the door of opportunity to the mammals: their warm blood enabled them to withstand the cold, their very smallness and insignificance was now a help when climate cut down the world's vegetable food supply.

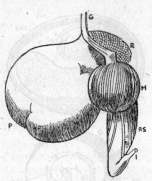

FIG. 236. The complications of chewing the cud; a calf's stomach.

The food comes down the gullet (G) into the paunch (P), and thence into the small reticulum (R), whose secretion partially softens it. After this it is passed up the gullet as the cud to be re-chewed; and descends again, now semi-fluid, by a special groove in the gullet into the manyplies (M). After further digestion it is passed into the rennet stomach (RS) and thence along the intestine in the usual way.

Climate may have more direct effects. The increasing dryness of the later Cenozoic led to the shrinking of the forests, the extension of grassy plains. This encouraged the evolution of grazing, as opposed to browsing, forms, and put a premium upon swift running, both in pursuer and pursued. It was also without doubt a decisive factor in the evolution of the ruminants. These not only include many grazers and rapid runners, but all have developed the complicated stomach which enables them to bolt their food hastily, and then grind it later, when they have retreated to some safer spot, by bringing it up and "chewing the cud." It is no coincidence that their rapid rise took place so late in Tertiary times: not until the dry open plains had grown and spread was there a premium on this strange digestive arrangement. And this same change of climate, as we shall see, in all probability decided the destiny of ancestral man.

The opening and closing of continents to migration brings in another element of chance. If the door into Australia had not been closed just when it was, the world would never have known the capabilities of the marsupial type: the pouched mammals would have remained as undeveloped and feeble as the egg-laying mammals. Nor would we have had the fantastic armadilloes and ant-eaters and sloths without the long isolation of South America. For competition plays a vital part in the evolution of efficient and varied stocks. It is in the great northern land-masses that the highest types of mammals developed: and this is through no reasons of climate, but because over their great extent the products of evolution of North America, of Europe, of North Asia, of India, and in later times of Africa, have all been commingled and forced into far more intense competition than prevailed in more secluded regions.

But finally climate comes in again to extinguish many of the strange and exciting creatures which the same blind agency, by removing their competitors, earlier started on their evolutionary career. Change of climate may cause extinction directly, as it did with so many of the larger herbivores during the Ice Age, or indirectly, as we have already seen in Book 4, Chap. 9, § 2, with the sabre-tooths, which died out because climate killed off the lumbering creatures to the slaughter of which they were so nicely adapted.

The skein of Evolution has many threads, and the threads are inextricably entangled with each other.

§ 5. Is There Progress in Evolution?

LOOKED at thus, Evolution would seem to be a chaotic affair, its changes dictated by one accident after another, each one the outcome of the chance advantage of the geological moment. And yet its movement as a whole does not seem to be haphazard. The whole of the history we have told in this Book is a witness to the fact that evolution is progressive. Life experiments and discovers; Nature selects; everlastingly the old is surpassed by the new and fitter. It is time, now that we have traced the main phases in our historical drama, to examine this idea of evolutionary progress a little more closely, to see in what precisely it consists and how it is brought about.

Specialization and progress in Evolution are both adaptive; they are both responses of evolving life to the pressure of competition, the struggle for existence, the need for more efficient living. Specialization is an improvement in life's machinery for a particular way of living. Progress in Evolution is an improvement in machinery for living in general. The latter is an all-round, the former a one-sided advance. We see the difference diagrammatically illustrated by the reptiles and the mammals. All the great lines of reptilian descent are specialized to particular modes of life; but the one line in which were evolved warm blood and milk and internal development of young achieved an all-round improvement in the art of living. The old type became subordinate, the new became dominant; and then the new could itself radiate out in every kind of specialization.

In this case we have judged that the mammal was more advanced because it succeeded in dominating its rival type in the course of evolution. But is this crude criterion the only one possible? Does it not sometimes happen that something really inferior gets the upper hand? In other words, is there any objective mark of progress, anything which enables us to say that one organism is higher than another? We all have some idea in our mind that this is so, for we all freely apply the words *higher* and *lower* to animals and plants. It would be difficult to find anyone who did not believe, even if he had not attempted to think the matter out, that a man was somehow higher than a wolf, a wolf than a salamander, a salamander than a jelly-fish, a jelly-fish than an amœba, an amœba than a bacterium.

But is there any objective justification for this belief? Some critically minded people would have it that there is none, that to call an animal *higher* or *lower* is unscientific, that there can exist no such thing as progress in Evolution. They point to the fact that survival is no criterion; bacteria and protozoa survive as successfully as birds or men, and have lived on while many elaborately organized types have been extinguished. Nor is successful increase and multiplication any badge of a "higher" type. There are innumerably more tubercle bacilli than human beings in the world. It might be imagined that adaptation was the key-property; but this will not work either. What could be more beautifully adapted to their own way of life than some of the world's most repulsive parasites? Yet these are always called degenerate animals. If survival and abundance and well-adjusted adaptation are no good in identifying a higher animal, what is? And the critics hint that we have been misled by our incurable vanity and merely call an animal higher because it happens to be more like ourselves.

But there is more in it than that. When we take a number of examples of what common-sense would call high animals and a number of what common-sense would call low animals, and reflect on the differences between them, we see that there is a real criterion of high as against low, of progress against standing still or degeneration. In a word, the higher creature has more control over the environment, and is more independent of it; it is in touch, through its sense-organs and brain, with more of the world about it—the world for it is larger and more varied and so far as we can judge from analogy with our own minds, its mental capacities of knowing and feeling, learning and foreseeing, are greater. A large part of Book 8 will be a running commentary on these last two clauses. Here we can confine ourselves to the first—which indeed in a sense includes all the rest.

The higher has more control and is more independent. To a certain extent this comes about through mere bulk. The microscopic animal is at the mercy of the wind or the current which the creature of any size can withstand. More definitely, it comes about through improvement of mechanical and chemical construction. The highly evolved reptile or mammal can do all sorts of things which the crawling amphibian cannot; the muscular stream-lined fish can swim in a triumph of accurate controlled motion when the medusa can only pulse blindly along. Partly the release from outer necessity comes through improvement in self-regulating machinery: the mammal carries about with it a special fluid environment for its tissues, the blood, far more delicately regulated to constancy of temperature and chemical composition than almost any environment, and made up so as to give optimum conditions for the work of the cells. Partly our freedom is due to a better co-ordination of parts. And very largely it comes about through improvement in the machinery for using the parts—the brain and its attendant mechanisms of sense-organs and nerves.

Without an efficient intelligence service of sense-organs most of the environment is a closed book; one has only to think in what a monotonous and tiny black box of a world a creature like an amoeba or Hydra is imprisoned—without either eyes or ears to give any knowledge of events at a distance, its whole experience apparently consisting of touches and tastes.

But even when the organs of information and the organs of action have been improved up to their biological limit, increased control and increased independence can be secured through a better use of these tools of life, by means of a brain better able to profit by experience. Increase of control and independence gained by this means has been the main feature of mammalian evolution, from the time of the archaic mammals to that of modern man.

If we make increase of control and independence our criterion of a higher organism, then when we look back over the course of Evolution we can say that there has been progress. For Evolution has gradually produced higher and higher types. But we also find that evolutionary progress is of a particular nature. In Evolution, all the lower types do not get changed into higher. On the contrary, not only do low types go on existing, and happily, side by side with the higher (as the horse, in spite of the advent of the automobile, goes on existing fruitfully in certain parts of the world and indeed performs certain functions of transport more economically than a mechanical vehicle, even in big

FIG. 237. The evolution of man and his upright posture—nine stages.

From left to right:—(1) Swimming: a lobe-finned fish from the Devonian (III D); (2) Crawling: a Stegocephalian from the Carboniferous (III E); (3) The body beginning to be raised from the ground: a primitive reptile from the Permian (III F); (4) Body fully off the ground: a mammal-like reptile from the Trias (IV A); (5) The beginning of tree-climbing: an opossum-like mammal; (6) Life in the trees: a primitive lemur from the Eocene (V A); (7) Brachiation in a primitive ape: a gibbon; (8) Brachiation in a higher type of ape: a chimpanzee; (9) The erect posture: man. (By courtesy of Dr. W. K. Gregory, from Proc. Amer. Philos. Soc.).

cities), but there has been a great deal of degeneration, active and independent types losing most of their independence to become fixed or parasitic. What evolutionary progress consists in is the raising of the upper level attained by life. The old and the simple types continue, but over great periods new types appear which attain some further degree of control and independence which life has up till then lacked. Evolution is very far from being all progressive; but it is shot through with progress.

Our critics may grant that control and independence, increased knowledge and intenser mental life, do, on the whole, characterize the animals usually called higher. But they will probably ask whether an evolutionary movement in this direction should be called progress, for *progress*, like *higher* and *lower*, implies some scale of values. The answer of course depends on one's attitude. If one wants to analyze purely objectively it is best to leave values out. On the other hand, if we want something fuller and more warming than detached analysis, if we want to build from the multitudinous facts a vision of Evolution as an imposing whole, there is nothing to stop our putting them in. The general results of this evolutionary tendency—increased physical efficiency, bigger brains and so on—happen to be results that in human affairs we call progressive. We do think independence and control, learning and foresight, nobler than their contraries. And so it is permissible to say that one animal is higher than another, and to give the general name of progress to this tendency of Evolution that raises the upper level of control and independence reached by life.

It is one of the most striking things in Evolution, this fact that the slow automatic movement of life, through ages before man ever existed, took a general direction which to our conscience commends itself as progress, and which, it seems, a dispassionate intelligence surveying earth's evolution from the outside would call progressive too.

Man happens to be the highest animal at the present time, as is evidenced, among other things, by the extent of the control he is exerting over the fellow-inhabitants of his world. It is this which takes the sting from the suggestion that we style creatures high or low merely on account of their greater or lesser resemblance to ourselves, irrespective of the evolutionary facts behind the resemblance or lack of it. If a tapeworm could for a moment be granted the power to think and philosophize, he would be bound to confess that the direction of his evolution did not agree with the main direction of Evolution, that his was a blind alley, while the opposite direction led on, apparently without limit.

It remains to ask whether evolutionary progress as we have defined it can be accounted for on ordinary biological principles. This is not an idle question, for many theologians definitely assert that it cannot be so accounted for. They have given up the belief that the apparent design seen in adaptation is evidence of a designer and of conscious purpose; they are willing to believe that Natural Selection will account for single adaptations. But they have not realized that it will equally account not only for the straight-line evolution of long-continued specialization, but for the still more general trend which we have called biological progress. Biological progress, in fact, consists in all-round adaptation, many-sided instead of one-sided specialization; and the struggle for existence will automatically make for progress as it automatically makes for long-continued specialization like that of the horses or detailed adaptation like the protective colour of a green leaf-insect.

For the qualities that we call progressive are very evidently of advantage in the struggle for existence. Clearness of eye, swiftness and sureness of foot, accurately standardized and regulated blood, memory, forethought, inventiveness—their desirability from the point of view of individual welfare is obvious. Manifestly they make for successful reproduction. And it is obvious also that anything making for independence from environmental influences and for control of environmental factors will help the race to survive. We must remember, too, how advance in one animal or plant automatically makes for corresponding advance in all those others whose lives are entangled with its life. The increasing of speed and skill in escape among herbivores was in the evolution of mammals as intimately bound up with increase of speed and skill in attack among their carnivorous enemies as was the increase of efficiency in the armour-plating of battleships with the increased range and penetrating power of big guns.

There will further be a premium upon progressive changes, since such a change will generally land the organism which adopts it in virgin soil, so to speak, where there is less pressure of competition—if not in an actually new physical environment, then in a biologically new situation. The modern mammals occupy the same dry land as did the wonderful reptiles of the Mesozoic. But constant temperature and embryonic existence within the mother provide delicately adjusted conditions which in their turn enabled a more elaborate and delicate brain-machinery to be developed, and so advanced their possessors on to new shores of control and independence.

Indeed, in a certain sense, biological progress is inevitable. It is inevitable provided that cosmic conditions do not alter so much that higher forms of life could no longer continue to exist; provided that helpful variations continue to be thrown up, and provided that the branch under consideration has not some initial defect of construction which will eventually set a limit to its advance (as insects, we saw, are robbed of all the advantages of bulk by the construction of their breathing-system).

In the size of the female human pelvis, for example, there may be a limiting condition to the size of human babies' heads, so to the expansion of the human brain, and so to the elaboration of the human mind.

We cannot say for certain whether this is so or not. But a check on the advance of *Homo sapiens* is not necessarily the end of progressive evolution.

In any case, when we look back at the past history of life, we see definitely enough that, through all the five hundred million years of adequate record, biological progress, though slow and sometimes devious, has been sustained. Progress, defined by human standards, has been the quality of the evolution of life, and there is no reason to suppose that it will not continue the quality of the evolution of life, so long as life continues to evolve.

VII

MAN DAWNS UPON THE WORLD

§ 1. *Why We Say "Man Dawns."* § 2. *The Remote Ancestors of Man.* § 3. *A Short History of Pleistocene and Recent Climate.* § 4. *Traces of Man Before and in the Pleistocene Period.* § 5. *The Advent of Modern Man.*

§ 1. Why We Say "Man Dawns."

MAN made no marvellous entry upon the drama of life. He did not appear suddenly upon the scene, strong, aggressive and prevalent, nor did he ascend slowly through a series of abundant and successful forms, as did the horses and camels. He dawns out of almost imperceptible premonitions. He appears first as a rare creature, and he seems to have been the descendant of ancestral species, perhaps as infrequent.

We know him now as a world-wide, abundant social animal but that has not been the case with him for more than twenty or thirty thousand years. We will discuss the why and wherefore of that later. But his remains before these recent dates testify to an animal almost as lonesome as the great apes and as unlikely to be preserved and fossilized. Were it not that several of the species of varieties of early *Homo* resorted for shelter and other purposes to caves, it is doubtful if we should know anything about them at all at the present time.

The reader will better understand the point we are making if he will try to imagine how many dead gorillas or chimpanzees or orang-utangs have been preserved by nature in the last hundred years. Possibly not one. These creatures, frequenting forests, are not likely to find a grave in the sediments laid down by swift rivers in flood over broad plains, nor in lake deposits, nor to be floated down towards the sea and dropped in an estuary. And as a matter of fact no fossil remains of gorilla, chimpanzee or orang-utang have ever been unearthed.

Were it not for the actual living survivors of these rare anthropoids, our grandfathers would never have had a suspicion of how nearly a mammal can resemble a man. A few unfavourable geographical accidents of no great general importance might have blotted out these sad-faced, illuminatory cousins of our breed altogether.

And of all the early *Hominidæ* up to seventy or eighty thousand years ago, the bones we know could be packed carefully in paper and put into a small suitcase. What we get are chiefly lower jaws—for reasons we have already discussed when describing the Mesozoic (IV) mammals. Some teeth, a few shattered crania, and a long bone or so complete the meagre list. Yet all we find is consistent with evolutionary ideas. And if bones are rare, another sort of trace is not rare. The *Hominidæ* were unique among Vertebrata in making instruments—making them, not merely using them—and among other materials they employed hard stone chipped to an edge or otherwise adapted to human purposes. No doubt there were implements made of wood and other materials which have perished. But a chipped stone is an almost imperishable memorial. For every single bone of a primate of which we know of an earlier date than twenty-five thousand years ago, we know of hundreds of thousands of stone implements. These stone implements show a progressive advance from very rough to more elaborate forms. They have been grouped into periods. We classify them into cultures and types, but we cannot say of the earlier ones whether they are the work of many varieties or species of *Hominidæ* or of just one or two races. Slowly, perhaps, happy finds and inspired search will fill in the yawning gaps of this least perfect and most fascinating part of the geological record.

But before we describe this dawn of man, we must go back and ask how it was that he dawned at all.

§ 2. The Remote Ancestors of Man.

FROM what mammalian stock have man and the rest of the Primates ascended? What were the steps by which he mounted, and what the evolutionary forces which impelled him onwards?

The bodily structure of these creatures at once bars out a number of mammal groups from the list of possible ancestors. Such obvious impossibilities as whales or bats can at once be left on one side. Then the full complement of nailed fingers and toes rules out the Ungulates, the teeth rule out the Carnivores and Rodents. And so it goes on, until we are left with the sole group of Insectivores, that most primitive and unspecialized order of the Placentals. Just as the ancestor of man must clearly have been a creature to be classified in the same group with tailless great apes, and the ancestor of the apes as clearly must have belonged with the tailed Old World monkeys, so these in their turn, on every evidence of comparative anatomy and habit, must be descended from a creature of lemur-like construction. On this all zoologists are in agreement; and they are further in agreement that the lemurs are closely allied with the Insectivores.

The Insectivores are little creatures, runners on all-fours, with, on the whole, remarkably small brains. The outstanding characters of the higher Primates are their large brains and their fore-limbs converted into grasping, manipulating hands. What can have brought about the change?

Without doubt, there were many contributory factors, but equally without doubt, one was predominant—life in trees. As Wood-Jones has pointed out in his *Arboreal Man*, an active life in trees calls for a number of qualities not needed by earth-dwellers. It demands a much greater variety of movement in the limbs; no longer will mere fore-and-aft swinging suffice as it suffices for even the speediest horse or antelope. It demands a prehensile hand and foot. Hoofs or talons would interfere; a reduction in the number of fingers and toes is not called for. It demands a quick capacity for judging distance; and this means reliance pre-eminently on the sense of sight, instead of on the sense of smell, to which most terrestrial mammals pin their faith.

A first step in this direction, or rather in these joint directions, has been made by a few Insectivores, such as Tupaia and other tree-shrews, which took to arboreal life. Tree-shrews are more agile and have a distinctly larger brain than ordinary Insectivores, and in the brain the area concerned with sight is markedly enlarged, while that concerned with smell is somewhat reduced. Indeed, so lemur-like are they in some respects that many zoologists consider them as barely modified lineal descendants of the forms from which the lemurs sprang.

Among the lemur group there is one little creature called Tarsius, which is probably nearer the main human stock than the rest. The ordinary lemurs still rely mainly upon smell, and are therefore dog-faced; but Tarsius relies mainly upon sight, its nose has dwindled, making it monkey-faced. Lemurs generally push forward with their muzzles to take food; it uses its hands. It has also an extraordinary power of turning its head to look; and, a very important fact, it had acquired the power of stereoscopic vision, using its brains to blend the two flat pictures given by the eyes to a single solid one. In addition, together with the higher monkeys, or the great apes and man, it forms the only branch of the mammals to have achieved a yellow spot in its eyes, a patch in the retina permitting especially clear vision, a patch which the animal turns on to any object he wants to examine closely; and this concentrated sight is one of the bases of attention, that mind-concentration which underlies human success. In Tarsius the yellow spot is still diffuse, in process of formation; from the monkeys up it is fully developed.

The first steps once taken, the various acquisitions reinforce each other. The possession of prehensile, mobile hands leads to their being used for holding food, for grasping, for trying out things. The more chance there is of this manual examination, the more important it will be for the brain to be alert to every subtle difference between objects; and once the brain is improved, there will be advantage in new delicacy of handling. Meanwhile, the information given by the hand can be linked up with that given by the eye, to provide a solid knowledge, in terms of sight and touch combined, which must be quite unattainable by such mammals as have only hoofs or claws and rely mostly upon smell. And thus, one improvement opens the way for another; the different evolutionary trends are mutual stimulants.

So brain and hand and eye progressed and pushed the early Primate on from tree-shrew condition, through early lemur type with dog-like head, and Tarsioid monkey-faced lemur, up to a new state, a new condition of animal life, in the shape of monkey. Here the handling of objects and the brain's interest in them has progressed so far that the creatures are possessed of a new biological feature in the shape of a veritable overflow of curiosity and inquisitive manipulation; and they have retained and improved the stereoscopic and concentrated vision of Tarsius.

A side-branch of the monkey-stem colonized South America, there developing into the marmosets and the prehensile-tailed spider monkeys, woolly monkeys, and howlers. They are in many respects, notably in their teeth and their noses, less human than the Old World monkeys, and no new types seem to have originated from them. Quite recently a photograph has been published of what purports to be a further stage in their evolution—a New World monkey from Venezuela, which it seems has lost its tail and attained an unusually large size. There is no other tailless New World monkey. If this discovery is authenticated we have here a case of parallel evolution; just as Litopterna produced horse-like creatures and Marsupials evolved a mole-like type, so it seems the New World monkeys have taken a step parallel with that which converted the Old World monkey into an ape. But for the moment we have no actual proof of this animal's taillessness nor of its exceptional bigness; the skull was not preserved and the photograph shows only a front view. In any case, it is quite certain that man is descended from a true ape and this South American creature, this pseudo-ape, has no place in his pedigree.

Between the stages of monkey and of great ape there were important changes. Among tailed monkeys only those like the baboons, which have taken to ground life, have attained much of a size. But all the true apes, even the little light gibbons, are bigger than almost any arboreal monkeys; and, of course, they have lost their tails. This comes from their having adopted a new method of moving through the trees. Instead of running along the upper side of branches and leaping like a squirrel, using the long tail as a rudder and balancer, they swing themselves along by their arms. The professional anatomist will tell you that they *brachiate;* but all that brachiation means is swinging along by the arms. The tailed monkeys have a chest deeper than broad; but brachiation broadens and flattens the chest. It demands even more prehensile hands, and greater sensibility in them to the shape of things; it means that the eye-hand correlation must become the most important factor in the animal's life, for eye and hand must co-operate in every swing. It means that the tail can be dispensed with; and that, since the body is habitually vertical, it will still tend to be held upright when the animal comes down to the ground.

All these changes depend upon a specialization for arboreal life. It is perfectly true, not only that man would never have arrived at the familiar construction we know unless he had once been ape and monkey, but, much more radical, that so far as we can see, he never could have achieved the character by which he becomes human, the intelligence which alone marks him off from all other creatures, unless he had had this apprenticeship to tree-life. Nor, when the time came for him to take to ground-life again, could he have done so in the only way which would leave him human—upright, with hands released from walking, and reserved for handling—unless he had passed through the stage of brachiating ape.

The fossil remains bearing on this period of evolution are not many. There were various lemur-like creatures in the Lower Eocene (V A 1). Of Old World monkeys there are scarcely any remains. But there is a Lower Oligocene creature from Upper Egypt, Propliopithecus, which appears to be a very primitive type of tailless ape. By the end of the Miocene (V C), a divergence is manifest. Pliopithecus foreruns the modern gibbons; and Dryopithecus the rest of the group of apes.

Dryopithecus, though known entirely from jaws, teeth and bits of skull, is a fairly abundant fossil. It ranged over a great part of the Old World, and had diverged into species of very various size, from that of a small gibbon up to that of a man. It must have closely resembled our ancestral stock; its teeth and jaw are just what we should expect from the ancestor of man, gorilla, chimpanzee, and orangutang.

Of the existing great apes, there are no fossil remains, the only other ape fossil known is Australopithecus, of which the skull has been found at Taungs, in Bechuanaland. This belongs to an immature ape, but an ape that seems in various ways less simian and more human, especially in the teeth and chin, than any of its living relatives.

Probably, we may conclude, the transformation from Insectivore to Lemur-like animal came in the late Cretaceous (IV C 3), of Lemur to Monkey in the early Eocene (V A 1), of Monkey to ancestral Ape in the warm forests of the late Eocene (V A 2 and V A 3). The specialization of the apes went on during the Oligocene (V B) and Miocene (V C); and the divergence of man-like from ape-like branches cannot well have taken place before the late Miocene (V C 3) or after the middle Pliocene (V D 2).

But though man could not have become man, intelligent and erect, without passing through monkey-like and ape-like stages, the continued existence of apes to-day is proof that to pass through these stages does not necessitate the further evolution into man. What was it that drove our pre-human ancestor down from the trees, shook him out of his arboreal habits? Bulk may have had something to do with it. The gorilla is the biggest of the apes (old males may weigh over 400 pounds); and gorillas are to a considerable extent ground animals. But they still live in forests, and all but the old males who are too heavy take to the trees at little provocation.

From what we know of other evolutionary changes, of how the big steps, the new achievements, seem regularly to be taken by life under the stress of necessity, of suddenly changed conditions, rather than inevitably and easily—from what we know of other creatures, we may suppose that the radical and vital change from tree-life to earth-life was also taken under some compelling stress. One of the most plausible suggestions which has been put forward is this. We know that from the late Miocene (V C 3) on, the climate of the Cenozoic grew drier and cooler until the Ice Age. Forests diminished. In most places the forest inhabitants could retreat with the forests; but in the regions of Central Asia, north of the Himalayas (which were then still being elevated), the high plateaus and mountains would bar the way south, and the forests would disappear, leaving their inhabitants, deprived of their old home, on the open ground. As this happened, there would be intensive selection in favour of any creature that could change its way of life; and so, it is suggested, arboreal forest-living ape was transformed into man, habitually on the ground and in the open. The apes that elsewhere retreated southwards with the trees being under no such stress of necessity, remained apes.

This is why several eminent authorities suggest Central Asia as the region of man's origin; and, as we have said, the suggestion is reasonable and plausible. It is, however, quite possible that the ancestor of man may have been from early times a more terrestrial creature than most of his relatives, adapted to life in open glades and the forest margins, and that he simply pursued this line of specialization until it led to a creature terrestrial and big-brained enough to be called man. After all, the gorilla spends a good deal of time on the ground; and among the tailed monkeys, the baboons and their relatives have left the trees (and, be it noted, have become the most formidable and the most intelligent of the monkeys). If this was so, the evolution of man could have taken place whether or no his forest retreat was cut off; and we can place his origin in Africa, if we like, as many have done, or in south-eastern Asia. The truth is that we do not know, and only new facts will tell us.

We may here take the opportunity of mentioning one strange idea which has been put forward, that man has a multiple descent, the ancestors of the white races being supposed to be chimpanzee-like, those of the yellow races orang-like, and those of the black races gorilla-like. But this encounters so many difficulties, and runs counter to so many evolutionary facts and principles, that we need not trouble to discuss it seriously.

In any case, it would seem that the apes have continued to pursue their line of brachiating specialization since the time of man's origin, so that they have accentuated their differences from man, notably in enlarging their arms and lengthening their fingers to hook over branches. We must not forget that they have evolved as well as we. They are better apes than they once were; we are worse.

Whatever view we hold, it seems certain that our pre-human ancestor was hairy all over—probably black-haired, though red is a possibility—had longer arms and fingers than now, was shorter, with legs more bent, feet turned inwards to grasp the branch on which the creature might be sitting, a head still set well forward on the neck instead of pillared erect upon it, and a brain of ape-standard.

Increase of brain was doubtless the prime agency of further evolution; but this was permitted and encouraged by other factors. Ground-life, by releasing the hands from the duties of branch-to-branch locomotion, freed them for manipulation, for the carrying of sticks and stones as weapons, for tool-making. On the ground, with climbing powers reduced, vision, alertness and quickness of thought and action were at a premium. Ground-life encouraged a more omnivorous habit; and once the early man-creature ceased to be almost wholly a fruit-eater and took to hunting, that too encouraged skill and observation. Both hunting and the increased dangers of the ground would foster the need for communication and co-operation, and so would encourage the birth of speech and idea. Meanwhile, since the use of artificial weapons diminished the need for natural weapons, the teeth, and especially the great ape eye-teeth, dwindled. As consequence of this, upper- and lower-jaws would no longer interlock as in the apes, and a freer movement of the jaw was possible. Not only the teeth, but the muscles for biting could be reduced, and so the muzzle shrank into a face. This release of the jaws and neighbouring regions from the savage activities of offence and defence meant that they could be used for more delicate activities, and as the violent biting spasms of the jaws were less needed, the delicate rippling contractions of the speech-muscles could be developed.

All these changes again reinforce each other and stimulate the main central change, the growth of brain. This was connected also with a further notable step forward—the prolongation of the helpless stage of man. The main lengthening out of development had taken place earlier, between tailed monkey and tailless ape. The common macaque monkey is sexually mature at four or five years old; the female carries her unborn young for about five and a half months, the baby cuts its first set of teeth by three or four months of age (instead of about eighteen months, as in a human infant), and begins shedding them again at about a year. The rather meagre data we have about the great apes, on the other hand, show that they are very like man in the tempo of their development. Even the comparatively small chimpanzee has a gestation period of at least seven months, and quite possibly more, and weans its infant at about nineteen months; the baby begins to replace its milk grinders at five or six—almost exactly the same age as the average child. Puberty comes on, it seems, at ten or eleven, which is little less than in many tropical races. As to the age to which the creatures may live, we have very few facts. Monkeys probably live for several decades; and there is every reason to suppose that chimpanzees live almost as long as men and women. As regards gorillas, one naturalist is of the opinion that they may live to be a hundred.

This slowing down of development permits a greater brain-growth and a lengthening of the period of educability. But in man a further step has been taken. The infant is more undeveloped, its helpless period is longer. This would make it harder for the mother to run with the rest of the group, which in its turn would encourage temporary settlements and make for a division of labour between males that hunt and fight and females that keep the home. And it strengthens the tie between mother and child; it makes educability longer-lasting and education more important.

The way in which this final change seems to have occurred is interesting. Apparently it has been brought about by making various characters present in the fœtus of apes last on into independent life or even to the adult stage of man. The most notable of these is human hairlessness, which, as we have pointed out in Book 3, prolongs the transitory stage, naked save for the crown of the head, which the ape-fœtus passes through after shedding its first crop of downy hair. Bolk, the Dutch anatomist, has itemized a whole list of similar characters, and is so impressed by them that he speaks of the evolution of man as a "fœtalization" of the ape.

Such a process, which is in a sense the reverse of recapitulation, is not uncommon in animals. In Book 4 we have met with a striking example in the axolotl, in which the tadpole characters are thrown forward, so to speak, and have come to cover the whole of life (Book 4, Chap. 5, § 4). In human evolution it will obviously not explain the increased size of special regions of the brain, but it does seem to have played a very real part. If we may hazard a speculative guess, it looks as if the general slowing down of the process of early development was advantageous as a method of delaying the closure of the skull sutures and so providing for increased brain-growth after birth; and that the later "fœtalization" of humanity was then evolved because it lengthened the educable period and made the most of the brains. Whatever the operating causes they have had important secondary effects. One of these, hairlessness, has had consequences, both practical and æsthetic, of the most far-reaching nature. So man acquired his unique character of being the animal which never grows up. Not only infancy, but the whole period of dependence is lengthened; the period of learning takes a far bigger fraction out of human life, even the most primitive human life, than out of any other. Childish curiosity, desire for play, and delight in experiment is prolonged, and comes to persist, like hairlessness, throughout life. What the axolotl is physically, man is mentally—a permanent youthful stage; and to this he owes his humanity.

We have now given a description of the different stages through which life has passed on its road to become man. It is perhaps of interest to point out that not merely has man passed through this particular set of changes, but that, so far as we can see, an organism with the essential human characteristics of speech and tradition, which have enabled man to dominate the world, could not have come into being in any other way. It was necessary for man's ancestral stock to acquire size: so union of cells to form a many-celled organism was essential, and all single-celled animals are ruled out. The mouth and nerves of Cœlenterates and the crawling of Flatworms were essential steps towards a head and bilateral symmetry; blood and body-cavity had to be evolved to give further size and efficiency; segmentation, internal skeleton, and swimming tail were required for speed and activity in the water; jaws and teeth for efficient feeding. On land, however, insects and other arthropods developed along quite another line, were extremely efficient: but we have seen how their external skeleton and especially their method of breathing prevented them ever growing big, and tied them to a mainly instinctive type of behaviour. Thus the vertebrate path remains, it appears, as the only one which could lead to an intelligent reasoning creature. The Sponges, the Echinoderms, the Molluscs, and all the horde of Arthropods were, from this point of view, all essentially doomed to come to a barrier to progress at one level or another. Within the Vertebrates, land life was necessary for further advance; the more specialized fish are a blind alley. The evolution of leg, lung and shelled egg were the next essentials. Then came warm blood: without this, the equability of internal process needed as a basis for true thinking could hardly have arisen. This cuts off Reptiles. The Birds, though warm-blooded, set themselves a barrier to unlimited advance in turning their fore-limb into a specialized flying organ: so we are left with mammals. Among mammals, only arboreal life could have led to the subordination of smell to sight, the development of binocular vision, the formation of a true hand, the tendency to intelligent manipulation. Thus all other mammalian branches save the tree-shrew-monkey-ape line were excluded. Further, only the subsequent return to the ground could have liberated the hand from over-specialization and imposed the necessity of winning through by intelligence, and only a gregarious species could have acquired speech. Thus only a mammal, gregarious and of ground habits, with a long arboreal phase in its ancestry, could have evolved to the level of speech and conscious reason. All other groups and lines were excluded.

What is more, there would seem to be no chance, even should the human species be wiped off the face of the globe, for any other type to take the opportunity of evolving to the same level as that reached by man. All the other groups have been specializing on their own lines; even the more primitive modern types are specialized in various details. And by specialization, they have in all probability cut themselves off from the path of advance which was still possible to their more generalized ancestors.

Thus man, with his unique intelligence, could not have evolved by any other route; and in himself he holds, it would seem, all the possibilities yet remaining of advance to higher levels. He has the privilege and responsibility of sole trusteeship for life's further progress.

When we come to the actual fossil remains of man, it is interesting to see evolution making experiments, so to speak, in different directions. We have already spoken of some of the early types of human and half-human beings (Fig. 154). Of these, Piltdown man was still extremely ape-like in chin, jaw, and teeth, especially dog-teeth; but in brain and brain-case he was definitely on the human line. Sinanthropus, similar in face and teeth, had a smaller brain. Heidelberg man had an amazingly massive jaw, but his eye-teeth were already quite human. Neanderthal man and Rhodesian man, in spite of having a far larger brain and more human teeth than Piltdown man, possessed huge brow-ridges which he largely lacked. Neanderthal man was still very bent in the leg, while the half-way ape-man Pithecanthropus seems to have stood almost as straight upright as we do.

We are reminded of the horse stock, where some species evolved by specialization in teeth, others by specialization in toes, others in simultaneous advance in both. Teeth, brain, chin, straight leg, muzzle and brow-ridges—these were so many improvements to make in our simian ancestor; and in the few fossils of early man that we possess, we see fragmentary samples of the many evolutionary attempts made upon the problem, attempts that did not result in the successful type of modern man until a mere twenty thousand years or so back.

But these were all men, tool-making men, and for the history of their improvement we must go back and study the stone implements that they have left; and to do this satisfactorily it will be advisable to tell first of the profound changes of climatic conditions which preceded and accompanied human development.

§ 3. A Short History of Pleistocene and Recent Climate.

THE climatic history of the world since the Pliocene (V D) has been worked out in very considerable detail with much patience and inspiration by a number of brilliant and devoted workers. The preceding chapters have told of the climatic decline that went on from the Miocene (V C). The world became austere. Towards the end of the Pliocene (V D) the circumpolar ice extended itself towards the equator and the mountains became the nests of vast glacier systems. Glaciers

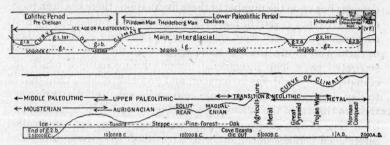

FIG. 238. Time diagram of the Ice Age (above) and the time since its end (below).

The wavy line represents temperature; when it runs below the horizontal line, ice-sheets were present in Europe and North America. The periods (Eolithic, etc.), the sub-periods (Chellean, etc.), and the types of men are given; in the lower diagram, the changes in flora and fauna are indicated. All events are dated with reference to Western Europe, except the discovery of agriculture and metal and the historical dates. In the lower figure the dating of Peake and Fleure ("Hunters and Artists") is followed; other authorities put the end of the Upper Paleolithic farther back. UP, Upper Paleolithic; C., civilization.

as they advance or shrink, pile up and leave fringes of mud and stones along their courses and at their ends, the moraines, and it is mainly upon the evidence of the moraines which the great ice-sheet built up at their margins and the boulder-clay and drift which their melting spread over the land that our history of past climate during the Pleistocene (V E) is based. It is based also on the evidences of change of climate revealed in the plants preserved in peat bogs, the shells of seas and lakes, and on the alterations of level between land and sea, which are traceable in raised beaches, and so forth. The annual rings of the stems of very old trees have also been helpful in indicating, by their openness or density, the nature of the seasons of long-past centuries. As this resurrected calendar approaches the present day it becomes more and more certain and exact. We must warn our readers that there is not full agreement among all workers on the conclusions here set forth; and that the actual dates we give, especially the earlier ones, may have to be revised in the light of fuller knowledge. But the picture has become a much more concrete and vivid one in the last few years, and its broad lines seem to have been drawn once and for all.

The Great Ice Age appears not to have been a single great wave of cold as the earlier geologists imagined, but to have struck the world in a series of cold spells with warmer times between. This was first suggested by James Geikie, and has been generally agreed upon since the epoch-making work of Penck and Bruckner, who explored the Alps from end to end, searching out the relations of every moraine and ice-deposit; but upon the number of these fluctuations there has been much dispute. Recent research, however, is gradually clearing up apparent contradictions. There seem to have been two main glaciations, both of them world-wide, the first more severe than the second. They were separated by a long interglacial period, during most of which the climate of the world was no harsher, and probably milder, than that of to-day. But each of the two main cold spells was itself discontinuous; the two thrusts of the advancing ice being separated by a retreat, and by a time of less severe climate. These retreats too are often called interglacial periods, but in order to stress their lesser importance and duration and to make our story clearer, we shall call them here *intermissions*, as opposed to the long main interglacial period. For this alone we will use the word *interglacial*.

Thus, we have two major glaciations, each consisting of two periods of severe glaciation separated by a milder period, and these two great glaciations are separated by the great interglacial. The whole Pleistocene Period (V E) seems to have lasted rather more than half a million years—a trifle, geologically speaking. Following a practice we hope our readers have found useful we propose to distinguish these main periods by letters: g 1 and g 2 for the first and second glaciation and i-g for the interglacial period.* Or if the reader wants the full style and title of the interglacial period he can mark it (V E i-g). The dates for the first half of the first glaciation are rather conjectural; but the others are based on solid calculations and on actual measurements of time.

We can say that the Ice Age ended either at the moment when the last great ice-sheet began to shrink, or when its retreat had gone so far that no ice was left on any European lowlands, and the climate had become no worse than that of to-day. For both these moments we can assign a date, based on De Geer's method of counting the annual layers of clay deposited by the summer meltings of the retreating ice. For the final amelioration of climate the date is close to 5,000 B. C., for the beginning of the retreat, about 20,000 B. C. As we shall

---

*The names given by Penck and Bruckner to the four glaciations as revealed in the Alps are Gunz, Mindel, Riss, and Würm; each after an Alpine valley where its moraines were well shown. The names were deliberately chosen to be in alphabetical order, with room for others between if the number needed adding to—a too-rare example of forethought in scientific nomenclature. Gunz and Mindel are our g 1 separated by its intermission. Riss and Würm together constitute g 2.

see later, the retreat was not by any means uniform. It is as if the ice fought rearguard actions, paused and even re-advanced; but it never recovered all its lost ground after 20,000 B. C.

The later of our two main glaciations (V E g 2) which thus ended close to historic times, lasted from about 130,000 B. C. It was divided into an earlier Ice Age of at least thirty thousand years, an intermission of about twice as long, and a second Ice Age, less severe than the first, of about twenty thousand years. Over the north temperate regions of the world, the climate of the intermission may have occasionally been warm, but for the most part it seems to have been cool and arid, like that of the Asian steppes to-day. The vegetation was sparse, and there was much deposition of wind-borne dust to form the thick covering of fertile material known to geologists as loess. After the final cold phase came the phase of alternation in which we are living.

Before this second glaciation (g 2) came the main interglacial (i-g); and this endured for nearly a quarter of a million years. Thus, be it noted, about half the so-called Great Ice Age was passed in a warm climate. During this period the ice must have been confined to the tops of the mountains and to the high arctic. The bulk of Europe and America supported a rich vegetation, with an abundance of animals, and animals of a temperate and even semi-tropical type—the southern elephant, a form of hippopotamus, Merck's rhinoceros, the sabre-toothed tiger, hyenas, lions, deer. The mammoths, meanwhile, found the main continent of Europe too warm, and browsed among the Scandinavian forests of birch and pine.

But before this interglacial period the first main glaciation (g 1) had had the world in its grip. It had been far more severe than the second. Its second culmination alone had lasted some sixty thousand years, longer that is than the two cold phases of the second glaciation added together. During this earlier time, the European ice-sheet had reached east to the Urals, covered all Ireland and Scotland, all England as far south as the Thames, and left only a corridor of about two hundred miles wide between its ice and the ice descending northwards from the local ice-sheet of the Alps.

In America the second phase of this first glaciation (g 1 b) seems to have been the worst, and the huge ice-sheets covered not only all Greenland and all Canada, but pushed south to St. Louis and Louisville and New York. They seem to have been over a mile thick in regions such as New England. Interestingly enough, most of Alaska seems to have escaped; there were local glaciers in its mountains, but the great bulk of it was ice-free. One of the most notable products of an ice age are the lakes that it leaves behind it, though these are destined to silt up and disappear in a comparatively short space of geological time. The Swiss lakes, the lakes of North Italy, the English lakes, and the Scottish lochs, are all presents from the Ice Age. But the most striking glacial lakes are the Great Lakes of America. The complicated history which they have undergone, joining up and separating, enlarging and shrinking, draining northwards and draining southwards, can be read in such books as Schuchert and Le Vane's *The Earth and Its Rhythms*. What is of interest to us here is the fact that these great bodies of water, so important for American commerce, owe their very existence to the ice and will diminish as the Ice Age recedes.

The ice has had other effects on our present-day economic life. Where it passed over hard rocks, it left behind it huge quantities of boulders; sometimes, as in parts of New England, the abundance of boulders makes agriculture all but impossible. When the ice ground its way over softer material, it generally left behind it a soft drift, often rich and tillable, and sometimes, as when it brought drift from a limestone region to one of sandstone, it markedly improved the soil. The yield of the driftless areas in Wisconsin, for instance, is only two-thirds of its drift-covered fields.

There are as yet no exact dates for the intermission of this first glaciation and for its earlier phase of severity. But together they cannot have fallen far short of one hundred thousand years, and this takes us back to 530,000 B.C.

Now, throughout this half million years and more, the map of the world was undergoing changes as remarkable and important as its fluctuations of climate. The geographical and climatic histories are indeed two aspects of a single story of planetary change. In the first great glaciation (g 1) it is to be noted that the shape of the ice-area was very different from that of the second (g 2). For example, although the main ice-sheet over Scandinavia spread as far east and south as it did in the last phase of all, yet Britain was not touched by it, and lacked even a local ice-cap. And though the Alps had their full portion the Pyrenees were very poorly glaciated. The explanation of this is to be sought in a change of level of the lands. It was the great emergence of the continents which helped to bring on the Ice Ages. Now, in the first glaciation (g 1) this probably went so far as to bring up into the light of day a mighty ridge, a sort of submarine hog's-back, known as the Wyville-Thomson ridge, which extends right across the North Atlantic from Scotland and Norway through Iceland to Greenland. If so, then the waters of the Arctic Ocean, chilly with ice, could no longer flow south to cool the Atlantic. In all probability it was the comparative warmth of the Atlantic which kept Spain and Britain unglaciated during this first age of severity, just as the Gulf Stream waters keep Norway habitable to-day.

In any case, the boundaries of land and sea were wholly different during most of the Pleistocene (V E) from what they are to-day. We hear, for instance, of man reaching Britain during the later Pleistocene, and we think heedlessly of an invasion of canoes crossing the English Channel. But the English Channel, as we know it, did not exist until 5,000 B. C. or even later, and Stone Age man had no need of boats to reach Britain; he travelled on his feet, as did the four-footed beasts that now live there with him.

In general, the glaciations seem to have been always times of elevation, the interglacial and the intermissions times of submergence. But throughout the whole of the Pleistocene Period England was connected with the Continent, not only over what is now the Channel, but also across the flat plains that are now the bottom of the North Sea. Across those plains a great river ran northward, the united Thames and Rhine, whose submarine bed can still be traced.

The straits of Gibraltar seem to have been closed at each time of elevation (g 1 and g 2). In addition, a second passage of dry land extended from Europe to Africa by way of Italy and Sicily, and Asia was cemented to Europe by a broad land-connexion across from Asia Minor to the Balkans. The Mediterranean was thus represented only by two inland lakes, the Black Sea by a third. And once the great ice-sheets began to form they locked up water in themselves, and lowered the seas still farther. The change of water-line thus produced was considerable; it probably amounted to about 400 feet at the maximum glaciation. But once the ice-sheets were formed, their weight began to press the crust downwards locally, and when their melting took the extra weight off, the crust began to warp slowly up again. There was thus an oscillation. Elevation produced ice, ice produced depression. These oscillations still continue. To-day the world is still recovering from the second great glaciation (g 2), and as it does so, countries like Northern Norway and Spitsbergen rise steadily out of the sea.

What with these major movements of elevation and depression, and the local warpings of the crust as the vast weight of superincumbent ice was put on or taken off, the geography of the Pleistocene is a kaleidoscopic thing. We find it difficult to realize these movements, so stable do the lands seem to our brief lives and observations, so hypnotic are the outlines of the countries in our atlases and so ineradicable the effect of that foolish early teaching of geography that begins with countries and "capitals." But let us not forget how extremely small the movements really are in comparison with the globe on whose surface they take place. To bring even the Wyville-Thomson ridge above water and make a northern connexion between Europe and America, needs a movement of five hundred fathoms—far beyond the reach of any diver. But this is less than 1-60,000th of the earth's radius: we could not touch an orange ever so gently without making a bigger deformation on its surface, for it would be a deformation of only about 1-500th part of an inch.

We are living now within a few thousand years of the second glaciation; the world's climate is still oscillating—one hopes towards mildness, though perhaps we are only living in the fool's paradise of an intermission or an interglacial—and modern research reveals the traces of ages in which humidity has increased and waned, and ages of comparative dryness and cold. In the early stages of the last shrinkage of the ice, towards the end of the Pleistocene that is, the temperate zone of the world lay across North Africa, and the Sahara was green and fertile. Northward treeless plains reached towards the ice. Southward was hotter and drier country. *Homo sapiens* was already in existence. The early true men of whom we shall presently tell drifted with the amelioration of world conditions northward. The Sahara began to dry. It became gradually the desert we know to-day; but it still preserves in some of its oases, though they are now entirely without connexion with any big rivers or lakes, dwarf crocodiles and occasional catfish as tokens of its moister past. Meanwhile, the green zone with its grass, its forests, its game and ample food crept across the great Mediterranean valley and extended from Spain, through Mesopotamia to Turkestan. The belt of steppes, the warmer belt of green, the heat-belt all quivered their way northward. We write "quivered" for there were constant setbacks. No two years were alike then any more than they are now. There would be periods when there was a phase of elevation, when the ice increased again and all the zones would be pushed back south. Somewhere in the history, but when we do not know, the barriers at Gibraltar were broken through and the Mediterranean became one sea. For a century or so about 5000 to 4000 B. C., there seems to have been a strong reaction, much snow upon the head waters of Mesopotamian rivers and great floods—perhaps the original of Noah's legendary Deluge.

By that time civilization was dawning. This phase of elevation, it is supposed, made the Nile valley possible for civilization. Before that time, it seems, it had been marshy and uninhabitable. Now it was drained and lay irrigated by a great river and open to the sun. And so in Mesopotamia and down the fertile strip of Egypt appeared a new and strange biological phenomenon, man the cultivator. But of him we must tell later. Unsuspected by him until quite recently and yet moulding the broad outline of his fortunes, the climatic fluctuations have gone on. The fourth and third millennia B. C. were ages of warmth and increase. Man multiplied in the fertile valleys and also in the belts of forest and steppe to the north. The second and first millennia were in comparison periods of elevation and cooler, drier weather. The zones of steppe, of green fertility, of heat and dryness moved towards the equator again and there was less rain upon the world. The men of the forests and steppes turned their faces southward with that southward shifting of the zones. In *The Outline of History* the reader will find the story of a great series of invasions and conquests of the ancient civilizations, which this southward swing of the climatic zones produced. But we will not run on into written history here.

We must first say a word on the effects which these violent changes of climate and level produced in the world of animals and plants, and then turn back to the story of man.

No life can exist in the centre of a great ice-cap, and thus, during the times of glaciation, great tracts of the world that in other ages were not merely habitable but comfortable, were swept wholly bare of living things. Later, as the ice melted away, enormous territories were thrown open again. But the conditions in these reclaimed areas are still not what they were. Both in Jurassic (IV B) and early Cenozoic (V) times, Spitsbergen was richly wooded; during the great glaciations there was no soil exposed over the whole archipelago—nothing but ice and snow; to-day the tallest "trees" that can grow there are the various creeping willows, that never rise more than four or occasionally six inches into the air. The climatic zones of the world are much more sharply marked to-day than in equable periods like Eocene (V A) or Jurassic (IV B) or Carboniferous (III E), when sub-tropical or warm temperate belts often crept up to within the arctic circle, and the cold temperate and sub-arctic zones, with their pleasant summers but harsh winters were reduced to insignificant rings round the poles.

The new territories exposed by the melting ice, therefore, though admirably adapted to support life in summer, were still hostile and barren in winter. And animals that could migrate to and fro with the seasons would have great advantages. Such seasonal migrations are undertaken by some of the larger northern mammals. The migrations of the caribou are the best known—mighty herds of animals moving north with the spring and south with the autumn, accompanied by wolves that prey on stragglers over the great spaces of northern Canada. But the method has been most thoroughly exploited by the mobile birds. In the tropics there is no irruption of birds that come to breed. In a north temperate region such as Britain, about a third of the breeding species are winter absentees, coming only for the summer; and among many of the other two-thirds, the so-called residents, there is really a great deal of migration, the individual birds that are with us in winter moving on northwards to breed, and having their places filled in spring by immigrants which have wintered farther south. And in arctic Spitsbergen, among about thirty breeding species of birds, all but the Spitsbergen ptarmigan and possibly its enemy the snowy owl, leave the country in winter. In every case, migrating birds breed in the highest latitude which they visit during the year, showing that competition for breeding-places is at the root of the migratory habit.

A wonderful description of the sudden change from the poverty of winter to rich and abundant summer is given by Seebohm in his *Birds of Siberia*. The tundra plains are snow-bound for over eight months out of the twelve. A couple of fine weeks melt the blanket of snow, and the low bushes appear still bearing the fruits they ripened the autumn before, conserved through the winter in cold storage. Flies and mosquitoes hatch out in such millions that they may obscure the ornithologist's view as he attempts to watch some rare bird through his binoculars. The birds arrive in swarms, feeding on the berries and the insects; they begin to nest as soon as the snow has left the ground; but in three months they and their young are on their way south.

It is almost universally found that the farther north a bird breeds, the farther south it winters; on its migration, it skips over the less mobile species of milder latitudes. This holds good even for the varieties of one and the same species. The sub-species of fox sparrow that breed in Alaska winter in California, while those that breed in Vancouver only move down to Washington and Oregon. This again is what one would expect. When the retreating ice leaves a new breeding-ground vacant, the birds that take possession of it in summer must support themselves in winter, too. The colonists of less northerly regions will have already established their winter quarters as near as possible to their summer home, and fresh competitors would put too great a strain on the food supply. So the new invaders of more northerly breeding-grounds push on in autumn towards the inexhaustible resources of the tropics, and are repaid for a more arduous journey by more abundant winter food.

Some kinds of arctic-breeding birds seem to have got so tied to cool conditions that they shun the warmer regions of the world altogether; and when the autumn comes they fly south, skipping all the tropics, to another spring and summer in the southern hemisphere. The American golden plover, for instance, breeds along the northern coasts of arctic America. In autumn the birds of this species move southeast, make a great ocean flight from Nova Scotia to South America, and then travel on to winter quarters in the Argentine and Patagonia. In the spring, curiously enough, they return by Central America.

No doubt there has been some breeding migration ever since birds were birds; but equally without doubt it was the advance and the retreat of the ice that have made migration an important part of bird life. For in an equable period migration need only be of small extent; and without the great equatorward thrust of the ice-sheets and their subsequent retreat, the skipping-over of the northernmost breeders to winter quarters nearer the equator than those of migrants breeding farther south would never have arisen.

The coming and the passing of the Ice Age had also very marked effects on the general population, both of animals and of plants, in temperate and arctic latitudes. For instance, in the temperate regions of the Old World to-day, the flowers and trees are less like those of pre-glacial times than is the case in America. In America, the vegetable army could simply creep south, and then creep north again when the ice allowed. But in the Old World there are ranges of mountains running east and west—Alps, Caucasus, Atlas. The plants were caught between two jaws of ice, and many were exterminated, so that after the Ice Age a more radical recolonization was necessary. In the most diverse regions, however, there was a widespread extinction of types, chiefly of large animals; but of this we have already spoken.

The changes of level, too, have played their part. We will give one example of the complicated business of post-glacial colonization—how that part of the European continent which has now been cut off as the British Isles was re-stocked after the Ice Age. The first main element in their population comes from the arctic and sub-arctic animals and plants, which in the Ice Age lived in what was then the northernmost inhabitable belt, from South England and Northern France along through Central Europe. When the ice retreated these migrated northwards. Some have died out, others survive in the mountains of Wales and Scotland. On the tops of a few Scottish mountains, such as Ben Lawers, little rock-gardens of glacial-relict plants are found, which are otherwise confined to the arctic or to the high Alps. The red grouse and the ptarmigan are examples from birds, the arctic hare from mammals. The arctic hare is also found on the Alps, the Pyrenees, and the Caucasus—islands of cold to which it could retreat, like the plants on the tops of the Scottish mountains, as warmth invaded Europe.

The most southerly region, apart from the high mountains, where the arctic hare exists is Ireland. Apparently it was unable to leave Ireland for the north because, as Scharff puts it in his interesting book on distribution, *European Animals*, "Ireland had become an island before the arctic hare had become aware of the fact." If the common hare had got into Ireland, its competition might have made life hard for its arctic relative; but it was prevented from entering in the same way as the arctic hare was prevented from leaving.

Then there was an invasion along the westernmost Atlantic coast, an invasion of forms that love the moist and equable oceanic climate. This invasion connects Portugal and Spain with Ireland and Cornwall and occasional patches of the west coast of Wales and Scotland. These plants and animals must have crept up from the Portuguese coasts, across the shore of the western bay that then joined England and Brittany, and round to the west of Ireland. The strawberry tree, so familiar in Mediterranean countries, is one of the plants that has reached Ireland in this way; the Mediterranean heather and the pretty saxifrage called London Pride are two others. The Lizard Point in Cornwall is covered with another kind of heather that is found nowhere else in Britain, but abundantly on parts of the west coast of France and in the Mediterranean, and there are plenty of other examples. Among animals there is a claret-coloured woodlouse and a spotted slug, both found in the Spanish Peninsula, and in Ireland, but not in Britain or Northern France.

Meanwhile, however, the greatest invasion was coming from the southeast and east. The mole, the roe- and the red-deer, the common sand-lizard, and the stag-beetle are examples; indeed the bulk of Britain's animals and plants, like the bulk of its human population, arrived from the east.

But the spread of a species is a slow process; and while this secular migration was going on, the land was sinking. Ireland was first cut off from Britain, and then Britain from the Continent. Many of the eastern army that had got to England found their way to Ireland barred because they had been too slow. Most people are aware that there are no snakes in Ireland. This, however, is not St. Patrick's doing, but due to the too early appearance of the Irish channel. Many other creatures besides snakes failed to reach Ireland. The common hare is not found there, nor the mole, nor the roe-deer; nor were there ever any Irish beavers. There are great gaps, too, among the flowering plants, Ireland possessing only about two-thirds as many kinds as Britain.

But just the same thing happened again when the English Channel was formed; the slower-spreading among the land-animals from the

MAN DAWNS UPON THE WORLD

east were held up at the frontier. Scandinavia was not cut off from the rest of Europe; and so, in spite of its inhospitable climate, it has almost sixty species of land-mammals, while Britain has only about forty, and Ireland about twenty-five. Belgium, just across the Channel, has been colonized since the Ice Age by twenty-two species of reptiles and amphibians; only thirteen of these have reached Britain, and only four have got to Ireland.

We have said enough to show what interesting problems in the migration and distribution of animals and plants the Ice Age has left us to unravel. Now we must take up our main story again, and go back to the first glaciation and the intimations of the dawn of mankind that are scattered throughout the deposits of the Pleistocene Period (V E).

§ 4. Traces of Man before and in the Pleistocene Period.

WE HAVE already given some account in Book 3 of the earlier types of the Hominidæ of which modern man, *Homo sapiens*, is the sole surviving species. Here we propose to add a little more colour and substance to that first account. As we have explained in the first section of this chapter, our material is at first almost entirely implements of stone.

The earliest are of so rude a construction that for a long time it was doubted whether they were artificial. It is now generally admitted that certain of them have been chipped purposely. They are called Eoliths. They are found in the earliest Pleistocene and even back in Pliocene deposits (V D). Over a period of hundreds of thousands of years, it seems, some creature was chipping stones to a rough but serviceable edge, probably in order to hack wood more conveniently. It is possible that the Piltdown man, of the two known individuals of which we have told in Book 3, was the maker, or one of the makers, of Eoliths. But we have no absolute proof of that.

Succeeding the Eoliths in the geological record are a series of big and clumsy implements known (from the spot in France where they were first abundantly discovered) as the *Chellean* type of implements. They pass by gradations into a still massive, but far better worked type, known as the *Acheulean* type. These implements were made by chipping bits off a flint which was already more or less of the desired shape. The bits were thrown away. These shaped flints are often called hand-axes and seem certainly to have been held, unhafted, in the hand; but it is more probable that they served, like the not dissimilar implements of the modern Tasmanians, for the peaceable purpose of grubbing up roots. We may call them "fist-flints." Quite manifestly they were not all or most of the equipment of their makers. But all their tools of wood have long since rotted away and only one bone implement of the time has survived. One very early wooden spear, from Clacton, has been ascribed to this period, but probably belongs to the next Age.

These tools are hardly ever found in caves (there are exceptions in the caves of Britain, such as Kent's Hole), but mostly in the gravels laid down by the rivers of that time. Their possessors must have lived in the open. In all probability they inhabited Europe in the periods of intermission during the first and second glaciations (g 1 and g 2) and during the entire interglacial period (i-g). We may imagine these submen hunting along the banks of the well-wooded rivers of those long ages of mitigation, camping under a windbreak of boughs at the edge of the wood, occasionally settling down for a spell (as we know from sites whose accumulated débris testifies to long occupation), stalking the deer that came down to drink, grubbing themselves roots in the glades.

Still there remain question marks against their physical form. We have told of the remains of Heidelberg man in Chapter 3. Was he, or was Eoanthropus, the fabricator of the Chellean and Acheulean fist-flints? We would give a great deal to know what were the relations of the Piltdown creature, with its huge dog-teeth but small forehead, with the Heidelberg men, definitely human in their teeth, but amazingly massive-jawed, and probably slow-spoken. The two species were certainly very different in construction, and yet the types of flints we find are remarkably uniform. Were the flints made by only one of these two kinds of submen, or did both live the same kind of life? Or was neither type responsible for these artifacts? We do not know; here is another secret that may yet be revealed to some fortunate worker. There may be a score of genera or species of Hominidæ still hidden unsuspected in the Pleistocene deposits.

However that may be, the fact remains that throughout the larger part of the Pleistocene Period, for anything between 300,000 and 400,000 years, the only human or quasi-human life of which we have evidence was of these Chellean and Acheulean implement-makers. Age after age they lived, content with the same unchanging pattern of implements and the same invariable life these implements indicate.

It is only when we come to the period of intermission of the second glaciation (g 2) that the first intimations of a new and much more human type of *Homo* appear. The Acheulean culture ends and a different one begins. It is not as if the Acheulean men or submen developed into the new race; it is as if a new kind of man replaced the Acheulean altogether. And now we find skeletons in close relation to the tools their owners made and we are no longer in any doubt of their authorship.

There is an abrupt break at this point. The Chellean types develop age by age by imperceptible degrees into the Acheulean. But the Acheulean types do not develop with any such gradualness into their successors. They end and something new begins. The new stone tools of the later Pleistocene (V E g 2 b) are of quite a different pattern altogether, and made in a different way. When you work a flint, you chip off flakes from a central core. All the fist-flints had been shaped from that core, and typically the flakes were waste; though in Acheulean times there was some utilization of flakes as well. But almost all the new tools were made from the flakes, and it was the cores that were rejected. The implements were accordingly flat on one side, more rounded on the other. They were never nearly as massive as the earlier tools, but they could be made much more quickly and in greater numbers. The particular type of implement of this age is very characteristic, and it is called *Mousterian*.

The men who made them had also discovered how to add deadliness to their spears by tipping them with flint points, and it was perhaps this which allowed their race to spread over the whole of Europe. For newcomers they undoubtedly were. Not only was their method of flint-working and flint-using new, but the men of the new culture were of a new and different species, *Homo neanderthalensis*. We know more about these Neanderthalers than about any other men of the Old Stone Age, and this for two reasons: they lived chiefly in caves, and they (often at least) buried their dead.

If we could revive one to-day we should see a short man, with big body carried a little forward on short and rather bent legs. We say "short" not "little," he was a decidedly massive creature. The head, with its powerful jaws and big face, hangs a little forward in the neck, so that both in attitude and in bodily proportions he is definitely more ape-like than any human being alive to-day. The face is not only big, but the jaws protrude so as almost to merit the term "muzzle," and the chin is still very poorly developed. The eyes look out from under a great pony pent-house, over which tumbles shaggy hair. The brain-pan is big, but big in the wrong place, for the forehead is low and receding. He takes a step towards us, and we notice that he puts most of the weight on the outer edge, instead of flat on the sole, in this, too, revealing his intermediacy between ape and modern man.

There is a good deal of evidence that these extremely unprepossessing savages had been established for some time in eastern Europe, from the Rhine eastwards, before they spread into the west. Of their original home we know nothing certain, but we may hazard a guess that it was in Asia. The fact that they lived in caves is enough to suggest that the climate of their time was cold, and this is confirmed by the kinds of animals found with their remains.

Each glacial thrust was preceded by a rise of the land and a worldwide change of climate. The Neanderthalers' appearance followed the intermission of the second phase of glaciation. In all probability it was the last return of severe conditions, the second push of the ice during the second main glaciation (g 2 b), which drove them from their original eastern homes. Some migrated westwards, to the milder seashore climate of France and Spain, ousting the men who made the Acheulean fist-flints. Whether these latter were killed in fight or, as is more probable, driven by competition into mountainous regions, there to die out as the last cold intensified, we do not know. We know as little of their fate as we do of their build and appearance. These earlier men remain tantalizingly hidden from the constructive imagination.

The ascendant Neanderthal stock of the late Pleistocene inhabited Syria and Palestine; a well-preserved Neanderthal skull comes from Galilee. The invaders must have crossed over one or other of the landbridges to Africa, for typical Mousterian flints have been found both in Algeria and in Egypt. Perhaps some of them spread southwards over the African continent. At any rate, the celebrated Broken Hill skull from Rhodesia, though it probably dates from much later, perhaps even after the end of the Ice Age, has certain Neanderthal features, such as the huge brow-ridges. The Neanderthal man was a hunter. We find the bones of the animals he ate, which include the bison, the mammoth, the reindeer, and the wild horse; we know he cooked over a fire; we know he used flint saws to cut up his joints of meat and flint knives to scrape the flesh off the bones. He lived in Europe for the tens of thousands of years of a long, cold period, but soon after its close and within some twenty thousand years from now he gave way to another type of the genus—*Homo sapiens*, modern man.

§ 5. The Advent of Modern Man.

AFTER the cessation of the Mousterian flints we find no more remains of Neanderthal man with his bent thighs and heavy jowl and great brow-ridges. All subsequent skeletons, however different from each other in detail, differ no more than do the existing races of man; they can all justifiably be called Modern Man, *Homo sapiens*.

This age of returning warmth and of a steady northward extension of life, was an age of comparative change and progress. In its first period men still hunted with weapons of Old Stone Age type; stones chipped and unpolished. Then, in some parts of the world at least, came a period with implements intermediate between the Old Stone Age type and those of what is called the New Stone Age, now generally called the Mesolithic or Middle Stone Age; and then followed the actual Neolithic Age with implements of polished stone, with agriculture and a settled life, and metal and civilization fast on its heels. For the purpose of this chapter we will restrict ourselves to the first of these three periods. We will not carry on into the beginnings of actual history. The men whose skeletons remain to us from this time, though modern men, were often primitive in this or that respect when compared with men of to-day. Their jaws were usually more powerful, they sometimes had brow-ridges (though never nearly so marked as in the Neanderthalers), the chin was not always so definite as to-day, and they were generally longer-headed. But almost without exception they had straight leg-bones and walked fully erect, not with bent knees; they had high foreheads, and the size of their brain was well within the limits normal to man to-day. Their great variability indicates that evolution was in rapid progress among them. That evolution has led on in the most advanced types of modern men to a smoothing away of the brow-ridges, a reduction of the jaws, an accentuation of the chin, and a general broadening of the head.

These late Paleolithic men, like the Neanderthalers, frequented caves and rock-shelters. They were savages still, but much cleverer savages than their predecessors. They passed through three main phases of culture. The first is called *Aurignacian;* in which period the flint-flake was brought to a new perfection and we find the first remains of definite human art. Before this, we have traces of the æsthetic impulse in the unnecessary fineness of chipping sometimes bestowed on flint tools; from now onwards we find personal ornaments, carved and patterned tools, drawings, paintings, and sculpture. The second period is called *Solutrean;* here a fresh type of flint instrument appears, called laurel-leaf, from its shape, in which the technique of flint-working reaches its highest triumphs, and then comes the *Magdalenian,* in which bone largely replaces flint as material for implements, and the flint technique degenerates. It is from this time, however, that there date the most remarkable drawings, paintings and sculptures which prehistoric man has left us. Many of the best-known dwelling-places of these men are in southern France and northernmost Spain; but quite recently caves full of relics of Aurignacian and Magdalenian people have been found in Palestine.

Aurignacian man seems undoubtedly to have come to Europe from

Africa across the land-bridges of which Sicily and Italy are to-day the remains. He must have found the Neanderthalers still there, and the difference between modern man and his more brutish predecessor was not so great as to prevent some traffic between them; for the earliest Aurignacian flints found in France show definite traces of the Neanderthal type of workmanship. It is an open question whether the two species mixed and inter-bred. If they did, then we have Neanderthal heredity among our genes to-day.

The Solutrean Period was, it seems, an interlude, an irruption of a race of hunters from the eastern steppes following the trail of great hordes of wild horses, whose bones are found in vast numbers in some of the caverns they inhabited. It seems to have been a change of climate that drove them and the wild horses westwards, and a fresh change that drew them eastwards home again. The men of the Magdalenian culture seem to have been the old Aurignacian tribes who returned to their old haunts with new methods and habits when the Solutreans withdrew.

The climate of the world was still cold in the Aurignacian and Solutrean periods and the reindeer was the characteristic animal of the central belt of Europe, with the mammoth and the woolly rhinoceros and the great cave-carnivores: cave-bear, cave-lion, cave-hyena. The land was largely open and treeless, oscillating between tundra and steppe; what forests there existed were dark forests of pine. The green belt was still away to the south. But in the Magdalenian Period, a steady amelioration of climate set in. Oak forests spread over Central Europe, driving the pines northwards. The reindeer moved north with the pine, the mammoth and the woolly rhinoceros and all the cave-beasts gradually died out, giving place to more familiar creatures. The age of the hunters in Europe was passing, and the period of cultivation and settlement was at hand.

But here we do not propose to follow up the extraordinary expansion of our species that now begins. In twenty thousand years or less, that not very abundant prowling animal made itself lord of the world and covered the earth with its habitations. With this expansion of *Homo sapiens* from the Magdalenian Age onward we shall deal in our concluding book upon the special biology of mankind. We end this strange eventful history of life here at the point when the sun of humanity rises upon the world and the dawn of man gives place to day.

BOOK SIX

THE SPECTACLE OF LIFE

I

HABITATS

§ 1. *Ways and Worlds of Life.* § 2. *Habitats and Their Inhabitants.* § 3. *Ways of Getting a Living.* § 4. *The Adjustment of Inhabitant to Habitat.*

§ 1. Ways and Worlds of Life.

THE home of Mr. Everyman is a house in the suburbs, between city and country, with a little garden attached. It is the average home of the civilized man. This small portion of the earth's surface is the abode of the multitude of living creatures, each fulfilling its own biological destiny as best it may. Each has its own way of life, each inhabits its own private world. Each lives self-centredly—the other existences with which its own intersect are for the most part not even suspected. And yet the whole assemblage of lives is biologically entwined; it forms an interlocking whole.

In this little domain there live Mr. and Mrs. Everyman, with their son, Master Everyman, one domestic servant, a tabby cat and a fox-terrier. Mrs. Everyman looks after the flowers, while Mr. Everyman makes himself responsible for the small vegetable plot at the far end of the garden. Here also Master Everyman has a couple of hutches with tame rabbits.

Mr. Everyman is an active member of a large economic community. He goes off every morning to work or business. Mrs. Everyman is, as the phrase goes, "economically dependent"; but she is the head of the little family world. The domestic life of the couple is on the whole exemplary—harmonious and agreeable. But Mr. Everyman finds it all but impossible to make his wife take an interest in business, while she finds him somewhat impervious in matters of dress and in local church affairs. The two get on very well together, but every now and then one of them is arrested before a gulf of incomprehension of the other's secret being.

As for Master Everyman, he is a source of mingled pride and trouble.

Both his parents have really quite forgotten what it was like to be eight years old, and their idea of his private world, consisting as it does of a few meagre recollections, stuffed with rose-coloured retrospective sentiment, inflated with adult morality and with parental ambition for their offspring, is very far from tallying with the reality. He pursues his own way of life as best he may.

So does the maid. Mr. and Mrs. Everyman treat her kindly, but they do not make much of an effort to understand her peculiar inner life nor to discover how she spends her time on her evenings out. Her life is interlocked with theirs in a hundred ways; but it remains intensely separate.

And when we come to the sub-human inhabitants the separateness and the mutual incomprehension increase. The Everyman fox-terrier is an affectionate dog, with the strong sympathy for his human master which so strangely characterizes the domesticated canine mind. He cringes to a reproof, is seized with tail-wagging at a cheerful word, and is thrown into a paroxysm of wriggling sentiment by forgiveness after a misdemeanour. But when he escapes from this human liaison, into what queer diversity of existence he plunges! An orgy of smells, a delightful rummaging in ordures, the strange but rigorous canine social life with its olfactory etiquette.

The cat is even further removed from its human masters. It likes being stroked, but that is almost its only rapport. It is an alien being, with its slit-like eyes and nocturnal habits. It is perhaps most alien when, impelled by love, it howls and caterwauls upon the roof. But it exists always in a private world which only includes the Everymans in an objective, physical way.

A great part of its interest is concentrated upon Mr. Everymouse and his family, who for some years have been established behind the wainscoting. Their life is a timid, twittering thing—pattering expeditions, lured on by good smells, out into the open kitchen or larder with scuffles and rushes back into holes and the dark safety within. Puss meanwhile is wound up by the smell of mouse into a special feline activity—long periods of intent watching at holes, twitchings of tail, alertness to pounce.

The rabbits in the garden lead the most subordinate lives of all Mr. Everyman's hangers-on. Their very matings are controlled. What does Buck Rabbit think about it when he is ignominiously lifted by the ears and put, scuffling and kicking, with a doe he has never seen? We do not know. Probably he does not think about it at all. But he fulfils his biological duty, and presently there is a litter of little rabbits to continue the nose-twitching acceptance of cabbage-leaves in the back-garden hutch.

The cabbages are grown by Mr. Everyman in the plot near the rabbits. They are organisms, too. But if they have an interior world it is so dim as not to be worth bothering about, so vegetable as to be meaningless to an active animal organism like one of the Everymans.

Master Everyman was much excited last summer by finding a chrysalis on the fence near the cabbages. His father, quite correctly, told him it belonged to a Cabbage-White butterfly, and it was carefully put into a box and looked at every day. But instead of producing a butterfly, the chrysalis became studded with a lot of little white cylinders, and out of each of these there hatched a lean and unpleasant-looking fly. Master Everyman was bitterly disappointed. He wanted to see the butterfly. It was no consolation to him that he had witnessed a remarkable case of parasitism. The flies were ichneumon-flies. Their parent had laid its eggs in the white butterfly's caterpillar and the grubs into which they hatched had devoured it from the inside.

Some years ago Mr. Everyman planted four little apple-trees at the end of the flower garden, and now he is very proud because he gets seven or eight nice apples off each tree every autumn. He congratulates himself, but he forgets to thank the bees. If it were not for these pertinacious little creatures, which visit his garden every fine day from a hive over a mile away, his apple blossoms would not have been fertilized and he would have had no apples.

He is also quite oblivious of his other garden allies. He knows, of course, that there are plenty of fat, juicy earthworms in the soil he digs over; but, if he gives the matter a thought, he supposes that the benefit is all on one side and that the worms ought to thank him for the provision of a home so admirably suited to their needs. Had he, however, read Mr. Darwin's delightful book on the subject he would realize that the benefit is mutual. The worms cannot thank him; they do not and cannot know of his existence. Even should he cut them in half with his spade, all they can know is the fact of the bisection. However, they pursue their own existence, and in the course of that existence they aerate the soil with their burrows, they help to drain it. They are all the time breaking the earth up into the finest soil as they eat their way through it and bringing material from the deeper layers to the surface in their castings. Mr. Everyman is at least aware of the existence of earthworms, but he does not suspect that in every ounce of soil there exist literally hundreds of millions of bacteria, and that on their chemical activities depends the fertility of his garden. Nor does he know that this microscopic flora has its microscopic enemies. Prowling through the soil are innumerable small amœbæ which live on the bacteria. They are all aquatic, but they are so small that they can travel about comfortably in the invisible film of water which clings to every grain in ordinary moist soil: if the soil dries up, the bacteria and their enemies alike pass into a passive resting stage. He is quite certainly unaware of the fact, established by the Rothamsted Experimental Station, that the partial disinfection of the soil, or its partial sterilization by heat, will leave most of the bacteria alive but kill most of their protozoan enemies, and that this will promote fertility. If he had a greenhouse this fact might be of considerable importance; greenhouse soil often goes "sick," and it used to be the practice to throw it away and get fresh soil in. But we now know that the condition is due to too many protozoa and that killing them by heat will very cheaply restore the soil's fertility.

We must not forget Mrs. Everyman's pretty flowers. They charm the senses by their colour and smell—an apparently unnecessary gift of beauty; they appeal by their tender green and their punctual growth, their reachings upward to the light. The Everymans, however, have probably never reflected that the green is life's badge of factory labour, the outward and visible sign of an inward chemical grace; that the plant's ways of sprouting and growing have been imposed on them during evolution by the unceasing struggle for moisture and light, in which millions that did not come up to the standards of their environment have been ruthlessly massacred; and that the beauties of their flowers have a purely commercial basis as advertisement to insects. Taking this as a basis, man has stepped in and constructed biological monstrosities. Generations of gardeners and seedsmen have laboured to produce double flowers that are sterile because all their reproductive parts have been converted into mere showy petals; to manufacture plants, like many roses, that can only be continued by the unnatural process of grafting; to bring into existence delicate strains that would never have a chance in nature. In garden flowers man has taken beauties generated by hard necessity and distorted them to serve his own sensuous and emotional ends. Biologically speaking, garden flowers are parasites on the Everymans' æsthetic longings.

Besides the flowers there are the weeds. Ill weeds grow apace, and Mrs. Everyman often wonders why such nasty plants were created. But the weeds' extraordinary capacity for sowing themselves and coming up where they are not wanted is no less a product of struggle and selection than the colours of the flowers, and what are weeds in a garden are essential elements in natural vegetation. A bare patch on the face of earth is in a few decades covered with rich natural vegetation again; and what we call weeds are among the most important colonizers of unoccupied soil, preparing it for finer types, paving the way for the full climax of plant-life.

Nor let us omit the ubiquitous bacteria and moulds and other microbes. Their spores float in every breath of air Mr. Everyman breathes, lie settled in every dusty corner. They turn his meat bad in hot weather, they sour his milk without a by-your-leave, they turn his bread mouldy if he leaves it too damp. A new strain of influenza microbes, started maybe in north-west Canada, or in Central Asia, sweeps across the world. The minute specks of disease-producing life infect Mr. Everyman as he travels to business; he brings them home and they pullulate in triumph through the bodies of his wife, son, and servant.

And we had almost forgotten to tell you that the dog has worms; that introduces another large category of animals to add to this suburban menagerie.

§ 2. Habitats and Their Inhabitants.

ALL this variety of ways of living exists in one little patch of earth's surface. We must multiply it many thousandfold if we take in the whole earth, with all its innumerable habitats, from pole to equator, jungle to desert, high mountain to deep sea. The result is overwhelming: our minds cannot hold its abundance without the aid of some principle of arrangement. Confronted with the same difficulty when we set about the descriptive cataloguing of the many hundreds of thousands of living things, we found that a classification based on resemblance in structural plan brought order into the chaos. With this to guide us, we could pigeon-hole our creatures, could brigade them into groups, could systematize the crowd into an evolutionary army in which each had its definite place.

This we did by concentrating our attention on constructional plan and leaving way of life out of consideration. The fish-like whales and porpoises were put with the mammals, the snake-like slow-worm with the lizards. But every organism, if it must have its own plan of construction, must also live in its own way. And now the opposite aspect of the variety concerns us; we are interested in function more than structure, and are seeking a principle to help us classify creatures not by their blood-relationship but by their ways of life. What interests now about whales is not their past derivation from land-mammals, although this has left its impress indelibly upon their construction, but their marine way of life, and the fact that some are adapted to straining off tiny crustacea and molluscs from the sea-water, while others are fiercely and frankly carnivorous.

The simplifying idea which serves us here is also an evolutionary one. It is the idea of the moulding force exerted, directly or indirectly, upon the organism by its environment and its method of gaining a livelihood.

Life is often thought of as insurgent, a rebel rising against the limitations imposed upon it by outer nature and surmounting them. In our preceding Book, we stressed that aspect of life. But if progress and the overcoming of difficulties by what we may metaphorically call biological invention are the cardinal aspects of life when looked at in the perspective of geological time, quite other aspects loom largest when we survey it as it is spread over the surface of the globe to-day. To such inspection, the main types of life—phyla, classes, orders—appear as definitely established and fundamental things; the rare "biological inventions" of life have taken place in the past and we take them for granted; what chiefly strike us in the present are the various ways in which these leading patterns have been adapted to different detailed conditions, the extraordinary plasticity of each of the main types of life's construction under the influence of different habitats.

In considering this plasticity and its results, there are two moulding forces which have to be taken into account. One is the effect of the organism's habitat, the other the effect of its way of life within that habitat. All animals and plants that live in the surface zone of the sea must in some way or other be able to keep themselves from sinking; all cave-animals must be adapted to darkness; all intestinal parasites to a shortage of oxygen. But in the surface layers of the sea one animal swims, another floats; one sifts and eats microscopic plants, another catches and devours large animals. In the intestine one parasite anchors itself, another wriggles freely about; one absorbs ready-digested food, another eats it when only half-digested. Our first and main simplification will be to divide the realm of life into a number of habitats, each with its own distinctive conditions; and then to study the way in which the animals and plants have become adapted to these conditions—in brief, how the habitat moulds its inhabitants.

Besides this, we shall also have to take some account of the different ways of life possible within each habitat, and see how they too mould the creatures which adopt them. But different ways of life are linked together. Carnivore eats herbivore, herbivore eats plants, plants live largely on the products of animals and their decay. Brought back from a contemplation of the variety of life to its inter-relations and to its unity, we shall find ourselves devoting the concluding chapters of this Book to what we may call the Economics of Life, the science of vital interconnexions which is called Ecology. But our first business is to set forth some of the varied spectacles of life that are revealed as we pass from habitat to habitat.

It is true that the biosphere, as the life-inhabited zone of earth is sometimes styled, is a mere skin. About ninety-nine hundredths of living things are crowded into a "life-skin" not more than a thousandth of the earth's radius in thickness, and occupying about one-third of one per cent of its volume. Such a skin on a regulation-size Association football would be less than a two-hundredth of an inch thick. Even if we take in the rare extremes this thickness need only be multiplied by four or at most five. In spite of this, the biosphere skin is extremely varied in the homes it offers to life. There are first the differences in medium—air, earth, water; differences in salinity from almost pure water to the Dead Sea's more than twenty per cent of salts; differences in temperature from hot springs that are nearly boiling down to many degrees below freezing; differences in pressure from well below half an atmosphere on high mountains to several hundred atmospheres in the deep sea; differences in light from the intense tropical sun to the utter darkness of caves, of the oceanic abyss or of an animal's gut. And all these various differences of temperature, light and pressure, of climate and situation, may be combined in an almost bewildering multiplicity to give the actual habitats in which animals and plants live out their lives. Of these we need not here give any detailed or formal classification; but before we go on to description, it will be as well to remind ourselves of the chief kinds of habitats available for life.

The original home of life was water—sea-water. The whole of the sea is inhabitable, both the firm bottom and all the vast succession of layers of open water in which life must float or swim.

As the shore is approached, and the influence of the tides and the waves is felt, conditions change, and a multiplicity of new habitats are provided. Climate too makes itself felt; there are no coral reefs in the arctic. And here and there special conditions make special habitats, such as that weedy Atlantic slackwater known as the Sargasso Sea.

Then there is brackish water, bridging the gap between the habitats of the sea and of inland waters. Inland waters, in spite of their small extent in comparison with the sea, provide an amazing diversity of habitats. There are salt lakes far saltier than the sea, there are almost saltless rivers. There are hot springs, and arctic waters that spend most of their time as ice. There are running waters of all degrees of turbulence, and there are deep and quiet rivers. There are lakes big enough to be oceans in miniature, with their own deep-water unilluminated zone and their various layers of free-floating and free-swimming life. And from these huge bodies of water there is every gradation down through ponds and pools, to temporary puddles, to the Lilliputian lakes that collect in the hollows of old trees and to the mere films of water on leaves or sticks between grains of soil, that can still harbour microscopic swimming life.

As lagoons and estuaries connect the worlds of sea- and fresh-water, so the worlds of water and land are connected by the transition zones of shore, of temporary pools, of swamps and marshes and mud-flats; and all these linking habitats will be differently populated according as they are fresh or salt.

The habitats provided by land are perhaps the most varied of all. There is the strange world of the soil itself, comprising the animals and plants wholly or mainly confined to a life below ground, embedded in or burrowing through the earth. But the greater number of land-habitats differ profoundly from those of water in being practically restricted to two dimensions only, spread over the surface that divides earth and air. Aquatic life exists in three dimensions; land-life is like a film. The inhabitants of this surface zone differ according to climate far more than do those of the sea, even of the sea's surface. So far as land-plants are concerned, climate is the overruling factor, and any broad classification of their habitats must be drawn on climatic lines. We have arctic habitats, sub-arctic, temperate, sub-tropical and tropical; besides these zonal divisions, there are divisions according to altitude—plain, hill and mountain. Finally, these habitats can be further classified according to soil conditions.

With animals, on the other hand, habitats are more an affair of the environments provided by plants; the influence of climate upon them is at one remove.

One important animal habitat, for instance, is the arboreal; the modifications induced by living in trees are more striking than those related to the climate in which the trees happen to grow. In forests, especially tropical forests, animal as well as plant life is stratified in horizontal layers, almost as in the sea; there is the tree-top layer, several layers in the less well-lit habitat of the region below the tops but above the ground, and the ground-layer between the trees.

The characteristic vegetation of steppe, savannah, and tundra is the chief agent in making of each of these a distinct animal habitat; and the very absence or limitation of plant-growth in deserts is one of the characteristics most important for their animal inhabitants. Caves afford a minor but interesting habitat to land and to fresh-water life; and another peculiar habitat is that now provided by man, in his buildings and yards, his fields and gardens, to special types, like sparrow and cockroach, that can take advantage of the opportunities so richly provided there.

Air as a habitat is in a certain sense less important than either earth or water, since no organisms inhabit it permanently, but only between terrestrial or aquatic interludes. Nevertheless it exerts a potent moulding force upon the creatures who have taken, however temporarily, to existence in its medium, and may properly rank as one of the larger habitat divisions of the earth. The less temporary and occasional, however, is the stay of a creature in air, the less does any subdivision of aerial habitat become possible for it; and it is symptomatic that the greatest mobility and range over the earth's surface found in any non-human species occurs in the birds. This is, if you like, the obverse of the fact that the only satisfactory classification of flying creatures is by the habitats they frequent when *not* in the air—water-birds and water-insects against land-birds and land-insects, and so on.

There remains one further major type of habitat, which is neither earth, air nor water, but rather fish, flesh or fowl; it is the habitat provided by the living bodies of other creatures and occupied by the horde of parasites. It is a habitat within a habitat: none the less, its moulding effect on its parasite inhabitants is striking in the extreme. And in variety it does not yield to any other main kind of habitat. A parasite may be external or internal: it may live in or on an animal or a plant: it may inhabit blood or muscle or intestine.

The common frog is an excellent example of their variety and abundance: it is a little zoological garden of parasite life. Besides various kinds of bacteria in its gut, every specimen harbours a swarm of big ciliate Protozoa of several kinds in its rectum, a roundworm and a fluke in its lungs, more flukes in its bladder. Sporozoan parasites are common, flagellates occur regularly, fungi may attack the skin, and there are literally dozens of rarer parasites of various kinds.

The world of parasites is a major world of life, worthy to rank with the worlds of sea, of fresh-water and of land in the number of its inhabitants and the variety of their ways of living.

§ 3. Ways of Getting a Living.

IN EACH habitat, different organisms live in different ways. One takes advantage of one set of opportunities which the habitat provides, another of others. Different creatures surmount the same difficulties by different methods. Accordingly, even in one and the same habitat, there will be many modes of life. The modes of life hinge first upon food, and then upon reproduction. And since the verb *to eat* is conjugated by life as much in the passive as in the active, modes of life in relation to food will include not only modes of feeding but modes of avoiding being fed upon. In this section, however, we shall illustrate our point with reference solely to ways of feeding.

The first great division in regard to feeding is that between green plant and animal. The green plant, strictly speaking, does not feed at all; it makes its own food inside the living factories of its cells, taking in the simple raw materials it needs from the inorganic medium in which it lives. The animal, on the other hand, can only utilize carbon and nitrogen in ready-made organic form: it profits by the green plant's labours.

Other modes of plant nutrition have been already referred to, such as that of most fungi, which need organic compounds, but of a much lower complexity than those required by animals; and the special modes of nutrition possible to nitrogen-fixing and other forms of bacteria. Every group of plants has also its parasitic representatives, and in the flowering plants there are some which are "carnivorous."

All animals, on the other hand, get their food either directly or at one or more removes from green plants. The usual division is into herbivores and carnivores but perhaps the best classification is into what may be called micro-feeders, which live upon relatively minute particles, engulfing them without any selection, and macro-feeders, which usually select their food, and in any case take it in relatively large portions.

The word *relatively* is used of set purpose. A whalebone whale swims its devouring way through swarms of little crustacea or butterfly-snails, taking ten thousand at a gulp. It is obviously using a micro-feeding method; yet one of these identical molluscs or crustacea, if captured individually by a small fish or medusa, would be a victim of macro-feeding.

The most general method of micro-feeding is to produce a current of water, usually by means of cilia, and then to sift out the contained food-particles from the current by some mechanical device. This is adopted by all sponges, sea-squirts and bivalve molluscs, many worms, Polyzoa, and other forms, including Amphioxus. The other main method is based upon chemical instead of mechanical sifting. The animal, instead of passing a current of water over its tentacles or through its gills, eats its way through its sandy, earthy, or muddy surroundings, and forces a column of the unpromising material through its tubular gut; the digestive juices dissolve any nutrient matter present, and this is then absorbed, while the remainder is passed out at the anus. This is the method of earthworms, lugworms, Balanoglossus and heart-urchins.

The essence of these micro-feeding methods is their automatic nature; the animal does not in any way select its meal, but its food is simply filtered or digested off from the liquid or solid medium in which it happens to lie. Occasionally, however, some selection enters in; herrings and whales choose patches of sea rich in plankton, and some of the sand-eaters appear to reject certain types of particles.

In general, however, all is grist that comes to the micro-feeders' mill, and they are essentially omnivores. If some, like whales, are predominantly animal-eaters, and others, like the minute Tunicates known as appendicularians, predominantly vegetable-eaters, this is an accident, determined much more by the size of the organisms eaten than by their vegetable or animal nature.

Macro-feeders, on the other hand, often show more specialization in their food. Since they take large mouthfuls, they need different kinds of mouths and teeth for animal- and vegetable-food, and, still more important, very different arrangements are required to catch and hold an active animal from those needed to browse upon stationary and unresisting grass or trees.

Vegetable macro-feeders, however, if less interesting in regard to their methods of securing their food, show numerous adaptations for its proper utilization. In the first place, the nutritious parts of green plants, the cell-contents, are all shut up in their little cell-wall boxes, made of cellulose (or even of wood), which is as indigestible to almost all animals as it is to us. A vegetable-feeding animal with no device for breaking open these microscopic boxes would be just as helpless as we are when all we have to eat is a box of sardines and we have lost the opener.

The usual method of box-opening is mechanical trituration; this may be done by rasping organs, like the radula "tongue" of snails, or by regular grindstones of teeth, as in a cow or an elephant, or by grinding gizzards, whose grinding power often depends on sand or stones deliberately swallowed, as with many birds. In addition, bacteria may be enlisted to break up the cellulose chemically, as in the gut of many hoofed mammals.

Owing to the bulk of cellulose, and the need for time to dissolve out the contents of the cell-boxes, the gut of herbivores is, almost without exception, relatively longer than that of carnivores, and often provided with voluminous outgrowths (like the cæcum in a rabbit, for instance; see Fig. 31 in which the food may be stored while it is exposed to bacterial action.

Most herbivores eat the green parts of plants, and may be roughly divided according to their method of eating. There are grazing types, whose green food, often relatively small, is spread out over a surface; examples of these are sheep, fresh-water snails, and those small caterpillars which merely eat away the surface of the leaves over which they crawl. The browsers, on the other hand (though there is every intermediate gradation), in general take portions from larger plants. There are fishes that bite pieces off the large seaweeds; giraffes and most deer browse off trees; many caterpillars take respectable mouthfuls out of the edges of leaves. Finally, there are the miners and borers, so small in relation to their food that they live not merely on but in it. The leaf-mining caterpillars and grubs of various moths and beetles are the best known examples. These eat their way through their environment just as do the sand- or mud-eaters; but what the herbivorous miners eat is all food, while the food of the soil-eater must be digested out of a mass of unnutritious material.

Besides the main type of green-eating herbivore, there are vegetable-feeders adapted for eating other parts of plants—fruits, seeds, bark (including even cork), wood, roots, tubers, the sap, special secretions like nectar, and even the pollen. In many cases the smaller animals live *in* their food, the large come and bite it off. There are, for instance, plenty of fruit-inhabiting and grain-inhabiting insects, as well as animals which eat fruits or seeds in a more ordinary way, such as toucans and fruit-bats on the one hand, finches, nuthatches and harvester-ants on the other; and there is even a Malayan squirrel which, after gnawing a hole in a coconut shell, gets right inside for the business of eating.

Among carnivore micro-feeders, there are as many adaptations as among herbivores. The tearing and cutting teeth of wolf or tiger are but one type. The teeth of some rays are turned into shell-fish crushers; of various bony fish into beaks for biting off coral. Whelks bore holes in the shells of their bivalve prey. Snipe and woodcock have beaks converted into sensitive and flexible worm-detectors and pincers; those of herons are converted into fish-spears, of hawks and owls into flesh-tearers. Ways of life are as multifarious as habitats; and both set their stamp upon living creatures.

§ 4. The Adjustment of Inhabitant to Habitat.

It is an obvious fact that on the whole animals and plants fit their surroundings; there is a definite correlation between the peculiarities of inhabitants and those of habitats. A few examples will illustrate the point better than pages of generalities. In the high northern latitudes, white birds and mammals occur in far larger proportions than elsewhere; while in deserts the preponderance of buff, fawn and sandy-coloured animals is equally noticeable. Or again, an unusual percentage of the inhabitants of the surface layers of the ocean possess one or more of the following peculiarities, which are rare elsewhere—either glassy transparency or else blue colour; long, projecting spines or flaps; bell-shaped construction. Below about a hundred fathoms, on the other hand, the majority of animals are either black or red (both colours being uncommon in most other parts both of sea and land) and possess luminous organs; and their eyes are either abnormally large, or else abnormally small or absent.

The preponderance of deciduous, non-coniferous trees in the temperate zone, of evergreen conifers in higher latitudes and on mountains, and of evergreen non-coniferous trees in tropical forests is a good example of correlation of flora with environment; and the unusual abundance of succulent plants in deserts and near the sea is another.

Half a century ago, almost every one would have unhesitatingly accepted such characteristics as adaptations which in some way conferred biological advantage upon their possessors. But there has been a reaction against a too facile application of Darwinian principles, and nowadays it is realized that there may be other interpretations for such facts as these.

As an example, let us take the similarity of the colour of desert animals to their surroundings. This is often very marked; and the older naturalists presumed that it had arisen through its conferring on its possessors a cloak of comparative invisibility. Later students of the problem, however, like P. A. Buxton in his *Animal Life in Deserts*, point out a number of difficulties. The most striking, perhaps, is that a number of desert animals which are wholly nocturnal, and therefore, one would think, can receive neither benefit nor the reverse from their coloration, are yet of this same sandiness of colour. He therefore concludes that one or other of the conditions of desert existence must act more directly upon the desert's inhabitants, forcing them to become sandy whether it is advantageous or not. There is in fact, as J. A. Allen long ago pointed out, a frequent geographical correlation of a more general nature than that between deserts and sandy colour, namely, a gradual darkening of the colour both of mammals and birds with humidity, a lightening with aridity; and it might be supposed that the amount of water-vapour in the air directly influenced colour. As further complication, however, desert creatures closely related to sandy-coloured species (and this occurs both among mammals, birds and insects) may be very conspicuous, black being a frequent colour; and yet such animals seem to be just as successful as their inconspicuous relatives.

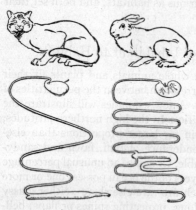

FIG. 239. Adaptation to different kinds of diet.

A cat and a rabbit of the same weight, with their stomach and intestines on the same scale. Note the big blind gut or cæcum of the rabbit.

There the problem stands. Upholders of the theory of adaptation will urge against Buxton's views that, in the brilliant atmosphere of the desert, it is just as important for an animal to blend with its surroundings on a moonlit night as it would be by day. And the supporters of the "direct action" view must, of course, admit that even if light sandy colours were first produced as the result of aridity (or some other condition of desert climate), with no reference to their biological value, yet when produced they would be likely to have biological value for many species, and would therefore be perpetuated. Their concealing qualities could later be perfected by Natural Selection if appropriate mutations turned up.

The temptation to interpret the facts in terms of the inheritance of acquired characters is a strong one. It is true, as Beebe showed, that in some species humidity and aridity have a direct effect upon the colour of individuals. This is so, for instance, with the dove, *Scardafella inca,* which when kept in a very moist atmosphere gradually assumed the dark colour characteristic of the sub-species found naturally in moister climates. But in other examples, such as the American deer-mice of the genus Peromyscus, the case is different. The various sub-species of the common Californian deer-mouse show a considerable degree of parallelism between their coat-colour and the prevailing colour of their surroundings, the sandy tone of those from arid habitats being very noticeable. But when these were bred by Sumner under experimental conditions, he found that the characteristic colours of the fur persisted unchanged in spite of quite new conditions of temperature and moisture. In other words, we are presented with the same phenomena already discussed in Book 4, Chap. 7, § 4, of a visible character having sometimes to be produced afresh in each generation as a response to environmental conditions, in other species being produced in all kinds of conditions owing to hereditary factors. At first sight this looks rather like the inheritance of acquired characters. But, as we concluded in our previous discussion in Book 4, facts of this kind are really not evidence at all.

The problem can only be solved by new observations and new experiments. The question whether such habitat-correlated characters are advantageous to all or to some of their possessors can only be settled by intensive work in the animals' natural surroundings; while, whether they are useful or not, the question as to their method of origin—by modification in each generation, by Lamarckian means, by induced mutations in the germ-plasm, or by random variation guided into certain channels by Natural Selection—can only be settled by a painstaking combination of physiological experiment and breeding tests.

Moreover, there is an important theoretical consideration that has often been overlooked. As we pointed out in Book 4, an adaptive character may give its possessors a definite advantage over other members of the species, and so in the course of generations automatically become a character of all the members of the species; and yet it may confer no advantage upon the species as a species. This principle of intraspecific selection is very possibly applicable to the persistence in deserts of sandy-coloured and conspicuous animals side by side. If variations crop up in the direction of sandiness and consequent concealment, they will gradually oust the other colours of the species. But if they do not, some species may persist quite happily in spite of conspicuous colouring.

In general, however, it must be admitted that probably the great majority of these correlations with habitat are adaptive; the most reasonable explanation at present is that they have arisen under the guiding action of Natural Selection. The existence of the same type of structure or habit in a number of different species in one habitat is not only an interesting fact, but it also constitutes *prima facie* evidence for the adaptive nature of the structure or habit in question; and when we can reasonably interpret the structure or the habit in an adaptive way, though we are not thereby exempted from the duty of putting our interpretation to more decisive tests, we are justified in so doing as a working hypothesis until evidence to the contrary is forthcoming.

Accordingly, what we shall do in the following pages is in the main to illustrate the intense variety of living things by pursuing life through a variety of habitats. We shall also point out those peculiarities which seem to fit the inhabitants to their surroundings; but we shall leave the origin of these adaptations to be interpreted in terms of our discussion in Book 4.

II

LIFE IN THE SEA

§ 1. *Life in the Sea.* § 2. *The Surface Life of the Sea.* § 3. *The Deep Sea.* § 4. *On the Floor of the Abyss.* § 5. *Sea-shore Life.* § 6. *Coral Reefs and Islands.* § 7. *Holes and Corners of Sea-life.*

§ 1. Life in the Sea.

THE sea provides a vastly greater space for life to inhabit than does the land. Not only does it extend over more than two-thirds of the surface of the globe, not only does it lack all blank lifeless areas such as are found on land in the Antarctic ice-cap or the tops of the great mountain ranges, but it is inhabited in three dimensions. The average depth of the sea is somewhere about 12,000 feet, and all of this vast body of water (save possibly a few of the deepest pockets) has its inhabitants. It is true that below the limits to which light penetrates the sea's population is sparse; but even so it is richly inhabited to a depth of a hundred and fifty or two hundred feet, and inhabited by a wonderful multiple population, layer below layer, each layer different from its neighbours.

And yet the number of kinds of creatures that live in it is very much inferior to the number of kinds that live on land. The last time that a detailed analysis was made (in 1898) only 85,000 species of aquatic animals—and this is including the fresh-waters with the sea—were on record, as against 327,000 land-animals. The difference was doubtless due in part to the greater attention paid to the inhabitants of the land; but in spite of this factor the comparative richness of the land-fauna cannot be denied. This paucity of species is apparently a direct result of the much greater uniformity of conditions in the sea. Nor is there any isolation of one part of the sea from another as there is between bits of land or bits of fresh-water; we have seen in Book 4 that isolation helps to generate new types.

But this numerical poverty is offset by the greater variety of the main types of construction, and life may be said to ring more variations but on fewer themes. Not one of the phyla of animals but has some marine representatives. A few classes of animals, notably the amphibians and centipedes, are altogether absent from the sea, and the biggest class of all, the insects, is extremely scarce in salt-water. But there are whole phyla of animals which are only found in the sea, like the lampshells and the echinoderms, and others, like the sponges and the cœlenterates, which are almost all marine. Besides this there are many classes that include sea-animals only, such as the cephalopods, in some ways the most highly developed of invertebrates, the sea-squirts, the radiolarians, amphioxus and its relatives—indeed over a third of the number of animal classes that are recognized by zoologists.

The explanation of this variety of type is doubtless a historical one: life originated in the sea. In this spacious home, life, evolving through long epochs, branched out into a great number of types, some lower, some higher; but only a few of these succeeded in making the advance into fresh-water or on to land.

. . . . . .

Sea-life is unlike land-life not merely in the strange and varied types of creatures that enjoy it, but in other more fundamental ways. Shelley vividly pictured its lovely profusion. Listen to him as he speaks of the West Wind:—

> "Thou who did'st waken from his summer dreams
>   The blue Mediterranean, where he lay,
> Lulled by the coil of his crystalline streams,
>   Beside a pumice isle in Baia's bay,
> And saw in sleep old palaces and towers
>   Quivering within the wave's intenser day,
> All overgrown with azure moss, and flowers
>   So sweet the sense faints picturing them! Thou
> For whose path the Atlantic's level powers
>   Cleave themselves into chasms, while far below
> The sea-blooms and the oozy woods which wear
>   The sapless foliage of the ocean, know
> Thy voice, and suddenly grow gray with fear,
>   And tremble and despoil themselves: O hear!"

Sea-life is as intense and as beautiful as he imagined it. But there are no flowers and no woods: save in parts of a zone of shallow water round the coasts, the sea-bottom grows no plants. The great bulk of the "meadows," "shrubberies" and "forests" of the ocean are animal in nature, incapable of making their own nutriment. Even when rooted, stalked, branched, and to a casual glance completely plant-like,

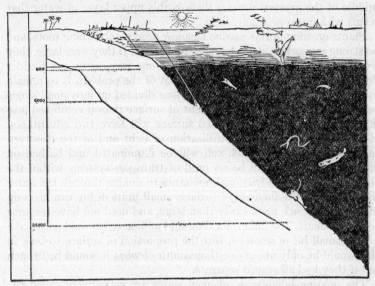

FIG. 240.   A diagram of the zones of the sea.

The sun helps the seaweeds and microscopic sea-plants to build up living substance. They feed the sea-animals. Light sufficient for abundant plant-life penetrates to about 200 feet; practically no plants are found below 600 feet. The creatures in the blackness of the deep sea are all animals dependent on a food-rain from above. The sea-bottom usually slopes gently at first to the edge of the continental shelf, then dips steeply to about 6,000 feet, then more gently to the abyss, little of which is over 24,000 feet deep. In the main diagram, the upper zones have had to be exaggerated; the true depth scale is shown in the inset.

the growths on the sea-bottom are without leaves or flowers. Their branches and stems are hollow, so many stomachs; they are studded with greedy mouths: and they owe their flower-like appearance to the cruel tentacles which bring about the capture of prey by snare or microscopic swirl.

None the less, the green plants are there: they must be there, or the sea could not support life. If green plants are to exist in the sea, they must live in the light: and light is gradually absorbed as it passes through sea-water. Not only will nearly all the ocean floor be dark, and uninhabitable by green plants, but even in shallow waters, where light can reach the bottom, there will be more light at the surface. The great vegetable-garden of the sea is its top layer of water, some fifty yards deep. All over the watery surface of the globe, even in the middle of the greatest oceans, there is this upper layer of water that teems with green productive life.

Plants growing in this marine meadow need no absorbing roots, and no strong supporting stem; but they must float. If they were large, they would need to be kept from sinking in some way, as by gas-bladder floats. But the most economical solution of the problem is to remain microscopic. A given bulk of plant-tissue divided up into small single cells will possess a far greater extent of surface than if combined into large plant-bodies. The increased surface will have two advantages. It will allow the maximum utilization of light and of the dissolved salts of sea-water, for each cell will be illuminated and bathed on all sides, and there will be no need of transport-systems within the tiny body; and, since frictional resistance to sinking through the water increases with the increase of surface, small units of life can keep up in the water much more easily than large, and need not have recourse to special floats. The union of a thousand million cells would only make a very small bit of seaweed. But the proportion of surface to bulk in this would be only about one-thousandth of what it would be in each cell if they had all stayed separate.

The floating inhabitants of open water are collectively called the *Plankton*—"that which is drifted about." Plankton is another of those technical terms which sooner or later will fix themselves into common speech; for it constitutes the main food-supply of the sea. Almost the whole of sea-life is nourished by the plankton. Plankton is the base of the sea's vital pyramid, on which are supported almost all our food-fishes, and even the great whales.

The microscopic plants of the plankton are mostly of two types, diatoms and flagellates. The larger flagellates belong for the most part to the curious-looking group called Dinoflagellates, while many of the smaller belong to the equally curious group of Coccolithophoridæ.

In certain spots at certain times, the crop of plant-plankton is so dense that it discolours the sea over large patches. But for the most part the plantlets remain invisible and unsuspected until tow-net and microscope are brought to bear. Even with the aid of the finest silk tow-net, however, many of the smaller floating plants are never captured; and only in the last few years has Lohmann discovered the extraordinary abundance and importance of this dwarf plankton or *nanno-plankton*. In part he obtained his knowledge by centrifuging large volumes of surface water, and examining with a microscope the fine sediment thrown down; in part he utilized the collections made by some of nature's tow-nets—the filtering and straining apparatus of small ciliary feeders like Appendicularia (a free-swimming relative of the Sea-squirts), compared to which the meshes of any tow-net would look like wiring-netting, for they catch objects down to 3 μ across, and exclude everything over 20 μ.

A great deal of work has been done in the last fifty years upon these tiny creatures; and as a result we are beginning to know something about their distribution, their habits, and their multiplication.

There is an orderly succession of different kinds of sea-drifting creatures through the year, just as there is an orderly succession of the growth and flowering of plants in a wood. In spring, the diatoms and other tiny plantlets begin to multiply; following on their heels come swarms of small animals, mainly larvæ which bottom-livers send up to take advantage of the diatom harvest; and their abundance is the cause of multiplication of larger and carnivorous animals. The plankton is at its richest in late spring; by then the burst of plant-growth has exhausted most of the available nitrates and phosphates, and the surface-zone must wait for a new quickening until the winter cools the top layer of water. The cold water is heavy and sinks: unexploited water, rich in nitrates and phosphates, rises from the depths to take its place; and the cycle can begin again as soon as the temperature rises high enough.

And there is an orderly distribution of the plankton over the roof of the sea. For chemical reasons, certain salts needed as plant-food are more abundant in cold than in hot water, and so plankton and surface life in general is more plentiful at either end of the globe, less plentiful round its middle. This regularity is interfered with by currents; the Gulf Stream flows on the surface right up to the coasts of Spitsbergen, but then sinks below the polar water, which is less salty because of melting ice, and plunges downwards; thus we may find warm-water forms descending to an inevitable death deep below the pack-ice of the arctic sea. The antarctic current brings cold water and rich life far up the west coast of South America. One day the life of the world's seas will be properly explored; and then we shall be able to chart them, season by season, according to the abundance of their basic food-supply, the plankton. A promising beginning has been made with the Atlantic, as the annexed figure shows.

Naturally this rich marine meadow is pastured by swarms of

FIG. 241. An animal forest—a scene at a moderate depth on gravel bottom.

Colony-forming cœlenterates studded with polyps are the most prominent forms of life: sea-fans on the left, tall Alcyonarians centre and right, sea-pens in the foreground. In the middle distance are sea-anemones and corals. Fish and a spider-crab complete the picture. (Based on an exhibit of animals collected by the "Discovery" Expedition, in the Natural History Museum, London.)

animals, but, owing to the microscopic size of its constituent plants, no large animals can browse directly on the vegetation; there are no creatures of the open sea corresponding in their diet to cow or deer, elephant or hippopotamus, or even to rabbit or prairie-dog. All the herbivores here are small, often indeed minute, and constitute only the first links of the food-chains which culminate in whale and dolphin, bonito and mackerel, argonaut and giant jelly-fish. As we have already seen in Book 5, this absence of large plant-eaters from the sea meant that all the ancestral land-vertebrates were carnivores. So the moulding effect of the environment radiates out from its original centre, affecting one remove of creatures after another, and extending its influence even into another habitat. Who would have imagined that the predominance of flesh-eaters among amphibians and the earliest reptiles was a consequence of the necessity that sea-plants should be microscopic in order to float the better? Yet so it is.

§ 2. The Surface Life of the Sea.

When Mr. Everyman takes his family to the seaside, and they all adventurously go out in a row-boat, they little suspect, as they look over the gunwale, how full of life are the blue-green, choppy waters around them. They could gain an idea of this abundance, if they provided themselves with a tow-net—a long conical net of fine muslin, or better, of bolting-silk of the sort used by millers to sift flour, to whose end is tied a glass or metal container. If this be towed patiently behind a boat for half an hour or so at slow rowing speed, it will collect a fair sample of the sea's surface inhabitants. Take the sample home and look at it under a low power of the microscope; a new and strange world of life is revealed. If you have chosen locality and season well, the variety of creatures will be extraordinary. Here are abundance of crustaceans, rowing themselves along by means of their long antennæ. Some of them are adult and spend all their lives thus constantly active in the roof of the sea. Others are the babies of bigger bottom-dwelling creatures—crabs, lobsters, prawns, barnacles—which spend their larval existence here before sinking to the bottom. It is these immature larval forms which make up the great majority of this population.

For the surface zone is the sea's main nursery. Drawing their nourishment either directly or indirectly from the minute plants of the plankton, the tiny glassy creatures can feed and grow here until they are big enough to cope with the different methods of feeding imposed by larger size and crawling life. There are fish eggs buoyed up with floats of oil, which presently hatch out into glassy fishlets, often amazingly different from their future adult selves. There are larvæ of sea-urchins and brittle-stars, looking rather like a painter's easel turned upside down, larvæ of starfish and sea-cucumbers, perhaps of sea-lilies; the "wheel-bearer" larvæ of bristle-worms and of molluscs, often curiously alike, with a wheel-like girdle of big cells carrying strong cilia. Later on some of the mollusc larvæ grow tiny shells and a rudimentary foot. There are larvæ of sponges and polyps and jelly-fish, of sea-mats and other Polyzoa, and the microscopic tadpoles that later degenerate and grow into sea-squirts.

The sea's nursery has its drawbacks. Of these millions of marine babies, only a minute fraction survives. If you send forth your new-hatched young unprotected into the world, you must expect a massacre of the innocents. They fall a prey to all kinds of small carnivorous creatures, glassy like themselves. Among these are the arrow-worms, rapid swimmers with cruel biting mouths; battalions of jelly-fish of every size, trailing their paralyzing net of tentacles; and Ctenophores whose tentacles capture not by their poison but by their adhesiveness.

There are swifter and larger creatures too who profit by the abundance of surface life, like the mackerel who strains off the plankton from the water-current through his gills, but these of course will elude your tow-net.

Then of single-celled creatures there is a great abundance. Our temperate waters on summer nights may come alive with pin-pricks of light; these are produced by the swollen spherical protozoan Noctiluca—"Shine-by-Night." Of the innumerable single-celled plants we have already spoken. Single-celled animals are not so abundant inshore; but far out in the ocean the open water is full of Radiolarians of strange and delicate construction, together with some floating Foraminiferans. And these creatures are mostly so small and so transparent that men pass through the midst of them without even realizing their existence. Yet it is they which make our fisheries possible.

In this section we shall consider mainly the well-lighted surface zone of the open sea, which is the prime generator of food for the whole ocean. But first we must introduce a couple of technical but necessary terms. All the open waters taken together make up the *pelagic* or open-sea zone; here life must either float or swim. All the bottom constitutes the *benthic* zone, where crawling or burrowing or fixed attachment becomes possible. Both zones are to be divided again according to depth, into the well-illuminated zone and the deep-sea or *abyssal* zone of darkness, with an intermediate twilit layer between. And both can be divided, according to their approach to land, into the *littoral* zone round the shore, and the great bulk of the waters, the high seas, outside.

The first need for the inhabitants of the surface zone is not to sink, not to lose contact with the light. Some creatures float passively. Others, which we may call the passive swimmers, make movements

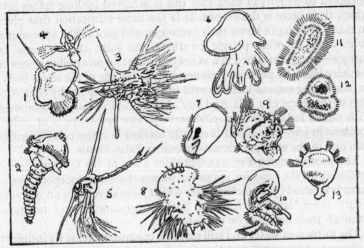

FIG. 242. Some free-swimming larvæ of the sea's surface zone.

(*1*) *The Pilidium larva of a Nemertine worm.* (*2*) *The Trochophore larva of a primitive segmented worm, with several segments already formed; the mouth is to the left.* (*3*) *The Nauplius larva of the goose-barnacle.* (*4 and 5*) *Two views of the Zoœa larva of a crab.* (*6*) *The larva of Phoronis.* (*7*) *The larva of a sea-cucumber; its gut with arms below, is seen by transparency.* (*8*) *The fully developed larva of the sandworm Nereis; it has three segments with bristle-bearing swimming-organs or parapodia.* (*9*) *Young larva of a limpet.* (*10*) *Older limpet larva (veliger stage) with a shell and rudiment of a foot (to right).* (*11*) *The solid planula larva of Clytia, a hydroid polyp resembling Obelia.* (*12*) *The larva of a lamp-shell.* (*13*) *The same just after fixation and metamorphosis.*

not to get anywhere in particular, but merely to keep up in the water. The more active swimming of still others serves not only to prevent sinking but to generate a food-current to be sifted. And finally there are the creatures which swim actively and deliberately seek their prey. Some of these, like the arrow-worms, are microscopic, but the larger, like whales and most fish, are big enough to be able to set currents at defiance and migrate from place to place. These last are sometimes spoken of as the true active swimmers, while all the rest make up the drifting plankton.

The plants of the sea are not nearly so passive as those of land. The diatoms of the plankton merely float; but the dinoflagellates and many others use their flagella actively and incessantly to help keep up in the water. The dinoflagellates and the diatoms also illustrate a common anti-sinking device, the increase of the amount of surface proportional to weight; in their case this is achieved by long spines into which their body is drawn out. It is the same adaptation that gives so many crustacean larvæ their grotesque and gnome-like appearance (Fig. 242). In other pelagic creatures, the same end—increase of surface—may be attained in a variety of ways: by protruding spines or feathery hairs, by projecting planes, by a flattening of the whole body, or (combining surface-increase with muscular movement) by construction in the shape of a bell. And any of these devices may be combined with others for reducing specific gravity. The bell-shape is of course common in jelly-fish. But it is equally marked in many pelagic cuttle-fish, and in the only pelagic sea-cucumber Pelagothuria.

The accompanying Fig. 243 will show some of the fantastic forms produced by the need for spines or planes. In some copepods, resistance is effected by branched hairs which have developed an extraordinary resemblance to feathers, both in appearance and function, although their branches do not interlock.

The saltness of water and still more its coldness increase its incipient stickiness or viscosity, and therefore its resistance to objects sinking through it. This physical peculiarity of the environment is reflected in the most delicate way in the anti-sinking adaptations of floating life. In dinoflagellates, the projecting spines become more elongated in summer and in brackish water. Similarly, feathery hairs are only found in the copepods of warm seas, not in polar species. When one and the same species of plankton-animal is found in polar and tropical regions, it is often found that it lives at the surface in high latitudes, but in low latitudes floats deep, where not only temperature but viscosity is suitable. For the same reason, it would appear, floating animals inhabiting the same depth-stratum are smaller in warm than in cold water, for if they grew larger in warm seas they would sink deeper.

The same fact often brings about a depth-stratification of related forms according to size. For instance, the different species of the Radiolarian genus *Challengeria* get larger as we descend, apparently being mechanically sorted out by their ability to float, and the same is true of arrow-worms and certain prawns and fish. A good example is the Atlantic fish *Cyclothone microdon*. Its average length is about one-and-a-quarter inch at 500 metres depth, double as long at 1,500 metres.

This anti-sinking trick of increasing surface is, it will be noted, often combined with some power of swimming. If the swimmers wanted to get anywhere in particular, the friction of the extra surface would hinder them; but as their swimming is only meant to keep them up in

FIG. 243. Adaptations to a floating life.

(A) The phosphorescent protozoön Noctiluca is lightened by containing comparatively fresh water. The other organisms are (B) Glaucus, a sea-slug; (C) Calocalanus plumulosis, a copepod crustacean with a huge tail; (D and E) Two kinds of flagellate. (F) A chain of diatoms. (G) Another species of Calocalanus. (H) The larva of the rock-lobster or langouste, Palinurus. (I) the larva of the common angler-fish, Lophius. All these have spiny or feathery outgrowths to prevent sinking. In addition the last two have flattened anti-sinking planes.

the water, the spines and feathers save some expenditure of muscular or ciliary energy. The same is true for most of the other passive antisinking devices of pelagic life.

As protoplasm is a little heavier than sea-water, any dilution of living bodies with water will lighten them and make them sink more slowly; and if, as is usually the case, the water is bound within a jelly, this can help support the animals—it can act as a primitive skeleton.

Among the ranks of water-swollen jellified pelagic sea-beasts are the horde of jelly-fish, in some of which all but one per cent of the body is water; all the comb-jellies, and many Siphonophores; swimming snails, and even octopuses and cuttle-fish; a few transparent fish; and the remarkable leaf-like larvæ of the eels. In the pelagic sea-squirts, the Salps and Pyrosomes, the same result is achieved by having the tunic of cellulose swollen with water.

Other organisms do not merely dilute their weight but counteract it by accumulating lighter substances inside themselves. Sometimes, as in the common phosphorescent Noctiluca, the main flotation material may be merely water which is less salty and therefore lighter than sea-water. But in most cases, the much more efficient method of using fat or oil is adopted; the creature is buoyed up by the microscopic liquid balloons scattered through its tissues. For instance, most small pelagic Crustacea have abundant fat in their tissues, as does the pelagic clam, Planktomya. It is no accident that the food reserves in the liver of Selachians and the abundant cod tribe are stored in the form of oil instead of the heavier glycogen of most vertebrates; it is interesting to reflect that, since vitamin A is soluble in oil or fat but not in other substances, the desirability of a low specific gravity has led to this vitamin, produced in abundance by the diatoms of the sea, being stored in easily accessible form in cod-liver oil, whereas had the specific gravity of the sea been a trifle greater, the cod might have stored its reserves as glycogen, and the bulk of the vitamin would have been destroyed in its body or excreted as waste.

Numerous free-floating fish-eggs, like those of mackerel, also owe their capacity of floating to oil-globules. The sun-fishes (*Mola*) and gigantic basking sharks (*Cetorhinus*) lounge lazily at the surface; they are only enabled to do this by the thick layers of fat below their skin, which reduce their specific gravity to that of sea-water.

The layers of blubber found in whales, dolphins, seals, and penguins, though doubtless indispensable for the minimizing of heat-loss by their bad conduction, serve also a secondary purpose in lightening the animals, and so rendering more of their muscular energy available for directed activities.

The most efficient lightening mechanisms are balloons of air or other gas; but the construction of these involves a high degree of specialization, and they are not common. Many Siphonophores have gas-glands which secrete gas into a special bladder; and the Pearly Nautilus and its strange relative Spirula still store gas in the chambers of their shells, as did all the Ammonites and Nautiloids of bygone eras. In one pelagic snail, Glaucus, the generation of gas by bacteria in the intestine has been turned to account, and the animal owes its capacity for floating to a flatulence made normal and physiological.

Gas-bladders are best developed in bony fish. The great majority of their abundant species owe their success to speed and directed activity;

to secure this they must dispense with all such aids to flotation as jelly or protruding spines or flaps, and concentrate on muscle and skeleton, and purity of stream-lined form. But as a result their tissues in general are a good deal heavier than sea-water, so that without a gas-bladder much of their energy would have to be expended in the never-ceasing task of fighting gravity. A few pelagic fish succeed in this task, such as

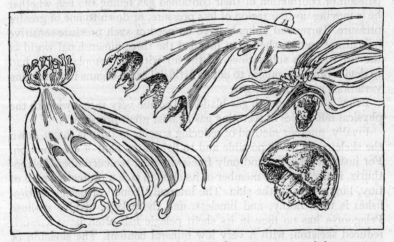

FIG. 244. Parallel evolution towards the bell shape in creatures of the open seas.

Above is an octopus, Cirrothauma, which has become bell-like by growing a membrane between its arms. Below it are two individuals of the pink pelagic sea-cucumber, Pelagothuria, with a fringe of tentacles held together by webbing. The one on the left has just made a downward swimming stroke with its tentacles. Below on the right is a large jellyfish, Stomolophus.

the immensely powerful sharks. The common mackerel, too, like its close relative the tunny, has also lost its air-bladder, presumably to achieve greater speed; as the result of its unceasing activity during the feeding season, it has to recuperate for several months of the year resting on the bottom without eating.

Apart from such rare exceptions active pelagic fish possess gas-bladders and are thereby enabled to regulate their specific gravity to that of the water in which they live, and so to utilize every ounce of their muscular power for pursuit or escape. If the gas-bladder had not been evolved, fish must have remained a predominantly shallow-water group because of the necessity of resting upon the bottom. As it is, the possession of a gas-bladder by almost all bony fish and the lack of it in all gristly fish like sharks and rays is reflected in the fact that a far greater proportion of the former are animals of the open water, while a far greater proportion of the latter spend most of their time near the bottom, alternately resting and indulging in sharp bursts of active swimming.

Gas-bladders may also act as sensitive pressure-recorders, the expansion or contraction of their contained gas telling the fish whether he is moving up to a region of less pressure, or down to one of greater pressure. Terrestrial animals have no need of such pressure-sensitive indicators; and so when man invaded the three-dimensional world of the sea with his submarines, of the air with his aeroplanes, he had to devise pressure-gauges to act as artificial sense-organs for recording variations in vertical height.

Below a certain depth gas-bladders become very rare, owing to the physical difficulties of secreting gas against great pressure.

Finally, another method of reducing specific gravity is to cut down the skeleton as far as possible and to lighten its plan of construction. For instance, the one and only free-swimming sea-cucumber, Pelagothuria, is also the only member of its class to lack a protective mail of tiny, limy plates in the skin. The internal "bone" of pelagic cuttlefishes is thin, horny, and limeless; and the pelagic bivalve mollusc Pelagomya has no lime in its shell; pelagic fish often have a much reduced skeleton, with a very low mineral content. The skeleton of Crustacea shows a progressive diminution of mineral matter as we pass from permanent bottom-dwellers to permanently open-water forms. Free-swimming Copepods of the plankton only have six or seven per cent of mineral matter in their constitution, and are lightened by having five or six per cent of fat. The heavy crawling shore-crab, on the other hand, has less than three per cent of fat, and over forty per cent of mineral matter.

Before passing to the deep sea, mention must be made of the strange life of calm regions of the ocean, of which the Sargasso Sea is one of the largest and best known. The currents which flow past such calms abandon into them a proportion of the floating animals and plants which they are bearing along. In the Sargasso Sea the most important of such flotsam is the gulf weed, *Sargassum bacciferum*. This grows on the coasts of the Caribbean Sea; pieces of it, broken off by the waves, float along in the Gulf Stream, supported by their gas-bladders, and accumulate in dense masses in the calm zone. Eventually their bladders decay, and they sink and die without posterity. But meanwhile, before their death, they support an abundant crop of strange animals. Some of these exist elsewhere; but some live only in this curious accumulation of doomed plant fragments, constantly passing from one temporary home to another. The animals often resemble the weed in an extraordinary way, both in colour and form. The weed is golden-brown, patched white with colonies of Polyzoa and with little tube-living worms. This coloration is frequent among its inhabitants; and in addition they are often beset with fantastic ragged lappets and membranes which mimic the leaves and branches of the weed. These devices appear to have been developed as a protection against the attacks of the sharp-eyed sea-birds which hover above this seaweed garden of the calm.

§ 3. The Deep Sea.

The inhabitants of the deep sea were to all intents and purposes unknown until the voyage of H. M. S. "Challenger" in 1872 to 1876. One of its objects was to discover the primitive ancestral forms of life which were suspected at that time to be lurking in the depths. This expectation, however, as we mentioned in Book 2 (Chap. 7, §1) was unrealized. When the deep-sea creatures were eventually brought by human ingenuity to the light of day, they were often more grotesque and strange than any imagination had dared to picture, but they were not particularly primitive. All sorts of families of normal surface-living creatures, including some of the most modern and specialized groups, have in fact contributed immigrants to this strange region; and among fish almost all the deep-sea creatures belong to the less primitive group, the bony fish, very few to the more primitive sharks and rays.

No living Trilobites or sea-scorpions or extinct types of Echinoderms, no Ammonites or Ostracoderms or primitive armoured fish have been hauled up to the light of an upper world that had outgrown them. It is possible that the marine abyss was such a difficult region to colonize that it stood largely untenanted through most of geological time, and that its invasion only began on a large scale when the more specialized types of sea-animals had come into existence—the prawns and crabs, the cuttle-fish and octopus, the bony fishes. Representatives of other groups had doubtless invaded it before, so that it was not as barren and empty as the land: but it offered no such encouragement and variety as the land, and so there was no blossoming of one or two special invading stocks into flourishing new groups, as happened with land-plants or vertebrates or insects, but a sporadic fitting of a number of isolated types to the queer and specialized conditions, as has happened in other queer and specialized habitats like caves or salt lakes.

Most authorities believe that the invasion of the great deeps by fish began no earlier than Cretaceous times (IV C).

If all marine life depends on the plants which build themselves out of inorganic matter in the narrow illuminated zone, then obviously

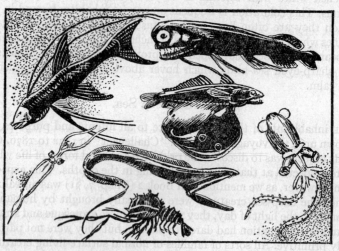

FIG. 245. Deep-sea creatures: four fishes, two squids, and a crustacean.

Top left, Bathypterois, with degenerate eyes and long sensory feelers on its head. Top right, Pachystomias, predaceous, with a sensory whisker on its chin, and two luminous organs close to its huge eye. Upper centre, Chiasmodus, which is carrying a fish bigger than itself in its distensible stomach. Lower centre, Gastrostomus, with enormous gape; it also has a distensible belly. Lower left and right, two squids, Toxeuma and Bathothauma, with huge stalked eyes. Below, centre, Eryoneicus, a blind prawn.

only this zone is biologically self-supporting, and the life at all greater depths must somehow exist at its expense. That is in fact so; deep-sea life is an unproductive assemblage of types nourished by scraps from the banquet spread above in the light of day.

Not every surface organism finds its fate in another's stomach. Thousands die from other causes, and their corpses, once their swimming movements cease, or their floats decay away, must sink towards the unilluminated depths. Over the whole of the ocean the rain of corpses proceeds without cessation, and the bottom is in large part covered with thick deposits made of their skeletons. This rain of death, however, gives life to all the deeper layers; for it is their only source of food. It is a constant stream of manna from above—often, it is true, in process of decomposition, but none the less nutritious.

The inhabitants of the regions passing down from the surface to the lightless abyss do not live haphazard at any depth. They each have their appointed place, each living out its life in its own particular layer of cold and silent blackness, often thousands of feet above the bottom, thousands of feet below the surface. The depth-range of some is comparatively large, while others are confined to a single level of the sea only a few hundred yards thick; and the zoning is complicated by the vertical movements which so many of the creatures nearer to the surface execute periodically every twenty-four hours, ascending at night and descending by day. None the less, the most striking feature of the life of the open sea is this stratification. Most pelagic animals live in definite storeys just as do the inhabitants of a skyscraping block of flats. We do not see the floors and roofs which limit the different zones. They are none the less perfectly real, and are constituted by the barriers set by degrees of temperature, pressure, and salinity and by the kind of food available.

For the food-rain will, naturally, not be the same near the surface as it is in the depths. Each layer will contribute its quota of dead vegetation and corpses to the layers below; and each layer will take its toll of the corpses from the layers above, altering not only the quantity which finds its way farther downstairs, but also the quality, since not all kinds of corpses will be eaten in the same proportions. The total amount of the food-rain will grow progressively less, since at each level there is a wastage, due to bacterial decay, with consequent dissipation of solid food into unavailable solution. On the whole, the smaller particles, of which the tiny plants of the surface make a large proportion, will be snapped up or decay away before the larger, so that the average size of the drops of the food-rain will increase with depth. They will be the biggest particles, the quickest sinkers, and the less rapidly decaying lumps. They will be bigger and rarer. We should therefore expect to find, as we go downwards, first a decrease in the amount of life in general; then a decrease in the proportion of current-feeding organisms, and an increase not only in carnivores, but especially in animals adapted for taking advantage of relatively large but infrequent mouthfuls. And these expectations are realized. On the other hand, the scarcity of food is so great that no large animals can exist in the great depths. The monstrous and fantastic deep-sea fish are almost all under a foot long. They are all extraordinarily voracious, some with jaws capable of grasping and swallowing bodies larger than the normal size of the elastic devourer. Their lives are a prolonged fast, varied by some swift mouthful and repletion.

It is worth while anticipating our story a little, and reminding ourselves that essentially the same conditions hold in forests, especially evergreen forests, and most particularly in the tropical rain-forests. Here again light, coming from above, is life's prime mover. Seen from above, the forest has an exuberant surface of green dense leafage, inhabited by a thousand birds, mammals, and insects which never descend to the terrestrial depths. Below the bright surface is a dim twilight, with a much-diminished life, the plants consisting mainly of parasites sucking nourishment from the trunks of the trees, or vines and lianas climbing up them towards fuller light, or epiphytes seeding themselves on the great trees' leaves and branches; and the space from roof to floor is divided into definite zones of life. The floor of the forest, like that of the ocean, is covered with the débris of the upper layers. Leaf-corpses are continually floating down, twigs and fruits falling, with an occasional animal body. The most essential difference is that in the forest the food-rain is not so slow-falling, and thus cannot be raided by layers of creatures on its way to the bottom. On the bottom it accumulates as mould, and gives nourishment to many fungi and bacteria and a horde of scavenging animals.

The tree-trunks support the illuminated surface. Competition between all the species and all the individuals of each species has finally resulted in a nearly flat sea of foliage, pushed heavenwards to the limit of the trunks' mechanical possibilities. The trunks are also supply-pipes bringing the necessary nitrogen, sodium, chlorine and other mineral constituents of plant-protoplasm up from the soil to the life-factory of the leaves. There is nothing like this in the ocean; but even there, as we have seen in § 1 of this chapter, the depths provide essential mineral supplies to the plant-life of the surface. In any case, it is as well to be reminded of the fact that the forest, like the sea, has its productive region at its upper surface, and that the solid ground on which we men are compelled to walk through it is a dependent parasitic zone as truly as are the ocean depths. The dependence upon a slow food-rain from above is the most important of the conditions influencing deep-sea life; and then, in order of descending importance, come the darkness, the cold, and the pressure.

The pressure at great depths increases by almost exactly one atmosphere for every ten metres of water. At the bottom of the great deep near the Philippines (9,788 metres) the pressure would be over 960 times as great as the ordinary pressure of air at sea-level—equivalent to the weight of a column of mercury nearly half a mile high! No animals have yet been brought up from over 7,000 metres (about 23,000 feet), but this is probably due more to the enormous technical difficulties of dredging at greater depths than to an actual absence of animal life.

It used to be supposed a century ago that no animals could exist at depths over one or two thousand feet, owing to the pressure. As a matter of fact, these great pressures make singularly little difference to animals, for the simple reason that the pressure of the water is externally the same on all sides, and is equalized internally by the pressure of the blood and the fluid contents of the cells. The animal exposed to a pressure equivalent to a quarter of a mile of mercury feels it no more than we feel the atmospheric pressure—15 pounds on every square inch of our bodies. The popular belief that the high pressure increases the density of the water so much that sunken ships would remain suspended when they reached a certain depth is quite unsound. Water is so incompressible that the increase in density at a depth of say 1,800 feet is only about $2\frac{1}{2}$ per cent; salt water with a specific gravity of 1.028 at the surface would at this depth have one of only 1.054.

Even the evil effects of rapidly bringing animals to the surface from the great pressures of the deeps have been much exaggerated. It is perfectly true that they are usually dead or dying when brought to the surface; but this is due much more to change of temperature than to change of pressure. This is shown by deep-sea hauls in the Mediterranean. The waters of this sea, owing to its excess evaporation, are more salty, and therefore heavier, than those of the Atlantic. Accordingly, the outflowing current in the Straits of Gibraltar sinks, and the inflowing current is near the surface. The Straits of Gibraltar make a sill nowhere more than 1,500 feet deep; and as a result only the highest and therefore the warmest layers of Atlantic water flow in. The Mediterranean thus receives no water colder than 12.9° C.; the whole of its contents from about 500 feet down to its greatest depths of nearly 13,000 feet are uniformly of this temperature. And we find that deep-sea animals can be brought to the surface in the Mediterranean, cuttle-fish for example, without showing any of the ill-effects which are suffered by kindred Atlantic animals brought up from the same depth.

The one exception to this innocuousness of pressure-change is afforded by fishes with closed gas-bladders. In spite of the arrangements which exist for resorbing gas from the bladder into solution in the blood, the expansion caused by rapid release of pressure is too great, the bladder swells up enormously as the fish is hauled to the surface, and forces the entrails out at the mouth, or even blows the fish into fragments.

That rapid but considerable pressure-change need have no ill effect is shown by the large daily vertical migration of various creatures. Many plankton organisms and pelagic fish migrate up to near the surface by night and down to 400 or 500 metres by day. In spite of the forty-fold change of pressure to which they thus expose themselves twice in every twenty-four hours, they suffer no more than do Swiss goats and goatherds in their customary vertical wanderings, up by day and down to the villages at night.

Sperm-whales dive down to great depths in search of their prey, the giant cuttle-fish; and all whales do so when harpooned; they may traverse over a thousand feet of vertical height in a minute or so, and some may be capable of diving to half a mile. A diver would die if he were hauled up from such depths so rapidly, through the appearance of bubbles of nitrogen in his blood (just as bubbles of carbonic acid gas appear in soda-water when it is released from the pressure within the siphon), which would then stop his circulation. Whales do not suffer in this way at all, largely, it would appear, through the presence of so-called *retia mirabilia* on their veins, "marvellous networks" where the vein breaks up into a great number of small vessels for a short section of its course. The gas-bubbles are probably caught in some of these small vessels and held there until they can be redissolved.

In all seas save the Mediterranean, temperature sinks steadily with depth. The average temperature for the great oceans is about 16° C. at 600 feet, only 10° at 1,200. At 3,000 it is only 4.5° C., while at 12,000 feet it is down again to 1.8° C.—only just above freezing. What is more, the deeper we go, the less variation is there in the temperature; at over 600 feet down, the seasonal change of temperature is trivial or nil. Nowhere else is life's environment, whether as regards light, temperature, or chemical and physical conditions, so uniform as in the marine abyss.

The darkness in the depths is complete, save for what light is generated by their living inhabitants. Water rapidly cuts off light. Light pouring directly down penetrates much deeper than oblique light; thus beyond a certain depth only the noon-day sun will penetrate, and so the length of the day decreases with the depth. In the harbour of Funchal, Madeira the 12-hour March day is reduced to 11 hours at 20 metres depth, to 5 hours at 30 metres, to a mere 15 minutes at 40 metres.

In general, light sufficient for the needs of green plants extends down, even in the purest water, only to about 100 metres. The great majority of marine plants live in the first 50 metres; at 75 metres there are only half as many plant individuals as at 50 metres.

Below about 200 metres, the only plants which exist are sinking downwards dead or damaged. Below this, the deep sea begins, with no living green plants and no true vegetarian animals, but only decay-producers, scavengers, and predatory creatures.

The longer wave-lengths of light are more quickly absorbed by water than are the shorter. All the red rays are absorbed soon after 100 metres, and all the green before 500, but some of the violet penetrate deeper than 1,000 metres. Even at 30 metres, a diver cannot see red objects as red; there is no red light for them to reflect and so they look black.

Most animals from the illuminated surface zone are of crystal transparency (sometimes, however, with a few coloured internal organs revealed). When colour is present, it is generally blue, as in the Portuguese Man-o'-war and other Siphonophores, some Copepods, and the lovely floating snail Ianthina. Many surface fish are bluish or greenish above, silvery below, like the mackerel. These colours, like transparency, would appear to have some protective value in the blue sea-water, though transparency seems to take its origin as a mere accident of jellification.

But in the depths, to reflect no light and so to fade into the surrounding blackness is the best means of protection from all those enemies which hunt by sight. And as red then looks black, red shares predominance with black and other dark colours such as deep brown or purple-black. Some groups, such as the fish, seem to find it difficult to produce red pigment, and in the deep sea incline to black; other deep-sea groups, such as Crustacea, are almost without exception red. There are also red Foraminifera and polyps, red sea-anemones and starfish, red octopuses and cuttle-fish, and even red arrow-worms. An interesting case is Velella, the Siphonophore. This, when adult, floats at the surface, with a sail projecting up into the air, and both animal and sail are a lovely blue. Its larva, however, lives in the deep sea and is red; it only becomes blue when it ascends to adult life at the surface.

Although some deep-sea gelatinous animals retain their crystalline appearance, yet jelly need not necessarily retain transparency, as is shown by the fact that many deep-sea jelly-fish and arrow-worms are dark-coloured or red: some, such as the jelly-fish Atolla, are transparent at the surface, and grow darker and darker with depth.

At the depths where these red prawns and black fishes live, whatever light is visible is produced by the animals themselves. Phosphorescent organs occur in almost every group of deep-sea animals, both bottom-living and pelagic, in polyps and corals, starfish and bristle-worms, cuttle-fish, Crustacea and fish. These organs are of various kinds. Most of the fixed bottom-living forms—sea-pens, sea-fans, and so forth—shine all over, diffusely, often with a marvellous radiance. It

FIG. 246. Luminosity in the deep sea.

Top left, the squid Pterygoteuthis, with big eyes and a pattern of luminous organs on head and trunk. Below it, the prawn Heterocarpus puffing out clouds of luminous secretion to baffle its pursuers. Centre, the gastropod mollusc Phyllirrhoe, with many small light-organs. Below, three fishes: from left to right, Vinciguerria, with light-organs looking like illuminated port-holes; Gigantactis, with a luminous nose which serves as a lure; and, Aulastomatomorpha, with its whole face and head luminous.

may be that phosphorescence of this type is a mere accident, serving no useful purpose, and that for unknown reasons conditions in the deep sea are especially favourable to its production. There are phosphorescent animals in plenty at the surface of the sea, and the luminosity of some of them, such as the abundant Noctiluca, seems to be as functionless and accidental as that of fungi on rotten wood. On the other hand, animals of this diffusely phosphorescent type are more abundant and produce more brilliant light in the deep sea than elsewhere. It has been suggested that the light serves to lure small fish and other prey within range of the polyps and their paralyzing tentacles.

This is made more probable by the extraordinary luminous organs of deep-sea angler-fish, which undoubtedly act as lures for prey. The shallow-water angler (Lophius) lies flat and well concealed on the bottom, with its huge predaceous mouth facing upwards. It grows a fishing-rod and bait on its head. The foremost ray of its dorsal fin has suffered a curious transformation; it is elongated, detached from the rest of its fins, and bears at its end an enlarged flap of membrane. The jerking of this attracts inquisitive and hungry fishlets, which in a moment are snapped up by the unsuspected jaws just below. As Aristotle observed: "They are often caught with mullet, the swiftest of fish, in their interior. Furthermore, the frog-fish is usually thin when he is caught after losing the tips of his filaments." In the deep sea, numerous members of this same family are found (though they are all much smaller in size): they also all attract their prey by means of lures, but these lures are always luminous. Perhaps the most extraordinary of these animals is Lasiognathus, whose fishing-rod in addition to possessing a luminous lure has a hinge in the middle and hooks at the tip. It seems to be thrown forwards when prey has been attracted near by the light, and then jerked back so as to impale them on the hooks.

It is as recognition marks that the luminous organs of deep-sea animals have their most frequent significance. The prowling population can never be dense; and without special advertisement it would often be difficult for the sexes to find each other. The advertisement takes the form of a pattern of luminous organs, each species having its own characteristic arrangement. Sometimes there is a single row of small lights down the body; sometimes tiers of lights, so that the fish looks like a liner at night; deep-sea cuttle-fish generally have a pattern dotted over their body and arms; some animals, both fish and cuttle-fish, make play with lights of different colours, or with different-sized lights. In any case, each species thus carries its own identity written upon it in letters of cold fire, can know its own kind at a distance, can recognize such or such a pattern as the badge of its enemy, this other as the badge of its prey.

But there are other uses of luminosity. Shallow-water cuttle-fish escape from their enemies under cover of a smoke-screen of ejected ink. This would be of no service where all is already black; and accordingly we find that some of their deep-sea relatives squirt out a luminous cloud to confuse pursuers.

Finally there are the luminous organs designed, like head-lights, to facilitate the animal's own seeing. The most curious example of this is the living torch carried by a hermit-crab from the Indian Ocean. Like many other kinds of hermit-crab, this animal carries on its snail-shell house a sea-anemone as partner. Doubtless the sea-anemone profits as do its fellows in similar situations in shallow water, by catching crumbs that fall from the crab's table. The benefit conferred in return is normally that of protection, by means of the serried batteries of stinging cells on the anemone's tentacles and the formidable stinging threads, which it can protrude through the portholes in the wall of its body, or its mouth. But in this case the anemone is phosphorescent, and confers the boon of light as well.

Frequently, in deep-sea fish, cuttle-fish, and Crustacea alike, there is a concentration of especially large light-organs near the eye, and in some cases these have evolved into veritable projectors. Cells which generate a luminous secretion are placed in front of a curved reflector made of a glistening membrane backed by black pigment, and the outward-streaming light is concentrated into a beam by means of a lens.

Where light is so scarce and so dim, eyes may be expected to evolve in one or other of two opposite directions. Either they may degenerate, wholly or in part, and the animal make up for its lack of vision by an excess development of other senses such as touch; or else they may be enlarged and improved to the pitch of extreme sensibility so as to catch every faint glimmer of light. Both of these methods are found in deep-sea animals. On the whole, the balance has been in favour of keeping and improving the eye. But bottom-livers leading a crawling-feeding life show eye-degeneration more frequently than do open-water creatures, and in both situations the more active kinds of creatures tend to enlarge their eyes, the less active to lose them.

In general, those deep-sea animals which have well-developed eyes include those which have definite arrangements of luminous organs, and, of course, those which possess searchlight organs. Recognition marks would be useless to a blind animal. But the deep-sea anglers are often themselves blind. Their light is aimed at other species; and so long as the inquisitive prey is lured alongside, it does not matter whether sight, touch, or any other sense is used to detect its presence.

Sometimes the eyes are positively enormous; in the crustacean Cystosoma, the eye occupies at least two-thirds of the surface of the head. In other deep-sea creatures (and nowhere else in nature), the so-called "telescopic eye" is found, which has been independently developed in eight sub-orders of deep-sea fish and in one cuttle-fish, while a somewhat similar arrangement is found in some deep-sea Crustacea. This type of eye, to be precise, is not "telescopic" at all;

it is an adaptation to save space. The bigger the surface of the lens, the more light does it collect; the more it approaches the spherical in shape, the more does it concentrate the light that falls upon its surface. With a lens as large as exists in many telescopic-eyed fish, an eye of the usual spherical type would be as big as the whole head. The so-called telescopic eye is simply the central cylindrical portion of an eye with a spherical lens. All the light that falls on it is concentrated upon a very small patch of retina; and range of vision is sacrificed to the pressing need for light-collection and concentration.

Even with such eyes, sight in the deep sea may well need to be supplemented; and accordingly it is common to find extraordinary developments of touch-organs, such as barbels on the head, or elongated spines or fin-rays acting like cat's whiskers, and most deep-sea fish have the mysterious lateral-line sense-organs better developed than their surface relatives.

One unique result of deep-sea life is the development of parasitism in the males of certain angler-fish. In these species, until recently, only females were known. Eventually some of these females were discovered to have growing on their bodies strange objects not unlike miniature and deformed fish. Further investigation showed these to be indeed fish—the long looked-for males of the species. When quite young they apparently bite on to the female, gradually embedding their snout in her skin. Both skin and blood-vessels of male and female grow together, and the male becomes a true parasite, nourished entirely at the female's expense. His heart and digestive system degenerate and the bulk of his body becomes filled with testis. Presumably when the female lays her eggs, some chemical stimulus acts upon the male and causes him to eject his fertilizing sperm at the same time.

The difficulty of male and female finding each other at mating-time in the dark and sparsely populated waters of the deep pelagic zone must be very great. This strange state of affairs—the only case of thorough-going parasitism in vertebrates—is without doubt a method for circumventing this difficulty.

§ 4. On the Floor of the Abyss.

NEAR inshore the bottom of the sea is covered with deposits derived from the land, eroded bits of land carried out to sea by rivers. Naturally the heavier particles settle first, so that in general the farther from the land, the finer-grained are the deposits. Such land-derived deposits, however, are rare at depths of over six hundred feet. Below that depth the floor of the ocean is covered with layers of material derived either from the skeletons of the animals and plants that live above, in the upper storeys of the sea, or else from volcanic dust. Some of the organic materials are predominantly limy, such as globigerina ooze

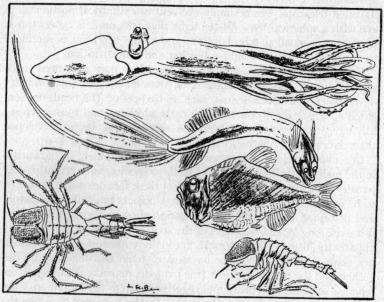

FIG. 247. "Telescope eyes" and other adaptations for seeing in dim light.

Above, the cephalopod Amphitritus. Center, two fish: upper, Gigantura; lower, Argyropelecus. Below, two views of the crustacean Cystosoma, with eyes so big that they cover all the upper surface of the head and meet on the crown like a dragon-fly's. The eyes of Gigantura look forwards; all the others look upwards.

largely made of the skeletons of that little foraminiferan, or the much rarer pteropod ooze in which the shells of those active swimming mollusca predominate. Others are mainly flinty, such as diatom ooze and radiolarian ooze. The finely powdered volcanic accumulations are known as red clay.

The calcareous deposits are poor or absent in high latitudes, owing to the difficulty of secreting a calcareous skeleton at low temperatures. They also never occur below a certain depth, usually about fifteen thousand feet, since under great pressures the calcareous matter passes into solution. But they have an enormous extent: globigerina ooze, especially abundant in the North Atlantic, covers nearly thirty per cent of the whole ocean bottom. Siliceous deposits can go a little deeper without being dissolved. They are especially abundant in colder seas, and in regions where the water contains much siliceous matter, for instance in many parts of the Indian Ocean, but cover less than ten per cent of the ocean bottom. Finally, in increasing quantities from a depth of thirteen thousand feet downwards, comes the red clay, covering over a third of the whole sea-bed. Only over very small areas is the ocean floor clear of ooze—here and there where deep currents flow powerfully, perhaps here and there where a rare slope occurs steep enough to shed its accumulations.

All these deposits have two things in common—they are soft, with a consistency something like that of butter on a very hot day; and they contain all the débris of the food-rain which has escaped the population above.

FIG. 248. Life on the muddy floor of the deep sea.

Sea-squirts (left) and glass-sponges (right) are carried clear of the mud on long stalks. Sea-anemones grow on the sea-squirts, and Pycnognoids crawl over them. A prawn walks over the mud on enormous stilt-like legs, and flattened sea-woodlice and starfish crawl on its soft surface. (Based on an exhibit in the Natural History Museum, London, of animals collected by the "Discovery" Expedition.)

It pays many organisms to sift and scavenge in and on and close above this rich ooze; but precaution must be taken by the surface animals not to sink in, and by the pure current-producers not to have their works clogged with the fine ooze. The surface scavengers and predaceous animals almost all show devices for spreading their weight over a large area. Some of the spider-crabs and other Crustacea have enormously developed legs, often with bristles or feather-hairs at the tips; some of the flat, cake-like sea-urchins of the deep sea have an expanded disk-like surface, and the sea-cucumbers have a special flattened sole. One or two of the current-producers have similar devices, such as the sponges which possess projecting collars of long spicules to prevent them sinking in the mud. But the great majority are stalked. Stalked sponges, stalked sea-lilies, stalked corals, stalked single polyps, stalked polyp colonies like the sea-pens with their strange bulb-like base thrust into the slime for anchorage, even stalked lampshells, sit balanced above the suffocating mud, on whose surface precariously crawl and forage the tribes of active scavengers.

The Hexactinellid sponges are a group almost entirely confined to the deep sea. They differ from all other sponges in the open lattice-work of their construction. The best known Hexactinellid is the lovely Venus's Flower-basket from the deeps off Japan. We marvel at its delicacy and beauty; but we generally fail to ask ourselves the biological reason for the construction which confers such properties. A simple reason, however, does exist. Since warm water has a lower specific gravity than cold, any currents flowing polewards from the warmed surface of tropical seas will, like the Gulf Stream, float upon the top; and the necessary return flow of chilled water from the poles will flow along the bottom. The Hexactinellids grow in the path of these slow but constant bottom-currents. They can afford to dispense with all the current-producing machinery of the ordinary sponge and can simply spread a lattice-net in the path of the débris-laden stream.

As Dr. Bidder writes of the group as a whole: "Food is brought to them, waste is taken away. For them in their eternal abyss, with its time-like stream, there is no hurry, there is no return. Such an organism becomes a mere living screen between the used half of the universe and the unused half—a moment of active metabolism between the unknown future and the exhausted past."

§ 5. Sea-shore Life.

AND now let us return from the extreme specialization of the deep sea to the primitive cradle of life—the shore, the intertidal region. It is so familiar to most people, and has been well described in so many admirable books, from Charles Kingsley's *Glaucus* to Flatteley and Walton's *Biology of the Sea-shore*, that we, with space pressing, can afford to treat it briefly.

LIFE IN THE SEA

We can best define the seashore as the strip of earth's surface between the highest land wetted by the sea and storms and the lowest low-tide mark. It has a very great biological interest. It is a zone of sharp transition from dry-land to salt-water; and, though a zone of transition, it is a permanent feature of our world. The outlines of sea and land may shift over the globe; but there is always a shore. Twice a day most of it is covered by water, and twice a day uncovered to the air as the tide ebbs off it. Thus, of all the habitats in the world, the shore is exposed to the greatest fluctuations and has the most variable conditions of life, in the completest contrast with the seasonless, dayless, changeless depths.

As we step across the shore from high-water mark to bathe at low tide, each point that we cross is exposed to a longer daily dose of water, a shorter dose of air. Accordingly, the shore is split up into parallel belts of zones of life according to the capacity of its animals and plants to resist exposure to air. At extreme high-water mark, or even above it, flicked only by occasional spray, there are still barnacles and the dwarf species of periwinkle, there are green seaweeds in the brackish pools, and the sea-slater Ligia runs about among the stones, a giant woodlouse that is tied here between sea and land. Between tide-marks brown seaweeds predominate, a number of kinds of them zone by zone as we descend. Most have gas-bladders on their leaves to lift them up towards the light directly the waters cover them, and to help minimize the danger of being smashed against the rocks by breakers.

Below them comes a zone of great ribbon-weeds (Laminaria), that are only uncovered at the lowest spring tides; and these are mixed with many red seaweeds, which grow rapidly scarcer as we return landwards.

The creatures of the shore wake to activity not once but twice in every twenty-four hours, when the tide comes up and they can expand in the salt embrace of sea-water, the primeval fluid that bathes and supports, brings food and distributes marrying cells and spores. Many of the obvious characteristics of sea-shore creatures are devices for living through the periods between the vivifying visits of the water. The barnacle has its doors uncompromisingly shut while it is in air. The door, however, is on the chain rather than barred and bolted tight. The slight sizzling noise you may hear on a still day on barnacle-encrusted rocks is the multitudinous tiny bubbling of the thousands of fixed and shell-imprisoned crustaceans. Then, for an hour or so each day, when the waters solicit attention, the panels slide back and the microscopic casting-net of feathered legs begins its rhythmic task. The limpet, under water, makes journeys a few yards long, to browse on weed; when the tide falls he returns to his familiar home, a slight depression in the rock which he has worn, and sits upon it with tent-like shell just raised to admit air. The mussels change as the tide goes out from busy sifters of débris to passive lumps of flesh shut in a strong blue shell. The tube-worms under the daylight air are all deep in their tubes, waiting for the water's renewal before they spread their flower-like tentacles; many of them have little doors to close the mouths of their homes.

The flattened shape of shore-crabs is adapted for retreat into crevices when the tide goes down; there to stay, audibly bubbling as they breathe, until the water rises again and they can sally forth on their scavenging jobs. In the moist miniature caves under rock ledges, sea-squirts and brilliantly coloured soft crusts of sponge wait with their apertures tight closed; sea-anemones are converted from carnivorous flowers to mere jelly-blobs. Some sponges that live in this situation differ from their permanently submerged relatives, which are barrel-shaped, in being flat and leaf-like. When the tide ebbs away, their two sides come together and retain a film of water.

Out on the sand and the mud, life disappears when the tide goes down. The buried cockles and clams and razor-shells withdraw their siphons—water-inlets and outlets—far beneath the surface, and the sand-eels burrow deep. Many creatures seek refuge from the air under stones—crabs and other small crustaceans of many kinds, little star-fish, long green or black Nemertine worms coiled up like skeins of ribbon. Even some fish and an occasional young conger can be uncovered by turning over the rocks between tide-marks. Among the colonies of hydroids and polyzoa that grow on rock and weed the tiny polyps are all withdrawn into their sheltering cups, which are often provided with some form of lid; and the periwinkles, withdrawn into their strong shells, close their homes with the horny door they carry attached to their foot.

But hard upon the retreat of the waters comes another fauna—a multitude of birds and mammals (man the chief among the latter) searching the pools and sounding the mud for the hiding water-feeders. Here and there in this world of waiting, hidden creatures are pools left by the retreating sea; and in these the varied population is a revelation to the ordinary landsman. There are so many patterns of life among them which he has never met before, so many unfamiliar ways of living. The rock-pools are seas in miniature. There you may find animals getting their food by ways unknown on land—current-sifting tube-worms and molluscs, plant-like tentacle-spreaders like hydroids and sea-anemones (it is an unforgettable sight to see an unwary fish, venturing too near one of these animal blossoms, suddenly trapped and paralyzed by a tentacle and drawn down to the gaping mouth);

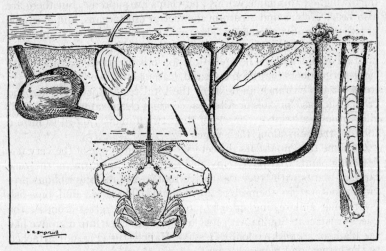

FIG. 249. Adaptation to life in sand.

From left to right: A heart-urchin eating sand in its burrow; it keeps the entrance passage open by means of a special long tube-foot. A clam with its inlet-tube collecting water and fine particles from the surface of the sand, its outlet-tube discharging upwards. A tube of Balanoglossus (Fig. 77). The tube is about eighteen inches deep; the sand which the animal eats is pushed out as castings. Between the inlet and outlet is an Amphioxus with only its mouth protruding; another mouth is seen farther away. A razor-shell clam with inlet and outlet tubes above and the distensible foot with which it burrows protruded below. Inset below, the crab Corystes breathing by a tube formed of its antennæ.

animals which browse, not on plants, but on other animals, like the lovely eolid sea-slugs that nibble hydroid polyps (and store up the nettle-cells of their prey, unexploded, in the tentacle-like projections, full of outgrowths of liver, with which their upper surface is beset, thus arming themselves with borrowed weapons). There, too, you may see animals which still reproduce by discharging a whole population of cells into the ambient water; animals with no head, like starfish; and animals with no mouth, like sponges. To watch and cogitate upon the life of a rock-pool near low-tide mark is to take a broad elementary course in marine biology.

There are three main types of shore habitat—rocky, sandy, muddy. On rock sea-life is adapted for clinging tight, for resisting the pounding of the waves, for retreating into cracks. Hence limpets, barnacles, dog-whelks, tough brown seaweeds, flattened crabby creatures, encrusting sponges and sea-squirts. On sand and mud the urge is to burrowing. At this game worms and clams are supreme, but there are also heart-urchins and a certain number of crustaceans, and even one or two peculiar sea-anemones. The animals often have remarkable anti-clogging adaptations, so that they can breathe in spite of sand and mud all round them. The crab Calappa has its claws converted into tight-fitting doors that keep out sand, leaving only two pairs of holes in front for the entrance and exit of the vital respiratory current. And others, such as Corystes or Albunea, conduct clean water down to their gills through a "siphon" made of their two antennæ joined to form a tube by the hairs along their edges.

On some tidal mud-flats the eel-grass can grow—one of the very few flowering plants that have reinvaded sea-water, and so one of the very rare sea-plants with true, absorptive roots—and its long ribbons provide a habitat in themselves for animals. There we find pipe-fish (swimming and resting upright, to simulate the grass fronds); the queer crustacean Squilla with its claws built to close into a groove like the blade of a knife; cœlenterates like Haliclystus (Fig. 99) attached to the leaves; and many sea-snails.

Below the range of the tides, the inhabitants of sand, mud, or rock continue to show the same general differences as between tide-marks. Creatures with special adaptations for temporary exposure to air are, of course, no longer found; but a greater abundance of fully marine animals takes their place. The number of free-moving animals especially increases below tide-marks, notably the fish and crustaceans. On rocky ground the fish tend to resemble their surroundings protectively by their blotchy colouring and projecting tags of skin that break their outline; the swimming crustaceans, such as prawns, are a transparent green. On sandy bottoms the most typical crustacean is the shrimp. Shrimps, like flat-fish, are strikingly protective in their colour, and both have the instinct to burrow down and lie half-buried in the sand.

The abundance of life in this well-lit zone beyond low-tide mark is extraordinary; our figures give an idea of the populations on and in different kinds of sea-bottom. This bottom-supply is of the greatest importance to many kinds of fisheries. Plaice, for instance, are bottom-feeders, much the most important item on their menu (in the North

Sea, at least) being a little clam called Spisula. Spisula is exceptionally abundant on the Dogger Bank, and this is why the Dogger is such good fishing ground. Indeed, experiments with marked fish show that on the Dogger Bank plaice grow almost twice as fast as off the Dutch coast. Between tide-marks, the zones of life were dependent upon times in and out of water. Below the reach of the tides, zoning of this sort disappears, and its place is taken by a zoning dependent on the rapidly decreasing light-intensity. But this we have already discussed (§ 3).

§ 6. Coral Reefs and Islands.

No ACCOUNT of the world's habitats would be complete without some mention of corals and their work. The formation of whole islands for the habitation of man by the unceasing industry of the tiny polyps (or "coral-insects," as they are still sometimes called in popular books —a survival from certain pre-Linnæan attempts at zoological classification!) has always fascinated the human imagination; and the sheer beauty, the strangeness and the variety of a coral reef have captivated all lucky enough to set their eyes on one.

Corals, broadly speaking, are polyps, belong to the cœlenterate phylum (Book 2, Chap. 4, § 3), which secrete a skeleton of lime. Most of them, but by no means all, belong to the same group as the sea-anemones. Most of them, too, are colonial, but there are plenty of solitary "cup-corals." The variety of their growth and construction is very great. Some, like the organ-pipe coral, grow as parallel tubes; others are branching and tree-like, like the precious coral or the stags-horn; others, again, like the brain-corals, produce great rounded masses of limestone; still others grow into flat, ridged disks, which look like the top of a mushroom turned upside down and converted into stone.

There are two very important facts about coral biology which help us to understand how reefs are made. One is that most corals are not really just animals; they are compound creatures like hydra and the many others discussed in Book 2, Chap. 6, § 5, consisting of an animal living in an obligatory partnership with a plant. The animal partner, as we have seen, is a polyp. The plant, as in most of these partnerships, is a single-celled green alga, which swarms in the tissues of the polyp. In one or two species of corals the polyps have lost their mouths and feed entirely by digesting some of the green cells, which on their side multiply so fast that they can afford to pay this tribute in return for the shelter and the supply of nitrogen which they receive. But in most cases a smaller toll of green cells is exacted, and the polyp supplements this by the ordinary sea-anemone method of catching animal-prey. In all cases, however, this army of corals which has gone into partnership with plants is definitely dependent on its partners: without their aid the polyps cannot exist. And accordingly all such types are restricted to shallow water where enough light penetrates for the green plants to utilize their energy-trapping powers. Reef-building corals are scarcely ever found below thirty fathoms.

The other point is one of more general application. It is that, for chemical reasons into which we need not enter, animals find it much harder to extract calcium from cold than from warm sea-water. And since calcium is the main ingredient of limy skeletons, animals with such skeletons flourish much better in warm seas. In almost all groups the number of kinds of animals that scaffold themselves with lime decreases rather rapidly as we pass polewards, and in polar seas the horny substance chitin largely replaces lime as the chief material for skeletons. Corals are one of the best instances of this rule. There do exist solitary corals in north temperate waters, but reef-building corals will not grow where the temperature of the surface water falls below seventy degrees Fahrenheit.

These two peculiarities restrict coral reefs to the shallow water of an equatorial girdle rather broader than the tropics. And this has, as further consequence, that, of all the calcium dissolved out of limestone rocks and poured into the sea by rivers, a great mass is being built up in solid form in this girdle which the world wears round its middle, and very little anywhere else.

Three chief kinds of coral reefs exist. There are fringing reefs which hug the coast and have no deep-water channel between them and the land. Then there are barrier reefs, which enclose a lagoon, usually deep enough for ships, but never over fifty fathoms, between themselves and the coast; and finally there are atolls, which do not border land at all, but consist merely of a coral platform rising out of the sea, its central cup covered with water and forming a lagoon, its rim raised to make a more or less continuous above-water circle.

Fringing reefs one can easily understand; they are what we should expect to find where conditions off a coast allow the growth of corals. But how explain the barrier reef and the atoll, with their wide lagoons and their steep outer slope, often extending down hundreds of fathoms into deep water? How can the deep foundations of these dams and pyramids have been laid, if the polyps cannot live more than a few fathoms below the surface?

Darwin sought to explain their origin by supposing a subsidence of land fringed by such a reef. As the land sank, new coral growth would be built up on the old. The reef would grow in height, and it would extend inwards as the land area shrank. Blocks of coral, piled up by the waves and cemented together by lime, would make a rampart on its seaward side, while the scour of currents in some places and the deposits of fine mud and sand in others would check the coral growth within this raised edge, thus giving us both barrier and lagoon. If submergence were continued until the last vestige of the original land had disappeared beneath the water level, the barrier reef would be converted into an atoll.

This theory, while it undoubtedly accounts for some of the facts, will by no means explain them all. It has since been pointed out that atolls might come into being in other ways. Submarine eruptions are not infrequent, geologically speaking, and may throw up volcanic cones above sea-level; but the erosive action of the waves is so great that such piles of débris, unless very large, are quickly eaten away and their tops eventually smoothed off flat well below the surface, thus providing an ideal platform for a future atoll. Indeed, wherever land is rising, but rising slowly enough for the waves to plane its head or its shoulders down to such a platform, atolls or barrier reefs may grow.

Still another view suggests that, however originated, the coral reefs that we know to-day have made most of their growth since the Ice Age. For during the main glaciations so much water was locked up on land in the ice-sheets that the sea's level was considerably lowered; and when this solid water was melted and restored to the sea the resulting submergence would mean rapid upward growth of corals. In support of this we have the calculations of Dr. Mayer, who showed that for Samoa, at the present rate of growth, all the existing reefs could have been formed since the Ice Age. And from other regions comes evidence of the decay rather than the growth of reefs, as if the rapid growth favoured by the post-glacial submergence was now choking itself.

But whatever be the origin of coral lagoons, there is no doubt that they provide a peculiar, varied and very lovely habitat. Mr. Beebe, in his *Beneath Tropic Seas*, has described the beauties of this submarine world as seen from a diving-dress; and Mr. Pritchard has even sat under water in a diving-dress to paint them for us in colours and media that are unaffected by sea-water. The coral pillars shoot up from the floor of coral sand over coral rock; there are arches, doorways, caverns of coral. Fish of fantastic brilliance dart in and out, the crevices are full of strange worms and crabs, here and there the lagoon floor is scattered with star-fish and sea-urchins and bright-coloured shells. And the lives of all these creatures are centred upon the dominant growths of coral, these strange compounds of flower-like polyp and microscopic green vegetables, as the lives of all the animals and plants of a tropical forest are centred upon the trees, or the lives of all the prairie fauna upon grass.

§ 7. Holes and Corners of Sea-life.

ONE could continue telling of sea-life through all this work and never be done, so rich and various it is. But we have the life of the land before us still, and we must be brief. In this last marine section we will bring together some of the sea's interesting minor habitats. One of the most remarkable of special seashore habitats is the mangrove swamp, found so often in tropical estuaries. Here the strangest mixture of creatures from land, fresh-water and sea meet each other. The mangroves and other kinds of trees invade the mud-flats, using their roots as stilts to keep their stems high. Millions of crabs dodge in and out of the roots; the mud is often as full of holes as a sieve—the work of armies of fiddler-crabs, the males with brilliantly coloured claws as big as their bodies, who scurry over the surface and pop sideways in and out of the holes; oysters grow on the mangrove-roots; the half-terrestrial fish Periopthalmus (Fig. 223) skips over the mud with the aid of its fins; at low tide the ants, descending from their nests in the trees, may be seen as they forage over the mud round pools in which sea-anemones still spread their tentacles; and birds, lizards, and monkeys come down to steal what they can from the sea's harvest.

The mangroves live by turning their roots into stilts; and they reproduce themselves by making dibbles of their seedlings. An ordinary seed or fruit simply dropped onto the mud would be washed away by the tide. Accordingly, the mangrove fruit is never shed, and the young plant actually germinates within it. In the course of six or eight months it develops a long, solid spike-like outgrowth, a foot or eighteen inches long and up to an inch or more thick, which forces its way out through the fruit-wall. At last the whole embryo plant, now weighing about three ounces, breaks loose from its parent, and falls, heavy and spear-like. Thus it stabs itself into the mud, penetrating thither even at high tide through a foot or so of water. And at once it expands its leaves and begins to grow.

A widely distributed way of under-sea life is that of the boring animals, some tunnelling in wood, some in the shells of other creatures, some even in solid stone. The wood-borers are still the cause of great damage to wharfs and piers, and the worst of them, the much-dreaded shipworm, was in the days of wooden hulls the cause of serious loss among ships. Drake's *Golden Hind* was infested with shipworms; and at one time their burrowings in the timbers of Dutch dykes threatened to let the sea in over Holland.

The shipworm, Teredo, is a bivalve mollusc, whose body has been elongated till it looks much less like a mollusc than a worm, while its shell is reduced to a pair of hard chisels at one end, which are moved up and down by powerful muscles and serve to excavate the animal's burrow in the wood. The shipworm is further remarkable in that it is one of the few highly organized animals capable of digesting wood-fibre. Its stomach secretes a digestive ferment which converts some of the wood into sugars. It is the only boring animal in the sea which is entirely dependent for food as well as for shelter on the material in which it bores; and it is so strangely specialized that it can only begin a burrow at one very early stage of its existence. If an adult shipworm is taken out of its burrow, it cannot make a new one, but dies helplessly. British shipworms are rarely more than eighteen inches long, but there is a tropical giant which may grow as thick as a man's arm, and six feet in length.

Crustacea take to wood-boring as well as molluscs; among them the gribble (*Limnoria*) is the worst offender. These are tiny creatures, related to the wood-lice, which rasp themselves tunnels with their jaws. Usually they live by pairs, a male and a female in each burrow, the female apparently doing most of the work. The damage they can do may be realized from the fact that up to four hundred of them have been found under one square inch of wood surface.

Among shell-borers the sponge Cliona is the best known. Its boring activities (carried out apparently by chemical means) are the cause of the holes and tunnellings so frequently seen in oyster shells. It burrows also into limestone. This sponge is the cause of considerable loss on some oyster-beds, for when it is really flourishing it may kill oysters by completely rotting and disintegrating their shells.

Another borer which uses chemicals is the date-mussel of warm seas; it burrows into limestone by dissolving it with acid from a special gland. The familiar Piddock or Pholas, however, rasps holes in rocks of various kinds by means of its shell, as the shipworm does in wood. Then there are sea-urchins which by unknown means excavate little hemispherical chambers for themselves in rocks; and rock-boring worms; and even seaweeds which use acid to excavate homes for themselves in limestone.

The boring habit is by no means confined to water-animals. There are no rock-borers on land, but wood-borers are numerous enough, almost all of them being insects. Though many of these feed on the wood they eat, yet none do so unaided; they all depend on the chemical activities of bacteria or protozoa, which they carry about with them inside—a specialist kitchen-staff. Many other insects once reputed wood-devourers have now been shown to feed on the moulds and other fungi which grow on the damp walls of their burrows. But, whether they use the wood or no, they one and all inflict damage, whether they be termites or beetles, moth-caterpillars or wasp-grubs, and the damage is sometimes serious.

But we are getting away from the sea, and before we leave it for good we must speak of one curious group of creatures—the lazybones of marine-life that get carried about free of charge and effort. Floating objects will naturally be covered with the same sort of creatures as grow fixed on more stable hard surfaces; yet the landsman is always amazed when he sees the mass of creatures that succeed in growing on a ship's bottom within the space of a year or two, so dense that they cut down her speed by ten or twenty per cent.

Prominent among these foulers of ships are the goose-barnacles; these, though they grow on piles, too, seem to have been specially evolved to hang down in the water from floating bits of wood. Big fish and turtles and whales are, from the point of view of such hangers-on, merely bits of floating substance which have the advantage of moving quickly; and they, too, are often covered with uninvited guests. In particular, the habitat provided by whales is sufficiently different from other floating habitats to have some unique inhabitants. There are certain kinds of giant acorn-barnacles which live exclusively there, anchored in a special way in the leathery skin.

But the most famous of all these passengers is the sucker-fish or remora. This carries on its head a powerful sucker, produced in the course of evolution by the transformation of its back fin. With this it sticks itself on to big marine creatures, usually sharks, and is whisked from place to place, detaching itself when the shark makes a kill to feed on the fragments. So tight does the remora stick that in several parts of the world man makes use of it to catch fish and turtles for himself. Professor Haddon has given an interesting account of this method of fishing among the natives of Torres Straits. First you catch your remora; you attach it to your boat by a string through the base of the tail, and then proceed in search of turtle. When you see one, you creep softly up, and throw your remora towards it. The remora, prompted by its violent instinct of attachment, darts for the turtle and fixes itself to the shell. If the turtle is smallish, you simply haul it in; if bigger, you may yet have a chance of spearing it.

One final point about the sea as habitat: it is singularly bare of insects. A few spend their larval lives in it; and there is one solitary species of fly which spends its whole life in salt-water. It is the only region in the world of life, except for a few cold and snowy wastes, where insects are not one of the major groups.

III

LIFE IN FRESH WATER AND ON LAND

§ 1. *Fresh-water Life.* § 2. *The Life of Flowing Waters.* § 3. *The Life of Standing Waters.* § 4. *Land Habitats.* § 5. *The Desert.* § 6. *The Tropical Forest.* § 7. *Regions of Rock, Snow, and Ice.* § 8. *Island-dwellers.* § 9. *Cave-dwellers.* § 10. *Out-of-the-way Modes of Life.*

§ 1. Fresh-water Life.

MASTER EVERYMAN, when he visits the country, likes to lean over the sides of bridges and watch what is going on below the surface of the moving stream. And his interest in brooks and ponds sometimes distresses his parents. But he is fascinated very powerfully by the unfamiliar world that exists under water. As he grows up, the pressure of earning a livelihood, the growth of other interests, the restrictive hedgings of convention, are more than likely to snuff out this healthy and natural curiosity, which is the necessary basis for science. Let the boy indulge in it while he can.

Meanwhile, what does he see from his bridge? He sees plants that trail and swirl instead of standing firm and erect; he sees occasional fish dart from one hiding-place to another, or use their tail-propeller to poise stationary in full current. There are water-snails, different in shape from the snails of land; in the quiet backwaters and eddies near the bank the whirligig beetles whirl and the water-skaters precariously skate on nothing but the water's own surface-film. On the bottom are little caddis-worms, slowly crawling about with their houses on their backs—houses they have made for themselves out of sticks or shells or little stones. If he is lucky, he may see the larva of a dragon-fly, or a may-fly crawling out of water on to a rush, bursting its skin, and escaping, a winged and aerial creature.

And then he goes to his favourite pond, armed with jam-jar and little muslin net. The surface is half-covered with a green sheet of duckweed—a plant that has reverted to the lowly condition of a mere plate of green tissue—and starred with white flowers of water-crowfoot. With his net he picks out a stickle-back or two, and some handsome newts with their black-and-orange bellies. Then there are magnificent great water-beetles, smooth and shiny, stream-lined like a submarine, and provided with oar-like legs; and unpleasant creatures half-way between a grub and a small Chinese dragon, with formidable sickle-jaws, which, though he does not know it, are the larvæ of these same water-beetles. He catches plenty of water-boatmen, which he notices (he is an observant lad) prefer to swim upside down, and is lucky enough to find a water-scorpion. Snails' eggs in transparent jelly-masses there are in plenty on the water-plants; the newt-tadpoles with their feathery gills are growing up; and the place is teeming with just-visible water-fleas.

It is indeed a strange world, and a varied world, this world of fresh-water life; and Master Everyman could go on exploring it for long years without exhausting its surprises and its interest. And yet it is but a pale shadow of marine life. The chief reason for the lesser variety of the life of inland waters is that many aquatic phyla and classes have never succeeded in quitting the sea. Only a few groups of water-dwellers like the aquatic insects, the ciliates and the rotifers, are more abundant in fresh than in salt-water, and one only—the amphibia—is wholly non-marine.

Life in fresh-water demands certain special adaptations which not every group of animals has been able to produce. The restricted size of most bodies of fresh-water is unfavourable; whales or giant jelly-fish or the most active type of pelagic fish could not very well thrive even in large lakes. The small amount of salt in the water creates another difficulty. Blood contains salt as one of its most necessary ingredients, and wherever blood comes in close relation to fresh-water, as in the gills of water-animals, there is a tendency for salt to leak out and for water to leak in. In amœba we saw a special structure, the contractile vacuole, everlastingly baling out water from the microscopic inside. In frogs and toads, the kidney has undertaken this duty; if a frog's ureter is blocked, and the animal is then immersed in water, it swells prodigiously. Bony fish have the power of keeping their blood-concentration almost constant in spite of variations in the salt content of the water; and hence are well represented in rivers and lakes; but in the gristly fish the blood alters with the surrounding medium, and, accordingly, sharks, dog-fish, or skates are almost unknown in fresh-water. The absence of salt also means a lower density, and this (as any bather knows) makes it more difficult for free-swimming and free-floating creatures to keep up; as a result, we find that the average and especially the upper limit of size of fresh-water plankton is much below that of marine plankton.

FIG. 250. Fresh-water carnivores.

Three water-beetles which use the hind pair of legs as oars, eating a small fish. Inset, the larvæ of two kinds of water-beetles; left, the Hydrophilus grub feeds on pond-snails and eats them out of water; right, the Hydrous larva catches its prey under water but lifts it into the air to eat it. Both obtain air for breathing by pores at their hind end. (Inset from "Concerning the Habits of Insects," by Prof. F. Balfour Brown, Cambridge University Press).

Changes of temperature are in general more extreme in fresh than salt-water, so that no animals which are intolerant of wide alterations of heat and cold can hope to leave the seas. Again, the danger of being carried away and down to the sea is ever present in rivers and in most lakes; and when that possibility is absent, there is often the danger of drying up or of rapid rise in salinity. In all these respects fresh-water is a more exacting environment than the sea, and many of the sea phyla have altogether failed to spread from their original home up the rivers into inland waters. But in contrast to this comparative poverty of primitively water-living creatures, we find that the so-called secondary aquatics—animals or plants which have gone back to water after a period of land-life—are much more abundant and varied in fresh-water than in salt. Any land-creature has learnt to overcome difficulties even more severe than those we have just considered, and after its arduous training, finds fresh-water life comparatively easy. Moreover, that such secondary aquatics should prefer fresh-waters to the sea is perhaps to be expected, since for one thing they are not so overcrowded with life, and for another the extent of shore-line inviting the evolutionary plunge is much greater, in spite of the total volume of water being so much less. There are plenty of secondary aquatics in the sea—whales and dolphins, seals and turtles, plants like the abundant eel-grass; but they make only a poor show compared with those of fresh-water—all the cohorts of fresh-water insects, spiders and mites, fresh-water snails, fresh-water lung-breathing vertebrates such as newt and crocodile and terrapin, and all the innumerable flowering-plants of fresh-water—water-lilies and water-crowfoot, lotus and arrowhead, pond-weed and bladder-wort.

The main reason why a number of land vertebrates have taken to fresh-water (the reason holds for hardly any which have taken to marine-life) is to escape terrestrial heat and terrestrial enemies. Such creatures often have the shape of their head modelled in adaptation to their amphibious life; nostril and eye are raised on protuberances so as to be above the water-line when the animal is floating—one has only to think of the silhouettes of frog, crocodile, and hippo. The conning-tower and periscope of submarines are analogous devices of man's contriving.

The smallness of most bodies of fresh-water is an important factor, for it leads to their being frozen when the inexhaustible sea would be merely cooled a trifle, dried right up when the sea would evaporate a fraction more from its surface. And this has imposed remarkable powers of self-protection on many fresh-water creatures. In temperate regions, they usually tide over the winter by means of a resistant resting-phase, which may be egg or gemmule, seed or pupa. Frequently, the need for a resistant winter stage together with that for the utmost possible utilization of summer's warmth has led to a remarkable type of life-cycle, in which generations of nothing but females, all parthenogenetic (Book 4, Chap. 2, § 5), succeed each other throughout the summer, while autumn brings males and sexual females, which produce fertilized "winter eggs," hard-shelled and cold-resisting, which will only hatch out in the following spring. This type of life-cycle is found among rotifers and in two separate groups of crustacea; but it occurs only among their fresh-water forms. Certain full-grown rotifers can be frozen solid and remain for a very long time in cold storage without taking harm. Some which were thawed out by the Scott Antarctic Expedition could not have been less than five years in this state of suspended animation.

In many small crustacea, inhabitants of small pools in dry climates, the fertilized egg is not cold-resistant but drought-resistant. In fact, in some cases, a preliminary thorough desiccation of the eggs appears to be necessary if they are to develop at all. Many rotifers, on the other hand, as well as most tardigrades (bear-animalcules), and of course many Protozoa, can be desiccated entire even when adult, and will yet revive again on being moistened. These remarkable resistances to freezing and drying are not present in any marine animals—they are not wanted in the sea, there is no selection in favour of them, and so they have not been evolved.

The smallness of fresh-waters has another important result, for it means that (except for a few vast lakes) they do not provide enough room for considerable waves to develop. This and the shelter afforded by banks means that the surface of fresh-waters will often be smooth, and this in its turn has encouraged the development of animals and plants of a type almost unknown in salt-water, which live in relation to the surface-film. Some of these actually live wholly in air and support themselves on this fragile watery skin, some suspend themselves from it down into the water, and a number live actually in it, their top-half out of water, their bottom-half in. Among plants, the duckweeds are the most completely adapted to this borderland existence, but all the forms like water-lily and lotus, in which the upper surface of the floating leaves is dry, belong to the same general type. Among animals, the whirligig beetles live most of their active life in the film, so much in it that their eyes are divided into two parts, the upper adapted for seeing in air, the lower for seeing down into the water; and the water-measurers actually treat the surface pellicle as a floor, and can skate over it dryshod, merely dimpling it with their feet. Many water-snails, on the other hand, even quite large ones, can use the surface-film as a ceiling, clinging to it with the surface of their foot and crawling along it; and Hydra may sometimes be seen doing the same thing. The eggs of many gnats and mosquitoes float in a raft in the surface-film; their larvæ, while they breathe, suspend themselves from it by means of a star-shaped set of plates at the end of their breathing-tube; and when their pupæ moult into winged adults, these are only saved from drowning by being able to support themselves on the film as they struggle out. Almost the only creatures of the sea's surface-film are the siphonophores, with sails sticking up into the air, and polyps and tentacles dangling downwards into the water, and a marine water-measurer, the sea-strider Halobates, which may skate over the waves to quite a distance from land.

LIFE IN FRESH WATER AND ON LAND

In connexion with this life between air and water, we meet with further adaptations. The South American Jaçana, a bird related to the moorhen, is adapted to walk over lily-pads to find its food. For this purpose, its toes are enormously elongated, to spread its weight over a greater surface of leaf, as skis or snow-shoes spread weight over a greater surface of snow. And there is Toxotes, a tropical fish, which stealthily stalks insects sitting on water-plants, and then shoots at them with a rapid jet of water from its mouth. The insect, thus brought down to the water, is held struggling there by the surface-tension of the surface-film, and is promptly swallowed.

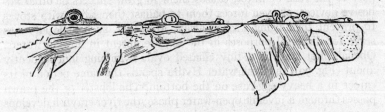

FIG. 251.   The fresh-water profile.

Frog, crocodile, and hippopotamus all raise their eyes and nostrils on protuberances to be above the water-line when they float. (Modified from R. Hesse.)

These creatures again remind us of the fact we stressed at the beginning of this section—that fresh-water animals and plants are highly specialized. The original home of life is the salt sea; to become adapted to living in saltless waters is itself an achievement.

§ 2.   The Life of Flowing Waters.

ONE general characteristic of all running waters is that they have no true or permanent plankton of their own. Plankton is constantly entering rivers from below at their mouths, and from above out of lakes; but it cannot colonize them. Plankton from the sea cannot swim upstream; plankton from the lakes is all, sooner or later, swept out to sea. The open water of rivers is not in itself in any way unfavourable to the growth and multiplication of plankton—the small crustacea and especially the quick-generationed rotifers from lake-plankton may grow and reproduce through several generations when swept into rivers; but all this multiplication is pure waste as regards the species, although, of course, it may serve in feeding other creatures. This source of food may be important; where the waters of the

Elbe are slowed down on entering its estuary, over ten million plankton crustacea may be found in every cubic yard of water. All these millions of creatures will eventually fall from the tree of the species, to be replenished by fresh accidental immigrants from above. The one fact of flow has prevented the rich open water of rivers from possessing its own fauna.

Indeed, it is rate of flow more than any other single factor which moulds river-life. Fast-running water hides animals under stones, robs them of the hairs and bristles by which small limbs become swimming paddles, flattens them, stream-lines the contours they expose to the rush of water, causes them to grow suckers or other adhesive devices, or even forces them to ballast themselves with stones to prevent their being washed away. Moreover, such little larvæ as serve for dispersal purposes in the sea are absent in flowing waters. Obelia's first stage is a tiny ciliated oval, swimming independently about (Fig. 99), the fresh-water Hydra spends the same period of its career in a heavy egg-case on the bottom. The lobster or the prawn passes through a juvenile open-water phase; the river crayfish develops as an embryo clinging to its mother's abdomen.

On the other hand, while swift water encourages animals to temporary attachment and wedgings away into chance crevices, it discourages all forms of permanently fixed habit, for the reason that the bottoms of swift rivers are largely composed of loose stones which, tumbling over and over, afford no permanent abode themselves, and would crush most organisms growing on rock; and, further, food is so scarce or so rapidly carried past (or both) that animals cannot afford to be tied to one spot. In the swift streams of temperate Europe, only thin crusts of one kind of fresh-water sponge and occasional colonies of a polyzoan are found in place of the sea's horde of stalked animals. Since plants cannot develop temporary anchors, and yet are exposed to the same violences of current and down-driving stones, they too are only poorly represented in rapid streams.

That it is the speed of the water, not the size of the stream, which is chiefly responsible for these adaptations is shown by the fact that when giant rivers, like the Congo or the Essequibo, have stretches of rapids on their middle course, the fauna and flora show adaptations of precisely the same nature as those found in a Westmoreland beck or a Swiss torrent.

In streams with gentler flow, these adaptations are not necessary, and are not found. When the flow is very slow, slow enough to allow the deposition of fine sand or mud, a special fauna arises to exploit this bottom layer, and may show some remarkable adaptations. Aquatic bristle-worms like Tubifex, Limnodrilus and Lumbriculus, which resemble miniature earthworms, may occur in vast quantities in such localities. In the Elbe below Hamburg, about 27,000 of such worms may inhabit one square foot; at low tide in central London great patches of the mud can be seen coloured dingy red by Tubifex. These

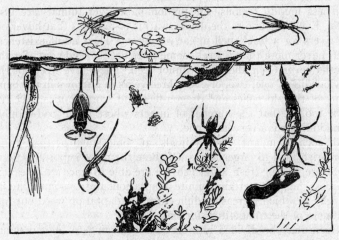

FIG. 252. Pond-life at the surface-film.

Water-lilies send their leaves to rest on the surface; duckweed plants float there, with roots hanging down. Water-measures skate on the film; a water-scorpion, several gnat pupæ and larvæ, and a water-beetle larva (seizing a tadpole) poke their breathing-tubes through it. A small water-beetle carries down air under his wing-cases, a water-spider (centre) transports air with which to fill its waterproof nest. Whirligig beetles swim in the film; a pond-snail crawls on it as if it were a ceiling.

worms often construct tubes for themselves. They bury their heads at the bottom of these pipes, and eat the slime; the tubes stick up like chimneys above the mud-surface and from them project the posteriors of the worms, waving rapidly to and fro to facilitate respiration, which is carried on, not only by the surface of the body, but also by rhythmically passing water in and out of the hind-gut through the anus. They are, like earthworms, constantly eating their habitat, digesting the nutritious bits from it, and ejecting the rest as castings. Their activity in this respect is prodigious; a Limnodrilus an inch and a half long may produce five feet nine inches of fæcal castings in the twenty-four hours! The huge numbers of worms thus continually bringing up new material to the surface, where it can be readily oxidized, play valuable rôles in purifying foul water.

Other slime-feeders, which may even exceed these worms in bulk per unit area, though not in numbers, are little bivalve molluscs. In other parts of the lower Elbe one of these, Sphærium by name, abounds to the extent of nearly 7,000 per square foot.

In less muddy regions, the predominant river-animals are insect larvæ. Everyone knows the dense swarms of mayflies that hover over our rivers for a brief spell in early summer. Their adult existence is brief, never more than a week, sometimes only a single day, for their mouths never open and they cannot feed. They have lived all their previous life of one, two, or even three years in burrows in the banks. Many of the alder-flies and dragon-flies and scorpion-flies have river larvæ too, so that the number of insects whose main food-supply is found in rivers is a very large one.

All such abundant small animals, at some stage of their career, afford rich food to larger forms. Of these the most important are the fish. Here, among fresh-water fish, we are able to trace with beautiful clearness, not only certain definite adaptations to fresh-water life, but the way in which increasing difficulties of adaptation weed out more and more of the competitors for existence.

Rivers may be divided into sections according to their dominant fish. In Europe, the head-water region, with swiftest flow and clearest water, is characterized by the trout, with minnows, miller's thumb (*Cottus gobio*) and loach (*Cobitis barbatula*) in attendance. Below this is often a zone with dominance of the grayling (*Thymallus vulgaris*), and then the barbel region; or this latter may immediately succeed that of the trout. Then comes the quiet-flowing domain of the bream; and down-river from this the estuarine region, with stickleback and smelt as most prominent members, pike and eels in its upper, less salty zone, and flounders and, in some rivers, sturgeons in the zone nearest the sea. These regions do not always succeed each other in this precise order, since streams may for instance have a region of rapid flow and pure water, and therefore of trout-fauna, intercalated on their middle course. But the order holds as a general rule.

Now as we go up from the river-mouth, in the first place the habits and the body-form of the fish change. Flat-fish never go far up the rivers, since food-rich bottoms of clean sand are absent. The chief estuarine types are those, like sticklebacks, adapted to rapid changes in salinity, or, like eels, to a muddy life. Above this stretch, within the true river, the shape of the body changes steadily with increasing rapidity of current. In the lower reaches are forms with poor musculature and sideways-compressed body. In more rapid currents, more muscle is needed to keep up, and the body must be more rounded so as to expose less surface to the flow of water. At the same time, the proportion of mud-rooters diminishes, the proportion of those which prey on tit-bits floated down in the current increases.

And finally the number of species diminishes. The number of fish species in the Rhine is as follows. In Holland, 41; in the upper Rhine just below the falls, 28; above the lake of Constance, 25; up to 700 metres above sea-level, 11; up to 1,100 metres, 5; to 1,900 metres and over, only 3 (trout, minnow and miller's thumb). And similar figures may be found in other rivers all over the world.

Perhaps the most remarkable of river-fish are the migratory members of the salmon family, which feed and grow in the sea, but for their reproduction migrate up rivers, often for hundreds or even thousands of miles, to the shallow head-waters, where in "redds" scooped in the gravel, under the well-oxygenated water, the eggs are laid and fertilized. Sometimes the great fish crowd into such shallow waters that they are scarcely submerged. In Alaska even bears and their cubs come to the brooks in salmon time and scoop the creatures out of water with their paws. In some species of salmon the spent fish may drop down to the sea again; but in others, not one of all the swarms which enter fresh-water survive; they make the great journey but once in their lives, and die when its biological purpose has been achieved. During all this time the fish seem to eat nothing, and their stomachs are contracted and secrete no gastric juice. All the energy for the journey and much of the materials for the development of sperms and eggs are obtained from the reserves of fat and muscle-sugar stored in the tissues, and even at the expense of the living flesh itself, so that the fish are left in a very emaciated state after spawning.

Considerable dispute has taken place as to whether the salmon family were originally fresh-water fish which have taken to marine feeding because of the greater abundance of food to be obtained in the sea, or marine fish which have taken to fresh-water breeding on account of the greater security provided to the eggs and young fry. But the latter view is in all probability true; so that permanently fresh-water species like the trout are degenerate or at least specialized forms which have secondarily ceased to migrate.

The incredible concentration of these fish in the early spawning season at the entrances to rivers makes the salmon fishery the most important of all river fisheries. In Canada alone, the value of the Pacific salmon caught in 1922 was thirteen million dollars.

Sturgeon and their relatives behave exactly as do the salmon, and like salmon, they are captured while on their long pilgrimage of reproductive destiny and turned to human profit. Fresh-water eels on the other hand execute a breeding migration in the opposite direction, as Kingsley has vividly described in his *Water Babies*. In early autumn the mature eels, leaving the quiet ponds where they have lived and fed for years, and, often crawling a mile or more across wet grass in the cool night to reach the nearest stream, make their way down to the sea. In preparation for this migration, they change their whole appearance, and from what fishermen call "yellow eels" become "silver eels." Their bodies grow silvery, their eyes enlarge enormously, their snout becomes pointed, their reproductive organs swell, and they cease to feed.

On reaching the sea, the eels are led by some mysterious instinct towards their distant breeding-places in the depths. English and Scandinavian eels, and even those from the recesses of the Baltic, travel across the whole Atlantic to the borders of the Caribbean Sea, where, at unknown depths, they spawn. The eels of Eastern North America, though of a different species, spawn close by. Those of Mediterranean countries spawn in the deeps off Sicily. The object of their pilgrimage thus attained, the creatures all die.

The eggs hatch into transparent leaf-like creatures, at one time thought to be a distinct genus of fish, and described under the name of Leptocephalus—"thin-head." These migrate slowly homewards, feeding and growing all the while. Johannes Schmidt has mapped them by their size, and by this means was able to show where the breeding-place must lie. Eventually, when about three years old, and two inches and a half to three inches long, they metamorphose rapidly into a miniature eel or elver, shrinking considerably as they do so, and swim in dense hordes into the mouths of rivers, where they are often caught by the bucketful for food. The American eel is more of a hustler: it metamorphoses before it is two.

There can be no doubt that eels were originally marine. The conger eel is wholly marine, and it too breeds in the deeps and has a similar leaf-like larva. So that the fresh-water eel is the reverse of the salmon. It has kept to its original deep-water breeding-place and feeds in fresh-water alone. Both salmon and eel have taken advantage of their adaptability to penetrate from the sea to waters where there is less competition. but they exploit fresh-water for different purposes.

LIFE IN FRESH WATER AND ON LAND

A lamentable feature of our industrialized world is the way in which it pollutes its rivers. Sewage; the poisonous run-off of certain kinds of tar on roads; chemical effluents of every description—they turn our streams filthy and turbid and destroy the best of their life. The prevention of pollution is a complex problem, with its economic as well as its biological sides, and we cannot enter into it here. But we can drive home the facts with the aid of a single example. The sewage and the chemicals in the Thames to-day would kill any salmon that tried to swim up or down its lower course. But less than two hundred years ago, the Thames was a good salmon river. And in the thirteenth century, the Thames at London was so full of fish that when Henry III was presented with a polar bear by the King of Norway, it was kept at the Tower of London and allowed to swim about at the end of a rope and supplement its allowance with the fish it could catch.

§ 3. The Life of Standing Waters.

THE life of fresh-water that is not moving is very different. There is no hurrying flow to inhibit plankton and larvæ or to insist on firm anchorage, and the size of the bodies of standing fresh-water may be so great as to permit a regular stratification of open-water life by depth, in faint imitation of the sea's vaster economy.

Lakes, owing to the physical properties of water, rarely freeze solid. Pure water is densest at 4° C., not (like almost all other liquids) at its freezing-point; thus, in winter, after the whole mass of water has been cooled to 4°, further loss of heat from the surface will leave a cold layer of temperature lower than 4° floating on the rest; this will freeze, and, owing to the poor heat-conductivity of water, further chilling of the air will cool the lower layers but slowly. If, as with most liquids, the density of water steadily decreased to its freezing point, a circulation of the lake-water would take place until all was at 0°, and then the whole mass could freeze almost at a bound into a solid block of ice.

As a result of this peculiarity of water, organisms can survive the winter through under the lid of ice, doubtless in a torpid condition owing to the low temperature, but without having to develop special adaptations against damage due to the freezing of their tissues. Since no vertebrate animals are known which can survive long-continued freezing solid, it is probable that if it were not for water's peculiar property of being heaviest at 4° C., there would be no fresh-water fish or amphibia in existence, or at least none whose adult life surpassed a single season. This property of water is one of the central themes in L. J. Henderson's interesting book, *The Fitness of the Environment*.

In all deep lakes in temperate regions the bottom water will be permanently at 4° C. Twice a year, in spring and autumn, the whole of the lake will be at this temperature. But the surface water will be the warmest layer in summer, the coldest in winter.

The plant-inhabitants of a lake margin are for the most part an invading army, taking possession of this vacant but difficult territory under stress of terrestrial competition. As the water deepens, the invaders change their tactics. Near the edge they are merely paddlers, with only the roots and a little of the stem below the surface. Farther out they must wade, and (to keep the animal metaphor) must grow longer legs, like any heron or flamingo. Some become amphibious, with part of their foliage below water, though the plant as a whole is still constructed on principles taken over from land-life. Deeper yet, and the whole plan is altered. Still clinging to air, the drowning plant sends up leaves and flowers to float at the surface at the end of stalks that are no longer supports but mere flexible mooring-cables and transport-conduits. Such are the water-lilies. Finally, the deepest venturers discover that submersion need not mean drowning, and like the pond-weeds, alter the whole structure of their leaves so as to be capable of food-manufacture and breathing under water.

It is more difficult to re-adapt reproductive methods to a submerged life; the dry dusty pollen, evolved during æons to be floated on the wind or to powder the hairy heads and backs of insects, does not take kindly to a wetting. Some plants whose vegetative life is all submerged, like the pond-weed Potomogeton, still send their flower-shoots up into air; the famous Vallisneria detaches its male flowers while still closed; these float up to the surface and open; the sepals bend back so as to raise the anthers to the level of the stigma of the female flower, which, meanwhile, has been paid out to the surface at the end of a long thin rope of a stalk. The male flowers, drifting hither and thither, may come to rest in the little harbours between the sepals of the female flowers, and some of their pollen gets rubbed off against the stigma. Once the female flower is pollinated, its stalk contracts into a spiral and pulls it down to safety near the bottom.

A few plants, however, have adapted their reproductive as well as their vegetative life to water. Naias, a relative of the marine eel-grass Zostera, forms pollen-grains without the usual outer coat, and elongated and thread-like instead of rounded in form. These are discharged under water; their large relative surface helps to float them, and some drift into contact with the stigmas, which, too, are submerged. Accordingly Naias and its relatives extend to greater depths than other secondary aquatics among plants, being released from all need of contact with the air; they are only limited in their downward invasion by the diminution of the light as it filters through the water.

Besides these secondary water-plants, there are, of course, others which have been aquatic throughout all their history. None of these, however, attain in the fresh-water community anything like the size or the importance of their marine congeners the seaweeds.

The animals of lakes are also perhaps best classified into the secondary invaders of water and its permanent inhabitants. The latter, save for one group, are less important than the former in size and variety. They are largely composed of small creatures belonging to groups from the lower crustacea downwards—water-fleas, shrimplets, rotifers, the little cœlenterate hydra, fresh-water sponges, flatworms and protozoa. The secondary invaders, on the other hand, include all the thousands of insects which live in water either all their lives, like the great Dytiscus beetle and the water-boatmen and water-scorpions; or through all their growing larval existence, like the caddis-flies, stone-flies, and the great dragon-flies, falcons of the insect world. They include also frogs, toads, salamanders and newts during their larval life and a certain season each year of their adult life; also most water-snails, and all the water-birds. These last are more bound to land and air but, like the plants, throw out representatives to ever more aquatic life, from paddlers like the sand-pipers to waders like the heron, and so to swimmers; and the swimmers become more and more bound to their medium as we pass from surface-feeders like moorhens and gallinules and bottom-grubbers like the swan, on to divers like the diving ducks and the grebes.

The single exception to the dominance of secondarily aquatic forms in lakes is an important one—the fish. Whereas the economically important river-fish are mostly wanderers to or from the sea, those of lakes are home products, and some lakes, like the Caspian and the Sea of Galilee, provide vast quantities of food to the neighbouring countries.

In general, the plankton of lakes is much less well-developed than that of the sea, and as direct consequence of this, active pelagic fish are less abundant, bottom-feeders more abundant than in the sea. Among the few plankton-feeding fresh-water fish are various members of the salmon family, such as the different species of white-fish (Coregonus) and some lake-trout. As most of the bottom of lakes is covered with fine slime, in whose nutritious lap live abundant insect larvæ, worms, and bivalves, many of the bottom-feeding fishes are provided with long whisker-like feelers, with which, as a matter of fact, the fish can not only feel but also taste; and also adaptations in the shape of the tail (as in the sturgeon) or the backward prolongation of the ventral fin (as in some of the catfish) which serve to force the head down in its rootings among the mud.

The large plant-plankton, as in the sea, consists largely of diatoms and peridians; but again as in the sea, the dwarf-plankton, which passes through the meshes of the finest net, is even more abundant.

The animal-plankton of lakes comprises for the most part small crustacea and rotifers; and besides that a few water-mites and protozoa; but very few insects have become so completely re-adapted to aquatic life as to have taken to open water and to the peculiar feeding methods there needed. Though the plankton may be abundant, yet even in quantity it falls far below that of cool seas; and in variety it is infinitely inferior. One or two interesting plankton animals deserve mention, notably the little glassy carnivorous crustacea, such as Leptodora.

In the great majority of lakes, in consequence of their smaller depth and their smaller geological age, there is practically no special deep fauna or flora. It is only in a very few large, deep and comparatively ancient lakes, like Baikal in Siberia and Tanganyika in central Africa, that a real deep-water world of life exists. In other lakes there may be deep-water forms, but they are ubiquitous creatures found also in the shallow zones, and have not had time—since most lakes are geologically very short-lived—to become specially adapted to the peculiar habitat. But in Baikal, for instance, at about 600 metres depth there are gammarid crustacea which are blind and have antennæ and limbs elongated as feelers, and fish like Comephorus, which has lost almost all its pigment and is a shimmering pink. In Baikal, the gammarids are the dominant group of crustacea and have launched out into all sorts of peculiar ventures. The same sort of thing has occurred with the fish of Tanganyika: over half of its one hundred and fifty species of fish belong to one family, the Cichlidæ, and the great majority of these are products of Tanganyika evolution, found nowhere else in the world.

But we have devoted our attention long enough to the inhabitants of watery worlds. It is time we turned to the creatures of land and air.

§ 4. Land Habitats.

IN BOOK 5 we have set forth in some detail how certain groups of animals and plants succeeded in invading the land. Here perhaps we may add a note on the reasons which prevented animals of certain other types of construction and ways of life from succeeding. First the net weight of an organism in water is negligible. In sea-water, protoplasm weighs only about a thousandth of its total mass; even in fresh-water, only about a two-hundredth. But in air, weight counts. And so all animals and plants without some means of mechanical support are debarred from terrestrial life. No jelly-fish can conceivably be imagined which could successfully invade a land habitat. There are a few land-animals that lack a skeleton. But of these, earthworms can only exist *in* the soil; and land-planarians are confined to hot, moist places. Slugs and land-leeches are the most successful of such creatures; but they both need a dampish atmosphere. Each is, however, in its own way, successful. The slugs are serious enemies of the gardener; and in some parts of the tropics the armies of land-leeches, attached to the vegetation and constantly waving their bodies about in search of possible prey, are among the most unpleasant pests with which explorer and pioneer have to contend.

And, secondly, no creature whose whole life is adapted to current-feeding can become terrestrial. Even if (as is probably impossible) the current-producing cilia could become adapted to beating in air instead of in water, there is scarcely any floating body of living things or their débris to be captured from the air, since, weight in air being what it is, all particles above a minimal size sink at once to the ground. For this same reason no sedentary animal which merely spreads its tentacles for prey can live on land. The tentacles would collapse and the abundance of prey is not there. Only much later, by new devices specially adapted to air and its inhabitants, have creatures like spiders exploited this catchment method of gaining a livelihood in terrestrial surroundings. This difficulty rules out from land-life the whole group of hydroid polyps, the jelly-fish and the corals; and the impossibility of current-feeding debars all the sea-squirts and other tunicates, the bivalve molluscs, polyzoa, lamp-shells, and lower chordates like Amphioxus.

No echinoderms have ever left the sea, although they possess a skeleton, and the free-moving types are not dependent upon current-feeding. On the other hand, their whole locomotion depends upon hydraulically-operated tube-feet, which apparently could not be made over so as to work in air.

The substitution of protected stages of seeds and embryos for free living stages, for larval forms that is, in the life-history, is one general rule of aerial adaptation—to which only the insects furnish exceptions.

The habitats open to life on land are more varied than in any other medium. But many of them are already familiar to Mr. Everyman, in virtue of the fact that he is himself an air-breather. Though he does live in a town, he has visited various parts of the country, and so (like Monsieur Jourdain, who found he had been speaking prose all his life without knowing it) has unconsciously imbibed a knowledge of the chief land habitats of temperate regions.

He has walked over the close-cropped downland, and the rich water-meadows. He has seen the hares (the mad March hares) chasing each other in spring over the open fields, and noticed that they avoid the woods. He knows the difference between a pine-forest and an oak-wood—the one permanently dark, carpeted with little save a thick brown layer of pine-needles, the other lighter and with a richer undergrowth, especially in the spring when the roof of leaves is not yet grown, and the primroses and wood-anemones and wood-sorrel cover the floor of the wood with blossom.

He knows the barren sandy heaths, alive with rabbits, bright with purple-flowered heather in July and yellow gorse in spring; the meadow-pipits sing there, and the nightjars lay their eggs on the bare ground. If he has visited Dartmoor, or the Lake District, or Scotland, he knows what a moor looks like, and has seen a peat-bog; he realizes that there are habitats so poor that trees will not grow on them, and can readily take the step in imagination to the more barren tundra of the arctic or the craggy regions of high mountain ranges. And yet these places, too, are full of beauty and of life; there are ravens and falcons in the crags, dippers by the streams, sometimes red-deer and mountain-hare, grouse, or ptarmigan. The stones are covered with lichens; between them grow mountain flowers, smaller but often brighter than those of the lowlands.

Or he may happen to live in the Middle West of America; and then he will know the great sweep of the prairie, covered with flowers in spring, growing into a sea of grass later in the year.

If he has any curiosity and interest in natural history, he will know something about the commoner land plants of his country, the commoner land beasts, birds, insects, and other animals, and where and how they live. Comparatively few people know by personal experience what a jelly-fish or a spider-crab, a sponge, a sea-cucumber or a cuttle-fish look like in their natural surroundings. But they inevitably have some acquaintance with rabbits and robins, bees and beetles, spiders and slugs, snakes and frogs and earthworms.

We cannot deal with all the diversified habitats of land in detail; many of these more familiar ones we shall leave out, or will deal with certain aspects of them only, in later chapters of this Book. Here we will take a few of the more striking of the contrasted habitats of dry land, and those that are likely to be unknown by personal experience to most of our readers, in order to bring out the salient facts of life's adaptation to land.

§ 5. The Desert.

WE MAY begin with the desert, since this possesses in exaggerated form all the characteristics which made the land hard to colonize. Botanically, the desert is at the other end of the scale from the tropical forest. In the equatorial forest, all that plants can desire is provided; the difficulties arise not from the harshness of the surroundings, but from the fierce competition due to the very ease of growth. In the desert, on the other hand, the soil is at the best of times not fully exploited; the desert plants provide an extreme example of what ecologists call an open formation, with great stretches of barren plantless environment between the scattered units of life. There is little struggle between plant and plant, but intense struggle of organism with inorganic nature.

On our globe, a double desert zone exists, girdling the tropics on both sides. The great Palearctic desert embraces the Sahara and all of Egypt save the coast and the Nile valley; Sinai, Arabia, the Syrian desert; and so across parts of Persia and Afghanistan to Turkestan, Tibet and the Gobi, with a side-branch into western India. The North American desert, including the celebrated Death Valley, reaches across some of the southwestern United States and northern Mexico. On the other side of the equator there is a corresponding band of deserts in South America; in South Africa there is the Kalahari; and Australia can boast of the earth's second largest area of completely desert land.

It is difficult to define a desert meteorologically. In general, it is, of course, a place of heat and drought. The rainfall is usually low, very variable from year to year, and, if it does not fall in minute amounts which evaporate again before reaching the roots of the plants, is often confined to a few heavy bursts which run off before they can soak deep. Deserts are not only hot places; they are places with very great fluctuations of temperature, both yearly and daily. This is doubtless in the main due to the clearness of the sky and to the absence of vegetation and of water-vapour. With a clear sky, heat radiates unbrokenly in from the sun, out again into space; but a blanket of vegetation impedes the passage of heat into or out of the soil. Where plenty of water exists, much of the sun's heat is used up, without rise of temperature, in transforming water into water-vapour; the restoration of this latent heat during night-cooling again acts as a damper on rapid temperature-change. In a desert, this temperature-buffer is almost absent. In addition, water has a high "specific heat"; if we put the same amount of heat-energy into a pound of water on the one hand and a pound of dry earth on the other, the second will get hotter than the first. That is another reason why wet clayey soils are "cold," why the waterless soil of deserts heats up and cools down with such surprising rapidity. The result is that the night temperature may drop below freezing point after a day maximum of 80° or 90° F. In many deserts the minimum temperature usually keeps just above freezing. But then, one year, there is a frost; it can be imagined what damage this will do.

In passing, it may be noted that merely irrigating a desert with water will not convert it to fertility. It works for a short time, but owing to the intense dryness of the air the water evaporates very quickly, leaving behind it any salts it may have contained. These gradually accumulate in the soil, until the last state is worse than the first. This problem of the caking of irrigated desert soils with salt has already become acute in various regions, including parts of Egypt, and soil science is being hard put to it to find a way out of the difficulty.

Deserts are violent places. As there is no mist and hardly any water-vapour in the air, the intensity of sunlight is greater in deserts than anywhere else save perhaps the tops of high mountains. Not only temperature and rainfall, but also relative humidity and windiness show huge fluctuations. A North Sea gale is nothing compared with the gale experienced by the explorer Augieras in the western Sahara, which lasted nine days, and was so violent that he could not stir the whole time from out of the lee of a sheltering rock.

We are accustomed to the two great cycles of day and year. Near the poles, the diurnal cycle drops out over a large part of the year. In some deserts, on the other hand, it is the yearly cycle which ceases to be important. Whole years may pass without rainfall, and there may be no regular relation between weather and season. Biologically speaking, the cycle of the year may cease to count in deserts; and only the long-range periodicities, such as the eleven-year cycle, of which we shall speak in a later chapter, with their disturbing effects upon earth's weather, make much difference to desert plants and animals.

In less extreme deserts, however, the seasonal cycle remains; there is a brief moist spring, and then indeed does the desert blossom like a rose. The bare soil comes to life, green shoots push up, burst into flowers that carpet the land with colour and bring with them as their biological attendants the hum and flutter of swarming bees and butterflies; the seed is set: and a few weeks later all has disappeared and the bare soil, dotted with a few perennial thorns, alone remains. Cinderella's coach has been turned back into a pumpkin, her lovely dress into dead leaves—only the change seems even more miraculous, for at first sight it looks as if the desert's beauty had vanished into nothing.

In reality, one half of the plant-life—the annuals—has retired into drought-proof safes, in the shape of seeds. The other half, the perennials, have their permanent being subterraneously, in living storehouses that we call bulb or tuber or fleshy root. In the few spring weeks they do all their year's work, of manufacture and reproduction alike, and then retire to vegetate invisibly below ground for the rest of the year.

But the most striking desert plants are the big fleshy water-storers. The most familiar of these are the cactuses, but the spurges (*Euphorbias*) and many of the composites of deserts have been modified in the same direction, often so thoroughly that all save experts mistake them for cactuses. The agaves and aloes and spanish bayonets show similar but usually less striking modifications. In such plants, the roots are generally deep, the tissues fleshy and built so as to be able to store large quantities of water in their cells. All plants must be constantly passing water from root to stem and leaf and out into the air as water-vapour; the current serves for vital transport; but in these desert plants its amount is, by one means or another, cut down. This enables them to draw water out of the soil during the rains and to store it for the hard times ahead. A hibernating bear or hedgehog lives through the time of food scarcity on its piled-up stores of fat; a cactus lives through the time of drought on its stores of water.

The usual method for achieving this is by a great reduction of the surface from which water can be given off as vapour; very often, as in cactuses, the leaves are small and temporary, or even absent, and their work is handed over to swollen stems, which expose a much smaller surface to the air in proportion to their bulk. The climax is reached in the barrel-cactuses, which stand about in the desert like casks, and like casks are full of water; but their surfaces are beset with wicked spines to protect the precious fluid inside from browsing animals. In passing, it may be pointed out that the plants that retire underground for the dry season and live as bulbs and the like are really doing something very similar, though more extreme, for they cut their transpiration current down to zero during the drought; and those which bridge the dry season in the shape of seeds go one better, by getting rid of all their water and passing into a desiccated state of suspended animation.

In the perennials, which do not die down above ground, mere reduction of transpiring surface is aided by other adaptations—the cuticle is varnished over with wax or resin, or a thick coat of hairs checks diffusion, a layer of cork protects the stem-tissues from evaporation, or the microscopic plant-mouths through which the water-current vaporizes out are tucked away in deep pits on the under side of the leaves. Often, too, especially in salt and "alkali" deserts, the amount of salts in the cell-sap is very high; this holds water and makes it harder for it to evaporate away, while at the same time helping the roots to pull water out of the soil.

Spines and thorns are a frequent accompaniment of desert-life. Hedges of prickly pear are more impassable than barbed-wire entanglements, and are often used for military purposes in protecting forts and camps; and the spines with which the agave's leaves are tipped are really formidable stilettos. In part, spininess seems to be the direct outcome of dry surroundings; but the tendency in that direction has undoubtedly been improved upon by Natural Selection to protect the succulent tissues—often the only source of moisture for miles—from animals. In certain North African agaves the long leaves are tipped with black-brown, and suggest the same hardness and sharpness as the leaves of ordinary agaves. But in reality the tips are perfectly soft and flexible—dummy spines. Apparently we have here a case of mimicry among plants, analogous to that of the dead nettles which look like stinging nettles. The harmless plant gains an advantage by looking like a dangerous one.

The desert has the same dual effect upon life as have other habitats with a combination of special and very unfavourable conditions. It exercises a rigid selection upon immigrants; and it calls forth a number of special adaptations in those which are successful in entering in. No true aquatic animals can live in deserts, save in the sparse oases. Amphibia, with their moist skins, are almost all turned back at the frontier; and so are those other moist-skinned animals, land-leeches and land-planarians, and those groups of insects, such as may-flies and dragon-flies, which have returned to water for their growth-stages. Animals confined to a purely insect diet and incapable of flight are almost unknown in deserts, probably because insects are abundant only at one season. Owing to the general scarcity of life, large animals are absent; the maximum size diminishes as we penetrate from the half-desert or desert-steppe to the true desert heart.

As a result of this selection, the types of animals which make up the bulk of the desert fauna are those which were already provided with a dry skin and a wholly terrestrial life-history—beetles, butter-flies, Orthoptera and Hymenoptera among the insects; centipedes, scorpions; and among vertebrates, birds, mammals and notably reptiles.

Those of other groups which have passed the immigration tests have done so by means of special aptitudes. Land-snails are not unsuccessful desert forms; they owe their success to their power of closing the opening of their shells with a door of hardened mucus, behind which they sleep away the driest time. A few land-crustacea in the form of woodlice live in deserts; most of them have a high dome-shape, which reduces their relative surface. No tailed amphibia exist in any desert; but there are a few frogs and toads. In the Australian desert, four species are not uncommon. They are without any annual breeding cycle, and can spawn immediately rain falls, whatever the season of the year. One species makes burrows down to the just-moist soil of torrent beds, and lays its eggs in a foamy mass at the bottom of its burrow; development proceeds within the egg to the early tadpole stage, and the eggs then wait for rain; when the rain comes, the eggs swell up, burst, and release the tadpoles, which grow rapidly and change into frogs. One form, called Chiroleptes, has taken a leaf out of the cactus' book, and stores water. An Australian observer writes: "If you put a lean dry herring-gutted Chiroleptes into a beaker with two inches of water, in two minutes your frog resembles a somewhat knobbly tennis ball." Absorption goes on all over the skin, and the water is stored not only in the bladder but in the subcutaneous tissue and the body-cavities, making the frog nearly spherical. The black-fellows, if hard pressed, use it as a source of drinking water. This is the only species which can live for long periods in completely dry conditions.

Another and very different adaptation to drought is seen in the sand-grouse. Those powerful fliers nest in the steppe-deserts of Asia, and the adults can quench their thirst by flying off to the nearest oasis, though this may be miles distant; but this the young cannot do. The problem of their drinking is solved by the parents soaking their breast-plumage in water, flying straight back, and allowing the nestlings to suck the moist feathers. However, a not infrequent adaptation to water-shortage is seen in the capacity which many desert animals possess of going without drink at all. They get all the water they require out of their food. Various desert mice and jerboas, and even some larger animals, such as gazelles, possess this valuable physiological power.

We have already spoken of the prevailing matching of the desert's sandy colour (Book 6, Chap. 1, § 4). Then there are a number of adaptations concerned with wind: we will mention two. Many ground-nesting birds build semi-circular ramparts of stones to protect their nests from the prevailing winds. Without these defences, the nest and eggs would be in frequent danger of being buried by the sand which the wind drives along. An even stranger case is that of a small butterfly in the Syrian desert. If it lived in the open, it would risk being blown clean away by the gales that sweep over the bare country; accordingly it spends the whole of its life fluttering about inside one of the rare desert bushes!

Another problem is temperature. In many deserts the temperature of soil exposed to the midday sun is enough to roast eggs—a good deal higher than anything which ordinary protoplasm can stand; and even in the shade, the thermometer will often go up to 120° or 130° F. The most obvious adaptation is to go about your biological business by night: and, as a matter of fact, a high percentage of desert animals are nocturnal. Of the creatures of the day, almost all avoid the sun as much as possible. All mammals and birds can regulate their temperature, and in deserts they can, if need be, keep it below that of their surroundings, just as in colder places they can keep it above the outside temperature. But many desert reptiles, notably lizards and snakes, have special adaptations tending in the same direction, by means of which they can at least keep themselves a few degrees cooler than the oven-like world outside, and so save themselves from sudden coagulation and death. The usual method is for the animals to have a mouth-cavity richly supplied with surface blood-vessels, and to pant rapidly with open mouth. This, as in a dog that has been heated up by running, causes evaporation and consequent cooling.

There is one final aspect of desert life that should be touched on. The scanty, open vegetation means a shortage of food for animals, and they must adjust their population to the supply. Sometimes they lay up stores against the time of drought and scarcity. This habit is what attracted Solomon's attention in ants. The ant he held up as a model to the sluggard was an ant of dry climates, a grain-storer. Our temperate-zone ants do not have the storage habit. Some desert mammals also store food: for instance, the kangaroo-rat *Dipodomys* of the American semi-deserts. This little agile, jumping creature makes big mounds, full of tunnels and chambers, and in them amasses huge stores of grasses, flower-heads, and the like. In one mound no less than $12\frac{1}{2}$ lbs. of hay-stores, mostly valuable forage-grass, was found. These depredations make *Dipodomys* a pest; for though in normal seasons there is enough to go round, in dry years the rodents take a serious proportion, and there is not enough left on the range for stock. Systematic extermination of the little beasts has led to a marked improvement of the grazing and the number of cattle it can carry per acre.

Hundreds of mounds belonging to this animal have been excavated after their inhabitants have been gassed; and in all of them there has been found either a solitary male, a solitary female, or a mother with her family: there are no couples to be found keeping house together. This, it seems, is a further consequence of the sparseness of the vegetation. Each mound is the centre of a largish area from which its owner draws supplies. If two were to live there instead of one, the supply-area would have to be uneconomically large, the animals' journeys in search of food unprofitably long. So the desert has imposed upon our kangaroo-rat this curious semi-bachelor life, in which adults visit each other for sexual intercourse, but never know a family existence.

And the same impression of the desert's poverty is brought home by the description of the Tibetan sheep given by a member of the Everest expedition. The tufts of herbage on these high and arid uplands are so few and far between that the sheep can only get enough to survive by running between mouthfuls. They chase their grass.

§ 6. The Tropical Forest.

THE tropical rain-forest has been the wonder of all naturalists. Alfred Russel Wallace tells us that it was von Humboldt's description which lit in him the desire for tropical exploration; but that the reality exceeded his expectations. This rain-forest girdles the tropical lands. It still covers about half of the South American continent—a forest over two thousand five hundred miles from west to east, over fifteen hundred miles from north to south; man's inroads into its green fastnesses are still negligible; the only way to penetrate it is along its rivers and streams. Africa's forest is equally celebrated; but it is considerably smaller in extent, owing to the height and dryness of much of the continent. And the belt of rain-forest continues round the world, through Ceylon and Malaya and New Guinea.

Naturally there is the greatest variation from place to place; but everywhere this huge expanse of green chlorophyll-machinery raised high on supporting trunks into the hot, steamy tropical air, has certain features in common. The variations in type occur especially where unfavourable conditions prevent the majority of species from flourishing. In tropical swamps, for instance, the ordinary forest gives place to a theatrical but much more uniform growth of palms; and only mangroves and a few other trees that use their roots as stilts have managed to grow over the mud of tropical estuaries.

Under conditions that are altogether favourable for vegetable life the struggle for light is never-ceasing. Here is no dead season when plants shed their leaves and have their roots frozen as they stand. Growth, activity, competition, continue year in year out. The most striking result of this is the huge variety of species that make up the forest. In contrast with the dozen or so kinds of trees in our temperate woods, the Cameroon rain-forest numbers close on five hundred. As well as the trees, there are the creepers and the parasites. In this same Cameroon forest, more than three hundred species of woody-stemmed plants occur which are incapable of standing on their own trunks but cling to the trees for support. In the great Amazonian forest, these lianas and creepers reach their maximum profusion and their greatest beauty of flower.

Miss Haviland in her *Forest, Steppe and Tundra*, describes the Amazonian forest as seen from one of its rivers: "On either side the banks are veiled by a wall of green foliage between one and two hundred feet high, towering above its own inverted image in the water. Here and there its splendid sameness is broken by a patch of coloured blossoms. The branches of the scarlet 'rose of the forest' are thrust out over the river, and sprays of Bignonia and other flowering creepers, yellow, purple and red, hang over the trees. The creepers cover the whole roof of the forest as with a canopy, and fall to its foot at the water-side like a curtain. In fact, the forests of the whole Amazonian region may be compared to a series of tables with many legs, separated by waterways, and each spread with a cloth which dips to the ground on every side. The table-legs are the upstanding trunks of the trees; the cloths are the tangle of vines and lianas which cover them with a close network. This mass of creepers is not altogether the suffocating burden or host of parasites that it appears to be. In exchange for support, it affords shade which is essential to the well-being of the forest; and it has been shown that when the veil has been torn aside so that the sun can beat down on the roots, the giant trees perish. For this reason an artificial clearing is usually fringed with dead trees.

"Here and there dark caverns yawn in the wall of foliage at the water-side. These are the mouths of creeks and streams, shut in by over-arching branches from which long aerial roots hang down like stalactites. To enter these caves by boat is like passing from the open air into a vast dim hall, supported by immense columns. The trunks of the trees rise up for seventy or eighty feet without a branch, and the undergrowth is thin and straggling. The ground is strewn with dead leaves, though it may be remarked that the accumulation of leaf-mould is not very great, owing to the rapidity of bacterial action."

The roof of the forest is the prime source of all its biological wealth, the scene of its greatest activity. As the same authoress says: "In some respects the roof of the forest may be compared with a prairie or savannah. There is the same wide green expanse, strewn with flowers and open to sun and rain and wind. Butterflies hover and grasshoppers skip over the surface, and its denizens are exposed to the unrestricted view of birds of prey—vultures, kites and harpy-eagles—which soar over the forest, ready to seize any bird or monkey which is not alert enough to dive under the foliage."

But, alas! we know tantalizingly little about this zone of brilliant light and rich life. Two-hundred foot trees are not easy to climb, especially when the attempt brings out armies of stinging bees and biting ants from their nests on every branch. We want a modern St. Simeon Stylites to set up a pillar in the rain-forest and use it not for meditation and prayer, but for the acutest observation and description of the unique life about him. We have already compared the forest to the sea—productive layers atop, with débris of light and food filtering and drifting down. Man on the solid ground of the forest is like a mere flat-fish on the sea-bottom. The tangle of life is layered according to its distance from the roof: in von Humboldt's words: "Forest is piled upon forest."

Those who come fresh to the rain-forest are generally disappointed at first by the apparent paucity of animal-life. This is partly because it is dwarfed by the fantastic luxuriance of the plants, partly because the greatest abundance, activity and brilliance of rain-forest animals is up in or near the tops. In the upper zones of the Amazonian forest, for instance, there are squirrels and sloths, tree-porcupines and tree-anteaters, bands of monkeys making the forest resound with their howling, tree-raccoons, climbing cats; there are innumerable birds—toucans, parrots, parrakeets, cotingas, barbets, frog-mouths, forest pigeons and nightjars, curassows, bell-birds. There are the marvellous tree-frogs and tree-toads, many of which never descend to earth but brood their eggs in pouches on their backs, or deposit them in foamy masses on the high leaves, to go through the tadpole stage in this pretence of a pond; while others put the egg-masses on leaves above pools, whence the tadpoles emerging slip into the water below. And there are the incredible hordes of insects—tree-nesting ants, and bees, and wasps, and termites; huge butterflies that never come down to ground-level, beetles and crickets and cicadas and plant-bugs that the human collector never sees unless a tree is felled.

What are the chief adaptations in this dense envelope of life? Luxuriance of growth is the first: no tree will ever succeed whose seedlings cannot shoulder their way up to the never-broken green canopy above. The next most striking fact is the abundance of plants that support themselves on others' shoulders in the race for light. Every observer of the tropical forest has commented on the extraordinary way in which the trees are beset with woody cables, thick and thin, like ropes and cordage carelessly and meaninglessly spread among a forest of masts. These are the lianas. Often their stems coil round each other like the strands of a cable, or are provided with a flattened spiral wing. They corkscrew it over the ground, hang in low curves, are pulled taut by the upward growth of the tree to which they cling. Here they make festoons of green, there hang in thick curtains or spread as sloping carpets.

Almost all of them have relatively huge conducting pipes to speed water and its dissolved salts up the long thin stem to the leaves. But their actual methods of climbing are very various. Some insinuate their growing tip through the interstices of bushes and branches, later weaving themselves firmly into place by sending out side-branches. The most wonderful of these are the Rotang palms, which have an additional support in the shape of wicked thorns on outgrowths from the tips of the leaves. Up in the light on the roof of the forest world, these thorny anchors wave round in empty air; but the old leaves die, and then the smooth stem slips until the young growth engages with the tops of the trees. In this way the old stem is continuously being paid out downwards onto the ground, and the total length of stem produced by one plant may reach the prodigious figure of three hundred yards.

Then there are the lattice-formers, that make a living trellis from branch to branch of their support; and the twiners, which climb by thrusting their stem spirally round the support. There are the tendril-bearers, which hook on by prehensile tendrils; and the root-climbers, whose stems produce clasping or adhesive roots wherever they come in contact with the support. A number of tropical species of fig have this form of climbing-iron well developed. In addition, many of them, once they have established themselves, let down long aerial roots which penetrate the soil and begin absorption. As a result, the supporting tree is often killed; but the climber, now strong enough to stand on its own legs, remains erect and independent.

Aerial roots are another striking fact of the tropical forest: and many of them are permanently aerial, never striking earth, but hanging like so many bell-ropes in the green gloom. They are the hall-mark of another set of plants that exploit the strength of trees—the so-called *epiphytes*. These are plants which do not even take the trouble to climb, but settle aloft as spore or seed, and begin their growth far above the soil. The aerial roots tap the water in the air, catching the rain before it reaches the ground, drinking in dew, or even sucking moisture from fog or direct from damp air. In this they are aided by an outer layer that greedily imbibes water. Their main difficulty is their supply of mineral salts. Some are dependent on the precarious supplies washed down in the débris of the stems on which they sit; but others have evolved remarkable adaptations for collecting little private gardens of soil up aloft. Some do this by producing a network of roots which grow upwards, nest-like, from the base of the stem and catch the dead leaves and twigs and other rubbish falling from the tops; some collect their humus in a nest of leaves into which the roots grow inwards and upwards; Dischidia, a Javan epiphyte, produces a set of pitcher-shaped leaves in which water and débris collect; into each of these leaves a special little root-system grows and absorbs what it needs from the soup therein contained. In certain species of Tillandsia (a genus related to the pineapple, other species of which are familiar in the southern United States as "Spanish Moss") roots have been dispensed with altogether; their leaves are arranged to make water-tight tanks, holding up to half a gallon, and are beset with tiny absorptive hairs that take the place of roots.

Some of the most beautiful orchids are epiphytes; and so are various figs. A number of these latter are only epiphytes for half their lives, for their aerial roots eventually grow down to the soil; and sometimes, as in the liana-figs we have described, these grow columnar and trunk-like and the plant, after killing its host, becomes an independent tree. This is the history even of the huge banyan-fig.

Besides flowering plants there are abundant epiphytic ferns and club-mosses; and many mosses and fungi, lichens and algæ, have found for themselves a station not too far from the light by adaptation to growing on leaves.

But we must not delay too long over the plants of the rain-forest, though it would be easy to fill a book with their beauties and peculiarities. We must pass to the animals. Among the animals the prevalence of the climbing habit is the first and most obvious feature. The forest is inhospitable to man, man an enemy of the forest. He forgets how much of the earth is still covered with trees, how much more was once under forest, not merely before he came to fell and clear, but in the great stretches of geological time when the moister, more equable climate spread the forest zones over much more of the world's surface. And so he is surprised at the abundance of arboreal animals, the importance of the climbing habit in the evolution of life. Not only has he himself descended from a tree-living ancestry, but the foot-structure of kangaroos and related marsupials makes it certain that they too were once arboreal, and many authorities believe that the ancestors of one great branch of Dinosaurs passed their apprenticeship in the trees.

Be that as it may, arboreal life is common enough to-day. In the Guiana forest, for instance, more than half the known species of mammals are climbers. It is an interesting fact that in the Amazonian forest, the greatest stretch of tropical forest in the world, more mammals than anywhere else have evolved that fine flower of tree-life, a prehensile tail. Only here do monkeys boast this fifth limb; and the tree porcupines, tree ant-eaters, coatimundis and kinkajous also possess it.

Life is so intense and competitive in the rain-forest that adaptations to escape enemies by utilizing colour and pattern are more numerous than elsewhere. The abundance of protectively coloured insects is astounding; the number of creatures, notably caterpillars and plant-bugs, which have evolved some form of terrifying device to bluff their enemies is far greater than elsewhere; so is the development of nauseous taste, combined with light colours to advertise the unpalatability; and the tropics is the chief home of that mimicry of nauseous by other species or by one another which we shall discuss in a later section.

Devices for enabling carnivorous creatures to deceive their prey are also much commoner. In the East Indies, to choose but one example, there lives a spider which is coloured black-and-white; after spinning a thick whitish web over part of a leaf, it lies on its back in the centre. In this attitude it looks precisely like a bird's dropping, and the web simulates the liquid draining away from it. Many butterflies have a curious partiality for sipping such excrementitious fluids; and H. O. Forbes actually saw one come down to take a drink, only to be captured by the spider.

§ 7. Regions of Rock, Snow, and Ice.

AT THE other extreme from the fantastic luxuriance of the tropical forest come the polar regions and their isolated counterparts, the bits of mountain-chains that protrude above the snow line. At both extremes, the severity of the struggle for existence is at a maximum; but while in the rain-forest it is the struggle between one creature and another which counts, in polar and mountainous regions it is the struggle with the elements. In the one case there is over-abundance of food; in the other, the extremity of scantiness.

The scarcity of food which besets land-animals in the polar regions may be illustrated by an incident which befel the Oxford Expedition to Spitsbergen in 1921. The sledging-party had surmounted the huge ice-fall, nearly 3,000 feet high, of the Nordenskiold Glacier, prior to setting forth across the inland snow plateau. They had collected a number of rock specimens and fossils, and cached them, all nicely labelled, to await their return. When they came back, the specimens were safe; but all the labels had been eaten off by desperately hungry arctic foxes.

That was in summer. In the long night of the polar winter the foxes must be still harder put to it, for all their possible prey has left the country, save only ptarmigan, which live in tunnels excavated in the snow, many feet below the surface, and subsist on the frozen shoots of the plants they find there. Some of the foxes are then forced out by hunger on to the sea-ice, where they play jackal to the polar bear when he kills a seal; others remain and try to catch the buried ptarmigan. The few men who stay in Spitsbergen through the winter help beguile their time in the first twilight of spring by trapping. They set traps baited with ptarmigan-heads. If they see fox-tracks going dead straight across the snow, they know the fox has scented the trap. One trapper traced such a straight fox-track five miles to a baited trap. What a natural sharpness of nose, sharpened still further by what an extremity of hunger, it must need to smell a bit of dried and frozen bird at five miles!

The two polar regions stand in sharp contrast; the high arctic is mainly sea, the antarctic is centred on a huge and mountainous continent. Accordingly the antarctic is far more rigorous and barren than the arctic. There are no land-animals whatever on the antarctic continent save a few wingless insects that scavenge round the shores. Its only vertebrate frequenters are the penguins—sea-feeders all—the skua-gulls which batten on them, and the petrels. And there are no flowering-plants—only a few patches of mosses and lichens.

Perhaps the most striking of all antarctic creatures is a temporary visitant—the emperor penguin. All penguins must breed where there are no predaceous land-mammals, or they would be exterminated; besides offshore islets and oceanic islands, the shores of the great antarctic continent are available for them, for no four-footed beasts exist there. The emperor penguin is one of these antarctic breeders. It is also the largest species of the group; and the young take so long to grow to full size that if they are to be ready to accompany their parents to the open sea and its rich food supplies in the late summer, they must begin their life in winter. Accordingly these great birds nest, in the total darkness of the antarctic winter, on snow-covered land, with temperatures often falling to 60° or 70° F. below zero. One of the achievements of the *Discovery* expedition was an arduous sledge-journey undertaken to study their breeding habits. The eggs and new-hatched young are protected from the cold by being held between the feet and the belly; and so violent is the birds' incubating urge that they will fight each other for the privilege of brooding young —often injuring and killing the chicks in the process—and will even satisfy their desires by incubating lumps of ice instead of eggs. Thus the penguins, though slow and awkward on land, nest unmolested on the mammal-free continent of the south.

Within the arctic regions, on the other hand, there are to be found musk-oxen and reindeer, the wolf, the lemming, and the arctic hare, and some valuable fur-bearing creatures, like arctic fox and ermine. Birds are very abundant round the coasts, but only during the breeding season; very few species (like ptarmigan and snowy owl) live there all the year round. Spiders and mites, mosquitoes, midges, and sawflies are to be found, often in comparative plenty, and even a few beetles, moths and butterflies. But the abundance of insect-life, so striking in the tropics, is absent. Insects, like reptiles and amphibians, get progressively less important as the climate grows colder, since their activities fall with the fall of temperature. If there were warm-blooded insects, they might well be as abundant in the arctic as birds; but insects cannot be warm-blooded: the limit of size imposed upon them (Book 5, Chap. 5, § 7) means that they have, compared with their bulk, too great an area of surface out of which heat can leak away. One of the few attempts at warm-bloodedness is made by humble-bees, whose comparatively large bulk and thick, hairy coat hinder the heat generated in their muscles from escaping. While they are moving, their temperature is several degrees above their surroundings: and this property enables them to penetrate much farther north than the smaller and less hairy hive-bees. The lower southward-facing slopes of the Spitsbergen mountains, only seven hundred miles from the pole, have a rich carpet of bright flowers during the short summer season; and even in the northernmost regions of Greenland there are great stretches of flattish tundra bare of snow in the summer, with a vegetation capable of supporting a population of shaggy musk-oxen.

There are some interesting adaptations to be found both in plants and in animals. Many of the plants, in response to the fewness or absence of bees, have changed over from reproduction dependent upon insect-fertilization to some other method. Some have become self-fertilizing; others, though they still produce flowers, never set seed, but rely on some form of sexless reproduction. In response to the shortness of the summer, many prepare all they can beforehand, and burst into flower and leaf the moment the sun thaws the snow off them. The same thing happens in Alpine regions. Here, however, Soldanella, the ice-flower, goes one better. Its flower begins to grow while the snow-blanket still lies over it; and the heat generated by the chemical activity of its growth actually helps to melt a tiny chamber over the bud and bring the plant to the air a few days before it could otherwise have escaped.

A very interesting effect of the arctic food-shortage is seen in some of the birds. The skuas, for instance, and probably the snowy owls, do not breed every year. In years of scarcity they make no attempt at nesting; their systems probably respond automatically to low temperature and lack of food, their ductless glands are not set in the direction needed if the ovaries are to pile up yolk in their eggs, and the reproductive impulse is never felt. In a somewhat similar way, butterflies which in warmer climates may have several broods in a year, in the high north take two or even three years to grow from egg to adult, the caterpillar hibernating between his summer feeding-periods.

But in the polar regions, arctic and antarctic alike, the great contrast is between land and sea. For if the land is poor in life, the sea is rich. For one thing, polar life, so long as it remains in the water, is not exposed to the extremes which it must suffer if it emerge into air. The sea may freeze over; but there is always water below, and this must be above freezing-point. But there is more than this; the life of open waters is actually more abundant near the poles than in the tropics, largely owing to there being more nitrogen salts available in cold than in hot waters. Diatoms discolour the polar seas for miles. Supported by the diatoms live hordes of crustaceans and other small creatures. It is a strange sight to lean over a ship's side in a Greenland fjord, with rock and ice all about one on the land, and see a constant procession, perhaps one to every square yard, of black specks of life, each provided with a pair of flapping sea-wings—millions upon millions of the pelagic snails called Pteropods.

As result of this richness of the polar plankton, the dominant life of both the polar regions lives either in the water, or, at least, upon its products. We have spoken at the end of Book 5 of the hordes of birds which take advantage of their winged mobility to visit the arctic in summer, take toll of the riches of the sea for themselves and their young, and leave for the south before ice and darkness cover their feeding-grounds.

Of the arctic mammals, aquatic forms are pre-eminent. The walrus browses on the shellfish which he rakes out of the arctic mud with his pick-like tusks; the seals swarm in both polar zones, and have a great range of habits, from the inoffensive crab-eater seal to the leopard seal, the tiger of the antarctic, which loves a penguin if it can get one. Their swimming powers are wonderful. Where there is a narrow passage into a lagoon, with a strong current through it, there the seals delight to play. They will test their powers against the stream, and dart forward in the face of a current against which a ship's boat with four oarsmen can scarcely make headway. Like land-carnivores, they are alert and intelligent—the real dogs of the sea. In the arctic summer, the seals love to bask contentedly on the slow-drifting ice-floes. When they are busy fishing they have to come up to breathe from time to time; this they generally do at particular holes or cracks in

LIFE IN FRESH WATER AND ON LAND

the ice; there the Eskimo hunter waits, harpoon in hand, for their emergence. Seals are the staple food not only of the Eskimo but of the polar bear, which is amphibious, and divides its time between water (in which it swims excellently) and ice. The land it has almost entirely forsaken.

Seals keep warm by means of the thick layer of blubbery fat which blankets them under their skin. It is noticeable, by the way, that wherever fat is employed as a heat-retainer as well as a store of food, it is spread uniformly all over the body; this characterizes not only seals, porpoises and whales, but also such creatures as reindeer, bears, and many other northern land mammals. But in hot climates, where it is needed only as a reserve of food, and the need of the warm-blooded animal is to lose heat, not to retain it, the fat is stored in local accumulations, leaving most of the body-surface unblanketed, as in the hump of the zebu or camel, or the tail of the fat-tailed sheep.

Seals, too, like sea-birds, come inshore to breed. The "rookeries" of some species rival those of the penguins in population and bustle.

Whales do most of their feeding and breeding in high latitudes. The inhospitable shores of antarctic islands, such as South Georgia, now hum with activity; and the whaling industry grows rich. The biological basis of this prosperity is the whale's need to keep warm; for this he grows his juicy under-garment, sometimes a foot thick, of blubber; and he comes to the antarctic to do it because of the richness of its waters in plankton. But there is a real danger that intensive fishing may bring whales as near extermination in the southern hemisphere as they have already been brought in the northern. World-regulation of the whale-fishery is the only hope.

The bottom life of polar waters is rich, too. Captain Scott's antarctic expedition secured wonderful hauls of such creatures as sea-urchins, sponges, and the queer sea-spiders or Pycnogonids—all leg and no body.

Recently Stefansson, the well-known explorer, has sought to persuade us that the arctic is not so bad as it is painted. *The Friendly Arctic* he calls it in the title to his book. He points out that the cold is much more intense in north-eastern Siberia than in the arctic, that the tundra supports a rich vegetation far beyond the arctic circle, that what with reindeer, musk-oxen and seals, the explorer need never want for fresh meat, and that we may hope to solve the world's meat problem by introducing reindeer into arctic Canada and breeding them there on a scale to fit those vast bare spaces.

There is much truth in this; but it is not all the truth. There are the winters to contend with; months of total darkness, even if enlivened by the displays of the Aurora, are hard to face. And the arctic which he speaks about is only the fringe of the arctic. The high arctic, save where local conditions keep it relatively fertile, is desolate enough. The west coast of Spitsbergen is full of flowers and birds in summer, because the climate is kept mild by the Gulf Stream. But the east coast is exposed to the polar current from the north and is barren and forbidding in the extreme. Then again, though it is true that seals abound all round the shores of the polar sea and for many miles northward, they do not seem to penetrate within several hundred miles of the pole. Our world is capped with a flat expanse of ice, barren of all life of mammal or bird, save rare storm-blown stragglers or the still rarer human explorers. On the other hand, as knowledge of weather conditions grows and aircraft improve in reliability, it is on the cards that this desert may see a busy traffic overhead; for it is by far the shortest route from Europe to Japan, or New York to China.

Over the life of the high mountains, interesting though it is, we cannot stay long. Broadly speaking, it is like that of a mountainous polar land, but without any compensating riches like those provided by the polar sea. Just as all land-life ceases long before the poles are reached, so the tops of the highest mountains are absolutely barren of life, plant or animal alike.

The summit of Mount Everest reaches 29,000 feet. On the Himalayas a few plants grow up to 19,000 feet. Most animals stop where the plants stop; but the climbers on Mount Everest saw the tracks of mountain-sheep at 20,000 feet; of hares and foxes at 21,000 feet, and wolf-tracks at 21,500 feet; a vulture was seen flying at 25,000 feet, and a few choughs visited the camp at 23,500 feet and followed the climbers out of curiosity to 27,000.

At lower levels mountain-life, if not rich, is full of beauty and interest. The low-growing plants, tufted and cushiony, often have flowers of a brilliance denied to those of lower altitudes; and there are grazing-animals like the ibex and other mountain-sheep, the mountain-goats and the chamois, all with an astounding agility and sure-footedness; and beautiful beasts of prey like the snow-leopard. The biggest of all flying land-birds are mountain-dwellers—vultures like lammergeier and condor; and there are smaller birds like cliff-swallows and mountain-choughs, and the rock-creepers, that search the faces of cliffs for insects, at each upward jerk displaying a crimson flash of wing-feathers.

Perhaps the most impressive thing about the mountain is that life, the insurgent, with all the pressure of millions of years of over-reproduction behind it, has not been able to scale their tops. The highest spot on earth's surface is only five miles and a half above sea-level; but life has faded out far below.

§ 8. Island-dwellers.

CERTAIN special aspects of island-life have already occupied us. We have seen in Book 3 how their inhabitants afford a proof of Evolution by resembling those of the nearest mainland; and in Book 4 how the isolation which they provide has brought new types into existence. But we have said nothing of the peculiar stamp which island-life sets upon many of its creatures.

The most obvious characteristic of island-life is the high percentage of flightless forms to be found among the great winged groups of birds and insects. There is a flightless cormorant on the Galapagos. New Zealand still possesses several flightless birds, such as the kiwi, the owl-parrot, and two or three rails; but once it also harboured the huge moa, a giant flightless goose, a giant flightless duck, and a flightless hawk, all now extinguished, probably by man. The most celebrated flightless bird of them all, the dodo, was an inhabitant of the oceanic island of Mauritius. Its anatomy shows it to have been a ridiculous and overgrown pigeon which had grown small in wing but large in body. The solitaire of the neighbouring island of Rodriguez was another overgrown flightless pigeon. This creature was not so plump, but had evolved a little way in the ostrich direction. On the same island lived a nearly flightless heron, also now killed off.

Then there are a large number of flightless birds which live on one or other of the South Sea Islands. Many of them are rails, which is natural enough, seeing that an ordinary rail spends most of its life without using its wings at all as it skulks through the marsh herbage. And others are moorhens, which also skulk.

New Zealand, though such a number of its birds have lost the power of flight, is not characterized by a very high proportion of flightless insects. It is only the smaller oceanic islands where these abound. Kerguelen and Crozet Islands, down south towards the Antarctic, between South Africa and Australia, are excellent examples. Of seventeen genera of Crozet insects, fourteen are flightless; of eight species of Kerguelen flies, only one lives up to its name and can fly. Nearer home, in Madeira, nearly half the beetles have lost the power of flight, and the same sort of thing is found in Hawaii, the Falklands, and elsewhere.

The adaptive meaning of the winglessness is clear. The insects of small, isolated spots of land will perish if they are caught on the wing by a wind and blown out to sea. Every mutation favouring shorter wings or a lessened inclination towards flight will benefit its possessors, until finally the race loses the capacity and the instinct of flying. For the larger and more powerful birds, however, with their better sense of vision, this will not be so important, though it may contribute. What will count with them is the fact that an island lacks predaceous land-animals, especially the active mammals. A rôle is thus vacant, waiting to be filled, in the economy of the place, for which the bird may fit itself if it grow wingless. For by sacrificing its wings it can put more strength into body and legs—as does the kiwi —and can escape from the mechanical limitations to size which flight imposes and grow enormous, like the dodo or solitaire or moa.

This lack of predaceous enemies may be revealed in other ways. Since island-birds are not under the necessity of escaping notice by matching their surroundings, there will be little selection against the albino and other colour mutations which in all birds occasionally crop up. And, as a matter of fact, white quails are common in the Azores, and pied blackbirds much less scarce than on the mainland, while pied and albino ravens are more frequent in Ireland and the Faroes than elsewhere; and the same is true of many New Zealand birds. The tameness of the birds of oceanic islands, which we have already mentioned, is due to the same lack of enemies.

So it comes about that flightless land-birds are typical of larger islands, but only if these lack mammals; while the percentage of flightless insects goes up with the smallness and the storminess of their island homes, till it reaches its maximum in Kerguelen, where a calm day, or indeed a day without a storm, is a rare exception.

There is one other characteristic of island-faunas. When a small island harbours representatives of some large mammal these are almost always small. One of the most striking examples is the dwarf elephant which lived in Malta in the Pleistocene (V E); and the islands of the Mediterranean still possess diminutive races of red-deer. Such creatures first inhabited what is now the island when it was still connected with the mainland. As seas encroached it was cut off, and grew smaller; and as it grew smaller subsistence grew less easy to find, and the less bulky creatures survived where the bigger ones starved. For a similar reason, it would seem, the reindeer on Spitsbergen are of a semi-dwarf race; but here the difficulty of finding subsistence depends more on the barrenness of their land than on its physical smallness. We can imagine the appalling struggle for existence that goes on as such an island shrinks to nothing and disappears below water. For instance, as the plain between England and the Netherlands became converted into the North Sea, what is now the Dogger Bank remained for long centuries as an island. On this a sample of the late Pleistocene fauna seems to have remained for a time, but the whole menagerie was eventually drowned.

§ 9. Cave-dwellers.

UNDERGROUND, in hidden streams and caverns, there lurks a specialized fauna that recalls in certain ways the inhabitants of the abyss. It is for one thing a parasitic world, bare of green plants, since light, the prime generator of all life, is absent. Cave-dwelling creatures either resort there for shelter and protection, as bats or breeding cormorants or rock-pigeons do, and seek their food outside; or, if they are permanent inhabitants, they live on the scanty scraps brought in by the shelterers and casual visitors, blown or drifted in by air, or floated through by subterranean streams. Among these permanencies there are a few beetles, grasshoppers, centipedes and spiders, but these live mostly near the mouths of caves; the aquatic forms are more interesting, especially the crustaceans, fish, and amphibians.

It is a curious fact that no cave-dwellers possess phosphorescent organs. Accordingly none have developed large eyes specially adapted for dim light, like some of the deep-sea dwellers, but a large number have become sightless, often developing long feelers or legs with sensitive hairs to make up for their blindness. Many also have lost their pigments and become white, since in the absence of searching eyes there is no need for the blacks and invisible reds of the deep sea which enable their wearers to blend with the darkness.

Amphibians which have found a refuge from competition in underground water have had to suppress their adult phase, living all their life as gill-breathers, and have become colourless and blind. The best known of these is the strange Proteus of the great limestone caves of Carniola, white (save for its pink gills), and blind, with tiny, sightless eyes. If when young it is exposed to white light, it develops abundance of dark pigment, and the pigmented skin even covers the eyes; but if it be exposed to red light the skin over the eyes remains transparent, the eyes develop, and the animal can be made to see, though its ancestors have been sightless for thousands of generations.

Another such blind and permanently gilled amphibian is Typhlomolge, which inhabits underground water-courses in Texas and is sometimes hurled surprisingly into the light of day from an artesian well.

In some caves, like the Mammoth Cave of Kentucky, large lakes lie underground; and here cave-fishes are to be found—they, too, exhibiting various stages in loss of colour and degeneration of eyes. And there are blind and pallid cave-crayfish and cave-prawns and well-shrimps, albino cave-snails and milk-white cave-flatworms. In the underground channel which leads off water through the mountain-side, from Thirlmere to supply the city of Manchester, numbers of the common green Hydra have established themselves, feeding on the débris brought down by the slow current. But in the darkness the green alga-cells which normally live as partners within the Hydra's body cannot exist, and have died or emigrated, leaving the race of subterranean polyps white and transparent.

The cave-fauna presents some of the perennial problems of Evolution in particularly clear-cut fashion. Are cave-animals blind, for instance, for Lamarckian reasons, because of the accumulation of the direct effects of generations of darkness, or has the inheritance of acquired characters had nothing to do with the matter? And if the Lamarckian view be ruled out, have they become slowly adapted to cave-life by the selection of smaller-eyed varieties? or was there what is sometimes called pre-adaptation, in that animals which happened to have poor eyes or to be blind, sought the shelter of caves and found there an environment suited to their constitution?

Lamarckian views we have already ruled out as improbable on various general grounds; but such examples as that of Proteus are additional evidence against them. If it has taken thousands of generations for the effects of disuse to make the eyes shrink to their present size, how account for the fact that a single life-time in red light will bring them all the way back to normal? There seems to be no doubt that long-continued disuse often leads to inherited degeneration; but the Darwinian would assume that, since selection no longer operates to keep the eyes up to the mark, animals with any mutation leading to incomplete eye-development could survive as well as those with normal eyes, and would indeed be at a slight advantage since there would be less material and energy of growth employed in building up an organ that no longer had any value.

There exist cave-animals which would help us to test the rival views. In the old mines of Clausthal in the Hartz, abandoned now for centuries, varieties of water-shrimps and water-slaters (Gammarus and Asellus) occur, almost indistinguishable from the common forms of the illuminated world above, save in their lack of pigment and their half-degenerated eyes. If these were crossed with normal types, and it was found that the difference was inherited in Mendelian fashion, we could be reasonably sure that the degenerate eyes owed their existence to mutation. There remains the possibility of pre-adaptation. It seems clear that this alone will not account for the blindness of cave forms. For one thing, some cave-animals are found without even a vestige of eyes, even in the embryo, although no such eyeless varieties have been found in their above-ground relatives. For another, all the evidence at our command indicates that eye-degeneration is a slow process, taking place step by step. This is well shown by the progressive reduction of eyes seen in various related species of cave-fish. Pre-adaptation, however, might play a part in starting a species of cave-life. Creatures that shun the light and do not rely much upon their eyes are much more likely to take to caverns to live in, than are light-loving species with good vision. But once established in their new habitat, further evolution will be needed to put the full cave-stamp upon them. This process we may call post-adaptation. On the whole (and this question of adaptation to cave-life is obviously only a corner of a more general problem), our knowledge indicates that pre-adaptation, though usually slight, may often be decisive in fixing an organism in a particular environment, while post-adaptation is the more important in working up the detailed correspondences between constitution and surroundings that are so striking to the naturalist in the field.

So far we have dealt only with permanent cave-dwellers; the part-time cave-inhabitants are often of the greatest interest. Bats, for instance, may hang in incredible numbers from the roofs of some caves, and their droppings may accumulate like those of sea-birds to make valuable deposits of guano. But we have no space to enter into the biology of bats, and will conclude this section with a single remarkable example from the Antipodes.

In the Waitomo caves in New Zealand, there are underground lakes whose roofs are studded with tiny points of emerald light. These stars are lights made by insect larvæ—the caterpillars of the fly *Arachnocampa*. Each spins for itself a number of glutinous threads which it lets down from the roof. Whenever any flying creature, attracted by the light, is caught on a thread, the grub above wriggles along and eats its catch. The caves communicate freely with the outer world, so providing abundant prey. When they are full-fed, they metamorphose, and fly out of the caves to the world outside. Another species is found in Australia; both kinds may live in dark crevices in shady moist spots as well as in caves. A strange feature of the larvæ is their sensitiveness to sound: even voices above a whisper cause all the lights to be extinguished. So abundant are these creatures in some of the caves that they make a subterranean Milky Way whose light is strong enough to see by.

§ 10. Out-of-the-way Modes of Life.

PERHAPS the most severe conditions which life has to endure, more severe even than the barren darkness of caves, are to be found in hot springs, for protoplasm simply coagulates like white of egg when heated above a certain point. The common grass-frog stiffens and dies thus at temperatures well below that of the human body; and there are very few animals or plants of normal habitats, even in the tropics, whose tissues can stand a temperature of over 40° C. Yet all but the hottest springs have some life in them.

Interestingly enough, the tolerance of heat goes down as we rise in the organic scale. For many-celled animals 45° C. (113° F.) is about the limit; and there are no vertebrates that can stand this, but only a few snails, beetles, worms, and crustacea, while some wheel-animalcules may tolerate a little more. Some single-celled animals on the other hand can stand up to 55° C.; and there are primitive algæ that live in water at 80° C. (175° F.)—nearly hot enough to make coffee with!

These creatures are interesting from the point of view of evolution; for their heat-resisting capacities lie well above anything to be found in more normal habitats; the adaptation to this hot-bath life must therefore be a new and special acquisition, brought about by selection acting on rare and lucky variations.

The same transgression of life's normal power is seen in the inhabitants of salt-lakes and lime-pans. Nowhere else has life ever encountered a salt-content greater than about forty parts per thousand by weight, a figure which obtains to-day in the surface waters of the Red Sea. But in the waters of the Dead Sea and the Great Salt Lake the salt-content is over two hundred parts per thousand, and even this concentration is exceeded by certain other salt lakes.

The Dead Sea contains no life; but in the Great Salt Lake there is a typical set of salt-tolerant creatures—a few algæ, the grubs of the salt-fly Ephydra, a water-boatman, the brine-shrimp Artemia, and

LIFE IN FRESH WATER AND ON LAND

various protozoa. The brine-shrimp and the salt-fly are the only two many-celled animals that are really successful exploiters of these highly salty waters. The brine-shrimps are sometimes so abundant that they colour the water red, and round some of the Californian salt-marshes there is in summer a black rim, visible from several miles away, consisting of millions upon millions of the little salt-flies that have been produced by the grubs in the brine.

Here again these abnormal powers of resistance must be the result of a special evolution; but the rarity of salt-specialized creatures shows how few must be the mutations out of which such almost un-natural tolerance can be built. The salt-fly grubs are by nature altogether tough. They can live half an hour in absolute alcohol, which will kill most creatures in a second or two, and in 4 per cent formaldehyde, a standard killing and pickling fluid for animals, they live over twenty minutes.

Transitory pools are another queer habitat. The beautiful fairy shrimp, Chirocephalus, may be found living and breeding in the water of cart-ruts. When the water dries up, it dies; but leaves its drought-resisting eggs behind to be blown away and perhaps to colonize some other cart-rut in the future.

In deserts and semi-deserts, the transitory pools of water brought into being by the rains are almost at once filled with the life that has hatched out of waiting eggs and cysts; sometimes amphibians manage to carry on their species in such districts by laying in the pools as soon as formed, the eggs and tadpoles hurrying desperately through their aquatic life to turn into frogs before the pools disappear.

Aquatic life too manages to exist in tiny hollows in tree-trunks, and even in the moisture absorbed by cushions of moss from showers of rain. In this latter case the drying-out of the microscopic swamps may be a matter of hours instead of days or weeks, and it will not be enough for the animals that live there to possess drought-resistant eggs, since their home will often dry up before even a microscopic creature has had time to get through its life-history: they must be able to become drought-resistant even when adult. The importance of the time-factor in animal affairs could not be better illustrated.

The chief inhabitants of these tiny moss-marshes are bear-animalcules (which mostly suck the juices out of the moss-cells, and cling on with hooked claws to prevent themselves being swept away by rain), small roundworms, wheel-animalcules and a few tiny crustacea; and they can all at any time of their life-history respond to drought by shrivelling up and banking the fire of their life so that it merely smoulders; in this state they can wait for months or even years until moisture again swells them to sappiness and activity. The seasons do not exist for these creatures; every now and then their existence is broken into by periods of suspended animation, out of which they emerge to take up their life at the point where it was interrupted.

The nests of ants and termites have their own highly specialized group of inhabitants; but of them we shall speak in a later Book. Even dung may provide a definite habitat; a distinct and specialized fauna and flora lives in and under cow-droppings, for instance. And there are a host of queer habitats provided directly or indirectly by man. Not only is there a rich fauna on sewage-farms, but a study of its habits is proving vital to scientific sewage disposal; and the inhabitants of waterworks are equally interesting and important. Sand-filters are widely employed in modern water-supplies; they are of the greatest value, since they will prevent the passage even of bacteria. But this is not due to the mechanical filtering power of the sand, but to the much finer filter formed on its surface by microscopic plants, mostly diatoms. This guards our water-pipes against two dangers—first against bacteria dangerous to health, and secondly against invasion by the reproductive bodies of animals and plants that could find shelter there.

When the water is not filtered, surprising results follow. Hamburg in 1886 was supplied with water from the Elbe, unfiltered, and stored in reservoirs for a quite inadequate length of time. Water shrimps began to pop out of the taps, and the pipes to get blocked with growths of Polyzoa, and even occasional eels. In 1886, Professor Kraepelin, by means of a wire-gauze cage which could be screwed on to the mains as desired, investigated the fauna of the water-pipes. The smallest creatures, such as Rotifers, passed through the meshes of the cage; but even so he secured specimens of fifty different genera of animals. Sponges encrusted the pipes, with huge mossy growths of Polyzoa and hydroid polyps, which often came away in tangled masses. There were plenty of worms, segmented, flat, and round, and of bivalve molluscs; shrimps and other crustacea, both fresh-water and marine, were not uncommon, notably "loathsome swarms" of the common water-slater, Asellus. Once a small flounder was captured; there were a fair number of sticklebacks; and eels (up to a foot in length) infested the pipes in thousands. In spite of the revelation of this huge fauna under the city streets, nothing was done for years. The population was easy-going, and the manufacturers of domestic filters, who did a roaring trade, succeeded in putting off reform. Then one day came the crash; in 1892 the waters of the Elbe became infected with cholera germs, and over eight thousand of the people of Hamburg died of that dread disease. Sand-filters were installed; the mixed fauna was barred from the biological Eldorado it had found in the pipes; and the people of Hamburg no longer ran the risk of cholera.

But we must bring our chapter to a close, although the catalogue of strange lives could be continued almost indefinitely. Let Mr. Everyman stroll round his garden and look (to give him a few suggestions) under stones, among dead leaves, on the branches of living trees, in the rain-butt, on the mouldering wood of an old fence, to see how many different habitats, each with its own peculiar zoo, he can find.

IV

SOME SPECIAL ASPECTS OF LIFE

§ 1. *Partnership and Parasitism.* § 2. *The Scale of Living Things.* § 3. *Colour and Pattern in Life.*

§ 1. Partnership and Parasitism.

IN THE previous chapters we have tried to show how life presses and crowds into every nook and corner of the earth's surface that is habitable. We have found life exuberant and various in hot springs, deep caves and dunghills, on remote oceanic islands, on frowning mountain crags; in all these places the same drama of hopeful germination and bitter struggle. But we have said little so far of how life jostles life, how, in virtue of that same life-pressure which injects organisms into the tiniest cracks on our world, living things impinge on each other. And yet a large part of the effective environment of any given kind of animal or plant consists of other animals and plants. The struggle for existence forces them into mutual relationships, often of the most intimate and extraordinary kind. In this section we shall describe some of the most striking of these, and show how, as in the more plastic changes of human relationships, casual association may pass over into mutually helpful partnership, or the undercurrent of competition may transform partnership into parasitism; how no sharp line is to be drawn between parasitism and the straight-forward relation between devourer and devoured; and how difficult it may be to distinguish between service and slavery.

Of the larger animals and plants, perhaps the majority live in intimate association with humbler organisms. Our own mouth-cavity and intestines harbour millions of bacteria, spirochaetes, and protozoa. Most of these are neither harmful nor the reverse to their giant host; they simply take advantage of the peaceful warmth and food provided by his tubular interior. But in herbivorous mammals the bacterial flora is part of the machine. Without bacteria, the mechanism of a horse would not work; for it is they which act as chemical tin-openers to all those microscopic boxes, the plant-cells, in which the digestible part of his food is locked up. To such mutual arrangements by which both partners gain there is generally given the name of *symbiosis*—a joint life. And then there are bacteria, no different in general appearance from any other sort, which penetrate into blood or tissues and there grow and multiply at their host's expense: they are parasites.

The abundance of parasites is extraordinary. Most species of animals harbour several different kinds of parasites, and though most of these are common to several hosts, the total number of parasite species must be almost as great as the total number of free-living creatures. We have already described one or two examples—the tapeworm, with its never-ceasing production of jointed living ribbon, each segment, when mature, being impregnated by one of its neighbours, and later dropping off to disseminate the embryos with which it is packed for the invasion of a second host (Fig. 91); the liver-fluke with its cycle, now in sheep and now in snail, always changing its form, but always a destroyer (Fig. 93); Sacculina, ramifying through the whole organism of some unfortunate crab, eventually bursting out like a hernia before killing its host, a bulging bagful of sperm and eggs (Fig. 82); the guinea-worm a living thread up to six feet long, making abscesses in human flesh through which to discharge its eggs, hardly to be got rid of save by winding it out, perhaps an inch a day, on a slip of wood which between whiles is fastened to the limb (Book 2, Chap. 3, § 5); the dreaded Trichina of measly pork; the roundworms of our own and our dog's intestines.

Here we will add a few more to our list. The hook-worm is another roundworm. Its eggs leave human bodies in the fæces; its young stages live free in moist earth, and gain entry to human body again by scratches in the skin. Bare feet and bad sanitation keep it going. Once inside, it is carried by the blood to the lungs, there forces its way into the air-spaces, and then, exploiting for its own use the normal human protective mechanism by which alien particles are expelled, is carried up to the mouth by the beat of the ciliated cells. Then it is swallowed; arrived at the intestine, it bites on to the wall with its powerful jaws, and remains there sucking blood until full-grown. A heavily infected man may contain many hundreds of hook-worms, each about half an inch long, all draining his life-blood. No wonder that the population of hook-worm areas are anæmic, listless and unprogressive.

Insects, if parasitic, are usually parasitic during their larval stages only. One of the most familiar examples is the ox warble-fly which ruins thousands of hides every year; the eggs are laid under the skin, and the maggots cause the "warbles"—large swollen lumps—through holes in which the fly eventually emerges. The horse botfly, on the other hand, penetrates farther. The eggs are laid on the limbs; they hatch within a day or so, and the tiny larvæ cause such itching that the horse licks the part and, without knowing it, swallows the maggots. Thus obligingly brought inside, they fix tight on to the stomach lining and grow there for the best part of a year.

Other flies lay their eggs in the noses of men and beasts, and the growth of the maggots may cause much suffering and even death. And there are a number of blow-flies which have departed from their normal habit of breeding in decaying flesh to deposit their eggs in open sores on living animals. The maggots live on the decaying flesh, and their presence ensures a continuance of the food-supply by aggravating and enlarging the wound. There are many millions of sheep in Queensland; and of each million nearly fifty thousand die a fly-blown death every year.

A rather different form of parasitism is found in many hymenoptera. Some of these are so small that their grubs are parasitic only on other insects' eggs; but most of them parasitize other grubs. The female ichneumon-fly, by means of her long sharp ovipositor, lays her eggs deep in a caterpillar's body. The young hatch out and devour the animal from within, as if a brood of rats were to eat their way through a living sheep. At first they spare the vital organs, confining themselves to fat-stores, connective tissue and the like; so that their host can still continue to feed and supply them with nourishment. No killing of the goose that lays the golden eggs! But the last golden egg produced for them by the caterpillar is that it should transform itself, when full-grown, into a pupa with a tough protective shell. Once safe within that, the ichneumon larvæ eat all that remains, themselves attain to full growth and bore their way out to pupate.

One does not at first know which to be more impressed by—the admirable delicacy of the adaptation, or the refinement of cruel exploitation. Such facts are indeed a real difficulty for those who believe in the creation of the world as it stands by a beneficent deity. But they offer no such moral problem to the evolutionist. Natural selection is a blind agency and knows no human values. It is as unjustifiable to ascribe moral qualities, such as cruelty or beneficence, to animals or the processes of their lives, as it is to ascribe purpose to the wind or consciousness to a mountain. The processes of Nature know no values; all values arise in the mind of man.

Microscopic parasites can be equally deadly. Those single-celled flagellates, the Trypanosomes of sleeping-sickness, are even beautiful with their flagellum attached to a transparent fin-membrane, as they undulate their way through a crowd of blood-corpuscles. But they multiply in the blood and then invade the cerebro-spinal fluid; the patient grows drowsy and wastes away, a ghastly bag of skin and bone, to an almost inevitable death. This microscopic flagellate killed over 200,000 men in Africa in seven years; and a near relative, which causes the nagana disease of domestic animals, makes it impossible to use horses, donkeys, cattle or mules over several million square miles of the same continent.

Animals are not parasitized solely by other animals. The bacteria are the most important of any parasites. They are all commonplace in appearance; their virulence is chemical. The bacillus of plague is a microscopic rodlet like any free-living bacillus; but it kills men like flies. The bacteria are as giants compared with the filter-passing viruses. But size is no criterion of deadliness; such virulent diseases as yellow fever and small-pox and foot-and-mouth are the result of parasites so small as to be ultra-microscopic.

Plants too may be parasitic. The spores of the fungus Cordiceps invade the tissues of caterpillars; then they germinate into a mass of filaments, which eventually permeate the host. Finally the caterpillar creeps down to the ground and dies, while the fungus sends out its fructification in the form of a long horn-like structure from the host's head. The caterpillar may be almost wholly vegetalized in death, and remain converted into a woody mummy for several months. Ringworm is due to another fungus which flourishes on living human skin.

Then there are animals which exploit plants. The most striking cases are those of the gall-producing insects and mites, the most abundant being midges and tiny hymenopterans. Their strange power it is to make their plant-host produce both food and shelter for their young. They lay their eggs on leaves or flowers, stems or roots; and the plant-tissues grow up round the developing grub to make a gall—a structure often quite unlike anything normally produced by the plant. The round fleshy oak-apples on oak leaves, and the brilliant red feathery robin's pincushions on roses are familiar examples. Sometimes the gall's structure is eminently adaptive—for the needs of the parasite; on sweet-gum trees you may find galls consisting of a central chamber lodging the grub, connected by a system of radiating struts with a protective shell outside.

It is not known how the insect forces these wonderful transformations on the plant-tissues, whether by substances injected with the egg, or by some stimulus emanating from the grub. If we could but discover their secret, we should be another big step forward towards the control of growth-processes.

And plants may parasitize plants. One of the most interesting as well as economically important of plant parasites is the fungus known as rust of wheat. We will speak of some of the rusts in a later chapter. A strange case is that of the flowering plant Rafflesia, a native of Malaya. This lives the whole of its growth-period within another plant. It has lost every vegetative organ which its ancestors possessed. Its host provides roots and root-hairs for anchorage and for sucking water and salts from the soil, leaves and chlorophyll for tapping carbon from the air, bark for protection, wood for support, and all the microscopic pipes for transportation; and so Rafflesia has need of none of these things and has come to consist merely of a tangled web of white filaments, very like those of most fungi, penetrating the host in every direction. All it has to do is to provide an absorptive surface; its host does the rest. However, the host-plant cannot very well provide for its parasite's reproduction; and so when the time comes, this degenerate mass of microscopic fibres gathers itself together here and there, breaks out to the exterior, and there grows out into blooms as elaborate and finished as those of any self-supporting plant. Indeed in one respect they outdo all others: for *Rafflesia arnoldsii* from Sumatra holds the world's record for flower-size, its blossoms sometimes exceeding a yard in diameter and weighing up to twenty-five pounds apiece.

This is a close parallel with the animal parasite Sacculina (Fig. 82), which at an early stage of its career is altogether inside its host, and consists wholly of a net-work of absorbing rootlets; and only later, for the purpose of reproduction, bursts through to the exterior and organizes a body with a definite anatomy.

It is worth emphasizing that parasitism and independence grade imperceptibly into each other. In the matter of food all animals are exploiters, either of other animals or of plants. Usually, however, the exploitation involves an elaboration of structure rather than a degeneration: one need but think of carnivore lion or herbivore horse or the beautiful sifting machinery of mussel or oyster. Some small caterpillars eat their way through the interior of fruits or stems or leaves. We do not generally consider them parasites for that: they are internal herbivores. But internal carnivores also exist, and in all degrees. Lampreys bore holes into the living bodies of their victims, and their relatives the hagfish may eat their way right inside. Miss

Worthington writes that on the Pacific coast of America many of the fish caught on night-lines are found to "have been entirely eaten away, nothing but the skin and bones being left. The Hagfish has bored inside the skin and eaten all the soft parts, and is sometimes caught in the very act of wriggling away at the close of its meal." Up to three or four may be found eating out a single fish. Such animals penetrate their victims only temporarily, but from this condition to the continuous internal carnivorousness of ichneumon-grubs is not a large step.

Among green plants, the same sort of series may be traced. Even within a single tribe of the family Scrophulariaceæ all gradations from independence to obligatory and complete parasitism occur. Many of them are always respectable green plants, independently manufacturing their nourishment out of air, water and soil-salts. The yellow rattles (Rhinanthus) have their feet on the first rung of the descending ladder: they *can* grow and reproduce in independence, but they often tap the roots of other plants with their own; and the pretty little eyebright Euphrasia, which no one would take for a subterranean vampire, behaves in the same way. The cow-wheats (Melampyrum), on the other hand, though they can germinate in isolation, die later unless they can parasitize their neighbours. Special side-roots are developed which, if they touch the root of another plant, grow round it, penetrate it here and there, and establish direct connexion between their own water-conducting tissue and that of their host. By this means a single plant of cow-wheat may simultaneously tap the roots of several hosts belonging to different species.

Even here, however, the parasitism, though obligatory, is only partial, since the plants still possess green leaves and manufacture their own carbohydrates. The mistletoe is in a similar category. The mountain-plant Tozzia is a rung lower. It germinates below-ground and for several years lives as a mere subterranean stem, entirely parasitic on other plants. Then it sends up a flowering shoot with pale-green leaves, poor in chlorophyll but still capable of making some contribution to its nourishment, and lives the brief rest of its life as a partial parasite. Finally, in the toothworts parasitism has become complete. They never produce any chlorophyll at all; they draw nourishment from other plants all their life long, penetrating right to the woody pipes of their hosts with special sucker-organs; and their seeds seem incapable even of germinating unless they happen to lie against the right kind of root. The Broomrapes have independently evolved into equally complete parasites.

These are all one-sided arrangements. Now take the opposite category, true partnership or symbiosis, in which the partners benefit mutually from their association. In the most extreme cases neither can exist without the other. We have already mentioned what is perhaps the most remarkable of these—the partnership between fungus and alga which has given the world the whole new group of compound organisms, the lichens, and enabled plants to push farther into the barren places of the earth than they could ever have done otherwise. The most important of such partnerships to us is that between leguminous plants and certain kinds of bacteria. Already in Roman times legumes like lupin and vetch were grown to enrich exhausted soils. We know now why the soil is thus enriched. If you pull up a plant of this sort, you find irregular swellings on its roots; and these swellings are inhabited by countless millions of bacteria. These bacteria have the power, denied to higher plants, of seizing and using free nitrogen from the air. They can do this to a limited extent when living on their own in the soil. But their nitrogen-fixing capacity is enormously multiplied when they are living with their big leguminous allies. These supply an abundance of carbohydrate materials; the bacteria obtain energy by breaking these down to simpler substances, and use this energy to conquer the chemical inertness of the atmospheric nitrogen. The green plant is good at fixing carbon, the bacterium at fixing nitrogen; in the partnership the two advantages are pooled.

Knowledge of these facts is of considerable practical importance. To get lupins or lucerne to grow well on new soil, poor in nitrogen bacteria, either they or the soil must be inoculated with the bacteria. Large areas of barren sandy lands in East Prussia have thus been reclaimed through lupins; and a new method of inoculating lucerne with its microscopic partner is now proving of great value in extending the area of that useful forage crop through various regions of Britain where it was hitherto not grown.

Another important symbiosis is that between termites, the so-called "white ants" of warm countries, and the bizarre microscopic flagellates which crowd their intestines. The flagellates can do what the termites, like ourselves, are helpless to achieve—they can break down cellulose and even wood into digestible substances. By exposing the termites to high temperature and by other methods, such as treatment with oxygen under pressure, the protozoa can be killed without damaging their partners. But termites thus deprived of their internal flocks and herds are powerless to digest their ordinary diet: abundant wood is now as useless to them as a bare table would be to us, and, although they go on eating, the chewed wood is passed out unchanged and they perish of starvation as surely.

The termite with its powerful jaws provides the raw food-material for the joint household partnership, and also the protected kitchen for its due preparation. The flagellates alone can achieve the culinary transformations there, and digest for both. As a result, the termite-flagellate partnership is the great scavenger of the tropics, before whose attacks leaves, twigs, and the fallen trunks of great forest trees melt and are destroyed, and their substance brought back into the vital circulation far quicker than by the ordinary processes of slow decay. As further result, however, white ants are among the most inveterate destroyers of houses and furniture, books and papers, in low latitudes. If the alga-fungus partnership has extended the range of humble plant-life in the arctic, that of termite and flagellate is hindering the spread of human civilization in the tropics.

Then a great many animals are green, owing to the presence of single-celled green algæ within their tissues. Among these are many of the beautiful Radiolaria that float in the sea, the green Hydra and the great majority of reef-forming corals. The animal profits by utilizing some of the carbohydrates built up by the alga's chlorophyll, and therefore need not trouble to find so much food; and the alga profits by utilizing some of the nitrogen which the animal is continuously giving off in its excretions, for there is in most habitats a shortage of available nitrogen, and an alga within an animal finds itself in an exceptionally favourable situation as regards this element. We have seen that the chief reason why reef-building corals cannot live save in a narrow surface layer of water is that they starve at greater depth, since their algal partners need light to make their carbohydrate contribution to the partnership.

During the last fifteen years, Buchner has been making a systematic study of symbiosis between animals and plants, and has discovered many interesting facts. Partnerships between animals and green algæ are confined (with the one exception of the sloths) to transparent and water-living creatures, for only here can the food-producing activities of the green cells be turned to advantage. We find them in protozoa, sponges, corals, and other cœlenterates, flatworms, wheel-animalcules, and possibly in a few pelagic snails.

When the plant-partners are algæ, they help in food-manufacture; when they are fungi or bacteria, they help in food-utilization. All arthropods that suck plant juices have such partners—plant-lice, scale-insects, plant-bugs, cicadas and the rest. So do all those that suck the blood of vertebrates—lice, bed-bugs, ticks, mites, tsetse-flies, probably mosquitoes and fleas, with leeches too; but their relatives that suck invertebrate blood have none; apparently the colourless plant-partners help in breaking down that typical vertebrate structure, the red blood corpuscle.

Hair- and horn-eaters like the bird-lice are also dependent upon symbiotic partners; and so are all the wood-eating animals and those whose diet is very rich in cellulose. Many beetles and beetle-larvæ, termites (though here the partner is a protozoan), ants, wood-wasps and gall-midges, ruminant and rodent mammals, and certain birds all fall into this category. In these cases, the partner-plants help to make an unusual type of food digestible. Where animal food is the staple, we find no bacterial or fungal partners. All cockroaches too are really partnerships, though in this case we do not know exactly what rôle the intestinal bacteria play.

Finally, partner-bacteria may play quite a different rôle—they may be used to generate phosphorescence. Many animals shine with their own light, but there are a number—most of the luminous cuttle-fish, all the brilliant Salps and Pyrosomes among the sea-squirts, and a few fishes—which cultivate luminous bacteria in special pockets evolved for the purpose. Symbiosis is thus not a haphazard occurrence. It has its rules and its meaning; each kind of partnership between an animal and a plant is helpful in a particular way of life.

A partnership less essential and less intimate, although no less interesting, is that between a common species of hermit-crab, *Eupagurus bernhardi*, and a sea-anemone. The mollusc-shell inhabited by the crab is always covered by a pink sea-anemone of the genus Adamsia. When the crab has grown too big for its house, and must change into another, it carefully detaches its partner and fixes it on the new shell. The hermit-crab with very little trouble and expense obtains a defence-force in the shape of the anemone's white stinging threads or acontia, covered with batteries of poisonous nettle-cells, which it shoots out when alarmed through little port-holes provided for the purpose in its flanks. The anemone gains transport and crumbs from the crab's table, as its active jaws break up a piece of carrion. The anemone's sole adaptation to a life of partnership seems to be its readiness to be detached by the crab, whereas to any other detaching force it offers the greatest resistance. Other crabs live with evil-tasting sponges; or carry sea-anemones in their claws as living knuckle-dusters.

From such examples one would imagine that symbiosis and parasitism were clear-cut biological categories. But consider a few others.

The common ling, or Scotch heather, *Calluna vulgaris*, is one of those numerous plants which have entered into partnership with a fungus. It is a real dual organism, like a lichen; the fungus extends its web of filaments right through the plant, most abundant in the cells of the roots, but reaching through stem and leaves and even into the ovaries. As the seed ripens, filaments extend on to its coat; and when it is shed and begins to sprout, these penetrate the living cells of the root. They secrete a ferment which dissolves cellulose; and by means of this, the filament spreads from cell to cell. After a short time nearly every cell of the root's outer layers contains a coil of this microscopic living thread; and thence it spreads up the stem. Finally it bursts out on to the outside of the root, there to form a microscopic feltwork, and also, though in less quantity, on to the outside of the leaves. So far this looks like a case of parasitism. But, surprisingly enough, it has been proved that the fungus feltwork round the roots is necessary for the life of the ling. If heather seeds are sterilized, so as to rid them of all traces of the fungus, and then sown, they begin to germinate; but, though stem and a few leaves at first grow normally, nothing of the root-system is formed save a few feeble stumps, and the rootless seedlings soon die. The fungus, on the other hand, can grow apart from its larger partner; and when thus isolated, it has been proved to have the power of fixing nitrogen from the air. Moreover it has been shown by Rayner that the heather-fungus partnership—in other words fungus-infected seeds—could germinate and grow quite well in spite of the absence of all nitrogen supplies, organic or inorganic, showing that the fungus was obtaining the necessary supplies of nitrogen from the atmosphere while in double harness just as well as when on its own.

Here, one might now suppose, is a perfect case of mutual benefit. The fungus receives carbohydrates made by the chlorophyll-machinery of the heather, the heather appropriates some of the nitrogen made available by the peculiar and specialized chemical activity of the fungus; and so long and close has been the partnership that the heather is now entirely dependent upon the fungus, and even needs its presence as a formative stimulus to its normal development.

But it is only in certain conditions that the partnership works thus smoothly and with mutual benefit. We have seen that heather seedlings grown without any fungus will come to nothing. But even if they are inoculated with some of the fungus, the partnership is only successful when they are supplied with little or no nitrogen salts. If more nitrogen is provided, the fungus grows too vigorously. It takes the upper hand, becomes a parasite instead of a partner, and kills the heather seedling. Through some subtle chemistry we do not yet understand, the balance is tilted, the regime of mutual help is destroyed and gives place to a regime of exploitation. Ling will only grow in certain kinds of soil. Possibly this is due in part to similar upsets of the precarious partnership.

Nothing could better illustrate the precariousness of life's adjustments. The phrase "hostile symbiosis" has been used to describe the state of our own tissues—all of the same parentage, all thriving best when working for the common good, and yet each ready to take advantage of the rest, should opportunity offer. There is a profound truth embodied in the phrase. Every symbiosis is in its degree underlain by hostility, and only by proper regulation and often elaborate adjustment can the state of mutual benefit be maintained. Even in human affairs, partnerships for mutual benefit are not so easily kept up, in spite of men being endowed with intelligence and so being able to grasp the meaning of such a relation. But in lower organisms, there is no such comprehension to help keep the relationship going. Mutual partnerships are adaptations as blindly entered into and as unconsciously brought about as any others. They work by virtue of complicated physical and chemical adjustments between the two partners and between the whole partnership and its environment; alter the adjustment, and the partnership may dissolve, as blindly and automatically as it was entered into.

In Calluna, beneficial symbiosis may pass over to true parasitism. Other partnerships partake of both qualities simultaneously. Take for example the little flatworm *Convoluta roscoffensis*, whose partnership with a single-celled green alga has been so graphically described by Keeble in his *Plant-Animals*. This all but microscopic worm-alga partnership is so abundant in some regions, as at Roscoff in Brittany, that it colours the sand green over patches many yards in extent, its hordes coming up with the inflowing of the tide and burrowing below the surface, out of reach of the pounding waves, as the beach is covered. Over two thousand million wormlets may exist in each square yard of such a patch. Convoluta, unlike the heather and the fungi of the most specilized kinds of lichen, does not take steps to carry over a supply of its microscopic partner in reproduction; its eggs are colourless, free of algæ. But the algal partner is abundant as a free-living creature in the environments where Convoluta is found; and every egg is soon surrounded by a little band of these green flagellates, attracted in all probability by the nitrogenous substances it gives off as waste. The egg develops into a larva, which one day swallows one or two of the green free-living cells. These multiply in its body and produce the large quantity of green tissue. A worm kept uninfected in sterile water not only remains colourless but fails to grow and after a short time dies.

The adult Convoluta thus contains a whole band of algæ. These find themselves in a very favourable situation for growth, since they snap up the animal's waste nitrogen before it diffuses out into the sea; and Convoluta, by its periodic migration to the surface, puts the algæ into the best conditions of light for splitting carbon dioxide. The worm, on the other hand, draws on its partners for food. While it is growing, it still eats in normal animal fashion, but once it is full-grown, it no longer uses its mouth, but nourishes itself entirely by digesting away the surplus of its green cells. So we have the strange spectacle of a worm that grows vegetables within its own flesh.

As the worms grow old, they mate and lay eggs. Then somehow the balance is upset and the whole internal population of algæ is digested; and when this has been accomplished, the Convoluta must die of starvation, since it is incapable of finding food for itself. Every worm of these millions on the beach, if it escapes a more violent end, is destined to starve, as a result of this incontinence of its own appetite.

During most of each worm's life-time, the partnership is a symbiosis; for both worm and alga benefit; but in its old age the partnership changes its character and the worm becomes a parasite, or rather a straightforward herbivore, destroys its plant-partners, and with them both the partnership and itself. From the point of view of the race of algæ, the partnership is never a symbiosis, since every alga within the body of the worm is in the long run doomed to destruction; the algæ are lured into slavery by the chemical bait dangled before their nitrogen-hunger, and are cared for only to be destroyed. The relation is like that between man and some animal which he would take into captivity only to fatten and kill, leaving the supply to be renewed by natural reproduction. Some kinds of oyster-culture approach this, although man here usually in some degree controls the reproduction of the subordinate partner; and luckily man is not inexorably tied to dependence upon oysters as Convoluta is to dependence upon its algæ.

Then both symbiosis and parasitism grade off imperceptibly into more casual relationships. The termite and its flagellate are indissolubly tied together—neither can live without the other. Ling cannot live in nature without its fungus; but the fungus can live without the ling. In many lichens both partners can exist separately, but they cannot grow under anything like such a range of conditions as can the partnership of both together. In other cases the relation is less obligatory. Almost all the familiar forest-trees of the temperate zone normally possess fungus in and on their roots like ling; such root-fungus is called mycorrhiza. Often the infected roots are covered with a visible feltwork of fungus, and this growth may be different from the normal. No root-hairs are formed when the mycorrhiza is present, the fungus threads apparently taking their place. The fungus is found actually inside many of the tree's cells. In some cells it is actively alive and growing; elsewhere it may die, and the tree-cells digest it. There is little doubt that in certain soils the proper development of the tree is bound up with the presence of the right fungus; but in certain conditions the trees can grow independently, without assistance from their fungi. At any rate, they are not, as species, inevitably tied down to the partnership.

The fungus-partners of many forest-trees lead a double life. Like so many fungi, they spend most of their existence as a mycelium, a thready feltwork of tiny white threads, penetrating the rich soil in all directions. If the threads continue their independent life, they eventually throw up a large toadstool to disseminate the race by means of spores. But if they chance to meet a tree-root, they find in it a habitat which is equally propitious but quite different. They change their whole mode of growth. The invasion of the very cells of the root at once suggests that the fungus is a parasite on the tree; but the digestion of a certain proportion of the fungus equally suggests that the tree has lured (metaphorically speaking, be it well understood) the fungus within itself to enslave and exploit it, as Convoluta enslaves and exploits its algæ.

Our simple human categories of exploitation and mutual benefit, although useful, are artificial and break down when confronted with the complex and inhuman, or at least non-human, realities of other life. Fungus and tree-root penetrate where they can and increase their substance blindly at the expense of whatever material they find; selection sees to it that beneficial variations are appropriated as they turn up. The fungus invades, the root resists; there is a balance between the invasion and the defence; there are some advantages on both sides, but damage on both sides also. The advantages reaped by either of the competitor-partners may cancel out its losses, and the net result be nil; or one may achieve a net benefit and to a certain degree exploit the other; or adjustment may be made so that in both the benefit outweighs the damage, and thus a true symbiosis arises.

We have seen that parasitism grades into the mutual dependence of symbiosis. It also grades into independence*—in animals generally the independence of the carnivore or the scavenger, in plants that of the green plant or the chemically scavenging fungus. It has usually been customary to call any organism a parasite which both lives on or in another and lives at its expense; but by so doing, we lump together a number of really quite distinct ways of life. Whatever our definition of parasitism, it is clear that such forms as a tapeworm, a Sacculina, or a Rafflesia, must be called parasites. They not only live at the expense of other organisms, but in so doing have sacrificed their own independence of life and lost the organs by which it was once achieved—the animals their limbs and stomach and sense-organs, the plant its leaves and stem and roots, and even its very greenness.

But what of the bird-lice or Mallophaga, which find shelter under the plumage of birds and only live by scavenging the débris of the feathers? What of the ichneumon-grubs that eat out the inside of their caterpillar-host and could just as well be called internal carnivores as parasites? Is there enough difference between the habits of a flea and a mosquito to justify us in calling one a parasite, the other not? What about the Remora or sucker-fish which obtains free transport from sharks and turtles and other large sea-beasts by clinging to them with the sucker on its head, but takes no food from them; or the gall-forming crab Haplocarcinus, which robs the corals among which it lives of no nourishment, but by the irritation of its presence distorts the growing tip of a coral branch into a bulbous protective case for itself? What of the little crustacean which inhabits the hollow interior of the barrel-shaped Salps, and scoops up its host's food as it is being slowly propelled by cilia along the food-groove inside its pharynx? What of the cuckoo, which lives at the expense of other birds when it is young, but then fends for itself?

The truth is that animals or plants can no more live in "splendid isolation" than can nations. Every organism is part of a network of relationships, biologically tied to a host of others. There are relationships concerned with feeding, with shelter and protection, with support, with reproduction, sometimes with such minor matters as toilet and recreation. Within every sort of relationship, one party may not only come to exploit the other, but to exploit it in such a way

---

*The independence of animals and fungi is a relative independence only: they are in the long run dependent on green plants. See Chapter 5.

that it loses some of its own independence and becomes, in the biological sense, degenerate. Or the two parties may come to benefit each other. But all relations of any intimacy, between lower organisms as between men and women, are precarious, supported, as it were, on a knife-edge. They may so readily over-balance and change into something different and even opposite.

§ 2. The Scale of Living Things.

IN THE preceding chapters we have set out something of the diversity of plans of structure and ways of existence. Life's purely quantitative diversity is no less striking.

It may interest the reader to give a brief scale of sizes. The biggest plants are nearly ten times bigger than the biggest animals. The biggest animals are whales, some of which must run to well over a hundred tons, possibly over a hundred and fifty; contrary to popular belief, the biggest extinct reptiles weighed only some fifty tons, or about a third as much as the giants among these biggest mammals. These giant reptiles were undoubtedly semi-aquatic; and water-animals can reach much greater weights than land-animals. Of land-animals, some extinct forms exceeded ten tons, but among those alive to-day, six or seven tons is the maximum.

All these are vertebrates. The biggest invertebrates are molluscs, some of the giant squids reaching two or three tons. The cœlenterates, startlingly enough, come next; a jelly-fish from arctic seas may weigh about half a ton. This group is followed at some distance by the arthropods, with giant spider-crabs of perhaps fifty or sixty pounds. Among groups which exceed two pounds, but fall short of ten, are snails, lamp-shells, echinoderms and also—a surprise to most people—earthworms and polyps. Perhaps, too, the largest tapeworms, with their huge lengths of seventy feet and more, just reach a kilogram.

Far below come insects and spiders, which never weigh more than two or three ounces. Most of them are much smaller. Even ants half an inch long are well below a gram in weight. The biggest ant-colonies have about a million inhabitants; all this population together weighs no more than a large man. As for fleas, three average fleas go to a milligram. If you bought an ounce of fleas, you would have the pleasure of receiving over eighty thousand of them. Even the solid queen hive-bee weighs much less than a gram, and the workers only about one-seventh of a gram—two hundred to the ounce. (An ounce is 31.1 grams.)

FIG. 253. The sizes of organisms.

(A) *The biggest living things. All the organisms are drawn on a uniform scale.* (1) *The largest known animal (sulphur-bottom whale) with African elephant, man, dog, and giraffe superposed;* (2) *Tyrannosaur;* (3) *Diplodocus;* (4) *Largest flying reptile (Pteranodon);* (5) *Largest flying bird (albatross);* (6) *Largest extinct bird (Aepyornis);* (7) *Ostrich;* (8) *Hen;* (9) *Largest extinct snake;* (10) *Length of largest tapeworm found in man;* (11) *Sheep;* (12) *Largest living reptile (West African crocodile);* (13) *Largest living lizard ("Komodo dragon");* (14) *Largest extinct lizard;* (15) *Largest single polyp (Branchiocerianthus);* (16) *Largest jelly-fish (Cyanea);* (17) *Horse;* (18) *Largest bivalve mollusc (Tridacna);* (19) *Large tarpon;* (20) *Largest crustacean (Japanese spider-crab);* (21) *Largest lobster;* (22) *Largest sea-scorpion (Eurypterid);* (23) *Largest fish (whale-shark);* (24) *Largest flower (Rafflesia);* (25) *Largest mollusc (deep-water squid, Architeuthis);* (26) *Bottom 105 feet of largest organism (Big Tree), with 100-foot larch-tree superposed.*

At the bottom are the rotifers, whose greatest giants fail to turn the scale at ten milligrams. The smallest many-celled animals are male rotifers, some of which weigh considerably less than one-millionth of a gram, so that it would take over a thousand to outweigh a single one of our muscle-fibres.

On the whole, the lower limit of size among the various groups is a good deal more constant than the upper. It is interesting to find that here again the limit for vertebrates is far above that for any other group. There is clearly a limit set to a many-celled animal in the number of cells out of which it is constructed; the smallest worm must have several thousand cells to build its organization. But it seems to be impossible, or unprofitable, to build a vertebrate out of less than several million cells.

The small scale on which such a complicated beast as an insect can be constructed is amazing. There are insects, complete with compound eyes, three pairs of jointed legs, wings, striated muscles, brain and nervous system and all the rest, which weigh less than the human ovum, and much less even than the nucleus of some large protozoa! A list of animals in order of weight would be full of surprises for most readers—a polyp as tall as a sheep, a frog as big as a terrier, a snail that lays eggs as big as a sparrow's. But we must leave mere curiosities on one side, and see whether we can find any thread of principle running through the tangled facts.

In previous books we have discovered some of the reasons which limit the size of different groups. Insects, as we have seen, are kept tiny by the fact that they breathe by branching air-tubes, not by lungs. Birds that fly cannot go above a certain moderate size for aerodynamic reasons; ostriches and moas show that birds can grow big if they sacrifice their wings. (Angels, by the way, are a biological impossibility. To obtain the motive force for flapping his wings, a normal-sized angel would have to have a breast-bone and breast muscles projecting four feet from the front of his chest.) Land-vertebrates are limited by the leg-bones. The supporting power of a pillar of bone is proportional to its cross-section. A little thought will show that the weight of an animal's bones will have to go up faster than its total weight if its limbs are to continue to support it. The bones of Swift's Brobdingnagians would have snapped under their weight if they had been mere enlarged scale-models of normal men and women. And there is clearly a limit to the amount of mere supporting material which living muscle and blood can economically move and nourish. In water, however, almost all an animal's weight disappears; bones are needed as scaffolds

SOME SPECIAL ASPECTS OF LIFE

and levers, no longer as supports. And so water-vertebrates can grow to a far greater size than can their terrestrial cousins; their size is limited only by the exigencies of food-supply and digestion.

Crustaceans, though aquatic, never grow really big, judged by vertebrate standards. In this case it is the habit of moulting which seems to be responsible. For, if you are a lobster or a crab, the bigger you grow, the longer it takes to produce an adequate new outfit of armour; a crab as big as a cow would have to spend most of its life in the retirement of the moult, hardening its hundredweight of shell before it could venture out and feed again.

Far down the scale, all creatures that move by cilia are kept extremely small. Not only can these hair-like rudiments of limbs never be more than microscopic themselves, but they are confined to the animal's surface, and cannot be massed in solid three-dimensional organs like muscles.

When ciliary-swimming marine larvæ seek to take advantage of the rich food at the sea's surface to grow to a larger bulk than usual before transforming into bottom-living creatures, they have to increase their cilia-carrying surface. In sea-urchin larvæ, this surface is a band of tissue bearing especially large cilia, which is braided, as it were, over the creature's protruding arms. In this case new surface is provided to cope with increasing size by growing new arms. In sea-cucumber larvæ there is a similar band of cilia, but they have no arms. And in these increase of total bulk is accompanied by the throwing of the band into lappets and folds, often of extraordinary complexity.

The limit to the employment of cilia comes in creatures only a few millimetres long, like young tadpoles. When these are first hatched out they have not yet the use of their muscles, but are covered all over with cilia. By the aid of these they can glide very slowly over the bottom, but are completely incapable of swimming, or of any rapid movement. (The speed of ciliated animals under the microscope is apparent only; the instrument magnifies their speed as well as their dimensions.) So it comes about that no free-swimming ciliate can grow heavier than about a milligram. Sedentary creatures that use cilia for feeding can grow much larger. The limit comes with Tridacna, the giant clam; but very few reach a pound weight.

Efficiency of circulatory, supporting, digestive and excretory organs—all are needed for any great size to be attained. Flatworms, with their lack of blood, can only grow more than microscopic by spreading out flat like a leaf; and even then their size depends on the degree to which their type of gut can be made to branch.

Then we must remember some curious consequences of smallness and bigness. The very smallest living things are so tiny that they are not out of the range of influence of mere molecules. Even if you are down to a bare ten-thousandth of an inch across, so many molecules are hitting you that the bombardment is equal on all sides. But reduce

FIG. 254. The sizes of organisms.

(B) *Medium-sized creatures. Nos. 3 and 8 are the same as those shown on Fig. 253. Nos. 8 and 15. The dog can be seen at the bottom right-hand corner of Fig. 253.* (1) Dog; (2) Song-thrush with egg; (3) Hen with egg; (4) Smallest bird (humming-bird) with egg; (5) Largest egg (Aepyornis); (6) Largest stick-insect; (7) House-mouse; (8) Largest polyp (Branchiocerianthus); (9) Queen-bee; (10) Largest beetle (Goliath beetle); (11) Common cockroach; (12) Largest frog (Goliath frog); (13) Common grass-frog; (14) Smallest vertebrate (a tropical frog); (15) Common herring; (16) Largest land-snail (Achatina) with egg; (17) Common snail; (18) Smallest mammal (flying shrew); (19) Human hand; (20) Largest starfish (Luidia); (21) Largest free-moving protozoan (Nummulites: extinct).

your bulk a hundred times further, and the rush of molecules may impinge with varying intensity, now on one side, now on the other, and what to all bigger creatures becomes mere uniform pressure keeps the tiny particle all its life long in incessant jerky motion. Filter-passers are so small that they no longer sink in water; the force of gravity has become negligible compared with the molecular bombardment.

Many units up the scale, we come on a very different consequence of relatively small size. Have you ever seen an insect drinking? Probably not. Insects do not drink, or at least they do not drink like larger creatures, by going down to a pool and lapping up the water. And yet they may be thirsty; butterflies may not infrequently be seen settled in crowds on damp ground, sucking moisture through their trunks. The reason for this in all probability lies in their small size. They cannot drink from open water surfaces; they seek a film on solid particles. They are most of them so small that the surface-tension inherent in the surface-film of water—that skin of molecules which is strong enough to support a water-skater and even a needle—is more powerful than they. Everyone has seen little moths and gnats struggling vainly against the grip of the surface-film; and one of the recognized ways of ensuring a peaceful night in flea-haunted corners of the world is to stand naked on a tray in the centre of the room; let the fleas jump on to you, and then sluice them off with water. For all their jumping powers, the surface-tension holds them tight; and to get away once they are wetted, they have to lift many times their own weight of water. As J. B. S. Haldane puts it: "An insect going for a drink is in as great danger as a man leaning out over a precipice in search of food."

Thus to insects and spiders the dangers of drinking are very real; and most insects get their drink in their food, or sop it up from wet earth as we might suck moisture out of well-wetted blotting-paper, or at most suck at very small drops of it; and then generally through a long proboscis. It is possible, too, that a curious fact in insect anatomy is to be explained by their difficulty in getting drink. Their excretory organs are unlike those of all other animals in opening into their intestine instead of direct to the exterior. Our own large intestine absorbs water from the undigested food. If this is the case with insects, too, they waste no water in their urine.

But size, though it enables land-animals to laugh at surface-tension, makes them susceptible to the dangers of gravity. The smaller you are, the greater is the proportion of your surface to your weight, since surface goes up as the square of your length, weight as its cube. A big African elephant weighs about a million times as much as a mouse, its linear dimensions being about a hundred times greater. A little easy calculation will demonstrate that if an elephant could be re-cast, so to speak, into the form of a million mice, the same weight of material would now have a hundred times as much surface.

In a perfect vacuum, all objects, big or small, would fall at the same constantly accelerated speed. But in nature, the resistance of the air comes in; and this depends on surface exposed. The huge proportion of surface to weight in the average bacterial spore means that it falls with incredible slowness; and even an average ant, though over a million times heavier, can hardly fall fast enough to achieve a real bump. A mouse can be dropped down a mine-shaft, however deep, and arrive at the bottom dazed but unhurt. A cat or a dog will be killed, however.

FIG. 255. The sizes of organisms.

(C) *Small, naked-eye creatures. Nos. 1. (Queen-bee) 9 (smallest humming-bird and egg), and 10 (smallest vertebrate) are those shown in Fig. 254, Nos. 9, 4, and 14. (2) Smallest fish (Lebistes); (3) Common brown Hydra, expanded; (4) House-fly; (5) Medium-sized ant; (6) Common flea; (7) Smallest land-snail; (8) Largest ciliate protozoan (Bursaria); (11) Egg of common grass-frog in its jelly; (12) Common water-flea.*

A man will be not only killed but smashed; and if a pit pony happens to fall over, the speed at the bottom is so terrific that nothing is left but a few of the hardest bits of its bones and a splash on the walls.

Size is also important in temperature-regulation, since the escape of heat is proportional to surface. For every gram of its weight, a mouse as big as an elephant would radiate away only one per cent as much heat as a normal-sized mouse. Accordingly, a small animal has to eat much more in proportion to its size to make up for this extra rate of heat-loss; and when the temperature gets very low, it cannot manage to stoke up sufficiently. A big dog, weighing forty-five pounds, for instance, for each pound of its weight needs only about half as much food to keep its temperature constant as does a small dog of seven pounds. A honey-bee would need 500 times as much: that is why bees cannot be warm-blooded. This explains also why small mammals and birds do not penetrate so far towards the poles as do larger ones. The smallest mammal in Spitsbergen is a fox; and there are scarcely any humming-birds that manage to exist outside the tropics or sub-tropics.

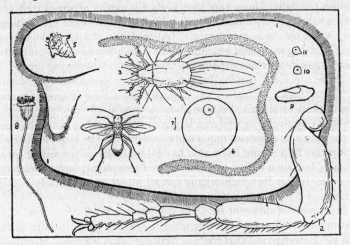

FIG. 256. The sizes of organisms.

(D) *The smallest naked-eye creatures and some large microscopic animals, and cells. No. 1 (Bursaria) is the animal shown in Fig. 255, No. 8; No. 2 is the fore-leg of the flea shown in Fig. 255, No. 6. (3) Cheese-mite; (4) Smallest insect (Elaphis); (5) Smallest many-celled animal (male wheel-animalcule); (6) Human ovum; (7) Human sperm; (8) The bell-animalcule Vorticella, a ciliate; (9) Another ciliate (Paramecium); (10) Human liver-cell; (11) Dysentery amœba. The dotted band in the body of Bursaria (No. 1) is its nucleus. It is much bigger than the smallest insect.*

For a similar reason, ears and other projections that increase surface are cut down to a minimum in cold climates; but in hot countries, where the need of the warm-blooded animal is not to conserve but to lose heat, they are often much enlarged, and provided with many blood-vessels. This is so, for instance, in the African elephant; when he holds his ears out from the side of his head, he is providing an extra fifteen per cent of surface through which heat can be lost.

Still another bearing of size on structure is to be found in paddles and wings. As we have seen, with very small water-creatures, microscopic vibratile cilia will serve. In moderately tiny creatures that swim with limbs, the limb need only be a mere stump fringed with hairs or spines. As the animal grows, the main limb itself must be flattened and expanded, though a border of hairs may continue to be of service; and with further increase the limb grows more and more of a flat paddle, the hairs get relatively less and less important. The change is often made within the individual life-cycle of many crustacea. The same applies to flight. Many tiny insects fly by means of wings that are a mere rod fringed with hairs. The lovely plume-moths are a little bigger; they work on the same principle, but multiply the hair-fringed rods. And all flying creatures weighing more than a fraction of a gram must have a flat impervious expanse as their main or sole flying-organ.

There are some curious facts about the proportional size of parts in big and small animals. Eyes in particular always increase far more slowly than total bulk, because above a certain minimum size a small eye sees practically as well as a large one. The number of touch-organs in the skin also increases much more slowly than the bulk of the body. It matters to a mouse to be able to deal with things the size of breadcrumbs, but such trifles do not concern an elephant; and the elephant accordingly has its touch-organs spread much more sparsely over its surface.

In animals of different size, but comparable intelligence, brains, for reasons we need not go into, increase roughly in proportion to surface and not to weight; so that mere percentage brain-weight is no criterion of intelligence.

§ 3. Colour and Pattern in Life.

HITHERTO in this Book we have been studying the adaptation of whole organisms to their habitats. It would be possible to embark upon an equally full and equally interesting study of the modification of single parts, isolated functions. Sight or digestion, the care of young or the capacity of producing sound, weapons of offence or organs of excretion —on any of such topics one could write an illuminating chapter of biological history. But space forbids us. We will here take the sole example of colour, that enrichment of the world, and deal with some of the biological uses to which colour and colour-pattern have been put. And we must leave it to our readers, with this sample to aid them, to use their imagination in reconstructing for themselves the richness and variety of life's functioning.

In this section then our topic is colour. We shall often use the word rather loosely to denote colour and pattern combined. We must also bear in mind that most animals are colour-blind, and that what to us is a pattern of colour to them becomes degraded into mere black, white, and grey; and also that most of the colours and patterns of organisms have been developed in relation to the eyes of other creatures, so that the colourings and patternings are not only not understood by their possessors, but may even be invisible and therefore non-existent to them.

A baby ringed plover fresh from the egg, which has never seen its mother, will, at the approach of danger, crouch and flatten with outstretched neck. In this position it becomes almost invisible in its native environment of parti-coloured shingle. But it cannot see itself or know what obliterative effect its actions have—as is indeed further demonstrated by the fact that it will crouch just as readily on grass or on a carpet, although there its behaviour merely makes it conspicuous. The egg out of which the baby plover hatched was equally protected by its coloration in the nest, although the hen-bird could neither know what colour her first egg would be, nor, even if she did, could she influence the way the glands of her oviduct would blotch on their pigments.

Some colours are, in their own right, useless. The most familiar case is that of our blood. Hæmoglobin happens to be so made that it looks red; but the redness has no value *qua* redness—it is a consequence of a particular chemical composition. Once there, the redness of blood may be used for the sake of its colour-effect, as in the comb and wattles of the barndoor fowl, or in our own cheeks and lips, or in the more lavish decorations of the baboon and the mandrill, who, as Mr. P. G. Wodehouse flippantly but expressively puts it, "wear their club colours in the wrong place." But in origin it is accidental. Other colours may have a physiological but not a biological meaning—the colour may help in achieving some function of the body. The most obvious example here is the green of plants. They are green because of their chlorophyll; and their chlorophyll is green because it absorbs the red and the violet parts of the spectrum to obtain energy for pulling the carbon out of carbon dioxide. There must be some reason why it does not absorb green light as well, but we do not yet understand it. If it did absorb all wave-lengths, the prevailing colour of our landscapes would be not green but black; and it is probable that we should have found in black the same qualities of freshness and restfulness that we now find in green. But the colour is green, and it has meaning in relation to its own chemical functions, not in relation to other organisms.

The pigment of the skin in the tropical races of warm countries is equally physiological; it prevents the entrance of ultra-violet light in the excessive and harmful amounts poured down by the tropical sun.

Finally, we come to colour with biological function. The two prime functions here are either to reveal or to conceal. The former are usually called *sematic* colours, colours which point you out, the latter *cryptic*, colours which hide you. In the earlier chapters of this Book we have talked of many concealing colours, especially those which harmonize the general tint of the animal with the general tint of its background (white in the arctic, green in the forest, sandy in the desert, blue on the surface of the sea, dark red or black in the abyss). Besides such general resemblances, animals may show protective resemblance to particular objects. We have the stick-insect, drawn out in fantastic fashion; the stick-caterpillars of many geometrid moths, which hold themselves rigid all day in amazing likeness to a twig (complete with buds), only moving in search of food when the sun goes down; the leaf-insect, which not only has its body transformed into the likeness of a green leaf, but bears little green leafy frills on its legs; the famous Kallima butterfly, which at rest not only is coloured brown like a dead leaf, and shaped complete with stalk and pointed tip, but has mid-rib and veins drawn on it, and in some species blotches that simulate mould spots, and even transparent patches bare of scales to look like holes; the little moths that escape unwelcome attention by a resemblance to a small bird's white dropping; the sea-horse whose strange garb of rags and tatters makes it look like a piece of the Sargasso weed among which it lives.

One of the most remarkable cases is that of a South American species of nightjar. All nightjars and nighthawks are protectively coloured, they and their eggs and their young; the common European species has, in addition, the habit of roosting crouched down lengthways along a branch, when its attitude and its mottled plumage make it seem one with the tree. But this South American nightjar goes one better. It nests in hollows on the top of tree-stumps or fence-posts. All day it broods in this exposed situation, but is very rarely discovered owing to its delicately pencilled grey coloration and its extraordinary habit of sitting bolt upright with head pointed up in air, seeming to be only a projecting bit of dead wood. If you approach, it shifts imperceptibly round, always facing you. When the young is hatched and grows too big to be brooded, it too adopts this same pose, and lives all day in this perpetual rigidity until it is able to fly. A combination almost as striking is seen in the bitterns, which, when alarmed, lift their beak and neck vertically in air, and by always fronting the intruder with their striped breast, manage to fade into invisibility in the reed-beds which are their home.

Sometimes concealment is achieved by colour-change. The chameleon is the most famous example, but his powers have been much

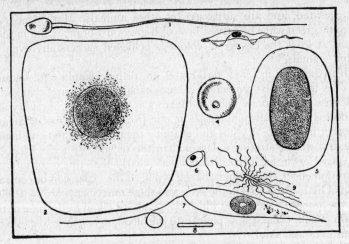

FIG. 257. The sizes of organisms.

(E) *Small microscopic cells and organisms.* No. 1 (human sperm) and No. 2 (human liver-cell) are the same as Fig. 256, Nos. 7 and 10. (3) Sleeping-sickness trypanosome; (4) Human red blood corpuscle, with a malaria parasite inside it; (5) Red blood-corpuscle of frog, with nucleus; (6) Smallest free-living protozoan (Oikomonas); (7) A green flagellate (Euglena); (8) One of the largest bacteria (anthrax bacillus); (9) Typhoid bacillus, with flagella.

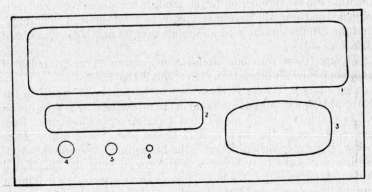

FIG. 258. The sizes of organisms.

(F) *The smallest organisms.* All are shown in outline. No. 1 (anthrax bacillus) is the same as Fig. 257, No. 8; (2) Tubercle bacillus; (3) Plague bacillus; (4) Bacterium of Malta fever (Micrococcus); (5) A particle at the limit of microscopic vision; (6) A filter-passing organism.

exaggerated, and are surpassed by other animals. Many tree-frogs become green when on leaves, brownish-grey on a branch; and in some cases at least the change of colour is achieved successfully even by blind animals.

But the facts in lower vertebrates are rather complicated, because besides apparently protective colour-changing, there are changes in response to such purely physical factors as the warmth, humidity and darkness of the surroundings. Put one frog in a dark moist jar and another in a white dry jar, and examine them after a couple of hours; the first will be strikingly darker than the second. Professor Hogben, of Cape Town, has demonstrated very neatly that this is an infernal secretion effect. The colour-change is due to the expansion or contraction of little black cells in the skin, and these movements are controlled by the amount of secretion which the pituitary gland at the base of the brain (Book, Chap. 4, § 5) pours into the blood, in response to messages from eyes and skin.

But flat-fish take the prize for concealment by change, for they can change pattern as well as colour. In the Plymouth Aquarium is a tank with two communicating compartments, one floored with fine sand, the other with boldly-variegated shingle; in the tank are flounders, and visitors are invited to drive the animals from one compartment to the other and watch the results. In the shingly half, the fish are coarsely blotched with dark brown, sandy-yellow and pale cream; but if one of them be chased through on to the sand, in less than five minutes its blotched pattern will have given place to a fine-grained uniformity, and the fish presents as good an imitation of its new background as it did of its old.

We must now turn our attention to pattern. A vast number of animals are lighter below than above; this is the simplest case of a concealing pattern. Its use is illustrated by an admirable model in the Natural History Museum in London. In a box, against a light-brown background, is a rough model of a duck, strongly illuminated from above. Although the duck is dark-brown above, and nearly white below, it blends wonderfully with the background, because the shadowed white of the belly becomes of the same tint as the illuminated brown of the back. But the model has a handle attached to it. Turn it upside down, and the duck stands out strikingly, the white made brilliant, the brown turned almost to black. Essentially the same thing is seen in nature, whenever a fish swirls over on to its back; all but invisible before, it gleams into brilliant conspicuousness.

When an animal habitually lives in an unusual position, its shading is often correspondingly adjusted. There is a Nile cat-fish, *Synodontis*, that normally swims upside-down. Though this habit had already attracted the attention of the ancient Egyptians, we do not yet understand why it has arisen; but whatever its meaning, the shading of the fish has changed with it, and its upper side, though the belly, is darker. Still more curious is the case of the crustacean Anilocra, an external parasite on the body of small fish. It always clings parallel with the fish's body, and one longitudinal half of it is dark, the other light, to match the shading of its host.

In animals living against an irregularly-patched background this irregularity is sometimes wonderfully simulated; one has only to think of the lizards, moths, beetles and other insects that imitate to perfection the rough and lichen-encrusted bark of the trees on which they crawl or sit; or the brooding woodcock whose brown plumage streaked freakishly with yellow makes her almost undiscoverable among the sun-flecked herbage and dead grass where she nests.

The woodcock introduces us to a new principle—the principle which underlies the art of camouflage. It is interesting that it was first fully stressed by a student of animals—Thayer, artist and naturalist, whose book, *Concealing Coloration in the Animal Kingdom*, if often over-enthusiastic (we cannot follow him, for instance, in believing that the scarlet of the flamingo is of use as a concealment when flying against a sunset sky!), is yet, in its way, a classic. He was the first to realize the value of what may be called *ruptive coloration*—bold pattern which breaks up the form and outline of an animal into irregular and meaningless pieces. Very conspicuous in a museum case, such animals in nature are among the best disguised. The same principle, in exaggerated degree, was adopted for concealing guns, ships, and so on in the War. As an anti-submarine measure, an interesting variation was adopted. Ships were painted in bold patterns, usually in black, white, and blue, so as to give, through false perspective effects, a wrong idea of their course; and this made it much harder for the submarine to take up the correct position for attack. For a crouching animal, falsification of outline is valuable; for a ship threatened by submarines, falsification of direction. The dappling of many deer, the spotting of the leopard and ocelot, and the pattern of the great boas and pythons may help in this way as well as by simulating sunflecks in the forest. Even the Malayan tapir may benefit by his white hinder-half.

It is worth mentioning that, as one would expect, all spotted forest-dwellers show light flecks on a darker ground (simulating sunflecks), while any spots in desert-animals are always dark on a lighter ground (simulating pebbles or local shadows against the glaring sand). But the most striking examples of ruptive coloration are afforded by bars and stripes, which cut across the animal's true form and take the eye from its outline. A familiar example is the ringed plover, so common on the beaches of western Europe. Its black colour and head-stripe disrupt its lines and make it blend with the background in a most unexpected way. Its American relative the killdeer is very similar; and in the turnstone the trick is played for all it is worth, the bird in the hand looking a regular harlequin, but achieving invisibility on a rocky shore.

Camouflage may be either protective or aggressive. Of aggressive disguises the most wonderful examples are the flower-spiders and flower-mantises, which frequent flowers in order to prey upon the insects which visit them, and escape notice by being dressed in the same brilliant colours as the flower.

It is but fair to say that many naturalists and sportsmen have severely criticized the whole theory of concealing coloration, and while some, like Thayer, would make all coloration concealing, others would go to the opposite extreme and deny that any colour or pattern ever concealed its possessor. The truth, as so often, is between the two extremes. Many colours do not have a concealing but a revealing function; many others seem to have no function at all. But one need only have been an amateur bird's-nester or a collector of butterflies and moths to know what protection colour may afford, while such marvellous adaptations as those of leaf-insects or bark-beetles or weed-like sea-horses may be best left to speak for themselves.

We now pass to our second main type of coloration—the coloration which reveals instead of concealing. First, there are the colours which help in recognition. The most familiar is the tail of rabbits, useless for all ordinary caudal purposes, but so white and conspicuous that in America it has earned for its possessor the name of cotton-tail. Its functions seem to be to act as a danger-signal and to guide others to safety. A band of rabbits is feeding as dusk begins to fall. One or two scent danger; out flashes the white signal and, as they run to their burrows, marks the path to home. At once the rest are on the alert; and the young and inexperienced run no risk of growing confused or forgetting which way to run. It is sometimes argued that the white tail is a handicap instead of an advantage, since it gives a nice mark to the sportsman. But this is unbiological. The while tail was evolved millennia before guns or even men were thought of. That it renders its possessors temporarily more visible to a pursuing fox is no matter, provided it helps them to disappear into burrows or brush before the enemy has caught them.

A case where colour links parent and young in a special relation is seen in the great majority of birds with helpless nest-fed young. The inside of the young bird's mouth is almost invariably of brilliant colour, often yellow, though some, such as ravens, show a rich crimson. As soon as the parent returns with food, up go the skinny necks, and open fly the large and brilliant gapes, making the destination of the food very conspicuous. In hole-nesters, like tits or wrens, this may be of real service by preventing waste of food and time. But in most birds we may hazard the guess that the adaptation is one of those which are of no service to the species, though a *sine qua non* to the individuals of which the species is composed. Watch any small bird feeding its offspring; it does not waste any time seeing that all take their turn, but jams packets of food down the throats that most obviously present themselves. Imagine one baby blackbird or robin of a brood unprovided with a yellow maw; it would not be so conspicuous, and would not be fed so often. It would thus grow more slowly and (since the rule that "unto him that hath shall be given," holds for birds as well as men) would so become still more handicapped in the struggle for prominence when food was brought; it would run the risk of being suffocated, of being the first to starve if there were not enough food to go round; it would get a poorer start in life than the rest. The bright mouths of nestlings have been developed through competition between brothers and sisters; they are the result of selection, not merely within the species but within the family circle.

Then, oddly enough, there are animals which are conspicuously coloured because they actually want to get eaten. A certain percentage of the black-caps, sparrows and other passerine birds that we see around us, are infested in their bowels with the fluke known to biologists as *Distomum macrostomum*. It is not a violently harmful fluke. It uses some of the bird's food and slightly damages the lining of its intestine; it is a nuisance rather than a danger. The birds were infected with the parasites when they were still nestlings, and we propose to note the manner of their infection as an example of what we may call appetizing coloration, and an illustration of Nature's impartiality.

The flukes, when they are mature, lay copious eggs, after the manner of internal parasites. The eggs pass out of the host bird with its excrement, and thus they are scattered over the leaves of bushes and herbs. Most of the eggs are wasted and come to nothing; a few of them, however, have the good fortune to be swallowed, with part of the leaf on which they rest, by a snail. The snail, we may note, must belong to a particular species, *Succinea putris:* in any other kind the eggs are digested and killed. But in *Succinea putris* they hatch out into microscopic, active larvæ. The larva bores through the snail's stomach and makes its way into his body tissues. There it comes to rest and undergoes a curious development. It grows enormously and becomes a shapeless radiating web of living tissue without brain or eyes or any of the usual marks of living animals. It spreads through the snail, growing at his expense, and becomes so mixed up with his tissues that it is difficult or impossible for a dissector to separate it completely away. But in the front, by his head, it sends out special shoots. It sprouts out into lobes which grow up into his horns and distend them, and are very conspicuous from outside—partly because they are gaily coloured, with green and white bands and vermilion tips, and partly because they wriggle and pulsate actively just under the skin of their host. They are designed, in short, for being eaten; conspicuous as they are, they catch the eye of insect-eating birds, and since they distend the horn they prevent its withdrawal into the shell. Wherefore, sooner or later, they are pecked off by the next host. Each of these gaily-coloured, attractive morsels is really a hollow bag containing hundreds of fluke eggs.

Now comes a curious development. If the bird swallows his mouthful, the bag with its contained eggs is simply digested away; it comes to nothing. But if the mouthful is carried back to the nest and given to the young birds, the fluke eggs, which can resist the milder gastric juices of the nestlings, hatch and infect their new hosts. So back to the beginning of the cycle again. And meanwhile the snail grows a new horn, and the parasite within him grows a new protrusion into it to attract and invade another bird.

The curiosity of animals is often played upon by man. When we bait for mackerel with a bit of red flannel we emulate the fishing-frog; and in past decades the hunters of the American "antelope" or pronghorn often lured it within gunshot by gesticulating and behaving in other odd ways. There are also well-authenticated stories of small carnivores such as stoats or weasels playing strange antics and so luring inquisitive rabbits and other small deer within reach.

But the majority of revealing colours have the opposite intent; they advertise distasteful qualities and fall under the head of *warning colouration.* An everyday example is afforded by the wasps and hornets, with their conspicuous banding of black and yellow; black-and-yellow too are the caterpillars of the cinnabar moth, which are rejected incontinently by almost all birds. The black-and-yellow fire-salamander is equally conspicuous; and he is equally nauseous, owing to the milky secretion of special glands in his skin, which is actually fatal to small animals. The poisonous coral snakes are ringed with brilliant black, white, and red; a number of butterflies which experiment has shown to be rejected on account of their acrid taste or their tough consistency are large, slow-flying, with showy and definite patterns.

FIG. 259. (Left) The fitting of mimic to model.

Left, the model, a distasteful African butterfly, Amauris niavius. Above is the sub-species found on the West Coast and across to Uganda. Below, that on the East Coast, which has much more white on its wings. The two in the middle are intermediate forms from the transition zone in Eastern Uganda and Western Kenya. Right, the mimic, a Swallowtail butterfly, Papilio dardanus. From above down, female and male of West Coast sub-species; male and female of East Coast sub-species. The males are not mimics, and, although they become blacker on the East, the females become whiter, like their models. (Courtesy of Prof. Poulton and the Hope Department of Zoology, Oxford.)

FIG. 260. (Right) Adaptation in mimicry among African butterflies.

Left, models. The top four are Planema epæa; first a male and female of the West Coast sub-species, then a female of the Uganda sub-species, then a female of the Abyssinian sub-species. Below, the Abyssinian sub-species of Amauris niavius (cf Fig. 259), a model belonging to another family. Right, mimics, all Swallowtails of the species Papilio cynorta. Above, male (not mimetic); next, West Coast female (mimics the West Coast Planema); third, female of Uganda sub-species (mimics Uganda Planema). As the female's normal pattern is nearer to the Amauris than the Planema model, in Abyssinia selection has moulded the mimic to resemble Amauris (below).

The meaning of these devices is clear enough. Lloyd Morgan made experiments with cinnabar caterpillars and chicks fresh from the incubator-drawer, which showed that the chicks have to *learn* that the caterpillars are nasty. A single lesson is often enough. The burnt child dreads the fire; ever afterwards the sight of the beast is sufficient reminder to prevent the chick from trying it again. All enemies who can learn must be taught by experience; and a certain number of the nauseous or noxious creatures are sacrificed for the good of their brothers and cousins, in order to educate their enemies. The individual dies for his cognates.

Warning coloration cannot afford to be subtle or refined. The conspicuous colours serve to make the process of education as easy and rapid as possible, and, by sharply distinguishing the inedible from the edible, make subsequent mistakes less likely. In the domain of sound, the rattle of the rattlesnake and the hiss of other serpents serve the same warning function.

The skunk is the typical instance of warning coloration, as it is certainly the closest relative of man to practise the device. Everyone knows the power of the skunk's weapon; how dogs bedewed with the vile liquid run off in howling despair; and how no wild creature, however large, ferocious, or hungry, will touch a skunk. Skunks most effectively combine appearance and behaviour for warning purposes. They have not adopted the ordinary principle of shading; white above and dark below, their form stands out sharply against the background. The animal knows no fear, but walks slowly, refusing to give ground or to run from man or beast. The white and bushy tail is held aloft as a warning banner. This fearlessness, justified by past conditions, leads at times to grave inconveniences in the face of modern inventions. A skunk will not make way even for a train, and a train is unable to step aside for a skunk; the resultant casualty may cause serious distress to the passengers for many hours.

But if enemies can learn the association of certain patterns with danger or distastefulness, and let their owners alone for the future, what is to prevent other animals from practising a simple bluff, and achieve immunity from attack by imitating the warning colours, without the trouble of developing the weapon or the offensive taste? The answer is that there is nothing to prevent it; natural selection can achieve any device, provided it is within the resources of life and of service. The only special condition is that the species that bluffs shall be much less common than the species with true warning coloration; if their proportions were the other way round, the lesson of the warning colour would never be properly learned. Such imitation of the outward appearance of a protected by an unprotected animal is called mimicry, and the species mimicked is styled the model.

Some of the best-known examples are the clear-wing moths, which closely resemble bees and hornets. Another is afforded by the two well-known American butterflies, *Danais plexippus*, the Monarch, and *Basilarchia archippus*, the Viceroy. The nauseous, flaunting Monarch is the model. It is geographically an invader, and apparently a recent invader, of America. The Viceroy mimic is a member of a quite different family, the Nymphalines, which are strongly represented in Europe and North America; it is so closely related to an American species of White Admiral that the two, in spite of their entirely different pattern, will interbreed and produce hybrids. The likeness to the model only holds for the adult stage; egg, caterpillar and chrysalis are all of regular White Admiral type. If mimetic likeness were the result of local conditions of life, we should have expected the invading Monarch to have become modified in the direction of the native forms rather than vice versa; but if it is an adaptation on the part of the model, the facts are at once intelligible.

It is worth while studying the methods by which the mimic's likeness to its model is achieved; for they throw light on the processes of Evolution. The salient facts are, first that the resemblance may be produced in all sorts of ways; and that so long as the enemy's sense of sight is deceived, it does not matter what tricks are employed. In Brazil, for instance, there is found a sand-wasp, *Pepsis*, together with two mimics, a plant-bug and a long-horned grasshopper. The wasp's antennæ are short and thick; the mimics have the base of their antennæ thickened to look like the whole of the wasp's antennæ while the rest is reduced to a filament so thin as to be invisible at any distance. Again, there is a distasteful group of South American butterflies which have prominent transparent areas on their wings; they are mimicked by a number of other butterflies and day-flying moths. In the models, the transparency is produced by the scales being reduced to minute vestiges. In some of the mimics, the same device is adopted, in others the scales are of full size, but enormously reduced in number; in others they are of full size and full number, but are themselves transparent; while in still others, they stand up on edge and so let the light pass through. In some clear-wing moths, on the other hand, the scales are loosely attached, so that they come off during the animal's first flight.

Examples such as these show how impossible it is to suppose that the conditions of life directly bring about the mimetic resemblance; any kind of Lamarckian explanation is ruled out of court, and as the only possible agency for the origin of mimetic adaptations, we are left with natural selection guiding random variation. The same reasoning holds good for the highly original method by which the plant-bug *Heteronotus* achieves its mimicry of ants. The roof of the front segment of its thorax is prodigiously expanded so as to cover all the rest of the animal, and modelled into the colour and shape of an ant!

The Membracidæ, the family of bugs to which Heteronotus belongs, provides in itself a remarkable object-lesson in deceitful adaptations and their origin. All of them have their first thorax-segment expanded into some sort of shield, often with no obvious adaptive significance. But this convenient expansion, once it is present, has been used as the raw material out of which all sorts of resemblances, both for concealment and for mimicry, have arisen. In one species, it is prolonged fore-and-aft to look like a grass-seed; in another, it is converted into a hollow semblance of a thorn, below which the little bug crouches against a stem; in another, it looks like a hooked burr; in Heteronotus, as we have seen, it takes the likeness of an ant; in Oeda it is much swollen and coloured bright orange, to mimic very exactly the cocoon of a distasteful moth. The whole group is defenceless and palatable; they all possess the thoracic shield which can be moulded into the most fantastic shapes without change in any of the vital organs; and so resemblance of one sort or another has become their main method of defence. In every case the resemblance is a hollow sham, and the general structure and habits of life are left unaltered.

In some cases, a single individual may pass from one sort of mimicry to another during its growth. The grasshopper *Eurycorpha* when young mimics ants, thinning its antennæ to invisibility, painting-in a waist, and having the instinct to run about restlessly, ant-like, among its models. But as it grows up, it becomes too big to look ant-like: and finally moults into a green creature with expanded wings, which escapes detection by its leaf-likeness and its immobility. To pass from the likeness of an ant to that of a leaf, however, it must go through an awkward intermediate stage when it resembles neither the one nor the other. At this age it is provided with yet a third set of instincts, and tries to escape from its enemies either by hiding or by "shamming dead."

The cuckoos provide us with one of the few examples of aggressive mimicry, in which a resemblance serves for the direct exploitation of the model. The common European cuckoo exists in a large number of strains or races, each of which lays its eggs almost wholly in the nests of one kind of dupe; the meadow-pipit cuckoos and hedge-sparrow cuckoos are the commonest strains in Britain, while in some parts of Europe there are redstart cuckoos, and in Japan, bunting cuckoos. In some cases the cuckoos' eggs are not particularly like the dupe's eggs; but usually the resemblance is striking. Recent investigation makes it almost certain that the differences are due to differences in the severity of selection. Sometimes the dupe is easy-going and stupid, like the hedge-sparrow, and will brood almost anything; then there is no need for the cuckoo's eggs to resemble their hosts, and no resemblance is achieved. But redstarts and buntings are stricter and will turn suspicious-looking objects out of the nest; and the cuckoo-strains that parasitize them have been rigidly selected for egg-mimicry.

The resemblance is protective so far as it concerns the eggs themselves. But it is aggressive from the point of view of the race; the egg, once preserved from expulsion, produces the ugly little nestling which in the common cuckoo destroys all its foster-brothers, in other species appropriates a share of their food.

Finally, we cannot pass over two extraordinary cases which show into what strange paths variation and selection may lead an animal. The palatable West African moth *Deilemera* spins a cocoon covered over with little white ovals. Any entomologist seeing this would say at once that these were the cocoons of Hymenopteran parasites, which after devouring the interior of the just-transformed caterpillar, emerge to pupate on its exterior. These parasites' cocoons are very tough and resistant, and are despised by most insectivorous animals. The average bird appears to be as readily deceived as the average entomologist, and so the Deilemera profits by its likeness to its deadliest enemy.

Then there is the case of the alligator-bug, *Laternaria lucifera*, one of the large insects of tropical America which are known as Lantern-flies (Fulgoridæ). From the front of its head there projects forward a huge hollow structure, which has been turned into a mask admirably representing a miniature alligator. The "eye" and "nostril" are raised on knobs, as in a real alligator's head; they are painted on in black, and the eye has even a white patch simulating the reflection of light. Along the side is a dark line representing slightly opened jaws, and from this there stand out in relief the most convincingly white and flashing teeth. At first blush the difficulties in the way of accounting for such a resemblance in terms of natural selection, or indeed on any rational basis at all, seem insuperable. The resemblance is not to a whole alligator, but only to its head, and to the head of an unnaturally diminutive alligator; and even if the resemblance be useful now, what advantage could have been secured by the early and incomplete stages in its evolution?

This last question is not nearly such a hard nut to crack as it seems, for all lantern-flies have an enormous hollow expansion in front of their head. We do not know in the least what function the expansion may have, but there it is; and in some forms it has roughly the general shape of the imitation alligator-head in Laternaria. It needs little change to convert this projection into a rough animal mask, and if that be of advantage, then the final touches can be easily added later. As to advantage, that too is not so difficult. A bird or a little monkey picking over the foliage suddenly comes on an object which looks like a grinning reptilian head. Is it likely to go on with its exploration and taste and try?

There are, as a matter of fact, other lantern-flies with less perfect animal masks, and masks which resemble not a particular creature, but serve their purpose by appearing simply vertebrate and ferocious. The effect here is based not on specific resemblance, but on general association. This particular form of bluff we may call associative mimicry or terrifying coloration. It is found in many insects, notably in their larval stages. A number of large caterpillars have prominent eye-like markings on either side of their body: when frightened, they draw in their head, puff out the eyed segments, and so provide themselves with a sufficiently vertebrate and alarming appearance to scare off some at least of their enemies. The caterpillar of the common pussmoth adopts a more curious method. It, too, puffs out its fore region, showing a prominent pattern; at the same time it erects its forked "tail" and from each fork protrudes a red and wriggling filament. In this guise it has startled many a juvenile collector, and doubtless many a bird. This particular appearance is not wholly bluff, for the pussmoth larva secretes formic acid, which it ejects as a last resort. The lobster-moth caterpillar assumes an even more grotesque appearance when alarmed.

There still remains, however, another type of mimicry to discuss. There are seven British species of wasp of the genus Vespa, and every one of them is banded with black and either yellow or orange-yellow. Black and yellow recurs as a warning colour in many other wasps, in several bumble-bees, in cinnabar caterpillars, in the fire-salamander. What is the reason for this recurrence of one colouration? Partly no doubt that it is a conspicuous one; it is not for nothing that the Automobile Association have chosen yellow for their motoring signs, and

SOME SPECIAL ASPECTS OF LIFE

that yellow signals are beginning to oust the old red and green on some railways. But might there not be advantages also for a species to be hall-marked dangerous with a widespread and therefore already widely known pattern? The possibility is converted into a certainty by what we find among butterflies. Here warning patterns of a complexity which rules out the possibility of chance coincidence are often shared

FIG. 261. Protection by bluff.

Left, the Puss-moth caterpillar in normal attitude (above) and in the terrifying pose it adopts when alarmed. Its front view simulates a face and from its tail protrude red, waving filaments. Right, the Lobster-moth caterpillar in normal attitude (above) and terrifying attitude (below).

by a number of distasteful species. The species may belong to different genera or to different families, whose patterns are characteristically quite unlike, so that their ancestry would be expected to make them differ, not to resemble each other.

Bates had been the first to put forward a reasoned theory of our first type of mimicry—mimicry by bluff; but he was puzzled by this likeness, often very close, between many distasteful species. It was reserved for Fritz Müller, following in Bates' footsteps in South America, to point out its meaning. The efficacy of warning colour in an insect depends upon the education of insectivorous animals; and to accomplish this education a certain number of individuals must be sacrificed each year for the good of the rest. If a number of species adopt a single hall-mark, the prescribed number of victims will be no greater,

960 THE SCIENCE OF LIFE

but the sacrifice will be spread over all the species instead of falling entirely on the shoulders of one.

The two types of mimicry are distinguished by the names of their discoverers. Mimicry by bluff is called Batesian, mimicry by pooling of warning colours is styled Müllerian. As Poulton writes: "A Batesian mimic may be compared to an unscrupulous tradesman who copies the advertisement of a successful firm; Müllerian mimicry to a combination between firms to adopt a common advertisement and save expense."

V

THE SCIENCE OF ECOLOGY

§ 1. *Ecology Is Biological Economics.* § 2. *The Chemical Wheel of Life.* § 3. *The Parallelism and Variety of Life-communities.* § 4. *The Growth and Development of Life-communities.* § 5. *The Grading of Life-communities.* § 6. *Food-chains and Parasite-chains.* § 7. *Storms of Breeding and Death.*

§ 1. Ecology Is Biological Economics.

WE COME now to a fresh way of regarding life, by considering the balances and mutual pressures of species living in the same habitat. We have studied the forms of life, we have traced what we have called, perhaps rather clumsily, "biological invention" and the evolution of existing forms, and we have considered the adaptation of these forms to the exigencies of this or that habitat. In every habitat we find that there is a sort of community or society of organisms not only preying upon but depending upon each other, and that certain balance, though often a violently swaying balance, is maintained between the various species so that the community *keeps on*. In this new chapter we are going to study this swaying balance.

The particular name given to this subject of vital balances and interchanges is called Ecology. Ecology is a term coined by Haeckel, the celebrated German biologist, in 1878; its root is the Greek οἶκος, a house, which is also the root of the kindred older word economics. Economics is used only for human affairs; ecology is really an extension of economics to the whole world of life. Man is always beginning his investigations too close to himself and finding later that he must extend his basis of inquiry. The science of economics—at first it was called Political Economy—is a whole century older than ecology. It was and is the science of social subsistence, of needs and their satisfactions, of work and wealth. It tries to elucidate the relations of producer, dealer, and consumer in the human community and show how the whole system carries on. Ecology broadens out this inquiry into a general study of the give and take, the effort, accumulation and consumption in every province of life. Economics, therefore, is merely Human Ecology, it is the narrow and special study of the ecology of the very extraordinary community in which we live. It might have been a better and brighter science if it had begun biologically. Here, until we come to our last Book, we shall say nothing further of economics, and then we shall write only in the most general terms. And all we shall say will be based on general ecology. Man is so peculiar a creature that a really satisfactory treatment of the science of work and wealth demands a companion work upon the scale of this *Science of Life*. But here we hope to show that ecology lays the foundations for a modern, a biological and an entertaining treatment of what was once very properly known as the "dismal science" of economics.

§ 2. The Chemical Wheel of Life.

EVERYWHERE, except in phases of extreme geographical and meteorological change, the world presents in every habitat this swaying balance of diverse species, and everywhere we find that these communities which prey and depend upon one another present many resemblances in pattern. It is not altogether fanciful to compare all these various communities, each one to a sort of super-organism, and in practice much of the thought and work of the ecologist involves that idea.

We have already made it manifest that except for the green plants (and possibly certain bacteria) no living thing can live upon any food except the substance of other once living things. The leafy green plant on the earth, the smaller green plants in the plankton of salt- and fresh-water, are the primary sources of all living substance. Every biological community exists on that as its basis. And as we have explained, the chlorophyll does its work by utilizing the energy of the sun. If animals and fungi are food-parasites upon green plants, green plants are energy-parasites upon the sun. The whole of life upon earth depends entirely upon solar energy. The sun's energy is the physical source of all life.

The great bulk of living substance is composed of carbon, hydrogen, oxygen, and nitrogen. But a number of other elements, though present only in traces, are equally essential. There must be sodium, potassium, calcium, magnesium, iron, phosphorus, sulphur, chlorine, iodine, and perhaps silicon, copper, and zinc.

Life's carbon is derived almost exclusively from carbon dioxide, which is present in the air to the average amount of 0.03 per cent and is held in solution in all waters, but in very varying proportions.

Nitrogen enters green plants in the form of nitrates dissolved in the soil water or in the water in which they float. All the other substances which plants need for construction—phosphorus, sulphur, calcium, magnesium and the rest—are brought to them dissolved in water in the shape of inorganic salts.

Hydrogen enters the green plant in the form of water, which is utilized with carbon dioxide in the manufacture of sugar. As we have described in an earlier Book, the oxygen of the carbon dioxide is split off and returned in elemental form to the air or water round the green parts of the plant.

So that, so far as the building up of carbon into organic substances goes, oxygen is a waste-product, there being sufficient in the water absorbed by the plant to cover all that is needed for the construction of sugars and other carbohydrates and of proteins. But, of course, it is necessary to take in oxygen at another part of the cycle to sustain the slow fire of living oxidations that generates the energy required for the business of living. The oxygen to feed this slow flame is obtained by green plants in elemental form, either direct from the air or dissolved in water. The atmosphere serves the oxygen needs of the great bulk of animals and of non-green plants; there are, however, a few parasitic animals and a fair number of bacteria and fungi which manage to extract the oxygen needed for their energy-production from the organic substances on which they feed.

The green plant having effected this synthesis of organic carbon and nitrogen compounds, of which it alone is capable, the rest of life steps in to seize its share of the spoils.

Plant-carbohydrates and plant-proteins are the only vehicles by which carbon and nitrogen can enter the animal kingdom. They must be taken to pieces by the enzymes of the herbivore's gut before they can pass through into the real interior of its body, but the pieces must not (save in some protozoa) be of simpler chemical nature than sugars and amino-acids. Carbon or nitrogen presented to an animal in any simpler form is as useless as a pile of banknotes to a man on a desert island.

Once the carbon and nitrogen have entered upon their animal career, they can continue to circulate in the animal kingdom as carnivore preys on herbivore and one carnivore on another. In the process, however, the complexity of the carbon and nitrogen-carrying compounds never falls below the level of simple sugars and amino-acids.

The handing on of these organic substances does not go on indefinitely. Some animals have no natural devourers; others escape devouring or die from disease or old age. Not all plants are devoured. The dead substance which would otherwise remain a locked-up accumulation of complex compounds, of no further use to life, undergoes dissolution. Some part of it undergoes purely chemical degeneration, some part of it is simply oxidized; but the greater portion is broken up by *decay*—that is to say, by the action of special bacteria. Decay is a living process; it is part of the cycle of life; it is not a mere inorganic decomposition. A great variety of bacteria are concerned in it, each species playing its own part in the decomposition.

If decay, helped by the natural oxidations due to the oxygen in the air, proceeds to its limit, the carbon that once was life's reappears in the form of carbon dioxide, its nitrogen enters air or water as ammonia, and the other elements for the most part become converted into water and salts. In poorly aerated situations, however, the carbon may be given off as the marsh gas or methane (CH_4) that we see bubbling up from the bottom of stagnant waters, and the sulphur as hydrogen sulphide (H_2S) with its familiar stench of rotten eggs.

The materials in this course of decay may be utilized by still other organisms before reaching the final stage and they may be once more directed upwards to greater complexity. Most moulds and fungi do this; they are in effect parasitic upon decay. Though they vary considerably among themselves as to the details of their nutrition, certain general features are always present. They cannot tap light-energy to win the carbon from carbon dioxide; and they find it inconvenient or unnecessary to utilize simple inorganic salts as source of nitrogen. But though food on the low chemical levels from which it can be hauled up into life by green plants is unavailable for them, they do not require it in such a complex form as is necessary for animal assimilation. They do not insist on sugars and amino-acids; they can utilize lower grades of chemical complexity.

Some fungi may fall a prey to animals and so lift the disintegrating compounds back to the full animal level, but in spite of this occasional upward return which fungi render possible, the vast bulk of living materials do at last deteriorate back to the extremity of decay. The carbon makes full circle and is once more available in carbon dioxide for the synthetic activities of green plants. The ammonia or ammonium compounds, in which form the nitrogen mostly returns to the air, (and also the far less abundantly generated sulphur compound hydrogen sulphide) form the basis for new activities on the part of bacteria. A series of bacterial species attack this ammonia and through their vital processes convert its nitrogen into the form of nitrates, which again are available for the nutrition of green plants. Once more these elements enter upon an upward phase in the wheel of life.

The rôle of the bacteria is absolutely essential in this cycle. Without the agency of a horde of microbes belonging to a great number of kinds, and each restricted to one chemical job, to one step only of the whole transformation, there would be an unbridged gap in the vital circulation. Doubtless, given time enough, inorganic oxidation would decompose proteins and carbohydrates, just as it succeeds in rusting iron; but the turnover would be so slowed down that only a negligible fraction of existing life would be able to carry on, and there would be a much greater leakage of nutrient elements into wholly unavailable forms.

Thus life depends for its primal supply upon green plants, and for its sustained supply upon bacteria. The whole of life considered chemically is one cyclic process from green plant to bacteria and so again to green plant.

The cycle does not run smoothly. It runs with various leakages and blocks, whereby some of the food elements are changed into forms, useless to the great majority of living things, or deposited in the solid state and so taken out of the reach of all life until inorganic solution or oxidation has made them again available.

Shortage of available nitrogen is a chronic state in life's affairs, and the formation from organic compounds of elemental nitrogen gas, chemically inert and so unavailable to almost all organisms, constitutes a serious leakage. Such a leakage is occurring all the time. Wherever insufficient oxygen is present, certain of the bacteria which normally convert ammonia into nitrates, can reverse their activities and use nitrates as a source of oxygen, restoring the nitrogen into the air as a useless by-product. If this leak were to continue indefinitely, the wastage of available nitrogen could not be made good, and life would gradually perish of nitrogen-want, though in the midst of an unavailable plenty of nitrogen.

However, the wastage due to such denitrifying bacteria is made good by means of yet another class of bacteria, the nitrogen-fixers, who are aided by the mycorrhiza fungi which we have already discussed (Book 6, Chap. 4, § 1). By chemical means at present unknown, these plants can seize hold of the nitrogen of the air and transform it into vitally-available compounds—a feat imitated by man in recent years, but only with the expenditure of great quantities of energy, derived from water-power or other mechanical sources. The nitrogen-fixing bacteria live for the most part in an association of mutual benefit with the members of the leguminous order of plants, the pea tribe. It is a symbiosis of great practical significance. The more leguminous plants, like beans, peas, lupins, clover, and alfalfa, man cultivates, the more nitrogen will divert from the air back into vital circulation.

There are other types of leakage or block in the rotating wheel of life. Through a lack of sufficient oxygen during decay, carbon may be locked up in its elemental form, unavailable to green plants. Many cubic miles of carbon have been thus solidified in the form of peat, lignite and coal, and much of this food capital has been idle, locked out of circulation, for hundreds of millions of years. It is not yet certain how the mineral oils which give man petrol and paraffin oil arose; but whether their origin is from plants, or, as is possible, from marine life, they again constitute a locked-up store of carbon that was once in circulation. The fireplace, the factory, and the automobile are doing all they can to restore this deposited carbon to a state of gaseous accessibility.

Carbon may also be deposited uselessly in combination with the important metal calcium. Considering that all chalk and limestone is entirely or almost entirely derived from the skeletons of living things; that an unknown depth of calcareous deposits, in the shape of Globigerina ooze, covers nearly thirty per cent of the whole sea bottom, and that corals are steadily imprisoning tons of calcium in coral reefs every day, we might anticipate a general calcium shortage. The element is, however, so abundant that only in certain restricted areas of the land is calcium-shortage of serious import to life.

Guano, a deposit consisting of the excrement of bats and birds, is another example of locked-up material. Most excrement serves to enrich the soil on which it falls, but on the Southern islands where guano is found the manure has been unused and has merely accumulated. Here again man comes to the rescue, and, by using guano as a fertilizer, restores its nitrogen and phosphates to the soil.

All these diversions of material into idleness are directly effected by or through life. In addition, purely inorganic processes may abstract quantities of this or that element from the rotation. Evaporation is the most potent of these agencies; evaporation has imprisoned many million tons of salts in the sterility of salt deserts and rock salt. The high prices given for fertilizers containing magnesium and potassium, sulphates and phosphates, indicate how serious to men such diversions of vital material from the wheel of life may be.

With such digressions and losses, with fluctuations and uncertainties, the wheel of life turns on. The driving force of the wheel is solar energy. By virtue of that energy and of that energy alone, the elements are drawn into the wheel, pass from lower to higher complexities of combination, pass from green plant to animal, from one animal to another, live, die, live and die again and again; some fall by the wayside as waste masses of substance, some stay unutilized for millions of years, others are caught by a fungus or a bacterium and turned back to the higher levels again; so sooner or later they return to the state of simple and stable combination at which they began, to be caught up once more by the sunshine and the chlorophyll and once more sent round the cycle.

§ 3. The Parallelism and Variety of Life-communities.

THIS fundamental Wheel of Life turns in a multitude of places and under a vast variety of conditions, and each one gives us a different sort of life-community. The Wheel of Life is indeed rather like the work of some extremely popular playwright who has only one leading idea in his head, which he repeats again and again with no great originality or invention. The scenery varies—it is not his work—the costumes and the names change, new fashions soak into him, the cast of the actors is different and now one personality dominates and now another, there is a lack or a superfluity of supers for the minor parts or some performing animal or other novelty has to be worked in, but beneath these changes we detect the same old plot and very much the same rôles.

We will consider now something of the variety and of the fundamental similarity of the life-communities thus evoked. They are, so to speak, organizations of species of organism. Life-communities develop and evolve as wholes. They might be called super-units of life. Always in a life-community there will be green plants as producers, herbivore animals (and parasitic plants) as exploiters of green plants, and decay-bacteria exploiting both and breaking down the substance of their dead bodies. Without the greenery, production would cease and the whole community come to an end. Without the decay-producing bacteria the return of plant-food substances would be so much slowed down that the whole drama would stagnate through the arrest of material in corpses.

Animals as well as bacteria speed up the circulation, for the urea and carbon dioxide into which they eventually turn the plant-food they eat are much more readily and quickly available for plants than the mere dead leaves and stems, full of wood and cellulose, that would be the main end of green plant substance if there were no animals to eat it. And even as corpses, animals are more speedily brought back into availability by decay than are the remains of dead plants. So, while green plants could exist by themselves as an independent life-community, carrying out a quite independent drama of transformations, yet animals are of the greatest importance as accelerators to such a cycle. They actually benefit the species of green plants whose individual bodies they devour alive or break down when dead, for without them the rate of growth of green plants would be enormously slowed down, and both their abundance and their variety would be infinitely less.

The life-community on land differs widely and necessarily from the more primitive life-communities of the sea. It has been said that "all flesh is grass and all fish is diatom," and while the life-community of the waters has its microscopic food-basis nearer the sunlight and carries out its interchanges in three dimensions, the life-communities of the land arise upon the soil and vary much more widely because of the diversity of rock and exposure, upon which they live and have their being.

Among the green plants of every land-community there are usually one or a few dominant species much more abundant or conspicuous than all the rest—the oaks in an oak-wood, the grass in a field, the heather on a moor, the great trees in a tropical forest, the bulrushes in a swamp. But besides these, there are always plenty of other less abundant kinds which find rôles for themselves among the margin of opportunity which the dominant form leaves over. Each life-community thus comprises a vegetable hierarchy.

This fact that there must always be a dominant vegetable type or types in any life-community comes about through the dependence of plants upon light and moisture. The kind of plant which, in the local conditions of moisture, can win in the struggle for light will in the long run kill out its main rivals or prevent them from reproducing and so rule the community. The climate may permit of forests; then the dominants will be trees. Or a dry soil may support nothing more than a steppe or a prairie; then the dominants will be grasses. In all cases the dominant is the kind of plant which can succeed in getting the biggest share of light-energy streaming down upon the area.

Light being the main necessity of plants, the dominant plant of a community is the tallest member, which can spread its green energy-trap above the heads of the others. What marginal exploitation there is to be done is an exploitation of the dimmer light below this canopy. So it comes about in every life-community on land, in the cornfield just as in the forest, that there are layers of vegetation, each adapted to exist in a lesser intensity of light than the one above. Usually there are but two or three such layers; in an oak-wood for example there will be a layer of moss, above this herbs or low bushes, and then nothing more to the leafy roof; in the wheat-field the dominating form is the wheat, with lower weeds among its stalks. But in tropical forests the whole space from floor to roof may be zoned and populated.

The plants of a life-community in their quest for light may become differentiated in time as well as space. In the lower layer of most woods, for instance, there are a few shade-loving plants that can grow and flower even in the leafy summer when nine-tenths of the light or more is being intercepted by the crowns of the trees. But if the early bird catches the worm, the early plant catches the light. Accordingly, another and the more numerous section of this layer of the woodland-community is made of specialists in vernal growth, which shoot up into activity to catch the early sunshine of the year before the trees have time to spread their light-traps across it. The primroses and wood-anemones and bluebells of English woods, the hepatica and the blood-roots and the spring beauty of the woods of America, all flower while there is still light in the lower storeys of the forest. Most of them are active for about three months of the year, and sleep away the rest.

In prairies and still more in deserts this seasonal specialization is very strongly in evidence, but it is concerned with water rather than light, of which latter necessity there is here more than enough for all. The studies of the Desert Laboratory at Tucson have shown how the desert plants divide up the year's rainfall of the Arizona desert. The first warmth of the year, combined with the slight winter rainfall, brings up a crop of small annuals, not markedly different from the small annuals of less extreme climates. They shoot up in January, flower in February, fruit in March or April, dry up and live through the rest of the year as seeds. And there is also a crop of perennials which have the same short period of active life, and survive the rest of the year as bulbs or leafless stems or underground root stocks. But as April passes, the temperature becomes very high and the rainfall declines. The tender vegetation dries up. This is the season of the plants we are accustomed to think of as most typical of deserts—the cactuses, the agaves, the yuccas. They can accumulate stores of water in their stems or leaves, and live and flower and fruit through the drought at the expense of these stores. They hold the stage until June is over. Then comes the main rainy season, with even higher temperatures, and the whole landscape changes marvellously. Millions upon millions of seedlings spring up and make the desert fertile. Speed is the keynote of their existence, for they must finish flowering and fruiting before, in less than three months' time, the next season of drought falls upon the land. And during this second drought, from late September on to the end of the year, with no rain and with increasing cold, the plant-life of the desert is almost at a standstill.

In the sea other seasonal factors come in. The researches of Atkins have shown that the reason for the sudden drop in the abundance of surface life during the summer months is that the diatoms and other basic food-producers of the sea have exhausted the available phosphates in the surface waters. Not until the winter, when cooled surface water sinks and phosphate-rich water rises from the unexhausted depths to take its place, can the full stream of life's abundance flow again.

When we pass from the fundamental vegetable hierarchy of a life-community to its animals we find a much greater amount of specialization and variety. We find not merely the herbivores that eat the plants, but the carnivores that eat the herbivores; we find parasites, a special and intimate sort of herbivores or carnivores, and we encounter scavengers which live upon the decaying remains of creatures of any possible grade in the scale. The scavengers of animal remains play a very different part from the scavengers of plant remains. The former (such as jackal or blow-fly grub, or the dung-beetle, though his actual food be vegetable remains extracted from the dung) are hangers-on of the purely animal part of the organization; they are exploiters of exploiters. But the scavengers of plant remains (such as the worms that eat dead leaves or the termites that eat dead wood) must be grouped together with parasitic plants like mistletoe and many fungi, and the few animal parasites of green plants, with the herbivores. They are all to be classed among primary exploiters of green-plant activity, middle-men between green plants and other animals.

Of these intermediaries, every community will have a few basic kinds which are the foundation for most of the rest of the community's animal life. From the point of view of biological economics, they are the animal counterparts of the dominants among the plants: but they differ in being often (though by no means always) small and inconspicuous, their importance unrealized by the casual naturalist. The tiny copepods of the sea, the sap-sucking plant-lice, the earthworms of the soil, carry on such decisive rôles, determining as they do the character of all the higher forms that prey upon them, quite as much as do such familiar herbivores as rabbits, sheep, or deer.

The carnivores are usually organized in what are called food-chains, each link of the chain serving as food for the next. If we eat sheep, we eat plants at one remove; if we eat snipe or woodcock, we are eating them at two removes, with worms and such small creatures making another link in the chain. Parasites, too, may be organized in chains, parasite upon parasite.

Again each of the main groups of exploiters, herbivores, carnivores, and the rest, can be sub-divided into well-marked minor rôles, which will be found, played by one actor here and another actor there, in all well-developed life-communities. There is, for instance, the rôle, whose importance in biological economics is not very commonly suspected, filled by the suckers of plant juices. This rôle is almost always undertaken by insects, almost all of them small insects—plant-lice, cocids, plant-bugs and the like—but the sum total of their pumpings is enormous, and converts a vast bulk of plant substance into a condensed animal form which then becomes available to carnivorous insects like ladybirds, and so passes, directly or indirectly, to bigger animals such as birds and mammals. All these higher forms are obviously conditioned by these juice-suckers.

Among the carnivores, there are carnivorous insect-eaters, some of them insects themselves, with spiders too, and frogs, reptiles, birds, and mammals. These insect-eaters fall into many sub-groups. There are the hawkers of flying insect prey, such as dragon-flies, swallows, swifts, nightjars, and some hawks, the ant and termite specialists like ant-thrush and flicker, ant-eater and ant-bear, the devourers of wood-boring types, such as the woodpeckers and other birds, the pickers-off of parasites like the tick-birds, and so forth. And then in a grade above the insect-eaters, and conditioned by the insect-eaters just as these latter are conditioned by the insects, are the carnivores that go for bigger prey, the hawks and owls, stoats and cats, wolves and boa-constrictors.

Among the scavengers there are the eaters of dead and decaying plant remains, like earthworms; the hangers-on of bigger carnivores, like the proverbial jackals with the lion; the undertakers of small corpses, like the burying-beetles; the funeral specialists of larger creatures, like vultures and marabou storks; and the devourers of dung, like the sacred scarab.

This life-drama, we have seen, is stereotyped in plot and construction, with the same main rôles to fill wherever it is played. But the caste varies. A rôle is played here by one kind of organism, there by another. It is interesting to note a few examples of how the different actors fit themselves to their parts.

What we may call the earthworm rôle, the soil-making rôle, is filled in arctic regions like Spitsbergen by hordes of the tiny insects called Collembola or spring-tails; in tropical forests it is partly filled by termites. On various coral islands it is filled by land-crabs; the most important source of humus there is rotting coconut husks, and these the land-crabs burrow into and consume, much as earthworms burrow into soil and consume the most important source of humus in temperate countries, decaying leaves.

Wherever there is abundance of sedentary and edible food, whether it be plant or animal, there is a rôle to be filled by browsers. Sea-slugs and sea-snails browse on seaweeds, and many coral-reef fishes browse on coral, just as land-animals like rabbits and sheep and deer browse on grass and bushes. Carnivores often take to supplementary diets and to scavenging. In Africa, the spotted hyena lives largely on the remains of lions' kills; in the high north, the arctic fox is kept alive in the winter by the remains of seals killed by polar bears. The resemblance goes further, for the arctic fox supplements his diet with sea-birds' eggs, the hyena with the eggs of ostriches.

The pursuit of earthworms in the soil is undertaken in Europe and North America by the common moles; in Africa by the quite separate family of golden moles, and in Australia by the pouched moles—marsupials, not placentals at all; and there are some rodents which have taken to a not dissimilar life.

These examples show how the plan of very different communities tends to repeat itself even in the details of its arrangement. Manifestly a life-community is an organization, with a definite unity of its own, an individuality of its own for the exploitation of natural resources.

Different kinds of plants and animals do not occur together haphazard; they are sifted out, selected, mutually adapted until a working organization results. Thus the life-community, if not an organism, is an organization of species which fill definite rôles, just as the body is an organization of parts and organs each with its special function. Certain kinds of species must be there for the life-community to live, just as certain kinds of organ must be there for the body to live. The life-community must have first its manufacturing side, the green plants; the different members of this department will be specialized to utilize light, air and moisture to the fullest extent, the dominant kinds doing most of the work, but the others subsisting either upon the surplus of raw materials left over by the dominants, or getting their chance during parts of the year when the dominants are inactive. Then when the raw material has been lifted to the organic level, a new series of forms, the animals and parasitic plants, play their part. Though animals do none of the primary production and are not self-supporting, yet in them life attains its most varied forms and its greatest intensity. They are arranged in successive grades. There are first the more basic grades which attack plants or their remains directly. These in their turn give sustenance to others—carnivores, parasites and the scavengers of animal corpses—which radiate out in linked chains from the basic plant-consumers. All these must be present in sufficient quantity and vigour if the life-community as a whole is to carry on its rhythms without catastrophic change.

§ 4. The Growth and Development of Life-communities.

WHETHER by Nature's agency or man's, bare stretches of land, devoid of life, are sometimes produced in the middle of fertile regions. A forest fire leaves nothing but dead trunks and charred soil behind it; on the shores of seas and great lakes, the wind and the waves may pile up great ranges of barren sand-dunes; a landslip or a rock-slide may strip a mountain-side of vegetation; mining or drainage or reclamation projects may leave big patches of untenanted soil.

These stripped areas do not stay bare. Life invades them again, and the invasion is a regular and orderly affair. There is a progression of inhabitants, one set of animals and plants succeeding another in sequence, until finally a stable state is reached. In a state of nature, the animal and plant life of this stable phase is the same as the original life on the area. The life-community has reproduced itself.

This community-reproduction was seen on the grand scale after the great eruption of the volcano Krakatoa, in the East Indies, in 1883. In three terrific August days the island blew itself in half, and threw such vast quantities of fine dust (more than four cubic miles of it) into the air that it floated round the world and for many months coloured European sunsets a richer red. On the island itself, and the two smaller neighbouring islands, Lang and Verlaten, every vestige of life was destroyed. The nearest land from which life could come was another pair of small islands about fifteen miles off, but they themselves had been three-parts devastated by the eruption. Java and Sumatra lay each about ten miles farther distant.

Less than three years after the explosion, a Dutch botanist visited the island. The ashy soil, from sea-level to peak, had been covered with a gelatinous layer of blue-green algæ mixed with diatoms and bacteria, in which a number of mosses and eleven kinds of fern were growing, some of them abundantly. These had all been brought by the wind in the form of tiny spores. In addition, there had appeared a few species of flowering plants, some from windborne seeds, some from seed floated in to shore by ocean currents. The ferns and the flowering plants were all growing as isolated scattered individuals, and there were no shrubs or trees.

Ten years later, a second visit showed the flowering plants in the ascendant: fifty species of them had arrived. In various places the ferns and flowering plants had closed their ranks to cover the soil. Inland were stretches of a regular jungle of tall grasses; and along the shore a characteristic community of straggling beach plants. Shrubs were scattered here and there, and a few rare trees had taken root. But there was no indication of anything that could be called forest.

Finally, in 1906, twenty-three years after the eruption, a party headed by Professor Ernst visited the islands. They were now the home of a rich vegetation. All along the shores a strand-forest had grown up, with several kinds of figs, and abundant coconut-palms, many of them with ripe fruit; and there were patches of a different kind of forest here and there in the ravines of the interior. Over ninety kinds of flowering plants and fifteen of ferns were discovered; and not only were there abundant mosquitoes, ants, wasps, birds and fruit-bats, but lizards had reached the island, including a three-foot monitor.

In spite of its luxuriance, the plant-life had not yet developed into a series of typical communities, each with its own few dominant species. New species were still arriving, and many kinds of plants to be found on neighbouring islands had not yet reached Krakatoa.

Ernst estimated it would take another half-century before the full richness of tropical climax vegetation had reproduced itself on the island; but considering the total annihilation of life, and the island's isolation, the progress made in twenty years is remarkable enough.

The lowest type of plants, wind-borne as minute germs, were needed to prepare the island soil, save round the level shores. This first primitive community paved the way for mosses and ferns. Each new addition helped to stabilize the soil and made it easier for higher plants to germinate; and so herbs and grasses, shrubs, and finally trees, came in in their due order. It would be difficult to find a more striking example of long-range colonization and slow successional development.

After the grand, the little. The same colonization, the same succession, will be seen in a half-pint of boiled hay infusion. First come bacteria, their ubiquitous spores settling into the liquid from the air. Then, once these have turned the organic matter of the hay into food fit for animals, there appear tiny infusorians like Paramecium and finally predaceous protozoa that eat their vegetarian brothers as tigers eat deer. This succession has no permanency about it; it is living on the food-substances put into the water by man, and all teeming life comes to an end in starvation, death and decay unless green plants (in the form of single-celled algæ) manage to gain a foothold. This they can only do at a particular stage: but if they succeed, they pave the way for a new phase of development, with quite different sets of inhabitants, which finally leads up to a balanced, self-supporting community of microscopic plants and animals.

Recolonization like that of Krakatoa is seen whenever a jungle clearing is left to itself, though here the reproduction is partial and regenerative, not complete, since by no means all the normal life of the area is destroyed, and only the later stages of the community's development have to be reformed. In some parts of Burma, the people clear and cultivate patches of forest for a few years, and then, when the soil's fertility is exhausted, move on to new regions. In a brief space, the clearings are swallowed up again, silently and inexorably, by the primeval jungle. Air-photographs show various stages in the process. In one place, a few trees are invading the edges of the clearing like dots of mould on a piece of bread; in another, a clearing is half devoured; in yet another, it is covered entirely, but the forest has not yet assumed its normal character; the full typical rain-forest eventually resumes its sway as if man and his works had never been.

The speed with which development of this sort proceeds, especially in its early stages, is truly remarkable. Let us take a recent example. During the War, the sea was deliberately let in over large stretches of Belgian land near the river Yser (including some of the richest agricultural land in the country), in order to prevent the German advance. By the time peace came, every land-plant had been killed off, and the sea-life had made great progress in taking over the territory thus made available for it. But in 1918 the land was drained again, and the lost district restored to land-life. The wet soil began to dry; meanwhile, salt-marsh plants colonized it and helped to loosen it with their roots and to fertilize it with their dead bodies. Only a year later, most of it was covered with a rich crop of grasses, asters and other plants, and after three years only the skeleton shells of barnacles and mussels here and there on fences and posts made it possible to believe that the whole countryside had so recently been covered by the sea.

The actual steps by which the normal world of life is re-established have been carefully worked out in a great many cases, and we know now almost as much about the details of ecological succession as we do about the development of individual plants or animals. Those who are interested in the subject can pursue it in such books as Tansley's *Types of British Vegetation* or McDougall's *Plant Ecology*. We must confine ourselves to a few illuminating instances.

Here and there on the shores of the Great Lakes of North America, the accumulation of dry sand provides bare areas for life to colonize. How it does so has been carefully studied on the Indiana shores of Lake Michigan. Sand brought by the waves dries in the sun, and its surface layer is blown off by the wind. Great strips of white lifeless sand accumulate, and may be heaped up to form dunes, dry on the surface, but moist below. Only a few exceptional plants can colonize such an area; they must be able to do with a miserably low allowance of the mineral ingredients needed for plant growth; and they must be perennials and able to bind the sand with their roots so as to prevent the dune moving slowly along with the prevailing wind, and overwhelming its plant inhabitants in its march. The sand-grass Ammophila and the wormwood are among the few plants that can manage this. Sometimes the dune grows too fast and runs away from their control, and all the pioneers are overwhelmed. But they may keep their home fixed. Then other grasses and plants like the wild rocket are able to come in and thrive now that the original deficiencies of the sand have been to some extent corrected by manuring with the decayed leaves and roots of the pioneers. After them bushes like junipers and sloes can get a footing.

When dunes get out of hand, their movement gradually slows down as they get farther from the windswept lake shore, and as soon as it drops below a certain speed they are "captured" and immobilized by plants, the capture usually proceeding up from the base of the sheltered lee slopes. The capture of dry dunes is effected by the same sand-grass and wormwood series of plants which we have just mentioned;

but where there are damp depressions in the sea of dunes, the pioneers are rushes, willows, and cottonwoods.

Eventually, however, the dry parts of the dunes grow moister, and the wet depressions grow drier, as plants live and die in them; the pioneer vegetation then gives place to another set of plants—Solomon's seal, horsemint, golden-rod and other familiar flowering plants, with shrubs like dogwood and bearberry, and soon a few pine-trees. The soil is now rich enough for more exigent kinds of trees, often black oaks. New shrubs and herbs follow the trees; the soil grows richer. Red oaks succeed the black oaks and, finally, when a thick layer of humus has been formed over the dune, sugar-maple and beech gradually replace the oak, and persist indefinitely, unless man interferes. And, of course, each stage in plant succession has its own animal inhabitants.

When the first white men came to Indiana they found this beech-maple forest in possession of the country wherever conditions were favourable. It was the fullest final expression of vegetable-life in the region. To-day we see this same life-community reproduce itself, in the space of a few score years, by the conquest of new-formed and barren land.

In this story there are two or three points of special interest. The first and most fundamental is that each life-community is something which develops. What we have been talking about hitherto are for the most part only the final stages of development, called *climax* stages by ecologists, and corresponding closely enough with the adult phase of an individual's life. But for a climax phase, such as a forest or a prairie, to come into being, it must have passed through a whole series of developmental stages. And just as in an individual reproduction must be by simple structures like egg or spore or gemmule, whose development into the adult state is through steps of increasing complication, so the pioneer group of organisms by which alone the life-community can reproduce itself is much simpler than the climax phase. It contains many fewer species; they are never big like trees, and are often, indeed, very small; and they make few demands upon Nature and so can get on in spite of poor soil.

In the developing single organism, each phase is its own executioner, and itself brings a new phase into existence, as when the tadpole grows the thyroid gland which is destined to make the tadpole stage pass away in favour of the miniature frog. And in the developing community of organisms, the same thing happens—each stage alters its own environment, for it changes and almost invariably enriches the soil in which it lives; and thus it eventually brings itself to an end, by making it possible for new kinds of plants with greater demands in the way of mineral salts or other riches of the soil to flourish there. Accordingly bigger and more exigent plants gradually supplant the early pioneers, until a final balance is reached, the ultimate possibility for that climate.

Another point is that whether the sand was colonized when it was too dry or too wet for most plants, the eventual result was the same—beech and maple forest. This is but an example of a general rule—that the general course of community-development makes dry environments moister, and moist environments drier. Even where development starts in the water, it is headed towards forest or whatever the normal climax of the region may be; for water-plants are all the time choking up their watery homes. We have already given an account of the zones of vegetation that characterize a large pond—floating plants, often microscopic, in the centre, then wholly submerged plants like pondweeds, then plants like water-lilies and water-crowfoot, that send their leaves and flowers up to float on the surface, then the plants that, as it were, merely wade or paddle with their feet in the water—bulrushes, arrowheads or pickerel weeds—and often a marshy zone with sedges and irises making transition to the real dry land. The plants grow and die, their remains accumulate, they help to silt up the pond. Thus, the water shallows, and the zones move in towards the centre. Presently the central zone is crowded out altogether, then, in the course of a few years, or scores of years, the next, and so on—unless man interferes—until the whole pond has become dry land, at first wet, then drier and eventually fit for trees to grow on.

We have already seen that the most unpromising dunes of bare, dry sand are made progressively richer and wetter by the accumulation of humus, until they too will grow trees. The same is true even of bare rock. Lichens are here the pioneers—first the ones like thin crusts, and then, when these have disintegrated the surface a little, the leafier ones, together with a few mosses. These plants catch a little dust and débris and so start the rudiments of a soil. Other mosses now come in, and a few grasses; and flowering herbs soon follow. Each addition to the population accelerates the rock's disintegration and helps in forming soil. Bigger herbs like golden-rod and shrubs like blackberries can now invade the place, and eventually the coating of humus is thick enough for tree-seedlings to take root. And so a rich woodland has taken the place of dry, bare rock-surface.

The way in which quite different developments from wet and dry beginnings may converge to produce the same adult phase is well shown in some of the middle-western States, such as Indiana, where all vegetation is destroyed in the process of surface coal-mining. The mining operations convert the area into a series of ridges and furrows, the ridges anything up to thirty or even forty feet high; and these bare "striplands" are usually left to themselves. The bottom of the furrows are wet, and may contain standing water; the tops of the ridges are drier than the neighbouring country. Although the wet furrows begin their development with bulrushes and water-plantain and cocklebur, and the dry ridges with white melilot and asters, sunflowers, and ragweed, both will end up, often in the short space of twenty-five years, in woodland typical for the region, consisting mainly of sycamore, elm, and honey-locust.

Many other examples could be given; but the principles remain the same. Life-communities develop; they develop from slight beginnings, and become progressively richer because each set of invaders paves the way for another, which grow more richly because they make greater demands on the soil. And however they begin, they tend in any one region towards the same stable climax.

This stable climax may not always be reached. It is the potential end which the climate of the region permits to its life-communities. But other factors may step in and prevent the potentiality from being realized. Though a pond will silt up in a comparatively few years, a lake may take immeasurably longer. Steep slopes and unfavourable soils, too, will retard or prevent the appearance of the climax stage, as absence of iodine in water will prevent newt-tadpoles from turning into newts. Perhaps most important of all, animals in general and man in particular may prevent the normal climax from appearing. In a later section we shall see how rabbits may turn woodlands into close-cropped grass-heath. Darwin found that cattle, by eating down the seedlings, prevented the heather commons of southern England from achieving what ought to be their last stage in development and becoming pine-woods. And it seems almost certain that the lovely close turf of the English chalk downs is not Nature's climax, but a substitute, induced by man's clearings, and his sheeps' croppings, for the more natural final phase of beech-forest.

We have spoken mainly of plants in this section. This is because animals usually follow the lead of plants. Plants must be the pioneers in exploiting the environment; and the final climax generally takes its main character from the plants which succeed in becoming dominant. One has only to think of the difference between a pine-wood, a prairie and a peat-bog. The animals of a community are not only dependent upon the plants for food, but their whole existence is modified by the character of the vegetation. It is hardly too much to say that while the effective environment of plants is provided by inorganic nature, the effective environment of animals is provided by plants.

An equatorial rain-forest owes its very existence to the intense light and heavy rain of the tropics. But the animals which live in its shelter receive very little light—less, indeed, than those of the arctic tundra; and though the air is damp, there is often a shortage of liquid water. It carries a special climate within itself. Furthermore, the forest provides a special mechanical environment. It encourages animals which can climb and swing from branch to branch; while the prairie and the pampas encourage burrowers and hoppers.

In the sea, where the main store of plant-life is floating and microscopic, and animals often fixed and plant-like in their growth, this primary importance of trees and vegetation does not hold; in coral reefs, for instance, the main lines of development are based upon the succession of different kinds of corals and the character of the whole community is given by the coral-climax, which provides effective environment not only for other animals, but for many seaweeds, large and small.

The facts of succession help to make clear a point of some general interest. The biologist is often asked why primitive types survive alongside of those that are more advanced, why they did not also evolve. The answer is really clear enough. During the course of evolution the more advanced types came into being in a world prepared by the life that had gone before. But they still need that preparation. They can only be large, strong, efficient, by making great demands upon their environment—the higher plants upon the soil, the higher animals directly or indirectly upon the higher plants. A tree cannot grow on bare rock, nor a sheep subsist on blue-green algæ. Many of the primitive forms of life survive by taking advantage of the less favourable kinds of environment that are always being brought into existence. Their own activity paves the way for their own supersession by higher types; but meanwhile their ubiquitous spores and germs have colonized other poor environments. Life exists by exploiting its environment. Competition forces a division of labour upon it. The environment is most fully exploited when each nook and aspect is worked by its own particular kind of life. And so besides creatures adapted to secure the riches of Nature in the most specialized and efficient ways, we should expect to find types which are adapted to her poverty—to barren and out-of-the-way environments like caves, to unconsidered nooks and crannies in rich environments, to times and seasons when the dominant plants are not at work, and to temporary opportunities in the environment, such as those we have been describing, when the feeble can thrive but the powerful not yet. And on the whole the creatures adapted to these poor holes and corners of space and time are creatures of primitive types.

Another fact of succession has some practical importance. In general, the natural vegetation of a region is the most luxuriant that it it can support. There are exceptions on oceanic islands, because they may never have received the best complement of plants. There are also exceptions due to the fact that Nature has hitherto not produced the most efficient kind of plant for the environment. The Lower Devonian lands (III D 1) had no forests because no trees had yet been evolved. This sort of thing may still happen. Of recent years, a remarkable grass has appeared which can grow on the fresh mud-flats of estuaries and harbours and cover them with rich vegetation. Hitherto such mud-flats have remained bare.

The grass in question is called the rice-grass, *Spartina townsendii*. It appears to be a natural hybrid between a native European species and one brought accidentally to Europe from America about a hundred years ago. However produced, it spread, slowly but steadily, from bay to bay and estuary to estuary, converting mud into meadow. The botanically remarkable fact about it is that it takes the ground it covers to an ecologically higher stage of development. As Professor F. W. Oliver says: "These bottomless muds, though they stood empty of vegetation and invited colonization, probably for thousands of years, found no plant capable of solving the problems of invasion and establishment till *Spartina townsendii* came and made light of the task." And the practically remarkable fact about it is that it provides the only case known of a plant which spreads rapidly and is not a pest, but a benefit, to man—up to the present at least, for it is on the cards that it will eventually begin to choke up harbours.

In the first place, it makes new land; and in the second, it provides a valuable crop. Round its plants the level is raised, and in place of mud, in which men may get bogged and even disappear, firm saltings develop. In place of a bare surface there grow fields of tall green grass, three feet or more in height, eagerly sought after by all stock and capable of being made into excellent hay. It may even prove possible to use it for paper-making.

The English, maritime and with no great land-hunger, do not particularly encourage it; but the Dutch are using it on a large scale for "poldering"—making new land by banking. Single slips are planted in the mud three metres apart. In three years the tufts have covered half the mud, in six years they all coalesce to make one meadow, with a surface about two feet above that of the original flat. *Spartina townsendii* will help Holland to accelerate very considerably her struggle to reclaim land from the sea.

Could man but find a plant which would do for fresh-water bogs what the rice-grass can do for sea-shore muds, enormous areas of land, especially in the tropics, could be quickly and cheaply made fertile. With the rice-grass before our eyes, we can go on experimenting with reasonable hope of success.

The rice-grass thus takes maritime muds to a new stage of development. This is an exceptional type of human interference in a life-community. In general the normal wild state of affairs is the highest, and when man interferes with vegetation he usually keeps it from attaining its natural climax. The major part of temperate Europe and North America should by right of nature be forest; the great stretches of grass and heath that exist there to-day owe their existence to man's activities in tree-felling and cultivation and in encouraging grazing animals. The grassland of the English shires is only waiting to become woodland again, but the grassland of the American prairies is grassland in its own right—the ground is not wet enough to develop naturally into forest. Only when man supplements nature, as when he irrigates the desert, does he carry the climax a stage higher than normal; and even then he may find the task harder than he imagined.

§ 5. The Grading of Life-communities.

IN OUR last section we have traced out the way in which a typical life-community develops from small beginnings up to the climax which is its normal completion. But in different regions the kinds of climax differ. In many parts of the world's land-surface the adult stage, so to speak, towards which all other arrangements of living things are tending, even if they are tending thither so slowly that they never actually reach it, is forest. In other words, the dominant plants of the climax are generally tall trees. Where the climate is not suitable for trees to dominate all smaller plants and form a forest, other types of climax occur: and in this way life is zoned over our planet's surface. The broadest zoning is the zoning by latitude. The normal climax of equatorial life is the typical rain-forest, green all the year round, with a multifariousness of splendid trees instead of one or a few dominant species. In monsoon districts, where rainfall is heavy but intermittent, the forests cease to be evergreen; they drop their leaves in the dry season. As we pass north and south towards the sub-tropics the amount of rain falls off, and the trees, unable to find sufficient moisture if they close their ranks, have to relinquish part of the soil to scrub and grasses. Dominance must be shared between them, and the result is the beautiful climax we call savannah-forest, where clumps of trees are dotted about with broad spaces between, as in a park.

With increase of latitude, we enter the world's two desert belts. We come into a region not only of lesser heat, but, under existing geographical conditions, of desiccation. The trees grow sparser, smaller, thornier, often with tough leaves. The African thorn-scrub is of this zone. As regions of less and less rainfall are approached the vegetation finds it more and more difficult to suck up the moisture it needs, and eventually the small plants, obeying the same necessity which overtook the trees long before, are forced to break their ranks. No longer is the surface of the ground completely covered with vegetation. The life-community, in the language of ecologists, is no longer "closed," but "open"; the lack of moisture has become a limiting factor and prevents life from fully exploiting the other resources of the soil.

The desert often passes over into steppe-deserts, steppes, veldts, and bad-lands, where the rain—small in amount and falling mainly in winter—usually only suffices to support an open community of grasses and shrubby plants like wormwood and sage-brush. As we pass farther polewards we reach, as a general rule, wetter country again, and the open communities of the true steppes give place to the full grassland of the grass plains, the prairies and the pampas, where vegetation once more covers the earth. All gradations occur from the short-grass plains that are only just closed communities to the tall-grass prairies whose plants are bigger and have deep-penetrating root systems.

These grasslands, open and closed, are for the most part found inland. Near the coast in the same latitudes climate is rather different, and generally allows a different climax, a climax of scrub or even forest. The rain falls mainly in the winters, and the plant's growing season is dry. Accordingly the trees of such sub-tropical and warm-temperate forests are adapted to low rainfall by being evergreen and having tough, leathery leaves. The chaparral in Texas, the maquis scrub and the forests of Aleppo pine and other conifers in various parts of the Mediterranean region, the Spanish woods of cork-oak and holm-oak, the forests of southern California and south-west Australia are examples. One of the tragedies of history is the cutting down of these warm-temperate forests over great areas of the Mediterranean basin, as in Dalmatia; green hillsides have given place to naked stony rock, perennial streams have turned into intermittent torrents, and the climate itself has been altered for the worse. In the last few years, however, a determined effort has been made to restore this coast to its pristine wooded state.

North and south of these dry, warm-temperate regions we come generally to a moister warm-temperate climate. Here deciduous trees like beech and oak, maple and ash, locust and chestnut, walnut and sycamore are the usual dominants. Where soil is poorer, however, or rainfall less abundant, evergreen conifers can do better, and they give us pine forests such as those of Scots pine or Austrian pine in Europe, long-leaf and loblolly pine along so much of the south-eastern coastal plain of the United States.

Conifers reappear again polewards of the main temperate zone. They encircle the whole of the northern hemisphere. An almost unbroken belt of them, four hundred to eight hundred miles wide, stretches for over five thousand miles from Scandinavia to the Pacific. In Siberia it is called taiga; but elsewhere, strangely enough, it has earned no special name. And a second huge forest of the same type covers the North American continent from Labrador to Alaska. Sometimes such a forest is all of fir, dense and gloomy; but deciduous trees like birch and alder may break the inhospitable monotony. Firs, however, are the dominants. Yet this does not mean that they find themselves here in the most favourable conditions, for spruce from sub-arctic North America grows much larger and better when transplanted to the milder climate of Scotland. It only means that, in the struggle for existence, they can survive where other trees with greater demands fail. They are not in any perfect adjustment with their not very attractive environment, but they are better adjusted than any other trees.

Towards the north (there is no corresponding belt in the south, owing to the disposition of land and sea) the forest thins out, just as the very different tropical forest thinned out towards the desert region. Trees can no longer grow where the subsoil is frozen all the year round; they dwindle, grow dwarfed and stunted, and eventually disappear, leaving only the arctic barren-grounds, with their permanently frozen subsoil, to which the name tundra is given. Dwarf and creeping willows, grasses, and sedges, are the chief among the tundra's higher plants, though bright flowers are by no means absent, and saxifrages and buttercups, campions and pale yellow arctic poppies and many other lovely plants, often crouched on the soil in cushions and rosettes, star the sunnier places during the short two or three months' summer. Lower types of plants like mosses and lichens are here relatively more abundant than anywhere else in the world, and may actually dominate the tundra's more vigorous regions. The "Reindeer moss," for instance, which is really a lichen, may extend almost unbroken over great regions of the American barren-grounds, often growing six inches or a foot high.

And finally, before life is made wholly impossible by perennial snow-fields and ice-caps, cold does what drought did in the desert belt: it breaks the closed ranks of the plants. When conditions become too extreme it is only here and there that plants can grow at all—where a slope faces south, where a patch of good soil has managed to accumulate, where snow is off the ground long enough for plant-life to run a segment of its course. Between the tundra and the ice-fields in Greenland and other arctic regions there may be such a zone, with a few poor plants in a desert of stones; and the antarctic continent nowhere knows any richer vegetation than this.

We have so far spoken of the zones of life on land alone. This is not because the grading of life is less important in the sea, but because we know less about it. We know that floating life is in general less abundant in warm waters than in cold, owing largely to high temperature reducing the amount of carbon dioxide in solution in the surface waters. But recently, thanks largely to the invention by Professor A. C. Hardy of an automatic plankton-recorder—an instrument which can be towed through the sea behind a ship, and which traps and preserves all the little floating organisms in its path and rolls them up neatly in order on a scroll of gauze for the investigator to examine at his leisure—we are making a beginning with the quantitative mapping of the sea's vegetation. A century hence the maps of the world will in all probability have diatom-belts marked on all the seas as they now have vegetation-belts marked on the lands; and a knowledge of these will be of great help to fishermen, and to whalers—if human improvidence and the absence of a cosmopolitan control of the sea has by that time left any whales.

This is grading on the grand or planetary scale. But there are other kinds of gradients. Altitude, for instance, proves a very good substitute for latitude. We can start at the equator in the Congo rain-forest, and by the vertical ascent of less than three miles to the peak of Ruwenzori, reach the same lifeless cap of snow and ice that we should have had to travel more than five thousand horizontal miles northward to find in the mountains of Spitsbergen. And in our ascent we should have passed through belts of vegetation almost as diverse as in that long, horizontal poleward stretch. Above the tropical forest, park-like savannah-forest, with richer, greener vegetation in the stream-gorges; and here and there bare and grassy stretches, treeless. Then, as the cool and the mountain rains begin, a queer forest of tree-heathers, all festooned with hanging moss. Above that, again, a still queerer forest of groundsels and lobelias grown to the estate of trees, looking not like any familiar terrestrial trees, but trees produced by some other planet. This gives place to mountain meadow, and this gradually to a true alpine flora, nestling in isolated tufts among the rocks. And above this, rocks and ice and snow, untouched by life.

Gradients in moisture may occur within a single climatic region, according to the lie of the land. There is such a gradient from every marsh or pond to the solid land around it, in every valley from the moist bottom up to the drier slopes. In the margins of salt lakes there is a gradient in salt-content of the soil, just as there is in salt-content of water between sea, estuary, and river, or, in some inland seas like the Baltic, between one end of the sea and the other. There is a gradient in the amount of sunlight as we pass round the shoulder of a hill from south to north aspect, or as we pass down from shore-level to deeper water in lakes and oceans. And there are other gradients in the environment—gradients in amount of oxygen dissolved in water, in acidity or alkalinity of soil, and so forth.

Each and all of these minor graded changes in the environment are reflected in its living inhabitants, just as definitely as the vaster change between equator and pole. And the change is always reflected in the same way—by a gradation of life-communities. Round the shores of the Great Salt Lake, for instance, there is first a barren flat, too salt for any plant; then a zone of one species of glasswort, capable of tolerating two and a half per cent of salt in the soil, then of another, more luxuriant species. Two or three further zones of shrubbier plants follow, until at a sufficient distance from the lake for the salt-content to have fallen to about one-tenth of one per cent, sage-brush becomes luxuriant. The most familiar examples of such gradations are that between water and dry land on the seashore or round the margins of a shallow lake or pond.

In every case, whether each grade extends for several hundred miles, like the prairie or for a few feet, like the belt of bulrush round a pond, a striking fact is to be noticed. However gradually and continuously the environment may change, the change in life is always more or less abrupt. Each belt of life is on the whole uniform, often remarkably uniform for most of its extent, and then in a short distance it makes a sudden transition to another and often markedly different belt. The cause of this is competition. If one kind of plant can do better than others in reaching up to the light, it will succeed, and the others will fail. This is why in each community there are one or a few dominant kinds of plants. Success in this kind of competition leads to dominance; but failure means not mere subordination, but complete or almost complete banishment. As long as the dominant species can hold their own, they can dominate. But when conditions change a shade further, they must give place to other and altogether different kinds of dominants. That is why there is generally such an abrupt change from woodland to prairie, or from one kind of seaweed to another on the sea-shore, why marsh-plants give place so suddenly to the bulrush zone round a pond, and why the timber-line is so sharply marked on a mountain. And in this way, since animals follow the vegetation, the rivalries of plants translate the sloping gradient of the lifeless environment into a staircase of life-communities.

Thus, with sufficient knowledge and patience, we could make a map of the whole world showing the distribution of life-communities as it was at a particular instant. But, as we were at pains to point out in the last section, it is not enough to think of life-communities in this fixed and static way, for they are continually changing. For one thing, they themselves are often reacting on their own environment, and so bringing about that community-development we call succession, up to the climax which is in equilibrium with itself. And for another, the environment itself is changing. Sea and land shift their boundaries; erosion levels mountains and builds plains; the belts of temperature move slowly up and down over the earth's surface as a glacial period or a dry spell comes and goes; and with such changes the life-communities must shift their boundaries too.

As Elton says: "If it were possible for an ecologist to go up in a balloon, and stay there for several hundred years quietly observing the countryside below him, he would no doubt notice a number of curious things before he died, but above all he would notice that the life-communities appeared to be moving about slowly and deliberately in different directions. The plants round the edges of ponds would be seen marching inwards towards the centre until no trace was left of what had once been pieces of standing water in a field. Woods might be seen advancing over grassland or heaths, always preceded by a vanguard of shrubs and smaller trees, or in other places they might be retreating; and he might see even from that height a faint brown scar marking the warren inhabited by the rabbits which were bringing this about."

If he stayed up long enough and reflected sufficiently hard on what he saw, he would begin to draw some interesting conclusions. He would realize, for one thing, that the transition from one life-community to another in space usually corresponded to an actual replacement of one by the other in time. The mapping of the gradation of life-communities gives a spatial picture of succession. That is clear enough with the belts of vegetation round a pond; they move inwards, each one replacing its more central neighbour, until all but the outermost, now spread over the whole area, have disappeared. And he would realize that there is no essential difference whatsoever between the narrow rings of life round a pond and the broad rings of life round the globe. They both depend on gradients in outer conditions. The sole difference is this. The one is not merely a gradient on a small scale, but one which natural forces, both those of life and those of lifeless matter, tend to roll out flat; the other, however, is not merely on a large scale, but is determined by agencies outside the scope of any changes in itself—by the very shape of the world. The one is essentially temporary, the other essentially permanent. Even so, the succession of stages round the pond may be interrupted and set back, as when a succession of wet years floods the margins and reverses the normal development. And if you like, you can think of equatorial forest as the climax community of the planet as a whole, towards which the life of all other regions is disposed to tend, though checked in its approach by limitations of warmth, of water, or of soil.

If our ecological observer could have stayed aloft during a geological period or two at the beginning of the Mesozoic Era, from the cold dry beginnings in the early Trias (IV A 1) to the warm moist uniformity of the late Jurassic (IV B 3), he would have witnessed the world's life-zones narrowing inwards round the poles, just as the bordering plants round the pond encroach upon the water; and he might have been pardoned if he had thought that the process would continue until the whole world, poles and all, was one climax forest.

However, there must always remain a sufficient difference between pole and equator to keep life zoned by latitude; and there must always remain land, high and low, and sea, shallow and deep, to maintain the complex set of gradients which arranges life in a series of belts, from mountain to plain, across the sharp transition of the shore and down more slowly again to the abyss. These two gradients are permanent features of our world; all the rest are temporary, and tend sooner or later to obliterate themselves.

There is another way in which the little mirrors the big. The same competition which results in the comparatively speedy development of ecological succession results also in the portentous slow development of evolutionary succession. A landslip or man's destructive hand uncovers a patch of the bare earth, or impounds a body of barren fluid; it is colonized by a succession of communities, and in a few decades is tenanted with rich life again. The whole world, both land and sea, was once free of life; and æons later all the land was still one great bare patch of earth and rock. First the seas, and then the lands were colonized. In both there has been a succession of faunas and floras, each one on the whole exploiting the environment a little more effectively than the one before. Evolution is a slow succession of a series of ever new and ever improved communities towards a still unrealized climax. The most up-to-date of life's existing communities are still very wasteful in their exploitation of the world's resources. It remains to be seen whether man, with his deliberate aim at a higher efficiency, his replacement of the hitherto dominant tree by his own cultivations and devices, will make a mess of things and fail, or will succeed and hold on from climax to climax. If he fails the forest will return.

§ 6. Food-chains and Parasite-chains.

Now that we have discussed the development and distribution of life-communities, we can return to the details of their interplay. Green plants draw their supplies from lifeless and universal sources; animals must live on other life. So it comes about that different kinds of animals will be at different removes from the prime source of food; and one of the characteristics of the animal part of every community will be its organization in the form of subsistence-chains. A subsistence-chain is a series of creatures, each living on its predecessor in the series. There are two main kinds of them, food-chains in the ordinary sense, in which the predecessor is devoured, and parasite-chains; and of course there may be subsistence-chains of mixed type, of which both devourers and parasites are members. The general rule is for the members of food-chains to get bigger and bigger as they get farther from the chain's original starting-point, while in parasite-chains we find the opposite tendency, each link is likely to be smaller than the one before.

The starting-point of a food-chain is normally among green plants, but from one and the same starting-point many food-chains may radiate out in different directions. In an English wood, for instance, plant-lice suck the juices of the twigs; these, either before or after having fallen a prey to spiders, are eaten by small birds like tits and warblers, and these in their turn by hawks. The same trees may contribute to the hawk's upkeep in another way. They drop their leaves upon the ground, the leaves are eaten by earthworms, the earthworms by blackbirds scratching among the underbrush, and the blackbirds by hawks. A different line is started through the seeds of the trees. Acorns and beech-nuts are nibbled by mice, and the mice are the chief support of tawny owls; and there are other chains running through woody branches or stem, to wood-boring insects and woodpeckers, through seeds and squirrels, through leaves and caterpillars or gall-insects, and so forth.

In the sea, where the single-celled diatoms are the main food-producers, the first link in the animal food-chain is invariably supplied by tiny creatures, very largely the little crustacea we call copepods, but also the microscopic larvæ of many larger animals like crabs and sea-snails and starfish. These are generally the prey of little jelly-fish, arrow-worms and small carnivorous fish, which in their turn usually fall victim to bigger fish.

A carnivore can only cope with prey within certain limits of size. Animals above a certain upper limit it is not strong enough to tackle; the most powerful spider cannot kill rabbits. Animals below a certain lower limit it cannot economically make a living off: a lion could not live, like a cat, by catching mice, nor an ostrich off insects as small as those that satisfy a tom-tit or a swallow. It is for this simple reason that each link in a food-chain is usually bigger, but not enormously bigger, than the one before. There are exceptions. The food-chain from the diatoms of arctic seas which passes through tiny crustacea, free-swimming pteropod snails, and fish may end in the body of a gull. But it may have a further link tacked on to it; this is the skua or jaeger which, though actually lighter than the gull, pursues and terrorizes it until by mere bluff and pertinacity it forces its victim to disgorge its last meal.

The skua, however (save for some depredations on eggs and young birds), is not strictly a carnivore though it is a link in the food-chain. But stoats feed largely on rabbits, which are both much heavier and much speedier than they. This they do by paralyzing them with fear, in some way which is as yet not fully explained; it seems, however, that it is the smell of the stoat which has this extraordinary effect upon the rabbit. Other exceptions are apparent only. Wolves are less than a quarter of the weight of the deer they kill; but they hunt in packs, and the pack is the unit that counts. The extremest case of numbers thus making up for size is found in the Driver ants; it needs hundreds of them to weigh an ounce; but their columns number millions, and they will eat anything, however much bigger than themselves, that cannot run away, from puppies and babies to tethered cattle.

The best example of an animal feeding on creatures vastly smaller

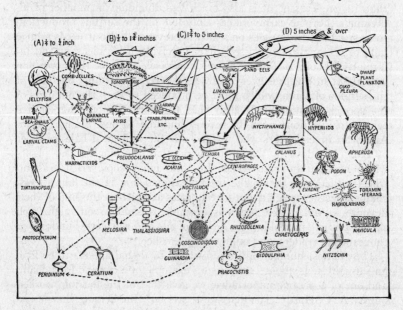

FIG. 262. A picture of the food-relations of the herring at different stages of its life-history.

The solid lines point to food eaten directly by herrings; other links in the food-chains are dotted. (A) The young herring feeds mainly on single-celled plants, nauplii and mollusc larvæ, and small copepod crustaceans. It is eaten by jelly-fish, comb-jellies, arrow-worms and bristle-worms (Tomopteris). (B) During its next stage it lives almost entirely on small copepods and no longer takes any plants or small larvæ. It is still eaten by jelly-fish and comb-jellies, but not by worms. (C) From 1¾ to 5 inches, its diet is exclusively crustacean but more varied—several kinds of copepods and large crustacean larvæ, and Mysis. (D) From 5 inches and upwards it eats copepods, large crustaceans, the tunicate Oikopleura, pelagic snails, and large numbers of sand-eels. The animals it eats support themselves, either directly or at one remove on various single-celled organisms, mostly diatoms (Melosira, Thalassiosira, Guinardia, Chætoceras, Nitzschia, Navicula, Coscinodiscus, Rhizosolenia, Biddulphia), flagellates (Prorocentrum, Peridinium, Ceratium), and algæ (Phæocystis). (Drawn from the data of Prof. A. C. Hardy.)

than itself is the whalebone whale. It has skipped several links in the chain by means of its special sifting device; but even so its usual prey is about an inch long, and the first links escape it. All big current-feeders have similar devices for accumulating particles which would be no use to ordinary eaters of the same size; the outstanding case is Tridacna, the giant clam, which in spite of living on microscopic débris, may grow a shell five feet across.

Not only is each animal link in a food-chain generally bigger in bulk than its predecessor; it is also much less abundant in number of individuals. For the carnivore can only live on the surplus production, so to speak, of the species on which it preys; and also it needs a huge bulk of material to keep itself alive. Percival, in his pleasant and informative book, *A Game-Ranger's Notebook*, tells us that one lion kills about fifty zebras every year. As this represents only one part of the surplus of zebras which must be produced to keep the numbers of the species constant, it is clear that the normal proportion of live zebras to live lions must be several hundred to one. Similarly, two American investigators who worked out the biological balance-sheet of Lake Mendota in Wisconsin found that the single-celled plants which nourish the many-celled but still microscopic crustacea and wheel-animalcules that make the first animal link in the lake's food-chains, together weigh about fifteen times as much as the sum of their devourers. And in our previous example, a small wood might well shelter but one pair of hawks, dozens of tits and warblers, hundreds or thousands of spiders, and millions of plant-lice.

The greatest number of links in such a food-chain seems to be five, and usually it is about three. As an example of a marine food-chain, and one of economic importance, we give that of the herring, worked out by A. C. Hardy. The diagram shows how the herring changes its food as it grows bigger. But it also shows how the starting-point always consists of single-celled plants, and the first animal-link almost wholly of one or another kind of copepod. And finally it shows what our readers may have been suspecting, that the various food-chains in a community need by no means be separate, but are linked up together in an interlacing meshwork like the chains in a coat of chain-mail.

To work out all this web of interrelations in detail for a whole community is all but impossible even in our temperate regions—let alone in the richer tropics—on account of the hundreds of kinds of animals and plants involved. To achieve anything like such completeness, the ecologist must turn to unkinder zones, such as the arctic, where the numbers of different species are so few that the food-cycle is reduced to a diagrammatic skeleton. Even a skeleton, however, still has a certain intricacy, as we may see by a glance at Fig. 262.

One or two interesting points in the integration of Spitsbergen life may be noted. There are no herbivorous mammals, but several herbivorous birds. Otherwise fresh plants are eaten by midges and sawflies, while the place of earthworms as the first animal link from dead plant-tissue is taken by the tiny wingless insects called Collembola, together with mites. All these tiny arthropods, before or after going through a spider link, are eaten by birds, and the birds are eaten by arctic foxes. This store of fox-food, however, does not seem sufficient, and the hard-pressed carnivore is driven to supplement it by living at the expense of polar bears, either eating their dung, or scavenging the remains of the seals they kill.

Another point is that the land is continually being enriched at the expense of the ocean. All the sea-birds—gulls, auks, guillemots, and ducks—get some or all of their food from the surface stores of the arctic sea. Their dung, rich in nitrogen thus extracted from the ocean, manures the ground and helps plant-growth. Below one of the great bird-cliffs where (for protection from foxes) the nesting sea-birds congregate in thousands, plants that elsewhere grow as miserable stunted things, one or two inches high, shoot up to a foot, the whole aspect of the vegetation is changed, and one may fancy oneself back in a temperate country.

Parasite-chains usually link secondarily on to some animal in a food-chain, though there are abundant examples which take origin directly in plants, such as the insects which cause and inhabit oak-apples, robins' pincushions and other plant-galls, or the trypanosomes—single-celled protozoa rather like the ones that cause African sleeping-sickness—that live in the milky juices of plants such as spurges.

As a matter of fact, the majority of parasites are not linked up ecologically into chains at all; or, if we like to put it so, most parasite-chains have but one link. There are no parasites living on tapeworms or on malaria germs. But sometimes there are parasites of parasites; and then the parasite-chain is a reality.

In such cases the size of each link diminishes rapidly instead of increasing, as in the ordinary food-chain. Everyone knows Swift's lines:

> "So, naturalists observe, a flea
> Has smaller fleas that on him prey;
> And these have smaller still to bite 'em;
> And so proceed *ad infinitum*."

This is, however, an obvious impossibility, since a few links would bring us down to a size smaller than a single molecule; and as a matter of fact, there seem never to be more than three or four links in such dwindling chains.

A two-link chain of sinister importance begins with the rat-flea and ends in the parasite which it harbours and may transmit to man—the bacillus of bubonic plague. In the same way parasite-ticks have as secondary parasites the spirillum of relapsing fever. The secondary parasites may of course be perfectly harmless; the intestines of many kinds of fleas, for instance, teem with perfectly innocuous single-celled flagellates called Leptomonas. In all these cases, numbers go the opposite way to size, as they did in food-chains. A common squirrel is quite likely to support hundreds of fleas, and each flea to shelter and nourish thousands of Leptomonas, though of course the total bulk of flea is only a tiny fraction of the bulk of squirrel, and the total bulk of flagellate again only a fraction of the bulk of flea.

Some of the most remarkable examples of parasite-chains are afforded by the parasitic hymenoptera, insects which, as we saw in our section on parasitism, can equally well be styled internal carnivores, their eggs, laid in the grubs or even the eggs of other insects, hatching out to maggots which consume their prey from within. Many kinds of these parasitic hymenoptera are themselves victimized by secondary parasites (sometimes styled hyper-parasites) belonging to the same group, and of the same unpleasing habits; and these in their turn may sometimes afford a livelihood to tertiary parasites of the same kind. Since each link may nourish parasitic protozoa in its intestines, the protozoa of the tertiary parasites make a fourth link; they are quaternary parasites. The insects which are tertiary parasites are all fabulously small.

Parasites and food-chains are sometimes tangled together in an interesting way. When one link in a food-chain is always being eaten by the carnivorous next link, any parasites that happen to be aboard are generally eaten, too; and so it is very frequent for parasites of a carnivore to become adapted to pass part of their life-cycle in its prey, so as to make certain of reaching the carnivore's interior again when the time comes. So the tapeworm of the dog and fox has the rabbit for its secondary host, the trypanosomes parasitic in tsetse flies have become adapted, unfortunately for us, to living out part of their cycle in the blood of men and other large vertebrates, while Aggregata, a common protozoan parasite of squids and octopuses, divides its time between them and their favourite prey, crabs.

Sometimes there are three stations on the journey. The enormous broad tapeworm begins life by infecting a tiny fresh-water crustacean, passes on to a second stage inside a fresh-water fish when this eats the crustacean, and so on to a mammal like an otter, when it eats the fish. But there may be other effects. The food-chain is like a railway with one-way traffic; and parasites will always be moving down the line. Squirrel-fleas often hop off on to the squirrel's enemy, the pine-marten, and have been known to become more or less acclimatized to life in this new environment; most of them, however, seem to die here. Sometimes this transference has unfortunate results for mankind. When human beings take a leaf out of the otter's book and eat raw fresh-water fish (smoked instead of cooked) they may receive a consignment of broad tapeworms. A knowledge of this one-way food-traffic along food-chains may thus be of considerable service in narrowing down the field of inquiry when tracing out the missing parts of a parasite's life-history.

It is rather rare for parasites to be important independent links in a food-chain: they are for the most part mere extras on the menu. But there are one or two rather interesting cases where they are eaten for their own sake. The most familiar example is that of the ticks of so many herbivorous mammals, which are eagerly devoured by birds. They may be merely a casual titbit, like sheep-ticks in the varied diet of starlings. Or they may be a staple, or even the only, article of diet: the African tick-bird seems to live solely upon the ticks of big game such as rhinoceros, zebra, and antelopes. One of the oddest cases is that old one, recorded by Herodotus and long dismissed as a traveller's tale, until re-established by the evidence of many nineteenth-century naturalists, of the little plover of the Nile that enters the mouths of crocodiles (held gaping wide to facilitate the bird's task) and picks off the leeches that suck blood from their gums. In the Galapagos Islands a similar rôle is played by a scarlet land-crab, which picks ticks off the big lizards that feed in the surf and come ashore to sun themselves. And there are the white Paddie-birds that live in the antarctic and in

Fig. 263. Primitive man observes a parasite chain.

A carving upon basaltic rock from South Africa, reputed to be between 25,000 and 50,000 years old. It represents a white rhinoceros with tick-birds in attendance, even as now.

some places subsist mainly on the roundworms which they pick out in huge numbers from the droppings of nesting penguins. In winter, the penguins move off to the open sea, and the Paddies grow progressively thinner, until the return of the migrants provides a new supply of worms.

§ 7. Storms of Breeding and Death.

IN THE last section, we pointed out that the abundance of different kinds of animals had a definite relation to their station in life, and their position in a food-chain. In general, the first animal links in such a chain are small in size and abundant in numbers, and each further link is marked by an increase in bulk and a marked reduction in abundance. But we said nothing as to the way in which this regulated equilibrium of numbers was achieved. It will be our business in the present section to give some account of the checks and counter-checks by which the swaying balance is kept within limits and of the consequences, often spectacular and sometimes disastrous, which follow when it is upset.

The first thing to realize is that the idea of a balance, albeit always a swaying balance, is a true one. The numbers of any species depend, on the one hand upon its rate of reproduction and growth, and on the other upon its death-rate from accident, enemies, and disease, just as the amount of moisture in the air is a nicely adjusted balance between the number of water-molecules that leave the liquid state every minute in vapour form, and the number that condense again in water.

Were it not for these two opposing forces at work, multiplicative and destructive, life's power of increase would be overwhelming. Before game in Africa was much interfered with by man (indeed up to less than thirty years ago), settlers in South Africa used periodically to be witnesses of the results of over-multiplication of that little antelope, the springbok. Trekking from the north, the springbok used to pass for days, and several hundred thousand might be in sight at the same moment. One migrating horde was estimated to be fifteen miles wide and a hundred and forty miles long. As a result of the innocent and unrestrained play of their natural instincts they were trekking to misery and death.

In tropical and semi-tropical regions the red single-celled plants called Peridinians occasionally multiply so as to turn the sea to the semblance of blood for miles, and may even be so abundant as to remove most of the available oxygen from the water, thus causing the death of thousands of fish. A few bacteria introduced into the body may in a ten-days' space have multiplied to a population more numerous than all the men and women in the world. A few prickly-pears introduced into eastern Australia as a botanical curiosity (and for a time propagated and spread by a kindly Society who thought that cactuses in pots might brighten the homes of immigrants' wives) covered thousands of square miles in the course of a few years. At the height of its multiplication the prickly-pear was invading a new acre of Australian land every minute of the day, until, as Dr. Tillyard says: "the vision arose of eastern Australia becoming in about a hundred years' time a vast desert of prickly-pear, with a few walled cities alone holding out against it."

But animals and plants very rarely find a chance of multiplying like this. For most species, the two great checks on increase are enemies and disease: enemies are generally the first line, so to speak. Epidemic disease rarely steps in unless the species has already multiplied abnormally. It is the outpacing of enemy checks which accounts for the extraordinary plagues of herbivores in different parts of the world. Let us give an example or so of the pitch that this abnormal increase may reach.

In a mouse plague which occurred in Nevada in 1907, three-quarters of the alfalfa acreage of the State was destroyed. The whole ground, for square mile after square mile, was riddled with mouse-holes till it was like a sieve. It was estimated that the several thousand mouse-eating birds and mammals busily gorging on mice in the affected district were killing over a million mice a month; and yet in spite of this toll, the numbers of the mice continued to increase. And in the Australian mouse plague in 1917, Hinton, in his booklet on *Rats and Mice as Enemies of Mankind*, records that seventy thousand mice were destroyed in one afternoon in one farmyard.

Such a state of affairs cannot continue for long. The multiplying species has escaped from the control of its carnivore enemies; but it must eventually run up against other controls. In the last resort, there is the control exerted by its food: if its multiplication is too excessive, there will not be enough for it to eat. But this control by starvation is a rare event in nature; in most cases, before the increase in numbers has brought the species within sight of food-shortage, a third kind of control steps in—control by disease. Almost all the vast outbursts of rodents end in appalling epidemics, which kill off the great majority of the teeming animals and leave the population far below its average abundance. We may quote Soper, a Canadian observer, who is describing the sequel to a great over-multiplication of snow-shoe rabbits: "Empire among the rabbits as elsewhere has its rise and fall, and then is swept away. A strange peril stalks through the woods; the year of death arrives. An odd rabbit drops off here and there, then twos and threes, then whole companies die, until the appalling destruction reduces the woods to desolation. One year (1917) in the district of Sudbury, northern Ontario, the signs of rabbits were everywhere, but not a single rabbit could I start. It seemed incredible. Local inquiries disclosed that a little over a year before the rabbit population was beyond count. Now, as if by magic, they were gone. Needless to say, however, a few individuals survive the epidemic. These now, because of their paucity, are seldom encountered."

Another reason why over-multiplication so rarely leads to starvation is that the first pinch of food-pressure is often the signal for great migrations, the animals crowding away from the area of shortage in search of new supplies. Locusts are the classical examples of this behaviour. Their inclusion among the Plagues of Egypt is proof of the impression their appalling visitations made in earlier ages: "The Lord brought an east wind upon the land all that day, and all that night; and when it was morning, the east wind brought the locusts. And the locusts went up over all the land of Egypt. . . . They covered the face of the whole earth, so that the land was darkened; and they did eat every herb of the land, and all the fruit of the trees, and there remained not any green thing in the trees, or in the herbs of the field, through all the land of Egypt."

To-day their visitations continue unabated in spite of all our civilization. Palestine was recently invaded by crawling hordes of the wingless immature form; in 1925 a plague of locusts threatened Egypt again, but prompt action by entomologists suppressed it; in Algeria and Persia, in South America and South Africa and Russia, serious plagues of them recur every few years. In February, 1929, it was announced that Kenya had had to institute a food-rationing system, so formidable have been the inroads of a sudden invasion of winged locusts.

Uvarov has recently discovered a number of interesting facts about the life-cycle of the East European locust. Its main breeding-grounds are in the huge, reedy deltas of the rivers that drain into the Caspian and Aral Seas. Bands of the immature and still pedestrian hoppers leave these swamps nearly every year, sometimes in great numbers. But it is only periodically, and, it would appear, only after a succession of dry years, that the hordes of adult winged locusts set out. These fly off in all directions; and when their reserves of fat are nearly exhausted, they settle down and fly no more, but lay their eggs wherever they chance to be. If this should be in the middle of crops, immense damage may be done by their offspring. In 1926, for instance, no less than 80,000 acres of wheat, maize, and millet were thus utterly devoured in northern Caucasus alone.

Then comes a strange fact. If the migrating swarm has chanced upon a reed-bed like its own original home in which to lay its eggs, its young develop into locusts of the same type as their parents; but elsewhere most of the young grow up into another type of locust, originally considered a different species. This type is not gregarious, and spreads slowly and individually over the countryside. If it or its young finds a reed-bed, the migratory type is once more produced. Thus from the permanent foci in the big deltas, the species is being disseminated by the armies of hoppers, the periodic winged hordes, and the slow and individualistic spread of the solitary form. And this existence of two forms, one solitary and the other gregarious, has been since shown to hold in several other kinds of locusts.

The ideal method for ridding the world of locusts will be to destroy their breeding-grounds. Failing this, we must learn to understand and foretell their cycles of abundance and scotch the beasts when they first appear, instead of waiting until their abundance has grown really formidable. In any case, their wide powers of dispersal and the irregular direction of their flights makes the locust problem eminently one for cosmopolitan control.

Similar outbursts of unbridled reproduction happen with lemmings, the little rat-like creatures that inhabit the moors of the Scandinavian mountains and the lower-lying tundras farther north. Periodically the lemmings, enormously multiplied, invade the lowlands, their huge migrating swarms moving mainly by night. So surprising are these sudden hordes, appearing as if from nowhere that Olaus Magnus, writing in the sixteenth century, was convinced that they fell from the clouds. They climb walls and swim rivers, losing many of their number every day. Finally the survivors reach the sea: apparently they take it for another river to be crossed, for they plump in and swim on until they drown. Collett records one case in which a ship steamed for a quarter of an hour through miles of swimming lemmings, and they have been discovered in the stomachs of cod. After such drownings in mass, the winds and currents may heap their bodies in thick drifts on the shore. In any case, of those that leave the mountains, not one returns alive.

In migrating locust swarms there seems to be no disease. But with lemmings, migration and disease go hand in hand. Not all the lemmings leave their homes. Of those that stay behind, the great majority sicken and die, and even of the migrating animals enormous numbers succumb to the epidemic. In some parts of the world squirrels show similar cycles, which terminate in a combination of epidemic and migration. A huge army of migrating grey squirrels swam the Ohio River in 1829; and even bigger hordes are recorded from Russia.

Violent epidemic disease seems to be the natural and inevitable result of overcrowding. Professor Topley, of Manchester, has demonstrated this experimentally in artificial mouse-populations which he has kept at different degrees of crowdedness; and the fact is a matter of common medical and veterinary observation.

This seems to be mainly a mere matter of chance and time. As animals are crowded together the chances of infection passing from one to another are increased, until, when a certain density of population is reached, the disease, hitherto a smouldering and sporadic thing, becomes a fulminating epidemic—spreading with maximum rapidity throughout the entire population.

Conversely, if the density of its victims falls too low, an infectious disease may die out. Malaria, as we shall see in Book 7, can only perpetuate itself by travelling to and fro in regular alternation between the digestive tube of mosquitoes and the blood of men. Sir Ronald Ross has demonstrated that if the population either of mosquitoes or of men falls below a certain density in a given area, the proportion of malaria-infected individuals will decrease, slowly but progressively, to nil. In a not dissimilar way, gun-cotton will burn harmlessly in the air, and remains unchanged altogether when left to itself at ordinary temperatures, when its molecules are relatively calm; but when detonated in a closed space, the violent movement of each molecule reinforces that of every other, and a formidable explosion is the result.

The enhanced rapidity of infection comes not only with artificial crowding, as on overstocked grouse-moors, or in densely packed human cities which have not yet learnt sanitary precautions, but in wild nature too. The abundances of rabbits and muskrats and gerbils and lemmings, even of deer and zebras, the mouse-plagues that the efforts of owls and hawks and men can scarcely palliate, are terminated in a month or so by pestilence; and the few survivors begin the cycle over again.

The next point is naturally to ask what is the cause of the occasional burst of increase? Here the statistics of trade first put science on to the right track. For over a century, the Hudson's Bay Company have kept records of the number of pelts and skins of different kinds of animals brought in to their posts. When these figures are plotted on a curve, they reveal a strange regularity of fluctuation. For almost every species, periods of great scarcity alternate with waves of great abundance; and the peaks of the waves succeed each other in a regular cycle.

The number of lynx skins brought in every year fell below 5,000 (and sometimes below 1,000) nine times between 1830 and 1914; and in the same period rose above 30,000 (and sometimes above 60,000) the same number of times. The oscillations of snow-shoe rabbits are precisely similar, but even more remarkable, since this species is more subject to disease; in some years epidemics may damp their numbers down to such an extent that only a few dozen skins are brought in. The two curves run parallel with each other, the peaks for the lynx tending to lag behind those for the rabbits. This is what we should expect, since lynxes feed mainly on rabbits.

The increase from the lean years to the crowded years is not a uniform progress, but an acceleration. In the years of great abundance the rabbits will have two or three broods, with eight or ten young in each brood, while in bad years there will be but one brood, with only two or three young to it. The rate of increase itself is thus almost twenty times as great in the favourable years. The numbers of young seem to increase with some regularity, for the Indian trappers are

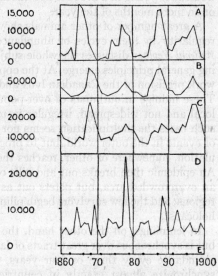

FIG. 264. The periodic ups and downs of northern mammals.

(A), (B), (C), (D) show the number of skins brought in to the Hudson's Bay Company from 1860 to 1900. The snow-shoe rabbit (A) has regular peaks about every eleven years; the lynx (B) is similar, but the peaks are a year or two later; the red fox (C) shows similar main peaks, with irregularities due to minor oscillations about every three years; the arctic fox (D) shows the three-year oscillations only. (E) Years of lemming migrations in Southern Norway. The abundance comes about every three years. (Modified from "Conservation of the Wild Life of Canada," by G. Hewitt; and "Animal Ecology," by G. Elton.)

said to prophesy the prospects of next season's rabbit crop by counting the number of embryos in this season's rabbits. The same thing, with a difference, occurs in field-mice. In favourable years, though the number of young in a brood is not increased, the breeding of mice goes on in more months of the year.

A great number of other animals show a greater or lesser degree of regularity in their cycles of abundance and scarcity. Elton in his *Animal Ecology* discusses the whole subject, and makes some interesting general principles emerge. At the opposite pole to the almost clockwork precision of the Canadian lynx and rabbit we have French mice. These indulge in outbursts of over-population, but the outbursts are local and not widespread, irregular instead of regularly recurrent. In such cases the multiplication seems not to be regulated by any cycle of events in the outer world, but to progress irregularly until the population, somewhere or other, reaches the saturation-point for disease. An epidemic then breaks out and kills off the majority of the mice in an overcrowded area, but peters out as it spreads into less populous regions; and the few survivors begin piling up numbers again for a later holocaust.

In lemmings, on the other hand, the variation is not only regular but is synchronous over great tracts of land. Lemmings have a peak of abundance every three or four years, and the years of abundance synchronize almost exactly in countries as thoroughly separate as Norway, Greenland, the North-Canadian mainland and the islands of the arctic archipelago. It is as if they were keeping time to the beating of some cosmic pendulum. And once the time is set for them, they pass it on to the arctic fox, whose staple food they are. Regularly, every three or four years, the number of arctic fox skins brought in by the Hudson's Bay Company trappers falls to 3,000 or under, while in the peak years in between, the number as regularly rises, usually to 10,000 or over.

British mice are rather more regular in their cycles than their French relatives; and they, like lemmings, have cycles of three or four years. The snow-shoe rabbit and the lynx have an even more regular but a longer cycle, with peaks and depressions about every ten years. And the red fox, which is bigger and lives farther south than his arctic cousin, lives partly upon rabbits but partly upon mice. Accordingly, his cycle is a double one, with main peaks corresponding to those of the rabbit, and minor ones superposed, corresponding to the ups and downs of mice.

Something outside the animals' own lives is imposing this regularity upon them; and that something, it seems certain, has to do with the weather. But what precise factors in the weather thus affect the herbivores is not always easy to say. The readings made by meteorologists, though of the utmost value in abundant ways, are not always very relevant to the life of animals. Temperature, for instance, is usually recorded at a height of four feet in the open and "very few animals live in the open at that height, except cows and zebras and storks and children and certain hovering insects."

Furthermore, what an animal or its food-plant responds to in the way of weather conditions is not likely to be the maximum or the minimum of any one factor, such as temperature or rainfall or sunlight, but especially favourable combinations of several varying factors. To take an example from ourselves the optimum geographical zone for white men is one of moderate temperature, moderate rainfall, moderate sunshine, and changeable weather; no extremes are involved in it, and it cannot be defined save as a complex meteorological combination.

The organism integrates the outer forces acting upon it. In the abundances and catastrophes of animals that fluctuate with a regular period we have in reality a new kind of instrument, more subtle than the thermometer or the rain-gauge, which will, we can feel sure, set the meteorologist on the track of new discoveries in his own science.

Sometimes, it is true, the weather does get to work in an obvious way. Very hard winters (which tend to recur with more or less regular periodicity) kill large numbers of the smaller birds. This is apparently due to starvation and not directly to cold. If birds can store up sufficient food, they can withstand astoundingly low temperatures; the little American Junco, even

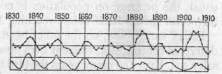

FIG. 265. The relation between solar disturbances and terrestrial life.

The upper curve shows the amount of growth made by trees in Germany, as determined by the thickness of their rings of growth. The lower shows the number of sun-spots recorded by solar observers. There is a considerable agreement between the two curves: (Modified from "Earth and Sun," by E. Huntington after Douglass.)

though it usually migrates south in winter and is no bigger than a sparrow, can withstand a blizzard with a temperature of 52° F. below zero, if well fed. The winter of 1928–9 was particularly severe on European bird-life. The hard winter of 1916–7 killed off the longtailed tits so thoroughly over large parts of England that many areas were not restocked up to the normal level of longtailed tit population until two, three, or even four years later.

Many of these animal-cycles seem to have a regular periodicity. The recurrent irruptions of unfamiliar birds are a case in point. The year 1927 witnessed a remarkable invasion of England by that extraordinary bird, the crossbill, which has its mandibles crossed over each other for the purpose of feeding upon pine-cones. These irruptions come westward from the pine-forests of central Europe, and occur at more or less regular intervals. One in the sixteenth century brought prodigious numbers of the birds, which did great damage by discovering that their beaks were admirably adapted for slicing apples in half as well as for obtaining the seeds from pine-cones. The dates of crossbill irruptions, however, have not been so well recorded as those of two other kinds of birds, the Siberian nutcracker and the sandgrouse. The nutcracker is an inhabitant of the vast coniferous forests of Siberia. It has invaded western Europe at intervals of about ten years, with what would be extreme regularity if it were not for the fact that now and again one of the invasions is "skipped." Although observations on the spot in Siberia are not forthcoming, it appears almost certain that the migrations are due to over-population in the bird's natural home, coupled with a bad harvest of the pine-cones upon which they feed. Doubtless, when the failure of the pine-crop is less extreme than usual, the pressure on population is not so great, and the wave of migration spends itself before reaching Europe.

Pallas' sandgrouse, on the other hand, is a bird of the steppes and deserts of central Asia, where it lives upon the scanty vegetation of the salty soil. Every so many years the bird leaves its home in huge flocks, migrating both eastwards into China, and westwards into Europe, even as far as the British Isles. Sometimes the migrations are continued for two or three years. Here a cycle of eleven years is pretty closely adhered to, with the additional fact that the alternate migrations are much bigger. As the records go, we seem safe in prophesying another imminent large invasion. The cause of the emigration again seems to be relative over-population, or what comes to the same thing, food-shortage, owing to their food-plants being covered by snow or glaze-frost.

A connexion has been suspected between the eleven-year cycles and the cycles of sun-spot numbers, which also have an average period of about eleven years. The sun-spots are a sign of increased activity and energy-radiation from the sun's surface; and this causes magnetic storms on our earth, ninety million miles away. Another fact of terrestrial climate which seems to be definitely correlated with sun-spot number concerns the tracks of thunder-storms. If the tracks followed by heavy storms are plotted on a map, it will be found that, in North America, for instance, there is in any one year a zone along which the majority of storms travel. Now this zone shifts up and down, with considerable regularity, from year to year, returning to the same position about every eleven years. Such a shift in the storm-tracks will obviously mean a slight shift of the margins of all the great climatic zones. It will mean that there will be cycles of rainfall, some areas getting more than the average every eleven years, while other zones in the same years will be getting less than the average; and this, according to the careful investigations of O. T. Walker, is what actually occurs. The autobiography recorded by trees in their annual rings of growth shows that they, in some situations, are under the influence of this eleven-year cycle. Not only does this hold for the giant sequoias of western America, but a fossil Canadian spruce from the Pleistocene (V E) shows that the Canadian climate in those days, certainly over 100,000 years ago, was oscillating with this same eleven-year period.

Such changes are likely to have the most noticeable effect upon plants and animals where conditions are difficult for life. For instance, a small change in rainfall in a semi-desert region will have much more effect than the same change in a well-watered country; and quite small temperature-changes in the arctic will have disproportionately large effects on the animals and plants which live there. Another interesting point that is now emerging is that the most important cycle in warm-temperate regions seems to be the eleven-year one; in regions farther north this gives place to a ten-year cycle, and in the far north, this again to one of about three-and-a-half years. But what may be the explanation of this strange fact we do not yet understand.

The various weather-cycles will have quite different effects on different kinds of animals, according to the length of the animal's own life-cycle. The short-period cycles of three years and a half would only be expected to affect small animals which reach maturity in a year or less. Larger animals have lives which are too long to be upset by such small cycles. In precisely the same way, the choppy little waves which are so unpleasant to the inmates of a row-boat have no effect upon the bulk of a liner. Even the eleven-year cycles will have little effect upon animals like deer or zebras. But deer and zebras and others of the larger herbivores do have recurrent plagues in wild nature, and these plagues recur at much longer intervals than those of rabbits; the very length of the cycles makes it more difficult to collect accurate information on them. Some idea of the times involved may be gained from the following rough calculation. If a single pair were to increase with no severe checks, an uncomfortable density of population would be produced by mice or lemmings in about three or four years, by squirrels in about five years, by rabbits or hares in about ten, sheep in twenty, by buffaloes in thirty, and by elephants in fifty or sixty years.

There is one fur-bearing animal which, as the Hudson's Bay Company's records show, seems to be exempt from these periodic fluctuations. It is the beaver. The beaver has made itself independent of all the short cycles of weather. It lives almost entirely upon the bark of the trees it fells, which themselves are so long-lived as not to be affected by ten-year or even thirty-year cycles. Then it constructs remarkable dams and canals which make it independent of all ordinary fluctuations in water-supply. And during the summer it stores great food piles of trunks and branches in its pond; thus it can get access to food under the ice, and is almost independent of the severity of winter.

When it has cut down too many of the trees in the neighbourhood of one pond, a beaver-community apparently just moves on in search of another; and at the end of summer any surplus population scatters on the same quest. It is on these treks that beavers seem most exposed to the attacks of beasts of prey; and in a state of nature it is then that most of the overplus of the beavers' natural increase is wiped out (just as most of the overplus of migratory birds' increase is wiped out on their migrations). But with their feeding and storage habits, and their normal immunity, safe in their ponds and houses, from most enemies, they have no need of the rapidity of breeding which a mouse or a rabbit must keep up to repair wastage; and so their breeding never runs away with them, so to speak, to lead to sudden huge increases of number. Then, too, their habits compel them to live in isolated colonies; there can never be a dense continuous beaver-population over a large area, and so there can never be an explosive outbreak of epidemic beaver-disease. And so the beaver-population (apart from man's inroads upon it) is never subject to the wild and rapid fluctuations that beset other rodents. It is regulated within much narrower limits and there is less wastage of lives.

There are other animals in which there is a kind of natural population control. It is found, for instance, among various kinds of birds. Eliot Howard has described and analyzed the system in his *Territory in Bird-Life*. In the breeding season, practically all small song-birds have the instinct to stake out a claim, so to speak, to a definite area or territory of considerable extent. This territory they defend against intruders of the same species, and often of allied species as well; they build their nest in it, and they confine themselves mainly to its boundaries in searching for insects and grubs with which to feed their young.

This "territorial instinct" doubtless had its origin in the nearly universal impulse to defend the nest and its immediate neighbourhood against intrusion. It takes a great many insects to supply the rapidly-

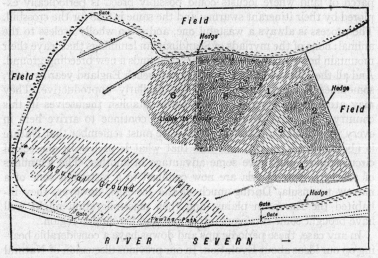

FIG. 266. The territorial system in birds.

Sketch-map of the nesting-territories occupied by six pairs of green plover in 1916 in a meadow in Worcestershire. The field measures about a quarter by an eighth of a mile. (From "Territory in Bird Life," by H. Eliot Howard. John Murray.)

growing naked young of a warbler or a finch; and any extension of the nest-defending instinct to cover a wider area would be of biological advantage to its possessor by reducing the infant mortality of his or her young in times of food-shortage.

When one kind of bird is unusually abundant in any given region, the pressure on space may force down the size of territories. But this process has a limit; a breeding pair will not tolerate another pair within a certain distance of their headquarters, and fighting will go on until one pair or the other are forced to leave. In years of exceptional abundance, some birds never find breeding-territory at all. They may penetrate to the northern limits of the species' range and drift about there in bands; or they may remain celibate in the middle of the breeding population. But they do not breed. So here again an upper limit is set to the population; and we do not find among small birds the same violent cycles, culminating in over-abundance and disease, that we do in small mammals.

When population-pressure seeks relief in migration, opportunity is given for the colonization of new areas. In this way, for instance, every patch of land where locusts could possibly breed is periodically explored by their itinerant swarms; and the same is true for the crossbill. The process is always a wasteful one, and often wholly useless to the animal; none of the myriads of Scandinavian lemmings that leave their mountain home in emigrating armies ever finds a new breeding-ground. And all the Painted Lady butterflies that reach England year by year, sometimes in abundant swarms, are similarly unproductive. They may attempt to breed, but they never establish themselves in this country. Yet these scouts of the species continue to arrive here in every year of high multiplication. But we must remember that Nature is abominably wasteful; and also that what is useless in one set of circumstances may have some advantage in another. The lemmings of southern Scandinavia are now confined to a restricted zone of a narrow peninsula. During much of the Ice Age, however, they inhabited the European plains; and then mass-migration might well have brought some survivors into a new and favourable region.

In any case, these periodic ups and downs have a considerable bearing on our ideas about Evolution. In our previous discussion of Natural Selection, we assumed that the struggle for existence exerts a more or less constant pressure. This is not true for species with violent cycles of abundance. When the survivors of a rabbit epidemic, for instance, are restocking the country, the struggle for existence will be very much lightened; as abundance increases, pressure on food-supplies will begin, and will slowly grow. Finally, when the inevitable epidemic breaks out, there is an intense selection at its hands for hardiness and disease-resisting qualities. In the same way, in a very hard winter only the lucky or the resistant birds survive.

In other words, selection itself is a fluctuating thing; and many qualities of plants and animals have been brought into being only by the intense selection of exceptional years. Most of the creatures are not fully tested most of the time; they have a reserve of biological adequacy and could dispense with this or that adaptation during ordinary seasons. But then comes the pinch, and all the reserves are called into play. Periodically the species has, so to speak, to pass special examinations. After a very favourable year, its members are put through a competitive examination first as regards their competency to secure food and breeding-space, and then as regards their disease-resistance; in very unfavourable years they are examined on their power of resisting hunger or thirst and the extremes of temperature. But in between they have a comparatively easy time.

The relaxation of selection after a catastrophic killing-off may also have important effects. Characters useless by themselves may be useful if they are combined together; and in these times of comparative ease the species is able to preserve and shuffle its mutations and get the most out of them. A concrete case in point is that of the Greasy Fritillary butterfly. This is a local species, its various stations being widely separated. At one place in the north of England where it has been abundant for a number of years, it suddenly grew scarce. In 1912 it was rare, from 1913 to 1920 very rare. In 1921 it began to increase rapidly again, and has since 1924 remained at its old abundance. During the period of its rapid increase, 1921, 1922, and 1923, it showed a remarkable outburst of variation, in size, colour, and pattern. From 1924 on, the range of variation decreased And now it is as constant as it was before the War—but not quite the same: it is darker, with a coarser-mottled pattern. It looks very much as if, while rapid increase was going on and selection was presumably relaxed, all kinds of recombinations made by the genes of the few survivors of the previous thinning-out came into being. Then, when population-pressure had brought selection up to the mark once more, most of these were weeded out, and only one main type was left—but it was not the original type. A better combination had been found.

But these cycles have more than theoretical interest. There are, for one thing, commercial advantages in knowledge. Fur-trading companies can regulate their staff of trappers according to the prophecies of the ecologist, and can guard themselves against periodic gluts and scarcities of pelts. Much more important is the medical significance of the facts. It is well-known that rats act as a reservoir of bubonic plague, transmitting it to human beings by their fleas; and the same is true for other small rodents such as gerbils. It has already been established that in Central Asia and South Africa the incidence of plague in man fluctuates with the abundance of these small mammals. The year 1910, when a small outbreak of human plague took place in the eastern counties of England, was also a year of plague and apparently of unusual abundance among English rats. In modern conditions, rats are not so ubiquitous as they used to be, nor do they come into such close contact with man; probably this fact saved England from a much more serious visitation of human plague in 1910. The early inhabitants of Palestine seem to have had some inkling of the connexion between rodents and disease. In I Samuel, chapters 5 and 6, we are told that the Philistines, afflicted with a grievous pestilence which seems to have been bubonic plague, were recommended to make and offer up golden images not only of the swellings or buboes, characteristic of the disease, but also "of the mice that mar the land." Modern biology has verified this connexion in detail and shown the real nature of the relation between the fluctuations of the species rat and the danger of human infection.

A recent application of this knowledge probably saved South Africa from a serious visitation. The gerbil is a common rodent of this and other warm-temperate regions. In 1924–5 the gerbils over a large extent of the Union were plague-infected and the area of gerbil-infection was still increasing rapidly. In a belt of country south of the infected area a war of extermination was waged against gerbils and all other plague-carrying rodents, and the epidemic passed away before reaching Cape Town and the populous coast owing to the natural dying down of the disease-stricken gerbil population to a density at which plague would no longer spread through it.

And it may well be that other epidemic diseases whose comings and goings are still mysterious will prove to be linked with the abundance or scarcity of some obscure rodent; but here only laborious research can enlighten us.

These facts help us to realize the real nature of the ordinary Struggle for Life. The careless thinker about things biological is apt to fall under the sway of military ideas and think of it as a war between one species and another. He envisages it as a regular battle between an inoffensive herbivore and its enemies, or a sort of athletic competition between a carnivore and its prey. In both cases he thinks of the struggle as something in which victory is to be achieved as it is achieved in war or sport. As a matter of fact, it is nothing of the kind. A herbivorous species without carnivorous enemies would tend to over-populate its territory, would become diseased and undernourished, would condemn itself to starvation by eating down its own food-supply; a carnivorous species again which was restricted to one kind of prey, and a kind it could too easily catch, would inevitably bring its own race to extinction by eating itself out of existence. To multiply and replenish the earth unchecked may be only the prelude to decay and extinction. Both of these eventualities have, through the interference of man, been realized. When red deer were introduced into New Zealand, they throve on the succulent forest and bush and multiplied exceedingly owing to the absence of all carnivorous enemies. But after a few decades they had changed the face of the country where they were abundant, and to-day the fine heads and heavy beasts are found only on the outskirts of the deer's range, where they are still advancing into virgin country. Elsewhere the herds, living on scarce and inferior food, are full of stunted specimens with malformed antlers, and the authorities have been forced to play the part of natural enemy, and to adopt a rigorous policy of periodic thinning-out to save the stock.

For a carnivorous instance of the evil of easy living we may quote from Elton the curious case of Berlenga Island, off the coast of Portugal. "This place supports a lighthouse and a lighthouse-keeper, who was in the habit of growing vegetables on the island, but was plagued by rabbits which had been introduced at some time or other. He also had the idea of introducing cats to cope with the situation—which they did so effectively that they ultimately ate up every single rabbit on the island. Having succeeded in this, the cats starved to death, since there were no other edible animals on the island."

We are often told that it is very important for children to select their parents wisely. It is becoming clear that a wise choice of enemies is an asset to an organism! One can hardly, perhaps, speak of an animal's enemies as part of its adaptations, but at least they are vital to its survival. In almost every case the word *enemy* is only applicable when we are thinking in terms of individuals: as soon as we think of species, the individual "enemy" often turns out to be a racial benefactor.

Unrestrained breeding, for man and animals alike, whether they are mice, lemmings, locusts, Italians, Hindoos, or Chinamen, is biologically a thoroughly evil thing.

VI

LIFE UNDER CONTROL

§ 1. *The Balance of Nature.* § 2. *Pests and Their Biological Control.* § 3. *The Beginnings of Applied Biology.* § 4. *The Ecological Outlook.*

§ 1. The Balance of Nature.

THE fluctuations in animal numbers we have been discussing give us new insight into the tangled web of interrelations summed up in the phrase "The Balance of Nature." A few further instances of the swaying of that balance may be interesting and profitable.

Change in one member of a life-community may transform the whole community into something else, as surprisingly as an increase of thyroid secretion will transform an exolotl into a land salamander. A classical case is that described by Ritchie in his interesting book, *The Influence of Man on Animal Life in Scotland.* It concerns a small stretch of moor in southern Scotland. When the story begins, this was covered with heather, and tenanted by typical heather-moor creatures such as the red grouse. In 1892, a few pairs of black-headed gulls came to nest there. This was probably the result of a general increase in the numbers of their species, but what produced this increase we do not know. Whatever the reason, the fact was the starting-point for an intricate chain of cause and effect. The owner liked the gulls and protected them, with the result that by 1905 there was a nesting colony of over three thousand. The trampling of the birds was bad for the heather, while their constant manuring of the ground changed the character of the soil. The result was that the heather vanished altogether, its place being taken first by rushes and later by coarse docklike plants. Pools of shallow water formed here and there. With the heather, the grouse disappeared; while the pools attracted teal and other duck. In 1905, protection was withdrawn from the gulls, their nests were robbed, and they decreased rapidly until by 1917 there were only a hundred or so of them left. The heather had re-invaded most of the ground, the pools were drying up, the teal had gone, and the grouse were returning. Thus in twenty-five years the ground and all its plant and animal inhabitants had changed completely, and then changed back again.

Another well-analyzed case comes from the Brecklands of East Anglia. The natural dominant vegetation of this strange barren country seems to be low pinewood. Wherever the trees for one reason or another fail to grow, dry heather-moor takes its place. Patches of thick-growing bracken-fern and low grass-heath also exist. The most abundant of the vertebrate inhabitants is the rabbit.

Rabbits are not indigenous to England. After the Ice Age they failed to reach it before the Channel had put a bar between it and the rest of Europe. They were certainly not introduced before the Neolithic Period, and many authorities believe that they were not brought over until after the Norman Conquest. However introduced, during the Middle Ages they were protected in warrens for the sake of their skins, but eventually spread and multiplied as they have in other countries. Their attacks on the pine-seedlings, together with the clearing and felling due to man, have swept the natural pinewood off the plateau, and left heather to take its place. A natural equilibrium was soon established between rabbits and their enemies such as stoats and weasels, and lasted for centuries. Of recent years, however, civilization in general and game-preserving in particular have enormously reduced the number of these carnivores, and the rabbits, relieved of this drain on their numbers, are increasing towards a new equilibrium of much greater abundance. The resultant "rabbit-pressure" is in its turn having striking effect upon the vegetation. Wherever the rabbits have access to pinewood, it fails to reproduce itself, and its place is taken by heather. But the heather, which could stand a certain density of rabbit-numbers, itself melts away when their concentration rises above a certain point. It is eaten down, and given place either to rush or to grass-heath. If rabbits are not too abundant, the outcome is decided by the nature of the soil; but once more the animals have the casting vote. If the pressure for subsistence is great, they attack the rushes, and these too are replaced by grass-heath.

The grass-heath itself is badly attacked by the rabbits; it is stunted and nibbled down close; and yet it can survive where the rushes and heather die out. As Farrow says: "The grass-heath owes its very existence to an extremely injurious influence which nevertheless greatly benefits it because it injures its competitors slightly more." Increasing rabbit-pressure, it will be noticed, progressively reduces the height of the vegetation. Pine-trees yield to bushy heather; heather to scrubby rushes; and rushes to the mere carpet of the grass. In normal circumstances the advantage given by height in the struggle for light and air causes a succession from low to tall plants, culminating in forest; and this is what happens in Breckland wherever rabbits are fenced out. But with intense rabbit-pressure, the contrary is the case, and the plant which can live and reproduce though cropped down to a mere inch or so will survive.

A complication is introduced by the bracken; for this, though tall and juicy, is distasteful to rabbits, and they leave it alone unless very hard put to it. As a consequence a miniature jungle of bracken is spreading rapidly over the landscape. Should it come to cover most of the country, the rabbits would be confronted by a new problem; they would have to eat bracken or starve. They would eat it; and so a new tilt would be given to the ever-unstable balance.

Thus the destruction of weasels and stoats has set the different kinds of plants advancing and retreating, marching and counter-marching, over all Breckland, and it is only because they march by yards in a year, instead of miles in a day, that their movements do not strike us as immediately and forcibly as the manœuvres of troops or the migration of birds or lemmings.

Both these cases provide good examples of a general principle—that change in life-communities goes by jumps, even though the change in conditions alters slowly and gradually. A change in the number of rabbits does not merely alter the proportion of pine-trees and heather-bushes and rushes and grass plants, but causes the total replacement of one kind of plant by another over a stretch of country. This is simply a particular aspect of the familiar fact of dominance, which we meet with in every plant-community.

Of the subtlety of the web's weaving, whereby a twitch on one life-thread alters the whole fabric, many writers have told us. A very simple case is the connexion of ravens with sheep-farming. In early spring the staple diet of ravens on the Scottish hills is afforded by the afterbirths of the ewes that have lambed. If sheep-farming ceased to be practised in Scotland, the number of ravens would go down with a bump.

A more curious example is the Box and Cox habits of mongoose and gerbils. The gerbil, a social creature, is a little burrowing rodent of the South African veldt; the yellow mongoose is a stoat-like carnivore inhabiting the same regions. Both retreat underground for safety; and it frequently happens that they live side by side, or even share some burrows and runways in common. Usually, however, their two streams of life do not come into more intimate contact, for the mongooses come out to feed by day and only use the burrows to sleep in at night, while the gerbils sleep through the daylight and are purely nocturnal feeders. However, when the gerbils are smitten, as is the fate of small rodents, with epidemics, they often crawl miserably out of their burrows by day, and then are caught and eaten by mongooses. One of the most frequent diseases of gerbils is bubonic plague. The ecologist accordingly examines the excreta of the mongoose; when he finds gerbil fur in them, he knows the gerbils are dying of some epidemic, and that this is more likely than not to be plague; and so he can either pursue his investigations further to make sure, or he can at once, though with a chance of being mistaken, recommend human precautions.

Many people have heard of Darwin's celebrated example of cats and clover. He pointed out that red clover, an important forage-crop, was absolutely dependent for its fertilization upon the visits of humble-bees, hive-bees not having a long enough proboscis to reach the nectar. Humble-bees make underground nests in banks and slopes; and these nests are often raided and destroyed by field mice, one observer estimating that this fate befalls over two-thirds of all the humble-bees' nests in England. The number of field-mice, especially near villages, is partly controlled by cats. And in this way, said Darwin, a decline in the number of cats would bring about a reduction in the amount of seed set by red clover.

Modern ecology is inclined to criticize this statement so far as cats are concerned. It would need a vast multitude of cats to affect the mouse population very seriously. On the other hand, there is undoubtedly a close connexion between the mice and the bees; so that the ups and downs of the mouse population will certainly affect the crop of clover-seed.

Miss Turner, the well-known student of bird-life, has pointed out the connexion between the growth of motor traffic and the decline of certain species of bird. Sometimes, as we all know, this connexion is obvious enough. The sparrow has almost vanished from the central parts of many American towns now that he is deprived of horse-droppings and scattered grain from nose-bags. But here is a subtler chain. In those remote days (a quarter of a century ago) when horse-buses were the Londoner's main means of transport, much of the horses' fodder was supplied from Norfolk. The rank marsh-grasses were regularly cut, ground into chaff, and sent off to the omnibus companies in London. The marshes that were cut in any one year provided ideal nesting-grounds for small waders and plovers the next season. To-day there is no market for marsh-grass. It grows dense and tall, and often is replaced, in natural process of ecological succession, by thick sedge. The snipe and redshank and plover can no longer force their way through this coarse tangle; and so fewer of them can breed in Norfolk and their races decline there.

Sometimes these obscure linkages have important practical results. As we have pointed out, the blackberry, imported into New Zealand, has there changed from a harmless weed which compensates for its thorns by its contribution to jam, to a real pest. But it is doubtful if it would ever have done so but for the introduction of European birds into the country. Some of these, notably the starling, devour its fruits, pass the seeds out undigested, and thus multiply its power of dispersal. Without this aid the invasion of new territory would probably have been so slow that the plant could easily have been kept in check.

§ 2. Pests and Their Biological Control.

ALL over the world man has been busy making difficulties for himself. The crowding of human beings into cities, like the crowding of animals in their times of over-multiplication, give new openings to disease. The city is the fosterer of commerce and architecture, of learning and the arts; but until it is disciplined and controlled, it is also the opportunity of the bacterium. Freedom of intercourse and communication stimulates both trade and thought; but it gives disease-germs new facilities for rapid spreading, as when the opening up of Africa brought sleeping sickness across from the West Coast to the East. Thus the growth of civilization has been marked by a trail of plagues, more explosive and more widespread than anything which primitive man can have experienced.

Agriculture brings similar difficulties. City life crowds and agglomerates human beings. Agriculture crowds and agglomerates single kinds of plants and animals; and the more thoroughly and intensively it is practised, the denser the agglomerations and the more unnatural the massing. This crowding not only gives new opportunities to the fungi and bacteria and protozoa that are the causes of most animal and plant diseases, but is an open invitation to insects. This is notably so in the tropics, where insects are more abundant and their life runs quicker. Tropical agriculture, though it gives promise of huge additions to the world's supplies of food, is not merely an invitation but an incitement to insects. And the greater the excellence of communications, the more chance of introducing lurking and often unsuspected pests from one part of the world to another—rats and earwigs, forest-devouring caterpillars and crop-choking weeds.

Again, the bringing in of the products of one region of the globe to supply the natural deficiencies of another is an obvious way in which man may improve upon nature. One has only to think of the food and recreation now abundantly provided in many previously barren mountain streams of the Rockies and the Bighorns by the introduction of trout, or the beautification of Europe or American gardens by the flowers of China and South Africa. But here almost more than anywhere else it behoves the would-be benefactor of humanity to proceed with caution; if he is not careful, he will do infinitely more harm than good.

Of late years, considerable progress has been made with a new and difficult art—the biological control of pests. Almost invariably, a pest is an animal or a plant which has been introduced, whether deliberately or accidentally, into a new country. Among the few exceptions, one is so interesting that we must cite it. The kea of New Zealand is a large and more or less omnivorous mountain parrot. Some time after the introduction of sheep into New Zealand, it was found that the kea in certain regions had taken to sheep killing. They settled on the sheeps' backs and tore away with their powerful beaks until they exposed the wretched animals' kidneys, which they devoured, killing the sheep in the process. Presumably they must have originally mistaken the sheep for moss and lichen-covered rocks, and in scratching for insects, have found warm meat. In this case it was the introduction of new food which turned a harmless creature into a pest. A similar though less striking example comes from Africa. The birds known as ox-peckers were adapted to picking parasites from the tough hides of rhinoceroses. When domestic cattle were introduced, the birds turned their attention to them too. But here they often penetrated the skin. When the flesh is thus exposed, they seem not averse to it, so that they too are on the way to become a nuisance.

When new species are introduced into a country, few will find themselves in the same balance as in their old home. For the majority, things will be unfavourable; they fail to gain a footing, and some disappear. Now and again, however, the introduced species chances to be better suited; and then its numbers will go up, often far beyond anything possible to it in its native country; and not infrequently its abundance will force it into changed habits. The starling in America has spread steadily since its introduction, and is reducing the numbers of many hole-breeding American birds by occupying so many of the limited supply of nesting sites. And once its population-density oversteps certain limits it is forced to change its food habits, and does a good deal of damage. In moderate numbers, starlings (and the same is true for a number of other creatures) do good, on balance; in great numbers they do harm. The earwig, a mere nuisance at home in Europe, has become a voracious and serious pest in New Zealand and the Pacific Coast of America, where some States have even set up special Bureaus of Earwig Control. The thistle was introduced into California by a Scotsman who wished to have his native emblem growing on his land; but it multiplied and infested the lands of everyone else. Another patriotic Scot in New Zealand built a fence round the first thistle that appeared on his farm, to protect it from possible enemies; but it was the advance-guard of a formidable invasion. Thompson's interesting book, *The Acclimatization of Plants and Animals in New Zealand*, is full of similar examples of misguided zeal. English sparrows, for example, were introduced so that their matutinal chirpings might help the early colonists to forget their homesickness.

Not infrequently, notably with insects, the devastating increase of an introduced species is due to its having arrived without its proper parasite enemies. If the right parasites can be found and turned out in quantity, the missing control is resumed and in a very short time the pest is reduced to harmlessness, or at least manageability. A good case is that of the Gipsy Moth, *Lymantria dispar*. This is a terrible enemy to trees, and even in its native Europe it will from time to time enter upon a period of over-multiplication and do enormous damage to forests. But in America, where it was introduced in the late nineteenth century, it threatened to develop into a new Plague of Egypt. Over wide stretches of country it stripped every tree of its leaves, and when the trees were finished, the hungry armies of caterpillars came down to earth and took to eating the herbs and flowers. An extraordinary sound fills the forest when the caterpillars are at the height of their abundance. Even on the stillest day there is a continual rustling patter; it is the sound made by their innumerable droppings.

It was found that the moth had succeeded in entering the country without any of its insect parasites; when three of the commonest of these were imported they imposed a new equilibrium on the population of the species and it became no more of a pest in America than in Europe.

LIFE UNDER CONTROL

One of the most striking examples of biological control comes from the Fiji Islands. Here, as on so many of the islands of Oceania, the coconut palm is one of the most important of vegetables, yielding not only many products for local use, but also the valuable coir fibre and the still more valuable copra, in which there is an extensive trade. Towards the end of the last century, the coconut plantations of Viti Levu, one of the two large islands of the Fiji group, began to fail. All sorts of soil investigations were made, but it was not for some years that anyone thought of looking for an insect enemy. It disclosed itself as soon as looked for—a lovely little purple-winged moth whose caterpillars devoured the leaves. The pest grew worse and worse, until in some plantations the trees were reduced to bare poles. So far the pest had been confined to the one island; then suddenly in 1922 it appeared on two small islands on the way between Viti Levu and Vanua Levu, the other big piece of land in Fiji, whose annual coconut crop was worth half a million sterling; and in 1923 took a further step to a new island. The planters now began to feel desperate. They offered a prize of £5,000 for a cure for the pest; but on its being pointed out to them that such a discovery would inevitably be the result of many men's brains, wisely changed their plan for one of deliberate research.

It had been discovered that the coconut moth in Fiji was exempt from parasites. Three entomologists were set the task of finding a parasite for it. They searched the coasts of the Pacific; and one of them in Malaya found a related moth which was parasitized by various enemies, the most important being a certain kind of fly. The next step was to get the parasites to Fiji. This was not so easy, as they do not hibernate. However, by chartering a steamer to make a special voyage direct from Malaya to Fiji, 300 flies were brought over in 1925. By twelve months later over 32,000 flies had been bred and set free, and by 1928 the fly had not only established itself wherever the moth was to be found, but was attacking between 75 and 90 per cent of the caterpillars. From Java two more parasites were introduced later, a second fly to prey on the caterpillar stage and a tiny wasp-like insect which is a parasite of the eggs. The result, three years after the first parasites were liberated, was that the moth had become quite rare, and that at a total cost of a few thousand pounds an important industry had been made safe, from one fatal enemy at least, in perpetuity.

This kind of work has its difficulties as well as its triumphs. The sugar-cane borer is a little weevil that was doing a vast deal of damage in Hawaii in the early years of this century. Muir set out to find a parasite, and eventually, after over two years of hunting round the Pacific, discovered one in Amboina, off the coast of New Guinea. But Amboina is 4,000 miles from Hawaii; and the fly has a short life-cycle and is very difficult to breed in cages. Eventually, after a number of failures (for instance, Muir developed typhoid at sea, when travelling with his flies, lost them all, and was forced to go back to Amboina and begin all over again), the fly was brought to Hawaii by stages—first to Queensland, where a new generation was bred, then to Fiji for a second generation, and so to Hawaii. Once it was introduced in Hawaii it soon reduced the sugar-cane weevil from serious pest to minor nuisance.

Biological control of plant pests is also possible, though both more difficult and more risky. You have to find an insect which will eat your weed and preferably nothing else; at any rate it must not eat anything of use or value. By the aid of this ally you may arrest the spread of your pest and can then proceed to measures of destruction—uprooting and the like—which are of no account whatever when the plant is in the full tide of its unnatural increase.

Considerable progress has been made by this means towards checking the onward march of the prickly-pear in Australia. It was at one time suggested that the spines should be burnt off and the plants used for feeding cattle; but it was pointed out that the annual increase of prickly-pear was considerably greater than the eating capacity of all the stock in Australia! It was eventually decided that the only hope lay in biological control. A well-financed scheme of research was brought into being in 1920. Entomologists scoured the United States, South America, and the West Indies for enemies of prickly-pear and related kinds of cactus. For these a breeding-station belonging to the Australian Commonwealth was set up in Texas, and special methods of transport were devised. Among the dozens of insects tried out, four main kinds have been liberated on a large scale—caterpillars of the moth Cactoblastis that tunnel through the plant; plant-bugs and cochineal insects which suck its juices, and the "red spider" (really a mite) which nibbles its surface. These are all confined to prickly-pear, and actually starve to death on any other plant, so narrowly specialized are their feeding-habits.

With the aid of these arthropods, the progress of this unpleasant vegetable has now apparently been checked. Australian land is no longer being covered with impenetrable prickly scrub to the extent of a thousand square miles (an area over the size of Warwickshire) every year; and Australian civilization has a breathing space to look round for other insect weapons to complete the pests' destruction.

To help in this work of biological control, special laboratories have been established in many countries. Perhaps the most remarkable is one near London attached to the Imperial Bureau of Entomology. In this "Parasite Zoo," biologists work out the methods of rearing all manner of insect parasites, and ship them in bulk to all parts of the British Empire as they are required.

But biological control is not always practicable. It is rarely possible, for instance, for man to employ vertebrates as his auxiliaries in this way, for the simple reason that they, with their more plastic nervous systems, will not consent to remain tied to one kind of food after the manner of so many insects. They have the habit of switching over even from their favourite diet, should it grow scarce, and taking to another which happens to be more abundant.

This habit has obvious dangers where a mammal, for instance, is introduced to cope with a pest. The most celebrated case of this kind is perhaps that of the mongooses introduced into the West Indies to cope with a plague of rats. They reduced the rats to a certain extent, but as the rats grew scarce, turned their attention to other creatures, especially wild birds and poultry; and speedily became a pest almost as bad as the one they had removed.

For pests with a complicated life-history, increase of knowledge may sometimes reveal unexpected methods of control. There is, for example, a disease of the white pine, known as blister-rust, which may inflict great economic loss. This, like other rust diseases, is caused by a fungus which requires two hosts to complete its cycle of reproduction. The white pine is the first; the second is wild gooseberry. If we can extirpate wild gooseberries, we can get rid of blister-rust as surely as we can stop men and women having malaria by extirpating certain kinds of mosquito (Fig. 271). This can be undertaken by direct methods. But a study of ecological succession has shown how good forestry will help on the extirpation. Where a clearing in the forest (the matter has been studied in New England and the Adirondacks) is left to itself, the first stage of weeds and shrubs gives place after a year or so to shrubs and bushes, among which the wild gooseberries find a place. And these are succeeded by a forest stage, with white pine, maple, and other trees. Now the seeds of the gooseberry are dispersed in the droppings of fruit-eating birds; and it so happens that when the forest stage is reached, these gooseberry-eaters no longer find the place to their liking, and depart for other clearings. The gooseberry plants still manage to survive under the shade of the trees, but in the absence of their natural disseminators, they do not spread and multiply. It is only in open shrubby clearings that they can increase. The proper reforestation or cut-over parts of the forest, with a little judicious weeding among the young trees, will help reduce the gooseberry bushes, and so the rust.

§ 3. The Beginnings of Applied Biology.

In these and many other ways, man is beginning to turn his all too scanty knowledge of the ecological web to good account. Apart from economic difficulties, most of the problems which agriculture has to face, and many of those which beset medicine, are problems in applied ecology. This is especially so in new countries and tropical climates. Knowledge is already so diversified that we partition up the task among a panel of specialists—soil chemists, entomologists, experts in moulds and fungi, agronomists, foresters, bacteriologists, public-health experts. But the problems interlock and shift from one field to another. The entomologist, faced by a disease of crops, will make it his business to look for insects; he may find them all right—and yet their undue abundance may be only the symptom of a weakened resistance of the plant, due to an attack of fungus, or to wrong methods of cultivation. The control of sleeping sickness, malaria, and yellow fever, we now know, depends upon a knowledge of the numerous species of tsetse-flies and mosquitoes and a full understanding of their habits. Medicine here would be helpless without the museum systematist with his vast collections, and the field entomologist, busy observing the insects' ways.

But even ecology, wide though it be, is not wide enough. Physiology and genetics, embryology and bio-chemistry and other sister sciences must also join in the counsels of applied biology if she is to rise to the level of her opportunities.

As J. B. S. Haldane has pointed out in his *Daedalus*, biological inventions have up to the present been few, and most of them were made before the dawn of history. Of these early achievements there is the domestication of animals, and the domestication of plants that we call agriculture. There is the utilization (doubtless not made without the overcoming of much sacred repugnance) of the milk of other creatures; there is the harnessing of yeasts and bacteria to make alcoholic drinks and vinegar and curds and cheeses. Perhaps, as he suggests, legends like that of the Minotaur hint at widespread and startling essays in hybridization; whatever the truth of this, certainly the discovery that stocks of animals and plants could be improved by crossing and selection, however unconscious the methods may have been at the start, is to be reckoned as another great biological invention. So was the idea of the rotation of crops; so was the practice of grafting; so was irrigation; so was the employment of castration to render domestic animals tamer and fatter, and so, too, was the deliberate practice of surgery. All these date back to prehistoric times; and from then until quite lately the tide of biological invention stagnated; any progress lay almost wholly in the improvement of what already existed.

The eighteenth and nineteenth centuries saw the tide begin to flow again. There was the discovery of artificial insemination by the Abbé Spallanzani; the use of chloroform as an anæsthetic by Sir James Simpson; the invention of artificial manures by Liebig; Pasteur's discoveries about immunity, which made it possible for some diseases previously thought intractable, like rabies, to be cured, and others, like typhoid, to be deliberately prevented; the utilization by Lister of Pasteur's discovery that putrefaction was caused by living bacteria, to give the world antiseptic and then aseptic surgery; the invention of new methods of controlling diseases, such as yellow fever and malaria, made possible by the discoveries of Manson, Ross, and Grassi as to the rôle played by insects in their transmission; the discovery of how to isolate and bottle up the active principles of the organs of chemical control, such as thyroid and adrenal, for use whenever needed—these are some. Another biological invention of this period must be mentioned, and that is the invention of safe and simple methods of preventing conception; for whether we approve or disapprove of their use we cannot but admit that their invention opens the door to momentous consequences.

Matters are moving a little more quickly in this twentieth century. Neither Loeb's great discovery of how to make unfertilized eggs develop, nor the equally remarkable discovery of tissue-culture, has as yet received any practical application. But we have made a beginning with this business of biological control of pests; we have begun to supplement the empirical practices (often admirable in their way) of the plant and animal-breeder with the application of Mendelian principles; and in a few places we have made a timid beginning in applying our knowledge of heredity to the improvement of our own species. We are making steady progress in the task of finding a drug which will produce healthy sleep without evil after-effects. The discoveries concerning vitamins and mineral salts and food-balance are making possible the invention of a healthy diet for city-dwellers; those concerning the effects of ultra-violet rays and radiant heat are on the way to give us healthy houses, and in time, let us hope, fogless towns. And we have had the invention of a method, however imperfect as yet, for rejuvenation.

The list, it will be perceived, is not a long one; and it is out of all proportion to the biological imaginations of mankind. Man has dreamt of prolonging his life; of controlling the destinies of society as he can now control a business or a machine; of eliminating pain; of building a new race, all of whom should be strong and beautiful, clever and brave and good; of harnessing the forces of life to work for us as effectively as we have harnessed the forces of lifeless matter; of creating living matter anew; of getting rid of disease; of making synthetic food and drink and substances which should stimulate and enlarge this or that faculty without being followed by depression or injurious effects; of fashioning new kinds of animals and plants as easily as he fashions clay or wood or metal; of painless, quiet and happy dying; of the abolition of fear and worry, cruelty and injustice; of an intensification of human capacity for living—the abolishing of fatigue, the enhancement of vigour and enjoyment; of making life yield happiness, or if not happiness, then joy and divine discontent.

Those are dreams that depend for their realization on the sciences of life; and what a paltry beginning we have as yet made with their realization! This is in part due to our refusal to use the knowledge already available to us; but to a far greater extent it is due to a lack of knowledge. Without a thorough knowledge of the abstruse and apparently academic principles of physics and chemistry not a single motor-car or wireless set, let alone an aeroplane or a television apparatus, could ever have been built; and we must get to know much more about the chemistry and physics of living matter, its psychology, the laws of its heredity, the mode of development of its body and its mind, before we shall be able to satisfy man's biological ambitions.

We may give one or two examples of the way in which the problems of applied biology are opening out. Let us take first the problem of the world's grass. The story begins with the veterinary surgeons. They told of diseases which mysteriously affected cattle, pigs, sheep, and horses, their growth and condition, their fertility, and their yield of meat, milk, or wool. Eventually these diseases were traced to a deficiency of diet; the animals were not getting enough mineral salts in their food. Sometimes it was iron that was deficient, sometimes iodine; not infrequently calcium, and most often phosphorus. Vast tracts of land in Africa, in Australia, in the west of Scotland, in the United States, are short of one or other of these vital elements.

Wild animals could thrive and reproduce in these regions because in nature a balance is automatically struck; the country carries what it can carry. Moreover, when the animals die, the materials of their bodies return to the soil.

When man comes on the scene, matters are altered. He crowds the country with animals. He hurries up their growth and increases the demands they make on the soil. A modern cow gives about a thousand gallons of milk at one lactation period, and produces her first calf at about three years; the native cattle of Africa do not breed till they are six, and yield at most three hundred gallons of milk at one lactation. And too often he ships off the meat, bone-meal, cheese, leather, and wool without putting anything back in the soil. He forgets that all their mineral ingredients have come out of the soil. A country that is exporting grassland products is also exporting grassland fertility. There are large areas which are naturally deficient in minerals; but man has been creating mineral-deficiency over other and vaster areas.

In untamed country, animals, wild and domestic, may make up for mineral defects by making periodic journeys to salt-licks and storing their systems with the elements they need. But when the lands grow settled, fencing interferes with these pilgrimages, as it has, for instance, in Kenya.

Once the diseases of cattle had drawn attention to this problem, research pursued clues in new directions. It was found that the amount of calcium or other mineral elements in the soil which was enough to prevent disease was not nearly enough to allow animals to give their maximum yield. Most pastures can have their stock-carrying capacity materially increased by adding mineral fertilizers. Then it was discovered that different grasses were by no means equal in their demands and their performance. In some dry countries, if you provide the right brand of salts and the right breed of grasses and clover, you can without irrigation turn a mean, scrubby pasture into a rich sward. You can breed grasses which will grow twice as fast as ordinary wild grasses. You can import new breeds of grass as you now introduce new strains of maize or wheat. In New Zealand, for instance, there are no indigenous animals that graze; and when cows and sheep are introduced the native grasses fade out under the unaccustomed nibbling. The New Zealand pastures can only continue productive if the right sward-plants are imported. Science is now making a resolute attack on the problem. Co-ordinated work like that of the Grass Research Station at Aberystwyth is making a good beginning; from these Welsh uplands new varieties are destined to be sent all over the world.

We have already bred animals that can build meat and milk and produce new meat-and-milk machines like themselves twice as fast as the wild representatives of their species; if we take half the trouble with the genetics of grass and clover which we have already taken with wheat and corn, we can make pasture that grows twice as fast as the average pasture of to-day; and if we pay attention to the elementary chemistry of the soil, we can ensure that this doubled demand shall be satisfied. The value of products which come out of grassland is enormous—as much as that of all our cereal crops together. If we like, we can double or treble this enormous yield.

From grass we may pass to wheat. The wheatfields of the world (we are citing Sir Frederick Keeble's *Life of Plants*) cover about 400,000 square miles, and the average yield is about thirteen bushels to the acre. A bushel of wheat weighs some sixty-three pounds, so that this amounts to just about one hundred million tons. There are one thousand six hundred and twenty-five calories of energy-value in a pound of wheat, and the average number of calories needed to keep a man going for a day is about three thousand. So, if we translate our wheat into terms of energy, and "if man could and did live by bread alone, the wheat crop of the world would each year provide sustenance for wellnigh three hundred million men."

No wonder that the world's wheat-belts are important. They can be made more important in various ways—by improved agricultural practice, by breeding disease-resistant strains, and so forth. Here we will only consider one way. They can become more important by being made to grow larger—through the breeding of special strains which will creep up towards the pole by growing and ripening earlier. Every day taken off the average time needed for a wheat to ripen means so many more miles advance of the wheat-belt northward.

In the early years of the present century the three Saunders, father and two sons, bred a new wheat called Marquis. It ripened a week to ten days earlier than Red Fife, which had been for years the staple Canadian strain. Between 1911 and 1916, Marquis superseded Red Fife, and the limit of wheat-farming was pushed fifty miles to the northwards. Since then other wheats have been invented which live at an even quicker rate—Ruby, Garnett, and Reward; and wheat has been brought another forty miles nearer the north pole. There must be a limit to the process; but it has not yet been reached.

§ 4. The Ecological Outlook.

Let us in conclusion summarize the ecological outlook of our species. The cardinal fact in the problem of the human future is the increase in the speed of change. The colonization of new countries, the change from forest to fields, the reclamation of land from sea, the making of lakes, the introduction of new animals and plants—all these in pre-human evolution were the affairs of secular time, where a thousand years are but as yesterday; but now they are achieved in centuries or even decades. One cannot estimate such changes exactly, but we shall not be far out if we say that man is imposing on the life of the world a rate of change ten thousand times as great as any rate of change it ever knew before.

In the second place, the change is becoming deliberate. What before was achieved by slow shiftings of balance due to unconscious competition is now being forced on nature at the point of human consciousness. And man is envisaging new methods of dealing with the old problems. He is tapping new sources of chemical supply and new sources of energy. He may even succeed in dispensing with green plants as prime producers, and himself obtain the manufacture of food-stuffs direct from their elements.

In all this there is promise; but there is also danger. The disadvantages of pre-human methods of evolution are their appalling slowness, their equally appalling wastefulness, and the fact that what is achieved is simply something that will work, and not something planned to work in the best and smoothest way. It is, humanly speaking, stupid that each year three-quarters of all the young that singing birds produce must come to nothing, and perhaps ninety-nine per cent of all the seeds that are made by flowers. It is stupid that the life-community in its task of utilizing the resources of nature, should be hampered by unnecessary middlemen and by creatures that short-circuit the vital circulation, but that is what the unrestricted competition of life leads to, as it leads to the sufferings of bacterial disease, and to parasite cruelty. It is stupid, again from our human standpoint, that the world had to rotate on its axis some fifty thousand million times after the reptiles began to dominate it before their brainless ascendancy was brought to an end in favour of the mammals.

Yet these disadvantages involve certain countervailing advantages. Such wastage evokes enormous reserves. The exuberance which most living things must possess to survive at all in such a wasteful world, is one of their beauties; and the reserves of energy, of leisure, of reproductive capacity which life must possess against the time of struggle, have been the soil out of which precious and unexpected advances have blossomed. If progress has been slow, it has been steady; if competition seems to have generated an unnecessary variety, yet it has ensured that when one type perished, new types were always present to take its place; if the communities of life are slow growths, they are adjusted and balanced growths; and epidemic disease and parasites, however cruel and wasteful, are among the checks and counterchecks by which this adjustment is maintained.

And conversely, the advantages opened up to man by the possibility of conscious quick attack upon his problems have their dangers. He can colonize a new country in record time and bring in his own appurtenances in the way of domestic animals and crop-plants; but, as we have seen, he almost inevitably upsets the balance of nature in the process and introduces devastating pests. He can make the soil produce a life-community which, like a wheatcrop, or a combination of grass and cattle, shall be most efficiently adapted to the one particular purpose he has in view; but he will be upsetting the chemical balance by removing the crop from where it grew without replacing its mineral constituents. He can tap new sources of food and energy; but too often lives on capital without putting by anything for the future. He can eliminate economic waste; but he runs the risk of creating a life without exuberance, with all the reserves of vitality thrown into the daily struggle. He can reduce disease and the wastage of human life; he is brought up against the danger of perpetuating weakly stocks that might better never exist at all. In a word, man can see what he wants; and because he sees what he wants he can make an immediate bid for it, and change the face of things with unbiological rapidity. But he will be very unlikely by the light of nature to see all the multifarious consequences of his bid; and too often the consequences will be quite different from what he wanted, and will turn to his harm instead of his help.

What makes it possible for man to go fast is his conscious mind and deliberate purpose. But it is not enough that merely his aim and his main effort towards it should be conscious; the whole process must be conscious. He cannot leave details to Nature and expect her to be on his side. He cannot mix the new process and the old. That is why the formidable apparatus of organized research and applied science is necessary if civilization is to continue; for it is the only possible substitute for nature's clumsy sequences of secular struggle, the sacrifice of the many for the few, broadcast waste to ensure the rare lucky survival, ruthless pruning, adaptation through strife and death.

From the standpoint of biological economics, of which human economics is but a part, man's general problem is this—to make the vital circulation of matter and energy as swift, efficient, and wasteless as it can be made; and, since we are first and foremost a continuing race, to see that we are not achieving an immediate efficiency at the expense of later generations.

To this end, man, with the aid of scientific breeding and selection, can produce organisms which are quicker and more efficient transformers of matter than anything found in nature; but he can only do so if he helps nature to satisfy their increased demands. A mere truism? Not by any means. It is true that since before recorded history cultivators and stock-raisers have used natural manure and extra fodder; it is true that in the last century, since the work of Liebig, Lawes and Gilbert, the employment of chemical manures has become almost universal. But up till quite recently man has taken little thought for the morrow beyond the single crop. It is true again that he has been forced by the demands of his wheat and corn to let his land lie fallow from time to time, or to introduce nitrogen-catching crops, like clover or lupins, into his rotation; but that is only a beginning, and the depletion of grassland we have just been considering is a reminder of the seriousness of the problem in other fields.

At last his difficulties are driving him back to first principles. In the last couple of centuries he has accelerated the circulation of matter—from raw materials to food and tools and luxuries and back to raw matter again—to an unprecedented speed. But he has done it by drawing on reserves of capital. He is using up the bottled sunshine of coal thousands of times more quickly than Nature succeeds in storing it; and the same rate of wastage holds for oil and natural gas. By reckless cutting without re-afforestation, he has not only been incurring a timber lack which future generations will have to face, but he has been robbing great stretches of the world of their soil and even of the climate which plant evolution had given them. This stripping the land of trees, soil, and moisture has gone on both in East and West. It is serious on the Mediterranean mountains; but it has reached its climax in China. That land is so densely populated that trees have given place to food-plants; there are often no trees at all, or only in gardens, over great tracts of country. Wood for fuel is in such parts almost unknown; the people burn straw and dung and refuse.

By over-killing, man has exterminated magnificent creatures like the bison as wild species. Less than a century ago, herds numbered by the hundred thousand covered the Great Plains. Buffalo Bill killed 4,280 bison with his own rifle in a year and a half; and that was far from being a record. The United States Government detailed troops to help in the slaughter, in order to force the Indians, by depriving them of their normal subsistence, to settle down to agricultural life on reservations. To-day there remain a few small protected herds.

By over-killing, he has almost wiped out whales in the northern hemisphere, and unless some international agreement is soon arrived at, the improvement of engines of destruction is likely to do the same for the antarctic seas. If he is not careful, the fur-bearers will go the same road; and the big game of the world is doomed to go, and to go speedily, unless we take measures to stop its extinction. By taking crop after crop of wheat and corn out of the land in quick succession, he exhausted the riches of the virgin soils of the American west; and is now doing the same for the grasslands of the world by taking crop after crop of sheep and cattle off them. To make good these losses of the soil, he has crushed up the nitre of Chile, the guano of Peru, the stores of phosphate rock in various parts of the earth's crust. But these too are capital and the end of them is in sight. The recklessness of the nineteenth century was appalling. Linnæus gave man the title of *Homo sapiens*, Man the Wise. One is sometimes tempted to agree with Professor Richet, who thinks that a more suitable designation would have been *Homo stultus*, Man the Fool.

Man's chief need to-day is to look ahead. He must plan his food and energy circulation as carefully as a board of directors plans a business. He must do it as one community, on a world-wide basis; and as a species, on a continuing basis. In the first place, he must learn to adjust population to supplies, and not be always and only thinking of the adjustment of supplies to population. Population may soon need to be controlled as urgently as war or unrestricted individualism needs to be controlled now. In the matter of supplies he must make provision for the future; the species must have its reserves of nitrogen and phosphorus, of timber-growth and soil-fertility, of useful animals and of sources of energy, just as surely as the Bank of England must have its reserves of gold and credit, or a factory must allow for the depreciation of its plant.

As a matter of fact, the situation is not so bad as it looks at first sight. We are using up our coal and oil; but water-power is always with us, and there are tide-power and sun-power and wind-power for us to tap. We are using up our oil; but sooner or later we shall replace it satisfactorily by power-alcohol made from plants. All over the world scientific forestry is beginning to replace irresponsible lumbering. The Peruvian government's regulations for their guano islands have ensured that each year the birds shall contribute to the needs of the future as much as man removes for the needs of the present. The nitre-beds will be finished up, but humanity need no longer worry now that a way has been found for bringing nitrogen from the air's inexhaustible reservoirs into a form available for plants. International agreements have not only saved the Alaskan fur-seals from imminent destruction, but restored them to abundance.

In the 70's of last century, the herd of fur-seals on the Pribiloff Islands off Alaska numbered about two-and-a-half million individuals. So long as animals were only killed on the islands, the number taken each year could be regulated. But private enterprise began to kill them on the high seas—with the result that by 1896 there were less than 600,000 left, and in 1911, only just over 200,000. In that year, however, all killing of fur-seals at sea was prohibited by international agreement, and by 1924 the herd numbered 700,000, and is still continuing to increase. If the whalers and the trappers and the big-game hunters are not too stupid to take similar precautions, whales and fur-bearing creatures and big game can be saved too.

In these fields man has only got to take a little trouble and he need not fear the future. But in the matter of phosphorus, the prospects are not so bright. Phosphorus is an essential constituent of all living creatures. It is, however, a rather rare element in nature, constituting only about one seven-hundredth part of the earth's crust. In the ocean, the proportion is infinitesimal, only five parts of phosphate to one hundred thousand of sea-water. This is because phosphorus is the limiting factor for marine life. To exist at all, living matter in the sea must contain, per unit of weight, almost one hundred times as much phosphorus as the surrounding water. Most of the sea's phosphorus is imprisoned in living bodies. Of all the phosphorus in the sea and in its animals and plants, some gets back to land in guano, some in phosphate deposits made of fossil bones and shells, some in fish that man catches and brings ashore; but all this is a trifling fraction of the whole, and for the most part slow-accumulating. On land, meanwhile, the soil is losing phosphorus all the time, partly by leaching out into rivers, partly by crop depletion. From the soil of the United States alone the equivalent of some six million tons of phosphate is disappearing every year; and only about a quarter of this is put back in fertilizers. Meanwhile, the store of fertilizers is being depleted, and man (*Homo stultus* again!) is sluicing phosphorus recklessly into the ocean in sewage. Each year, the equivalent of over a million tons of phosphate rock is thus dumped out to sea, most of it for all practical purposes irrecoverable. The Chinese may be less sanitary in their methods of sewage disposal, but they are certainly more sensible; in China, what has been taken out of the soil is put back into the soil. It is urgently necessary that Western "civilized" man shall alter his methods of sewage disposal. If he does not, there will be a phosphorus shortage, and therefore a food shortage, in a few generations. But even if he does that he will still have to keep his eye on phosphorus; it is the weak link in the vital chain on which his civilization is supported.

BOOK SEVEN

HEALTH AND DISEASE

I

INFECTIOUS AND CONTAGIOUS DISEASE

§ 1. *Is Man Particularly Unhealthy?* § 2. *Microbes.* § 3. *Insects as Microbe-carriers.* § 4. *Immunity.* § 5. *Avoiding and Killing Microbes.*

§ 1. Is Man Particularly Unhealthy?

IT WILL have become evident to the reader that much of what is written in Book 6 bears very directly on the problems of human health and disease. We have seen that all living things are continually engaged in an exacting struggle with their environments. For every single healthy animal or plant which attains maturity and hands on its genes to future generations a multitude suffers and fails. Human ill-health is no more than a special case in that great spectacle of defeat and failure. We find throughout the rest of life parallels to the diseases that haunt our own. Birds, for example, are affected by their own special form of malaria; mice have cancer and bubonic plague; plants suffer extremely from contagious diseases. Naturally enough crippled and unhealthy individuals are not so noticeable in wild creatures as they are in our own species, for the greater severity of their struggle for existence removes them speedily from our observation. But the diseases are there, and besides the more striking diseases to which we have just alluded there are the subtler derangements produced by unsuitable environmental factors; every species has, ringing the area in which it lives and to which it is fairly adapted, a margin of less suitable habitat, where existence is barely possible, where food is short, or the climate too hot or too cold, but into which the pressure of population drives a certain number of individuals who can just hold out there. Few of any living species are completely happy in their habitat, for exterior circumstances are always changing and are generally ahead of adaptation. A consequent stress and malaise rests therefore upon the surviving minority of each generation even if such species are holding their own in the struggle.

Man has evolved along unique lines because of the extensive grey matter of his cerebral hemispheres; to a quite unparalleled degree he controls his environment; and so his struggle for health and bodily and mental vigour assumes special features of its own. He makes a longer fight for it. One hears a lot of nonsense about "natural health"—about savages and wild things being almost uniformly fit, while only civilized man is disease-ridden; the suggestion generally follows that by turning one's back on civilization and taking to the woods one could achieve a savage healthiness. In truth what would be achieved would be a swifter death. No one who has taken the pains to look carefully into the matter will carry out this idea as to the freedom of savage races from disease. What one sees among savages is not the vigour of prolonged life but the brief precarious vigour of youth; the ailing have already been removed from the sample. Our flint-chipping ancestors were not exempt, as their skeletons show, from gouty disorders, from abscess and caries. They probably suffered from most sorts of human disease. "Perfect health" is a fantasy. The unstable health of man is only a case of the universal instability of life, and we are more vividly aware of unhealthiness in human communities because it is less swift in its action. We are, paradoxically, more aware of human unhealthiness because man is one of the most resistant of living animals.

Man's body, as we have shown in our opening Book, is a complicated and delicate structure, liable to derangement in endless ways. Its continual equilibrium exceeds the feats of a Japanese balancer. Yet man takes it into the strangest surroundings, into extremes of heat and cold, subjects it to unwonted pressures and the greatest desiccation and humidity that any living thing can endure. He is continually varying his dietary. He is continually exposing himself to strange infections. Yet by means of skin, hair, sweat, shivering and so forth, by the perpetual vigilance of liver and kidneys, by the antiseptics of its gastric juice, its tears and other protective secretions, its armies of white corpuscles and anti-toxins, this astounding body of his sustains itself in effective action.

We are now going to review the general struggle of man to keep healthy and fit, to give a brief summary of medical principles at the present time. But first we may say a word or two about the great individual variability of human beings in these matters. Because man has so great a power of artificial resistance to the assaults of his environment, a number of congenital peculiarities present themselves, which would probably not have survived under more stringent conditions. There are numbers of individuals who are not so much diseased or unhealthy as abnormal. They must avoid things that most of us need not avoid, or do things that most of us need not do. There are people, we noted in a previous chapter, whose blood will not clot of its own accord and who may therefore lose quite considerable volumes of blood from a small cut. There are people whose ductless glands are defective; their thyroids, for example, may not produce enough secretion and they may remain infantile and idiotic instead of growing up. Recently the attention of medical workers has been increasingly focused on the study of "diathesis"—i.e., of these constitutional peculiarities. Other kinds of inherited tissue weaknesses do not manifest themselves at birth but appear later in life and then, it may be, simulate diseases due to the action of external causes. There may be various symptoms; the nerve-cells or muscle-cells wither away slowly and are replaced by useless fibrous tissue; or the red or white blood-cells may be abnormally formed; or the bones may be unusually brittle, especially in later life; or a bony stiffening may be developed in the connective tissue round the muscles. Gout is a disease in which hereditary factors play a large part. Then there are various chemical abnormalities, such as the regular excretion of an aromatic substance homogentistic acid in the urine, and the presence of a pigment in the blood which makes the victim painfully sensitive to light, developing a pox-like eruption and perhaps deeper, graver symptoms if too brilliantly illuminated.

These are rare conditions. But in addition to such quite obvious abnormalities there must be smaller, less conspicuous inherent differences which in conjunction with outer circumstances may inconvenience or even impair or injure, and there are all sorts of odd congenital variations in appetite, preference and response to particular conditions. We lay stress on the fact of idiosyncrasy here because it compels us to write very generally in what follows; there are very few rules of health of universal validity. What is one man's meat may be another man's poison. In quite a number of matters we have each of us, with such medical advice as we can get, to discover our own personal equations and work our own rules of life.

§ 2. Microbes.

As we all know nowadays infectious diseases are due to living parasites. Seventy years ago this was quite unknown. Bacteriology has existed for no more than a human lifetime. It was generally believed at that time that some sort of relation existed between disease and putrefaction and such changes as the souring of milk and the frothing alcoholic fermentation of sugary fluids, but the basis of the relation was entirely obscure. The processes were supposed to be purely chemical changes. And although it was known that microscopic "animal-cules" were present in fermenting and decaying things, and even in diseased blood—the anthrax bacillus was seen by a German worker in 1855—they were regarded as results, not causes of the phenomena; they were supposed to be generated by the chemical changes. Then very rapidly the work of Pasteur transformed men's ideas and revolutionized hygiene, medical and surgical practice.

In the following pages we shall follow current usage and employ the terms "microbe" or "germ" to cover any invisibly small disease causer. To some biologists these terms may be objectionable because they may designate members of very different groups—bacteria, protozoa, simple moulds and yeasts, or filter-passing viruses—but they are too convenient to be sacrificed on that account.

Not all microbes are harmful. That is a quite erroneous idea. As we saw in Book 2, there are neutral and beneficial microbes; as examples of the latter the cheese-makers and the nitrogen-fixers in the soil may be cited. Nor are the microbes which are definitely disease-causers all on the same grade of specialization and malignity. First there are highly specialized ones, which have adapted themselves completely for a parasitic life in our bodies. Examples of these are the organisms causing smallpox and influenza, which can only be caught by direct or indirect contact with an infected person—being transferred, for example, by a towel, or an article of clothing, or a droplet of sneezed-out sputum. The protozoon that causes malaria is another highly specialized parasite, carried from man to man by mosquitoes. Anthrax is another specialist, although it will invade the blood of any warm-blooded animal indiscriminately. It is a great destroyer of flocks and herds. The anthrax bacillus is able to turn itself, when conditions are adverse, into hard, resistant spores which can wait for years in a state of suspended animation. If such a spore should chance to be swallowed with a blade of grass by a grazing animal, or, dry and dusty, get on to a sore or scratched skin, it revives and multiplies in the blood.

Contrasted with these completely parasitic bacteria there is a series of disease germs which can live inside or outside human bodies as they choose, given suitable conditions of temperature and nourishment. The cholera vibrio and the typhoid bacillus can flourish and apparently multiply in fresh-water or in damp soil, and great explosions of these diseases may follow the entry of their microbes into the water supply of a town. The microbes causing tetanus (or lockjaw) and gas gangrene are essentially not parasites at all, they can go on flourishing if there are no big animals in the world, but if they get into a cut or scratch, then they increase, multiply and poison the whole system. During the Boer War, which was fought mostly on wild uncultivated land, tetanus was practically unknown, but during the Great War, fought on richer and infected soil, tetanus was very common, and led to a considerable mortality until suitable means of combating the disease had been adopted.

When it gets a footing in a human body, a parasitic microbe may harm its host in various ways and to various degrees, according to its nature. It may actively attack his tissues—thus the malaria parasite enters and destroys his red blood-cells. Or it may injure by shedding poisons into the blood-stream. The tetanus bacillus stays in the region of the wound but sheds a highly poisonous substance (or toxin) into the blood which upsets the nervous system and produces violent and painful muscular spasms—a poison so deadly that less than a ten-thousandth of a gram—about one three-hundred-thousandth of an ounce— is enough to kill a man.

The diphtheria bacillus lives in the throat of its victim, where there appear characteristic films of broken-down, inflamed tissue. Nevertheless the effects of diphtheria are felt all over the body. This is because the bacilli shed traces of a very powerful poison into the blood. If they are grown in artificial broth for some weeks and the broth is then filtered through a porcelain filter so that all the bacteria are left behind, their toxins remain and the broth will now produce diphtheria symptoms in the heart, kidney and nervous system, if injected into an animal. But that animal will not distribute diphtheria infection, because that requires the bacteria.

The actual process of infection, the entry of the microbes into their host, is an extraordinarily complicated and various one. It is affected by a great number of different factors, and round these questions of invasion and distribution the chief difficulties of bacteriology arise.

The central puzzle is the fact that one and the same microbe can be carried by some people without hurting them in the slightest, while on others it has the most destructive effects. There are people, for example, who can carry typhoid or cholera in their systems and excrete swarms of active, deadly bacteria in their fæces and urine without themselves showing the slightest sign of illness. One has only to imagine a typhoid-carrier with a job as cook or waiter in a restaurant to realize how large a part such people play in spreading diseases. Sometimes they have had a mild bout of the disease which has left them immune; it has been estimated that one person out of every twenty that recovers from typhoid becomes a carrier for the bacillus. Sometimes these "carriers" are congenitally protected against the particular microbe that they spread. Such phenomena are found in most microbe infections.

Diphtheria is a disease spread almost entirely by carriers. The microbe has often been found at the back of the nose of healthy people.

Fig. 267. Bacilli of typhoid fever, which swim by means of long flagella.

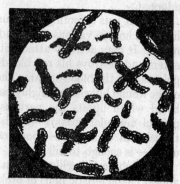

Fig. 268. Bacilli of diphtheria, which live in the throat, but shed a powerful poison into the blood, affecting especially the heart, kidneys, and nervous system.

It has even been estimated that in a crowded working-class district where diphtheria is about, one child in every ten may be a "carrier." The factors controlling the virulence of the diphtheria bacillus are complicated and by no means clear. The constitution of the host certainly plays a part. Some people seem to be naturally immune, and the percentage of these immune people varies with age in a curious way. For the first six months after birth almost all children are immune, but at the end of that time the immunity has fallen off. Then it appears again in an ever-increasing number of individuals as they grow up. But accidental influences may also affect the virulence of the diphtheria germ; the presence of another bacterium, *Streptococcus pyogenes*, seems either to increase the vigour of its attack or to lessen our resistances.

Somewhat similar relations exist with tuberculosis and pneumonia. Here the bacteria are very widespread, and commonly found in perfectly fit, active individuals. Whether they are dangerous or not seems to depend on the individual's condition.

As another example of the way in which the host's condition can affect his relation with microbes, we may take the *Bacillus coli*, a very common parasite normally present in the human intestine, and held in check by our organic defences. But if the bowel wall is perforated by some accident the bacillus may get out into the belly cavity and cause serious trouble. Or if through interference with the blood-supply the vitality of the bowel wall is weakened, the bacteria can make their way through as if there was a perforation.

This shows how the host's resistance to a microbe may vary. But the virulence of a bacterial strain is itself a changeable thing. Bacteria grown in artificial media are often much less virulent than those from living animals. Changes of this kind may underlie sudden outbreaks of infectious disease; thus smallpox in England usually occurs in a comparatively mild form, but occasionally sudden lapses into a more dangerous type occur. A good example of sudden change in character of a disease is afforded by syphilis, which smouldered obscurely in various countries during the Middle Ages and suddenly broke out as a violent epidemic in the fifteenth century. It may be that *Encephalitis lethargica*, or sleeping sickness, a disease which was first recognized in Europe ten years ago, is an example of the same kind of thing.

Even the weather may affect these bacterial attacks. A cold, muggy spell, by weakening people's throats, may hasten the spread of diphtheria.

Thus, by variation in the virulence of a parasite and the resistance of its host, and by other circumstances, the relations between a human population and the bacterial population inside it may be altered. A disease which has been locally confined and comparatively harmless may suddenly break loose and sweep through vast populations unaccustomed to its attacks. In the first condition it is called endemic; in the second it is epidemic, or, if very widespread indeed, pandemic. Influenza gives a good example of a pandemic; in 1918–19 it swept through the disorganized world and destroyed more lives than did five years of the Great War. In London alone it killed eighteen thousand persons; in India upwards of six millions. The Black Death is a classical example of an infectious disease thus breaking loose. Cholera is endemic in the warm, humid delta of the Ganges, and from there it raids from time to time in various directions, borne by actual sufferers or by the immune "carriers." On a rather humbler plane, measles breaks out with particular violence among the children of great cities about once every two or three years. Here the determining factor is the supply of raw material for the measle germ (a filter-passing virus) to work upon. Children in the first three years of life are most susceptible, so each wave sweeps through the children of that age and then dies away and the bacteria wait in obscurity until a new set of infants is ready to sustain their next phase of exuberance.

§ 3. Insects as Microbe-carriers.

IN DISCUSSING parasitism in Book 6 we noted that one of the chief problems confronting a parasitic organism is the spreading from host to host. In biological slang, we say that it lives in a "discontinuous habitat," and some method of infection must be a part of its life-history.

Most of the microbe diseases described in the last section are spread by more or less direct transference from person to person; the ordinary contacts and movements of humanity are enough to keep them going and to spread them about. But there are stranger, more elaborate life-histories among our parasites. It will be interesting to consider as an example of extreme parasitic specialization the life-story of the microbe responsible for malarial fever.

It has been said that malaria, ague, or marsh fever is the most destructive ill to which our species is subject. Certainly, seen in historical perspective, it assumes vast proportions. It has been estimated that in India forty per cent of the death-rate is due to malaria in one form or another—six times as much as cholera. In the past, marsh fever has wiped out armies and depopulated cities. It may have been responsible, in part at least, for the downfall of the Greek and Roman civilizations; it was certainly known as a scourge in those days. It is described unmistakably in the Hippocratic writings. Until recently it was very widespread. It was endemic in England, particularly in marshy districts such as the fens of Cambridge or Lincoln, and it hung over London until the second half of the last century. In 1850-60, barely a lifetime ago, over one-twentieth of the patients at St. Thomas's Hospital were ague cases. Its retreat from London was the result of the building of the Thames Embankment and the drying up and utilization of the land along the Thames; from other countries its recession was brought about by agricultural drainage. When the swamps went, malaria went too.

The fact that malaria brooded chiefly over marshy districts has been known for centuries, and it was inferred from this fact that the cause of the disease was damp marsh air. Hence the name, *mala aria*, which is Italian for "bad air." But other guesses had been hazarded. In the writing of the great Indian physician, Susruta, of the fifth century, malaria is attributed to the bites of mosquitoes. It is pointed out that violent malarial outbreaks sometimes coincide with plagues of blood-sucking insects.

We know now that the disease is due to a protozoan parasite belonging to the group called Sporozoa, and that it has a complicated life-

 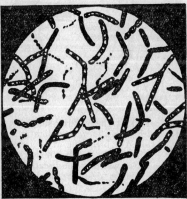

FIG. 269. Cholera vibrios, which swim with very active writhing movements.

FIG. 270. Bacilli of tuberculosis, which infect nearly everybody, but are generally held in check.

story, part of which is spent in human blood and part in the body of a mosquito. It cannot be transmitted directly from man to man or from mosquito to mosquito, but has to alternate regularly between the two. This rhythm is its only way of spreading. Incidentally, only the female mosquitoes are concerned, because the males, in this species at least, never suck blood.

The details are curious and interesting and we will give them at some length. (To make this rather complicated string of events clear, we punctuate our description with numbers corresponding to those on Fig. 272). Let us start with the infection of the man.

When a mosquito bites a man, she follows up the puncturing of his skin by injecting a drop of "saliva"—the name given to a fluid produced by two large glands that open into her mouth. This injection irritates the tissue of her victim and causes the blood-vessels to expand locally (hence the red spot at the site of puncture) and also it prevents the blood from clotting. Thus it doubly assists her meal. It is in this saliva that the parasite invades its human host. If one examines the saliva of an infected mosquito with a powerful microscope it is seen to contain a number of little spindle-shaped creatures, each with a nucleus in the middle. If they are injected into the blood-stream of a man, these parasites get busy at once (1). They swim actively by spiral wrigglings of their bodies and approach the red blood-cells. Each parasite sticks one of its pointed ends into a red cell and in about forty minutes of violent wriggling, works its way inside (2).

Once in the red cell the shape of the parasite changes. It becomes rounded, moving about in its little living habitation like an active amœba, and presently as it grows it develops a clear space in its middle, a vacuole, which gives it the appearance of a ring (4). Feeding by absorbing the substance of the red cell it gets bigger and bigger, until it nearly fills the latter (7); then its movements slow down and stop and it begins to divide itself up, parcelling out its whole body, after the manner of protozoa, into its progeny. In this manner it gives rise to some ten or fifteen little parasites (10). When the division has taken place the exhausted red cell breaks up and the parasites are liberated, to invade and consume other red cells as their parent did (11).

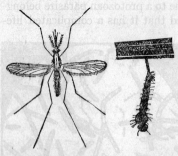

FIG. 271. Left, the malarial mosquito *Anopheles maculipennis*. Right, a mosquito larva breathing at the surface of the water. A thin film of oil on the water will clog the breathing-tube, and thus destroy the larvæ.

Now it is when the cells break up in this way that the malarial fever assails the sufferer. During the growth of the parasites inside the red cells he feels little harm. But when the red cells break down there is a liberation of poisons into his blood, which produce a general fever, and the parasites and broken-down red cells may actually plug up his capillaries, in the brain or heart or lungs or elsewhere, and give him various local symptoms in addition. And as the red cells are destroyed in ever-increasing numbers, he suffers from anæmia and another set of distressing symptoms. Moreover, there are three distinct species of malarial parasite, differing in the time relations of the life-cycle. One, *Plasmodium vivax*, takes just two days to grow in the red cells, so that the sufferer has acute feverish attacks every other day, starting with the second day after he received the poisoned bite. Another, *Plas-*

modium malariæ, takes three days to grow, so the fever comes on regularly every third day. The former rhythm is called tertian, and the latter quartan, malaria. The remaining species, *Plasmodium falciparum*, is irregular, some individuals growing faster than others, and therefore produces a continuous or irregularly recurring fever.

Thus, for a time, the parasites grow and proliferate in the red cells of their host. But in due course, perhaps warned by the damage that they have done that their food-supply is running out, they change their tactics. Instead of dividing up into batches of little ones the fully grown parasites convert themselves into sexual forms (12 and 13). The exact appearance of these sexual forms varies from species to species—our figure shows how they are seen in *Plasmodium vivax*—but they are always of two kinds, male and female. When they reach this stage, the parasites stop and wait.

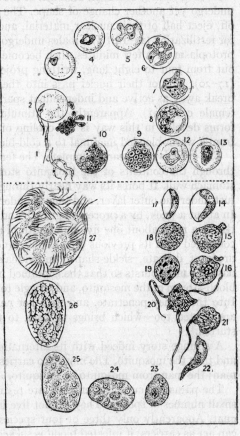

FIG. 272. The life-cycle of the malaria parasite.

The numbers show the successive stages in order, and are referred to in the text. All stages above the dotted line occur in human blood; those below occur in the mosquito which carries the infection.

Now these sexual forms cannot develop further unless a very improbable accident happens. Most of them never do get any further; in a few days the great majority perish. But one or two may have been lucky, a few among millions, and have found themselves sucked up with a drop of their host's blood by another mosquito.

As soon as they reach the stomach of the mosquito these lucky ones undergo yet another change of shape. The females round themselves off, eject half of their nuclear material, and thus prepare themselves for fertilization (14–16). The males undergo violent streaming of their protoplasm for a few minutes, then become still, then abruptly squirt out from four to eight long, writhing projections from their surfaces (17–20). Bits of their nuclei pass into these projections which then break away as active and independent spermatozoa, and fertilize the female cells (21). Apparently the stimulus that makes the sexual forms develop in this way is the cooling of the blood when it passes from a warm-blooded mammal to a cold-blooded insect.

The cycle is now nearly complete. The fertilized female cell wriggles through the contents of the mosquito stomach until it reaches the stomach wall. It bores its way nearly through the wall, coming to rest just under the outer layer, and then it undergoes further changes (23). In about a week, by a process of growth and proliferation, it has given rise to a mass about one five-hundredth of an inch across—enormous compared with its previous dimensions—and consisting of some hundreds of delicate, sickle-shaped individuals (27). In another three days the mass bursts so that the contained individuals escape into the blood-stream of the mosquito, and wriggle towards her salivary glands. Into these they penetrate, and with her next ejection of saliva they are ejected too—which brings us back to the point from which we started.

A strange story indeed with its alternation of hosts, first the man and then the mosquito. The mosquito carries the parasite from man to man; the man from mosquito to mosquito.

The parasites, curiously enough, are restricted for their hosts to a small number of species. They cannot live in any sort of mosquito or gnat. Apparently only three or four species of Anopheline mosquito can act as carriers; if infected blood is sucked up by any other kind of blood-sucking insect the parasites are digested at once. On the other hand another and closely related species of parasite that causes a malaria-like disease in birds can only be transmitted by the gnat *Culex*, and there are two other parasites causing redwater fever in cattle and malignant jaundice in dogs which are carried from beast to beast by ticks.

There are a number of other cases of this carrying of parasites by blood-sucking arthropods. This is why arthropods can be so dangerous to man. The harm that their infections may do is very much greater than the mere irritation of their bite.

We cannot list all the diseases which have intermediate hosts of this kind, but we may consider briefly one or two of the most important. The case of sleeping sickness, due to a flagellate protozoon carried from man to man by the African tsetse-fly, is well known. A curiously ingenious parasite is the spirochæte *Borrelia berbera*, causing African relapsing fever, which is carried by lice; here the spirochæte does not infect by way of the bite but simply swarms in the body-fluid of the louse; when the louse is crushed as it bites (as very often happens, because of the irritation it produces) the parasites are liberated and enter the new host through the puncture it made. Typhus fever also is borne by lice, the parasites being shed in the excreta of the lice and entering through cracks in the skin.

Plague is a flea-transmitted disease, and biologically rather a curious one. Like malaria, it has played a dramatic part in history; always it has been hanging about in the East, and time after time it has swept across the Mediterranean and into Europe. The Black Death was only one of its many onslaughts. Nevertheless, it is essentially not a disease of men but a disease of rodents. It is apparently endemic among the rats and ground squirrels in the East and in America, being carried from rat to rat by their fleas. Occasionally, for reasons that are at present obscure, it breaks out with particular violence, and the rat-fleas leave their dying hosts and climb on to men, or any other warm-blooded creatures that present themselves, carrying the plague bacilli to them. This is why onsets of plague are mostly preceded by a high mortality among rodents.

As a final example of an arthropod-borne disease we will consider yellow fever. This is not as widespread, geographically speaking, as plague or malaria; nevertheless it has been a grim scourge. Essentially it is a Central American disease, but it was carried to West Africa by the slave trade and has flourished and spread in that region. It was Yellow Jack that smote the ship that carried the Ancient Mariner and thus avenged the albatross. Sometimes it has raided northward from these tropical zones, in the Old World very devastatingly to Spain, Portugal, and Italy, and once even to England; in the New to Baltimore, New York, and even so far as Boston. But generally its ravages have been in warmer climates. And until recently it hit hard. Dr. Charles Singer in his admirable *Short History of Medicine* says that every year for many years the British garrison in Jamaica lost 185 men per 1,000 to Yellow Jack—nearly one in five!

The mystery of the transmission of this disease was solved by an American commission led by Walter Reed, in the opening years of this century. Theirs was an extraordinary investigation. Yellow fever cannot be given to animals, and it is still doubtful whether the responsible microbe has been seen and recognized. So the commission had to experiment on volunteers and on themselves, with microbes they had never seen. But they established the fact that the disease could only

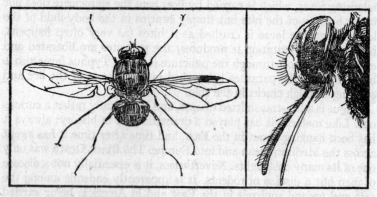

Fig. 273. The tsetse-fly, Glossina palpalis, which transmits sleeping sickness in Africa.

Left, the fly (in life, about a third of an inch long), and right, a close-up of its head showing the long, piercing mouth-parts.

be caught as the result of the bite of a particular species of mosquito, and their culminating experiment is well worth quoting for its pure physical heroism. When their preliminary experiments had shown, pretty clearly but not conclusively, that the mosquito in question was guilty, they went to a district where there was no endemic yellow fever, and there they built two wooden huts. One was horribly stuffy, the other pleasant and airy. Into the first hut went nine volunteers, three at a time. They tried in every imaginable way to infect themselves with Yellow Jack, except that they did not have any infected mosquitoes near them. They slept between sheets and blankets from the beds of men who had died of the scourge, and even wore the pyjamas of the sufferers. They wrapped round their pillows towels that had been soaked in the blood of men dead of yellow fever. Each set lived under these conditions for twenty nights. And not one of the nine got yellow fever! Into the other hut went a second batch of volunteers, and they lived under healthy, normal conditions except that they deliberately allowed themselves to be bitten by mosquitoes which had fed, a fortnight or so previously, on the blood of fever patients, and they caught the disease—at least, five out of six did. Then, a final touch to make the chain of proof complete, two of the men from the first house, the stuffy one, were bitten with the infected insects—just to see whether, by some coincidence, they were naturally immune! But they caught it from the mosquitoes. So there was no doubt at all, after this, that the fever was transmitted by the mosquitoes and in no other way.

Once this had been proved, war was declared on yellow fever, by systematically destroying the breeding-places of the mosquitoes responsible. The result has been tremendous; the disease is rapidly coming under control, except in the wilder parts of Central and South America. Soon, perhaps, it will be stamped out altogether. Similarly with the other diseases that we have discussed. Malaria left the Panama Zone with yellow fever; elsewhere, and notably in India, it is still rampant, although it is being systematically attacked along these lines. Plague can be controlled by wholesale exterminations of rats. Now that the necessary knowledge is in our possession the stamping out of these scourges is simply a matter of organization. We destroy the breeding-grounds and exterminate the tainted carriers, and the disease is ended. So we cure the race. The cure of the infected individual is a different question.

§ 4. Immunity.

ONE of the chief ways of combating disease-germs in individual cases is by aiding and intensifying the natural resources of the patient. In the event of bacteria making an entry—through a scratch in the skin, for example—phagocytes are rushed at once to the spot. They attempt to surround and devour the bacteria. Some kinds of germ are more readily attacked than others, and this is presumably an important factor in determining virulence. For example, guinea-pigs and mice are very susceptible to anthrax, while dogs are much less so; and it has been shown that for some reason the phagocytes of the former animals do not attack the anthrax bacillus, while those of the latter do. But besides this actual physical battling of invaders with defending cells, there is a subtler warfare of chemical products in the blood-stream which is of great importance from a medical point of view, and which we must now consider.

One of the most striking protective devices in the body is its ability to respond chemically to foreign substances introduced into the blood. Some aspects of this ability were discussed in Book 3. In the matter of bacteria, the body manufactures "anti-toxins" which neutralize the toxins which they secrete and rob them of their sting; it attempts, for example, to withstand the chemical assaults of the diphtheria bacillus by producing a diphtheria anti-toxin. Moreover, these anti-toxins are specific. Diphtheria anti-toxin is no good against tetanus toxin, and if after recovery from diphtheria the patient is attacked by tetanus he has to start the business of anti-toxin formation all over again. Where these substances are produced is at present not clear. There is some evidence that that versatile organ, the liver, plays a considerable part in their manufacture.

Not only does the body respond in this way to toxins; apparently the materials of which the surface membrane of bacteria are made can, in many cases, stimulate chemical retaliation of this kind. Various kinds of substances are turned out to combat whole bacteria. Some, called agglutinins, make them sticky so that they adhere together embarrassingly in clumps, others, called lysins, make them break up and dissolve, and so forth. The strongest substances are the so-called opsonins, which make the bacteria more attractive to phagocytes and thus assist the latter in their perpetual policing of the blood-vessels, the streets of the body-state. In this way, by living phagocytes and chemical anti-bodies, the invaders are resisted and very often overcome.

Now in many cases the chemical anti-bodies that we have just been considering do not disappear from the blood when the disease is over. Once it has made them the body keeps its weapons, for a time at least. This is why, after an attack of chicken-pox or measles, for example, a person is immune to that particular disease for some years. The anti-bodies are still circulating in the blood. At the price of suffering the disorder, immunity to its future assaults has been acquired.

Upon these facts one of the most important branches of modern preventive medicine is based. With many contagious diseases, an attack confers subsequent immunity to that same disease, often for a period of years; so that if we can give a person a mild attack, we can make him or her immune.

To give a simple example. Pasteur found that if anthrax bacilli were kept at a high temperature they lost much of their virulence. If these weakened parasites were injected into a sheep the animal underwent a mild attack of anthrax from which it easily recovered. After that it was immune to the attacks of anthrax bacilli even if they were of

INFECTIOUS AND CONTAGIOUS DISEASE

natural undiminished activity. The defensive substances were preserved in its blood.

A more important application of these principles is vaccination against smallpox. Until about a hundred years ago this disease was an extraordinarily widespread scourge. In many countries practically everybody went through it; some it killed, some it branded, some it left comparatively unharmed. In England, in the eighteenth century, probably three or four people in every thousand died of smallpox every year. But it was noticed very early that an attack of smallpox conferred immunity in future. If the sufferer recovered, he did not contract the complaint again. It used to be common in the East for people to infect themselves deliberately from a sufferer who had smallpox in a mild form, in order to avoid getting it badly—a practice that found its way into Europe in the eighteenth century. In 1798 the physician Jenner discovered a safer method of securing protection. There is a disease of cows that is rather like smallpox, and Jenner found that if a person is inoculated with this cowpox he gets a comparatively harmless disease, which is not infectious, and which leaves him immune not only to cowpox but to smallpox as well. Apparently the two diseases are due to strains, of different virulence, of the same germ (which belongs to the filter-passing group). That is the principle of vaccination (the word comes from vaccinia or the cowpox) as it is done today.

The desirability of universal vaccination is still in dispute. There are active groups of anti-vaccinationists in many countries. We do not propose to go in detail into the pros and cons of this involved discussion. We consider that the statistics of smallpox incidence and mortality prove quite conclusively that vaccination has the greatest effect in protecting against the disease.*

Smallpox is now to a large extent robbed of its terror. Occasionally a vaccination has ill effects, especially in people who were not vaccinated in infancy. There is a puzzling group of nervous diseases, such as a form of encephalitis, which in very rare cases appear after infectious diseases, such as influenza and whooping cough, and may follow vaccinia. But such sporadic accidents are heavily outweighed by the benefit that the spread of vaccination has brought about. For the statements we hear about vaccination being responsible for the increase in cancer and in infantile syphilis there is no foundation whatever.

---

*The figures are to be found in a pamphlet, "Smallpox and Vaccination," published by the Ministry of Health, London, 1924.

For some diseases it is not necessary to inject living microbes; dead ones may stimulate the body to produce the necessary anti-bodies, and thus be used for immunization. Vaccination against typhoid fever is done in this way, by injecting dead cultures. It has been carried out very extensively in the armies of various nations, with admirable results.

In those cases that we have just considered the body is forced, by inoculating it either with microbes or with their products, to arm itself with anti-bodies against the invaders. This is called "active immunity." There are cases, however, in which it is impossible to make the body protect itself in this way. There may be no time. Thus, in preventive inoculation against tetanus, horses are made immune to the tetanus toxin by injecting it in gradually increasing doses, and then a little of their serum, containing anti-toxin, is transferred to the human patient. Thus the man is protected by having in his blood the anti-bodies that the horse made. This kind of immunity, produced by injecting ready-made anti-bodies, is called "passive immunity." Anti-tetanic serum has already saved countless lives. In 1914 upon the European War front there was a terrible mortality from tetanus among the wounded, owing to their wounds becoming infected with bacilli from the soil. But after that year all wound-cases were treated with anti-tetanic serum as a matter of routine, with the result that the incidence of tetanus fell from sixteen per thousand to two per thousand. Nowadays such injection is a routine procedure at hospitals whenever an accident case with wounds is brought in. Diphtheria is a second example of a disease whose course can be very considerably modified by the injection of anti-toxins.

These are but a few of many successful applications of the facts the study of immunity reactions has brought to light.

§ 5. Avoiding and Killing Microbes.

BESIDES these methods of strengthening our bodily defences, bacteria can be attacked directly by various chemical and physical agencies. An enormous amount of sickness and suffering has been saved by the working out of reliable methods of disinfection.

Less than a hundred years ago a surgical operation was a horrible risk, a thing to be undertaken only as a last resource. Death was common after such comparatively simple operations as the amputation of a limb; the cut surface was a door flung open to the bacteria whose very existence was still unsuspected, and the patient got gangrene in the wound with various kinds of blood poisoning; often the ligatures came unstuck in the rotten tissue and he bled to death.

A pioneer in the reform of surgical technique was Semmelweis (1818–65), a young Hungarian physician who worked in the maternity wards of the General Hospital at Vienna. The mortality in those wards was appalling; at least six per cent of the women died after childbirth of the disease known as puerperal fever, sometimes for months at a time the rate was two or three times as great. Semmelweis noticed first that the mortality was very much higher in a ward where students were instructed than in one which was only attended by midwives. Then he was suddenly shocked by the death of a friend, who died from blood-poisoning resulting from a cut received while he was making a post-mortem examination. The symptoms were very like those of puerperal fever. Semmelweis concluded that there was something produced in decomposition that caused these diseases, and that the students carried it on their hands from the dissecting-room to the wards. He noticed that the usual soap-and-water wash did not remove the smell of putrefaction from the students' fingers. Also, he thought, it could be carried from diseased to healthy women in the same way. So he gave instructions that before any patient was examined the examiner must wash his hands in a solution of chlorine or chlorinated lime-water, substances which were widely used for deodorizing putrefying things at the time. His experiment was successful; as a result of this precaution the mortality from puerperal fever fell very markedly in his wards.

Two or three years earlier, in 1843, Oliver Wendell Holmes, in America, had pointed out independently that puerperal fever was apparently contagious and carried in some way by the hands of the operator. He suggested strict cleansing of the hands, with a routine like that adoped by Semmelweis.

These pioneers were feeling their way towards what is now called asepsis. But until it was demonstrated that the suppurative diseases that generally followed surgical operations in those days were due to bacteria, they could not get a real grip on the matter. They were fighting an unknown enemy in the dark.

The importance of Pasteur's results to the surgeon was first realized by Lord Lister (1827–1912), who had been studying the unpleasant results of surgical operations for several years when Pasteur's writings began to reach him. He had concluded, as Semmelweis and Holmes had concluded, that suppuration is essentially like decomposition. Pasteur showed that decomposition was due to the action of microbes, and presently it became clear that several infectious diseases were due to microbes, too. Evidently, then, the thing to do was to kill the microbes before they could get into the wound.

At first Lister was too violent in his methods. He dabbed the wound, as he worked, with lint soaked in crude carbolic acid, and he had a spray of the same fluid playing over the operation. This, he found, sometimes killed the tissues as well as the bacteria. Then he tried milder antiseptics, and he began to sterilize his instruments and dressings by heating them instead of by soaking them in strong chemical solutions. Soon he got striking results. Before his adoption of these antiseptic methods forty-three per cent of his amputations had proved fatal. After, the number fell to fifteen per cent. We have gone a long way since then; fifteen per cent would be considered appallingly high nowadays; nevertheless the drop in mortality to one-third of its former value showed that the new methods were bearing fruit.

From this foundation of Lister's the modern school of aseptic surgery has sprung. Nowadays heat has largely replaced Lister's crude chemical antisepsis. The operators and nurses wear white, sterilized garments and masks, and sterilized rubber gloves; they work in a clean, white, sterilized operating-theatre. As a result of this advance surgery has increased enormously in scope and efficiency. It is possible now to perform operations that would have seemed ridiculously audacious a life-span ago. In 1850 it took a bold surgeon to open the abdominal cavity—and a bolder patient; nowadays a bullet can be extracted successfully from the beating heart.

At the same time there has been a great increase in our knowledge of microbe-killing agencies. By boiling, heating in air or treating with appropriate chemicals any bacterial contaminations on foodstuffs or articles of clothing can be destroyed. Moreover, it is possible to a certain extent to disinfect the inside of our own bodies, for substances have been discovered which will poison particular bacteria or protozoa without seriously affecting our own tissues. This, we may note, is by no means easy. The ordinary antiseptics like formalin, calomel, iodine, chlorine and carbolic acid are general protoplasmic poisons; they kill anything living and can only be applied to our skins because of the horny imperviousness of our dead outer layer. On the other hand one or two recently discovered substances are selective. The dyes trypan red and trypan blue are lethal to certain trypanosomes but harmless to ourselves. Quinine (as has been known for centuries) kills the malaria parasite without injury to our tissues, and various synthetic substances of similar constitution, such as "optoquine" are being tried with varying results. The most spectacular and successful selective antiseptic is salvarsan, a complex substance containing combined arsenic, which is very effective in killing spirochætes (such as the syphilis spirochæte), but so harmless to our own tissues that in most cases its injection into the system causes no ill effects. The story of this discovery illustrates the extreme difficulty of the problem of killing bacteria without hurting the surrounding tissues. Paul Ehrlich and his colleagues, working systematically with the technical assistance of a great German dyeworks, tried over six hundred compounds on their experimental animals before they found a satisfactory one—the name 606, commonly attached to salvarsan, is simply its serial number in the books of the investigators. Recently, continuing their investigations, they have discovered another compound, neo-salvarsan—number 914!—which seems to be even better than number 606.

II

THE NOURISHMENT OF THE BODY

§ 1. *Mr. Everyman at Table.* § 2. *The Six Vitamins.* § 3. *Some Possible Poisons.* § 4. *Drugs, Their Uses and Dangers.*

§ 1. Mr. Everyman at Table.

BUT Mr. Everyman cannot lay the whole responsibility for his disorders and discomforts to the account of microbes. To a large extent his health is his own affair. He must keep up his resistance. If he eats and drinks unwisely, or if he over-smokes, if he over-muffles himself and takes no exercise, he is weakening himself, so that for some at least of his troubles he has only himself to blame. Consider, for example, his meals. Three or four times a day he stokes his inward engines. Is he fuelling them correctly or is he fouling the flues and choking the furnace with ash?

The body needs food for two purposes—firstly as fuel, secondly as material for building up its own living machinery. Food, therefore, should fulfil these two requirements. It should have enough energy bottled up in its molecules to keep our wheels going round, and it should include those chemical materials which are required for our multitudinous chemical operations. That, in brief, is why Mr. Everyman has to eat. His choice of food, however, is governed by influences of a very different kind; by custom and pocket and the whims of a more or less erratic palate. Let us see whether physiology has any words of guidance for him as he hovers over the menu.

So far as energy is concerned, that useful abstraction, the average man, needs about three thousand calories per day in a temperate climate. He needs more if he does strenuous muscular work, and less if he lives in a hot country. Roughly speaking, these three thousand calories are contained in a pound and a half of cheese or sugar, or in three pounds of bread, or in four pounds of herrings or eggs or lean beef, or in five pounds of apples, or in seven pounds of potatoes. They are contained amply and to spare in the diet of people who are fairly

THE NOURISHMENT OF THE BODY

comfortably off—in those, for example, who can afford this Book. The chief trouble in their class is the tendency to eat too much. Man is still born with the organic constitution and instincts of an improvident savage, of an ape-man living from hand to mouth. Such a creature snaps up what it can get—berries, roots, insects—the limiting factor is not appetite but supply. If it finds a windfall—if, for example it kills a large mammal—it gorges itself and lays down stores of fat, which are slowly reabsorbed and consumed. Suddenly, in a few thousand years, this creature has civilized itself and finds itself with food enough and to spare; and the instinct to gorge and fatten, now made useless by security, has become a dangerous factor.

This man-ape, our reader, enjoys eating, and if he can afford it he eats to repletion—not once in a while, which is, so to speak, what Nature intends—but two or three times a day. In extreme cases, of gourmets who take little exercise, the body has great difficulty in getting rid of the fuel thus generously supplied to it. It makes enormous deposits of fat. It burns the stuff extravagantly, and, since most of the body is covered by clothes, the face and neck become flushed and purple with distended blood-vessels, so that as much heat as possible may be lost to the surrounding air.

Mortality statistics throw an unflattering light on human appetite. In the well-to-do a much higher proportion of deaths is due to disease of the digestive system and to disease (such as a form of diabetes) caused by over-nutrition than in the poor. It would seem that those in a position to do so over-strain their digestive machinery. At the other end of the scale, among the poorest classes of the community, there is under-nutrition, manifesting itself as a lowered resistance to a number of more definite diseases; but this, it would seem, is due not so much to too little being eaten as to the ingredients in the diet being unsuitable. It brings us to the second aspect of nutrition—to food as a source of the various chemical substances that the body requires.

Here again the reasonably well-to-do, taking an average common-sense mixed diet, will in the long run get all they need. The human inside can handle and survive a great variety of diets ranging from vegetarianism to the rich oily and meaty diet of an Eskimo. What is best for any individual depends on his age and condition, and it depends also on his idiosyncrasies, for these are matters in which people vary enormously. For these reasons we do not propose to deal systematically and in detail with all the different foodstuff that may appear on the menu. Those who wish, for economic or other reasons, to pursue these questions further, will find a clear treatment in Barbara Callow's admirable little book *Food and Health*. We will confine the present discussion to a few points of general interest, bearing in mind as a sort of guiding spirit that picture of the dawn-man, a snatcher of fruits, nuts, and invertebrate animals, who has civilized himself abruptly and finds himself in an environment to which his native equipment, anatomical and instinctive, is quite out of adjustment. He has a hitherto unapproached variety of edible things at his disposal; how and what is he to choose?

One danger which he runs, and it is a very grave one, is that he may not get enough *fresh* foods among his dietary. As civilization advances and complicates itself, the town dweller comes to rely more and more on preserved articles. Instead of growing his cabbages in his garden he takes them out of a tin. True, he gets more variety, and there is a great saving of trouble; he or she can come home tired from a day's work and make a meal by simply opening a tin. But all of this stuff has suffered, even if its suffering has been very slight, by some method of artificial preservation. Even milk from the surrounding country, in a great city, contains a little preservative. How far are the various methods of preservation detrimental and how far are they good?

Cold storage is the method of preservation that involves least alteration of the composition of a foodstuff. Biological processes, like chemical processes, go on slower at low temperatures; so that merely chilling or freezing food delays the normal changes that it would undergo if left about in the warm. Drying also is comparatively harmless—at least it does not involve the addition of dangerous substances to the food. It may, however, mean the loss of good things—of appetizing flavours, for example, or of those vitamins which are readily oxidized. Of smoking and salting the same may be said. Chemical preservations such as boric acid or benzoic acid or sulphur dioxide are more dangerous. They are preservatives because they poison bacteria, and in excess they may be harmful to ourselves. Their strength in any article of food is now controlled by law, but in a breakfast including, say, fish, butter, bacon, and jam (all of which may legally contain preservatives) they might well accumulate and cause indigestion. Salicylic acid sometimes accumulates in the body to harmful amounts. Moreover, there is always the chance of a preservative neutralizing or decomposing some of the substances which are present in the merest traces in our food and nevertheless play important parts in our bodily processes.

In recent years there has been a rapid increase in the proportion of tinned and bottled foods, which become more and more numerous, varied and attractive. The tinning or bottling process consists essentially in heating the food sufficiently to destroy any decay bacteria in it, but without spoiling its appearance or flavour, and then sealing it up. Preserved foods in glass containers have been similarly treated —(as a matter of fact much of it is tinned food that is transferred to glass pots before it is sold to make it more saleable). But during the heating process there is always a certain amount of chemical alteration of the actual foodstuffs—vitamins, for example, are destroyed. Moreover, canned foods are often touched up and coloured in various ways so that it is difficult or impossible to tell by merely inspecting it what one is eating. With the responsibility of tinned things for food-poisoning we shall deal in a later section.

Quite aside from preservation, a great number of the articles of food that are now set before Mr. Everyman are not natural articles at all. They are substances extracted from the natural plant and animal tissues upon which his uncivilized ancestors subsisted, and for which his gastric apparatus is presumably prepared. Sugar is a case in point. The sugar cane has been cultivated in Asia from great antiquity, but in Europe the common availability of sugar is only a few generations old. Before the eighteenth century it was a costly luxury and an article of medicine in Europe and it spread with tea and coffee. Now sugar is not in itself a necessary substance; starch does all in our bodies that sugar does, and indeed it is turned into the simple sugar, glucose, by our digestive juices before it is absorbed into the blood. Nevertheless, under primitive conditions, a sweet tooth is a very useful thing, for its indulgence involves the consumption of such natural plant products as fruit or honey, which are rich in vitamins and other healthful ingredients. But nowadays the appetite simply leads to the eating of refined sugar and sweetmeats. It has been said that excess of sweet things can do as much harm as excess of alcohol, for it cloys the appetite and takes the place of more natural foods. Also, since sugar is an admirable pabulum for bacteria, sweet-sucking is bad for the teeth. This is a desire that Mr. Everyman should watch and control, since he has such unprecedented opportunities for its indulgence. His desire is an entirely misleading desire in this respect.

Another foodstuff, also chiefly a carbohydrate one, that may fairly be called unnatural, is white flour. Wheat before it is milled is a richly varied substance. Each individual grain has a central mass largely consisting of starch, which is a store of food for the young growing plant just as egg-yolk is a store of food. Pressed against this "albumen" is an embryo plant. Surrounding the whole is an apparatus of envelopes, of which botanists distinguish five. Our sub-civilized ancestors mashed up the whole thing to make meal, and as a result they got a nutritious bread. Nowadays milling processes have been perfected until it is possible to sift away everything except the central food-store, and to make bread only from that. Such highly refined bread is much whiter, cleaner looking, and therefore more attractive than wholemeal bread and it has the same fuel value, but it is poorer both as regards vitamin content and as regards something whose importance has only recently been realized—the amount of rough, indigestible matter that it contains.

It is not possible to say with any certainty on what an average apeman fed before civilization got to work upon him. Probably he ate whatever he could get. The greater apes to-day are largely vegetarian, taking roots, tender shoots, seeds, and fruit, but they vary their diet with eggs or nestlings, and such invertebrate animals as insect larvæ. But whatever the details were, the primitive diet must have been a chemically varied diet in that it consisted of fresh natural products, whether animal or plant. Any such product contains all sorts of chemical ingredients. The current tendency is to over-simplify foods; the cases of sugar and white flour are good illustrations. The natural man, if one may use the term, flung all sorts of stuff down his gullet—and left the job of selection to the powers within. They took the rough with the smooth. They got a good deal that they didn't want, but on the whole they were well protected. They had an elaborate apparatus of tonsils and lymphoid patches and vermiform appendix to defend them against living undesirables, and a liver to censor the products of their digestive activities. But at least the body could be sure of finding all the ingredients that were necessary in the miscellany of stuff with which it was supplied.

Apparently the human abdomen is made to be treated in this somewhat unceremonious way, and much of the prevailing digestive incompetence is due to its task being made too easy by modern separated and refined foods. A "natural" foodstuff, especially a vegetable foodstuff, contains a certain proportion of fibres and husks that our digestive machinery can make nothing of. These act as a stimulus to the bowels. They insist on being thrown out. If one eats only highly digestible food, this "roughage" may fail and the processes of evacuation slacken accordingly, with the most insidious and pervading effects on one's general vigour. Robert W. Service, in his entertaining and lively book, *Why Not Grow Young?* speaks of a handful of bran tempered with skimmed milk as the best lunch-time sweet. Tastes vary; some of us have more easily stimulated insides than others; but if one does not eat enough of this precious rubbish called roughage one has to choose between constipation and a variety of artificial chemical aids—substances whose unpleasantness, acting as a substitute for the natural stimulus, hurries up the movements of the bowels, or lubricants which assist the languid intestinal muscles in their task.

So Mr. Everyman, in taking the onus of choice upon his own conscious shoulders and consuming separated products instead of plant or animal tissues, is running several risks. His belly can go slack and out of training, just as his muscles can, if he gets too easy a time. Also his refined dietary may be too refined. He may be leaving out essentials. Just how he may be leaving out essentials we will next proceed to explain.

§ 2. The Six Vitamins.

ONE hears so much nowadays about vitamins and their importance that it will repay us to devote a section to these elusive substances. Their importance is very real. What exactly are vitamins and what do they do?

Let us go back for a moment to the sixteenth century, when the navies and commercial ships of the world were being ravaged by a painful and dangerous disease known as scurvy. Nobody knew anything of the cause of this disease. It first appeared on a ship when she had been at sea for some weeks, and its onset was gradual. First the crew became weaker, less able to make efforts of any kind, easily put out of breath, and easily depressed. Then the condition became more serious. Sunken eyes, fœtid breath, constant pain in the muscles, mental and physical exhaustion, a strange upset of the blood-vessels so that the gums and eyes and nose oozed blood, and even a gentle pressure on the skin might cause local bleeding into the spaces between the cells—these symptoms led up to graver and graver conditions, which were often fatal.

It was early suspected that the disease was in some way due to the limited and monotonous ships' diet. The sailors had to nourish themselves largely on salt meat and the biscuit called "hard tack"; any kind of fresh fruit and vegetable was unheard of on board. Soon the suspicion was confirmed. Already in the sixteenth century there were writings on the importance of fresh fruit, and in 1593 Sir R. Hawkins cured a ship's crew of scurvy by making them drink lemon-juice. Gradually other captains followed the lead. There is a credibly reported story of four of the East India Company's ships that sailed together in 1600; the captain of one made his crew take lemon-juice and thus kept them in good health, while the other ships were so crippled by the scurviness of their crews that he had to lend them his men to set their sails. Presently the value of lemon-juice became widely realized, and in 1795, over two centuries after the essential facts were known, the British Admiralty made its use compulsory in the ships under their control. The Board of Trade followed with regulations for the mercantile marine in 1863.

We now know that there is a substance contained in minute traces in fresh fruits and in actively living plant-tissues (as distinct from resting seeds or storage organs like potatoes), but particularly plentiful in oranges and lemons, and necessary for normal animal-life. If it is absent or deficient in the diet there is a rapid loss of weight, followed by scurvy. If the disease is not too advanced, the symptoms can be reversed very rapidly by the mere consumption of lemon- or orange-juice.

A second and closely similar case was worked out during the latter half of the nineteenth century. The Japanese navy at that time was plagued by a disease known as beri-beri—a disease known in tropical countries since a remote antiquity—just as the European navies had been plagued by scurvy a century before. Beri-beri is essentially a disturbance of the nervous system. It begins gradually, as scurvy does, with languor and exhaustion. Then there is a spreading paralysis of the limbs and a burning or tingling feeling in them, followed by numbness. The nerves to the heart and diaphragm are also disturbed, causing characteristic irregularities of the breathing rhythm and heart-beat. The complaint is often fatal. It was noticed that it tended to break out among people whose diet was artificially restricted. The Japanese navy at the time under consideration was fed largely on polished rice, i.e., rice-seeds from which the husks had been taken off. Moreover, alterations in the diet, such as substitution of barley for part of the rice, diminished the frequency of the disease very considerably. We now know that beri-beri is another deficiency disease, just as scurvy is; if the rice used by the Japanese sailors had not been polished they would not have suffered, for the husk (like most fresh vegetable tissues) contains the necessary substances.

Now, as we have told them, the stories of beri-beri and scurvy are straightforward enough. The interpretations seem obvious. Sailors deprived of fresh vegetables and fruit get scurvy; what more natural than to assume that the vegetables contain some substance necessary for life, whose absence causes the disease? But as a matter of fact it is less than twenty years since that interpretation was firmly established.

THE NOURISHMENT OF THE BODY 1061

For there are other possible interpretations. The vegetables might neutralize some poison that appears when salted meat is stored; or they might be disinfectants and keep some microbe in check. It was only as a result of painstaking experimental work that the true relationships were made clear.* If bacteriology has existed for the space of a full human lifetime, the study of vitamins is still adolescent and unformed; it has just about reached the university age.

The classical experiments which brought the new branch of science into the world were made by F. Gowland Hopkins at Cambridge in 1906 and 1907, and published in 1912. He found that rats fed on a pure synthetic diet, containing all the proteins, fats, carbohydrates, and salts, and so on, that were at that time known to be necessary, gradually lost weight and died. On the other hand, the addition to their diet of a few drops of fresh milk daily made all the difference; in this the rats thrived and grew. Why should this be? The increase in amount of fat, protein and carbohydrate is negligible, and so is the increase in energy content of the diet. Hopkins inferred that there must be something in the fresh milk, present only in minute quantities so that it was undetected by the analysts, but traces of which in the diet are in some way necessary for life. The analogy with scurvy and beri-beri was obvious; it was pointed out at once, and put to experimental test. These mysterious agencies, present in minute traces in certain articles of food and necessary for vital processes, are called vitamins. We say "mysterious" advisedly. At present nobody knows what exactly their chemical constitution is, or how they act. Nevertheless our knowledge, as far as it goes, is sound. There is no doubt at all that these substances exist. Their relative amounts can be roughly estimated by exploring the physiological activity of extracts containing them. It has even been shown that vitamin A (see below) gives a characteristic colour reaction. But the strange thing about these vitamins is that they work in astoundingly small traces, which have defied analysis for nearly twenty years.

Since in the early years of vitamin work so little was known about the substances, the problem of naming them was a difficult one. They had no popular names and it was not possible to give them names like those of other organic compounds, which would describe their chemical constitution. So, instead of being named, they were simply catalogued and given index letters. Vitamin A occurs in fresh milk, in

---

*The reader who doubts this should consult the articles on scurvy and beri-beri in the 1911 *Encyclopædia Britannica*, which review the state of our knowledge a year or two before the publication of Hopkins's work.

butter, in cod-liver oil, and in various vegetable foods, and is particularly necessary for young, growing animals. The substance occurring in rice-husks, whose lack causes beri-beri, was labelled vitamin B, and the ingredient of lemon-juice which prevents scurvy, vitamin C.

Soon complications began to turn up about vitamin B. The extracts containing it were found to have effects which differed puzzlingly according to their source and their manner of preparation and administration. Recently it has been made clear that there are at least two different substances concerned, which can be separated by appropriate chemical methods; they are called vitamins B_1 and B_2. Both these substances, which occur plentifully in yeast, in the germs and outer layers of plant seeds and in eggs, are of the utmost importance to our nutritive processes. Lack of B_1 produces beri-beri; lack of B_2 produces pellagra, another distressing disease that was entirely mysterious until very recently.

A fifth, and very important, vitamin—vitamin D—has been added to the list. Deficiency of this substance causes the widespread disease of children known as rickets. In rickets the deposition of calcium phosphate, which normally infiltrates and stiffens our bones, is interfered with, so that the skeleton is softer than it should be; in grave cases the weight of the body or the pulling of the muscles may curve the enfeebled spine and limbs and produce the characteristic malformations seen in rickety children. Together with this failure of bone formation there is general digestive disturbance and loss of weight. So vitamin D which occurs in fresh natural oils and fats and is richly contained, together with vitamin A, in cod-liver oil, is certainly essential to young, growing people; probably it has at least a beneficial effect in adults as well.

Apparently vitamin D has an important effect on the structure of the teeth. If it was deficient at the time when the teeth were being formed, their protective coating of enamel will be badly made, containing little rifts and flaws. Recent work has made it very likely that this is one of the chief causes of dental decay. This effect is a delicate one; a deficiency of the vitamin which is too slight to produce noticeable rickets symptoms may lead to little defects in the dental armour, and thus to decay in later life. Moreover—and this shows how complicated the activities of these factors in the food may be—it appears that in some cereals there is an "anti-vitamin" which neutralizes vitamin D. Oatmeal, Mellanby has shown, contains such a substance; if the staple carbohydrate is oatmeal, more vitamin D will be needed than if it is wheat-flour or rice. (Scotsmen can eat quantities of porridge and yet have good teeth, because they eat it with plenty of fresh vitamin-containing milk.) So that the state of one's teeth at thirty years of age is more than a sign of one's assiduity at tooth-brushing; it is a judgment on the way one was fed in one's youth. It may be that we are only beginning to decipher the complicated story of the interplay of these minute but important ingredients in our food.

Recently vitamin D has been the subject of another series of exciting investigations. Apparently its lack can be compensated for, to a large extent at any rate, by sunlight. If young rats are fed on a diet containing no vitamin D, but otherwise adequate, and if they are kept in the dark, they develop rickets in three or four weeks. But if they are put in direct sunlight for as little as twenty minutes a day, the symptoms do not appear. Moreover, if the sunlight, instead of being direct, has passed through a flint-glass window, it is ineffective and the disease manifests itself—a sign that it is in the ultra-violet end of the spectrum that the curative rays are located.

What is true of rats is true also of men. Sunlight tempers rickets; that is why the disease is comparatively rare in hot countries. With the importance of this work in light therapy we shall deal in a later section. Here we simply stress its interest as an extension of our ideas about these substances. It has now been shown that the ultra-violet rays act on a substance called ergosterol, traces of which are present in all animal tissues, and actually turn it into vitamin D. Ergosterol is a substance whose molecular formula is known, so probably by following up this discovery it will be possible to isolate the vitamin and find out exactly what it is. In illustration of the exceedingly small amounts in which vitamins act, it has been said that half an ounce of pure ergosterol, treated with ultra-violet light and then jealously stored and paid out, would be more than enough to supply all the vitamin D needed by a man in his whole lifetime.

Finally, another vitamin, E, has recently been recorded and indexed. It is found in various vegetable oils, particularly in wheat-germ oil and in the cells of green leaves, and in small traces in animal tissues. Absence of E causes sterility in rats—in the male, destruction of spermatozoa in the testis and ultimately of the whole germ tissue; in the female, death of the embryo in the uterus and its absorption by the maternal blood. But it remains to be proved that the substance is necessary for effective reproduction in our own species.

Here, then, we have a group of substances present in our diet in the minutest traces and yet playing parts of the most fundamental importance in our lives. They are not so much parts of our food as necessary impurities. Probably more will be discovered. It is worth noting that, to a large extent at any rate, the diseases produced by their lack are diseases of civilization. They are extraordinarily dependent upon our social and economic organization—upon the methods of preparation and distribution of foods and their price. Obviously if an artificial cereal product like polished rice or white bread is plentiful and cheap, and if fresh untreated animal and vegetable products are harder to get, the poorer sections of the population will be prone to those diseases which result from vitamin lack. Man does not live by bread alone. Dark, sunless houses and air greyed with smoke will help in producing rickety children. So that the effective combating of these deficiency diseases calls for more than individual effort.

Nevertheless, the essential facts about vitamins are well worth noting from the point of view of individual hygiene. Even if the reader is not visibly scorbutic or rachitic he or she may be suffering slightly from a vitamin lack. There is, of course, a sufficient amount of vitamins in the diet above which further consumption does no good; but between that value and deficiency grave enough to cause the unpleasant diseases that we have had to consider, there is a range over which effect fluctuates with supply. A slight vitamin shortage may produce a lack of physical stamina, mental slackness, a somewhat lower resistance to bacterial invaders, very like the earliest stages of a deficiency disease.

In a word, these mysterious vitamins are a warning against excessive sophistication of the diet. They are a warning against "pure" food. Man suffers less by adulteration than by refinement of his food substances. We are brought back to the conclusion of the last section, that natural foodstuffs are in many ways preferable to artificial products. Plenty of green vegetables, fresh fruit, and dairy produce will supply the necessary substances; but if the appetite is satisfied largely by white bread, elaborately cooked meats and cakes and other sweet confections, the possibility of a deficiency of these essential constituents is considerable.

§ 3. Some Possible Poisons.

AND now, having considered the possibility of vitamin deficiency, let us take up the question of certain dangerous poisons that may contaminate our food.

One often hears of "ptomaine poisoning" following the consumption of potted meat, sausages and similar articles, and it is worth noting that the name is a very inaccurate one. A ptomaine is a substance pro-

THE NOURISHMENT OF THE BODY

duced by bacteria of decay in rotting things, and, although it is poisonous, a foodstuff would have to be aggressively bad to contain dangerous amounts of it. These cases are really due to the products of a different group of bacteria, which live in the intestines of rats, mice, farmyard animals and birds. Sometimes they come to contaminate carcasses or they may infect eggs. The commonest of these microbes is *Bacillus œrtrycke*, which produces a very resistant poison, unaltered either by cooking or by our digestive juices. If food with this bacillus in it is cooked, the organisms are killed, but their toxin remains. The poisoning caused by eating it may be severe, but it is rarely fatal. But if the same food is eaten uncooked, then the bacteria may live and breed for days and produce their toxin in the bowels, with altogether more damaging results. Even in the last worst event, however, the majority of sufferers recover.

A more serious, but happily rarer, kind of food-poisoning is known as botulism. The organism which causes it, *Bacillus botulinus*, has the distinction of making the most poisonous substance known to man. Moreover, it is insidious, for its presence makes no difference to the appearance or taste of the food. So active is the toxin that even if a piece of food containing it is taken into the mouth and immediately spat out a fatal dose may be ingested. J. B. S. Haldane has estimated that about sixty pounds of this substance would be enough to kill the entire human race. Although it can resist the digestive juices, this toxin is destroyed by cooking. Its bacillus can only live in the absence of oxygen, so it is chiefly tinned or bottled things which were inadequately sterilized before being sealed up that carry it. In England only one outbreak has been recorded, but in America hundreds of cases of botulism have been caused by canned vegetables.

Besides these poisonous by-products of bacterial activity, food may be dangerous as the vehicle through which parasitic organisms get a footing in our insides. The most important case nowadays is milk, which, if it has not been properly pasteurized, may contain the bacilli of tuberculosis; this is one of the chief ways in which young children get infected. Oysters may contain typhoid bacilli through lying in estuarine water that contains human sewage to which typhoid patients or typhoid "carriers" have contributed. Green vegetables may be similarly contaminated, or they may carry the eggs of parasitic worms. Often meat and other foods which have been exposed for sale in shops are badly infected with bacteria, partly because of microbe-bearing dust settling on them, and partly by the agency of flies. The fly is a great disseminator of bacteria. It passes from a dung-heap to the dining-table and back, and wherever it crawls over food it leaves bacteria in its excretions and vomit—for flies have a habit of vomiting before they settle down to a meal. Food containing living bacteria can be disinfected by thorough cooking.

Finally there may (albeit very rarely) be metallic poisons in food. Beer is sometimes artificially contaminated during manufacture with arsenic; so also may cheap sweets. Recently a number of people in Newcastle-on-Tyne suffered temporary discomfort through consuming lemonade that had stood overnight in a cheap enamelled pail, from which it had taken up a little antimony. At one time a kind of lead poisoning, caused by cider that had been kept for a while in leaden vats, was so common in Devonshire that it was called "Devonshire colic." Copper sulphate has been used to colour vegetables (such as green peas), but its use is now prohibited by law.

It is evident that these various kinds of poisoning, bacterial or otherwise, have to be guarded against by state supervision of food production and distribution, and by systematic testing, analysis and grading of foodstuffs. The problem is quite beyond the selective powers of the individual household. Experience has shown (the confession has to be made) that the people who produce and handle food will not take adequate hygienic precautions, which they regard as unnecessarily bothersome or unnecessarily expensive, without some measure of official compulsion and control. Mr. Everyman, in his civilized world, can help to safeguard himself by co-operating with public health authorities whenever possible—for example, by paying due regard to the grading of milk and purchasing only Grade A.

There remains, however, another question. May it not be that among the commonly marketed and consumed foodstuffs there are some that we would be better without, quite apart from their cleanliness and freedom from bacterial and other contamination? We have already said something about the unnaturalness of sugar as an article of diet; there are some who would banish it altogether from the menu. Others have written that milk is a pernicious food, at least for adults. One or two extremists see in it a cause of cancer. And numbers of healthy and vigorous people never touch meat. In this very general review of nutritional problems we cannot contribute to the already voluminous mass of controversial writing on these questions; we owe it to our readers to mention them, but we could undertake to produce for every healthy vegetarian a healthy meat-eater, for every healthy milk-shunner a healthy milk-drinker. Let the reader remember the singular elasticity of the normal human constitution.

FIGS. 274–275. Model farms where Grade A milk is prepared.

The milk is untouched by hand, coming direct from the cow and being automatically filtered, sterilized, cooled, and bottled in a few minutes.

Indeed, if one takes care not to over-eat, to indulge in refined and highly seasoned food with restraint, and to include plenty of fresh garden and dairy produce in the dietary, one has done about all that is necessary in the way of personal selection. It takes two to make a meal—the eaten and the eater—and what Mr. Everyman eats may not matter so much as how he eats it. The most hygienic meal in the world can weigh on the system like lead if it is badly cooked and served on a dirty cloth, hastily chewed, bolted down like posting letters, and then handled languidly and sluggishly because of a traffic-block farther along the digestive tube.

Abdominal discomfort does not necessarily mean wrong food. The delicate machinery can be upset in other ways. Bad teeth or gums, or a septic focus elsewhere in the body, can poison it, and as we shall see in the next Book, the health of all our organs is most intimately bound up with the well-being of the nervous system. A worried or excited man digests badly; a serene, unruffled man digests well. There is something very appealing to the human mind in the dietetic fad; from its first beginnings, as far back as history takes us, medicine has had to resist and overcome a tendency to crystallize out into materia medica, to treat by simply answering symptoms with special herbs or drugs, such-and-such a herb for such-and-such a symptom. Similarly many people ascribe their general slight ill-health to wrong feeding, and believe that if they knew what the particular error was everything would be all right. If only it was as simple as that! In many cases at least bad nutrition is not a cause but a symptom, not the fault of the food but the fault of the system that receives it. The system is living lazily or irregularly.

§ 4. Drugs, Their Uses and Dangers.

WE MAY close this chapter on diet by discussing certain chemical substances which many of us are in the habit of absorbing, although strictly speaking they are not foods. Very often, in a difficult situation, one can get considerable nervous and physical support from such drugs as alcohol and tobacco, just as Sherlock Holmes solved his most baffling problems with the help of cocaine. These substances buck us up, quite apart from any nutritive value they may possess. How far is it possible to extend our natural powers, to raise the general level of our output, by the use of such drugs? Can we imagine any drug which would stimulate and sustain without the harmful effects that usually threaten the excessive smoker and drinker?

THE NOURISHMENT OF THE BODY

First, before we turn to this more general question, we may examine the three or four comparatively mild drugs that are so commonly used as to be part of normal life to-day.

Tobacco, as is well known, owes its soothing properties to a drug called nicotine. But it contains other ingredients too. The yellowing of the teeth and fingers of confirmed smokers is due to other substances, and the flavours of different tobaccos to aromatic compounds of various kinds. Moreover, cheap tobaccos are sometimes treated with chemicals such as saltpetre, an ingredient of gunpowder, to make them burn more readily, and the paper of a cigarette is very often so prepared. As one smokes, little bits of leaf often get detached and swallowed. So there is always the chance that a bad result of smoking—on the stomach, say—may be due to substances other than the nicotine itself.

The following simple experiment, which the reader may perform forthwith, is instructive. Fill the mouth with smoke; put the lips into the kissing position; hold a reasonably clean handkerchief taut across them; eject a thin jet of smoke slowly through the handkerchief; examine same.

Excessive smoking fouls the mouth, acidifies the stomach and dulls the appetite. It may also interfere with the sight; sometimes heavy smokers see faint hazy clouds near the middle of the field of vision, and have difficulty in distinguishing colours in that region, although their sight is unimpaired round the edges. This effect is due to an action on the optic nerve. Smoking also reddens the throat and makes it sore; herein such adulterants as saltpetre play an important part by producing irritating fumes. Nicotine, absorbed into the circulation, raises the blood-pressure and accelerates the heart-beat and breathing movements, so that in people whose blood-pressure is already on the high side it may be dangerous (but the stories one hears about tobacco actually promoting arterio-sclerosis are probably exaggerated). Another danger resulting from over-smoking, especially with incessant cigarette-smokers, is that carbon monoxide—the dangerous ingredient in ordinary coal-gas—may be produced and act as a stupefying drug. Generally speaking, and with the exception of the gas just named, a large amount of the nicotine and irritants produced when tobacco burns accumulates at the end of the cigarette, cigar, or pipeful nearest the mouth, so that the latter end of a smoke is far more potent than the first. Also, of course, these agents act much more powerfully if the smoke is inhaled into the lungs.

Against these disadvantages, various arguments have been put forward in favour of tobacco-smoking. It has been very plausibly suggested that by depressing the appetite smoking lengthens life, for most of us eat too much! The smoke is also said to disinfect the mouth by killing bacteria, but its power so to do is probably not considerable and almost certainly outweighed by its irritant action on the natural defences of the throat. The chief argument in favour of smoking is that it pleases and soothes. It allays "nerves," and moderate indulgence in the weed may save one from falling into worse vices. One can live perfectly happily and healthily without smoking at all—indeed, giving up tobacco altogether is easier than rationing one's consumption—and if Mr. Everyman has any cause to suspect that he is overdoing it he would be well advised to lay off for a while at any rate. Under ideal conditions of health and mental happiness one need never smoke. But in the bustle and worry of modern life it is not by any means a bad thing to have a comparatively harmless drug at one's command—provided that one keeps the upper hand and controls the habit instead of being controlled by it.

People who give up smoking often become confirmed sweet-suckers. This also, as we have said, may be a harmful habit. The tobacco manufacturers fulminate against it in their advertisements, and the sweet makers retaliate against tobacco. One is reminded of the pot and the kettle, although neither vice, in the great majority of cases, can be called black.

The question of alcoholic drinks is more complicated. To begin with, alcohol can be taken in a great variety of different forms. Wine contains up to 10 per cent of alcohol together with a number of ethers and other complex organic compounds which give it flavour and colour. Most wines of reasonable price are touched up with such improvements as sugar, dyes and clearing agents, ethers, alum, and sulphate of lime. Strong wines like port and sherry are, of course, fortified to 10 or 15 per cent of alcohol by the addition of spirit. Beers, brewed by fermentation of starch and sugar and flavoured with various vegetable bitters, may contain from 3 to 7 per cent of alcohol, together with various other substances such as sugars. Spirits, made by distilling over and thus concentrating the alcohol fraction of a fermented liquor, contain from 40 to 66 per cent of alcohol. Of these, whisky comes from malt and grain, gin from malt and rye with additional flavoring, brandy (theoretically) from grapes, rum from molasses. Moreover, the more concentrated alcoholic drinks may have other important active principles besides straight alcohol. Thus raw whisky often contains fusel oil—hence the deadliness of bootleg liquor—and

THE NOURISHMENT OF THE BODY

absinthe owes its powerful action on the nervous system chiefly to oil of wormwood.

Alcohol itself may be said to act in three capacities—as a food, as a drug, and as a poison. It acts as a food because it can be burnt in the tissues, although it is of no value for body-building. It is a food of the fuel type. But it differs in certain essential respects from other fuel foods. First, its molecules are comparatively simple, and these pass through the wall of the digestive tube without any preliminary digestion. So that if, in treating some ailment, one has to administer a food which is rapidly and easily absorbed, alcohol may be useful. Moreover, it is a completely artificial food, for handling which our cells are not naturally prepared. They have no way of storing alcohol and are quite unable to cope with an excessive supply. So that if we drink a large amount the alcohol simply washes round in the blood until it is all burnt up. In a chronic drunkard this state of affairs may last from one bout to the next, and the blood may always be slightly vinous. The actual food value of an alcoholic drink, we may note, is never very great; and even in the case of sweet wines and beers, which contain such additional nutrients as sugar, there can be no doubt, since fermentation is necessarily a wasteful process, that the drink contains less food than did the fruit or grain from which it was made.

In its capacity as a drug, alcohol chiefly concentrates its attack on to the nervous system. It is commonly said to act as a stimulant, but one should realize that in this particular case the word is very loosely used. There is not a shred of evidence for a direct stimulating action of the drug on any vital process (with the dubious exception of speeding up the breathing rhythm). When one brightens after a glass of wine, speaking and laughing more freely than before, one does so simply because restraining influences are narcotized. One is bucked up by the paralysis of one's conscience. In this manner alcohol can bring valuable relief from excessive worry, for the little gnawing part of Mr. Everyman that watches and doubts and criticizes his every action is the one most easily put to sleep. It is pre-eminently the worried, overconscientious type of person who is genuinely braced by alcohol and rises to a difficult situation all the better for a nip.

A somewhat similar illusion about alcohol is its warming effect. It makes one feel warm, but without producing any increase at all in the temperature of the body. This it accomplishes by a local action on the blood-vessels of the skin, which it dilates, thus bringing about an increased flow of warm blood to the surface, where most of our temperature organs are placed. Indeed, by diverting blood into our natural radiators, alcohol may actually cool our bodies while warming our skins with this deceptive glow.

The poisonous action of alcohol is hard to separate clearly from its drug action, since the latter, depending on a depression of vital processes and not on a genuine acceleration, is in a sense a poisoning process. In chronic alcoholism, the same effects are exaggerated. The nervous centres become permanently impaired, first the highest and then down to the lower levels. The self-control is weakened, leading to such complications as syphilis. Alcohol taken in too concentrated a form (such as neat spirit on an empty stomach) may harden and inflame the stomach walls, and this, by interfering with the production of gastric juice, upsets assimilation. The first steps in digestion being clumsily performed, the whole process is thrown out of gear. At the same time the microbe-killing action of the gastric juice is interfered with. Thus, and by a general devitalizing action on the system, the barriers to infection are lowered. The liver, because of its exposed position, is often involved in these consequences and may become chronically inflamed; sometimes it shrinks and acquires a wrinkled, warty surface, becoming the so-called hobnail liver. The reproductive glands, too, may be affected, resulting in sterility. So alcohol undermines the bodies and minds of its too faithful devotees. But in moderate drinkers there is no evidence at all of any really permanently harmful effects, although the possibility even for them of incidental impairment of function cannot be disputed.

The ethics of this question—for example, whether it is right to prohibit any consumption of alcohol by a whole population in order to prevent its abuse by some of the members—are complicated, and we do not propose to enter into them here. They lie outside the province of biological science. As far as moderate, restrained drinking is concerned, we can see no organic harm, and if it soothes the mind at a difficult time, or pleases the palate, and thus helps the gastric juice, it may be temporarily convenient. But the facts that some people are more susceptible than others, and that the general accessibility of alcohol leads to its grave and damaging abuse in many cases, have been amply demonstrated. There, as biologists, we leave it.

Even tea and coffee constitute a drug problem, for they are taken as stimulants and not as foods. They contain small amounts of an alkaloid, caffein, which has an invigorating and restorative action on the nervous system. But even these mildest of vices can have their bad effects; too much tea leads to indigestion, constipation, trembling, and insomnia, while strong, black coffee can also upset the stomach. With tea, many of the symptoms are due to substances such as tannin which are present in the leaves in addition to caffein. Tannin diffuses out into the water more slowly than caffein, so if tea is not allowed to stand in contact with the leaves for more than five minutes, but poured into another vessel at the end of that time, it has all its invigorating principles and little of its deleterious elements.

Looking over the substances that we have discussed in this section one gets the idea that from a scientific point of view they are prepared in extraordinarily untidy ways. Alcoholic fluids are natural juices in which whole populations of organisms have thrived. The smoke of burning tobacco or the infusion of dried bits of the tea-plant are equally complicated chemically. Our meals should, of course, be varied; one should beware of relying for his nourishment on artificially extracted and purified foods, for reasons that we have already noted. But, when we are deliberately taking a drug because of some special effect it produces on body or mind, surely the cleaner, more straightforward way would be to use a definite dose of a purified product? For example, to take one's caffein pure and without contaminating tannin? In this case at least the harm is in large measure due to an accompanying substance and not to the drug itself.

We do not wish to advocate the consumption of any of the hundred-and-one stimulating pills, offering iron, phosphorus, ozone and what not, that are on the market to-day. There is at present no extracted drug of proven efficacy which can safely be taken without medical advice. But we do suggest that as modern methods of analysis and purification improve, the possibility of using tonic aids with more precision than heretofore presents itself; and the idea is well worth some further discussion.

The chief danger with any stimulating drug is that it may worm its way into one's life and become a habit. This is, of course, seen most clearly with such evidently harmful drugs as cocaine and morphine, which provide a considerable clearing and enlivening of the mind when they are first used, but leave, when their effects wear off, a longing for further indulgence which is wellnigh irresistible. The morphine maniac becomes less and less sensitive to the characteristic influence of the stimulant, so that his craving for that influence drives him to ever-increasing doses. Slowly but infallibly he goes to pieces. It is perfectly possible for such a person in an extreme instance to become so dependent upon his drug that sudden and complete cutting off of his supply may kill him. But tobacco can spread its milder, more insidious dominion over one's life in the same way. It is easy for a sedentary worker who gets into the habit of smoking as he considers the papers on his table to slide into an almost incessant use of the drug. The craving for alcohol, also, may become imperative. So Mr. Everyman, who is probably a little self-indulgent, is warned to be on his guard. In all these cases his motto should be "Less anyhow, none if possible."

Habit-forming drugs are substances for which the body is constitutionally unprepared. One can imagine a considerable stimulation resulting from the use of more natural excitants, such as refined preparations of some of the internal secretions. The physiological basis of our changes of state and mood is becoming gradually intelligible. May it not be that, just as we can revolutionize the mentality of a guinea-pig by modifying the internal secretions from its sex glands we shall some day be able, by taking now this extract and now that, to put ourselves into the frame of mind most appropriate to the job in hand? That a Mr. Everyman with greater self-control and self-understanding will take a teaspoonful of one kind of extract before he sits down to work, and a teaspoonful of another before he goes on to the tennis-court, and crowd sixteen or so hours of vigorous activity into his waking day?

The beginnings are somewhat unsatisfactory, it must be confessed. Thyroxin, the active principle of the thyroid gland, has a tonic action on the metabolism of the whole body and lends vigour to growth and to physical and mental activity. Here, at first sight, is a promising suggestion for further improvement. But unhappily the good effects of thyroxin administration are only shown in people who are naturally under-thyroided; one can have too much of the stuff, just as one can have too little. So that thyroxin-taking is a matter of compensatory medicine rather than an extension of human ability. The same is presumably true of our other endocrines. Even vitamins, it appears, can be taken in excess. Recently, rats fed on large amounts of irradiated ergosterol have been shown to suffer from it. Massive doses are even fatal. We hasten to reassure the reader that the amounts used in the investigation were about one hundred thousand times the normal consumption, and quite unlike anything that he is likely to meet in his own adventures.

An ingenious suggestion for tonic improvement was made during the War by Professor Embden, of Frankfurt. He was studying the important part played by phosphates in muscular activity, and he tried the effect of dosing men with acid sodium phosphate (a salt that normally circulates in small amounts in the blood). He found that their output of muscular work was markedly increased, and many of his human subjects felt mentally invigorated as well. Apparently the substance has no harmful effects, and can be taken at the rate of a quarter of an ounce (dissolved in water) a day for considerable periods of time. It can be bought very cheaply from chemists, but the experimentally-minded reader is warned against the similar-looking potassium salt, which acts as a purge of great violence and thoroughness.

Perhaps in the cleaner, healthier cities of the future, the use of such comparatively simple substances as this will be found to provide refreshment enough, and the more elaborate and dangerous vegetable narcotics will pass away. We cannot tell. Perhaps, when we know more of the laws that govern our being, and apply our knowledge to the regulation of our daily lives, the natural tone of nerve and muscle will be sufficient without any artificial improvement at all. Tobacco and alcohol are, to a large extent, refuges from the physical and nervous frustrations of modern life.

This gesture towards the future may seem a disappointing conclusion to our inquiry. But it is the best we can do. Let Mr. Everyman seek consolation, as he reaches again for the decanter or the tobacco-jar, in the thought that his near descendants will probably fortify themselves, if they need fortification, without the faint threat of sore throat or hobnail liver that hangs over his own indulgences.

III

FRESH AIR AND SUNLIGHT

§ 1. *Town Air and Country Air.* § 2. *The Air of a Stuffy Room.*
§ 3. *Sunlight as a Tonic.*

§ 1. Town Air and Country Air.

THE body's everlasting battle with its environment goes on at three main fronts. With the digestive tube we have already dealt. We turn now to the other two—the lungs and the skin—for Mr. Everyman is exposed to fluctuating influences besides those of the substances he absorbs.

One often talks about the desirability of plenty of fresh air. It will be worth our while to look a little more closely into this idea. What exactly is fresh air? To put it in another way, what are the factors that can make air foul? How does the air of towns and human habitations differ from the cool forest winds that filled the earliest human lungs?

The most obvious alteration, to a town-dweller at any rate, that human activities produce in air is when they charge it with soot. Here the American citizen has the advantage of the British. No American town can show anything like a London fog. The rapid industrial development of America followed after that of England and profited by riper experience; the pollution of the air by smoke was reduced by consuming fuels other than coal, in many cases under the urgency of legislation. But in an English industrial district, where coal is burnt more or less inefficiently in factories and domestic grates, the amount of soot in the air is appalling. In Oldham, a big Lancashire cotton town, the amount of soot deposited in the year ending March, 1915, reached nine hundred and sixty tons per square mile. London nowadays has a deposit which may in certain districts exceed a third of that value. In London it is domestic fires that are chiefly responsible.

A large amount of this urban smoke is carbon, but there are various other chemical ingredients; it is often acid with sulphurous acid, and therefore does an enormous amount of damage to masonry, and especially to old buildings. The dense fogs that occur in London, for example, are due entirely to smoke pollution. They are produced under appropriate atmospheric conditions by the condensation of droplets of water round the soot particles as they float in the air.

Besides its immediate unpleasantness and the general depression it causes, soot has a very harmful effect on the health of town-dwellers. It holds up ultra-violet light (see § 3); it enters and fouls the respiratory passages. Time after time it has been shown in various towns that a severe fog-period is accompanied by a rise in the death-rate from disease of the lungs and air passages. Moreover, it is not only the town-dwellers who suffer; the smoke may drift surprisingly far on a gentle wind over the surrounding countryside. Belgian smoke is sometimes carried over England. As it travels the smoke settles on fields and gardens, clogging the stomata of plants and poisoning them with acid and thus indirectly interfering with the supply of fresh foods, both vegetable and dairy produce, to the towns. Here at least the Briton can be grateful for his climate, for rain is a great cleanser of the air.

The smoke nuisance again is a matter for social organization rather than of individual hygiene, although the individual can help by watching his own grate and chimney. In New York it is a punishable offence to emit visible smoke from a chimney. The replacement of the old-fashioned coal fire by gas or electric heating, or its adaptation to a more efficient smokeless fuel, would have a profound effect on the health, not to speak of the comfort, of city dwellers in Great Britain.

A variety of other kinds of suspended matter are grouped together as dust. To a large extent ordinary dust is organic in origin; it is made up of particles of decaying plants, manure, pollen, fragments of insects, and their eggs, spores of microbes and moulds, and so on. Under artificial conditions it is complicated to a greater or less extent by inorganic particles. These latter are the most destructive of health. Workers whose occupations involve the handling of quartz, granite, flint, sandstone, china-clay and the like are particularly prone to pulmonary diseases because of the flinty particles they inhale; grinders and polishers are also liable to suffer from the injurious effect of dust from emery wheels and sand grindstones on the membranes of the lungs. Dust from cotton and other plant fibres can also be harmful. But happily these injurious dusts are not as widespread as smoke is.

The organic elements in dust are altogether less damaging. For one thing, we are prepared for them. Any animal, living under normal conditions, inhales a fair amount of plant spores and pollen and fine organic débris, and we find that mammalian noses, even our own degenerate noses, are well equipped to filter them out before they reach the lungs (Fig. 27). And the lungs are provided with special cells which act like phagocytes, taking up and removing any such particles as penetrate to the alveoli. The flinty particles just discussed are dangerous partly because they are so numerous and partly because, for chemical reasons, the phagocytes cannot deal with them properly. They do not take them up, but leave them about to irritate and inflame the delicate tissues.

But the dust nuisance was enormously increased when men began to make roads. A modern tarred and concrete highway is almost dustless, but a white country road may seriously strain the cleansing devices of the nose, eyes, and throat of those who journey along it. For centuries the road has evolved side by side with the domesticated horse, as a special artificial strip of open country along which that ungulate could use his hoofs to our advantage, and the horse, by dropping balls of finely minced hay to dry on the roads, is responsible for much of the irritation and infection that human travellers endure. Nowadays a revolution is occurring in road transport. The pedestrian may fail to recognize any advantage in the change as he watches the passing automobile from the roadside. He notices only the foul blue breath of the engine and the dust-clouds stirred up by the wheels. But there is another side to the question, for the automobile is diminishing the multitude of horses that pollute the roads, and at the same time bringing about a great improvement in road surfaces. There may soon be a day when there will be no more dust on a country road than there is in a country wood. It is our misfortune to live in a period of transition, when the roads of the old type are being torn up and flung about in powdery handfuls by the traffic of the new.

Some people have a peculiar sensibility to particular kinds of organic dust, to which they react by skin-rashes or by the distressing asthmatic and catarrhal symptoms of "hay fever." The physiology of this complaint is rather curious. We noted in the last chapter how the body protects itself by forming anti-substances when certain foreign matters intrude into the blood-stream. But sometimes, under conditions that are not at present properly understood, the introduction of a foreign protein produces, not immunity-reactions, but a strange sensitivity to future administrations of that particular protein. For example, if we inject a little white of egg under the skin of a guinea-pig, nothing much happens. The animal is apparently undisturbed. But it acquires a sensitivity to egg-white; if a month later the same substance is injected, even in a minute dose, the animal has violent convulsions and respiratory distress, and may die. Hay-fever, it seems, is a parallel case; the disturbing substance being usually the pollen of some plant or other, which acts as an irritant when it touches the mucous membrane of the nose. At first the reaction is slight, and then, later, the sensibility increases to a distressing pitch.

Perhaps this phenomenon of anaphylaxis explains another curious fact. It is known that some people cannot bear the presence of cats in the room, even unseen and unheard cats cause them definite physical distress. Or they may be sensitive to dogs, to horses, or to some other common species. According to Shylock:

> "There is no firm reason to be render'd
> Why he cannot abide a gaping pig,
> Why he, a harmless necessary cat."

But it is at least a plausible suggestion in the light of modern knowledge, that some kind of organic particles emanate from the cat or the pig (such as specks of dandruff), which set up reactions like the hay-fever reactions.

But we are getting away from the subject of our discussion.

Far and away the most dangerous particles that can be carried in the air are the living ones—microbes and their spores. The number of these creatures vary enormously. They increase with dust (since they can be carried on dust-grains) and they are more numerous in towns than in the country; they are particularly abundant indoors in stuffy, warm, ill-ventilated apartments. Hill and Campbell have described a staircase in which there were 750 bacteria in every cubic metre of air before sweeping, and 410,000 after the carpet had been brushed for ten minutes. In places like crowded picture-galleries, where, apart from the possible presence of variously infected people, hundreds of feet bring in dust and keep it stirred up, the numbers of floating microbes are enormous; in the Paris salon five million per cubic metre have been counted on a Sunday afternoon. On the high Alps, over fifteen thousand feet, there are only four or five microbes per cubic metre; in pure air in mid-ocean a single bacterium has twice this amount of elbow-room. There is little need to stress the meaning of these figures. Mr. Everyman breathes about a third of a cubic foot of air per minute; this means that if he is high in the Alps, a microbe will be drawn into his lungs every twenty minutes or so, while in the picture-gallery he inhales about sixty thousand per breath. Some of these organisms are bound to be disease-carriers. Such complaints as influenza and pneumonia are spread exclusively by bacteria carried in tiny drops of saliva coughed or sneezed out by infected people.

We look back with disgust on the water-supply of a century ago. If the town-dweller of those days chose to put a drop of water from the mains under a microscope, he could entertain himself for hours by watching the water-flies and gnat larvæ, and other curious creatures that abounded. And we read of subtler and more sinister contaminations. But may it not be that our grandchildren will look back with equal disgust upon our air-supply a century hence? That they will listen almost incredulously to the descriptions of the soot and acids and microbes and powdered dung that polluted the air of an early twentieth-century town? We use water to drink and to wash with. But we admit the air we breathe to an intimacy at least as close as that of the water we swallow, and the whole surface of our bodies is continually being wiped with air, with a vigorous pressure of fifteen pounds per square inch. Surely, then, the cleanness of air is a matter that deserves our serious consideration?

§ 2. The Air of a Stuffy Room.

WE TURN now from the gross particles of dirt that may defile air to its subtler physical and chemical influences. Why is it that the town-dweller, spending most of his time indoors, is thin, weedy, and often snivelling, while the country-dweller, or, to a lesser extent, the townsman who gets into the open for a few hours every week, is ruddy and robust? Let us see how far we can analyze the causes of this healthiness of the open air.

We will start with a simpler problem. You come into a room, full of perfectly good, clean air; you shut the door and windows and light a fire; you sit down with a book. Presently the room gets stuffy. The air has become dull and oppressive. Your face flushes, and you feel slack and drowsy, with perhaps a slight headache. Why is this?

As you sit and read, you burn, just as the fire burns. You use oxygen and exhale carbon dioxide. It was thought for a long time that the gradual loss of freshness in the air was due to these chemical processes —to oxygen scarcity or to carbon dioxide excess. But the matter is not as simple as that. Normally slightly over one-fifth of the air is oxygen, and it is very rare to find an atmosphere in which that value is seriously

FRESH AIR AND SUNLIGHT

departed from. In a badly ventilated school-room or theatre or law-court the percentage of oxygen in the air may fall from twenty-one to twenty; in your stuffy sitting-room it has probably not fallen even as far as that. Changes of this magnitude are altogether too trivial to affect the body noticeably. With carbon dioxide also it is perfectly safe to bet that no significant change has occurred. Normally 0.03 per cent of the air is carbon dioxide; in badly ventilated meeting-places and school-rooms, it may rise to 0.5 per cent, or even in rare cases to 1.0 per cent. But so accurately does our respiratory machinery respond to these changes that they barely affect the composition of the blood. They merely produce an imperceptible quickening and deepening of the breath. Only when the carbon dioxide percentage rises to 3.0 does it produce troublesome symptoms—violent breathing and slight headache—and this is six times as high as anything the reader is likely to encounter. So that neither oxygen nor carbon dioxide can be made responsible for the stuffiness of our sitting-room.

Another idea that has been widely held, is that human bodies exhale traces of organic poisons of complicated but unknown composition, and that these are responsible for the unhealthiness of crowded, stuffy rooms. Here again the evidence is negative. The odorous molecules that emanate from bodies are not directly harmful to health. The objections to them are purely æsthetic. Moreover, it can be proved that stuffiness is really not due to the chemical composition of the air at all. It is due to its physical properties.

In order to explore this question of indoor air with more accuracy, experimental rooms have been fitted up in physiological laboratories, in which the composition of the air can be controlled by adding or absorbing any of its ingredients, and in which such physical properties as warmth and dampness can also be regulated. It is found that if people are crowded together in such rooms, the oppressive feelings of stuffiness can be relieved simply by keeping the air cool and dry. Chemical factors have nothing to do with it. Moreover—and this is a conclusive proof that stuffiness is not due to "impurity" of the air—a tube can be fitted, opening to the exterior; under these circumstances people inside the chamber are not refreshed by breathing outside air through the tube, nor are people outside oppressed by breathing air from inside it. So that the unhealthy oppressiveness of a stuffy room does not concern the lungs at all. It concerns the skin.

As we saw in Book 1 (Fig. 34) the skin is an organ of many and varied functions. Most interesting for our present purpose is the fact that it regulates the body temperature—partly by acting as a radiator, in which blood is cooled, and partly by perspiration. Now this regulation is a continual and very important process. Heat is always being generated in our muscles and other organs, and it is always being lost through the skin to the surrounding air. There is, so to speak, a ceaseless flow of heat outwards through the skin. Our temperature is a sort of compromise between the speed at which heat is generated and the speed at which it flows away; it is rather like the amount of water in a bath with the taps turned on and the plug open, a balance between inflow and outflow. Now the processes that generate heat in our body are difficult to regulate. Just as a petrol-engine warms as it works, so heat appears whenever our muscles contract, and, indeed, whenever any energy-changes occur in our bodies. Plainly the source of the heat-river is intimately associated with our vital processes. It is a measure of vigour. A healthy, robust man produces and loses more heat than a weakly man*; his heat-river is fuller and swifter. And it fluctuates with his ever-varying activity. To cool the body by interfering with heat-production would involve depression of all sorts of important activities. So temperature regulation occurs at the other end. When heat-production is excessive, or when the surrounding air is too warm, the flowing away of heat through the skin is speeded up as much as possible. The blood-vessels dilate so that plenty of blood may come and be cooled, and the sweat-glands keep the skin moist so that there may be extra cooling as a result of evaporation. If, on the other hand, the rate of heat-production within the body is low, or if the air is cold, the heat losses are checked; the blood-vessels in the skin contract and the surface of the body is kept dry. Thus the skin is a weir-gate across the stream of heat, which can control its speed and flow.

Now it is evident that hot damp air will obstruct heat-loss. Warm air will not take up heat as readily as cold air, nor will our sweat evaporate as readily if the air is already humid. It is upon these facts that the stuffiness of a warm, close room chiefly depends. The heat-stream is dammed, and in order to avoid over-heating, the warming processes are delayed and interfered with—which has the most widespread depressing action on our organs. We suffer, so to speak, from calorific constipation.

Moreover, it is clear that moving air will enable the skin to discharge its cooling function better than still air. If the air is motionless,

---

*But there are unhealthy kinds of heat-production, as we saw in § 2 of the preceding chapter.

the part in contact with the skin soon warms up and saturates with water vapour. This is why a slightly open window, by making a current of cool air, is so immediately refreshing, and why even an electric fan, though it makes no alteration in the chemistry of the air, nevertheless has a grateful effect in a stuffy atmosphere. The facts that we have just set down explain the feeling of oppression one gets in the warm, still, damp air of a thundery summer day.

FIGS. 276, 277. The evolution of clothes. Tennis champions in 1908 and 1928.

In addition to the interference with the normal cooling of the body, produced by moist and motionless air, there is apparently a nervous effect on the sense-organs of the skin. Slight changes in coolness of the air produce gentle stimuli which in some way keep the mind refreshed. How these sensations should affect a busy brain is by no means clear, but it does seem as if our cerebral machinery goes all the more smoothly for this atmospheric caressing.

So that if we want to keep the air in a room fresh and stimulating we must not let it get too warm, nor too moist, nor too still. We ventilate rooms for freshness and variety in the atmosphere, not for purity. And while we are on this subject of stuffiness, let us remember that, from the point of view of most of Mr. Everyman's skin, it is the climate under his clothes that matters. Thick heavy garments lead to perspiration, and if they are not loose-fitting they imprison a layer of warm damp air next to the skin.

We may add a note here on the common cold—of all diseases the one which causes the greatest loss of time and energy. Nearly everybody has one or two colds a year and many of us have them almost continually. Apparently the common cold is infective—that is to say, due to a microbe which can be passed from person to person. Just what the responsible microbe is, we do not know. It may be one of the filter-passing viruses, or it may be a group of microbes of different species. Whatever it is, it gets into the nose and irritates the mucus-secreting cells, producing a grossly excessive secretion. The nasal membrane gets inflamed. The trouble may spread into the air-sinuses opening out of the nose, giving the headaches and other discomforts of a severe cold in the head. Or it may spread to the throat and farther down the respiratory tubes. A severe throaty cold produces general feverish symptoms by blood-poisoning.

Our susceptibility to this annoying complaint can be affected very markedly by external conditions. Cool air, in itself, is not a cause of colds. Fishermen do not get colds, nor do polar explorers. Shackleton recorded that in his antarctic expedition there were no colds except for a single outbreak, which occurred after a bale of clothes from London was opened. The idea that chilly air may cause a cold is due to the fact that in the early stages of the disorder, before the nose runs, the temperature-regulating mechanisms of the body are upset and we feel chills more sharply, and notice them before we know we have got a cold.

Agencies which enfeeble or otherwise trouble the mucous membrane are great helpers of the cold microbe. They prevent the mucuos membrane from defending itself. It has been shown that in warm, badly ventilated rooms the flushing of the skin is accompanied by a congestion of the nasal membranes, which become swollen and boggy. Consequently one feels "stuffy" in such places. In this condition the nose relaxes its normal defensive vigilance against disease-germs, and the cold is "caught"—or rather the cold catches us.

Susceptibility to colds varies from person to person. A nose may be especially prone, for structural and other reasons, to invasion. But in one individual, the frequency with which colds are caught is on the whole a very fair measure of the general toughness of the body, of the condition in which one is keeping oneself.

Perhaps a healthier and more vigorous generation will cut out the cold altogether, either by toughening themselves or by identifying and combating the microbe. They will look back on our sniffing, nose-blowing population with the feelings we experience when we look back at the pock-marked population of barely a century ago.

§ 3. Sunlight as a Tonic.

EVERYBODY knows how mind and body rejoice when the sun breaks for a little while through the mists and clouds after a spell of overcast weather. Muscles become tauter and the step livelier; the mind works more swiftly and sweetly; the digestive secretions seem all too perfect. Sunlight is a very real and active tonic; of that there can be no doubt. Of recent years there has been a great amount of active research, based on the discovery that some, at least, of the beneficial effects of sunlight are due to the rays called "ultra-violet." These, as everybody has heard nowadays, are rays to which our eyes do not happen to be responsive; they lie outside the range of the visible spectrum, and it would be possible to have a room that looked pitch dark but was nevertheless full of this kind of radiation. We may call the ultra-violet rays invisible light.

In the excitement that followed the first fundamental work on these rays all sorts of physiological action were attributed to them with more or less justification. The clearest, best attested and most striking story was of how ultra-violet light can delay or altogether prevent the onset of rickets in a diet deficient in vitamin D. (Book 7, Chap. 2, § 2.) It converts a substance called ergosterol, present in the cells of the skin, into the vitamin. This is not, be it noted, a justification in itself for treatment of weakly or rickety children with ultra-violet light; the light simply compensates for the lack of the vitamin, and does nothing in this respect that cod-liver oil or any other vitamin-containing substance could not do.

Then there is a group of actions depending on the fact that ultra-violet light is harmful to protoplasm, and acts as an irritant of the skin. It is the chief cause of sunburn, and artificial exposure to lamps specially made to emit ultra-violet light is followed by a similar train of responses—first redness or erythema, developing about twelve hours after exposure, then, if the burning has been severe, soreness and blisters and peeling, then the bronzing of the skin with the pigment malanin as a protection against further exposure. The most powerful oil lamps conceivable, those of a great lighthouse, for example, might play upon a man's skin for ever without producing sunburn, because they do not emit a sensible proportion of ultra-violet rays.

A number of practical applications depend upon this action. Ultra-violet light has been used very successfully against lupus or tuberculosis of the skin, because it irritates the tissues and stimulates the defensive process called inflammation. It spurs on the normal attack against invasive organisms. If a large area of the skin be mildly irritated by this means there is a temporary increase in the vigour with which the leucocytes throughout the body go about their scavenging duties. But other skin irritants besides ultra-violet light can have results resembling these. According to the authors of the report of the Medical Research Council for 1927–8 "there is no present reason to know that artificial light can do more in this respect than a mustard plaster, which is indefinitely cheaper." Ultra-violet light simply happens to be an irritant whose dosage can be easily and accurately controlled.

Another aspect of its irritant properties is that ultra-violet light acts as a disinfectant and will kill bacteria, just as in heavy doses it will kill human cells. But this is probably of little importance in the physiology of sun-bathing. Ultra-violet light will only penetrate the skin a very short distance—it has been said that the skin of a blister is amply enough to block the rays and so protect its nasty contents from disinfection—and bacteria in the ducts of the sweat-glands and in the hair-follicles will also escape. A beam of direct sunlight, however, falling through an open window into a room will disinfect the specks of dust that float across it.

Then there are a number of special actions—for instance, ultra-violet irradiation has been shown to hasten notably the healing of ulcers of the cornea—the glassy front part of the eye.

But aside from these instances, all of which are matters of special medical treatment rather than of general hygiene, there is a widespread belief that ultra-violet light braces and stimulates the whole system. Many people who have undergone courses of ultra-violet treatment, not for special complaints but because they felt run-down, are enthusiastic in their recommendation, and the monkeys and reptiles in the London Zoo are all the healthier for the provision in their cages of lamps emitting this form of radiation. But in this field there is a regrettable lack of clear scientific evidence. It seems to be established that the number of red cells in the blood increases after irradiation— perhaps this is another after-effect of its inflaming action—and there is a toning up of the apparatus of temperature control; people who are being treated commonly cast off mufflers and wraps. One worker claimed that he could detect a change in the gastric juice after treatment. But in all these claims there is an element of uncertainty. There is, for example, the possibility of auto-suggestion; our next Book will show how subtle and powerful is the influence of the nervous system on such other parts as the digestive organs. As for the monkeys and snakes, the light may simply be compensating for the lack of vitamin D in the Zoo diet.

We are not denying here that ultra-violet light treatment has beneficial effects that cod-liver oil and auto-suggestion between them could not produce. In the present somewhat confused state of medical knowledge it would be premature to deny. But the uncritical nature of the evidence should be kept in mind; there is a lot of exaggeration and a certain amount of sheer fantasy in the ultra-violet boom.

At present special lamps are being manufactured and put on the market which allow of ultra-violet treatment in the home. Most artificial illuminants are, of course, quite useless from this point of view, but by fitting a small arc-lamp or a mercury vapour lamp it is possible to strip and bask in one's bedroom in a sun that turns on and off to order. But those who may be tempted to get and use such lamps must remember that ultra-violet light is a drug that should be cautiously handled, especially in the highly concentrated form in which it is emitted by these special lamps. Ultra-violet light can cleanse and stimulate but it can also irritate and blister. Fair people in particular, who have little protective melanin, should be chary of its use. A little too much exposure may make one itch devilishly, and there are cases of people who have done serious injury to their skins by falling asleep under one of these artificial suns and exposing themselves too long. Unless goggles are worn the eyes may be badly and even permanently hurt by these dazzling light-sources. The only clearly established virtues of artificial ultra-violet light concern its use in the treatment of certain special afflictions. It is a useful new weapon in the doctor's armoury. Leave it to him.

It is a curious fact that just as a piece of blue glass may let through the blue rays and hold back the red, yellow, and green, so most kinds of glass can let through the visible rays and hold back the ultra-violet. Light which has passed through an ordinary window-pane is useless medically, although it looks the same as unfiltered light. It may warm and brighten, but it has lost its real "kick." In research work or medical practice in which ultra-violet light is used, the transparent parts of the apparatus are made of quartz or other substances which let the ultra-violet light through. Various kinds of glass are being manufactured for window-panes and the like which admit these rays to a considerable extent. More important from the health point of view than the opaqueness of glass is the opaqueness of the atmosphere to ultra-violet light. Pure dry air lets the rays through, but hazy or foggy air obstructs them, and smoky air may stop them altogether, as glass does. This, according to ultra-violet enthusiasts, is why cold mountain sunlight is so much stronger and more stimulating than the sunlight of plains which has filtered down through more or less water-vapour, and why it is so incomparably superior to the smoke-emasculated sunlight that reaches a city-dweller. But we should not forget the beneficial effect of clean, dry air on the lungs and skin.

Whether these particular rays are the responsible agents or not, there can be no doubt as to the comparative healthiness of a smoke-free atmosphere. The excitement with which ultra-violet treatment has been received is a confession of our social imperfections. With cleaner air and less smoke pollution we could dispense with artificial aids altogether. They are the complement of bad town-planning and bad combustion.

While great towns remain what they are, Mr. Everyman is well advised to go into a purer atmosphere for at least two or three weeks in the year, to the seaside or the mountains, and there expose as much of his body to clean air and the tonic rays of the sun as his ideas of decorum and his liability to sunburn will allow.

IV

THE PRESENT HEALTH OF *HOMO SAPIENS*

§ 1. *The Control of Epidemic Diseases.* § 2. *The Heart and Lungs.* § 3. *Cancer.* § 4. *Tuberculosis.* § 5. *Mr. Everyman and the Fight Against Disease.*

§ 1. The Control of Epidemic Diseases.

AND now, in the light of what we have already learnt, let us look round at the general health of the world-community to which we belong. Comparing the modern world, and especially its most intensively civilized parts, with the world of a hundred years ago, the most striking difference is the enormously reduced importance of pestilence—of all forms of microbe-carried diseases. The main ways in which this reduction has been effected have already been sketched out in Chapter 2. In England, for instance, leprosy, sweating sickness, and plague have vanished. With the general improvement in public hygiene, and particularly in the disposal of sewage and the purification of the water-supply, cholera has ceased to make periodic raids on the population and the typhoid mortality has fallen to about one-thousandth of what it was ninety years ago; isolation and preventive vaccination have reduced smallpox to a rare and comparatively mild complaint; and such still-prevalent infections as measles and whooping-cough are also retreating, although more slowly. In many other parts of the world this eliminating process has barely begun. Pestilence still broods in hot, uncivilized countries and in the disorganization of the Great War it broke loose with devastating results. Malaria, plague, and influenza are the chief infectious scourges nowadays. But influenza presents problems of its own. It is apparently followed by immunity of the most transient kind, and the mildness of many cases, hardly distinguishable from the common cold, makes isolation impracticable. But for malaria, plague, and the rest, the cleaning-up of Europe shows what can be done. It needs only similar efforts on a world-wide scale instead of a national scale to drive these pestilences altogether off our planet.

Meanwhile, in those countries which have been purged already of the more serious infectious diseases, there are less urgent but nevertheless serious problems to solve. Measles presents a definite parallel to the smallpox of a hundred years ago. Although its ravages are much less severe, it is so universal in Great Britain that few escape it; it is pre-eminently a disease of children, as smallpox was; as with smallpox, it may be followed by a train of damaging consequences, and open the door to other more fatal complaints. That measles, in common with other still-prevalent infections, will go as smallpox, plague, and malaria are going can hardly be doubted.

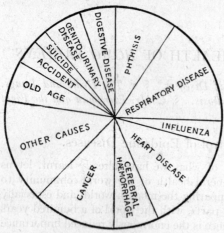

FIG. 278. A diagram to show how the total death-rate in England is divided up between the main groups of causes.

Here we have been forced to deal very summarily with the battle against epidemic disease because of the limitations of our space. A number of minor engagements and triumphs we have regretfully omitted. The interested reader will find further information in such books as Paul de Kruif's spirited *Microbe Hunters*, or Dr. Charles Singer's *Short History of Medicine*.

§ 2. The Heart and Lungs.

TO-DAY the centre of interest has shifted somewhat from these dramatic scourges. A consideration of recent mortality figures will show very clearly where the principal problems now lie. In this and the following sections we shall base our discussion on the deaths occurring in Great Britain in 1928; the picture presented applies in its essentials to all civilized countries at the present day.

In that specimen civilized community during the year under consideration, 11.7 people died out of every thousand. What were the causes of their deaths, and how far could they have been avoided?

First among the various causes of death as set forth by the Chief

Medical Officer of the Ministry of Health, we find diseases of the circulatory system, accounting for 22.9 per cent of the total mortality; more than two-thirds of these deaths are due to diseases of the heart. Here, of course, we are not dealing with a simple disease like cholera or malaria, but with a complex group of conditions. Generally speaking, heart-disease is primarily caused by some kind of microbial invasion of the body; rheumatic fever is often responsible, although a number of other infections may play their part in weakening the vital pump. Besides parasites, there are various secondary factors. Physical strain, blood-poisoning due to inadequate elimination by the kidneys and excessive smoking or drinking are examples.

The next group, responsible for 12.9 per cent of the total mortality, is disease of the respiratory system. Mostly these deaths are due to pneumonia and bronchitis. Tuberculosis of the lungs is left out and included with other kinds of tuberculosis; we shall discuss it later. Here again we are dealing with a complex group of conditions. Lobar pneumonia—pneumonia primarily affecting the alveoli of the lung (Book 1, Chap. 2, § 5)—is due to the presence of an infecting organism, usually a bacterium called "pneumococcus." But the conditions determining invasion are very complicated. Pneumococci are to be found in the throats of many perfectly healthy people, especially in winter and early spring; the number of such people has been variously estimated at from twenty to fifty per cent of the population. Apparently the normal, healthy body is able to resist the further penetration of the invaders. But fatigue, exposure to cold, or some other invasion such as measles or diphtheria, may lower its resistance and so cause a dangerous attack. On the other hand the pneumococcus itself varies in virulence; there are a number of recognizably different strains; and we find that pneumonia occasionally breaks out in epidemic form. Besides the pneumococcus there are two or three other microbes which may be waiting about in the throat and which join in the assault. Bronchitis means inflammation of the membrane lining the bronchial tubes and it may be caused in various ways—by microbes, or by direct irritation by chemical fumes or dust. In some ways intermediate between the two is broncho-pneumonia, in which the fine end-branches of the bronchial tubes, just where they enter the alveoli, are affected.

As a cause of mortality, we noted, respiratory disease is second only to diseases of the circulation. As a cause of ill-health and disablement it has no rival. An idea of its extension can be obtained from the figures of sickness insurance; in 1928 more than twenty-two per cent of the cases that insurance practitioners had to deal with were bronchitis, catarrh, colds, or pneumonia. The next group, abnormalities or irregularities of the digestive system, totalled only thirteen per cent. We are confronted here by a tremendous complex of disorders of the respiratory system, varying in severity from a slight inflammation and over-activity of the mucous membrane of the nose to acute and fatal disease of the lungs, and together accounting for a very great proportion of human discomfort and death.

Strictly speaking, pneumonia is an infectious disease. The pneumococcus can only be caught in the neighbourhood of an infected person, for it cannot survive for long away from the moist dark corners in which it lives. But, except in the case of particularly virulent outbreaks, it is hardly practicable to attack it by using the methods employed against epidemic diseases because the number of people carrying the microbe and feeling no obvious discomfort from its presence is so much greater than the number of actual sufferers.

Some vividly interesting figures, which suggest an altogether more profitable line of attack, are to be found in the *Registrar-General's Decennial Supplement* for 1921, in the part dealing with Occupational Mortality. The death returns for England and Wales over a period of three years, from 1921 to 1923, are analyzed with regard to the professional occupations and economic status of the dead. The first striking fact that emerges is that workers in certain industries have a very high mortality from bronchitis and pneumonia because of the damaging action of dusts on their lungs; thus potters, china kiln and oven men, and various groups of metal grinders suffer by inhaling particles of silica, while certain groups of workers in the cotton and woollen industries are injured by vegetable dusts. But aside from factors peculiar to special occupations there is a marked and regular grading in all forms of respiratory disease corresponding with economic status. We cannot give full details here; the reader who is interested should consult the report; but we may note the difference between the two extremes of the scale. The mortality from lung disease (excluding tuberculosis) among unskilled workers is two and a half times that among the upper and middle classes. Here, therefore, we are dealing not with any simple cause, but with the general conditions under which people live.

The mortality from influenza shows a similar grading. This is because in itself influenza is not a killing disease; but in people whose resistance is low it helps to open the door to other microbes, and so leads directly to fatal respiratory complications.

In a previous discussion of the common cold we found a similar story. We are dealing here with a group of very prevalent germs, too uniformly prevalent to admit of direct frontal attack. But the facts that we have just given show that a large proportion of their toll on human life and energy is preventable. The road lies through the control of the secondary causes that assist the invasion of the microbes, through the building up of a healthy, resistant population.

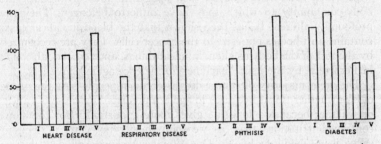

FIG. 279. A diagram to show how certain causes of death affect different sections of the community.

The numbers I to V represent five social classes, grading from the well-to-do (I), to the unskilled workers (V). The height of each column shows the comparative death-rate in that class from the disease in question during the years 1921–3, expressed as a percentage of the average death-rate from that disease. It will be seen that heart-disease, respiratory disease (bronchitis and pneumonia) and phthisis fall most heavily on the poorer classes, and show a regular grading with social circumstance. But diabetes, like most diseases of the digestive system, grades the opposite way, indicating that those in a position to do so overstrain their nutritional machinery. (Figs. 278, 279, 280 are after the Registrar-General's Decennial Supplement for 1921, Part II, Occupational Mortality.)

§ 3. Cancer.

THE next cause of mortality, after diseases of the circulation and diseases of the lungs, is that outstanding puzzle, cancer; 12.2 per cent of the deaths in 1928 were due to this cause. As everybody has heard, the death-rate from recorded cancer increases slowly but steadily from year to year; in 1928 it was nearly three times as great as fifty years ago.

Cancer is a rebellion of the tissues of the body—its alternative name, malignant growth, is a good one. It starts locally. Suddenly some of the cells lose their specialization and begin to grow like an independent parasite. For some months the revolt spreads slowly into the immediate neighbouring tissues. Then it begins to diffuse itself more extensively; some of the cancer cells detach themselves and float along the lymph vessels, lodging ultimately in the filtering glands through which those vessels pass every now and then. Thus secondary cancers are set up. If nothing is done the spread continues for two or three years until either by direct interference with the functions of some vital organ or by sheer exhaustion, the sufferer is killed. The remarkable thing about the picture is the thoroughness with which these cells fling off their allegiance and turn against the body-community to which they have hitherto belonged. They are profoundly altered. It has been shown that the blood in cancer cases contains anti-bodies opposed to the cancer cells. They are recognized by the rest of the body as alien, inimical things, and resisted, although ineffectively, by the usual machinery of defence against invaders.

The great majority of cancer deaths occur in people past the prime of life; whether this is because old age is a predisposing cause of cancer or whether because a long life gives more opportunity for the slow, continued irritations that, as we shall shortly note, are a frequent cause, is not definitely known. One often hears cancer called a "disease of civilization," so we may point out here that its occurrence has been recorded in primitive races of man, and even in other vertebrate animals living both in a state of nature and captivity. In the lower animals, cancer presents the same microscopical features and the same higher incidence in old age as it does in our own species. Manifestly in a species where individuals die early, deaths from cancer will be less abundant proportionally than in one, such as our own, in which the average life has been greatly prolonged. A part of the increased ratio of cancer deaths to the total death-rate is due to the fact that people have escaped earlier deaths and so survived to the cancer age.

What are the causes of this strange and terrible rebellion among the tissues?

There is as yet no clear evidence that human cancer, at any rate, depends on the presence of a microbe of any kind. One or two types of malignant tumour in fowls have been shown to be transmissible from bird to bird, using a clear, filtered fluid, which indicates that ultra-microscopic viruses are at work. But the conditions are very complicated. It is apparently necessary that the virus should be accompanied by a delicate and easily destroyed chemical factor. The researches are continuing; but it is not yet possible to speak of practical human application.

On the other hand it is abundantly proved that cancers can be started by the local application of irritants of various kinds. Thus skin cancer is usually common among chimney sweeps and certain groups of cotton-workers, owing to their exposure to irritating dusts, among gas-stokers, and coke-oven workers because of the continued action of tarry and intense heat, and among iron-puddlers and glass-house workers, where it is caused by heat alone. Cancer of the tongue, œsophagus, and stomach, on the other hand, is common among barmen, cellarmen, and waiters, whose occupations enable them to consume unusually high amounts of strong alcoholic drinks. Parallel

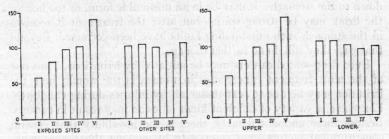

FIG. 280. Cancer and social position.

Left: A diagram constructed in the same way as Fig. 279, and showing how cancer of different organs varies from class to class. Cancer of sites directly exposed to the environment, such as the skin, mouth, and stomach, grades sharply with economic positions; cancer of deep, sheltered sites shows no grading. Right: A parallel diagram, but dealing with cancers of the digestive tract only. The upper part of the tube, from mouth to stomach, is exposed to irritation by food and drink; and here cancer grades from class to class. The lower part, intestine and rectum, is only exposed to digested food, and here there is no grading.

data have been obtained from experiments in which cancers were produced in animals by the local application of irritants to the skin. It is worth noting that these irritants have to act over long periods of time; it is chronic, not acute, irritation that is effective.

Apart from the risks entailed in these special occupations, there is a very interesting and suggestive correlation between cancer and the general conditions under which people live. In the report on occupational mortality to which we have just referred, it is shown that cancers can be divided into two groups, according to the original site. The first group includes cancers of the skin, tongue, larynx, œsophagus, and stomach, which show a very pronounced and regular grading with economic position, their incidence being more frequent in poorer sections of the community. Just over half of the total cancer mortality falls in this group. The second includes cancers of other parts, not exposed to fluctuating conditions, and these show no grading at all, but afflict the poor and well-to-do in equal measure. It would seem that the exposed surfaces of the poor are subject to irritation by ill-fitting clothing, bad food, and drink, and the like; the figures suggest that a large proportion at least of cancer deaths could be prevented by more comfortable living. The most interesting figures concern the different parts of the digestive tract. Cancer of the tongue, of the œsophagus, and of the stomach, all show a pronounced grading from class to class. Cancer of the intestines, of the liver and of the rectum do not. Apparently food and drink may act as an irritant on its way down to the stomach—it may be in an unsuitable form, or too hot, or the drink may be strong spirit—but after the treatment it receives in the stomach such outstanding faults have been corrected. Beyond that part they will do no further damage.

Even deep-seated cancers may be initiated by irritations. Thus one of the most frequent cancer-sites in women is the womb; here the irritation may be due to scar-tissue left by injuries during childbirth. Elsewhere, a variety of different kinds of lesion may result in cancers. Besides these undoubted facts, there is the possibility that general constitutional factors may co-operate by making the tissues more sensitive to local irritation. It is imaginable, for instance, that chronic poisoning of the kind that damages the heart and arteries might be a contributing factor. In mice, it has been shown that heredity plays a part; certain strains are more liable to cancer than others; but there is not a shred of evidence at present for the common belief in the inheritability of a cancerous disposition in our own species.

As for the treatment of cancers, the only hopeful course at present is to destroy the primary centre before it begins to spread and give rise to secondary growths. If it has spread to that extent, there is no hope as yet of a cure. The primary centre, in early cases, is generally removed surgically; but recently a number of brilliant successes have been obtained by killing it with radium. This new method was first used successfully in cases where surgical operation was dangerous and inadvisable; nowadays it is spreading even to operable cases and becoming a substitute for the knife. The chief trouble about treating cancer in the present state of our knowledge is that the disease is often not noticed until it has progressed too far to be curable. Perhaps intensive research will reveal yet other methods of attack; the fact that the cancer-cell is sufficiently different from other cells to stimulate the formation of anti-bodies suggests that there may be chemical ways of getting at it.

There we must leave the question of cancer with this practical maxim; avoid chronic chafings and irritations whenever possible and have the scars of old operations examined from time to time.

THE PRESENT HEALTH OF *HOMO SAPIENS*

§ 4. Tuberculosis.

But we must continue our review of the chief causes of mortality in a civilized community to-day.

Next after cancer, in the official report, comes a group called "Diseases of the Nervous System," accounting altogether for 9.1 per cent of the whole mortality. But this is a very mixed collection indeed. Nearly half of the deaths are due to cerebral hæmorrhage, which is really a local failure of the circulation rather than a defection of the nervous tissue. The rest are brought about by a miscellany of conditions, such as special infections, complications of syphilis, and so forth; none of them by itself accounts for any large number of deaths, and we need not discuss them further here.

Next comes tuberculosis, responsible for 8 per cent of the total mortality; here again those vulnerable organs the lungs come into prominence, for phthisis, consumption, or tuberculosis of the lungs accounts alone for 6.5 out of that 8 per cent. Besides the actual mortality, there is an enormous amount of illness and enfeeblement due to lesser degrees of tuberculosis. The disease is world-wide in its extent; it does not occur in epidemics but is constantly taking its toll of the species; probably no other single microbe is so destructive. Moreover, it affects other mammals besides man. Experimenters have even succeeded in infecting fishes and worms.

Tuberculosis is caused by a highly characteristic bacillus, which may invade any organ or tissue, causing the latter to form tiny lumps. Its course depends on a number of circumstances. The infected tissues may soften and die, or the germs may be hemmed in successfully by white blood-cells and by the growth round them of fibrous tissue and their spread may be checked. Apparently the majority of people contain, somewhere in their bodies, the remains of such a battle; it is said, from experience at post mortems, that the signs are recognizable in at least ninety per cent of adults. Moreover, in most cases at least, the infection seems to occur in childhood. Most children of fourteen have already acquired tuberculosis scars.

It is obvious from these facts that the presence of the bacillus in the body does not necessarily mean disease. Normally the natural resistances are enough to hold it in check. But, as with pneumonia, anything that lowers one's vitality may upset the balance in favour of the dormant germs. Fatigue, other infectious diseases, malnutrition, or life under unhealthy conditions may set it off. Here, the report on Occupational Mortality already referred to shows a striking parallelism between the death-rate from tuberculosis of the lungs and that from pneumonia and bronchitis; it increases rapidly and regularly as we pass down the economic scale.

The fight against the tubercle bacillus involves two campaigns—preventive and curative. The preventive steps are largely directed against the spread of the more damaging kinds of infection; the most notable is the systematic testing of milk for the bacillus, and the certifying of uncontaminated sources. A large proportion of abdominal tuberculosis, especially in children, comes from cows via their milk. The prompt isolation of people with seriously consumptive lungs, or with tuberculosis sores on the skin, helps to check dissemination. Recently considerable interest has been awakened by the work of Professor Calmette, who claims that by inoculating with a special harmless strain of living bacilli immunity can be produced, lasting for about a year—the principle being the same as vaccination against smallpox. But the completeness of the resulting protection is still doubtful.

As far as the treatment of sufferers is concerned, there is no direct attack on the bacillus; the method is to isolate the patients in special sanatoria, where everything possible is done to increase their natural resistance. They are given plenty of exposure to the open air—pure air—and to sunlight; they are given a generous but carefully chosen diet; their exercise and their personal hygiene are watched and regulated. In early cases the treatment is often strikingly successful. But tuberculosis is a very insidious disease; it produces a general weakening rather than arresting local symptoms; and often the treatment is not undertaken until it is too late for repair.

It is consoling to set against the alarming steady increase in the cancer mortality the fact that as the general standard of life of the community rises, tuberculosis retreats. It is still a tremendously injurious and wasteful disease. But in England the death-rate is now less than one-fifth of what it was seventy years ago, and in other civilized countries it has undergone a parallel decline. Direct preventive means have certainly played their part; but the most important factor of all is undoubtedly general public health. Sharp reductions have been found to follow the substitution for back-to-back houses of houses with through ventilation, and the draining away of sub-soil water. During the Great War the mortality from tuberculosis increased again, but except for this interruption its retreat has been steady and continuous. If the present rate of progress can be kept up, if the improvement in housing and general hygiene can be continued, tuberculosis may be a rare disease in two generations from now. The clean, open-air life of the sanatorium points the way.

§ 5. Mr. Everyman and the Fight Against Disease.

CIRCULATORY disease, respiratory disease, cancer, nervous disease and tuberculosis—those, in descending order, are the main causes of death in Great Britain at the present day. Together they make up 65 per cent of the total mortality. Next on the list comes disease of the digestive system—5.8 per cent—and so on to a number of other groups. We need not pursue the analysis further. The chief problems have already been stated; to continue would be to multiply details without any important addition of principle.

There remains the question of how Mr. Everyman should face the present position. To re-state our percentage in a slightly different way, the odds are almost exactly two to one that he will sooner or later die of one of the ailments that we have discussed. What should he do, from the point of view of his own personal welfare? How can he help the species in its collective struggle for healthier living?

The salient fact, to which we were brought time after time in our review, is that these main diseases depend enormously on one's general power of resistance—on a host of, at first sight, minor matters of every-day personal hygiene. To a large extent, at any rate, they are preventable maladies. Or rather they are delayable. Sooner or later as age creeps on death must come in some form or other, but there seems little reason why it should ever come before threescore and ten, or why it should take a painful form. But this tale of deaths coming for the most part before life has played itself out, impresses us not so much by mortality as by the enormous mass of unsatisfactory living, of malnutrition and under-exercise, of bad housing and clothing, of carelessness, ignorance and fatigue that it makes evident. That is the fact; the mass of minor physical inefficiency is the main thing; the fatal diseases are only its culminating symptoms. It is not merely that people do not live long enough, but that they are not living well enough.

We have seen that much of this prevalent ill-health is due to circumstances outside the immediate control of Mr. Everyman. Such matters as the control of supplies of food and water, the systematic checking of the spread of infectious disease and the regulation of conditions of industrial housing and employment are for public health authorities. The raising of the general economic status of the population—here let us remind ourselves again of the Report on Occupational Mortality, which shows how much ill-health and disablement is caused by poverty—is a problem for governments to handle. But there remains the individual problem; the course that Mr. Everyman should adopt if he wishes to live as long and as vigorously as his circumstances allow.

First, then, let him keep himself clean and fit. Let him brush his teeth regularly, evacuate his bowels regularly, take a reasonable amount of exercise so that his muscles are firm. Let him think his way of living over, and avoid any habit—such as the sort of over-smoking that leads to a perpetual sore throat—that may open the door to grave disease. Above all, let him not get too slack; let him be interested and active.

Second, let him consult the doctor frequently, whenever he is in doubt. There is a sort of heroism that keeps a man at his work when he ought to be in bed; when he has a severe cold in the head that might be mild influenza, for example. It is a dangerous form of heroism. To neglect any slight trouble in its early stages may have serious consequences; if the complaint is an infectious one, delay will assist in its spread.

Moreover, healing is a specialist's job. No busy man would think for a minute of purchasing an expensive automobile and running it for years without ever calling in outside assistance and advice; the running of a body for its seventy or eighty years of life is an altogether trickier affair. True, books can be bought which explain all about cars down to the minutest detail, so theoretically the first proposition might be possible. But bodies vary much more than motors. One's state of health is a resultant of so many interacting factors that every individual case presents peculiar features of its own, demands special expert consideration. Therein lies the inadequacy of home doctoring, of attempting to cure oneself by simple old-fashioned remedies or by some attractively advertised patent medicine. The balance of health is too delicate and too intricate for that kind of stereotyped adjustment.

What we have written here sounds very elementary and almost platitudinous; and yet it has to be stressed. Many people seem to live anyhow, without a thought for their physical welfare, until suddenly some troublesome or painful derangement sends them to the doctor. They look upon the medical man as simply a patcher-up of damage done. But the doctor should be more than that. He should be a regular medical adviser. So much preventable suffering has its root in causes which could easily be remedied if they were detected early enough and promptly attacked. Mr. Everyman would be well advised to see his doctor periodically, just as he sends his automobile to be overhauled once in a while, not so much for the correction of distressing symptoms as for the detection and rectification of minor errors in his way of living. Then he can be sure that the hygienic rules and disciplines he has adopted suit his own particular case.

Naturally enough, vigorous health cannot last for ever. As the years go by, Mr. Everyman's power of resistance will gradually fail, and sooner or later something will get him. But by living a careful, reasoned life he can postpone that moment as long as possible; he can get as vigorous and fruitful a life as his inborn constitution and accidents of circumstance permit; he stands the best chance of putting off his ultimate defeat until his declining appetites and interests and the natural resignation of age rob death of its last elements of frustration.

BOOK EIGHT

BEHAVIOUR, FEELING AND THOUGHT

I

RUDIMENTS OF BEHAVIOUR

§ 1. *The Three Elements of Behaviour.* § 2. *Receptivity.* § 3. *Response.* § 4. *Correlation: the Origins of the Nervous System.* § 5. *Vegetable Behaviour.* § 6. *Instinctive and Intelligent Behaviour.* § 7. *The Behaviour of the Slipper Animalcule.* § 8. *The Different Worlds in Which Animals Live.*

§ 1. The Three Elements of Behaviour.

WE HAVE now studied a living body in some detail, and reviewed the whole spectacle of life. We have studied in particular the resistance of the human body to disease. We are now turning to the spontaneous activities of living things. They act of themselves, and they act discriminatingly and not merely according to the laws that govern the movements of lifeless matter. And we shall find creeping in by almost imperceptible degrees the persuasion that living things do not merely move spontaneously, but feel. As we ascend the scale of being we shall find movements more and more suggestive not merely of feeling but of perceptions, observations, and thoughts as we know them in our own minds. A point will come when we shall have to abandon exterior observation and turn our minds inward for light upon the problems of animal behaviour.

All living things respond to their surroundings. Even the simplest microscopic animals will pursue their prey and avoid hostile influences; even a plant pushes its roots towards moisture and its leaves towards the sun. All through the spectacle of life we perceive that adjustment of activities to circumstances of which our own triangle of feeling, thinking, and doing is only the culminating case. To this apparently spontaneous adjustment we are now to give our attention.

It may be well to insist upon a possible danger in our treatment of this question, the assumption, which may be either premature or quite wrong, that amœba really feels the influences, such as harmful chemical substances or excessive heat, which it avoids, or that a plant is aware of moisture and light, and responds to them in a way which involves some sort of parallel to our own consciously appreciated feelings and responses. There is only one conscious mind (the point is too often made in abstract discussions to need further elaboration here) of whose existence a man may be sure, and that is his own. His belief in the existence of other similar minds in his friends and acquaintances, rests only upon inference. Their acts and speech, and general similarity to himself, force him to that conclusion. When he considers creatures of other kinds the same inference follows the same observation; the behaviour of his dog at a pleasing or a disappointing event recalls in many ways that of his child under similar circumstances, and presumably betokens the same joy or sorrow. Perhaps he reads a similar mental life, albeit at a lower level, into frogs and fishes, and even invertebrate animals. That is the tendency. But let us be on our guard. To read a consciousness like our own into a creature so elementary as an amœba simply because it shrinks at a harmful stimulus is intellectually dangerous; to read a mind like our own into so differently organized a creature as an insect, or a mollusc, is dangerous, too. They may have minds, but as different from our minds as their bodies are from our bodies. In the past the study of the lower creatures has too often been complicated and impeded by such insidious temptations.

Presently we shall study in some detail a little living automaton called the slipper animalcule. It is one of the protozoa. Can such a creature be said really to feel? We do not know; that depends on whether it has a conscious mind. But external influences, such as warmth of the water, or chemical substances dissolved therein, can act upon the creature and modify its behaviour. Does a slipper animalcule think? Apparently not, since, as we shall see, it lives entirely in the present, and shows no evidence either of memory or forethought. But there is a certain evident adjustment of its acts to its circumstances. The sudden impact when it runs up against a hard obstacle is met by an avoiding reaction, but at the different touch of a mass of the bacteria upon which it feeds the animal stops and begins to eat.

To review the whole range of behaving life we shall find it best to consider our matter under three aspects. First, there is receptivity; organisms can be influenced by external circumstances. This in ourselves is often (but not always) accompanied by sensation. Second, there is activity; organisms swim, or run, or grow, or blush, or perspire, or perform any of a multitude of apparently spontaneous acts. Third, there is correlation; the acts are more or less accurately suited to the circumstances, and the whole of the machinery of this fitting of the one to the other, ranging from the simplest automatism to Mr. Everyman's conscious and deliberate thought processes, we treat under this third head.

It will make the matter plainer if we take these heads one by one, and see how in each case the first organized beginnings lead up to the state of affairs found in ourselves.

§ 2. Receptivity.

The frog is so constructed that its brain and spinal-cord can be destroyed with a needle in less time than it takes to hang or electrocute a man. For this reason, besides its commonness and general availability, it is one of the most-used laboratory animals.

Suppose that we have a frog which has been "pithed" in this way—the brain and mind destroyed, the heart, abdominal organs and so forth still living—and that we remove from it the calf-muscle with its attached nerve (Fig. 39). They lie limply on the experimental table. The muscle is pink and spindle-shaped and about an inch long. Before its removal it was firmly attached at its upper end to the bones at the knee, while below it tapered away into a glistening tendon—a tough, flat ribbon of tissue which ran round the heel to the sole of the foot. Most of the tendon and a fragment of bone from the knee have been taken out with the muscle. The nerve is a white, slender, living string, and an inch or so of it is still attached to the muscle.

These organs can retain their vitality for hours if they are properly tended. They are kept reasonably cool, and moistened from time to time with a little of a properly mixed salt-solution.

If we pinch the muscle or press on it, it will suddenly quiver and shorten; the same effect may be produced by giving it a weak electric shock, or by adding a drop of an appropriate chemical solution. In biological language, the muscle is said to be "irritable" or "excitable," and the pinch or electric shock or chemical agent by which we can make it move is called a "stimulus." A fraction of a second after the stimulus the contraction passes off and the muscle lies limp as before, so that we can repeat the experiment again and again.

The nerve also is irritable, but its activity manifests itself in another way. Its rôle in the organism is not to contract but to conduct. So, when we pinch it or otherwise stimulate it at any point, it shows no outward sign of change, but the muscle to which it is attached abruptly shortens, as if in response to direct stimulation. Our experiment simply sent impulses flying along the nerve from the point we stimulated, and those impulses, reaching the muscle, stimulated it in its turn to activity.

The impulses in the nerve are invisible, but they can be detected electrically. If we connect to some part of the nerve a delicate apparatus for recording very slight electric currents, and then stimulate another part of the nerve, we should see a little tremor of the recording-apparatus a fraction of a second after the stimulus, as the impulses flashed by on their way down the nerve.

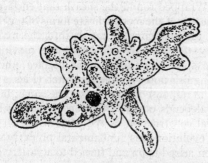

FIG. 281. The microscopic, shapeless, fluid body of Amœba, which, although it has no special sight-organs, shows a primitive sensibility to light.

FIG. 282. Stentor, another protozoon which is sensitive to light.

It can anchor itself to a fragment of pond-weed, as shown, or swim about in the water.

Here we have a very simple demonstration of irritability, which is one of the most fundamental properties of living substance; and if it has been set forth in a manner somewhat recalling the elementary class-room, the triplex author pleads the importance of the phenomenon as his excuse. Irritability is a universal property of protoplasm, and without it there would be no sensation, no consciousness, no nervous activity. The whole harmony of the body, the whole of the matter with which this Book is concerned, depends upon it.

The rich variety of sense-organs with which our bodies are provided have already been reviewed in Book 1, Chap. 3, § 5. We saw a very considerable diversity of structure and working. The man in the street speaks of the evidence of his five senses; to the biologist this is undue modesty, for in actual fact (allowing for the different kinds of skin sense, internal sense, and so on) he can give a list of about twenty. We watch the isolated muscle as it twitches at our bidding on the table before us; at first sight the relatedness of the two processes, its twitching and our watching, is not apparent. But related they are. The working of a sense-organ is no more than a refinement and a specialization of the irritability of an isolated muscle or nerve.

Consider what we have already learnt. The frog's muscle can be stimulated by a pinch or a light jolt. In our bodies there are various kinds of cells (or parts of cells) which specialize in extra sensibility to such mechanical stimulation. Wrapped round the sheath that encloses the root of a hair, they underlie one of the most delicate forms of touch. Grouped at the openings of the semicircular canals into the main cavity of the inner ear, they tell us how our heads are moving. Arranged in a row along the spiral chambers of the cochlea, they enable us to hear and to distinguish notes of different pitch (Figs. 45-46). The essential nature of the response is in all cases the same; but its meaning to the organism depends on the situation of the touch-sensitive organs, on the special construction of the parts where they are placed. In the course of evolution this fundamental property of mechanical irritability has been seized upon and turned to a variety of uses.

To make the idea plainer, let us take another example; let us watch Nature at work and trace stage by stage how out of a generalized irritability she builds that marvel of organic engineering, the human eye.

Light can be shown to have a direct stimulating action on undifferentiated protoplasm. A beam of intense light focused on to one end of an Amœba causes, first a local contraction of the stimulated region, and then movements of the other parts, which lead to the animal's crawling out of our experimental limelight. Very strong light suddenly flooded on to the whole animal makes it contract and draw itself together into a rounded blob, and may paralyze and even kill it. From such crude light-sensitivity as this we start our story.

The second step is found in other protozoa, where part of the body is specially irritable. The front end of the ciliate Stentor, for instance, is considerably more sensitive to light than the rest; and the animal shows the curious reaction that whenever the illumination of this delicate front end is suddenly brightened, it abruptly reverses its direction and swims backward for a short distance, then twists round a little so that its front end points a new way, then goes forward again. This behaviour has a definite value. If we put a number of Stentor into a glass dish, fairly brightly illuminated, but with a patch of shadow in it, the animals will be found in a little time to have collected in the shade. If now we watch them through a microscope as they swim about, we shall see that every now and then one of them gets to the edge of the shadow; but as soon as its sensitive pole feels the bright direct light of the sun, the animal gives its "avoiding reaction," backing abruptly and changing direction, and thus keeps in the shade. Moreover, when it is outside, in a bright light, the animal only swims straight when its front end lies in its own shadow, i.e., when it has its back to the light; if it is pointing in any other direction it twists about (apparently at random) until this position is achieved. This reaction will evidently guide it automatically out of open, sunlit waters to shady corners.

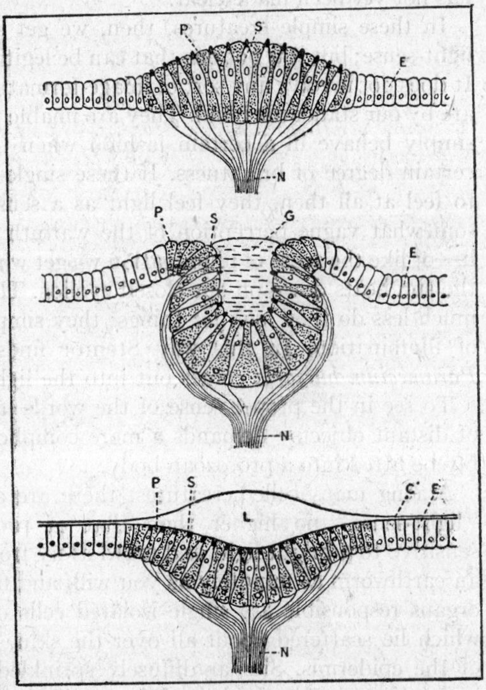

An opposite reaction to light is shown by

FIG. 283. Three simple kinds of eye, as described in the text.

(E) *Cells of skin.* (S) *Cells sensitive to light.* (P) *Cells full of black pigment.* (N) *Nerve fibres.* (G) *Glassy substance exuded by the cells.* (C) *Cuticle.* (L) *Lens.*

those protozoa which have simple plants living symbiotically inside them (Fig. 117). The common slipper animalcule, *Paramecium caudatum*, a creature found in stagnant pools containing decaying leaves and the like, neither seeks nor avoids light unless it be so very intense as to be injurious. But its close relative *Paramecium bursaria*, which contains algæ, swims towards the light and therefore collects near the surface of the water, where its green partners can build up starch and sugar most effectively.

In many protozoa there are special structures called eye-spots—usually red in colour—and it would appear that they are involved in some way in the animal's responses to light. But just how they act has not yet been made clear.

In these simple creatures, then, we get something we may call a light-sense; but it is not one that can be legitimately described as sight. It does not involve any sort of image-formation whatever. The animals are by our standards blind. They are unable to see other objects. They simply behave in a certain fashion when they are illuminated to a certain degree of brightness. If these single-celled bodies can be said to feel at all then, they feel light as a sensation more like our own somewhat vague perception of the warmth of a fire than our seeing it—or like the sense of illumination we get when we lie sun-bathing and the sun beats down on our closed eyelids. They do not look *at* things, much less do they look *for* things; they simply respond to the degree of illumination. In this way Stentor finds a shady corner, while *Paramecium bursaria* comes out into the light.

To see in the proper sense of the word—to focus and form images of distant objects—demands a more complicated eye than can possibly be fitted into a protozoan body.

Among many-celled creatures, there are a number which show a "light-sense" no higher than that of protozoa. Earthworms are sensitive to a bright light and crawl away from it, but you can dissect an earthworm as minutely as you will, and find no trace of eyes. The organs responsible are single isolated cells of a rather curious kind, which lie scattered about all over the skin, in among the other cells of the epidermis. Similar diffusely sprinkled light-sensitive cells are known in a great number of forms. Tadpoles apparently feel light over their whole skins; if after a tadpole has been blinded it is put in the dark and then suddenly illuminated, it begins to swim actively about—which shows that the light must act as a stimulus on its skin. The lancelet has light-feeling cells in a curious place—buried in the nervous substance of its spinal cord; because of its thinness and translucence the whole of its anatomy is illuminated, and a light-cell inside is pretty nearly as useful as one on the skin.

In these cases, of course, we are no more justified in speaking of sight than we are in the case of the protozoa. It is doubtful if we may even speak of feeling. The earthworm lies out of its burrow before dawn, feeding on dead leaves and the like; as the sun rises and shines on it, it retreats into its burrow. That is about all its light-cells can do for it. It does not see the early bird approaching; but with reasonable luck (and with its great sensitivity to even the slightest jarring of the earth, as stand-by), its light-sense will guide it to safety as its enemy begins to stir.

The next great step in the evolution of sight is the collection of the scattered sight-cells together into primitive eyes. Fig. 283 shows three types, such as are found in a great number of invertebrate animals. In each drawing the skin of the animal is supposed to be cut across and magnified; the conspicuous palisade of cells being the epidermis, separating the rest of the animal below from the outer world above. The sense-cells are gathered together into patches, and at their inner ends they send nerve-fibres down to a brain. In the uppermost eye the sensitive bit of skin (which is thicker than the rest) bulges slightly outwards, and there is a copious deposit of black pigment (shown stippled) between the cells. In effect, this pigment forms a black tube round each sensitive cell, so that the latter is only strongly excited by light that falls upon it in the direction in which it points.

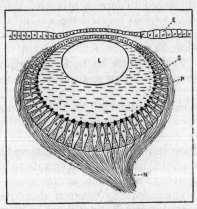

FIG. 284. A further stage in the evolution of the camera eye (see text).

The parts are lettered to correspond with those in Fig. 283.

So in this very simple kind of eye there is a sort of crude image formation. The creature distinguishes when there is something pale or luminous in one direction and something dark in another, and is perhaps aroused by conspicuously moving objects—although this sort of vision must be at the best hazy and without definite outlines. Such eyes as this are found in a number of worms and other lowly organized invertebrates. They are, of course, very small, and appear simply as little specks of colour on the head, horns or other parts of their possessors.

The eyes in the middle and lower part of our figure show a different tendency; here the sensitive surface is curved inwards instead of outwards. But pigment between the cells serves the same purpose as formerly. Here, however, we note the first rudiments of an additional refinement—a device for concentrating and perhaps focusing light. In the middle eye the elementary retina has been dimpled in to form a definite cup, and inside the cup there is a special glassy substance secreted by the cells themselves. This state of affairs is found in many jelly-fishes, star-fishes, and worms. The eyes of a limpet are of this kind, but those of its more familiar relative, the snail, are more complicated. In the eye at the bottom of our figure the retina is less markedly curved, and here part of the cuticle (a thin, hard, transparent layer of secretion that covers the body in most invertebrates) is thickened to form a primitive kind of lens. This last arrangement, with various minor complications, is found in other members of the groups just named, and in molluscs and the arthropod Peripatus (Fig. 84).

In these three kinds of sense-organ, primitive and inefficient though they be, we can discern the two fundamental features of true eye-construction. The first is the presence of protoplasm which is specially sensitive to light*. This localization first appeared in our account with Stentor, and we trace it through the light-feeling cells of earthworms to these elementary eyes. The second is the grouping of a number of these cells into a working unit; and it is this which makes true seeing possible. Each cell corresponds to a particular direction from which light-rays may come. It corresponds, so to speak, to a sector of the environment. So between them, taken together, the cells form a spatial image. Not until such an arrangement as this had been evolved could the shapes and movements of distant objects be perceived and influence an animal's behaviour.

There we have the essentials of the evolution from mere light irritability to sight. To attain to the clear vision of a bird or a higher mammal is only a matter of perfection of detail. We can note some of the main steps, although it would not be worth our while to go into all the minutiæ of the process. The story is complicated by the fact that the invention and gradual elaboration of the eye has gone on independently in a number of groups, sometimes along parallel lines and sometimes in strikingly different ways.

From the hollow kind of eye shown in the middle and at the bottom in Fig. 283, we can lead up to eyes like our own. First the sensitive dimple of skin pinches itself off altogether from the rest and comes to lie a little way below the body surface in the shape of a hollow ball.

---

*Our own retinal cells are not directly stimulated by light, but by a chemical substance produced when light falls on the retina (Fig. 47). How far this is general is not yet clear, for the discovery is a recent one, but in the diffuse light-cells of a sea-squirt (which are like those of earthworms discussed above) it has been shown that stimulation has also a chemical basis. Whether this is also true of protozoa and of the light-sensitivity of undifferentiated protoplasm is not yet known.

Then the optical apparatus is improved and extended. In some molluscs and worms this is done by specializing part of the glassy substance inside the eyeball as a more or less spherical lens; in this case only the deep half of the eyeball is sensitive to light, the side lying against the skin, and the skin itself being transparent to admit as much light as possible. Fig. 284 shows an eye of this kind. In other

FIGS. 285, 286, 287. The three climaxes of the camera eye.

The eyes of birds of prey are remarkable for clear vision at a distance. Those of monkeys and men are unique in having a "yellow spot" on the retina; also they are placed side by side to allow of stereoscopic vision. Those of octopuses and cuttle-fish, although independently evolved, show striking parallels with the eyes of vertebrates.

cases, such as the third eye in the middle of the forehead with which our reptilian ancestors were provided, the outer wall of the eyeball itself forms a lens. Or the lens may be made out of the skin over the eyeball; this is what happens in ourselves and in a number of the higher molluscs.

Thus the retina and the lens arise. Similarly, we can trace further elaborations—the cornea and iris, the appearance of muscles to focus the eye on objects at various distances, the gradual refinement of the structure of the retina itself, culminating in the yellow spot of monkeys and men. It is fascinating to note that a parallel evolution has taken place in those highly organized molluscs, the octopuses and cuttle-fish. A glance at Figs. 288 and 47 will show how closely the cephalopod eye, with lens and cornea, iris and focusing-muscles, resembles our own. Nowhere else among the invertebrates has such perfection of detail been evolved—at least, not in an eye of this kind.

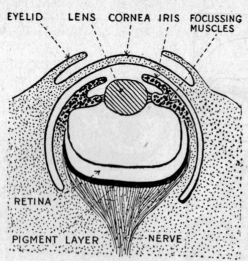

Fig. 288. The eye of a cuttle-fish cut across, for comparison with the human eye shown in Fig. 47.

But there is another, very different, direction in which the eye has evolved. We may call the type which culminates in octopus, vulture, and man the camera eye. But the primitive convex type shown at the top in Fig. 283 contains other possibilities which are worth tracing as they develop stage by stage.

As we noted earlier (Fig. 90), there are marine worms which live in tubes they build out of sand or fine gravel, spreading a crown of feathery tentacles to catch minute particles of food. Many of these worms have simple eyes on the tips of their tentacles, and Fig. 289 shows one of these eyes cut across. It is like the uppermost eye of Fig. 283, but it shows an improvement in that each of the sensitive cells is provided with a tiny light-concentrating lens of its own. Each is in fact a miniature unit eye. Because of their convex grouping each of the eye-cells looks out in a slightly different direction from the others, and between them they build up a kind of mosaic picture of the surrounding world. Each by itself is unable to make an image, since it has only a single nerve-fibre; it simply telegraphs to the brain: "I am receiving light; I am not receiving light," with additional comments, perhaps, on the intensity and colour. The image synthesized in the brain is a mosaic of all these separate units of information, much as a weather-chart is built up from the reports of local meteorological stations.

This is how we may suppose the eye of the tube-worm to act. But, oddly enough, there is no direct evidence that the creature forms images in this way. So far as we know it only shows one kind of reaction to light, and that is a very simple one. Whenever the light falling upon the worm is suddenly dimmed, it flicks back with lightning speed into its tube. A passing shadow will do it. The value of this "shading reaction" to the organism is too plain to need stressing; one has only to think of a fish browsing over the sea-bottom in search of attractive morsels to snap up. Moreover, the reaction is obviously a very simple one—as simple as the avoiding reaction given by Stentor when suddenly illuminated. Similar reactions to a passing shadow are seen in a great number of sluggish and sessile shore-creatures. While these worms snap back into their tubes almost too fast to see barnacles close their shells abruptly; many sea-squirts contract up into gelatinous blobs; burrowing bivalves withdraw their soft, protruding siphons into the sand. Sea-urchins cannot withdraw into safety; in some of them, however, sudden shading causes them to bristle up their spines, and if the shading is only partial the spines turn towards the shaded spot. Thus, as our browsing fish prowls along with his tell-tale shadow sliding beneath him, the invertebrate world hides or arms itself at his approach.

Many of these creatures are eyeless; and whether the comparatively well-formed eyes of a tube-worm are really used only for this purpose, or whether they provide the owner with a further crude visual power that has not yet been detected, we cannot say. But we must return to our evolutionary story.

Eyes like this one that we have been describing, but with various extra complications, are found in one or two bivalve molluscs, where presumably they have been independently evolved. But the type reaches its highest development in the astonishing compound eyes of arthropods. If magnified some eight or ten times, the head of a dragon-fly shows most of its surface covered with an enormous pair of faceted eyes—each eye a multitude of crowded eyelets, each eyelet a somewhat more complicated version of the single eyelets seen in the tube-worm.

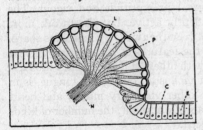

FIG. 289. An early stage in the evolution of the compound eye—the eye of the tube-worm lettered to correspond with Fig. 283.

Thus, stage by stage, and in various directions, the eye has evolved; thus a primitive vague awareness of light has been elaborated and refined into sight as we know it. We could trace other stories like this one; for in the same sort of way the sensibility of the lowest animals to warmth or cold, to a touch or a jar, to various chemical substances has been specialized in higher creatures to serve a multitude of purposes. By insensible degrees we pass from a primary sort of receptivity that is like the purely mechanical irritability of an electric bell-push, to a discriminating reaction that is indistinguishable from our own seeing. Vitalists like Bergson have made a tremendous difficulty about the evolution of the eye, declaring it is too complicated a process for unassisted natural selection. But is it after all so inexplicable?

§ 3. Response.

As every reader knows, Mr. Everyman's movements are brought about by muscles. Of these he has several kinds. There are the "voluntary muscles," by means of which he performs deliberate movements; there are the "involuntary muscles" in his intestines, heart, and so forth, that behave independently of his will. These two kinds of muscles can be distinguished by their microscopic structure and by their physiological properties. The voluntary muscles only come into action at the call of impulses from the central nervous system, but when they do so they contract with great promptitude and vigour. The involuntary muscles are independently active, but they work more slowly and feebly. But the line between the two is by no means sharp. Heart muscle combines many of the features of such muscle as that found in the walls of the stomach with others of the voluntary muscles. Muscle from that strange, almost molluscan organ, the tongue, is in some ways intermediate between other voluntary muscle and heart muscle. When we take other animals into account, the number of kinds of muscle becomes very great indeed.

Now, just as the responsiveness of our sense-organs is no more than a specialization of protoplasmic irritability, so the shortening and relaxation of our muscles is no more than a specialization of a general power which living substance has of changing its shape. We have already dealt with the slow, flowing movements of Amœba, which can bulge out its shapeless body into a protruding lobe in whatever direction it pleases. Related to this, it would seem, are the restless, flowing movements of other cells—such, for example, as the endless circulation of the protoplasm of many plant cells round the tiny boxes of cellulose in which they are imprisoned. In recent years Pantin has shown, at Plymouth, that the movements of an Amœba have many properties in common with the movements of a muscle-cell; the two are similar in their relation to oxygen-lack, to temperature, and to a great number of chemical substances. In brief, a muscle-fibre is a cell specialized in such a way that this primitive contractility of living material is made vigorous and efficient.

In another direction, we can trace a series of intermediate stages in one creature and another, leading from the flowing, temporary lobes of Amœba to the delicate vibrating filaments called cilia or flagella—found, for instance, in many protozoa, in spermatozoa, and in various other kinds of animal and plant cells. Here, again, a recent piece of work by Gray, at Cambridge, has shown that cilia have the same fundamental mechanism as muscle.

But movement is not the only way in which an animal can respond to environment. Mr. Everyman may change colour at a sudden surprise; he may even break out into a cold sweat. His colour change is, of course, a change in the amount of blood in his skin, and is really only a disguised kind of muscular movement, for it depends on the contractility of muscles or muscle-like elements in the walls of his blood-vessels. But perspiration depends on glands, not muscles. Similarly, if something has delayed his meal, his mouth may water at the sound of a dinner-gong from a neighbouring house. Here, also, he responds to the stimulus by secretion.

We, ourselves, like most vertebrates, are predominantly muscular animals, and glands play a comparatively small part in our behaviour. For clear examples we must turn elsewhere. If you take a newt out of water and handle it, it will exude a clear, slimy secretion over its whole skin to make itself slippery and help in wriggling out of your hand. If you treat it too roughly or hold it too long it will exude another kind of fluid—less transparent and less oily to the touch, but containing an irritating poison. A lady who tried the experiment of gently biting the tail of the crested newt—the larger of the two common English newts—reported a bitter, astringent taste, irritation of the throat, a strong flow of foamy saliva, violent automatic movements of the muscles of the mouth, and a headache that lasted for several hours.

Many other amphibians show a similar power of responsive exudation; indeed, the highly glandular skin is as distinctive of amphibians as is the scaly skin of reptiles, the hairy skin of mammals, or the feathered skin of birds. The poison glands cluster very thickly in the warts with which the skin of a toad is beset; and anybody who has seen a dog take a toad in its mouth knows that they can be used to advantage. Their action on ourselves results in symptoms not unlike the early stages of a cold in the head. According to report, the Colombian tree-frog, *Dendrobates tinctorius*, has so venomous a secretion that the Indians use it for poisoning their arrows; one frog held over a fire (and this species seldom exceeds an inch and a half in length) pours out enough for fifty arrow-heads. In this case the poison acts chiefly on the nervous system. Snails provide a well-known example of defensive secretion; everybody knows how an irritated snail will retire into its shell and exude a quantity of frothy slime.

We have already said something of the gas-bladder of fishes (Book 5, Chap. 4, §5, and Book 6, Chap. 2, § 2). Apparently fishes have the power, by means of a special gland, of secreting gas into the bladder and so regulating their density according to the depth at which they are swimming.

A special kind of secretion is the formation of luminous substances —a property which is surprisingly widespread in the animal kingdom. As anybody knows who has taken a row-boat or a canoe on the sea on a summer night the surface waters abound in tiny organisms which give out flashes of white or greenish light when they are alarmed; the sea seems to light up at every stroke of the paddle. In a number of nocturnal animals light-producing organs are used to attract the notice of the other sex. In fire-flies the various species have characteristic rhythms which they flash out as they fly, much as grasshoppers have distinctive rhythms in their chirping.

Sometimes the light-producing organs are simple glands, no more complicated than the skin glands of newts and toads; but in other cases (such as deep-sea fish) there may be elaborate accessories— reflectors behind, and even lenses in front, so that the light shines out in a definite beam.

Clearly, the variety and degrees of organization of different kinds of responsive tissue is considerable. In what follows we shall need a word to cover all these various responsive tissues and structures—muscles, cilia, glands, and the other organs that we are about to discuss. The most convenient word for the purpose is "effector." A muscle-fibre is one kind of effector; a gland-cell is another.

As a further kind of response, we may note the changes of colour of chameleons, frogs, and flat-fish and the even swifter changes of octopuses and their relatives (Fig. 87). In the latter case the colour-change is really only a kind of muscular movement. If you look carefully at the skin of a living octopus in an aquarium, you will see that the pigment is not diffused uniformly but gathered into definite specks. These specks can expand or shrink, and thus darkening or paling of the whole skin is brought about. Each speck is really a tiny elastic bag containing pigment, and it is attached to a number of muscle-fibres which radiate away from it through the skin. When the muscle-fibres contract they pull on the bag from all sides and it expands; when they relax it shrinks again because of its elasticity.

But the colour changes of other creatures are less prompt and vigorous, and due to effectors of a different kind. Here the pigment-cells are branched, radiating structures, and the pigment inside them can either spread through the whole cell (producing darkening of the skin), or collect into a tight little speck in the middle (producing paleness). Recent experimental work suggests that these cells are modified contractile cells, and fit into the movement-producing group that we discussed above.

But perhaps the strangest effector of all is seen in certain fish, which can give electric shocks to any other animals that happen to come into their immediate neighbourhood. Many of the common skates and rays have living batteries at the roots of their tails; if you grasp them at that point they can give you a shock of about a volt, or about half the voltage of an ordinary pocket flash-lamp battery. In the aquarium at Monaco you can, for a small fee, make the experiment of touching the Mediterranean electric ray, *Torpedo marmorata*, which has stronger batteries at the sides of its body. If it is feeling vigorous, and if you have not had too many predecessors, it will give you some twenty or thirty volts. One or two other kinds of electric fish can do better than this. Strongest of all is the electric eel, a fresh-water fish from South America. This animal, which may be five feet long, has a powerful battery running the whole length of its body, and accounting for more than a third of its weight; it can give you a shock even if you handle it in a surgeon's rubber gloves, its potential reaching several hundred volts (more than the voltage of domestic electric light). Frogs and little fish can be electrocuted at once by the electric eel.

In most cases at least, these electric organs are modified muscles. The normal activity of a muscle is accompanied by slight, incidental electric changes; and in the electric fishes, Nature has seized upon this fact and modified the muscle-cells until their power of shortening is lost altogether, while the electric currents they produce are intensified.

Each of these modified muscle-fibres has the shape of a thin plate of tissue, and an electric organ consists of a great number of these plates, put together like a pile of pennies, with their separate electrical effects summing up, just as ordinary dry-cells sum up if they are connected in series. The electric eel has about seven thousand of these living electric cells in series with one another on each side of its body, all ready to work at the bidding of the central nervous system.

These electric organs afford one of the best examples of parallel evolution, for they have been evolved independently in several groups of fishes, now as a modification of one muscle and now of another. In the electric eel and in the common skates and rays part of the column of muscle that runs along the whole trunk and tail, and performs the swimming movements, has been converted to this new use. In the electric rays, or "torpedoes," a muscle that generally lies close to the skin and wraps round the gill region has been "electrified," and lies as a large, white cake-like body on each side of the fish. In a recently discovered electric fish, *Astroscopus*, which is related to the perch and stickleback, the battery lies just behind the eye, in a cavity that pouches off from the orbit. Also—oddly enough—it is supplied by a branch from one of the nerves to the muscles which move the eyeball; so it looks as if the battery itself is a specialized part of one of the muscles that roll the eyes.

But the strangest case of all is found in the electric cat-fish, *Malapterurus*, from African rivers. Here the batteries lie actually in the skin and form a blanket round most of the body. Moreover, they are supplied by two nerve-fibres only, one on either side of the animal. In this case, the organ is apparently not a modified muscle at all, but a modification of the glands of the skin. It has been shown that in ordinary glands (such as the salivary glands), activity is accompanied, as in muscles, by slight electrical changes. So Nature can make her living batteries from two quite different kinds of raw material.

§ 4. Correlation: the Origins of the Nervous System.

As WE ascend the scale of life we discover a more and more elaborate co-ordination of receptivity and response.

As we explained in Book 1, a nerve is a sort of living telephone wire, and the central nervous mass made up of brain and spinal cord is like a telephone exchange. Nervous messages flash up from the sense-organs along nerves, and, also along nerves, outward messages flash down to muscles, glands, and so on. In the central nervous mass the sensory messages are weighed and integrated; decisions are made; appropriate commands are dispatched that result in action.

Now the simplest living creatures have none of these organs. How, in them, is action suited to sensation? Can we discern the rudiments in an Amœba of our own nervous activity, as we have already discerned the rudiments of our sensibility and muscular movements?

The whole surface of an Amœba's body, we noticed, is equally sensitive to various stimuli such as heat, strong light, harmful chemicals, or the neighbourhood of food.

The whole surface is contractile, and every part can be protruded as a temporary limb, or can form an improvised mouth or anus. Similarly its whole body can conduct impulses from a stimulated spot to other parts at the rate of a small fraction of an inch per second. This slow, sluggish communication between part and part is the primitive type of co-ordination from which the vastly more rapid and efficient conduction along our own nerve-fibres has been evolved. It is, like irritability or contractility, a fundamental property of protoplasm. Just as the one property has been refined and specialized in a sense-cell and the other in a muscle-cell, so a nerve-cell is simply a cell which is pulled out into a long, thin thread running from one part of the body to another, and in which this primitive protoplasmic property of transmission of impulses is picked out and made into the primary function.

Amœba, then, has all three of the fundamental elements of behaviour—the reception of stimuli, the transmission or conduction of impulses, and the final action of an effector organ, but all are indiscriminately blended, diffused through the whole body. There lacks the definition of special parts, and since any bit of an Amœba's body may be now a sense-organ, now a nerve, now a muscle, the protoplasm manifests that general inefficiency that one expects from a jack-of-all-trades.

Other protozoa possess a permanent form, and permanent effectors such as cilia or flagella. Much of their behaviour is determined by this structure, for they move with one end forwards, and in particular ways, swiftly or slowly, smoothly or in jerks, in straight lines or spirals, as determined by the arrangement of their cilia. A number have also sense-organs or the beginnings of sense-organs. As we have seen, there may be a zone round the mouth or at the front end which is more sensitive than the rest of the body, and quite frequently there are such special organs as eye-spots.

Finally, in some of those highly organized protozoa, the Infusoria, there are special nerve-like organs, in the form of fibrils which radiate away from the sensitive parts and co-ordinate the movements of the cilia all over the body. In one or two cases there is even a central mass of this substance, acting perhaps like a very simple brain, to which the sensory portions of the system run in, from which the motor fibrils run out. These fibrils have apparently the same function as nerves, namely that of rapid conduction from one spot which is excited to another where action is carried out; but, of course, they are of quite a different origin from real nerves, being all differentiated inside a single cell, while each of our millions of nerve-fibres is an outgrowth from a whole cell. Thus in this respect as in so much else the protozoa, within their minute and single-celled bodies, have anticipated many of the inventions made later and independently by many-celled animals. These devices are therefore not directly ancestral to ours, but interesting parallels at a lower level of organization.

The bodies of the higher animals consist of republics of co-operating cells. In surveying the whole range of living things known to-day (Book 2) we noted a number of forms on the border-line between the single-celled and the many-celled—there are the flagellates that live in colonies, the simple filamentous algæ and fungi—and these lead up to more highly organized forms. During early evolution there must have been a transition stage when life was experimenting with variable success with such aggregations of cells. On the cellular level there are surprisingly highly organized creatures like the infusorians that we have just discussed; the first many-celled aggregations were very much less individualized than these. A sponge is a multitude of cells, specialized into several co-operating kinds and growing as a whole into a definite and characteristic shape; but the cells still retain a very considerable degree of independence; cut away a bit of a sponge, or pass it through a fine sieve, and it will re-organize itself without difficulty. A polyp, or a flatworm, or a sea-squirt, although showing a much higher degree of specialization of organs and tissues, still retains the power of reconstitution to an astonishing degree. It would seem that its parts specialize themselves, but not whole-heartedly, keeping a certain potential independence in reserve. Finally, in the most elaborately organized creatures the subordination of the cell-life to the life of the whole is well-nigh complete.

As one would expect, the development of nervous communications between part and part shows a parallel growth from a stage of preliminary experimental inefficiency to its final elaboration.

We have a primitive kind of communication between cell and cell in the microscopic green spheres, each about one-fiftieth of an inch across, of the flagellate Volvox, which consist of hundreds of individual cells embedded in a common mass of jelly. Each of these cells has a pair of lashing flagella, but they all beat in a disciplined way, so that the colony travels with one side always in front. Here the cells are connected together by fine strands of living protoplasm, and we may guess that, nervelike, these strands convey impulses from cell to cell, and so make possible their co-operation.

The somewhat similar flagellated cells of a sponge (Fig. 103) are less well disciplined. In normal conditions each beats away, its stroke quite independent of that of its neighbours; and when conditions are unfavourable it stops and draws in its flagellum. Through the gelatinous tissues of the sponge other cells crawl about like Amœbæ and transport food from part to part. As the various cells multiply the sponge as a whole grows. The only reactions which are directly and rapidly adjusted to changes in the outer world are performed by muscle-like rings of specialized cells round the water-canals and the main mouth through which the current is ejected. Under unfavourable conditions (such as shortage of oxygen in the water) these contract and the water-system is temporarily closed. Also there is a crude kind of transmission in the sponge that foreshadows true nervous action. Parker has shown, working with a tube-shaped sponge that grows on the coral-reefs of Bermuda, that an injury near one of the great outward openings is followed after a little time by the closure of the latter. Apparently an impulse of some kind has crept sluggishly through the tissue from injury to opening—not through special nervous tissue, for such is lacking, but through the general living flesh of the sponge. To such primitive forerunners of nervous activity Parker has applied the term "neuroid transmission."

True nervous tissue appears in the cœlenterates, where it is associated with specialized sense-organs and with sheets and bands of muscle. But it is curiously primitive in its arrangement, which differs strikingly from that of our own nervous systems. Let us choose a jelly-fish to typify this stage.

The body of a jelly-fish, as is well known, is a transparent jelly-like bell, fringed at its edges with stinging tentacles. In the middle of the lower side of the bell is the mouth, surrounded by larger tentacles and generally at the end of a longer or shorter tube, that dangles down like the handle of an umbrella.

Except for feeding movements of the tentacles and mouth, the jelly-fish has only one form of muscular activity, and that is its swimming.

A living specimen in an aquarium tank can be seen to beat rhythmically almost like a heart. Each beat consists of a simultaneous contraction of the whole bell, and it has the effect of jerking the animal upwards through the water. After the beat the bell expands again because of its elasticity; the contraction is brought about by a layer of muscles in the bell.

In his admirable *Science from an Easy Chair*, Sir E. Ray Lankester describes the behaviour of a little fresh-water jelly-fish. First it drives itself upwards through the water with a series of powerful contractions of its whole body; then it rests and floats slowly down like a parachute. But as it descends it captures water-fleas and other animals with its tentacles. Because of its transparency, they do not notice the living snare that is sinking down over them. Then it jerks its way up again, then floats down again, and so on.

That upward jerk, a sudden convulsive contraction of the whole bell, is the main activity of the jelly-fish. Sometimes it beats rapidly and sometimes slowly. Sometimes it rests altogether for a while. But these changes of rhythm are the only variations it can ring on its single swimming movement.

Spaced at equal intervals round the margin of the bell, are eight "marginal bodies," which contain the chief sense-organs of the jelly-fish. Each includes several different structures. First there is an organ which resembles in its design a simplified version of our own inner ears and perhaps underlies a primitive sense of balance. Then, also, there may be a more or less elaborate eye; and it is believed that the marginal bodies are sensitive to chemical stimulation.

The nervous system does not lie in definite tracts, or nerves, as ours does, but it consists of a diffuse lacework of nerve-fibres which pervades the whole bell. Round the rim there is a denser ring of nerve-fibres; that is the only special aggregation of nervous tissue the jelly-fish possesses, the only thing about it that even remotely suggests a brain. As a matter of fact, it is more like a circular nerve than a brain, for it is only a region of especially efficient conduction, and, as far as we can tell, performs none of the peculiar brain functions.

The simplicity of this nervous system obviously harmonizes with the simplicity of the movements that the animal performs. We ourselves can make a thousand different movements at the bidding of our brains; the reader has only to think of all the possible ways in which his or her wrist and fingers can be moved. Now these hand movements involve the action of thirty-nine different muscles, and to each muscle there must run a separate bundle of nerve-fibres from the brain so that it may be properly controlled—thirty-nine separate and distinct nerve-paths from brain to hand! But for the one movement of the jelly-fish such an elaborate telephone system would be unnecessary. The diffuse network suffices, spread over the whole bell and securing only that it beats together as a single unit.

The nervous system of a jelly-fish may be compared to a nightmare telephone system in which there is no proper exchange but only a tangle of wires; where the raising of one receiver would call up hundreds of subscribers, or indeed, if you spoke loud enough into it, every subscriber in the system.

The absence of definite nervous paths in the jelly-fish is easily demonstrated. It so happens that in many jelly-fishes (including the commonest kinds) the impulses for movement emanate from the marginal sense-organs. If these are cut off, the whole animal is motionless unless it be directly stimulated, by a touch, for example, when it responds by a single contraction of the bell. But if only seven of the eight marginal organs are cut off, the remaining one suffices to keep the jelly-fish active; rhythmical impulses radiate out from it along the nerve-net to the muscles, and so the normal swimming movements are produced. Now in such a jelly-fish it is possible to cut the bell into any shape, into strips or spirals or other patterns, as Fig. 290 demonstrates, and in spite of such mutilations the contraction still spreads rhythmically from the marginal organ over the bell. So obviously there are no special conduction paths; the nerve-net conducts perfectly well in all directions.

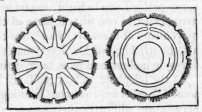

FIG. 290. The bell of the jelly-fish Aurelia can be cut up in various ways and will still conduct impulses.

In these examples, pulsations spread rhythmically from the remaining marginal organ (see text) over the whole bell. (After Romanes.)

The jelly-fish, then, presents an interesting stage in the growth of the apparatus of behaviour. It shows the three essential kinds of tissue. Some of its cells are specialized for the detection of stimuli (some for light, some for chemical substances, some for balance, and so on) and others are turned to muscle-fibres. Yet other cells, the nerve-cells, specialize in rapid conduction of messages from part to part. But this nervous system is astonishingly diffuse and decentralized, compared with our own.

In the evolutionary passage from cœlenterate to higher mammal, three important changes have taken place. The first is the supersession of a diffuse, directionless, all-pervading network of nerve-fibres by definite nerves, running, as telephone wires run, from one point to another without side-branches or wayside entanglements. As the range of possible sensations and movements is extended, these improvements in the telephone system become more and more necessary. The main steps in this transition still survive in the more primitive animals, where an extensive nerve-net is found side by side with a few special nerve-paths. Even in highly organized creatures the primitive network may persist in one or two parts of the body. The wave-like creeping movements of a snail's "foot" are controlled by a nerve-net, and so are the writhings of Mr. Everyman's intestine. But most of the nerve-fibres, in mollusc and man, lie in definite tracts or nerves.

As the telephone web of the body becomes organized in this way, a second parallel improvement is brought about in the speed with which nerve-fibres will carry their messages.

Here we have to break away for a moment from our telephone analogy. In a telephone wire the electric currents travel with an extreme velocity from mouthpiece to receiver. But the "impulse" in a nerve is not an electric current; it is a change that runs along more like a ripple on the surface of a pond than an electric current in a wire. The first beginnings of nervous activity are by our standards impossibly slow. Even in surprisingly highly-organized creatures it may be hardly swifter. In the nerves of a fresh-water mussel, an impulse takes two or three seconds to travel an inch. In slugs it runs forty or fifty times as quickly. In king-crabs the messages from the brain to the muscles travel at about ten feet per second; in a frog they do seventy or eighty; in ourselves four hundred feet per second—which is more than two hundred and fifty miles per hour.

Perhaps the meaning of these facts will be most forcibly brought home by imagining a man with the nerves of a fresh-water mussel. Imagine that he carelessly let his cigarette slip between his fingers, so that it began to burn him. In something over two minutes he would feel the pain, and, if his brain reacted at once, the impulse to make him drop the cigarette would arrive at his fingers in another two minutes. But he would have suffered extensive organic damage in the interim.

Thus the telephone-cables of the body become more and more efficient as we ascend in the scale of Evolution. The third important advance is the development of a central exchange, the central nervous system, to which the nerves converge, where the information from the various sense-organs is received and collected, and whence harmonious

RUDIMENTS OF BEHAVIOUR

impulses are sent to the various effectors. Only thus can behaviour be knitted together, and the different parts be made to work as a single whole.

To go into the details of the nervous system in the various phyla of animals would take us deeper into biological technicality than we need to plunge. A few points of importance may be summarized very briefly.

In the echinoderms (Fig. 89) the nervous system is very decentralized; the same may fairly be said of their behaviour. The main centre, a loop of nervous tissue, runs round the mouth; from this five nerve-trunks depart, one along each of the rays of the body. But these five nerves are not simple telephone-cables, as our nerves are: they are governing exchanges like our spinal cords.

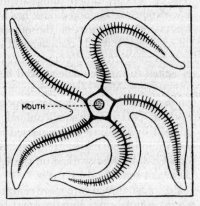

Fig. 291. A diagram of the nervous system of a starfish.

Five main nerves run along the arms from a central ring round the mouth.

Each of the five arms of the starfish behaves with remarkable independence; the creature is in many ways more like five individuals springing from a common centre than a single person with five fully subordinated limbs. A single arm, cut away from the rest, will crawl about on its tube-feet and turn over again if put the wrong way up (Fig. 88). When a starfish is thrown on its back, each arm tries to right itself by twisting over at the tip, gripping the ground and pulling. When one arm gets a secure hold it sends nervous messages to the others, which then stop their endeavours and stay passive, so that the first arm pulls the whole animal over on to its normally lower face. These nervous messages are sent by way of the ring round the mouth. If the ring be cut between each pair of arms, there is no co-ordination between them: if the animal is upside down, each arm goes on struggling to right itself, and, as they work against each other, the starfish simply ties itself into knots.

Moreover, as we saw in Book 2, the starfish is covered with little stalked pincers, with which it defends itself against aggression and keeps itself clean. But these are best studied in its relatives, the sea-urchins, where they are larger and more numerous.

A sea-urchin is a hollow, bony ball bristling with spines. Each spine can be moved about; it has a ball-and-socket joint at its base, and a special set of muscles that move it. In among the roots of the spines is an undergrowth of little stalked pincers. If an enemy approaches, the spines at that side of the sea-urchin turn like lances towards him, and the jaws below stand up, ready in case he should get to close quarters. Now none of these things are due to impulses from a central brain. Each spine, each stalked beak has a little reflex system of its own, and reacts independently of the others. Each is so constructed that it directs itself towards the chemical stimulus of an enemy. A tiny chip broken away from a living sea-urchin's shell, with only a single spine attached to it or a single stalked beak, will show the same alarm and preparation. Most of the behaviour of the creature is due to a similar summation of the automatic responses of its thousands of separately working parts. Any small animal touching the urchin is suddenly gripped by the tiny beaks, passed by the tube-feet and spines to the mouth and there eaten. Here again, the parts are working separately. Von Uexkull, who worked on the behaviour of sea-urchins, speaks of them as "republics of reflexes."

But most animals are more centralized than this, and with the lower worms—flatworms, roundworms, bandworms and so on—the head comes into our story. Here for the first time we have animals which regularly progress with one end always in front; and as the head evolves with its special sense-organs and mouth, the brain appears and evolves with it. In these worms there is a simple brain in front, sending off two main trunks, one down each side of the body.

In annelid worms (Book 2, Chap. 3, § 4) and arthropods (Book 2, Chap. 2, §1) there is a characteristic type of nervous system. The brain is two in parts, one just in front of the mouth and one just behind it; they are connected together by thick nerves, and the whole apparatus forms a dense ring round the animal's gullet. From this a single great trunk runs away, corresponding to our spinal cord, but lying along the belly instead of the back. In these creatures, as we have noted, the body consists essentially of a chain of more or less similar segments, and in each segment the nerve-trunk swells out to form a little local brainlet, from which lesser nerves radiate out to the sense-organs, muscles and other tissues. The nerve-trunk of an earthworm may have over a hundred such subsidiary centres, one in each ring of its body.

That these swellings are indeed brain-like in their working was shown by some very suggestive experiments by Yerkes, subsequently confirmed by Heck. A narrow tube is made in the shape of a T, and earthworms are put into the stem of the T so that they creep along towards the cross-bar above. When they come to the parting of the ways, they can go either right or left. Should they go right, they get a gentle electric shock from a pair of electrodes concealed in the wall of the tube. Should they go left, they escape without this punishment.

Even a worm will learn. At first the choice of route is made at random, and the animal goes right as often as it goes left. But, very slowly, the fact that the right-hand path is dangerous is forced upon its elementary intelligence. After a hundred trials or so, it goes left definitely more often than it goes right. After a hundred and fifty, the lefts were found to outnumber the rights by about ten to one. Then the electrodes were moved from the right-hand tube to the left-hand tube; slowly the worm unlearnt its first lesson and mastered the new one.

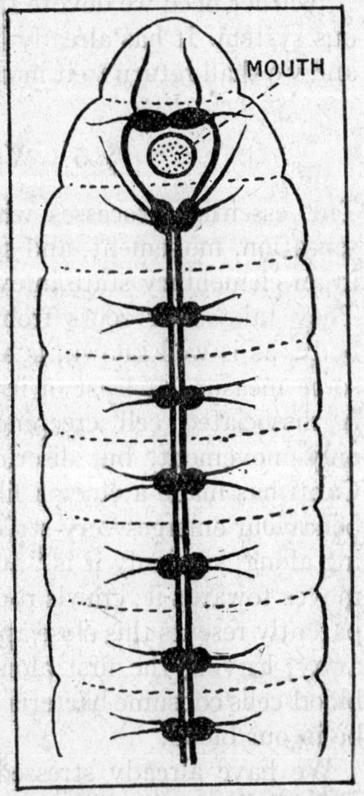

But the most interesting point is this. It was found that worms could learn in this way if the brain-mass round the mouth was removed altogether; and even if, to make sure, the whole central nervous system of the first six segments was absent. So undoubtedly these swellings all along the worm's nerve-cord can perform a function which is confined to the brain in ourselves.

Fig. 292. A diagram of the nervous system of an earthworm.

From a main nervous ring round the mouth, a double cord runs down the lower side of the body, with a swelling in each segment.

Sometimes this segmental arrangement may be modified. The most noteworthy alteration is the pulling together of a number of these subsidiary segmental brains into a single nervous mass. Thus the house-fly has, between its legs, what we may call a belly-brain, which seriously rivals its superior centre, the head-brain, in size. The same is true of crabs.

The state of affairs in molluscs (Fig. 294) is rather complicated. As in the arthropods and segmental worms, there is a ring of brain round the gullet which dominates the rest of the nervous system. From this ring two main pairs of nerves run off. One pair goes to a pair of subsidiary brains which co-ordinate the movements of that highly characteristic molluscan organ, the foot. The other pair goes to other subsidiary brains which supervise the working of the digestive apparatus, gills and so forth. But into the details and variation that the group presents we need not penetrate.

Neither need we devote space in this section to the vertebrate nervous system. It has already been considered in a single type (Book 1), and we shall return to it more fully in a later chapter.

§ 5. Vegetable Behaviour.

THE essential processes which underlie behaviour—nervous action, sensation, movement, and so forth—may be traced, as we have seen, in an elementary state in even the simplest animals, such as amœba. They take their roots from general properties of protoplasm. The body, as is well known, is a vast community of cells; any cell shows some measure at least of irritability, contractility and co-ordination. A dissociated cell creeping about in a tissue culture shows not only movement, but discriminative movement. Recently Dr. R. G. Canti has made a cinema film of such cells, speeded up so that their behaviour emerges very strikingly. One sees, for example, a cell creeping along; suddenly it is "attracted" by another cell, turns aside and moves towards it; crawls round it and over it. But the second cell apparently resents this close approach, for it struggles and moves rapidly away, leaving the first alone. With similar independence our white blood-cells consume bacteria and other things that have no business to be in our blood.

We have already stressed the difference between specialized and unspecialized cells. We can all add two and two together, or run fifty yards without excessive discomfort, or sing the first few notes of the National Anthem. An Einstein, a Nurmi, or a Chaliapine is simply a person who can do one of these things outstandingly well. Similarly a muscle-cell, a nerve-fibre or a sense-cell surpasses the general cell population of our bodies. But just as Chaliapine would lose most of his value were there no stages or opera-houses, or an Einstein if there were no books, publications and learned societies, so specialized cells require an appropriately organized body-community, whose elaboration increases with their specialization. When we turn from animal to plant behaviour we find the most remarkable differences due to the different directions in which specialization has gone.

One often sees accounts in the newspapers of how in many ways plants parallel ourselves. They have beating hearts, we are told, and nerves like ours; they can be shown to tremble at an unpleasant stimulus, or to squirm in a death agony. Generally these accounts centre on a series of ingenious experiments conducted by Sir J. C. Bose, of Calcutta. Now, since contractility, conduction, and sensibility are

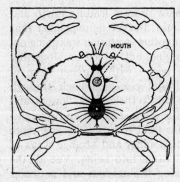
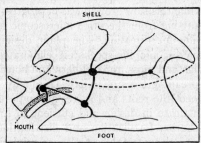

FIG. 293. A diagram of the nervous system of a crab.

It is built on a segmental plan (compare Fig. 292), but most of the "subsidiary brains" behind are gathered together into a single mass between the legs.

FIG. 294. A diagram of the nervous system of a mollusc.

From a main ring round the mouth, cords run back to "subsidiary brains" controlling the viscera and foot.

general protoplasmic properties, one would expect to find something of the sort in plants. Irritability to different kinds of stimuli exists undoubtedly in plants, and in one or two cases a kind of slow transmission of impulses has been demonstrated which seems to parallel the "neuroid" transmission of an amœba or a sponge. Moreover, many plants, as is well known, exhibit various movements: opening and closing of flowers, turning of leaves towards the light, and so forth. These are, however, due in most cases to a rather different kind of mechanism from the contractile tissues of animals. The plant-effector is usually a pack of cells that can swell and stiffen by absorbing water and becoming turgid.

Here then plant behaviour parallels our own, but no more than does the behaviour of an amœba or a sponge. It is ridiculous to overstress the analogy; to speak, for example, of a plant having nerves like ours simply because it shows crude nerve-like transmission. In another instance Bose has detected rhythmical pulsations in the central tissue of stems, and interprets them as the motive-force that drives sap upwards. They may be—although it has been denied by several botanists —but it is certainly misleading to speak of such pulsations as heartbeats. The writhings of Mr. Everyman's bowels are a hundred times more closely akin to the beating of his heart than are these alleged pulsations in plants.

Evolution is commonly presented as a process leading up to and culminating in the crowning human brain; but let us remember that while this slow growth was in progress another evolution was going on, starting from the same microscopic origins, but taking an altogether different direction. The plant design began with a green scum of microscopic life-specks; it elaborated and perfected itself through alga and moss, through fern and horse-tail, to the grasses, flowers, and spreading trees of to-day. Gradually the stem stiffened and straightened, the roots and leaves and pipes for fluid transport became more intricate in design and efficient in functioning, and strange ways of securing fertilization and distribution came into being. Yet all the time the vegetable stock was minimizing the nervous and muscular phenomena that we prize so highly. The rudimentary powers of sensation, conduction and response, latent in its protoplasm, remained unexploited. Plants are green; they have no need to hunt or forage; so the vegetable stock evolved upon fundamentally different lines.

The main activity of a plant is not discriminating movement, but discriminating growth. It spreads its leaves towards the sun and its roots down into the soil. In this it shows a certain parallelism with our own behaviour; for it grows in this direction and avoids that much as an animal moves in this direction and avoids that. Anyone who sees a speeded-up film of plant behaviour will at once recognize this parallelism.

Take a particular case: the growing root of a nasturtium seedling is photographed every fifteen minutes. The resulting film is run through the projector at the ordinary rate of fifteen pictures a second. The rate of movement has thus been magnified 13,500 times, and the behaviour of the root-tip now seems that of an animal instead of a plant. It pushes its way like a white worm through the soil; like a worm's head, its advancing end waves from side to side as it moves onwards; like a worm, it avoids obstacles by bending and crawling round them.

The same apparent acquisition of the attributes of an animal as a result of mere change in the tempo of behaviour is seen in nearly all such speeded films. The nasturtium bud, which takes all a morning in its slow expansion into a flower, becomes alarmingly like a lion suddenly throwing open great jaws; the snowdrop folding its petals and drooping its blossom during the hour before sunset, has the air of a man relaxing his limbs and hanging his head on his breast in sudden grief. The horse-chestnut twig, emerging so slowly from its winter dormancy that its movements, its onward growth, its putting forth of successive leaves one by one, its alternate slow rotation, a little to one side and then to the other, are as imperceptible as those of the hour hand of a watch, suddenly acquires all the life and grace of a dancer.

This appearance is not deceptive. The reactions of plants are indeed not dissimilar to those of many lower animals; and the fundamental difference is precisely a difference of rate. The plant, rooted in the soil, cannot escape danger by flight, or capture food by active pursuit; rapid reactions are therefore, with rare exceptions, pointless, and its movements are almost wholly self-centred, devoted to attaining the right position for itself and its organs, unrelated to enemies or to prey or to any rapidly moving object whatever. That being so, they can be executed quite well through the slow agency of growth, and no special rapid machinery such as muscle has ever been needed or called into existence.

Two main outside forces influence the direction of plant growth. The first is gravity; leaf-bearing stems tend to grow upwards (as is very easily shown by growing seedlings in the dark), while roots tend to grow downwards. The second is light; stems grow towards light and roots (for instance, in a seedling suspended in the air and illuminated from one side) grow away from it. A third factor, moisture, may also play a part; in most cases roots grow towards moist regions in the soil. Responses of this kind are called *tropisms;* and, as we shall very shortly see, the word has a similar application in animals. The response to light is called *phototropism*, that to gravity *geotropism*, and that to water *hydrotropism*. As a further verbal complication, if the plant grows towards the stimulating influence its tropism is called positive, while if it grows away it is called negative. Thus a leaf-bearing shoot is positively phototropic and negatively geotropic; while a root is negatively phototropic, positively geotropic, and positively hydrotropic. The utility of these phrases will soon become apparent.

§ 6. Instinctive and Intelligent Behaviour.

WHILE plants specialized in directional growth, animals took the other road and developed an apparatus for directional movement, with the more refined sensitive structures that this faster method of response necessitates. But for all their muscles, sense-organs, and brains, animals may be ruled in their activities by external influences as slavishly as plants are.

While these lines are being written, a little zoo of insects is flying in, one by one, through the open window from the night outside. As they enter, the pilgrims go straight for the object which has fascinated them and drawn them in—the acetylene lamp on the table. They flutter round it and crawl over the shade; some are held by the white circle of light it throws on to the ceiling. Sometimes one of them finds the way round under the lamp-shade and flies straight for the hypnotizing glare—only to fall, burnt or dead, a second later. One fat moth fell with a flop at the heat, then pulled himself together, rose and, undaunted by his experience, made straight for the white glare again.

Sir E. Ray Lankester wrote of a house in Java, in which an open lamp was lit every night. "Regularly two sets of animals . . . arrived on the scene. Swarms of moths and flies dashed in and out of the flame and fell, maimed by the heat, to the ground. There a strange group had already assembled. Gigantic toads and wall-lizards crept from their holes in the masonry and woodwork and awaited the shower of injured insects, which they snapped up in eager rivalry, as the infatuated flame-seekers dropped, hour after hour, to the floor." (*Diversions of a Naturalist.*)

A kindred case was observed by Loeb. If you take a number of prawns and put them in a dish through which a gentle electric current is continuously passing, they will show no sign at first of awareness of the current. But in a little time every one of them will be at the end of the dish which is connected with the positive pole of the battery. The prawns seek the anode as slavishly as the moth seeks the lamp. Various other creatures can be similarly influenced by a weak electric current. The slipper animalcule, for example, always swims to the negative pole—the opposite direction to the prawn.

How are we to interpret these facts? Here, obviously, the bias is of no value to its possessor, because weak electric currents of this kind never occur in nature. It is a purely accidental phenomenon. In the case of the prawn the nerve-fibres in the central nervous system lie in different directions, according to the muscles with which they are connected; and it is known that a constant electric current will affect the activity of a nerve-fibre according to the angle at which it happens to cross it. In a prawn which happens to be facing the positive pole, the nerves to the muscles which push it forwards are made more active, while those which carry it backwards are enfeebled. In a prawn facing the other way the reverse occurs. So that the first prawn progresses somewhat more vigorously and easily than usual, while the second goes with labour and difficulty, until it happens to turn round. It is as if they were living on the side of an invisible hill. Just as in the latter case gravity would be constantly urging them down the slope so the current constantly urges them to the positive end of the tank.

Something of the same kind happens in the case of the insects that fly into the light. They are so made that the angle and intensity of the light that falls into their eyes affects their muscles; and they tend to fly towards the light as infallibly as a prawn tends to make its way towards the plus end of its electrified aquarium.

Gravity can act upon animals in a similar way; there are earth-seekers which tend always to direct their course downwards, and earth-shunners which tend to travel upwards.

The parallelism between these slavish responses to physical factors and the tropisms of plants, which we described at the end of the last section, was first stressed by Loeb. He extended the words photo-tropism, geotropism, and so forth, to animals. Movement directed by an electric current is called galvanotropism. Moreover, Loeb performed a number of pretty experiments to show the automatic, mathematical way in which the animals were driven by their inner urges.

For instance, he tried the behaviour of phototropic animals in the presence of two lights. If both were equally bright, the animals generally went, not to either light, but in a direction that led halfway between them. This shows that they seek the light because of a simple automaticity and not because they like light and deliberately seek it out, for in the latter case they would go to one light or the other. If one light was stronger than the other, the animals took a course with a slight bias that way, and its angle could be calculated very accurately by means of a simple mathematical formula. In another series of experiments rapidly flashing lights were used, and here it was found that the pull a light exerted on the creature depended on a straightforward relationship—it was proportional to the product of the light's intensity and its duration—which is parallel to what we find in certain simple light-produced chemical reactions, like the changes that occur in a photographic plate during its exposure.

These experiments were, of course, prepared under simplified conditions; in nature the action of a tropism is often counteracted and delayed by a host of other stimuli and reactions. The insects that make for the light do not go altogether blindly and directly; they may pause on the way to chase or avoid each other, or to rest for a little while. But sooner or later they get there, just as a feather floating in the air will sooner or later come to earth, though on the way it may indulge in all sorts of devious twists and flutters. Under natural conditions a tropism is a sort of directional urge that runs through the behaviour of a creature, and gives it a pervading bias, perhaps conspicuous, perhaps hardly noticeable, but always in a given direction.

Galvanotropism, we noted, is useless to its possessors; the positive phototropism of a night-flying insect is not only useless but sometimes, in the neighbourhood of human habitations, definitely destructive. But often a tropism plays an important adaptive part in the life of a creature. Some animals, like the roots of a plant, are negatively phototropic; a blow-fly grub will crawl away from the light in a straight line, and, as it lives buried in decaying things, the helpfulness of this tropism is manifest.

Moreover, in many cases the sense of a tropism can be reversed by other conditions. Thus barnacle larvæ seek light in the cold and avoid it in warmth; similarly many small aquatic crustaceans, such as the water-flea Daphnia, tend to swim downwards in a bright light and upwards in darkness.

There is a very pretty case in which such a change of tropism underlies a complicated instinct. The caterpillars of the goldtail moth *Porthesia chrysorrhœa* hatch in autumn and spend the winter hibernating in nests near the ground on the stems of the shrubs on which they feed. In early spring they leave the nest and crawl straight up the shrub to the tops of its shoots, where they find the first buds beginning to open, and these they devour. The coming out from the nest is apparently a response to warmth, for they can be made to leave it at any time in winter by warming it artificially. But why do they crawl upwards—how do they know that the only place where they can find food is at the tops of the branches? If they went downwards they would starve.

They do not know. This upward creeping can be shown to be a simple positive phototropism, directed by the light reflected from the sky. If some of the caterpillars are taken as they are leaving the nest, and put in a glass tube lying near a window, they will all collect at the end of the tube nearest the light, and stay there. If the tube is turned round they will crawl to the other end, again towards the light, and there wait. Most surprising of all, if a few young leaves from their food-shrub are put at the other end of the tube, the end farthest from the light, they make no attempt to reach the meal; the light-rays hold them captive at the end nearest the window, and here they stay until they starve.

Under natural conditions the tropism guides them up to the only place where they can find their food. But there is a further complication here. For only an unfed caterpillar is positively phototropic; in some way nourishment removes the tropism, and after eating light has no effect. The utility of this is clear. The caterpillar quickly clears up the leaves at the top of the twig, and if its slavery to light persisted it would have to stay there and starve. But having eaten it is freed and can creep in any direction; the experiment with the glass tube and the window now has no result. So it is able to work its way downwards and find the lower buds as they begin to open.

A tropism, then, is a blind and unreasoning drive within an organism; it may lead the creature to destruction or to salvation, but in either case it is a simple result of the interaction between the inborn structure of the nervous system and stimuli from the environment.

Another kind of response, equally blind and unreasoning, is the reflex. To reflex we introduced ourselves in Book 1, Chap. 3, § 6, taking as an example the automatic movements of the hand when the fingers touch the red end of a cigarette. We shall note many other examples in later chapters. The differences between a reflex and a tropism are two. Firstly, a reflex usually only concerns a part of the body (in this case, the hand), while a tropism affects the position and movement of the body as a whole. Secondly, in a reflex an abrupt reaction is elicited by some sudden change in the environment, while a tropism is more a steady, underlying bias in behaviour, brought about by a constant stimulus. But the distinction between the two will become clearer in later sections, as we handle the words.

Between them, tropisms and reflexes underlie most of the phenomena usually loosely called "instinct." The flying of the moth into the flame is often spoken of as an injurious instinct. Young chickens when they hatch have the instinct to peck at any small object that they notice lying on the ground; but soon experience teaches them to distinguish the edible from the inedible. In this case the first instinct is a reflex.

Instinct is rather a dangerous word because it has been used in a variety of senses in ordinary speech; any unconscious impulse may find itself labelled instinctive. But here we are going to restrict the term to those elements in behaviour which are inborn in the organism, or which develop in later life (as the beard and deep voice of a human male develop) as a simple result of the organism's own constitution.

Instinct is congenital behaviour. Contrasted with instinct we have all those elements which depend upon individual experience—upon memory and learning. These, for want of a better word, we may call intelligent behaviour.

Every animal comes into the world with a certain inherited endowment of congenital behaviour. Some go through life with that alone. Before we carry our analysis any further, it will be well to consider an example of that kind. But most animals learn as they live; as we ascend the animal scale we find them extending and improving on their original outfit, sometimes only slightly and sometimes very considerably, by personal adaptability and experience. Finally, in ourselves intelligent behaviour outweighs and largely supersedes instinct. Our next chapters will trace the inter-relations of the two kinds of behaviour, first in the highest and most interesting invertebrates, and then in the group to which we ourselves belong.

But first we must introduce ourselves, by means of a microscope, to a creature whose behaviour is entirely congenital.

§ 7. The Behaviour of the Slipper Animalcule.

IF WE take a few bits of water-weed from a pool in summer—preferably a stagnant pool—and leave them in a glass of rain-water to rot, in a few days we shall very probably see a multitude of white specks in the glass, on holding it up to the light. These specks, the largest ones about a third of a millimetre long (an eightieth part of an inch), are slipper animalcules. The rotting weeds have first nourished a host of bacteria, visible perhaps as a faint turbidity in the water, and the slipper animalcules, tiny but ravenous beasts of prey, are now living on their very much smaller neighbours.

The appearance of a slipper animalcule (Paramecium is its technical name), when magnified with a low power of the microscope, is shown in Fig. 295. It is covered with cilia—the minute, vibrating, hair-like projections that we first met in the lining of our windpipes—and by means of their almost ceaseless lashing it swims rapidly along. But the cilia are not all alike; some are stronger than others and the direction of their lashing is not quite the same as the long axis of the body. Because of these facts, and also of a curious twist in the front end of the creature (its body has been likened to a leaden rifle-bullet twisted by being shot through a grooved barrel) it swims in a spiral course, spinning round and round on its own axis as it travels forward.

About half-way along the body is a wide funnel-shaped pit, which sinks in towards the middle of the body. This is the mouth; like the rest of the body surface it is covered with cilia, and their lashing keeps a whirl of water flowing into the funnel, where such nourishing particles as bacteria are seized and swallowed.

Paramecium is one of the biggest of the better-known protozoa, and as with the rest of the group it has an organization like that of a single cell from our bodies. Most of its interior consists of the viscid fluid protoplasm; in the centre somewhere is a large nucleus with

FIG. 295. *Paramecium caudatum*, the slipper animalcule.

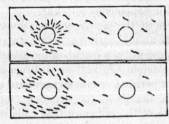

FIG. 296. A number of slipper animalcules, imprisoned between two slips of glass, show their liking for faintly acid regions.

Above, they collect round a bubble of carbon dioxide (left), which dissolves in the water, forming an acid, and neglect a bubble of air (right). Below, some time later, the water round the carbon dioxide has become too acid for comfort, and the creatures now lie in a ring where the acidity suits them better.

a smaller one (or perhaps two) lying beside it. Any food particles that are taken in at the mouth pass into the interior contained in little water-filled cavities and thus drift round the body, being slowly digested.

By our standards, the ways of a slipper animalcule are as strange as its structure. It multiplies by tearing into halves after the manner of protozoa, and it is neither male nor female, for its equivalent of the sexual process, divorced from the reproductive function (Book 4, Chap. 2, § 3), is an affair between exactly similar individuals. But here we are not concerned with these rarer adventures; it is on the humdrum daily business of feeding and self-preservation that we shall concentrate our attention.

As we watch a paramecium bustling its spiral way across the field of the microscope, we may chance to see it give a highly characteristic and important reaction. Suddenly it stops, darts swiftly backwards, turns slightly, and then starts swimming forward again in a new direction. This is called the "avoiding reaction," and it means that the animalcule has come up against something that it doesn't like.

Suppose that some kind of obstacle, such as a bit of plant débris, blocks its path. As soon as it touches the barrier the creature gives this avoiding reaction—starting back, turning, and trying a new line. Perhaps this still does not clear the obstacle; if not, there is another collision and another avoiding reaction, and so on until the paramecium gets round. There is no system in this exploration. It is a perfectly random affair; the animalcule turns first in one direction and then another until it happens to get clear.

Exactly the same reaction is given if very strong light suddenly falls on the animal, or if it gets into water which is unsuitable—too hot, too cold, too salty, too acid, or too alkaline. To all these unpleasant stimuli paramecium has but one answer; and it repeats this answer until it either perishes or else escapes into better conditions.

The details of the avoiding reaction may vary according to the nature of the influence that calls it forth. Sometimes the backwards movement is considerable while at others it is slight; sometimes the change of direction is greater than at others; and so on.

This "try, try again" of a single form of response may be called the Method of Trial and Error. In paramecium, we see it in its simplest form. But it continues to be employed in one way or another on every level of behaviour up to the human. A dog with a stick in its mouth is trying to get through a set of railings; it turns its head now at one angle, now at another, it shifts the stick in its teeth, it tries over and over again without any evident system in its investigation, until finally one position allows it to pass—that too is trial and error. So, too, on a more elaborate level, is the behaviour of a man seeking frantically to escape from a room in which he has been locked, as he rushes round and hammers the walls, in the hope that by chance he will hit on a weak spot. So, on a different level once more, is the behaviour of the scientist when confronted with a strange fact to whose solution his previous experience suggests no obvious line of approach, he tries what Darwin called "fool's experiments" in the hope that one or other of them may yield a clue. But as we go up the scale, the method of trial and error ceases to dominate behaviour, it becomes a tool in the hands of this or that instinct or desire, as when the dog with the stick desires to get to his master through the railings; it becomes modified by experience and intelligence, as in a prisoner trying to break out of his confinement and examining his cell systematically; and, finally, in man it becomes supplanted, to a certain variable extent, by deliberate planning. But the idea of "have a shot at it" is always there, from Paramecium to Plato, and if it was indispensable as the master of the protozoan activities, it remains equally indispensable as the exploratory scout of man.

As Jennings (an American investigator) showed, almost all the apparently purposive behaviour of paramecium depends solely on the repetition of this piece of automatism. Normally the stagnant waters, full of decaying vegetation, which it inhabits, are slightly alkaline. Where, however, a multitude of the bacteria on which it chiefly feeds are collected, the carbon dioxide they give off will make the water more acid. It is, therefore, an advantage for paramecium to find and stay in any region of slight acidity. That it does so can be easily demonstrated. If a number of paramecium be put in a drop of water between two slips of glass, and then, by means of a fine pipette, a bubble of air and a bubble of carbon dioxide be separately introduced into the tiny aquarium, the carbon dioxide bubble will soon be surrounded by a dense ring of animalcules, while the air bubble will be neglected.

As more carbon dioxide passes into the water, the ring of paramecium will move slowly out from the bubble, the water immediately against it being now too acid for comfort. If during this latter phase, we watch our little experimental world through a low-power microscope, what we see is this: if a paramecium happens to swim from the general drop into the region of the crowded ring, it gives no reaction, but swims on with its normal spiral. When it reaches the region of the stronger acid on the inside of the ring, however, it gives an avoiding reaction, and changes direction; and it does the same if now, swimming back again, it reaches the neutral water on the outside of the ring. So it continues, giving the avoiding reaction every time it gets to the edge of the ring, and the acidity becomes too high or too low. Thus by means of its sensitiveness to high or low acidity, and its single answer to both, the paramecium becomes trapped, but trapped in the middle zone where the conditions are best.

Precisely similar behaviour can be seen when drops of weak acid, or of various salt solutions are used; in each case the animalcules wander by chance into the most suitable regions, but, once there, refuse to leave them; and we have the paradox that paramecium is guided in its positive behaviour by a negative reaction.

The creature does, however, exhibit a positive piece of behaviour which is directly advantageous to it. Its bacterial prey has a habit of swarming together and forming raft-like collections of great numbers of individuals. If a paramecium is guided to the faintly acid zone round such a raft by its avoiding reaction, it will obviously be advantageous for the animal to stop swimming and anchor itself directly to the bacterial raft. This is done by means of a characteristic reaction to solids. Whereas strong mechanical contact (such as is produced by the creature suddenly bumping its front end against a hard obstacle) calls forth the avoiding reaction, light touches, especially with objects of irregular and thready surface such as the bacterial raft or a scrap of paper, brings about a total stoppage of the cilia on the side touching the object. The other cilia also slow down, except for the ones round the mouth, which continue actively sucking food towards the mouth. The result is that the paramecium stops and browses over the surface it has discovered. It is interesting to note that when the animal is doing this its sensitivity to other agencies is diminished—its attention, so to speak, is occupied by its meal—a phenomenon which is of course paralleled in our own behaviour.

This second reaction, in spite of its simplicity, is obviously of service to paramecium under natural conditions. The rough, solid objects which elicit it will, in general, be rich in nutritious particles. But the experimenter can show how mechanical and unreasoning a thing it is; Jennings found that paramecium would stop and browse eagerly but emptily over such objects as scraps of paper, fine fabrics, threads, or heaps of carmine powder, in spite of the fact that "the cupboard was bare."

In addition to these two reactions, paramecium may respond by a directional bias in its swimming to one or two simple physical agencies. For example, it is negatively geotropic; it tends to swim upwards, against gravity. In the pools it inhabits it does not feed on the bottom but swims about in the upper layers of water, finding out and browsing over floating particles in the manner we have described. The sense-organ by which it feels its way upwards is, apparently, its digestive apparatus. Paramecium has of course no permanently marked-off alimentary tube, but when it eats food-particles it takes them into little vesicles of water, in which they drift round its body. Each of these little vesicles, or "food-vacuoles" is in effect a stomach and intestine combined. Digestive ferments ooze into them from the living protoplasm around; the products of digestion are absorbed out of them. Now the food is in general heavier than water, and so it comes about that the food-particles rest on the lower sides of these temporary stomachs in which they are enclosed. It would seem that the paramecium can feel this, much as we might feel a small but heavy object in our stomachs pressing downwards, and that this is the only way in which it gets any sense at all of up and down. If the reader can stretch his imagination sufficiently, let him be mentally floating in water (not on it), with eyes closed, with the balancing organs in the ear temporarily paralyzed, and only the pressure of a small hard object in the stomach to tell him which way up he is. It is possible to trick paramecium because of this fact. By inducing it to swallow grains of finely powdered iron or nickel and then holding a powerful magnet over it, the animal can be persuaded that up is down, and its reactions can be reversed accordingly.

But this response to gravity only takes place if the water is faintly acid. In neutral water there is no reaction. Apparently it depends for its manifestation upon the physiological state of the animalcule. This typifies a set of phenomena of very great importance; a number of reactions which occur with machine-like regularity in one set of conditions are changed or fail to appear altogether when the conditions are altered. We met a similar case in the caterpillars of *Porthesia chrysorrhœa*. A great deal, perhaps all, of the variability and apparent spontaneity of the behaviour of lower organisms is due to such alterations of physiological state; and we have only to think of our own feelings and reactions to food when hungry, just after a good meal, and when on the verge of sea-sickness, to realize that in higher forms also they play a leading part.

Thus, by means of a few simple and perfectly automatic reactions, the slipper animalcule swims about, seeks and consumes its prey, and avoids hostile influences. To us it is almost incredible that so small an equipment should suffice. If an animal like a pig, say, or a deer were to live with such slender resources, it would normally bustle blindly along, bumping into trees and rocks, and only able to get round obstacles by backing away for a few steps, turning into a new direction at random and starting off again. Put in a small pen with an open gate, it might give ten or twenty such reactions before it hit the way out. When it smelt a suspicious smell, it could not make off in the opposite direction, or wait warily and find out just what kind of potential enemy was the cause of the smell, or what it was doing. It could not see or snuff its food from afar, much less learn its way about its locality and remember where to find the best grass and juiciest thickets, but would be reduced to careering along, brought to a standstill now and again by the feel of herbage or leaves on its legs; and indeed, it would not seek for its food at all, but would simply happen upon it by luck. But for paramecium, swimming about in a bacterial soup, such reactions are enough, and thus these tiny, single-celled creatures conduct their lives.

Can we imagine so limited a creature as having a conscious mind? Let us assume so for a moment, and see what kind of mind it must be if it exists at all.

The first important difference between its experience and ours is that it has no special sensory apparatus, such as the eye or the ear, for determining the direction from which such agencies as light arrive, nor for determining their relation to each other in space. It knows nothing but its own body, and the things that touch its body. Neither has it, as far as we can judge, much power of discriminating between different kinds of influence; it gives one reaction only, the avoiding reaction, to such diverse conditions as a hard obstacle, too-acid water, too-alkaline water, salt water, hot water, chilly water, and so forth. Presumably it experiences but one kind of sensation for all these things, since it gives but a single response. So that we can read into the mind of our paramecium no variety of qualities, colours and tones; no images, no sense of near and far; at most, nothing more than monotony of faint pleasure and displeasure.

Secondly, paramecium has no memory. It gives no sign of profiting by experience; even the dodging by which it tries to get round an obstacle is random and unsystematic. Sometimes indeed it shows quite transitory changes of intensity or kind of reaction, after long repetition of the same external influence, but these are merely fatigue effects and not expressions of a memory faculty at all. It reacts differently because it is tired. Paramecium's life is forgotten as soon as experienced. The animal recalls no past and anticipates no future, but lives life after life, as it were, in a perpetual chain of "nows."

Specialists in the study of protozoa rank paramecium as a highly-organized member of the group. One could find even simpler lives. Amœba, for instance, does not swim by rippling cilia, but oozes along by a kind of viscous flowing of its whole body that we have already described. It lacks a special mouth, and embraces and takes in its food at any point on its surface. Its behaviour is as simple as or simpler than that of the slipper animalcule. Going lower down the living scale, we find kinds of bacteria which do not move at all; which do not eat, but simply absorb their food; whose only activities are growth, sexless reproduction, and the automatic exudation of various chemical materials. We find the life, like the structure, is at or below the cellular grade; there is as much behaviour—sensibility and activity—in one of our Mr. Everyman's white blood-cells, or in a connective-tissue cell in a tissue-culture, as there is in these elementary free-living creatures.

§ 8. The Different Worlds in Which Animals Live.

THERE are two great mistakes into which everyone, it would appear, falls by some kind of inevitable tropism when beginning to observe and think over animal behaviour, and it may be well before we go further to devote a brief section to their discussion. One is to ascribe to animals the higher faculties of human mind, such as intelligence, purpose, elaborate emotional states like hopes or fears or aspirations, powers of rapid learning, of imitation, of recalling an idea. The other is the assumption that the world in which animals live is that world of objects and events in space-time in which we pass our lives. The ascription of human faculties to animals underlies all the familiar stories of dog intelligence, all the pleasant anecdotes of dozens of books on popular natural history. We shall have occasion to prick that bubble, regretfully but decisively, in several of the later sections of this Book. But this is perhaps the place to discuss our other point.

When we say that the world of a beetle is different from that of a dog, and both from that of a man, we naturally do not mean that the external reality in which a dog lives its life differs from that in which a beetle lives. But the nature of external reality—that is a problem for the philosophers. What concerns us as biologists is the nature of the external world as it touches and appears to the animals, the world as it effectively is for them. The world of a slipper animalcule does not consist of a number of objects as it does for us, each object, like a tree or a dog, possessing a number of properties, some concerned with shape, others with hardness and heaviness, others with temperature or smell or taste or colour. There are no *things* in its experience, only separate stimuli; it apparently has no capacity for perceiving two kinds of stimuli joined up into one compound experience, for thinking them together, as we do when we think of the yellowness and roundness of an orange. There is no space in its experience, no right nor left; it has no capacity for telling where anything is in relation to anything else; nor can anything it experiences have a shape—all the stimuli that beat upon it are as formless as smells are to us. And there is no time in its experience; it lives only in the narrow boundary between the past and the future. Once an experience is past, it is blotted out for ever; past and future have no meaning and indeed, no existence for paramecium.

Most of the lower animals live their dim and windowless existence in a world of this limited kind. Let us think of one or two.

A jelly-fish bell is an arrangement of nerves and muscles which has only one answer to all the possible questions which our varied universe can put to it: it can pulse. It can vary the emphasis of its answer by pulsing more strongly or less strongly; that is all.

A sea-urchin is a feudal system of reactions, but a feudal system without a king. The barons occasionally combine, though they enjoy a great deal of local independence; but there is no brain-parliament where the local lords can pool their ideas. There is in the sea-urchin almost no apparatus for putting together the experiences of the parts and making a single experience for the sea-urchin as a whole.

The worm, the mollusc, and still more the arthropod, have the makings of such an apparatus: yet their sense-organs and their brains often do not permit their experience to be anything very elaborate. Even for such a complicated creature as a snail, for instance, the sun does not exist; there are only degrees of light and warmth. And it cannot see things. It only becomes visually aware of objects when they are between it and the light; and then they are merely shadows of more or less intensity. The world of crustacea begins to acquire more of a framework and a greater richness within the framework. A crab scuttling over the shore at low tide can see on which side of him you are approaching; and objects begin to exist for him, because he can distinguish something of sizes and of flat shapes, if not of solidity. But, all the same, the shapes are wretchedly blurred and dim; he sees men as trees, walking; his visual world is little but a world of dark dangers of different extent, between which he can draw no further distinctions.

To such creatures as hermit-crabs, objects with solid shapes begin to exist. This is more or less of a necessity for them with their shell-inhabiting propensities; and experiment has shown that they can distinguish spheres from cubes and flat from pointed cones.

With the perfection of the eye as an image-forming camera instead of a mere light-perceiving organ, and with brains capable of linking up impressions from other senses like touch or smell with those from sight, the world of evolving life grows rapidly richer; it comes to have some resemblance to the world we know, by consisting of solid objects in space. When a bee is flying to and from its hive, across our garden, it sees the same objects as we do. It may not know that this is a chair or that a tree; but at least it sees them and distinguishes them.

Even so, the world of such a creature may differ from ours in many ways. It may, for instance, be a world of black and white, since the

RUDIMENTS OF BEHAVIOUR

animal has no colour-sense; and most insects are deaf, so their world is soundless.

There are other frameworks in our human world besides that of space; there is the framework of time, and the framework of cause and effect. These evolve long after that of space. The story of the evolution of mammalian intelligence, which we shall give in a later section, is in large part a story of life making groping experiments in the direction of these new frameworks. A dog is just beginning to put two and two together; but his powers in that direction bear about the same relation to our human capacity for digging out causes and drawing deductions as the power of a crab's eye to distinguish the shape and pattern of things does to a dragon-fly's or a bird's. So with time; the non-human animal does not have its life fitted to a framework of time. The past may be alive in the present for it; but so far as we know, the past does not exist in its own right, as it does for us, as something to which we can have access when we wish. The length of time for which an animal can hold an image in its head is very short; the image speedily gets crowded out by the insistent throng of new sense-impressions. It is probable that no animals below apes can call up images of past events as we can; a dog probably is incapable of remembering and reflecting about his absent master, although he recognizes him again at once on his return, even after years. This lack of imagery and recall too makes the animal's time framework a poor one.

With man and man's greatest invention, language, the world once more becomes richer: it becomes an orderly whole with at least the possibility of having all its aspects related one to the other.

Cell-colonies acquire a purely physical unity; they are marked off in space; cell-colonies then become many-celled animals, and the nervous system confers on them a unity of behaviour—they act as wholes; the human cerebral cortex provides men with an inner unity of experience—their world of thought becomes a single whole. Looked at from a slightly different angle, we see the aggregates of cells we call higher animals acquiring a physical unity quite early in their evolution; but only at the very end, in man, do they come to possess an individuality of the inner conscious life. Before, any inner life there may have existed has been a mere aggregate of shreds and incidents of consciousness; now it becomes organized—a personality.

There are thus three main kinds of worlds in which animals live. There are spaceless, timeless worlds consisting of mere stimuli. There are worlds consisting of stimuli put together to make objects, things with shapes and sizes. And there are worlds of space and time, of objects held together in bonds of cause and effect to make orderly constructions.

These are the three main stages in the evolution of the world as it is known to life. They are connected by every possible transition, and the evolution is still proceeding. Our sun is a very different sun from Abraham's; it is much bigger, much farther away, and much hotter. Even our framework of space has recently changed.

We are often apt to think that our capacity for changing our thought-constructions—as when we abandon the flat earth for the round, the central for the merely planetary globe, the first chapter of Genesis for Evolution—is something specifically human. So in a sense it is; for it is only we men who have got any such elaborate thought-constructions to change. But it is also the culmination of the same slow process which was begun when life first began to enlarge its world by increasing the range and number of the bare stimuli to which it could respond.

II

HOW INSECTS AND OTHER INVERTEBRATES BEHAVE

§ 1. *The Arthtropod Mind as the Culmination of Instinct.* § 2. *An Anatomy of Instinct.* § 3. *Solitary Wasps.* § 4. *Insect Societies.* § 5. *Ways of Life Among Ants.* § 6. *The Parasites of Ant-colonies.* § 7. *Termites.* § 8. *Bees.*

§ 1. The Arthropod Mind as the Culmination of Instinct.

AS WE have pointed out in Book 3, two only of the various streams of evolving animal life have attained outstanding success. These are the vertebrates and the arthropods. In Chapter 3 of Book 2 we were at some pains to stress the radical differences between their two plans of construction. In Chapter 5 of Book 5 we pointed out some of the differences in their evolutionary fate. We shall here discover an equally profound divergence between them in behaviour and in plan of mind.

The mammals, with man at their head, and the insects, with ants, bees, and termites as their highest specialization, are the culminating branches of these two divergent stocks. The divergence began with obscure details of bodily chemistry and construction, such as the preference of insect tissues for secreting chitin and using it for a skeleton on the outside of their bodies, of vertebrate tissues for making cartilage and bone and using it as an internal scaffolding. The vertebrates went in for tails and for limbs restricted to two pairs, the arthropods for taillessness and a whole battery of appendages. There were many other differences, but we need only stress once more the fact that land-vertebrates breathe by lungs, land-arthropods by air-tubes. This fact is not so unconnected with our present subject as it seems, for it was air-tube breathing which limited the size of insects. Their small size condemned them to comparatively few brain-cells; and this limitation of the number of brain-cells made it impossible for them to provide all the myriad alternative brain-pathways which are needed for any elaborate process of learning and for the plastic behaviour we call intelligent. (It is perhaps no coincidence that several observers have remarked on the "insect-like" behaviour of the smallest warm-blooded vertebrates, the humming-birds.)

The construction of vertebrates, on the other hand, was admirably suited for strength, size, and power. All the biggest animals, both of land and water, are vertebrates. Thus when they had exploited to the full the mechanical resources of their bodies, there remained the resources of behaviour of which they could take advantage; for they could grow big brains, they could learn, they could adapt themselves individually instead of having to leave adaptation to the slow variations of the germ-plasm. Their size necessitated a longer growth; and this too favoured learning—and in its turn favoured longer life, so that experience should not be wasted. So the arthropods and the vertebrates represent two different lines of mental development. Chitinous armour and air-tubes limited their possessors to a reliance upon instinct, the vertebrates' tail and backbone, their general size and muscularity allowed them to develop their intelligence.

We ourselves are almost destitute of instincts. When people talk of human instincts they usually mean either habits or intuitions; and both habits and intuitions are based on years of previous experience. We find it very difficult to imagine the mental life of a higher insect. What can it feel like to be born with a nerve-machinery that ensures you shall react without prior experience and without education to quite elaborate situations; to react to them in a finished and apparently purposeful way; and yet to be baffled by quite small variations in the situation?

What does it feel like, for instance, to be very hungry and yet starve to death rather than try any unaccustomed food outside the one species of animal or plant normally eaten, as many insects (but not a single vertebrate) will do? Or to be impelled, like some kinds of solitary wasps, to hunt for one kind of spider only, to sting it in a particular way so as to paralyze it, to wall it up with an egg as food for the grub which will hatch out of the egg—and then never pay any more attention to your offspring at all? What does it feel like to be able to build a honeycomb—a double plate of deep cells, the cells accurately hexagonal, the bases of all the cells of one plane elaborately dovetailing with those in the other? And all this without the least instruction in geometry or in wax-modelling? What does it feel like to be a worker-ant and know without being told exactly what to do with the babies of your own species; but to be so incapable of profiting by experience that if you are provided with ant-babies of another species which require a different food and different treatment, you never make the least effort to alter your routine and continue in your own ways although all your charges without exception sicken and die after a few days of your care?

The answer is that we do not know what it feels like; but that in all probability the level of accompanying consciousness is not high. Probably the insects do exceedingly little thinking; one object releases an elaborate train of behaviour, while another is paid no heed to, in the same automatic way that light of a certain wave-length hitting our eyes has to give rise to a reaction which we feel as a sensation of red, while light of only slightly longer wave-length cannot be perceived as light at all. The instincts of insects, however extraordinary, are for the most part nothing else but reflexes. The animal is turned out complete with the possibility of them, as a musical-box is turned out complete with the possibility of playing a definite but limited repertory of tunes. Its behaviour is part of its inheritance; and just because it is so automatic, no more demands thought than does our withdrawal of a pricked finger from a needle, or the secretory activity of our pancreas when stimulated by secretin in the blood .

The insect has its repertory of inborn tricks. It does not have to learn them, however elaborate; but in revenge the repertory is limited, and the tricks are so automatic that they easily fail in any unusual situation. The vertebrate has the trouble of learning by experience (which may be bitter), but in return it has a far greater range of possibilities open to it, a much more adaptable behaviour.

It is interesting to find that the same rigid and narrow specialist adaptation to a particular way of life is seen in other departments of insect life. This is particularly well marked as regards hearing. The ear of all higher vertebrates is sensitive to tones of a wide range of pitch; and from the enormous variety of sounds which the animal is thus enabled to perceive it picks out those which have meaning for it. Among insects, comparatively few have hearing-organs at all, and these, in some cases, are adapted to perceive only the particular note emitted by other animals of the species; their world of sound is confined to one or two tones.

We may reasonably assume that when an insect cannot satisfy one of its instinctive impulses it feels some sort of displeasure, and if it is thwarted, experiences something like anger at whoever is doing the thwarting; that it is somehow excited by the sight of its prey or the smell of its mate, and feels some kind of satisfaction in the proper accomplishment of the behaviour its instincts force upon it. But of anything like a train of thought, rational or not, it cannot be capable. And experience, even in the highest insects, can do little except to put a polish, so to speak, on the machine-made instincts already conferred on it by its genes.

We can sum up this divergence between insect and vertebrate from another and rather different point of view. Behaviour and mind, no less than bodily structure and chemical physiology, have all in the long run been evolved in relation to their environment. We need not beg a philosophical question and say they are determined by it, but at least they are conditioned by it. The very sense-organs, on whose information we depend to build up the highest flights of intellect or imagination, show this conditioning very clearly. No animal, for instance, possesses any sense-organ for detecting whether a wire or a rail is carrying an electric current or not; and yet such a knowledge is sometimes a matter of life or death. The reason for this gap in our repertory is doubtless the fact that powerful electric phenomena (apart from lightning, which cannot be avoided) do not occur in life's normal environment; they have only begun to exist in the last few decades. But had electric currents dangerous to life been running through the landscape during geological time, then animals, we can confidently assert, would have evolved sense-organs to detect them.

And the sense-organs which life does possess are narrowly conditioned by the facts of the lifeless environment. Sugar is abundant in nature, and sugar-containing substances are nutritious. Hence we not only possess sense-organs capable of detecting a sweet taste, but we find sweet things agreeable. Had the nutritious sugars been rare in nature, and saccharine, which is useless for food purposes, been abundant, the sensation of sweetness would doubtless not have been pleasant; while if lead acetate or sugar of lead, which is sweet but poisonous, had been the common sweet substance, sweetness would of necessity have been disagreeable to the higher animals, for only those with natures that found sweetness nasty would have escaped being poisoned. As final correspondence between sense-organs and environment let us mention the fact that of the energy which the sun radiates upon the surface of our earth, the maximum intensity is of the wavelength which gives us a sensation of greenish-yellow; and this is right in the middle of the narrow range of radiations (only a single octave of them) to which our eyes are sensitive. If the sun were in a somewhat different stage of stellar evolution, and was sending out most of its energy in the form of orange or red rays, it is likely that the

INSECTS AND OTHER INVERTEBRATES

range of radiation we can see would centre on red instead of on yellow. We should see other colours beyond the red, and we should be blind to blue and violet.

Organisms, in fact, are relative beings; they have meaning in relation to their environment, no meaning apart from it. And this relativity of life is just as pervading in regard to the senses and the mind as it is in regard to the mechanical construction of a limb or the adaptive significance of a colour-scheme.

But arthropods and vertebrates differ in the way in which mind and behaviour are fitted to the environment. Your arthropod is born adapted. Turn back to Book 2, Chap. 2, § 1, and remind yourself of the battery of tools that grow out of the lobster's body—walking legs and pincers, jaws of various kinds, swimmerets—each capable of doing some special thing. Get hold of a lobster and see them for yourself if you can. The arthropod mind is rather like that. Contrast your own hand with the lobster's organic tool-kit—it has no sharp edges or crushing forces, none of the special structures with which a lobster's limbs are so richly and variously provided. But it can move, plastically and adaptably, and be turned to a great variety of uses. Much the same can be said of your mind. It starts off comparatively unencumbered by special instinctive furniture, and adapts itself as you grow up by learning what to do. Your butterfly creeps out of the pupa-case in which it has shaped from an almost unorganized pulp. It brings with it two pairs of wings, six pairs of legs, two antennæ, and a highly specialized battery of mouth-parts; most remarkable of all, it brings a brain finished and complete. It knows from the outset what it is to do and how it is to do it. It never grows up or learns; it enters the life drama in fully adult form. Contrast your own first appearance.

Those two types, butterfly and man, present the contrast of arthropod and mammal in its extremest form. All arthropods are mechanical as the butterfly is, but few vertebrates have a plasticity that even approaches our own. In studying the behaviour of arthropods we shall see how much is possible on the level of purely mechanical behaviour. Later, when we turn to vertebrates, we shall trace the gradual supersession of inflexible instinct by a new and more effective kind of mental organization.

§ 2. An Anatomy of Instinct.

THE complication of the actions which insects perform untaught, and in the absence of all experience, has struck observers from the earliest times; and it was the study of insect behaviour which played the dominant rôle in generating the concept of instinct which prevailed both in biology and in general thought for the best part of the eighteenth and nineteenth centuries. The concept was also modified by philosophical and theological notions, it being almost universally held, before the evolutionary idea came into prominence, that the actions of man were guided by reason, those of animals by a wholly different faculty called instinct, directly conferred upon them by God. How erroneous is the first part of this antithesis will become apparent in full force when we study human behaviour; and the last half-century's detailed studies of animals have made it increasingly dubious whether the very term instinct should not be discarded.

Addison reflected the general view of the eighteenth century when he wrote in the *Spectator:* "I look upon instinct as upon the principle of gravitation in bodies, which is not to be explained by any known quantities inherent in the bodies themselves, not from any laws of mechanism, but as an immediate impression from the first Mover and the Divine energy acting in the creatures."

To us, with the evolutionary idea as background, the facts of nervous physiology as foundation, a radically different approach is necessary.

Let us take one or two examples of what are commonly called instincts. One of the most illuminating is the spinning of a cocoon by caterpillars about to pupate. They have never seen their parents or a cocoon belonging to their species; they cannot know by any form of experience that they are suddenly going to pass over into a passive pupa, with all their old organs in a state of dissolution; nor that they will emerge from this second period of embryology as a winged insect which must escape from the cocoon and will have only a tongue, no jaws to bite its way out.

And yet, what elaborateness of apparent precision! The cocoons of the moth *Saturnia pavo*, for instance, protect the pupa by being exceedingly tough; to make it possible for the winged adult to get out, the substance is fitted with an arrangement of spikes like that in a lobster-pot, but leading the other way, allowing egress and not ingress. Addison, confronted with such a cocoon, would doubtless have said that God had conferred upon the caterpillar a knowledge of the mechanical principles involved in lobster-pot construction.

In an attempt to become more scientific, many naturalists later abandoned the view that the knowledge had been implanted by the Creator, in favour of the idea that instinctive actions were inherited habits. Previous generations had once done them consciously, deliberately, with effort; the actions became gradually perfected, but at the same time more automatic, until they were done with the same effortlessness as that with which we achieve such difficult but habitual tasks as talking or writing.

But this view is untenable. For one thing, it demands the inheritance of acquired characters if it is to work, and that, as we have seen in Book 4, is something which in all probability does not happen. But even if acquired characters were inherited, matters would be little better. For no one can seriously suppose that a caterpillar, for instance, can possibly know that he or she is going to wake up one day as a winged butterfly, and can have the foresight to make provision for this future event. It can never have done its spinning work consciously and deliberately, because it can never have possessed the facts on which to deliberate.

The impossibility of there being knowledge behind instinct is perhaps most prettily illustrated in the well-known case of the yucca plant and its moth, *Pronuba*. The yucca, or Spanish bayonet, with its lovely spikes of white bells, can only be fertilized by the help of this particular moth. The female moth visits the yucca flowers (each of which stays open only for one night), scoops up the pollen, which is rather sticky, and kneads it into a pill, which she holds in a pair of specially shaped appendages. Then she pierces the ovary of the flower with her long ovipositor and lays three or four eggs just between the future seeds. And then, flying up to the pistil, she pokes the pollen-ball down into the cup which it conveniently bears at its tip, and leaves it sticking there. She then repeats the process with another flower. The pollen germinates, sends down its pollen-tubes, fertilizes the plant's eggs, and the caterpillars that meanwhile hatch from the moths' eggs begin to eat the seeds. As there are about two hundred seeds in one flower, and as each of the three or four caterpillars needs about twenty-five seeds, almost half are left over for the needs of the plant. The full-grown caterpillars eat their way out of the seed-capsule, let themselves down to the ground by a silk thread, and make themselves a cocoon there; without eating any more, they await the transformation to a pupa, which does not overtake them until the next summer; and later in the season they emerge as moths, to mate and repeat the cycle.

The association is one of mutual benefit, a reproductive symbiosis; the action of the female moth in putting the pollen-ball on the pistil seems admirably purposeful, just as her care not to kill the goose that lays the golden eggs, by only introducing three or four future grubs into each flower-capsule, seems admirably calculated. But when we reflect that the mother moth dies before the seeds mature, and that the moths of the next generation have never seen a yucca in flower before they begin their business of pollen-gathering and egg-laying, it becomes obvious that foresight and reason can play no part in the instinct—quite apart from the fact that experiments have decisively shown that no insect is capable of drawing such conclusions as the moth would have to draw if it were really being intelligent on the facts presented to it. We have no more right to suppose that the moth

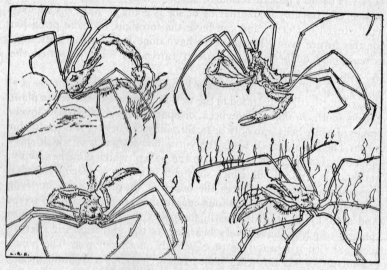

FIG. 297. An elaborate instinct.

The spider-crabs disguise themselves by picking bits of weed, biting them off to the right length, and attaching them on to the little hooks with which their backs and claws are covered. It has been shown by experiment that the action is purely instinctive.

is being purposeful and intelligent in its actions than that the yucca is being purposeful and intelligent in growing a pistil with a cup at its tip to receive the pollen; or, to confine ourselves to the moth, we have no more reason to find proof of intelligence in its actions in putting the yucca pollen in the proper place than in its growing the special appendage with which to manipulate the pollen.

There we have the gist of the matter; an instinct is, like a leg or a gland or an adaptive colour-pattern, one of the tools of the species; it merely happens to be a bit of behaviour-machinery instead of a bit of engineering or chemical machinery. It is the outcome of the animal's nervous construction, as the leg and its working is the outcome of its mechanical construction. It is a bit of nerve-clockwork.

As such, instincts can be polished up, altered, specialized by the same evolutionary agencies as limbs or glands. If we believe that natural processes, such as variation and selection, can account for the existence of adaptive structure or colours, we need have no difficulty in thinking that the same processes can account for adaptive instincts.

In the same way, the leaf-insect does not say to himself: "Every day in every way I will look more and more like a leaf"; or the bug that grows a sham ant on its thorax deliberately pretend to be an ant. That was all worked out for it by the forces that mould the species, no more and no less than were the levers of our arm-bones or the light-focusing apparatus of our eyes. In fact, the ant-bug has got such poor eyes that it is probably quite incapable of forming an image of an ant nearly as accurate as the imitation he grows on his thorax. It is the sharp eyes of the birds that have ensured the bug's likeness to the ant, not any volition or intelligence on his part. Nor is there any more proof of intelligence in the behaviour that goes with the colour to make it protective. The young plover that at the sign of danger crouches down on the shingle into invisibility will do the same on a carpet or a lawn where it is extremely conspicuous. The stick-insect will stiffen into the resemblance of a twig whenever it is illuminated, and stay thus rigid so long as it is illuminated—and this whether it is among twigs, or on the laboratory bench, or exposed to the eyes of its enemies on an open path. The mother bird that trails one wing as if wounded in front of an intruder is not deliberately simulating injury; she is carrying out an inborn method of response.

Here the curious fact (already mentioned in Book 8, Chap. 2, § 6) that prawns will creep to the positive pole if a weak electric current passes through their tank, is illuminating. Obviously there is no advantage in their going to that pole. Under natural conditions such circumstances never arise; it may be said with certainty that no more than a few hundred of prawns out of the myriads that have lived on earth have ever had an opportunity of exhibiting their galvanotropism. Yet the phenomenon depends on a property common to all prawns—upon the way the nerve-fibres happen to be arranged in their central ganglia. It is a necessary consequence of their anatomy. The instincts that we are discussing are really only more complicated instances of the same kind of thing; the animals are so made that given a certain combination of circumstances they will—and must—behave in a certain way. Just as the colouring of a leaf-insect or a young plover depends on the way its pigments are disposed, so the actions of a cocoon-spinning caterpillar or of a female yucca moth depend on the way their inner protoplasmic telephone system is planned.

The absurdly automatic nature of insect behaviour is well illustrated by the mating reactions of certain moths in which the males smell out the female by the aid of their huge feathery antennæ. So acute is their sense of smell that males will come from more than a mile away to flit around an empty box in which a female has been confined. If a male be put with a female in a cage, mating will follow in due course. If the male's antennæ be cut off, however, he is incapable of recognizing the female as such—or rather, his antennæ are the one channel through which the mating reaction can be stimulated, so that without them his sex-behaviour cannot be set going. Most remarkable of all, however, is his reaction to the two little scent-organs which the female possesses near the tip of her abdomen. If these be cut out (an operation which does not seem to incommode the female) and they and the operated female are put in a cage with a normal male not deprived of his antennæ, he will pay no attention to the female, but will make vain attempts to mate with the two little scent glands. It is all so different from human reactions that we can scarcely grasp it. But if we want to understand the world of insects, we must try to grasp it. We have to realise that the male moth has no idea of a mate; he merely possesses mating reactions; and these are fired off by one stimulus and one only —a particular smell.

When a bee stings, a similar automatism is revealed. The actual mechanism of stinging is purely reflex. If a worker-bee stings a man, it leaves its sting in the wound; and the isolated organ there continues to execute the same movements as during the first piercing of the skin, so piercing deeper into the flesh. When the experienced bee-keeper is stung, he therefore makes haste to extract the sting at once. This reflex-machinery is, however, normally held under higher control. A bee whose brain has been extirpated, has had its chief inhibiting centre removed; and it puts itself without ceasing through its various reflex paces, cleaning itself, protruding its sting, walking restlessly about, fanning its wings, and so forth.

Higher vertebrates have instincts just as much as insects. But in them, the instinct is rarely so machine-made; it is flexible at its two ends, both as regards the situations which call it out, and the methods adopted to execute it. And the process continues until in man nothing is left but the most central part of the instinct, the instinctive impulse like fear or attraction or anger which pushes one on to act in a certain general way when confronted with a certain general kind of situation. Nothing approaching this flexibility is seen in any insect. Men have trained performing fleas; but the fleas are merely crawling or hopping along in the ordinary way while harnessed to a miniature carriage. No one could train a flea or any other insect to do anything so different from its ordinary activities as shamming dead or jumping up at the word of command, like a dog, or bicycling, like a performing bear, or sitting up at table and politely passing the plates like the young chimpanzees at the London Zoo. Insects have elaborate instincts because they have no elaborate brains to be intelligent.

This lack of intelligence is most clearly seen in those numerous instances in which an instinct, beautifully adapted to the ordinary conditions of the possessor's life, makes default and even acts wastefully or harmfully, when the conditions are changed—even though the least glimmer of intelligence would have set matters right.

In one species of solitary bee, each mother builds a mud cell which she fills with honey and pollen up to a certain level; then she lays an egg on the food; and then seals the cell up. An observer broke open a cell while the bee was away. On her return, she noticed the hole, for she explored it thoroughly with her antennæ. But she had not the sense to mend it. She proceeded with her business of food-gathering, load after load of food being deposited in the cell only to fall through the open hole. When she had brought the regulation number of loads, she laid an egg, and sealed up the foodless cell. As with the solitary wasps we shall shortly describe, the normal behaviour is a chain of actions, each determined by the one before, each determining the one after. It is adaptive, but not purposeful.

An even more obvious example of clockwork instinct is provided by certain cocoon-spinning caterpillars. If the animal is interrupted in the middle of its task and the half-finished cocoon removed, it will not begin again from the beginning, but will spin only what remained for it to do, in spite of the fact that the half-cocoon it produces is completely useless for protection.

Forel describes how, when a battle is raging between different kinds of ants, you can decapitate two of the combatants, and the heads will go on trying to fight each other for some minutes. That perhaps depends as much upon the fact that insects get their oxygen by air-tubes and not in blood, as upon the reflex nature of their behaviour. But what are we to think of the worker-wasp who was imprisoned with a grub of its own species, but without food? It wanted to feed the young wasp, but was in evident perplexity how to do so. Eventually it bit off the grub's hind end, and offered it to the front end!

There we may leave our anatomy of instinct. We shall describe a great variety of insect instincts in subsequent sections; but they are all variations on the same theme, they are all behaviour which is in the main the rigid outcome of inherited nerve-structure. We should find the same machinery of behaviour at work in other groups. There are the hermit-crabs which have the instinct to find shells as houses for their unprotected abdomen, the other crabs which go about, one sort with stinging sea-anemones held as knuckledusters in its claws, another sort with a distasteful garlic-smelling sponge. There are the octopuses which have the instinct to build themselves a little den of stones, behind which they can lurk unseen; but if you give them bits of glass, they will build their rampart as readily with them, although it is transparent and lets its occupant be seen. There is the elaborate flight-instinct of the squid, with its ejection of ink and a right-angled turn at the crucial moment, which we have already described (Book 2, Chap. 3, § 2). There are the instincts of spiders, which rival those of insects in their elaborateness. The combination of structure with appropriate instinct is beautifully seen in a borrowing spider, whose abdomen ends behind in a hard flat plate; and it uses this as its front door to block the entry to its burrow. There are the web-spiders, which build their wonderful snares without ever being taught; the common garden spider usually builds two new nets a day, one in the morning and one in the afternoon, for its many weeks of life.

But it is in the Hymenoptera, the ants, wasps, and bees, and in the termites that rigid instinct attains its highest levels of intricacy. First we will consider the very interesting mechanism of the solitary wasps and then pass to the complex interplay of the social insects.

§ 3. Solitary Wasps.

MOST wasps are solitary; only a few are social. You may read an excellent account of their ways in G. and E. Peckham's *Wasps*, or (together with many other vivid and illuminating descriptions of insect behaviour) in the works of Fabre.

The almost universal characteristic of solitary wasps, out of which the social habit has evolved, is their provision of food for their young. And though the adults live mainly or solely upon fruit or the nectar of flowers, the growing grubs are universally carnivorous. In solitary wasps there are no neuter workers. The males have but one biological

INSECTS AND OTHER INVERTEBRATES

duty—to impregnate the females. The females, once charged with a store of sperms, make burrows in the ground or in walls or tree-trunks. They then hunt for prey, which may be spiders, caterpillars or ants, flies, moths, plant-lice or other arthropods, and sting them through the nerve-centres in their nerve-cords, a procedure which kills in some species but which usually paralyzes without killing. The paralyzed prey is then put in the burrow, and an egg laid on it or beside it. In most cases the burrow is then sealed up and abandoned, so that the parent makes provision for an offspring it will never see. Each species of solitary wasp—and there are many hundreds—chooses a particular kind of situation for its nest, hunts only one kind of prey, and makes and seals its burrow in its own characteristic way.

All the wasps of the genus Ammophila, for instance, confine themselves to caterpillar-hunting. They first make a nest in the soil. Then they seal it up with sand, take a flight round to fix the neighbourhood in their memory, and depart to hunt for caterpillars. This is not a light task. Although the wasps spend most of their time running up and down likely plants examining the under-side of leaves, they rarely find more than one victim a day. When the victim is found, it is stung and paralyzed. The stinging is an elaborate business; we will quote the Peckhams' account of one wasp they watched in the act: "The wasp attacked at once, but was rudely repulsed, the caterpillar rolling and unrolling itself rapidly and with the most violent contortions of the whole body. Again and again its adversary descended, but failed to gain a hold. The caterpillar, in its struggles, flung itself here and there over the ground, and had there been any grass or other covering near it might have reached a place of partial safety; but there was no shelter within reach, and at the fifth attack the wasp succeeded in alighting over it, near the anterior end, and in grasping its body firmly in her mandibles. Standing high on her long legs and disregarding the continued struggles of her victim, she lifted it from the ground, curved the end of her abdomen under its body, and darted her sting between the third and fourth segments. From this instant there was a complete cessation of movement on the part of the unfortunate caterpillar. Limp and helpless, it could offer no further opposition to the will of its conqueror. For some moments the wasp remained motionless, and then, withdrawing her sting, she plunged it successively between the third and second, and between the second and the first segments.

"The caterpillar was now left lying on the ground. For a moment the wasp circled above it, and then, descending, seized it again, farther back this time, and with great deliberation and nicety of action, gave it four more stings, beginning between the ninth and tenth segments and progressing backward."

After this she gave herself a thorough toilet; and then proceeded to bite at the neck of her victim, pinching it repeatedly between her strong mandibles.

The paralyzed caterpillar, though much heavier than the wasp, is then dragged across country, through jungles of grass-stems, to the nest, which has been previously sealed up. One wasp was watched dragging its prey nearly a hundred yards, which took over two hours. When the nest is reached it is opened, the prey pushed in, an egg laid on it, and the nest-opening carefully sealed up again. A second caterpillar is usually brought later to the same nest, as one is not enough for the young grub.

Fabre had described his French Ammophilas as stinging their prey with an uncanny precision which could only mean that they possessed in some mysterious way a knowledge of caterpillar anatomy. The wasp's sting, he maintained, always pierced the caterpillar's nerve-centres (of which there is one in every segment) in such a way that the victim was completely paralyzed, and thus could neither wriggle and crush the young wasp-grub, nor die and decay. And his statements are still often quoted as examples of the unerring and supernormal revelations of instinct. But the careful observation of the Peckhams has shown that Fabre was here at fault, or at least that the conclusions he drew are not in the least general. In the American Ammophilas the stinging is much more variable, both in execution and results, than he records. The instinct of the wasp, it seems, is simply to thrust its sting into the lower surface of the caterpillar (the sting goes in most easily at the joints between the segments of the body), and to go on doing so at different places until its victim becomes more or less limp. Sometimes the wasp is content with a single sting, sometimes it stings the caterpillar between all of its thirteen segments.

And not more than half of the stung caterpillars observed by the Peckhams lived on in a paralyzed state. Some of them reared and rolled about violently when the wasp-grub began to take mouthfuls out of them; others died quite soon and became decomposed before the wasp-grub had finished its growth. In both cases, however, the grub's development was not prevented. So long as the victims are not fully active, and so long as most of them do not decay prematurely, the method will work. The quasi-miraculous accuracy of the wasp's surgery turns out to be a rough and ready reflex of no great complexity or regularity: the wasp has no surgical knowledge of the insides of caterpillars, but merely displays a reaction to the feel of their bodies. The same simplicity of behaviour is seen in those species that provide spiders for their grubs; they react to captured prey by thrusting their sting into the lower surface of the spider's fore-part. As the nerve-centres of spiders are concentrated in a single mass surrounding the gullet, the sting will in nine cases out of ten pierce some part of this compound centre and so achieve some degree of paralysis.

The instinctive and machine-like quality of most of their behaviour was clearly shown by some experiments of Fabre on the wasp Sphex, which hunts crickets. When the Sphex has brought a paralyzed cricket to her burrow, she leaves it on the threshold, goes inside for a moment, apparently to see that all is well, emerges, and drags the cricket in. While the wasp was inside, Fabre moved the cricket a few inches away. The wasp came out, fetched the cricket back to the threshold, and went inside again—on which Fabre again moved the cricket away. He repeated the procedure forty times, always with the same result; the wasp never thought of pulling the cricket straight in. Drag cricket to the threshold—pop in—pop out—pull cricket in: the sequence of actions seems to be like a set of cog-wheels, each arranged to set the next one going, but permitting of no variation. The Peckhams repeated the experiment with an American Sphex. This creature was not quite so automatic, for after her prey had been moved a number of times, she did drag it straight into the burrow.

These wasps have a certain power of learning, as shown by their memory for the surroundings of their nest: and some of them at least can modify their behaviour in a way which shows the beginnings of intelligence. But the main flow of behaviour is automatic. All that experience does is to change a detail here and there; and the change is made as it were under protest, and only after many repetitions.

The same cogging up of behaviour in a rigid sequence of actions was found by the Peckhams in *Pompilius quinquenotatus*, one of the spider-hunting wasps, which has the instinct to hang its prey in the crotch of a grass-plant. This species only digs a burrow after catching its spider, and the hanging up of the prey serves to keep it from being stolen by ants while the wasp is busy digging.

If a paralyzed spider is removed from the crotch where the wasp has hung it, and an uninjured spider substituted, this latter will usually remain quiet for a time, as it reacts to handling by "shamming dead." The wasp comes back in due course, and in some way notices the difference. But instead of stinging the new spider, she refuses to have anything to do with it (this is not due to the spider having been handled, as was readily shown by experiment), and flies off to find another. When she comes back with the new prey, instead of using the burrow she has just made, she digs another. Here again an invisible sequence must, it seems, be preserved. Only a paralyzed spider in a grass-crotch releases the reaction to drag the prey into the nest. An unparalyzed spider in a crotch is not a normal phenomenon, and interrupts the whole chain. Similarly, the hanging up of a paralyzed spider provides the stimulus for the next action in the sequence, the digging of the burrow; and this is called out, even if a perfectly good empty burrow is waiting.

Among individuals of a single species there is a good deal of variation in instinct, just as there may be in colour or size; but in a single individual the instinct is, within very narrow limits, fixed and rigid.

§ 4. Insect Societies.

WE COME now to the most interesting aspect of arthropod behaviour, the formation of co-operative communities of insects which store food and have a real economic life.

The incipient stages of sociality, in which parents remain with and look after their young, are to be found here and there in many orders of insects, such as earwigs and some beetles. But true social life, in which the young stay on with their parents and help with further broods, is known in two groups only. One is the Hymenoptera, which includes ants, bees, wasps, and ichneumons. The other, the Isoptera, a very different order allied to the cockroaches, consists of the termites, often miscalled white ants.

But there is a still further degree of insect social life. Some of the young may be transformed into unsexed neuters, who take upon themselves all the duties of the colony except that of reproduction.

The ants are the most specialized and the most successful of all social units. We will begin with some account of them, and then pass to the termites as a remarkable example of parallel independent evolution. We shall next deal, very briefly, with some interesting facts about bees. The social wasps, a further variant of the social theme, will demand no more than incidental mention.

All ants, like all human beings, are social animals. Many people seem to imagine that ants are like miniature men—that there are a number of ant races, some practising one mode of life, some another, but all part of one big single group. This is very far from being the case. There are over three thousand five hundred separate species of ants already known to science, each one a biological unit pursuing its own independent path, incapable of interbreeding with any other.

These various species show an extraordinary diversity of size and structure. The largest worker-ants weigh about a fiftieth part of an ounce—four or five thousand times as much as the smallest. There are some worker-ants with formidable grinding mandibles for chewing up grain, others with sabres for piercing their enemies' heads, others with leaf-cutting scissors for jaws. There are those with huge heads and those with tiny heads, those with squat bodies and those with slender bodies. For we must not forget that, in striking contrast with men, ants have to be built for their jobs. They do not make tools; they grow them as parts of their bodies. Almost all ants for instance have an antenna-comb, an ingenious gadget for holding their sensitive antennæ and combing them free of dirt, which is built into one pair of their legs. Two or three kinds of ants have huge heads expanding to end forwards in a round, flat, hard surface. These creatures station themselves at the entrances to the nest, which they block completely with their heads, moving aside when another worker taps them with its antennæ. They are porters and doors in one.

Each species of ant is thus built specially for its own particular kind of life and is quite unadaptable to any other. Even within the single community there is the same kind of specialized physical diversity. Only the males and females have wings; the neuters grow up wingless. The neuters have much bigger brains than the males or the queens; but, as they never have to fly, their eyes are smaller. In an ant called Carebara, the neuters are minute; they are only about one-thousandth of the bulk of the queen. It is as if Lilliput were not only sober reality, but as if Lilliputians lived side by side with normal-sized men and women to make up a single human community.

Even within the neuter caste there may be markedly different sub-castes. Contrasted with the more ordinary-looking workers, there may be soldiers, larger, with portentous heads, and battle-axes for jaws. In such creatures as Atta, the leaf-cutting ant, there is a huge range of neuter size and structure. In this genus we can if we like distinguish soldiers, worker majors, worker minors, and the tiny worker minimæ; but the different forms are actually connected by every possible gradation. In more specialized cases, soldiers and workers may be sharply marked off from each other, with a gap between.

This physical diversity goes hand in hand with diversity of behaviour. The males do nothing but fertilize the queens when the time comes. The queens lay eggs eternally. The workers have the instinct of tending the young, the soldiers are impelled to bite and snap in defence of the colony. The workers of one kind of ant keep ant-cows, but never look at grain or make raids on other ants. Those of a second are only graminivorous, those of a third live by slave-labour. Thus the division of labour in an ant-community, unlike the division of labour in a human community, is based on marked, inborn individual differences of structure and instinctive behaviour between its members.

The way in which the ant-stock exploits the resources of the world is equally different from the human way: it does so by splitting up into thousands of separate species, each with its own inborn peculiarities, instead of remaining one species with branches differing only in their acquirements of tradition and culture.

The neuters of ants, bees, and wasps are not a third sex; they are sterilized females, which develop with rudimentary ovaries and oviducts. The difference between them and the fertile females, the queens, seems to be brought about solely by feeding. This we know for certain in bees. Female bee-grubs fed on "royal jelly," which contains more pollen and, therefore, more protein, grow into queens, those fed on a less protein-rich diet grow into workers. The constitution of the female bee is so devised that it has these two possibilities of expression to be unlocked by the keys of two different diets.

Bees have only one sort of worker, but among ants further differences obtain, giving soldiers and workers of different types. While the difference between queen and neuter seems to be due to a difference in the quality of food, that between worker and soldier depends apparently on differences in quantity. Where there exists a gradation between worker and soldier, we find that the proportionate size of the head goes up with the absolute size of the whole body, just as we found for the big claw of the fiddler-crab (Fig. 188). The developmental machinery is so set that keeping the grubs underfed and making them pupate when quite small will lead to a creature with relatively tiny head, while fattening them to the greatest possible size will mean that the neuter into which they metamorphose must not only be big but bigheaded.

The difference between female (whether neuter or queen) and male, however, depends on heredity. But the method of sex-determination is different from the normal. Instead of the male having one chromosome less than the female, he has a whole set of chromosomes less; for he arises from an unfertilized egg.

The eggs of these social insects are unique in the animal kingdom, for they will develop equally well whether fertilized or no; they can be parthenogenetic, but do not need to be. Unfertilized eggs always turn into males, fertilized into females. Whether an egg shall be fertilized or no is controlled by the queen as she lays it. The queen, in bees and wasps as well as in ants, mates once only in her life—during her "nuptial flight." The population of spermatozoa she then receives are stored in a little purse, opening off the oviduct, with purse-strings of muscle round its neck. In this they live for all the rest of the queen's life —a period that may extend to several years; seventeen years is the longest recorded. If an egg is to grow into a female, the sphincter is relaxed an instant as the egg passes across the bagful of sperms, a few sperms escape, and one fertilizes it. If it is to grow into a male, the sphincter is kept tight shut, and the egg pursues its parthenogenetic way.

There is a reason for this. The ordinary methods of sex-determination inevitably give equal proportions of males and females. But the states of ant, bee, and wasp are based on the labour of sterilized females. What should they do with a huge population of useless males, when a few score are ample to perpetuate the race? The problem has been neatly solved by the adoption of this other method, in which the proportion of the sexes can be varied as desired.

The way an ant-community develops is as follows: Swarms of winged males and virgin females fly out of their nests (leaving an excited population of workers round the door-ways) and up into the air for the only flight of their lives. Nuptial hordes issue at the same time from all the nests over a large area; this simultaneous flight is prompted apparently by special weather conditions, and certainly promotes cross-fertilization between different nests. Queen ants differ from queen bees in permitting several males to mate with them, one after the other. When this polygamous flight is over, the males drift off to perish of starvation. Of the females, some return to their old nests (for in ant-societies many fertile queens may live and lay in mutual tolerance within a single nest); others fail to find their way home, and proceed to found new colonies. They find a sheltered spot, creep in and break off their wings at a preformed plane of weakness. For weeks they take no food, but live, ripen their first eggs and feed the first batch of young grubs at the expense of the material produced by their degenerating wing-muscles. The first grubs pupate and turn into workers. These at once, without any instruction, take over all the duties of the nest-building galleries, keeping the place clean, and tending the young. From now on the queen does nothing whatever for the community but lay eggs.

In bees and wasps the nests are often beautifully regular, and always contain rows upon rows of developmental cubicles for the young. Every member of the colony lives boxed up in his or her or its private cell from first egg-hood to adult emergence. But in ants the nests, often subterranean, consist of an irregular network of galleries and chambers, and they never contain special cells for the grubs. Ant-homes thus lack a certain picturesqueness to be seen in the hive, but in both respects the ant is really more advanced than the bee or wasp. The irregularity of the nest allows much greater plasticity, more freedom in adjusting means to ends; and the young, not being confined to one spot, can be taken from chamber to chamber as conditions demand, or even given an airing or sunning out of doors.

What interests us most of all, however, is to find what keeps the ant-community together, how its social and economic life is organized and unified. In human societies there is an economic nexus, of goods and services and money; there are sentiments of patriotism and social devotion; there is education at home and at school, to fit the growing child for his life as a social being and to train him for a particular career.

It is wholly different with the ants. Among them there is no education. The workers, it is true, can learn to modify their behaviour to cope with certain unaccustomed situations, but the modification is slight, and when the shroud-like covering of their pupal life is stripped off them by their nurses they emerge fully equipped with the instincts needed to carry on with the work of nest-building and food-gathering and nursing. The division of labour, again, which in human communities is the result of special training, is, among ants, the immediate result of inborn differences in structure and instincts.

Then ants have a sort of patriotism; but it seems to be based entirely upon smell. Ants which by some means or other have acquired the characteristic nest-smell are tolerated and treated as fellow-citizens; those which have not this shibboleth, even if they belong to the same species, are attacked and killed. By putting a number of pupæ of different kinds of ants together, an artificial mixed ant-state can be produced. The animals on emergence all acquire the same smell and live together in amity, though in nature they would fight to the death. Blood is thicker than water; but for ants at least smell is more powerful than blood.

Ants have an economic life, too; this is based entirely upon direct exchanges of food. If you watch ants in an observation nest, you will often see one ant go up to another and stroke it with its antennæ.

INSECTS AND OTHER INVERTEBRATES

This is a solicitation of food. If the second ant is well-fed, it will stop; the two will raise the fore-parts of their bodies until their mouths are close together, and the second ant will produce from its mouth a drop of liquid which the other will swallow.

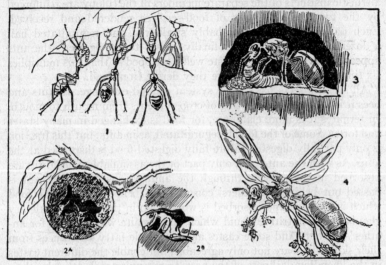

FIGS. 298-299. Scenes from ant-life.

(*1*) *Workers of the ant Œcophylla repairing their nest. The nest is made of leaves. One set of workers is pulling the two sides of the rent together, others are using grubs as secotine-tubes to stick the leaves together. (2A) Colobopsis at home. This Texan ant makes its nest in live-oak galls, 1 to 1¼ inches across. One gall is here represented as if cut across. A soldier is using its head to plug the entrance. (2B) A "porter" soldier enlarged to show how its head is modified to serve as a door. (3) The beginning of a colony of wood-ants. A queen after her nuptial flight has broken off her wings, imprisoned herself in an underground cell, and laid her eggs. Of the grubs that hatch out she gives one much more food than the rest, in order that she may have a worker to assist her as soon as possible. (4) A queen Carebara, whose workers are less than a thousand times her weight. She is leaving on her nuptial flight, and a number of workers are clinging to the hairs on her legs.*

This liquid it pumps up from its crop, a large reservoir in the front of its abdomen, between gullet and true stomach. Forel calls it the "social stomach," because any food which it contains is part of a common store, available for any other ants which come begging. Only when food passes the valve from crop to true stomach is it digested and utilized by the private individual. When honey coloured blue with a harmless dye is given to ants, the blue crop can be seen through the thin parts of the abdominal walls. If a few ants are fed on such dyed honey, within a couple of days practically every ant in the community will show a pale blue tinge, showing how vigorous is the exchanging of food. There is in fact a circulation of food within the colony very nearly as essential as the circulation of blood in the human body; and the altruistic instincts of the separate members of the colony are reinforced by the tangible economics of food-services rendered and received. Each particle of food is probably swallowed and regurgitated half a dozen times or so before it finally passes to be digested; the ants appear to have the sense of taste well developed so that this multiplies six times the collective pleasure they derive from food.

All social insects have some system of food-exchange. In ants and bees it is confined to a passing of crop contents from mouth to mouth. In termites it is more elaborate, for food is exchanged in many states and forms. Some of the food is regurgitated, as in ants; but this fraction is only partially digested. More fully digested food is dispensed at the anus. As with the ant-cows, only part of the assimilable substances are absorbed in the passage through the animal's own gut; the rest is passed out behind for others' consumption. Then some of the food which is digested and absorbed is worked up in the so-called salivary glands into a nutritive liquid which one termite will produce for another's benefit. And some castes at least exude fatty substances from their skin, which are not only agreeable, but enable the different castes in the unending darkness of the nest, to recognize each other by their taste. The difference between the food-exchanges of ants and termites is like the difference between a monetary system with only one kind of currency, like cowrie shells, and one with copper, silver, and notes.

In wasps we get an interesting variation. There is no exchange between adult and adult, but there is between adult and young. After a worker has fed a grub, it taps its head and the grub gives out a drop of slightly sweet liquid from its salivary glands, which the worker eagerly licks up. This difference from ant or bee practice is connected with wasp diet, which is mainly flesh. The growing grub needs protein for its growth; the adult wasp needs chiefly sugars and such-like fuels for its muscles. The grub, by returning part of its food in the form of sugary liquid, not only provides a bribe to keep the worker at its nursemaid work, but helps in distributing the available food economically. Sometimes the workers cheat, and try to get the grub to produce the sweet bribe without feeding it. Occasionally they succeed; but for the most part the system works as an economic system should, to encourage legitimate business.

§ 5. Ways of Life Among Ants.

THE ways of ant-life are various in the extreme. Most of their species are general foragers, picking up what miscellaneous small animal prey they can find. But there are many which have specialized. Certain ants, unique among sub-human animals, keep domestic animals. Man, a vertebrate, domesticates other vertebrates; ants domesticate other insects. The commonest ant-cows (as Linnæus first dubbed them) are plant-lice or aphids, but coccids or scale-insects are also kept. We keep cattle in order that they may transform grass, inedible by us, into edible milk and meat. The ants keep plant-lice and coccids to tap the resources of the plants. No ant has developed sucking mouth-parts; their jaws are all built on the biting plan. They cannot therefore get at the rich currents of plant-sap. Aphids and coccids, however, have a proboscis designed for this purpose; they sit tight, anchored by their tongues, and pump themselves full of nutritive liquid. They are, however, wasteful in their internal arrangements. Much of the nutriment remains undigested, and is passed out in the form of little drops of sweet liquid, so-called honey-dew. Ants have always had a reputation for thrift, and they do not tolerate this waste. They lick the sugary mess off the leaves; or they catch the drop as it emerges from the aphis; or, last stage of all, they "milk" their cows, caressing them abdominally with their antennæ, upon which the ant-cows void their liquid contents.

The more specialized among pastoral ants really domesticate their insect-cattle. Some build little wood-pulp stables over them, and may connect the stables with the nest by covered passages. Others excavate underground chambers and set their cattle to exploit roots. And some tend the aphis eggs through the winter and put them out on plants when they hatch in spring.

Then there are the agricultural ants. In these, the larger workers climb trees, cut bits out of leaves, and bring them home held aloft like green umbrellas; in the nest they are chopped up fine and turned into regular beds of leaf-mould by the smallest workers. On these beds, which are suspended from the roofs of special subterranean chambers, the ants grow a fungus, a white meshwork of filaments. This, if left to itself, fruits in the form of a large toadstool. But so long as the ants have it under cultivation, it never fruits. They plant it on the chewed-up leaves, they weed the beds of other moulds, they manure them with their own excrement, and they treat the precious vegetable so that it grows in a peculiar way, with little knobbed heads. It is these knobs which the ants especially fancy as food, and on which they almost entirely subsist. These underground fungus-gardens are connected with the outer air by ventilating shafts, which are closed or opened to regulate the temperature and the moisture.

When the queens of these agriculturalists are preparing to leave the nest on their nuptial flight, they take a good meal of the fungus. A mass of the filaments and of the leaf-mould in which they grow is collected in a little pocket in the floor of the mouth. This pocket is present in all ants, and serves to collect dirt and solid particles out of the food (for no ant ever swallows any food that is not in the liquid state); when the pocket is full the contained pellet is ejected on a rubbish-heap outside the nest. After the agricultural queen is fertilized, she excavates a little earthen chamber, and snaps off her wings. She voids the fungus-pellet on to the floor of the chamber, manures it with her own dung, and keeps the fungus going until the eggs she lays hatch out into grubs; she feeds them with bits of fungus-heads. The grubs pupate and turn into workers. These, all untaught and without a previous glimpse of a leaf, sally out, cut leaves, chop them up, and add to the garden.

Fungi grow readily in the humid atmosphere of an ants' subterranean city. Doubtless these fungus-growers at first supplemented their ordinary diet with a few casual fungi, and only gradually were the instincts evolved which led to the perfection of fungus-agriculture which we see to-day. The transference of the staple vegetable from one nest to another seems at first sight the most difficult evolutionary step; but we see how utilization of the already existing débris-pocket in the mouth, together with a trifling change of behaviour as regards its contents, suffice to bring it about.

The third type that we will choose is that of the grain-collectors. Perhaps we should have put them first because of their literary celebrity. For to this tribe belonged the ants which impressed King Solomon. They have been celebrated in fable from Æsop's day to our own. However, their achievements, remarkable as they are, are not so extraordinary as those of the cattle-keepers and the fungus-gardeners. It was at one time supposed that these little creatures not only stored grain, but planted and cultivated it. This has now been shown to be a myth. They are in the stage of food-collectors, the stage through which our own forebears must have passed in their transition from hunting to farming. They all inhabit dry countries, and the store of grain is gathered against the season of drought. Sometimes neglected seeds germinate near the nest when the next rains come, and it is this which naturally enough gave rise to the legend of deliberate cultivation.

The soldier-caste of the grain-ants has been demilitarized. Their swords have been beaten into plough-shares by evolution—in point of fact, their militarist jaws have been converted into heavy grinding and crushing tools, which are able to achieve what the workers' slighter jaws cannot do—break up the hard grains. In some species the workers then chew up the flour thus produced, moisten it into a kind of paste, and put cakes of it out to bake in the sun.

Another remarkable ant of dry regions is the Honeypot, Myrmecocystis. The bees are the only insects which make store-chambers for liquid food. But these ants have got over the difficulty by turning some of their own number into living honey-jars. The ordinary workers collect the honey-dew voided by plant-lice; arrived back in the nest, they hand over most of the contents of their crops to a special kind of neuter appropriately call edthe Replete. These have the capacity of distending their crops until their abdomen swells to the size of a pea, perhaps a hundred times its original bulk. At the close of the wet season the repletes hang themselves up in rows from the roofs of underground cellars, and in the dry season of scarcity they are, so to speak, taken down and tapped, supplying the whole community for months from their superfluity.

This moulding of the individual's structure so that it becomes a tool of the community is equally well illustrated by the leaf-nest ant, Œcophylla, which inhabits nests made of green leaves glued together. These ants share with man the doubtful honour of being the only known organisms to employ child-labour. If you tear a hole in a leaf nest, you will see a gang of workers swarm out and pull the edges of the gap together. Meanwhile, on the inside of the breach, another gang appears, each member of which has a grub in its jaws. When the outside gang have got the leaves in place, the interior workers begin their job. They squeeze the grubs; the grubs exude threads of extremely sticky secretion from their so-called salivary glands, and with these threads the workers repair the damage, dabbing the grubs from one side of the rent to the other. This is child labour, if you will, but with all kinds of fundamental differences from its human counterpart. The children in the mills of Lancashire, three-quarters of a century ago, did not secrete thread like spiders or possess hands like shuttles or carding-combs. But in the ants, the constitution of the grubs has been altered during evolution to fit them for this work. Their salivary glands produce this very adhesive substance, which is different from the secretion of the corresponding glands in ordinary ant-larvæ, and the glands are much larger than usual.

Another remarkable ant variation is the slave-maker. The term *slave-making* is firmly engrained; but in reality the relations of slave-maker to slave are often more like those of parasite to host. The workers of the slave-making species set out and raid the nests of other kinds of ants. The best-known are the Amazon ants, Polyergus. With their long sickle-shaped jaws, they pierce the brains of the defenders, and carry away a store of cocoons. Forel gives some wonderful descriptions of these expeditions, the organization and the tactics involved, the fierce fighting, the occasional repulse of the attackers. The pupæ hatch out in their new home, and set about what their instincts impel them to—the care of ant-babies, even though these are not of their own kind; the building of the home, albeit an alien one; the foraging of food. The Amazons themselves do not stir leg or jaw in any such domestic duties—for which, indeed, the build of their mandibles altogether unfits them.

The red slave-maker, *Formica sanguinea*, is not so specialized. Its neuters can still work as well as fight; and old colonies may give up their raids and allow all their slaves to die off. Their queens are unable to found a colony independently; apparently they have not enough reserve materials stored in their bodies. After the nuptial flight the queen either returns to her old home or to another colony of the same species, or else invades a small nest of a related species, the brown ant *Formica fusca*. There she excitedly seizes a number of pupæ and mounts guard over the pile, killing any of the *fusca* workers (who are much smaller than she is) if they try to recapture the cocoons. Sometimes, it seems, the *fusca* queen is killed by the intruder; but sometimes the workers adopt the stranger and turn against their own queen.

The Amazon queen is even more warlike and invariably founds new colonies by invading a small community of the slave-species (again usually *Formica fusca*) and killing the queen with one bite of her sickle-shaped jaws through the brain; the *fusca* workers then adopt her and care for her and for the eggs she lays. So the mixed colony of warriors and alien slaves is begun.

This way of life must clearly have arisen out of the general foraging habits of ants. Alien cocoons of some related species, carried away in a raid, aroused the nursing instincts of their captors, and were tended instead of being devoured. The workers that emerged from them would automatically take their place in the life of the colony, which thus would receive an accession of working strength without having incurred the expense of feeding the grubs. And so it would thrive, and any further strengthening of the instinct to raid and to look after the raided cocoons would be preserved by selection.

The path of biological dependence, however, is a slippery one. In many species dependence has gone so far that the slave-makers are the

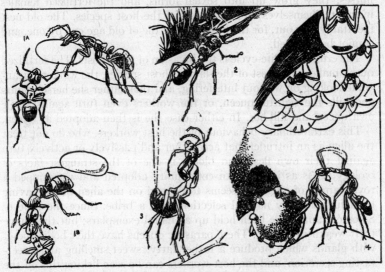

Fig. 300. Scenes from ant-life.

(*1A*) *A worker of Polyergus, the Amazon slave-maker, carrying away a cocoon of Formica rufibarbis. One of the Formica workers is trying to retrieve the cocoon.* (*1B*) *The Amazon has dropped the cocoon and has pierced the brain of the Formica with her mandibles.* (*2*) *A grain-crushing neuter (modified soldier) of Messor, one of the harvesting ants, carrying a seed home in her jaws.* (*3*) *A worker of the common field-ant, Formica pratensis, defending the colony. Holding on by her middle pair of legs, she is squirting a jet of formic acid at the enemy.* (*4*) *Honeypot ants. A number of replete neuters, with their abdomens enormously swollen by their crops distended with honey-dew, are hanging from the roof of an underground store-chamber. Below, one of the repletes is regurgitating honey-dew to an ordinary worker.*

merest parasites. Ordinary parasites, like the tapeworm, tend to lose their organs of food-finding and of protection against the outer world; they become masses of tissue focused upon food absorption and reproduction. The only department of their life in which there is elaboration above the normal is in their means of dispersal, to enable them to make the difficult passage from one host to another. These parasitic ants are no exception. But the biological unit in ant-life is the colony, not the individual, and it is the colony which shows degeneration. In these species the colony has lost all its workers; it has come to consist of nothing but reproductive units, male and female. The queens are found in alien nests, entirely incapable of looking after themselves. Their offspring are the only young ants in the nest. Tended by alien nurses, they grow up into sexual forms, and the fertilized females insinuate themselves into new nests of the host species. The old nest inevitably dies out, for the host workers die of old age one by one, and cannot be replaced.

The crux of the life-cycle is the invasion of fresh hosts. The fertilized queen makes for a nest of the slave or host species, the workers meanwhile looking on without interfering, and then either she herself fights and kills the rightful queen, or the workers even turn against their own queen and kill her. In either case she is then adopted as queen.

This extraordinary behaviour of the host workers, who do not treat the alien as an intruder, but accept her and passively or actively turn against their own flesh and blood, is one of the strangest facts of biology. It is as if a human community adopted and worshipped a royal family of aliens. It seems to depend on the alien queen having been furnished by natural selection with a bribe. Since the time of Solomon, ants have been held up as moral exemplars; but they have their weakness—greed. These parasitic queens have their bodies beset with glands which produce a rather fatty, sweet-smelling and sweet-tasting secretion; and the host-workers can be seen licking this off the tufts of hair on to which it exudes with the greatest avidity.

§ 6. The Parasites of Ant-colonies.

THE individual ant has its own internal parasites; but the ant-colony, that amorphous beginning of a new and compound individuality, is the real unit of the species and harbours many more hangers-on. Already over two thousand species of animals, mostly insects, but with a fair sprinkling of spiders and mites and a few crustacea, have been discovered living in the nests of ants, and incapable of existing anywhere else. That is about half as many kinds of colony-parasites as of ants.

In this respect, the communities of ants and termites have attained a complication beyond that of human societies. The relations of the ants to these ant-guests, indeed, are very different from those between man and his domestic animals and pets. Ants have no other domestic animals but aphids and scale-insects. These are their milch cattle; but they have no beasts of burden like horses or camels, no animals yielding them clothing like sheep or vicuna, none whose eggs they take, like poultry, or whose flesh they eat like ox or pig, no companion and guard like the dog, no vermin-catcher like the cat. Nor do they have any pets deliberately domesticated, like canary or goldfish or fancy mice, for their attractiveness.

For in the insect-world the whole mechanism by which things happen is different. The guests and parasites of the ant-colony have been moulded to their present queerness of structure and behaviour by natural selection, not by artificial selection. No conscious purpose on the part of the ants themselves has intervened in their making. As we shall see, the tastes and habits of the ants themselves have contributed to the evolutionary shaping of the guests' germ-plasm. But the shaping has been always indirect; the ants are part of the guests' environment, and if the guests are to survive, they must be adapted to life among ants as tapeworms have had to become adapted to life in vertebrates' intestines. Some of the smaller hangers-on of humanity, like the cricket on the hearth or the house-martin, have become adapted in this way to their human environment; but all such creatures are very small compared with man, while ant guests, almost without exception, are about of a size with their hosts.

If we imagined that in England our houses were, against our wills, inhabited by cockroaches as big as wolves and house-flies like hens, and that there were also crickets to whose presence we were indifferent, although they were the size of our own children, and pet-like creatures whom we liked because they rendered us some agreeable service, as it might be parrots which had the instinct of scratching our backs for us; and monstrous animals which we allowed to eat our babies in their cots because they secreted hot rum-punch or some equally fascinating liquid, and that in France, say, and the United States, there were similar sets of animals, willy-nilly sharing men's houses and reproducing there, only all belonging to different species from those in England; and finally that they were all incapable of existing permanently anywhere outside our houses; then we should begin to get some idea of the ants' menagerie of guest-animals.

The simplest relation between a colony-parasite and host is one of simple thieving on the part of the parasite, unmitigated hostility on the part of the ants. One of the most amusing and impudent of such thieves is Lepismina, a wingless silver-fish insect. When one of the worker-ants is engaged in regurgitating a drop of food to a sister, Lepismina steals up, crawls below their upraised bodies, make a sudden snap at the drop of liquid as it passes, and bolts. Regurgitation is a ticklish matter, and demands rather a delicate pose on the part of the two ants engaged upon it. Before the aggrieved couple can disengage themselves for pursuit, the thief is generally well away towards safety. But Lepismina has always to keep out of its hosts' way if it does not want to be attacked and killed. There are some parasites, however, of which the ants take no notice. The maggot of the little fly Metopina, for example, is found in the nests of an ant called Pachycondyla, clinging round the necks of the ant-grubs. In these ants, the grubs are fed by a primitive method; pieces of insect-meat are placed in a natural trough which the grub grows near its mouth. When a grub is fed, round whose neck a Metopina is wrapped like a boa, the parasite uncoils, helps itself, and coils up again. The ant-nurses pay no attention to it at all; they clean it as if it were a part of the ant-grub. Apparently it has acquired the authentic nest-smell, and this is enough to make the nurses overlook the very abnormal shape of those of their charges which are being exploited. Then there are other parasites which, though the ants are perfectly aware of their presence, are tolerated. One of the queerest of these is the mite, Antennophorus, which rides about on the heads and bodies of its host. Although it is as big in comparison with its host as a small monkey in comparison with a man, up to four or even six may be found on a single ant. To secure food, these creatures employ a trick found in many ant-guests; they stroke their hosts so as to simulate the food-begging caresses of one worker soliciting another to regurgitate. And they get what they ask for. In all ant-guests which solicit food by stroking, the organs for soliciting and caressing are moulded into passable imitations of ants' antennæ. Sometimes these caressing organs are the parasites' own antennæ; but often, as in Antennophorus, one pair of legs is made antenna-like. It is as if a badger or a squirrel were to grow imitation human lips on the end of his tail to kiss us into giving him his dinner.

Still other parasites are fed for services rendered. One little beetle, for instance, called Oxysoma, seems to perform various agreeable toilet duties for its hosts, licking them clean and pleasantly stroking them.

Finally come the strangest gang of all—those parasites of the colony which are not only tolerated, not only paid with food, but looked after and protected by their hosts, sometimes before the ants' own flesh and blood. And the reason for this unnatural behaviour is the ants' boundless greed. The parasite exploits the cravings of its hosts by offering them a delicious secretion to lick. Worker-ants and termites thus share with man the unenviable distinction of being the only animals with vices.

INSECTS AND OTHER INVERTEBRATES

Most of these vice-exploiting parasites, these perambulating bars of the ant-world, are beetles. There are several hundred species of them; they all produce the special sticky secretion which their hosts like so much, and this always oozes out on to bunches of special hairs, which act as teats from which the ants can suck the juice. The beetles usually have had their antennæ altered so that they are like ants' antennæ. Thus the beetles earn a living by providing luxuries in return for necessities. Their relation with their hosts is a perversion of the normal food-exchange on which the life of the ant-colony is based. The ants, like many human beings, put luxuries first; at a threat of danger, the beetles are generally carried off to safety before the ants' own grubs and pupæ.

The most remarkable of all guest-beetles is Lomechusa, a parasite of the Red Slave-maker, *Formica rufa*, whose history has been unravel-

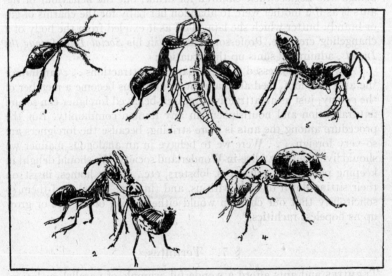

FIG. 301. Some guests and parasites of ant-colonies.

(1) *A beetle, Mimeciton, which closely mimics the Driver-ants with which it lives.* (2) *The beetle Atemeles soliciting a worker Myrmica to regurgitate food for it by caressing the ant with its fore-legs. In return it will allow the ant to suck the secretion from the tufts of yellow hairs on its abdomen.* (3) *The silver-fish Atelura rushing up to snatch the drop of liquid food that one Lasius worker is regurgitating to a second.* (4) *Another Lasius worker which is carrying three mites, Antennophorus, one under its head and one on either side of its abdomen.* (5) *A Honeypot worker with two little beetles (Oxysoma) attending to its toilet.*

led by the Jesuit entomologist, Wasmann. Not only does the adult beetle live by pandering to the ants' gustatory desires, but its grubs appear to exude some equally desirable bribe. The ants look after the beetle-grubs better than their own brood, and even allow them to eat the ant-grubs. As a result of the neglect of their beetle-infatuated nurses, a large proportion of the ant-larvæ perish; others are wrongly dieted and grow up as useless intermediates between workers and queens. A heavily-infected colony of the Red Slave-maker is doomed to extinction; but such heavily-infected colonies are rare, and the parasite itself sporadic and infrequent.

This perversion of the normal nursing instincts of the workers argues overpowering attractions in the beetles' secretions, comparable to those possessed for human beings by spirits or opium. But there is always some striking difference between insect and man. There are human beings who neglect their children for drink; but the behaviour of the ants is as if a mother were to abandon her baby for the charms of gin or brandy-butter which she lapped off as it exuded from the body of a changeling creature. Professor Wheeler, in his *Social Life Among the Insects*, admirably sums up the situation:

"Any insect possessed of these glandular attractions . . . can induce the ants to adopt, feed and care for it, and thus become a member of the colony, just as an attractive and well-behaved foreigner can secure naturalization and nourishment in any human community. But the procedure among the ants is more striking, because the foreigners are so very foreign. . . . Were we to behave in an analogous manner we should live in a truly Alice-in-Wonderland society. We should delight in keeping porcupines, alligators, lobsters, etc., in our homes, insist on their sitting down to table with us, and (in some cases) feed them so solicitously that our children would either perish of neglect or grow up as hopeless rachitics."

§ 7. Termites.

TERMITES and ants afford a wonderful example of parallel evolution. Termites, though sprung from a stock related to the cockroaches and differing fundamentally from ants in having no grub-stage and no metamorphosis, but hatching from the egg as babies of the same general build as their parents, have evolved independently a social life strikingly similar to that of ants. They, too, have sexual and neuter castes; the neuters are often divided into workers and soldiers; their cities are an irregular network of galleries and chambers; the community's economic existence is based on an elaborate system of food-exchange; they have numerous colony-parasites, many of which exploit their hosts' greed just as do the guest-beetles of ants' nests; and some species even cultivate fungi in a way almost identical with that of the fungus-gardeners among ants.

But when we look into the organization of the termite state, we find interesting differences of detail between it and the ant community. The neuters are not all neuter females; there are neuter males as well, in equal abundance with their sterile sisters. Then when we find soldiers in addition to workers in a colony, the two castes are not merely unlike in their proportions, as with ants, but qualitatively unlike. This and other evidence has led many biologists to believe that the difference between sexual individual, worker and soldier, in termites, is not an affair of feeding, as with ants, but is due to some elaborate hereditary machinery, such as that which controls the production of the two sexes in most higher animals. There is a further probability in this. Since the young, growing termites are free to feed where they will (in the gardener termites the young have been described as browsing like miniature white lambs on fungoid herbage of equal whiteness), it is difficult to see how their diet could be regulated. But full proof is as yet lacking.

Then not only are there neuter males, but the royal, sexual males help in founding the colony, unlike the male ant, whose services to the race are at an end with his single act of fertilization. The kings and queens, as in ants, fly out into the world when the time comes to mate; however, their flight itself is not a nuptial flight, but an opportunity for the young people to meet. A male and female join fortunes, descend to earth, both shed their wings, both set to and excavate a nest, and only then is the marriage consummated. The fully fertile queen usually grows to an unbelievable size, several inches long, her original slimness revealed by the patches of dark chitin which once made a continuous skeleton, but now are separated by areas of white, egg-distended skin. Some of these queens lay eggs almost continuously, at the rate of about one egg every two to four seconds, for many years.

Besides these fully-developed queens there may be numerous "second form" and "third form" females, which have reproductive organs, but are in other ways intermediate between the true queens and the wingless workers; these always remain unbloated. And there may be a similar gradation in the males. The function of these creatures in the colony is obscure; we know, however, that in some species, instead of soldiers, there is a special caste of *nasuti*, or "nosy ones," who constitute a sort of chemical warfare corps. These have long snouts on which open the ducts of special glands that secrete an unpleasant liquid which is extremely sticky, and possibly corrosive as well. With this the nasuti immobilize their enemies by gumming them up—legs, antennæ, and body—in a helpless mass. The size of these defence glands in some of these creatures is enormous, much bigger than all the other organs put together—a visible proof of the extraordinary degree to which division of labour has specialized the various castes.

One of the most curious and uncanny things about termites is the fact that all of them, save the winged sexual kings and queens, are subterranean creatures that shun the light. The workers forage for hundreds of yards round the nest, but always at night or else under cover of tunnels which they build for themselves as they go. When they reach a supply of food, whether it be a log in the forest or a piece of human furniture, they eat it out from the inside. Finally it becomes a mere shell and collapses at a touch. The workers are all pigmentless, fleshy, white creatures; the winged kings and queens, which will one day have to brave the light, are dark brown. The soldiers have huge, cast-iron heads, but soft and defenceless bodies; the hard heads are brown, the soft bodies white.

Parallel with this shunning of all direct contact with light and the outer world by termites has gone a specialization of their nests. Only in the more primitive termites do we find diffuse nests, underground or in dead wood, like those of most ants. In tropical forests there are many tree-nests, usually made like wasps' nests of chewed-up wood cemented by saliva. Some are as big as a good-sized barrel; others are plastered on to trunks like a cluster of sausages, and are protected from rain by a series of inverted carton V's built above them.

Tropical ants build tree-nests, too; but no ants have nests resembling those of some African termites—stalked objects protruding from the ground with a mushroom-like top acting as an *en tout cas* against both rain and sun. The highest developments of termite architecture are the giant termitaries (wrongly styled "ant's nests") of African and Australian tropical scrublands, which may harbour over a million individuals, reach a height of fifteen or twenty feet, and are so concrete-like in their strength and hardness as to be almost indestructible save with the aid of dynamite. In some parts of Africa the abundance of these nests makes the clearing of aerodromes a costly and laborious business.

It is in their food-economics that the termites are most extraordinary. In the first place, they live almost exclusively upon wood. This is entirely indigestible by most animals, so that the termites here enjoy a notable advantage in the struggle for existence. But, as we have seen, they are only enabled to do this by the aid of the swarms of symbiotic protozoa in their insides.

Maeterlinck, in his book on termites, holds up the termites' capacity to digest wood and the presence of neuter castes in their societies as proof of more than human intelligence. This is uncritical fantasy. The two types of existence stand on a different level. The termites, we can be completely certain, do not even know that they possess flagellates in their insides, or that these digest wood for them; and even if they did know, the presence of these admirable aids to digestion would be no more due to cleverness on the part of the termites than is the presence in our insides of that remarkable fluid, the pancreatic juice, due to our cleverness. And similarly with the castes. Workers and soldiers are what they are because natural selection has moulded the termite germ-plasm in a particular way, not because of any forethought or conscious planning by the insects themselves.

FIG. 302. The ravages of termites.

Only the workers and soldiers contain the precious flagellates, because only they eat wood. The rest of the colony receive their food from these wood-eaters. The details of their food exchanges we have already mentioned. Even their concern for their precious, swollen-bellied queen is in return for food. Hundreds of workers are generally to be seen in attendance upon the queen. It used to be supposed that they were actuated by the motives which prompt human worshippers. But once more the gulf between man and insect is revealed. They surround the queen because she exudes a specially rich and fatty secretion; and their apparent attentions consist in licking her to get something for themselves—sometimes so violently that they rasp holes in the royal side.

A piece of wood excavated by white ants at Singapore (Courtesy of the British Museum, Natural History.)

Termites do not spread so far out of the tropics as do ants, but where they are abundant they are of more economic importance. They are extremely destructive, not only to wooden fences and buildings and furniture, but to paper. And since ink and the written or printed word are no deterrents to their hunger, books and documents are continually being destroyed by them. Von Humboldt noted that in tropical South America books more than forty or fifty years old were great rarities; before attaining such an age they generally go to feed termites. Some authorities, indeed, believe that the termites' destruction of books is one of the chief reasons which have hitherto prevented tropical civilizations from reaching any pitch of progress comparable with that of more temperate nations. Our descendants will be able to judge whether this suggestion is true or no, for some tropical countries are now constructing libraries and archives with concrete foundations to make them termite-proof.

On the other side of the account must be entered the fact that termites (unlike ants) are beneficial to plant growth. Cellulose and wood are among the most resistant of organic substances. The termites, in combination with their indispensable flagellate allies, are among the few agencies which rapidly break them down. And thus much of the capital of life, which would otherwise be locked up for years, is speedily brought back by termites into the vital circulation.

§ 8. Bees.

IN ANTS all stages in the evolution of communal life have been lost, but in bees there survive many stages in the process. There are bees which are entirely solitary, where the females simply store up food for their grubs and leave them to hatch out untended. There is here no trace of social life, and yet it is the first step towards it, since the social life of insects has evolved in connexion with the care of the young. A further step is seen in bees like Halictus, where the female stays by her little nest and waits until her children emerge as winged creatures as big as herself. The children then add to the nest and throughout the summer there is a true colony-life. But autumn breaks up the community, and the females that survive the winter must each found a separate nest again. Here is no continuity of social life, nor is there any division of labour between castes.

The humble-bees illustrate the next step. The community still endures for one season only, but it shows the beginnings of a worker-caste. The females that were fertilized in the autumn before and have successfully survived hibernation, dig themselves an underground chamber in which they build a few irregular rounded cells. In the first cell they lay perhaps half a dozen eggs, put up with them a supply of honey and pollen, and seal the cell up. From time to time they open it again and give the grubs fresh food. But the solitary mother can only manage to provide the bare minimum of food, and the weight of the bees that hatch out is only some fifteen or twenty per cent of their parents'. Further, the machinery of humble-bee development is so arranged that the ovaries only grow properly when food is abundant. Accordingly these first stunted bees have rudimentary ovaries, and are sterile; they are in fact neuter workers. They help their parent in the duties of the nest; with their aid, food becomes more abundant, the state's new children bigger. As the season goes on, all transitions are thus produced between sterile worker-females and fertile queen-females; and by late summer queens are plentiful. Males also are produced in late summer; and though they do not help at home, they at least support themselves by gathering their own food; they are thus not so specialized as the males of ants or hive-bees, which have become merely the male gonads of the colony, nourished by the efforts of other members of the community. Winter kills off all the males and workers, since the reserves of food are but scanty.

In several ways the characteristic features of the hive-bees' advanced polity are foreshadowed in a half-and-half condition by the humble-bee. It possesses neuter workers; but these differ from the queens only in their sterility and their size, not in their construction and instincts. The queens still undertake other duties than that of a mere egg-laying machine, the males do not deserve the name of drones, for they still feed themselves. The females, like worker hive-bees, secrete wax, sweating out plates of it from between the crevices of their chitinous armour; but wax is not yet used as sole building material, being mixed with resin and pollen before use. The humble-bee does store up food; some of the empty cells from which workers have hatched out are used as honey-pots. But they build no special combs of cells all devoted to honey-storage, and the amount of stored food is small.

In other bees, further gradations to the fullness of social life are to be found; but we will pass straight to the consideration of the most specialized of bees and the only insect to have been domesticated by man, the hive-bee.

The bee-hive, like the ant-nest, is on the way to become a super-individual. It has the same division of labour; the workers work for the community, the queen lays eggs for it, the males are sacrificed in autumn for its good. It is in certain ways more of a unit than the ant-nest—it has one queen, not many. It shows its incipient individuality in another interesting way. Although its separate members cannot maintain a constant temperature, the community can and does. The bee-hive, at least in winter and spring, is for most purposes a warm-blooded organism. In winter, if the temperature falls too low, the bees set the heating machinery in motion. They assemble hanging from the roof in a dense bunch; when this cools to a temperature of about 55 degrees Fahrenheit, the bees get restless, they take a meal of honey, come back and crawl actively within the cluster. The heat produced by their movements cannot readily escape from the dense mass of bodies; and the temperature rises rapidly—within an hour it may jump from 55 degrees to 75 degrees. The outer bees which make the skin of the cluster cool off first; then they crawl into the interior and others take their place, until finally the whole cluster is cooled to 55 degrees again. This takes the best part of twenty-four hours; and then the heating process is repeated.

In spring and summer, however, when the development of the young brood is going on, the hive becomes fully "warm-blooded"; the bees keep it almost at blood-heat (actually 93 degrees to 95 degrees F.), whatever the outer temperature. At lower temperatures, the grubs not only develop more slowly, but abnormally; and temperatures much higher than this would be as fatal to bee-protoplasm as to our own. On cold days the workers who are looking after the brood crouch and crawl in a thick layer over the brood-cells, so constituting both a protection against heat-loss and a source of warmth. On hot days they fan their wings, drawing cool air through the hive.

It is often asked how any theory of inheritance can explain the facts of bee-life. The workers alone have the elaborate instincts which keep the business of the hive going; and yet they leave no descendants. The drones do nothing but fertilize the queens, the queens nothing but lay eggs; and yet they give rise to the gifted and industrious workers. The same question, of course, applies also to ants and termites. The question would perhaps not be asked if it were not for the unconsciously held Lamarckian beliefs about heredity which most people cherish: they are surprised that the workers can do things which none of their ancestors ever did. The difficulty is not so great as it seems; it disappears as soon as we think in terms of the modern idea of the germ-plasm. The case of worker-bees is precisely the same as the case of specialized tissues within our own bodies. Our muscle-cells, our blood or glands, our brain-tissues—none of these leave any descendants: they do their jobs in our economy in spite of the fact that they are descended from a long line of germ-cells which have never done anything of the sort, but have merely divided, become gametes, fertilized one another and divided again. So with bees. The queens and drones are the germ-plasm of the community, they alone are part of the immortal racial stream; the workers are its soma, doomed to work and die for it without direct posterity. As the same human chromosomes in two different parts of the embryonic environment will set to and help build up tissues as different as muscle and nerve, so the same bee-constitution with one kind of diet will generate egg-laying queens, with another busy neuters.

The instincts of the workers can be kept up to the mark by natural selection. Those fertile females whose genes under worker-diet do not develop into workers with proper instincts, will produce inefficient hives; such communities will go under in the struggle for existence, and so the defective genes will be eliminated from the bee germ-plasm. Here again the process is analogous to what happens in our own bodies. Those germ-cells whose genes give rise to inefficient thyroids or brain-cells will be eliminated from the race, although as mere germ-cells they may be adequate enough. Once more we see the community as the true unit of the bee or ant species, the single bee or ant as a subordinate member of this super-individuality.

There are numerous books, both scientific and popular, on the hive-bee. To them we must refer the reader for the picturesque details with which bee-life abounds. Here some interesting and very recent work of the Bavarian zoologist Von Frisch must suffice. His patience and skill have elicited many new facts, and have set many old ones in a new light.

The essence of his method has been intensive observation of individual bees leading their natural life. As there may be thousands of bees in a hive, and as they all look alike even to a trained eye, the first requisite was some means of identifying individual workers. To this end Von Frisch modified and improved the marking method which Lord Avebury was the first to use. He caught bees, and by means of combinations of dabs of bright colour on different parts of the body, gave them labels which made identification easy, even in flight. If a white spot on the thorax is taken to mean 1, the same colour on the abdomen to mean 6, red on thorax is 2, on abdomen 7, and so on, a glance at a marked bee enables you to write down its number. With the aid of this method, of specially-devised observation hives, and infinite patience, Von Frisch discovered how the numerous duties demanded of the workers—nectar-gathering, grub-feeding, comb-building, and the rest—were distributed. The reception of such paint-marked bees by their fellow-workers is worth noting, as it throws an interesting sidelight on bee-mind. No difficulty is made about old bees; but ones that are taken and painted immediately after emergence are roughly handled and usually thrown out of the hive or stung to death. The unfamiliar smell of the paint is reacted against unless the familiar nest-smell is superimposed. And if the young bee is smeared

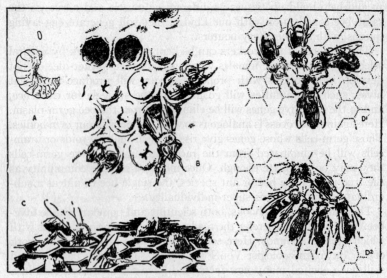

FIG. 303. Scenes from the life of the honey-bee.

(A) A bee's egg and a full-grown grub on the same scale. (B) Worker-bees emerging from their cells. Some cells are unopened; others are just being gnawed open; from others bees are emerging. (C) The workers' first duty: cleaning out cells for the queen to lay eggs in. (D1) A bee returned from honey-gathering regurgitating food to three other workers. (D2) Before flying off again she executes a dance over the comb, followed by other workers, who smell her with their antennæ.

with a little honey as well as being painted, the honey is licked off, and after this she is accepted without more ado.

In human communities, division of labour comes about by different people learning to do different kinds of work. You are not born a lawyer or a colonel; training and experience are needed. In the more specialized kinds of ants and termites everything, or almost everything, depends on heredity; different jobs are performed by creatures who are born different—soldiers, workers, and so forth. But in bees a third method is adopted. Queens, males, and workers are born different; but the workers' instincts develop and they take on different jobs

INSECTS AND OTHER INVERTEBRATES

as they grow up—as if human beings should automatically and without training change from nurse-maids into bricklayers, then into hall-porters, and wind up in business.

A queen-bee may live five years; but workers die of old age after about five weeks of winged existence. (To this must be added three weeks' development—three days as an egg, six days as a growing grub, and twelve days as a resting pupa.) This short span of adult life falls into three main periods. The first, which the bee spends entirely inside the dark hive, guided by smell and touch and hardly ever employing her eyes, lasts about ten days and is concerned with the care of the young. The second, from about the tenth to the twentieth day, is concerned with building, cleaning, and acting as guard at the main gate of the hive, with occasional short excursions. And the third, from three weeks old till death, is spent mainly outside the hive, and is concerned entirely with collecting pollen and nectar.

We may follow an individual bee through her life-cycle in a little more detail. On emergence from her pupa-skin and cocoon, she gnaws open the thin wax lid of her cell, which was sealed over her when she had ceased to grow and had turned into a pupa. Then she dries and cleans herself, and within an hour, all uninstructed, may have begun work. She begins cleaning out cells from which other bees have hatched, removing débris and licking their walls with her saliva. Only when cells have been thus treated will the queen lay new eggs in them.

But in the first three days the young bee sits about a good deal doing nothing in particular (the idea that bees are always busy is a delusion, which springs from the fact that we notice them at work, not when they are resting; to someone stationed in a central street of a great city, it would appear that human beings were always busy). Apart from cell-cleaning, the only other duty of the young bee is keeping the brood warm on cold days by blanketing it with its body.

When the young worker is about three days old, however, the nursing instinct stirs within her, and she begins feeding the older grubs with honey and pollen from the communal stores. The younger grubs, however, cannot digest raw pollen, and have to get their nitrogenous food from a secretion made in the worker's "salivary" glands. This is closely parallel with the feeding of the human infant on milk, and is one of the few cases of an invertebrate animal feeding its young on specially-secreted fluids. On about the sixth day of a worker's life, these glands suddenly begin to swell and grow, just as the milk-glands in the breasts of a woman grow in preparation for suckling her child. With the growth of the glands, there develops the instinct to use them; and from about the sixth to the tenth day the worker does little but nurse the young grubs. Then the nurse-glands shrink again, and the bee's impulses prompt her to new activities. She takes nectar from the older bees returning from their work among the flowers, and pumps it into honey-cells; packs the pollen tight in the pollen-cells; and removes dirt and débris a short distance outside the hive. In place of the shrunken salivary glands, her wax-glands have enlarged, and sheets of wax grow out and protrude between the plates of her abdomen. With this, her building instincts come into play, and she spends a good deal of time on the construction of new comb. Besides the ordinary cells which may serve either for honey, pollen, or developing brood, workers of this age may build cells of the same shape, but larger, whose greater dimensions stimulate the queen, as she pokes her abdomen into the cell, to lay an unfertilized instead of a fertilized egg; these are the drone cells, since unfertilized eggs grow into males. If the workers decide to bring up a female grub as a queen, they build a protruding and very large "queen-cell."

All through this period, the workers, prompted by a new restlessness, will occasionally leave the hive and take short flights to get to know the surroundings. These explorations are pushed farther and farther afield, until before she is three weeks old, a worker has a good knowledge of all the neighbourhood within a few hundred yards.

Before she goes off collecting from flowers, however, she spends two or three days on guard-duty—work that is on the dividing line between her previous life inside the hive and her future activities outside it. There are always a few guards or door-keepers near the entrance to a hive, some just within it, some just outside, and these are always eighteen to twenty days old; so accurately do the different instincts succeed each other in a worker's life. They investigate the entering bees with their antennæ, and will drive off any members of a strange hive. They attack marauding wasps or other creatures who try to gain entrance to make free with the community's honey-stores, and it is they who will fly out and sting big animals or human beings who carelessly approach the hive.

It is a well-known fact that if a bee stings a man, the sting sticks in his skin and is torn out, and the bee dies. But this does not happen when a bee attacks a creature its own size; the deadly sting (which is an ovipositor or egg-laying tube converted into a weapon, with certain of the attached glands made over into poison-secreting organs) can be used over and over again on another insect. This particular sacrifice of the worker in the interests of the community is only demanded on the rare occasions when large animals are to be stung.

Only after about her twentieth day does the worker begin food-gathering. All her work is now to visit flowers, suck up nectar, and collect pollen. The nectar she brings back in her crop or honey-stomach, vomiting the contents up again on her return, either into store, or usually to other younger workers, who themselves either feed others or pump the nectar into cells. Her honey-stomach is about the size of a pin's head; she must fill and empty it over fifty times to get a thimbleful of honey. And it takes the contents of over a thousand clover flowers to fill it once! When we remember that a single hive may store over two pounds of honey in a day, in addition to what it eats, we begin to realize the efficiency of the bee as a device, evolved late in the history of life, for exploiting the secretions of flowers.

On cold, rainy days, the food-gatherers do not go out much, but sit about on the combs (thus unconsciously helping to keep the hive's temperature up). Eventually they grow old, and spend much time doing nothing, even on fine days; and at last, if they have met with no accident, they die of old age, and their corpses are transported out of the hive by their sisters, and thrown away.

The most novel and interesting of Von Frisch's discoveries concerns what has been sometimes called the language of bees. It is so far a language in that it is a means of communication, that by its aid they convey information to each other about the available sources of honey and pollen. But as always, there is a radical difference between insect and human methods. The so-called bee-language is really only a chain of stimuli. It is based on a strange and elaborate system of instinctive or reflex actions, not learnt by a painful process of associating symbols with meanings as ours is, and the result, though effective for its purpose, is achieved without the aid of any powers deserving to be called intellectual.

Before the experiments on this means of communication could be planned, it was necessary to find out a good deal about the senses of bees. It was shown that bees' eyes, like our own, distinguish colour. But they do not distinguish it in quite the same way as we. For one thing, they see farther up into the ultra-violet than we do, but are quite blind to most of the wave-lengths which to us appear red. (That is why flowers visited by bees are very rarely pure red.) For another, though they distinguish between different intensities of illumination, different grades of black, grey, and white, as well as we do, their discrimination of colour-shades is a good deal less accurate. They distinguish readily enough between the whole group of colours included between blue and violet and the other whole group between orange-yellow and yellowish-green; but they cannot see any difference between the tints within each group—yellowish-green and orange are alike to their eyes. They can also distinguish one form from another, but not so thoroughly as we; interestingly enough, objects whose shapes recall the shapes of simple flowers are more readily learnt and remembered. They see quite well, but a good deal less well than birds or men.

Bees, again like other insects, have their sense of smell, as well as their most sensitive faculty of touch, in their antennæ; and experiment shows that they can distinguish various smells readily enough. Once bees have been taught to associate the smell of peppermint, for instance, with a supply of sugar-water, a drop of peppermint oil will attract them from a considerable distance. On the other hand, bees are only about as sensitive to smell as men, and fall far below dogs or certain moths in their olfactory capacities. They are even excelled by the professional perfume-experts employed at scent-factories when it comes to distinguishing between two very similar smells.

They have a brain-machinery which enables them to use these senses in quite a varied and effective way. They can be trained to associate particular colours and forms and smells with food, and to remember the association apparently for the rest of their short lives. By pretty experiments it has been shown that when both colour and scent have been associated with food, it is the colour which they look for and recognize while still at a distance, while the effect of the scent preponderates when they are close. If, for instance, they have been used to find food inside a small box painted yellow and scented with oil of roses, and then a number of similar boxes are set out, one unscented but painted yellow, another not painted but smelling of roses, they will begin by flying straight to the yellow box, but will check and refuse to enter it when they miss the familiar scent; however, any which then happen to come within range of the smell of roses will enter the scented box, though it lacks the familiar yellow colour. As we shall see later, this is what happens in nature; different flowers are recognized far off by vision; but smell is the final means of identification.

The bees' powers of finding their way are remarkable. If you catch a bee as it is entering the hive, imprison it in a box, mark it with a dab of colour, and liberate it a mile away, within a few minutes it will be at the hive again. Some observers thought that bees and other insects possessed a special "sense of orientation" through which in some un-

INSECTS AND OTHER INVERTEBRATES

explained way they were drawn back to the hive, but this is a baseless and needless assumption. The bee learns the lie of the land round the hive by exploration flights. The flights grow longer and longer, until a radius of two or three miles is thus learnt by heart; but if you take your captive bee beyond this radius, she is hopelessly lost.

Even when there are a number of similar hives in a row, the bees usually find their way back to the right hive. But mistakes are by no means infrequent, as you can convince yourself by marking a series of bees from one hive. Indeed, we should expect mistakes, for finding the right hive must be as hard as finding the right door in a long, drab street where the houses had no numbers. When a queen makes a mistake on her return from her marriage-flight the matter becomes serious; for she will get killed in a strange hive, and her own hive will be queenless and will peter out unless the workers manage to raise a substitute-queen. Accordingly, for years past, many bee-keepers have painted their hives different colours; but opinions were long divided over the value of this measure. The discovery of the limitations to bees' colour vision has cleared everything up. Colours do help the bees to find their way home—but, naturally enough, only if the bees can see them. Black, white, yellow, grey, blue—the bees can and do profit if these colours are used to identify their separate hives; but if you put bright-yellow next to yellowish-green or orange, or blue next to purple or violet, or red next to dark grey, it is no help to them, because to their eyes those colours look alike. Von Frisch records that in 1922, the monks of the Bavarian monastery of St. Ottilie, who have a very large apiary, begin to paint their hives in accordance with these biological discoveries. In the two previous years, sixteen out of twenty-one young queens had been lost; in the next five years, only three out of forty-two.

In the absence of colour-differences, the bees go by differences of shape and by surrounding landmarks, big and small. In addition, smell is employed to help. Bees have a special scent-organ near the hind end of their bodies; and it has been shown that the bees of different hives exhale scents which their sense of smell can distinguish. As with ants, home and patriotism are linked with smell. A number of them may often be seen near the entrance of the hive, using these organs to dab their special scent on the alighting platform, and then fanning with their wings, so as to drive the scented air out as a recognition-mark for their compatriots. And this is done especially in early spring when memory is dim after the winter, or a swarm has started life in a new situation, when the landscape is still unfamiliar.

Two other remarkable faculties help the bees to find their way home.

One is their power of guiding their flight by the direction of the sun. If you imprison a bee while it is feeding on a sunny day, and then liberate it after two, four or six hours, it will at once start off on a straight course, as if it knew the way for home; but the path it chooses is not the right one; it makes an angle with the right direction, and that angle is precisely the angle through which the sun has moved in the intervening time. This curious faculty of steering by the sun is not uncommon in insects; ants in particular, since they cannot fly, have to rely on it to a great extent.

We can thus prophesy the direction-line a temporarily-imprisoned bee will take on liberation. If we mark her, and post observers along this line, we discover a further remarkable fact. She does not simply go on and on in this direction; when she has covered a distance about equal to that between the point where she was captured and the hive, she ceases her straight flight and begins questing around in bigger and bigger circles until, if she is lucky, she finds the hive. The distance which she traversed on her outward flight is in some way recorded in her brain. This faculty of automatically recording distance (which probably depends in a bee on the number of the wing-beats she has made) is possessed by many animals, vertebrates as well as insects; not only is it prominent in ants, but when rats learn to run a maze, it is this on which they chiefly rely.

The bees' dependence on this "compass sense" and "sense of distance" can be dramatically demonstrated by imprisoning a feeding bee, at once walking with it to a spot some way on the other side of the hive, and setting it free. Guided by the sun, it now flies directly away from the hive; but stops and searches when it has flown the right distance. Bees in nature use any and every one of these methods of finding the way home—sight, smell, compass-sense, and sense of distance; the emphasis falls on one or the other according to the nature of the country round the hive.

With this knowledge about the capacities of their senses and their memory, Von Frisch could go in to the real kernel of his experiments. If you put out a source of food—a strip of paper smeared with honey, or (better for experimental purposes) a little dish of sugar-water—not too far from a hive, sooner or later it will be discovered. It is almost always some hours, and often several days, before this happens; but once one single worker has found it, you can be sure that within the next hour or two it will be visited by scores or hundreds of bees. Somehow the hive has been told.

Marking, watching and waiting enabled Von Frisch to unravel

the story. A bee that has found a new source of food flies back when her crop is full. In the hive she pumps up the shining drop of liquid and gives it to younger workers. Once rid of her burden, she begins a peculiar little dance, running with rapid steps in narrow circles over the comb, first in one direction, then in another. The dance, which always takes place in a crowded part of the hive, creates a good deal of excitement among the bees near the performer. They run after her, trying to touch her abdomen with their feelers. After anything from a few seconds to a minute of this excited dance, with its comet-tail of interested bees, the dancer stops. She may repeat the dance somewhere else, but soon goes off again to fetch more nectar, repeating the performance each time she returns with a full load.

It is easy to convince yourself that the dancer does not act as a guide when other bees follow to the source of food; and yet in a short time there will be plenty there. It looks as if the bees must have a real language, and that the dancer can tell her fellows that there is a good bed of clover under the hedge two hundred yards south-south-east— or a dish of sugar-water in the angle of the garden wall—information which they then act upon at their leisure.

Experiment dispels such illusions. If you provide a dish of sugar-water, mark the first few bees that visit it, and then put out half a dozen more dishes in different directions and at various distances from the hive, then although all the marked bees return only to the original dish, within an hour all the dishes will have been discovered by unmarked bees. The dance is simply a sign that food has been found; it stimulates honey-gatherers who have been sitting quiet at home, and impels them to go out and search; but it gives them no information as to where they should search. Von Frisch found that bees from quite a small hive, in hilly country when stimulated by the dance to go searching, would find his little sugar-water dishes put out in the meadows more than half-a-mile away from the hive. It is known that bees from larger hives in flatter country will visit flowers two or three miles off; and doubtless this represents the distance to which they will search. Take the sugar-water away, soon after it has been first found, and the bees behave as they would when a natural source of nectar fades; they dance no more on returning to the hive, and sit at home; and no fresh bees are stimulated to the search.

If instead of a dish of sweet liquid we put out one containing layers of blotting-paper moistened with sugar-water, the bees have great trouble in sucking it up. Apparently the poorness of the supply reacts on their feelings; at any rate, such bees no longer dance when they get home, but simply give what they have got to their younger sisters, and return straight away to the dish. The same happens in nature when a source of supply is drying up and the bees have to visit an unduly large number of flowers to get a load. It would be uneconomical for all the battalions of the hive to be mobilized for a skimpy supply of nectar. And through this simple arrangement—that the bees dance when they have had an easy time getting food, and do not dance when they have had trouble or difficulty—a rough proportion is ensured between the richness of the supply and the number of workers that turn out. So on a cold day, or at a season with few flowers, only occasional scouts will be going out from the hive. But if they find anything, their return sets a number of others in motion.

This, however, is only the A B C of the bees' code. Now come the subtleties. We proceed to attract a few bees to a dish of sugar-water as before; but we stand the dish on a piece of paper on which we have put a few drops of some strong scent, like peppermint-oil. We then put out a number of other dishes, two or three also perfumed with peppermint, others with other strong scents (essential oils are the best, such as oil of thyme or jasmine, bergamot or lavender), still others without any addition of scent. And now we find that only those scented with peppermint will attract any visitors; the rest are neglected. The experiment works equally well with natural flower-perfumes. In place of the artificial dish, we use a bunch of flowers, say phlox, each flower filled with a drop of sugar-water to ensure abundance of supply. Then, when a few bees have found this, we set out other bunches of phlox not provided with sugar-water, together with bunches of cyclamen. Some bees are busy on every phlox posy, but leave the cyclamens strictly alone, even if they are stood close alongside the phloxes. This is all the more remarkable since phloxes are adapted only to moth visits, and the short-tongued bees cannot reach the nectar in their deep corolla-tubes. Yet once the artificial treatment of one bunch of phloxes with sugar-water has associated phlox-smell and food-supply, the bees go on visiting all the phloxes they can find, searching busily for the nectar which ought to be there by all the rules of the game but isn't. After this has been going on for a time, we substitute for the sugar-watered bunch of phloxes a similarly treated bunch of cyclamens; gradually the number of bees on the other phloxes diminishes, and the cyclamens begin to be visited, until in an hour or so all are searching cyclamens. You can vary the experiment as you like; any scented flowers will do— beans or thistles, gentians or vetches—the result is always the same. The scent of the flower clings to the bees' hairy abdomen; the trail of workers behind a dancing bee returned from honey-gathering is sniffing at her (if the word sniff is permitted of antennæ) to get her adherent flower-scent.

Now we see that the apparently inconvenient arrangement by which the dance of a successful nectar-gatherer sends out other bees in all directions and not only to the original source of supply, is seen to be admirably adapted to the realities of the situation. The recruits are sent out in all directions; but they are only searching for flowers of one particular kind. In nature many plants of one kind will begin to bloom all over the place at about the same time. Thus the discovery of a single plant freshly in bloom will set large numbers of bees searching the countryside for others of the same sort. The method is as good a one as could be devised by exploiting as rapidly and fully as possible the different sources of honey which succeed each other during the summer season.

Finally, a further subtlety ensures that there shall be no undue waste of energy. We put out two dishes of sugar-water at the same distance, but in different directions from the hive. Each is soon discovered by a few bees, which then remain constant to their own source of supply. In place of one dish we then substitute a poor supply, in the shape of blotting-paper barely moistened with sugar-water. The bees from this dish, as we have seen, will not dance when they return to the hive. But those from the other dish will dance, and their dances stimulate new contingents of workers to fly off and search. We should expect that equal numbers of these fresh workers would visit both sources of food; but this is not so—the visitors to the moist blotting-paper number not more than a tenth of those that come to the easily available food. By close observation, followed by detailed experiments into which we cannot go, the reason for this was found. The different scents of different flowers act as so many signals in this code. But the worker-bees themselves also have a say in the matter. They, too, have a "scent-sign"; they can add this scent to that of the flowers, or they can withhold it. When present it emphasizes what the flowers have to tell; its absence weakens their appeal. The worker-bees have two scent-glands near the tip of their abdomen. These are usually tucked away inside the body, but they can be extruded, and then exhale a scent which to human nostrils smells like the plant called balm. When they are feeding at a rich source of supply, be it dish or flower, they keep on dabbing with these at their surroundings, and before settling they usually fly about for a little impregnating the air with this scent of balm.

This scent is the emphasis-note in the bee code. When bees are on the search for a flower with a particular scent, they will be moderately attracted by the scent alone, but they will be strongly attracted when the flower-scent is thus underlined and reinforced by bee-scent. This fact ensures that when a group of flowers has been thoroughly exploited by bees, it shall not have the same attraction for newcomers as flowers already discovered but not yet fully spoiled of their nectar.

This seems to exhaust the bees' code about nectar. But nectar contains no nitrogen, and nitrogen is as necessary for the repair of bee bodies as of our own. Flowers also produce pollen; and pollen is rich in nitrogen. Accordingly pollen-gathering is just as important a part of a bee's activities as honey-sipping. Nectar and pollen between them make up the whole of a bee's dietary. When a bee is going out after pollen, she first takes a little honey in her crop from the stores at home. Arrived at a flower, she scratches the pollen off the stamens with jaws and front legs, meanwhile bringing up a little honey from her crop to make the dusty yellow powder sticky. Her feathery body-hairs also catch pollen, so that she soon gets floury all over. With the elaborate combs on the much-enlarged first foot-joint of her hind-leg (as in other insects, her tools are part of her body), she rakes this pollen-flour off; then she rubs her hind-legs together below her body, and by the aid of a wonderful bit of machinery at the joint just above the combs, the pollen they have collected is pushed through on to the outside of the joint next above, which is hollowed out into a shallow trough bordered with long springy incurved hairs, and acts as a pollen-basket. With her middle pair of legs she pats the accumulating pollen into shape, until eventually she is carrying two great pollen-masses, each nearly as big as her head. So home to the hive, where she strips off the pollen-masses into a pollen-cell, leaving them to be pounded tight by younger bees; and then back for another load.

The division of labour in the hives goes deep. Bees that go pollen-gathering rarely search for nectar at the same time. And, like the honey-gatherers, they usually restrict themselves to one flower at a time. The methods by which their activities are regulated turn out to be extremely similar to those we have just set forth for the honey-gatherers. When they come back to the nest after a successful trip, they execute a dance (which is as distinct from the honey-dance as a waltz from a one-step), and this stimulates others to go pollen-hunting. The surprising new fact here discovered by Von Frisch is that each kind of pollen has its distinctive scent; this is different from the scent of the flower as a whole, which is usually exhaled by the petals. While the nectar-gatherers are guided by flower-scent, the pollen-gatherers go entirely by the smell of the pollen. This was prettily proved by cutting off the stamens of two different flowers, like campanula and rose, and putting the stamens in the wrong flowers. When a campanula with rose-stamens was provided and a few workers came pollen-gathering at it, their dance stimulated the bees at home to go out looking not for campanulas but for roses. Campanulas were not visited; the smell of rose-pollen prevailed over that of campanula-petals. Curiously enough, although the honey-dance and the pollen-dance are quite different, it is the scent adhering to the dancing bee, not the type of dance she executes, which decides the activity of those she stimulates. This Von Frisch proved by catching a bee who was drinking sugar-water and sticking a pair of pollen-bags of rose-pollen on to her legs. She flew home and, being agreeably full of nectar, executed the honey-dance; but the bees that were stimulated to go abroad went hunting pollen in roses. As with nectar, the pollen-gatherer who finds but scanty pollen does not dance on her return; but if pollen is abundant, she underlines the fact by scenting the flower and its neighbourhood with her own scent-organ.

The bees' code thus consists, first of twice as many "signs" as there are kinds of flowers in the neighbourhood adapted to being visited by bees; each flower contributes one sign through its general scent, another through the scent of its pollen. The colours and forms of the flowers also have to be learnt by the bees; they enable another sense to be used in the search. This mere signal-list is converted into a means of providing information by the bees themselves—the dancing after a successful trip, and the imprinting of their own scent on plentiful sources of supply.

Doubtless, as Von Frisch says, there are many other things to learn about the signalling of bees. The gathering of pollen and nectar is but one of their many activities; and there must, it seems, be methods of conveying information about comb-building, grub-feeding and other things which are done inside the hive. Here a fruitful field for experiment remains.

Certain central parts of the comb are reserved for the developing brood, with a zone of pollen-cells round them, and the honey-cells filling up the rest of the space—a lucky arrangement, since it enables the bee-keeper to provide us with pure honey-comb without admixture of grubs or pupæ.

When the hive's population has increased to a certain size, the workers begin raising a half-dozen or so grubs to be queens. A few days before they are due to hatch, the workers take a further decision; after a time of unaccustomed restlessness, about half of the thousands, filling their crops with honey from the stores, fly out in a wild cloud of circling, buzzing life, together with the queen. Soon the queen settles on a branch, and all the flight settles round her in a dense cluster, pounds of bees in a solid mass, constituting a swarm. A few scouts are busily flying about looking for suitable spots for a new nest, and if the bee-keeper does not speedily come and coax the swarm into one of his own hives, he will have the chagrin of seeing the cluster resolve itself once more into a cloud and fly away out of sight.

Among the population left behind, a new queen hatches out and after a week or two flies out on her nuptial flight, is fertilized once and for all by a single one of the pursuing drones (the race for reproduction between the rival males has led to their possessing powerful flight and very large eyes to keep the mounting queen in sight), and returns to the hive, which she will never leave save at swarming-time. If the population is only moderate, the workers kill all the other queen-pupæ. But if it is still large, the first-hatched of the young queens will depart with a new swarm, and then the workers keep the other queens in their cells, making a little slit in the roof through which the prisoners stick out their tongues to be fed, and keep the first queen away from them; otherwise she would sting them to death before their emergence, for unlike the queen ant, the queen bee tolerates no rivals.

As autumn draws on, the attitude of the workers towards the drones undergoes a change. As long as new swarms were possible and there might be new virgin queens to fertilize, the drones are carefully tended and fed (they are incapable of getting food for themselves from flowers), even if they chance into a strange hive. But now they begin to be bullied and nipped, and the workers pull them roughly about and throw them out of doors. They try to make their way in again to food and shelter, but are greeted with more bites and even stings. And so gradually, some by starvation and some by cold, some stung to death, this one-sided sex war ends with the death of all the drones, and the community, now solidly female, settles down for the winter.

In some ways the hive-bee communities are not so highly organized as those of ants; they have only one neuter caste, and their rigid dependence upon flowers for their food has prevented any such marvellous radiation into many ways of life as took place during ant-evolution. But in their buildings they are pre-eminent and their storage-system is better developed than in any other invertebrate animals. Their remarkable method of giving information so as to exploit to best advantage a great number of kinds of flowers also stands alone, though we can feel sure that fresh observations will reveal something analogous among ants and termites. But, as with ants and termites, there is no real comparison to be made between their communities and those of man. The two organizations are of different kinds, and rest on different foundations. The successes of human communities have been gained by abandoning as far as possible all the things to which the successes of insect communities are due. The power of automatically performing unlearnt actions, which is the basis of the acts of bees or ants, in man is nearly absent; individual flexibility and indefinitely growing community-experience take the place of rigid behaviour fixed by inheritance.

III

THE EVOLUTION OF BEHAVIOUR IN VERTE-
BRATES

§ 1. *The Vertebrate Nervous System.* § 2. *The Mind of a Fish.* § 3. *The Amphibian Mind.* § 4. *The Brain in Reptile, Bird, and Mammal.* § 5. *Courtship in Animals.* § 6. *The Evolution of Mammalian Intelligence.* § 7. *Education in Animals.* § 8. *Play.* § 9. *The Behaviour of Monkeys and Apes.*

§ 1. The Vertebrate Nervous System.

WE NOW leave the arthropods and their elaborate instinctive life and proceed to consider the evolution of that type of brain which is characteristic of the vertebrata and which leads up to and culminates in our own.

We shall have to go somewhat more deeply into anatomical detail than we have hitherto done. The human mind can be properly understood only if one has a knowledge of its material substratum, the brain. That brain is a very complicated organ, and though the triplex author will simplify his account as far as he possibly can, sparing his reader every technicality that can be spared, yet, even so, what follows will need close and attentive reading.

To begin at the beginning: all vertebrate animals are made on the same plan; all vertebrate brains consist of the same chief parts, although, as we shall note, the different classes vary enormously in their elaboration of detail and in the relative emphasis which they lay on the different brain-regions. We will start with the earliest and simplest phases of development.

In a young embryo, the first parts of the nervous system to appear are the great central exchanges, the brain and spinal cord, and, curiously enough, these originate as a tube of skin. We will choose that primitive vertebrate, the lancelet (Fig. 77), as an illustration of this point, for it lays eggs that contain very little yolk, and its development is therefore comparatively straightforward. In higher vertebrates

EVOLUTION OF BEHAVIOUR IN VERTEBRATES

there are various complications due to the presence of vast yolk stores, or to the growth of embryonic membranes, which we will do our best to ignore.

An early lancelet embryo, five or six hours after fertilization, is a roughly egg-shaped object, just visible as a whitish speck to the naked eye. If we were to cut it across, as cucumbers are cut across, the cut end would look like Fig. 304. The body is made of two layers of tissue, each only one cell thick. The outside layer (stippled) is the skin of the embryo. The inside layer (shaded) is the wall of its digestive tube; the cavity in the middle is the digestive cavity. Now it can be seen that the part of the layer which lies along the creature's back is a little thicker than the rest (shown by a darker stipple in the figure). This part is to become the nervous system. The skin is just beginning to grow up outside it, along its edges.

Fig. 305 is a similar slice of another lancelet embryo, about twenty hours older than the first. Several important changes have taken place in this brief interval. The one that most concerns us is that the rudiment of the nervous system has been overgrown altogether by the skin, and it has rolled up to form a tube which runs along the whole creature, from end to end. But while this special bit of skin has thus been turning into a spinal cord, a rather similar change has befallen the part of the wall of the digestive tube which lies nearest the creature's back. It has separated off from its fellow-cells and now lies as a cylindrical rod, seen in the figures as an oval shaded area, underneath the nerve-cord. This is the notochord—the elastic rod that serves as a backbone for these primitive animals (Book 2, Chap. 1, § 3g). We also note a third change. A new layer of tissue has appeared between skin and digestive tube (it arises as an outgrowth from the latter), and from this middle layer, shown cross-hatched in the figure, the muscles, blood-vessels, and connective tissue of the lancelet will develop.

In the higher vertebrates, the fact that the nerve-cord arises as a strip of embryo skin which separates off from the rest and rolls up to form a tube, can be demonstrated, although, as we have already said, there are complications. The hollowness of the central nervous system persists throughout life. If we cut across the spinal cord of an adult man or woman, the cut surface looks like Fig. 306. On examining it closely, a tiny hole can be seen in the middle. This narrow channel, which runs right along the spinal cord, corresponds to the bore of the nerve-tube in the lancelet. In the head it opens into a series of spaces, to which we shall return.

The channels and cavities inside the central nervous system are filled with lymph-like fluid, and they provide one of the ways by which nourishing substances can reach the brain-tissue. But apart from their utility in that direction they are of considerable interest, for they are a unique characteristic of our phylum. No other animals but vertebrates have hollow brains. The brain of an insect is a solid lump of nervous tissue. It is at least possible that this fact accounts for the superior mental development of vertebrates, for, as we shall note, this primitive tube has a way of blowing itself out here and there into pouches, with thin sheets of nervous tissues as its walls. This is a point of importance.

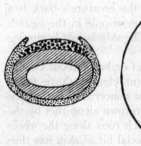
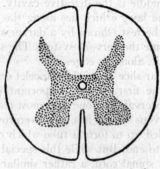
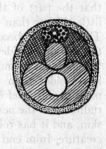

FIG. 304. The first stage in the growth of the nervous system.

FIG. 306. The human spinal cord cut across, magnified.

FIG. 305. The second stage in the growth of the nervous system.

Wherever brain-tissue collects together into dense, solid masses, whether in our own brains or the brains of insects, its working is stereotyped and automatic; apparently an efficient apparatus for plastic, educable behaviour can only develop if the grey matter is spread out thinly. So by making such spreading possible, the primitive hollowness of the vertebrate nervous system enabled its possessors to dominate the world.

There are one or two other significant facts about the spinal cord of an adult vertebrate. First, its substance is of two kinds. Most of it is white, but there is a central core of grey matter (shown stippled in Fig. 306), which has roughly the shape of an H in our cross-cut. As with the brain, the grey matter of the spinal cord is the more vital part; this is where the living telephone exchanges are situated. The white matter simply consists of the telephone wires—of bundles of nerve fibres running from point to point, and mostly up and down along the length of the cord. Moreover, the H-shaped arrangement of the grey matter is characteristic. The two horns of the H which point

EVOLUTION OF BEHAVIOUR IN VERTEBRATES

towards one's back (upwards in the figure) are receiving stations; it is here that the nerves from sense-organs in the skin, muscles, and so on, deliver their information. The two horns which point towards one's belly (downwards in the figure) are transmitting stations, whence nerves run away to muscles, glands, and so forth, and control their activities. Another point seen in the figure is that two deep clefts, front and back, divide the spinal cord almost completely into halves, one to each side of the body.

This central tube, then, is the first part of the nervous system to appear in an embryo, and from it the nerves presently sprout, growing and branching out through the body as the roots of a young seedling grow and branch out into the soil.

We have examined the spinal cord in some detail because the brain is really only the front end of the spinal cord made bigger and more intricate; and we can understand it more clearly by studying its derivation from this simpler part. There was once a medical student who described the brain as "a bit of spinal cord with knobs on." The lancelet has hardly any brain; the front end of its spinal cord barely differs from the rest. This is because it has few special sense-organs on its head; it lacks eyes, ears, and nose. Moreover, it is jawless and feeds by means of a filter in its throat, which is largely automatic and independent of nervous control. There would be little for the brain to do.

From an ancestor not unlike this we are descended. First special sense-organs developed, and then came the jawed mouth with a definite, discriminating bite instead of a continual automatic sieving of the water. Thus arose the need of a brain to control the jaws and to receive the reports of the sense-organs. As the latter became better and better developed, the brain became more and more dominant over the rest of the central nervous system. It grew to be the best informed part of the whole organization; gradually the other parts came to subordinate themselves, to send their own sense-information up to this main centre, and to take instructions from it.

In a vertebrate embryo, the brain first appears as a widening out of the front end of the nerve-tube. Strictly speaking, there are three swellings, not one, and they lie just behind each other. They are called the fore-brain, the mid-brain, and the hind-brain. Fig. 307 shows in profile the head of a chick embryo, about two days old. The front end of the nerve-cord curves forward like a crook, for the chick is hanging its head on its chest. The fore-brain is a hollow swelling, and the right eye can be seen lying over it. The mid-brain is another hollow swelling at the bend of the crook. Behind it is the hind-brain, and that melts gradually into the spinal cord. All these divisions run into each other without sharp boundaries and their cavities communicate freely, for they are only local inflations of the primitive nervous tube.

That is the vertebrate brain in embryo. In later development the three parts take different courses, and each grows in its own characteristic way. To illustrate the state of affairs in the adult vertebrate we choose the brain of the frog, which is drawn from above and from the left side in Fig. 308. It is a fairly primitive brain, and shows all the more important parts very clearly. A mammalian or human brain, although perhaps of more immediate interest to us, would be much less suitable, because in these higher types the two cerebral hemispheres grow enormously and overlap the other parts, so that their relations are hard to make out. We shall consider them later on. Let us take the three main divisions of the frog brain one by one.

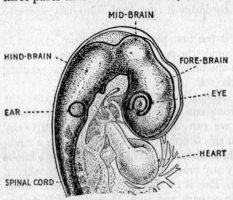

FIG. 307. The head of a chick, two days after fertilization.

The hind-brain is not clearly marked off from the spinal cord; the latter gradually broadens out into the former. Just as the spinal cord is a subordinate controlling centre with nerves running to the trunk and limbs, so the sides and floor of the hind-brain are subordinate centres with nerves running mainly to the face and throat. The movements of the jaws (and of the gills in a fish) are controlled from here. But there is a curious nerve called the Vagus ("The Wanderer") which differs from the rest; it runs back into the chest and belly and helps in the regulation of the viscera.

The roof of the hind-brain is, however, very characteristic (by "roof" we mean the part lying uppermost in a quadruped or a fish in the natural posture). In front, just behind the mid-brain, it sprouts out into a special mass of brain-tissue, the cerebellum. This, as we have seen, is an organ concerned with balancing the body and keeping it poised. In that flat, automatically stable creature, the frog, the cerebellum is small, as our figure shows; but in ourselves, poised precariously on two legs, it is large and complicated (Fig. 53). It is no accident that such an organ should have developed in this part of the brain, for here the nerves from the ear arrive, and the ear includes the most important sense-organs of balance in the body.

Just behind the cerebellum is another peculiar region, where the central canal comes up to the surface and spreads out as a wide, shallow trough. Over this the roof of the hind-brain is not nervous at all, but an extremely thin film of living tissue. Outside it is a dense network of blood-vessels. The reason for these dispositions is not far to seek. Oxygen and food substances diffuse from the blood through the membrane into the fluid inside the cavities of the central nervous system, and this helps considerably in fuelling and victualling the brain. The thinned-out part of the roof of the hind-brain is lightly stippled in Fig. 308.

The mid-brain has thickened, nervous walls, so that its central canal is in most animals narrowed like the canal in the spinal cord. The most important centres lie in its roof, and bulge up as the solid "optic lobes," of which there is usually one on each side of the brain. In most vertebrates these optic lobes are the chief centres to which the nerve-fibres from the eyes take their course, and they serve for that important function, visual reception. They are very conspicuous prominences in the frog. But in mammals they are superseded by the hollow cerebral hemispheres, and suffer a corresponding diminution in size. Moreover, the mid-brain is the seat of origin of nerves to the muscles which move the eyeballs, and is the main controlling centre of their movements. These are its best understood duties, but there are others which still await full analysis. In particular, the hind-brain, like the spinal cord, seems to be concerned in essentially unconscious and mechanical activity, but with the mid-brain we come to a higher plane. Much of the activity of this region also is simple automatism, at least, in ourselves; but in fishes and frogs the mid-brain seems to do many things that we relegate to our cerebral hemispheres, and even in mammals it may perhaps be the seat of conscious sensations of pleasure or pain.

So far the parts of the brain are fairly straightforward, and similar, except for very minor variations of detail, in all back-boned animals from fishes to ourselves. But the story of the fore-brain is more difficult. Here the organs of intelligent thought arise; here are found the greatest divergences between the members of the vertebrate series. And the fore-brain has a somewhat confusing way of sprouting out into hollow outgrowths at various points, which are not always nervous in function.

Very early in the development of an embryo it can be seen that the fore-brain is swelling out into two additional hollow bulbs, one on each side, much as the sides of the reader's face will swell out if he shuts his lips firmly and slowly puffs out his cheeks. These bulbs sprout out more and more distinctly and their connexion with the middle part of the fore-brain narrows down to a stalk like the neck of a flask. These two bulbs are not going to be parts of the brain at all; they are the first rudiments of the eyes. They grow nearer and nearer to the skin, and as they approach it the skin changes curiously and gives rise to the lens and cornea (Fig. 47). The outgrowths from the brain become the retina and the black pigment-layer outside the retina; their stalks grow solid and become the nerves which carry sight-impulses from eye to brain.

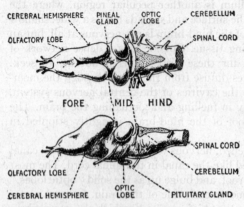

FIG. 308. The brain of the frog, seen from above and from the left side, to illustrate the chief parts of the vertebrate brain.

The cross-hatched area shows where the brain has been cut away from the spinal cord. The numerous little stumps are nerves, cut across a short distance from the brain.

Meanwhile, two smaller bulbs with thicker walls grow out from the extreme front end of the fore-brain. These ultimately will be connected to the nose by nerves, and they are the brain centres at which the impulses of the sense of smell first arrive. They are called olfactory lobes, and are seen at the front end of the brain in Fig. 308.

Another and even stranger growth springs from the floor of the fore-brain. A hollow protrusion, rather like the finger of a glove, sprouts out and grows downward and backwards towards the roof of the mouth. As it does so a bit of the embryo's palate separates itself off from the rest and moves up to meet it. The two organs, the bit of brain and the bit of mouth, unite together to form a compound organ, the pituitary gland (Book 1, Chap. 4, § 5). That the brain should participate in this adventure is altogether strange, for the pituitary gland has nothing to do with nervous activity, neither is it a sense-organ; as we saw in

Book 1, it is an organ of internal secretion. It influences growth and the general chemistry of the body; in frogs it plays an important part in the animals' changes of colour.

An interesting parallel case of entanglement between the nervous system and the ductless glands is found in another region of the body. The nerves running out from the central nervous system divide into branches of two kinds. Some run to the muscles of the limbs, trunk, face, and so forth which are under immediate voluntary control, and also to sense-organs in the skin and in deeper parts concerned with the position and movements of the body. These nerves execute our ordinary conscious movements. But others go to our more unconscious regions, and supervise such activities as the writhings of the digestive organs, the speed and force of the heart-beat, the tension of the muscles in the arteries, and so forth. Moreover, the nerves of this second group swell out here and there into knot-like swellings—subsidiary exchanges, on the telephone analogy—where a certain amount of routine work can be done without bothering the central nervous system.

Now, in a developing embryo, the rudiments of the nerves of this second group can be traced; and it can be shown that as they grow the knot-like swellings on them bud out tissue of a different kind. Besides giving rise to nerve-tissue, these germs give rise to a set of little ductless glands. In later development, most of the glands thus formed are gathered together into two main masses, one lying on each side of the backbone just in front of the kidney, called the adrenal glands. (As a matter of fact, the adrenal gland is a double structure with a main central mass surrounded by a distinct peel; the centre arises as we have described, while the peel has a different origin and different functions. Why they thus come together nobody knows; and the fact is a complication which does not concern us now.) This case is, however, not so puzzling as the case of the pituitary gland, for, as we saw in Book 1, the adrenal glands help in regulating the organs of digestion, circulation, and so forth. They are a labour-saving accessory of the nervous system. In a crisis when the organism has to fight or fly for its life it is obviously advantageous to suspend the operations of the digestive organs so that all the bodily energies can be concentrated on the effort; at the same time the circulation must be speeded up and made more efficient. This is done generally by sending a nervous message to the adrenal glands. The glands pour their secretion into the blood and it circulates everywhere, mobilizing the resources of the body as it goes. In the absence of such a device we should have to send special nervous instructions to every organ concerned, so these glands effect a great economy of nervous effort.

But we must return to the fore-brain, and take note of the fate of its roof. A large part of this roof is thin and richly supplied with blood-vessels; like the thin part of the roof of the hind-brain, it assists in supplying the central nervous exchanges with nutriment. But out of it there grows a little stalked knob, the pineal gland, showing very clearly in Fig. 308. This, like the pituitary gland below, is an organ of internal secretion. But it has a curious history. At one time the vertebrate stock had at least one eye—probably two—staring upwards from the middle of the head; to this day traces of those eyes can be made out in many amphibians and reptiles and in the lamprey, that relic from a remote past when our stock was jawless. Apparently the pineal gland is a forehead eye which first became blind and useless and then (at least in the higher vertebrates) was turned to another purpose, and made into a ductless gland.

The brain, like the rest of the nervous system, starts off in the embryo as a specialized bit of skin. Besides nervous tissue, this rudiment gives rise to two sense-organs, the eyes, and to a whole series of ductless glands. We will conclude our preliminary survey of the vertebrate nervous system by noting some vitally important, dominating exchanges that appear in the walls of the fore-brain as it grows.

In most vertebrates the fore-brain has only one direct nervous connexion with a sense-organ, and that is the nose. We have already seen how the primitive smell-centres push out in front as the olfactory lobes. True, the eye and its nerve grow from the sides of the fore-brain; but the nerve gets connected with the optic lobes in the mid-brain, and to them it sends all its information. Moreover, there are no outgoing nerves from the fore-brain to muscles or glands. All the effectors receive their nerves from farther back along the nervous tube. A glance at Fig. 308 will show that the hinder half of the brain (like the spinal cord) is thickly beset with nerves, while the front half is comparatively nerveless. In a word (except for its nerves to the nose), the fore-brain is free from the immediate routine business of receiving information and dispatching replies.

And now, in the light of the things we have already learnt about the brain, let us return to the story of how it evolved. We start from brainless creatures like the lancelet; as the vertebrate stock evolves, a biting, tasting mouth and a set of highly specialized sense-organs appear, and the front end of the nervous tube enlarges and reorganizes itself to meet the responsibilities thus put upon it.

EVOLUTION OF BEHAVIOUR IN VERTEBRATES

The mouth and throat come under the supervision of the hind-brain. Here the nerves to their muscles arise; here the reports from the taste-organs are considered. The ears also, and the lateral-line organs that we shall consider in the next section, send their information to this region. The mid-brain enlarges itself primarily in order to receive and analyze information from the eyes; it also regulates the eye-movements. At the front end of the fore-brain a centre grows up where impulses from the organs of smell arrive and are dealt with. Each of these departments of the brain is dominated by a particular sense-organ. Naturally they do not work in isolation and mutual independence; they intercommunicate; they are connected together by nerve-cables running within the substance of the brain, so that their various activities can be brought together and harmonized.

In the lower vertebrates, one or other of the executive departments may stand out over its fellows and become the leading part of the brain. Thus the trout is predominantly a visual fish. Its optic lobes are very large and they receive nerve-fibres from the other sense-centres; all such reports are considered but the final decisions are made in the optic centres, and impulses from the eye naturally play a dominating part in the conference. The dogfish, on the other hand, relies more on smell; the olfactory centres at the front end of its brain are enormous, and hither the other regions send their reports. As any keen angler will tell you, a trout has a critical eye for the appearance of the lure. But fly-fishing for dogfish would be a very different game; instead of tinsel and jay's feathers the lure would have to be ornamented but judiciously chosen scents. To take a third example, the carp (which has the gait and general expression of an incorrigible epicure) is a gustatory fish; it has very well-developed taste-buds in its mouth, and here the leading part in determining its behaviour is played by the distended taste-centre in the hind-brain.

But none of the highest vertebrates allow one sense thus to dominate over the rest. They cultivate them all, and keep a balance between them. Moreover, to guard against any unfair weighting of the scales they hold the conferences in neutral territory. They have developed special parts of the brain which are not under the sway of any one sense-organ, and to these new centres all the others send their reports. The obvious place for these new developments was the comparatively unoccupied territory of the fore-brain.

These new centres may be called centres of correlation, for they are not immediately concerned with reception from sense-organs, or with the dispatch of instructions to effectors. The sensory information is digested by other centres before it comes on to them, and their decisions are carried out by subordinate centres. To draw an analogy with a business organization, the new fore-brain centres are managers and dictate the policy of the body; the routine business of opening envelopes and writing letters, and even of making automatic reflex replies in easy cases which do not demand special consideration, is done by secretarial staffs elsewhere.

Crude rudiments of these centres of correlation are found in fishes, but they do not attain to their full dominance until we reach the higher land-vertebrates—the reptiles, birds, and mammals. They fall into three groups. First, there is a set developed in the side walls of the fore-brain near its hinder end. Chief among these is the thalamus. This structure cannot be seen in Fig. 308, for it does not protrude conspicuously from the surface of the brain; but it lies at the side between the optic lobe and the hinder end of the cerebral hemisphere. The second set appears on the floor of the fore-brain in front of the thalamus; it is known collectively as the corpus striatum. As we shall see, this set is the dominating part of the nervous system in birds. The third set is a series of sheets of grey matter which appear in the roof of the front part of the fore-brain. These constitute the cerebral cortex, which attains its highest development in ourselves. A frog or a fish has no cerebral cortex and we find its first beginnings in reptiles. The part of the brain labelled "cerebral hemisphere" in Fig. 308 consists largely of the centres which receive smell-impressions and of the somewhat rudimentary corpus striatum. But of all these parts we shall learn more in the following pages.

§ 2. The Mind of a Fish.

THE first vertebrates lived in water; we ourselves show in our own embryonic development that our bodies are the bodies of fishes turned and twisted about to fit them for a life on land. This is true of our brains, and correspondingly of our minds. Some of our sense-organs are still much the same as they were in our finned ancestors. The smell-cells in the nose, for example, are definitely fishy and will only work if they are immersed in water; so we find in an out-of-the-way corner of the cavity of the nose a special set of little glands, evolved when the vertebrates came on to dry land, whose business it is to secrete a film of moisture over the smell-cells—a tiny vestigial sea for them to work in. So let us throw our imagination back through the eras, and try to get an idea of what it was like to be a fish.

The most arresting difference between the behaviour of a fish and that of a land-vertebrate is the manner of muscular response. Instead of long jointed limbs the fish has a sinuous trunk and a powerful tail; instead of our standing, running, walking, gesticulation, manipulation, and so forth, it has a series of undulatory movements. It does not know the feeling of planting its feet on firm ground, for its weight does not greatly exceed that of the surrounding water; instead it swims through a medium somewhat heavier and more resistant than air, and its muscular energies largely consist in a rhythmical sideways pressing of body and tail against that enveloping substance.

Running along the sides of a fish's body is an important sense-organ that we lack completely—the lateral line. This is noticed as a white stripe along nearly the whole length of the dogfish, from the tail to the back of the head, in Fig. 68. When it gets to the head it divides into two or three branches, which run about in the cheeks, chin, and snout, and make patterns which vary in different kinds of fish.

On examining the lateral line closely we should see a row of fine pores all along it, but larger and more conspicuous on the head than elsewhere. These pores open into a common canal, which lies just under the skin. Inside the canal there are bunches of sensitive cells rather like the sensitive cells of the inner-ear. As a matter of fact, there are grounds for believing that the ear is really only a bit of the lateral line enlarged and glorified for its special functions—much as our cerebral hemispheres are enlarged and glorified bits of spinal cord. But that is by the way.

Just what this lateral line does is by no means clear. Some writers believe that it detects currents, others that it is sensitive to pressure, others again that it responds to slow pulsing vibrations of the surrounding medium. Perhaps the most plausible suggestion is that it feels the force with which the fish's flanks press against the water. The animal progresses by means of a series of wave-like sideways pushings of great strength and delicacy—for they have to be nicely controlled if it is to go straight. We may assume that it has a delicate sense of the contact between itself and the water, responsive to every slight variation in intensity, and that this sense is one of the regular pervading elements in the background of its mental life.

The eyes of fishes are less well developed than our own, and this is only to be expected, for water is not so favourable a medium for vision as air. Generally it is more or less turbid, either because of suspended mud or sand, or because of the very intensity of minute floating life in it; even in the exceptionally transparent water of coral islands it is not possible to see very far. To a fish distant objects are perceptible as looming shadows—even with our own refinements of the eye they would be little more—and clear vision is only possible at close quarters. Accordingly we find an interesting difference between a fish's eye and our own. We ourselves, when we relax our eyes, focus them on distant objects; that is the natural position of the parts to which they automatically tend, and to examine close objects, as in reading or writing, can only be done by exerting certain little muscles which alter the shape of the lens. But the machinery of accommodation in the eye of a fish is different. Here, in the relaxed position the eye is focused for near objects, up to a foot or so away, and the focusing-muscles work in the opposite direction. By straining their eyes big fishes can focus them on objects up to some thirty or forty feet in distance. But the figures depend, of course, on the size of the fish. Little trout, about six inches long, can see clearly for about a foot or eighteen inches. Outside that radius their world is hazy and fades away.

Can fish smell? The point is often disputed, for our noses detect perfumes borne in the air, and on the surface of things, at least, it is hard to imagine a fish as smelling in an analogous way. But the difficulty disappears when we remember the little bit of sea in our noses in which our smell-cells are immersed. Odorous substances from the air have to dissolve in that film of fluid before they can be perceived at all. As a matter of fact, it can be proved that in fishes the senses of smell and taste are as distinct as they are in ourselves. Such substances as we know as odorous stimulate the nose and the information is conveyed thence to the olfactory lobes at the front end of the brain. Substances which are bitter, sour, and so forth and stimulate our own taste-organs are perceived by similar organs in fishes, and in this case the information is received by a centre in the hind-brain. But in a fish there is no reason why the taste-buds should be confined to the mouth, for the whole skin is moist and suitable for their operation; and we do indeed find that many fishes taste with other parts of their anatomy. In cat-fishes, and in the carp and its relatives, taste-organs are found over the whole body, being especially plentiful in the little "feelers" and barbels that often project downwards from the chin and are used to examine the mud and to detect any smaller animals or other appetizing morsels. If you quietly drop a bit of meat close to the flanks of a cat-fish, the animal will twist round and snap it up at once, in virtue of its diffuse powers of taste.

Moreover, there is a third chemical sense in fishes, distinct from these two. This so-called "common chemical sense" is stimulated by

EVOLUTION OF BEHAVIOUR IN VERTEBRATES

various irritating substances and by alterations in salinity, and it is apparently much less discriminating than taste or smell. It has no specially constructed sense-organs but depends, as pain depends (Fig. 41), on free nerve-endings in the skin. But it is a sense distinct from pain. Newts show the same common chemical sense. In the higher land-vertebrates it is largely lost; the dry, thick skin that they find necessary to prevent excessive evaporation of water makes such a sense an impossibility. But just as smell and taste linger on in special moist corners of the body, so the common chemical sense lingers in such places as the inside of the nostrils and the cornea of the eye. The smarting of the eyes, nostrils, and mouth, produced by ammonia vapour, is an instance of its operation.

For the chemical senses, then, fishes are well equipped. Their ears, on the other hand, are primitive. They have, of course, the apparatus of three semi-circular canals at right angles to each other with which all vertebrates balance themselves, but the part which corresponds to our cochlea is so poorly developed that it is doubtful whether they can hear sounds at all. The fish ear lacks the apparatus of outer-ear, ossicles, and ear-drum that collects vibrations and transmits them to the inner labyrinth in ourselves; embedded in the skull, the latter can only be stimulated by vibrations that have passed through the tissues of the head. Fishes can be trained to come for their food when a shrill whistle is blown or when a bell is rung; but Parker has shown that similar reactions can be brought about by jarring the side of the aquarium with a pendulum as stimulus. In both cases the response is abolished by cutting the nerve to the ear. It may be that a loud sound affects the fish like a sort of fluid jolt, a sudden disturbance of the water, and no more than that. Their ears are, perhaps, more tactile than auditory.

In the matter of other senses—touch, pressure, pain, and so on—fishes are probably very much as we are, save for an outstanding exception. Even if they are sensitive to temperature they can seldom experience such sensations as the sudden warmth of a ray of sunshine or the chill of a breeze, because of the very uniform temperature of the water in which they live.

As for the intelligence of fishes, most people who have studied them carefully are agreed that they are stupid and learn but slowly. One or two instances of their educability are worth quoting. Triplett kept a pike in an aquarium with a number of smaller fish, but he separated the two by means of a glass plate. Soon the pike learnt that to leap at the other fish was to get a sharp blow on the nose—although it could not possibly understand the reason, for the glass plate was perfectly invisible. Presently the glass plate was removed, and the pike swam round with the other fishes. But it never made any attempt to seize them.

Thorndike put fish at one end of a glass tank, with an attractive shady corner with food in it at the other. In between were a series of glass partitions, each with a hole in it in a different place. The fish felt their way to the other end of the tank by bumping up against the glass plates and, after a considerable amount of trouble, finding the openings and getting through. But after a number of trials they learnt where the invisible openings were situated, and swam directly to the goal. Recently Bull has begun to analyze the learning-powers of fishes by means of Pavlov's method of conditioned reflexes (of which we shall learn more hereafter), and he is getting surprisingly good results.

A fish, then, can learn a certain amount, albeit with labour and slowly. But it lives in a curiously circumscribed world. Light soon loses itself as it travels through the water; sound-waves also travel badly in the heavy medium. Even smell is a short-distance sense in a fish and not a long-distance sense as it is in a dog or a deer. Water currents are very different from breezes, and fish sniff at things and thus find their prey, but do not scent it from afar. One of the most important things that happened to the vertebrate mind when our stock came on land was an enormous extension of the radius of the perceptible world.

§ 3. The Amphibian Mind.

BEFORE we go any further we must take up again our account of the vertebrate nervous system, and note an important difference between the fore-brain of all land-vertebrates and that of most fish. The rest of the brain need not detain us, for it is very much the same in all vertebrates; they all have underlying their conduct the same basis of automatic reflex behaviour. As we proceed the fore-brain will force itself more and more exclusively on our attention.

Fig. 309 is a rough diagram to show the essential points at issue. On the left is the fore-brain as it appears in most fish. It is drawn from above. The front end of the mid-brain (M) is seen behind; this leads into a hollow swelling (F) which is the fore-brain. At its extreme front two pouches (O) have grown out; these, as we have seen, are the olfactory lobes. The chief nervous exchanges are indicated as stippled areas in the figure. First there is a mass of tissue concerned with the reception and analysis of impressions from the nose. This appears as a U-shaped area coming back from the olfactory lobes. In most fish

EVOLUTION OF BEHAVIOUR IN VERTEBRATES

it lies as a conspicuous swelling on the floor of the fore-brain, and near the middle line. At the sides of the fore-brain some other centres can be seen, especially near its hinder end. These are not intimately bound up with any of the sense-organs; they are the rudiments of those superior centres whose evolution we discussed at the end of §1.

Contrast with this the drawing on the right, which is a general scheme of the fore-brain in land-vertebrates. Instead of being a single chamber the fore-brain has nearly divided itself into two, so that now it has the shape of a Y. Behind is a region (T) called the between-brain, which remains undivided through life. In front of this is the divided part; its two halves are called the cerebral hemispheres. (They are clearly seen in Fig. 308, but the frog differs from most land-vertebrates in having its olfactory lobes joined together in the middle line.) Properly speaking, a fish has no cerebral hemispheres, although it has the undivided parts corresponding to them.

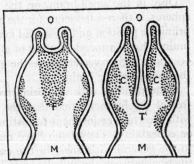

FIG. 309. The evolution of the fore-brain (see text).

(M) is the mid-brain, (F) the fore-brain, (O) the olfactory lobes, (T) the between-brain, and (C) the cerebral hemispheres.

Just why the front part of the brain should bisect itself in this way is not known; but the resulting Y-shape of the fore-brain runs through all the animals with which we now have to deal. Oddly enough, it first appears in lung-fishes, those strange links between water-life and land-life.

Notice how the main nervous exchanges come to lie when the partial bisection has been done. Those concerned with the sense of smell have been divided by the incision and now lie in the inner wall of each hemisphere, and rather towards its floor. In that position what we may call the "nose-brain" is found in all vertebrates from the amphibians up. The superior exchanges are enlarged and elaborated in the brains of land-forms, and in our diagram we can distinguish two pairs of these. The first pair lie one in each of the cerebral hemispheres; they are thickenings of the outer halves of the floors of the hemispheres. Each of these is called a corpus striatum. The second pair lie at the sides of the hinder, undivided part of the fore-brain; each is a set of three or four centres, of which the most important is the thalamus.

This brings us to the fore-brain as it occurs in the most primitive land-animals, the amphibians. The cerebral hemispheres are largely occupied with the sense of smell. Over most of the walls of the fore-brain, where they are nervous in structure, the grey matter lies up against the internal cavities, and outside it is a layer of white matter. That is to say, most of it is organized on the same plan as the spinal cord.

Only in the smell-brain, on the inside walls of the cerebral hemispheres, is there a tendency for the grey matter to move away from the primitive position and to spread evenly through the whole wall of the brain. This is a foreshadowing of a most important development that occurs in higher vertebrates—the growth of layers of grey matter on the surface of the brain.

Corresponding with this primitive arrangement of the highest centres of the brain we find that the amphibians are stupid creatures, with a certain endowment of instinct but with very little intelligence or educability. Those who have made pets of newts or frogs testify to the general witlessness of the creatures. There is, however, a distinction between the tailed and tailless amphibians. Newts and their relatives still wriggle precariously on the border-line between water and land; their tails are not very powerful for swimming compared with the tails of fishes, and their limbs are small and feeble. Frogs and toads are altogether more competent creatures. Except for their larval history, they are true land-animals with powerful legs. They are not nearly so dull as newts; and their brains (especially their optic lobes and thalami) show a better development of the leading nervous exchanges. But even frogs are morons compared with reptiles.

Nevertheless, a study of amphibian behaviour illuminates the contrast that we have drawn between life in water and life on land. The lateral-line organs, for example, linger on in aquatic amphibians—in tadpoles and many newts—but disappear in the definitely terrestrial frogs. Again, true hearing and the voice appear as the vertebrates move into a light gaseous medium which can convey sound-waves adequately. Most newts are deaf and dumb; but frogs have ear-drums, and a single ossicle connecting each drum with the labyrinth of the inner-ear; and frogs, as anybody knows who has lived near a frog-populated pond, have vocal cords of which they make good use. Some such chorus of croaks, trills, grunts, and metallic clicks must have echoed in the warm Coal-measure marshes when, with the amphibian, the voice came into the world.

Eyes are small or absent altogether in newts, but in frogs and toads

EVOLUTION OF BEHAVIOUR IN VERTEBRATES

they are well developed, and are the chief sense-organs by which the creatures hunt their prey. Even so, the amphibian eye at its best seems to give a very blurred image. Frogs and toads will not touch their prey unless it moves. A hungry frog sees a fly crawling along; he sits up alertly and takes a few steps towards it. If the fly stops still, nothing further happens; but as soon as it moves again, out flicks the frog's sticky tongue, and the fly disappears. Not only is a motionless insect invisible to the frog, or at least not recognized as possible food, but any small moving object will elicit the reflex snapping movement. Frogs in a glass jar can be made to snap at the end of a pencil pulled along the glass from outside, and will do so again and again, learning nothing by the fruitlessness of their efforts.

Indeed, although amphibians can be induced to learn under experimental conditions, their powers in this direction are comparatively poor, and cannot play a very important part in their normal life. They live almost entirely on instinct, such as the impulse to snap at any small moving object, which keeps them fed. Sometimes they have surprisingly elaborate instincts.

There is a Brazilian tree-frog called the "ferreriro" or smith (because its voice sounds like slow hammering with a mallet on a copper plate), which builds special little nurseries in the shallow muddy water at the margin of ponds to keep its tadpoles safe. The female, who does the work, plunges down and brings up an armful of mud, and out of a series of armfuls constructs a little circular wall. Observers have described the careful way in which she smooths the inside of the nest with her hands. Another Brazilian tree-frog lives in willows near the water and never comes down even to breed. It lays its eggs in jelly-like masses wrapped up in a sort of nest made by sticking two or three leaves together. As the tadpoles develop the jelly softens and liquefies until finally they plop down into the water below.

The courtship of newts is an interesting display of instinct which any country-dweller may witness by improvising an aquarium. In spring the male develops a magnificent crest on his back and tail and his colouring becomes more brilliant, with patches of black, orange, iridescent blue and white that vary from species to species. Putting himself across the path of a female, he "dances" and waves his crest in front of her, with a curious sideways bend of his body so that his fringes hang almost over her head. Then he turns his back on her, takes a few steps away, and, with violent wavings of his plume-like tail, he very conspicuously extrudes a little white packet of sperms. If she has been affected by his courtship she steps towards it as it floats in the water, seizes it in her hands, passes it to her feet and inserts it into the opening of her oviducts.

But these fine elaborations of instinct are behaviour climaxes, occurring only at special times of the amphibian life. Usually, their daily round of feeding and so forth is comparatively simple. They survive individually not so much because of their activities as because of their general timidity and furtiveness. Usually they live obscurely hidden in damp corners; most land-living frogs and toads come abroad at night, and day-frogs often have striking powers of protective colour-change. As a result of their marginal existence on the fringes of water and land, it is interesting to note that less than three thousand species of amphibians are known, as opposed to over five thousand reptiles, thirteen thousand mammals, twenty thousand fishes and twenty-eight thousand birds. The amphibian is on the whole an ineffective creature, neither truly aquatic, nor truly terrestrial. Partly in virtue of its slimy venomous skin, and partly in virtue of its prolific spawning, its race manages to muddle along.

§ 4. The Brain in Reptile, Bird, and Mammal.

THE first land-brains must have been very like the brains of our modern newts and salamanders, the most primitive of living amphibians. In lung-fishes we find very similar brains. It is a simple, inefficient type, as we have seen, and corresponds to a general dullness and limitation of the possible modes of behaviour. In many respects amphibians and lung-fishes are less well equipped mentally than the majority of modern fish. From an evolutionary point of view, they represent a stage when the attention of the stock, to speak very metaphorically, was focused on those alterations which made land-life possible—on the shift from gill-breathing to lung-breathing, with its consequent revision of the whole layout of the throat, the heart, and the main arteries. It may be their very mental inadequacy which, by putting them at a disadvantage, compared with their competitors in the waters, forced them to make these adjustments and come on land.

But when the vertebrate body had been properly fitted for a terrestrial life, when the dry impermeable skin and the protected egg, which characterize all vertebrates above the amphibian level, had been evolved, there arose a new era of intense competition on land. Terra firma ceased to be a refuge for the witless, and, except for a few out-of-the-way corners, became a hard school in which only the nimble and vigorous, or the exceptionally well protected, could hope to survive.

The whole apparatus of behaviour was improved. The vertebrate body was raised from the ground; instead of crawling on their bellies, reptiles began to run, leap, and fly. Sense-organs, and especially the eye and the ear, got progressively better and better. There was a general speeding up of the metabolic processes of the body; the heart and arteries were made more efficient; the lungs became more elaborately honeycombed to increase the rate at which oxygen could be absorbed; the kidneys were reorganized and their ducts were disentangled more completely from the reproductive apparatus than had previously been the case. This general revision culminated in the birds and mammals. Meanwhile there was a necessary correlated increase in the elaboration and efficiency of the central nervous system.

The most essential difference between the brains of amphibians and reptiles is the appearance in the latter of a new way of arranging grey matter in the walls of the cerebral hemispheres. It is a departure of the profoundest significance. In an amphibian fore-brain the grey matter is mostly massed next to the inner cavities in the same primitive position as it occupies in the spinal cord. But in reptiles, and in all higher vertebrates, thin grey sheets appear in the roofs of the hemispheres, separated from their cavities by layers of white matter of varying thickness. In ourselves the sheets lie right at the surface of the cerebral hemispheres (Fig. 54), and they are therefore spoken of as the cerebral cortex ("cortex" means "peel").

It is important for us to realize the distinctive features of this cerebral cortex. The first is its general design—it is laid out, as we have already stressed, in thin sheets and not in dense masses, and this is necessitated by its highly characteristic microscopic structure. To penetrate into the minute details of that would be laborious and take us further than we need to go. The second point is its position. We saw that the primitive vertebrate brain consists of a series of centres intimately connected either with sense-organs or with groups of muscles, and of tracts of fibres running telephone-like from centre to centre, and enabling them to communicate with one another. But soon special centres of a superior kind are evolved, not dominated by any immediate connexion with a sense-organ or an effector, but communicating only with other nerve-centres, not occupied with immediate reception or control, but supervising and regulating the rest of the brain.

Now evidently it is of advantage for such presiding regions to be somewhat aloof anatomically from the subsidiary centres they control —just as it is a good thing to have parliament in a separate building of its own and not mixed up with one of the executive departments such as the War Office or the Ministry of Labour. Herein lies the importance of the roof of the fore-brain as a site of the cortex.

Let us recapitulate very briefly the general layout of the brain to make the point clear. At the front end of the fore-brain lie the nerve-centres connected with the nose. To return to the parliamentary analogy (for the central nervous system is indeed the government of the body state) this is the Ministry of Smells. Further back, in the mid-brain and hind-brain are the Ministry of Vision, the Ear Office, the Ministry of Position and Equilibrium, the Ministry of Respiration, the Ministry of Abdominal Contentment, and so forth. These are all routine executive organizations. They communicate with each other by massive tracts of nerve-telephones; and in particular we may note tracts that run up along the floor and sides of the fore-brain to keep the other departments in touch with the Ministry of Smells.

In the simplest brains, the roof of the fore-brain is not primarily nervous; as we saw in the first section of this chapter, it is mostly thin and nutritive in function. The first superior regulating centres develop along the tracts between the nose-brain and the rest. Here they find space enough, unoccupied by the executive centres; here, also, they are able to inform themselves and to regulate, because of the tracts of nerve-fibres. This is how the thalamus and the corpus striatum arise. But when they have appeared, on the sides and floor of the fore-brain, a new possibility follows. Hitherto the development of the roof into a controlling centre has been impracticable; but now fibres can grow up into it from the corpus striatum and thalamus, and the cortex can develop out of the way of the main lines of communication between ministry and ministry, using the corpus striatum and thalamus as intermediaries for its information and decisions. Once this step has been taken, to develop the human brain becomes only a matter of elaboration and refinement; all the essential zones are present.

The cerebral cortex is found in reptiles, birds, and mammals. Its appearance is a tremendous stride away from simple automatism towards intelligent behaviour. But, curiously enough, its possibilities are only fully exploited in the mammals. For the brain in land-living vertebrates evolved along two different contrasting lines. One leads through the dinosaur-like reptiles to the birds; the other leads to the mammals. For a time they evolved together, but soon they separated.

In the bird development, the principal evolutionary change was the

EVOLUTION OF BEHAVIOUR IN VERTEBRATES 1221

enlargement and complication of the corpus striatum. That part of the brain may be said to culminate in birds. Nowhere else is it so elaborately organized; nowhere else does it outweigh the other parts so markedly in size. But while the floor of the fore-brain was thus undergoing improvement, the roof remained thin and the cortex small. In the mammal direction, precisely the opposite occurred. The corpus striatum did, indeed, undergo a certain amount of revision and reconstruction, but the main progressive change was the enlargement and

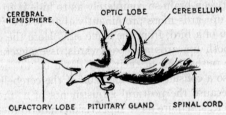

Fig. 310. The brain of a lizard.

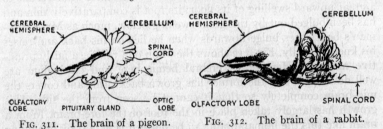

Fig. 311. The brain of a pigeon.   Fig. 312. The brain of a rabbit.

The brains are drawn from the left side. (Compare Fig 308.).

elaboration of the cerebral cortex. As we shall see in a moment, this difference underlies a very striking contrast between the behaviour of birds and that of mammals.

Modern reptiles are more bird-like than mammal-like in their brains. They are, of course, a mixed bunch, scattered and often only remotely related survivors of an enormous group which at one time dominated the world. But our own stock broke away very early indeed from the main trend of reptilian evolution, and took a different direction. Both from the brains of existing reptiles and from the skulls of fossil ones we can see that all through the Mesozoic Era the dominant vertebrates were enlarging the floors of their fore-brains and neglecting the cortex. This occurred most markedly in the dinosaur stock which led up to the birds, and to which the crocodiles are closely related. The tortoises and turtles diverged at an early date, and in their brains we find the same tendency, but in an altogether less exaggerated form.

Let us illustrate these points concretely, by considering three actual brains. Fig. 310 shows the brain of a lizard seen from the left side; on comparing it with the frog's brain in Fig. 308 the increase in importance of the cerebral hemispheres will be obvious. Already in the frog there is a perceptible tendency for the hemispheres to bulge backwards over the other parts; but in the lizard they have reached to the optic lobes and overlap the between-brain so completely as to hide it from view. Also they protrude upwards more prominently than in the frog. Fig. 311 shows the brain of a bird, from the same side. Here the growth of the hemispheres, both upwards and backwards, is exaggerated, and the optic lobes are pushed out of their way down on to the sides of the brain-stem. Also a considerable expansion of the cerebellum is noticeable; this is because the postural adjustments of a bird, flying and walking on two legs, require more supervision than do those of a lizard. The expansion of the cerebral hemisphere in both these brains is due to a great upward swelling of its floor; its roof is comparatively thin and has been bulged out by this great swelling below, much as Mr. Everyman's bedclothes bulge upwards when he lies on his back and raises his knees. Finally, Fig. 312 shows the brain of a rabbit, a fairly primitive mammal. Here both cerebral hemisphere and cerebellum are well developed, and the former has grown backwards and covers the mid-brain completely so that it cannot be seen from the side. This growth has largely taken place in the roof of the fore-brain. From a brain like this the human brain can be derived (Fig. 53) by imagining this growth of the cerebral hemispheres greatly exaggerated, and their surface becoming wrinkled to increase the area of the now essential cortex. As a complicating factor man walks upright, so his brain is twisted in such a way that the spinal cord runs downwards from it instead of coming straight out behind.

How does this structural divergence between the brains of reptiles and birds on the one hand and mammals on the other reflect itself in their behaviour?

The outstanding difference between mammals and other vertebrates lies in the adaptability of the former. They can remember more surely and learn more quickly than any other living things. Moreover, they tackle a new problem more competently. Suppose you put a fish, or a frog, or a reptile, in some experimentally devised strange situation— such as a maze in which the animal has to follow a definite path if it is to reach a comfortable nest and food, while other possible paths lead nowhere. It will simply muddle about among the unfamiliar surroundings until it gets to the goal. Put the animal in again, and it will muddle its way through again; but gradually as the lessons are repeated, it will get less and less dilatory and make with more and more directness and certainty for the goal. This kind of learning is simply trial-and-error, like the random trial-and-error of a slipper animalcule, but with a certain element of memory attached to it.

Yerkes, an American investigator, devised a maze for some tortoises on which he was experimenting. Part of the path to the comfortable nest, to which they were supposed to find their way, led down a steep inclined plane. It so happened that one of the tortoises, put in the maze and not liking the look of it overmuch, began poking aimlessly about and tumbled off the side of the sloping plane. Surprisingly, it found itself near a pleasant nest. Thereafter this animal invariably solved the problem by repeating his accident—by making for the edge of the plane, deliberately tumbling over, and then calmly walking into the nest.

To a large extent, all animals learn in this way—by muddling about more or less at random, and by remembering which activities are followed by pleasing results and which by harmful. But mammals do it very much more quickly and less laboriously than other vertebrates. Moreover, in the higher mammals at any rate there often appears a somewhat different kind of solution when the animal comes up against anything new. The creature supplements this profiting by chance events by means of inner resources of its own. It stops to think; then it "gets an idea"; then it tries the new idea to see whether it will work.

How far this reflective pause is really a process that differs in kind from the simpler learning, and how far it is merely an extension of that process made possible by the greater extent and better organization of the store-rooms of the mammal's memory—an overhauling of past experience to find whether there is any illuminating precedent—is a question that will come up for review in a later section, when the intelligence of mammals will receive a more thorough examination. Here it is enough to note, as a fact of observation, that mammals are altogether more competent to deal with novel and puzzling situations than any other vertebrates.

As we noted in Book I, a mammal—or at any rate a higher mammal —thinks by means of its cerebral cortex. In ourselves, everything that is not simply automatic is determined from that dominating centre. But this does not mean that learning is always confined to the cortex. Indeed, in frogs and fishes this structure is absent, and yet a certain slow learning is possible, although their adaptability is slight compared with our own. Their somewhat rudimentary counterparts of our higher mental processes are apparently performed mainly by the thalamus and other incipient correlation centres; perhaps in some cases by the mid-brain also. The growth of the cerebral cortex represents an extension and elaboration of this ability. Cortical tissue is grey matter arranged in such a way that it can do these things much more effectively. There is evidence that the power of the thalamus to participate in the formation of habits persists in such comparatively lowly mammals as rats; but even here the chief mental processes are carried out by the cortex.

The evolution of the mammalian brain is largely a matter of the improvement of this all-important grey film of tissue. It increases in comparative size and comes to overlap more and more of the other regions of the brain, until in ourselves (Fig. 53) only the cerebral cortex and the cerebellum are visible from the side, and only the first from above. At the same time it increases greatly in complication. As in the early mammals the new centre began to prove its worth, the older brain regions became more and more subservient to it. Its fibrous connexions thickened enormously as it grew, and it took over more and more of the responsibility of running the body state.

Here let us note the meaning of two further words which are making their way into popular biological writing. The cortex of a tortoise— the most primitive cortex we know—is divisible into three distinct fields. At the outside and inside edges of the roof of each hemisphere it is supplied chiefly with fibres from the edges of the corpus striatum below. The information supplied to these lower regions, and through them to the marginal field of the cortex, is mainly concerned with smell, taste, and the condition of the viscera. But the third strip of cortex, lying along the middle of the roof of the hemispheres, is mainly connected with the thalami in the side walls of the between-brain, and thus indirectly receives information from the eyes, the ears, the skin, and the tendons, joints, and muscles involved in position and voluntary movement. Naturally these three regions do not work in isolation, but as a single unit. Nevertheless, all through the mammalian series, it is possible to make out a similar distinction between the parts of the cortex which are mainly concerned with smell and taste and the parts mainly concerned with sight, hearing, and touch. But their importance varies. The part concerned with smell and taste is well developed in the most primitive mammals; it was the first of the two to assume its definitive elaborate form, and is known as the

EVOLUTION OF BEHAVIOUR IN VERTEBRATES

archipallium. The other part is small in primitive mammals, but acquires more and more relative importance, until in ourselves it is enormously expanded and overshadows the archipallium completely. This part is called the *neopallium*. The shift of responsibility from archipallium to neopallium corresponds to a gradual alteration in behaviour, to the shrinking of the nose and the improvement of the eyes, and (since the neopallial tissue is differently and more efficiently disposed than the archipallial) to an increase of intelligence and adaptability.

Thus the mammalian brain has evolved. The primitive amphibians are spawned in great multitudes; they are stupid, inadaptible creatures, but out of the thousands thus launched a few survive to carry on the species. The higher mammals reproduce altogether more sparingly, but they are better able to take care of themselves and to survive unusual combinations of circumstances. A large share of the responsibility for carrying on the race has been handed from the genital organs to the brain, from the proliferating germ-plasm to the thinking individual soma. But meanwhile the other vertebrate line that leads up to the birds took a somewhat different direction.

Birds can learn. Anybody who has seen performing pigeons in a music-hall knows that their brains can hold one or two tricks. But performing pigeons generally come on the stage in great flocks, for the number of tricks that any single bird can do is much less than the repertoire of a good performing mammal. Indeed, the educability of a bird is hardly greater than the educability of a reptile, and certainly less than that of a rat. Where birds excel is in the performance of elaborate but purely instinctive acts. An incubator-hatched bird will build its nest as neatly and pick its material as carefully as any other member of its species; at the first appropriate occasion it will go through the routine of courtship with perfect correctness. In a word, the great development of the corpus striatum in birds means a tremendous endowment of inherited instinct, rivalling that of insects in its complexity and far exceeding our own, but without any notable individual intelligence or adaptability to modulate it.

An admirable example of blindly mechanical instinct is afforded by the so-called thermometer bird or bush turkey of the Solomon Islands. The eggs in this species are laid in heaps of mixed plant material and sand, and incubated for some six weeks by the heat of the rotting vegetable matter. Moreover, they all lie with the blunt end upwards. The chicks hatch out from this blunt end, and their feathers all point stiffly backwards; so by wriggling and struggling in the heap they can force their way upwards. As soon as they get out they shake themselves, and then dash into the shadow of the nearest undergrowth. Here we have an admirably adaptive chain of reactions. But if you take a chick which has just liberated itself and dig it into the heap again, it is quite incapable of coming out once more, but stays there struggling ineffectively until it dies. Its movements are now of the wrong kind. The reactions follow one another automatically, a mechanical chain of instinct, and the creature can no more go back to the beginning and start again than can the cocoon-spinning caterpillars described in Book 8, Chap. 2, § 3.

One of the present authors (J. S. H.) has given another illustration in his *Essays of a Biologist* of how limited the behaviour of a bird may appear when some unforeseen accident disturbs the usual routine of its instinctive acts. The scene described has been filmed by Mr. Chance, and the chief actor is a pipit in whose nest a cuckoo has laid her egg.

"When, after prodigious exertions, the unfledged cuckoo has ejected its foster-brothers and sisters from their home, it sometimes happens that one of them is caught on or close to the rim of the nest. One such case was recorded by Mr. Chance's camera. The unfortunate fledgling scrambled about on the branches below the nest; the parent pipit flew back with food; the cries and open mouth of the ejected bird attracted attention and it was fed; and the mother then settled down upon the nest as if all was in normal order. Meanwhile, the movements of the fledgling in the foreground grew feebler, and one could imagine its voice quavering off, fainter and fainter, as its vital warmth departed. At the next return of the parent with food the young one was dead.

"It was the utter stupidity of the mother that was so impressive—its simple response to stimulus—of feeding to the stimulus of the young's cry and open mouth, of brooding to that of the nest with something warm and feathery contained in it—its neglect of any steps whatsoever to restore the fallen nestling to safety."

Clearly we are dealing here with a far more mechanical type of conduct than occurs when the mammalian cortex is at work. One cannot imagine anything to parallel this in the behaviour of a bitch or a ewe.

§ 5. Courtship in Animals.

On a fine spring day go to the Zoo, and stand in front of the Amherst pheasants' cage. The hen is a dowdy creature; her dowdiness is protective. But the male is brilliant in the extreme, with his fabulous colours, gorgeous tail and frill. If you are lucky, you will see him go through his courtship. He makes a little rush past the hen; as he passes, he droops his wing, shows off his tail and fans forward the frill on his head. Then he will repeat the process from the other direction. You are struck by the way in which the display is directed at the female. It is only executed at the moment of the rush past, the body and the tail are sloped towards the hen, and only on the one side is the frill fanned out. The frill on the far side is kept closed; he wastes no beauties where they will not be seen.

Most people have seen a peacock displaying his train—perhaps the most sheerly beautiful sight in nature. The bustard inflates his throat, throws back his head, and everts his wings so that he may strut before the hen, like a surprising animated giant white chrysanthemum. The argus pheasant is perhaps the most striking of all. The long brown wings, patterned with a series of light spots, wonderfully shaded so that they look like solid spheres, are spread and thrown forwards like the bell of a great flower. The long tail plumes are waved up and down behind, and from below one wing an eye peeps out to keep the hen in view.

Here we have in its fullest development the phenomenon of "courtship" which plays a large part in the life of a great number of the higher animals. It is not confined to the vertebrates. The lowest types in which the process can be traced are certain marine bristle-worms. At the breeding season, the males may be seen writhing and contorting themselves in a frenzied dance among the females; this appears to stimulate the females to shed their eggs, whereupon the males emit their sperms. The masculine excitement has been turned into a stimulus for the females' egg-discharging instincts.

In the alert and half-terrestrial fiddler-crabs (Fig. 188) there is a rudimentary but illuminating form of display. When, in the mating season, a female crab appears, the males in her path stand tiptoe, with their single huge claw held aloft—a statuesque attitude. If she is indifferent, they may run along a little and strike their attitude again, but will not pursue her far. As Dr. Pearse writes, after watching them carefully, "they seem to be advertising their maleness." There you have the clue. There are not many possible situations in a fiddler-crab's life. There are food-situations, danger-situations, and sexual-situations. The male's upreaching pose, with his masculine appendage so prominently displayed, is an advertisement to the female that a sexual-situation exists, and so is not only an advertisement, but a stimulus.

Similar excitation occurs in the case of snails, octopuses, and cuttlefish, but good studies of these processes have still to be made.

This apprising the female of the existence of a sexual-situation is of vital importance in many spiders. Here a smallish moving object is normally the sign of a food-situation—a stimulus for the female to pounce and kill. Accordingly it becomes essential for the male to distinguish himself sharply from other smallish moving objects. Among the almost blind web-spinning spiders, he does this by approaching the web and vibrating one strand, producing a movement quite different from that due to the struggles of captured prey. The female may rush at him once or twice (on which he very hastily makes himself scarce), but eventually this message of love vibrating along the web attunes her to the sexual-situation. Even so, however, after mating has been accomplished, he may again become a mere moving object to her, and she may then kill and eat him.

In scorpions, Fabre has described a sort of set-to-partners of courting male and female; but here it seems to be the invariable rule for the female, once one instinct is satisfied, to change to another, and eat the male after his biological duty to the race has been carried out. And the female of the predaceous praying mantis may begin eating her mate while the act is still being accomplished.

The instrumental music of the grasshoppers and crickets (leg on leg, or wing-case on wing-case) serves a sexual function, but it is rather one of bringing the sexes together than of courtship; and the same holds for the flashing of fireflies and the glowing of glow-worms. But the little carnivorous flies called empids have a true and very interesting courtship.

In some species, the males present the females with insect prey which they have captured; and usually the gift is either wrapped in silk spun by the male, or embedded in a large glistering "balloon" of bubbles which he has secreted, as the gift of a necklace might be enclosed in a striking casket. Carrying these embellished gifts in their legs, they fly to and fro over pools and streams. Here the mere advertisement of a sex-situation is reinforced by the stimulus of a gift. But in other species, the place of the prey is taken by a flower petal or a blade of grass, and the males will put coloured paper in their balloons if you strew it for them; this appears to be the only case outside warm-blooded vertebrates in which objects are picked up and used in courtship for what we may call their æsthetic appeal.

Many female insects cannot ripen their eggs without a rich protein meal; the female mosquito, for instance, needs blood. This may account for the female mantis or spider devouring her mate; and so the gift of prey may save the male empid's life.

Throughout the vertebrates there is a steady development of this business of courtship. In fish it is exceptional, and is found mostly where internal fertilization is necessary. But where it does occur, it may be quite elaborate. The male dragonet assumes the most brilliant colouring in the breeding season, and flutters round the female like a gorgeous butterfly. The little sword-tails owe their name to the male's long-bladed tail fin, with which he prods and pokes the female into interest as he swims about her. And in the humble stickleback of country ponds, the male becomes a lovely iridescent creature in the spring, bright with rich red, which kindles with his excitement or pales if he is beaten and chased away by a rival.

Among amphibians there is a marked difference between the courtship of the tailless frogs and toads and the tailed newts and salamanders, a difference connected with their methods of reproduction. In the frogs and toads, the male waits, tightly clasping the female, until she lays her eggs, when he ejects his sperm upon them. There is thus no need of courtship, but only of the sexes finding each other, and accordingly the spring behaviour of the males is limited to vocal performances which advertise their whereabouts. But, as we have described in Book 8, Chap. 3, § 3, the fertilization of most newts and their allies is more elaborate. It is necessary that the female should be roused to play her part. To this end the male newts have a magnificent nuptial dress and courtship-dance, in which with arched back they prance in front of the female, while the tail wafts the presumably odorous secretion of certain glands towards her. And it has been proved by experiment that she will not pick up a sperm-packet unless a courting male is about—in the absence of such stimulation she treats it with complete indifference.

The courtship of reptiles has been very little investigated. Among our fragmentary knowledge about them is the fact that various snakes, including the common European grass-snake, will assemble in writhing, intertwined masses in the spring, apparently for the purpose of choosing mates; and that the males of some of the Australian lizards will get up on their hind legs and wrestle in the most human-looking way for the possession of the female.

It is in birds that courtship-display reaches its greatest elaboration. We will begin with some account of the usual type of courtship in songbirds, which Eliot Howard has observed with such patience and analyzed with such care in his *Territory in Bird Life*.

In migratory song-birds, such as the Old World warblers, the males in spring arrive at their breeding quarters a week or so ahead of the females. They then proceed to find suitable sites for nesting, and each appropriates a more or less definite area or territory for himself. He may have to fight to obtain it, and is almost sure to have to fight to keep it. There he awaits the arrival of the hens, spending almost all the time left over from feeding and sleeping in song. When the hen-birds arrive, they search for the songsters. One may simply take possession of a male and a home without any ado. Or she may have to fight another hen for her prizes; in such a quarrel the male, though often an interested spectator, will not take sides, and becomes the associate of the victress. There appears to be no courtship display on his part at this juncture, although from this moment onwards the pair are mated.

They are mated for the duration of one brood, but not for life. For though many birds, like storks and parrots, eagles and ravens, do mate in permanency, the recent practice of bird-banding, by which wild birds are temporarily trapped and have light metal identification rings fastened round their legs, has revealed an unexpected inconstancy in the marriage systems of small song-birds. The most systematic study has been made on an American wren. Mr. Baldwin had a large number of pairs of wrens occupying nest-boxes in his orchard. During their first brood of the season he banded all of them. Then while the second brood was being raised, he caught the birds again, and found to his surprise that the great majority had changed partners between broods! The single brood or the single season appears to be the usual time for which song-bird matings last; but during this period the marriage is definite and monogamous.

It is after the mating that the males begin their display. They have yet to excite the female to actual intercourse. It is to that end that most courtship is directed. Although the warblers are mostly very soberly coloured, they display as many gaudy birds do—they spread their wings, fan their tails, bristle up the feathers on throat and head. At the same time they often seize on some piece of potential nest material, such as twig or leaf, and advance towards the female with it in their mouths. There seems to be a strong association in their minds between the two emotionally-coloured activities of courtship and of nest-building. In addition, a frequent form of courtship is the pursuit-flight, in which the cock darts, twisting and turning, after his mate. The hen warbler may be interested in the male and his displays, but it is not until after an "engagement period" of a few days that she permits the consummation of their mating; and shortly after this the first egg is laid. The hen in most song-birds does all or the greater part of the work of brooding the eggs. But once the eggs hatch, both parents share in collecting enough food for the prodigiously rapid growth of the fledglings.

In song-birds that are not migratory, but resident all the year round (as in the buntings studied by Howard), the males often begin to stake out their territorial claims in January or February, spending an increasing part of each day on their territory as individualists, a decreasing time with the flock as gregarious creatures. The females leave the flock and follow them on to the territories a few weeks later, but still much earlier than the beginning of egg-laying; so that the "engagement period" is here very much prolonged.

The "engagement period" probably has a physiological basis. We know that in pigeons at least there are two stages in the growth of the eggs in the ovary. At first, yolk is laid on very slowly; but a few days before the egg is ready the rapidity of yolk-growth suddenly increases about twentyfold. And this change is associated with profound alterations in bodily chemistry; the amount of sugar in the blood increases, the adrenals enlarge, and other ductless glands alter their activities. It is probable that the end of the "engagement period" coincides with this change in female metabolism. In any case, while the males have but two main phases in their sexual life—unsexed in winter, fully-sexed in the breeding season—the females have three; they pass from winter neutrality to full breeding activity through a third, intermediate phase, in which they are enough sexed to take an interest in their territories and future mates, but not sufficiently so to be ready to pair. The familiar and rather ludicrous spectacle of a cock sparrow hopping love-sick before a hen who repels his advances by violent peckings is due to the hen being only in the half-way stage, while he is fully ardent. Often not one but many male sparrows will gather in fruitless and disorderly courtship round one hen. In such cases the sight of one bird courting has excited the rest. The same spread of excitement at the sight of display (or of combat) may be witnessed in other birds; but in the sociable sparrow, with his dense urban population, it is more frequent and leads to bigger gatherings.

The biological function of display seems to be the emotional stimulation of the female; display prompts to the act of mating rather than to the choice of marriage partner. This is demonstrably so in newts. Even if two rival male newts deposited their sperm-packets before a single female and simultaneously executed a courtship dance, it would be asking too much of amphibian mind to suppose that she would remember which packet belonged to which male, and would pick up the packet deposited by the one whose dance pleased or excited her most.

Where we should expect male competition and sexual selection to have fullest scope is in species that, like the ruff, are polygamous, or promiscuous in their mating habits, so that each female has the choice between many males. The facts confirm our expectations. Edmund Selous spent a whole spring watching a ruff assembly-place in Holland. In spring the males repair to definite areas called "hills," since they are generally a few feet above the marsh. On one such hill there may be anything from half a dozen to twenty or thirty males. They whirl round like dancing dervishes in sheer overwhelming excitement, or spar viciously with each other. Occasionally the females, or reeves, visit the hill, and it is there that actual pairing occurs. This is the only contact between the male and female, for the males never visit the nests or look after the young.

When a reeve visits the hill the scene changes. The ruffs squat or prostrate themselves in strange attitudes—wings spread out, beak pointing down at the ground. They will remain thus, as if hypnotized, for a considerable time, occasionally shifting round to face in a new direction. Sometimes the reeve will just fly away again; sometimes she will walk up to one of the ruffs and touch him with her bill, on which mating will take place. The choice lies entirely with her.

Selous soon distinguished all the male frequenters of his particular hill by their appearance, for the ruff is a variable and highly individualized bird, and he found that their success in obtaining mates was very unequal. One bird (interestingly enough, one whose appearance was to human eyes very striking) mated more often than all the other males on the hill put together; and there were several males who never mated at all. Here is Darwin's sexual selection in diagrammatic form. The females can be seen exercising a choice of mates, and some males are chosen much more frequently than others; and so some have many offspring, some few, some none. The premium upon whatever promotes feminine choice is therefore much higher than where monogamy prevails, and, accordingly, we find that the polygamous birds usually have more elaborate display-characters than their congeners.

It is harder to understand the biological value of courtship display in monogamous creatures like song-birds. However, as Eliot Howard has shown us, display undoubtedly helps to regulate the emotional reactions of the pair, and so ensures that both male and female shall at intervals be brought simultaneously into a state of readiness to pair. Female doves reared in solitary captivity can be made to lay eggs (though the eggs will, of course, be infertile) by keeping them in the next cage to a male; the emotional stimulus of his bowings and scrapings will set going some change in their ductless glands which will lead to egg-production. Even caressing a virgin pigeon's head with the finger, so imitating the billings of the male, may have the same result.

In creatures like grebes, where courtship is a joint affair of the two sexes, the stimulation is doubtless mutual. But here it looks as if other functions had been added on. No one who has watched grebes, or other birds with elaborate mutual courtship-ceremonies, can doubt that their performance is extremely pleasurable; and in all these kinds of birds the ceremonies continue right through the season, from the moment of pairing-up until the young are well-grown. And both parents must share the duties of incubating the eggs and feeding the young if the full complement of offspring is to be reared. It looks as if the elaborate display-ceremonies constituted an emotional bond between the pair, which, by helping to keep them together through the season, cemented a union which was of biological value. This is no fantastic suggestion, for something of the same sort is at work in human family life.

This brings us to the relation between kind of courtship and way of life. Courtship and display in general are seen to have as their main function the stimulation of the other sex's emotions. But the means by which courtship achieves this end differs according to the habits of the species. Need for protection will work against the tendency of sexual selection to produce brilliant colours and striking display ornaments. Where sexual selection is very strong, as when polygamy is the rule, extra weight is put into the scale against this need for protection, and we get the fantastic beauties of the male pea-fowl, ruff, or black game. But when monogamy is practised, and the birds are in danger from hawks and other enemies, sexual selection is weaker, the need for protection greater; and so we have the state of affairs seen in larks or Old World warblers, or the common European quail and partridge in which the male is almost as dull-coloured as the female, and courtship consists merely of exciting actions.

Often a compromise is struck between the exigencies of sexual and natural selection by having the bright colours tucked away out of sight during ordinary existence, but brought suddenly into play at courtship. The great bustard is an excellent example of this, and so is the prairie hen, where the males during courtship ruffle up tufts of feathers on either side of their neck, and so reveal brilliant patches of bare skin which they then inflate into the semblance of half-oranges. Such displays have as additional advantage the stimulus of novelty and strangeness.

Birds which have few or no enemies to fear, such as the gregariously-nesting herons and spoonbills, are often very conspicuous—snow-white, for instance—since there is no special need for protection, while the conspicuous colour helps the birds to recognize each other from far off.

Courtship is also subtly entangled with family life. In almost all song-birds the food-territory round the nest is the basis of reproductive success. And the fights of such birds are concerned with property rights rather than with sex. Male fights male to secure territory; female fights female to secure a male—but only if he is already in possession of a territory; and pair fights pair over boundary disputes. Song is also in the main concerned with territory. When at its most intense it is serving a double territorial purpose. It is an advertisement to any female within hearing that here is a male in possession of an eligible nesting site; and it is a warning notice to other males—trespassers keep out, or else fight.

Song, being thus essentially an advertisement, is generally given from as conspicuous a position as is consonant with safety. Where there are trees most birds sing from a top branch; where there are none, the singers often make up for their absence by aerial song—as larks over fields and heaths, or pipits over moors. Many wading-birds, too, have songs or their equivalents, and they also give them aerially; we have only to think of redshank or curlew or godwit. In snipe, the aerial "song" is mechanically produced; the strange, far-carrying bleat is made by wind vibrating the spread tail-feathers. Many woodpeckers also have a mechanical substitute for song; they drum their beaks with incredible rapidity on dead branches to make a resonant note; and some, like the American red-headed woodpecker, select their instrument, making trial of tin roofs, the metal arms of telephone insulators and so forth, until they find something to their liking. Ruffs and such-like birds have no territory, and therefore no song.

To say that song is a territorial advertisement does not imply that it is never given save with some territorial meaning. Nature does not work in that way; she employs machinery which ensures that on the whole the biological functions of song shall be carried out, but does not worry if the bird sings on other occasions, too. Male birds are constructed so that they shall sing when in certain physiological states.

For most song-birds, one condition is that the gonad shall be pouring its hormone into the blood; another is that they shall enjoy a sense of well-being and not be cold or starved. A third is that they shall not have additional calls upon them. Almost all male song-birds stop singing when the eggs hatch, and they have to begin foraging for their young; but if now the young are destroyed, song will begin again almost at once. When family cares are past, and moulting is over, many birds have a period of autumnal song before their gonads shrink and turn them into neuters again for the winter.

Similarly, the sex activities of ruffs or black-cock depend merely upon sex-hormones and superabundant vigour. The males will continue dancing and sparring on their assembly-places for weeks after all the hens, now fully occupied with their eggs and young, have ceased to visit them. Some observers have for this reason concluded that the assemblages can have nothing to do with sexual selection; but this is to misunderstand the blind and blundering way in which selection works.

We often find even the details of display determined by the bird's mode of living. Diving-birds may use their powers to disappear below the surface and re-emerge close to their mate in what must assuredly be an agreeably stimulating manner; this occurs in the grebe and some diving ducks. Or they may raise fountains of spray with their feet, as does the golden-eye duck, or rush erect over the water, in apparent defiance of the laws of hydrostatics, like the loon. Birds that are masters of flight use air as their medium; the peregrine tercel will dash down-wind at the falcon as she sits on a rocky ledge, and swerve up when only a foot or two from her. The crepuscular, insect-hawking nightjar of Europe flies moth-like above his mate, and now and again claps his wings startlingly above his back; an African species trails strange whip-like plumes in his display-flight; the related American bull-bat lets himself drop from hundreds of feet aloft, and swerves up with a "zoom."

Fulmar petrels have no bright colours of plumage; but for their display they use "interior decoration," in the shape of the delicate mauve lining of their mouths; two fulmars languishingly waving their open mouths at each other is a sight not to be forgotten. Male Adelie penguins present their mates with nesting material—in this case stones—and the pair indulge in a mutual ceremony which Dr. Levick, in his delightful book *Antarctic Penguins*, calls "going into the ecstatic attitude." Mutuality in courtship and display was hardly known in Darwin's time, but is actually quite common. We find it not only in herons, grebes, petrels, penguins, but also in divers and pelicans, in cranes and terns, in the dull-coloured stone-curlew, and in the flightless but well-crested kagu, and in the great albatrosses of the Laysan Islands.

The males of the bower-birds, an Australian group, construct peculiar bowers which are quite different from the nests. The bower often consists of a short tunnel of twigs, at the entrance to which the bird deposits shells, bones, berries and other bright objects. This museum of specimens appears to be a substitute for courtship decorations, and the "display" of the male consists in his driving the female through the bower and drawing her attention to his collection. In another species there is no actual bower, but the male clears a space a few feet across, and on it lays a certain kind of leaf with silvery underside. He always puts the leaf with its silvery side showing, and if the wind blows it over, turns it right way up again. As the leaves wither, he clears them away and brings new ones.

The Phalaropes afford another interesting case. The females are rather larger and more brilliantly coloured, and the whole duty of incubation and feeding the young devolves upon the males. In courtship, too, it appears that the females take the more active part. Sexual selection has here been reversed.

Sometimes, instead of presenting nest-building material to his mate, the male bird may present food. The male harrier, returned from his hunting, gives a special call when near the nest; the female leaves her eggs and flies below him. At the right moment he drops his prey; she, back-somersaulting almost completely, catches it in her claws, and returns to her duties. Male terns give their mates fish, male titmice present theirs with caterpillars. In many such cases the female while being fed adopts the same attitude and gives the same call as does the half-grown young when begging food from its parents—a simple example of association. More subtle is the cross-association with sex, which may be witnessed in various birds, notably the common house-sparrow, where this same attitude and note is used by the female as a symbol of readiness to pair.

Psychologists, indeed, have in the courtship-behaviour of birds a rich field to explore. We have already mentioned the association between the presentation of nest-material and courtship ceremonies. A good example of the transference of this attention to an unusual object is found in the Adelie penguin. When there are men near their rookeries, these brave but comic little creatures will sometimes approach them instead of their mates, and solemnly deposit a stone at their feet. Dr. Levick records that he was quite embarrassed the first time that he was a recipient of this tender attention. They may also make these gifts to sledge-dogs; but then matters may end disastrously, for the dog will often snap at them and kill them. It would appear that this action is an expression of special interest. The only way in which it can be normally expressed is to a bird of opposite sex; but strange, impressive beings provide fresh outlets.

Another queer transference of instinct, recalling certain traits of abnormal human behaviour, was noted in a male argus pheasant in the Amsterdam Zoo. As no mate of its own kind was available, the bird was put with a female of a related species. Now in the courtship of the ordinary argus pheasant the hen usually stands and looks on while the male shows off his plumes. But matters must be different in the species to which this hen belonged, for she absolutely refused to stay still. The male grew more and more discouraged. He rushed about the cage, throwing up his wings and tail spectacularly; but when he peeped under his wing, the hen had moved on. Eventually he gave it up; but the urge to display was still on him, and he indulged it by performing before the dish in which his food was put out! The dish might be inanimate, but at least it was associated with some form of gratification. In mammals courtship is less frequent and, when present, less spectacular than in birds. Often, as in deer and wild sheep, bisons and sea-lions, there is no courtship, but the males fight for the possession of the females, who passively fall to the lot of the victor. Such males are naturally characterized by weapons and prowess in fighting, and display-decorations are absent.

Deer size each other up from afar, and battle is not joined unless the rivals are on fairly equal terms; for each encounter that comes to fighting, there are a dozen where one beast slips away. But once the fight has begun, it is fierce; the combatants shove with antlers interlocked, and whichever gives up first has to beware lest, as he disengages, the other's brow-tine pierce his neck. The break-away is thus the most exciting moment of a fight; the vanquished unlocks his antlers and is round and away in a flash.

In sea-lions and sea-elephants, the bulls, who are generally enormously bigger and stronger than the cows, come ashore on the breeding-beaches and stake out mating-territories there. As the females land, the males fight for them and establish harems. In some species they may seize the cows by the neck and pitch them over their shoulders into the harem. The old bulls are so busy defending their sexual rights that they take no food at all during the breeding season, which lasts for several weeks. In spite of this, their vigilance is sometimes deceived; in one species at least the cows may occasionally slip off into the sea, and there mate with the half-grown bulls who are too small to fight and establish harems of their own. But doubtless the official husband hands on his genes in a sufficiency of cases to make strength and valour pay in the struggle for reproduction.

But in many mammals there is little of either fighting or display in connexion with sex. In dogs, for instance, the male is only attracted by the female when she is ready to receive him; and at such times her scent stimulates all the dogs within a mile. The reproduction of most mammals, indeed, is on a more physiological, less psychological plane than that of most birds. Emotional stimulation has no influence upon the development of their eggs, but an automatic cycle of the ductless glands controls the female's mating rhythm. The sexual situation is brought into being from within, and so needs no stimulus from without. In polygamous mammals, capture of females is enough. When the time comes, they will accept the victor. And in monogamous species, some apprisal (usually by scent) of the female's physiological state is often sufficient.

Matters are not always quite so simple as this; there are many cases on record of female mammals—mares or bitches, for instance—who refused to have anything to do with particular males; and it is probable that careful observation would reveal many examples of discrimination. But the broad fact remains that in mammals, as compared with birds, courtship is rare, and display rudimentary; and the explanation undoubtedly is to be sought in the difference in their reproductive physiology.

The monkeys and apes alone among mammals have begun to subordinate the chemical control of reproduction to a control by the higher centres of the brain. The fish or amphibian, reptile or bird, has a fixed breeding season. At other times of the year it is indeed, physiologically speaking, castrated; its ovary or testis shrinks to a tiny fraction of its breeding size, and the amount of sex hormone in the blood falls until it has no effect. The creature oscillates each year between two phases of existence, a neutral and a sexual being. Most female mammals, in addition to the annual cycle of breeding season and neuter season, have a cycle within the breeding season, of readiness to mate and the reverse, determined by the state of the eggs in the ovary. But in monkeys and apes not only will the females mate at other times of this ovarian cycle, but the neuter part of the annual cycle also shrinks. Some human races (like the Eskimo) conserve traces of a restricted breeding season, but in civilized man it has disappeared. In the higher primates, the chemical control of mating is only slight, the emotional control predominant.

And parallel with this change, display-characters come into prominence, although never so strikingly as in birds. Many male monkeys have facial adornments, sometimes bizarre to human eyes, sometimes alarmingly reminiscent of human whiskers or mustachios. Many also have developed adornments at the hinder end of their persons, in the shape of coloured patches of bare skin—rose, turquoise, golden yellow, scarlet, green. There are traces of these left in the gibbons, but otherwise the tailless apes are without them. The abandonment of this posterior display was a very important step in human ancestry; for it meant that interest could be chiefly concentrated on the face, and features and expression could become the main appeal of love.

In man, the predominance of mind has allowed a more deliberate choice of mates; and with this courtship enters on a new phase. Feminine beauty becomes, almost for the first time in the history of life, more important than masculine decoration. Dress and finery, gifts and poems, can excite in the place of special plumage or inborn colours. Physical prowess, brains, wealth, can all become linked up with courtship. Sex and mating cease to be in a compartment of life by themselves; it is a prime characteristic of human mind that it enables different aspects of existence to be brought into relation with each other, different departments of life to interpenetrate.

The varied and individualized love-making of humanity differs fundamentally from the courtship of birds. The human lover woos with the cerebral cortex, he (or she) is plastic and responsive, and adapts the means to the occasion. The impassioned bird woos ardently but automatically with the corpus striatum. In one case the individual of the opposite sex is a problem; in the other an exciting situation. The human lover may do a thousand things; the courting bird is an elegant determinate machine.

§ 6. The Evolution of Mammalian Intelligence.

The intelligent mind is a mammalian invention; compared with any of the higher mammals, a bird is little more than a highly complicated and highly emotional instinctive machine. But before we turn to a fuller analysis of the growth and meaning of intelligence, let us warn ourselves against the tendency to exaggerate this contrast unduly, and to overrate the mental powers of mammals because of their many other resemblances to ourselves. As a matter of fact, the reputation of mammals in this respect has sunk considerably during the past half-century. Nowhere has the rigour of the scientific attitude of mind been more destructive of loose and tolerant ideas than in this field. Fifty years ago, books on animal psychology were largely occupied with the mammals, and nearly all that space was taken up by anecdotes illustrating their wisdom. We are all familiar with the elephant who cherished a grudge and squirted his old enemy with dirty water after thirty years; or with the lion who refused to eat his benefactor Androcles. We have all been assured of the exceptional cleverness of our neighbour's dog or cat. The laws that govern social intercourse have perhaps forbidden any explicit expression of disbelief in these marvels; yet, inwardly, a certain saving scepticism has reminded us that anecdotal evidence is often untrustworthy. For one thing, it picks on occasional striking events for narration, and, for all we know, they may be accidents or coincidences; for another, it usually depends on the report of a casual observer, untrained in what to look for and prone to read his own private ideas and interpretations into what he sees.

So let us go into this matter of mammalian intelligence with care and circumspection; and let the reader beware of too hasty indignation if in what follows we seem to be disrespectful towards some particular four-footed favourite of his or her own.

The lower of the two main branches of the mammalian stock, the marsupials, are on the whole stupid creatures, above the reptiles in respect of their warm blood and their mode of reproduction, but hardly in their behaviour. They seem to learn with difficulty, and this is helping towards their extermination in Australia; they are no match for the more enterprising and plastic placentals introduced by man. The few marsupial carnivores, such as Tasmanian devil and marsupial wolf, are almost untameable. They are fierce bits of flesh-eating machinery that go their way almost regardless of change in outer circumstance. In zoos, for instance, they do not get to know their keepers in the personal and intimate way of lions or wolves or seals. The placental wolf has been tamed into our friendly and serviceable dog; it would probably be impossible to do anything of the sort with the marsupial wolf. Nor do the instincts underlying family life, which are at the basis of so much of the latest developments of mind, seem to be well developed in marsupials; at least, a mother kangaroo, when hotly pursued, has been seen to eject her well-grown young from her pouch so as to help her own get-away.

The marsupial type, as we have seen, is a survival from the Cretaceous world of life, preserved by the accident of Australia's isolation.

And in that isolation they seem to have made as little progress in brain-power as in method of reproduction. In other parts of the world the early placentals ousted and supplanted them; but the superiority of the victors seems to have rested in their new method of nourishing their unborn young rather than in their superior intelligence. Their brains in the early Cenozoic were no larger or more elaborated than those of modern marsupials, and we may reasonably take the wallaby and marsupial wolf as giving an approximate picture of the behaviour of all the early Eocene mammals. Afterwards, the intenser competition in the broad spaces of the northern hemisphere put a premium on intelligence; and, as we saw in Book 5, increase in brain-size was one of the salient features of mammalian evolution between Eocene and Pliocene.

Even so, some creatures with poor brains survived, either in out-of-the-way corners of the world, like the armadillos or sloths in South America, or in humble or obscure ways of life, like the shrews or moles. Such animals, it would seem, are almost on a par with lizards or birds as regards the automatic quality of their behaviour.

But as, in evolution, the proportionate size of the cerebral cortex was increased (a change visibly reflected in the foldings and convolutions of this outer layer of brain-tissue), the faculties of association and memory improved. In higher mammals all learn well and rapidly. A dog knows his master, his friends, and his enemies; circuses would be impossible without the cleverness of elephants, horses, sea-lions, and land-lions; the bears at the Zoo will gladly give you an exhibition of how quickly they can learn to manipulate such an unaccustomed object as a tin of treacle; the foxes could not survive in the European countryside if they were not adaptable. But all this learning, however rapid, is, by our standards, of a restricted kind. It makes behaviour supple rather than truly intelligent. The performances of trained animals, and the tricks and much of the knowingness of our household pets, are deceptive because they are due very largely to the activities, deliberate or unconscious, of human teachers. The animals have been helped in their learning by man. Evidently the real test of an animal's intelligence lies in the things that it can find out of its own accord, without any assistance, and this is a matter which has only been investigated with scientific rigour in the last few decades. The method is to set a problem for the animal to solve, and then to see what it will do, taking the greatest care not to prompt or guide it in any way. If possible, the investigator should be outside the room and watch through a peep-hole.

A favourite type of experiment is as follows: Food is put inside a shut box, to which some latch or spring or hidden passage will give access, and the unfed animal is left outside the box; or vice versa, the animal is imprisoned in a puzzle-box, and food and liberty are outside. If practicable, the first method is the better, as the animals are not flustered or frightened at their imprisonment. Rats and squirrels can do little but find hidden weak spots through which they could dig a tunnel; but creatures with considerable skill in manipulation, like dogs and raccoons, are capable of quite elaborate feats. They learn how to slide back a bolt, to push down a lever, to pull the dragging-loop of a string which opens the door, and so forth. They even learn, all by themselves, combination fastenings in which several operations have to be performed in a definite order before the door will open. But— and this is a very important point—they never learn their tasks with even an infusion of what one might call reasoning; the process is entirely one of trial and error. They claw about until a happy accident produces the right result; and then, as the test is repeated, by what seems a fundamental law of learning, the successful movements are gradually "stamped in" to their behaviour, the unsuccessful one gradually "stamped out." They improve steadily in their performance until they do the trick quickly and automatically. But what they have mastered is a trick of movement, not an understanding of how the lever or the catch or the string works. They have not so much learnt a lesson as formed a habit. For alter the arrangement of the fastenings, without any change in the mechanical principle involved, and they have to begin again at the beginning with random scrabblings and build up a new movement-habit as laboriously as before.

A dog, for example, learns of its own accord to get into a large box for food, by pressing a little lever. One day the box is turned through a right angle. The animal is completely put off; it goes up to where the lever used to be and scratches away fruitlessly; and it takes as long to learn how to get in as it did the first time.

Just the same automatic habit-formation without any insight is brought about when animals are trained to run a maze to get food or liberty. Rats, for instance, will quite soon learn the secrets of even complicated mazes, such as a miniature of the celebrated one at Hampton Court. But this is a blankly unintelligent habit. A human being who has learnt a maze fairly well, will never go wrong at the entrance; the first turning is the first thing he learns. But rats, even when they have reduced their mistakes to very few, are just as likely to make one at the first turning as anywhere else. Similarly, if a man is set to learn

EVOLUTION OF BEHAVIOUR IN VERTEBRATES

a maze which is a mirror-image of one he has already learnt well, he quickly recognizes the fact, and by always substituting right turn for left turn, succeeds very soon with his new task. Not so the rats; such a maze is no easier to them than a completely new one. They are unable to grasp its analogy with the first.

We ourselves form motor-habits of much the same kind. Your house is, in a sense, a problem box; you have to learn your way about it; and the mechanical automaticity with which you proceed from the bedroom to the dining-room to get your breakfast, perhaps while thinking of something quite different, is parallel to the mechanical way in which the animals solve their problems when they have been learnt. The animal sees its box, presses the lever, gets inside, and thus earns its meals; but it no more thinks out the train of actions as a logical sequence when it performs them than you think out the motions of your legs and the turnings of your door-handles on your way to breakfast. The process by which your motor-habit was established was rather different from the trial-and-error method of the animal, but the final result is strictly parallel. Another example of a human motor-habit is afforded by the actor who has repeated the same line night after night for hundreds of nights, and now goes through his performance mechanically; if the habit, the automatic flow of words, should fail for a moment, he is helpless and breaks down completely.

It is not by any means easy to tell what part automatic motor-habits of this kind play in the lives of wild mammals. We may guess that a rabbit runs round its burrow as automatically as we run round our houses. In one or two cases, we have more definite information of animals whose actions have been stereotyped into a cast-iron routine. Thus there are animals which deposit their dung in the same place day after day, with extraordinary pertinacity. Then there are the rounds made by animals in search of food. The Malayan rhinoceros, for instance, has a regular round of feeding-grounds, which he covers in about a month. Dr. Ridley describes well-worn tracks through the forest, about three feet across, along which the animals travel, generally by night. There is a story of a temporary hospital hut, which had been hastily run up, and happened to be exactly on a rhinoceros track; when the beast came round again, he went in at one door, down the central passage between the beds, and out at the other, much to the alarm of the patients.

But we must get back to our puzzle-boxes. When a habit is established, its automatic execution is very similar in man and beast. But there is a great and important difference in the way in which habits are acquired. The animals in the experiments, in most cases at any rate, solve their problems by scrabbling about aimlessly and remembering the movements which happen to work. But put a man in a strange situation, and the odds are he will behave in an altogether more reasonable way. He will think things over, and not begin to try solutions until he gets some kind of idea to work on.

We would make it clear here that the essential contrast does not lie between human behaviour and the behaviour of other mammals. It lies between two different methods of attacking problems. One method is to try all sorts of movements in the hope of muddling through; the other is to attempt to understand the problem before actually performing its solution. On the whole, most mammals in these experiments employ the first method, and most men, in the affairs of everyday life, employ the second.

Let us take a simple example. We rig up a little piece of wire netting at right angles to the walls of a house; after three or four yards we put a right angle bend into it, so that it runs parallel with the house for another couple of yards. For our experiment with this simplest of "mazes" we choose three organisms—a hen, a dog, and a child of five or six. We lead them up to the wire netting and throw a tit-bit (the tit-bit is, of course, suited to each subject) over it. The problem is successfully solved if the subject of the test without hesitation sums up the situation and makes off round the backwardly-projecting bit of wire-netting to the prize. You may say that the problem is so stupidly simple as to be no problem. Not at all. The hen never solves it properly. So long as it sees and wants the food, it will dash and flutter vainly against the netting. If it does succeed in getting round to the tit-bit, it will be because it has abandoned the problem and started to go away, and then accidentally turned so as to see the food from a more favourable position. The little human creature, on the other hand, will never fail to trot round. And the dog is intermediate. If the food is thrown well over the wire, the dog may take a few ineffectual jumps towards it, but then will, it seems, suddenly grasp the problem, and run round to get it in a purposeful and single sweep. But if the food be dropped just over the wire, so that it is within a few inches of its nose, it behaves as stupidly as the hen. The stimulus is now too potent; the dog is, as it were, magnetized by it and cannot acquire the detachment needed to run round; and so he remains scrabbling and barking stupidly at the unattainable food.

The differences in method of attack are impressive; they represent an important step in the evolutionary process that has led up to the human intellect. Let us summarize now the main stages in that process, as we have seen them.

First, is the completely unintelligent stage, when behaviour is inborn and stereotyped, and the individual has no power at all of profiting by experience but reacts like a machine to whatever stimuli may be acting at the moment. That stage we illustrated by considering in some detail the behaviour of the slipper animalcule. Confronted with a problem—such as a barrier across its path—this little automaton simply changes its direction at random and pokes about until it happens to get round.

Then comes the power of remembering which of a number of random responses happened to work in a given situation. This brings about a considerable economy of reaction; when the same problem confronts the organism again, it is solved in less time and with less expenditure of energy than before. Learning of this kind appears in a very crude form in various invertebrates—we noticed it in earthworms (Book 8, Chap. 1, § 4), and it runs through the vertebrate series. But it is always subordinated to instinctive reaction in animals below the mammalian grade.

Stage three is a further move in the same direction. The capacity of the cortex is extended; memories are accurately stored, and the power is acquired of comparing and contrasting different situations, of noting their resemblances, and of putting two and two together. Many mammals even have never reached this stage. It represents a further saving of effort and time; the dog which has learnt to open a puzzle-box and is then completely beaten when the box is turned through a right angle has to start again with random movements and gradually worry out a new solution. But a man who can look at the box and realize its change of position performs the appropriate movements as soon as that one thought-process is accomplished.

The performances of sheep-dogs show what the dog species can do when what is demanded is skilful adjustment of behaviour to a situation where essentials are already grasped; the experiments with wire-netting and puzzle-boxes show how limited is their power of thinking out the principles of wholly new situations.

Thus, with all their capacity for learning, most mammals do not seem to dispose of anything that merits the name of an idea. In the main, their learning powers are of our second grade above; they enable them to be steered into useful habits by a mnemonic selection of random movements. Their actions may look deceptively like our own—until suddenly some unexpected incident shows the profound difference. A cow, for instance, whose calf has been taken away from her is wonderfully comforted by its stuffed skin. Cases are on record where the mother, happily licking away at such a dummy calf, has licked its seams open, so that the hay with which it was stuffed has protruded; then it has proceeded to eat the hay. Ewes know their own lambs —a remarkable feat of discrimination, one would think. But in reality, the recognition seems to depend on simple smell. Once the mother has licked her offspring, it is recognized as hers; if you substitute another lamb before her own has been licked, the changeling is licked and then treated as if it were the legitimate child. A rat, if the nerves of its foot have been severed, no longer recognizes this foot as part of its body, though it is still alive and pulsing with blood, but treats it as an extraneous object and proceeds to gnaw and eat it. It has no "idea" of its own body; and is normally prevented from maltreating its toes only by the pain which arises when they are nibbled.

And what of the dog, one of the more intelligent mammals, who still persists in burying bones however well-fed, and in turning round and round before going to sleep as if the mat were herbage in which a bed had to be made?

This last example is a reminder of the importance which straightforward instinct still plays in the life of most mammals. Their instincts are much less fluid and command their lives much more directly than do ours; ours influence each other, become incorporated with ideas and modified by traditions, in a way which is new in life's evolution.

Miss Pitt, in her various books on natural history, has described how fox-cubs and young otters, growing up in captivity and free from the need of finding their own food, will suddenly reveal their latent instincts. Particular smells pull the trigger, and the hunting, pouncing, and worrying reactions, till then never exercised and never seen in others, are automatically released in full force.

The most remarkable mammalian instincts are undoubtedly those of beavers, who build dams and dig canals untaught. The dams serve to make pools in which they store food (bits of trees with the bark on them) and where there will be water under the ice in winter for them to swim to their food-pile. The canals are to facilitate their lumbering operations; when they have felled a tree and cut it up into segments, it is easier to float them to the pond by water than to pull and push them overland. Dams may be up to a fifth of a mile long; and canals even longer. Both kinds of works seem to show engineering skill; the canals especially are run so as to ensure a gentle but unfailing flow of water. It is in the highest degree improbable that beavers have any comprehension of the principle of gravity or the fact that water finds its own level; exact observation will probably show that they regulate the course of their canals by some simple method of trial and error, such as only continuing to dig if the water flows, or is more than a certain depth. It is often stated that they fell their trees intelligently, gnawing at them so that they shall fall free and not be wasted by remaining entangled with other trees, but this appears to be a myth.

It would be a fascinating study for a student of animal behaviour to rear a group of young beavers by hand and discover exactly how much their unaided instinct was capable of; and still more fascinating if the group of untaught beavers could be compared with a group of untaught human children. But both experiments remain to be done.

It is a curious fact that the scientific mind and the activities of the reasoning faculty are so frequently written down as "inhuman." Actually, this "cold" power of abstraction, this "inhuman" reason, is the one emergent property which the human species possesses, while our warm "human" emotions we share with the brutes. There can be no reasonable doubt that other mammals are subject to the same kinds of passions, feel the same sorts of emotion, as we ourselves. Our sheep can be frightened; our dog is glad when we come home, feels something closely akin to shame when caught in some misdeed; our cats can experience anger and disappointment. But the capacity to subtract eleven from twenty-four, to grasp that the earth is round and the sun some ninety million miles away, to understand general statements such as that Honesty is the Best Policy; to attach any meaning to abstract terms such as Space or Truth—this is all distinctively and exclusively human. There are only the barest germs of such capacities in other creatures.

§ 7. Education in Animals.

THE higher vertebrates, we see, can learn, mammals much more than birds. The development of the brain and especially of the cortex means the progressive replacement by flexible responses of the fixed responses of an instinctive system. Putting it compactly, the purely instinctive animal is born complete; the higher animal is born incomplete and learns. And it is advantageous therefore that the protective association of parent and offspring so characteristic of the higher types should involve a certain assistance in the learning process. This assistance is the beginning of education.

Not all animals that learn educate. For education, family or community life is one necessary pre-requisite. Birds like the bush-turkeys of Australia, which bury their eggs in mounds of earth and leave them to develop by themselves, hatch out away from all parental care, and have to rely on their own innate machinery of response from the very start of their lives. Obviously no education is possible to such a type. And in many other birds, even though family life is developed and the parents feed and protect the young, nothing one can really call education occurs. Mechanical responses of offspring to parent are not education. The crouching of young birds at their mother's warning call does not have to be learnt; it is an automatic and innate response, like the blinking of our eyelids to a threatening hand. According to Hudson, unhatched birds that are squeaking while still inside the egg will stop their noise at their parents' alarm-call. The small part which training plays in bird life is brought home forcibly by the familiar fowl with her foster-brood of ducklings, who, one fine day, all untaught and very much to the distress of their acting parent, proceed to swim away across a neighbouring pond.

Flying again is an activity as innately given as swimming; when the time comes and the muscles and nerves are developed to the proper pitch, a bird will fly; it will fly a little awkwardly at first, and practice will be needed to give it full mastery of the air, but it has no need of any training to be able to use its aviation machinery without danger of crashing. Most young birds take their first flight quite independently of their parents; but in some species the old birds, though they are in no way concerned with helping their young to greater skill, as a golf or a skating instructor helps his pupils, still assume the definitely educational task of stimulating the young bird to take the plunge at the earliest possible moment. Some eagles, for instance, when the young have been five or six weeks in the nest, no longer feed them so regularly; after a time they may withhold food altogether, but will sit about in the neighbourhood of the nest, calling, until the young birds grow bold through hunger and launch themselves into the air. It has even been stated that old birds will take their young on their backs and then swoop downwards, leaving them to fly as best they may; but this is probably a traveller's tale. Curiously enough, although ducklings will go to water of their own accord, young gulls and swans have to be brought there and induced to enter the strange element.

Many birds of prey, however, plainly educate their young in hunting. Falcons catch and cripple prey and then leave it to be finished off by their offspring, just as many beasts of prey bring home wounded animals for their half-grown young to try their claws and teeth upon. Grebes let their children pursue fish which they have caught and liberated again in a damaged state. In many cases the mother helps to teach the children what kind of food to look for; the clucking of a hen attracts her chicks to come and see the morsel she has found. Even if the parents do not impart instruction so deliberately, the mere existence of a family group constantly puts the young in situations where they must profit by their elders' experience of the world. Lion-cubs, when they are big enough, are taken on their parents' hunting expeditions, and it is stated that they do not become independent hunters until they are over eighteen months old.

Just how far these cases represent a true handing-on of actual experience from old to young is difficult to determine. We have seen that in most mammals and probably all birds, learning consists in making more or less random movements, without any definite purpose behind them, and remembering which of them are useful. The parents, in the examples which we have discussed, do not so much present their young with the results of their experience as help them to get experience for themselves. They hurry the normal process up; that is all. The kitten confronted with an injured mouse examines it, plays with it, claws at it, and so learns how to handle it in much the same way as it would learn how to solve the riddle of a puzzle-box; the mother cat, by bringing the mouse, has provided opportunity but not direct instruction. The education is essentially self-education.

One commonly hears it said that parent animals teach their young by doing things in front of them, which the young try to imitate. But in actual fact, it is very doubtful whether any mammals, other than monkeys and apes, have a sufficiently well-developed imitative faculty for this. In experiments with puzzle-boxes it was often noticed that if animals saw one of their fellows press the right lever, or had the human experimenter repeat the trick a dozen times before their eyes, they learnt nothing from it. Even if they were held and passively put through the trick themselves, it was rarely of any avail; the movements, to be learnt, must be their own. In our own species, any modern educationalist will tell you how much better children remember what they have done or been persuaded to do of their own initiative.

In monkeys and apes and ourselves, the instinct to imitate brings about a great abbreviation of the educational process, by leading the young to make only those movements that they will find to be profitable. But how far this occurs in other mammalian species is still somewhat doubtful.

The greatest extension of the educational process came with the invention of language; for this made it possible for the actual experience of one generation to pass on to the next. Not only this, but contemporaries share their experience. "I don't advise you to buy such-and-such a make of car," says Jones to Brown. "Robinson got one, and it's rotten." By means of the written and spoken word, the joint experience of millions of living beings rolls up into a single whole.

Nowhere else in the mammals is this handing of experience from the individual who has experienced it to one who has not, even paralleled. But there is a somewhat analogous case in birds.

It has been known from time immemorial that not only parrots but many birds will imitate the sounds they happen to hear. To this, the mocking-bird owes its name; the American blue-jay, the European starling and sedge-warbler imitate other species in a state of nature, while raven and mynah can readily be taught new notes in captivity. Bullfinch fanciers train their prize birds by the aid of a pipe to whistle special phrases, and then the birds are made to sing against each other in public.

Of late years, critical investigations have revealed the surprising fact that whereas in some birds the song is entirely inherited, so that young males raised out of earshot of their own kind (and even forced to hear the songs of other birds) will in due course give the characteristic song of their species, in others the full song must be learnt, and isolated males never get further than a few notes or feeble phrases. An experimentally-minded German fancier has even succeeded in grafting the nightingale's song on to a strain of canaries. The young canaries were reared in a soundproof room where the only singing birds were cock nightingales, and the nightingale's short song-period was supplemented by gramophone records through the winter. The canaries picked up the alien song; and now, after a few years, this has become self-perpetuating, since the cocks sing only nightingale-wise, and their offspring learn from them. The song is not exactly like the nightingale's, for the tone-quality is shriller, the phrasing not quite perfect; but it is far more nightingale than canary. In such species, song (like human language, but unlike the croaking of frogs or the music of grasshoppers) has to be learnt anew in each generation.

But in general, education, even in the warm-blooded birds and mammals, plays only a minor rôle. It is one of the latest tools of life, whose more elaborate possibilities, latent from the time when men began to speak, are only beginning to be exploited.

This is, perhaps, the place to speak of the various animal prodigies,

EVOLUTION OF BEHAVIOUR IN VERTEBRATES

the calculating horses and conversational dogs, who have made their bow before the public in the last thirty years. (It is interesting, by the way, how such phenomena come and go in bunches; they seem to be catching, or at least there are fashions in them. This has been true for mesmerism, for table-turning, and for various aspects of spiritualism.) The claim is made (large books have been written about it) that horses, for instance, can be trained to perform quite complicated arithmetical calculations, and that Airedale terriers can discuss morality and a future life with their mistresses. Since animals lack the power of vocal speech, their utterances have to be conveyed by movements, generally of the feet. The commonest method is that employed in table-rapping; so many taps with the right foot mean such and such a letter or number; with the left foot, they mean some other letter or number.

The calculating horses of Elberfeld attained such a celebrity that solemn commissions of university professors were appointed to consider their case. Their portentous arithmetical capacities and the equally super-normal (but rather banal) communications of the philosophical terriers have provided much helpful material for theosophists and spiritualists.

But in sober truth, these thinking horses and dogs do not reflect at all deeply. What the creatures invariably do is to notice and respond to little signs given by their masters or mistresses. Sometimes the signs have been given deliberately; but in the great majority of cases they are unconscious—slight gestures, a shift of the head, a twitch of a finger, unconscious movements of the same kind as those which are utilized by the professional "thought-readers" who, though their eyes are bandaged, find a hidden object by holding the hand of someone who knows where it has been hidden. They may be all but unnoticeable by the human observers—movements through a twentieth or a fiftieth part of an inch; but the animals who do not have to keep their attention on the talk, can spot them. The dog or the horse goes on tapping the ground with his foot until he sees the movement which he takes as the sign to stop; then he stops. It is worth mentioning that the horses often do not even look at the blackboard on which the problem is chalked up, before beginning the tapping. Also that the errors which they most frequently make are not in the least the sort of errors one would expect, like a failure to carry ten; usually the mistake consists in being one wrong in the answer—tapping twenty-two instead of twenty-one, for instance; or else in transposing digits and tapping twenty-seven, say, instead of seventy-two. In the former case they have not noticed the movement until just too late; in the latter, they have used the right foot in place of the left—that is all.

§ 8. Play.

PLAY is so engrained in human life that we rarely trouble to ask ourselves how it originated and what its meaning may be. The children of Greece and Rome had their dolls and toys; so did those of Ancient Egypt, twice as far back in time, and so, we may safely hazard, did the children of the later Old Stone Age. And if the children had their playthings, the parents had their games; one of the recent discoveries at Ur, dating from over five thousand years ago, was a board for some sort of game resembling draughts.

But play is rare in other than human life. With the possible exception of ants, play is unknown outside the vertebrates; and among vertebrates, it is not certainly known outside the two warm-blooded groups, the birds and mammals. If the bird attains the highest level of courtship, it is the mammal which best knows how to play.

What is the biological function of play? Let us consider a few examples to clear our minds about this question. We are standing in the bows of a steamer in the Mediterranean. Some distance away we see a series of leaping forms, one behind the other, each curving over in a semicircle to dive below the surface and re-emerge a few seconds after. They are a file of dolphins. After a time they sight the ship, and alter their course to meet it. Then the real fun begins. They play around her stem, never actually achieving contact with her sides, yet always on the verge of it. Now one of them bores down into the blue water until he is a mere dim shadow; now he comes twisting up again. Sometimes they circle right round the ship; but usually they are content to gambol around the bows, effortlessly keeping up with the power-driven machine. Only after hours will they leave her.

No fish would ever behave like this. Fish will leap out of the water, but only to avoid their enemies; they will keep poised in the current of a stream, but only because that is the business of their lives. The dolphin and the porpoise are mammals; and they prove it by their playing. Those who prefer everything to be sensible and simple have suggested that porpoises really frequent the bows of ships to rub barnacles and other encumbrances off their backs; but the unanimous verdict of those who have watched them is that this is not so—the porpoises are not being reasonable, they are being playful. Sometimes this play-impulse may be linked up with others; porpoises may play round ships not only because they like playing, but because they anticipate scraps being thrown overboard.

Kittens are perhaps the most playful of all young animals. Everyone knows how a kitten will spend minutes intently patting a cotton reel or a rolled-up scrap of paper across the floor, how a dangled string or watch-chain will lure him on, how he will pounce in pretence at hands moving under the coverlet, or even pursue his own tail in ludicrous gyration. Gradually this playfulness will die out, and the kitten give place to the staid cat, who prefers dignified fireside purring to the frivolities of play. Most adult cats preserve only one play-activity; they play with the mice they have caught, letting them go to recapture them a hundred times before they finally kill them.

Kittens, as befits the offspring of a solitary species, are perfectly contented to play alone; but puppies, though they are by no means averse to solitary sports, such as worrying an old boot to bits, play best in company. On such occasions the make-believe and pretence which characterize so much playing, animal as well as human, is vividly revealed. Two half-grown pups are playing at fighting; they rush at each other, roll over and over, snap, growl, worry. And yet neither ever hurts the other; they always know when to stop. They are not angry; but they are thoroughly enjoying themselves playing at being angry. And in just the same way, they will indulge in mimic chasings, first one and then the other taking the rôle of pursuer.

Birds, like mammals, will play. The snake-bird or darter is a fresh-water cormorant with amazingly long and flexible neck, by whose aid it pursues and catches agile fish under water. A female snake-bird has been seen sitting on the swamp-cypresses in a Louisiana pond and playing at catch, all by herself. Reaching down with her long neck, she picked a small twig, then threw it up in the air and caught it in her beak. This was repeated until she misjudged matters and missed the twig; she cocked her head at it as it fell, and proceeded to pick another with which to continue the play.

In birds, the commonest form of play is flying play. Ravens, in spite of their size and staid appearance, have various aerial sports. A favourite one is to turn almost completely upside-down in the middle of ordinary flight, giving a hoarse croak at the critical moment; this may be repeated over and over again. Or they may dive and somersault together in the breeding-season. The small egrets and herons that nest here and there in protected rockeries along the coast of Louisiana and Texas return every evening to their breeding-pond from feeding on the marshes. A steady concourse pours in, along various flight-lines, about two hundred feet up. Arrived above their home, they simply let themselves drop. Their plumes fly up behind like a comet's tail, they scream with excitement, and when not far above the ground, spreading the wings so as to catch the air again, they skid and side-slip wildly before alighting.

Now we can begin to say something more definite about play—what it is and what it is not, its occurrence, its meaning. Play occurs in adult as well as in young animals. Usually it is described somewhat obscurely as an outlet for surplus energy. But such activities as the cat's playing with mice are definite responses to particular situations and the downward plunging of home-bound egrets and herons could equally well be described as a way of achieving a necessary action in the most pleasurable and exciting way. Whether or not it serves to excrete surplus energy the play may take very various forms. It may be a direct imitation of some regular activity of the species, but playfully carried out, like the sportive fighting of dogs, or the cat's behaviour with captured mice. Though not a complete imitation, it may be obviously connected with some such special activity, as is the pouncing of the kitten; or it may serve to use some special activity in a new and interesting way, as in the snake-bird's twig-catching play. Or again, it may serve the same end for some more general activity like bodily movement; and in such case, the surplus activity may simply overflow in sheer exuberance, as in the romps of young lambs or puppies, or be guided into more special and fixed channels, as in the somersaulting of ravens. The same differences are to be seen in our own play. Many games are more or less accurate imitations of serious activities, others are mere romps; and there are sports which, like ski-ing, we practise mainly for their thrills of new and violent bodily motion; there are adults' games which go back to the playfulness of childhood. There are many children's games which anticipate the business of later life:—

> "Behold the child among his new-born blisses,
> A six years' darling of a pigmy size . . .
> See, at his feet, some little plan or chart,
> Some fragment from his dream of human life
> Shaped by himself with newly-learned art."

From the point of view of its evolution and its biological meaning, play seems to have a double origin. There is the play which is biologically useful as a preparation for adult life; this is play in the strict sense. And there is the play which results from a mere surplus of energy being directed into pleasurable or exciting outlets; this, if we want to distinguish it, we can call sport. The former is more characteristic of young animals, the latter of adults; the former predominates in the mammals, the latter among birds. Karl Groos in his classical book, *The Play of Animals*, stressed the view that play is essentially useful to the species; and that, since it was useful, its origin could be accounted for on ordinary selectionist principles. Others have preferred to believe that play was wholly useless, and could be accounted for altogether as an aimless overflow of energy. But our distinction between preparatory play and sportive play enables us to take a middle course.

Preparatory play is found almost exclusively in mammals. It depends on the fact that a mammal comes into the world as a singularly unfinished product. Most of us do not realize how much we had to teach ourselves in our childhood. Avoiding objects as we walk about, judging distances, picking things off a table—all such acts are so automatic to us that they seem trivially simple. Yet we had to learn them all, and the learning of them was one of the most elaborate of our acquisitions. As a reminder of what that learning meant, we may describe a striking experiment carried out by Professor Stratton on himself. He had spectacles made which inverted everything: when he put them on, the world was upside-down to him. He wore them continuously, and naturally had to learn how to fit his movements to this new picture of his surroundings.

"When I saw an object," he writes, "near one of my hands and wished to grasp it with that hand, the other hand was the one I moved. The mistake was then seen, and by trial, observation, and correction the desired movement was at last brought about."

It took him several days before he could begin to work smoothly in the new conditions. Even more difficult (perhaps because it looked less queer) than the inverted world was a world produced by another set of glasses in which right and left were transposed, as in the world we see in the looking-glass. With these glasses, "At table, the simplest act of serving myself had to be cautiously worked out"; and it took him a week before he had adjusted himself fairly adequately to this mirror-image world. On the eighth day he removed the glasses—and by then the new habits had become so well learnt that it took him over twenty-four hours to get back to normal. The effect of a week's reversal of right and left was strong enough to override the habits of a lifetime, and for a day to keep him very literally "out of touch with reality."

An insect, because of the stereotyped elaboration of its nervous system, can fly as soon as its wings are dry and stiff, and not only so, but it can direct its flight in relation to other objects. But the mammal has to learn how to control its limbs, and how to correlate movement with sight. Its mental peculiarity necessitates a period of playful immaturity.

This is why the higher mammals, with their greater brain-power and capacity for learning and their longer infancy, excel in preparatory play, while the birds, which become adult in a few weeks, and are less intelligent, play very little when they are young. But in sportive play the birds, with their high temperature, ceaseless activity, and wonderful bodily powers of flight, excel the mammals, who tend to become more lethargic as they grow up; only a few mammals, like some seals and dogs and monkeys and apes, go on playing all their life long. Whales may leap bodily out of the water; and probably this is a form of sporting thrill. The two forms of play, of course, grade into each other. The exercise of twig-catching skill by the snake-bird helped, no doubt, to keep her hand in (or rather her neck) for the business of catching fish, and the flight-sports of birds must keep their powers in trim against the serious tests of gales and storms. None the less, the distinction is a useful one.

To what a pitch pure sport may go in birds is shown by the community flying-games of rooks. Here is a typical observation. A large gathering of these gregarious and intelligent birds was seen in a field in February. About half the birds were excitedly walking about and cawing; others were mounting in steady spiral flight, all fairly close together. When they had arrived at a height of four or five hundred feet, one after another they folded their wings and dropped. They whizzed down like plummets; when only forty or fifty feet from the ground they spread their wings and began braking with them. As a result they skidded and swerved through the air in the wildest way, eventually alighting to walk about and caw a little before repeating the sport. The sport itself is like that of the egrets; here, however, it is not merely an embroidery of the necessary return to their homes, but is organized for its own sake, as we laboriously plod up hill with toboggan or skis in order to enjoy the downward rush. The same sort of sport has also been observed in early autumn; and once, it is recorded, the rooks were seen to mount so high that they disappeared from sight; only then, at several thousand feet up, did they turn, dropping out of the blue to attain alarming velocities before they put on their brakes. As sensational and as clearly practised for sheer enjoyment is the behaviour recorded of snipe—not the species as a whole, but a few

EVOLUTION OF BEHAVIOUR IN VERTEBRATES 1257

sporting individuals—in flying upside-down. In the most remarkable case, a snipe which had been "drumming" up aloft came swooping down, and when only a few feet above the ground turned on its back, and continued, thus inverted, in a horizontal course for several hundred feet.

Perhaps the most human of all cases of animal play is one recorded by Levick for Adelie penguins. Floes and little bergs of ice were all the

FIG. 313. Adelie penguins at play—taking a ride on a drifting ice-floe.

time drifting in a strong current past the edge of the land where their rookery was situated. The penguins loved to take joy-rides on these. They swam out and leapt up, until the ice-platform was sometimes too crowded to take another bird. They drifted down, contentedly excited, for about a mile; and then swam up again to take a fresh ride.

Probably similar to the rooks' aerial sports are the dance-gatherings of some other birds. Jaçanas, according to W. H. Hudson in his *Naturalist in La Plata*, assemble in bands and dance and scream, meanwhile waving their wings whose brilliant yellow is usually hidden. And our stone-curlews assemble, after the breeding-season is over, to indulge in the strangest antics, flapping, rolling sideways, racing about. From these primitive dances it is not a long step to the choral performances of another bird studied by Hudson on the pampas, the crested screamer. Although these are big, heavy creatures, a large flock of them often circles up until lost to sight, and there, in the upper air, the birds begin a melodious trumpeting, which floats down to the hearers below with more than the mystery of the hidden choir in Wagner's *Parsifal*.

Such community-singing is not uncommon, although in a less romantic setting, among song-birds. Although that migratory thrush, the redwing, only breeds in the far north of Europe, yet before they leave their winter quarters in England, considerable flocks of them may assemble in trees and give a concert, which, since the full song is not yet developed, is strangely subdued in its effect. These breed gregariously; but even in solitary nesters like the European goldfinch, such concerts may be given in early spring when the urge to song has begun to be felt, but the birds are still in flocks and bands.

Birds that roost communally often carry on the most animated "conversations" morning and evening at their roosting-places. Rooks are a familiar example: starlings are still more striking. Of late years starlings have taken to sleeping in huge numbers not only in the trees of London squares, but on the City buildings. It is one of the most remarkable experiences to hear the thousands of starlings on the cornices of St. Paul's Cathedral begin to sing and chatter at dawn in autumn, long before the City's population has arrived. The volume of sound is enormous; for a time the birds merely converse, but soon a few begin taking little flights from place to place on the building; the flights become longer and more numerous; and after twenty minutes or so bands begin to leave in all directions for the business of the day.

An even stranger preface to the day's work has been observed with swallows and martins. If you get up at dawn on a fine morning, somewhere where there are a number of house-martins' nests still unfinished under the eaves, you will probably see no birds about. They have all flown up many hundreds of feet into the air, as if to greet the rising sun as early as possible. When the rays of the sun touch them, their little twittering band sinks slowly down with the sunlight; and they begin their ordinary activities when they are down again to earth.

Certain animal habits are more like obsessions than play. Creatures that collect stores for the winter, for instance, often amass much more than they can ever use. The passion for collecting takes them, and they lay up food like a miser hoarding gold. Such animals as squirrels bury their food-stores in little caches here and there; and perhaps the majority of these hiding-places are permanently forgotten. The squirrel who makes many food-caches and forgets most of them is in its way like the fish or the sea-urchin who produces huge numbers of young only to have the great majority die early. The system is a wasteful one, like so many of nature's systems, but it works because enough of the caches are re-discovered, just as the sea-urchin's reproductive methods work because the survivors, though few, are enough. The squirrel's habits, however, may benefit the plants whose seeds he stores; he often provides for their dispersal and by burying and then forgetting them, he puts them into the best condition for germination.

The passion of certain birds, notably those of the crow family, for bright objects is well known; though magpies are really the most thorough-going performers, literature has made the Jackdaw of Rheims the best-known example. We do not know how this behaviour has originated; probably it is a combination of a hoarding instinct with an appreciation of brightness and striking colour. It is at any rate certain that quite a number of birds have a simple fondness for things that to us are striking or pretty, and use them to embellish their nests in the same way as we call in the decorator as well as the utilitarian builder when we are building houses. Birds of prey, such as buzzards and eagles, break off branches of greenery to put round their nests, renewing them as they fade. A number of wading birds put shells or bright pebbles round the depression which serves for their eggs; an interesting point in many of these species is their variability—some individuals decorate their nests abundantly, others sparsely, still others leave them bare. A queer case is that of some of the American flycatchers, who always hang a cast-off snake-skin on their nest—whether to decorate it or to protect it is quite unknown.

Levick found that Adelie penguins loved bright colours. The sitting birds are always stealing stones from each others' nests; Levick accordingly painted a number of stones with different colours, and put them within reach of the birds at one edge of the rookery. They were much coveted, and, by repeated thieving, travelled steadily across the colony. He made the interesting discovery that red stones travel the fastest. Although red is a colour which the penguins can seldom see in their normal environment, it tickles their senses as it does ours.

§ 9. The Behaviour of Monkeys and Apes.

THE construction of apes is so like our own that their actions constantly remind us of familiar human doings. We see an ape mother fondling her baby, and it seems to us that she must be experiencing the feelings appropriate to a human mother; we see a sad-faced orang-outang in a cage, and his expression convinces us that he is thinking about his past life, free in the Bornean jungle, as a human prisoner would think of his lost liberty. But then something happens that gives us pause. The orang mother wants to travel from one end of the cage to the other. The baby that her arm has been encircling at her breast is shifted to her prehensile foot and is bumped over the floor as she swings herself arm over arm along the roof-bars; she does not look nearly so human

FIG. 314. Experimenting with chimpanzees.

Mrs. Kohts testing her chimpanzee Ioni in the matching of colours.
(Courtesy of Mrs. Kohts, Moscow.)

now. Or the melancholy philosopher in the corner—if he is thinking, why does he never talk? If he is so human, why does he suddenly break off into some unrepressed obscenity? Is the mind behind the actions really so like our own?

Curiously enough it is only in very recent times that any systematic study of ape behaviour has been made. Just before the War, the German psychologist Koehler made a notable beginning on this problem, studying the behaviour of a group of young chimpanzees, four to seven years old, some fresh from the wild, and none previously trained in any way, in the warm climate of Teneriffe. The Americans, headed by Yerkes, have since then made intensive studies of various apes, and so has Mrs. Kohts in Moscow.

How like are these creatures to ourselves? That is the fascinating problem in all these researches.

EVOLUTION OF BEHAVIOUR IN VERTEBRATES

On the emotional side, there is a very strong resemblance. One need not be an expert observer to tell by looking at its face what a chimpanzee is feeling; its series of emotional expressions is almost identical with our own. Weeping, perhaps, has a rather unfamiliar look, and the

FIG. 315. Anthropoid expressions.

Mrs. Kohts' chimpanzee Ioni, moved by various feelings. (Courtesy of Mrs. Kohts, Moscow.)

pursed and protruded lips of excitement so much exaggerate the thrilled child's gesture when it says "Oo-ooh," that the grimace seems unnatural and grotesque. But they fondle their babies and kiss their friends, human as well as simian, to show their affection in an entirely human way. They obviously like play for its own sake, especially when young, and are human enough to enjoy teasing other and stupider creatures like fowls.

When it comes to more complex feelings, the resemblance continues. Jealousy they share with many of the tailed monkeys, such as baboons. Madame Abreu, a Cuban lady, whose large collection of apes and monkeys on her estate in Havana has been studied and described by Yerkes in his book *Almost Human*, had a baboon who always tried to hide his mate when any male human being came near the cage; women, on the other hand, he did not worry about. Madame Abreu once brought a Catholic priest as test-object to see if the baboon would take him for a woman on account of his long cassock; to judge by his behaviour he was not deceived.

Sympathy is strongly developed in most chimpanzees. If one of their group is ill, he will not be teased or disturbed; now and again one will come over and caress the ailing companion. Actual whimperings or moans of pain will almost invariably bring the other chimpanzees round a sufferer, often with touching manifestations of concern; but Koehler found that ape-sympathy, though so readily expressed, needs stimulation. Out of sight, out of mind; when he removed sick chimpanzees to a distant hut, the others betrayed no recollection of them— they neither searched for them nor showed any signs of sadness or feeling of loss. Much the same is true of their affection for their babies. When a baboon or a chimpanzee baby dies, its mother generally refuses to part with its corpse, but carries it about until it becomes dry and mummified. Tricks have to be resorted to get the body away. But once it is gone, memory is short-lived; after a little searching, the mother shows no sign of remembering her loss. In this as in so much else, the ape's behaviour differs from ours chiefly in respect of its immediacy.

The chimpanzees are among the most sociable of animals. The greatest punishment you can inflict upon a young chimpanzee is to put him into solitary confinement. He pines, mopes, grows listless and sickly. They are almost as ready to accept companionship from human beings as from their own kind; but now and then explosive happenings remind the human that he is dealing with a different species of organism. Koehler, for instance, in his *Mentality of Apes*, forcibly describes the working of their herd instinct. "The moment your hand falls on a wrong-doer, the whole group sets up a howl, as if with one voice. . . . At times the most insignificant episode between man and ape, which arouses a cry of anger against the enemy and springing against him, is sufficient for a wave of fury to go through the troop; from all sides they hurry to a joint attack. In the sudden transfer of the cry of fury to all the animals, whereby they seem to incite one another to ever more violent raving, there is a demoniac strength. It is strange how full of moral indignation this howling of the attacking group sounds to the ears of man; the only pity is that every little misunderstanding will call it forth as much as a real assault; the whole group will get into a state of blind fury, even when most of its members have seen nothing of what caused the first cry, and have no notion of what it is all about."

Fear, too, can be induced. If Koehler went into the cage, simulated terror and looked fixedly in one direction, the chimpanzees ran together and looked fearfully in the same direction, although of course there was nothing whatever to be frightened of.

In such a socialized animal, the instincts of self-suppression and self-assertion are of great importance. One outcome of the self-suppressing tendency is seen in the trustful way in which many chimpanzees submit themselves to medical treatment. An outcome of the urge to self-assertion is the immediate advantage which they take of the least weakness or timidity in other apes, or in human visitors. And the interplay of the two opposed tendencies results in the establishment of a regular order of precedence. After a group of them have lived together for a week or so, each will have found his social level. One (who may be of either sex) will be head of the gang; and of the rest each will boss certain individuals and let himself be bossed by others, without matters ever coming to a head in a fight.

Human beings may be drawn into the network of precedence. De Haan records a remarkable instance of how this system works with tailed monkeys. He was experimenting with a mangabey and a macaque. The mangabey entirely dominated his companion; he had only to lift his eyebrows or show the whites of his eyes for the macaque to throw down his food and retreat hurriedly into another compartment. The mangabey, on the other hand, felt himself inferior to the man. He offered no resistance to De Haan's threatening gestures or looks, but reacted to them in a regrettably human fashion (very like that of the clerk who has been hauled over the coals by the manager and vents his spleen on the office-boy), by looking viciously at his companion, who immediately slunk off into the small compartment.

One day, De Haan wanted to demonstrate this procedure to the keeper. But now when he threatened the mangabey, it sprang at him instead of turning on the other monkey. To the monkeys, the keeper was a much more important personage than the man of science; and the fact that he was there and remained benevolently neutral was sufficient stimulus to the self-assertive side of the mangabey to make him react directly to De Haan's threats; but as soon as the keeper left the room, the mangabey gave up his bold face. If the keeper threatened the monkey and De Haan stayed quiet, the beast turned on De Haan; but if De Haan came with a visitor and made threatening gestures, it was the visitor who was treated as an inferior and snarled and screamed at. If De Haan pretended to have a scrap with another man, the mangabey always took sides; and he always sided with the "higher in rank"—with De Haan against a casual visitor, but with the keeper against De Haan. As our author says: "It was certainly surprising and somewhat disappointing to see how this feeling of the rank of the different persons predominated over sentiments of personal affection."

We could run on and fill a chapter with details of the emotional behaviour of apes, so fascinating are its odd resemblances and differences to our own. But the reader can go to the admirable books of Yerkes and Koehler; for our present purposes a few more anecdotes must suffice.

A young orang-outang whom Yerkes was testing would show his discouraged puzzlement with a problem that was too much for him by repeatedly bumping his head none too softly against the floor, just as a man might hit his head with his fist in despair at his own stupidity.

An adult male chimpanzee at Madame Abreu's place began to evince an embarrassing interest in a fair-haired girl in the kitchen, whom he could see at her work from his cage. The door into the kitchen was accordingly screened; and the chimpanzee saw the screen being put up by one of the men attendants. Before this, the man who had been instrumental in depriving the chimpanzee of the sight of the blonde kitchenmaid, had been on very friendly terms with this chimpanzee; but a few days later the ape, seizing his opportunity, made a vicious attack on him. The complex feelings engendered by this incident prevented the ape ever afterwards from looking his former friend in the eyes, or even accepting food or caresses from him.

Large and unfamiliar animals produced panic in Koehler's chimpanzees. The sight of two big oxen so terrified them that it acted like a purge; and the passage of a camel made it impossible for any experiments to be done for a considerable time.

Koehler then undertook some experiments with crudely-stuffed toys. The realistic school of animal psychologists would have us believe that animals are only likely to react strongly to an object which is familiar, or at least pretty similar to something in their natural environment. "But," he writes, "the chimpanzee's reactions were in almost comic contradiction to this view. Almost any representation of an animal, even if small and friendly-looking, is treated as uncanny; and larger and more grotesque toys are the occasion for paroxysmal terror." When Koehler came into the cage carrying a goggle-eyed quadruped, about eighteen inches high, with some resemblance to a donkey, "in a second a black cluster, consisting of the whole group of chimpanzees, hung suspended in the farthest corner; each tried to thrust the others aside and bury his head deep in among them," and when he suddenly put on a Cingalese devil-mask, the apes were equally terrified.

It was the combination of resemblance and difference that inspired their terror. Purely geometrical constructions had no such effect as animal toys, and merely concealing the face in a sheet would not act as did the mask. This encourages us to believe that the lantern-bugs which look like miniature reptilian heads do profit by the resemblance; and also gives us an insight into human nature. The uncanny, that which inspires with awe and sacred terror, is not the completely strange; if it is too far beyond ordinary experience it is simply not grasped. To be effective, it must combine the familiar with the unfamiliar; it must be strange, but recall what is well known.

Let us come now to the intellectual development of the ape. Here the difference between apes and men is more marked—and this in spite of the apes' conspicuous superiority over other mammals. They have no true language; that is their first and greatest deficiency. They have a rich and varied vocabulary of sounds, and these are often used for communication; but they are always expressions of their feelings, never descriptions of objects. If a chimpanzee has a banana taken from him, he can express the fact that he is angry; if he wants a banana, he can express the fact that he is hungry; if he gets a banana he can express the fact that he is satisfied. But he cannot say anything about the banana itself. No ape has any words for things. Nor is it easy to train apes in language habits. With great difficulty, one chimpanzee has been taught to use a few real words; other experimenters have failed to achieve even this much.

Their mental life extends very little either into the future or the past. Like young children, they live mainly in the present. There are many stories, apparently well-authenticated, of apes attacking the author of some injury after months of absence; and we are apt to think that this implies a memory like our own. But there is no evidence that the creature has any power of mental recall. The offender has generated resentment against himself; when he shows himself again, resentment is again aroused. This is an example of a long-enduring effect of past experience; but that is no evidence that apes (or any other animals) can be human enough to brood over their injuries, or think revengefully of an absent enemy.

In what way, then, are the apes on a higher level of behaviour than cats or dogs or horses? In considering this question, we must bring in the tailed monkeys, since in brain and behaviour they are halfway between the tailless primates and the other placentals. The first characteristic of the higher primates' behaviour is their manipulative restlessness. They are always inquisitively exploring their environment, always enjoying themselves by doing something with their hands. They are much better acquainted with the varied objects around them than are any other creatures. These monkey tricks are the foundation on which human science and industry was finally erected. Confronted with puzzle-boxes and similar problems, monkeys seem little more intelligent than other mammals. But they have a greater range of movements, and they are undoubtedly more imitative, though not so imitative as popular belief would have them.

Their quickness and restlessness seem not to be particularly useful to them; it is rather an overflow, a by-product of their active arboreal life. Their lack of concentration prevents them turning their capacities to service.

The tailless apes are a stage higher. For one thing, they are more imitative; for another, they have more insight. The surprising tricks which apes can be deliberately taught are only of secondary interest. It is remarkable that chimpanzees can be trained to have as good table manners as many children; that they will learn to dress and undress themselves; that they will even sign their names (Consul, the famous performing chimpanzee, had his own banking account and signed cheques on it). But all these are only evidence of the apes' manipulative skill and docility. Consul, we can unhesitatingly affirm, had no idea that the marks he made on the paper were his name, or the least notion of what a cheque was; he had been taught a trick, like a dog that does "*Trust and Paid For.*"

What really interests us is their level of intelligent insight. Koehler tried to find out what untaught chimpanzees could achieve by putting food out of their unaided reach, at the same time providing simple implements by which, if they were clever enough, they could reach it. Throughout, the greatest care was taken that no human being should give the apes any hint of what implement to use or how to use it.

The chimpanzees succeeded in accomplishing the following feats.

EVOLUTION OF BEHAVIOUR IN VERTEBRATES 1267

They realized that they could use sticks to beat down a banana hanging from the roof, or to reach one on the ground outside the cage. Once they had learnt this, they learnt to break off branches to use when no sticks were handy. They employed bigger sticks as vaulting-poles to leap at suspended food. If a string which they could reach was tied to food which they could not reach, they at once pulled the thread towards them (a feat of intelligence that sounds simple, but is probably quite beyond a horse or a cat). To catch ants (which they liked eating

FIG. 316. An anthropoid engineer.

One of Koehler's chimpanzees has discovered how to pile three boxes on top of one another to secure a banana. The construction has proved adequate, but is rather insecure. Note the sympathetic movement of the spectator's left hand. (From Prof. W. Koehler's "Intelligenzprüfungen an Menschenaffen" ["The Mentality of Apes"], by permission of Julius Springer of Berlin and Kegan Paul, Trench, Trubner & Co., Ltd., of London.)

FIG. 317. The use of implements by apes.

A young and untrained chimpanzee, using two packing-cases to stand on, employs a long pole to knock down suspended food. (From Prof. W. Koehler's "Intelligenzprüfungen an Menschenaffen" ["The Mentality of Apes"], by permission of Julius Springer of Berlin and Kegan Paul, Trench, Trubner & Co., Ltd., of London.)

for their acid flavour) outside their cage, an ape would poke a straw among the ants until they crawled on to it, then pull the straw back and lick the ants off it. They used a hanging rope to swing on and so clutch a suspended banana. They used packing-cases, big stones, or even people to climb on and reach up to fruit. Most interesting of all, they could combine implements. We have mentioned in Book 3 the chimpanzee who saw how to fit one stick into another to make the long implement he needed (Book 3, Ch. 5, § 1). They also piled two, three, and sometimes even four packing-cases on top of each other when more height was necessary.

Thus they have, it is clear, a considerable power of solving simple mechanical problems; and they usually solve them, not by blind trial and error, but with the aid of what we have called insight—either solving them outright, or by trial and error illuminated by some understanding of the situation.

But their limitations were quite as striking as their achievements. They had only the feeblest insight into mechanics. Their towers of boxes were usually unstable. They could never see that a ladder was safe when it made an angle with the ground, but persisted in pushing it with one side right up against the wall, and its rungs accordingly at right angles to the wall—with the result that it fell over when they began climbing up it. If their rope was wound round a beam in three turns, not even overlapping, they never could see how to begin unwinding it, but pulled haphazard at it like a man confronted with a vast insoluble tangle of string. When given the choice of pulling two strings, one of which was attached to a banana, while the other merely lay close to it, they almost invariably pulled the string that ran straight towards the fruit, irrespective of whether it was fastened to it or not; it seems very dubious whether apes have any understanding of the mechanical situation which we call connection of string and object. They very rarely thought of using an implement which was out of sight. Usually the stick and the food had both to be visible for the ape to think out the connection between them.

Many other facts could be cited to show that their insight, though it takes them further than any other animals, fails before situations which to us appear laughably easy. Their limited association-centres will not allow them to grasp many elements in one act of thought.

And their average powers as a species are much lower than we have implied. For it is by no means all of them which can solve such problems. Individual chimpanzees differ as much, both in temperament and intelligence, as human individuals. Many remain baffled, or sulky, or complacently unsuccessful; the more difficult problems were never solved, or even repeated, by any save rare ape-geniuses. Perhaps the most interesting of all the many interesting things that wait to be done in biology would be to take a group of clever chimpanzees and see what could be accomplished by fifty generations of selective breeding for intelligence. They are so near the critical point at which language and abstract thought begin; could one help them across it?

IV

CONSCIOUSNESS

§ 1. *Objective and Subjective.* § 2. *Is Consciousness Passive or Active?* § 3. *What Is the Range of Consciousness?* § 4. *Body-Mind.*

§ 1. Objective and Subjective.

AND now, before we can go very much further with *The Science of Life*, we must bring into consideration a whole new system of aspects which so far we have avoided. We warned the reader at the outset (Introduction, § 5) that we were going to do this. As long as we could, we have viewed life as visible, tangible, material fact external to our conscious minds, and ignored any other possible point of view. "Feeling" we have left out of account. We have studied life *objectively*, using that word as it has been used since the days of Kant. It has been the spectacle of life, the spectacle of its evolution and behaviour, that has engaged our attention. We have avoided any element of introspection in our view.

But as we have studied the behaviour of creatures, the questions of feeling and knowing and thinking and willing have come nearer and nearer to us, and the fact that we feel and think and know and will begins now to force itself upon our attention. The contrast and the relations between the world of feeling within, the subjective world, and the world of exterior reality, the objective world, can no longer be disregarded. They must now be discussed.

They have to be discussed, they have to be stated, but let us say clearly they cannot be explained. This duality of all our individual universes, this contrast of objective and subjective, is an inexplicable duality. So perhaps it will always remain. It is a fundamental condition of life as we experience it. It is possible that a day will come when all the processes which go on in the brain when we think or fall in love, will be described fully in the physiological terms of matter and energy. The explanation may be complete in its own sphere—but the experience of thinking, or of being in love, will not even have been described, let alone explained. That applies with equal force to simple sensation. When we have the sensation of redness, light of a particular wavelength is stimulating a certain kind of cell in our retina, and there it sets going nerve-impulses to certain centres in our brains. But no amount of knowledge of wave-lengths and retinal cells and nerve-centres will make a blind man understand the unique quality of redness as opposed to greenness or blueness; we can describe and explain the machinery underlying sensation, but not the sensation itself. Material processes cannot explain consciousness any more than consciousness can explain material processes; they are different qualities of being.

We can, in general terms at least, explain the physical mechanism of brain; we cannot explain how its working makes us feel and know—or indeed, why we feel and know at all. There we come to a riddle that smiles away any completeness from a purely physiological, mechanical account of life. Nor has philosophy or theology any answer to this riddle. Theology stumbles upon this opposition in its dilemma of predestination and free-will. From one point of view, the point of view from which we contemplate the outside world, there seems to be a definite flow of causation from one moment to another, so that if we knew all the facts about anything at any one moment, we could predict exactly what would happen to that thing. That is our normal way of regarding exterior events. They are, we assume, fated. They are predestined. If something unexpected happens we infer that there was some cause we overlooked in operation, and we seek it, and generally find it. But from the point of view of looking inward, the things within have none of this certainty. We feel free to make up our minds in any way. We choose continually, but it does not seem to us that our choice is conditioned. To other people it may seem that we cannot help but choose so and so, but not to ourselves until the instant of decision. Quite conceivably minds are very complex weighing-machines and this freedom of choice may be a delusion, but it is a delusion woven into the very stuff of the weighing-machine. The mental weighing-machine can never realize it is a machine. A theologian might put it that predestination is the universal law, but free-will the law of the individual moral life within the frame of that universal law. Or, returning to the language of eighteenth-century philosophy, predestination is objectively true and free-will subjectively true. Each is true in its own sphere and there is no way of synthesizing these two into one.

§ 2. Is Consciousness Passive or Active?

NOWADAYS these perennial questions present themselves with superficial differences and essential resemblances to the paradoxes of our forefathers. There is the old incompatibility, clothed in new phrases. There is re-statement but no explanation.

And apart from the insoluble riddle of this duality, there are questions of the relationship between these modes of experience that are only one degree less refractory. Men still differ profoundly upon the question whether this dualism is, so to speak, a dualism on equal terms, or whether consciousness is dependent upon objective reality, as the picture in a mirror is dependent upon the things that pass before that mirror. Is consciousness merely a reflection in a mirror, or is it associated with other kindred powers or qualities, so that it can be not only affected by objective things, but active and able to react upon them? Or, what is very nearly the same question, is consciousness merely an aspect of reality or a separate independently active factor in living realities?

We cannot answer that briefly here. Indeed, we cannot answer it decisively at all. But the rest of this Chapter and the next two Chapters following it will be contributory to the reader's judgment on the matter. Perhaps these questions, so widely asked and so inadequately answered, are difficult because they are still badly put. Perhaps they assume too complete an opposition between the objective and subjective. Perhaps we treat body and mind as opposites in kind, when in fact each is one face of a single two-faced reality. Let us consider first what we know about consciousness at the present time. Let us ask how far it extends in the world about it? And in the course of answering that we may find these profound questions take on a different and less difficult appearance.

§ 3. What Is the Range of Consciousness?

WE KNOW by direct knowledge of no consciousness but our own. We know that individually we think and feel. Or rather, *I* know. But that people about me feel, I assume and infer. I have no direct knowledge of that. We infer that other people feel from their behaviour—from the movements of their facial muscles that gives smiles or frowns; from their actions that imply a conscious purpose; from the words they use, which we interpret through long familiarity so wholly in terms of *meaning* that we are apt to forget they are only symbols, themselves mere air-vibrations of peculiar and arbitrary form. But no sane person hesitates to infer that all normal human beings are as capable of conscious thought and feeling as he himself.

But there many thinkers have stopped short. The great Descartes (1596–1650), whose views exerted the profoundest influence on the thought of the next two centuries, held that animals were automata, while man alone had a true conscious soul. "The animals," he writes, "act naturally and by springs, like a watch"; or again, "The greatest of all the prejudices we have retained from our infancy is that of believing that beasts think." The soul, thought Descartes, communicated with the body through that central and unpaired organ of the brain, the pineal gland; the soul operated in the pineal gland like a captain in a conning-tower, in some way it directed the activities of the animal and automatic machine constituted by the rest of the body. Apparently Descartes was unaware that other vertebrata also possess pineal glands.

Such a restriction of consciousness to humanity became impossible when the facts of reproduction, development and evolution were fully realized. It was only credible—and even then with difficulty—during the age of creationist dogmas. In each one of us we are now free to recognize there has been an unbroken development from fertilized egg to adult conscious human being. Yet no one will maintain that the ovum or the early embryo can be conscious in the same way that the man is conscious. None the less, it is impossible to draw any sharp line in development and to say, "Here consciousness enters the embryo or the infant." There is an imperceptible sliding into conscious life. The same difficulty greets us when we look at other animals. The higher vertebrates have so many points of similarity to ourselves in their behaviour and the construction of their bodies and brains that we all, in spite of Descartes, ascribe to them almost intuitively a consciousness akin to ours. Evolution only confirms this most natural impulse. It seems obvious that apes and dogs and mice and birds have consciousness.

But there is a limit to the ready recognition of fellowship. What sort of consciousness has a frog or a fish, for all its kindred brain? We may find it tempting to read consciousness into the behaviour of an insect; but the interpretation has to bridge a still wider gap of inference. We speak of an angry wasp. But may not a wasp's anger be as different from ours as is her way of seeing or hearing from ours? What about the consciousness of a crayfish or an earthworm, in which the brain is rudimentary? Of a sea-urchin, in which there is no brain? Of a jellyfish, in which there is no centralized nervous system? Of a sponge, in which there is no nervous system at all? A cabbage—an amœba—a bacterium? And yet the facts of biology have forced Evolution upon us, and we must believe that all these creatures are part of the one stream of life, that we with our consciousness have developed imperceptibly out of ancestral fish, out of still more remote ancestors no more elaborate than worms, than polyps, than amœbæ.

And pushing our exploration further back we have seen every reason to suppose that life has evolved from not-life, living matter from matter that had never been alive.

What conclusion can we draw from this array of facts and ideas? First, if we are not to break the principle of continuity that is at the root of any connected thinking about the world, and is revealed with diagrammatic clearness in the material side of development and evolution, we must suppose that consciousness such as we possess has evolved and developed, and that just as our muscles and nerves and stomachs and eyes and their activities have gradually been produced out of undifferentiated protoplasm and its less specialized yet not alien activities, so our consciousness and our capacities for feeling and thinking have gradually evolved out of some capacities of the same general nature as consciousness, which are the properties of protoplasm even in such simple form as ovum or amœba. There is no escape from this unless we believe, as many theologians do, that consciousness has suddenly appeared out of nowhere, so to speak, during the course of evolution, and during the course of each man's or woman's personal development. Psychology shows us clearly enough that processes like those of consciousness, but of which we are not fully or intensely conscious, can go on in us—our visceral activities for example—and doubtless the mind-like properties of embryo or amœba are so rudimentary that we can scarcely begin to imagine what they may be like. We can only say that they must be of the same nature as our mind—a dim awareness, a dim striving.

An analogy will make the point clearer. There are fishes which can give violent electric shocks. There seems at first sight to be no bridge between this power and any capacities of other animals. But the invention of delicate instruments and the investigation of life-processes by their aid has shown that this was due merely to our ignorance. As a matter of fact, there is not a single activity of life which is *not* accompanied by electrical changes—only these are so minute that we need special instruments for their detection. Every time a muscle contracts, a gland secretes, a nerve conducts an impulse, an electrical change takes place which can be measured by the aid of a sufficiently sensitive galvanometer. In the electric eel, certain of the muscles have been transformed in such a way that their several small electrical effects are enlarged and summed together to produce a considerable discharge. In the normal muscle or gland the electrical effects that accompany its working are inevitable results of the nature of protoplasm; living matter is so made that when it suffers physical or chemical change an electrical change takes place as well. In the average animal these tiny electrical changes are of no special biological value; they are only necessary by-products of its nature. But in the electric eel a special machinery has been built up which intensifies the electrical happenings and makes them of direct value to their possessor.

May this not be what has happened with mind? Something of the same general nature as consciousness, we suppose, accompanies the activities of all living matter, it may be of all matter; but it is generally beyond comparison feebler than ours, and, like the electric properties of ordinary nerve or muscle, is undetectable by ordinary inspection and is of no specific use to the animal. It is an unavoidable part of the process, but not an important object of the process in such instances. In the course of evolution, however, special machinery is built up—brain—through which these mind-properties of life have been utilized, intensified, and harnessed, and finally in ourselves made the most important single biological property of the organism. The brain is the organ bringing mind into effective activity, as the intestines bring fermenting and digestive reactions into effective activity, or the electric organ of the electric eel, the electric concomitants of the vital process.

§ 4. Body.-Mind.

IF WHAT we have suggested in the preceding section is so, then many still current questions about the relation of mind and matter cease to have a meaning. The brain does not *generate* mind any more than the mind generates brain. But the human organism is so constructed that when it is thinking particular and complicated happenings are going on in its brain, and when certain happenings are going on in its brain it is thinking. Consciousness, on this view, is how the organism experiences the brain-happenings, which are all that the external observer could detect with the methods of physical science.

Professor Whitehead, in his *Science and the Modern World*, takes this view. Reality, as he envisages it, is a flow, a series of events; and "the private psychological field is merely the event considered from its own standpoint." Or, as Lotka puts it: "To say that a necessary condition for the writing of these words is the *willing* of the author to write them, and to say that a necessary condition for the writing of them is a certain state and configuration of the material of his brain, these two statements are probably merely two ways of saying the same thing." The evolution of a complex responsive and selective apparatus and the appearance of a mind, each imply the other.

To regard a conscious being in this fashion is quite incompatible with the older idea of him as a "soul" imprisoned in a "body." He is, on the contrary, a portion of the stuff of reality organized so that it is intensely conscious; he is not mind *and* body, but body and mind in one. Body is one aspect of this unity, mind is another. The matter of physics and chemistry and the conscious spirit of the human mind are two aspects of the organisms we call men and women. In the light of such a conception the old question, whether mind determines the actions of matter or matter determines those of mind, ceases to have any meaning at all. If our thoughts and our brains and bodies are only two aspects of one reality, we cannot think of our living brains and bodies apart from our minds. If the world-stuff is organized in a particular way, in the form which develops into a human being, it will be both a body and a mind. Man, in this hypothesis, is not Mind plus Body; he is a Mind-Body.

The survival value of this complex choosing organization which involves consciousness is obviously very great. By means of its enhanced powers of association and analysis, the newly evolved human mind enables life to grasp much larger situations than it was ever capable of before, and to hold in thought two alternatives and their consequences in a way impossible to any lower animals. On this power of deliberately balancing two alternatives simultaneously present in thought depends our persuasion of free will. We may never be able to untie the knot of predestination, but we can assert that the sense of freely willing is an essential part of this new machinery evolved by life for choosing between different ideas or courses of action.

The world-stuff, it seems, could not have reached this new level, where it can for the first time balance elaborate alternatives in one act of mind, without experiencing the feelings of effort, indecision, arbitrary power, or flow of will in one direction, which contribute to our conception of free will. To spare oneself these feelings by substituting the idea of predestination in the excessively crude form of Kismet or Fate imposed on us from without is not only poor philosophy but a retrograde step in evolution. We are each of us a part of Fate, and the sense of free will is one of the methods through which our destiny is achieved.

This conception of the body in space among objective things and consciousness which apprehends space but does not seem to occupy it, as being merely two distinct and infusible aspects of one substance, one mind-body, is called and has been called since the time of Spinoza, Monism. Spinoza's monism is the flat opposite of the extreme dualism of Descartes. It is the conception most prevalent among biological workers, and it dominates the thought of the three-fold author of this present work. But it must be clearly stated that this is not the common way of looking at these things. For many centuries a very emphatic dualism has ruled human thought and impressed itself upon language. We still talk of body, soul and spirit; we put physical and psychic into antagonism and treat them habitually as systems of reality separable not merely in thought, but in fact. But from these time-honoured established ideas modern biology is steadily breaking away and moving towards this newer conception of a single universal world-stuff with both material and mental aspects, of which, so far as we know, life is the crowning elaboration, and human thought, feeling, and willing the highest expression yet attained.

V

THE CULMINATING BRAIN

§ 1. *The Expansion of the Cortex*

§ 1. The Expansion of the Cortex.

BEFORE we come to the human mind we must say something of the organ in which the highest mental processes take place.

The human brain has roughly the size and shape of a half-melon. It is pinkish and very soft to the touch. Its outstanding peculiarity, when contrasted with the brains that we considered in the early sections of Chapter 3, is the great size of its cerebral hemispheres and the way their surface is increased as much as possible, within the limits set by the size of the bony cranium, by being thrown into a maze of wrinkles.

As their wrinkled surface indicates, these relatively enormous cerebral hemispheres of ours are largely made up of cortex and of the white matter supplying the cortex. The centres in the floor of the forebrain which achieve such distinctive pre-eminence in birds, are comparatively small in ourselves. Herrick calculates that the grey matter of the cortex accounts for about half of the total weight of the human brain; it is twice as massive as that of an ape of equal body-size. During their colossal growth the hemispheres have overlapped the other parts so completely that nothing else shows when we look at a brain from above, and only certain parts of the hind-brain in the profile view. To see where the other centres have got to we must adopt a different method of approach.

Let the reader imagine himself a homunculus on the borderland of the visible, apparent only as a speck to the naked eye. Let him be placed inside the central canal of the spinal cord (Fig. 306) and, inspired by curiosity, let him creep upwards into the cavities of the brain. What could he discover?

As a convenient map of his peregrinations we submit Fig. 318, which shows a brain cut into equal halves. It is as if the brain had been

THE CULMINATING BRAIN

sliced neatly along the groove that runs down its middle and the right half were viewed from the cut surface.

First, then, as he nears the hind-brain, our tiny adventurer finds that the canal along which he is creeping widens out, and at the same time the roof gets thinner. This is the part that was described in our account of the frog's brain, where blood-vessels press up against the

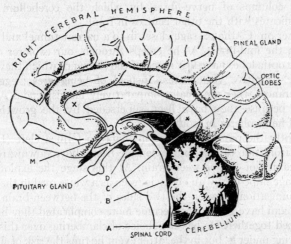

Fig. 318. The right half of a bisected human brain; for full description see text.

Parts cut through are shown as uniform grey surfaces. The thick, white, slightly arched band XX is the "Corpus callosum," a communication between the two hemispheres present only in the higher or placental mammals.

thin roof and where nourishing substances soak through into the fluid within. But note a complication. Most animals go on all fours, while humanity goes upright; the human spinal cord is vertical, while most spinal cords are horizontal. So our homunculus is clambering upwards like a chimney-sweep, and the "roof" of the hind-brain—the part that forms the roof in the great majority of vertebrates—rises as an upright wall behind him. Perhaps as he climbs past this thinned-out region (which is marked A in the drawing) he hears a faint drumming of arteries on the other side of the wall.

But soon it thickens again, and he gets to the wide cavity marked B. Here, in the middle of the hind-brain, he finds himself surrounded by important nerve-centres. Behind him bulks the massive cerebellum;

beside him and in front of him a series of slight bulges of the wall show where a number of reflex executive departments are located. Here the movements of the muscles of the face and throat and of those that bring about the breathing movements are controlled; hither the ear, the taste-buds, and the sense-organs of touch and so forth from the head region send their reports. At the two sides of this chamber are great bulging columns of nerve-fibres by which the cerebellum behind communicates with the other centres in front.

Farther on (C) the passage closes in to a narrow tunnel and plunges through the mid-brain. As he creeps through, our explorer observes that the optic lobes have lost much of the eminence which they possess in more primitive vertebrates. Instead of being great, egg-shaped protuberances, they are small, comparatively inconspicuous swellings. This is because they have been superseded by the growth of the cerebral hemispheres in front.

Abruptly the passage changes in shape to form a cleft (D), very narrow from side to side, but extensive from above downwards. This is the cavity of the between-brain. At this stage the explorer finds his passage impeded by an obstacle (seen as a white oval in the figure), a communication between the two sides of the between-brain. As the two thalami have grown and become more complicated they have met and joined together. He has the choice of clambering over this barrier or of diving under it, but in the latter event he runs the risk of slithering down the funnel that leads towards the pituitary gland. So he takes the safer course. As he goes over the barrier he leaves behind him the pineal gland—a little, stalked knob about the size of a cherry-stone—and he notices that most of the roof overhead is thin, and for the second time he hears the pulse of circulating blood.

But now forward progress is completely barred; the slit-like cavity of the between-brain ends in an impassable vertical wall (shown pale in the figure). This corresponds to the front end of the brain of an embryo, or of a primitive vertebrate. However, the two cerebral hemispheres have grown out of the sides of the fore-brain, and into these he may venture if he chooses. On either hand there is a round opening called the Foramen of Monro (M), and through one of these he may clamber into the great, rambling cavity of the cerebral hemisphere.

Here he finds so complicated a series of chambers, and so much inside them to arrest his attention, that he could spend an hour or two in observant prowling. He could crawl over the deeper parts of the forebrain that are concerned with smell-impressions, or over the inner surface of the corpus striatum. But because of its superficial position—it is outside and he is inside—he could never get near the grey matter of the cortex; and that is the part to which we must now turn our attention. So let us leave him to his investigations.

One of the most fascinating developments of recent neurology has been the discovery that different parts of the cortex have different duties to perform. This branch of study received its original impetus in the early years of the nineteenth century from the work of a German physician, Franz Joseph Gall. He was the first to stress the importance of the grey matter in the brain; he showed that the white substance was fibrous and consisted merely of communicating strands running between the grey centres; and he succeeded in dissecting out a number of the most important tracts of fibres. But he misconceived the significance of these intercommunicating grey centres. He believed that the cortex was a patchwork of areas, each the special seat of some sentiment or faculty, and that by observing the shape of a person's head you could infer which of those areas were well developed, and thus what his character and disposition were like. He noticed, for instance, that several of his more quarrelsome acquaintances had prominent bumps behind the ears, and therefore placed a centre for "combativeness" in that region; "amativeness" was put at the back of the neck because of the heat of that part in a hysterical widow; and there was a centre for "tunefulness" on the temple, because of a bulge that Gall noticed on the head of a musical prodigy of five. It is now known that the whole of this theory is nonsense, and although practising phrenologists may

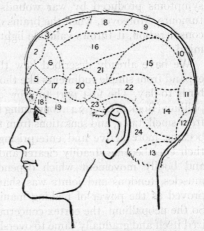

FIG. 319. Where Gall believed some of our faculties to have their controlling stations.

The centres are: (1) Sense of size; (2) Sense of causality; (3) Disposition to imitate; (4) Sense of colour; (5) Sense of time; (6) Wit; (7) Sense of wonder; (8) Optimism; (9) Firmness of character; (10) Vanity; (11) Constancy in friendship; (12) Philoprogenitiveness; (13) Amativeness; (14) Aggression; (15) Caution; (16) Poetry; (17) Sense of melody; (18) Sense of order; (19) Mathematical aptitude; (20) Mechanical sense; (21) Acquisitiveness; (22) Cunning; (23) Love of food; (24) Cruelty.

still be found, their profession ranks with palmistry and the casting of horoscopes as far as its evidential basis is concerned.

The subdivision of the hemispheres that we are going to note rests upon much sounder grounds—partly upon observations of the minute structure of the grey matter itself, and of the course of the fibres that supply it, partly upon the results of systematic experiment and the symptoms produced by war wounds, accidental injuries, and local tumours. Moreover, when the brains of other creatures are taken into consideration, it throws valuable light on the evolution of the human mind.

We have already described how the mammalian cortex began to expand from simple rudiments like those that are found in reptiles and birds to-day. The first part to grow and differentiate was the archipallium (Book 8, Ch.3,§ 4), concerning itself chiefly with matters arising from smell, taste, and sensations from the viscera. But as the mammals became more active and enterprising they developed other senses; their eyes became steadily clearer and stronger; the sense of attitude and bodily movement which depends on organs scattered in the muscles, tendons and joints was sharpened; the sense of touch improved as the power of making manipulative movements increased. So the neopallium, the cortex concerned with these activities, specialized itself and gradually came to overshadow the older parts concerned with smell.

Now just as the mid-brain and hind-brain contain a number of distinct centres corresponding to various functions, so the neopallium began to subdivide itself into a series of specialized but interacting regions as its responsibilities increased, and it took over more and more of the business of control.

Fig. 320, which we reproduce with the permission of Professor Eliot Smith, shows a series of mammalian brains; the types are all fairly primitive and unspecialized, and they are picked to give as near an approximation as possible to the line through which we have evolved. The cerebellum and the rest of the hind-brain are represented as solid objects with a light stipple. The archipallium is white. The neopallium is variously shaded so that it looks rather like a patchwork quilt. The gradual supersession of the archipallium by the neopallium is brought out very forcibly by this diagram.

Even in the simplest of these brains—that of the jumping shrew—the neopallium is divided into a number of distinct zones. Let us note one, shown black, concerned with the reception of impulses from the eyes; another, shaded with T-shaped marks, responsible for hearing;

THE CULMINATING BRAIN

a third, cross-hatched, for the sense of touch; and a fourth, coarsely stippled, whence fibres run down to the movement-centres in the hind-brain and spinal cord to carry the controlling instructions of the cerebral hemispheres. In a mammal even lower down the scale, such as a rat or an opossum, these same areas would be present, but very much less sharply defined. It is interesting to follow their development upwards and to see the gradual improvement and extension in the behaviour of the animals closely reflected in the structure of the brain.

The jumping shrew is a small gerbille-like African creature, living furtively, generally concealing itself by day and coming out at night to feed on insects and the like. Its name is due to its hopping kangaroo-wise on its hind feet. As one would expect from its brain, its behaviour is elementary and mainly dominated by the sense of smell. Some of the jumping shrews have such long and mobile noses that they are known as elephant shrews.

The tree shrew is a close relative of the jumping shrew, but it is an altogether more agile creature. It is able to scuttle very nimbly along surprisingly narrow branches, and for this clear eyesight and accurately controlled movements are necessary. Further, in an arboreal mammal,

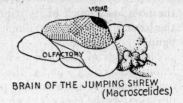

BRAIN OF THE JUMPING SHREW
(Macroscelides)

BRAIN OF THE TREE SHREW
(Tupaia)

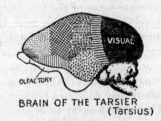

BRAIN OF THE TARSIER
(Tarsius)

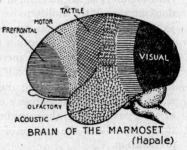

BRAIN OF THE MARMOSET
(Hapale)

FIG. 320. Four stages in the evolution of the mammalian brain; for full description see text.

The "prefrontal areas" and those marked with small circles and horizontal lines are the "association areas"; their extent corresponds with the intelligence of the animal. (Courtesy of Prof. Elliot Smith, from his "Essays in the Evolution of Man.")

the nose loses much of its importance. How profoundly this affects the balance between neopallium and archipallium can be seen in the figure.

The brains of the tarsier—a creature on the border-line between lemur and monkey—and of the marmoset—one of the most primitive living monkeys—show further progressive specialization in the same direction. As the eyes get clearer and the hand becomes a more sensitive, plastic instrument the corresponding areas of the neopallium expand until the old smell-brain is dwarfed altogether.

If the reader turns back to the section on the Evolution of the Mammalian Intelligence, he will see how important these changes have been. A creature like the jumping shrew, snuffling over the ground and with dim eyes and little adaptability or delicacy of movement, must live in an extraordinarily limited world. Of the smells of neighbouring objects it is acutely aware; probably its mobile nose gives it a more accurate sense of the localization of odorous things than we can well imagine; but anything more than a very short distance away is presumably only a blurred form, alarming perhaps or simply uninteresting, but never very clearly perceived. A rat or a mouse, sniffing its way along, is at this level; and the experiments with rats in miniature mazes taught us that although the animals can form automatic motor habits they do so in a random way, without any sort of understanding. Their world seems to be olfactorily varied, but lacking in shape. The clear eye and the skilful hand of a monkey bring with them a tremendous increase in the ability to explore and understand the forms of objects. They enable their possessors to become less intensely self-centred; to shift some of their attention from their immediate neighbourhood to distant objects. They increase the radius of the accurately perceptible world, and in broadening the animal's world they broaden its mind, for, since reflective ability is obviously useless in a creature which can only clearly perceive events that are either happening to it or closely impending, the introduction of the remote into its scheme of things introduces the chance of study, forethought, and appropriate preparation.

Thus the growth and differentiation of the neopallium enabled one of the great steps in the evolution of the human mind to be taken. It made possible the supersession of the old method of problem-solving by the new, of aimless poking about by the reflective pause, of chance discovery by invention. We traced this supersession in Chapter 3, and saw the new method appear occasionally in mammals of intermediate grade, such as dogs, and more often and strikingly in apes. But what of the final step, from ape to man? Can we trace a corresponding change in brain-structure in this case also?

Fig. 321 shows the chief subdivisions of the human cortex. As with the marmoset, there is a sight-area at the hinder end, a hearing-area low down on the side, a touch-area and a motor-area above the hearing-area. In the motor-area, the figure shows more detail than Fig. 320. It is seen that different parts of the body are under the control of different parts of the area, its top end being responsible for the legs and the lower parts of the back and its bottom end for the head. But the striking thing about this brain when compared with the others that we have been discussing is the enormous development of the areas which are not so allocated. It will be noticed that in the jumping shrew, the whole of the hemisphere falls into one or other of the sensory or motor areas. In the tree shrew a new area appears, very small, at the front end of the brain; it is indicated by horizontal shading. This area expands in the marmoset and monkey and another appears in front of the visual cortex. In man, these new regions (white in Fig. 321) are enormously developed; and there is a third, at the side of the hemisphere below the hearing-area. Undoubtedly the growth of these new zones of the neopallium underlies and determines the growth of the thinking mind.

These new parts of the cortex differ from the old in not being directly connected with any of the sensory or motor centres in other parts of the brain. All their fibres communicate with other parts of the cortex. They serve to co-ordinate the other parts of the hemispheres. Their appearance is a curious echo of something that happened very early in brain evolution. The first brains, as we have seen, consisted of automatic reflex centres connected with sense-organs and special groups of muscles. Then correlation centres arose to supervise and regulate the activities of these administrative departments, as we termed them. The cerebral cortex was such a co-ordinating centre. But in the mammalian stock the distinctive plasticity of the cortex proved such a valuable asset that more and more responsibility was shifted up from the lower centres. The cortex was itself subdivided into regions of higher grade. Once more arose the need for controlling centres unoccupied by immediate executive activity, and the new areas arose in response to this necessity.

The steady, progressive development of these new areas in the neopallium characterizes the primate stock. They are the last parts of the brain to reach their full development in the child. In the jumping shrew the neopallium consists entirely of areas immediately concerned with sensation or response, and its mental life is also largely or entirely occupied by proximate activities. In the higher mammals and man the "association areas," as the new zones are called, expand as intelligence grows. The difference in the size of the brain between a man and a gorilla is due almost entirely to the enormous expansion of the association areas in the man.

As far as behaviour is concerned, man stands out from all other animals in being able to speak and think in words; and this power underlies so much of the difference between man and ape that we may give some attention here to the parts of the brain concerned in language.

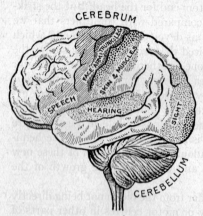

FIG. 321. A diagram of the human brain, seen from the left side (compare Figs. 318, 319, 320) to show some of the special regions discussed in the text.

Areas concerned with sensations are stippled, those which control movements are shaded.

It has been known for a long time that disturbances of speech —either of utterance or of understanding—may follow injury to the left cerebral hemisphere. Oddly enough, such disturbances very seldom result from injury to the right. It is a curious and quite unexplained fact that the nerve-fibres cross over on their way from the lower centres to the cerebral hemispheres, so that the left side of the brain controls the right side of the body, and vice versa. But why the mental processes concerned in speech should be confined to one hemisphere—as apparently they are—is completely mysterious. We all have a certain asymmetry of the apparatus of behaviour, being able to perform more delicate movements with the right hand than the left; the same is said to be true of gorillas. Perhaps the fact that we learn to write with the right hand determines the fact that our language-associations are built up primarily in the hemisphere which controls the right hand. However that may be, the fact remains; and the opinion is current among neurologists to-day that we do most of our thinking with the left halves of our brains, the other side being comparatively vacuous.

Now any injury may cause a disturbance of speech if it falls within a wide strip lying along the side of the hemisphere, from just in front of the sight-area by the hearing-area to below the motor-area. Moreover, the nature of the disturbance varies according to the part affected. Local disease at the hinder end of the strip near the sight-area produces "word-blindness"—an interference with the interpretation of writing or print; the patient can see the letters perfectly well, but has difficulty in making out what they mean. Disease in the neighbourhood of the hearing-area produces a similar complaint called "word-deafness," when, although sounds can be heard the meaning of spoken words is imperfectly apprehended. Finally, an injury at the front end of the strip, near the motor-area, produces disturbances of articulation, although the patient may think clearly and understand both spoken and written words perfectly well.

These facts reveal a state of affairs contrasting strikingly with the mosaic which the phrenologists drew on the human cranium. It is evident that a large part of the cortex participates in the speech function, not any sharply delimited centre. The special areas for sight, hearing, and movement participate; and so do the association areas between them. The local injuries that we have discussed merely produce blockages at one part or another of the system; thus a tumour at the front end of the strip simply interferes with the power of framing words; the patient may still be able to laugh, to express emotions by grunts and other appropiate noises, and even to sing tunes. There are cases where in this condition the ability to write is retained.

Speech, then, is not the function of some special part of the brain that other animals lack; it is an extension of faculties, present in cruder form in other mammals.

In the early study of brain anatomy there was eager search for some part of the brain that would distinctively characterize our species—something that we possess, and that is absent in all other animals. Various discoveries were announced, but in every case it was found that the part could be traced elsewhere, at least as far as the higher apes. Now we know that no such distinctively human feature exists. The differences between the brain of a man and the brain of an ape are simply differences of proportion—of the extent to which the association areas are developed. And just as the human brain has evolved through a series of stages that we can trace quite closely to-day, so with the human mind; it is no more than a culmination of possibilities that can already be detected far down the vertebrate scale.

VI

THE CORTEX AT WORK

§ 1. *Pavlov.* § 2. *What Is a Conditioned Reflex?* § 3. *The Dog as a Simpler Man.* § 4. *Inhibition and Control.* § 5. *The World of a Dog.* § 6. *Boredom, Alertness, and Sleep.* § 7. *The Dog Hypnotized.* § 8. *Temperament in Dogs.*

§ 1. Pavlov.

WE SHALL start our analysis of the working of this supremely important thin layer of grey matter, to which our examination of the culminating brain has led us, by first using what is called the "Behaviourist" point of view, and studying mental phenomena objectively and from outside. Only afterwards shall we resort to those more introspective methods which the Behaviourists treat as negligible or relatively unimportant.

At present there are two chief schools of Behaviourists—one in America and one in Russia. The name "Behaviourist" is often confined to the former. But the Russians, who have been working now for nearly thirty years under the leadership of Professor Pavlov, at Leningrad, are Behaviourist in principle. They show the same bias for purely objective methods of research. The Russian school works mainly on dogs, and, as we shall soon see, this has its advantages; the dog is simpler in mind than we are, but it evidently acts according to the same fundamental laws; and it enables us to get at those laws in a less complicated and more understandable fashion than if we approached them first through man. The American work has a more intimate human application, and we shall find it best to leave it for a time until we have learnt what we can from our humbler, simpler, canine material.

The full importance of Pavlov's work has only been realized in the last few years, as its results have been made available in translations. Its principles have still to find their way into current psychological teaching and popular literature. That they will play a fundamental part in the future development of the science of mind cannot be

THE CORTEX AT WORK

doubted. Much of the following account may strike the reader at the first onset as artificial and remote from his or her own thinking brain; some of it may seem to savour of pedantry, with its rather tiresome new terminology. That is how the growing experimental sciences—physics, chemistry, biology—impressed many educated minds a century or so ago, before they had begun to work their revolution in practical affairs. This work that we are about to examine is still in the experimental stage, and the reader must not look for the complete logical integration of a riper science. Beyond occasional very illuminating hints, it has no immediate practical application. But as a pure analysis, as an indication of how mental operations can be studied and described in terms of simpler conceptions, it is of the greatest interest. We make no apology for giving a fairly long account of it here. It is and it will remain of fundamental significance in mental science.

§ 2   What Is a Conditioned Reflex?

It was not Pavlov's original intention to explore the machinery of the brain. He happened upon that field of work in the course of other investigations. It appeared suddenly and challenged his attention while he was studying the digestive glands and digestive processes. He was like a man exploring the upper valleys of a mountain-range, who happens unexpectedly upon a pass into hardly suspected vast regions beyond.

Pavlov, in his earlier work, had noticed that the salivary glands are delicately controlled according to the needs of the organism. If dry food is taken into the mouth a lot of saliva is produced, so that it may be properly moistened; with watery food the flow is less copious. Moreover, the texture of the spittle varies besides its quantity. If the mouthful is going to be swallowed, a slimy, lubricating saliva appears; but if it is nasty and must be ejected, the saliva becomes thin and watery to help in rinsing the mouth. These facts do not at first seem likely to throw much light on mental processes, but let us continue.

These variations in the activity of the glands are in themselves simple, inborn reflex responses. No mental activity is involved. The substance in the mouth acts on special taste-organs on the tongue and palate and the action of the glands is automatically regulated in accordance with the information thus supplied.

But now the mental factor creeps in. The salivary glands are not merely automatic in this straightforward way. They can be controlled in a less direct fashion. Not merely the contact of the food, but the

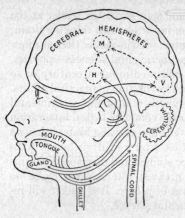

FIG. 322. A diagram to show the difference between inborn and conditioned reflexes.

If food is put in the mouth, impulses flash up to the brain along a nerve from the taste-organs on the tongue. In the brain, owing to an inborn nervous connection, they travel to the centre controlling the salivary glands and so out, along another nerve, to the glands themselves. Thus the mouth waters and the food is lubricated for swallowing. That is an unconditioned reflex. When we learn things, we do so by forming similar nerve-connections (shown dotted) in the cerebral hemispheres. The man shown has learnt that a particular cooked dish is good to eat. When he sees the dish, impulses travel from the eye to the visual centre (V) in his cortex. As he has formed a connection (dotted arrow) between this centre and the motor centre (M) controlling his salivary glands, the impulses follow that route and his mouth waters. Similarly, when he learns that a particular gong heralds dinner he forms a connection between the hearing centre (H) and the motor centres concerned. Compare Fig. 321.

mere sight or smell of food, or even signs of preparation of a meal, can make the mouth water. Approach a dog with a dish that it loves, and slimy saliva will flow in its mouth at once. Approach it with a nasty medicine that has to be forced down its throat, and a defensive, watery secretion will appear. But here we have something more than the inborn response of a simple reflex action. None of these things will happen the first time the dog sees that particular dish or medicine. An element of memory comes in; the animal reacts as it does because of past experience. A dog which has had its cerebral hemispheres removed will produce the right kind of saliva to suit a substance thrust into its mouth—it has the simple reflexes—but it will not react to the sight or smell of such substances. It does not recognize them. Obviously these are subtler responses and we are dealing with a more complicated state of affairs than before.

This was Pavlov's point of departure. He asked and sought to find out as exactly as possible, how these "psychical responses" of the salivary glands worked. He was not a psychologist, he knew next to nothing of psychological terms, and so he set about the problem as a physiologist should—by experimenting, and by speculating no further than the observable facts would carry him.

To begin with, he simplified the conditions of the experiments as far as possible, so as to get the problem into a clear, diagrammatic form.

His work on digestion had been done on dogs, so with dogs he continued. Let us give an example of the sort of experiment with which he started, and one which embodies his first underlying idea.

A dog is taken and put by itself in a quiet room, with as few distracting excitements as possible. Presently it hears the regular click of a

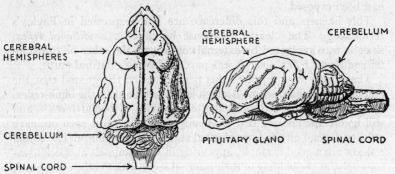

FIG. 323. The brain of a dog, seen from above and from the left side.

The cerebral hemispheres are better developed than those of a rabbit (Fig. 312), but not so well as those of a chimpanzee or a man.

metronome (a thing it has never heard before), and a few seconds after this has started, a plate of food swings down on the end of a string from a place of concealment. The dog sniffs, looks, and begins to eat; when it has finished, the plate swings away. Presently, after a suitable lapse of time, the clicking begins again, and, as before, the plate swings down almost at once, bearing a new meal for the animal. And so on, time after time. Soon, as we watch the progress of the experiment through a peep-hole in the wall, we notice that the dog's behaviour has changed. At first the sound of the metronome meant nothing to the dog, but after a few repetitions it calls forth signs of expectation—excited movements, licking of the lips, watering of the mouth and a similar, but invisible, watering of the inside of the stomach. In a word, the clicking is now recognized as a signal of impending food.

One might say that the dog has "learnt" that the metronome means food, or that it has associated the two together. But Pavlov was distrustful of such phrases. He wanted to make a new analysis, using terms of his own invention and definition, and to bring the phenomena as far as possible into line with simpler physiological facts already known with precision. First, then, he remarked that when the new habit was established it became indistinguishable in its action from an ordinary reflex as far as an outside observer could see. The clicking now evoked signs of appetite, precisely as a sudden movement near one of the eyes evokes an automatic blink. The difference was simply that while the latter reflex is inborn in the organism, the former had been acquired and depended upon the peculiar conditions to which the dog had been exposed.

This likeness and this difference are both expressed in Pavlov's terminology. The "learnt" response he called a *conditioned reflex*, since it was implanted by external conditions; the other he called an *unconditioned reflex*, since it was part of the dog's natural outfit.

The reader may object to this parallelism on the ground that the reaction to the metronome is much less infallible than the blink-reflex. An animal will only register appetite if the idea of food attracts it, and will ignore the clicking if it is already well filled. But even ordinary reflexes depend on the physiological state of the organism. The sexual reflexes are a good example, appearing in most animals at special seasons only, at the bidding of their internal secretions. On the other hand, a learnt reflex based on physical punishment instead of feeding may become quite as powerful and regular as an inborn reflex. The distinction is not a sound one. These are differences not of kind but of degree. A conditioned reflex is usually less powerful than an unconditioned one. In general a conditioned reflex is slower than an unconditioned one, because it involves a more complicated mechanism in the brain.

It will make this idea clearer if we put it in terms of brain structure. An ordinary reflex depends, as we have learnt, on the way the nerve-fibres are arranged, just as the course of the electricity in a circuit depends on the way the connexions are arranged. Take the reflex of producing a watery saliva whenever an unpleasant substance is put into the mouth. Impulses flow up special nerve-fibres from the sense-organs in the mouth to a centre in the brain. Thence they run away along other fibres to another centre, and thence out to those parts of the salivary glands which make watery saliva. At the same time some of them will be conducted along other appropriately arranged fibres to the muscles which enable us to spit things out. All this is a natural and necessary consequence of the way in which the nerve-fibres are arranged in the brain. But the conditioned reflex involves a more tortuous journey for the nervous impulse. Let us consider an example. Suppose that a conditioned reflex is built up in the following way—a whistle is blown and a nasty substance is popped into the mouth, and the procedure is repeated until the sound of the whistle is alone sufficient to call forth watery saliva and expulsive movements of the tongue and throat. How has this come about? We know that when the whistle blows impulses stream from the ear to a special part of the brain, and we know when the expulsive response appears that impulses are streaming out from other brain-centres to the appropriate glands and muscles. Manifestly the formation of the conditioned reflex must consist in establishing a connexion of some kind between the two centres. A track has been opened between these two centres. The simultaneous excitation of these centres has brought about their mutual connection. As a matter of fact the cerebral hemispheres have played the part of a switchboard, and the two centres have been put into more or less permanent connexion with each other. It is this intervention of the cerebrum which is the essential distinction of the conditioned from the unconditioned reflex, and it was the realization of this fact which opened to Pavlov the new vast region of research into cerebral function and mental action he has so ably exploited in the past quarter-century.

§ 3. The Dog as a Simpler Man.

LET us try to realize the implications of this conception of the cerebrum as a switchboard for the conditioning of reflexes. If Mr. Everyman runs through his daily routine in his mind and tries to calculate how much of it is determined by his own personal circumstances and experiences, and how much of it is inborn, how much is learnt and how much instinctive, he will get an idea of the importance of conditioned reflexes in his own mental life. But he will probably underestimate that importance. He will be disposed to count many things as unconditioned which are in fact conditioned. For example, if little puppies are taken away from their mothers at birth and fed on milk, they do not recognize bread or meat as food the first time they see it. They sniff up to it curiously and taste it experimentally; behaviour that contrasts very strikingly with the mouth-watering and lip-licking of a more experienced dog who sees food. Only when they have eaten a substance once or twice do they give food-reactions at the sight of it. So that the very recognition of food depends on conditioned reflexes. Modern psychology is revealing more and more how responses which seem to us to be instinctive and automatic depend on events of early childhood, events that the passage of years has blotted altogether from our memories.

Because of its very importance and multiplicity of detail the thinking machinery of the human brain is difficult to analyze. Here our remote but sympathetic cousin, the dog, is of great help to us. The mind of a dog is built on the same plan as ours, about as much as its body is built on the same plan as ours. But the mind of the dog is very much less intricate in detail, less fine in its working than the man's mind. In a dog's mind we can see many processes at work that are practically identical with those we observe in ourselves, but we see them in a simpler, more readily apprehended form. There is less danger of confusing incidentals with essentials in this simpler mind. This is why these experiments on dogs give such valuable indications of the nature of our own mental machinery.

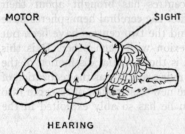

FIG. 324. The brain shown in Fig. 323, with some of the more important regions of the cerebral cortex indicated. Compare Figs. 320 and 321.

In the matter of speech, for example, we can discern in the dog's mind the same essential processes that we use when we learn a language. The animal can be taught that particular combinations of sounds have special meanings, and to associate simple responses with them. Everybody knows that a dog will learn its own name and one or two simple commands such as "heel" or "down." Sometimes one hears of dogs which have exceptional ability in this respect. A couple of years ago a report appeared signed by Drs. Warden and Warner of the Department of Psychology at Columbia University. It concerned a male German shepherd dog called Fellow, owned by Mr. Herbert, of Detroit, Michigan, who had been at great pains to teach it to respond to a number of verbal commands. The tests were performed with the master out of the room, to exclude any involuntary gesture which might convey to the dog what was expected of it, and nevertheless it would obey such shouted commands as "turn your head," "look up high" or "roll over." The psychologists report that "there would seem to be no doubt that scores of associations between verbal stimuli and definite responses have been well fixated by the patient teaching of Mr. Herbert during the past several years."

The second element in the dog which foreshadows our own power of language-building is the formation of chains of conditioned reflexes— but this is a more complicated story. It depends upon the fact that one originally neutral stimulus can be "associated" with another which has already acquired a conditioned meaning, as the following example from Pavlov's laboratory illustrates.

A dog is taught by the method described above that the clicking of a metronome means food; and now that sound is itself enough to evoke typical expectation signs. After this state of affairs has been established, the following sequence of events is arranged: first a large black square appears in front of the animal and then it vanishes again. Ten seconds later the metronome begins to click, and at once the dog salivates and licks its lips—but this time no food appears. This new sequence is repeated several times, being interspersed with the more normal procedure of metronome followed by food. The ultimate result is that the black square gets "associated" in the dog's mind with the metronome; it is a signal that fore-runs the metronome, and since the latter is a food-signal the former becomes a food-signal, too. So presently the black square by itself comes to evoke the appetite-signs.

But this is a tricky experiment to perform. If the interval between black square and metronome is too short, of if they are applied together, a different thing happens; the dog discriminates, and comes to respond to the metronome by itself but not to the metronome heralded by the black square. With this complication we shall deal more fully shortly. When the experiment is successful as we first described, it is evident that a conditioned reflex can be built up not only on the basis of a natural reflex but on the basis of another conditioned reflex, too. We can condition a conditioned reflex; we can set up a conditioned reflex of the second degree. And we can proceed to condition that. In the brain-switchboard an arriving impulse may be made to pass not merely along a single connexion between two points but along a whole chain of connexions between different points before it is sent down to the lower centres again. In this special case of the square and the metronome two connexions have been made, and the impulses sent up from the retina when the black square is seen are sent from the eye-centres in the brain to the ear-centres which would be stimulated by the metronome, and thence to the areas from which fibres run to the salivary and gastric glands and muscles of lip-licking, tail-wagging, and so forth.

Now this is, in embryo so to speak, what we do when we learn a language. At first we associate particular noises with particular experiences; these are simple conditioned reflex mechanisms. Then we begin to build word on word, to learn them not by actually experiencing their meaning but by description; and here chains of conditioned reflexes are built up and come into play. The process is, of course, a vastly complicated one, enormously more complicated in our case than anything a dog can do, and the inhibitions that we are about to discuss play an important part in it. Nevertheless, what happens is fundamentally the same thing; it is no more than an elaboration of the cruder, simpler "learning" that was studied by Pavlov in his dogs.

So we begin to realize the conditioned reflex as the unit of which our higher mental activities are built up. When we perform such complicated automatic acts as standing erect or walking, we do so as a result of the interplay of a great number of inborn reflexes. We might call such an activity a symphony of simple reflexes. When we think, the same sort of thing happens on a higher level; instead of running along the inborn nerve-paths in the spinal cord and brainstem, impulses are circulating through the labyrinth of tracks that education has laid down in the cerebral hemispheres. In this sense our thinking is a symphony of elaborately associated conditioned reflexes.

But how is the actual thinking done? When we have equipped ourselves with these conditioned reflexes, how do we use them? To this problem we must now turn, with our humbler cousin the dog as a simplified diagram of ourselves.

But first a word about the experimental methods.

As a result of the assiduity with which the case against vivisection has been put before the public, many people have come to think that any experiment on a dog is necessarily a horrible, agonizing performance. We can reassure the reader at once that this at any rate does not apply to the experiments that we are now considering. Let us note as exactly as possible how the animals were treated in the ordinary routine experiments, so that the humane reader can proceed with our brief summary of a brilliant piece of pioneer work without any qualms on the score of cruelty.

First, then, if the laws controlling conditioned reflexes are to be worked out properly, it is necessary to study some response that can be measured with a reasonable degree of accuracy. Then any fluctuations in its intensity can be followed. This requirement is fulfilled by that simple response, mouth-watering. All the animals used in the work had undergone a minor operation, conducted under full anæsthesia and with proper surgical precautions for the welfare of the patient, in such a way that one of the salivary glands discharged on the outside of the cheek instead of into the mouth. As a dog has several salivary glands, this fraction of the secretion could well be spared. Recovery was rapid, for a dog is more elastic than a man is in this

THE CORTEX AT WORK

respect, and thereafter the animals showed no sign of discomfort at their treatment.

During the experiment, the dogs had to be held still. This was done by soft loops passing under the arm-pit and groin, in such a way that they could stand comfortably and shift their position, but could not move away. Moreover a little hollow capsule was put on the cheek, which communicated through a long thin rubber tube with a gauge outside the experimental room. The saliva from the operated gland was shed into the capsule, and its volume could be accurately read off on the gauge. This apparatus involved an amount of discomfort about equal to that produced by a pair of spectacles or a well-fitting denture. That the experience was not a terrifying one is amply evidenced by the fact that several of the animals used to run into the rooms of their own accord, and jump up on to the table where they stood during the experiment. When off duty, they were well cared for.

The rooms in which the experiments were conducted (except in the earliest stages of the investigation) were in a specially designed building, as sound-proof and smell-proof as possible, so that all distracting stimuli was avoided. These details may seem tedious, but it is because of the rigorous experimental methods that the results are so illuminating. Obviously if one is studying the responses of the animal to its environment, the first thing to do is to get an environment that one can regulate with precision. The experimenter himself was outside the room, controlling the experiment by means of electrical and pneumatic controls. By means of a peep-hole he could see how the animals were behaving, and by means of his gauge he could measure their salivary secretion with considerable delicacy and almost watch the thought-processes going on in their brains.

And now for some of the more important results that have emerged, and are still emerging, from this almost meticulous analysis.

§ 4. Inhibition and Control.

THE essential idea running through Pavlov's work is that the operations of the cerebral hemispheres are controlled by the interplay of two antagonistic processes, excitation and inhibition. This is not a peculiarity of the higher nervous centres. All through the nervous system a similar antagonism can be traced.

We noted in Book I that the heart is an automatic pump. It will continue to beat after its removal from the body if it is properly cared for. But in the body its activity is controlled. It quickens cr slows and thus regulates its output to suit the needs of the organism. Running down to the heart are two pairs of nerves; one pair is the bearer of messages which hurry it up, and the other of messages which curb its activity and slow it down—the first excite the heart while the second inhibit it. One is reminded of the dilemma with which poor Launcelot Gobbo was tormented when at one elbow the devil pressed him to take to his heels and run away from his master, while at the other his conscience urged restraint. In the same way as the heart is controlled, the churning movements of the stomach and bowels can be excited or inhibited by appropriate nerves, and the muscle in the walls of the arteries can be tightened or slackened. To the importance of these and many other similar adjustments (and to the participation therein of internal secretions) we gave our attention in Book 1. In moments of violent exertion the heart and arteries are excited, so that the circulation speeds up and delivers blood more rapidly and at higher pressure, while the digestive organs are inhibited, so that the resources of the body can be concentrated on the voluntary muscles and nerves. In moments of relaxation, on the other hand, the circulation is calmed down, while the bowels churn happily away. Here we remind ourselves that our activities are reflected in this delicate interplay of excitation and inhibition in our chests and bellies because, as we shall shortly see, there is a similar interplay in our brains.

In the very flesh that clothes our limbs and bodies the two processes are continually active. Our voluntary muscles are always braced up and partly contracted in order to support the body and maintain its posture (except of course when we are limply asleep), and whenever a muscle tightens to produce a special movement the opposing muscles must slacken correspondingly. Both these changes are brought about by messages from the central nervous system. When we bend the knee, the bending muscles are excited and those which straighten out the knee-joint are inhibited. Strychnine upsets the nervous system in such a way that these compensating inhibitions are turned into excitation; in strychnine poisoning, whenever a muscle contracts its opponent contracts too, with the most agonizing convulsions as the result.

In the central nervous system, inhibition can also be traced. The reflex centres can be excited or put temporarily out of action. As with our muscles this is a vitally important fact, for there is always a powerful undercurrent of reflex activity flowing through the nervous system, and if it were not for mutual inhibition the spinal cord would be a hopeless chain of jostling, conflicting reflexes. It is because one reflex can inhibit and temporarily suspend another that our organized automatic activity is possible. To take a simple example, when a dog feels an itchy stimulus on one side of its trunk it lifts the nearest hind-leg and scratches the stimulated spot. Also, if one of the feet is hurt (for instance, by a cut, or the sting of an insect) it is reflexly raised, just as we would automatically withdraw our fingers under the same circumstances. Now suppose that while one hind-foot is busy scratching, the other is suddenly hurt—if both reflexes operated at once the animal would evidently fall down. But this never happens. Actually the "scratch reflex" is inhibited by the "flexion reflex" of the hurt foot, so that the scratching leg is put to the ground to support the animal as the other is raised, even though the stimulating itch continues. The more urgent of the two reflexes has taken precedence over the other. All this is inborn and automatic; it does not involve the intervention of the brain.

Imagine what the streets of a great city would be like if there were no control over the movements of motor vehicles. It is because of the policemen at the more congested corners, now moving the traffic along and now holding it up, now exciting and now inhibiting the flow, that road-transport can take place at all. So with the nervous flow and circulation of our body-city. Governing and integrating all the more mechanical of our activities we have these two tendencies—excitation and inhibition. Pavlov, already familiarized by his earlier work on the control of blood-pressure and the digestive organs with this balance of excitation and inhibition, extended the idea to the higher centres of the cerebral cortex. The conception of a balance in nervous impulses is by no means new; it is suggested already in such common expressions as "concentration," and "self-control." But it went no further than a half-metaphorical statement before the work of Pavlov. He applied it definitely and precisely, as the statement of a physiological fact. He brought it to the test of measurement in his experimental animals. And following it up, his work developed in the most surprising and illuminating way. As an instance of the part played by inhibition in the cerebral cortex, let us consider what he termed the "extinction" of a conditioned reflex.

Suppose that a conditioned reflex has been established in the brain of a dog, so that whenever it hears the sound of an electric buzzer it shows the usual signs of expectation of food. This is done, as we have already seen, by regularly sounding the buzzer and giving food immediately afterwards. Now suppose that the buzzer is sounded several times in succession (at intervals of a couple of minutes), but without the usual meal. At first the dog responds to the sound, but gradually the reactions become feebler and less definite, until finally, after half-a-dozen consecutive disappointments, they cease altogether. This is the process of extinction. But it is not, he found, a simple fading-away or rubbing-out of a trace left in the brain; it is something more complicated than that. It is not even a permanent extinction. If the original conditioned reflex was firmly established the extinction is found to be merely temporary; the conditioned reflex has not been destroyed at all, it has only been covered over by something else; two or three days later (or even two or three hours, in some cases) the buzzer will call forth a full expectant response, in spite of the previous disappointment.

Actually, during the extinction, a process of active restraint was going on in the dog's brain. One might imagine the animal was thinking, "Hallo—what's wrong? The food isn't turning up as it should—better ignore the buzzer for a bit until things become normal again." But to do that is, of course, to read more lucidity into the animal's mind than we have any right to presume. We shall find it more satisfactory to say that the conditioned reflex becomes inhibited, much as the movements of the bowels are inhibited when occasion demands. By counting the drops of saliva produced by the animal each time the buzzer sounds, the progress of such an inhibition can be measured, and a curve can be plotted showing its steady intensification.

But there may be complications. First, if, after a conditioned reflex has been thus extinguished and allowed to recover, the experiment is repeated, one finds that the second time the extinction occurs, it takes place more rapidly than the first. What took half-a-dozen buzzings to accomplish may now take three or four. And a third extinction will be swifter still. An inhibition, therefore, like any other kind of mental activity, can be improved by practice. Moreover, if extinction is accomplished, the buzzings being repeated until there is no response, and then immediately after there is a return to the normal procedure—i.e., a buzz followed by food—it will be found that the conditioned reflex is at once restored, and in the future, as in the past, buzzing will produce copious salivation.

Clearly here is a very delicate system of adjustment between the organism and its circumstances. Its practical value is evident enough. Consider a wild dog born into the world with its few necessary primary and unconditioned reflexes and with its wide capacity for making associations and conditioning these reflexes. Such a dog in forming conditioned reflexes learns to anticipate and prepare for the special accidents of its individual life. Its behaviour adapts itself to the fact that in some special part of the forest there is plenty of food, while in another there are painful thorns; that men and danger are about at certain seasons or at certain times of day and not at others. But in a changing world there must obviously be some means of revising this knowledge. There is no point in haunting a warren for rabbits if it becomes deserted, or in shunning a human dwelling-place if it has ceased to be inhabited. A signal that has lost its meaning must not be responded to; it must be taken out of current use; and it is by means of "extinction," the inhibition of reflexes that no longer "work," that the guiding collection of conditioned reflexes is overhauled and revised.

And now for another aspect of this same tidying-up, adjusting process in the mind. Imagine that a fresh dog is taken into the laboratory and put into one of the usual quiet experimental rooms. From time to time a tuning-fork is sounded, and whenever this happens food appears. Soon, as we have learnt, it forms a conditioned reflex and regards the sound of the tuning-fork as a food-signal. But what we have not yet discussed is that, other things being equal, the stimulus is at first "generalized." The stimulus is not very precisely apprehended. Any tuning-fork—or, indeed, any similar noise—will produce signs of food-expectation, and the reaction will be the greater the more closely the noise resembles the original tuning-fork. But the animal can be taught to discriminate more precisely between the noises. If a number of different notes are sounded pretty frequently, and only one of them is followed up with food, the others soon cease to evoke any response at all. The animal's brain finds out which one of the noises matters. Obviously the brain-process here is analogous to the extinction of a conditioned reflex, since it consists essentially in suppressing the responses to those stimuli which, though similar, have no consequence of importance. This is another manifestation of inhibition in the cerebral cortex. Quite similarly in the equivalent human process, whenever we learn that although X has a certain meaning, something else rather like it has not, we undergo what is at bottom an inhibitory process, like these processes that we are studying in dogs.

As another variant of this discriminating education, the animal can be taught to distinguish between combinations of stimuli. For example, it can learn that a whistle by itself means food, but that the same whistle accompanied by the simultaneous flash of a light means nothing. This evidently corresponds to our learning that what does in a particular set of circumstances will not do if certain added conditions are present. It may be tempting to skate on a frozen pond, but not if a danger-board is added to the scene. The same fundamental process of selective blotting-out is at work.

An inhibitory process, fundamentally identical with the others but having rather a different effect, appeared in the so-called "delayed conditioned reflexes." A dog has learnt to expect food at some signal—say, at the flash of a bright light. Then the interval between the flash and the feeding is slowly lengthened, from day to day, until it reaches two or three minutes. At first the dog begins to salivate and lick its lips as soon as the flash appears, but gradually it learns to bide its time. When the lesson is completely learnt it shows no sign of response at all for about a minute and a half or two minutes after the flash—that is, until shortly before the food appears—and then it begins to show the usual signs of expectation. The pause between the flash and the signs of expectation here is really a period of suspense, of internal conflict; the flash has produced a tendency to respond, but a simultaneous inhibitory process has been evoked which holds the nerve-centres concerned in check for most of the regular interval between signal and meal.

The most obvious human parallel is the learning of movements which have to be properly timed—such as changing the gears of an automobile, or judging the bounce of a tennis-ball and striking it at the correct moment.

In the experiments that we have just discussed the inhibition concerns condition reflexes. But even inborn reflexes can be inhibited as a result of training. Boxers learn to keep their eyes open, even when an opponent's glove jerks towards them, by inhibiting the natural protective blink. A striking demonstration of this power of education was given in Pavlov's laboratory when a dog was taught to regard an electric shock as a dinner-gong. At first the shock was so weak as to be a barely perceptible tickle, but as the training progressed it was stiffened up stage by stage, until an astonishing state of affairs was reached. The application of a surprisingly powerful shock produced no sign of pain or displeasure, not even the quickening of pulse-rate and breathing that usually accompanies an unpleasant surprise, but was followed by mouth-watering, tail-wagging and the other physical accompaniments of expectant appetite. In other dogs a prick of the skin, deep enough to draw blood, was robbed of its terrors and became an apparently welcome signal of approaching food. We see the same kind of thing in the keen cricketer who does not notice the sting of the ball he catches, or the medical student who overcomes his or her natural aversion to dissection.

These experiments of Pavlov's show very clearly how the reaction of the brain to the stimulations that are constantly streaming in from the sense-organs are controlled by a set of delicately adjusted inhibitory processes. One of the most striking peculiarities of these inhibitions is their extreme sensitiveness. They are anything but blunt and brutal restraints. They are fragile. If, for instance, there is any extraneous disturbance during one of these experiments (such as the noise of people talking in the corridor outside the room, or a faint whiff of some strange odour in the air), the inhibitions are the first things to suffer. If the animal so disturbed happens to be undergoing an extinction experiment, and has already got some reflex partly or wholly extinct, this disturbance will upset all the newly-acquired restraint. The animal may revert to the old responsiveness even when its reactions have almost fallen to zero. Or if a discrimination experiment is in progress the dog will go back upon the discrimination it has learnt and respond to the wrong notes or combinations of stimuli as well as the right ones. Or again, this effect of disturbance in diminishing inhibition is very clearly shown in the case of a "delayed" conditioned reflex, in which two or three minutes intervene between signal and meal. Suppose that during this period, in which the animal waits and withholds its response, there is some slight distraction, its effect is to produce immediate salivation and appetite signs. The pause, as we have seen, is a period of internal struggle between excitation and inhibition, and the distraction weakens the inhibition so that the response breaks through. A very strong extraneous disturbance, however, has the opposite effect to a moderate one. It does not simply extinguish the inhibition. It goes deeper. It interrupts the whole conditioned reflex and so abolishes the positive response.

Besides such adventitious distractions as noises and stray whiffs, there are physiological ones, such as a sore on the skin or a full bladder (for Pavlov's dogs had been trained to propriety on this matter, as men are). These also upset the inhibitory side of the balance before they disturb the positive responses. The name given by Pavlov to such an interruption of inhibition, whether it springs from external or internal causes, is "disinhibition."

These experimental results are evidently of the first importance. They are providing us with a new language, in which we can express all sorts of phenomena of everyday mental life in terms of analyzable and measurable processes. The distracting effect of a barrel-organ in the street, the faulty timing of a tennis player who is "put off his game" by the barracking of the crowd, the general irritability and weakening of control that a toothache produces—a hundred kinds of restlessness and worry can be interpreted in terms of disinhibition.

Perhaps this difference in sensitiveness between excitation and inhibition, so that inhibition is the first to go, explains the action of small doses of alcohol, which elates by weakening the restraining processes. It has already been shown experimentally that caffeine, the essential drug of coffee and tea, sways the balance in favour of excitation. It thus, but by a different method, releases impulses from inhibition and extinguishes discriminations. In a dog taught that a touch on the paw meant food while a touch on the back meant nothing, injection of caffeine weakened the differential inhibition so that the animal responded by salivation to a touch in either place.

A complicated thing, this thinking brain of ours! We can get a rough, very metaphorical picture of its activity by returning to the architectural details which were set forth in the last chapter. The fibres which stream up to the hemispheres spread out, we saw, in such a way that various parts of the body are connected to and represented by various parts of the cortex. There is a sight-centre, a hearing-centre, a movement-centre, and so on, and these main areas are more minutely subdivided. We all come into the world with a plan of our bodies mapped out on our brains in this fashion. Over the surface of that plan, while we are alive, there is a continual play of excitation and inhibition, of activity and restraint, like the shimmer of light and shade on the underside of a bridge. Some minds, we shall note, are naturally brighter than others; they have a preponderance of excitatory over inhibitory processes. Others are darker, being more reserved and controlled. But we live and learn, and as we do so the first crude plan with which we started gets marked and written over. As our conditioned reflexes develop, an intricate net-work of connecting paths between point and point spreads over it, and as we inhibit and discriminate a further pattern condenses on this web, some groups of cells becoming permanently excitatory in tendency and some inhibitory, according to the meaning of the impulses which come along their fibrous connexions. Pavlov likens the cerebral cortex in this stage to a patchwork, and speaks of it as a "mosaic of functions." This mosaic is the record of our individual lives, the form in which our experience and our little share of wisdom are engraved. But it is a living mosaic. It can be revised and altered, its detail can be made finer and more precise, as it both guides and is moulded by the rippling excitations which run over its surface and well up from below.

Pavlov's investigations into these two interacting processes of our brains go much deeper than does this descriptive introduction. The fundamental empirical laws governing their interactions are being worked out, and already generalizations are emerging which explain many of the known facts. Thus it seems (and the cases of inhibition already described illustrate the point) that whenever an experimental stimulus falls alone on the cortex, unsupported by any natural reflex such as feeding, it will sooner or later become inhibitory. That is an example of the simple laws into which our mental processes are being translated. But the details of the analyses are too complicated to be set forth here.

We may, however, mention very briefly the following fact. If any point on the cortex is in a state of excitation, its activity tends to inhibit the activity of other parts of the brain-surface; and, conversely, an area of inhibition tends to create excitation and more active response elsewhere. (This phenomenon, which is called "induction," interacts with and opposes a tendency of excitation and inhibition to spread about in the brain, from part to part, that we shall presently discuss.) In this, the experiments point towards an explanation of a number of states in which one part of the brain is excited and holds others in check—such as concentration, attention, and obsession.

§ 5. The World of a Dog.

BESIDES the light thrown upon our own mental processes, this work on conditioned reflexes enables us to probe into many matters that were once mysteries of the dog-mind. It is possible to learn how much this animal hears, smells, and feels, and to realize how far its world is like our own. Does it see things as we see them and hear what we hear?

The method of study, which developed out of the discrimination experiments just described, is not difficult to understand. In one case a dog was taught to expect food whenever a black screen was held before it, but not when a white screen was similarly exposed. After the lesson had been learnt and the distinction mastered, a pale grey screen was presented instead of the white one. This was found to make little difference to the animal; black was still a food-signal and grey had no effect. Day by day, the grey screen was made darker and darker, in order to find out at what stage the animal would begin to confuse the two and mistake the gradually darkened grey for a food-signal. The result was surprising. It was found that the dog had far better powers of distinguishing shades than the experimenters could claim. Ultimately there were two screens, both of which looked black to the human eye, but which were obviously perfectly distinct to the dog; one of them produced copious salivation whenever it appeared, while the other had little or no effect. But in other respects the canine eye was found inferior to the human. It cannot distinguish shapes as clearly as the human eye can, and (when they had excluded differences in sheer brightness) the investigators were unable to get any unequivocal evidence of colour vision in their animals. The dog, it would seem, looks out on a black-and-white world, rather like the world that the cinematograph presents; the projector is rather out of focus so that the shapes are blurry, but there is a much more delicate appreciation of luminous and shadowy patches than we can ever experience.

When the experimenters turned their attention to the dog's hearing they found a marked superiority to that of humanity. The animals could distinguish variations in the loudness and softness of a note far more delicately than any man, and they had a truly wonderful power of judging rhythms; one of the dogs discriminated without difficulty between a metronome ticking at one hundred beats to the minute and another at ninety-six. Dogs can also hear notes which are so high in pitch as to be beyond the range of the human ear, as Galton showed many years ago. Equipped with a Galton whistle, you can go for a walk, whistle for your dog, and he will come running, though neither you nor any other human being can hear a sound. Pavlov has built up conditioned reflexes on the basis of these apparently inaudible notes.

It would be fascinating to use these methods for exploring the canine nose—for our powers of smell are as nothing compared with a dog's—but unhappily, little has been done in this direction. The difficulty in this case is a technical one; the problem of how to generate smells of known and constant intensity has yet to be solved.

But these suggestive investigations into the world as it appears to a dog are only a side-branch from the main trend of Pavlov's work. Let us return to the study of inhibition, for it is there that the most illuminating hints are found to help in our study of the human mind.

§ 6. Boredom, Alertness, and Sleep.

WHILE the experiments on inhibition were being conducted, it was noticed that the animals often became curiously dull and drowsy. Perhaps in the early stages of an investigation all had gone briskly and well, the dog being alert and mastering its lessons without difficulty; but when experiments were begun which involved considerable inhibition—such as the learning of a subtle distinction between similar stimuli—the drowsiness made its appearance. (Can we not remember

THE CORTEX AT WORK

as a human parallel how Master Everybody is bright and happy at his play, how quickly he gets drowsy at his lessons, how quickly he wakes up again when he is freed, and how suddenly he is sobered when his conditioned or natural reflexes are corrected by parental "don'ts"?

Some dogs showed this effect more than others; one or two bad learners even went right off to sleep during the work. This somnolence, we may note, was one of the chief obstacles to that exploration of the dog-mind that we have just discussed. At first it was simply a nuisance that held up the experiments. But gradually it gathered an interest of its own.

Another way in which the dogs could be made sleepy was by means of unfamiliar stimuli. Some unusual noise, let us say, is sounded at fairly frequent intervals without any painful or pleasurable accompaniment such as feeding. At first, the dog's curiosity is aroused; it pricks up its ears and turns its head in the direction of the new sound. But as time after time the novelty is unattended by exciting consequences the animal loses its interest; the responses get weaker and more fleeting until they vanish altogether. Now this, it is worth noting, is a process of inhibition, like the "extinction" of a conditioned reflex. The pricking up of the ears and turning of the head towards a sudden sound is an inborn reflex, just as is the withdrawal of the hand when it touches a burning cigarette; the reader has the same reflex as part of his or her make-up, and tends to glance automatically in the direction of any abrupt, unfamiliar noise. A dog whose cerebral hemispheres have been removed will do it too. But the loss of interest in a persistent and meaningless stimulus is a higher phenomenon, an active inhibitory process that occurs in the cortex; and the most striking proof of this is that the dog without cerebral hemispheres never gets used to the interruption but gives a vigorous "curiosity reflex" time after time, no matter how often the meaningless noise is repeated.

To put it in another way, the boredom that a tediously reiterated stimulus inflicts is *a sign of activity in the highest part of the brain.* It is a protective device. Were it not for this gift of selective indifference, the organism would be everlastingly distracted by meaningless stimuli and everlastingly giving the "curiosity reflex." It is not enough to note what is important; one must also ignore what is inessential.

Another common fact of everyday experience fits in here, for, like other forms of inhibition, this contempt bred of familiarity can be disinhibited. The dog has ceased to be interested in the experimental noise; but now if it is suddenly distracted by some further novelty—

a photographer's flash, let us say—it will respond again to the old stimulus as long as the disturbance produced by the new endures in its brain. The reader has probably noticed that when one is startled one becomes aware of all sorts of little things—the crepitation of a fire, the ticking of a clock, the sound of one's own movements—which had previously been set down in the hemispheres as unimportant.

But we have strayed somewhat from our argument. It was with the sleepiness that appeared in inhibition experiments that we were concerned, and the reason why this particular inhibition of a natural reflex interests us now is that it also has a soporific effect. The dog, like the reader, gets drowsy when it is bored. If the occasional noise is continued after the time when it ceases to call forth a "curiosity reflex," and if the other surroundings are quiet, the animal first becomes dull and still and then falls fast asleep. Even conspicuously alert and lively animals can be thus depressed. We may note here, as another fact which presents an obvious parallel to our own species, that puppies are far more easily bored to the drowsy stage than grown dogs.

It was soon made clear that this sleepiness was not a simple exhaustion effect. It was not that the dogs were tired out by the problems presented to them. In all the cases that we have described the process of getting drowsy, strange though it may seem, was an active process, just as inhibition is. That is a novel result which runs counter to our common preconceptions about sleepiness. Gradually Pavlov's investigators became more and more convinced that the two phenomena, inhibition and sleep, were closely related to each other. Sometimes, in exceptional instances, they were even interchangeable. There was an animal which had been taught to expect food at a particular signal. Then the interval between signalling and feeding was lengthened by degrees to three minutes. In most dogs, as we have already seen, the result would be as follows: the signal would be recognized, but it would initiate an inhibitory process in the brain so that there would be no visible response for, say, two minutes; then the mouth-watering and other signs of appetite would appear. But in this particular animal sleep took the place of inhibition. Whenever the signal was given the animal went promptly to sleep, its eyes shut, its muscles relaxed. It even emitted an occasional snore. Then, after a couple of minutes, it woke up suddenly and gave the usual signs of confident anticipation of the coming meal.

The Pavlov investigators have accumulated a wealth of detailed evidence to show that inhibition is a process which spreads. It radiates out into the adjacent cortex from its centre of activity. If a conditioned

THE CORTEX AT WORK

reflex is extinguished, or if a discrimination is applied, the inhibitory process is not confined to the particular cell-groups in the brain that are immediately concerned, but tends to invade other neighbouring cell-groups, too, and even to diffuse itself widely through the brain. But it has to fight its way. Opposing this tendency there are antagonistic forces, so that the extent to which the spread occurs and the time for which it lasts depends on the interaction of a number of circumstances. A concrete example of the effects observed will make the idea clearer. Five points were chosen on the hind-leg of a dog and given numbers in order from the paw to the thigh—o, 1, 2, 3, and 4. (Fig. 325). The places 1, 2, 3, and 4 were used as food-signals, one of them being touched whenever a meal was impending, and soon the animal responded by signs of appetite to such a stimulus. But the place o was differentiated from the rest, and the dog learnt not to expect anything when touched on the paw.

This differentiation, like any other discrimination, depended on inhibitory processes in the dog's brain. Now suppose that o was touched, and immediately afterwards one of the other places, what happened? Touching o set up an inhibition; and it was found that the second touch, on one of the positive places, did not bring about such a marked response as usual. The inhibition had spread in the cortex from the cells connected with the paw to the neighbouring cells connected with the leg. Moreover, by comparing the results of a long series of experiments, the spreading out of this wave of inhibition through the brain could be followed in some detail. It was found that the inhibitory process weakens as it diffuses itself, being strongest at 1 and weakest at 4, and that it is opposed and finally overcome by antagonistic forces, so that in a few minutes it recedes and is swept back to its point of origin.

Fig. 325. To illustrate the experiment on inhibition described in the text.

In other animals events were followed up in other ways. It was found that the effect might extend even to quite remote parts of the brain-cortex. After extinguishing a conditioned reflex to a flashing lamp it was possible to detect a slight and fleeting weakening of another to a musical note, so that the inhibition had spread from the sight-centre to the hearing-centre (Fig. 321).

The neighbouring areas of the brain-surface can be affected after only a single application of an inhibitory stimulus. But suppose the thing repeated, suppose that one starts a number of consecutive inhibition-waves, what will result? The spread will be stronger and wider, the inhibition of the activating forces more and more effective. Finally the whole cortex may be inhibited, and even the lower reflex centres concerned with maintaining the position of the body may be invaded. The animal will then become motionless, limp, unresponding —in a word, asleep.

Herein is the answer to a question which has probably arisen already in the reader's mind. If inhibition is such an important process in the active brain, responsible for the most delicate adjustments that it performs, how is it that in these experiments inhibition so often brings about a drowsy, sluggish state of the animal? To reply in Pavlov's own words, "To my mind all the facts which have been given in the preceding lectures dispose at once of the apparent contradiction. Internal inhibition during the alert state is nothing but a scattered sleep, sleep of separate groups of cellular structures; and sleep itself is nothing but internal inhibition which is widely irradiated, extending over the whole mass of the hemispheres, and involving the lower centres of the brain as well."

Inhibition is sleep controlled and localized. Sleep is inhibition at large. Under the artificial conditions of the experiments the dogs were subjected to exceptionally uniform surroundings, to a deliberately simplified environment, for reasons that we have already noted, and there were less stimuli than usual to keep their brains busy. Accordingly the excitatory processes, which normally prevail against inhibition, were lessened. The balance between inhibition and excitation was upset; the former got out of hand and spread itself over the brain much farther than it would in the normal varied doggy life.

This identity is one of the most suggestive ideas that has emerged from Pavlov's work, for it seems to foreshadow an analysis of all sorts of subtle fluctuations in the vigour and acuteness of our minds. Mr. Everyman probably imagines that he divides his twenty-four hours between wakefulness and sleep, but in fact, as these investigations demonstrate, he is never completely awake. There are always bits of his brain that are calmly slumbering. Whenever he decides not to perform and defers some natural act, and so inhibits a conditioned or unconditioned reflex, he is literally putting a part of himself to sleep, so that in a sense sleep is a necessary and integral part of his waking activity. Conversely, when most of him sleeps, little bits of his brain may stay awake. Let him note as an illustration how his sleeping wife, although she cannot be roused by a host of accidental noises, may wake up at once at the feeble cry of her sick child. Sleep is everlastingly ebbing and flowing in the brain. With the coming of night it wells up, with the coming of day it recedes, but only in very rare moments, if at all, does it drain away altogether from the hemispheres.

§ 7. The Dog Hypnotized.

And now we turn to certain strange half-way states between sleeping and waking that occasionally appeared during the experiments. In rare cases, the dog would sink into a "trance"; it would stand in an erect, alert posture but perfectly motionless, sometimes for minutes, sometimes for hours. In this condition there was no sign of the limpness that accompanies true sleep, but there was also no sign of active control. If the limbs were moved into a new position by the experimenter the animal would "stay put" in the new attitude.

The condition is apparently intermediate between the strictly localized inhibition of normal activity and the complete diffusion of sleep. All the control centres in the cortex are asleep, but the lower reflex centres which keep the body erect and balanced (Book 1, Ch. 3, § 6) are still in full activity.

Or the inhibition may be even more local, and affect a part only of the cortex. Sometimes the dog is immobilized as far as its muscles are concerned, but still in control of its glands. Then a strange spectacle may be witnessed by peering in through the window of the experimental-room. The dog stands "entranced"; presently a metronome begins to click (or whatever conditioned stimulus has been used for the particular animal is brought into action) and at once there is copious salivation—but without the slightest sign of any movement to interrupt the fixed poise of the "trance."

These experiments throw light on that puzzling department of human psychology, hypnotism. One of the first expressions of hypnosis is the "cataleptic state"—a state of immobility in which voluntary control is lost, but the power of balancing and holding up the body retained. Evidently this corresponds to the partial sleep of the brain that is sometimes seen in dogs. The subject, like a dog, may have part of his cortex awake and part inhibited; being unable to control his limbs and remaining in any position in which he is put, but all the while fully realizing his own helplessness. As a further parallel between dog and man we may note how closely the passes and reiterated formulæ with which the hypnotist "puts his subjects under" resemble the monotonously repeated stimulus which was found to be so powerful an initiator of inhibition in the animals.

Throughout the animal kingdom, even in invertebrates, similar phenomena can be traced. If a crayfish is held in one hand and stroked somewhat firmly along the back of its carapace, from tail to face, it will pass into a cataleptic state, which bears a suggestive resemblance to that seen in dogs or men. It becomes motionless and its muscles have a stiffness that has been compared to the stiffness of wax. In this state it can be put into most grotesque attitudes, and will hold them until the inhibition passes off. Frogs can be thrown into the same condition by stroking, or by holding them between the two flat hands so that they cannot move, and then suddenly turning them on to their backs. Indeed, a great variety of creatures can be similarly immobilized —goats and pigs, ducks and hens, crabs and water-scorpions. The Indian snake-charmer knows that if a cobra is suddenly grasped behind the head and pressed on the back of the head when it is in the "threatening attitude" it becomes cataleptic and wax-like; it would seem (Exodus VII) that these facts were known to Aaron and to Pharaoh's sorcerers.

In many cases these curious tricks bring no benefit that we can see to the animal in which they appear. They may be looked upon as an accidental result of the way in which its nervous system works. A sudden strange stimulus falling on the cortex will temporarily inhibit its other activities—in more familiar words, the creature concentrates its attention to the new event. And so, as we have seen, a monotonous, meaningless stimulus is inhibitory too. Both these kinds of inhibition, normally controlled and purposive, may under artificial conditions become exaggerated and lead to complete immobilization of the organism.

But there are special cases in which the same tendency has been built upon and utilized. Variation and natural selection between them stretched the neck of the giraffe and the little finger of the pterodactyl to astonishing proportions; they seized on the accidental fact that when muscles or glands work they produce slight electric currents, and built from that the powerful batteries of the electric eel and the torpedo. So with these obscure nervous processes.

It is well known that many kinds of spiders and insects will "sham dead" if they are alarmed, becoming stiff and motionless; and the same thing can be seen when a speckled fawn cowers flat on the ground, or when we startle a nestling ring-plover (a bird which nests on the ground) and it flattens itself out and becomes all but invisible. Here the stimulus that frightens the creature sets up an inhibition which is allowed to sweep through the brain; as a result the creature becomes paralyzed and motionless, and therefore inconspicuous.

Perhaps the most striking case of all is that of the stick-insect *Dixippus morosus*, which prowls about by night, but falls at once into a rigid catalepsy whenever it is illuminated. It spends the day thus hypnotized, and its stillness conspires with its admirable protective architecture to merge it with the twigs on which it lives. In this state it can be put into strange attitudes, and will hold them, just like a hypnotized man. And just as hypnosis is in ourselves a phenomenon that happens in the highest parts of the brain, so with these insects: a Dixippus whose brain ganglion has been removed will prowl restlessly and aimlessly around, and never show any signs of "shamming dead," even when brightly illuminated, touched, or jarred.

Thus, under a variety of strange appearances, we trace this process of inhibition-spread. In a spider shamming dead, in the sudden stiffening and insensibility to surrounding things of a man struck by an idea, in the docility of a charmed serpent, we can see the same process at work, now as a mere curious accident and now turned to this end or that. Like the molecules of a crystallizing salt the facts arrange themselves, and the science of the mind takes form.

Naturally, Pavlov does not claim that his experiments explain all the phenomena of hypnotism. They simply show how the onset of the hypnotic state can be interpreted. Such facts as suggestion, which involves the use of language, are untouched. But it is worth noting that Pavlov has put forward a theory of suggestion (although to give an account of it here would involve a deeper and more technical analysis of the fundamental properties of excitation and inhibition than we can embark upon) and declares in his latest book, "I hope to be able to produce a phenomenon in animals analogous to 'suggestion' in man during hypnosis."

Other interesting phenomena were observed while the dogs were waking up and the flood of inhibition in their brain was receding and localizing itself. There was, for example, a "paradoxical phase," when all the reactions of the animal seemed to be upside-down. Stimuli which usually produced very weak conditioned reflexes now acted very strongly, while stimuli which were powerful had no effect. The retreat of the inhibition was not a simple ebb: a release here meant a passing intensification there. We cannot go into details with the space at our disposal. We may, however, pause and point a moral.

We have seen how the deliberately simplified conditions to which the animals were subjected led to an unusual extension of the inhibitory processes in their brains, and what peculiar results this spreading produced in several cases. Put a dog—or a man, for that matter—into a darkened room, with strange and unfamiliar and on the whole, meaningless things going on, and heaven knows what disturbances of the normal mental balance may not take place in his hemispheres. This is a hint with various implications; we commend it especially to amateurs in spiritualistic science.

§ 8. Temperament in Dogs.

FINALLY, the dogs were not simple infallible machines but living things; they had their weaknesses and individual peculiarities; and here also the experiments proved suggestive from the human point of view.

It was found that in the presence of too difficult problems the normal relations between excitation and inhibition could be upset. Thus in one case experiments were being made to see how clearly a dog could distinguish shapes. The method was like that of the grey and black squares (§ 5). First a conditioned reflex was built up by heralding the food with a luminous circle, projected on to a screen in front of the animal by means of a lantern. Then a rather flat ellipse was similarly projected, but without feeding; and soon the animal learnt that the circle meant food while the ellipse did not. When this had been done, the shape of the ellipse was gradually altered, and it became more and more circular from day to day, in order to find out at what point the dog would begin to confuse the two shapes.

Suddenly, when the limit was nearly reached—the ellipse being about eight-ninths as broad as it was long—the behaviour of the dog changed. It became very excitable, and struggled and yelped in its stand; its inhibitions disappeared, and it gave positive responses even to the first and flattest ellipse. Even outside the experimental room it remained nervous and irritable. The problem had been too much for it. So the experimenters started again with the first ellipse and the circle, and re-taught the old differentiation; and as this was done the animal recovered from its nervous trouble and became quieter and more tractable. Then, even more slowly than before, they began to work up to rounder and rounder ellipses. But as before, and at the same point, there was a breakdown when the problem became too difficult.

THE CORTEX AT WORK

A number of other experiments were made of this kind, in which neurotic disturbances were produced by a difficult "collision" between excitation and inhibition. In another case the animal had learnt to discriminate between two different rates of rhythmic touching of the skin, one being inhibitory, and the other excitatory; then it was found that if one rhythm followed immediately after the other without any interval the machinery of the brain gave way. Or, in animals with "weak" nervous systems, experiments in which powerful natural reflexes were inhibited had similar results.

Very powerful strange stimuli could also produce neuroses. The reader who wishes to follow this matter up will find in either of Pavlov's books* a full account of the behaviour of a dog upset by a great flood which swept through Leningrad in 1924, and in which the animals were nearly drowned.

The most interesting thing about these results is that the nervous breakdowns thus produced—which were often serious, and might last for months—were of two main types. In some animals the conflict produced a great predominance of excitation over inhibition—the example just quoted is one of these; in severe cases the hyper-excitability was even more marked, and all of the inhibitory processes in the brain were weakened or destroyed by the clash of experimental circumstance. But in other animals the same kind of conflict unbalanced the mind in the opposite direction—i.e., in favour of inhibition. First of all the conditioned reflexes disappeared; then after a long time they slowly recovered, but in so doing they passed through various "paradoxical" stages, like those seen in hypnotic states, when originally weak stimuli acted strongly and originally strong ones were weak. But each stage lasted in this case for several days.

It is obvious that these disturbances present a parallel to the nervous breakdowns or minor nervous troubles that we ourselves may experience as a result of shock, worry, overwork, or internal conflict. The human cases are discussed more fully later. Pavlov likens his two types of breakdown, in which excitation and inhibition predominate, to neurasthenia and hysteria respectively. It is interesting to note that he succeeded in curing one or two cases of the first condition by means of bromides, which, he believes in some way assist the inhibitory proc-

---

* Pavlov's work is available in English in two books—*Conditioned Reflexes*, published by the Oxford University Press, and *Lectures on Conditioned Reflexes*, published by Martin Lawrence. The first is a systematic presentation of the whole field and it is written in a compressed, rather difficult style. The second is a collection of lectures, with a personal account of Pavlov. Both are for the serious reader only; they are not light reading.

esses, and that, as with ourselves, a period of rest, with gentle stimulations only, always proved beneficial.

In severe cases the disturbances were serious illnesses, demanding special treatment of this kind. But, as one would expect from the human analogy, they vary in intensity from case to case, and grade by intermediate stages into slight everyday fluctuations in irritability and inhibitability.

Some dogs proved weaker, more easily upset than their fellows. Corresponding to the two types of nervous breakdown, two tendencies were visible in the normal animals. We have seen in the active, properly adjusted brain a balance between excitation and inhibition; in some animals the scales are congenitally weighted in favour of one or the other. At one extreme was the excitable animal, in which positive conditioned reflexes could be established readily enough, but inhibitions (e.g., in discrimination experiments) with difficulty if at all. Such an animal would be naturally confident, inaccurate, and aggressive, and might easily lose control when punished even by a well-known master. At the other extreme was the inhibitable animal, in which positive reflexes were hard to establish and easily disturbed by distracting stimuli, but in which the inhibitions were strong and ready. This kind of dog is cowardly in general behaviour—shrinking and tucking in its tail at any loud word or sharp gesture.

Between the two are more balanced animals; but even here an excitable group and an inhibitable one can be distinguished. The dogs of the first are lively and curious, running and sniffing about, almost impossible to keep in order; those of the second are stolid, reserved, sedate. Paradoxically enough, it was the first type which was most readily sent to sleep by the monotony of the experimental room; apparently the lively animal needs plenty of stimulation to keep it going.

Pavlov compares his grouping of dogs with the age-old Hippocratic classification of human temperaments. The extreme types are choleric and melancholic; the central types sanguine and phlegmatic. It seems at least that his analysis has the merit of simplicity; it puts the main facts in order and states them in terms of measurable processes, even if there are many complicating subtleties that for the present at least it leaves unexplained.

But here our account of the dog-mind must end. In 1930 the investigation has been going on for less than thirty years; and we have summarized the main facts already brought to light. Who can guess what facts and laws will not be added if it continues with the same vigour for another thirty? We have seen the dog-mind as a very much simpler, more elementary mind than ours, but one with the same basal processes at work in it. Most of the threads of which our daily mental life is woven—learning and inhibiting, waking and sleeping, alertness and drowsiness, temperamental fluctuations, and even the mysterious disturbances commonly spoken of as "nerves"—appear in a more diagrammatic, more readily analyzable form in this less elaborate brain.

Let us turn now to the much more difficult problems that we ourselves present.

HUMAN BEHAVIOUR AND THE HUMAN MIND

§ 1. *Human Behaviourism.* § 2. *The Mind as the Thing That Knows.* § 3. *Emotion and Urge.* § 4. *Hypnosis.* § 5. *The Unconscious.* § 6. *The Splitting of the Self: Multiple Personalities.* § 7. *Hysteria.* § 8. *Exaltation.* § 9. *Automatism and Mediumship.* § 10. *Neurasthenia.* § 11. *Repression and the Complex.* § 12. *Psychoanalysis.* § 13. *Minds Out of Gear and In Gear.* § 14. *Differences Between Minds.*

§ 1. Human Behaviourism.

PARALLEL with this mainly Russian work on the behaviour of the dog, to which we have devoted the last chapter, there has gone on a considerable amount of American research, which can be explained most conveniently and lucidly here in direct sequence to the account of Pavlov's work. Its leader has been Dr. J. B. Watson, and he has a very considerable following, especially in America.

At the outset, it may be well to anticipate a possible confusion about the word Behaviourism. As used by Watson and his disciples, it has a double meaning; it denotes not merely a method of work but a theory of mental realities. In its narrower sense, Behaviourism signifies the study of behaviour—plant, animal, or human—in a strictly objective way, by the same methods as are used in such sciences as chemistry or physiology—the observation, comparison, and measurement of physical facts. In that sense it is used by both Pavlov and Watson. But the school of Watson makes Behaviourism more than a method; it makes it an exclusive "doctrine." The American Behaviourists tell us that the study of consciousness or mind, as distinguished from brain, has no place in science. They go even further and deny the very existence of conscious processes. If they allow consciousness to exist, it is without processes of its own. Hardly do they tolerate the passive-mirror idea.

We will quote from Dr. Watson himself. Here is one passage: "According to the opinion of many scientific men to-day, psychology, even to exist longer, not to speak of becoming a true natural science, must bury subjective subject-matter, introspective method and present terminology. Consciousness, with its structural units, the irreducible sensations (and their ghosts, the images), and the affective tones, and its processes, attention, perception, conception, is but an indefinable phrase." Or again: "Whether there are ten irreducible sensations or a hundred thousand (even granting their existence), whether there are two affective tones or fifty, matters not one whit to that organized body of world-wide data we call Science." Or finally: "If Behaviourism is ever to stand for anything, it must make a clean break with the whole concept of consciousness. Such a clean break is possible only because the metaphysical premises of Behaviourism are different from those of structural psychology."

We have in the Fourth Chapter of this Book given our reasons for adopting a different attitude. If consciousness exists, it is, we hold, as worthy of scientific study as any other phenomenon. It is as legitimate to use subjective evidence as objective and to deal with subjective and objective realities on a common monistic footing. But our disagreement with Behaviourist philosophy has nothing to do with our attitude to Behaviourist method. That method, we agree with Watson, has remarkable virtues, and we differ from him with respect. Behaviourism permits of an accuracy of measurement, a scientific standardization and comparison of results impossible by any method which trusts solely or mainly to introspection and description.

Watson's most interesting and suggestive experiments have been concerned with the "conditioning" and "de-conditioning" of children's reactions. He has shown that the human species is so poor as to be almost destitute of the sharply-defined "pattern-instincts" of so many lower animals, in which particular situations evoke particular kinds of behaviour with as much definiteness as response follows stimulus in a reflex action. We have already described such instincts in ants and bees and other insects. We have told how there are solitary wasps which will attack only one special kind of prey; insects that will starve to death on any plant but their usual habitat, not because it is unwholesome to them, but because they have only the instinct to eat one particular kind; ants which have the instinct to make fungus-gardens; and so forth. Birds are similarly equipped, and for a long time, while great reptiles dominated the earth, this was the direction which vertebrate behaviour took as it elaborated and improved itself. Even such a close relative of man as the cat has instinctive reactions towards mice. Take a kitten a few weeks old, that has been brought up away from all other cats and fed solely on milk, and put it and a mouse together in a box; it will show all the signs of feline hunting excitement, growling and bristling, and will try to catch and kill its natural prey. But this sort of thing is less well seen in kittens than in birds, for mammals tried a unique mental experiment as they evolved; they made for individual adaptability, and therefore cleared the mind as far as possible of fixed reaction-systems. In man it seems that there is scarcely anything of the sort. He starts with no such equipment of directive dispositions.

The new-born infant will suck, and it will grasp and hold on to a finger or stick; and it will show signs of fear if support is suddenly taken away from it and it feels itself dropping. In addition to this it manifests a very distinct fear reaction at loud noises, but it has no other specific fears at all. No other avoidances, and no other impulses but this simple grasping and suction. Careful experiments show that babies have no instinctive fear of particular kinds or sizes of animals. A five-and-a-half months infant was tested with a black cat, a pigeon, a rabbit, a white rat, and a large dog; the smaller creatures she tried to get hold of, the dog she looked at with unafraid interest. When taken to the Zoo a week or so later, she showed no fear when brought quite close to camels, bears, zebras, ostriches, nor even when a monkey made threatening gestures through the wires. Similar tests on a seven-and-a-half months child gave the same result. In the same way, it can be shown that babies have no fear of flame. Nor is there any specific instinct for catching any particular kind of animal, as with the kitten.

The child has to learn what to avoid and what to be afraid of. It begins by a positive grasping and mouthing reflex to almost everything that is presented to it. It reaches out, grabs it, manipulates it, and brings it up to its mouth. If these positive reactions lead to a painful result, a selective inhibition will be set up.

Sometimes it takes a long time to effect this. A five-months baby, for instance, which was tested by Watson, reached for a lighted candle and followed it about with her eyes. If allowed to put her fingers in the flame, the hand was reflexly withdrawn; and the withdrawal was quick enough to save the child from pain, for she did not cry. If the fingers almost but not quite touched the flame, they were bent, but the whole hand was not snatched away. On the first occasion, the baby was tested twenty-five times, the painful stimulus being usually enough to cause this finger-bending, and two or three times enough for the hand to be pulled away. One week, and again two weeks later, the experiment was repeated. On neither occasion was there any noticeable inhibition of the impulse to reach for the flame. On the fourth trial, a further fortnight later, the reflex had begun to be modified; the baby, now six months old, sometimes checked her hand half-way to the candle, sometimes would not reach at all but sucked her fingers instead. But sometimes she still reacted as positively as at the beginning. Six weeks later, however, she only reached out once in the primitive unconditioned way; and after this the reaction became habitually negative.

In this series, about a hundred and fifty stimuli, slightly or moderately painful, were needed to inhibit the original seizing reflex, though doubtless if the experimenters had been willing to let the child burn itself more severely, fewer trials would have been needed. As we have already mentioned, the human infant is instinctively afraid of nothing but loud noises and loss of support. But by utilizing this innate unconditioned reflex, it can be made afraid of any object you like. Watson, for instance, showed a baby a furry toy animal, and at the same moment his assistant made a loud noise by hitting a steel bar out of the baby's sight. After this had been repeated three or four times, the baby had acquired a conditioned reflex in the same way as did Pavlov's dogs; on seeing the toy animal, it showed "conditioned fear," even though no noise was made, and from now on it was afraid of the animal. What was more, it did not differentiate the toy clearly from other similar objects; as with Pavlov's dogs, any stimulus resembling the conditional stimulus worked, and the baby was frightened of other toy animals of about the same size, and things like a seal-skin coat which shared the property of furriness. The conditioning had put a widespread twist into its mental structure.

But such impulses can be de-conditioned, and the fear again taken out of the objects into which the conditioning experiment put it. Watson, for instance, gave one child of eighteen months a conditioned fear of a bowl of goldfish. The goldfish had been associated with painful stimuli, and whenever the child saw the bowl, it said "bite" and refused to come within a couple of yards of it. But we will let Dr. Watson tell his own story.

"If I lift him by force and place him in front of the bowl, he cries and tries to break away and run. No psycho-analyst, no matter how skilful, can remove this fear by analysis. No advocate of reasoning can remove it by telling the child all about beautiful fishes, how they move and live and have their being. As long as the fish is not present, you can by this verbal organization get the child to say: 'Nice fish, fish won't bite'; but show him the fish, and the old reaction returns. Try another method. Let his brother, aged four, who has no fear of fish, come up to the bowl, and put his hand in the bowl and catch the fish. No amount of watching a fearless child play with these harmless animals will remove the fear from the toddler. Try shaming him, making a scapegoat of him. Your methods are equally futile. Let us try, however, this simple method. Get a table ten or twelve feet long. At one end of the table, place the child at meal time, move the fish bowl to the extreme other end of the table and cover it. Just as soon as the meal is placed in front of him, remove the cover from the bowl. If disturbance occurs, extend your table and move the bowl still farther away, so far that no disturbance occurs. Eating takes place normally, nor is digestion interfered with. The next day, repeat the procedure, but move the bowl a little nearer. In four or five such sessions, the bowl can be brought close to the food tray without causing the slightest bit of disturbance."

By continuing this process of substituting a pleasant conditioning stimulus for an unpleasant one, the identical object was finally changed from a source of irrational fear to an agreeable reminder of dinner-time. This is in entire harmony with Pavlov's work.

The learning and unlearning of reflexes can go on equally well in adult human life, just as we have seen it go on in the case of dogs. Everyone who has been in the reptile house at the zoo knows how the spectators will jump back and blink every time a snake strikes in their direction, even though there is a stout bit of plate-glass between them. This blinking and starting back is an automatic reflex defence-mechanism. But it can be inhibited. An apparatus was rigged up in which a heavy wooden hammer, faced with rubber, could be made to hit a piece of plate glass just in front of a man's face. At first blinking and shrinking occurred at almost every stroke of the hammer. Later, as nothing happened to hurt him, the reflex was more and more thoroughly inhibited. In a series of trials, each of which comprised 400 blows of the hammer, one subject raised the percentage of times in which he refrained from blinking thus: one per cent, three, nine, sixteen, sixty-seven, eighty-eight per cent. This recalls the fading away of the dog's secretion of saliva when a stimulus which usually brought feeding in its train was repeated over and over again without any food after it.

Watson and his school have made a real contribution to psychology in showing how plastic the mind of the child is, and what a huge part conditioning plays in the building up of much human behaviour that looks at the first glance simple, characteristic, and instinctive. The human mind, they have demonstrated, is a mass of acquisitions and has a smaller proportion of innate dispositions than that of any other creature. But they have asserted this with a strange lack of balance. They have over-asserted it. Because human minds are built up, so to speak, upon a practically blank sheet by an accumulation of conditioned reflexes, they have run on to the absurdity that any system of conditioned reflexes can be built up in any infant. They have leapt on to the assumption that every human being starts with the same blank sheet of the same texture and capacity for receiving and carrying impressions. They deny that heredity counts for anything in determining personal quality.

In Watson's own words: "The Behaviourist no longer finds support for ... special abilities which are supposed to run in families. He believes that given the relatively simple list of embryological responses, which are fairly uniform in infants, he can build (granting that environment can be controlled) any infant along any specific line—into rich man, poor man, beggarman, thief." He can, in fact, train the offspring of two certified defectives to carry on the work of Jeans or Einstein.

But this is manifestly absurd. It is only the romantic doctrine of Rousseau and the believers in human perfectibility brought up to date. Because environment is important in making men of different sorts it does not follow that heredity is unimportant or less important. Pavlov gives no countenance to this extraordinary heresy. The distinctive quality of the nervous tissue, the area of the grey matter, and the precision of the sense-organs, must all play their part in assisting or impeding the task of the educator.

We point this out as plainly as possible here. These really very wild exaggerations that Dr. Watson has imposed upon his otherwise valuable results have had a widespread diffusion, more especially in America, and they may be applied and are being applied very mischievously to condone the careless breeding of inferior and defective social types.

§ 2. The Mind as the Thing That Knows.

IN SPITE of the dogmatic prohibitions of the American school of Behaviourists we shall now begin to look inward. In what follows, the reader will find we are introducing a considerable amount of introspection. We shall call in the evidence of subjective impressions, evidence subject to many risks that objective observation escapes and evidence we can check only by comparison with what other people choose to tell us about their interior worlds. In the earlier Chapters of this Book we have studied behaviour as objectively as possible in relation to the developing brain. Now we are turning about, so to speak, and looking at the same activities from within. We are entering upon psychology properly so called, which Dr. Watson insists is no science at all.

It has to be admitted that so far as psychology has been purely introspective it has been a very barren and unprogressive study. It has associated itself with metaphysical and philosophical discussions. It has analyzed and over-analyzed mental processes; it allowed itself to be brought into relation with the realities of hypnotism, of nervous physiology and mental disorder, only very reluctantly and protestingly. It is difficult to the point of impossibility to say where psychology ends and where physiology begins. What is the subject of psychology? Consciousness; the thing that knows and feels. Is it also the thing that thinks? The Behaviourist will say, No. Is it the thing that wills and sets outer impulses going? We are back again at the insoluble duality we discussed in Chapter 4.

The mind I possess, and into which I look, and which is, I gather from information and observation, not profoundly dissimilar to the minds of other people, presents itself to me as an active process of which, strangely enough, I can recall no beginning; an active process which undergoes intermissions of which I am only subsequently aware, such as sleep, insensibility, and forgetfulness. By noting the development and action of other minds I conclude that my mind is the outcome of a process of synthesis, elaboration and, at last, remembrance, giving continuity to the transitory feelings and responses of myself as a baby, and that at the end of all its activities and intermissions comes a final intermission that, for all I know, may lengthen out into an endless cessation, death. That for other people may be the end of my mind, but manifestly my mind can never know of my final cessation. My mind thus viewed by itself is a very paradoxical thing indeed, without either a definite beginning or any end that I shall ever perceive, and yet with an effect of continuing process.

In its higher phases my mind is either apprehending vividly or thinking. Or it is giving a direction to a flow of will. I may be resolving to do something or actually doing it. To me all these processes have an immediacy and a simplicity that only the most elaborate analysis will persuade me is a false simplicity. I cannot at the same time perceive anything as what it is and simultaneously carry all its analyzed factors in my consciousness.

I sit at my dinner-table reflecting upon this section, and incidentally and almost heedlessly take and eat an orange. That seems to me a perfectly simple thing to do, but yet hear what an eminent modern psychologist, Professor McDougall, can find in it. He gives an analysis which shows how many elements, some conscious, some once conscious but now long forgotten, some entirely unconscious, enter into my mere perceiving of that orange. A Behaviourist might say that he gives a remote and shadowy description of the interplay of a vast concourse of nervous impulses which worked out to a resultant and pictured itself on my mind as the casual eating of an orange. But we are exploring now beyond the present boundaries of Behaviourist method.

"Now note," writes Professor McDougall, "that, when I perceive the orange, there is somehow operative in the act of perception not only my previous experience of oranges, constituting the act one of recognition, but also a vast amount of other experience, experience of material objects of many kinds—experiences of seeing, of handling, of tasting, of hearing, and so forth. And all this mass of past experience contributes to determine the 'meaning' of the object, or, more strictly, my 'meaning' as I perceive the orange. I know it, not merely as something that may yield me sensory experiences of taste and odour and touch, but also as a solid object out there in space, with all the properties that I have learned to expect of solid objects. If I put out my hand to raise it to my lips the strength of my innervation of my muscles is nicely adjusted to its weight, without my reflecting on, or being explicitly aware of, its weight; as I realize, if the object proves to be a mere hollow shell, of less weight than a normal orange; for it flies up in the air in a way which surprises me.

"Similarly, if you push it off the edge of the table and it remains suspended in the air instead of falling to the ground, I realize that its gravitational property was implicit in my perception. Or if I take a knife to cut it, and the knife meets with no resistance; or if I throw it forcibly against the wall, and it is not shattered. In a multitude of such ways I may be brought to realize the extreme complexity of my act of perception, to understand how much is implicit in my recognition of the orange, and how much previous experience has gone to the formation of this mass of knowledge of material things which is implicitly operative in my act of perception, guiding my behaviour in relation to the object."

Whatever one may think of this minute analysis of orange-taking, there can be little dispute about the complexity of the process by which I have conveyed this orange of mine—which, by-the-by, is a quite imaginary orange—into the mind of the reader. Seeing, sniffing, recognizing, and eating an orange is a thing any lemur or gibbon might do. But this orange we are discussing and imagining was never grown on earthly tree. It was evolved originally by Professor McDougall out of his private wealth of orange-impressions some time ago to body forth certain ideas about perception, and I (one of a trinity of authors) have received it with its intimations of colour, shape, scent, and so forth, by reason of an extract from the professor's book, which one of my collaborators has had typed for my information. I have written these words with that orange rolling quite vividly through my mental being. I have transferred it to an imaginary moment when I sat after dinner and ate it as I meditated on this section. And now the reader has it in mind quite distinctly.

I will not elaborate the picture of my communion with the professor's thought-orange which is going on as I write this passage. I will not dwell on the printing, proof-reading, publishing, selling, nor on the buying and sitting down to read, that has intervened before that same imaginary orange has been evoked in the present reader's mental being. There you have this orange now in your mind, and so far as you are concerned it was conjured up by the sight of a group of certain little black shapes on the white page that need the most subtle discrimination on your part to know them for what they are. The key to it all is these shapes which make the word "orange." In the act of recognizing them you are, as we have learnt from Pavlov, actively inhibiting the associations of quite a multitude of kindred words. Orange (with a capital O and a Roman arch), orage (the French word). Orage (the Yorkshire writer, who is now in America), orang (utang), and so on. The impression started by the impact of this group of shapes arises very rapidly, so rapidly as to be scarcely perceptible, and passes through, a similarly grouped, similarly etched out and defined sound record, the sound "orange" reverberates faintly in your mind, very probably your vocal cords and mouth make a faint intimation of an approach to vocalizing it, and that sets going a carefully and definitely selected and associated group of impressions of shape, scent and other qualities, and so to your general orange revival. From this momentarily excited agglomeration which is your idea of orange, associations radiate in every direction; you might have gone on to marmalade, Seville, Florida, orange-groves or the Hesperides, but until these have appeared before you here, you have probably ignored these associations in favor of those to which the trend of our argument has directed you, a trend giving you the supposedly tangible ripe orange produced for dessert and used as an instance of perception.

If indeed those threatened elaborations of this process were pursued as Professor McDougall has pursued the mere recognition of his orange as orange—it would fill a large volume to trace them all—you would find that all the time we were dealing with a directional selective flow through continually radiating associations, a complex journey that would make the course of a message through the most elaborate switchboard of a telephone exchange seem the simplest of processes. Or putting it in language the preceding chapter will have made familiar, you are dealing with a stupendously intricate migration of nerve impulses, of which your idea of "orange" is not so much a terminus as an important junction and shunting place, starting fresh trains of impulse as soon as one arrives.

The difference between adult human thought and the mental reactions of a dog or monkey, or a very young child, seems to be a difference not in kind but in complexity. A new stage above these animal levels is established by the introduction of words, and a still further stage by the introduction of writing. To evoke the group of associations that constitute orange in a monkey's mind, one must present some tangible element of the group, the smell of it, or the bowl in which oranges are put, or the sight of someone who is in the habit of giving oranges, or else the creature must feel a hunger and craving that oranges have at some time satisfied. It has our orange junction and shunting yard but on a much smaller scale. The human mind has developed an unprecedented system of symbols, *words;* and by means of these tags it can call up and bring together the groups of associations we call ideas in a way that seems to be beyond the power of any other creature. And it can transfer its mental activities to a kindred mind as no animal can do. By superadded systems of symbolism, by drawing and by writing down, the human mind can at the same time fix its ideas for subsequent consideration and free itself from the immediate necessity of attending to them in detail while it explores or develops further association-groups.

We have already discussed the behaviour of the higher mammals and shown that, with the probable exception of some of the higher apes, their mental action hardly ever goes beyond the establishment by trial and disappointment of *habits* of reaction to the stimulus of particular circumstances. All their learning is the acquisition of habits. The dog or monkey rarely seems to pose the situation, correlate it with memories of previous experiences and bring out of these experiences an idea of appropriate conduct. It is not at that level of mental lucidity. We have, however, described instances of greater deliberation in apes (see Figs. 316, 317). But when we come to the developing human being we find the floundering habitual responses of the dog or monkey giving place more and more to an acuter perception and an exacter response. Which is often less prompt. The higher animal has formed not a habit which goes off almost as automatically as a reflex, but a directive idea.

The development of language has conduced greatly to this process of deliberate and measured response. Words and phrases have come to play the part of handles by which reality can be held off a little and examined. Where the needed responses are invariable the human being will indeed form rigid habits exactly like the responses of a well-trained animal, but there is a great field wherein it is advantageous to vary the response, and then the apparatus provided in verbal and visual symbolization gives the man a versatility altogether beyond the powers of any beast.

The grouped associations, the ideas of things which are gripped together and defined in the human mind by names and other words and symbols, can be subjected, because of these handles, to a searching analysis and can be viewed from this aspect or that. One aspect can be considered, exclusive of others, and so the object can be brought into relation with other things which share some common characteristic. An orange can be thought of as a fruit along with other edible fruits, or as a round thing, along with the earth and a pill and any other round things, or as a yellow golden thing, or as a softish thing, and so on. The human mind can then make general propositions about round things or about yellow golden things. We cannot conceive of any parallel process in a wordless mind.

These general ideas give men quite distinctive powers of producing reactions between this and that group of associations. A young child or an animal has no such general ideas. When a child calls every man it meets "dada," or when a dog barks at every stranger, that is not because child or dog possesses general ideas. The child has not generalized from its "dada"; it has not yet learnt to distinguish its dada, which is a very different thing. The general ideas of an intelligent adult are the result of a clear discrimination of distinct qualities common to all the items brought under that general idea. The pseudo-general ideas of a child or animal are of an altogether different order; they are vague unanalyzed ideas due to a want of discrimination. They are a lower grade of grouped associations altogether.

Out of the general ideas which words enable men to use, they weave a coherent picture of the whole world about them and develop a consciousness of themselves in definite relationship to it. They find themselves with standards. From such abstractions as "good" or "bad" they can classify things into a system of values. They can attain to the idea of "truth" and of knowledge or delusion. They can frame their thoughts so that they seem to mirror the whole possible universe. They can make a picture of the universe which is generally valid.

So far as habit and custom rule the lives of men, they are all very much alike, but within they may have developed the most extraordinary differences. The natural aptitude of human beings to use general terms and to apprehend and react clearly to circumstances varies widely; their acquired aptitudes vary still more widely, both in extent and nature. The mental differences among men in range, depth, sensory quality and arrangement, are as wide as the differences of one animal phylum from another. There are men with minds like earthworms and men with minds like soaring eagles. A great number, perhaps a majority of human beings, seem hardly to think at all, except in the vaguest way, of anything outside individual concerns. Others think occasionally of more general things. Some think continuously as far as thought will take them. Some mentalities are no more than the picture of a few squalid relationships to which the sky and stars play the part of a theatre backcloth. Some mirror the whole of being. There seems to be no internal limit to the infinite patterning of associations that are possible. But as a controlling power there is the limiting influence of natural selection which will not tolerate a picture of the universe that betrays its possessor to danger, injury, and destruction. But any picture that *works* will be tolerated. Subject to that limitation of survival, nature will permit a man to develop whatever picture and map of the universe may shape itself in his mind.

Education (using the word in its widest sense) plays a very large part in determining the form and layout of the widening general ideas of the expanding human mind. The education of a human being comes mostly from the human beings about him, and so we find most human communities have distinctive pictures of the universe as the common basis of their moral culture and general conduct. No two languages have identically the same general terms and their grammatical structure also gives a preference to this or that method of grouping and correlating words. There is no such thing as exact translation, and so the framework of the mind of an adult Frenchman or German is subtly different from that of an English speaker. Every language and every tradition carries its distinctive tendencies towards this picture of the universe or that. These pictures evolve from age to age, they come into contact and react upon one another. When we discuss conduct we shall have more to say of the rôle of this picture of the universe in Mr. Everyman's life. This work, *The Science of Life*, is an attempt to give some of the most important aspects of the picture of the universe as Science is redrawing it to-day.

§ 3. Emotion and Urge.

THE older psychology, psychology before experiment, had much more to say and said it more connectedly, about ideas and the association of ideas, than about the phenomena of feeling, emotion, urgency, and willing. What it had to say may conceivably have been erroneous or misleading, but it did have a certain system and order. The passage from perception to urge and initiative remained however an uncultivated jungle of difficulties. The older text-books degenerated into a discursive description of the expression of the emotions and suchlike subjects. The study of dreams under the old regime also remained slight, disconnected, anecdotal, and unilluminating. Until the present century there was an extraordinary ineffectiveness before the facts of hypnotism, mediumistic seances, and mental aberration. Mental disorder was left to the doctors and hypnotists were treated as charlatans.

It would be outside our scope here even to attempt to trace the story of how psychology has broadened its contacts with reality during the past half-century. We will not even attempt a complete list of such names as Wundt, Fechner, Ebbinghaus, Brentano, Höffding, and the like, who played significant rôles in this renaissance.

One dominant liberating personality was William James. His animated catholic receptivity did much to draw the diverse phenomena of hypnosis, mediumistic phenomena, and psychotherapy towards convergent reaction. His treatment of the emotions was a very complete anticipation of the Behaviourist line of thought. Before his time it had been customary for psychologists to assume that the physical changes that make our emotions visible to others are the result of our experiencing these emotions. When we turn pale with fear, when our hair stands on end and our heart beats fast and we suffer other visceral disturbances, the bodily change had been supposed to follow upon the mental. But James argued that matters were the other way about. It is not that "we meet a bear, are frightened and run" but "we meet a bear, run and are afraid." Complex reflexes move us; when the stimulus does not overpower us altogether, these reflexes are generally in the nature of preparations for the emergency; the quickened circulation, the withdrawal of blood from the surface and so forth; and the emotions are merely the shadows of these preparations upon the mirror of the mind.

A very fertilizing stream of influence upon psychology which rose to its maximum twelve or fifteen years ago, has been what is generally known as psycho-analysis, the work of Freud, Adler, Jung, and their followers. (We say "generally known" because there are disputes as to the use of the word, which we shall use here in its less restricted sense.) Their work has gone on with a certain independence of the researches of the Behaviourists and it would be premature as well as presumptuous for us, in this popular summary, to attempt a synthesis of these different systems of thought and inquiry. A certain miscellaneousness must rule therefore in the sections that follow this. We shall deal first with the phenomena of hypnotism and the integrity of the individuality. Then we shall go on to the essential ideas of psycho-analysis. Throughout all this discussion and leading up to our treatment of conduct in the subsequent chapter, the reader will find that current conceptions of motive, individuality, and personality and therewith the reader's own egotism are undergoing an intensive illumination. We are exploring "selves" and "will" here, of the same constitutions as your own self and will, and it is impossible to avoid the personal application.

§ 4. Hypnosis.

IN THE previous chapter we described how various animals can be hypnotized, and explained the relation of hypnosis to the general process of inhibition. Here we will deal with the much more elaborate phenomena of the hypnotized human being. They throw a very penetrating beam of light into the self under consideration.

There are few things more striking than a successful experiment in hypnotism. The subject is told to lie back, relaxed, in a chair, with his gaze directed up at some bright object held near him. The hypnotizer, or operator, as we will call him, talks quietly and firmly to the subject, suggesting various appurtenances of sleep—drowsiness, heavy eyelids and limbs, a sense of warmth, numbness. At the same time, he will generally stroke the subject's limbs and make "passes" in front of his face. It used to be supposed that these encouraged a flow of "animal magnetism" from operator to subject; but this mysterious "fluid" has no existence, and the strokings and movements of the hands, like the flow of soothing words, act it would seem merely by their rhythmic monotony.

After a time, the subject finds his limbs beginning to feel heavy; he does not want to move them, and eventually cannot move them, even when challenged by the operator. His limbs then become rigid, and later, as the hypnosis becomes deeper, pass into a state of plasticity, as it is called, in which they remain in any position they are put, however unusual or fantastic, for a more or less indefinite period. If the subject is now awakened, he will be able to give a fairly complete account of his experiences; but if the experiment be continued further, he passes into a state in which, as in dreamless sleep, there is no remembering.

His condition differs from sleep, however, in one important particular. The subject is still in touch with the operator, or as it is generally called, *en rapport* with him. He is cut off from the rest of the world. Loud noises, the exhortations or commands of the bystanders—these have no effect; but he will obey the whispered suggestions of the operator.

When hypnotism is practised as an entertainment the operator generally orders the performance of some ridiculous action—the subject is told to scrub the floor with the imaginary water in an empty basin; it is suggested to him that a raw potato is an apple, and he is told to eat it; or he is informed that he is a sheep—and he begins to do what he is told, to act in keeping with the suggestion.

When experiments are scientifically carried out, it is found that the operator's powers of suggestion and control are enormous. He can suggest that the subject's thumb shall be insensitive to pain, the first finger hypersensitive; and then the subject will draw away his first finger in agony from the tiniest prick, while he will let his thumb be jabbed with a needle without a sign. Blindness of one or both eyes can be successfully suggested, or paralysis of a limb. The subject can be made to believe that one of the party is not present; if told to sit on the chair which this man is occupying, he will sit on his lap. Most extraordinary of all, in some subjects blisters can be raised by suggestion, or the temperature can be made to go up or down—bodily activities not under the control of our volition at all.

Another remarkable set of facts is provided by what is called posthypnotic suggestion. The subject, while still in hypnosis, is told that, after he comes out of his hypnotic condition, he is to perform a particular action, but that he is not to remember the command to do so. The action may be some absurd one, such as putting a footstool on the middle of a table; it may be commanded to take place at a given signal, or after a definite lapse of time. In the great majority of cases, it will be performed. Often the subject is uneasy for a little time beforehand; some sort of a struggle is going on in him between the conscious self which knows that putting a footstool on the table is absurd, and the hidden part of his mind which has been in subjection to the operator's bidding. But eventually he will do as he has been told. When he has been told to execute the action after the lapse of a definite period of time—so many minutes, hours, days, or even weeks—the instruction will often be carried out with extreme punctuality. Post-hypnotic commands of this sort have been effective after a whole year.

Not infrequently the subject will throw an ingenuous ray of light on human rationality, by making up some reason for what he is impelled to do. If he has been told to open a window, he will say, "Very stuffy in here, don't you think?" although it may actually be rather fresh; or if he is to put a vase on a book-shelf, he may suggest that it really looks much better there. This is what is generally called *rationalization*, the making up of reasons to justify an act whose motive is not rational. The rationalizer may be consciously or subconsciously ashamed of the motive; he may simply not know what his motive is but merely that the impulse is there. We realize here the gulf between Reason and reasons.

One of the salient facts revealed by such experiments (which have now been carried on by psychologists in all parts of the world for over half-a-century), is that in hypnosis the personality is as it were split or, to use the technical term, *dissociated*. The part in rapport with the operator is in some peculiar way isolated from the rest. The case of post-hypnotic suggestion we have just cited shows that this cleavage may remain after waking. The two dissociated parts of the mind are dissociated in the sense that they can act separately and even antagonistically, and that there is a barrier between the consciousness of the one and of the other. Yet they act upon one another in spite of this mutual unconsciousness. As we have already said, after waking from hypnosis, the conscious self may grow restless and uneasy when the time comes for carrying out a post-hypnotic suggestion, and is almost always aware of some extraneous impulse to action. Usually the subject experiences the same sort of uneasiness that he feels in normal life when he knows he ought to remember to do something but has quite forgotten what the something is. At other times he knows precisely what the impulse is urging him to do, but cannot ascribe a reason or a motive to it. Between these extremes there are all gradations.

From the point of view of the normal waking self, the impulse to post-hypnotic action is real enough; but its origin and often its very nature are hidden. And yet the command is carried out. The command has been given by word of mouth, received by the ordinary channels of sense. Though the subject may have no recollection of it whatever, yet when he is again hypnotized, he will remember all about it and tell you; and if he be told to remember it instead of forgetting it when he wakes up, he will remember.

Here again, in the domain of memory, there is a complete splitting.

§ 5. The Unconscious.

IN OUR description of the hypnotized dog we seemed to have a case of restricted activity, due to the inhibition of wide regions of the brain. These regions are held down while others remain in control. In these human cases, and especially in the cases of post-hypnotic suggestion, there is a parallel division into regions that are operative and regions that are restrained. There is a barrier to connexion and recollection between the two systems. One which is in abeyance during hypnosis resumes as the main stream of awareness, the personal consciousness. The other system "goes under" at the end of the trance. But though it is not apparent in consciousness, it is still somehow active. It becomes apparent by its influence as an urge of which the sources are hidden. Possibly it bears with it a faint consciousness of its own which has no relation with the consciousness of the general self.

In the language of psychology, a language repudiated by the Behaviourists, the phenomena of suggestion are a proof of the existence of an underworld of mind, a welter of activities, of which the personal self is unconscious and which are yet of the same nature as consciousness. This underworld is called by the psychologists "the Unconscious" or "the Subconscious," using the article "the" and a capital letter to indicate that the word is used in a special and definite sense for this world of masked, hidden, disconnected and unremembered activities. In the work we shall prefer the former term, the Unconscious. It is a region of events out of reach either of direct observation or direct introspection. We can, as we have just seen, get reactions from it through hypnosis and we shall find that in many other ways its activities well up and affect the conscious sphere. The realization of the existence and activities of the Unconscious is the fundamental idea of this modern psychology.

Generally the Behaviourist avoids the use of any particular name for the restrained, sleeping, or inactive regions of the cortex outside the system concerned in the behaviour of the subject. Clearly he will find no difficulty in recognizing that the inhibition of any region may be incomplete and that a secondary system of nervous reactions may arise and at last come to mingle or conflict with the ruling system in progress. Or that a ruling system of reactions may be split into divergent systems with a decreasing and disappearing amount of intercommunication. In such terms all this world of phenomena we are now describing, all this splitting of the personality, and all this appearance of barriers between process and process, may perhaps be stated. But for the last thirty years it is the language of psychology that has ruled this field of research, and in that language it is that we must deal with its facts here.

It is a language very liable to certain forms of misconception that have to be borne in mind. In the first place the name (with its capital letter) of "the Unconscious" for this outer welter of "mental" activity "below the threshold of consciousness" gives a false unity to what is really a vast and vaguely defined miscellany. Indeed at times it personifies this miscellany. The tendency to think of it as a second self, a self in the dark, equivalent in its complexity and homogeneity to the conscious self, is a very powerful tendency and few escape it altogether. But it has no such cohesive factor as the consciousness of self provides. Its groupings and systems are capable of limitless inconsistency. Selfhood is precisely what it does not possess. It is not indeed a mere confusion, a chaos of unrelated things. It has its systems; its groupings of emotions and ideas, and some of them may be very complicated systems. But they gather to no head until they enter the conscious field.

In no science perhaps is terminology so metaphorical and inexact as in this field of psychology. Consciousness as consciousness seems to be nonspatial and the Unconscious is of the same nature, yet, for want of anything better, we are continually driven to use spatial metaphors in talking about its activities and to speak of "parts of the mind," to distinguish between "superficial" and "deep" mental processes and the like. So long as the loosely metaphorical quality of such statements is remembered, we may not be greatly misled by them.

And having made these explanations we can proceed to explore this "stirring and teeming wilderness," this underworld of the Unconscious, as the psychologist sees it. It has many "levels" and many "regions," and hypnosis enables us to make a general survey of its extent. Many hypnotic results are concerned only with the superficial unconscious.

When a post-hypnotic command, such as shutting a door at a given signal, is concerned, we seem to be altogether at the superficial level. There is a barrier, which has somehow been inserted and left, between shutting the door and knowing why we want to shut it, that a simple verbal suggestion can remove. The unconscious element here is quite of the same stuff as normal conscious thinking and doing.

But hypnosis can carry us down from such surface effects as this to regions normally inaccessible to conscious activities. During hypnosis, the operator, as we have already noted, can tap regions of the subject's being which in healthy waking life are partially or wholly out of conscious control, and can cause blisters and changes in the blood-supply. This power can be used for therapeutic purposes. Control of the bladder and the bowels may be much disturbed in neurotic people, and here suggestion under hypnosis may often achieve a regulation impossible to the patient's conscious will. Professor Delboeuf carried out a remarkable experiment in which he made equal and symmetrical burns on the arms of two subjects. The burn on one arm of each patient he left to nature; but he suggested during hypnosis that the other should heal better. And he found that this burn healed quicker and with less inflammation than the other.

Now these bodily processes upon which hypnosis opens a door are under control from the autonomic nervous system, that subordinate organization of nerves to which, as we hinted in Chapter 3 of Book 1, the control of the viscera is largely delegated. It controls most glands, and many of the "involuntary" that are not attached to the skeleton and concerned with locomotion—all the muscles that squeeze food along our intestine, that expand or contract our blood-vessels and the pupils of our eyes, that make our hearts beat. In normal life, these activities are regulated automatically, or with only a moderate degree of control from our will; they are things that ought to happen of themselves, leaving the conscious mind free to deal with the unexpected changes of the outer world. But in hypnosis it appears that a more effective connexion is established (or released) between the still active part of the surface mind that is in rapport with the operator, and the deep-buried processes of bodily life.

Hypnosis may enable us to go deep in another direction—in memory. The hypnotized subject has access to all sorts of recollections (which can then often be verified as recollections of actual facts), which to the waking self are lost and unavailable. And the deeper the hypnosis into which he is plunged, the greater may be the number of these lost and buried recollections that are exhumed, and the further back into past life may they penetrate. In many subjects, if the hypnotizing process be often repeated and carried on for a longer and longer time on each successive occasion, memories of very early childhood can be brought out. It is as if there were a sort of stratification. Some experimenters state that certain subjects can even be transported back to infancy and that there in the lowest layers of the Unconscious still linger lost memories of sucking and crying.

To the Unconscious, too, belong all those impressions which enter the mind without being attended to—the ticking of the clock on the shelf, the beat of the heart in your own body. There is here a complete gradation between an intense focus of consciousness on which attention is concentrated, a zone of less acute consciousness where attention and interest are not aroused, a fringe of events of which we are barely conscious, and an outermost zone of events that do not enter consciousness at all and yet may be recorded in the Unconscious.

Superficial mental matter from which we are temporarily shut off, visceral nervous activities, unobserved stuff, and the lumber-room of memories—these are four leading realities in our unconscious worlds. But there is another of much greater practical significance. There is an accumulation of ideas and feelings that we have deliberately thrust out of our consciousness because we found it undesirable to be aware of them. Forgetting is not always, perhaps not often, a mere passive fading out. It can be a very active process. There are things we cannot bear to think of, do not dare to think of, will not think of. Out they go. This wilful banishment of thoughts and impulses is called *repression*.

We have seen how in hypnosis the suggestion of the operator can bar one part of the mind from access to the rest. Here something wilful within plays the part of the operator. Many lost memories are really of this type—they have been forcibly forgotten. Many of our more primitive impulses have been normally repressed in this way. Lusts, appetites, hatreds that we "control" are thus driven under. But the thing is done against resistance; impulse is in conflict with impulse, our self-control against a rebel. Very frequently when the impulse itself is repressed in its naked reality it will push back towards consciousness in a disguised form. Our repressive powers are limited. Our most nude and cloven-footed impulses, which we should be horrified to admit into the polite society of our conscious ideas, still manage to exert an influence over us. The same may be true, however, for other more angelic impulses. It is not only evil which is repressed by self-control. Most of us are uncomfortably conscious that the angelic does not go very well with our business and the practical life of every day. Hence such impulses, too, are often partially or wholly repressed; and we do good by stealth or entertain angels unawares, the good impulses exerting their influence from the Unconscious in the same disguised way as the bad ones. As Freud himself has acutely remarked (in *The Ego and the Id*), "The normal man is not only far more immoral than he believes, but also far more moral than he has any idea of."

Hypnosis is one way of getting into touch with the Unconscious, but it is not the only way. At moments between the unconsciousness of sleep and the full consciousness of waking existence come those flashes of something neither real nor absolutely non-existent which we call dreams. We have opened to us a world of emotions and images, undisciplined, uncontrolled by any logic, regardless of criticisms. Sometimes dreaming may bring before us things in our mind we never imagined were there; hates, desires, fantastic suggestions. Sometimes it may reveal unsuspected connexions. Sometimes it may diagnose disease. As we come awake, logical thinking and self-control assert themselves. We try to lay hold of these strange intimations before they sink back into the darkness. We attempt to rationalize the strange vision. It is rare that the Unconscious produces an abstract idea or a logical process—perhaps it never does so. It thinks—if we can say it thinks—not in abstractions, but in sensuous symbols. We do not find the idea of power in our dreams, we find giants and monsters. We do not find truth, but balances and rectilinear figures. We do not find love, but endearments and embraces. This is no longer the method of adult human thought; it recalls the fantasies of childhood, the mythologies of savages and a more primitive stage in the development of our race. The psycho-analysts have gone far in codifying of the symbols the Unconscious uses, and connecting them with the intricately symbolic art of the early world. Early man had to express and work out his ideas in complex images, because the methods of abstraction and logical thought had still to develop. He was incapable of abstractions, and so it seems is the ordinary Unconscious, the artist of our dreams, to this day.

The conception of the Unconscious as, among other things, a realm of repressions, it must be understood, is a relative one. Relative to the conscious self. When dissociation is at work, the dissociated mind-system is, from the viewpoint of the rest of the mind, part of the Unconscious; but the processes going on may be conscious within its own boundaries. There are numerous well-authenticated cases where two part-systems within one mind can be made to do different things at the same time, without one being conscious of what the other is doing.

Some people, for example, with the gift for automatic writing, can be engaged in conversation, and then, if a pencil be put softly into their hand and questions whispered to them, will write perfectly intelligent answers—without the main self knowing anything about it. The system that writes behaves just as if it were conscious; to believe it is not conscious raises more difficulties than to believe it is.

§ 6. The Splitting of the Self-Multiple Personalities.

THIS concurrent activity of entirely separate mental processes is our first introduction to a series of phenomena, which in their complete development amount to the existence, side by side, in one brain, of two or more active mental systems so disconnected and so highly developed as to be in effect separate persons.

A classical example of multiple personality, as this type of derangement is called, has been described by Morton Prince in a book far more absorbing than most novels, *The Dissociation of a Personality*. Let us tell the story as briefly as we can. Miss B. had been a nervous child, solitary and addicted to day-dreaming, living in an unhappy home. She became a hospital nurse, and when eighteen formed a romantic attachment to a young man but soon afterwards went through an emotional crisis, apparently because the young man's expression of his affection was not so idealistic as hers. During the next six years Miss B.'s character changed considerably. She went to college and became very studious; but remained neurotic with poor general health, and eventually called in Dr. Prince.

Then came another emotional shock, again connected with her love-affair; she sent for Dr. Prince, and in his presence actually underwent a sudden change of personality. From now on, for over a year, this new personality alternated with the old; first one and then the other was in possession of Miss B.'s bodily equipment.

Dr. Prince speaks of these two personalities as B_1 and B_4. Analysis showed that they represented complementary phases of character. B_1 was the girl Dr. Prince had known before the crisis. She remained studious, submissive, retiring, religious, self-sacrificing, and highly suggestible. B_4 was quite a different creature—self-assertive and irascible, sociable but worldly and vain, impatient of things intellectual and religious, self-opinionated. B_1 was devoted to children and the old; B_4 frankly disliked them. Though both were unusually emotional, B_4 reacted against her emotions, while B_1 let herself be carried away by them. Perhaps the most curious fact is that, although they shared the same body, they differed markedly in health. B1 was easily tired and had poor general health; B4 was energetic and robust.

The two personalities had no knowledge of each other's inner life. Their memories were different. Both had knowledge of all Miss B.'s early years, up to the first emotional crisis six years before, but B1 alone could recall anything of the happenings of the six years since then. After the appearance of B4, each remembered only what happened during the periods when they were "top dog." Circumstances and tastes decreed that they had different friends; and sometimes the change would come suddenly, and B1, for instance, would find herself talking to a friend of B4's whom she, B1, did not even know, or if she knew, knew only slightly and disliked.

They could not communicate with each other directly, and so had to obviate some of the difficulties which their situation entailed by writing notes and leaving them where the other half-self would find them. The two alternating selves had to practise tolerance, but really disliked each other. B4 despised B1's physical weakness and her idealism, and could not understand her wasting her time over books; B1 shrank from B4's worldliness and found her selfish and hard. And yet here they were tied together, forced to share a single bodily frame, and by that fact often forced to commit each other in all kinds of disagreeable ways.

By the aid of hypnosis, Dr. Prince was able to dig deeper into the meaning of the case. To cut a long story short, he concluded that B1 and B4 were really complementary parts of Miss B.'s personality, organized round two sides of her nature which had long been in conflict—her altruistic impulses on the one hand, the more practical but egoistic side of her nature on the other. From earliest childhood the girl had been passionately attached to her mother; but the mother had never cared for her, and had treated her unkindly. The conflict thus set up in the child's mind between the unattainable happiness she longed for and the harsh world of actuality, found expression in her day-dreaming, where she could give free rein to her aspirations and suppress all thought of the practical reality which was always so unkind. The expansive altruism of adolescence ranged itself with the other battalions of idealism; and doubtless any compromise with the worldly world of everyday was coloured by the suppressed conflict and made to appear despicable and hateful. The conflict must have been acute, with the forces of altruism and romance on the whole winning, when Miss B. devoted herself to nursing. She undoubtedly gave full rein to her idealism in her thoughts of the young man to whom she grew attached; and when he failed to live up to the level of unreality which she had ascribed to him, and revealed himself as an inhabitant of the flesh-and-blood world which demands kisses as well as nebulous romance, the shock was so great that, in attempting to push him out of her thoughts, she automatically intensified the old-standing push against all the worldly, selfish, unromantic impulses and ideas, with which his behaviour was now associated. And under the intensified onslaught, this whole complex of ideas was pushed clean out of consciousness—dissociated from the rest of her mental life.

But though repressed, these tendencies were not extinguished. Everything that could feed them was switched away from the daylight of consciousness into the limbo of the Unconscious. The buried half-self could be stimulated, but could never act or express itself, save in obscure influence, upon the other half-self, exerted below the level of consciousness. Then came the second shock, the mechanism of the mind was jarred, the repressive force weakened, and the repressed organization of ideas and desires sprang into conscious being as the new personality B_4.

This story, which recalls Stevenson's famous tale of Dr. Jekyll and Mr. Hyde, would be remarkable enough, but it is made more so by the emergence of yet a third "personality" of quite a different type— childish, puckish, irresponsible—who christened herself Sally. "She" appeared on the scene before B_4, but at first only when B_1 was in hypnosis. Later, during the period of alternation between B_1 and B_4, she too sometimes managed to obtain possession of the waking body to the exclusion of the others. B_4, on learning of Sally's existence from Dr. Prince, and by letters from B_1, made vigorous attempts to suppress her. Sally in return attempted her tricks on B_4, but was less successful than with B_1, partly because B_4 was a stronger nature, partly because Sally had direct access to B_1's thoughts but not to B_4's.

Later, Dr. Prince got Sally to write her autobiography. This unique document claims to be the record of a soul suppressed for all its early life, at long last finding access to the world of human beings and human communication. In it Sally maintained that she had had a separate conscious existence right back to Miss B.'s infancy, and that even when the little girl was learning to walk, Sally's tastes and ideas were different from those of the main personality. It is, of course, impossible to say how much of this autobiography is true in detail; but it seems clear that Sally had existed as a more or less distinct and organized mental being, but subordinate and suppressed, for a long time.

Then came the denouement. Dr. Prince, by the aid of suggestion in deep hypnosis, was able gradually to fuse the two antagonistic personalities B_1 and B_4 and weld them together into one. The more successful the fusion, the more thoroughly was Sally suppressed. Her own description was that she felt "squeezed." The rebuilding of a single healthy personality was much delayed by Sally. She had at last tasted the joys of full existence. She had been given the privilege of controlling the body in which she had lived for years as an irresponsible but helpless passenger—a transformation more dazzling than Cinderella's. And she was not going to surrender without a struggle. Repeatedly she would force herself up into control to prove that she was not going to be squeezed out of existence; and each time this happened, the amalgamation of B_1 and B_4 was hindered. But eventually she began to grow discouraged, and at last, as the result of appeals by Dr. Prince to her better nature, consented to give up the struggle. From that moment the restoration of the single and unitary personality went rapidly forward. As Dr. Prince writes, "The resurrection of the real Miss B. was through the death of Sally." Since the synthesis, it has been impossible to communicate with her; but we may presume she lives once more an obscure and powerless existence, suppressed by the maturer organization in control.

The final outcome was a restored Miss B., who commanded the memories both of B_1 and B_4, but not of Sally. She was a well-balanced personality, who "lived normally ever after."

So far Dr. Prince's account—we hope not stripped of too much of its absorbing interest in the process of boiling down. It remains to ask what are its general bearings.

Dr. Prince interprets it, as we have seen, in terms of dissociation of different parts of the personality. B_4 and Sally were both mind-systems organized round tendencies which were more or less antagonistic to the dominant tendencies of the growing girl, and which were finally excluded by her from any participation in the control of the body. But while Sally seems to have been in the nature of a left-over, a set of childish tendencies which never got incorporated in the main organization, B_4 was organized round maturer ideas and desires which had to be definitely fought against. As result, Sally retained some degree of independent and simultaneous consciousness—she was "co-conscious" with the main personality; but B_4 was driven out of consciousness, though the system still existed in the Unconscious and could still be stimulated there. With much of her energies thus devoted to repressing half her own nature, B1 grew weak and neurotic; and first one and then the other of the repressed systems managed to burst up into full consciousness again.

The main criticism which has been made of Dr. Prince's views is that, since he used hypnosis and suggestion as his main method, he himself suggested, without intending to, many of the symptoms which he afterwards sets out to explain.

Some critics, for instance, maintain that the very sharpness and distinctness of the three alternating personalities, and their long-continued separate existence, were due to the treatment employed and to the unconscious bias of Dr. Prince in favour of an interpretation in terms of split personality. When psychological analysis, Freudian or otherwise, is used, we are told that such marked and long-continued dissociation of personality is not seen. There is perhaps an element of truth in this criticism. It is possible that had Dr. Prince adopted other methods of treatment, he would have been able to restore Miss B. to a balanced mental existence without her passing through a long period in which her body was the prize of rival personalities. But under the conditions of the experiment she *did* pass through such a period, and antagonistic personalities *did* alternately take control of her organism. We are for the moment naturalists of the mind; and the case of Miss B. assures us that the strange mental creatures called multiple and alternating personalities can in certain circumstances exist.

There seem to be all gradations between such a complete splitting of the personality as we have here described and a mild degree of dissociation between two sets of tendencies—the evasion of conflict between two unreconciled systems of ideas. In some types and temperaments dissociation evidently happens much more readily than in others. In what we may call the easily-dissociated type, real alternations of personality may come into being without treatment by suggestion, although the abnormal state of affairs rarely lasts so long as in Dr. Prince's case, and the two personalities are rarely so sharply defined or so balanced in antagonism.

Many cases reported in the papers as "loss of memory" are really due to the emergence of a dissociated and repressed part of the personality which suddenly comes to the surface and ousts the normal self. Such happenings are called "fugues" by psychologists. Professor Janet reports on one such case. A boy lived with his mother who kept a little shop in a city not far from the sea. He hated the routine of the life, and used to spend much of his spare time sitting with sailors in the bars and taverns and listening to their stories of adventure. One day he disappeared from home. Some months later, a travelling tinker in another part of the country treated himself and the lad who worked for him to some wine, mentioning that it was such and such a feast-day. At this the boy exclaimed, "It's my mother's birthday," and remembered that he was the boy who used to sit with the sailors and hate his home-life. But with the revival of these memories came an effacement of the memory of the intervening months. Later investigation showed that he had worked his way to the coast on canal barges and, after various hardships, had taken up with the tinker. His repressed dreams of romance had taken shape in this adventure.

There are plenty of other well-authenticated cases, but we will avoid vain repetitions and go on to other aspects of our subject with those two important facts established—the fact that minds can be split, and the fact that such splitting is usually at least the result of mental conflict.

§ 7. Hysteria.

AND now we will take up an instance or so of hysteria. The term hysteria itself has a curious origin; it is derived from the Greek word for womb, since the ancients (and the moderns also until after the sixteenth century) regarded it as a disease of women only, caused by disorders of the womb, or by its getting loose from its anchorage and moving about within the body so as to press on various vital organs! It is now used to cover certain types of mental disorder, in men as well as women, which we can best define by illustrative description.

The symptoms are extremely varied. At one extreme we have what is usually known as "hysterics," when the sufferer (generally a woman, and more often a woman of Latin than of Teutonic race) indulges, to quote Dr. Gordon, in an "ebullition of emotional expression, accompanied by diffuse and purposeless motor activities." And this may terminate in a regular convulsive fit, closely resembling the fits of epilepsy. Apparently we have here a release of energy, a physical ebullition, but the reality of the case is that there has been an inhibition of the controlling centres in the higher part of the brain and the lower centres have broken loose. At the other end of the scale to "hysterics" are cases which simulate those of some real complaint—hysterical blindness or dumbness, or the paralysis of a finger or a limb, or pain—without there being anything organically wrong with the affected part. In the past, hysterical symptoms were often put down to mere malingering; and many laymen still hold this view, not only about hysteria, but about the complaints usually referred to under the name of "neurasthenia," which we will presently describe; they think them moral lapses.

But true hysteria is as much a disease as measles, though the one is caused by a germ, the other by a conflict in the mind. The popular idea has, however, a grain of truth in it; the hysterical symptom is the work of our Unconscious; and our Unconscious often employs it to keep us from doing what some part of our nature does not want to do. The difference is that in hysteria we, as conscious beings, may have no notion of the cause of our disability.

Let us take one or two cases. Dr. Gordon, in his book, *The Neurotic Personality*, quotes one of a young man who came back to England after the War suffering from a paralysis of one hand; the fingers were held stiffly out, and he could not work—a fact which gave him the greatest anxiety about his future. He had been a prisoner in Germany, and the paralysis dated from a day when a heavy piece of iron fell on his hand. Five minutes' explanation and suggestion, without even hypnosis, cured him—he moved his fingers again as well as ever he could. The explanation appears to have been as follows: The man was reacting against his work in the prisoners' camp—he hated working for the enemies of his country; the work itself was hard and unpleasant. Then came the accident. His hand was quite severely hurt, the fingers stiffened with pain. Here steps in the Unconscious, seizes the heaven-sent opportunity of finding an excuse for stopping work, and commands the fixing of the hand in the outstretched attitude, just as the operator in a hypnotic seance might do. The conscious self knows nothing of all this; all he knows is that the fingers are paralyzed. (We are manifestly taking the utmost liberty of metaphor here. We are personifying the Unconscious outrageously.) The paralysis was real, but hysterical. If the subject had been consciously malingering, a doctor could have detected the fraud in a few minutes, and he would have found it physically impossible to keep it up. Only suggestion acting through the Unconscious can control bodily functions in so perfect a way.

In many such cases the paralysis of movement will persist even in deep sleep. As complete proof that this patient's inhibition was genuine, comes the fact that his paralysis did not disappear when he came back to England and wanted to work. The Unconscious is blind; it operates in a dark realm and in very different ways from the Conscious. A reaction once established through it may work even against the desires of consciousness. The paralysis continued because the patient believed in it; but once he realized that he need not believe in it, it disappeared.

Another case of Dr. Gordon's was a man suffering from blindness—purely hysterical blindness; there was nothing wrong with his eyes, but he could not see. Here the accident which set off the blindness was an attack of conjunctivitis. The man's mind had been distressed by certain hidden conflicts and these were waiting for some chance of pacification. This conjunctivitis served its purpose excellently. Symbolism is a primitive way of thinking which preceded abstraction. In the Unconscious, as in dreaming, cymbolism is the normal method of interaction. The man, it turned out, was feeling exaggerated remorse over some private misdemeanours, and, as he himself put it, "could not bear to look himself in the face." He could not bring himself to look squarely at the situation and face up to it practically, either by coming to some worldly compromise with his conscience, by cutting loose entirely from certain habits he felt were wicked, or at least by not crying over spilt milk and tormenting himself; and the Unconscious, with its incapacity for logical solution, solved his distresses by this device of mental blindness. He was cured of his hysteria by a few explanations and a course of waking suggestion which enabled him to master his moral problem.

The hysterical symptom may be a loss of memory just as well as a loss of some power of movement or a loss of sight; and then we have conditions which grade imperceptibly up into the "fugues" and double personalities described in a previous section. In War cases, the peg which the Unconscious used to hang the symptom on was often some sudden physical shock, like a shell-explosion. The man recovers consciousness; he is dazed, and cannot remember properly what has happened. He cannot remember——

This is precisely what half of him has been wanting to do for months; he would like to forget all about the War and the horrors he has seen, but his sense of duty has repressed his wish. Now, while he is suffering from the shock, the repressed wish has its chance. "I can't remember," thinks the conscious self. "Quite right, you can't," respond the repressions in the Unconscious, "and you shan't." And the memories are dissociated from the rest of the mind.

Here are two simple War cases from McDougall. In one, only the memory of the incident was thus split off. A soldier, brought to hospital, dazed after an explosion, was about to be discharged as fit, when it was discovered that he walked in his sleep. He got out of his bed and went over to the bed of the only sergeant in the ward. When hypnotized, memory returned, and he re-lived the scene. A shell had killed several men in his party; he was running off to report to his sergeant when a second shell exploded near by and left him half unconscious. With the recovery of the lost memory, the sleep-walking came to an end.

In the second case the loss of memory lasted longer. A sergeant carrying dispatches had, as it turned out later, been blown off his motor-cycle by a shell-burst. A few hours after, he found himself on the sea-coast, over a hundred miles from the front, with no recollection of the shell-burst or of anything that had happened later. Distressed at his dereliction of duty, he surrendered himself to the military police. Much later, under hypnosis in hospital, he recovered the lost memories; he recalled how he had asked questions and studied the sign-posts; how he was actuated by a violent desire to get away from the zone of danger. Here natural fear, long honourably repressed, had got the upper hand, while the personal self was dazed and had split off all the memories of past existence, all the ideals of duty. But once he was well out of danger there was no longer the same urge to the instinct of self-preservation, and the normal self could again get the upper hand. The case is very like one of double personality, save that one of the "personalities" is a very minor and low type of mental being, with only the one impulse of self-preservation as its mainspring instead of a whole system of complex motives. It might well have been more powerful and more complete; if so, it might have held its own more or less permanently, and we should have had a typical case of total and long-continued loss of memory.

Thus the mechanism of hysteria, as of hypnosis and multiple personality, is a splitting, a dissociation; in cases of hysterical blindness or paralysis just as much as in loss of memory. Let us add two further illustrations. A soldier was stooping to pick up a bomb, when it exploded. He was luckily not wounded; but he found himself with mouth fixed open, tongue stuck out, and unable to move his jaws or utter a sound. He had begun to give vent to an instinctive cry; but in the very act his control of the machinery of voice was cut off, dissociated, probably as a natural reaction to the extremity of fear and shock. The tongue and mouth came under control again in a few hours; but the hysterical dumbness persisted. The Unconscious had used the paralysis of fear as its peg on which to hang the symptom which would withdraw the man from further danger.

One final case of great interest from McDougall. An Air Force cadet —keen, clever, athletic—fell and bruised his right arm as he was running to mess, and the arm was put in a sling for a few days. When the sling was removed, the arm was rigid, paralyzed. The boy denied that flying had "got on his nerve" and said how keen he was to go to the front. It turned out, however, that he had recently had a narrow escape from a serious accident while flying. McDougall put him into hypnosis; but when he was then commanded to move his arm, a strange conflict was revealed. "The arm moved" (we are quoting McDougall's words) "but, instead of obeying my suggestions, it performed the most extraordinary contortions with tremendous energy. Very soon my 'spell' was broken; the patient came suddenly out of hypnosis; and then, in place of his usual cheerful friendliness, displayed a fierce resentment, told me he hated me, etc. On the following day he was his usual cheerful friendly self; but he reported that, when he woke from sleep, he had found his right arm interlaced among the bars at the head of his bed and had had the greatest difficulty in extricating it. These vents were repeated with some variations. It was as though he were possessed by a devil.

"There is no room for doubt that this was a case of strongly-repressed fear, generating a strong subconscious aversion from return to duty. When I tried to force him back to duty by removing the disability, the repressed fear broke through and (as in an animal brought to bay) generated anger against me. The significance of the contortions of the arm is less clear; we may fairly suppose that they expressed a conflict between the repressing forces and an impulse to strike me."

Thus even in the finest characters the Unconscious may be busily at work as the conscious self's antagonist. And the more strongly natural instinct is repressed, the deeper-buried is the conflict and the more difficulty is experienced in bringing it to the light of day. The repressed fear had used the accident and the sling for its own ends. The anger of the Unconscious, as it felt its unworthy drift resisted, was forced temporarily upon the whole personality. This throws a flood of light upon much of our ordinary experience. And finally the independent action of the arm is noteworthy. We have again a resemblance to dual personality; but the minor personality here is in control only of a single limb. McDougall himself likens the symptoms to those of demoniac possession, and we cannot doubt that those who in other ages were described as possessed by evil spirits were in reality the sport of repressed parts of their own selves acting through their Unconscious.

Another interesting form of escape from the demons of an overwhelming situation is what is called Regression. The patient, so to speak, drops adult self-direction, abandons all memory of recent years and relapses towards childhood or infancy—and irresponsibility. That is the fundamental wish of the worried spirit, childish irresponsibility, the dear safety of the nursery again. McDougall gives a case of a wounded man, further distressed by the fear of air raids, who relapsed to childish crawling, childish ways of feeding, fretful crying and complete dependent helplessness. Slowly, as fear ceased to press upon him, he grew up again. He is the type case of a whole series of such instances of retreat from adult distresses.

From such strange wholesale reversions, there are all grades and variations up to apparently normal people who merely prefer to look back instead of pushing forward with life. Regression in one form or another is one of the commonest symptoms of maladjustment, and one of the most potent causes of wasted human energy.

§ 8. Exaltation.

THE hysterical exaggeration of inhibition does not always recall the ideas of our great-grandparents about diabolical possession. It may sometimes produce saints. The normal personality may be in conflict not only with tangles of suppressed desire of a lower moral type. Its aims and discretions may also be in conflict with excessively noble and generous impulses. And when the insurgent factor gets loose and is able to inhibit the normal motives of the personality, it may be that the inhibition is not of good, sane and high-minded activities, but of commonplace and lowly desires. There is a higher as well as a lower hysteria.

Many of those who are called mystics are the exalted outcome of a distressful mental conflict. They may believe themselves, and other people may believe them, to be not "possessed" but "inspired." An inspiring force may be interpreted and personified in very different ways according to the atmosphere of ideas about the individual. The religious mystic finds it is his Divinity with whom he has become united and identified. The exalted naturalist, like the late Richard Jefferies, will find he is "at one" with Nature.

We cannot go into all the fascinating questions involved in a full exploration of mystical experience. Sometimes it is the product of a special kind of training and concentration, just as is the power of following and appreciating a difficult mathematical theorem or a difficult piece of music; and then its machinery has little in common with hysteria. But when there is repressed conflict in the mystic's mind and a sudden illumination, then the machinery is similar, and then may show the completest parallelism with what we have identified as operating with demoniac possession.

In passing, let it not be supposed that the value of mystical experience is destroyed if we find out that it has a psychological cause within our own minds and is not due to a spirit or spirits outside and separate from us. To avoid theological entanglements, we can best illustrate the point from the experience of poets and artists; though with the proviso that there seems no difference of principle between this and the experiences of religious mystics. When Wordsworth wrote of

> "a sense sublime
> Of something far more deeply interfused,
> Whose dwelling is the light of setting suns,
> And the round ocean and the living air,
> And the blue sky, and in the mind of man"

he was describing an experience which, however he or anybody else may interpret it, can become to any man, as it did to him, one of the mainsprings of life.

And whether we believe that *Kubla Khan* (which came to Coleridge in a dream) or Shelley's *Ode to the West Wind* were inspired by some spirit or being other than themselves, or were the products of their own minds in a particular state of exaltation, the poems remain great poems, sources of joy and delight, things of permanent value.

The phenomena of religious experience may parallel those of morbid hysteria in the closest way. The most remarkable among such parallelisms is undoubtedly what is called stigmatization—the appearance on the worshipper of marks like the wounds inflicted on Jesus at crucifixion. The most famous man to receive the stigmata was St. Francis of Assisi, but there are now several score cases (mostly women) on record, many of them quite recent, and authenticated without any shadow of doubt. In addition to the marks in hands, feet, and side, some have developed marks which simulate those made by scourging, by the crown of thorns, and so forth. In every case the stigmatization has only appeared after prolonged meditation upon the passion and crucifixion. In some people, once initiated it was repeated at more or less regular intervals; in Louise Lateau, a Belgian peasant girl who received the stigmata in 1868, they bled regularly every Friday. In those cases which have been at all properly looked into, it seems that the region affected becomes tender and red, with blood leaking out of the capillary walls. Almost always, there is acute pain in the region of the stigmata; and sometimes only the pain is felt without any visible marks being left.

The Roman Catholic Church herself has pronounced some of the cases as impostures; there is every reason to believe that others have been unconscious impostures, the worshipper inflicting the injuries on herself—or rather having them inflicted on her body by her Unconscious—a not infrequent happening in hystero-epileptic persons. But there remain a number that are apparently genuine. In these it would seem that the Unconscious is again showing its amazing power over bodily functions that are beyond conscious control, just as it does in hysteria by producing mental blindness or lack of memory, in hypnotism by raising blisters or causing temporary paralysis. Stigmatization is the most startling of all such phenomena and can probably only occur when a man or woman is in that exalted mental state called ecstasy by the mystics, or in some approach to it. It is noteworthy that many of the stigmatized (though not all) have been hysterical or have suffered from epilepsy. They have turned their weakness into something which from the religious point of view is of inestimable value.

§ 9. Automatism and Mediumship.

OCCURRENCES such as automatic writing, which play so large a part in psychical research, appear to be only phenomena of the split mind. Some people with a gift for automatic writing find that they get their best results when, after taking a pencil in their hand, they occupy their conscious attention with talk or a book. Others, on the other hand, find it best to relax, and may pass almost into a trance. In these circumstances the hand with the pencil may write not only voluminously, but often coherently and interestingly. What it writes would appear to be the product of some system of ideas which is denied full access to the normal main consciousness—more or less thoroughly dissociated, repressed, or buried.

An interesting case of this sort was recently reported from the United States. An active stockbroker, who rather despised literature, found that lines of poetry came into his mind in the state between sleeping and waking. He was interested and wrote them down. They seemed quite good, so he sent them to a magazine, which accepted them. He now earns regular money by subconsciously composed poems —a product, doubtless, of a normal romanticism which his business life has suppressed.

There is no ground for believing that automatic writing is the work of extraneous "spirits"; the recesses of the mind have a sufficient population of partial personalities to account for all the automatic scripts that have been published. The same is true for the well-known phenomenon of "control" of a so-called medium. The medium relaxes and sinks into a trance-like state, in many ways resembling that produced by hypnotism; and after a time he (or, generally, she) begins to speak in an altered voice and manner. The new voice asserts that it belongs to some spirit which has entered into the mind of the medium and so has obtained control over her brain and actions during her trance. The "control" will give answers to questions or may volunteer statements of its own accord; it is, of course, through these channels that the bulk of spiritualistic information is obtained. There are doubtless cases where fraudulent mediums simulate trance and "control"; but when all these have been discounted there is an abundance of genuine cases in which a personality distinct from the medium's waking self does speak and act, and the medium has no knowledge of its words and actions when she comes out of trance.

A well-tried rule in science, as in practical affairs, is what was known to scholastic philosophy as William of Occam's razor, which, being translated into modern terms, lays down that unnecessary causes should be avoided; if you can explain your facts with the aid of well-tried principles, do not drag in new ones.

This rule is very much to the point as regards mediums and their "controls." All the facts, remarkable as they are, can be explained perfectly well as being due to the activity of secondary systems in the medium's own mind—repressed ideas, split-off personalities of varying degrees of completeness, deep layers of the Unconscious. Those who doubt this assertion should read the account of the celebrated Hélène Smith, given by Flournoy, Professor of Psychology at Geneva, in his book, *Des Indes à la Planête Mars*. Hélène Smith was not only an outstanding medium, who was perfectly genuine and believed in the literal truth of her control and its utterances, but one of the very few mediums who have consented to a thorough scientific investigation. She had visions, automatic writing, auditory hallucinations in which she heard messages or even whole poems, and trances in which a romantic and benevolent "spirit," who called himself Leopold, was the control. Among other revelations made by him was one of the conditions of life on Mars, including a detailed account of the language of the inhabitants.

Flournoy analyzed this language and showed that it was all composed of European word-roots, mostly French—the language of Mlle. Smith herself. He was also able, by a careful analysis, to trace the genesis of Leopold to an incident in Hélène's childhood, when a strange man of striking appearance came on the scene in time to save her from the attacks of a big and savage dog; and his memory was invested with all sorts of ideal qualities by the girl. And the stories which were heard by Mlle. Smith, or told by "Leopold"—some of them long and florid romances—were just what one would imagine as the products of the romantic side of a rich personality which the exigencies of life (she was employed in a business) had forcibly repressed.

§ 10. Neurasthenia.

LET us turn now from these aspects of dissociation to another type of mental disorder, which occurs where dissociation is not achieved. All minds are not equally prone to dissociation. Some are so close knit that they are unable to keep apart even the most violently discordant groups of ideas. And then open conflict goes on.

During the War there appeared a great number of cases of mental disorder of a non-inhibitory, non-dissociative type, which were at first classified as "shell-shock." In pre-War days the psychological theories of nervous disorder were at a discount; orthodox medical science preferred to look for its cause in physical damage or disorder in the nerve-cells. When a shell bursts close to a man the concussion may cause actual tiny hæmorrhages from the blood-vessels of the brain, with consequent damage to the nerve-cells; and at first cases of shell-shock were almost without exception set down to organic injury of this kind. But soon even the keenest upholders of this view found that the facts were against them; in many patients no trace of hæmorrhage could be found, and some severe shell-shock cases had not even been rendered unconscious by concussion.

The main cause of the nervous collapse was then sought within the mind; the shell-burst had set things off, not by means of its mechanical effects, but by working on the hidden springs of emotion. The old-fashioned medical men, robbed of the comfort of a tangible cause, flew to the other extreme and accused all shell-shock patients of malingering. It is on record that one medical officer with the rank of general in the British Army declared roundly that every such case should be shot as a warning to other shirkers and malingerers.

This view was not tenable for long—indeed, it was really not tenable at all by anyone acquainted with the facts. The next stage was the admission that most shell-shock was mentally caused, but this was coupled with the assertion that it only occurred in men who were somehow below the normal level—in degenerates or defectives, in those of unbalanced mental constitution or unstable temperament. It was a stigma of inferiority. But this idea, too, had to be given up. Many of the sufferers were men of the highest type of character and the most unblemished record. Whatever their weaknesses may have been—and everyone has some weak point in his mental organization—they were not in any way inferior.

And meanwhile the misleading word "shell-shock" itself had dropped out of the medical vocabulary. It was recognized that these cases were essentially failures of inhibition. Their essence was continued unabating conflict because inhibition had no decisive power, and the ease with which excitations spread and circulated through the cortex meant that the antagonistic ideas and urges could be continually struggling with and chafing against each other. The neurasthenic types are the worriers, the anxious souls. Their conflicts were of the same general nature as those which lead to hysteria, but they do not end in hysteria. In part, this difference is due to a difference in the circumstances; a sudden shock helps the development of hysterical symptoms, while a less violent but chronic rousing of the conflicting impulses to renew their never-decided battle favours a neurasthenic outcome. But in the main the difference is due to inner causes, of temperament or of training.

To the subject of innate differences of temperament we shall return; it merits a section to itself. Here we will only mention the significant fact that, while the percentage of mentally-caused disorder in the War was about the same among commissioned officers as among other ranks, there was more neurasthenia among the officers, more hysteria among the N.C.O.'s and privates. Apparently the closer knitting of the mind which higher education and public school training brings about helped to counteract the tendencies to segregation which a more naïve and more rough-and-ready mental life encourage.

The neurasthenic patient grows thin, tired and, above all, worried. He knows there is something gravely wrong with him, but has not the satisfaction of the hysterical patient of being able to show any definite symptoms which would justify his conviction of illness, and so he continues to reproach himself. The tradition of not giving in aggravates this feeling of self-reproach, and so does the prevalent belief, openly expressed or tacitly held by most laymen, that neurasthenia is really all the patient's own fault (while smallpox or a cold in the head are not, even if you have contracted them by neglecting obvious precautions). Depression, even frequent dallying with the idea of suicide (though serious suicidal attempts are rare among neurasthenics), an awful conviction of sin or uselessness or inferiority, of difference from normal, healthy people, a wearisome difficulty in getting through one's work, and especially in coming to a decision, a feeling that the mind is wearing thin and that the familiar self may fall through into a pit of mingled nothingness and hatefulness below, and with it all a longing to escape from the invisible prison in which the spirit finds itself, but a feeling of utter helplessness because the prison is one's self—these are some of the common symptoms of neurasthenia.

The neurasthenic is a living battle-ground. Two sides of his nature are in conflict; but instead of his being able to separate the combatants by suppressing one altogether as in hysteria, the dog-fight goes on and on. Sometimes a change of circumstances, tonics and tangible hope may suffice to dispel the conflict, but more often an analysis of the issue and a new outlook is needed.

The War cases of neurasthenia were usually simpler and more clear-cut than those of peace-time. The conflict often went on in the full daylight of consciousness, as in the earnest and religious young stretcher-bearer who, in spite of constant prayer for strength, was seized with uncontrollable fear—not fear of death, for he came to desire death as his only solution; but just *fear*—at every shell-burst. He came to sleep less and less; grew thinner; became nervous and jumpy and distressed. The conflict continued—it became one between his religious convictions and this irrational fact of fear.

Much more frequently there is a perpetual but unavailing attempt to be rid altogether of one of the rival sets of impulses. The conscious self is repressing them, holding them down, but circumstances are always stimulating them, and they arise again, taxing the strength of the repressive ego; and so the fight worries along at the boundary of Conscious and Unconscious. Often repression is successful during the day, but fails when the fatigued higher centres are asleep. That was the case with those numerous men in the War who repressed their fears and their fear-arousing memories of battle, only to have them rush up again during the night in the shape of nightmare battle-dreams.

A fact which throws a remarkable light upon the hidden springs of the mind is that severely-wounded soldiers very rarely showed any symptoms either of neurasthenia or hysteria. Why? Because the fact of being wounded relieved the Unconscious of all need to invent excuses for getting away from danger; because a wound, and especially a severe wound, brought to a sudden end the conflict between, on the one hand, repressed fear and worry over wife and children, and on the other, self-respect, the honourable claims of duty and the grip of military discipline.

In peace-time, the problems that face men are neither so violent nor so crude as those of war; and when the neurasthenic state develops, it usually does so by a slow, creeping and complicated growth, with its roots in a hundred incidents of different periods of the past.

§ 11. Repression and the Complex.

IN OUR last two sections we have been speaking of hysteria and neurasthenia as if they were sharply distinct. So they are, when we are confronted with typical cases. But there are all gradations between the extremes, and many cases that combine symptoms of both sorts.

This is what we should expect, since both kinds of disorder are the result of the same cause—namely an unresolved conflict. Hysteria and neurasthenia are two symptom-complexes, which differ from each other more because of a difference in the patients' constitution and temperament than because of differences in the prime cause of the upset. In a similar way the same germ may cause no symptoms in one person, but may kill another.

When the nervous system with all its complexity is involved, symptoms can shift about in an extraordinary way. For instance, symptoms of eye-strain can manifest themselves as pain in the eye; as headache; as general tiredness; or in the form of dyspepsia or other bodily disorders. The last is usually the case in men of strong powers of will and concentration; they refuse, as it were, to allow themselves to grow tired and headachy—but the strain continues, and throws the nervous control of the intestines or blood-vessels out of gear.

In the present section we will consider certain problems of neurotic disorder in general, paying attention more to the causes than to the various resulting symptoms; and here we hope to unearth ideas of value not only as regards nervous disease, but in the conduct of everyday life.

The centre of our interest will be the initiating conflict. Conflict is always there, but the conflicting impulses will differ from case to case. Sex trouble in one form or another is in peace-time at the bottom of most neurotic cases, as it is the basis of most peace-time novels and plays. The crude sex-impulse may be in conflict with romantic idealism, or with morality; mature passion may be fighting convention and the desire to keep on good terms with the world; the lack of outlet for a normal instinct may warp the old maid or the old bachelor; every nuance of complexity may be touched in an unhappy marriage, where hate and love, physical attraction and repulsion, remorse and hope, selfish and social motives, the claims of personal life and of parental love and parental duty may all be mingled in a shifting battle.

But any other powerful impulse may be in conflict with its opposites. In war, repressed fear was at the root of most neurotic cases. In peacetime the purely egoistic desires for success and prosperity may come into collision with the social impulses of altruism. The resulting conflict may be very severe; an outlet from it which is not infrequently taken advantage of, especially by sensitive and high-minded young men, is the cutting loose from the world and its claims, provided by a religious vocation.

So, too, the will-to-power may come into opposition with the rest of the self. And one impulse may be mixed or seasoned with another in every degree and every combination. The struggle that leads to a religious vocation is usually not with worldliness alone, but with sex as well. The will-to-power may take sides in a sexual conflict, or fear and idealism may be blended with remorse.

If the conflict stays unresolved and causes neurotic symptoms, then the more the sufferer is of the hysterical type and inclined to dissociation, the more likely are definite, specific physical defects to be substituted for mental discomfort; while if he is of the neurasthenic type, the existence of the conflict (in addition to causing diffuse physical symptoms like weariness and irritability, giddiness, sweating, queer sensations in the pit of the stomach, or irregularities of heartbeat) will issue chiefly in mental symptoms such as sense of guilt, fear of sin, feelings of self-disgust and inferiority, of depression and lack of energy, of gloomy, helpless struggle and humiliation.

In general, psychologists agree that the more an unresolved conflict is shoved out of sight into the Unconscious, the more serious are its resultant symptoms likely to be. The skeleton, when put away in the cupboard, comes to life and acquires new power. The reason is not far to seek. By being made unconscious, the conflicting impulses are debarred from access to the higher conscious levels of mind, cut off from reflection and deliberation, from rational purpose and will. But they are still seeking an issue; and as they are cut off from reason, they adopt the more violent and more primitive methods of the irrational, which resemble those of the animal or the child much more than those of the adult human being.

We have already seen how the Unconscious of hysterical types seizes swiftly and cunningly on chance accidents and turns them to its own ends; and we have also seen how unsatisfactory in the long run are the solutions thus brought about. We will give one or two further instances of neurotic symptoms to show how a repressed conflict may work out.

A common symptom is irrational fear of particular situations—what the psychologist calls a phobia. It may be a fear of enclosed spaces or, precisely the opposite, a fear of open spaces or of getting away from a wall into the open; a fear of water, of loud noises, of everything connected with the physical side of sex, of large animals, of cats. In practically every case, such uncontrollable and irrational fears can be traced back to some incident, usually in early life, the memory of which has been repressed. W. H. Rivers gives an account of an interesting case of his. A doctor in the early prime of life suffered occasionally from stammering and nervousness; and whenever he was in a narrow enclosed space he had a feeling of unpleasantness and discomfort, often of actual fear. As he had experienced this ever since he could remember, he thought it was a normal trait which he shared with the generality of the human race. In the War, living and sleeping in dugouts brought his horror of narrow spaces to a head, and often forced him to spend his nights in the trenches rather than sleep in the safety of the dugout. At last he realized that this fear was something abnormal. He grew more and more nervous, and he had to be sent to hospital, suffering from severe stammering, headache, and depression; he either could not sleep, or if he slept, had terrifying battle-nightmares.

Rivers got him to write down his dreams and try to recollect any incidents of early life which they recalled. He eventually recalled an incident which took place before he was four, and seems to have been the origin of his morbid fear. He had gone alone to the house of an old rag-and-bone man who used to give the children a half-penny for various objects they picked up; and on his way out, which lay through a dark, narrow passage, found the street door shut. He was too small to open it; and at the same moment a large dog at the other end of the passage began to growl.

As he recollected the incident, the memory of his childish terror came vividly back; but from that moment on the irrational fear of closed spaces became very much less, and his general condition improved. Probably he had been forbidden to have anything to do with the old man, and his guilty conscience saw to it that the memory of the fear was repressed. He remembered that ever afterwards he had been afraid even to pass the house.

The feeling of guilt is the commonest cause of repression. An act is over and done with; it cannot be undone; and yet we know it was wrong and wish it undone. Instead of facing things out, we try, in the terms of the moralist, to stifle the pricks of conscience—to forget the heavy sense of guilt. But the sense of guilt may be associated with any kind of action. In the case just quoted, the guilty act was an act of disobedience to parents. It may be murder, and spring from the lust for power, as with Lady Macbeth's crime. It may be sexual transgression, real or imaginary, as is very frequently the case with boys and girls of a rather idealistic type during their adolescence. Such repressions are apt to work out in uncontrollable motor impulses, either irrelevant to their cause or symbolically associated with it. Shakespeare seems to have heard of such a case and used it in his invention of Lady Macbeth's sleep-walking gestures.

Some mental patients are constantly washing their hands and displaying an exaggerated regard for cleanliness which may become so excessive that they have to be certified insane; this, as with Lady Macbeth, symbolically expresses the desire to be rid of guilt's uncleanliness. And as sexual offences are commonly regarded as "dirty," this ritualistic washing is frequently the result of sexual escapades.

Besides such direct symbolizations, the Unconscious may make use of other mental mechanisms, such as simple association. An impulse which is thought of as guilty or shameful seeks and finds expression; we strive to repress its further activity; we succeed in blotting out the memory of the guilty incident—but only by switching the impulse, so to speak, from its normal outlet on to something else which happened to be associated with that moment of intense experience. This is especially common with the sex impulse, and results in what is generally called *sexual fetishism*, where the most heterogeneous acts or objects are found to give some degree of sexual excitement and pleasure, and the normal attraction of the opposite sex is correspondingly weakened. There are men who form collections of women's boots or, a common police-court case, find themselves driven to cut off the pigtails of young girls.

These facts throw a vivid illumination upon human motives. For one thing they show how futile it is to interpret human acts in terms of reason or advantage, or of a balance-sheet of pleasure and pain. This sort of mental accountancy, which the utilitarian school of Bentham and Mill brought into vogue in the middle of last century, is too rational; indeed, the whole philosophy of utilitarianism is one gigantic rationalization—it invents polite and logical reasons to account for the workings of obscure and irrational impulses. In our conduct we may be partly guided by reason, but we are not *impelled* by it. We are pushed and shoved by impulses and strivings which have nothing to do with logic, whose end and aim is not pleasure but expression in action.

It is widely believed among psychologists that many repressions may, so to speak, make a deal with the repressive forces and find relief in an ennobled form. This they call *sublimation*. Artistic and literary gratifications are considered by these authorities to be sublimated repressions. The sublimated sexual impulse they think may have given the world many acts of devotion as well as poems and works of art; the will to power may have been sublimated to help in scientific discovery and the creation of great works of art. And so on. But this attribution of intellectual and æsthetic achievement to sublimation may easily be carried too far. There has been extraordinarily little repression in the lives of many great artists and thinkers to account for their achievements. It is questionable if sublimation, as thus defined, can be treated as anything but an extreme type of symbolization, or whether it can be credited with more than a little amateurish art, writing, research, or devotion.

So far we have made very little use of the term "complex." Its definition will introduce nothing really new to the reader of §§4, 5, and 6 of this chapter. It is typically a mental system coloured with emotion, connected with some instinctive urge, linked up with an organized set of associated ideas and in conflict with the more dominating system of the mind. So it is more or less repressed. It differs in degree rather than in nature from a secondary personality on the one hand and a simple repressed impulse on the other. It never has the importance of a secondary personality; the main personality is affected by it but remains in control.

Among all the troubles of mental development, the formation of repressed complexes is the commonest. One might almost call it the normal experience, since different sides of the self are almost bound to come into conflict during development, and unreasoning repression seems often the simplest way of dealing with such a conflict. It is the method to which children or young people resort very readily. The more effective the repression the less the normal conscious self knows about the repressed side of his nature, though it is still active, and still influencing him; but at the same time, the more independent and active the complex becomes. Furthermore, the more the complex is cut off from the main consciousness and reflective reason, the more devious, primitive and irrational the ways in which its pent-up energies force their way into action—until in the extremer cases we get hysterical paralysis; irrational fears, compulsions, fetishisms; the power of decision brought to a standstill; or a hidden sense of guilt undermining the entire fabric of self-respect and self-confidence.

The discovery that such repressed bits of mind exist and continue to act in the Unconscious is one of the great achievements of modern psychology; the resolution of insurgent complexes is one of the main tasks of psychological medicine, while their prevention should be a prime aim of education.

In concluding this section, we will draw attention to one property of the mind-systems out of which repressed complexes spring. They are one and all invested with a double quality of simultaneous attraction and repulsion. This is, if you like, another way of saying that they are fraught with conflict, for conflict will only arise when one part of the self desires, another part dislikes or is disgusted; but the fact is so important that it deserves stressing. This double quality, positive and negative in one, has been called *ambivalence*. Normally animals can suffer little from this double-edged attitude of mind, because they live so much in the present and have so little capacity of reflection or recollection. The germ of it is present in them, however, and can be brought into prominence artificially, as in Pavlov's dogs, which suffered from neurasthenia and hysteria because they were subjected to conflicting impulses. But in man, matters are far more elaborate; and when we begin to reflect, we realize what a part is played by ambivalent systems of feeling in the development not merely of every individual one of us, but also of the most deeply ingrained human traditions.

Whole tracts of our own nature may come to have this quality. The sex-impulse is the most obvious case in point. It is very common for adolescent boys and girls to combine passionate urgings of their sexual instinct with fear and disgust at the idea of its physical manifestations; and this attitude may continue into adult life. But in the same way the urge to power and self-expression may become tinged with ambivalent feeling, and an explosive mixture of self-assertion and self-depreciation result.

Or particular objects or ideas may become thus charged, as it were, with contradictory feeling. The attitude of children to parents is the most familiar example. The boy loves and respects his father. But the father is the source of punishment and prohibition as well, and the boy is frightened of his anger and galled by his authority. He loves, fears, admires, and dislikes all in one. Hence, frequent cases of difficult children who improve enormously on being sent to school. And there is every nuance according to sex and temperament and circumstances of parent and child. Wife or husband may become similarly double-charged with the electricity of mental life. Physical attraction and satiety, the desire for stability and irritation of being tied for life, real affection and the friction of daily domesticity—these and many other pairs of impulses may be at work charging up the ideas centred on the marriage-partner. In extreme cases, there may be the most violent oscillations between passionate love and wild hatred; or the positive and negative may be locked in a continuous, indecisive, and exhausting struggle.

Many religious ideas are similarly double-charged. In most religions love and fear are intimately blended in the attitude towards the god. In more primitive religions, it is rather objects and persons that are ambivalent, and they become ambivalent in a special way to which the epithet *tabu* is given. Something or someone—a king, a place, an action, an object—is regarded as both sacred and forbidden, often almost unclean—both revered and, in certain aspects, abhorred. As one anthropologist has put it, there are two kinds of holiness or sacredness to primitive man, good-holiness and bad-holiness; and both are combined in what is tabu. Many of us can recall a similar ambivalence in our own early struggles with religious propositions, and the history of Manicheanism seems to present that indecision on a gigantic scale.

§ 12. Psycho-analysis.

THE repressed complex in one or other of its protean forms is at the root of most neurotic disorder. Different methods of getting at the peccant complex and eradicating its symptoms have been used. Suggestion under hypnosis was for long the chief deliberate method of attack. Its usual alternatives were to send the patient a long sea-voyage or give him a rest-cure. But when the root of the mischief was in a deep-seated and long-continued conflict, the voyage or the nursing-home would often give renewed opportunity for the conflict to rage without interference from outer claims; and the last state of the patient was worse than the first. Even if the rest did benefit the patient, in most cases it only cured effects and left the cause unaltered. And the same objection applied to hypnotic suggestion; it often removed symptoms but did not touch the cause, which sooner or later broke out in new effects.

In the War, many cases of shell-shock were successfully treated by another use of hypnotism. When it became clear that the patient's neurotic state was due to some violent shock of horror, disgust or fear, the terrifying or shameful memory of which he was suppressing, hypnosis was employed to recover the lost memory. We have given several examples of this, and of how the recovery was often difficult and accompanied by deep-seated resistances; but once recovery was accomplished the symptoms generally disappeared. This "blowing off" of symptoms and cause in one blast of remembered emotion is sometimes called "*abreaction.*" There is no doubt as to the efficacy of the method; but it has its limitations. For one thing, abreaction alone may merely cure a particular incident of mental breakdown. The man became neurotic because of a terrible experience—but in most cases, also because of some inner conflict which was the opportunity of the experience. When the conflict has been long-continued and severe, abreaction alone will not cure it; the man goes back to the trenches apparently healthy, but receptive soil for the next violent shock that is to befall him.

In the neuroses of peace-time there is often no single experience from which the patient's illness dates; usually the conflict itself becomes gradually more acute; and the complex grows by feeding on the experiences of years.

Persuasion and waking suggestion have also been used; but they too often fail to reach the true cause of the complaint. During the present century, the method known as psycho-analysis has come into ever greater prominence.

Historically, the method of psycho-analysis grew out of the method of abreaction, which was discovered by Breuer (1842–1925) in 1880. Towards the end of the nineteenth century Sigmund Freud became associated with Breuer in developing the method. Later, Freud, pursuing the subject by himself, became progressively more dissatisfied with the results of abreaction alone, and came to place more reliance on analysis. The first method which he adopted was the method of "free association." The patient in this method is asked to make his mind as relaxed and blank as possible, and then to tell his physician everything that comes into it, following out the chain of association whithersoever it may lead him. If this be practised, it is found that association repeatedly leads the patient along certain roads of thought which are then suddenly barred by a resistance, a reluctance often so strong as to be a veritable compulsion to stop. Freud was the first to ascribe such resistances to their true source—in repressed shocks, memories and desires buried in the Unconscious. By encouraging the patient to go on or to start again, such resistances may be gradually broken down, and not only will abreaction be brought into play, but the patient will have had the root-causes of his disorder made clear to himself, the hidden workings of his own personality laid bare. Analysis, at its best, results in ruthless self-knowledge which should automatically lead on to conscious self-discipline. Freud soon came to supplement free association by the analysis of the patient's dreams. And in the subsequent quarter of a century various improvements, or at least elaborations of method and of the ideas behind the method, have been introduced.

In former ages dreams were considered to be of the utmost significance in life and simple people have always believed them to have a prophetic and warning quality. But the disposition of psychology up to the time of Freud was to belittle their importance. To him and his associates we owe our modern realization of their great symptomatic value. Essentially a dream is the appearance between sleeping and waking of an uncriticized and uncontrolled flood of associations. Repressed complexes get an opportunity in these unwary phases for more or less complete expression before the normal self is fully reconstituted and alert. They reveal themselves, albeit often disguised, distorted, and symbolized, to the trained observer.

The War brought home very forcibly the need for proper treatment of mind-disorders; and in the last ten years there has been an extraordinary development of psychological methods for the treatment of sick minds. The main trends in that development have been the recognition that in psycho-analysis and the ideas behind it Freud had placed a new tool of the utmost value into the hands both of science and of medicine. His initiatives have been followed up with passionate zeal by a number of disciples and rivals. It is a manifestation of the vigour and richness of this new department of inquiry that it should develop wide divergences of opinion and that the master should presently find himself under the searching criticism of his livelier followers. Heated controversies ensued and dissentient schools of psychoanalysis arose. The best-known of the seceders are Adler and Jung, who left the Freudian movement in 1911 and 1913 respectively.

Freud has consistently maintained that sex was the great motive force in all subterranean conflicts of the mind. The civilized human being has repressed and delayed the sexual impulse beyond any animal precedent in the interests of educational development and social peace. His social life and his sexual life are in necessary conflict. There Freud is borne out by contemporary anthropology (see Book 9, Chap. 1). But Adler would replace the sex-impulse very largely in his interpretations by the will-to-power. And there he also has anthropological support. The human animal has not only a retarded adolescence but a lost independence. Self-assertion rather than sexual desire is for Adler the main suppression. But after all, sexual exercise is one main form of self-assertion. It is evident in his writings that Freud uses *libido*, his term for the sexual impulse, in a sense much wider than most of us would attach to the word sexual. The quarrel with Adler may be to a large extent terminological. Freud's interpretation may be the juster for some races and types, and Adler's for others. Freud admits that the bulk of his cases are drawn from prosperous Viennese Jews.

Jung, however, goes further. He has enlarged our notion of the Unconscious. He finds in it not merely our conflicts and repressed desires, but also the source of our inspirations and higher impulses, to make it, in the coarse, regrettable, but illuminating words of an anti-Freudian critic, a well-spring, not merely a cess-pool. That is to say in decent language Jung finds it much more than a dump of repressions. There is, he asserts, an Unconscious "already there." That idea, as we shall see in a moment, is a very considerable extension of the Freudian Unconscious.

Let us define the conflicting views a little more exactly. The orthodox Freudian asserts that we all of us pass through a phase of infantile sexuality, in which every sensory gratification, be it suckling or tickling, is in essence sexual. This is manifestly stretching the meaning of the word sexual. He claims also that we all normally acquire an "Œdipus complex" in early childhood. Œdipus, it will be remembered, unwittingly killed his own father Laius, and married his own mother, Jocasta. The Freudian view is that every boy passes through a phase when he is stirred by a sexual attraction towards his mother, a sexual jealousy of his father, and that the reverse arrangement holds good for girls. This attitude may be complicated and modified and varied in many details, but the basic principle, Freud asserts, is always traceable. There is always a conflict and restraint. The passage through this phase in development and the conflicts which it inevitably engenders, leave indelible marks for good or ill on the personality. Out of this primary conflict emerges the normal self, shorn of the primitive possibilities of incest and parricidal violence.

The Freudians claim that the Unconscious, which, as we have pointed out, necessarily thinks in symbols, does so in forms that are almost universally the same for everyone. They give a list of things which symbolize sex, and it is an amazing list. Nearly everything in the world, it seems, from a church steeple to running water, possesses sexual significance in the human mind. With a Freudian dictionary of symbols, the practitioner, they claim, can, as it were, translate the contents of dreams and the ideas thrown up in free association and give them meanings which must remain unsuspected by the patient until he has conquered his resistances. There is an element of truth in this conception of symbolism, but it seems to the present writers that the Freudians—who frequently seem to outdo and caricature Freud—carry this too far and make it too definitive.

Latterly Freud has shown a disposition to define the operative aspects of the mind more exactly. He divides it up into an "Id" which is the whole reservoir of impulsive reactions, the Ego which is the superficial layer of the Id in conscious contact with reality, and the Super-ego which in man (and man alone) is developed out of the Id; it represents the repressions of instinct characteristic of man and dominates the Ego. This Super-ego or moral or ideal self is first evoked, he asserts, by the conflict of the Œdipus complex. It will be convenient to return later when we discuss conduct, to this question of the interacting parts into which the mind, for the purposes of an analysis of conduct, can be divided.

Adler has opened up a very valuable variation of psycho-analysis by stressing how the sense of inferiority, from which so many human beings suffer in their early years, produces by over-compensation an exaggerated or even abnormal desire for success, power, and accomplishment. The mere fact of being a child, forced to obey the prescriptions of its elders and confronted with the task of learning all about the world and the conduct of life in a few short years, in itself conduces to a feeling of inferiority; the failures to achieve must always be so many, the thwarted sense of helplessness in face of superior authority is inevitably so often aroused. And if to this be added strict and uncomprehending parents and teachers and an ambitious but over-sensitive nature, the conflict between self-depreciation and self-exaltation is almost inevitable. Many young men whom their elders stigmatize as conceited and pushing have been forced into this attitude by the conviction of their own inferiority which was imposed on them in early life and which they are trying to stifle in the depths of their Unconscious; it is a protective covering of their own inferiority-complex.

Jung has attempted to do fuller justice to the complexity of the mind than either of his rivals. For him, it is a general urge of life, rather than the particular urge of sex or of mere self-assertion, which drives us on towards finer adaptation and fuller satisfactions. He undermined much of the Freudian analysis by pointing out that, even if we discovered some infantile "cause" of adult conflict, it was rarely likely to be the whole cause. For, once conflict comes into being, any later clash of impulses is likely to be drawn into the existing battle; and any new repressed impulse, even if different from the old, will almost always form an alliance with its fellow rebel. In Dr. Prince's Miss B. (§ 6), when she fell in love, the repressed sexual impulse took sides with the worldly or selfish impulses repressed in earlier years.

And sometimes, Jung argues further, another step may be taken, and the new experiences may come to colour the old. Physical sex-impulse repressed in adolescence, for instance, may come to be projected back and colour some incident of childhood which had to do with mere childish curiosity about forbidden, but not necessarily sexual, things; or sex may come retrospectively to colour the child's inferiority-complex and project a tinge of sexual jealousy into his simple envy of the grown-up size and strength of his father. Thus when Freudian analysis discloses an infantile complex, it may have really discovered a mare's nest, a chimæra manufactured by the patient's own mind in maturer years.

For this and other reasons, Jung deprecates a too exclusive concern with the patient's individual past. Knowledge of the past is often necessary to understand the present; but it is the setting of the patient's mind at the present moment which is important for his future cure. One final point on which Jung has done great service is in stressing the innate differences beween human beings in their ways of thinking and feeling and reacting. If we would understand a man, whether as personality or patient, we must pay as much attention to these as to external incidents of life.

Finally, as we have noted, Jung claims that there is a racial as well as an individual Unconscious. Some of the Unconscious was always there. He has adopted a quasi-Lamarckian view of mind, maintaining that in the Unconscious there is a store of racial memories, laid up in the course of generations as what he calls Archetypes or primordial images. It is these Archetypes, he supposes, which form the basis of dream-symbolism and of mythology and religion as well.

With such secessions and rivalries among the schools of psychoanalysis, many of the acerbities that characterize sectarian controversy have appeared. Psycho-analysis can easily become a controversial weapon and the dignity of science has been waived for the amusement of the vulgar. Some of the Freudians have accused Jung of genteel dismay at the revelations of universal sexuality which psycho-analysis was making, and to have shrunk timidly and unscientifically from following out the conclusions to be drawn from his work; while they account for Adler's "Will to Power" by hinting that this is what one would expect in an ambitious Jew. On the other hand, Jung has ascribed Freud's constant preoccupation with the "father-complex" to *his* Hebraic ancestry and the attitude of his race to their jealous god, Jehovah. But the psycho-analysis of a thinker does not necessarily destroy the validity of his thought.

We have dealt thus explicitly with these rival schools of thought, because their differences still bulk very large in current discussion. But we may confidently expect that, when the current edition of *The Science of Life* appears twenty-five years hence, the whole controversy between Freudian, Jungian, Adlerian, and other brands of psychologists will have been relegated to the attics of scientific history. Each party is making its contribution to truth, and less partisan psychologists are already drawing impartially on all those divergent explorers in the field of psychological observation for a more solid edifice of theory. Let not our criticisms seem to be a depreciation of their work, or above all, a belittlement of Freud. Sigmund Freud's name is as cardinal in the history of human thought as Charles Darwin's. These psycho-analysts, under his leadership, have created a new and dynamic psychology, one that thinks in terms of activities and strivings, of impulses and conflicts, in the place of a flat and lifeless picture of mental states.

If the facts of hypnotism first demonstrated the existence of the Unconscious, it is the psycho-analysts who have shown its extent and its importance in our daily life, in sickness as in health. They have made the study of dreams a valuable and interesting branch of psychology, have shown the connexion between the methods of thinking found in primitive peoples and those adopted by the Unconscious. They have rightly stressed the frequency of regression, the mind's return backwards to the past and its simplicity, the need of attaching due weight to apparently trivial slips and to foolish symbols, the importance of hesitations and other impediments to the free flow of thought in unveiling the hidden complex hedged round with paralyzing resistances. It was they who first stressed the principle of abreaction; they drew attention to the important idea of sublimation whereby impulses directed to a lower activity such as physical sex-gratification can be harnessed to other and higher aims. If the Freudians have overstated their case in dealing with the Œdipus complex, they drew much-needed attention to the fact that the child's relations to his father and mother are amongst the most fruitful sources of troublesome repressions and conflicts in later years.

Last, but not least, they have largely helped us to understand the truth (so hard at first to grasp) that neurotic disorder is in its essence purposive. The symptoms, however much distorted by repressing forces, however poorly adapted to their aim, do represent an urge towards adjustment in the repressed Unconscious. It is a striving in perplexity.

§ 13. Minds Out of Gear and In Gear.

INSANITY is not a definite disease like measles; it is merely a convenient term under which are lumped the disorders of a great many sufferers, whose only common feature is that the workings of their minds are so different from those of other people, and so divorced from reality, that it is impossible or dangerous to try and carry on the business and the social intercourse of life with them.

Certain types of insanity recur over and over again; the science known as psychiatry has classified them, and the psychological and medical researches of the last half-century have helped materially in our understanding of them. In the Middle Ages the people we now label insane were thought of as possessed by good or, more usually, evil spirits. In the "Age of Reason," the eighteenth century, they were thought to be suffering from a disease of the reason which put them beyond the pale of ordinary men and women. The claims of society alone were recognized, and when the lunatics were loaded with chains and herded into filthy prisons, society congratulated itself on having done its duty. To-day, we realize that the insane person is a human being like ourselves; his insanity may be the result of deficient heredity or some physical disease, or due to the abnormal exaggeration of some quite normal trait of human mind. Our treatment of them has correspondingly changed; indeed, it is one of the ironies of our civilization that many lunatics, could they but have enjoyed all their life the attention and the healthy conditions of existence which they find in the asylum, would never have become mad.

Let us imagine that we are being taken round a big asylum, and that one of the medical staff is pointing out to us examples of the different main types of insanity and explaining their history and origins. Over and over again we shall be likely to think how easily we might have overstepped the border-line of sanity in the same way, how, but for accidents of heredity and circumstance, we might be the lunatics and they the normal human beings. Here we see a wretched man living the life of an imbecile and paralyzed baby.* He cannot move his limbs; his mind is a blank; control over his bodily functions has disappeared; little remains of his behaviour but the almost reflex action of taking food when a spoon is put to his lips. The doctor tells us that he will eventually lose even this faculty, and will die of starvation unless fed by a stomach-tube; but even so his nervous system will continue to disintegrate until he dies. He is suffering from a late stage of general paralysis of the insane. This is a sequel to infection with the germ of syphilis, and its symptoms are general decay of body, and especially of brain and mind. About one in every fifteen asylum patients suffers from it.

Some two years ago our patient was healthy; then he began to fail in the normal affairs of life; he forgot all the odds and ends he ought to attend to; his judgment grew impaired; he thought he was rich, and began treating everyone with indiscriminate generosity. His capacity for doing simple arithmetic failed (long before his powers of general reasoning). His control, as well as his judgment, grew impaired; he neglected all his duties in order to go to the races on every possible occasion, and began making the most absurd and ambitious collections. It was then his relatives had him put in the asylum. There he grew worse, until finally his delusions and his normal human desires disappeared one by one. You can watch his behaviour going to pieces as the spirochætes rot his brain. Finally, even the power of speech and the instinct to walk will disappear, and he will revert to the condition of a very dirty baby.

It is this condition which, as we have seen, may often be ameliorated and its further progress arrested by infecting the sufferer with malaria. It is an object-lesson in the interrelation of mind and body, and shows us how even when slow physical degeneration of nerve tissue is the prime cause of mental derangement, the mental architecture is broken down in the reverse order from that in which it was built up; the highest faculties go first, the simplest instincts and reflexes last. Higher control disappears early; and its loss manifests itself equally in every sphere of mental life—the logical intellect refuses to work properly, repressed desires can manifest themselves in delusions, the capacity for decision and action is impaired.

Near by is another case of a baby's behaviour in a grown-up body. And here, too, the prime cause is purely physical; the condition is very different from that of the general paralytic because the patient has

---

* These cases are taken from the words of Kraepelin, Kretschmer, Jung, Bleuler, McDougall, and Bernard Hart.

always been like this—his defects are congenital. He can understand a few easy words, but cannot speak himself; his behaviour is in every way grossly subnormal. If we were able to examine his brain-tissue after death, we would, in all probability, find that the outer layers of cells in his cerebral cortex were definitely under-developed. He is a congenital idiot; he has never had any mind worth mentioning.

Idiocy and imbecility may be caused by the most varied influences—any influences that retard or prevent the proper development of the cerebral cortex. Toxic poisoning before birth, injuries at birth, water on the brain, and especially the inheritance of certain defective genes, are perhaps the most important. Perhaps the most striking fact about the condition is that nowhere between the grossest idiocy and the brilliancy of genius is there any sharp line to be drawn. The two extremes are connected by an unbroken series, via talent, ability, normal capacities, backwardness, mental defectiveness, mild imbecility. Some combinations of genes over-equip us for life; others are a sad under-equipment.

Here is another patient, who looks miserable and depressed. If you talk to him you will find that he believes he is eternally damned, for he is a renegade from the Catholic faith. He would like to commit suicide, were it not that he is sure of going straight to hell if he did so. His will-power and his capacity for feeling normal emotion, he will tell you, have disappeared, together with his interest in the ordinary things of life. This feeling is the basis for his belief that he is damned; for his soul feels already lost. When left to himself he sits alone and broods all day. Fear, misery, and self-reproach make up his habitual state. Before he was brought to the asylum his wife had scoffed at his religious beliefs, which he cherished at the back of his mind long after he had given up being a practising Catholic, and in his gloomy state the thought of her unbelief fills him with fear and horror.

But this same man, the asylum doctor will tell you, will at other times be altogether different. There is then no sign of gloom, but he looks cocksure and confident; he is bright-eyed and active—too active —with a running spate of talk. And his talk is the antithesis of what it was. He is now a thorough-going sceptic, and one who feels it his mission to destroy the superstitions of the credulous believers in revealed religion. What pleases him most is to get into religious controversy with one of the doctors or visitors; the more distinguished his opponent the better—he likes a foeman worthy of his steel. But if he is thwarted, he flies into a violent passion; and then you must look out that he does not smash up his furniture, or attack one of the attendants. If he thinks of his wife during this active violent phase, it will be to fly into a jealous passion on account of her propensity for going out with other men; though while he was in the melancholy phase it was impossible to arouse his jealousy at all. And at still other times he will be a normal human being for months together, and can then be discharged from the hospital.

He is what the psychiatrist calls a manic-depressive, a man in whom the exaltation and excitement of mania oscillates with the gloom and melancholy of depression. In one or other of its forms, the manic-depressive disorder comprises a very large number of asylum patients. It is one of the kinds of insanity which most often comes on and passes off again, leaving the patient normal between attacks. In its most interesting types, the depression and the excitement regularly alternate; and then, as in true multiple personality, the actual physical make-up changes with the mental. In the melancholic phase, secretions are diminished, temperature is often a little sub-normal, and the "reaction-time," or time needed to perform a simple voluntary action at command, is slow. The reverse holds for the excited phase—secretions of all kinds are over-abundant, the pulse is rapid, the reaction-time is very short. The attitudes adopted in the two phases are, we should expect, very different; a queer feature is that the maniac generally shakes hands with arm outstretched from the shoulder, the melancholic only from the wrist, with arm drawn into the side.

In our round of the asylum we shall find other cases who show only one of the two abnormal phases. Some are always depressed, and never pass into the excited manic state, while in others this is the only symptom of madness, and they never become melancholic. The chronic cases of either sort are very troublesome—the maniac grows violent, abusive and destructive, the melancholic must be constantly watched to prevent attempts at suicide.

What is disorganized here? It is the faculty of self-criticism, which in its turn is based upon a proper balance of the impulses of self-assertion and its opposite. In the normal man these two opposites are continually at work; self-criticism and sober judgment are the outcome of their balanced hostility. The machinery whereby one of the two impulses gets the upper hand completely is not yet fully understood, though it seems akin to that by which dissociation of memories takes place. There seems to be no psychological method of curing established cases; beyond a certain point there is no return. Yet it is clear that mania and melancholia are only exaggerations of what we may see in normal people. We all, even the most equable of us, have moods of elation and energy, of depression and self-reproach; we may know them to be irrational, but they dominate us and our feelings. In the moody man these oscillations are more marked; and when circumstances or temperament aggravate matters still more, the man becomes "impossible" in daily life and must be shut up. Then there are people who all their lives long show abnormal energy and self-assurance; they are not maniacs, but the stuff of which maniacs are made. We envy them their untiring flow of hopeful activity—until we notice that their performance does not keep pace with it. Self-criticism is painful, but it is necessary for real achievement. The activities of such "sub-manics" are not fully logical. When this trait is exaggerated it passes over into the insane flow of ideas so characteristic of the chronic maniac, whose "conversation" is often a mere string of phrases held together by accidental associations, sometimes by mere associations of sound.

So, too, we all know the gentle, slow, patient men who are half-way to melancholia; they are no good where speed or bluff or quick decision is wanted, but they are often excellent at the routine work of the world.

The maniac breaks out in rage at each trivial thwarting and may maim and slay. But have we not heard of choleric business men who tear the telephone attachments from the wall? And has not the temper of the retired colonel grown proverbial? The normal man with tendencies to melancholy often commits suicide when circumstances go against him. He differs from the melancholic who must be watched all the time, but only because the melancholic catches at ridiculously trivial pretexts for his suicidal attempts.

Here is another patient of a different type. This woman has been in the asylum for many years; for years she has spent all her waking hours sitting huddled in one position, continuously moving her hands and arms in a peculiar stereotyped way. These movements turn out to be an imitation of a cobbler stitching boots. She cannot be made to utter a word, and never shows the least interest in anything that goes on around her. She is completely shut up in her self; her life has narrowed to one compulsive action.

Her past history is simple; she was engaged to be married, when her lover suddenly broke it off. As a result of the shock she soon passed into the state in which she has remained ever since. The young man was a shoemaker by trade; and the stereotyped movement as of cobbling shoes may be a symbol of her preoccupation with the hopes of love and marriage which circumstances so violently destroyed.

Usually, when inner longings come into conflict with hard reality, and the impulses from which the longings spring can find no other outlet, it is the unsatisfied desire which is repressed or split off. But in this case it is reality which has been repressed; the woman lives an inner life that is totally unreal.

This loss of contact with reality is the most constant symptom of insanity. In mania and melancholia it is the emotional tone which is out of contact with reality. In this woman, reality is simply shut out from making any contribution to mental life and growth; the particular way in which it is here shut out is by the sufferer turning inwards, as it were, and living a dream.

This turning inwards into self and away from outer reality is the essential feature in a great number of kinds of insanity which are now generally grouped together under the title of *schizophrenia* or mind-splitting—the splitting being not, however, that which we have discussed as dissociation, but a more radical divorce between inner and outer, self and reality, wish and experience. And the most common of such disorders is that which is called *dementia præcox*, precocious loss of mind, in which the symptoms usually begin to appear about the time of puberty. The disorder generally first manifests itself in moodiness and depression, reluctance to work, over-preoccupation with self, long fits of day-dreaming now and then broken by emotional outbursts. Gradually the sufferers grow less and less interested in the world about them, sit idle and refuse to work, and develop delusions, fixed ideas, or queer actions. Like manic-depressive insanity, *dementia præcox* is psychologically incurable once it is fully established. Once committed to an asylum, patients continue their progress towards imprisonment in self. The delusions, the dreams, and the actions, abraded, as it were, by constant repetition and not fed by new experience from without, tend to become more stereotyped and often degenerate into mere symbols or hints of their former selves; until finally a large proportion of the patients sink into an apparent stupor—"those cold, lifeless ruins," as Kretschmer calls them, "who glimmer dimly in the corners of asylums, dull-witted as cows." Minds can degenerate through disuse even more radically than muscles or tendons. But even in these unfortunates, there seems always to be a nucleus of inner life, atrophied by disuse, but still revolving round a trace of some self-centred idea.

We can follow a series of stages in this downward and inward progress. Over there is one of Kretschmer's "cow-like" cases—for ever silent and apathetic. Here is a second, who lives only to repeat endlessly all day: "Will that be all right if I walk up to the door and back again? Will that be all right if I walk up to the door and back again?

Will that be all right ....?" There is a third, in whom not speech but action is repeated and stereotyped; she stands for hours together in one spot, gesticulating with outstretched right hand.

This patient has inner life so divorced from outer that he lets his limbs stay in the position the doctor cares to put them. And another shows the same phenomenon in speech; asked a question, he repeats it, often just altering the pronoun. "How are you to-day?" says the doctor. "How am I to-day?" answers the patient. His interest is focused on his hidden thoughts, and the outwardly directed part of himself acts like an automaton. In other less striking cases, the patient will do what he is asked, but in a stupid and repetitive way. If you put a dustpan and brush in this woman's hands and tell her to clean the floor, she will begin to do so, but will go on brushing the same little bit of carpet for hours. More frequently the reverse is the case, and the patient is entirely "negativist"—he does exactly the opposite of what is suggested or commanded. Here is an extreme case. The doctor makes as if to touch the patient's hand; it withdraws. It moves about like one magnet repelled by another, as the doctor moves his own hand. But if the doctor pretends that he is anxious to avoid being touched, his hand is followed about by the patient's, as if the sign of the magnetism had changed. Here it seems probable that the strong wish of the Unconscious, not yet fully gratified, to cut itself off from the outer world, expresses itself in this symbolic reversal of all the suggestions which the outer world continues to inflict upon it. Sometimes the resistance is violent; this other patient needs three attendants to dress and undress him, must be taken by force to his dinner and to exercise, and will try to tear up every new suit of clothes.

There are many other stages and varieties, but all have in common this inward preoccupation. As a result, they are in a real sense living a dream. It is a waking dream, and they cannot wholly prevent the outer world from obtruding upon it, but it is a dream, and often their thinking is typical of dream-thought, where symbols and wishes take the place of ideas and rational will.

Sometimes inwardly-directed thoughts become focused, not on a personal desire or simple wish, but on some more impersonal construction of the mind. This happens when the man has an intellectual bent; and the result is the crop of "men of one idea," who believe that their idea is to save or revolutionize the world. When they and their ideas happen to be in harmony with reality and to fit in with accidental circumstances they become the great prophets, reformers, and creators of history. When their notions are less in touch with reality and the spirit of the times, we call them faddists, cranks, unpractical fanatics; they found sects, develop new theories of the universe without worrying too much about facts or verifications, preach at street-corners, or promulgate new "-isms." If their ideas are too flagrantly unreal, we say they are insane and shut them up. Don Quixote is the immortal embodiment of a borderland case of this type; and every asylum contains his like, though each will be focused on his own peculiar craziness.

Conflict followed by the triumph of wish and the repression of reality is the most frequent symptom of insanity, and sometimes its cause. But it is not always accomplished by a turning away of the self from the outer world. This woman here, for instance, has become insane through her repressed wishes bursting through, so to speak, and justifying themselves against reality by an elaborate set of delusions. She announces firmly that she is a descendant of Queen Elizabeth, and, therefore, the rightful monarch of England. She already boasts a title —"Rule Britannia" she calls herself—and has armies at her disposal which she is preparing to launch against various European nations. Unfortunately her plans are always being hampered by the rival claimant, who happens to be in possession; she will not accord him the title of king, she refers to him as "Mr. Guelph." It is he who has frustrated her schemes and eventually, alas, got her shut up! Even so he is not content, but is all the time trying to humiliate her still further by depriving her of the privileges due to her exalted station.

That is "Rule Britannia's" version. The medical officer informs us, however, that she is really a poor woman who had previously been a domestic servant, in which capacity she had experienced more than the usual amount of hardship and drudgery. Eventually, perhaps as the result of some more definite shock of which we have no record, the strain resulted in dissociation, and what had been repressed came to the top, while the painful reality see-sawed out of consciousness. The mechanism is like that at work in Dr. Prince's case of double personality. But there the split was between two warring parts of one nature, both of them compatible with the ordinary affairs of life; while here the split is between one part of her nature adapted to the harsh drudgery that was the only reality for her, and another part so incompatible with any reality open to her that it could only manifest itself as fantasies and wishes. And further, the wishes, being recognized as unreal, were repressed, and then, being out of contact with reality, could elaborate themselves in naïve, childlike, dreamlike ways. "Rule Britannia" is a wish-personality come into control of a body.

After the crisis the conflict still continued, but now reality had to be repressed. Hard facts kept on telling her that she was not in point of fact on a throne, or even at liberty; and so the Unconscious manufactures "reasons" to explain the anomaly, and she asserted that the machinations of a powerful persecutor were to blame for her present circumstances.

This is the familiar process we have already encountered and called rationalization—the manufacture of reasons to cloak with justification our unacknowledged wishes and impulses. The "reasons" here are glaringly false and pitifully inadequate to us proud possessors of normal judgment; but do not let us be too proud of our rationality. "Rule Britannia" is only exemplifying in extreme form a process which is at work in every one of us—though from its very nature it is always so much easier to detect it in others than in oneself!

Here is another lady suffering from delusions; but her delusions are of rather a different kind. She believes that Mr. X, a gentleman of her acquaintance, is passionately enamoured of her and has for years been scheming to marry her by force. Her record was until quite recently a blameless one. She was unmarried and had reached a considerable age without a blemish on her reputation, when she began hinting to her friends of Mr. X's unwelcome attentions. The hints grew broader; she spread stories that she was being followed. Finally she insisted that he was plotting to abduct her, and wrote to the police to demand protection. On investigation, it turned out that Mr. X, far from having any passionate interest in her person, merely knew dimly of her existence. But the lady continued to stick to her story; and eventually had to be removed to an asylum.

"Methinks the lady doth protest too much," says the Queen in *Hamlet*, when the Player-Queen reiterates her aversion to re-marriage. Our poor little old maid is the victim of similar over-much protestation, but carried to a pathological extreme. She has suffered for decades from the repression of her natural impulses of sex and affection. She has never acknowledged the existence within herself of anything so unladylike, and accordingly has developed a defensive mechanism of pious prudery. Then one fine day she found herself attracted by Mr. X; but she never knew it, for the attraction, naturally linked up with the repressed complex, was never consciously acknowledged. Her repressed wishes, however, were all the time trying to bring him into her thoughts; accordingly the frontier police of the conscious mind, in the shape of her inveterate prudery, not liking that a man should be admitted so frequently to the sanctum, only allowed him in after he has been disguised as a wanton male, assailant of female virtue. The repressed impulses succeeded in slipping the ideas of Mr. X, love and marriage into consciousness; and the repressing ego still satisfied its moralizing tendencies. These delusions grow like war-rumours; and even now the lady is able to keep everything coherent by explaining that her removal to the asylum was part of her wicked lover's schemes for bending her to his will.

Such a delusion is an example of what is called *projection*. The repressed complex, felt as something alien which the sufferer would like to get rid of, is projected outside, as it were, and attached to some quite innocent scapegoat in the external world. Here again, we find all gradations from insanity to everyday behaviour. The "protesting too much" of the Player-Queen is a frequent symptom. The repressed urge inside us makes itself such a nuisance that we attack it wherever we find it outside, thus acquiring an agreeable sense of virtue, without, in fact, becoming more virtuous. For this reason, as Bernard Hart rightly says, "whenever one encounters an intense prejudice, one may with some probability suspect that the individual himself exhibits the fault in question or some closely similar fault." The people who prohibit mixed bathing or protest against the immodesty of female dress are seldom those who have obtained the wise man's rational control of desire; usually they have merely repressed their violent instincts, and are dimly aware that these impulses, undisciplined by reason, are likely to throw them off their moral perch if unduly stimulated. The repressive attitude often taken up towards the dances of primitive peoples by missionaries is in part at least an outcome of the same machinery. Because of excessive repression, certain things which should be natural and could be beautiful, become painful, disgusting, or shocking; and in violently attacking the unfortunate natives' customs, the missionary is too often attacking his own repressed self.

The most striking characteristic of lunatics, after their loss of contact with reality, is their lack of logic. The lady who calls herself "Rule Britannia" is still a good scrubber of floors; and she does not let her belief in her regal state interfere in the least with a perfectly cheerful and thorough performance of her daily task of cleaning the ward. The two sets of ideas concerned with royalty and with floor-scrubbing might seem to be incompatible. So they would be if they met; but they do not meet. They are prevented from meeting by the machinery of dissociation and repression. Dissociation does its best to keep them in separate compartments and make it hard for them to achieve contact; should they begin to do so, repression steps in, forces criticism into the Unconscious, and substitutes absurd "reasons" for Reason. The mind thus becomes divided into what have been aptly called "logic-tight compartments," in which different systems of ideas can develop in splendid isolation. Sheltered behind such barriers, the lamb of a delusion can grow and flourish safe from the wolfish logic that would otherwise devour it.

Repression and dissociation are not simply morbid phenomena. Their morbid manifestations are exceptional; normally they are adaptive and useful. They are the protective part of the mind's machinery, for defending itself from disruption when confronted with two opposing impulses or two incompatible sets of ideas. The world is so complicated, conduct so difficult, and yet the need of firm belief and speedy action so vital, that it would be impossible to work out every problem of thought and morals on its rational merits. If we did try to do so, our minds would either disintegrate, or we should be immobilized like the ass in the fable, between the two exactly equal bundles of hay.

The arrangements by which the mind conserves its unity and its force in the midst of a chaos of warring facts and ideas are, we may recapitulate, of several kinds. There is first, the capacity for belief by suggestion, submission to authority, unquestioning loyalty, and obedience. This disposition, most evident in the earlier half of life, must have done much to facilitate gregarious tribal existence in the opening stages of human society. Next, there are the faculties of repression and dissociation, which parcel out the mind into compartments. Some repressing or inhibiting force is always needed to maintain dissociation, so that the two agencies generally act in conjunction. Sometimes the split is complete, the compartments become quite impervious to each other's ideas and impulses. More often, however, there is a certain leakage of ideas from one compartment to another. When this is so, another faculty of the mind comes into play—distortion. Repressed ideas which cannot be altogether repressed are distorted and disguised so that they can gain expression and enter consciousness without a disabling conflict.

We have already mentioned the chief methods of distortion. Rationalization is perhaps the commonest obsession, and compulsive expression, as in Lady Macbeth's washing of her hands, is another. Projection we have already explained. Unacknowledged transference of the repressed impulses to new aims and objects is yet another channel. The transference of repressed maternal instincts to pets is familiar; many of the hobbies of elderly bachelors and the hoarding impulses of misers and certain types of collectors are also examples of this process acting in relation to the will-to-power or sexual suppression.

All these ways of distortion occur. All of them are convenient and necessary protective mechanisms of the mind. But all can be dangerous. All are bounded on the other side by insanity. Some degree of repression and dissociation seems necessary to achieve action, some rationalization is often essential for self-respect. But it can be safely laid down that certain things are always bad—long-continued and ardent repression, extreme dissociation that turns the mind into a set of separate compartments, persistent self-justification by inventing "reasons" that undermine the sanity of the general outlook, persistent flight into the interior world of dream or wish or memory. And it can be further laid down that a solution of a conflict which is brought about with the aid of consciousness and reason is almost invariably more satisfactory than one accomplished with the aid of unconscious processes alone.

There is one irrational distortion, however, which is nearly always helpful, and that is the distortion provided by laughter. There have been many theories of laughter, and perhaps none of them are wholly satisfactory. But this explosion, this fit of gasping cheerfulness which aerates the blood and quickens the circulation, certainly has one useful function which must have promoted its evolution during man's rise from his pre-human ancestors, and that is the function of escaping conflict without calling in repression or leaving behind a sense of bitterness or failure. Man is the laughing animal; he "laughs off" endless contradictions. Early man, savage, puny, and ignorant, could never have reached his present state without laughter; his constant failures to achieve his desire, his hardships and disappointments, the incongruity between wish and reality, were all the time inflicting wounds on his mind. He laughed and had another go at it. If hope is a tonic and idealism a slightly narcotic salve, laughter is an antiseptic for mental wounds; a saving abandonment.

Yet even humour, like any other irrational solution of a conflict, can be exaggerated, turned to harm, or employed as an outlet for insanity. Humour is a frequent refuge for the inefficient. It can be a vulgar form of escape from the embarrassment of effort. In some forms of insanity, especially that brought on by alcoholism, the sufferer's Unconscious is all the time keeping him from realizing his true state and the misery he has caused, by cheap and silly jocularity.

Man turns out to be non-rational and non-moral much oftener than he supposed. Nor *can* he lead an existence that is purely rational or purely moral; the mainsprings of his life are the deep-seated impulses he inherits from the past. He cannot escape them; he sterilizes himself if he merely fights against them; he loses contact with all solid earth if he tries to soar too high above them. It is his affair to permit the development of his impulses and yet harness their energies.

§ 14. Differences Between Minds.

PAVLOV, as we have already described, has demonstrated that his dogs differ among each other in their reactions. They differ in their temperaments as much as in their intelligences. Mind varies as much as body—as, indeed, we should expect if both are aspects of the single reality, life. The breeds of dogs differ as much in mental as in physical attributes—there is the tenacious bulldog, the clever poodle, the stupid toy terrier, the savage mastiff, the intelligent airedale, the lazy pug. Koehler found that one or two of his chimpanzees were, relatively to their fellow-apes, geniuses; and some were sulky, some cheerful, some shy, some bad-tempered. Among rats, some individuals are quieter and more docile than others; and it appears that the docility and gentleness of tame rat-breeds is due to man's selecting the tamest animals to keep and breed from, generation after generation.

The most obvious of differences among human beings are differences in intelligence. Choose a thousand children at random; there will be a few who are brilliant, avid of knowledge, a few who are so slow of comprehension that, however well taught, they always plod on far behind the average children of their own age; there will be a number of distinctly able but not brilliant children, and a number who are distinctly stupid, though not deficient; and the majority will be just ordinary in their capacities. Of late years, methods have been devised for measuring intelligence. The criticism of such "intelligence tests" has always been that they do not distinguish between the innate capacity for being intelligent and the acquired capacities due to training—in fact, that they measure education rather than native intelligence. But they have been improved, until now it is generally agreed that they do in large measure fulfil their purpose. They work admirably when applied to a group of children (or grown-ups) all of the same nationality and the same social class; but they need to be applied with the greatest caution when we are comparing groups of different nationality, language, and tradition. They and similar tests for other capacities are already being used successfully and on a large scale in education and also in industry to minimize the wastage involved in trying to fit square human pegs into round occupational holes and vice versa.

Using these intelligence tests, we can compare the performance of a child with average performance. We find that one child of ten attains the average for twelve-year-olds, while another is only up to the eight-year average. Both children are ten years old; but we can say that one has a "mental age" of twelve, the other a "mental age" of eight. A child is generally classed as mentally deficient when its "mental age," determined by these tests, is less than three-quarters of its chronological age in years. The actual capacity for being intelligent does not go on developing, even in the ablest men and women, after about fifteen or sixteen years of actual age. What increases after that is judgment and concrete knowledge, so that the intellectual edifice continues to be built, but not the pure faculty of intelligence by which it is built.

If we take a set of children and test their intelligence year by year, we find that the tests are consistent—a child is not below normal one year, above normal the next. What we are really measuring is the rate

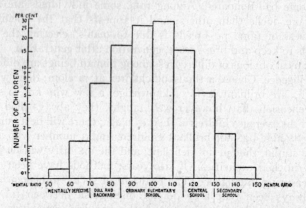

Fig. 326. The variation of intelligence in a group of 2,000 school children.

Carefully devised intelligence tests are applied on a large scale, and the average performance of children is found for each age. This is taken as a standard (100). If a child is ten years old and only attains the level of an average child of seven or eight, he is said to have a "mental ratio" of $\frac{7}{10}$ or 70 per cent.; children with mental ratio below this are definitely defective. If he had done as well as the average child of 13, his mental ratio would be 130. There is no break to separate the mental defective from the normal, or the normal from the genius. (From "The Measurement of Mental Capacities," by Dr. Cyril Burt. Oliver & Boyd.)

at which intelligence develops. And we find that intelligence behaves just as do physical characters like the shrimp's eye-colour, which we discussed in Book 4 (Fig. 185); the intelligence of different children grows at different speeds, and stops growing at different ages. Professor Spearman has given strong reasons for believing that what is measured by the intelligence tests is some general factor which underlies the performance of any and every kind of mental operation demanding intelligence. He prefers to speak of it by the non-committal symbol of "g." In popular parlance it could be called mental energy, but the word energy has a perfectly definite scientific meaning, and its use in this loose way should be avoided. There is assuredly a flow of nerve-impulses which is the physical basis of thought; and when we know more about the mechanism of this flow we shall be able to give Spearman's "g" a more concrete interpretation. Meanwhile it must suffice to know that we can measure something which is the basis for general intelligence.

There is one other conclusion that we can draw; and that is that differences in this capacity for intelligence are due mainly to differences in hereditary make-up. As always, nature and nurture are blended. Birth-injuries, childish illness, remediable defects in eye or ear, underfeeding, fear, and unhappiness—all these can hamper the development of intelligence just as a favourable environment can help it on; but the single cause which far outweighs all others in determining the degree of intelligence is the combination of genes with which the child is born. This is seen with especial clearness in those numerous cases— like the Cecils, or the Darwins—where intellectual ability runs in families. It is also well brought out by intelligence tests on children brought up from their earliest years in orphanages; in spite of the uniformity of their upbringing, they show as wide a range of capacity, as measured by intelligence tests or in other ways, as do other children.

Besides these differences in general intelligence, there are many differences in special capacities; and these are even more exclusively determined by heredity. The most obvious of such special gifts are the mathematical and the musical. Pascal as a mere child begged to be allowed to learn about numbers, but his father insisted that this must wait until his education in languages and history was more advanced. But his mathematical bent was not to be denied; when he was still a mere child, his father found him busied with problems in geometry. Unaided, he had already worked out for himself a series of the theorems of Euclid.

At the age of four, Mozart could play minuets; at five, he had already composed a number of little pieces, and gave his first public performance. So great was his passion for music that he slipped out of bed to practise in the middle of the night.

At the other end of the scale, we have men of the highest intellect who cannot tell *God Save the King* from *Pop Goes the Weasel*, and others to whom any mathematics beyond the simplest arithmetic is a closed book.

Another important difference is between brain and hand, intellectual and motor ability. Everyone knows how some children, stupid at book-learning, excel in carpentry or have an uncanny knowledge of mechanical toys and the insides of motor-cars; while others who are near the top of their class are both awkward with their hands and uninterested in handwork. Statistical tests seem to show that this is not exceptional. Highly intelligent children are, on the average, a little behindhand in their motor abilities, while those who are exceptionally gifted with their hands are, as a group, a shade below the average in their intelligence. Plenty of exceptions exist, but the tendency is there.

With this difference between brains and hands we are introduced to something new—a difference in quality of disposition rather than in quantity of some single endowment. Early in the present century, Jung, the psycho-analyst, formulated a broad distinction between two outstanding types of personality. Other attempts at classifying types of mind had been made before him—the most celebrated perhaps was William James's division of the human race into "tender-minded" and "tough-minded"—but Jung was the first to attack the problem systematically in the light of modern knowledge. One type he called *introvert*—turned in upon itself; the other *extrovert*—turned outwards towards the outer world. The one is more interested in his own inner being, the other in the practical give-and-take of life. To the one the outer world provides the necessary material out of which the inner constructions of thought may be built; to the extrovert, the inner construction is of value as enabling him to live more fully in the world of people and things around him. The introvert tends to be solitary, shy, given to speculation and imagination, often troubled about himself, his future, his relation to the universe at large. The extrovert, on the contrary, tends to a balanced expansiveness and sociability which contrasts with the somewhat fitful liveliness of the introvert; he makes friends easily, has little embarrassment about such public performances as acting or singing, which are often a torture to the introvert, is bored with anything abstruse or speculative, lives much more in the present.

There is no doubt that this picture has a great deal of truth in it. One can often see this turning of the mind inwards and outwards exemplified in two members of the same family, even at an early age. It is further likely that the lack of correspondence between ability of hand and of brain which we have just noted, is concerned with a difference in psychological type. The introverted child is more interested in things of the mind, takes less trouble with practical handwork; and vice versa with the extrovert. There is also the fact, elicited during the War, that dissociation and hysteria are the normal outcome of conflict in the extrovert, while in the introvert similar conflicts are likely to end in repression and neurasthenia. In fact, the fundamental basis of the difference between the two types would seem to be a difference in ease of dissociation. The extrovert is able to keep conflicting ideas apart—to use a popular phrase, in "watertight compartments"—and thus to assimilate cheerfully and with equanimity the most divergent experiences. In the introvert, on the other hand, the mind is more unified; its different parts cannot be kept apart, but are continually infringing on each other; and much of his mental life is occupied by their attempt to fit themselves together into logical, harmonious schemes. The more discordant elements are fiercely repressed and may become troublesome complexes, while in the extrovert they are simply walled off from the rest of the mind.

Recently, McDougall has hit on an ingenious method which throws some light on the constitutional differences between extrovert and introvert. The accompanying figure represents a cube; but as you look at it it shifts its perspective—now one of its two square faces is nearer you, now the other, now it is sloping up and to the right, now down and to the left. If you look fixedly at the figure without trying to force on it either of the two possible perspective interpretations, you will find that it changes from one to the other at more or less regular intervals. McDougall found that the interval varied very much for different people; and that it was much longer for extroverts, much shorter for introverts. Here evidently we have a model on a small scale of the way in which the mind reacts to the graver conflicts of life. The two interpretations of the figure are mutually exclusive, and equally satisfactory. When one of them is adopted, the other must be inhibited. In the extrovert, such an arrangement is comparatively stable; in the introvert the inhibitory process is feebler and the two rival interpretations are continually struggling for supremacy.

But the most interesting thing that emerged from this investigation was that the rate of alternation could be altered in one and the same individual by the use of drugs. Quite small amounts of ether or alcohol would lengthen the interval to double or quadruple, while coffee or strychnine would speed up the change. "The drugs of the two classes were perfect antagonists; for example, during the hastening produced by strychnine, which lasted several hours, a whiff of ether or a dose of alcohol temporarily antagonized and reversed its effect."

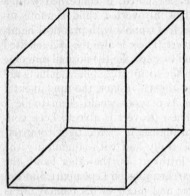

FIG. 327. The Deceptive Cube.

It is perhaps unfortunate that these terms extrovert and introvert have come to be so widely used, for they give a false impression of simplicity, as if the world was really just two sorts of human beings, "extroverts" and "introverts," as distinct as males and females. It would be truer to say that everyone fluctuates between extroversion and introversion. Some of us have a bias one way and some another, but a glass of wine may make us more extrovert and a cup of coffee swing us towards introversion.

Recently new suggestions have come from another quarter. Our knowledge of heredity, of physiology, of brain-structure and brain-function, has forced upon us the view that body and mind are not distinct entities, but two expressions of the unitary living organism. In actuality, they are inextricably entangled. On this view, minds and bodies (for we cannot help continuing to make the distinction between them for convenience sake) would not be expected to stand in a haphazard relation with each other, any kind of mind with any kind of body, but certain types of mind and body should go together. We have already maintained this view for male and female instincts (Book 4, Ch. 6, § 8); we have refused to believe that a fully masculine soul could inhabit a fully feminine body, or vice versa. Now we are extending this idea to cover the various temperaments. We are asserting, contrary to such pre-scientific notions as reincarnation, or the idea that a waiting soul somehow slips into a body when the body has reached a certain stage of development, that soul and body are both aspects of one whole, the mind-body, the living human being, who is always different from every other individual human being.

Kretschmer, in his book *Physique and Character,* gives some account of recent work in this field. In his psychiatric practice he noticed that certain kinds of insanity went together with certain kinds of body-build. Those who suffered from the manic-depressive type of madness were usually stout and stocky, big-waisted, well-covered, with rather barrel-shaped bodies, short neck and limbs and a squarish face. Those who suffered from *dementia præcox* and related kinds of madness, on the other hand, were rarely of this thick-set build. They mostly oscillated between the athlete type—well-developed muscles with little surplus fat, broad shoulders and narrow waist, oval face and longish limbs—and the scraggy or asthenic type—thin, narrow-chested, with little muscle and less fat.

We have already found reason to suppose that the maniacs and melancholics are of Jung's extrovert type, the *dementia præcox* patients more introvert. So, if the correlation holds for sane people, we would expect to find the extroverts and introverts among our acquaintance differing in body as well as in their mental attitude. Popular beliefs and everyday experience justify our expectation. The fanatics, the prophets, the spinners of theories, those who will spend their life in the service of a single idea—they tend to be thin, drawn out in length—

"Yon Cassius has a lean and hungry look;
Let him be watched; such men are dangerous."

Calvin, Savonarola, Robespierre are examples in real life. We have no record of what John the Baptist looked like, but there is a more or less unanimous conviction that he was not short and stocky. And who could ever depict Sancho Panza as anything but plump and stocky, and Don Quixote as anything but lean?

The medium-broad type is often the intuitively practical man, the man of affairs, the administrator. Mirabeau, Cavour, Lloyd George come to the mind. Among literary men Dr. Johnson is a good example. He was not interested in the working-out of philosophical systems, in close-knit imaginative constructions; the Dictionary typifies his practical discursive bent, and Boswell's Life is, above all, a record of his preference for the social interchange of ideas over solitary meditation. He is a complete contrast to a poet like Wordsworth, the lean lover of long, solitary walks, the constructor of elaborate systems of thought to bear the weight of his intuitive feelings about Nature.

There can be little doubt that these researches will soon link up with work upon the ductless glands. We already know that too much thyroid thins and excites, while too little bloats and damps down. The physical type produced by too much anterior pituitary resembles an exaggeration of the athlete type. Too much adrenal cortex masculinizes, too little pituitary fattens and makes greedy and sleepy.

On the more purely mental side people differ enormously in the way they think. Galton, in his classical book, *Researches into Human Faculty*, found that some people think predominantly with the aid of picture-images; others rely more on sound-images and, as it were, hear their thought; still others are tied to the motor side, so that their thinking runs along the lines of an incipient pronunciation of words. Besides these, there is the abstract thinker, exemplified most clearly by the pure mathematician, who has apparently left the solid ground of sense-images and words and floats among abstractions. What is strongly developed in him is the capacity for perceiving the relations between things and ideas; he moves among problems in higher algebra as securely as the visualizer among remembered scenes. The same word is to one man a picture, to another a sound-image, to a third a cog in a logical chain; small wonder that human beings find difficulties in the way of complete mutual understanding.

Recently Jaensch has brought together some interesting facts in this connexion. By specially devised methods he showed that a considerable proportion of children have the image-forming capacity developed in high degree; their mental pictures are so clear and sharp that if you tell them to project their mental image of a face, say, on a blank screen, they can, as it were, fix it there; and you can, guided by their answers to your questions, measure it or draw it. In most cases this capacity disappears during early adolescence as the power of abstract thought is cultivated, though in some people it persists throughout life. This antagonism of picture-thinking and concept-thinking was also revealed by Galton, who found that highly educated people had, on the whole, less capacity than the uneducated for recalling vivid images.

Jaensch found further that certain drugs such as an extract of the American cactus called Mescal or Peyote (Anhalonium) could increase the vividness of image-formation. It conferred vivid imagery upon many of those who normally lacked it; while in those in whom the faculty was already strong the pictures were converted into hallucinations: the image was projected into the external world and seemed to exist in its own right. Certain preparations of calcium, on the other hand, often reduced the vividness of imagery, and could be used successfully on children who suffered from violent nightmares.

Such studies are likely to throw a good deal of light on human faculty, for we know the chemical constitution of the drugs and can use them in accurate experiments. As knowledge progresses, we shall be able to classify the differences of humanity with increasing accuracy. We shall presumably find that individuals may differ in regard to many separate qualities; and in each quality there will be a scale from high to low, plus to minus.

It is not only that these drugs illuminate our capabilities. A time will come when they may be used to assist and enhance them. A time may come when we shall be able to supplement our normal mental powers with chemical assistance as nowadays we supplement our muscular forces with power machinery.

VIII

MODERN IDEAS OF CONDUCT

§ 1. *The Conduct of Life.* § 2. *What Is the Self We Conduct?* § 3. *The Primary Biological Duties.* § 4. *Of Self-knowledge and Moods.* § 5. *Candour.* § 6. *Restraint and Poise.* § 7. *Evasion, Indolence, and Fear.*

§ 1. The Conduct of Life.

WE HAVE already warned, and several times reminded our readers of the warning, that with the introduction of the subjective element into our account of vital phenomena we have departed from that atmosphere of clear, cold statement, proof, and certainty we were able to maintain so long as our method was still wholly objective. But to shirk questions of feeling and will because they did not admit of the hard precision of a purely objective treatment would be to rob our *Science of Life* of half its interest and two-thirds of its practical value.

We have already said something of the essential paradox in things. We have pointed out how comparatively easy it is for us to regard the whole world, including our fellow-creatures, as a fated system of cause and effect, as a system entirely mechanical and determinate, until we come to ourselves. But within ourselves we find it is at least equally true that we choose, that we will to do this and refuse to do that; that we are not fated, but free. Your sense of your own freedom is as primary as your sense of my complete subjection to controlling causes. You may deny this practically, but your every act will assert it. For all practical ends your liberty and your sense of your personal responsibility for what you do is ineradicable. Even if you fling yourself down and say, "I am the toy of destiny; I can do nothing that I am not compelled to do," you know all the time that you will be doing and saying this of your own free will. So soon as you have said it, you can say: "That was silly of me. I did not mean to say that."

"Kismet," says the good Moslem, and stands hesitating what next he shall say or do, still under the plainest necessity of making a decision.

So, whatever logical juggling we may do in this matter, in reality you and we, the present trinity of authors, and Mr. and Mrs. Everyman and the whole Everyman family find ourselves making up our minds, deciding what to do, forming our plans and occupying ourselves with hopes, determinations, reflections, self-congratulations or remorse, and the general conduct of our lives.

It must be poor biology, feeble psychology, and indifferent science we have been dealing out to Mr. Everyman if we cannot draw from this accumulation of fact some clear directions and helps and reinforcements for him and Mrs. Everyman, in this essential problem in their lives: What to do and how to do it.

§ 2. What Is the Self We Conduct?

THROUGHOUT the larger part of the preceding chapter we have been elaborating a picture of the mind as it is revealed by the introspective and comparative methods of psychology. We have talked in metaphors, about levels and strata, searchlight (of attention) and focus, images, complexes, repressions, threshold of consciousness, and the like. Thereby we have been able to convey and to bring to a considerable coherence an important mass of fact about divided personality, mental inhibitions, hypnotism, and impulse. But the reader will have been haunted throughout, as we, the writers, have been haunted, by the feeling of metaphorical vagueness and the dangers of misconception that accompany metaphor. The reader will have been moved to ask again and again: "Where precisely in this seething mass of mental activity does the self begin and end?" Or, "Am I all my mind or only some of my mind raiding about amidst the rest of itself?" At times it has been almost as if we described the coming and going, the conflicts, overlappings, and replacements of clouds in the sky. This sort of indistinctness is not confined to psychology. Has the reader ever troubled to ask: What *exactly* is the state? What *exactly* is the nation? What *is* public opinion?

The leading psycho-analysts have evidently been haunted by the same dissatisfaction as the reader. They have made, and they are still making, attempts to mark off the regions or activities of the mental stir with a clearer definition. We have already noted Freud's distinction of the Ego, the Id, and the Super-ego (Ch. 7, § 12). More recently Jung (*Two Essays*, 1926) has been making a parallel attempt to define the—

what shall we say?—parts, active forces, powers, rôles?—in the intricate mental interplay of *Homo sapiens*. His essay throws much helpful light upon the conditions and limitations of the adaptation of this strange animal Man to happy participation in a continually more complex and penetrating social existence, and yields us some convenient new forms of expression.

As the starting-point of the self-conscious life of a man or a higher mammal, he explains, is the realization of the "I," the "Ego"—the realization of oneself as pitted against the universe. To the very shallow and unthinking this ego is all that one is, but, as we now realize, a great undefined field of mental activity goes on in everyone, outside the conscious ego, and nearly everyone discovers sooner or later divisions of motive which are in a sort of struggle for the control of the ego. Before Mr. Everyman was in knickerbockers he had learnt that it would not suffice always to say: "I want," "I will," or "I won't." There were times when he found himself saying: "I want and I don't want," or "I will and I won't." From the start he has had this balancing going on in him.

Often but not always these conflicts resolved themselves into a struggle between what he was informed was his better and his worse self. Generally the worse self had a quality of being more flesh of his flesh and bone of his bone than the better. It was as if, under the pressures of external and internal experience, a ruler was arising within him to which his conscious ego in general had to be subservient. It was a part of his ego arising to dominate the rest of his ego. What is this second more *selected* self in Mr. Everyman and all of us?

Dr. Jung has devised for it the excellent name of the "persona." It seems to be in many respects the same thing for which Freud uses the term Super-ego. The persona is begotten of a need to escape from the complexity, the realized dangers, and the dissatisfactions of the ego. Material—but here Jung parts company from Freud—is drawn in to it from the Unconscious outside the ego. The persona is the self we want and do our best to imagine ourselves to be. It is how we present ourselves to ourselves and the world. It is the story we tell ourselves about ourselves. It is not a fixed thing. It varies with our audience and our mood just as the various performances of a play may vary with audience and mood. Now something is out, now something is added, now an understudy plays. But generally it is how we wish others to take us. It is a mask into which we weave very many extraneous and suggested things that we suppose people to expect of us. And it is always thrusting down incompatible impulses, desires, and cravings out of sight, out of our ego into the Unconscious. And also it thrusts down incompatible and disturbing memories. These suppressed elements, denied recognition and honour, are usually less lucid, more in the quality of emotion and blind feeling. They become like a dark shadow of the persona, they are everything it is not; they are, as it were, in necessary conflict with the persona for the ego. For this suppressed and insurgent underself in the Unconscious, Jung suggests the name of the "anima."

The difference between these two aspects of the self is not necessarily a moral one; the persona is not necessarily good, the anima not necessarily bad. The persona is the mask, and the mask may be, for example, that of a stern implacable judge upon the bench, a pig-headed, intolerant man, while the suppressed and imprisoned anima cries and worries for mercy and consideration. But the normal anima is something suppressed, something getting loose in dream values or unwary impulses, in the inadvertencies of hypnotism or drunkenness or fatigue. The tension of its upward thrust towards recognition and effect in the ego may lead to serious nervous lesions and mental disorder.

Jung makes a very interesting but very questionable distinction between this dark shadow of the persona in men and in women. The persona imposed upon the average woman is, by tradition, social and economic conditions and physiological necessities, different in its nature from that of the average man. There is in women generally more suppression of judgments and above all of initiatives and less of emotions and receptive response than in man. The suppressed elements of a woman are therefore commonly more wilful than those of a man. Jung would go to the length of making a difference between an *anima* in the case of a man and an *animus* in the case of the woman. His choice of these words is not we think so happy and unambiguous as his choice of *persona*. *Anima* and, even more so in English, *animus* are loaded with false associations. Moreover, these two things differ not in kind but in degree. For English use, we prefer to use *anima* for both sexes.

In the earlier psycho-analysis of both Freud and Adler, as we have said, the Unconscious was treated as largely or exclusively individual and its content was supposed to be merely personal experience, suppressed and forgotten. But Jung would insist upon a vast proportion of inherited material in the unconscious. We cannot follow him all the way in this Lamarckian view. But though it seems impossible for ideas to be inherited as such, we can clearly inherit predispositions to particular ways of thinking. At their first onset in a human being's life, many things are found to be attractive or repulsive in this way or that.

Ideas are grasped or refused. The mind, one must conclude, is ready to receive or repel them. Just as all our brains and bodies are alike with small variations, so this "general Unconscious," this readiness to receive or resist, is alike for most members of our species.

On this sea of general unconscious preconception and sustained by it, floats our individual ego, with the persona it has gathered for itself in the full light of consciousness and its anima hidden below the surface. By such a figure—and again we remind ourselves and our readers that all this field of science is still at the metaphorical stage—we may convey this conception of the "collective Unconscious," which all of us have practically in common. And from it also, just as much as from the suppressions of our own personal past, we draw values and dispositions for our dreams and imaginations and unwary actions; some of us much more than others; we draw the forms of expression, the shapes in which we find ourselves disposed to clothe our gathering and accumulating personal response to the outer world.

The persona, says Jung further, is capable of "inflation"—that is to say is capable of such an extension away beyond the natural ego into the collective Unconscious that the sense of personal identity becomes weakened and the individual confuses himself with his official functions or with the race or with the Deity. This inflation of the persona implies a process of devotion and self-forgetfulness. It may make a man a pedant or a saint. The anima may by a similar diffusion dissolve the boundaries of the ego in mystical ecstasies. "Inflation" may have its desirable and admirable side; it is an open method of escape from the distresses of individual frustration and it can be a method of reconciliation to honourable but contributory and inferior tasks. It has played and may still play a great part in the religious life. It may play an even greater part in the future development of our kind. A general inflation of the persona may be going on in most civilized communities.

So, very sketchily we give the main lines of Dr. Jung's account of the broad structure of our mental being. The reader must bear in mind that this is a speculative account. It is not an account of established and universally accepted views. But it marches with common experiences. It corresponds very closely to facts in the behaviour of people about us. What we are dealing with here is not so much science as thought on the borderland of science and literature. We give it because it is so richly suggestive, because it had at least the sort of truth that literature can convey, because it is interesting and helpful to record the vision of a particularly fine and powerful intelligence which has been devoted for many years to the task of peering into the teeming reflections and creations of the human mind.

It becomes manifest under such a scrutiny that, in the inner world, just as in the outer, the individuality comes and goes, that it changes, is now more and now less, now fainter and now intenser, that it assimilates and again rejects. We untrained people assert and believe so firmly in our complete unity and our unbroken identity because, when our identity weakens or changes or dissolves, we are by that very fact no longer there to observe and deny.

In the light of this shrewd rephrasing of conduct by Jung, we can speak of the conduct of life as the rule, development, and wary correction of the persona. When we talk of a man behaving well we mean that he is under the sway of an efficient and fine persona. A weak person is one with a fluctuating persona; an impulsive man with poor self-control has a too vigorous anima. It is the disposition of most of us to rationalize the impulses that come up to effectiveness from the anima. We try to square them with the persona and may make considerable changes in the persona to do so. Clearly moral training and education are essentially attempts to mould the persona upon socially desirable lines, and most of the graver internal conflicts that give us hysteria, neurasthenia, and mental collapse are conflicts of the persona with the anima. Most, but not all, for both the anima and the persona may have their own inconsistencies and internecine contradictions.

§ 3. The Primary Biological Duties.

SINCE Mr. Everyman, as we have seen, is from the biological point of view a mass of living cells detached from the rest of life and organized to act together as one individual, it seems reasonable to assume that the first duty for education to impose upon the developing persona is that he keep this self of his in a fit and proper condition to act. He has to eat, drink, and sleep; imperatives so plain that no system of conduct, however fantastic, has ever ventured to forbid them altogether. And our modern disposition is to carry out these manifest indications to their logical conclusion, and declare that health and fitness, having regard to the characteristics of the individual rôle and organization, are primary demands upon a human being's attention and energy. Mr. Everyman should keep his body and mind as a soldier keeps his weapons—ready for the utmost use.

That does not mean that he should become enslaved by an elaborate ritual of hygienic precautions. Health in itself is not a final human end; it is merely a condition for the easy and thorough attainment of more justifiable objectives. The health faddist who places himself outside the normal usages of the life about him, in pursuit of some imaginary super-exaltation of his bodily life, and the game-playing enthusiasts of "fitness" and exercise, who spend so much time keeping fit that they are ultimately fit for nothing, are warnings to us that in this field, as in most fields of human concern, excess can be as absurd and futile as disregard.

In Book 7 we have reviewed the conditions of bodily health, and in the preceding chapter of this Book we have indicated many of the ways in which the nervous and mental organization may be thrown out of gear. So long as Mr. or Mrs. Everyman feels well, keeps happily active, does not suffer from boredom, sleeps well, is capable of occasional effort and exertion without a sense of stress, is not gnawed by unsatisfied desires, and is neither over-sensitive to minor stimulations nor subject to fits of bad temper, he or she may rest content with the balance of his or her bodily and sub-conscious processes. But it is the business of everyone to note these little discomforts and discontents that intimate the beginnings of a jar within or without one's organism. Then comes the duty of finding out what is wrong, seeking outer advice, and adjusting routines, diet, habits, and circumstances to restore the smooth working of the body-mind.

That readjustment may be impeded by many conditions beyond Mr. Everyman's immediate control. He may find he is undernourished, ill-housed, cramped and restrained, living under vexatious compulsion, and urged to efforts beyond his powers. He may find it is not so much himself that is out of gear as the world of Man about him. His attempt to change his circumstances may bring him face to face with colossal adversaries. Social, economic, political forces, the pressure of other and more fortunate people may impede him in the performance of these primary duties. Then, if he is still to be content with life, his persona will adapt itself to a struggle in which hope, and sometimes disinterested hope, may have to be the substitute for fulfilment. Some sort of inflation of the persona is the normal response of the human being to frustration of the impulse to be free, vigorous, and mentally and physically satisfied. That impulse is thwarted because of fear and indolence; it is thrust down into the anima, and its failure is rationalized often in some quasi-religious form as resignation, as virtuous contentment, or the like. Compensation is sought in hopes and dreams of a future life, in which abstinence will be a recognized virtue and everyone who has meekly buried his inadequate talent will be disproportionately rewarded. Or the thwarted, hampered, and restricted human being may still keep the spirit of conflict alive, and then the inflation of the persona will take the form of self-identification with all the disinherited or oppressed in the community, with the Common People, with the Proletariat or what not, and lead to socially wholesome insurrectionary and revolutionary activities.

Here our interest in such activities is strictly biological. They appeal to a biologist as being more wholesome than submissiveness and self-effacement. They are more conducive to progressive evolution. They are attempts to readjust through inflation and not through repression, and they may open a way to more generous institutions and more favourable methods of social association. The body-mind of Mr. and Mrs. Everyman is the outcome of a vast process of evolution, and it exists now—so far as we can say it exists for any purpose at all—to try out its distinctive possibilities to the utmost and to reproduce and multiply its type.

Clearly, then, it is a primary duty for Mr. Everyman to see that he with his like and kin is fed properly, clothed properly, that he has elbow-room for the development of all his possibilities, and lives as fully as his nature permits him to do. Only through the inflation of his persona in such a fashion that he becomes—to use Aristotle's term—a political animal can he bring himself into an effective relationship with the general problems of the species. He can only do his primary duty to himself and struggle for sustenance, freedom and opportunity with any hope of enduring consequences, by merging his individual push and his individual hopes in some aspect or another of the collective struggle of his kind.

§ 4. Of Self-knowledge and Moods.

SINCE the individuality seems from a biological point of view to be the primary business of the individual, a certain intelligent egotism may be considered as among the more fundamental moral requirements. "Intelligent" we say, and not merely egotism. Modern morality differs from all preceding conceptions of goodness in its increased insistence upon the duty of knowing and understanding ourselves and others to the full extent of our capacity. The "simple faith," the "simple virtue" of former times we are now inclined to dismiss as intellectual sloth. There is no goodness in conscious conduct without criticism.

It is not sufficient for the individual to be generally well-behaved. In that case Nature might just as well have made us all alike. Each individual is an experiment; each individual has special qualities, can do some things better and more effectively and joyously than others. It is the biological obligation of Mr. Everyman to develop his distinctive disposition, to spend rather than bury his talent. Circumstances, lack of education or mis-education may render self-discovery and self-development perplexing or difficult. That is Mr. Everyman's personal bad fortune; that is his fight.

To measure ourselves against our fellows and define our differences is not an easy task, but we have all our lives to do it in, and a large part of our daily experiences to do it with. In the great development of psychology that has gone on since this century began, there has been a steady approach to a classification of temperaments and intellectual types. Of that we have written. It is the business of Mr. and Mrs. Everyman to learn about these things. There is no need now to live in that atmosphere of cloudy illusion about oneself which seems to have been the lot of our grandparents. It is our business to know what we can do well and what we cannot do well. The thing we can do well is what we are for. It is our duty to occupy ourselves with that if we possibly can and not with other things.

It is not only our duty. Doing the thing we can do exceptionally well is also the most agreeable way of living. We should not be deflected from that by snobbish or sordid considerations. It is better to be a good cook than a mediocre hostess, or a good instrumental musician than an unoriginal composer. To do what one can do best as well as one can possibly do it is success in life, and honours, rewards, praise, recognition, and record are mere incidental stimulants that have no part in the final reckoning.

Self-knowledge is not complete with only a realization of our gifts. Equally incumbent is it to define and measure our deficiencies. We have to keep a watch against those pleasant self-delusions for which our egos have an inordinate appetite. None of these things are easy, none can be achieved by a simple prescription, and the duty of modern men and women to keep the edge of self-criticism keen by a sustained and sedulous study of mental processes is manifest. A time may come when an ignorance of mental science will be scarcely distinguishable from moral imbecility.

One of the most important and one of the most neglected aspects of self-knowledge is a measure of the quality and extent of our moods. The stupid man thinks he is the same man always; children and immature minds cling to the same delusion. The trained observer of mental states knows that he plays the theme of his individuality with many variations; it may even be a fugue with interlacing themes. Many people, the majority of people perhaps, still go through life without realizing how their minds are coloured and diverted by, for instance, physiological states. Unaware of this, they strive to rationalize all they do, insist on a false consistency, and accept and make themselves not merely responsible but advocates for the righteousness of every escapade of the insurgent anima. They cling obstinately to the things they have done or to the attitudes they have adopted long after the moods that produced those attitudes have diffused away. It is the afterthoughts that are thrust back into the anima.

"If I had not *meant* to do this, I should not have done it," they say. "This is precisely what I have always intended to do."

Or they make rash vows to fix the impulses of some quite exceptional emotional storm. "Never more will I—do so and so. Never will I forgive. Never will I speak to him."

This is a condition of things that will change very rapidly with the diffusion of newer and clearer psychological ideas.

It is not only that these ideas will dispose us to be more wary of holding our persona to a rigid consistency and make us more discriminating in our definition of our own purposes, but they will also do much to mitigate the harshness of our judgments upon other people. We shall discover a new charity towards those about us when we realize how often their acts, attitudes, and failures may be the result of transitory and explicable irritations, fatigues and passing obsessions. We shall become more critical of ourselves and more tolerant of others. We shall define our fundamental ends with a greater clearness and pursue them with all the more effectiveness because we shall not allow our incidental aberrations to turn us permanently aside, and we shall encounter the resistances and opposition of others without bitterness or condemnation. The hates, the revenges, the scoldings, false accusations, campaigns of spite, sudden acts of violence that now embitter the lives of the great mass of human beings may be very largely eliminated as our realization of the mischief of our own moodiness and our comprehension of the inconsistencies of our fellows increases. Life may change. It is still for most people much more of a screaming, grabbing, quarrelling, lying, resentful, blindly impulsive and incoherently wasteful affair than it need be. But there exist now nearly three centuries of novels and descriptions of manners and behaviour, and nobody who studies these can doubt that in that space of time common everyday life has become more dignified, kindlier and more generous. After his primary duties to himself, the first duty of Mr. Everyman to others is to learn about himself, to acquire poise and make his persona as much of a cultivated gentleman as he can. He has to be considerate. He has to be trustworthy.

It is in our most intimate relations, in the relations between husband and wife, between parent and offspring, between directors and subordinates, between employers and employed, for example, that the erratic impulses of the mood work their worst and need most to be anticipated, averted, and when they break loose allowed and atoned for. Steadfastness is a strain and should be realized as a strain. Repressions accumulate below any line of conduct continually followed. Life is change; when change ceases life ceases, and many of these close and important relationships tend to become habitual in their manifestations and lose their stimulating quality. In spite of their great intrinsic value to us they cease to interest; they begin to bore and the anima gathers vigour. One of the commonest experiences in the lives of saints is the discovery that all the opening ecstasies of faith have faded, that the wonderful life of holiness has lost its light, that God has hidden himself away. Lovers can have the same dismaying experience. Then the way is open to irrelevant impulses and many sorts of self-contradiction.

Such lapses are psychologically inevitable and two-thirds of their harmfulness vanishes if this is recognized. The harm comes in when they are rationalized as permanent changes of direction, when the saint finds out he is damned, the lover that he never loved, or the three collaborators find each other "impossible." The sane modern man takes holiday and returns refreshed, and is not too greatly perturbed if his anima suddenly thrusts a needed holiday upon him. And what is perhaps a degree more difficult, he must realize the same necessity in others. He cannot exact from others a perpetual persistence he does not practise. The real value of conduct lies in its main tenor and in the vigorous resumption of that tenor after interruption, and not in its perfect coherence and inflexibility.

§ 5. Candour.

THE conception of the conscious individual that is built up by modern biological science and introspective psychology, differs in many respects from the older idea of a human self. The older idea was more compact and definite and altogether unified. John Everyman was just plain John Everyman and evermore should be so. In those unanalytical days people no more questioned that than they questioned that

Old England was Old England or that people were either good or bad. But the scalpel, the microscope, the study of the less familiar aspects of life and the searchlight of an intensive criticism have all combined to undermine our confidence in the simplicity and absolute integrity of individuals. We begin to apprehend the transitory, provisional, and fluctuating factors and aspects of a human self.

In those unsuspicious pre-scientific days John Everyman could really think of himself as boxed-up in himself, as absolutely separate from all the rest of things, able to keep his thoughts and motives to himself, free to do just what he liked with himself. He could make his own plans for his own ends. He could lie and cheat other people, but it did not occur to him that he could ever lie to himself.

But now objective science is bringing it home to Mr. Everyman that he is a product of organic evolution and that this conscious individuality of his is a synthesis of processes, which is directed rather to the survival of a collection of genes than to its own distinct and separate affairs. He has been boxed-up in himself for a particular end, and that boxing-up is not the primary verity he thought it was. And psycho-analysis shows him that all unawares his ego has been perpetually imposing interpretations upon the intimations of fact that come to him, exalting this in the persona or thrusting that down out of recognition into the anima. It dawns upon him nowadays that this working idea of conduct as a boxed-off separate game against the universe may be illusion. He is not entitled to all that privacy. It was a narrow and cunning idea of existence. He belongs, he now apprehends, to something greater than himself, something that modern science is gradually enabling him to realize. He is a part. He is not a cut-off unity. He is neither a beginning nor an end.

As this realization soaks into people's minds, it changes their attitude to conduct very profoundly. They develop what is called "the scientific attitude of mind"; the scientific style of behaviour. The first, most marked characteristic of this modern attitude is the unprecedented value it gives to candour. We are all experiments together, says biology; we are all serving in the education and growth of life, and all the plotting and planning and hiding things from each other for small private and personal advantages that constituted the bulk of human reactions in the past is seen suddenly for the waste and folly it is. The psycho-analyst has revealed to us the advantages of candour with ourselves, the advance of science in the last two centuries has opened before us the splendour of candour between man and man. A commandment, omitted strangely enough from the Ten delivered at Sinai, has become the dominant one for the new world: "Tell the Truth. Hide nothing."

As that has been realized as a primary duty the pattern of persona in fashion in our world has undergone enormous changes. We have no quantitative science of conduct, no systems of measuring changes in the moral atmosphere, but it will express something of what we are seeking to convey if we assert that the changes in the spirit and nature of conduct that have occurred in our modern communities during the past hundred years far exceed any of the changes in mechanical realities or any of the extensions of knowledge that have gone on during the same period. A hundred years ago conduct went on under a burthen of definite injunctions and definite prohibitions, under a burthen of repressions and concealments, that have become almost inconceivable to-day. To go back to the novels and plays before the new era began is to realize something but not all of that change. All life in these representations seems to be going on in blinkers, to be "make-believe" enforced in the most terrible and cruel fashion. Everyone pretended to be pious, patriotic, and genteel, and many succeeded in being so. The Unconscious was indeed a cesspool in those days, and the general decorum was relieved by outbreaks of dirty and nasty libidinousness. These lacked even the dignity of defiance. They were flushed and furtive in manner; they were made for the sake of their foulness rather than their freedom, like the stories of mediæval monks. There was plenty of shamefaced vice, but then so many things were vicious. The sexual suppressions of the genteel found release in a perpetual discussion and reprobation of "the nude." All mankind seemed hiding from itself. Souls and bodies were hidden even from their owners. Slowly throughout the past hundred years the long lost human form has been unveiled from the ankles upward and from the chin downward. The long lost human soul has dared an even greater unveiling. Could our great-grandparents be brought alive again to-day, the miracle would probably be wasted upon them. They would die of shock.

To them our atmosphere would seem to be one of blinding candour, of reckless revelation. If they did not instantly die they might come to realize and even admit that social existence is already cleaner, easier, happier, and better than it has ever been before.

The movement towards frankness, which marks the new confidence in reality that has been diffused throughout the modern community by scientific progress, is still only in its opening phase. Great sections of the human population remain untouched by it and whole regions of the world. So far the main initiatives of the dawning scientific period (for we are still only in its dawn) have produced little more than a conspicuous abundance of mechanisms, a more efficient hygiene, and these new qualities of the persona that are spreading so rapidly about the world. The exterior things were the easiest to achieve, steam and electric mechanisms, ships of steel, great guns, chemical dyes, aeroplanes, submarines. It is more difficult to undermine and replace mental organizations. They live out their lives, will neither change nor die, and offer great individual and collective resistances. They insist therefore upon a slower rate of progressive change.

It is easier to scrap a stage-coach than a public school, and to cover the world with a network of wireless messages than to change the framework of a system of ideas. The bitter experiences of great wars and much waste and suffering seem to mark an unavoidable phase in the adjustment of such serviceable ideas of the past as patriotism and loyalty to the ever-broadening necessities of the coming years. In social and political questions, and in particular in our loyalties and patriotism, we are still terribly insincere and repressed. There remains far more political than sexual repression. The affairs of this swiftly developing world of mankind are now being conducted, obstinately and tenaciously, in accordance with traditional pictures of the universe that are anything from fifty to a thousand years out of date.

§ 6. Restraint and Poise.

WE HAVE shown that the contemporary mind is a mind that has been throwing off traditional repressions to an enormous extent. The conscious life has descended into the Unconscious and flung open the gates of repression. Among the younger generation nothing seems to be hidden and nothing forbidden. But having established this wide liberty, the problems of the balance of motives and decision to act return in the new daylight and demand an open and conscious solution.

Nobody—the young least of all—wants a mere life of immediate impulses. Living for the moment brings its punishments swiftly. Some selection of impulses is unavoidable, some hierarchy of purposes and motives. Every generation in its turn seeks that definitive element for the persona that our Victorian ancestors used to call "a purpose in life," a conception of our general aim by which all our other impulses can be measured and to which they can be adjusted. To-day there is as urgent a need as ever for that guiding outline and framework of the persona, and the great advances that are being made in the clarity and reality of our psychological ideas make it far more possible than it has ever been before to effect an adjustment.

The first, most imperative needs of our organism have first to be assuaged. There is no possibility of a general scheme of conduct when one's mind is obsessed by overpowering hunger or thirst or fear. Until these are assuaged the rest of life's problems are in suspense. These provided for we have next to consider our sexual urgencies. To multitudes of people and to most of the young, morality is little more than a feverish struggle for sexual adjustment. In the everyday use of the word in English a moral man is a man who is sexually correct in his behaviour. Many people never get past this intense preoccupation with sex to any more general moral concepts.

The sexual life of a human being is a retarded one. It is thrusting to the surface of consciousness against many retarding and repressive forces from a very early age. There may even be innate restraining influences. It can be evoked prematurely and inconveniently, and there lies one of the most difficult of modern social problems: the protection of the immature from unnecessary stimulation. That is the reason for public decency. The sexual possibilities of the young may lead them to a concentration upon sensuous and crude emotional reactions, too vivid, exacting and exhausting to permit full mental growth. They can be "used up" by sex. But on the other hand excessive concealment and a decency carried to the pitch of inquisition and exaggerated horror may drive the young into secret cults and practices and may impose upon them quite monstrous suppressions. It is a pity that sexual curiosities should have been labelled for ages with such words as "dirty" and "filthy" and represented as shameful. Every healthy human being has passed through stages of curiosity, peeping, and experiment. The holiest saints, the noblest of women, could confess to indelicate incidents in childhood. It would be far better if the young understood the universality of their troubles. The first flutterings of desire are no more shameful than the first flutterings of a nestling. They are clumsy, but natural, and there is no need to taint and poison them with horror and moral dismay. The avoidance of unnecessary and premature sexual stimulation does not mean concealment. But it does mean the passionless presentation of facts. Clearly and calmly imparted knowledge, not too long delayed, is the first necessity for a wholesome adolescence.

And naturally as adolescence proceeds the sexual complex gathers power. We may present the facts as passionlessly as we like; but that will not prevent the growing individual making a personal application of them and feeling in due time the increasing power of desire. What has modern morality to say to that?

The rude morality of the past, preoccupied with problems of social stability, was all for prohibition and restraint except within the limits of a rigorously defined marriage. Within the marriage bond no amount of indulgence was excessive. The "new moralist," tremendously impressed by the psycho-analysts' revelation of an almost universal craziness in mankind due to sexual distortion and suppressions, has been throwing all his weight in favour of as free and abundant gratification as possible. He advocates the widest possible diffusion of birth-control knowledge, trial marriages, and temporary unions. The more play of sex there is, he asserts, the less we shall be bothered about it. His hostility to suppression is as great as the mediæval hostility to indulgence.

The spirit of the age seems to be with him. But it is arguable that he goes as far in one direction as the old school did in another. The old teaching was that complete continence for long periods of time or for always was quite possible for human beings and on the whole better for them than indulgence. A priori this seems improbable. Many of our new liberators, however, seem to go altogether beyond established fact in the opposite direction. They seem to insist upon unrestricted sexual indulgence as a condition of bodily and mental health. There is a sort of propaganda of indulgence and shamelessness now as mischievous as the old propaganda of disgust and repression. The fact seems to be that human beings are capable of a very wide range of behaviour in these matters. Little or none of the seasonal excitability of many lower mammals is observable. Many men and many more women practise complete abstinence from physical love-making for long periods or for a lifetime without any ascertainable mental or physical distress or injury. Many others seem under the sway of a real necessity and may be greatly troubled or disordered mentally by restraint. In women particularly sex may be latent for long periods and then begin to stir, often in very vague emotional forms. For normal men an occasional happy indulgence seems to secure a general tranquillity. The meaning attached to that "occasional" may vary widely for different types. If we had more exact facts to give we would give them here. But that is all we know.

Because of the tabus and repression that have hitherto prevailed and the consequent obstruction of research, we are still in the dark about the physiological facts underlying that extraordinarily variable and intricate thing, the sexual complex. Plainly it is a thing that bulks enormously in our mental life. We must admit that, even if we do not go all the road with the extreme Freudian. Its most perplexing aspect is the way in which it spreads its tentacles from the lowest to the highest strata of our minds and the rapid interaction between highest and lowest that it makes possible. If sex were a mere physical need it would present no problem of any difficulty to the moralist. Satisfy it, he would say, and take any necessary and obvious precautions that may be necessary so that it does not disorganize your population balance nor disseminate any infectious or contagious disease. So far as disease goes, prompt douchings and washings with such a substance as potassium permanganate in any case of doubt becomes a moral obligation, and any germicide that will kill the spermatozoa or any contrivance that will bar efficiently the access of the spermatozoa to the ovum, is manifestly sufficient to meet the needs of his second qualification. It is not within the purview of *The Science of Life* to discuss these matters in detail; suffice it here to say that what we may call the coarse control of sex, that is to say the easy elimination of its possibilities of undesired offspring or disease-dissemination, is quite within the reach of intelligent people.* The complete abolition of such hideous diseases as syphilis and gonorrhœa could be achieved in two or three generations if a world-wide observance of a few perfectly simple precautions could be imposed. But at present the mental and moral confusion of our species forbids any hope of such a feat of hygiene.

But this cleaning up of the sexual act which is now in progress in the more civilized communities so that it becomes socially and physiologically harmless, does not begin to give us a new sexual morality. It merely clears the ground for a new sexual morality. Except in possibly the case of a few imbeciles the sexual act is never merely physical. It is not even mainly physical. It is involved with a whole world of sensuous and æsthetic discriminations. The state of the genital secretions causes changes in the blood that affect the colour and stir of the whole mental life, and conversely mental and particularly æsthetic excitation affects the amount and vigour of the physical sexual activities. There is a quality of preference in most sexual advances and a sense of selection. Self-love is involved profoundly. There is a peculiar pride and glory in the successful exercise of choice in this field. Hardly any form of human love-making exists without some element of flattering illusion. The sense of personality is stirred to its depths. For most people it is

---

* Further information can be found in *Wise Parenthood*, by Marie Stopes; *Parenthood: Design or Accident?* by Michael Fielding; or *Happiness in Marriage*, by Margaret Sanger.

the culminating self-assertion. It is therefore usually and very easily competitive. It may give rise to intense rivalries, devouring jealousies, bitter resentments, and the extremest humiliations. As we have already pointed out, all our troubles to adjust ourselves in these matters are quite unaffected by the modern detachment of sexual intercourse from parentage and infection.

The sexual complex is the greatest power in our minds to bring our conduct into relation with our fellow creatures; there we incline towards the Freudians; but it is not the only one. There is, for example, the whole tangle of solicitudes about our personal power and position in life, a dread of being in difficult and possibly unsuccessful competition with other people, which the Freudian would explain as obscurely sexual or ignore, and which Adler would make our predominant concern. The problems and rules of conduct towards others differ only in intensity in these two and any other possible fields.

The individual has first to make up his or her own mind about physical sexual gratification. Is it necessary in his or her individual case? Is it oppressively necessary? What restraints are necessary to keep it within bounds and prevent injuries or evil accompaniments? What are to be its mental accompaniments? Does it concentrate on some real or imagined person? What possibilities of hatred or tormenting jealousy must be guarded against? How is the individual sexual life to be lived without injury, annoyance, or boredom to other people? All these are private questions, for which no general rules apply. The defect of most moral systems in the past has been their rigorous insistence upon universally applicable codes of behaviour. But now the contemporary disposition is to let everyone build up a distinctive persona. That persona is what we promise ourselves to be and what we promise others to be, and it is not only wise but an obligation to clip our more aberrant impulses to the promise of the persona we have adopted. We must not betray the expectations we have induced people to form of us. And though we have a right to expect a similar consistency in them, we must exercise charity if we find at times some repression has got away with them.

In its general trend biological science is at one with all these higher and more intuitive developments of religious thought that are called "mysticism." In all the more highly intellectual developments of mysticism there is a struggle to escape from too intense preoccupation with the "self" ("the body of this death," to quote Saint Paul) and an endeavour to identify oneself with some greater, more comprehensive, immortal being. Now as we have unfolded this general outline of biological fact we have found a constant dissolution of our ideas of the primary importance of individuality, and a growing realization of the continuity of life as one whole. All that has gone before in this work, the physiology, the comparative anatomy, the genetics, the psychology, has agreed in showing that individualities such as ours are temporary biological expedients, holding great somatic aggregations together in one unity. Our sense of the supreme importance and unbreakable integrity of our "selves" is, in fact, a dominating delusion with great survival value. We feel it most in youth and ignorance. Then our concentration upon self is most intense, and the thought of defeat, frustration or death, exquisitely intolerable. We fight for self-expression, for our own survival, for our reproduction, with extreme effort. But ripening knowledge and the progress of adolescence tempers this fierce self-concentration.

It seems possible that man is being evolved past this phase of extreme selfishness and self-concentration. The tendency of all moral teaching and of all progress in conduct throughout the development of civilization has been to replace selfishness by fellowship, to treat secretiveness, cunning and secret motives, greed, injustice, self-assertion, disregard of the feelings and good of others with increasing reprehension. Modern biology simply takes these generally recognized virtues of trustworthiness, frankness, fairness, willing helpfulness, charity, devotion to the commonweal and recognizes them as not only normal duties but as reasonable and natural ways of establishing and retaining mental serenity.

§ 7. Evasion, Indolence, and Fear.

BEFORE we conclude this very general review of modern ideas about conduct, we must direct attention to a group of phenomena which have this in common, that they concern a recoil from activity, a certain paralysis of will. In the Catholic system, based on a very huge amount of searching experience, there is, among the list of the seven deadly sins, accidie or sloth. It is the inability to live, as St. John put it, "abundantly."

In the contemporary civilized community there is a very large number of people who display this weak disinclination for life, who live by habit, who react feebly to stimulation, who seem to be inattentive, spiritless and joyless. They are what are often called "devitalized" individuals. They may have lost physical health, or that state of mental health, that confidence in things and themselves being "all right," which is called "faith." Effort never seems to be worth while to them. They accept second-rate things, are content to do things in a second-rate fashion, and to be second-rate people. Rodents and other creatures, when their population numbers have passed a periodic minimum and selection is not pressing them too hard, may display parallel types of listlessness, but most living species are too close to death to carry many such slack individuals. If most creatures do not want to live very strenuously, nature sees to it that they do not live at all. Natural selection does not generally tolerate the second rate. But, compared with all other animals, mankind is hardly subjected at all now to even the shadow of selective killing. There is a kind of low-grade security for nearly everyone. People are indeed enslaved, sweated, fed on broken crusts, but they drag on, and they can procreate. Humanity is certainly accumulating a substratum of these dull unkilled.

But it is not to be too readily assumed that all people who submit to lead dreary, unventilated, uneventful lives, are inferior and unnecessary people in their essential quality. If individuals inheriting the best genes in the world are born into grey conditions and ill-nourished and discouraged, they may never get stimulation early enough, or of the right quality, to develop their faculties fully. They may be caught early and cowed by pain before their courage develops; they may be trained to humiliations and submission. All conscious living is a balance of impulse and inhibition, and the balance may easily acquire a permanent tilt towards inhibition. Life can scorch and sting those who challenge it rashly, but, from the modern point of view, it is better to be scorched and stung and searched to the core of one's being, than to stand aside and watch the world go by.

These people whose hearts fail them may, and do, find consolation and compensation in a multitude of self-protective mental complexes. They represent their slackness as commonsense, as quiet modesty, as a mysterious subtle refinement that keeps them aloof from the brawling strain of vigorous vulgar life. They thank the gods they are not "pushers," not "highbrows," not the mad, restless followers of every novelty. They like to do things in their own quiet, unobtrusive way. They cannot but be amused at the exertions and the so-called knowledge of those they do not understand. There is much quiet laughter of a really malicious sort among the victims of accidie. They have a sociability of their own and are capable of immense passive obstructions in a progressive world. They line the streets for any passing show, and rather hate it as it passes. Then home to the den that is never lit by strenuous effort or any sense of the purpose that may lie before mankind.

There can be little doubt that in the new clearness that is coming to mankind, it will be realized that submissiveness to limited education, under-development, and under-employment of one's faculties, will be recognized as a cardinal sin. This cowering into ignoble but apparently safe niches in the social fabric, this burial of one's talent, this refusal to learn and understand and serve and live to the uttermost, this suicide of most of one's individuality in order to keep the rest of the body alive, is even less tolerable to the new morality than it was, in theory, to the old. The world is passing into a new self-conscious phase of economic and social organization, which has little use for acquiescent drudges, and may develop an active impatience with merely consuming parasites and commensals. Modern thought calls to everyone who discerns himself or herself to be cramped and restrained from vigorous self-expression to struggle out of that net, play a part and live.

IX

BORDERLAND SCIENCE AND THE QUESTION OF PERSONAL SURVIVAL

§ 1. *The Theory of Body-Soul-Spirit.* § 2. *Dream Anticipation and Telepathy.* § 3. *Clairvoyance, Table-tapping, and Telekinesis.* § 4. *Materialization and Ectoplasm.* § 5. *Mythology of the Future Life.* § 6. *The Survival of the Personality After Death.*

§ 1. The Theory of Body-Soul-Spirit.

FOR the larger part of this work we have confined ourselves to facts and generalizations that would be regarded as "scientific" by the majority of scientific workers. Only in the last three or four chapters have we invoked the introspective evidence of psychology, and even then we have done our best to keep the very metaphorical quality of its phrasing in view and to correct it by frequent reference to Behaviourist methods and ideas. Upon evidences universally acceptable and verifiable we have built up our picture of the co-ordinated realities we call "life" in which our own consciousness floats.

We have attempted no philosophical nor metaphysical "explanations." We have dealt with facts, and our picture is a presentation of fact. It is not in the sphere of science to offer fundamental "explanations." Science is simply a scrutiny and a putting together of scrutinized facts. Mind has come into our picture and we have traced its entry phase by phase. We have observed this new side of existence becoming more important in the scale of being until in ourselves it has the effect of an inner world, reflecting all the processes of the material world, and *conscious*. We have shown reason for supposing that our own consciousness, so elaborate, so intensely and clearly focused upon the realization of self, is only the culmination of a series of developments closely correlated with the material evolution of living things. We have suggested that the stuff of consciousness may be co-extensive with the material stuff out of which life has elaborated itself. As our material bodies have been evolved, so this stuff has been gathered together into the unity of our minds. But that is only a suggestion, a way of putting it. It is difficult to see how it can ever be more than a working hypothesis. Life we observe about us abundantly and freely; consciousness only in ourselves.

In this account of life we have avoided as far as possible certain ancient controversies and time-honoured theories. There is a very ancient theory that a human being consists of three distinct and separate factors, the body, the soul, and the spirit. We have troubled our readers very little with this concept of our fundamental triplicity, either to deny or discuss it, because it has been possible to render our account of life without it. It is not apparent in the ordinary facts of life. What we have said about it hitherto will be found in the Fourth Chapter of this Book, on *Consciousness*. And here we will make no attempt to give a definition of these three alleged factors nor say where one is supposed to end or the other begin.

Those who hold this theory of our threefold nature are apt to be loose in their use of the words "soul" and "spirit" and run them together very confusingly. But they seem to agree that in the case of human beings, and possibly also of dogs, cats, parrots, and other creatures, there is something detachable, a two fold or onefold something—the soul-spirit or ghost let us call it—which includes the consciousness, memories, and an indefinable multitude of characteristics constituting the individual difference or personality. This soul-spirit can operate beyond the body and by other than material means, and may possibly persist for an indefinite time or for ever after the material body, the body of fact, has been destroyed.

It is difficult to say whether this immortal part is supposed to exist before the conception or hatching or birth of the individual. If it does so, it must be without individual characteristics. So far as these go, it is clearly a synthetic product. It is what the assemblage of genes have made it. It arises and develops these characteristics, and once it has achieved this, then, from this ancient point of view, there is no ending it. We have traced the growth of mental life and seen how this arises from general material processes. Gradually the personality is gathered together out of its material sources and gradually it becomes detachable. At last it is supposed to float off from its material associations altogether; they have been merely its matrix and substratum and in the end they may be destroyed; this destruction liberating rather than injuring it.

In a summary of the Science of Life we are bound to consider how far this widely diffused and time-honoured theory of the possible disintegration of the individual into a material perishable part and a more refractory and enduring, if much less palpable, "spiritual" part is sustained by the body of biological knowledge we now possess.

§ 2. Dream Anticipation and Telepathy.

BUT before we come to issues in which this theory is involved, there are others leading to it and associated with it that can nevertheless be dealt with to a certain extent in a preliminary study. No one with anything beyond the most elementary knowledge of biological fact will imagine that in any direction our knowledge can be regarded as complete and final, and there is much to be said for the view that considerable regions of mental and physiological activity remain to be properly explored, and that their exploration may yield facts differing in quality and implication from any facts at present established. Facts may become incontrovertible that will modify even our fundamental conceptions of life. For example, orthodox science knows no way in which the experiences of to-morrow may cast their shadow on our thoughts to-day. Yet Mr. J. W. Dunne in *An Experiment with Time* has made very curious and suggestive observations upon his own dreams and the dreams of others that point to the possibility that such a foreshadowing occurs. And there is a copious—a terribly copious—literature, recording facts that seem to show that the mental states of one person may produce impressions upon the mind of another, without the use of any means of communication at present known to the biologist.

Mr. Dunne's observations, reinforced by the observations of various friends who have adopted his methods, are made at the moment of waking from sleep. The observer trains himself to write down, at the moment of awaking, all that he can recall of the content of the dream, if any, through which he passes from complete slumber to the waking state. This—in the case of the more conscientious workers—is typed and put on record. In quite a large number of instances, it would seem, subsequent experiences occur very strongly reminiscent of these dream expressions. A considerable part of our dreaming, Mr. Dunne notes, seems to be based on experiences in the immediate past; much of it is the misinterpretation of bodily states; the chill of an exposed limb, for example, suggesting bathing in cold water; much, as we have noted in our account of psycho-analysis, is ascribable to repressed complexes; but also, he believes, a considerable residue anticipates experiences of the near future. His statements and those of his associates are certainly striking enough to justify further experiment in this field.

Science has nothing to offer upon this matter. Here may be something to modify our idea of the relationship of consciousness to time, or Mr. Dunne may be the victim of coincidence and his remarkable facts may become attenuated by further inquiry. He has not attempted to filter out the anticipatory element of his dreaming for any practical purpose, but he has written some very engaging speculations upon the relationship of consciousness to the time-dimension in space-time. It is to be noted that there are well-authenticated cases of dreams charged with foreboding, which nothing occurred to justify. They have to be weighed in balance against dream-anticipations.

The possible action of one mind upon another so that a more or less exact parallel to a mental process in the one is induced in the other, is called *telepathy*. It is, to use a very clumsy parallel, a sort of mental wireless telegraphy. As exemplified by Professor Gilbert Murray and his daughter, for example, one observer concentrates his attention on a book or picture, while the other, out of normal sight and hearing in another room, waits with a vacant mind to record whatever impression is received. Results have been got by these two, of a quality and exactitude difficult to explain by any other hypothesis than that of a direct thought-transmission. Fairly complex drawings have been rudely reproduced, and scenes read by the one have been described by the other. Sir Oliver Lodge, among others, has produced very remarkable results with drawings. Nevertheless it remains quite impossible to formulate any explanation of this process in terms of established scientific fact. The analogy to wireless telegraphy is at best a very loose analogy; a human brain has no structural resemblances to a wireless station, and the mechanism of transmission, if there is any transmission except through the normal sense-organs, must be upon widely different lines.

That distinguished man of science, Professor Charles Richet, who accepts telepathy as a fact, hypotheticates a "Sixth Sense" which operates in these cases. This seems to be a sense without special sense-organs. Other critics resort to the phenomena of hyperæsthesia with which we have already dealt. Others again either challenge the good faith of the experimenters or minimize their agreement down to the level of coincidence and natural parallelism. Kindred minds they say are parallel minds; two people very intimately related and accustomed to each other's society may follow closely similar paths of mental association and may respond in precisely the same way to the same circumstances. In this way, too, by this insistence upon the parallel working of similar things, it may be possible to account for the remarkable unison in the flight of social birds and in the movements of gregarious herbivora.

It increases the difficulty of this discussion that every shade of credibility is to be found in the cases cited, from the unimpeachable integrity of Professor Murray to manifestly dishonest witnesses and observers. And there is hardly any form of telepathy that cannot be imitated by conjurers and other professional entertainers. There is an indefinable element of untrustworthiness in most of the witnesses and possibly in all. Exaggeration of statement for the sake of emphasis is a common error of the human mind, and most of us would rather have an over-accentuated story to tell than nothing remarkable to say.

Moreover, these mental experiments are dependent upon the moods and health of the experimenters, and all are complicated by the fact that as frequently as not the recipient draws or relates something entirely different from what was in the mind of the transmitter. When the transmitter thinks of a lion, it is perhaps explicable if the recipient thinks of a cat, the British Empire, a battleship, the zodiac or the map of Asia, but it becomes more difficult if he records a pair of slippers or the North Pole. It is often very difficult to define what is relevant and continuous, and what irrelevant and discontinuous in mental association, but when every concession has been made there remains a vast proportion of failure in telepathic endeavours. The public is too apt to hear only of telepathic successes.

It is to be noted that in chess tournaments and bridge clubs, where human brains intensely concentrated upon identical problems are brought into close proximity, telepathy is not observed, nor does the thought of it trouble us in the ordinary reservations, evasions and falsehoods of everyday intercourse; it is unknown in the jury-box and unused or not reported from the connubial pillow. The mind of the recipient must be lax; that is the claim; and where people are too vividly interested, that laxity of mind is unattainable. It is not high-pressure strained activity but low-pressure activity. For all practical purposes at least the human cranium remains opaque.

These are reasons for keeping our heads when we hear marvellous stories of thought-transmission, but there is no justification for an intolerant rejection of the idea. After the monstrous accumulations of half-evidence and pseudo-evidence in this field have been sifted and reduced, there remains enough to justify an attitude of critical indecision. Whether through Professor Richet's as yet unlocated and undefined "sixth sense," or in some other way or ways, there does seem to be sufficient evidence of some unknown reaction of the thought-process of one individual upon the thought-process of another to invite further inquiry. There is certainly nothing in the idea of telepathy that runs counter to the general scientific ideology. But also there is nothing to forbid a practical scepticism in the matter.

It is arguable that if this telepathic faculty or "sixth sense" is an actual possibility, it should have played an important and recognizable part in the evolution of the animal world. Has it done so? If it possessed a survival value for any particular species, we might reasonably expect to find it highly developed in that species. But do we find it highly developed in any species? It should have a use in social co-operation. A re-examination of the behaviour of wasps, bees, and ants from this point of view might shed new light on what Maeterlinck calls the "spirit of the hive." Something may still remain unexplained in the emotional infection of crowds. There are such things as undesirable possibilities and for the human type telepathy may be one of them. Man is an intensely "individualized" animal. Individuality is a biological device that, like all Nature's devices, has to be protected. At certain stages in evolution, telepathic faculty might easily prove a hindrance rather than a help to survival, and the opacity of the normal human skull to "thought-waves" may be a necessary condition for efficient action and intercourse, blinding us to some very exciting possibilities. It may have been an essential part in that separation and enhancement of individuality which has certainly gone on in the vertebrata and other animal phyla.

§ 3. Clairvoyance, Table-tapping, and Telekinesis.

CLOSELY associated with this experimentation upon mental reactions by unknown means at a distance, is another body of alleged phenomena of a much more questionable type. For the experiments and the generalizations arising from them Professor Richet, in 1905, adopted the word "Metapsychics," which is now in general use. It had previously been suggested by Mr. Lutoslawsky in 1902. The human mind is a necessary factor in these experiments and they demand concessions and limiting conditions known in no other field of research. They are admittedly associated with a network of deception and deliberate fraud. That is the misfortune rather than the fault of these inquirers. Their general procedure can be burlesqued or imitated by impostors with the greatest ease.

Possibly these "psychic phenomena" are obtainable from all of us, but it is only a limited number of people who can produce them at such a level of vividness and emphasis as to yield observable and recordable results. These special types, when they are induced to develop their peculiarities systematically, are called mediums. We have already had something to say of this class in § 10 of Chapter 7.

"Mediums," says Richet, "are more or less neuropaths, liable to headaches, insomnia, and dyspepsia. . . . The facility with which their

Fig. 328. Drawing from a flashlight photograph showing levitation of a heavy wooden table during a séance with the medium Eusapia Paladino.

(*From Lombroso's "Fenomeni ipnotici e spiritici" Unione Tipografico-Editrice Torinese.*)

consciousness suffers dissociation indicates a certain mental instability, and their responsibility while in a state of trance is diminished. . . . A powerful medium is a very delicate instrument of whose secret springs we know nothing, and clumsy handling may easily disorganize its working."

Geley insists that the medium must be in good health, in a good temper and not distressed by the experimenters. In italics he writes: "*The phenomena are the results of a subconscious psycho-physiological collaboration between the medium and the experimenter."* Not every sceptic may be admitted to these investigations. In a séance the observer is part of the exhibit.

Manifestly, in view of these considerations, "séances" demand a very special sort of inquirer; we need to combine the qualifications of the sympathetic alienist, the criminologist, and the professional conjurer with those of the psychologist, physiologist, and physicist, and to blend faith with criticism in a remarkable way, if we are to approach the experimental part of this work with a reasonable hope of competent observation. The ordinary citizen who plunges untrained into psychic and metapsychic experimentation is as likely to make a useful contribution to science and to profit by his self-confidence, as if he set himself without any special preparation to trying out new types of aeroplane or the investigation of high explosives. It is to be regretted that so many unqualified people, often people gifted and experienced in other directions, have lacked the modesty needed to refrain from casual participation in these experiments and have consented to the publication of their injudicious and practically worthless condemnations or confirmations.

The physical conditions under which these sensitives consent to display their gifts increase the difficulty of the inquirer's task. Light, it seems, is very inimical to psychic phenomena. It must be shut off more or less completely. Red light is least unfavourable and may be used with discretion. On the other hand some such distracting noise as the tinkle of a musical box is a great help. The inquirers' sight and hearing must be used with these handicaps. Moreover, the observers have to assist with their hands and often their feet embarrassed. A circle must be formed for the "influence" to operate. Hands must touch hands and feet feet. Indefinite periods of waiting ensue under conditions strongly conducive to boredom, fatigue, hypnotism, and sleep. Few dabblers in metapsychics can be at their critical best when "phenomena" occur.

Moreover, the investigator is usually obliged to acquiesce in a sort of question-begging mythology about the facts he sees. The medium, it is commonly alleged, is merely a medium for other minds and wills; through him or her, ghosts or other bodiless spirits are supposed to operate, either by taking possession of the central nervous system and working the larynx, limbs, and other parts, or by other less direct manifestations possible only in the medium's presence. Phenomena of various types are produced which are ascribed to some ghost or elemental spirit which is known as the medium's "control." The critical observer must tacitly accept this assumption, or, at any rate, he must waive his objections to it, if he assists at a séance. He may assume that the controls are partly dissociated personalities in the medium's brain, or that they are products of her subconscious mind, analogous to the creations of our dreams. But he will have to put his mind in the posture of recognizing these "controls" as realities.

"Psychic" phenomena are of various types. One group comes under the heading of "*telekinesis*" and includes all movements of objects due to the presence of the medium and not made by means of any forces known to normal science; tables tilt and are elevated, objects are thrown about, tambourines are rung. Another set of phenomena includes clairvoyance and clairaudience. The observer is dependent on the word of the medium, who sees and hears things unseen or unheard by the observer. In more elaborate manifestations the medium goes into a trance-state and then talks, often in strange voices which ramble on, give advice, answer questions, profess to be the voices of deceased persons, deliver messages from the spirit world. Certain mediums are said to have told people secrets known to no other living person but the hearer, and the controls, by the use of intimate names, old memories, and personal turns of thought and speech, are frequently able to satisfy those present that they are the spirits of deceased friends and relations.

The successes are more apt to be recorded than the failures. In some recent sittings in London Miss Rebecca West, the writer, was addressed by a control who announced himself as "Grandfather West," and hailed her as "Rebecca." Mr. Dennis Bradley, who records the incident, remarks that she became "emotional." This is the more understandable when we learn that "Rebecca West" is a pen-name taken from a character in a play of Ibsen's, and that the lady's real surname is Fairfield. The possibility of a horde of ink-relations, if one may coin a phrase, parents, grandparents, cousins, and so forth being added to one's blood-relations is surely enough to make anyone emotional.

Sir Conan Doyle received a long series of communications from a "control" named "Pheneas," who is declared to be a gentleman of prominence resident "thousands of years ago" in Ur, before the time of Abraham. Archæologists may yet discover his home. He is not, however, a Sumerian, as one might have expected, but an Arab—an anachronism. He has a penchant for predicting earthquakes. And he seems to have contrived a bicycle accident for Sir Conan in order to oblige him to take a needed rest; he was helpful to Sir Conan in finding a house in the New Forest and he compared Sir Conan's spiritualist book-shop to a "great flare in a pea-soup mist." In the spirit world he is known as the "Star of Hope" and "leaves a trail of joy behind him." In his childhood (at Ur) he was taught to *fear* God and that, he explains, was all wrong. In that pagan city before the time of Abraham, he was perhaps fortunate to hear of God at all—even with a Presbyterian flavour. Later he "won a great battle against odds and that was what first drew the people's attention to him." It would. Such feats confer distinction on the most obscure. Afterwards people remarked "his glorious character." At Ur he used to write "through pieces of leafy stuff"—disdaining, it would seem, the clay tablet commonly used in that part of the world for cuneiform writing. There is a bookful of communications of this type from and about Pheneas published by Sir Conan Doyle. A large part of this mass is vaguely platitudinous and uninteresting.

FIG. 329. Drawing from a photograph taken during a séance, showing the medium Kathleen Goligher raising a table.

The table is obviously propped up by a rod coming from between her knees, and from which a veil-like object is hanging. These structures are interpreted as being formed out of teleplasm exuded by the medium. (From "The Psychic Structures at the Goligher Circle," by W. J. Crawford, D. Sc. John M. Watkins.)

This Pheneas matter is an extreme instance of the type of material produced by a medium in the clairvoyant state. Generally the "control" represents itself as coming from less remote sources than Pseudonymia or Ur. More often than not, it claims to be the ghost of some recently deceased person known to some of the investigators. Sir Oliver Lodge has been the means of publication of various books about lost sons, including his own son Raymond. The Raymond "communications" are made up into a book together with Raymond's very vivid letters home from the Western front, where he was killed in 1915. The contrast between the interest and conviction of these letters and his essentially futile communications through the medium is very great. Certain recognitions and identifications are achieved, and beyond that the alleged Raymond is either weakly incredible or platitudinous. And it would seem that his messages have now faded out. Earlier communications professing to come from the late Frederic William Myers and the late W. Richard Hodgson, through such celebrated mediums as Mrs. Piper and Mrs. Verrall, have proved equally lacking in novelty or profundity and equally have they faded out. The alleged posthumous minds of these acute persons appear as if flattened and faded. And they desist. They cannot keep it up—important though the business must be to them. But occasionally in the minor matters used for identification, the grasp of the medium upon small intimate points has been remarkable. Such a grasp is no evidence whatever for this mythology about controls and spirits which seems to be necessary for the operations of most (but not all) mediums; in many instances it has been of a type to sustain a provisional belief in telepathy, and a case may be made out for further systematic investigation by competent persons of possible elements of an unknown nature still awaiting recognition in the clairvoyant state.

The talking medium is, however, only one way in which these abnormal responses and communications ascribed to ghosts and other spirits are made. Talking is not the usual initial stage. The development of mediumship often begins with rappings and the movement of articles of furniture. Then the rappings are codified. So many mean Yes, so many No. Next there is the spelling out of words by means of an alphabet of raps. As the communications continue, the gathering impatience of controls and mediums and observers alike is mitigated by the spoken word. Or communications may be made by automatic writing; the experimenter thinks of something else and his hand writes automatically. Planchette is a little mobile apparatus which runs about under the hands of the transmitter, writing down messages. There is indeed a great variety of transmission methods. The messages of the planchette, like those of automatic writing, are concrete enough; but the host of experiments carried out by Janet and others show that, far from being evidence of communication with a spirit world, they are normal channels for the outpourings of the operator's own subconscious mind.

A method of communication, once very popular but now out of fashion with mediums, was slate writing. In this type of manifestation the slates are tied and locked and sealed together, with a piece of chalk or so between them. After suitable incantations the slates are opened and communications, answers to questions and so forth, are found written upon them.

This slate work has been successfully imitated by conjurers. The classical case is that of the late Mr. S. J. Davey (set forth with the utmost particularity in the *Proceedings of the Society for Psychical Research*, Vol. IV). Davey asserted that he had mediumistic powers, and, under proper observation, produced all the characteristic phenomena. He induced those who were present at the séance to set down, as explicitly as they could, exactly what they had observed. He then gave his own account of the manipulations by which he had deceived them. A comparison of the records showed wide disagreement between the observers, and it was surprising to realize what they had missed and what they had misinterpreted. The brightest incident was the production of a concealed name by the operator. A Persian gentleman known as Mr. Padshah had asked him to ask the spirit for his name, Padshah being merely the name he found convenient to use in America. His proper formal patronymic was Boorzu. Davey had either forgotten this question or, failing to understand the drift of it, ignored it, but wanting to perform some wonder with books, he wrote the word "Books" between the slates. He wrote it indistinctly. When the slates were opened Mr. Padshah immediately read the word as his own name and explained the wonder to the company. It was difficult for Davey to convince him that a miracle of telepathy had not occurred. It was still more difficult for Mr. Davey to persuade many of the believers in mediumistic phenomena that he was not really a medium, basely representing his gifts as mere tricks. This was the attitude taken up by Dr. Alfred Wallace, the great biologist, who was also a spiritualist, towards the celebrated conjurers Maskelyne and Lynn.

§ 4. Materialization and Ectoplasm.

The word "materialization" first made its appearance in America in 1873 to denote what had been previously called "spirit forms." It was at a time when the rage for "occult" pursuits had become almost universal in Europe and America. In the States this development is usually dated from 1848, when the "Rochester rappings" first attracted attention. These were assumed to be signals from the dead; and although they were afterwards explained as sounds produced by snapping the big toe inside the shoe by the very women who had figured as the mediums, this belated confession was powerless to stop the wave of occultism which swept America and Europe, invading the humblest homes as well as the most brilliant Courts, and demanding official recognition from the guardians of science.

Thus we find names of the first magnitude like Faraday, Lubbock, Huggins, and Crookes on the list of persons seriously studying the alleged phenomena and hoping to find something which could be established on a secure basis. They found themselves on unfamiliar ground. Instead of dealing with material bodies and processes under conditions of their own choosing, they found these conditions arbitrarily varied by what claimed to be supernatural authority, an authority not amenable to the ordinary conventions and canons of evidence. It was obviously impossible to obtain anything in the nature of scientific evidence in the circumstances. Crookes alone chose to waive the precautions which would have made his results evidential.

His "success" marks the pinnacle of mediumistic achievements. The "spirits" gained such confidence in his readiness to observe the prescribed conditions that eventually a complete materialized spirit calling herself Katie King appeared, walked about the room on his arm, submitted to being photographed and even embraced, and finally disappeared by being in some mysterious way absorbed into the body of the medium, Florence Cook, a girl of about the same age.

This episode has led countless investigators to emulate Crookes. It has appealed to many instincts, combining as it did the mystery of Undine with the religious wonder of spirit made flesh. The subsequent history of occultism is largely the story of attempts to reproduce the phenomena related by Sir William Crookes.

But the development of mediumistic technique has not kept pace with our growing knowledge of its methods. The experiments of Crookes, both with Florence Cook and D. D. Home who "levitated" out of one window and back through another, have been submitted to a searching criticism, and it is now seen that they have no claim to be in any way scientific. Crookes himself abandoned the attempt to convince his scientific brethren and returned to his chemical work.

Eusapia Palladino, the well-known Neopolitan medium who died in 1918, mystified two continents for fifty years. She was repeatedly caught in the act of fraud, but never ceased making new converts, one of her most important captures being Cesare Lombroso, the renowned criminologist. He was convinced that his deceased mother had appeared to him with the help of Eusapia's mediumship. Reading the detailed accounts of the many séances she gave to committees of scientific men, one is struck by the marvellous dexterity with which she played her mother-wit and knowledge of human nature against the ponderous science of her judges, now refusing, now conceding, controlling her controllers, wearing out their patience, choosing the best

Fig. 330. Drawing from a flashlight photograph of a materialized head, formed of teleplasm that has exuded from the medium known as "Eva C."

(*Courtesy of Baron von Schrenck-Notzing, from his "Phenomena of Materialisation." Kegan Paul, Trench, Trubner & Co., Ltd.*)

means of deceiving them and carrying out her manœuvres with supreme skill and agility, while always ready to cover a critical situation by a display of temper.

Perhaps the most detailed account of a materializing medium is the story of "Eva C.," the French medium studied by Dr. von Schrenck-

Notzing and Madame Bisson. They took several hundred photographs of what were claimed to be partial materializations. One of the most

FIG. 331. Drawing from a photograph of Mme. Bisson of a materialized visitant produced by the medium Eva C.

To show the figure, the medium is holding aside the curtains of the cabinet which concealed her during the sitting. (Courtesy of Baron von Schrenck-Notzing, from his "Phenomena of Materialization." Kegan Paul, Trench, Trubner & Co., Ltd.)

striking of these we reproduce. The features of this figure, writes Mme. Bisson, who worked with the Baron in these experiments, "express earnestness and dignity, as in a conventional Christ-like head." It stands "with crossed arms and upward gaze and by its height and attitude gives an impression of solemnity." The reader may judge for himself. It is interesting for the reader to ask himself what he would have imagined from Mme. Bisson's description and to compare it with this flashlight record. Here evidently we have the will to see the thing in the best light and in a mood of exaltation put quite plainly on record.

Before the days of flashlight photography Tissot, the great painter, was shown what he believed to be the reincarnation of a woman friend in the company of a spirit guide. He painted the very beautiful impression he received and made a noble picture. The harsher methods now in use permit of no such sublimations of the vision. Our cameras show the actual things. Often these figures and faces begin small and are, as it were, inflated to a proper size.

Whatever these flattened and crumpled visages and figures may be, they are certainly not materializations in living flesh and blood; these chosen photographs demonstrate clearly that they never resemble anything more than queer flimsy stuff bearing the likeness of a painted face, whose eyes never move, whose eyelids never flicker, whose lips are fixed in one rigid expression.

In the great days of the past the alleged materializations appeared with a certain quiet dignity, but now they arise in a less agreeable fashion. They begin by being exuded by the medium as a formless stuff (styled *teleplasm* or *ectoplasm*). Usually but not necessarily, the exudation occurs by mouth and nose; it may sometimes come out of the head and neck, the ears, or from other orifices of the medium. It has a quantitative abundance like the foam of bottled beer when the beer is "up." It may have as little substance. This ooze presently takes on forms and, it is asserted, organic structure also. Possibly all the ectoplasm does not come from the same source or have the same nature. There may be a primary ectoplasm and secondary ectoplasm or ectoplasms produced in other ways. Hands, feet, grotesque bestial forms and at last these crumpled paper-bag faces emerge from the accumulating stuff. Jan Guzik, of Warsaw, materialized pet dogs. Parrots have also been recalled from the Great Beyond. There is no record known to us of canaries or white mice which have "passed over" returning to comfort their surviving owners, and no materialization of departed invertebrates, flies, spiders or the like which have solved the great mystery, has occurred. But this may come.

There is much to arouse prejudice in the literature of this ectoplasmic research. Many of its illustrations are ugly to the pitch of disgust.

They turn one back with regret to the age of faith and the romantic brush of Tissot. Yet we have to remember that these inquiries are pursued by men whose substantial honesty cannot be fairly questioned

FIG. 332. An impression by the painter Tissot of two figures who appeared during a sitting with the medium Eglington in 1885.

From "Clairvoyance and Materialisation," by Dr. Gustav Geley. Ernest Benn, Ltd.)

and it must be conceded that they take no pains to make their exhibits alluring. In the end these researches may turn back from the ectoplasm in order to illuminate many as yet unexplored subtleties in the psychology and physiology of mediums and investigators. We may ask no longer what they saw and produced but how they came to see and produce such things. It cannot be too insistently repeated that these are highly specialized researches in which the enthusiastic amateur will be welcomed only by experimentalists who are propagandists rather than scientific workers.

A considerable publicity has been given recently to the exploits of a lady in Boston, Mrs. Crandon, better known to the world as "Margery." Her principal control is her brother Walter who was killed in a railway accident in 1912, and most of the ordinary phenomena of séances are well shown in connexion with her. She is a pleasant lady with a humorous smile. She has been the subject of acrimonious disputes, challenges, alleged exposures by Houdini, the conjurer; she has her champions and her manifest enemies. Recently she has attracted the attention of Dr. R. J. Tillyard, an F. R. S., and a trained entomologist. He has been convinced of the survival of a human personality after death by his study of her, and his convictions have been given prominence in *Nature*, the organ of science in English throughout the world (August 18th, 1928). We will not enter into the details of this stormy tangle; we will merely state simply and plainly certain things we are asked to believe in accepting the good faith and soundness of the Margery displays. That statement goes to the root of the professed scepticism with which all this mediumistic business is regarded by the majority of educated people.

The personality of Walter is recognized by a number of characteristic things. In his life it seems there was a certain lack of glossiness in his manners, and in these séances he has used terms like "damned fool" and "bastard" (applied to the incredulous Houdini), showing the unmitigated survival of his distinctive quality. His voice is heard from the air from a point about ten inches in front of the medium's stomach, and is said to be recognizably his voice and no other. It is not produced by Margery. It is "masculine, fairly loud and slightly hoarse." Whistling is also one of Walter's gifts. In life Walter produced his voice by means of his lungs, his larynx and the cavities of his head, and when he whistled he used his lips. When they changed his voice changed. When his throat was dry, when he had a cold, when his lungs failed him the difference was immediately apparent in the sounds he made. The words he used depended upon the movements of his lips, tongue, teeth, and so forth. We know of no means by which a human voice can be produced, in the first instance, within a foot of a lady's waist, and say things without the activity of normal speech-organs. And Walter's speech-organs ceased to exist at his death. The instrument was broken. But the voice is still heard. The air is thrown in vibrations—by what? Why was it ever necessary to have that apparatus? Could Walter have talked distinctly without a larynx or a palate during his life? In some of these séances this difficulty is, as it were, noted and an ectoplasmic larynx appears. But how Walter gets his labial, dental and palatal sounds with this thing, or what winds from the pit stress its vocal cords remains to be explained. It would be easier to understand if he produced an ectoplasmic gramophone record.

FIGS. 333 and 334. Drawings from two flashlight photographs of teleplasm produced by "Margery" from her mouth.

Note also a connecting cord of teleplasm going to her ear.

It seems Walter can materialize a thumb also. When he died, his limbs, his fingers and thumbs fell into decay. But at the Margery séances fresh thumbmarks in wax are produced, and it is alleged that they are recognizably the same as one he left very conveniently on his razor just before his death. They are not all normal thumbprints such as are made by a real thumb. Some are the prints of thumbprints. And some are neither the one nor the other but mirror-prints, so to speak, of the thumb. They are not all alike; a few are of a quite different thumb, so that either Walter can summon other spirit thumbs to his aid or else he has become at least quadrumanous on the psychic plane. And Walter, though invisible and unsubstantial, exhales carbonic acid. Our world is richer for these material products of the combustion of refreshment, nectar and ambrosia perhaps, consumed beyond the veil. What sort of being is this invisible creature, which talks distinctly without a palate, which whistles without lips, which adds its waste products to the volume of plant food in our world and possesses positive, negative and supplementary thumbs? Is it a ghost? Never before was there a ghost after this fashion. It is something as different in its nature from Raymond or Pheneas as a knock on a door is different from a water-colour drawing or a dream about a butterfly.

Fig. 335. The well-known medium "Margery"—Mrs. L. R. G. Crandon, of Boston, Massachusetts.

(From "'Margery' the Medium," by J. Malcolm Bird.

It is just this variety in essentials which furnishes the strongest argument against the objective reality of these affairs. They differ enormously with the group of observers. One series of "manifestations" does not confirm another. Each series of manifestations contradicts some other in its quality and implications. It is not as if one consistent outer world, a spirit world or what you will, was really communicating with ours. It is not in the least as if something outside was communicating through different media—remaining itself the same. It is as if one group of mediums had one set of ideas and another another, and as if what they had in their heads found expression in "phenomena."

We cannot absolutely reject the evidence for these phenomena. The group around Margery has much to lose and little to gain from their publication. It lays itself open to irritating sceptical criticisms and many unpleasant imputations. They are evidently pleasant human people who can be very indignant with and disagreeable to an aggressive inquirer like Houdini. To disbelieve this conception of what happened is to risk an appearance of accusation against them. That we would eagerly disavow. Nevertheless, we have a right to incredulity here. Perhaps no individuals are quite so simple as our ordinary law and business customs assume. There may be a "will to believe" and make-believe in people more powerful and devious than is currently admitted. There may be a capacity for self-deception and collective hallucination greater than we have supposed.

To turn from the tangled "phenomena" of the séance as delusive is to turn to psychological speculations. Yet it is a lesser improbability to suppose that a charming lady, an eminent entomologist, some highly respectable Bostonians and a few privileged visitors have been mistaken in their impressions or inaccurate and imperfect in their accounts of what happened on certain obscure and secluded occasions, than that all the rest of our general ideas about life are wrong. These Margery thumbmarks and so forth remain, more or less, very curious material for inquiry even when we decline the spiritualist hypothesis. As Geley implies, the observer is part of the phenomena under observation. Which we would extend to this proposition, that the circle of observation includes all the reality of the affair.

It is well for us to recall that a century ago, a controversy very like that which now rages over metapsychics and spiritualism was raging over what was then called Mesmerism or Animal Magnetism. To-day we know that the phenomena of mesmerism were compounded from three sources. Some, the majority perhaps, were fraud and charlatanism; others were the result of exaggeration, self-deception, or misinterpretation; but there remained a residuum of facts which we now call the facts of Hypnotism. Under competent and critical investigation these were elucidated. And the study of hypnotism has now become an important aid to our modern deeper knowledge of the human mind. The metapsychic controversy may follow a similar course. Such bodies as the British and American Societies of Psychic Research will go on with their work, avoiding as far as possible the sensation-seeker and mercenary impostor on the one hand and the implacable sceptic on the other. It was Sir Oliver Lodge on one occasion who said to one of the present trinity of authors, "Save me from my friends," and the restraint of the credulous enthusiast is among the most necessary conditions for progress in this field.

So by degrees the grain, whatever grain there may be in this matter, will be sifted from the mass of chaff; recriminations will be forgotten; and science may add a new source of controllable power to the service of mankind. Even if we find the "other-world" phenomena dwindle to nothing, we may learn very much that is now scarcely suspected about joint and collective suggestibility and joint and collective hallucination. We may come to realize that our perceptions depend less upon the immediate fact before us and more upon the prepared matter in our minds than we are at present disposed to admit. We may find our memories less rigid than we imagine them to be. Events may often be less exterior to ourselves than we suppose.

"Impossible" is a word scientific men should never use. "Highly improbable" is as far as they are ever justified in going. We do not hesitate to find Walter "highly improbable." But, as Richet reminds us in his *Thirty Years of Psychical Research*, such a great scientific man as Bouillaud declared the telephone was ventriloquism and the still greater Lavoisier said conclusively that stones cannot fall from the sky because there are no stones in the sky.

§ 5. Mythology of the Future Life.

BEYOND the world of "occult" phenomena that claim recognition as material for scientific inquiry, there is now a vast, abundant literature of loosely authenticated "revelations" about the future life, beyond the scope of any exact treatment whatever. Remarkable and moving accounts of how it feels to "pass over" and the agreeable or disagreeable opening phases of the new state abound, and every month adds to their multitude and variety. A mawkish prettiness and unattractive poetry adorn many of these effusions.

Their disagreement is stupendous; apparently there is not so much one future life as a thousand thousand, varying in quality with the imagination and mental texture and equipment of the seer. These stories do not really support each other; they smash each other to pieces. A point upon which none of them insist but which is very manifest in most of them, is the very much lower intellectual level at which the departed spirits are living in comparison with normal worldly intelligence. And none of these revelations seem to be in precise accordance with that posthumous separation of the sheep and goats, which was formerly, at any rate, the teaching of the Christian churches. This new necromancy is a cult as far removed from orthodoxy as from unbelief. We do not wish to dogmatize about this literature; we owe it to our readers to mention it here, but we fail to see how we can square its fundamental assertions and implications with the main mass of *The Science of Life*.

§ 6. The Survival of the Personality After Death.

THIS obscure and often distressing and grotesque borderland of biological science would have demanded attention, if for no other reason, because it comes so close to another question we have all asked ourselves. Alone, in the silence of the night and on a score of thoughtful occasions we have demanded, Can this self, so vividly central to my universe, so greedily possessive of the world, ever cease to be? Without it surely there is no world at all! And yet this conscious self dies nightly when we sleep, and we cannot trace the stages by which in its beginnings it crept to an awareness of its own existence.

Mr. Everyman sets down the printed word and reflects. "I am I," seems to him the statement of a veracity beside which number and space and time seem flimsy abstractions. But then he reflects upon a number of things *The Science of Life* has brought before him. All the way through this work has been throwing light upon the nature of individuality. We have recognized grades of individuality, the cell individual centred on a single nucleus, the individual metazoon, the individual colony made up of individual zooids. We have found it impossible to define individuality in the case of many creatures; in the case, for example, of the sponges and Obelia and other colonial polyps. We have seen individuals melt together and become one, and individuals break up into many. In our study of mental life we have seen that in one single brain it is possible for separate and even antagonistic individualities to exist. Even in clearly defined human individuals we are constantly aware of a conflict of motives, a war between a better and a worse self, a divergence of loyalties and ends. Is the whole subconscious and conscious self the immortal part, or is it the persona only? Is it an inflated self that survives? Many of Mr. Everyman's intensest passions do not so much further his individual interests as they do those of the race. Sometimes he would rather love than eat.

Some of the best things in our lives are the least individual things. When a man is exalted by high aims, possessed by some exquisite effort or occupied by profound study, he becomes altogether self-forgetful. In moments of great passion he "forgets himself." These are no metaphors. The conscious self is not the whole of a man. It is the central bureau for his general bodily behaviour, but it is subjected to systems of motivation, rational thought, scientific curiosity, loyalties, mass-suggestions, which come into his being from without, as general instructions from headquarters come in to the semi-autonomous activities of a branch bank. Many of our sense-impressions undergo interpretation in the brain. Perhaps the collectivity of our sense-impressions is interpreted to suit the needs of our mind. It is possible that it has served the ends of survival that Mr. Everyman should think himself a much more independent being than he is. Personality may be only one of Nature's methods, a convenient provisional delusion of considerable strategic value.

Moreover, individual death is one of the methods of life. That we have already enforced in our comments on rejuvenation in Book 4. Every individual is a biological experiment, and a species progresses and advances by the selection, the rejection or multiplication of these individuals. Biologically, life ceases to go forward unless individuals come to an end and are replaced by others. The idea of any sort of individual immortality runs flatly counter to the idea of continuing evolution. Mr. Everyman makes his experiments, learns and teaches his lesson, and hands on the torch of life and experience. The bad habits he has acquired, the ineradicable memories, the mutilations and distortions that have been his lot, the poison and prejudice and decay in him—all surely are better erased at last and forgotten. A time will come when he will be weary and ready to sleep.

It is the young who want personal immortality, not the old.

Yet these considerations do not abolish the idea of immortality; they only shift it from the personality. In the visible biological world, in the world of fact, life never dies; only the individuals it throws up die. May there not be another side of existence of which our consciousnesses seem to be only the acutest expression we know, a perceptive side of matter, if one may strain a term, which also is more enduring than any individual experience? Just as our bodily lives stipple out the form of the developing species, so our mental lives may stipple out its dawning consciousness. Though we are mortal as ourselves, we may be immortal as phases and transitory parts of an evolving undying percipient continuity. When we philosophize in the stillness it may be not ourselves alone, but Man that feels his way to self-realization through our individual thoughts.

Apart from such speculations we may say this much: upon the continuity of any individual consciousness after bodily cessation and disintegration *The Science of Life* has no word of assurance, and on the other hand it assembles much that points towards its improbability. But so far as our lives go, as matters of fact apart from consciousness, *The Science of Life* has no doubts; it does not speculate, it states. Our lives do not begin afresh at birth and do not end inconclusively; they take up a physical inheritance, they take over a tradition, they enter into a set drama, they are conditioned from the outset, and each has a rôle to play, different from any rôle that has ever been played before or will ever be played again. And our lives do not end with death; they stream on, not merely in direct offspring, but more importantly perhaps in the influence they have had on the rest of life. According to the playing of the rôle the unending consequences are determined. They endure in the fabric of things accomplished for ever. That at least is not theory or speculation; it is as much a statement of fact as that every stream that flows upon this planet earth flows down towards the sea.

BOOK NINE

BIOLOGY OF THE HUMAN RACE

I

PECULIARITIES OF THE SPECIES *HOMO SAPIENS*

§ 1. *Fire, Tools, Speech, and Economics.* § 2. *Origins of* Homo Sapiens. § 3. *Primary Varieties of Human Life.* § 4. *The Development of Human Interaction.*

§ 1. Fire, Tools, Speech, and Economics.

IN THESE concluding Books we have returned to that point of maximum interest from which we started out upon our general survey of biology, the life of Mr. Everyman. But whereas hitherto it has been convenient and interesting to study Mr. Everyman as an individual, we now propose to study him collectively as Man, *Homo sapiens*, the last surviving species of the family of the Hominidæ, which differentiated from the great apes in the shrinking forests of the Miocene or Pliocene Age (V, C and D). We have described his body and its evolution; his brain and his mind. We have studied the operation of his mind until at last we had to stop short at that apparently insoluble mystery of life, consciousness. Our last Book culminated in a great interrogation: is there an immortal individual consciousness? Does the individual consciousness survive matter? We left that question open.

Now we are going to consider our Mr. Everyman very briefly as a unit in a rapidly developing species of animal. We are going to treat *Homo sapiens* as a species among other species. It is a species which now dominates life, which is producing a whole multitude of unprecedented problems and novel biological situations, and seems to be breaking more and more away from the control of the blind forces that have hitherto determined the course of evolution.

Homo sapiens differs widely in his action upon his environment from any other form of life. Let us examine in what that difference consists. In common with the rest of the Hominidæ, the apes and some monkeys, the human animal uses its hands to pick up sticks and stones, to make tools of them and supplement its forces and extend its range. This readiness to use tools is closely associated with the high development of the hand and with the unprecedentedly efficient use of that hand which the possession of a spot of distinct vision has made possible. But modern man carries this tool-using disposition to an extent immeasurably beyond the range of kindred types. All of the Hominidæ fabricated tools, chipping flint and wood to forms more convenient and effective, but the adaptation of tool to use is comparatively crude and primitive in all other species of Homo except our own. All other species of man reached their highest mechanical aid in a shaped flint or a pointed stake; *Homo sapiens* still soars on, with no sign of any finality, from the dynamo and the aeroplane engine. Other Hominidæ also supplemented their powers with artificially made fire. But their use of fire was incidental, while Modern Man now at a great rate burns his world for his ends. More definitely restricted, it would seem, to Man is the use of articulate speech. Great numbers of vertebrated species communicate by sounds, but all surviving races of Homo have conventionalized their vocal sounds to convey complex meanings in a way no animal does. They have language. Whether *Homo neanderthalensis* had, strictly speaking, a language is questionable.

The fossil extinct species and genera of men lived in small groups as the great apes do, small family groups usually about an old male; their remains never indicate larger social aggregations. But the true men (*Homo sapiens*) from the earliest appearance of their remains are seen to be going beyond that primitive assembly. Some antagonism has been allayed. Some kind of toleration has been established among the adult males; the family group has been expanded into the tribe. That, however, is the less important aspect of the matter. Many other animals have increased and decreased in their gregariousness. As the excavator, explorer and archæologist trace out man's history to us, we realize that in the last fifty thousand years or so he has not merely become rapidly more and more gregarious, but also, what is far more remarkable, he has left that life of haphazard which is the lot of nearly all the rest of the animal creation, that practice of eating your food where you find it and taking no thought for the morrow, and he has become an economic animal, preparing for the future, cultivating and storing food, domesticating other creatures for his own support. And he does not do this mechanically, by instinct, as the ants do. He does it by forethought.

Some other mammals store food—the beaver and squirrel, for example; the leopard will hide away a half-consumed corpse and return to it as a dog will bury a bone, but no other mammals cultivate. To find any creature that cultivates we must go outside the vertebrata altogether, to the ants and termites. In the social life of these insects we find a superficial parallel to the social life of man. When, however, we scrutinize the methods of co-operation, the parallelism disappears. In Book 8 we have made a study of these methods of insect organization, and pointed out the essential difference of a society based on instinct from one based on intelligence. One point in that comparison we may recall. A fundamental feature of the insect society is the differentiation of the individuals into a number of types. The individual is either a worker, a soldier, a queen or a drone. Sometimes there are several different types of worker. Each does its work as a machine does its work in accordance with the way it is made. A worker cannot become a soldier, or a soldier a male.

But in the case of human society, there is no such differentiation. Save for abnormalities and unimportant exceptions of usage, every individual remains a physiologically complete individual, every man is a "man and a brother," and queen and beggar girl are "sisters under their skins." Except for rare abnormalities every individual remains male or female and has at least the capacity for reproduction. That is in flat contrast with the condition of the social insects. There may be social considerations to prevent the profoundest philosopher or the richest business man marrying and begetting children by a Negro defective or giving his daughter in marriage to a jail-bird or a tramp, but there is no irrevocable physiological separation to prevent such things occurring. In various regions, class and caste restriction of intermarriage may have sustained slight differentiations of manners, intelligence and colour, the importance of which has been frequently exaggerated, but at present such ancient barriers seem to be weakening and dissolving. They have scarcely any relationship to current economic organization.

The complex economic life of mankind, infinitely more complex than that of any other creature, has been achieved without any such differentiation. Man is the sole economic animal undifferentiated for function, and he is the sole known user of grammatically arranged words, as distinguished from merely expressive and indicative sounds and signs. He is the sole surviving species which uses fire and deliberately shaped tools, and he has now carried his use of extraneous power and of accessory external limbs and organs (for that is what tools amount to) to such an extent that much of his exacter knowledge comes to him through such artificial sense-organs as microscope and galvanometer and the greater part of the energy of the economic life of his species is derived from other than his bodily sources, and a great part of its handling is mechanical.

The development of this abnormal and novel social life is the latest, greatest and strangest of the products of evolution. It is a new phase in the history of life. Essayists of the "smilingly thoughtful" school are apt to ascribe man's disposition to set himself apart as the head and centre of the story of life to the natural egotism of the species, and playful writers have "turned the tables" by writing accounts of the world from the point of view of an ant or a crocodile, in which man is spoken of with the pity and contempt we have for these lower creatures. But as a matter of fact there is no view of the world from the point of view of an ant or crocodile, their interests have no such range; and an abstract intelligence without the least prejudice in our favour would have to do just as we do here, and treat the collective life of *Homo sapiens* as the present culmination and most distinctive and wonderful phase of vital evolution.

The outer forms, the political forms of the rise of this unprecedented specific collectivity, mankind, can only be treated in an *Outline of History*, and an adequate review of its economic developments will need a work even more bulky and extensive than the whole of this *Science of Life*. Here we restrict ourselves as severely as we can to the more definitely biological aspects, to social origins, broad factors in social adaptation, religion, education, each regarded strictly as a biological force, and the possible increase, survival and suppression of types; leading up to a final consideration of the biological outlook of our kind.

§ 2. Origins of *Homo Sapiens*.

ADAM and Eve die hard. People are still apt to talk of "the first man and woman" and to discuss the claims of this or that restricted region to be "the cradle of mankind." They imagine, one supposes, a particular couple of some species of sub-man suddenly discovering themselves "different" and starting out upon a new way of life. "Let's found a new species, my dear," is the note of it. But the reader will have read this work but carelessly if he is still under the delusion that new species appear in any such fashion.

As we have spread modern concepts before him, he will have realized how species constantly produce mutations of type which, if they give advantages leading to their natural selection, may become prevalent in that species, locally or extensively. Most of these mutations are very slight in their effects. They turn up here and there over the range of the species; one individual appears with a change in colour-pattern; another with a difference elsewhere in its body; and because of the constant mixture by interbreeding of the population these mutations are brought together into more or less favourable combinations. Locally or extensively one combination may prevail altogether, and if it prevails locally and is cut off geographically, or produces in association with its other characteristics some change in its reproductive quality, it may become incapable of interbreeding with the original type and so become in the fullest sense a new species. All species present the phenomenon of local variation and some, like the domestic dogs, present extraordinary varieties of type which yet retain a common capacity for fertile offspring. It is possible that the Hominidæ, whose rare infrequent fossils are now coming to light in this or that region of the old world, and of whom we are continually discovering fresh types, may have been as are the domestic dogs, interbreeding varieties. It is pure guesswork whether *Homo neanderthalensis* in any region interbred or did not interbreed with *Homo sapiens*. Just as there never seems to have been a single sort of domestic dog from which all other sorts sprang, but an advancing interplay of various sorts, so quite possibly there was never a single primordial sort of *Homo sapiens*. The race varied widely locally, strongly marked types appeared under special conditions, and then with geographical changes other human types came flooding in, still capable of fertile interbreeding, to kill, to mix, to learn from each other by imitation and precept, and start yet further variations through genetic discord.

We cannot say that the primordial varieties and species of Homo were forest or bush inhabitants or dwellers on the border of steppe country. Much more probably they were all those and also other things. The early skeletons of true men found in the Grimaldi caves belong to two types, the Cro Magnon type and the Grimaldi type, as widely different as the Red Indian is from the Negro to-day. But certain characteristics the species of Homo already had in common. They were, for example, less highly specialized for the forest life and they were already far more gregarious than the great apes. The great apes are all rather solitary creatures. They live in little family groups, which are typically dominated by an old male; they are jealous of intruders upon their territory and mix freely only in their own group. Consequently, comparatively rare though these creatures are, they display considerable local variation. The West Coast gorilla is quite distinct from the mountain gorilla of Central Africa, not only in physical characters, but in temperament. Several varieties of orang-outang are known, some of which should probably be classed as distinct species. As we have previously mentioned, the old males of at least one variety have strange fleshy frills round their faces. And the chimpanzees from different regions of Africa are of markedly different types. Although chimpanzee coat-colour only ranges between brown, grey and black, the naked skin of the face varies almost as much as man's skin—it may be as black as a Negro's, or dark brown, or lighter brown, or almost as pinky-white as a white man's.

Some interesting speculations upon the social life of the primordial Hominidæ were made years ago by J. J. Atkinson in his Primal Law (*Social Origins and Primal Law*. Lang & Atkinson). His views, like Darwin's, have been revised and modified in several particulars, but they still remain of paramount importance and value in human biology. The whole theory of psycho-analysis rests upon such ideas and is in entire harmony with them.

Atkinson, arguing back from the practically universal tabus against various forms of incest, and the world-wide traces of the custom of exogamy (the custom which makes ordinary males capture wives from another tribe and abstain from marriage with the women of their own people) developed a remarkably plausible theory of the early constitution of human, or rather sub-human, society.

FIG. 336. A new link in the chain of human descent.

The brain-case of the sub-man, Sinanthropus, found in China by Mr. W. C. Pei, of the Chinese Geological Survey. The prominent brow-ridges over the eyes are seen at the right. This skull is like that of the Ape-man, Pithecanthropus, except that it housed a distinctly larger brain. This important discovery was only made in December, 1929.

He assumed that the normal social group among the earlier Hominidæ was, to begin with, like the more usual social group of the great apes, a small family herd under the leadership of a big male. As the other males approached maturity this head male drove them off, the young females for the most part remaining with him and bearing him offspring. Of course, such a normal type of group does not exclude, indeed it almost requires, that sometimes two or three brothers who had been chased off have wandered together and even have attracted a stray female or so. Such leaderless groups occur among the gorillas, but they are much rarer than the family group. The adult male's fierce jealousy for his females and for his territory ensured the prevalence of the family group, and the restoration of fresh family groups when the old ones were broken up. It is a grouping well adapted to forest conditions where food is scattered, and forthcoming only in sure and sufficient quantity for a limited number of individuals. It has consequently remained to this day the typical social method of the forest primates, though chimpanzees will sometimes band themselves together in larger groups.

But the Hominidæ, being less highly specialized for forest life, more inclined to eat animal food, and better adapted to bush, grass and rock country where fruits and roots are found less easily than prey best hunted in co-operation, would be advantaged by any mental or temperamental adaptations that would admit of the primitive families growing into larger social groups. This adaptation became possible, Atkinson suggests, through the interplay of certain natural dispositions on the part of the females and young males.

Briefly, the mothers, after the fashion of most mammals, were disposed to protect and cherish their male quite as much as their female offspring. But an adult male in the breeding season—and all round the year is the breeding season for the primates—is apt to be intolerant of competitive males. In order to keep their young sons by them, then, it was necessary for the mothers to inspire their young with awe for his seniors, and particularly for the ruling Old Man, and to make the juniors chary of infringing his rights and rousing his jealousy. By example and crude precept, the natural awe of the young male for his father's strength and possible rage was given form and direction. The young males grew up learning that his personal possessions and particularly the females of their group were tabu to them, that certain things must not be done in his sight or proximity. The fear of the Old Man was the beginning of wisdom and decency. Many of them retained their natural infantile disposition to propitiate beyond adolescence. The younger male deferred to the older male; so, said Atkinson, men learnt the elements of self-suppression, and the idea of sin, and particularly the sin of incest, was born in the human mind. So, say the psychoanalysts, the first repressed complexes arose.

Human society became possible through this primary suppression, and it is hard to imagine how it could have become possible in any other way. No other animal but *Homo sapiens* betrays the slightest objection to incest, and that the objection to incest is a tradition and not an instinct, the records of any provincial criminal court will show. A few eminent sociologists like Dr. Hobhouse are of the opposite opinion and believe that there is an instinctive objection to incest, but all the known facts point to an imposed and habitual avoidance as the real bar against this form of intercourse.

And now comes the next step in the history of the fundamental human institutions. As the young man, growing in strength and desire, wandered discontented upon the borders of the family territory, he discovered there were other women in the world, women who were not his chief's women, women unprotected by the tabu. He went for one of these women when he got the chance.

Conceivably if she was a neglected woman astray, a rather superfluous woman from another family group or the woman of a group whose head man was slain or enfeebled, she went in equal measure for the straying young man. If we are to suppose that the older males were in the habit of attacking and often killing the young ones, the probability of such superfluous women is increased. Atkinson represented the young man as always attacking and overpowering the stray women, but then he wrote a quarter of a century ago when a certain veil of modesty hung about feminine desire and enterprise. Attack may not have been necessary. The young man brought the strange woman home to the tribe or to the outskirts of the tribe, or she came with him without being brought, his own woman. Less typically she beguiled him towards her parental hearth—her man. If she was the stranger and came into his group, naturally she looked up to him; he was her chosen, and she did not yield to the Old Man. The family women did not want her as an equal and a rival; they were on the young man's side against any interference with this pleasant acquisition of his on the part of the Old Man. They were willing to set a tabu between her and the Old Man. If, on the other hand, the man went to the woman's family group, equivalent tabus would become necessary.

Now all this is very plausible theorizing indeed, and in support of it we find tabus of practically world-wide extent that are entirely consistent with it.

Out of such crude and obvious occasions, which probably presented themselves with wide variations among the Hominidæ and were repeated millions and millions of times in the course of tens of thousands of years, and out of these instances of the successful grafting of a stranger on the group, exogamy (marriage by the acquisition of strange mates) would have crept into existence almost imperceptibly, and a second fundamental tabu between mother-in-law and son-in-law, and between daughter-in-law and father-in-law have arisen.

Such in its essentials is Atkinson's guess at the method by which the extension of the primordial family group to the proportions of a tribe was attained. It is a guess, a theory, but there is a great mass of fact to keep it in countenance. It explains most plausibly a change in social habit which would have been extremely advantageous to man in certain phases of climatic change to which we will presently advert.

Throughout the world, everywhere tabus and customs occur that are strictly in accordance with this theory of Atkinson's, and which are explicable in no other way. First come the incest tabus that are woven into the fabric of every human society. Biologically they are unique, purely human and universally human. And they are not instinctive; they can fail to be established, or they can be broken down. There have been exceptions to these tabus. The Pharaohs could marry incestuously and so could the Incas of Peru, and there is no evidence that they made any difficulty about it. Indeed, only their incestuous offspring were legitimate and could succeed to the throne. But here, and in similar exceptions, we seem to be dealing either with extreme sophistications or with the heirs to the sexual freedoms of the Old Man and not with the normal marriage of the sons of the patriarch. Even into the twentieth century the sexual restrictions upon royalty have been different from those of other sections of the community. The scions of many royal houses were not bound by marriages contracted with women below their rank and could commit with impunity what was for lowlier men the crime of bigamy. The royal families of Europe up to the Great War constituted a very closely intermarrying system. Such exceptions confirm rather than disprove the general thesis. They show that the objection to incest is not innate and instinctive but a fundamental tabu, and that it is something less binding upon the ruler than upon the commonalty. The morality of the head man, the patriarch, was not that of the son.

Throughout the world now, even in the most isolated and savage communities, the incest tabu holds. No variety of human association is known that ignores it. The primitive man-ape family-tribe whose mental conflicts laid the foundations of all our present social organizations, has gone altogether, leaving only its traces in our minds, customs and institutions. In the English prayer book we recall the most fundamental of all human institutions when we read the prohibited degrees of affinity. In less civilized and presumably more primitive communities we find a far-reaching system of " avoidance" tabus which are clearly intended to hedge about and strengthen the essential tabu. The woman of the Siberian Ostyaks must not appear before her father-in-law nor her husband before his mother-in-law until they have children and the woman must muffle her face against her father-in-law throughout life; the Buriaks, Kalmucks, Altaian Turks, and Kirghis have similar restrictions. The woman must never use the name of her husband's father. Now let us leap to Ceylon and we find the Veddah must not speak alone to his mother-in-law nor speak with or take food from his son's wife. Again in Melanesia in the Banks Islands a man must not enter a house in which is his mother-in-law, and if he meets her in the bush he must turn aside to avoid her. The daughter-in-law must not name her father-in-law. Similar restrictions are found in New Guinea and Torres Straits. Parallel tabus prevail throughout Australia. In Africa a Zulu covers his face with his shield if he chances to come upon his mother-in-law, throws away a mouthful he is eating if she happens to pass by and must never mention her name. Frazer cites similar customs from the Bantu tribes and the Masai. They are found very widely among the American aborigines. It is impossible not to believe that here we must be dealing with a "fossil institution" of once universal importance, a common fundamental idea. That in many instances it should be lost and in many distorted or perverted is only what was to be expected.

The "fossil" reminders of exogamy in marriage ceremonies are equally widespread. The evidence of the marriage ceremonies lies parallel to the tabus and is similar in its world-wide extent. The primordial procedures have been complicated, distorted, varied enormously, but always at the root we can detect a barrier against free intercourse and the necessity of mitigating by special conventions the discord and harmful excitement of a stranger introduced to the close intimacies of a family group. The primordial marriage introduces while the primordial tabus protect the newcomer. The excitement of novelty is denied free play. The disturbance is damped down below the level of social disruption.

The sexual psychology of the apes and monkeys in their natural surroundings has still to be studied with any thoroughness. What is known of the behaviour of these creatures is entirely compatible with the suggestions of Atkinson. They have no breeding season, no rut, such as prevails among mammals; like man their breeding season is all the year round. And all the year round sexual intercourse goes on and most vigorously in seasons of abundance and well-being. Their sexual impulse is excessive. When they are social like the baboons, there is a perpetual bickering and scolding going on in the swarm, insults, pursuits and recriminations due mostly to their intense sexual impulses. The males are possessive and keep a sharp eye on their particular mates. In the London Zoological Gardens there were for a time only six females to many more males and all these were individual wives. When one "married" male happened to die, his widow was actually torn in bits by other males eager to acquire her now that her strong Old Man was no more. *Homo sapiens*, when he is thrown out of his ordinary social atmosphere so that the fundamental tabus are no longer in effective action, that is to say when his sense of sin is lulled or out of action, shows himself to be closely akin in these matters to the rest of the primates his cousins. His instinctive restraints, as distinguished from his reasoned ones, are slight or non-existent. He is possessive as a lover, but disloyal. It is the primary tabus that restrain him. It is these tabus that hold back our species from incessant sexual squabbling and make the disciplined tolerance needed for sustained economic co-operation possible.

The survival value of the tribe (that is to say, the group with numerous mated males, living in comparative mutual toleration) over that of the narrower family group under a jealous patriarch must have been particularly evident in these drier phases of climatic change when forests were giving place to more open country and shrinking in their area. Such phases, as we have shown in Book 5, Chapter 7, were recurrent during the age of human evolution. Life was being repeatedly squeezed out of the forests to try its luck on the steppes to the north or the deserts to the south. Whenever a diminution of forests occurred a dietary of varied small pickings would be replaced by a menu with quantitatively larger and less diversified items. The tribe of kindred, as distinguished from the mere patriarchal family, was able to hold larger territories and choose and hold the best territories against any smaller groups of competitors. It was also more able to attack larger animals for food than a lonely hunter or a one-man group could do, and to defend itself more effectively against hostile beasts. It could venture into open country where isolated men would certainly have been hunted down and killed.

It is idle to guess how early in the development of the Hominidæ this movement towards the co-operative gregariousness of a life in the open began. It may have begun first as the vegetation of the Miocene Age (V C) diminished and have marked the earliest stage of differentiation from the apes. It may have been a recurrent experience. Man may have fluctuated between a more and a less numerous assemblage and between intenser and lighter tabus. From his very first visible appearance upon the terrestrial stage *Homo sapiens* at any rate, in all his varieties, is a social animal becoming steadily more social, with suppressions, with tabus, with a sense of sin. He appears from his very beginnings as a being of incomparable mental power—divided against itself.

Human society is built up upon a balancing of motive against motive. Man is discovered from the first to be a moral animal. He fits in by individual self-suppression, a thing he has also imposed upon some of his domesticated animals. His suppressions are individual acquisitions, varying with his training, his social position, and the chances of his life. Through his facility for suppression he has been able to achieve this unprecedented miracle of an economic society without fundamental differentiation. Never before in the whole history of life has such a being appeared.

§ 3. Primary Varieties of Human Life.

FROM the beginning of our certain knowledge, *Homo sapiens* appears as a widespread number of types, the primary "races of mankind." There have evidently been profound local separations and modifications in the prehistoric period, but none so complete and enduring as to break up the species. As geographical conditions have changed— and, as we have shown, the last twenty-five thousand years have achieved immense modifications of the map of the world—these variations have resumed communication with others, have remingled with others, to produce mongrel groups, which, if they merit the name of race at all, can only be styled "secondary races." Much nonsense has been talked and written about racial purity. Except possibly in the case of certain very isolated peoples, the now extinct Tasmanians and the Andaman islanders, for example, racial purity is a myth. Men and women are all mongrels, showing in various proportions the characteristics of this imperfectly specialized type or that.

Man is, indeed, in respect of his variation, as in respect of his mental capacity and other characters, a unique organism. The fact that this now dominant animal type is represented, not by many families and orders as was the case with other types dominant in their time, such as the Reptile or the Mammal, but only by a single species or interbreeding group; the fact that this species has a greater range, geographical and ecological, over the face of the planet, than any other single species of animal or plant; and the fact that the extent of naturally-occurring variation (we say naturally-occurring to exclude the humanly produced variation of certain domesticated species) within the species *Homo sapiens* is much more considerable than in any other creature—these are all noteworthy facts of human biology.

They are largely, it seems, corollaries of his capacity for speech and tradition, combined with his irresistible migratory urge. Man was an incurable migrant, certainly from palaeolithic times, and he shows not only slow mass movements similar to those of other animals (though often more conscious and more rapid), but also, and on a constantly increasing scale, the migration of individuals or small groups, prompted by trade, persecution, discovery or conquest.

But whereas in animals, slight differences in instinct or structure or physiology can, and often do, keep varieties from interbreeding even if migration brings them together, man's plasticity of behaviour can bridge these gaps, and so re-cement groups, which had begun to acquire distinctness through isolation, into a single interbreeding whole, which is, we repeat, without precise parallel in other forms of life.

As Professor T. H. Huxley pointed out long ago, there is a central series of human races ranging from the Atlantic coasts on either side of the Mediterranean to the east of Asia, Malaya, and Polynesia. These various peoples are brunette in various degrees of intensity, dark or black-haired, with hair that does not frizz and which may be very hard and straight. South of them, in the old world, are the black peoples of very variable type, with frizzy hair. Their hair, their colour, the abundance of sweat-glands in their skins show them to be an adaptation to tropical forest conditions. To the northeast of the central series appears a more distinctly yellow type of race, with oblique eyes, broad cheekbones and very hard, black hair, the Mongolian group of races. These, with their scantier sweat-glands and protected eyes, witness to the long influence of drought and dust. To the northwest and north, that is to say, over the centre and north of Europe and once extending far into north Asia, are a series of fair peoples with grey and blue eyes, the "white" group of races more specially adapted to the deciduous forests of a temperate climate and insufficiently protected against very bright sunlight. *Homo sapiens* seems to have been a late comer in America, an immigrant always. It is doubtful if he has been there for more than ten or twenty thousand years. The American Indian peoples find their closest relations among the eastern Asiatics.

That in general terms is the fundamental shape of human ethnology. A rather more detailed account of the races of mankind and a diagram showing their relationships will be found in *The Outline of History*, and it is unnecessary to repeat these here. As soon as we attempt greater detail than such broad classifications give, we find ourselves in a tangle of kaleiodoscopic racial types, due to the indefatigable mongrelization that has been in progress. Whatever differentiations were going on ten thousand years ago, they have long since been defeated by the steadily increasing communication between one part of the world and another. Man is no longer differentiating locally. Some day, perhaps, ethnology will abandon its attempt to separate what is so obstinately confluent and may set about classing human beings in other categories. A classification of genetic units may become possible, and individuals may become definable for formulæ which will show the particular grouping of characteristics that has occurred in each case.

Meanwhile it is perhaps desirable in view of the present political emphasis on race, to stress a few special points. First, there is no such thing as an "Aryan race." There are only groups of peoples of very various stock who speak languages of Aryan type. Here a linguistic, cultural term has been quite illegitimately transferred to the racial, biological sphere.

Secondly, there is no such thing as a pure "Jewish race." The term *Jewish* denotes a community with a certain religious and semi-national tradition, in which some community of descent is also involved. But the Jews themselves are of markedly mixed origin, including for instance marked Armenoid as well as Semitic elements, and in their dispersal have interbred sufficiently with the surrounding peoples for their stock to contain a strong admixture of alien genes, and an admixture which varies from country to country.

Thirdly, the Nordic race, on which so much political stress has been laid, hardly exists anywhere in a state even approaching purity. In Germany, for instance, Nordic genes are very much mixed with those of Alpine, and, to a lesser extent, with those of Mediterranean provenance, and there has also been a certain infiltration of Mongolian characters from the East. In Britain and France and Italy, again, there is an equally thorough mixture, though the proportion of genes and resultant types are different.

Fourthly, the Nordics have not, as is often claimed, been responsible for all the great advances in human history. The greatest advance of all, from barbarism to civilization, by means of the inventions of agriculture, permanent building, and written record, was made in the Near or Middle East, probably by dark-haired people of Mediterranean type, certainly not by any tall and fair-haired, blue-eyed Nordic stock. Similarly, in modern times, the greatest men of the white race, writers, explorers, artists, scientists, musicians or statesmen, have more often than not been of non-Nordic type.

The present unfortunate state of thought on the subject is chiefly due to a confusion of the ideas of *nation* and *race*. Appeal is made to a mythical racial purity to bolster up the claims of national unity,—to a mythical racial inferiority to support nationalist claims to independence or expansion. The intelligent reader of this work will be able to detect the falsity of these appeals, and to realize that, by and large, race-mixture is more desirable than race-purity. True that the degree of race-mixture and the rate at which it is to occur must be adapted to political and cultural conditions. But race-mixture alone can give that variety on which the many-sided progress of nations and of humanity at large must be based.

Returning now to the primary varieties of our species, man is not only far less specialized and more versatile and migratory than the great apes, but also he is far more omnivorous. *Homo sapiens* varied his menu with his opportunities, and here subsisted on a wholly vegetarian diet, and here on fish, and here again was completely carnivorous. Difference of habitat means difference of habit and difference of physique and successful type. No human varieties survive to-day that we can call "primitive" in the proper meaning of the term. We can only speculate about the nature of the "primitive" life of man. But even as late as the nineteenth century, various savage races remained sufficiently isolated to contrast very vividly in their customs, physical types and social organization with the main masses of mankind. There were, for example, the aboriginal Tasmanians, a people who had still lingered at the early Paleolithic level and were out of all comparison inferior to the Solutrean and Magdalenian Europeans. There are still the Brazilian forest folk and the Pigmies of Central Africa. There are also the Veddahs of the Indian forests, the Australian Blackfellows, and the Bushmen of the Kalahari desert.

All these are divergent or retrograde forms of human life. Indubitably men, their little communities of at most a few score individuals, with their crude equipment of implements, their limited language, and their rigid traditions, keep before us a realization of the narrow limitations from which the mass of our species is now escaping. None of these folk seem ever to have lived in larger groupings than they do at the present time, but most of them have been influenced and sophisticated by some intercourse with less specialized humanity. They are not to be rashly taken as primitive. Rather they are the preservers of aberrant and less successful tabu experiments. Many of the more peculiar institutions of the vanishing savage societies of our time are the sociological equivalents of the platypus and echidna. They are not ancestral survivals but side branches.

§ 4. The Development of Human Interaction.

IN CHAPTER 7 of the preceding Book, we have considered the intricate system of associations and symbols by means of which three writers, working together in a country house in Essex in England, are able to arouse the thought, form and flavour of an orange in the minds of readers in America, Australia, Africa, China or anywhere else to which this work may penetrate. We pointed out how it is that the human mind is able and free to isolate qualities and pursue generalizations, to form, analyze and reconstruct concepts, because of its use of word-symbols. This does not merely carry the power and range of the individual human mind immeasurably beyond that of any other living creature; it also establishes a new and strange interdependence among human individuals. The development of that interdependence is one of the leading *motifs* in *The Outline of History*, and we refer to it only briefly here. Probably the first language was a language of gesture and mimicry. Imitative sounds were an obvious way of suggesting other animals, or any other noise-making objects, such as a stream or a breeze. With most gestures, as Sir R. Paget has recently pointed out, goes an associated movement of the vocal apparatus, and it was a convenient and labour-saving expedient to detach this concomitant as a symbol of the gesture it accompanied. Man also drew early. The aggregation of his sound symbols into grammatical speech and of his pictorial gestures into picture-writing and at last into script, once it had begun, may have progressed very rapidly. Symbolism must have involved a great economy in association. It was a process that probably took a few thousands rather than scores of thousands of years. Later Paleolithic man and early Neolithic man were already freely-talking and picture-writing animals.

With the development of speech and image and picture, and with the multiplication of instruments and constructions, man began to supplement his heredity in an entirely unprecedented fashion. We have told how with the evolution of parental care among the higher vertebrata, education appeared in life. All the Hominidæ were exceptionally educational animals, and with the development of speech, precept was added to example and memories began to be transmitted from old to young. *Homo* was the first living creature to form a picture of his universe that transcended individual experience. The elders supplemented their stories of what had happened to them and what they had been told by their predecessors with imaginations about the beasts and rocks and the sun and moon; myth and legend were added to tradition.

It is not so very difficult to imagine, once the process of symbolization was begun, once the point of crystallization was reached and language became possible, a very rapid development of the traditional element in human life. Men began to "explain" things, and particularly the tabus and customs, by telling stories about them. Man added tradition to heredity. He is the first and only traditional animal. There again he is separated from all the species. And now he ceases again to be traditional. He is supplementing tradition by science and analysis. Tradition has been a phase in his development which has lasted only a few score thousand years.

Primitive human thinking was like the thinking of children and uneducated people to-day. Something was imagined and either liked and sought, or disliked and avoided. Things were grouped in the mind to see how they looked and felt together. Countervailing ideas were evoked to alleviate, distort or suppress disagreeable realizations. Thinking was more like reverie and had little use for words until it had to be told. It has only been very slowly that an acuter observation, an exacter definition, a more logical process has come to the aid of these primitive methods, and now begin to supersede them.

The great period of Hellenic thought between the sixth and the fourth centuries B.C. marks the transition from what Jung, in his *Psychology of the Unconscious*, calls Undirected Thinking to Directed Thinking. Plato has recorded and immortalized for us the birth-cries of logical thought. Aristotle was the Father of Natural History and Philosophy. From that period onward, the earlier mythological method of expression, dream-like in its quality, gave way slowly but surely to philosophical analysis and open-eyed scientific classification. We are still in the closing centuries of that phase of transition. Only now does it become possible to present the ordinary human being with a picture of the universe that is generally valid and divested of fabulous interpretations. The bulk of mankind is still thinking mythologically. Only now is it possible to replace dogma by rational direction.

In *The Outline of History*, the expansion of man's picture of the universe is traced. Step by step we see how man passed from a picture of the universe centring upon his family and his tribe and having a radius of a few score miles, a little fear-girt picture, filled with the projection of his personal reaction to his father and his associates, to broader concepts, to the picture of the city, the nation, the state or the empire. His mythology, in that story of the past, retreats before the advancing realism of his thought, his sympathies expand, his sense of fellowship replaces an animal hostility to strangers and to unfamiliar types. Throughout that story there go on a concurrent improvement of his means of transport, and a steady development of his methods of expression, record and communication. In spite of hates and brutalities, of an inherent disposition to distrust, of the crazy egotism of the ordinary individual in a position of power, of a troublesome inheritance of greed, cowardice, sloth, and self-protective illusion—in spite of all these things, this advance continues steadily. We live in a clearer and cleaner light than the men of the past. The average person is more lucid and less obsessed. An ever-increasing proportion of human beings realize sane and comprehensive pictures of the universe. Loyalties grow wider and more rational. It is a process of mental personal expansion to which the only visible limit is our planet and the entire human species.

II

THE PRESENT PHASE OF HUMAN ASSOCIATION

§ 1. *The Religious Tradition.* § 2. *The Passing of Traditionalism.* § 3. *The Supersession of War.* § 4. *The Change in the Nature of Education.* § 5. *The Breeding of Mankind.* § 6. *The Superfluous Energy of Man.* § 7. *The Possibility of One Collective Human Mind and Will.* § 8. *Life Under Control.*

§ 1. The Religious Tradition.

WE HAVE considered how the Hominidæ became more gregarious not by a diminution of individual lust, combativeness and self-assertion through variation and selection, but by the suppression of egoistic passions by countervailing inhibitions. Man's social evolution has been a mental process, through the development of a tradition of restraint upon impulsive conduct. Possibly there has been a selective preference for inhibitory types, so that now he restrains himself with increasing ease, but that must remain a guess. Nothing has been subtracted from man in the processes of socialization but something has been added and imposed. He is not a fierce animal that has become a weak one, but he is a fierce animal bridled and trained. He has been "caught young," he is now trained from the plastic days of childhood to control himself, and this observance of self-control and the rules that determine it, is in most savage communities regulated by the system known, broadly speaking, as *tabu*. The same carefully cherished infantile awe more highly elaborated and directed, the same organization of an inner moral conflict for the good of the community, becomes in the larger, more complex societies that dawn upon us in the Mediterranean region and east central Asia at the very beginning of history, the core of religion.

There remains some wide gaps in our direct knowledge of the evolution of human society. It is only in a few regions of the world that any systematic search for traces and vestiges of early man has yet been possible. We have still to piece together the stages by which man under favourable conditions passed from the phase of a casual feeder and hunter, not very definitely fixed to a definite place, to the condition of a settled cultivator. In Egypt, in Mesopotamia, we find him already settled more than seven or eight thousand years ago. But it is not certain that the early stages of this change-over from casual to economic living occurred in these regions. Presently we may find remains of these phases in the Near East or North Africa or elsewhere, or they may have occurred in lands now submerged and inaccessible to us— the basin of the Mediterranean, for example. At present we have to fill the hiatus with speculative matter.

FIG. 337. Paleolithic and Modern Savage Art.

Two clay figures of bison, about 2 feet long, in the Tuc d'Audoubert cave, in France. (From le Comte de Begouen, in "L'Anthropologie.")

On one side of the hiatus we have the later Paleolithic men, such as inhabited Spain and the South of France some twenty thousand years ago. They seem to have had a tribal organization of unknown range. They were wandering barbarians who had made very considerable upward progress from primordial savagery. The artistic value of their sculptures and their drawings upon rocks (See Figs. 337-9) is well known. The South African bushmen who make similar drawings to this day believe they have magic power, and that they cast a favourable spell over the hunted game they depict. In addition to his tabu system, the later Paleolithic savage had also, it would seem, a practical science and art, a fetish or magic system. We find the similar mingling of awe and rather badly reasoned practical magic in surviving savage communities, and it is reasonable to conclude that in the later Paleolithic period over most of their wide range the human tribes had arrived or were arriving at about the same level of mental development. Then on the near side of the hiatus, without any satisfactory connecting bridge, we find these early agricultural communities in the Neolithic stage.

There is nothing to indicate that these Neolithic peoples and culture developed directly from the later Paleolithic. These two cultures may have developed divergently from a common origin at the lower Paleolithic level. Except for a certain want of artistic freedom, the Neolithic peoples have got practically everything the later Paleolithic folk have. In addition they have carried their implement-making to a much higher and more polished level, they have domesticated and use a number of animals the Paleolithic people merely hunted and they practise a most elaborate and religious agriculture. How they attained to agriculture is one of the most fascinating riddles of human evolution. It may have been a very gradual process.

We can offer here only contributory considerations to the answering of that riddle. Certain salient facts have to be noted. First and most striking is the fact that everywhere there is the closest association of agriculture with sacrificial religion. To a modern mind that is very perplexing. The Neolithic cultivator did not simply prepare the ground with hoe or primitive plough and sow the seed. That was not enough. He also sacrificed one or more living beings and in the most primitive cases he made a human sacrifice. He made it with an elaborate ceremonial and the act was performed by a special person, the priest. The victim's body was partly eaten and partly distributed over the ground to be cultivated. Seed-sowing and sacrifice seem to have been equally important in the mind of the Neolithic cultivator. Neither was much good without the other. The survivals and vestiges of this seed-sowing sacrifice are found all over the world, wherever there are crops. That copious erudite work, Frazer's *Golden Bough*, traces this connexion. How did this association arise? We may guess but we cannot feel sure.

It is so much a matter of common knowledge now that the seed and its consequence, the plant, are connected, we are told of it so early in life, it is so woven now into the hackneyed metaphors of thought, that it is a little difficult for us to imagine intelligent adult Paleolithic savages to whom the idea was unknown. But there must have been a stage when it was still as unknown as the science of electricity. There still exist to-day savage peoples who are unaware of the connexion between the union of the sexes and the birth of children, and the connexion of seed and plant is at least as remote. Maybe the predecessors of the first agriculturalists wandered after food and at certain places men died or were killed in hunt or combat, and were buried. Early man, some of the varieties of early man, may have had cannibal habits. They ate part of a body to share in its strength; they buried the rest of the body if it was the body of anyone they respected, with honour and with supplies of food, quantities of grain. The grain may also have been scattered about the dolmen, the tumulus they were making. They would return later and find an unusually dense crop of their grain-bearing plant upon the site. They might believe this to be a special bounty from the departed. There was nothing to direct their minds to the fact that scattering the seed was the essential thing to get such a crop; it was not irrational for them to think the burial was the essential thing. "Bury a body with proper respect," they would reason, "and you get a crop of grain." It was not absurd for them to suppose that if a well-treated youth or maiden was specially killed and buried, a crop of food would ensue.

Always we have to make great efforts before we can hope to approach the primitive savage's conceptions. It is not only in the case of seed and plant that we are obliged to clean out all our fundamental

FIG. 338. A spirited sketch of a bison in the Niaux cave.

Such representations undoubtedly were used in hunting-magic. Three spear points have been drawn, piercing the bison's lungs. (From Cartailhac and L'Abbe H. Breuil in " L'Anthropologie.")

ideas, so to speak, before we can imagine the early savage's thought-process. In regard to death we find our minds stored with the accumulation of ages of thought, speech and tradition. But does an ape, did the early Hominidæ, know of death? You killed and you ate. Your companion was wounded and got better, or he was wounded and became immobile. Would he get up again? You left him and presently you saw him alarmingly in a dream. He was still alive, you inferred, but queer! Uncritical people brought up with elaborate beliefs in immortality, are apt to say that the burial of the dead by various species of the Homo with their weapons and provisions and women, shows that these early people had a belief in "immortality." It is, we suggest, much truer to say that they did not believe the deceased was altogether dead. They were left uneasy by his enigmatical behaviour. It was wiser to respect and consider him still. So in the case of chiefs and masterful people, you bewailed their loss, you implored them to aid you, you did not dare share out their weapons and treasures. You waked them and buried them with all that was by tabu untouchably theirs. And killing the sacrificial victim was not thought of as extinction; he was simply released to work in a new way for his slayers.

These are matters of the purest guesswork and supposition, and so, too, are the mental processes by which men ceased to be hostile to many of the animals about them. We are not stating proven facts here; we are offering suggestions. The fabulous stage of a child's imagination may give us some inkling of what went on in the mind of man when he crept close to the beasts. One may understand with a lesser effort how tribes following herds of wild cattle may have developed a proprietary sense, protecting them from wolves and other hunters and penning them into convenient valleys, and how the dog, playing much the same rôle towards man as the jackal does to the lion, may have grown imperceptibly to companionship. It is the most natural thing in the world now for men to drink the milk of their cattle, but there was a time when it must have seemed a strange, unnatural thing, a "beastly" thing to do. Our ancestor in that past had no trained observation, no lucid language, no logical method of thought; he thought experimentally in images and myths. In the fantasies of children and dreamland, the species recapitulates its mental growth; in traditional mythology we have the mental fossils of man's intellectual evolution.

Moreover, in these Neolithic communities of cultivators which appear on the dawn of history, the stars and the seasons, *the fact of the year*, have been discovered. An immensely clumsy astronomy has come into existence. Pyramids, obelisks, great standing stones, are being used to measure the altitude of the sun and determine the cardinal points of the year, from which the propitious times for seed-time and harvest may be deduced. We may hazard the opinion that in some level land open to sun and stars, where a periodic inundation was the vitalizing power upon the soil, men first observed the midday shadows growing shorter or longer as the floods drew near.

Speculation apart, the facts remain that in the Neolithic phase we begin to recognize the development of settled human societies, with the main features that we still trace in our present communities. The central fact of social life has become the altar; the directive force is the priest. Tabu, that is to say primitive moral control, and magic, which is primitive science, are now grouped about the directive priesthood, and an elaborate astronomy, fraught with worship, links the plough and the labouring beast and the sacrifice upon the altar with the constellations. On these stellar and mystical co-ordinations the uncertain prosperity of community and individual is understood to depend. There is a great fear of disturbing the order of things by any strange act or any negligence. If things go wrong then someone must have sinned against tradition. The sinner must atone for his sin in order that the majestic order of seed-time and harvest should continue.

Whatever were the precise steps man took, or whatever elaborate mental suppressions and elaborations made it possible for him to take them, the fact remains that in a period of five hundred thousand years or less, and mainly through the efficacy of this internal conflict, this system of suppressions, in making co-operation possible, *Homo sapiens* ceased, over favourable areas of the world, to be a casually living animal like all the rest of the vertebrata, and became an economic animal, foreseeing, domesticating, cultivating, storing and toiling as no other animal species had ever done. Tradition, the new invention of Nature, guided and controlled him and gave him an unprecedented security. It guided and controlled him, but never completely subjugated him.

Fig. 339. A drawing by a modern Australian native, of a kangaroo hunt.

By a curious convention, the animal's interior anatomy is shown. (From Baldwin Spencer's "Wanderings in Wild Australia," Macmillan & Co., Ltd.)

He was subdued to toil but he developed no instinct for toil. The history of his social development until the dawn of our own time is largely the history of an intricate interplay of the desire to evade and thrust off the incidence of toil, with the traditions, training and suppressions that made submission possible. So intricate is that moral and economic drama even in outline, that it can be dealt with satisfactorily only in a separate work, at least as extensive as this present summary of *The Science of Life*. For that it must be reserved.

And settlement and civilization were not the lot of all the species. In suitable regions it became settled, but over great areas of land where the grass was intermittent it roved with its cattle, and became nomadic. And in forests, tropical uplands and mountain regions it developed minor systems of living, minor traditions and mythologies. Tradition never attained complete uniformity or complete stability. The nomad reacted upon the agriculturalist and the agriculturalist on the nomad. They swapped stories and imitated methods. They never ceased from interaction and interbreeding. Of this interplay *The Outline of History* tells more fully than we can do here.

Obscure and difficult as the story of social development still is, there can be no question of the immense survival value of nearly every phase in the process of settlement. Man, from being a not very abundant species, sparsely diffused, began to multiply extremely in the regions best adapted to primitive cultivation. The human population of the world, which before may have been no more than a few score thousand, soon mounted far beyond the million mark in the great alluvial areas. We have already said something in Book 5 of the effect of climatic fluctuations on the phases of man's early development. The time is almost at hand when it will be possible to correlate the broad movements of human population both in the old and new worlds with the extensions and retreats northward and southward of forest and desert conditions. The way in which the Aryans came down on the earlier Aegean and Semitic civilizations, otherwise so inexplicable, becomes understandable when we realize that the forests and their way of life was coming with them. Fluctuations of the grass on the steppe lands drew and drove the nomads, accumulated energy for centuries and then sent it trekking. Agriculture adapted to flooding lands may have responded with more and more efficient irrigation systems as the world grew dry.

§ 2. The Passing of Traditionalism.

THE development of human societies was a development of traditions. Usage, justified by mythology, was the method of human association for scores of centuries. But the different conditions under which our widely diffused species was living in different regions of the world forbade the establishment of any uniform usage and mythology. By wars, raids, and the clash of traditions, the spirit of comparison, disputation, and inquiry was fostered. Undirected thinking gave place here and there in a few minds to a more sceptical, sustained and efficient process. The climatic fluctuations that brought the Semitic and Aryan-speaking peoples down upon the primitive civilizations opened the human mind to the possibility of alternatives in tradition. Men asked "What is truth?" Inventions in thought and method no longer awakened the same effective distrust and opposition. Plato's Utopias display the full-fledged realization that tradition could be set aside and a social order still exist. Aristotle embodies the release of the human mind towards new knowledge and power. These are the pioneers of modern thought and effort. How their initiatives lost force and were renewed again in the birth of modern Science, the historian must tell.

There is a very able and too little known book upon the Fijians by Sir Basil Thomson—a "study of the decay of custom," which illustrates very admirably the tentative and intermittent way in which progress has been achieved in the human past, and the new spirit in which man now faces his world. We thank Sir Basil for the liberty of quoting a striking passage from the book, so well does it say what we wish to express here. It is all the better that it says it at an angle of approach and with implications rather different from our own.

"The law of custom was the law of our own forefathers until the infusion of new blood and new customs shook them out of the groove and set them to choosing between the old and the new, and then to making new laws to meet new needs. This happened so long ago that if it were not for a few ceremonial survivals we might well doubt whether our forefathers were ever so held in bondage. With the precept —to do as your father did before you—an isolated race will remain stationary for centuries. There is, I believe, in all the history of travel, only one instance in which the absolute stagnation of a race has been proved, and that is the case of the Solomon Islands, the first of the Pacific groups to be discovered and the last to be influenced by Europeans. In 1568 a Spanish expedition under Alvaro de Mendana set sail from Peru in quest of the Southern continent. Missing all the great island groups Mendana discovered the islands named by him Islas de Saloman, not because he found any gold there, but because he hoped thereby to inflame the cupidity of the Council of the Indies into fitting out a fresh expedition. Gomez Catoira, his treasurer, has left us a detailed account of the customs of the natives and about forty words of their language. And now comes the strange part of the story. Expedition after expedition set sail for the Isles of Solomon; group after group was discovered; but the Isles of Solomon were lost, and at last geographers, having shifted them to every space left vacant in the chart, treated them as fabulous and expunged them altogether. They were rediscovered by Bougainville exactly two centuries later, but it was not until late in the nineteenth century that any attempt was made to study the language and customs of the natives. It was then found that in every particular, down to the pettiest detail in their dress, their daily life and their language, they were the same as when Catoira saw them two centuries earlier, and so no doubt they would have remained until the last trump had not Europeans come among them. . . .

"In the sense that no race now exists which is not in some degree touched by the influence of Western civilization, the present decade"— (the book is dated 1908)—"may be said to be a fresh starting point in the history of mankind. Whithersoever we turn, the laws of custom, which have governed the uncivilized races for countless generations, are breaking down; the old isolation which kept their blood pure is vanishing before railway and steamship communication which imports alien labourers to work for European settlers; and ethnologists of the future, having no pure race left to examine, will have to fall back upon hearsay evidence in studying the history of human institutions.

"All this has happened before in the world's history but in a more limited area. To the Roman armies, the Roman system of slave-owning, and still more to the Roman roads, we owe the fact that there is not in Western Europe a single race of unmixed blood, for even the Basques, if they are indeed the last survivors of the old Iberian stock, have intermarried with the French and Spanish people about them. An ethnologist of the eighth century, meditating on the wave upon wave of destructive immigration that submerged England, might well have doubted whether so extraordinary a mixture of races could ever develop patriotism and pride of race, and yet it did not take many centuries to evolve in the English a sense of nationality with insular prejudice superadded. Nationality and patriotism are in fact purely artificial and geographical sentiments. We feel none of the bitter hate of our Saxon forefathers for their Norman conquerors; the path of our advance through the centuries is strewn with the corpses of patriotisms and race hatreds.

"Nor was the mixture of races in Europe the mere mingling of peoples descended from a common Aryan stock, for if that were so, what has become of the Persians and Egyptians, worshippers of Aeon and Serapis and Mithras, who garrisoned the Northumberland wall; of the host of Asiatic and African soldiers and slaves scattered through Europe during the Roman Empire; of the Negroes introduced into Southern Portugal by Prince Henry the Navigator; of the Jews that swarmed in every mediæval city; of the Moors in Southern Spain? Did none of these intermarry with Aryans, and leave a half-caste Semitic or Negro or Tartar progeny behind them? How otherwise can one account for the extraordinary diversity in skull measurement, in proportion and in colour which is found in the population of every European country?

"If we except the inhabitants of remote islands probably there has never been an unmixed race since the Paleolithic Age. Long before the dawn of history kingdoms rose and fell. Broken tribes, fleeing from invaders, put to sea and founded colonies in distant lands. Troy was no exception to the rule of the old world that at the sack of every city the men were slain and the women reserved to be the wives of their conquerors. Doubtless it was to keep the Hebrew blood pure that Saul was commanded to slay 'both man and woman, infant and suckling' of the Amalekites, the ancestors of the Bedawin of the Sinai peninsula.

"It may be argued that the laws of custom have been swept away by conquering races many times in the world's history without any far-reaching consequences—those of the Neolithic people of the long barrows by the warriors of the Bronze Age; those of the British by the Romans; those of the Romano-British by the Saxons; those of the Saxons by the Normans. But there was this difference: in all these cases the new customs were forced upon the weaker race by the strong hand of its conquerors, and as it had obeyed its own laws through fear of the Unseen, so it adopted the new laws through fear of its new masters. It was a rough, but in the end a wholesome schooling. We go another way to work; we do not as a rule come to native races with the authority of conquerors; we saunter into their country and annex it; we break down their customs, but do not force them to adopt ours; we teach them the precepts of Christianity, and in the same breath assure them that instead of physical punishment by disease which they used to fear, their disobedience will be visited by eternal punishment after death—a contingency too remote to have any terrors for them; and then we leave them like a ship with a broken tiller free to go whithersoever the wind of fancy drives them, and it is not surprising that they prefer the easy vices of civilization to its more difficult virtues. In civilizing a native race the *suaviter in modo* is a more dangerous process than the *fortiter in re*...."

There an exceptionally penetrating colonial administrator, who was specially engaged in the problems of Polynesian human biology, gives an admirable summary of the mutual destruction of traditions. What he says of Fiji applies with appropriate modifications and differences of intensity to the whole world of mankind. Amidst this dissolution of traditions Science draws the lines for a fresh material and moral organization of our race.

§ 3. The Supersession of War.

PALEOLITHIC man in most of his varieties may have been a very combative creature as an individual, he was probably as fiercely territorial as many birds, but it is doubtful if anything of the nature of war had a place in human experience until these later stages in the Paleolithic record, when the social group had grown to the dimensions of a tribe. Then probably the collective hunting of big game and tribal bickering and warfare developed together.

The settled population of the Neolithic period were already familiar with warfare. Early sculptures from the Sumerian cities and from predynastic Egypt show us disciplined fighting men. War became an integral part of the social tradition. Conquest and the interplay of the warrior and priestly tradition supply the main themes of recorded history.

In Book 4 (Chap. 8, § 3) we discussed the evolution through natural selection within a species of structures and habits inimical to the species as a whole, but giving an advantage to the particular individual over his fellows. Such are many conspicuous sexual colours and noises, enormous antlers, exaggerated size and the like. The war-making disposition seems in many cases to have been a variation of this sort, which gave particular human tribes and communities a distinctive or dominating advantage over more peaceful societies. In the past, wars and conquests have no doubt done much to accelerate human progress by breaking up tradition-systems that threatened to become rigid and facilitating the establishment of larger and more efficient unities, but that was a phase out of which we have passed, and there can be little question now of the biological disadvantage which rests upon our species through its present preoccupation with war and war organization.

War prefers the healthier and more vigorous males for possible destruction at an age when the chances are against their having produced offspring; it misdirects and wastes a grave proportion of the none-too-ample directive and organizing ability of mankind, and its consumption of material resources even during that preparatory phase which we dignify with the name of "peace" is disastrous.

Inseparably associated with the habitual idea of war as a normal feature in life is the idea of the independent competitive sovereign state.

The picture of the universe in the minds of a vast majority of men and women is distorted by this idea of a necessary hostility to foreigners and the fear of any relaxation of the disciplines of the state to which they are devoted makes them obstruct every effort to release the new generation from its obsession with belligerent ideas. For many people who are adult and set, such ideas have become incurably a part of the mental structure. They cannot think of political and social questions except in patriotic forms. Yet the independent sovereign state tradition, which is really inseparable from and in part identical with the war tradition, cripples education at the present time, limits human freedom, hinders the development of a sane conservation and exploitation of the economic resources of the world, and is altogether so patently evil that it is impossible to believe that it will maintain itself for many more generations against the accumulating commonsense of mankind. A great cultural effort is certainly necessary, and a thousand intricate problems of tactics and strategy must be solved before human education can be turned away from its traditional prejudices, but the experiences of the last hundred years of release and restatement give good grounds for confidence that the thing may be done.

§ 4. The Change in the Nature of Education.

WE HAVE said that man, like many of the higher mammals and much more than any of the others, is an educational animal. But education for him, as for them, has been until quite recent times the imparting of tradition, the building up of his systems of association upon traditional lines. In education the human young learnt the wisdom of its forefathers. Education was an entirely conservative force; it functioned to preserve the traditional state of affairs. So it is still over large parts of the world. So it is wherever it is under the direction of religious bodies who maintain a view of the universe which they believe to be final. So it is in the completely self-satisfied atmosphere of a typical English public school. But in quite a little space of years the conception of education in many progressive minds has undergone the most revolutionary developments. The introduction of scientific work has infected even the most dogmatic centres with a sense of intellectual incompleteness. Even the most traditional education glances now ever and again, almost unwittingly, towards the future. Instead of "forming minds" and "moulding character" to a certain pattern fitted to a definite rôle and then turning out the completed product to astonish the world, the educational machinery of to-day begins at least to think of its function as a preparation for adventure, experiment and learning that will continue throughout life.

Education, from the modern point of view, consists of four chief factors. There is, *one*, the training of all the individual faculties to as high a level as possible, speech, drawing, the full use of hands and body generally; *two*, the development of a persona and of the self-knowledge and the practical psychological commonsense necessary for happy personal conduct and the filling of a distinctive rôle in life; *three*, the establishment of a picture of the universe in accordance with reality, the realization of the great adventure of humanity and of a personal rôle in that drama; and *four*, the special technical training and experience needed for the due enactment of the individual rôle. Or inverting the order, education should aim to make of each individual a good versatile artisan—versatile, for conditions change—a good citizen consciously playing a part in a general scheme, and a well-disposed, considerate, amiable person in full possession of all his or her powers. And throughout life there should be a persistence in educational adaptation. The dismal old-fashioned idea that one learns all that one has to learn before adolescence and then works out the consequences, as many animals do, is fading out of human thought. Adult education and self-education become duties in the modern state, and sloth, as we pointed out in our chapter on Conduct, a deadlier sin than ever. The modern world has less and less use for men and women who have ceased to learn.

It is not within the scope of this book to deal with educational method and still less with educational organization. The work of the behaviourists is manifestly bound to have a profound influence on early training and we have shown by example how the problems of conduct change as the light of psycho-analysis is brought to bear upon them. General education has become a function of the modern state. The whole of education does not go on in schools even during the school years, but the purpose of state education is plainly to equip all its citizens, or all the citizens that matter, with the means of understanding and participating in a collective aim. It is natural for the state to seek to justify and establish itself in the minds of its future subjects. In a world of militant independent sovereign governments, it follows that the inculcation of patriotism and a military spirit will be a constant preoccupation with the directors of the people's schools. But as we have pointed out in the previous section, the division and wastage of human effort by a multiplicity of independent potentially belligerent governments may be a transitory phase in human affairs which is now approaching its end. The struggle to bring it to an end will be necessarily an educational struggle. For the next few centuries that struggle will determine the main forms of intellectual life. There will be much instinct, much vigour and passion on the nationalist side, but in every country the nationalist side will be telling a different story, while all over the world men imbued with the scientific spirit and a realization of historical values will be working for identical ends. Social, economic, political and intellectual progress will be different aspects of the same process.

§ 5. The Breeding of Mankind.

WE HAVE already called attention in this work to the fact that *Homo sapiens*, or at any rate, his civilized variety, is peculiar among other animals in the lesser pressure of natural selection upon him; he is not being killed off to the same extent as most creatures. He has so fenced himself about with security, so increased his food supply and defeated his enemies, that for a time at least he is relieved from those searching destructive tests which in the case of most species under normal conditions maintain a certain numerical stability. The human population of the world is increasing very impressively. In Book 6 (Chap. 5, §7) we wrote of the "breeding storms" that will occasionally upset the balance of this or that biological community, when conditions are favourable, and produce a vast excess of some particular creature. Mankind seems to be going through such a breeding storm now, the favourable conditions being provided by his new-found control over nature.

According to Professor Carr-Saunders in his excellent little book on Population, there now exist one thousand seven hundred million human beings. A century ago there was not half this number. The present rate of increase, he estimates, is about one per cent per annum. Such a rate would in five hundred years' time give a world population of two hundred and fifty thousand million. This means that all over the globe there would be five hundred times as many people as you now find in such congested regions as England or Belgium. And a not very remote prolongation of our arithmetic would bring us to a time when all the land and shallow water of the world would scarcely afford standing room for mankind, much less space for food growing. Manifestly something will happen long before that state of affairs arrives.

One thing indeed is happening now, a growing voluntary restraint upon increase. We have already mentioned what is called birth-control in Book 8 (Chap. 8, §6) and indicated the essence of its methods. Knowledge of them is being diffused and through the spread of this knowledge and of the ideas that promoted that spread, and through a rise in the age of marriage, an effective check on human multiplication is coming into play. The rate of increase has fallen and continues to fall in just the countries in which increase was most conspicuous half a century ago. From 1870 to 1880 the annual English birthrate averaged 35.4 per thousand living; in 1928 it was 16.7—less than half as great. The rising birthrate turned and began to fall traceably from the time that birth-control methods became reasonably effective and the knowledge of them became widely disseminated. This breeding storm has certainly passed its maximum in the United States and all the Western European countries.

There seems to be an irresistible seeping of birth-control knowledge now throughout the world, even where all outward discussion of birth-control and all facilities for its practice are forbidden. There is a direct correlation of the standard of life with the birthrate, and no population to which these ideas and methods have come has failed to respond to them. Even in blackest Italy, in spite of the most vigorous formal suppression of birth-control propaganda, the eloquent exhortations of the Duce to Italian womanhood and the threatenings and outrages of philoprogenitive Fascisti spread the suggestion, and the birthrate falls —under conditions of dingy concealment, no doubt, and with much mental trouble. For some generations, and more and more universally as the modern ways and conceptions of life that are known as Westernization spread about the world, the birthrate will probably continue to fall, and it is quite possible that there may even be a marked diminution of the total human population in the concluding phases of the process.

That we cannot certainly prophesy. But it seems at least probable that this present human breeding storm will not have those tragic consequences in pestilence and famine that follow naturally upon the breeding storms of rodents and birds. There is enough reason to suppose that our modern productive organization, running at full time, can, for another century or more, support an even denser world population than now exists at a fairly high standard of life. But if a final disaster is to be averted, the slowing down of the increase must continue until the increase ceases. Manifestly, long before man arrives at the pitch of crowding we have anticipated, other restrictive forces will come into play. As we have already suggested in our sections on

Ecology, he is living at present beyond his income. He is using up capital. Coal, many necessary metals, and above all phosphorus, are being used up and not replaced. High prices and hard times will come in an intensifying form long before that phase of "standing room only," and force this problem into the focus of human intelligence.

This breeding storm in which we are living now, unlike all other breeding storms in the world of life, may pass without a subsequent massacre. And here again the abnormality of human biology forces itself upon us. Man it seems is breeding now without selection. And if presently the problem is tackled and the increase checked effectively, the phase of severe selection that is the normal sequel to a breeding storm may never occur. In which case, with a cessation of selection there will be no further biological progress.

Confronted with this possibility, liberal thinkers are accustomed to reply that selection will still go on but that it will no longer be by killing but by relative breeding. That may or may not be so. Certain types, it seems natural to assume, will breed more abundantly than others. The most vital issue from the point of human biology is the question of what types will breed most, whether they will be the types most helpful in the progressive development of the world community and, if not, what measures are possible, advisable and desirable to replace nature's method of selection-by-killing by an alternative method of selective reproduction—Eugenics: that is to say, the preferential breeding of the best.

It is customary to speak of negative and positive eugenics. Negative eugenics is the prevention of undesirable births. Positive eugenics is the promotion of desirable births. Except in so far as the private judgments of young people about to marry are concerned, no attempts at positive eugenics are traceable in the world about us. Most young people about to marry seem to find an ounce of flattery, or a trustworthy investment list, more directive than any eugenic ideas. But negative eugenics is already in operation. In several American states surgical sterilization —a very slight operation, the ligaturing of the oviduct or the vas deferens—is performed upon various types of mental defectives incapable of self-control. Six thousand such operations have been performed in California alone and it would be difficult to find fault with the results. The reader will find an up-to-date account in Gosney and Popenoe's *Sterilization for Human Betterment*. That there is a pressing need for such negative eugenics in the Atlantic communities, due to the steady elimination of death selection from human conditions, is shown by the British Board of Education Report of the Mental

Deficiency Committee (1929). This records an increase of one hundred per cent in the defectives of Great Britain between 1906 and 1927, while the population as a whole has increased only fourteen per cent. At present there are close on ten thousand certifiable defectives in every million of the English population. The birthrate of defectives it seems has not risen, but the defective children have been better taken care of and have survived. So that they in their turn are capable of parentage.

Apart from this traceable (and easily controllable) increase of idiots and imbeciles, there is very little evidence that any change in the average human being is now going on. There is a considerable amount of talk about the rapid multiplication of the unfit, but underlying this there is an assumption of the eugenic superiority of the more prosperous classes over the artisan and labourer mass. It is doubtful if there is any such superiority. People talk of "the ladder of social opportunity" lifting brains and energy, and of stupidity and shiftlessness sinking to the slums. But on the other hand it may be argued that a quite stupid acquisitiveness and meanness may accumulate wealth, and that a large proportion of business successes are merely the lucky ones among a multitude of gamblers. Few great scientific discoverers have died rich, and the best human brains have quite other preoccupations than getting or getting on. Advocates of the universal diffusion of birth-control knowledge argue that shallow, self-indulgent and mean-spirited types will be tempted to eliminate themselves altogether. How far self-indulgent people fall into one or several general and distinguishable types is not known.

In a little while it may be possible to handle these issues with exacter definition and much more confidence. All those who have had experience of birth-control work in the slums seem to be convinced that there is a residuum, above the level of the definable "defective," which is too stupid or shiftless or both to profit by existing birth-control methods. These "unteachables" constitute pockets of evil germ-plasm responsible for a large amount of vice, disease, defect, and pauperism. But the problem of their elimination is a very subtle one, and there must be no suspicion of harshness or brutality in its solution. Many of these low types might be bribed or otherwise persuaded to accept voluntary sterilization.

But while there is no evidence of any alarming uncontrollable degeneration of *Homo sapiens*, there is still less indication of any modification to adapt him to his extraordinarily changed and changing circumstances. That is the most perplexing aspect of human reproduction. Positive eugenics remains a dream, a note of interrogation. We have the rapid development of novel political, social, and economic arrangements with which the ordinary man does not keep pace. Science has vastly increased the destructive possibilities of war, and there is no sign yet of any effective restraint upon war possibilities. The inventions and organizations that have produced the peculiar opportunities and dangers of the modern world have been the work so far of a few hundred thousand exceptionally clever and enterprising people. The rest of mankind has just been carried along by them, and has remained practically what it was a thousand years ago. Upon an understanding and competent minority, which may not exceed a million or so in all the world, depends the whole progress and stability of the collective human enterprise at the present time. They are in perpetual conflict with hampering traditions and the obduracy of nature. They are themselves encumbered by the imperfection of their own trainings and the lack of organized solidarity. By wresting education more or less completely from its present function of transmitting tradition, they may be able to bring a few score or a few hundred millions into active co-operation with their efforts. Their task will still be a gigantic one.

We are only able to guess at the amount of undeveloped capacity that goes to waste in each generation. There will certainly remain a considerable proportion of mankind, incapable it seems of being very much educated, incapable of broad understandings and co-operative enterprise, incapable of conscious helpful participation in the adventure of the race, and yet as reproductive as any other element in the world community. For a number of generations, at any rate, a dead-weight of the dull, silly, under-developed, weak and aimless will have to be carried by the guiding wills and intelligences of mankind. There seems to be no way of getting rid of them. The panics and preferences of these relatively uneducatable minds, their flat and foolish tastes, their perversities and compensatory loyalties, their dull, gregarious resistances to comprehensive efforts, their outbreaks of resentment at any too lucid revelation of their inferiority, will be a drag, and perhaps a very heavy drag, on the adaptation of institutions to modern needs and to the development of a common knowledge and a common conception of purpose throughout mankind. Obsolescent religious forms and plausible political catchwords will be used to rally and canalize their mental weaknesses. The brighter, more energetic types of stupidity and egotism will be constantly organizing and exploiting the impulses and uneasinesses of this universally diffused multitude. For here we are not writing of any social class or stratum in particular. The inferior sort is found in greater or less abundance at every level.

Possibly mankind will find that positive eugenics is unattainable and undesirable, and what we have said about the differences in minds in Chapter 7 (§ 14) of Book 8 seems to promise the possibility of an upward extension of negative eugenics by the recognition of definably inferior intelligences. If these can be detected and set aside so that, without humiliation or other cruelty, they can be debarred from breeding, then presently the mass of mankind will begin to follow its leaders up the scale of understanding.

The struggle of intelligent and energetic minds throughout the world to clear out their own lumber and get together for the conscious control of the affairs of the strangely mingled multitude of our kind, to develop the still largely unrealized possibilities of science and to organize a directive collective will, is the essential drama of human life. All other great human events, wars, epidemics, revolutions, strange fashions of living and the like, are by comparison either phantasmal or catastrophic or both.

§ 6. The Superfluous Energy of Man.

THE same forces that have lifted *Homo sapiens* for a time from off the grindstone of natural selection and allowed all types to multiply, give him also a redundancy of leisure and energy far beyond that vouchsafed to any other creature. In Book 8, we discussed the sportive play of animals, which goes on when they escape for a time from the urgency of life. Modern man has developed sportive activities to a quite distinctive degree. Here we cannot make more than an allusion to this development which manifests itself to Mr. Everyman on every hand, in his newspaper, in his conversation, in his participations. Essentially the race-meeting or the dance is the highly organized and regularized equivalent of those rare community games, the flying sports of rooks and the joy-rides of penguins we have already described.

Another outlet for human energy finds its anticipation in the decorative efforts of buzzards and wading birds described in that same section. Man's excess of energy over his material requirements finds a vent in a multiplicity of imitative and creative activities which we lump together as the Arts. It is not for three modest writers about biology to venture into the cloudy and stormy realm of art-criticism and decide what is Art and what is not; our interest here is strictly confined to the biological meanings and possibilities of the Arts. As we pointed out in our Introduction, biological science itself was created and revived by collectors and artists, and from our point of view there is hardly a dividing line between a man who vents himself in carving stone to represent lovely forms, or in arranging beautiful sounds in beautiful patterns, and one who uses his mind to experiment interestingly with living forms or to pursue thought into its remotest recesses. The impulse in all these instances is akin to sportive play. It is the sportive play of the brain and hands, eyes and ears. It may discover itself to be immensely important, but to begin with it was no more than surplus energy seeking an outlet.

And just as in that earlier section on the play of animals we found it almost impossible to say where sportive play ended and practical training and serviceable experimenting began, so here we have to recognize the stupendous and still increasing value of both sport and art in exploring the possibilities of human effort, feeling and desire. It is difficult to believe that as the ever-increasing productivity of the social organization increases human leisure, the already vast developments of sport, art, creative literature and scientific research will not continue to absorb a larger and larger proportion of the total output of human energy.

§ 7. The Possibility of One Collective Human Mind and Will.

THE progressive development of human inventions, the onset of power production and the present rapidly extending and unifying economic organization of mankind, are subjects that cannot be properly dealt with in a general review of Life; they belong in part to general history and in part to descriptive economics. But here we may say a few words upon the subject of the development of that conscious unification of the human species which is now going on very rapidly. And again we face something quite unparalleled elsewhere in the entire realm of biology.

Other species of animal seem to have an individual conscious existence limited strictly to their individual experiences, but with the dawn of tradition the human mind began to extend itself in time and space beyond the individual range. The difference between the human mind and the mind of a chimpanzee is infinitely greater than the difference of the bodies or brains of the two; it is as different as a bird is from a snake; it moves in more dimensions. At present a human mind, fully developed by education and inquiry, reaches so far and so wide that individual experience is a mere point of departure for its tremendous ramifications. What it has of its very own is altogether dwarfed by what it has in common with other individuals of the species. Just so far as a human mind is well informed and soundly instructed, so far is it able to understand, that is to say to identify itself with, other well-informed and soundly instructed minds. By means of books, pictures, museums and the like, the species builds up the apparatus of a super-human memory. Imaginatively the individual now links himself with and secures the use of this continually increasing and continually more systematic and accessible super-memory. The human mind neither begins nor ends therefore with the abruptness of an animal mind. As it grows up, it takes to itself more or less completely the growing mental life of the race, adds a personal interpretation to it, gives it substance and application, and in due course fades out as an individuality, while continuing in its consequences as a contribution to the undying flood. From the point of view of the species, the consciousnesses of men are passing trains of thought and impulse. They are now as much part of a larger life as the perception of the sunlight on this sheet of paper and of the singing of a bird outside are parts of the life of the writer of this sentence. They are material and enrichment. It is not a metaphor, not an analogy; it is a statement of fact that this larger comprehensive life is going on.

In this work we have traced a long process of synthesis from the single cell to the multicellular organism and from the coelenterate to the coelomate. We have seen the interdependence of individuals in space increase with the development of colonial and gregarious forms and of individuals in time with the growing care and intimacy of parent for young. The higher forms of interdependence have involved great extensions of mental correlation. We have shown how human social economy is based almost entirely upon the mental modifications of the individual and how little it owes to instinct. This mental modification is steadily in the direction of the subordination of egotism and the suppression of extremes of uncorrelated individual activity. An inflation of the persona has gone on, so that the individual has become tribal, patriotic, loyal, or devotee. *Homo sapiens* accommodates this persona, by which he conducts his individual life, to wider and wider conceptions.

The more intelligent and comprehensive man's picture of the universe has become, the more intolerable has become his concentration upon the individual life with its inevitable final rejection. No animal, it would seem, realizes death. Man does. He knows that before his individuality lies the probability of senility and the certainty of death. He has found two alternative lines of accommodation.

The first is a belief in personal immortality, in the unendingness of his conscious self. After this life, we are told, comes the resurrection—and all necessary rejuvenescence. This idea is the essential consolation of several of the great religions of the world. We have already discussed its credibility.

The second line of accommodation is the realization of his participation in a greater being with which he identifies himself. He escapes from his ego by this merger, and acquires an impersonal immortality in the association; his identity dissolving into the greater identity. This is the essence of much religious mysticism, and it is remarkable how closely the biological analysis of individuality brings us to the mystics. The individual, according to this second line of thought, saves himself by losing himself. But in the mystical teaching he loses himself in the Deity, and in the scientific interpretation of life he forgets himself as Tom, Dick, or Harry, and discovers himself as Man. The Buddhist treatment of the same necessity is to teach that the individual life is a painful delusion from which men escape by the conquest of individual desire. Western Mystic and Eastern Sage find a strong effect of endorsement in modern science and in the everyday teaching of practical morality. Both teach that self must be subordinated; and that self is a method and not an end.

We have already, if this account of mental processes is sound, the gradual appearance of what we may call synthetic super-minds in the species *Homo sapiens*, into which individual consciousnesses tend to merge themselves. These super-individual organizations have taken the form of creeds, communities, cultures, churches, states, classes, and such-like accumulations of mentality. They have grown and interacted in the history of the species very like the complexes of an individual human mind. They seem to have now under current conditions a ruling disposition to coalesce. They seem to be heading towards an ultimate unification into a collective human organism, whose knowledge and memory will be all science and all history, which will synthesize the pervading will to live and reproduce into a collective purpose of continuation and growth. Upon that creative organization of thought and will the continuing succession of conscious individual lives, drawing upon and adding to its resources, will go on. At the end of our vista of the progressive mental development of mankind stands the promise of Man, consciously controlling his own destinies and the destinies of all life upon this planet.

But note these words we are using, "seem" and "promise." This is no assured destiny for our kind. The great imperfect conflicting collectivities of to-day, swiftly as they have developed and wonderful as they are in comparison with all other animal life, may never become a unity. Man may prove unable to rid himself of the over-development of war; he may be hindered too long by the dull, the egoistic and the unimaginative, by the stupid, timid, and tradition-swayed majority, ever to achieve an effective unity. The dead weight of inferior population may overpower the constructive few. Or the incalculable run of climatic changes may turn harshly against him. Strange epidemics may arise too swift and deadly for his still very imperfect medical science to save him from extirpation. There is no certain assurance that rats and mice, dogs gone wild again, prowling cats, flies, and a multitudinous vermin may not presently bolt, hide, and swarm amidst the decaying ruins of his cities. Shoals of fish may dart in the shadowy encrusted wreckage of his last lost ships. We have no assurance that so *Homo sapiens* may not end. But such an end is hard to believe possible.

§ 8. Life Under Control.

ON THE whole we believe that our species will survive and triumph over its present perplexities. There is much in life that may make intelligent men impatient, but it is not reasonable to let impatience degenerate into pessimism. Vulgar fashions, false interpretations and decaying traditions make a vast show and noise in the world; the crowd is always about us; but we forget that these things are divergent and inconsecutive and accumulate no force, while scientific work and lucid thought are persistent and cumulative. They remain and reappear when the shouting dies away. The progressive development of the scientific mind may survive all the blundering wars, social disorganization, misconceptions and suppressions that still seem to lie before mankind. Until in due course the heir comes to full strength and takes possession. But he will survive only on one condition, and that is that he must take control not only of his own destinies but of the whole of life.

By that time the body of modern science will be enormously greater and more closely knit than it is to-day, and at the most we can only point out the drift of constructive thought and power as it manifests itself at present. We are almost driven to believe, from a brief contemplation of the advances made in the past third of a century, that scores of unsuspected fundamental new possibilities and hundreds of shortcuts to now almost inaccessible ends are bound to present themselves as the work goes on. The immense possibilities of mechanical reconstruction that open out to man are outside the scope of such a work as this, except in so far as they involve changes in environmental conditions. Such well-informed and ingenious writers as Mr. J. L. Hodgson (*The Time Journey of Dr. Barton*), or many of the little volumes in the *To-day and To-morrow* series, will supply the reader with plentiful food for his imagination in these matters.

It is fairly plain that the acceleration of transport and communication is still in progress and shows no certain limits. The facilities for every form of mental exchange and interaction increase. The thrusts that tend to interlock the economic life of the species into one system, and which strain the traditional political network to the breaking-point, increase. Traditionalism and its inseparable attendant, war, remain huge and dangerous powers in human affairs; they must be denounced, written about and fought and defeated, and they will claim their toll of martyrs, but so soon as they have been thrust out of action, if ever they are thrust out of action, men will find it difficult to realize how great and terrible they were. They may wax old and shrivel away. The plans for the exploitation of the earth may be released almost insensibly from the entanglement of frontiers. The deserts of the world may be irrigated and its changes of weather and season foretold. It may become a planetary farm and garden, a playing field, a fishpond, paddock, workshop and mine. It is possible that there will be a considerable shrinkage then of the areas devoted to food-production. Not that there will be any diminution of the food-supply; but there is no reason to suppose we have yet reached anything like the full possibility of yield per acre. Our chapters on Ecology supply the fullest justification for a forecast of an intensive cultivation demanding only a fraction of the space and toil now given to food-production.

Concurrently with this shrinkage of the food-growing areas there may be a considerable fall in world population. At present the full extent of the decline of the birthrate of most civilized countries is masked by the prolongation of the average life due to better hygienic conditions. If there is a limit to the latter process, and at present there seems to be one, then a real fall in total population will presently become apparent. This may go on for some time, and it may involve the elimination of types unwilling to bear and rear children. To have a world encumbered for a time with an excess of sterile jazz-dancers and joy-riders may be a pleasanter way to elimination than hardship and death. Pleasure may achieve what force and sword have failed to do. The world can afford it; it is not a thing to fret about. It is only a passing fashion on a grand scale, this phase of sterilized "enjoyment." The great thing is that it should be able and willing to sterilize itself. We may expect an increase in the gravity and sense of responsibility of the average sort of people even in the course of a few generations because of this elimination. The types that have a care for their posterity and the outlook of the race, will naturally be the types which will possess the future.

But these are only the opening sentences of the next chapter of human biology. The fall and recovery of populations, the politico-economic unification of human affairs, may present phases of intense stress and tragedy, periods of lassitude and apparent retrogression, distressful enough for the generations that may endure them, but not sufficient to prevent the ultimate disappearance of misleading tradition and the dominance of a collective control of human destinies. And by that time biological science will be equipped with a mass of proved and applicable knowledge beyond anything we can now imagine.

At present eugenics is merely the word for what still remains an impracticable idea. But it is clear that what man can do with wheat and maize may be done with every living species in the world—including his own. It is not ultimately necessary that a multitude of dull and timid people should be born in order that a few bright and active people should be born. That is how things have to be to-day, but it is an unnecessary state of affairs. Because at present our knowledge of genetics is too limited to do more than define certain sorts of union as "undesirable" and others as "propitious," it does not follow that we shall always be as helpless. There may come a time when the species will have a definite reproductive policy, and will be working directly for the emergence and selection of certain recessives and the elimination of this or that dominant. In our treatment of genetics we have given a few first-fruits of the science, which suggest what forms the practical eugenic work of the future is likely to take. Once the eugenic phase is reached, humanity may increase very rapidly in skill, mental power and general vigour.

And it is not only human life that human knowledge may mould. The clumsy expedients of the old-time animal and plant breeder will be replaced by more assured and swifter and more effective methods. Of every species of plant and animal man may judge, whether it is to be fostered, improved or eliminated. No species is likely to remain unmodified. Man's protective interference goes far to-day, and it may extend at last to nearly every life-community. Perhaps no man has yet imagined what a forest may some day be, a forest of great trees without disease, free of stinging insect or vindictive reptile, open, varied and delightful. The wilderness will become a world-garden and the desert a lonely resort for contemplation and mental refreshment. An enormous range of possibility in the selective breeding of plants and animals still remains to be explored. One may doubt the need to exterminate even the wolf and tiger. The tiger may cease to be the enemy of man and his cattle; the wolf, bred and subdued, may crouch at his feet.

Consider what man has learnt to do with plants—quite apart from mere selective breeding, the picking and approving of Nature's creative experiments. Muse for a moment on the idea of a graft. A bit of one kind of plant is joined, welded, one flesh with another. A thoroughly unnatural thing, which never happened before the human cultivator appeared, is brought into existence. Or think of the astonishing graft-hybrids; the living skin of a lemon tree enclosing the middle of an orange tree, growing together, one flesh.

It is perfectly possible that man will do these things with animal bodies in the near future. Already he is learning to handle the difficult, delicate material. The tissue cultivator takes bits of muscle or nerve or kidney and grows them for years in his incubators, quite isolated from the rest of the body. The plastic surgeon cuts a splint from the tibia and welds it into a broken lower jaw, or takes skin from the buttocks to mould an eyelid. Soon we shall be doing much more tremendous surgeries. It may even prove possible to operate directly on the germ-plasm, for the geneticist can already produce mutations by means of X-rays. Man has conquered the hardness of steel; he cuts and twists it and builds with it as he pleases; to-day he is learning a new art, with living protoplasm as his medium.

A quality of fantasia comes into our writing as we follow up these possibilities. Yet it was not fantasy but hard fact that brought us to this point. Arising out of the thought and effort of to-day, it is plain that human achievement marches on to fresh powers and fresh vistas— until our utmost imagination is strained and exhausted. We are dazzled by the conquests we deduce; we laugh; our minds gasp like newborn children when they first meet the free air.

And will the personal life in these coming ages of man's complete ascendancy be as happy and exciting as it can be to-day? In that great age the subordination of self will certainly play a part. But the subordination of self is not by any means the same as self-sacrifice. The individual life may be infinitely richer as a part than as a whole; the whole sustains and inspires its members; experiences we cannot dream of may lie before our descendants. Great and wonderful and continually expanding experiences lie before life, intensities of feeling and happiness we shall never share, and marvels that we shall never see. Yet we need not envy that ampler life. With a necessary but quite practicable effort of self-control and self-subordination it has been possible for many of us, in our own time, within our phase and in our measure, to live intensely interested and happy individual lives. Even to those to whom the scheme of things has turned an adverse face, courage can give its own high and stern satisfactions. For the stoicism of the scientific worker at any rate, there can be no complete defeat. And these mightier experiences and joys of the race to come will be in a sense ours, they will be consequence and fulfilment of our own joys and experiences, and a part, as we are a part, of the conscious growth of life, for which no man can certainly foretell either a limit or an end.

THE END

INDEX

Figures in bold-faced type indicate the pages on which illustrations may be found.

A

Aard-Vark, **384**.
Abdomen, 36, 37; of lobster, 194, 195.
Abyssal zone; life in, 846. *See* Deep sea.
Acartia, **991**.
Achatina, **940**.
Acheulean culture, 817-818.
Acquired characters, inheritance of, 581-592.
Acromegaly, 165.
Actinozoa, 247.
Activation, 150.
Adamsia, 930.
Adaptation to habitat, 835-838, **869**.
Adaptive radiation, 743.
Adirondacks, 1021.
Adler, Alfred, 1364, 1365, 1366, 1368, 1393, 1407.
Adolescence, 161-162.
Adrenal gland, 104, 162, 165, 166, 574, 1207.
Adrenin, 162, 166.
Æpyornis, 446, **937, 940**.
Afghanistan, 895.
Africa, 334, 376, 388, 390, 396, 398, 400, 733, 785, 812; as possible region of man's origin, 802.
African mammals, typical, **384**.
After-birth, the, 159.
Agave, 897.
Age, old. *See* Senility.
Ages of rocks, 419.
Agglutinins, 1048.
Agouti, 389.
Agriculture, and sacrificial religion, 1456; applications of modern biology to, 496-497, 501-502, 928, 1016-1032; importance of bacteria in, 303.

Ague, 1040.
Air, and sunlight, 1076-1088; as a habitat, 830-831; conquest of the, 748-756; of a stuffy room, 1081-1084. *See also* Respiration.
Alaska, 779, 809, 814, 887.
Albatross, **937**.
Albinism, 488, 500, 501.
Albunea, 870.
Alcohol, use and abuse of, 1068-1072, 1074.
Alcoholism and heredity, 591-592.
Alcyonarians, **844**.
Alder, 984.
Aleppo, pine, 983.
Alfonso XIII of Spain, **490**.
Algæ, 285, 287, 288, 292, 293, 661, 701.
Alimentary canal. *See* Digestive system.
Alligator, 183.
Alligator-bug, 957.
Alps, 328, 654, 808, 809, 815.
Alternation of generations, in animals, 228, 238; in plants, 268-274, 702 *sqq.*
Amauris niavius, **953**.
Amazonian forest, 902-903.
Amber, 330.
Ambivalence in psychology, 1361, 1362.
Amblypods, 782, **778**.
Amboina, 1020.
Amherst pheasant, courtship of, 1226-1227.
Amia, 735.
Amino-acids, 74-75.
Ammonites, 632, 692, 735, **693**.
Ammophila, 976, 1159.
Amœba, 275-280, 662, **1105**; and dysentery, **943**; co-ordination of, 1119; compared with body cell, 279-280;

1481

movements of, 114; reproductions of, 280–281, 439, **278**; sensitivity to light, 1106.
Amphibians, 184–186, 368; brain and behaviour of, 1214–1218; evolution of, 722–729, 775–776, **693**, **740**; in coal-measure forest, 732; metamorphosis of, 185–186, 526, 527; naked skin of, 184–185; number of species of, 1218; poison of, 1115–1116.
Amphidasys betularia, 595.
Amphioxus, 189, 190, 191, 369, **869**.
Amphitrite, **224**.
Amphitritus, **864**.
Anaconda, **365**.
Anaërobic life, in bacteria, 302; in yeast, 292.
Anaphylaxis, 1079.
Ancylostoma, 227.
Andrias scheuchzeri, **323**.
Anemone, plumose, **248**.
Angiosperms, 267, 764–765.
Angler-fish, larva of, **849**.
Anhalonium, 1388.
Anilocra, 949.
Animal forest, **844**; light, 192, 860–862.
Animal magnetism, 1431.
Animals, classification of, 15–16, 168–169, 378–379; contrasted with plants, 256, 292, 301–302; dependence upon plants, 962; origin of, 660–661; sizes of, 936–939.
Annelida, 221–222, **1127**.
Anopheles maculipennis, **1042**.
Antarctic, 288, 294, 658, 719, 733, 908, 913.
Ant-eater, 176, 178, **386**, **391**, **778**.
Antedon, **220**, **370**.
Antelope, 178, **384**.
Antennophorus, 1176, **1177**.
Anthracite, 720.
Anthrax, 1048–1049; bacillus, 1036, **947**.
Antibodies, 359, 1048.
Antirrhinum majus, 621–622.
—molle, 621–622.
Antiseptic surgery, 1050–1053.
Antitoxins, 1047–1049.
Antitoxin treatment, 359.
Ants, 1162–1178, **363**, **942**, **1167**, **1173**. parasites of, 1174–1178, **1177**.

Anus, evolution of, 667–668.
Ape-man, 411–412, *facing* **774**.
Apennines, 654.
Apes, 179; behaviour of, 408–409, 1259–1269; compared with man, 406–411; embryos of, **413**; evolution of, 796–805, **778**; habits of, 1238–1239, 1440–1442; local races of, 1440–1441; skull of, *facing* **775**.
Apherusa, **991**.
Aphrodite, **224**.
Apotettix, 579, 733.
Appalachian mountains, 654.
Appendicularia, 843.
Appendix, 89, **87**.
Aquatic life. *See* Sea, Fresh Water, Rivers, Lakes.
Arabia, 396, 895.
Arachnida, 204.
Arachnocampa, 917.
Aral Sea, 998.
Archæopteryx, 330, 350–351, 651, 752, **349**, **740**.
Archegosaurus, **730**.
Archeocyathus, 678.
Archeozoic Era, 322, 324, 677, **325**.
Archipallium, 1225, 1282.
Architeuthis, **937**.
Arctic, 294, 908; fox, 908; hare, 908; ocean, 658, 810.
Arenicola, **224**, **672**.
Argentina, 815.
Argus pheasant, 1237.
Argyropelecus, **864**.
Aristocystis, **687**.
Aristotle, 5, 15, 861, 1452.
Arizona desert, fertility of, 969.
Armadillo, 178, 388, **778**; nine-banded, **391**; Texas, 441.
Arrow-worm, 673, **676**, **991**.
Arsenic, sweets and beer contaminated with, 1066.
Arsinoitherium, 785, **778**.
Art, modern savage, 1455; of early men, 1455.
Artemia, 918.
Arteries, 35, 160, **54**, **57**.
Arthrodires, 700.
Arthropoda, 194 *sqq.*, 805; behaviour of, 1147–1151; contrasted with verte-

INDEX

brates, 194; evolution of, 669, 678, 693; relationship with worms, 209, 670.
Articulates, 717.
Artiodactyla, 178.
"Aryan race," 1449.
Ascaris lumbricoides, 225.
Asellus, 917, 920.
Asepsis, 1051.
Aseptic surgery, 1050–1052.
Asexual reproduction. See Reproduction.
Ash, 984.
Asia, 777, 789, 900, 1057; Minor, 340, 811.
Aspidium, **269**.
Ass, 178, 334.
Aster, 979.
Asterias, **220**.
Asterina, 218, **217**.
Asteroxylon, 706, **706**.
Astrapotheres, 394, **778**.
Astronomy, early, 1458.
Astroscopus, 1118.
Atelura, **1177**.
Atemeles, **1177**.
Atlantic Ocean, 810–811, 857.
Atolla, 859.
Auk, great, 624.
Aulastomatomorpha, **860**.
Aurelia, 247, **248**, **1123**.
Aurignacian culture, 821.
Australia, 176, 249, 388, 389, 392, 393, 395, 677, 733, 777, 781, 785, 786, 895, 899, 913, 918, 997, 1021, 1240; how the marsupials colonized, **394**; mammals of, **386**; mouse plague in, 997.
Australopithecus, 801.
Autolytus cornutus, 237.
Automatism and mediumship, 1351–1353.
Autonomic nervous system. See Nervous system.
Avebury, Lord (1834–1913), 1185.
Axolotl, 527.

B

Babirussa, **778**.
Baboon, 1262, **384**, **778**.
Bacillus botulinus, 1065.
— coli, 1039.
— œrtrycke, 1065.
Bacillus of bubonic plague, 994, 1009; of cholera, 1036, **1041**; of diphtheria, 1037, **947**, **1038**; of tuberculosis, 1038–1039, **1041**; of typhoid fever, 1036, **1038**.
Backbone, 32.
Bacteria, 48, 299–306; and agriculture, 303–304, 928; and circulation of elements in nature, 305–306; and decay, 304–305; and disease, 303, 1033–1053; and fermentation, 304–1036; and phosphorescence, 304, 930; economic importance of, 304, 928, **304**; in digestive organs, 83, 89, 834; in food, 1065–1066; nitrogen-fixing, 305, 928; number of, in air, 302, 1079; resistance of body against, 92, 1038–1039; sexless reproduction of, 301, 439, 443; sizes of, **947**; spores of, 301, 438.
Bacteriophage, 308, **652**.
Badger, 178.
Baikal, Lake, 397, 892.
Balance of nature, 1012–1016.
Balanoglossus, 192, 356, **190**, **869**.
Baldwin, Prof. Mark, 589.
Balkans, the, 811.
Balloons, heights reached by men in, 7.
Ball-urchin, 688.
Baltic, 888.
Baluchithere, **778**, **784**.
Baly, Prof., 651.
Bandicoot, rabbit-eared, 389, **386**.
Banyan tree, 443.
Baobab, 142.
Barnacle, 203; larvæ of, **991**.
Barnard, J. E., 309.
Basal Eocene period, 781–782, 785.
Basilarchia archippus, 955.
Bastian, 5.
Bat, 178, **359**, **361**, **778**; gamete of, **447**; vampire, 388.
Batesian mimicry, 960.
Bateson, Prof. William, 433, **474**.
Bathothauma, **854**.
Bathypterois, **854**.
Batrachiderpeton, **730**.
Beadlet, 247.

Bean, Florida Velvet, 578.
Beans, selection in, 470.
Bear, 178, 386, 778; polar, 911.
Bearberry, 977.
Beardmore glacier, 735.
Bear island, 720.
Beaver, 1006, 1246-1247, 778.
Bechuanaland, 801.
Beech, 977, 984.
Bees, 1156, 1182-1199, 1186; flight of, 1192-1193; queen, 940, 942; sense of color in, 1189-1191; sense of smell in, 1190, 1195-1196, 1197.
Beetles, 1177, 1177.
Behaviour, 408; essentials of, in animals, 104-105; human, 133-139, 1318-1389; in plants, 1128-1131; instinctive and intelligent, 1131-1136; of insects and other invertebrates, 1147-1199; of monkeys and apes, 1259-1269; of the lower vertebrates, of the slipper animalcule, 1136-1143; rudiments of, 1102-1146.
Behaviourism, 1318-1325.
Behring Strait, 340.
Belemnite, 694, 693.
Belgian Congo, 313.
Belgium, 817, 975.
Bell-animalcule, 943.
Ben Lawers, 815.
Bennettites, 764.
Benthic zone, 846.
Bergson, Prof. Henri, 428, 433, 435, 638, 1114.
Beri-beri, 1060, 1062.
Biddulphia, 991.
Biffen, Sir Rowland, 496.
Big Tree, California, 937.
Bile, 84.
Bilharzia, 230.
Biological duties, primary, 1395-1397.
— economies, 961-962.
Biology, beginnings of, applied, 1022-1026; history of, 17-18.
Biosphere, 829.
Birch, 984.
Birds, 175, 180-182, 805, 937; and pollination, 768; brain of, 1221; courtship of, 1226 sqq.; education in, 1248; evolution of, 693, 740, 751; flight of, 751; flightless, 913; fore limbs of, 359; instinct in, 1225; play in, 1253 sqq.; social life of, 1255, 1256; song of, 1250, 1258; territorial instinct in, 1007.
Birnbaum, 7.
Birth, 154.
Birth-control, 428, 1469, 1470.
Bivalves, 211.
Blackberry, 1016.
Black Death, the, 1039, 1045.
— Sea, 811.
Bladder-wrack, gamete of, 447.
Blastostyle, 240, 242.
"Bleeder's disease," 563; pedigree of, 564; in European royal families, 563.
Blister-rust, 1021.
Blood, 35-37; cells of, 38 sqq.; clotting of, 51; course of, 52-59, 57; evolution of, 667-674; functions of, 48-49; of lobster, 196-197.
Blood-pressure and nicotine, 1069.
Blood-test, the, 360.
Blood-vessels, 35-36.
Blubber, uses of, 850.
Boar, wild, 778.
Body-mind, 1275-1277.
Body-soul-spirit, the theory of, 1411-1413.
Bones, evolution of ear-bones, 353; human, 33; minute structure of, 47; See Skeleton.
Boring Animals, 875-876.
Borrelia berbera, 1045.
Bose, Sir Jagadis Chandra, 259; experiments with plants, 13.
Botfly, 924.
Bothriolepis, 699.
Botulism, 1065.
Bow-fin, American, 735.
Brachiopods, 233, 233.
Brachydactylous hands, 488.
Bradley, Dennis, 1419.
Brain, 35, 128-130, 159-160; development of, 1203-1207; evolution of, 666-667, 1215; gray matter of, 136-137; human, 133-137, 1278-1287, 1279, 1286; main parts of, in vertebrates, 1203-1205; of amphibians, 1214-1218, 1206; of extinct reptiles, 760; of invertebrates, 1122-1128;

INDEX

of reptiles, birds and mammals, 1218-1226. *See* Nervous system.
Branchiocerianthus, **937, 940**.
Branchiosoma, **740**.
Brazil, 380, 955.
Bread, 1057-1058; leavening of, 291.
Breathing. *See* Respiration.
Brecklands of East Anglia, 1013.
Breeding of plants and animals, applications of modern knowledge to, 496-498; superstitions about, 508-513.
Breuer, Robert (1842-1925), 1363.
Bristle-worm, 678, 885, **672**.
British Columbia, 385.
— Museum (Natural History), 948.
Brittany, 816, 932.
Brittle-star, 219, 355, 688, **220**.
Broken Hill skull, 820.
Bronchitis, 1091.
Brontosaurus, 757, 761, **771**; fabled African, 313.
Brontotherium, 631.
Broom, purple, **455**.
Browne, Sir Thomas (1605-1682), 5.
Bryophyllum calycinum, **257**.
Bryophyta, 272.
Bryozoa, 671.
Bubonic plague, 1009; bacillus of, 994.
Buccinum undatum, **213**.
Budding, 440.
Buffalo, 388.
Bullfinch, 1250.
Bulrush, 978.
Burbank, Luther (1849-1926), 497, 622.
Burchell's zebra, **384**.
Burma, 975.
Bursaria, **942, 943**.
Bush turkey, 1225.
Bustard, great, 624.
Butcher's broom, 365.
Buttercup, 261, 264, **262**.
Butterfly, 953; African Swallow-tail, 364, **953**; Monarch 955; Pierine, 626; Viceroy, 955; White Admiral, 955.

C

Cactoblastus, 1020.
Cactus, 897, 969.
Caddis-fly, 762.
Caisson disease, 9.

Cake-urchin, 219.
Calamites, 720.
Calanus, **991**.
Calf, abnormally developed, **518**; stomach of, **788**.
California, 329, 814, 1018.
Calluna vulgaris, 931, 932.
Calocalanus plumulosis, **849**.
Calvin, John (1509-1564), 1387.
Cambium, 705.
Cambrian period, 322, 678, 680-683, 684, 686, 688-692, 739, **325, 693**.
Camel, 178, 400, 401, **359, 401, 778**.
Cameroon rain-forest, 902.
Camouflage in nature, 944-948.
Camphor, 777.
Camptosaurus, **740**.
Canada, 675, 677, 678; Arctic, 911.
Canary, 1250.
Canary Islands, 619.
Cancer, 1093-1096, **1095**.
Candour, 1400-1403.
Canidæ, 379, 380, 387. *See* Dog.
Canis aureus, 380.
— lupus, 380, 381.
Cannel coal, 720.
Canti, Dr. R. G., 43, 1128.
Cape Verde Archipelago, 398.
Capillaries, 36, 53, 56, 57, 58; regulation of, 99-100.
Carbohydrates, 73-74.
Carbon, 73-74; circulation of, in nature, 305, 962-964.
Carbon dioxide, 29-30, 98-100; and plants, 254-255. *See also* Respiration.
Carboniferous period, 322, 684, 688, 719, 720, 724, 739, 742, 763, 813, **325, 693**.
Carebara, **1167**.
Caribbean Sea, 852.
Carniola, 915.
Carnivora, 178, 380.
Carnivores, aquatic, **778, 880**.
Carp, 142, 1209.
Carpel, 713.
"Carriers" of disease, 1037-1039.
Cartilage, 47.
Caspian Sea, 891, 998.
Castle, Prof. W. E., 610, 611, 612.
Catfish, 892, 949.

Catkins, 263.
Cats, 178; classification of, 382-383; digestive organs of, **836**; play in, 1253.
Cattle, 178; breeding of, 508-513; ecology and, 1024-1025; in Neolithic age, 17.
Caucasus, 654, 815.
Cave bear, 822.
Cave hyena, 822.
Cave-lion, 822.
Cave men. *See* Homo.
Caves, animal life in, 915-918.
Caytonia, 765.
Cells, 38-47, **41**, **49**, **64**, **81**, **92**, **119**, **123**, **125**, **143**, **251**, **440**, **947**; co-operations of, 665; germ (*See* Gametes); length of life of, 44-45; movements of, 41-42, 1128; multiplication of, 44, 286, 459—461; number of, in body, 46; of blood, 47-48; of plants, **255**; stages in aggregation of, 285-287.
Cenozoic era, 323, 324, 737, 761, 767, 769, 774, 775, 776, 777, 779, 780, 781, 788, 801, **325**, **693**, **778**; climate during, 776-779.
Centipede, 208-209.
Central Africa, 733.
— America, 268, 313, 340, 815, 1047.
— Asia as the region of man's origin, 802.
Central nervous system, 106-110.
Centropages, **991**.
Centrosomes, 461.
Cephalaspis, **696**.
Cephalopoda, 214, 691-694, **864**.
Ceratium, **665**.
Cercaria, 230.
Cerebellum, 137, **1206**, **1221**, **1279**.
Cerebral cortex, 137, 1219, 1220, 1222, 1278-1287.
— hemispheres, 134, 135, 136, 137, 1219, 1222, **135**, **1215**.
Cetacea, 179, **180**, **778**.
Cetorhinus, 850.
Ceylon, 497, 902.
Chætoceras, **991**.
Chalicotheres, 784, **778**.
Chalk, biological origin of, 283.
Challenger expedition (1872-1876), 312, 853.
Challengeria, 848.

Chameleon, 184, **946**.
Chamois, 912.
Chaparral, 983.
Charles II of Spain (1661-1700), **490**.
Charr, 620.
Cheese-mite, **943**.
Chellean culture, 817, 818, 819.
Chemical sense of fishes, 1212.
Chemistry of living substance, 71, 962.
Chemotherapy, 1051.
Chestnut, 984.
Chiasmodus, **854**.
Chicken-pox, 1048.
Childhood, 159-167.
Children of late marriages, 513.
Chimæras, grafts and, 454-457, **455**.
Chimpanzee, 388, 406, 408, 409, 410, 411, 1260-1266, 1381, 1441, **792**, **1261**, **1267**; embryo of, 154, **413**; hands of, **409**.
China, early man, in, 412, **1441**.
Chinese Primrose, mutations in the, 579.
Chirocephalus, 919.
Chiroleptes, 899.
Chiroptera, 178, **778**.
Chitin, 627.
Chlamydomonas, **286**.
Chlorodendron, **665**.
Chloroform, 1023.
Chlorophyll, 254, 627, 945, 962.
Chloera, 1039, **1041**.
Chordata, 191.
Chorion, 154-155.
Chromosomes, 431, 460-507, 531, 598, 599, **464**, **493**, **495**; and sex, 552-575; function of, 462-473; in plants, 466; mapping the, 491-496.
Chrysocyon, 380.
— jubatus, 380.
Chyme, 83.
Cichlidæ, 892.
Cider and poisoning, 1066.
Cilia, 63, 147, 189, 213, 284, **1115**.
Ciliates, 444, **662**, **943**.
Cinnamon, 777.
Circulation of the blood, 36, 54-56; evolution of, 667.
— elements in nature, 305, 962-967; of carbon, 962; of nitrogen, 305, 306, 962; of phosphorus, 1031.

INDEX

Cirrothauma, **851**.
Civet, 178.
Clacton, 818.
Clairvoyance, table-tapping, and telekinesis, 1416–1422.
Clam, **869**, **991**.
Classification of living things, 14–16, 168–170, 378.
Clausthal, 916.
Clavellina, 543, 544, **544**.
Cliff-swallow, 912.
Climate, changes in, 653–659, 733–737, 777–779, 806–817.
Climatius, **699**.
Cliona, 875.
Cloaca, 153.
Clothing and health, 1083; evolution of, 1083.
Club-foot, cause of, 488.
Club-moss, 272, 711, 716, 717, 718, 720, 730, 693, **709**.
Clytia larva, **847**.
Coal, cannel, 720; Brown, 720; Carboniferous, 720–721; Devonian, 720; Eocene, 720; household, 720; Jurassic, 720; rate of use of, 721.
Coal-measures, 719; life in, 729–733.
Cobitis barbatula, 886.
Cobra, 184.
Coccolithophoridæ, 842.
Coccyx, 152.
Cochineal insect, 1020.
Cocklebur, 979.
Cockroach, **206**, **940**.
Coconut moth, 1019.
— palm, 1019.
Cœlenterates, 244, 665, 666, 683, 805, 1121, **246**, **248**, **844**.
Cœlom, evolution of, 667.
Cœnolestes, 389.
Coffee, 1072.
Cold, common, 1083–1084; susceptibility to, 1084.
Cold storage, preservation of food in, 1056.
Collembola, 972, 993.
Collozoum, **665**.
Colobopsis, **1167**.
Coloration, concealing, 945; ruptive, 949; warning, 952.

Colour and pattern in life, 944–960.
— changes of, in animals, 946–948, 1115; in chameleon, 1116; in cuttlefish, 216; in frog, 1116; in octopus, 1116.
— of desert animals, how evolved, 835.
— of marine animals, 859
Columba livia, 376.
Comb-jellies, **991**.
Comephorus, 892.
"Comet starfish," **238**.
Comfrey, 378.
Complex, in psychology, 1356–1357, 1360.
Conditioned reflex, 1289–1293; inhibition of, 1297–1305.
Conduct, modern ideas of, 1390–1410.
Condylarths, 782, 783, **778**.
Coney, 388, **384**.
Congenital idiocy, 1371.
— physical defects, 1034–1035.
Conger, 725.
Congo, 884; rain-forest, 985.
Conifers, 267, 763, 764, 984, **693**, **709**.
Connective tissue, 47.
Consciousness, 12, 1270–1275; passive or active, 1272; range of, 1272–1275.
Constipation, 1059; and tea-drinking, 1072.
Consumption, 1096.
Continents, maritime outlines of, **655**.
Convergence in evolution, 744.
Convoluta roscoffensis, 932, 933, 934.
Copepods, 852, 859, 971.
Copra, 1019.
Corporomonas, stages in sexual process of, **442**.
Coral, 247–248, 439, 871–874, 929, 844; reefs and islands, life on, 871–874.
Cordaites, 716, 730, 731, 734, **693**.
Cordiceps, 925.
Coregonus, 620, 891–892.
Corizus, **560**.
Cork-oak, 983.
Cornwall, 816.
Corpus luteum, 157, 158.
Corpuscle, human red blood, 48, **49**, **947**; of frog, **947**.
Correlation. *See* Nervous system *and* Internal secretion.

Cortex, at work, 1288 *sqq.*; expansion of, 1278 *sqq.*
Corycella, **662.**
Coryphodon, 782, **778.**
Corystes, 870, **869.**
Corythosaurus, **758.**
Coscinodiscus, **991.**
Cotton-spinner, 220.
Cottus gobio, 886.
Cotylosaurs, 741, **740.**
Courtship in animals, 1226-1239; in newts, 1217.
Cousins, marriage of, 501, **502.**
Cowper's glands, 146.
Cow, quantity of milk given by a, 1025.
Cowpox, 1049.
Cow-wheat, 927.
Coyote, 380.
Crab, 201, 694, 868, **1129;** hermit, 201, 535, 862, 930, **201,** 869; larvæ of, **991;** nervous system of, **1129.**
Crandon, Mrs. ("Margery"), medium, 1428-1431, **1429, 1430.**
Cratægus, 457.
Creodonts, 782, 783, 785, **778.**
Crested Dinosaur, **758.**
Cretaceous period, 322, 323, 692, 694, 725, 742, 748, 752, 753, 754, 759, 764, 767, 769, 772, 773, 776, 777, 779, 780, 781, 782, 785, 786, 854, **325, 693, 747, 758.**
Cretinism, 163.
Crinoid, **693.**
Crocodile, 995, **740, 747, 883;** West African, **937.**
Cro-Magnon type of man, 1440; *facing* 774.
Crookes, Sir William, 1423.
Crossbill, 1004.
Crows, carrion, hooded, and hybrid, **382.**
Crozet Island, 913.
Crustaceans, 201 ff., **693, 849, 854, 864, 937.**
Cryptic colours, 946.
Cryptodifflugia, **662.**
Cryptogams, 268.
Ctenophores, 248-249, 846.
Cuba, 497.
Cuckoo, 956-957.
Cucumaria, 220.

Culex, 1044.
Cultivation of the soil, origin of, 1456. *See* Agriculture.
Curie, Madame, 419.
Cuscus, 389.
Cuttle-fish, 180, 215, 673, 691, 857, **215, 963;** eye of the, **1112;** smoke-screen of the, 216, 860.
Cuvier, Georges, Baron (1769-1832), 357, 385, 784.
Cyanus, **937.**
Cycadeoids, **693.**
Cycads, 713, 735, **693.**
Cyclops, 202, 227, 245.
Cyclostomes, 187-188, 695, 697.
Cyclothone microdon, 848.
Cynogathus, **353.**
Cypress tree, 776.
Cystosoma, 862, **864.**
Cytisus adami, 456, **455.**
— laburnum, 456.
— purpureus, 456, **455.**
Cytoplasm, 42.

D

Dalempatius, his diagram of human sperm, **530.**
Dalmatia, 984.
Danais plexippus, 955.
Daphnia, 578.
Darwin, Charles, 334, 395, 398, 428-430, 502, 600-603, 618, 632, 645, 767, 825, 1015.
Darwinism, meaning of, 428; vindication of, 603-604.
Davey, S. J., 1422.
Dawn-Man, 412.
"Dead man's fingers," 247.
Dead Sea, 918.
Deaf-mutism, 500; how inherited, **502.**
Death, causes of, 140-142; life after, 1433-1435; nature of, 140.
Death rates, in animals, 601, 996-1011; in man, 1089-1092, **1090, 1093.**
Death Valley, 895.
Decay, as a living process, 964; biological importance of, 305.
Dedifferentiation, 543.
Deep sea, life in the, 9-10, 853-875.

Deer, 178, **778**; red, 389, 1010.
Deer-mouse, America, 837; Florida, 618.
Deilemera, 957.
Dementia præcox, 1374.
Dendrobates tinctorius, 1116.
Dendrocystis, **687**.
Dendrosoma, **665**.
Denmark, 385.
Descartes, Réné (1596–1650), 1273, 1277.
Desert, belts, 983; colour of animals of, 835; life in, 895–901; nature of the early world, 701.
Desmids, 287, **662**.
Development of embryo, 150–155, 515, 533; abnormal, 518.
— life communities, 973–982.
Development, normal and monstrous, 514–517, **516**, **518**, **527**; of life-communities, 973–982; of the individual, 514–551.
Devonian old red sandstone, 684.
— period, 322, 683, 697, 698, 718, 719, 763, **325**, **693**.
"Devonshire colic," 1066.
Diabetes, cause of, 1055.
Diaphragm, 36, 60–62.
Diathesis, 1035.
Diatoms, 287, 843, 848, 892, 910, 920, **662**, **849**.
Dicynodont, 742.
Diet. *See* Food.
Digestion, 71–90; in amœba, 277; in lobster, 197–198; organs of, **78**; slackness of, 1057–1059; with the aid of bacteria, 834.
Digestive system, 71–90, **78**; absent in tapeworm, 230–231; cancer of, 1095, **1097**; evolution of, 666–668; in flat-worms, **226**; in herbivores and carnivores, **836**; in lobster, 197.
Dimorphodon, **754**.
Dinichthys, 700, **699**.
Dinictis, 636.
Dinoceras, **778**.
Dinoflagellates, 842, 848.
Dinosaurs, 184, 756–761, 772, 774, 781, **740**, **758**, **760**, **771**.
Diphtheria, 1037, 1038, 1048, 1050, **1038**.

Diplocaulus, **730**.
Diplodocus, 761, **740**, **937**.
Dipodomys spectabilis, 901.
Dipper, 608, **593**.
Diprotodon, 395, **396**.
Dischidia, 905.
"Discovery" expedition, 908.
Disease-carriers, human, 1037; insects, 1040–1047; rats, 1045.
Disease, epidemic, 1089–1090; infectious and contagious, 1033–1053; nature of, 1033–1034; the fight against, 1099–1101.
Disinhibition, 1303.
Distomum macrostomum, 951.
Distortions, 586–587, **589**.
Distribution of animals, as evidence for evolution, 388–404.
— of human races, 392, 1447–1450.
Dixippus morosus, 1313.
Dog, 170, 178, 374–377, **375**, **377**, **390**, **396**, **937**, **940**; brain of, **1291**; different species of, 379–381; hypnotism in, 1311; intelligence of, 1242; mind of, 1293–1297; senses of, 1305–1306; temperament in, 1314–1317.
Dogfish, 171–172, 187, 417, 1211, **171**, **174**, **367**.
Dogger Bank, 871, 915.
Dogwood, 977.
Doliolum, 236.
Dolphin, 179, 785, 850, **359**, **778**.
Dominance in heredity, 481.
Dorsal lip, in embryo, 519.
Dove, 836.
Doyle, Sir Arthur Conan, 1419–1420.
Draco, 750.
Dragon fly, 207–208, 1113.
Dragonet, 1229.
Dreams, prophetic, 1413–1416; psychology of, 1358–1364.
Drepanaspis, **696**.
Driver ants, **1177**.
Drosophila, 465, 475, 476, 491–496, 570, 579, 614, **495**, **579**, **652**.
Drugs, and the mind, 1386–1389; their uses and dangers, 1068–1075.
Dry land, as a habitat for life, 830–831, 893–895; invasion of, by animals, 722–729; invasion of, by plants, 701–716.

Dryopithecus, 414, 800.
Duckbill, 176, **386, 740.**
Duckbill Dinosaur, 772, **740, 758.**
Duckling, an abnormally developed, 518.
Duckweed, **885.**
Ductless glands. *See* Internal secretion.
Dugong, **778.**
Dungeness, 319.
Dunne, J. W., 1413, 1414.
Dust, and hay fever, 1078; and health, 1077; defences against, 63.
Dwarf-plankton, 892, **991.**
Dysentery, and amœba, **943**; and ciliates, 284.
Dytiscus beetle, 891.

E

Eagle, 142, 1248.
Ear, evolution of, 353; working of, 120–124. *See* Hearing.
Early Cambrian period, 690.
— Cenozoic period, 813, 1241.
— Devonian period, 724.
— Eocene period, 801.
Earthworm, 222, 971, 972, 1108, **1127.**
Earwig, 1018.
East Africa, 385.
— Indies, 907.
Echidna, 352, **386.**
Echinoderms, 217–221, 805, **220, 693**; behaviour of, 1125–1136; classes of, 219–221; evolution of, 686–688, **687, 693.**
Echinosphæra, **687.**
Echinus, **220.**
Ecology, 22, 961–1011; economics and, 961.
Ectoplasm, 1422–1432, **1424, 1429.**
Edentates, 178, 786, **778.**
Edrioaster, 687, **687.**
Education, in animals, 1247–1252; in man, 1293–1296.
Eel, 886, 888; electric, 697, 1117, 1118, 1275.
Eel-grass, 890.
Egg, evolution of the, 736; human, 151; of grass-frog, **942.** *See* Ovum.
Egret, 1253, 1254.

Egypt, 5, 820, 895, 896, 998, 1018.
Ehrlich, Paul, 1053.
Élan vital, the, 428, 433, 637–639.
Eland, **384.**
Elaphis, **943.**
Elbe, 884, 885, 886, 920.
Electric eel, 697, 1117, 1275.
— ray, 1117.
Elephant, 388, 786, **778**; African, 941, 937; evolution of the, 350.
Elephas. *See* Elephant.
Elginia, 741.
Elm, 979.
Elodea, 553.
Embryo, evidence of evolution from the, 366–373; formation of, 150–152, 514; human, 150–155, 405, 410–411, 416, **151, 153, 155, 174, 403, 413**; of chick, 1203, **174, 1204**; of chimpanzee, 154, **413**; of dogfish, 171–173, **174**; of gibbon, **413**; of gorilla, 154, **413**; of lancelet, 1201; of lizard, **174**, 367; of newt, 367; of rabbit, **174**, 367; organizers in, 518–520; symmetry of, 515.
Embryonic membranes, 155, **787.**
Emotion and urge, 1330–1331.
Empedocles, 428.
Emperor penguin, 908.
Empid, 1228, 1229.
Encephalitis lethargica, 1039, 1049.
Endocrine cells, 166.
Endocrines. *See* Internal secretion.
England, 809, 811, 815, 817.
English Channel, 811, 816.
Entelodon, 784, 778.
Environment, 6–11; heredity and, 467, 470, 472, 489, 582, 1383; response to, 110–114. *See* Habitat.
Enzymes, in digestion, 77–78.
Eoanthropus, 412, 818.
Eobasileus, 782.
Eocene period, 323, 341, 345, 348, 775, 776, 777, 780, 781, 782, 783, 785, 1241, **325, 344, 778.**
Eohippus, 341, 345, **333, 336, 344.**
Eoliths, 817.
Epidemic diseases, 1089.
Epidermis, 91.
Epididymis, 145, **147.**

Epihippus, 341, 342, **344**.
Epiphytes, 905, 906.
Epithelium, pavement, 40.
Equus. *See* Horse.
Ergosterol, 1063, 1074, 1085.
Ermine, 908.
Eryoneicus, **854**.
Eryops, **740**.
Eskimo, 911, 1055, 1238.
Essequibo, the, 884.
Eugenics, 1468–1471, 1477–1478.
Euglena, 284, 285, 287, **947**.
Eunotosaurus, 352.
Eupagurus bernhardi, 930.
Euphorbias, 897.
Euphrasia, 927.
Eurycorpha, 956.
Eurypterids, 684, **937**.
Eustachian tube, 152, **123**.
Evadne, **991**.
Evaporation, 966.
Evasion, indolence, and fear, 1408–1410.
Evening Primrose, 581, 582, 621, 766.
Everest expedition of 1924, 7, 8, 9.
Evolution, 316–424 (evidence for).
— 425–643 (mechanism of).
— 644–822 (story of).
— and geographical isolation, 617–621.
— by Natural Selection, 429, 600–617.
— chief theories of, 425–433.
— continuity of, 346–348, 580–583.
— convergent, 743–745.
— cosmic, 422, 644–647.
— crossing and, 621–623.
— Darwinian theory of, 428–430, 600–617.
— diagrams of, showing history of main groups, **676**, **740**, **778**.
— evidence from anatomy, 356–366.
— evidence from geographical distribution, 388–402.
— evidence from the variability of living things, 374–388.
— evidence from useless structures, 362–366.
— evidence of the embryo, 366–373.
— evidence of the rocks, 318–355.
— evidence summarized, 402–404.
— in action, **470, 471, 472**.

— Lamarckian theory of, 427, 428, 429, 583–593.
— mystical theories of, 433, 629–643.
— of adaptations, 835–838.
— orthogenetic, 432–433, 629.
— parallel, 348, 743–744, **851**.
— progress in, 789–795.
— purpose in, 639–643.
— random nature of, 639–643.
— slowness of, 347–348, 612.
— spurts in, 580–583.
— straight line, 432, 629–632, **636**.
— tangled nature of, 786–789. *See also* the names of the various groups.
Ewart, Cossor, 509.
Exaltation, psychology of, 1349–1351.
Excretion, 29–30, 66–71; organs of, 65.
Excretory system, **65**.
Exercise. *See* Physical exercise.
Exophthalmic goitre, 163.
Extinction of species, 623–624.
Extroversion, 1384–1385.
Eye, 1107, 1109, 1111, 1113; defects of, in man, 124–126; development of, in embryo, 1206; evolution of, 1106–1114, 1284; in deep-sea animals, 862–863, **864**; inheritance of colour in man, 487–488; of cuttle-fish, 1112, **1112**; of fishes, 1211; of insects, 1113; of scallop, 212; third, in primitive vertebrates, 1208; working of, 124–126. *See* Sight.

F

Fabre, Jean Henri (1823–1915), 1158, 1160, 1161, 1228.
Faeces, 88–90.
Falkland Islands, 913.
Faroe Islands, 914.
Fasciola, 230.
— hepatica, 228.
Fats, 73; storage of in body, 81–82.
Faults, geological, 328.
Feathers, 181, 351; evolution of, 752.
Feather-star, 221, 370, **370**.
Femur, **34, 169**.
Fermentation, yeasts and, 291.
Fern-plant, 269, **271**.

INDEX

Ferns, 268-274, 716-719, **269**, 271, **709**; evolution of, **693**; reproduction of, 269-270; sperm, **271**.
Ferreriro, the, 1217.
Fertilization, of the egg, 148, 465, 514, **464**; by artificial means, 449; cross and self-fertilization, 264; dispensed with, 448-449; in plants, 261-265, **273**.
— of the soil, 288, 1025, 1028, 1029.
Fetishism, 1359.
Fiddler-crab, 1227, **540**.
Field ant, **1173**.
Field-mouse, 1002; gamete of, **447**.
Fig, 776.
Figwort, 366.
Fiji, 1019, 1020, 1463.
Filamentous bacteria, 302.
Filaria medinensis, 226.
Filter-passing organisms, 306-310, **652**, **947**.
Fire-salamander, 953.
Fish, 170-174, 184-187, **937**; anatomy of, contrasted with man, 170-172; bony and gristly, 186; colour changes in deep sea, 948-949, **860**; earliest, 694-700; electric shocks given by, 1117; evolution of, **693**; eyes of, 1211; flat, **372**; fresh-water, 878-892; mind of, 1210-1214; senses of smell and taste in, 1208, 1209, 1210; swimming bladder of, *see* Gasbladder.
Flagellates, 283-287, 443, 842, **286**, **662**, **665**, **849**, **947**, **991**.
Flamingo, 949.
Flatworms, 227-232, 805, **226**, **668**; gamete of, **447**; regenerating, **447**.
Flea, 935, 936, 993, 994, **363**, **942**, **943**.
Flicker, 385, 622.
Flight, evolution of, 749-756; in archæopteryx, 351; in birds, 180-182, 751-753; in insects, 205, 749; in pterodactyls, 751.
Florida, 720.
Flounder, 886.
Flour, value of white, 1057, 1064.
Flournoy, Professor, 1352.
Flowers, 261, **937**; and bees, 1191-1199; evolution of, 713, 765, **709**; inheritance of colour in, 476; insects and, 262, 761-769; T. H. Huxley on, 358; variation in, 378; wind-pollinated, 263.
Flukes, 228, 951, **229**.
Fly as disseminator of bacteria, the, 1065.
Flying-fish, 187.
Flying-fox, **778**.
Flying-reptiles, 752-753, **740**.
Fogs, cause of, 1077.
Food, amount needed by man, 1054, 72; and cancer, 1095; artificial, 1056; chemistry of, 71 *sqq.*, **76**; choice of, 1054-1064; excess of, 1055; of different animals, 832-835, **836**; of plants, 254; poisoning by, 1064-1068; preservation of, 1056; uses of, in body, 29-31, 71-90, 1054. *See* Digestion *and* Vitamins.
Food-chains and parasite-chains, 989-996.
Foraminifera, 283, 439, **662**, **991**.
Fore-arm, structural plan of, 357, **359**.
Fore-brain, 1203, 1205, 1206, **1204**, **1215**.
Forel, 1157.
Fore-limbs, structural plan of, 359, **361**.
Forest, tropical, 901-907.
Formica fusca, 1172.
— pratensis, 1173.
— rufa, 1177.
— rufibarbis, **1173**.
— sanguinea, 1172.
Fossils, age of, 330-332, 419-424; evidence of, for evolution, 318-355; formation of, 328-330.
Fossil record, coherence of, 332; diagram of, **693**; gaps in, 326-332; sample sections of, 332-346. *See* Evolution *and* Geological time.
Four o'clock, 476, 481, 485, **478**, **500**; Mendelian explanation of breeding behaviour of, **478**.
Fox, Arctic, 378, 1002; common, 379, 380, 1241; cross, 378; Fennec, 380; red, 378, 1002; silver, 378.
France, 329, 777, 821, 822.
Frederick III (1415-93), Emperor, **490**.
Fresh water, life in, 878-883, **880**, **883**.
Freud, Prof. Sigmund, 1331, 1363, 1368, 1391, 1392, 1393, 1407.

INDEX

Frog, 185, 891, 1215-1218, **359**, **883**, **940**;
brain of the, **1206**; evolution of, **542**, **740**; gamete of, **447**; hand of, **409**; segmenting eggs of, **534**; tadpoles of, 185, 526, **529**.
Fruit-fly. *See* Drosophila.
Fulgoridæ, 857.
Funchal, 858.
Functional differentiation, 523.
Fungus, 289-291, 925, 928, 964; parasitic and beneficial, 929-932.
Fur. *See* Hair.
Fur seals, extermination of, 1031.

G

Galapagos Islands, 398, 618, 619, 620, 995.
Galilee, 820.
Gall-bladder, 84.
Gall, Franz Joseph, 1281; and controlling stations of the faculties, **1281**.
Galton, Sir Francis (1822-1911), 1388.
Galveston, 385.
Gametes, 161, 444-448; female, 145; how formed, 462; male, 145, 148; of various animals, **447**.
Gammarids, 397.
Gammarus, 916.
— chevreuxi, 532, **532**.
Ganges, river, 1039.
Garnett wheat, 1026.
Gas gangrene, 1037.
Gas-bladder of fish, 10, 850, 1116.
Gastric juice, 81.
Gastropods, 213, 690.
Gastrostomus, **854**.
Gavial, 183.
Gecko, 184.
Gemmule, formation of a, **441**.
General paralysis of the insane, 1370.
Genes, 473-498, 531-535, 552-575, **532**; dominant and recessive, 480-481, 486; in man, 485-491, 563; multiple, 478-491.
Genetics, 434, 468-513.
Genus, 169, 378-388.
Geographical distribution, and evolution, 388-402.
— isolation and origin of species, 617-623.

Geological time, diagrams of, **325**, **647**, **693**; how determined, 419-423; magnitude of, 419-424; main divisions of, 322-326. *See* Fossil record *and* Rocks.
Geotropism, 1131.
Gerbil, 1010.
Germ-cells. *See* Gametes.
Germ-plasm, 457-458; and soma, **585**.
Giant pig, **778**.
— Rush, 731.
Giants, 165.
Gibbon, embryo of, **413**.
Gigantactis, **860**.
Gigantura, **864**.
Gill-clefts, 171, 368; of human embryo, 152; of fishes, 152.
Ginkgo, 371, **693**.
Gipsy-moth, 572, 1018.
Giraffe, 178, 388, **384**, **778**, **937**.
Glaciers, 807.
Gland-grafting, 541.
Glands, Cowper's, 146; ductless, 162, 165, **105**; of common toad, 1116; parathyroid, 166; pineal, 167, 173; pituitary, 165, 167, 173, **1206**; post-pituitary, 164, 166, **164**; pre-pituitary, 164; salivary, 1289, 1290, 1292; thyroid, 157, 162, 165, 166, 173, 977.
Glass-sponge, 865.
Glaucus, 850, **849**.
Globigerina ooze, 283, 966.
Glossina palpalis, 1046.
Glossopteris, 735.
Glyptodont, 395, 634, 748, **390**, **778**.
Gnat pupæ and larvæ, **885**.
Gneiss, 328.
Gnu, white-tailed, **384**.
Goat, 178.
Gobi desert, 895.
Godunoff, 7.
Goitre, exophthalmic, 163.
Golden-rod, 977, 978.
Goldfinch, 1258.
Goldtail moth, 1134.
Goliath beetle, **940**.
— frog, **940**.
Gondwanaland, 733.
Goose-barnacles, 876.
Gooseberry, 1021.

Gorgonia, 248.
Gorilla, 388, 801, 1440, **410**, **413**; embryo of, 154, **413**; resemblance between man and, 406; skeleton of, **410**.
Graafian follicles, 146.
Grafting, trees, 451, **450**; yellow laburnum and purple broom, **455**.
Grafts in plants and animals, 454.
Graptolite, 666, 683, **693**.
Grass, 983, 1024, 1025, 1030.
Grass-frog, **940**, **942**.
Grasshopper, colour-patterns in wild, 579.
Grasslands, 983, 1030.
Grass-snake, European, 1229.
Gratiola, 366.
Grayling, 886.
Greasy fritillary, 1009.
Great ant-eater, **391**.
Great Dismal Swamp, Florida, 720.
Great Lakes, 976.
Great Salt Lake, 918.
Grebe, 1249.
Greenfinch, gamete of, **447**.
Greenfly, 449.
Greenland, 388, 675, 776.
Grey matter in brain, 137.
Gribble, 875.
Grimaldi type of man, 1440.
Gristle, 33, 47.
Groos, Karl, 1255.
Ground-sloth, 782, **390**, **778**.
Grouse, red, 619.
Growth, and ductless glands, 162–165, 536; discriminating, 1130; exaggerated local, 539, **540**; human, 159–160, 535, **538**; in Crustacea, 539–540; limitation and control of, 535–540; of the individual, 514–551.
Guano, 966, 1030, 1031.
Guiana, 906.
Guinardia, **991**.
Guinea-pig, 388.
Guinea-worm, 226, 923.
Gulf Stream, 811, 843, 912.
Gull, 1248.
Gwyniad, 620.
Gymnosperms, 267.
Gynanders, 563–565, 568.
Gypsum, 733.

H

Habitat, 823–838, 919; adaptation to, 835–838; aërial, 831; freshwater, 878–892; land, 893–895; marine, 839–877; parasitic, 831; terrestrial, 830, 893–918.
Habsburg lip, 488, **490**.
Haeckel, Ernst Heinrich (1834–1919), 368, 369, 430, 961.
Hæmocyanin, 221.
Hæmoglobin, 64, **652**.
Hæmophilia, 563; pedigree of, **564**; in European royal families, 563.
Hag-fish, 188, 927.
Hair, 95, 112, 175; length of, and temperature, 589–590, **591**.
Haire, Norman, 167.
Haldane, J. B. S., 496, 609, 651, 755, 1022, 1065.
Haliclystus, 247, 870, **248**.
Halictus, 1182.
Hamburg, 920–921.
Hands, **940**; brachydactylous, 488; comparison of, **409**.
Hardy, A. C., 985, 992.
Hare-lip, 416.
Harpacticids, **991**.
Hartebeest, 385.
Hartsoeker's diagram of human sperm, 530.
Hartz Mountains, 916.
Haviland, Miss, description of an Amazonian forest by, 902.
Hawaii, 913, 1020.
Hawk-moth, 767, **363**.
Hawthorn, 457.
Hay fever, 1078.
Health, 1033, 1097–1101, 1395–1397; clothing and, 1083–1084; food and, 1054–1068; fresh air and, 1076–1084; of animals and savage races, 1033–1034; sunlight and, 1085–1088.
Hearing, 120–121; in dogs, 1306. *See* Ear.
Heart, **36**, **39**, **55**, **57**, **61**; disease of, 1091; regulation of, 101, 1297–1298; survival of, 46.
Heart-urchin, 688, **869**.
Heather, Mediterranean, 816; Scotch, 931.

INDEX

Hedgehog, 178, **778**.
Heidelberg man. *See* Homo.
Helium, 419, 420.
Hen, **937, 940**.
Henderson, L. J., 650.
Heredity, and sex, 552-628; in man, 485-491; mechanism of, 459-467; Mendel's laws of, 473-485, 505-508; non-Mendelian, 505; superstitions about, 508-513.
Hermit-crab, 862, 930, 1158, **201**.
Herodotus, 995.
Heron, 1253.
Herring, 725, **940**; food-relations of the, **991**.
Hesperornis, **351, 740**.
Heterangium, 718.
Heterocarpus, **860**.
Heterocephalus, 590, **591**.
Heteronotus, 956.
Hexactinellid sponge, 866.
Himalayas, 654, 801; animal and plant life in the, 912.
Hind-brain, 1203, 1204, **1204, 1206**.
Hingston, Major, 8.
Hipparion group of horses, 345, **344**.
Hippidium, 345, **344**.
Hippopotamus, 178, 784, **778, 883**.
Hoatzin, 351.
Hogben, Professor, 948.
Holland, 887; and land reclamation, 982.
Holmes, Oliver Wendell (1809-1894), 1051.
Holm-oak, 983.
Holoptychius, **699**.
Homarus vulgaris, **197**.
Hominidæ, 797.
Homo, 168.
— heidelbergensis, 412, 806.
— jaw of, **414**.
— kanamensis, 412.
— neanderthalensis, 170, 412, 806, 819, 820, 821, 1440.
— rhodesiensis, 412.
— sapiens, 168.
Homology, 171.
Homunculus, 531, **529**.
Honey-bee, 1184-1191.
Honey-locust, 979.

Honeypot ant, 1171, **1173, 1177**.
Hooke, Robert (1635-1703), 39; microscope used by, **21**; observations on cork, 39.
Hook-worm, 227, 923.
Hopkins, F. Gowland, 1061.
Hoplophoneus, **636**.
Hormones. *See* Internal secretion.
Hornea, 710, **706**.
Horsemint, 977.
Horses, 178, **778, 937**; calculating, 1251; evolution of, 332-346, 458, 787-788, 333, 33**6**, 33**8**, 344.
Horseshoe crab. *See* King crab.
Horsetail, 717, **709**; evolution of, **693**.
House-fly, **363, 942**.
House-martin, 753.
House-sparrow, 389.
Howard, Eliot, 1232.
Huanaco, **391**.
Hudson Bay Company, 1000, 1002.
Hudson, W. H., 752.
Human body. *See* names of parts.
— association, the present phase of, 1454-1480.
— behaviourism, 1318-1323.
— interaction, development of, 1450-1453.
— sacrifice, 1456.
Humble-bee, 1182.
Humboldt's woolly monkey, **391**.
Humming-bird, **940, 942**.
Humour, biological function of, 1380.
Hunter, John (1728-1793), 454, 525.
Huxley, Thomas Henry (1825-1895), 358, 406, 430, 732, 1448.
Hyæna, 111, 972.
Hyænodon, **778**.
Hybrids, between species, 385, 386; graft-hybrids, 454, 456, **455**; intersexuality in, 572; sterility of, 384; vigour of, 497.
Hydra, 245, 247, 295, 452, 916, 929, **668, 942**.
Hydrophilus grub, **880**.
Hydrotropism, 1131.
Hydrous larva, **880**.
Hydrozoa, 244.
Hygiene. *See* Health.
Hyperieds, **991**.

Hypnotism, in dogs, 1311; in lower animals, 1312-1313; in man, 1331-1339.
Hypohippus, 634, **344**.
Hyracotherium, 337.
Hyrax, Cape, **384**.
Hysteria, in dogs, 1315; in man, 1344-1351.

I

Ianthina, 859.
Ibex, 912.
Ice Ages, 677, 733, 777, 779, 784, 789, 801, 807, 808, 809, 810, 815, 817, 820, 873, **647**, **734**; time diagram of, **807**.
Ice-flower, 909.
Iceland, 810.
— moss, 294.
Ichneumon-fly, 825, 924.
Ichthyosaur, 184, 319, 330, 746, 785, 359, **740**, **747**.
Iguana, 184.
Iguanodon, 329, **740**.
Immunity from disease, natural and artificial, 1047-1050.
Inbreeding, effect of, 500; its dangers and uses, 498-503.
Incest, origin of tabus against, 1441-1447.
India, 340, 401, 733, 734, 1039, 1047.
Indiana, 979.
Indian Ocean, 862.
Indigestion and tea-drinking, 1072-1073.
Individuality, in cells, 44; in the lower animals, 235, 241-244, 246; in plants, 259.
Indo-China, French, 619.
Infection, the process of, 1036-1037.
Inflammation, 50, 93.
Influenza, 1036, 1039, 1049, 1080, 1089, 1092.
Infusoria, 284, 1119.
Inheritance, of acquired characters, 583-593. *See* Heredity.
Inhibition, in the nervous system, 1297-1305; and hypnotism, 1311; and sleep, 1306-1311.
Inoculation, preventive, 358-359.

Inostransevia, **740**.
Insanity, 1369-1381.
Insectivora, 178, **778**.
Insects, 205-208; as disease carriers, 1040-1047; as pests, 1018-1021; behaviour of, 1147-1199; determination of sex in, **554**; earliest known, 685; effect of island life on, 913-914; evolution of flowers and, 763; faces, **363**; flight of, 749; flightless, 913; fossil, 329, 685, **693**; parasitic, 923; pollination of flowers by, 262-264; rivalry between land-vertebrates and, 769-770; societies, 1162-1168.
Insemination, artificial, 559.
Insomnia and tea-drinking, 1072.
Instinct, culminates in Arthropods, 1147; defined, 1135; in amphibians, 1216-1217; in birds, 1225-1226; internal secretions and, **158**, 567.
Insulin, 166.
Intelligence in birds, 1225-1226; in fish, 1213-1214; in mammals, 1239-1252.
— tests, 1381, **1382**.
Internal secretion, 52, 103-104, 157, 162; and growth, 162, 164-165, 526-529; and sex characters, 161, 566-569. *See* Adrenal, Pituitary, Thyroid.
Intersexes, 569-575.
Interstitial cells, 146, 161.
Intestinal worms, 223-226, 230.
Intestine, 87, **90**; large, 89; small, 84.
Intra-specific selection, 616.
Introversion, in psychology, 1384.
Iodine, 288, 529.
Ireland, 288, 809, 815.
Irritability of protoplasm, 1105.
Irvine, A. C., 7.
Islands, life on, 398-400, 618-620, 913-915.
Isolated organs, survival of, 45-46.
Isolation, as a species-maker, 617-621.
Israelites and manna, 294.
Italy, 385, 810, 822, 1040, 1468.

J

Jackal, Indian, 380.
Jaguar, **636**.
Jamaica, 1045.
James Island, 618.

INDEX 1497

James, William (1842-1910), 1330, 1384.
Japan, 288, 866, 957.
Java, 313, 412, 497, 498, 768.
Jaws, evolution of, 352-353, 695-696, 414.
Jeans, Sir J. H., 12, 645, 654.
Jelly-fish, 241, 247, 678, 859, 936, 1110, 1121-1123, 1144, **248, 458, 851, 937,** 991.
Jenner, Edward, 1049.
Jerboa, 178, 900.
"Jewish race," 1449.
Johannsen's experiment, **470**.
Junco, 1003.
Jung, Dr. C. G., 1331, 1364, 1366, 1367, 1384, 1387, 1391, 1393, 1394, 1395, 1452.
Jurassic period, 323, 324, 692, 742, 746, 747, 748, 752, 753, 756, 759, 762, 763, 764, 765, 775, 780, 813, 988, **325, 349, 693**.

K

Kala-azar, 284.
Kalahari desert, 895.
Kamchatka, 779.
Kangaroo, 176, 389, **386**.
Kangaroo-rat, 901.
Kea, 1017.
Kelp, 664.
Kent's Hole, 818.
Kenya, 1025.
Kerguelen Islands, 913, 914.
Kiang, 334.
Kidneys, 37, 66-71, **65**.
King-crab, 205, 684, 695, **693**.
Kittens, play in, 1253.
Kiwi, 913.
Koala, **386**.
Koehler, Wolfgang, 1381.
Komodo dragon, **937**.
Koumiss, 142.
Kraepelin, Prof., 920.
Krait, 184.
Krakatoa, 973.
Kudu, **778**; greater, **384**.

L

Laburnum, 456, **455**.
Lacerta viridis, **183**.

Lake Baikal, 892.
— Mendota, Wisconsin, U. S. A., 992.
— Michigan, 976.
Lakes, life in, 889; salt, 552.
Lamarck, Jean Baptiste (1744-1829), 427, 583, 584.
Lamarckism, 428, 431, 583-593.
Lamb, an abnormally developed, **518**.
Lamellibranchia, 212.
Laminaria, 288, 663, 867.
Lammergeier vulture, 912.
Lamprey, 188, 191, 926.
Lamp-shells, 233, 671, **233, 693**; larva of, 847.
Lancelet, 189, 1108, 1201, 1203, **190**.
Land, dry, as a habitat, 830-831, 893-918; invasion of, by animals, 722-729; invasion of, by plants, 701-716; originally lifeless, 701-702.
Land reclamation, 981-982.
Langouste, **849**.
Lankester, Sir E. Ray, 1122, 1132.
Lanugo, 154.
Larch, **937**.
Larvæ, in surface zone of sea, **847**; of insects, 207-208, 886, 1171, 1187; of sea-squirts, 192; of worms, **847**.
Lasiognathus, 861.
Lasius, **1177**.
Late Cretaceous period, 801.
— Devonian period, 700.
— Eocene period, 801.
— Miocene period, 801.
— Permian period, 779.
— Silurian period, 695, 723.
Lateral line, the, 1211, **171**.
Laternaria lucifera, 957.
Laughter, 1380.
Lawes, 1029.
Layering of plants, 451.
Lead, 420.
Leaf-insect, 1155.
Lebistes, **942**.
Leech, 223.
Leeuvenhoek, Anthony van (1632-1723), 275; microscope made by, **20**.
Lemming, 908, 999, 1008.
Lemonade and poisoning, 1066.
Lemon-juice as a cure for scurvy, 1059-1060.

Lemur, 179, **778**, **792**.
Leopard, 388.
Lepadocrinus, **687**.
Lepidodendron, 730.
Lepidonotus, **224**.
Lepismina, 1175.
Leptodora, 892.
Leptomonas, 994.
Leucocytes, 91, 92, 93.
Levick, Dr. G. Murray, 1257, 1259.
Levitation, 1416–1422, **1417**, **1420**.
Liana, 904.
Lice, 1045.
Lichen, 292–295, 928, 978, **294**.
Liebig, Justus von, Baron (1803–1873), 1023, 1029.
Life, after death, 1433–1435; before fossils, 660–674; conquers the dry land, 701–737; distinctive features of, 4–6; earliest, 660–661; evolution of, diagram, **676**; in fresh water and on land, 878–833; in the sea, 9–10, 839–877; length of, 141–142; limitation in space of, 6–10; meaning of, 2–6; of flowing waters, 883–889; of standing waters, 889–892; on floor of the sea, 863–866, **865**; on other planets, 11–12; on seashore, 866–871, **869**; on surface of the sea, 845–853; origin of, 649–653; oxidation and, 30–31; subjective side of, 13–14, 1270–1277; the ascent of, **676**; under control, 1012–1032, 1476–1480; vegetable, 253–274.
Life-communities, 967–1011; growth and development of, 973–982; grading of, 982–989; parallelism and variety of, 967–973.
Life force, 637–639.
Ligaments, 33, 47.
Light, and plant growth, 254–259, 1131; fascination of, for insects, 1133–1134; its influence on amœba, 1106–1108; on earth worm, 1108; on lancelet, 1108; its penetration into water, 858–859. *See* Animal light, Ultraviolet light.
Ligia, 685.
Lignin, 704.
Lignite, 720.

Limax amœba, **278**.
Limbs, evolution of segmental, 672; movements of, 36.
Limestone, 328; biological origin of, 283.
Limicolæ, 608.
Limnodrilus, 885.
Limnoria, 875.
Limpet, 1110; lava of, **847**.
Limulus, 684.
Ling, common, 931.
Lingula, 234.
Linnæus, Carol von (1707–1778), 210, 426.
Lion, 388, **778**.
Lister, Lord (1827–1912), 1023, 1051–1052.
Lithographic stone, 350.
Litopterna, 348, 394, 799, **390**, **778**.
"Little Joss" wheat, 497.
Littoral zone, 847.
Liver, 67, 86, 1072, **39**; and the manufacture of anti toxins, 1048; of lobster, 198–199.
Liver-cell, human, **943**, **947**.
Liver-fluke, 228, 238–239, 923, **229**.
Lizard, **937**; flying, 750; green, **183**.
Lizard Point, 816.
Lizards, evolution of, **740**.
Llama, 178, **401**.
Loach, 886.
Lobster, 194–196, **197**, **937**.
Lobster-moth caterpillar, **959**.
Lobworm, 223.
Lockjaw, 1037.
Locust, **363**; plagues, 998–999; tree, 984.
Lodge, Sir Oliver, 1420–1421, 1431.
Loeb, Jacques, 449, 1023, 1132, 1133.
Lomechusa, 1177.
London, 776, 1039.
— clay, 320, 776.
— Pride, 816.
Lophius, 861, **849**.
Louisiana, 1253.
Louisville, 809.
Lower Cambrian period, 688.
— Carboniferous period, 727.
— Devonian period, 685, 698, 718, **706**.
— Eocene period, 800.
— Oligocene period, 800.
Loxomma, **740**.

Lucretius, 428.
Lugworm, 221, **224, 672**.
Luidia, 219, **940**.
Lumbriculus, 450, 885.
Luminosity in the deep sea, 861-863, 860.
Lung-fish, 400-401, 700, 722-723, 726, 1215.
Lungs, **36, 39**; heart and, 1090-1093.
Lupin, 928.
Lyginopteris, 717, **354**.
Lymantria, 572.
— dispar, 1018.
Lymph, 58-59.
Lymphatic system, 58-59.
Lynx, 1002.

M

Macaque, 1263.
Machærodus, **636**.
Macrauchenia, **390, 778**.
Macro-feeders, 833.
Madagascar, 379, 446, 733.
Madeira, 858, 913.
Madreporite, 218.
Maeterlinck, Maurice, 1181.
Magdalenian culture, 821-822.
Magnolia, 776.
Magnus, Olaus, 999.
Magpie, 1259.
Maidenhair-tree, 371, **693**.
Malaria, 1000, 1021, 1037, 1040, 1041, 1047, 1052, 1089; parasite of, **947, 1042, 1043**.
Malaya, 313, 400, 401, 497, 733, 902, 926.
Mallory, G. L., 7.
Malta, 914.
— fever, bacterium of, **947**.
Malthus, Thomas Robert (1766-1834), 428.
Mammals, 175 sqq.; brain in, 1223-1224, **1283**; classification of, **175**; courtship in, 1237-1239; education in, 1247-1249; evolution of, 779-786, 805, **693, 740, 778**; evolution of intelligence in, 1223-1224, 1239-1241; extinct South America, **390**; length of the Age of, 422; of Australia, 176, **386**; periodicity of northern, **1001**; placental, **778**; play in, 1252-1254; pouched, 176; smallest, 179-180; typical African, **384**.
Mammoth, 329, 809, 822.
— cave, Kentucky, 916.
Man, anatomy and physiology of, 32-167; biological peculiarities of, 1436-1453, 1472; brain of, 1278-1287; embryo of, 150-155, **153, 367, 413**; evolution of, 405-424, 796-822, **778, 792**; fossil (*See* Homo Eoanthropus, Pithecanthropus, Sinanthropus), hand of, **409**; heredity in, 487-491, 562-563; physical health of, 1033-1035, 1089-1090, 1096-1099; place in nature, 405-411; place in time, 419-424; place of origin, 802; races of, 1447-1450; rate of increase of, 1467-1472; tail in, 37-38, 152; vestigial organs in, 415-417.
Manatee, 179.
Mangabey, 1263.
Mangroves, 874.
Mania, 1372-1373.
Manic-depressive insanity, 1372.
Manna-lichen, 294.
Maple, 776, 984.
Maquis scrub, 983.
Marble, 328.
Marmoset, **391**; brain of, **1283**.
Marquis wheat, 1026.
Marriage, 1441-1446; of cousins, 501.
Mars, life on, 11-12.
Marsh fever, 1040.
Marsupials, 176-177, 781, **740**; colonization in Australia, **394**.
Mastodon, 634.
Materialization, 1422-1432.
Maternal impressions, 510.
Matthew, Dr. W. D., 781.
Mauritius, 913.
Maximilian I (1459-1519), **490**.
Meals, 1063-1064, 1065-1066.
Measles, 1039-1040, 1048, 1090.
Mediterranean, 249, 812, 858, 914.
Mediumship, automatism and, 1351-1353.
Mediums, in spiritualism, 1351-1352, 1416-1432.
Medlar, 457.
Medusa, 241, **242**.

Megatherium, 395, **390**.
Melampyrum, 927.
Melancholia, 1373-1374.
Melanesia, 573.
Melanogaster, 496.
Melilot, white, 979.
Melosira, **991**.
Membracidæ, 956.
Mendel, Abbé, Johann Gregor (1822-1884), 21, 431, 434, 473-475.
Mendel's laws, 265, 474-485, 505-508.
Mermis, 227.
Merychippus, 342-343, 370, **333, 336**.
Mesmerism, 1431.
Mesohippus, 343, **333, 336**.
Mesolithic Age, 821.
Mesopotamia, 812.
Mesozoic era, 323, 324, 692, 702, 735, 737, 748, 755, 756, 761, 769, 774, 777, 779, 780, 788, 988, **325, 647**.
Mespilus, 457.
Messor, **1173**.
Metamorphic rocks, 328.
Metchnikov, Ilya (1845-1916), 142.
Metopina, 1176.
Metridium, **248**.
Metriorhynchus, **747**.
Mexico, 895.
Micraster, 348.
Microbe-carriers, insects as, 1040-1047.
Microbes, 275-276, 280-281, 299-301; and disease, 1035-1040; avoiding and killing, 1050-1053; carried in the air, 1079-1080; harmless, 1036; in crowded and open places, 1079-1080. See Bacteria.
Microbrachis, **730**.
Micrococcus, **947**.
Micro-feeders, 832-833.
Microscope of A. van Leeuwenhoek, **20**; of Robert Hooke, **21**; development of, 20.
Microscopists, early, 530.
Mid-brain, 1203-1204, **1204, 1206, 1215**.
Middle Cambrian period, 678.
— Carboniferous period, 685.
— Devonian period, 699.
— Eocene period, 785.
— Jurassic period, 765.
— Ordovician period, 697.

— Pliocene period, 801.
— Stone Age, 821.
Migration, of birds, 814-815, 1004; of butterflies, 1008; of caribou, 813; of fox sparrow, 814; of lemmings, 999; of locusts, 999-1000; of mammals, 392-393, 813, 814, 815, 816; of salmon, 887.
Milk, 175, 1026; and tuberculosis, 1065, and internal secretion, 157; contamination of, 1065; longevity and soured, 142; model farms for preparation of Grade A, **1067**; why it turns sour, 303-304.
Milkwort, 378.
Miller, Hugh (1802-1856), 684.
Miller's thumb, 886.
Millipede, 208-209.
Mimeciton, **1177**.
Mimicry, 955-960, **953**; Batesian, 959-960; Müllerian, 959-960.
Mimosa pudica, 259.
Mind, collective, 1471 sqq.; drugs and the, 1386-1389; human, 1318-1389; in gear and out of gear, 1369-1381; in insects, 1147-1158; individual variation in man, 1381-1389; of a dog, 1293, 1317; of amphibia, 1214-1218; of a fish, 1210-1214; of reptiles, birds and mammals, 1218-1226. See Behaviour, Consciousness.
Minnow, 886.
Miocene Age, 323, 324, 777, 779, 782, 800, 807, 1436, **325, 344, 778**.
Miohippus, 345.
Mirabeau, Count Honoré Gabriel Riqueti (1749-1791), 1387.
Mirabilis jalapa, 476, 481, 484, **478**.
Miracidium, 229, **229**.
Missing links in evolution, 348-355.
Mississippi, 318.
Mites, 205.
Mitosis, 461.
Moa, 913.
Mocking-bird, 619, 1250.
Modern Era, 774-795.
Mœritherium, **778**.
Mola, 850.
Mole, 178, **778**.
Molluscoidea, 233.

INDEX

Molluscs, 211-216, 805, **860, 937**; evolution of, 688-694; nervous system of, 1129.
Monarch butterfly, 955.
Mongolia, 340, 760, 781, 785.
Mongolian desert, 334.
Mongoose, 1014.
Monitor, 184.
Monkey, Humboldt's woolly, **391**.
Monkey-puzzle tree, 776.
Monkeys, 179; behaviour of, 1259-1269. *See* Apes.
Monotremes, 176, **740**.
Moods, self-knowledge and, 1397-1400.
Moropus, **778**.
Mosasaur, 184, 746, **740, 747**.
Mosquito, 1000, 1021, **363, 1042**; malaria-carrying, 1041-1042, 1043-1044, 1046-1047.
Moss sperm, 271.
Mosses, 268-274, 978, **255, 271, 709**.
Moth, cinnabar, 953; gold-tail, 1134; Poplar Hawk, **363**; Vapourer, 363.
Moths, and light, 1132; melanism in, 595; smell in, 1156.
Moulds, 964, **290**.
Mountain-chough, 912.
Mountains, life on, 7, 912.
Mount Everest, animal and plant life on, 912.
Mouse, 178, 997, **28, 940**; gamete of, 447, **480**.
Mousterian culture, 819, 820.
Mouth, digestion in, 78-79; evolution of, 666.
Mud-hopper, 726, **724**.
Mulatto, 489.
Müllerian mimicry, 690.
Multiple genes, 489, 610 *sqq.*
Multiple personalities, 1339-1344.
Multituberculates, 780, **740**.
Murray, Prof. Gilbert, 1414.
Muscles, 32-36, 109, 1104, 1114; of the forearm, **37**; of a frog, 111.
Mushrooms, 285, 290, 663.
Musk ox, 388, 908, **591**.
Mutation, 473, 577-580; in Chinese Primrose, 579; in Drosophila, 579, **495, 579**; in Evening Primrose, 581; in moths, 593; induced by X-rays, 593-596; steadiness of, 58.
Mycetozoa, 295.
Myers, Frederic William Henry (1843-1901), **1421**.
Mynah, 1250.
Myriapoda, 208.
Myrmecocystis, 1171.
Myrmica, **1177**.
Mysis, **991**.
Myxoedema, 165.
Myxomycetes, 295. *See* Slime, Fungi.

N

Nagana disease, 925.
Naias, 890.
Nanno-plankton, 842-843.
Naosaurus, **740**.
Natural selection, 429, 600-605; as a conservative force, 606-609; in action, 609; may be useless to species, 612-617; not acting on man, 1468-1469; under changing conditions, 609-614.
Nauplius larva, **847**.
Nautilus, 692, 694, 850, **693**.
Navicula, **991**.
Neanderthal man, *facing* **774**. *See* Homo.
Necator, 227.
Nectar, 261.
Nemathelminthes, 223.
Nematophycus, 707.
Nemertine worms, 232, **847**.
Neo-Lamarckism, 428, 432.
Neolithic Age, 17, 821.
Neo-Malthusianism, 428.
Neopallium, 1225.
Neo-salvarsan, 1053.
Nereis, 224, **672**; larva of, **847**.
Nerve fibres, 106-109, **105, 108**.
Nerves, 35, **105**; function of, 106-110, 1298; impulses in, 1104; structure of, 106-110.
Nervous impulses, 109.
Nervous system 35, 107-111, 133-139, **668, 1202**; central, 106; development of, in embryo, 1200-1210; evolution of, 666, 1117-1128; in lower animals, 1200-1211; in vertebrates, 1200-1226. *See* Brain, Nerves.

Nest of ants, **1167**; of termites, **1181**.
Neurasthenia, in dogs, 1315; in man, 1353-1355.
Nevada, mouse plague in, 997.
New England, 810, 1021.
— Guinea, 176, 389, 902.
— Hebrides, 573.
New Stone Age, 821.
— York, 809.
— Zealand, 379, 390, 392, 400, 913, 914, 1010, 1017, 1018, 1025.
Newt, 185, 516, 1216-1217, **516**; evolution of, **740**; poison of, 1115.
Nicotine, 1069.
Night-blindness, 563.
Nightjar, European, 1235.
Nightjar, South American, 946.
Nile valley, 812, 895.
Nipa palm, 776.
Nitre, 1030.
Nitrogen, cycle, 305, 962-966; fixation, 305, 928; in blood, 52; in proteins, 74.
Nitzschia, **991**.
Noctiluca, 846, 850, 860, **849**, **991**.
Nordenskiold glacier, 907.
Nordics, 1449.
North Africa, 340.
— America, 340, 380, 389, 393, 401, 734, 779, 789.
— Sea, 870-871, 915, **370**.
North-western India, 321.
Norton, Lt.-Col., 7, 8.
Norway, 677.
Notochord, 153, 190.
Nourishment of the body, 71, 1054-1075.
Nova Scotia, 815.
Nummulite, **940**.
Nyctiphanes, **991**.

O

Oak, 977, 984.
Obelia, 239-244, 245, 260, 683, **242**, **668**.
Ocean, depth of the, 11; life in. *See* Sea.
Occupational mortality, 1091-1093.
Octopus, 214, 673, 691, 694, 994, 1112, **693**, **851**.
Œcophylla, **1167**.
Oeda, 956.

Œnothera lamarckiana, 581, 582.
Oikomonas, **947**.
Oikopleura, **991**.
Okapi, 388.
Old Age. *See* Senility.
Old Stone Age, 17, 423, 819, 821.
Oligocene period, 323, 777, 782, 785, 801, **325**, **344**, **778**.
Oliver, Prof. F. W., 981.
Onager, 334.
Onychophora, 209.
Ooze, abyssal, life on, 863-866; globigerina, 283; radiolarian, 283.
Ophiothrix, **220**.
Oporabia autumnata, 609.
Opossum, 176, 378, **391**.
Optoquine, 1052.
Orang-outang, 796, 1260, 1264, **407**; late embryo of, 413.
Orchids, 767, 906.
Ordovician period, 322, 324, 683, 684, 686, 690, **325**, **693**.
Oregon, 814.
Organisms, sizes of, **937**, **940**, **942**, **943**, **947**.
Organizers, in embryo, 519.
Origin of species, 429. *See* Evolution.
— life, 649-653.
Ornitholestis, **740**, **758**.
Orohippus, 341, **336**.
Orophocrinus, **687**.
Orthogenesis, 432, 629.
Oryx, **384**.
Osborn, Prof. Henry Fairfield, 432.
Osteolepis, 726, **699**.
Ostracoderms, 695, 697, 699, **693**, **696**.
Ostrich, **937**.
— Dinosaur, 772, **758**.
Otter, 178.
Ovary, 144, 146, 157, 162, **149**; in plants, 263.
Oviduct, 146, 148, **149**.
Ovum, 144, 146, 149, 445, **145**, **943**.
Owl, Snowy, 814, 909.
Ox, 176, **778**.
Oxford clay, 320.
Oxidation and life, 31.
Ox-pecker, 101.
Oxygen, 29, 98; amount needed, 98; and plants, 253-255; as a poison, 301, 302;

INDEX

transport of, in body, 48, 64; urgency of need for, 53–55. *See* Respiration.
Oxysoma, 1176, **1177**.
Oysters, 210–212; as typhoid carriers, 1065.

P

Pachycondyla, 1176.
Pachystomias, **854**.
Paddie-bird, 995.
Paddle-worm, 223.
Pain, 116.
Palæomastodon, **778**.
Palæotherium, 345.
Paleozoic era, 322, 324, 679, 683, 686, 691, 692, 703, **325**, **647**.
— fish, **696**, **699**.
— vertebrates, **696**.
Palestine, 294.
Palinurus, **849**.
Pallas' sandgrouse, 1004.
Panama, Isthmus of, 397.
Pancreas, and diabetes, 166; and digestion, 84.
Pangenesis, Darwin's theory of, **585**.
Pansy, wild, **626**.
Panthera tigris longipilis, 382, 383.
— sondaica, 383.
Papaver nudicaule, 621.
— striatocarpum, 621.
Papilio cynorta, **953**.
— dardanus, 364, **953**.
Paraguay, 380.
Parahippus, 345.
Parallelism and variety of life-communities, 967–973.
Paramecium, 284, **943**.
— bursaria, 1107, 1108.
— caudatum, 1107, 1137–1139, **1137**.
Parasite-chains, 989–996.
Parasites, 284, 372, 831, 961, 963, **203**, **662**.
Parasitism, discussion of, 831, 922–936.
Paratettix, 579.
Parathyroid glands, 166.
Paratyphoid, 359.
Pareiasaurus, 741.
Parrot, 142.
Parthenogenesis, 448–449.
Partnership and parasitism, 922–936.

Passerella ilica (fox-sparrow), migration of, 814.
"Passive immunity" from disease, 1050.
Pasteur, Louis (1822–1895), 5, 291, 437, 1023, 1036, 1048, 1051; experiment by, **437**.
Patagonia, 815.
Patella, **169**.
Pattern in life, colour and, 944–960.
Pavement epithelium, 40.
Pavlov, Prof. Ivan Petrovitch, 1214, 1288–1293, 1296, 1302, 1303, 1304–1305, 1306, 1308, 1310, 1313, 1315, 1322, 1361.
Peacock, 1227.
Peat, 720.
Peccary, 388, 784.
Pedicellariæ, 219.
Pelagic zone, 846.
Pelagomya, 852.
Pelagothuria, 848, 852.
Pelican, **182**.
Pellagra, 1062.
Pelvis, **169**.
Pelycosaur, 742, **740**.
Penguin, 850, 996.
— Adelie, 1257, 1259, **1257**.
Pennaria, **243**.
Pepsis, 955.
Peridians, 892, 996.
Peridinium, **991**.
Periophthalmus, 726, 874, **724**.
Peripatus, 209, 222, 356, 1110, **207**.
Perissodactyla, 178.
Peristalsis, 80.
Permian period, 322, 684, 725, 733, 736, 739, 741, 762, 763, 775, **325**, **693**, **734**.
Persia, 340, 895, 998.
Peru, 1030.
Pests and their biological control, 1016–1022.
Petrel, 908.
Phaeocystis, **991**.
Phagocytes, 91, 93, 1048.
Phalanger, vulpine, 389, **386**.
Phalarope, 608–609, **593**.
Phanerogams, 267.
Phenacodus, **778**.
Philip IV of Spain (1605–1665), **490**.
Philippine Islands, snails of the, 619.

Pholas, 875.
Phoronis larva, 847.
Phosphates and muscular activity, 1074–1075.
Phosphorescence. *See* Animal light.
Phosphorus, importance of supply, 1031–1032.
Phototropism, 1131.
Phthisis, 1090, 1093.
Phyllirrhoe, 860.
Phylum, classes of the vertebrate, 174–175; meaning of, 170–171.
Physalia, 245.
Piddock, 875.
Picard, Professor, 7.
Pierine butterflies, 626.
Pig, 178, 784, 359, 778.
Pigeon, 1225, 1233; wild rock, 376.
Pike, 886.
Pilidium larva, 847.
Piltdown man. *See* Eoanthropus.
Pineal gland, 173, 1208, 1280, 1279; of frog, 1206.
Pine-tree, 984; spores of, 709.
Pipe-fish, 870, 248.
Pithecanthropus, 411, 412, 806, *facing* 774.
Pituitary gland, 142, 157, 158, 164, 165, 173, 529, 948, 101, 1206, 1279.
Placenta, 155, 159, 155.
Placental mammals, 177–178; archaic, 740; evolution of the, 778; modern, 740.
Plague, 1045, 1089.
Plague bacillus, 947.
Plaice, 870.
Planaria, 227, 237, 544, 226, 668.
Planema epœa, 953.
Planktomya, 850.
Plankton, 842–850; in lakes, 891–892; in polar seas, 910; in rivers, 883–884.
Plankton-recorder, 985.
Plant-animals, 284–288.
Plant-body, the building of the, 661–664.
Plant-bug, sex determination in the, 552–553.
Plant-carbohydrates, 963.
Plant-cells, 255.
Plant-lice, 971.

Plant-proteins, 963.
Plants, 253–274; and water, 968–970; aquatic, 841–845, 844; behaviour of, 1128–1136; breeding of, 496–503; cells of, 255; classes of higher, 267; cycad-like, 693; development of, 761–769; distinction from animals, 253–254; earliest known land, 706; evolution of true flowering, 693, 709; fertilization of, 273; flowering, 267; food of, 257–258; growth of, 129; importance to animals, 962; individuality in, 259–260; invisibly small, 284–286; light and, 253–259, 968–969; nutrition of, 253–259; of the ancient world, 716–719; organization of, 256; pollination of (*see* Flowers); regeneration in, 257; reproduction in, 261–274; single-celled, 286–287; Sir J. C. Bose's experiments with, 13, 1129; stomata of, 255; sunlight and, 257–258; variations in, 598–599.
Planulæ, 241.
Plasmodium falciparum, 1043.
— malariæ, 1042–1043.
— vivax, 1043.
Plato, 1452.
Platurus, 747.
Platyhelminthes, 227.
Platypus, 352, 356, 363, 389, 386.
Play, biological function of, 1252–1259; in man, 1472–1473.
Pleistocene period, 323, 324, 775, 777, 807, 817, 915, 1005, 325, 344, 778.
— and recent climate, 806–817.
Plesiadapis, 785.
Plesiosaur, 184, 746, 740, 747.
Plesippus, 343, 344.
Pleurobrachia, 248, 248.
Pliocene period, 323, 777, 806, 817, 1241, 1436, 325, 344, 647, 778.
Pliohippus, 342–343, 336, 344.
Plover, 1155.
— golden, 815.
— ringed, 945, 950.
Plumose anemone, 247, 248.
Plymouth Aquarium, 948.
Pneumococcus, 1091.
Pneumonia, 1080, 1091, 1093.
Podon, 991.

INDEX

Poisons, in food, some possible, 1064–1068.
Poisons, produced by amphibians, 1115–1116.
Polacanthus, **740, 758**.
Polar regions, 907–908.
Pollan, 620.
Pollen, **264**; and bees, 1196–1197; evolution of, 708–712. *See* Flowers.
Pollination, 261–265; by animals, birds, and insects, 765–769.
Polyergus, **1173**.
Polygalum, 378.
Polyploidy, 598.
Polyps, 244–245, **242, 248, 668, 847, 937, 940**.
Polyzoa, 233, 920.
Pompilius quinquenotatus, 1161.
Pond-life at the surface-film, 882, **885**.
Poplar, 442–443, 731, 776.
Population, 1467–1468.
Porcupine, 178.
Porpoise, 179, 180.
Porthesia chrysorrhœa, 1134.
Portugal, 816, 859.
Portuguese man-of-war, 245, 859.
Portunion, 372.
Potamogeton, 890.
Poultry, determination of sex in, 562; heredity in, 487, 560–563.
Prawn, 694, 870, 1132, **854, 860, 865**; larvæ of, **991**.
Praying mantis, 642, 1228.
Pregnancy, 156, 157, 158; extra-uterine, 148; influence of shock on, 511.
Preservation of food, 1056–1057.
Pressure at great depths, 856–858.
Pribiloff Islands, 1031.
Prickly-pear, 390, 997, 1020.
Primary era, 324.
Primates, 179, **178, 778**. *See* Man, Apes, *and* Monkeys.
Primitive streak, in embryo, 151.
Primula, 581, 621.
— sinensis, 579, 599.
Prince, Morton, 1339, 1340, 1341, 1342, 1343.
Proboscidea, 178, **778**.
Procerodes, gamete of, **447**.
Progeria, 541.

Projection in psychology, 1378.
Prokofieff, 7.
Pronuba, 1153.
Propagation, artificial, 449–452.
Propliopithecus, 800.
Prorocentrum, **991**.
Prostate gland, 146, **147**.
Protection, by bluff, **959**.
Proteins, 74, 308; in animals of different species, 359–360.
Proterozoic era, 322, 324, 677, 678, 679, **325, 647**.
Proteus, 366, 915, 916.
— amœba, 277.
Prothallus, 268.
Protoceras, **778**.
Protocrinus, **687**.
Protohippus, 345.
Protonema, 270.
Protoplasm, 40; irritability of, 1105, 1106; nerve-like condition in, 1119; origin of, 650.
Protozoa, 280, 444, 660–661, **440, 442, 849, 940, 947**; multiplication of, 280–281; sexual process in, 282.
Przevalsky's horse, 334.
Pseudocalanus, **991**.
Psychology and conduct, 1390–1410; comparative (*See* Behaviour, Mind); human, 1318–1389.
Psycho-analysis, 1362–1369.
Ptarmigan, Spitsbergen, 814.
Pteranodon, 753, **754, 937**.
Pteraspis, **696**.
Pterichthys, **699**.
Pteridophyta, 270.
Pteridosperm, 355, 717, 734, **693**.
Pterodactyl, 747, 774, 361, **754**.
Pterolepis, **696**.
Pteropods, 690, 910.
Pterosaurs, 754, **754**.
Pterygoteuthis, 860.
Ptomaine poisoning, 1064–1065.
Puberty, 161.
Puerperal fever, 1051.
Pulse, in man, 56.
Pure lines, in genetics, 469–473.
Puss-moth caterpillar, **959**.
Pycnogonids, 911, **865**.
Pylorus, 83.

INDEX

Pyramids of Egypt, composition of, 283.
Pyrenees, 654, 815.
Pyrosoma, 192.
Pyrosomes, 850.
Pyrotheres, 394, **778**.
Pyrotherium, **390**.
Python, 184.

Q

Quagga, 334, 624.
Quartzite, 328.
Quaternary era, 324.
Queen ant, **1167**.
Queen bee, 1184, 1197–1198, **940, 942**.
Queensland, 1020.
Quinine, 1052.

R

Rabbits, 178, 389, 979, 997–998, 1000, 1011, 1013; brain of, **1221**; digestive organs of, **836**.
Races, distribution of human, 392; do they grow senile? 633; mixture of, 621, 1462; of mankind, 1447–1450.
Radio-active elements as geological clocks, 419.
Radiolaria, 283, 295, **662, 665, 991**.
Radiolarian ooze, 283.
Radium, 419.
Radula, 689.
Rafflesia, 766, 926, **937**.
Ragweed, 979.
Rag-worm, **224**.
Rana. *See* Frog.
Rat, 178, 1009, 1045, 1242; black, 601, 624, **613**; brown, 601–602, 624; hooded, 611, 612, **613**; and plague, 1009, 1045.
Rat-flea, 994, 1045.
Rationalization, in psychology, 1333.
Rattlesnake, 184.
Raven, 1250, 1253.
Ray, 187.
Rayleigh, Lord (1842–1919), 420.
Recapitulation, in man, 152–154; theory of, 366–373.
Recent period in geology, 323, 324.
Recessiveness, in heredity, 481.

Record of the rocks. *See* Fossil record *and* Evolution.
Rectum, **149**.
Red deer, 914, **359**.
— Fife wheat, 1026.
— Sea, 918; origin of name of the, 276, 288.
Red spider, 1020.
Redi, 437.
Redshank, **593**.
Redwing, 1258.
Reed, Walter, 1046.
Reflex, 133–134, 135, 136, 137, 1291–1292, 1298.
— conditioned, 1291–1292; contrast of, with inborn, **1290**; inhibition of, 1297–1305.
Regan, Dr. Tate, 388.
Regeneration of lost parts, 236–238, 450–451, **238, 257**.
Reindeer, 822, 908.
— moss, 294, 985.
Rejuvenation, 142–143, 541; is it desirable? 550–551.
Religion and biological discovery, 18–19; and theory of Evolution, 314, 432–433; in man, 1454–1460.
Remora, 876.
Repression and the complex, 1356–1362.
Reproduction, 145–146, 236, 438–458; in plants, 261–274, 442–443, 702–718, 761–769; of our species, 1467–1472; rapidity of, 281, 600–601.
— sexless (asexual) 438–439; artificial, 449–452; in amœba, 281; in animals, 236, **238**; in bacteria, 301; in plants, 260; the most primitive method, 438.
— sexual, 144–145; a complication of reproduction, 443; absent in bacteria, 301; evasions and replacements of, 448–449; in fungi, 289; in Protozoa, 281; mechanism of, 459–467; reason for, 596–598. *See also* Fertilization of the egg.
Reproductive system, 144–150.
Reptiles, 182–184, 805, **747, 937**; beginning of the Age of, 737, 738–743; brain in, 1218–1227; evolution of, 738–761, **693, 740**; flying, **754**; length of the Age of, 423.

INDEX

Resin, 330.
Respiration, 60; of plants, 254; organs of, **61**; reason for, 29; regulation of, 97–99.
Respiratory system, 60–66; diseases of, 1091–1092; of fish, 172; of insects, 770.
Response, forms of, 1114–1118.
Retina, 124, 125, **125**.
Retrogression, 541.
Reversion to ancestral type, 503–505.
Reward wheat, 1026.
Rhabdopleura, 193.
Rhamphorhynchus, **754**.
Rheumatic fever, 1091.
Rhinanthus, 927.
Rhine, 811; species of fish in the, 887.
Rhinoceros, 101, 178, 388, **778**, **995**; Malayan, 1243; Merck's, 809; woolly, 822.
Rhizocrinus, **370**.
Rhizopoda, 283.
Rhizosolenia, **991**.
Rhodesia, 820.
Rhodesian man. See Homo.
Rhynia, 719, **693**, **706**.
Rhyniella, 685–686.
Ribbon-weeds, 867.
Ribs, **169**; action of, 61–62.
Rice-grass, 981.
Richet, Prof. Charles, 1030, 1414, 1416.
Rickets, 1062, 1063.
Rivers, early vertebrate evolution in, 699–700; life in, 883–884; pollution of, 889.
Rockies, 654.
Rock-lobster, larva of, **849**.
Rocks, age of, 419–424; diagram of the record of the, **325**; evidence of, for evolution, 318–355; formation of sedimentary, 319. See Fossil record.
Rodents, 178, 784–785, **778**.
Rodriguez, 913.
Rook, 1256.
Roscoff (Brittany), 932.
Ross, Alexander, 5.
Ross, Sir Ronald, 1000.
Rotang palm, 904.
Rothamsted Experimental Station, 826.
Rotifers, 232, 920, 938; evolution of, 670.

Roughage, in food, 1058.
Roundworms, 223–227.
Roux, Wilhelm (1850–1924), 523.
Ruby wheat, 1026.
Ruff, 1232, 1234.
Ruminants, **778**; stomach of, **788**.
Ruptive colouration, 949.
Ruscus, 365.
Russia, 1000.
Rust (of wheat), 926.
Ruwenzori, 985.

S

Sabella, **224**.
Sabre-tooth, **778**.
Sabre-toothed cats, **636**.
Sacculina, 203, 373, 923, 926, **203**.
Sagitta, 673.
Sahara, 388, 777, 812, 895.
St. Francis of Assisi, 1350.
— Helena, 399.
St. Hilaire, Geoffroy, 427.
— Kilda, 618.
— Louis, 809.
Salamander, 185, 727, 1229; evolution of, **740**.
Saliva, 79.
Salmo, 620.
Salmon, 725; migration, 887.
Salps, 850.
Salts, mineral, and fresh-water animals, 278, 879–880; as food of plants, 254; in blood, 49, 879; in diet, 71; in lakes, 918–919; in sea, 648.
Salvarsan, 1053.
Sand-dollar, 219, 688.
Sand-eel, **991**.
Sand-grouse, 900.
Sand-hopper, 397.
Sand-rat, 590, **591**.
Sandstone, 327.
Sandworm, **672**, **847**.
Saprophytes, 290.
Sargasso Sea, 829, 852.
Sargassum bacciferum, 852.
Sassafras, 776.
Saturnia pavo, 1152.
Sauropods, 757, 759, 761, **740**.
Saxifrage, 768.

Scale-worm, 224.
Scales, in fishes, 698; in reptiles, 183.
Scallop, 212.
Scapula, 169.
Scardafella inca, 836–837.
Schizophrenia, 1374.
Schopenhauer, Arthur (1788–1860), 428.
Science, dawn of modern, 17.
Scorpion, 204, 205, 693.
Scotland, 385, 809, 810, 815, 816.
Scottish Highlands, 321, 328.
Scrophularia, 366.
Scrophulariaceæ, 622, 927.
Scurvy, early treatment of, 1059–1060.
Scymnognathus, skull of, 353.
Scyphozoa, 247.
Sea, as original home of life, 648–696; deep, 853–866; life in, 10–11, 839–845; shore life, 866–871; surface life of, 845–871, 847; temperature of, 657–658, 857.
Sea-anemones, 247, 862, 930, 844.
Sea-cow, 785.
Sea-cucumber, 220, 678, 688, 852, 866, 693, 851; larva of, 847; pelagic, 851.
Sea-dahlia, 247, 248.
Sea-elephant, 1237.
Sea-fan, 844.
Sea-gherkin, 220, 220.
Sea-gooseberry, 248, 248.
Seal, 178, 850, 910, 911, 778.
Sea-lettuce, 287.
Sea-lily, 221, 678, 688, 735, 866, 220, 370, 693.
Sea-lion, 1237, 778.
Sea-mouse, 223, 224.
Sea of Galilee, 891.
Sea-pens, 866, 844.
Sea-pudding, 220.
Sea-scorpion, 684–685, 695, 700, 693, 937.
Sea-serpents, 311–313.
Sea-shore life, 866–871.
Sea-slug, 690, 849.
Sea-snail, 991.
Sea-snake, 747.
Sea-spider, 911.
Sea-squirts, 209, 236, 369, 543, 190, 544, 865.

Sea-urchin, 219, 347–348, 687, 1113, 1144, 220, 527, 693; behaviour of, 1126; gamete of, 447; segmentation of eggs of, 527.
Seaweeds, 284–288, 444, 705, 707; gamete of, 708, 447.
Sea-woodlice, 865.
Sea-worms, 223, 224.
Secondary Era, 324, 769.
Secretin, 166.
Sedge-warbler, 1250.
Seed, 266–267; distribution of, 266–267; evolution of, 709.
Seed-ferns, 355, 718, 731, 354, 693; evolution of, 693.
Seeds, 261–267.
Segmentation of the body, in animals, 200, 222, 669–670.
—— egg, 151, 527.
Segmented worms, 221–223, 237, 668.
Selaginella, 711, 709.
Selection, and change of race, 613; by man, 374–378, 468–473; in evolution (See Natural selection); sexual, 616, 1232.
Self-fertilization in plants, 264.
Selous, Edmund, 1232.
Sematic colours, 946.
Seminal vesicle, 147.
Semi-vertebrates, 188–189, 190.
Semmelweis (1818–1865), 1051.
Senility, 140–141, 160–161, 541; of evolving races, 633.
Senility and races, 632–633; precocious, 541.
Sensation, 110–119, 1103; internal, 117–118; of position and movement, 130–133.
Sense-organs, 112–133; evolution of, 1106–1114; irritability as common basis of, 1105; limitations of, 1149, 1150.
Sepia, 215.
Sequoia gigantea, 142.
Sertularia, 248.
Sex, and marriage in primitive man, 1439–1447; conduct in relation to, 1404–1408; control of, 554–555; determination of, 552–575, 554; evolutionary significance of, 596–598; impor-

INDEX 1509

tance in psychology, 1364–1365; intermediates, 569–575; normal causation of, 552–555; reversal of, 568–569. See Reproduction.
Sexes, proportion of the, 555–558, **556**.
Sex-hormones, 565–568.
Sexless reproduction. See Reproduction.
Sex-linked inheritance, 560–563, **561**.
Sex-mosaics, 563, 568; butterflies produced by shock, 565.
Sexual aberrations in fowls, 568.
Sexual fetishism, 1359.
— selection, 616.
Sexuality, evasions and replacements of, 448–449.
Seymouria, **740**.
Shark, 187; basking, 187, 850.
Shaw, George Bernard, 428, 430, 435, 638.
Sheep, 178, 228, 1017, **229**, **937**; an abnormally developed, **518**.
Shell-shock, 1353.
Shetland Islands, 618.
"Shine-by-Night," 846.
Shipworm, 875.
Shock on pregnancy, influence of, 511.
Short-sightedness, cause of, 488.
Shrew, flying, **940**; jumping (brain of), **1283**.
Shrew-mouse, 178.
Shrimp, 201, 870.
Siberia, 329, 385, 397, 892, 911, 984, 1004.
Sicily, 811, 822, 888.
Sight, 124–126; in bees, 1189–1190; in dogs, 1305–1306; in fishes, 1211–1212. See Eye.
Silurian period, 324, 683, 685, 688, 697, 698, 707, **325**, **693**.
Simocephalus, **672**.
Simpson, Sir James Young (1811–1870), 1023.
Sinai, 895.
Sinanthropus, 412, 414, 806; skull of sub-man, 412, **1441**.
Singer, Dr. Charles, 1045.
Single-celled life, 275–288, 291, 661, **662**.
Siphonophore, 246, 669, 683, 850, 859.
Sirenians, **778**.

Sivatherium, **778**.
Sizes, of organisms, 936–944, **937**, **940**, **942**, **943**, **947**; of ultra-microscopic particles, **652**.
Skate, 187.
Skeleton, development of, 523–524; human, 33–38, **34**, **36**, **39**; of dog, **169**; of gorilla, **410**; of lobster, **195**; of vertebrates, 173, 697–698, **364**, **365**; strength of, 941–942.
Skin, 91–96, 1081–1082; and temperature regulation, 94–96, 1082; of amphibians, 184–185, 1115–1116; of coloured races, 490–491, 583, 588; of fishes, 698; sunburn of, 588, 1085.
Skink, 184.
Skua, 908, 909, 990.
Skull, Broken Hill, 820.
Skunk, 954.
Sleep, nature of, 1306–1311.
Sleeping sickness, 284, 925, 1016, 1022, 1039, **947**.
Slime-fungi (slime moulds), 286, 295–299, 444, **297**, **298**.
Slipper animalcule, 1136, **1137**; behaviour of, 1136–1143.
Sloth, 178, 389, 634.
Slow-worm, 184.
Slug, 213.
Smallest living things, 307–308.
Smallpox, 1036, 1039, 1049, 1090.
Smell, 119–120; in bees, 1191; in dogs, 1306; in fishes, 1212; in moths, 1156; in primitive mammals, 1282.
Smelt, 886.
Smilodon, 635, **636**.
Smith, Hélène, medium, 1352.
— Prof. Elliot, 1282.
— William (1769–1839), 320, 327.
Snails, 211, 212, 1116, **940**, **942**.
Snakes, ancestral stock of, **740**; extinct, **937**.
Snapdragon, 622.
Snipe, 1016, 1234.
Snow-leopard, 912.
Solar disturbances and terrestrial life, relation between, **1003**.
Solar plexus, 107.
Soldanella, 909.
Solenhofen, 330, 350.

Solomon Islands, 1459.
Solomon's seal, 977.
Solutrean culture, 822.
Soma, 457; germ-plasm and, **585**.
Somerville, Dr., 7, 8.
Song-thrush, 940.
Soot, and health, 1077; amount deposited in modern towns, 1076; defences against, 63.
Soredia, 293.
South Africa, 998.
— America, 340, 345, 351, 387, 388, 389, 395, 396, 401, 733, 777, 781, 785, 789, 815, 895, 902, 955, 998; characteristic animals of, **391**, **399**.
— American ungulates, primitive stock of, **778**.
— Georgia, 911.
Spain, 811, 816, 821, 1045.
Spallanzani, Abbé, 1023.
Spanish Bayonet, 767, 1153.
— moss, 905.
Sparrow, English, 601, 606, 1018.
— migration of the fox, 814.
Spartina townsendii, 982.
Species, chemical basis of, 360; meaning of, 168, 378-388; origin of. *See* Evolution.
Speech, 1327, 1328, 1437; fundamental processes of, 1294; parts of the brain concerned in, 1286.
Sperm, human, **943**, **947**.
Spermatophytes, 267.
Spermatozoa, 144, 145, 146, 147, 148, **444**, **558**; of ferns and mosses, 268, 269.
Sperms. *See* Spermatozoa.
Sperm-whale, 858, **778**.
Sphærium, 886.
Sphenodon, 184.
Sphenophyll, 731.
Sphex, 1161.
Spider, 204, 941, 1158, **693**.
Spider-crab, 866, 936, **844**, **937**; gamete of, **447**; instinct of, **1154**.
Spinal cord, 35, 173, 1201, **135**, **1202**.
Spiracle, 171.
Spirochætes, 302.
Spirogyra, 287, 663, **665**.
Spirula, 850.
Spisula, 871.

Spitsbergen, 720, 729, 776, 811, 843, 907, 972, 986, 993.
Spleen, 101.
Sponge, reproduction of, 442, **441**.
Sponges, 249-252, 442, 805, **251**, **441**.
Spongilla, **441**.
Spontaneous generation, 5, 436, 438.
Spores, 442; evolution of, in plants, 706-716; of bacteria, 300, 301; of ferns and mosses, 267, 269, 270, 271, 272; of fungi, **297**.
Sporogonium, 272.
Sporozoa, 284, 1041, **662**.
Spring-tails, 972.
Spurge, 897.
Squat-lobster, gamete of, **447**.
Squid, 673, 694, 994, 1158, **854**, **860**, **937**.
Squilla, 870.
Squirrel, 178, 378, 834, 995, 1258, **778**.
Stag, 632.
Stamen, 263, 713, **262**.
Starfish, 218, 237, 687, 1110, **217**, **220**, **238**, **693**, **865**, **940**; behaviour of, **1125**.
Starling, 389, 1016, 1250, 1258.
Stefansson, Vilhjalmur, 911.
Stegocephalians, 728, 736, **730**, **740**.
Stegosaurus, 761, **740**, **760**.
Steinach, Eugen, 142, 167, 547, 548, 549, 566.
Steno, Nicholas, (1638-1686), 20.
Stensio, Dr., 697.
Stentor, 1106, 1108, 1110, 1113, **1105**.
Sterilization against bacteria, 300, 1052, 1057.
— of the unfit, 1469.
Sternum, **169**.
Stick-insect, 1155, 1313, **940**.
Stickleback, 886, 1229.
Stigma, 263.
Stigmatization, 1350-1351.
Still-births, 557.
Stomach, human, 80-81, **36**, **39**; of calf, **788**.
Stomata of plant, **255**.
Stomolophus, 851.
Stone curlew, 1257.
Straits of Gibraltar, 811, 857.
Strangeways, Dr., 43.
Strata, building of geological, 316-326; formation of, 318-320. *See* Fossils *and* Rocks.

Stratton, Professor, 1255.
Strawberry, 442-443.
Streptococcus pyogenes, 1038.
Structural plans as evidence for evolution, 356-362.
Struthiomimus, 757, **758**.
Sturgeon, 888, 892.
Stylonychia, 439, **440**, **662**.
Sublimation in psychology, 1360.
Succinea putris, 952.
Sucker-fish, 876.
Sugar, 497, 1057.
Sugar-cane weevil, 1019-1020.
Sugar-maple, 977.
Sumatra, 926.
Sun animalcules, **662**.
Sun-fish, 850.
Sunflower, 979.
Sunlight and plants, 257-258; as a tonic, 1085-1088; fresh air and, 1076-1088. *See* Ultra-violet light.
Surgery, antiseptic and aseptic, 1051, 1052, 1053.
Survival of the personality after death, 1433-1435, 1474-1475.
Swallow, 1258.
Swallowing, 79.
Swallow-tail butterfly, **953**.
Swamp, Great Dismal, of Florida, 720.
Sweat, 95-96.
Sweden, 421.
Sweets contaminated with arsenic, 1066.
Swimming bladder of fish. *See* Gas-bladder.
Sycamore, 979, 984.
Syllis, ramosa, 237, **238**.
Symbiosis, 294, 923-935.
Sympathetic nerves. *See* Nervous system.
Symphytum, 378.
Syncoryne, **458**.
Syndyoceras, **778**.
Synodontis, 949.
Syphilis, 302, 1039, 1049, 1053.
Syria, 396, 820.
Syrian desert, 895.

T

Table-tapping, 1416.
Tabus, 1441-1447, 1454, 1458.

Tadpoles, 185, **885**; transformation of, 526-530, **529**.
Tænia, **231**.
Taiga, 984.
Tail, 37-38; characteristic of vertebrates, 671, 672, 673, 674; in birds, 351; in human embryo, 38, 152, 416, **153**; prehensile, 906.
Tamandua, **391**.
Tanganyika, Lake, 892.
Tapeworm, 230, 232, 923, 935, 993, **937**.
Tapir, 178, 388, 400, 949, **399**, **778**.
Tarpon, **937**.
Tarsius, 799; brain of, **1283**.
Tartary, 294.
Tasmania, 388, 389.
Tasmanian devil, 389, **386**.
Taste, 119; in fishes, 1212, 1213.
Taungs skull, 801.
Taunus, 654.
Tea, use and abuse of, 1072-1073.
Tealia, **248**.
Teeth, in horses, 335-346, **336**; of sabre-toothed tiger, **636**; of Trachodon, 759; vitamin D and the, 1062-1063.
Telangium, 718.
Telegony, 508.
Telekinesis, 1419.
Telepathy, 1413-1416.
Teleplasm, 1426.
Temora, **991**.
Temperament, in dogs, 1314; in man, 1381-1389.
Temperature, and life, 8-9, 918; of deserts, 896; of the body, 30, 93, 94, 1081-1082; of the sea, 658, 857; sense of, 116, 117.
Tendons, 47.
Tenrec, **778**.
Terebratula, **233**.
Teredo, 691, 875.
Termites, 928-929, 971, 1178-1182, **1181**.
Terrapin, 183.
Territorial instinct in birds, 1006-1007.
Tertiary period, 324, 788.
Testes, 145, 166, **147**.
Testis-grafts, 549.
Tetanus, 1037, 1050.
Tetrabelodon, **778**.
Texan ant, **1167**.

Texas, 916, 1253.
Thalamus, 1210, 1215, 1280.
Thalassiosira, **991**.
Thames, 487.
Thayer, 949.
Theriodonts, 742.
Theromorphs, 352, **353, 740**.
Thirlmere, 916.
Thistle, 1018.
Thorax, 37.
Thomson, Sir Basil, 1461.
— Prof. J. Arthur, 10.
Thorium, 420.
Thorn-scrub, African, 983.
Thrush, 1258.
Thylacoleo, 395.
Thymallus vulgaris, 886.
Thymus, 166.
Thyroid extract, effect of, 526–530. *See* Thyroxin.
— gland, 142, 157, 163, 164, 165, 166, 167, 173, **105**; and growth, 526–530.
Thyroxin, 162, 163, 166, 1074.
Tibet, 895.
Tibetan sheep, 901.
Tick-bird, 995, **995**.
Tiger, sabre-toothed, **636**; species of, 382–383.
Tillandsia, 905.
Tillyard, Dr. R. J., 1428.
Time. *See* Geological time.
— idea of, in animals, 1143, 1145.
Tissandier, Gaston, 8.
Tissue culture, 40.
Titanotheres, 348, 631, 632, 783, **778**.
Titanotherium, **778**.
Tit, longtailed, 1003.
Toad, common, 185, **185**; evolution of, **740**; poison of, 1115–1116.
Toadstools, 288, 289, 290.
Tobacco, 578, 766; use and abuse of, 1069–1070, 1073–1074, 1091, 1100.
Tomopteris, **991**.
Tooth. *See* Teeth.
Torpedo, 697, 1117.
Torres Straits, 876.
Tortoise, 183, 756, **740**; gamete of, **447**.
Touch, sense of, 115–116.
Toxeuma, **854**.
Toxodont, 785, **390**.

Tozzia, 927.
Trachea, 60.
Tracheæ, 204, 770.
Trachodon, 330, 759, **740, 758**.
Traditionalism, passing of, 1460–1464.
Transformism, 426, 427.
Tree, method of grafting a, 451, **450**
Tree ant-eater, **391**.
Tree-ferns, 268.
Tree-frog, 904, 1116, 1217.
Tree-shrew, brain of, **1283**.
Tree-toad, 904.
Trembley, Abbé, 452.
Trepang, 219.
Triassic period, 323, 692, 702, 729, 735, 736, 741, 746, 748, 756, 757, 760, 762, 775, 780, 782, **325, 693**.
Triceratops, **740**.
Trichina, 923.
Trichinella, **236**.
— spiralis, 225.
Trichomonas, **662**.
Triconodonts, **740**.
Tridacna, 992, **937**.
Trilobites, 680, 683, 694, **700, 681, 693**.
Trinidad, 768.
Trinil, 411.
Tripe de roche, 294.
Trituberculates, **740**.
Trochophore larva, **847**.
Tropical forest, life in, 901–907.
Tropisms, 1131.
Trout, 886, 1017, 1212.
Trypanosomes, 925, **947**.
Tsetse fly, 1022, 1045, **1046**.
Tuatera, 775.
Tuberculosis, 1096–1099; bacillus of, 947, 1041.
Tube-worm, eye of the, 1112, **1113**.
Tubifex, 885.
Tulip, 627.
Turbot, **372**.
Turkestan, 895.
Turtle, 183, **740, 747**; an abnormally developed, **518**.
Twins, 554–555.
Tylenchus scandens, 227.
Typhlomologe, 915–916.
Typhoid fever, 359, 1050; and oysters, 1065; bacillus of, **947, 1038**.

INDEX

Typhus fever, 1045.
Typotheres, 394, 785, **778**.
Tyrannosaurus, 757, 772, **740**, **937**.

U

Uga pugilator, **540**.
Ultramicroscopic organisms, 306-310.
Ultramicroscopic particles, size of, **652**.
Ultra-violet light, 309, 1077, 1085-1086, 1087, 1088; and vitamins, 1063.
Umbilical cord, 155, **155**.
Unconscious, the, 1334-1339.
Unguiculates, 178, **177**.
Ungulates, 177, **175**, **778**.
Universe, scale of the, 644-646.
Unwins, Capt. C. F., 8.
Upper Cambrian period, 690.
— Carboniferous period, 719, 721.
— Devonian period, 718, 727, 729.
— Egypt, 800.
— Jurassic period, 752, **747**.
— Silurian period, 685.
Uranium, 421, 422.
Ureter, **65**, **147**, **149**.
Urinary organs, **65**.
Urine, 68.
Use, moulding power of, 523-526, **525**; and disease, in evolution, 583-584.
Ussher, Archbishop, 326.
Uterus, 148, 150, 156, 157, **149**.

V

Vaccination, 1049, 1050.
Vagina, **149**.
Vagus nerve, the, 1204.
Vallisneria, 890.
Vanadium, 627.
Vancouver, 814.
Vanua Levu, 10, 19.
Varanus, 748, 775, **937**.
Variation, 576-577, **591**; accidental and inherited, 469-473; in plants, 598-599. *See* Mutation.
— local, 617-621; in apes, 1440-1441; in man, 1447-1450.
Variation, failures of, 623-628.
Vas deferens, 547.
Vegetables, value of green, 1063; contamination of, 1065.

Vegetarianism, 1055-1056.
Vegetation, belts of, 988-989.
Veins, 53, 54, 58, 59, **57**, **59**.
Velella, 859.
Vendace, 620.
Venereal disease, 1053, 1369-1370, 1406.
Ventricles of heart, **55**, **57**.
Venus' Flower-basket, 866.
Venus' girdle, 249.
Vertebral column, 33.
Vertebrates, 170-192, 805, **696**, **754**, **940**; behaviour of the lower, 1200-1226; characteristics of, 171; classes of, 174-175; contrasted with arthropods, 194-199; evolution of, 671, 694-700, 722-729, 738-761, 769-773, 779-786, 796-822, **693**, **787**; flying, **754**.
Vespa. *See* Wasps.
Vestiges, 362-366.
Vestigial limbs, **364**, **365**.
— organs in man, 415-417.
—— as evidence for evolution, 363-366.
Vetch, 928.
Viceroy butterfly, 955.
Vienna, 1051.
Villi, 86.
Vinciguerria, **860**.
Vinegar bacteria, 304.
Vila tricolor, **626**.
Virgin reproduction, 448-449.
Virus diseases, 308, 1040-1042.
Viscera, 148.
Vital energy, source of, 28-31.
Vitamins, 850, 1059 *sqq.*; effect of lack of, 1061-1062; the six, 1059-1064.
Viti Levu, 1019.
Vivisection, 22, 1296-1297.
Vizcacha, 388, **391**.
Volcano of Krakatoa, 973.
Volvox, 664, 1121.
Von Baer, 366, 368.
Von Frisch, 1185, 1191, 1192, 1193, 1197.
Von Humboldt, Baron (1767-1835), 1182.
Voronoff, Serge, 142, 167, 541, 549.
Vorticella, **662**, **943**.
Vulpes, 380; vulpes, 380; zerda, 380.
Vulture, 1112.

INDEX

W

Waitomo caves, New Zealand, 917.
Wales, 816.
Wallaby, 389.
Wallace, Alfred Russel (1823-1913), 398, 428, 429, 901.
Walnut, 984.
Walrus, 178, 910.
War, the supersession of, 1464-1465.
Warble-fly, 923-924.
Warblers, courtship of, 1230.
Warfare, primitive, 1464.
Warning colouration, 952-953.
Wart-hog, 388, 784.
Wasmann, 1178.
Wasps, 958-959, 1158-1162; solitary, 1158-1162.
Water, as habitat, 829-830; contamination of, 1080; importance of, to organisms, 72-73; life in fresh, 878 sqq., 919; life in salt, 839-845, 918-919; supply, and disease, 920-921, 1036-1037.
Water-beetle, **885**; larva of, 880, **885**.
Water-flea, 578, **672**, **942**.
Water-lily, **885**.
Water-measurer, **885**.
Water-plantain, 979.
Water-plants, early, 663.
Water-scorpion, **885**.
Water-snail, 238.
Water-spider, **885**.
Water-weed, Canadian, 601.
Watson, Dr. J. B., 1318, 1319, 1320, 1321, 1322, 1323, 1324.
Weasel, 178, **778**.
Weather and bacteria, 1039.
Weights of living things, 936-944.
Weismann, August (1834-1914), 431, 457, 586.
West Indies, 249.
Whale, 10, 179, 180, 785, 850, 858, 911; Greenland Right, **364**; sulphur-bottom, **937**; whalebone, 832, **778**.
Whalebone, 180, 832.
Whale-shark, **937**.
Wheat, 496-497, 1026.
Wheel-animalcules, 232, 670, **943**.
Whelk, 213, **213**.
Whirligig beetles, **885**.
White Admiral butterfly, 955.
White ants, 1178-1182, **1181**.
— melilot, 979.
Wholemeal bread, value of, 1057-1058.
Whooping cough, 1049.
Wight, Isle of, 777.
Williamsonia, 764.
Willow, 977, 984.
Windpipe, 60.
Winkle, 213.
Winton, Prof. W. M., 421.
Wöhler, 6.
Wolf, 380, 908, **778**; European, 381; maned, 380, 387; marsupial, 389; range of the, 381; timber, 381.
Wombat, 389, **386**.
Wood-ants, **1167**.
Woodcock, 949.
Woodlouse, 680, 899.
Woodpecker, 626; North American, 385.
Word-blindness, 1287.
Word-deafness, 1287.
Worms, 221-232; evolution of, 667, 668, 669; eyes of, 1109; flat, 227, 805; intelligence in, 1126-1127; marine, **224**, 238; nemertine, 232; parasitic, **226**; round, 223; segmented, **668**.
Wren, common, 951, 1230.
Wrens, geographical variation in, 617-618.
Wyoming, Bad Lands of, 331.
Wyville-Thomson ridge, 810, 812.

Y

Yeast, 288, 291-292.
Yellow fever, 1022, 1045-1047.
"Yeoman" wheat, 497.
Ypsipetes trifasciata, 595.
Yucca, 767, 1153.

Z

Zebra, 178, 334, 388, 624, **384**.
Zeuglodont, 785, **778**.
Zoæa larva, **847**.
Zones of the sea, **841**.
Zoophyta, 244.
Zostera, 890.
Zygote, 445.